ELEMENTS OF FUEL & COMBUSTION TECHNOLOGY

ELEMENTS OF FUEL & COMBUSTION TECHNOLOGY

Om Prakash Gupta
B. Tech. (Chemical Engg.)
Ex DGM, HRD
Bokaro Steel Plant
Steel Authority of India Limited (SAIL)

KHANNA BOOK PUBLISHING CO. (P) LTD.
Publisher of Science, Technology and Engineering Books
4C/4344, Ansari Road, Darya Ganj, New Delhi-110002
Phone: 011-23244447-48; **Mobile:** +91-99109 09320
E-mail: contact@khannabooks.com, sales@khannapublishers.com
Website: www.khannabooks.com, www.khannapublishers.com

Price: ₹ 499.00

ELEMENTS OF FUEL & COMBUSTION TECHNOLOGY

O.P. Gupta

ISBN: 978-93-86173-32-4

Edition: 2018

Published by:

KHANNA BOOK PUBLISHING CO. (P) LTD.
4C/4344, Ansari Road, Darya Ganj, New Delhi-110 002
Phone: 011-23244447-48 **Mobile:** +91-9910909320
E-mail: contact@khannabooks.com

Printed in India by:
S.P.S. Printers & Binders, Delhi

PREFACE

Fuel & combustion is an integral part of the degree & diploma curriculum of various branches of engineering. Besides, it is a compulsory paper for various associate membership examinations conducted by professional bodies like Institution of Engineers (AMIE), Indian Institute of Metals (AMIIM), Indian Institute of Chemical Engineers (AMIIChE), Institute of Chemist etc. This book has been prepared strictly as per the syllabus of these examinations.

This book contains detailed description of solid, liquid & gaseous fuels, combustion and furnaces.

Short and multiple-choice questions & answers drawn from the examination papers of various engineering colleges and professional bodies examinations are given at the end of the book which further enhances its utility for the students.

I hope that the book in its present form will be useful for the students.

Constructive suggestion for the improvement of the book are welcome.

26th January, 2018
Ranchi

Om Prakash Gupta

CONTENTS

Introduction and Classification of Fuels

A substance which produces heat is called a fuel. The heat may be produced either by combustion or by nuclear fission & fusion. Those fuels which produce heat by the combustion are termed as simply *'fuels'* whereas those which release heat by fission or fusion are called *nuclear fuels.*

Fuels are classified mainly in four general classes namely *fossil, by-products, chemical and nuclear fuel.* Out of these, first three release heat by combustion in presence of air or oxygen. Nuclear fuels release heat by nuclear fission or fusion by converting mass into energy.

Fossil fuels are those which have been derived from fossil remains of plant & animal life and are found in the crust of the earth *e.g.* coal, petroleum, natural gas etc.

By-product fuels are the co-product of some regular manufacturing process and are of a secondary nature. As for example, coke oven gas and blast furnace gas are the by-product fuels in the process of coke & iron making respectively.

Chemical fuels are of an exotic nature and normally not used in conventional processes. Examples are hydrazine (a rocket fuel), ammonium nitrate, flourine etc.

Nuclear fuels which release heat by fission are uranium, plutonium etc. and those generating heat by fusion are deuterium & tritium (both are isotopes of hydrogen) etc. In both the situations, the mass is converted into energy.

Depending upon the state of occurrence, the fuels are classified as solid, liquid or gaseous fuels.

Fuels in each general division are further classified as *Primary and Secondary fuels.* Primary fuels are those which occur in nature *viz.* coal, wood, petroleum, natural gas etc. Secondary fuels are those which are derived from primary fuel *e.g.* fuel oil & kerosene (derived from petroleum), coke oven gas (derived from coal) etc.

Secondary fuels are further classified into *manufactured and by-product fuels.* Manufactured fuels are those which are made for some specific purpose *e.g.* coke (made for iron making), gasoline (made for internal combustion engines), producer gas (made for industrial heating) etc. By-product fuels are those which are a co-product/side product (unavoidable product) of a regular manufacturing process *e.g.* bagasse, tar, refinery gas etc. By-product fuels help industries in conserving primary fuels.

Table 1.1 gives the classified list of important fuels (excluding chemical & nuclear fuel).

Nuclear fuels are of two types. One type can release heat by its fission by neutron bombardment. Such fuels are called fissile fuel *e.g.* uranium-233, uranium-235, plutonium- 239. There are some nuclear fuels *e.g.* thorium-232 and uranium-238 which can not be fissioned directly, but they can be converted into fissile material *e.g.* uranium-233 and plutonium-239 respectively. Such nuclear fuels are called fertile fuels.

Still another type of nuclear fuels release heat by fusion, *e.g.* deuterium & tritium (both are isotopes of hydrogen), hydrogen etc. But in both the cases whether the fuel undergoes fission or fusion, conversion of mass into energy takes place to release heat.

Depending upon the usage, fuels are classified as domestic fuel, illuminating fuel, industrial fuel, rocket fuel etc.

Table 1.1: General Classification of Fuels

General Division	Primary Fuels	Secondary Fuels	
	Natural	Manufactured	By-product
Solid	Wood , Coal	Semi-coke , Coke, Charcoal, Briquettes Pulverised Coal	Charcoal, Wood refuse, Bagasse, Coke breeze, Waste material from grain
Liquid	Petroleum	Petrol, Kerosene, Alcohol Colloidal fuels, Fuel oil, Naphtha, Vegetable Oil	Tar Pitch Benzol Paper pulp mill waste
Gaeous	Natural gas	Producer gas, Water gas, Carburetted water gas, Coal gas, Oil gas, Gobar gas, Reformed natural gas Butane Propane Acetylene Hydrogen LPG	Blast furnace gas, Coke oven gas, Oil refinery gas, Sewage gas, L.D. converter gas

Domestic fuels supply heat for cooking and space heating. *Illuminating fuels e.g.* kerosene, town gas etc. supply light in darkness. *Industrial fuels* supply heat for process heating, steam and electricity generation etc. *Rocket fuels e.g.* hydrazine are used for producing enormous thrust for the propulsion of rocket.

As per another general classification system, the fuels which generate heat by combustion are called *'Chemical Fuels'* and the fuels which generate heat by nuclear fission or fusion are called *'Nuclear Fuels'*.

Characteristics of a good fuel

- It should be readily available at cheap rate.
- It should be easy & safe to handle, store and transport and at a low cost.
- Storage cost of the fuel should be low.
- It should have high heating value (Calorific value).

- Its moisture content should be low, as it reduces the heating value.
- It should have moderate ignition temperature as low ignition temperature fuel will catch fire easily and one with high ignition temperature will present difficulty in igniting it.
- Its non-combustible matter content should be low.
- It should not produce on burning harmful products of combustion.
- It should possess moderate rate of combustion. If rate of its combustion is low, the required high temperature may not be attained as part of heat generated will be lost. Too high combustion rate of fuel may be uneconomical and uncontrollable to achieve the desired effect.

DEFINITIONS & UNITS CONVERSIONS RELATED TO FUEL TECHNOLOGY

Definitions

Introductory concept & definitions of certain preliminary terms are required for the study of this subject. A first hand information is given below regarding them.

Rank of coal. It denotes the maturity of the coal. Coal is formed from vegetable matter which gets converted into coal *via* different stages of maturity of coal namely peat → lignite → bituminous → anthracite. Here, peat is the most immature coal hence it is of lowest rank. Whereas anthracite is the most matured coal hence is of the highest rank. *Rank* of a coal denotes its *degree of maturity* or the *degree of metamorphism.*

Coalification (or metamorphism). The process of conversion of lignite to anthracite is called coalification or metamorphism of coal.

Carbonisation of Coal. Heating of coal in absence of air at high temperature to produce a residue coke, tar and gas is called its carbonisation.

Gasification of Coal. Heating of coal with insufficiently less quantity of air plus steam to produce a gas rich in CO and H_2 is called its gasification. This is done to produce gaseous fuel from solid and liquid fuel.

Caking and Coking Coal. When powdered coal is carbonised (*i.e* heated in absence of air), it forms an expanded lumpy mass in some cases. Such coals are called caking coal and the phenomenon is called caking of coal. When the residue is very hard & strong it is called coke and the process is called coking of coal. The coal is called coking coal. All the caking coals do not form strong, hard and coherent residue coke. Hence all the caking coals are not necessarily coking coal but all the coking coals have to be necessarily caking in nature.

When coal is carbonised, the caking coal will first expand and then contract slightly leaving a larger, weak and porous residue. When coking coal is carbonised, it will expand first and then it will contract to a greater extent (than merely a caking coal) to form a very compact, relatively less porous and stronger residue. Caking corresponds to a slower rate of heating whereas coking corresponds to faster rate of heating. All the coking coals are essentially bituminous coal. Strongly caking coals are called coking coals.

Non-caking/Non-coking Coals. Those coals which do not fuse, form lumpy mass and a coherent/ compact residue are non-coking/caking coals.

Basis for reporting the analysis of coal. Following bases are in use :

(i) **Run-of-mine coal (R.O.M.).** The crude coal as obtained from mine is called run-of- mine coal. It involves no treatment of coal like size reduction, moisture elimination ,washing etc.

(ii) **As-received.** When the coal is received by a consumer from the mine then its analysis and other properties are reported on As- received' basis. During the transportation of coal from mine to the consumer, lot of physical& chemical changes occur during its transportation and processing *e.g.* size reduction, washing, weathering etc. Hence the analysis of coal, at consumers end is reported on 'As-received' basis.

(iii) **Air-dried.** Freshly mined coals gets air & sun-dried during its storage, hence its moisture content, varies depending on the humidity and temperature of the atmospheric air. In fact the data on As-received' basis and air dried basis may be same because the coal looses its moisture due to e×posure to the atmospheric air during transportation and storage.

However, for getting accurate and comparable results the coal is exposed to an artificial and standard atmosphere at 40°C & 60% relative humidity to get air-dried data. Even after air-drying, coal contains some amount of moisture.

(iv) **Dry-coal.** When the effect of moisture on analytical data is to be completely eliminated then the coal analysis is reported on completely 'dry' basis.

(v) **Dry-ash-free (d.a.f.).** Effect of ash and moisture in coal is eliminated by reporting the data on dry-ash-free basis. This data is suitable for comparing the pure coal with low ash (< 10%).

(vi) **Dry-mineral matter-free (d.m.m.f.).** For high ash coal (> 10%), as is the case with high ash Indian coal, the mineral matter content of the coal is around 10% more than its ash. Whereas for low ash coal, mineral matters can be considered to be equal to its ash content. Hence for high ash coal, the comparison of pure coal can be done if its analysis is reported on dry-mineral matter-free basis instead of dry-ash free basis.

(vii) **Mineral matter-free.** Effect of mineral matter present in coal is eliminated by reporting the data on mineral matter-free basis. It includes the moisture in the coal.

Proximate Analysis of Coal. Proximate analysis means finding out the weight percentages of moisture, volatile matter, fixed carbon and ash in coal. This analysis gives the approximate composition of the main constituents of coal and is useful in deciding its utilisation for a particular purpose.

Ultimate Analysis of Coal. Ultimate analysis means finding out the weight percentages of carbon, hydrogen, nitrogen, oxygen & sulphur of the pure coal free from moisture and inorganic constituents. This analysis gives the elementary constituents of coal and is useful to the designe of coal burning equipments & auxiliaries.

Combustion. It is an exothermic chemical reaction of a fuel with oxygen or air at high temperature to liberate heat.

Calorific value. The quantity of heat (kcal) liberated by the combustion of unit quantity of fuel is called its calorific value. Unit of calorific value is kcal/kg for solid & liquid fuels and kcal/Nm^3 for gaseous fuels. Nm^3 means volume of gas in m^3 at Normal Temperature and Pressure (NTP) which is 0°C & 760 mmHg. Since the volume of gases vary sensitively with pressure (Boyle's Law) and temperature (Charle's Law), hence their volume is expressed at NTP in Nm^3 to have a standard measurement.

Gross calorific value or higher heating value at constant volume is the quantity of heat liberated by combusting the fuel at constant volume in oxygen saturated with water vapour, the original material and final products of combustion being at a reference temperature (25°C) and the water obtained from the fuel being in the liquid state.

Gross calorific value at constant pressure implies that the combustion takes place at constant pressure and not at constant volume. In the laboratory determinations, solid & liquid fuels are burnt at constant volume and gaseous fuels are burnt at constant pressure. In the ovens and furnaces, however, the combustion takes place at constant pressure. The difference in the two corresponding values is small. For coal, the calorific value at constant pressure exceeds the calorific value at constant volume by about 5.5 kcal/kg.

Net calorific value or lower heating value at constant volume is the quantity of heat evolved when unit quantity of fuel is burnt at constant volume in oxygen saturated with water vapour, the original material and final products of combustion being at a reference temperature (25°C) and the water obtained from the fuel being in the vapour state. The net calorific value is therefore less than the gross calorific value by the amount corresponding to the heat of condensation of water vapours, which at 25°C is 583.5 kcal/kg of water. On the basis of hydrogen of water, this is equal to 5,252 kcal/kg or 468.9 kcal/Nm^3 of hydrogen. The following formula is used in calculating the net calorific value from gross calorific value of *solid and liquid fuels* approximately.

$$C_N = C_G - 53\,H$$

where C_N and C_G = net and gross calorific value in kcal/kg, respectively, and H = percentage of hydrogen of coal, including hydrogen of moisture and of water of hydration of minerals.

For *a gaseous fuel,* the formula is :

$$C_N = C_G - 4.7\,V$$

where C_N and C_G = net and gross calorific values in kcal/Nm^3 respectively, V = volume percentage (as H_2) of total hydrogen of the gaseous fuel, including the hydrogen obtainable from other combustible components.

Net calorific value at constant pressure implies that the combustion takes place at constant pressure and not at constant volume.

Flue gas. It is the gaseous product of combustion of a fuel.

Fuel gas. All gaseous fuels which are burnt to heat up the furnaces. Fuel gas when burnt produce flue gases.

Thermal capacity (heat capacity) and Specific heat. Thermal or heat capacity of a substance is defined as the amount of heat (kcal) required to raise the temperature of unit weight (kg) of the substance by 1°C (from 15 to 16°C say).

The specific heat of a substance is the ratio of the heat capacity of that substance to the heat capacity of water. Specific heat is a number. Specific heat of water is 1 at 15.5°C.

Specific heat of aluminium (at 0 to 100°C) is 0.226. It means that if it takes certain number of kcal. to heat a certain number of kg of water by a certain number of °C ; it will take only 0.226 times as many kcal to heat the same number of kg of aluminium by the same number of °C.

The amount of heat required to raise the temperature of equal masses of different substances to the same temperature level are different because their specific heats are different.

Specific heat increases with increase in the temperature of the substances. 'Mean specific heat' is the average value of the specific heat between two temperature levels.

For gases, there are two values of specific heats namely specific heat at constant pressure (C_p) and specific heat at constant volume (C_v). There is a difference between these two values ($C_p > C_v$) due to the heat equivalent of the work of expansion caused by an increase of volume resulting from a temperature rise.

Heat content. It is the heat contained at specified temperature above some fixed temperature (datum). It is calculated by multiplying the weight of a substance by the mean specific heat times the temperature difference.

Boyle's law. The volume of a fixed quantity of gas at a constant temperature varies inversely as the absolute pressure

i.e.,

$$\rho \propto \frac{1}{V} \text{ or, } P_1V_1 = P_2V_2$$

Charle's law. The volume of a given mass of gas varies directly with absolute temperature.

i.e.,

$$V \propto T \text{ or, } V_1T_2 = V_2T_1$$

Kilo mole. It is that quantity whose weight in kilogram is the same number as the number of molecular weight.

$$\text{kilo mole} = \frac{\text{Weight of substance in kg}}{\text{Molecular weight of the substance}}$$

Dalton's law of partial pressure. The total pressure of any gaseous mixture is equal to the sum of the pressure of each component (called partial pressure). Partial pressure of each component is proportional to its concentration in the gaseous mixture.

Saturated gas. In a mixture of water vapor and any other gas, each exerts a pressure proportional to its percentage by volume and since water has a definite vapor pressure at various temperatures, the concentration of water vapor in a gas is limited. When this limit of water vapor is reached, the gas is said to be saturated.

Dew point of gas. In saturated gas, any decrease in temperature (at a given pressure) will cause condensation (dew formation) of the water vapor present in it. The temperature at which the first drop of water condensate appears is the dew point of the gas. In fact, the condensation can occur by increasing the pressure also.

Humidity of gas. Amount of water vapor present in the gas is related to the humidity of the gas. Gas, which is completely saturated with water vapor has a humidity of 100 per cent ; if only 50% saturated, it has a humidity of 50%.

Ignition temperature. It is the minimum temperature at which the fuel ignites (catches fire) in an oxidising atmosphere without any External source of fire.

Flash point. It is the minimum temperature at which the liquid fuel gives enough vapor which produces a momentary flash when exposed to a flame.

Fire point. It is the minimum temperature at which the liquid fuel vapor produces a continuous flame when exposed to a flame instead of a momentary flame. Fire point is higher than the flash point.

Cracking. Decomposition of larger molecules of fuels into smaller ones by application of heat is called cracking.

Orsat analysis. Determination of percentage of carbon dioxide, carbon monoxide, oxygen and nitrogen in flue gas by Orsat's apparatus is called Orsat analysis.

Theoretical flame temperature. It is the temperature attained by the products of combustion of fuel when there is no loss of heat to the surrounding. Higher the flame temperature of the fuel, better it is as it provides higher thermal efficiency.

Octane number. It is the anti-knock rating of gasoline (Petrol). It is equal to the volume percent of iso-octane in a mixture of iso-octane and n-heptane which gives the same knocking as the gasoline under test when burnt in a standard test engine. Higher the octane number, better is the petrol.

Cetane number. It denotes the ignition quality and anti-knock rating of diesel fuels. It is the volume percentage of cetane in a mixture of cetane and alpha methyl naphthalene which gives the same ignition delay as the diesel under test when burnt in a standard diesel engine under standard conditions. Higher the cetane number of diesel, better it is.

Pour point. It is the lowest temperature at which an oil (say fuel oil) will pour or flow when cooled under specified conditions. If the oil is further cooled below pour point it will not flow and eventually freezes. This is important if the liquid fuel is to be handled during cold weather and at high altitude.

Units and Their Conversions

Length

[In-inch, ft-feet, m-metre, mm = millimeter, cm = centimeter, Å = Angstrom, km- kilometer, □m-micro-metre]

1 Å = 10^{-4} micron = 10^{-8} cm = 10^{-10} m, 1 μ m (micrometre) = 10^{-6} m, 1 mm = 1000 microns

1 micron = 10^{-4} cm ; 200 mesh size = 74 μ m (micro-metre) = 0.074 mm = 74 microns

1 inch = 2.54 cms = 25.4 mm = 0.08333 ft = 0.0254 m, 1 □m = 1 micron = 10^{-6} m

1 ft =12 inches = 30.48 cms = 0.3048 m = 0.3333 yard, 1 yard = 0.9144 m

1 m = 3.28 ft = 39.37 inches = 1.0936 yards = 100 cm = 1000 m = 10^{-3} km

1 mile = 1.609 km (i.e, 5 miles = 8 km) = 5280 ft = 1760 yards = 1609 m

1 Nautical mile = 1852 m = 1.852 km = 1.15 mile ; 1 mm = 1000 microns

Area

1 barn = 10^{-28} m^2, 1 mm^2 = 10^{-6} m^2 = 0.00155 in^2 = 10.76 × 10^{-6} ft^2

1 in^2 = 6.45 cm^2, 1 ft^2 = 0.0929 m^2 = 929.03 cm^2 = 144 in^2 = 0.093 m^2

1 hectare = 10000 m^2 = 2.47 acres, 1 acre = 100 m^2, 1 cm^2 = 10^{-4} m^2

1 acre = 4840 $yard^2$ = 0.4047 hectare = 4047 m^2 = 43560 ft^2 = 4840 $yard^2$

1 m^2 = 10.76 ft^2 = 1550 in^2 = 1.196 $yard^2$ = 0.000247 hectare = 10^6 mm^2

1 km^2 = 10^{12} mm^2 = 10.76 × 10^6 ft^2 = 10^6 m^2 = 1.196 × 10^6 yard^2 = 247.1 acre = 0.3861 mile^2

1 mile^2 = 27.88 × 10^6 ft^2 = 3.097 × 10^6 yard^2 = 2.59 × 10^6 m^2 = 640 acre = 2.59 km^2

Volume

[ml-millilitre, KL-kilolitre, cc-cubic centimetre (cm^3), dm-decimeter]

1 mm^3 = 10^{-9} m^3 = 16.39 cm^2, 1 ft^3 = 28.32 litres = 0.02832 m^3

1 litre = 1000 ml = 1000 cc = 10^{-3} m^3 = 0.22 U.K. gallon = 0.264 U.S. gallon = 1 **dm^3**

1 imperial gallon = 1 U.K. gallon = 4.546 litres (dm^3) = 1.201 U.S. gallon = 4.546 × 10^{-3} m^3

1 U.S. gallon = 3.785 litres = 0.833 U.K. gallon

1 U.S. barrel (bbl) = 42 U.S. gallon = 34.973 U.K. gallon = 0.159 m^3 = 0.159 KL = 5.615 ft^3

1 ml = 10^{-6} m^3, 1 pint = 0.5682 litre = 1/8 U.K. gallon

1 m^3 = 1000 litre = 1 KL = 35.3148 ft^3 = 219.97 U.K. gallon = 264.17 U.S. gallon

1.308 yard^3 = 10^3 dm^3 = 61024 inch^3 = 10^9 mm ^{3} = 6.29 U.S. barrel

1 ft^3 = 0.028317 m^3 = 6.023 U.K. gallon = 7.481 U.S. gallon = 1728 in^3 = 28.32 dm^3

Specific Volume

1 litre/kg = 1 c.c/gm = 10^{-3} m^3/kg, 1 litre/gm = 1 m^3/kg

1 ft^3/kg = 2.8317 × 10^{-2} m^3/kg, 1 ft^3/lb = 6.2428 × 10^{-2} m^3/kg

Volumetric Flow Rate

1 ft^3/hr = 7865 mm^3/sec = 0.0166 ft^3/min = 7.865 × 10^{-6} m^3/sec, 1 ft^3/s = 1 cusec = 2.8317 × 10^{-2} m^3/s

1 ft^3/min = 60 ft^3/hr = 471.9 × lO^{-6} m^3/sec = 1 cumin = 4.7195 ×lO^{-4} m^3/s

1 m^3/sec = 127.1 × 10^3 ft^3/hr = 2119 ft^3/min = 10^9 mm^3/sec

Velocity

1 in/sec = 60 in/min = 5 ft /min = 0.08333 ft/sec = 0.0254 m/sec = 25.4 × $10^{-6\text{km}}$/sec

1 ft/sec = 720 in/min = 60 ft/min = 0.3048 m/sec = 304.8 × $10^{-6\ \text{km}}$/sec

1 m/sec = 2368 in/min = 196.9 ft/min = 39.37 in/sec = 3.281 ft/sec = $10^{-3\ \text{km}}$/sec

1 km/sec = 2368 × 10^3 in/min = 196.9 × 10^3ft/min = 39.37 × 10^3 in/sec = 3281 ft/sec = 10^3 m/sec

1 km/hr = 0.6214 mile/hr, 1 mile/hr = 1.609 km/hr

Frequency

1 cycle/second (cps) = 10000 Hertz (Hz)

Mass

[lb-pound, oz-ounce]

1 grain = 0.0648 gm ; 1 mg = 10-6 kg ; 1 gm = 15.432 grains = 0.0353 (ounce) oz = 0.002205 lb ;

1 oz = 28.35 gm = 984.2 × 10~9 ton (U.K/long)

1 lb = 453.6 gm = 7000 grains = 0.454 kg = 16 oz = 446.4 × 10~6 ton (U.K/long)

1 long ton = 1 U.K. ton = 2240 lb = 1.12 short ton = 1.12 U.S. ton = 1.016 metric ton = 1.016×10^6gm = 1016 kg,

1 U.S. ton (short) = 2000 lb = 0.8929 U.K. ton (long) = 0.9072 tonne = 907.2 kg

1 tonne = 1 metric ton = 1000 kg = 2204.6 lb = 0.984 U.K. ton (long) = 1.1023 U.S. ton (short)

1 kg = 1000 gm = 2.205 lb = 984.2×10^{-6} ton (U.K / long)

1 slug = 14.59 kg, 1 oz = 28.35 gm, 1 tonne = 106 gm.

Mass Flow Rate

1 Ib/hr = 1.26 × × 10^{-4} kg/hr = 2.7778×10^{-4} kg/s ; 1 tonne/day = 1.1574×10^{-2} kg/s,

1 tonne/hr = 2.7778×10^{1} kg/s

1 ton (short)/day = 1.05×10^{-2} kg/s, 1 ton (short)/hr = 2.52×10^{-2} kg/s ; 1 ton (long)/day = 1.176×10^{-2} kg/s, 1 ton (long)/hr = 2.8224×10^{-1} kg/s

Density and Concentration

[ppm-parts per million]

1 ppm = 1 ml/m^3 = 1 mg/kg; 1% = 10000 ppm (for water, density = 1 gm/cc)

1 gm/ml = 1 gm/cc = 62.4 lb/ft^3 = 10.022 lb/gallon = 10^{-3}gm/m^3

1 lb/ft^3 = 16.02 kg/m^3 = 0.4536 kg/ft^3 = 578.7×10^{-6} lb/in^3 ;

1 kg/m^3 = 0.625 lb/ft^3 = 9.8 N/m^3 = 0.02832 kg/ft^3, 1 kg/litre = 10 kg/m^3

1 kg sec^2/cm^4 = 9.81 kg/m^3, 1 mg/ml = 1 kg/m^3, 1 mg/litre = 10^{-3} kg/m^3

1 m^3/kg = 16.019 lb/ft^3 ; 1 ft^3/lb = 0.062428 m^3/kg

1 kg/ft^3 = 35.31 kg/m^3 = 2.205 lb/ft^3 = 1.278×10^{-3} lb/ in^3 = 35.31×10^{-6} gm/mm^3

1 lb/in^3 = 27.68×10^3 kg/m^3 = 1728 lb/ft^3 = 783.8 kg/ft^3 = 27.68×10^{-3}gm/mm^3

1 gm/mm^3 = 10^6 kg/m^3 = 62.43×10^3lb/ft^3 = 28.32 × 10® kg/ft^3 = 36.05 lb/in^3

Absolute (Or Dynamic) Viscosity

[cp-centipoise, Pa-Pascal, s-second]

1 poise = 100 cp = 1 gm/cm sec = 0.01 kgf sec/m^2 = 1 dyne sec/cm^2 = 10^{-1} Pa s = 0.1 N sec m^2= 2.0885×10^{-3} lbf sec/ft^2

1 Pa.s = 10 poise = 0.02089 $lbfs/ft^2$

1 kgf sec/cm^2 = 9.81 N/m^2 ; 1 kgf sec/m^2 = 98 poise = 0.205 Msec $/ft^2$

1 lb_f sec/ft^2 = 478.8 poise = 4.88 kgf sec/m^2 = 47.88 Pa.s

1 kg/m.hr = 2.7778×10^{-4} $N.S/m^2$,1 lb/ft.hr = 4.1338×10^{-4} $N.S/m^2$

1 $mN.S/m^2$ = 1 centipoise = 10^{-3} $N.S/m^2$, 1 kg/M.S = 1 $N.S./m^2$

1 lb/ft.S = 1.4882 $N.S/m^2$

Kinematic Viscosity and Thermal Diffusivity

[cst—centistoke, rpm—rotation per minute]

1 stoke = 100 cst = 1 cm^2/sec = 0.36 m^2/hr = 3.875 ft^2/hr = 10^{-4} m^2/sec

1 m^2/s = 10^4 stokes = 10.76 ft^2/s

1 m^2/hr = 2.78 stokes = 10.76 ft^2/hr = 2.78×10^{-4} m^2/s

1 ft^2/hr = 0.26 stokes = 0.09 m^2/hr = 2.5806×10^{-5} m^2/s

1 ft^2/s = 0.09290 m^2/s

1 cst = 10^{-6} m^2/sec = 1 mm^2/s ; 1 rpm = 1.047×10^{-1} radian/sec

Force

[pdl—poundal, kN—kilo Newton]

1 pdl = 0.1383 N = 0.03108 lbf = 0.01410 kgf

1 N = 7.233 pdl = 0.2248 lbf = 0.1020 kgf

1 kN = 7233 pdl = 10^3 N = 224.8 lb = 102 kgf

1 lbf = 31.17 pdl = 4.448 N = 0.4536 kgf

1 kgf = 70.90 pdl = 9.803 N = 2.205 lbf

1 dyne = 10^{-5} N

Pressure

[mmwg—millimeter water gauge, inwg—inch water gauge, mmHg—millimeter mercury gauge, psi—lbf/in^2, (Pounds per square inch)

mB—mega bar, atm—atmospheric pressure, mmwc—millimeter water column, N—Newton, M—mega, Pa—Pascal, kPa—kilo pascal]

1 atm = 760 mmHg = 29.92 in Hg = 14.7 psi = 1.013 bar = 1.013 kgf/cm^2
= 34 ft water = 10033 mmwc = 10033 kgf/m^2 = 1.013×10^5 N/m^2
= 1013.2 mB = 2116 lb/ft^2 = 1.013×10^5 Pa

1 in wg = 1.87 mm Hg = 0.07351 in wg = 0.036 psi = 1.87 torr = 1.87 mm Hg = 249.1 N/m^2
= 249.1 Pa = 2.49 mB = 25.4 kg/m^2 = 25.4 mm wg = 5.202 lbf ft^2

1 Pa = 4.015×10^{-3} in water = 295.3×10^{-6} in Hg = 145×10^{-6} psi = 64.75×10^{-9} $tons/in^2$

1 kPa = 20.89 lb/ft^2 = 0.145 psi = 10^{-2} bar = 64.75×10^{-6} ton/in^2 = 1 kN/m^2 = 102 mmwc

1 in Hg = 13.6 in wg = 3386 mB = 25.4 torr = 25.4 mm Hg

= 0.0334 atm = 0.4912 psi = 0.03386 bar = 70.73 lb/ft^2 = 3386 Pa = 3386 N/m^2

1 ft water = 304.8 kgf/m^2 = 2986.1 N/m^2 ; 1 mB = 100 N/m^2

1 mm Hg = 1 torr = 133.3 Pa = 133.3 N/m^2 ; 1 μ bar = 10^{-1} N/m^2 , 1 millibar = 100 Pa

1 Pascal = 1 N/m^2, 1 bar = 1 kgf/cm^2 = 10^5 N/m^2 = 10^5 Pa = 10^6 $dynes/cm^2$

= 401.5 in H_2O = 14.5 psi = 29.53 in Hg

1 psi = 0.07031 kgf/cm^2 = 68.95 mB = 6.895×10^3 N/m^2

= 6895 P = 0.06895 bar = 2.036 in Hg = 27.68 in water = 144 lb/ft^2

1 kgf/cm^2 = 14.7 psi = 10000 mm wc = 0.1 MN/m^2 = 9.81 N/cm^2 = 98 kPa

1 kgf = 9.8 N, 1 lbf = 4.448 N, 1 MN/m^2 = 10 $kgflcm^2$, 1 kgf/m^2 = 1 mmwc = 9.8067 N/m^2

1 lbf/ft^2 = 47.88 Pa = 47.88 N/m^2 = 0.1922 in water = 0.01414 in Hg = 6.944×10^{-3} psi

= 478.8×10^{-6} bar

1 ata = 1 kgf/cm^2 = 9.8067×10^3 N/m^2, 1 N/mm^2 = 10^6 N/m^2

1 $ton_f/inch^2$ = 1.5477×10^5 N/m^2, 1 $Poundal/ft^2$ = 1.4881 N/m^2

Heat, Work, Power and Energy

[BTU—British Thermal Unit, CHU—Centigrade Heat Unit, J—Joule, kJ—kilo Joule, Cal-Calorie, Kcal—kilo calorie, kWh—kilowatt hour, W—watt, hp—horse power, min— minute, TNT—Tri-nitro toluene, G—Giga = 10^9]

Heat

1 J = 0.7376 ft lbf = 0.2388 cal = 947.8 × 10”6 BTU = 277.7×10^{-9} kWh = 9.478×10^{-9} therm

= 1 N.m

1 BTU = 252 calorie = 0.252 kcal = 778 ftlbf = 0.5556 CHU = 25030 pdl

=1055 J = 2.931×10^{-4} kWh

1 Therm = 2503×10^6 pdl = 105.5×10^6 J = 77.82×10^6 ft lbf

= 25200 kcal = 10^5 BTU = 29.31 kWh

1 calorie = 0.003968 BTU = 4.186 J = 4.186 watt. sec. = 3.088 ft lbf = 1.163×10^{-6} kWh

1 kcal = 1000 calorie = 3.968 BTU = 0.00116 kWh = 3087 ftlb

= 4186 J = 4.186 kJ = 2.205 CHU = 427 kg m =0.001560 hp

1 ft.lbf = 0.3238 calorie = 0.1383 kg.m = 1.356 J = 1.356 N.m

1 kgf. m = 9.8067 J

1 hp.hr (metric) = 2.6477×10^6 J

1 hp.hr (British) = 2.6845 times 10^6 J

12.85×10^{-9} therm = 1.285×10^{-3} BTU = 3.776×10^{-7} kWh

1 ft lb/min = 0.0226 W, 1 ft lb/sec = 0.00136 kW

1 ton of TNT = 4.2×10^9 J = 4.2×10^6 kJ = 4.2×10^3 MJ = 4.2 GJ

Work and Heat Flow Rate :

[sec—second, MW—mega watt = 10^6 W, kW—kilowatt, = 10^3 watt, hr—hour, N-Newton]

1 BTU/s = 1055.1 J/S, 1 BTU/min = 0.02356 hp, 1 CHU = 1.8 BTU = 0.453 kcal

1 BTU/hr = 0.2931 W = 0.2162 ft lbf/s = 0.00393 hp =0.2931 J/S

1 Therm = 100,000 BTU = 25200 kcal = 29.31 kWh = 2503×10^6 pdl
= 105.5×10^6 J = 77.82×10^6 ft lbf; 1 m.Kgf/S = 9.8065 J/S

1 ton of refrigeration = 3.5169×10^3 J/S

1 J = 1 W . sec. = 10^7 ergs = 9.5×10^{-4} BTU = 0.2388 calorie = 1 N.m

1 ft lbf/s = 4.628 BTU/hr = 1.356 W = 0.01819 hp = 1.356 J/S

1 kcal/hr = 1.163 J/s, 1 MW = 10^6 W = 1000 kW, 1 N = 10^5 dyne, 1 therm/hr = 2.9308×10^4 J/S

1 MW day = 8.64×10^{10} J

1 W/m^2 = 0.3170 BTU/ft^2 hr = 1 J/S.m^2 ; 1 BTU/ft^2 hr = 3.155 W/m^2 = 3.155 J/S.m^2

1 kcal/hr.m^2 = 1.163 J/S.m^2, 1 KW/m^2 = 1000 J/S.m^2, 1 kcal/hr.ft^2 = 12.518 J/S.m^2

Power and Energy

[W—watt, kW—kilowatt, MJ—mega joule]

1 hp = 0.746 kW = 746 W = 550 lbft/sec. = 632.34 kcal/hr = 2545 BTU/hr

1 W = 3.412 BTU/hr = 44.24 ft.lb/min = 0.2389 cal/sec = 0.7376 ft.lbf/s = 0.01341 hp

1 kW = 13.41 hp = 1000 W = 102 kg m/sec = 3412 BTU/hr = 737.6 ft.lbf/s ;

1 BTU/min = 17.58 W, 1 MW = 1000 kW, 1 Watt = 3.412 BTU/hr

1 kWh = 3.6×10^6 J = 2.655×10^6ft lb f = 859.8 kcal = 3412 BTU = 3.6 MJ = 0.03412 Therm

Joule's equivalent = 778.3 ft.lb

Specific Heat

1 cal/gm ,°C = 1 kJ/kg . °C = 1 BTU/lb . °F = 1 BTU/lb . °R = 4.1868 kJ/kg . °K

Enthalpy

1 cal/gm = 1 kcal/kg = 4.1868 kJ/kg, 1 BTU/lb = 2.326 kJ/kg

Calorific Value

[N—at N.T.P. *i.e.* 0°C and 760 mm Hg, S = at S.T.P. *i.e.* 15°C and 760 mm Hg]

1 kcal/kg = 1.8 BTU/lb = 0.0403 therm/ton = 4.187 kJ/kg

1 BTU/lb = 0.555 kcal/kg = 0.0224 therm/ton = 2.320 kJ/kg

1 BTU/Nft^3 = 8.9 kcal/Nm^3 = 0.038 J/N cm^3 = 0.038 MJ/Nm^3

1 kcal/Nm^3 = 0.1124 BTU/N ft^3, 1 therm/U.K. gallon. = 5543.3 kcal/litre

10000 kcal/litre = 1.804 therm/U.K. gallon

1 cal/m^3 = 0.11237 BTU/ft^3

1 J/kg = 4.29 × 10^{-3} BTU/lb ; 1 MJ/kg = 238 kcal/kg

1 J/cm^3 = 27 BTU/ft^3 ; 1 MJ/Nm^3 = 238 kcal/Nm^3

1 MJ/sm^3 (dry) = 26.81 BTU/sft^3 (dry); 1 MJ/sm^3 (saturated) = 26.79 BTU/sft^3 (saturated)

1 BTU/sft^3 (dry) = 0.0373 MJ/sm^3 (dry);

1 BTU/sft^3 (saturated) = 0.03732 MJ/sm^3 (saturated)

1 kg coal equivalent = 7000 kcal/kg = 2.9308 × 10^4 kJ

1 kg oil equivalent

Temperature

Conversion of degree Farenhite (°F) to degree centigrade/celcius (°C) and *vice versa* is done by the following formula:

$$\frac{C}{F} = \frac{F-32}{9}$$

From where, it follows that:

$$°C = (°F - 32) \times \frac{5}{9} = (°F - 32) \times 0.555$$

$$°F = °C\ \frac{5}{9} + 32 = (°C + 17.778) \times 1.8$$

°A (degree absolute) = °K (kelvin) = 273 + °C

°F absolute = °R (Rankine) = 460 + °F

Temperature difference : $1°K = 1°C = \frac{9}{5}°F = \frac{9}{5}°R$

Heat Transfer Rate

1 BTU/ft^2 .hr = 2.712 kcal/m^{2}hr, 1 kcal/hr = 1.163 W

1 kcal/m^2 . hr = 0.3687 BTU/ft^2 hr = 1.163 W/m^2

1 BTU/in^2 . hr = 390.6 kcal/m^2 . hr ; 1 kcal/m^3 hr = 1.163 W/m^3

Heat Transfer Co-Efficient

1 kcal/m^2 hr °C = 0.2048 BTU/ft^2 hr °F = 1.163 W/m^2 °K = 1.163 J/S.m^2 . °K

1 BTU/ft^2 hr °F = 4.88 kcal/m^2 hr °C = 20.44 kJ/m^2 hr °C = 5.678 J/S.m^2 . °K

= 5.678 Wm^2 °C = 135.6 × 10~^{6}cal/cm^2 sec °C

1 W/m^2 °C = 0.1761 BTU/ft^2 .hr. °F = 23.88 × 10^{-6} cal/cm^2 . sec. °C = 1 J/S.m^2 . °K

1 kcal/hr. ft^2 °C = 1.2518 × 10 J/S.m^2 . K

1 cal/S.cm^2 . °C = 4.1868 × 10^4 J/S.m^2 . °K

Radiation Co-Efficient:

1 kcal/m^2 .hr. °K = 1.163 W/m^2 . °K^4

1 cal/cm^2 . sec. °C = 41868 W/m^2 . °C = 7373 BTU/ft^2 .hr . °F

Thermal Conductivity

1 BTU/ft^2 .hr °F/in thickness = 0.124 kcal/m^2.hr. °C/m = 0.344 cal/cm sec °C

= 6.93 W/min. °K = 1.4423 × 10^{-1} J/S.m. °K

1 BTU in/ft^2 S °F = 519.2 W/m °C = 300 BTU/ft.hr.°F

1 BTU/ft^2 hr °F/ft = 1.488 kcal/m^{2}hr °C/m = 1.7308 J/S.m. °K

1 kcal/m hr °C = 0.672 BTU/fthr °F = 8.063 BTU/ft^{2}hr °F/inch

= 1.163 W/m °K =1.163 J/S.m. °K

1 cal/cm sec °C = 360 kcal/m hr °C = 242 BTU/ft hr °F

= 2902.9 BTU in/ft^{2}hr °F = 4.1868 × 10^2 J/S.m. °K

1 W/m °C = 0.5778 BTU/ft hr °F = 1.928 BTU in/ft^2 s °F

1 W/cm °C = 85.985 kcal/m.hr. °C = 57.779 BTU/ft.hr. °F

1 BTU/ft hr °F = 1.731 W/m °C = 3.333 × 10^{-3} BTU in/ft^{2}s. °F

1 kcal/hr.ft. °C = 3.8156 J/S.m. °K

SI (International System) Unit

SI units have been introduced to standardise the units throughout the world. India has also adopted this unit but still M.K.S. unit is in vogue here. The base unit in SI unit system is also metre, kilogram and second.

Some of the important *SI derived units* having special names and symbols are tabulated in table 2.1.

Table 2.1: Names and Symbols of SI derived units

Quantity	Name	Symbol	Expression in terms of other SI units	Expression in terms of base units
Force	Newton	N	—	kg m/sec^2
Pressure	Pascal	Pa	N/m^2	kg m/sec^2
Energy	Joule	J	N.m	m^2 kg/sec^2
Absolute viscosity	Pascal second	PaS	—	kg/m.sec
Heat capacity	Joule/kelvin	J/K	—	m^2 kg/sec^2.°K
Thermal conductivity	Watt per metre kelvin	W/m.°K	—	m kg/sec^3.°K
Molar heat capacity	Joule per mole kelvin	J/mol. °K	—	m^2 kg/sec^{2}mol °K
Moment of force	Newton-metre	N.m	—	m^2 kg/s^2
Heat flux density	Watt per square metre	W/m^2	—	kg/s^3
Surface tension	Newton per metre	N/m	—	kg/s^2
Specific heat capacity	Joule per kilogram kelvin	J/kg.°K	—	m^2/s^2.K
Energy density	Joule per cubic metre	J/m^3	—	kg/m .s ^{2}
Specific energy	Joule per kilogram	J/kg	—	m^2/s^2
Currents density	Ampere per square metre	A/m	—	A/m^2
Electric field strength	Volt per metre	V/m	—	m kg/s^3 .A
Magnetic field strength	Ampere per metre	A/m	—	A/m
Molar energy	Joule/mole	J/mol	—	m^2 kg/s^2 . mol
Molar heat capacity	Joule per mole kelvin	J/mol.K	—	m^2 kg/s^{2}mol.k
Angular velocity	radian per second	rad/s	—	rad/s
Angular acceleration	radian per second square	rad/s^2	—	rad/s^2

Abbreviations used for prefixes in SI units are tabulated in table 2.2.

Table 2.2: Abbreviations used for prefixes in SI units

Prefix	Symbol	Factor	Prefix	Symbol	Factor
deca	da	10^1	deci	d	10^{-1}
hecto	h	10^2	centi	c	10^{-2}
kilo	k	10^3	milli	m	10^{-3}
mega	M	10^6	micro	m	10^{-6}
giga	G	10^9	nano	n	10^{-9}
tera	T	10^{12}	pico	P	10^{-12}
femto	f	10^{15}	atto	a	10^{-18}

Table 2.3: Solid Fuel to Furnace Oil Conversion Ratio

Calorific value of solid fuel, Kcal / kg	Kg. of solid fuel required to substitute 1 litre furnace oil
3000	4.5
3500	3.9
4000	3.4
4500	3.1
5000	2.7
5500	2.5
6000	2.3
6500	2.1

Table 2.4: Approximate Conversion of Various Fuels to KWh

S.No.	Fuels	Conversion factor for converting into KWh
1.	Natural gas	m^3 × 10.6 KWh ft^3 × 0.3 KWh Therms × 29.3 KWh
2.	LPG (Propane)	m^3 × 25 KWh
3.	Coal	Kg × 8.05 KWh
4.	Coke	Kg × 10 KWh
5.	Gas oil	Litres × 12.5 KWh
6.	Light fuel oil	Litres × 12.9 KWh
7.	Medium fuel oil	Litres × 13.1 KWh
8.	Heavy fuel oil	Litres × 13.3 KWh

WOOD, CHARCOAL AND PEAT

WOOD

Wood is a domestic fuel used in tropical countries where forests are abundant and other fuels are not easily & cheaply available.

Main combustible components of wood are cellulose & lignin, which are compounds of carbon, hydrogen and oxygen. Other minor combustibles are resin and waxes. The major non- inflammable component of wood is water which is present upto 25-50% in freshly cut wood. Normally, it is used after air-drying to 10-15% water.

Though, the ash content of wood is very low (< 1%) but because of its very high oxygen content (upto 45%), its calorific value is very low (4000-5000 kcal/kg).

Composition and properties of a typical air dried hardwood :

Proximate Analysis :

Cellulose	= 50%	Lignin	= 30%
Resin + wax	= 2%	Moisture	= 15%
Ash	= 0.5%	Water solubles	= 2.5%
Calorific value	= 4500 kcal/kg	Density	= 650 kg/m^3.

Ultimate analysis of typical air dried wood:

Carbon = 50%,	Hydrogen = 6%,	Oxygen = 44%

Burning characteristics of wood:

- It can be ignited easily.
- Smaller pieces of wood burn more readily compared to large pieces. In large pieces layers of semi-fused ash form on the surface thus preventing its free burning.
- It burns with a long non-smoky flame when burned in excess air.
- Wood fines e.g. sawdust burns quite easily and readily. Moreover it can be made into binderless briquettes at high pressure.

Uses of wood. It is used:

- as domestic fuel on large scale in India and for furniture making.
- to produce wood charcoal by its carbonisation or destructive distillation.
- to produce producer gas by its gasification.

WOOD CHARCOAL

Wood charcoal which is called simply charcoal is made by carbonisation (destructive distillation, i.e. heating in absence of air) of wood at 600°C.

Physico-chemical changes during wood carbonisation. Following four stages are involved in the carbonisation of wood :

- When the temperature reaches 100-120°C, the moisture of the wood is expelled first.
- At 275°C, initial decomposition of wood takes place resulting in the formation of little distillat gas containing acetic acid and water.
- Active distillation of wood takes place upto 350°C till the process is exothermic producing liquid products (like acetic acid, methyl alcohol, pyroligneous acid, tar etc.) and gaseous products containing carbon monoxide, carbon dioxide, nitrogen, hydrogen, hydrocarbons etc.
- From 350 to 600°C, slow evolution of residual volatile matters (*i.e.* gases) from the wood/charcoal left in 3rd stage.

Products in wood carbonisation. Charcoal is the solid product left after the carbonisation of wood. Hot gases are cooled to separate wood gas and liquid into two layers. Upper layer of the liquid is pyrolignious acid and the lower level is wood tar. Pyrolignious acid is an aqueous solution of acetone, methyl alcohol, acetone and wood spirit mainly. Wood tar can be fractionated to separate many chemicals. Besides it is used as a supplementary plant fuel. Normally matured dense wood gives dense charcoal on carbonisation. Yield of products in wood carbonisation is tabulated in Table 3.1.

Table 3.1: Typical product yields in wood carbonization

Product	Yield,% air dried
Charcoal	30
Pyrolignious Acid	38
Acetic Acid	8
Wood Spiri	15
Water	15
(Acetic Acid + Wood Spiri + Water)	=38
Wood Tar	10
Wood gas	22

Scheme of wood carbonisation. Carbonisation of wood is done in open pits (primitive method, now obsolete), kilns or metal retorts. Pits and kiln are located in forests and retorts in factories.

In open pit carbonisation wood is burned in large heaps with restricted air. Yield of charcoal is

20% which is of inferior quality. Besides, gas and liquid by-products are lost to the atmosphere as they can not be recovered. Even the charcoal that we get from domestic wood burning ovens also comes in this category.

In charcoal kiln also, charcoal (of better yield and quality) is the only product as gases and liquid by-products are not recovered. Kiln is parabolic in shape having typical dimensions, radius = 3 metres, height = 2.5 metres and capacity = 30 m^3 of stocked wood.

Wood is stacked on the ground with one vertical central passage acting as the chimney and a horizontal passage at the bottom for introducing fire to the centre. The kiln is covered with thick layers of grass, leaves etc. and then plastered with a mixture of earth soil and charcoal dust. Initial firing is done with grass and twigs and then the wood is partly burned to supply heat for the process. Carbonisation time is 7-10 days. After that fire is extinguished with water and the kiln is allowed to cool for a week at least before the charcoal is taken out.

Low temperature carbonisation of wood :

Carbonisation in metal retorts is done at a low temperature of 350°C. Retorts may be of four types namely :

- Externally fired (heated) batch retorts
- Externally fired continuous vertical retorts
- Internally heated batch vertical retorts
- Internally heated continuous vertical retorts.

Most recent design is internally heated retorts which use forced recirculation of heated inert gases evolved during carbonisation. Besides, it employs efficient mechanical handling of wood and charcoal and has high thermal efficiency.

Characteristics of products of low temperature carbonisation of wood are :

- Charcoal yield is high (35%)
- Ash content of charcoal is low (< 2.5%)
- Volatile matter in charcoal is high (upto 15%), hence it can be easily ignited and burns at low rates.
- Heating value of charcoal is high (7500 kcal/kg)
- Charcoal produced has high vapor adsorption capacity.

High temperature carbonisation of wood. It is mainly carried out for the production of town gas and chemicals besides for charcoal. Carbonisation temperature is 1000-1200°C. Besides, town gas (which can be used for heating of domestic ovens) the valuable liquid chemicals like creosote (used as a wood preservative), turpentine, light & heavy oils are produced.

Product characteristics are :

- Charcoal yield = 25%
- Heating value of charcoal = 8000 kcal/kg
- Gas yield = 850 Nm^3 gas/ton dry wood
- Gross heating value of gas = 3000 kcal/Nm^3
- Composition of wood gas :

$CO_2 = 13\%$,	CmHn (unsaturated hydrocarbons) = 2%
$CO = 24\%$,	$CH_4 = 15\%$ $N_2 = 1\%$ $H_2 = 45\%$.

Uses of charcoal:

- Because of its large specific surface area (150-450 m^2/gm) and light & porous nature, it is used for removal of obnoxious and colouring materials from solutions, gases, vapors, petroleum products etc. by adsorption on its surface.
- It can be used as a feed stock for gasification to make producer gas which is used for domestic and industrial heating. During second world war, this producer gas was used as a fuel in road vehicles in many countries including India.
- It is used as a clean and smooth burning fuel in domestic heating ovens but it is a costly fuel.
- Previously, it was being used for metallurgical furnaces but now it has been replaced by coke. There was one blast furnace at earstwhile Mysore Iron and Steel Ltd. Bhadrawati (Now in SAIL) which was using charcoal instead of coke. The charcoal consumption was 1 ton charcoal/ton pig iron. Capacity of the blast furnaces was 250,000 tons pig iron per year.
- It is used very widely as a fuel for *blacksmith's* and metal worker forge furnaces/ovens.
- It is a raw material for the manufacture of carbon disulphide.
- It is mainly used as a domestic fuel in India.

Composition of charcoal. A typical composition of charcoal is given below :

C	= 80%	O_2 & N_2	=15%
H_2	= 2%	Ash	= 3%

Merits of charcoal as a fuel:

— It has a very high specific surface area compared to coal (20-200 m^2/ gm coal).

— Its ash content is very low (< 3%).

— Its calorific value is high (7500-8000 kcal/kg).

Demerits of charcoal as a fuel:

Its mechanical strength is very poor, hence it gets crushed to powder in operation which is easily swept away in a current of gases and also it may prevent the proper flow of gases in the furnace.

Wood gasification. Especially, the hard wood is gasified in presence of steam and air to make producer gas. This producer gas can be used as a fuel or for synthesis of ammonia (by removal of hydrogen from it).

Gas yield = 15000 Nm^3/ton of crude wood. Heating value of wood gas = 1500 kcal/Nm^3

Typical composition of wood gas :

CO_2	= 9%	CO	= 25%	N_2	= 48%.
H_2	= 15%	CH_4	= 3%		

PEAT

Peat is the first stage in the formation of coal from wood (cellulose). It is not strictly a coal or else it can be termed as the most immature coal. It has been formed by gradual decaying of vegetable matter in moist places.

It appears light brown and fibrous at the surface of peat deposits but with increase in depth the colour becomes darker and the vegetable structure disappears.

Composition and properties of peat. It varies from place to place depending upon the nature of original vegetable matter, depth in the deposit and age. Peats contain about 90% moisture and only 10% solids when freshly mined. Hence it can not be utilised unless air dried to reduce the moisture content to 30% or less.

Its heating value is very less due to high moisture content and is only slightly greater than wood.

Typical composition of freshly mined peat is :

Moisture	= 85%	Volatile matter	= 8%
Fixed carbon	= 4%	Ash	= 3%
Calorific value	= 650 kcal/kg.		

After this, peat is sun dried for final use. Its typical composition is (on air dried basis):

Typical proximate analysis :

Moisture	= 20%	Ash	= 5%
Fixed Carbon	= 25%	Volatile matter	= 50%%
Calorific value	= 5000 kcal/kg.		

Typical ultimate analysis (ash free-dry basis) :

Carbon	= 55%	Hydrogen	= 6%
Oxygen	= 35%	Nitrogen	= 3%
Sulphur	= 1%	Calorific value	= 5000 kcal/kg

Occurrence of peat. *Russia* has the *largest deposit* of peat (60% of the world reserve) in the world. Besides, it occurs in Germany, Finland, Poland, America, Sweden, Norway, Ireland etc. India has very small deposit of peat in Nilgiri Hills (Tamilnadu) and two sides of Hooghly river in and around Calcutta.

Peat is extracted by hand cutting and also by the mechanised excavation method. Air drying of peat may require 40-50 days.

Burning Characteristics of Peat

- It offers low furnace temperature and efficiency due to its low calorific value.
- Its bulk density is very low *i.e.* 300 kg/m^3 (compared to coal having 700 kg/m^3). Hence its transportation cost is high as it occupies large volume per unit weight. Besides, storage volume requirement is also more.
- Its friable nature causes appreciable losses on handling prior to its burning.

Tar produced can be further separated into paraffin wax, illuminating oil and creosote oil (similar to wood creosote). Water produced contains ammonia, calcium acetate and methyl alcohol. The gas may be used for heating the ovens.

A typical comparative analysis of peat charcoal and wood charcoal is given in table 3.2.

Table 3.2: Analysis of Wood Charcoal and Peat Charcoal

Composition (% by weight)	Wood charcoal	Peat charcoal
Carbon	85	84
Hydrogen	3	2
Oxygen + nitrogen	3.5	8
Ash	2.5	3
Moisture	6	3
Heating value, kcal/kg	7500	7000

Gasification of Peat. Peat is gasified in presence of steam and air to produce producer gas. Steam requirement is very low as compared to coal as peat itself contains enough moisture. The gas produced is very suitable for gas engines. Good quantity of ammonia can be recovered from the gas. A typical composition of gas is given below :

CO_2 = 13%, CO = 18%,	H_2 = 11%, CH_4 = 2.5%, N_2 = 55.5%.
Calorific value of gas	= 1000 kcal/Nm^3
Gas yield	= 2500 Nm^3/ton peat.

Utilisation of Peat. It is mainly used as a domestic fuel in Great Britain. Peat from Nilgiri Hills (India) is brought to Ootacamund for disposal as fuel.

Besides it is used for steam and power generation (particularly in U.S.S.R.). It is gasified to make producer gas and carbonised to get peat charcoal and by-products. It is also used as a fertiliser or for making fertilisers.

Other Non-Fossil Domestic Solid Fuels

These fuels are normally used in rural areas by low income group people as domestic fuels. These include wood wastes, bagasse, agricultural vegetable wastes, cattle dungs and dry leaves/ twings collected from forests.

Typical properties of some of these non-fossil domestic fuels (on air dried basis) is tabulated in table 3.3.

Table 3.3: Composition & Properties of Some Non-fossil Domestic Fuels

S.N.	Non-fossil domestic fuels	%Ash	% Moisture	Calorific value, kcal / kg
1.	Fresh bagasse	1.0	45	1500
2.	Cow dung cake	33	4.5	3150
3.	Cow dung	21.5	8.5	3300
4.	Mango leaves	18	10	3400
5.	Rice husk	19.5	11.5	3450

6.	Stored/dried bagasse	–	15	3800
7.	Date palm leaves	4.5	10.5	4050
8.	Ground nut shell	1.5	12	4200
9.	Palmyra leaves	5.5	9.5	4200
10.	Coconut shell	0.5	9.5	4300
11.	Casuarina leaves	1.5	8.5	4400
12.	Sawdust from sal wood	1.5	10	4450
13.	Sal wood	0.5	10.5	4500

Typical ultimate analysis of some agro residue is tabulated below in table 3.4

Table 3.4: Ultimate Analysis and C.V. of Some Agro-residues

S.No.	Ultimate analysis, %	Saw dust	Coconut shell	Paddy husk	Deoiled bran
1.	Moisture	37.98	13.95	10.79	7.11
2.	Mineral matter	1.63	3.52	16.73	19.77
3.	Carbon	48.55	44.95	33.95	36.59
4.	Oxygen	41.93	31.94	32.52	31.02
5.	Hydrogen	6.99	4.99	5.01	4.15
6.	Nitrogen	0.80	0.56	0.91	0.82
7.	Sulphur	0.10	0.08	0.09	0.54
8.	Calorific value (C.V.) (Kcal/Kg)	4801	4565	3568	3151

ORIGIN, COMPOSITION, CHARACTERISTICS AND SIGNIFICANCE OF CONSTITUENTS OF COAL

ORIGIN OF COAL

Coal is a complex mixture of plant substances altered in varying degree by physical and chemical processes. These processes which changed plant substances into coal has taken million of years and has been accomplished by bacteria, heat & pressure inside the earth's crust.

Two theories namely 'in situ' theory and 'drift' theory have been suggested by the geologists regarding the mechanism of formation of coal from plant substances (*e.g.* wood etc.).

In situ theory. According to this theory, coal seam occupies the same site (place) where the original plants grew and where their remains accumulated several million years ago to produce coal under the action of heat, pressure & bacteria.

Drift theory. According to this theory, the plants, trees etc. were uprooted and drifted (transported) by rivers to lakes and estuaries to get deposited there to form coal during the course of time after they got buried underground. Indian coals are formed according to drift theory in general. If sea water is the carrying agent or if the deposit is covered by marine inorganic sediments, then coal formed are rich in hydrogen and possess coking properties.

The stages in the formation of coal from vegetable/plant (wood) matters are as follows : Plant debris (wood) → peat → lignite → brown coal → sub-bituminous coal → bituminous coal → semi anthracite coal anthracite coal → graphite.

This process of coal formation may be complete or may be stopped at any stage giving rise to coal of varying maturity which is implied by the term 'rank' of the coal.

- In this series from peat to anthracite ; each member is more mature or of higher 'rank' than the one which precedes in the series. Thus anthracite coal is of higher rank than bituminous coal. Bituminous coal is of higher rank (*i.e.,* more mature) than lignite. With the progress of coal forming reaction, moisture and oxygen content reduces and percentage of carbon increases. Also the calorific value increases from *4500 kcal / kg for lignite to 8500 kcal/kg for bituminous coal.* There is a slight fall in the calorific value of anthracite coal because of removal of hydrogen.

Points in favour of 'in situ' theory :

- In the existing peat deposits ; the decayed vegetable matter had accumulated at the place of origin.
- Coal seams are found on the bed of fire clay. The constituents like lime, alkali, iron oxide, etc. which act as flux in such clays were found extracted by plants growing on it.
- Underclays of coal have been found to contain large quantity of fossil roots.
- Composition of a particular coal seam is generally constant over a wide area. Had the original decaying vegetable matter drifted from its original place, there would have been a great variation in the ash content of coal at various places in a seam.
- Aquatic fossils (marine remains) are not found in coal.

Points in favour of drift theory :

- Generally, there is no seat of earth below coal and the coal has been found lying on sandstone, shale etc.
- In a coal seam, percentage of inclined trunks of fossils is much more than the vertical position. Had the coal seam been formed at the same place where the vegetable materials decayed and embedded, the fossil trunks would have been vertical.
- To form 1 metre thick coal seam, 10 metre thick seam of peat is required. So, the formation of 10 metres thick coal seam would have resulted from 100 metres thick peat seam which occurs nowhere.
- Coal seams appear like sedimentary rocks. Seams of coal are made up of different layers separated by layers of shale, clay or sandstone which vary in thickness from a mere film to several metres.

As the plants accumulated under water, some silt also settled along with them. Thus at a later stage some mineral matter got mixed with the coal as it formed. The initial transformation of wood is due to bacterial action causing degradation of organic matters *(e.g.* cellulose, lignin etc. present in wood) and removal of oxygen. The bacterial action produced acidity which if accumulated prevented further action and coal formation could not proceed. In some cases, the soil was alkaline which neutralised the acidity and bacterial action continued to take place bringing about further degradation of vegetable matter. This explains why coal is found associated with certain types of rocks. Bacterial action takes place in the initial stage and breaks up organic matters into simpler molecules. The degraded organic matters, as time passed, gave rise to peat. Beyond this stage, heat, pressure and time became the chief factors for further conversion of peat into different coals.

Most coals found in the world are of the *'in situ' origin* ; *very little being of drift origin.*

Many coalfields consist of coal seams and sedimentary rocks in a sequence that is repeated several times and corresponds to a cyclic process of accumulation of plant debris and cover by *sedimentary rocks.* Such formations of stratified coal seams and sedimentary rocks are known as *'coal measures'.*

There are two important stages in the formation of coal from vegetable matter, *viz.* peat stage or biochemical stage, and metamorphism or dynamochemical stage. In the biochemical stage, the plant material underwent decay under moist conditions by bacterial attack. This could have been caused either by rapid and complete burial of the peat deposits under a cover of inorganic sediment or by complete flooding by stagnant water, followed later by a covering of sedimentary earth.

Changes in the metamorphism state are due to the pressure of overburden, *tectonic pressure* caused by severe earth movements, regional temperature increasing by 3-5° for every 100 m increase in depth and contact with molten rocks and lava. These were operative for a very long time, spread over geological ages. Thus resulting in coals of different maturity/rank.

The rank of coal rises with the increase in the age of the geological system of the coal deposits as shown in the Table 4.1. Bituminous coals are found in *Paleozoic & mesozoic deposits* and lignite in *tertiary deposits.*

Table 4.1: Normal Age of Coal Deposits

Geological System Era	Mean Age (x10^6 years)	Rank of Coal Formed
Paleozoic	300-245	Anthracite üüüüüüü Bituminous
Mesozoic	200-120	Bituminous Bituminous Bituminous and sub-bituminous
Tertiary	60-20	Bituminous and Lignite Lignite Lignite
Quarternary	< 1	Peat

The effect of depth of occurrence on the maturity of coal is obvious. Both temperature and pressure rise with the increase in depth and the rate of a chemical change doubles for a rise of about 10°C. Hence the coals in the lower seams of coal measures are generally more mature than those of higher seams. This variation of rank with depth is known as *Hilt's law.* Temperature gradient in earths crust in India is about *increase of 1°C / 30 m depth.*

The precise nature of the changes that took place in the gradual conversion of plant material into coal is not known but it may be inferred that the bacterial attack in the first stage of coal formation brought about chemical reactions under oxidising atmosphere, in which all the ingredients of plants took part and produced the humic acids and other components of peat. Later peat was buried under mineral rocks and pressed into more and more compact material. Chemical reaction (such as dehydration, decarboxylation, dehydrogenation etc.) took place with the result that H_2O, CO_2 CO CH_4 and H_2S were removed off and the humic acids of peat were converted more and more into neutral substances. The result was lignite. In contrast to the biochemical stage, the lignite formation passed through a reducing atmosphere. The aging of lignite into bituminous and anthracite coals was due to further reduction and condensation reactions.

Generally, with the *increase of vertical depth, the carbon content of the coal increases and the volatile matter decreases.* The process of coal formation (called *coalification or metamorphism)* may be complete or may be stopped at any stage, thus giving rise to coals of varying maturity represented by the term *"rank of coal".*

Different *stages in the formation of coal* from plant/vegetable debris (called peat to anthracite series) is as follows :

Plant / vegetable debris → peat → lignite → sub-bituminous coal bituminous coal → semi-bituminous coal → semi-anthracite coal → anthracite coal → graphite.

Each member is more mature *(i.e.* of higher rank) than the one which precedes it in the series. Thus, anthracite coal is of higher rank than bituminous coal.

Generally, as the *coalification (i.e.* maturity of coal) progresses, the moisture and oxygen percent in the coal decreases whereas the carbon content and the heating value increases. Heating value of anthracite coal may be slightly less than that of bituminous coal due to the removal of hydrogen.

Stages of coalification. Table 4.1:(a) below depicts the coalification process which has been divided into several stages and lists them along with their typical characteristics. Peat is the stage next to wood and shows very high percentage of oxygen and hydrogen. This is due to entrapped moisture. After that the major chemical change (from lignite to anthracite) is gradual decrease of hydrogen and oxygen and increase in the percentage of carbon as shown below in table 4.2.

Table 4.1: (a) Approximate Characteristics of Various Stages of Coalification

S. No.	Stages of coalification	Proximate Analysis, %			Ultimate Analysis, %			Gross calorific value, Kcal/kg
		Volatile matter	Fixed carbon	Moisture	Hydrogen	Oxygen	Total carbon	
1.	Wood	-	-	-	6.3	43.1	49.5	3200
2.	Peat	26	11	56	8.3	63.0	21.0	2000
3.	Lignite	35	23	35	6.6	42.1	42.4	4000
4.	Sub-bituminous coal	28	45	24	6.1	33.8	55.3	5200
5.	Bituminous coal	27	63	3	5.2	7.5	78.0	7700
6.	Semi-bituminous coal	15	75	2	4.1	4.2	80.0	7800
7.	Semi-antliracite coal	9	77	3	3.6	4.9	78.4	7300
8.	Anthracite coal	1	88	3	1.9	4.4	84.4	7400

The deposits of all the above modification of wood are found throughout the world. The factors that led to the difference in the stages of coalification of different deposits are as follows:

- Not all the plant constituents decay to the same extent.
- Differences in heat and pressure at various deposits also contributes towards the differences. At a given point in a coal field, the carbon content increases as we go down vertically. Some lignite deposits are located just at the surface.

- If the acid products of bacterial action accumulates then the decay does not proceed fast and may hinder coalification. A more alkaline clay removes the acid products more and thereby helps coalification.
- The time is also a factor. Some of the lignite deposits are of more recent origin.

Coals are extremely complex substances and are the mixtures of various complex chemical compounds produced from the constituents of initial vegetable matter. Some extraneous mineral matters also get entrapped mechanically.

Composition of Coals

Composition of coal is expressed in terms of its proximate analysis and ultimate analysis.

Proximate analysis of coal means determining the percentage of moisture, volatile matter, fixed carbon and ash. Proximate analysis with sulphur content and heating value of coal helps in deciding the mode of utilisation of a particular coal.

Ultimate analysis of coal means determination of its total carbon, hydrogen, oxygen, nitrogen, sulphur and ash content. It gives analysis in terms of the elementary constituents and is helpful in combustion calculations for design of furnaces and its auxiliaries.

The changes in average composition from wood → peat → lignite → bituminous coals → anthracite (on dry ash free basis) is given in table 4.2.

Table 4.2: Typical Composition Change from Wood to Anthracite

Fuels	% Composition on dry ash free (daf) basis				Calorific value (daf), kcal/kg	*Moisture 60% R.H. and 40°C, %
	Carbon	Hydrogen	Nitrogen	Oxygen		
Wood	50	6	0.5	43.5	4990	25
Peat	57	5.7	2	35.3	5490	25
Lignite	67	5	1.5	26.5	6495	18
Sub-bituminous coal	77	5	1.8	16.2	7210	11
Bituminous coal	83	5	2	10	8595	4
Semi-anthracite	90	4.5	1.5	4	8690	1
Anthracite	93	3	0.7	3	8500	1.5

*Equilibrium moisture content under saturation condition are considerably higher for wood to sub-bituminous coal. (R.H - Relative Humidity)

A typical composition of different types of coals of U.K. & U.S.A. and some other solid fuels is given in the Table 4.3.

Table 4.3. Typical Composition of Different Types of British Coal

Fuels	Proximate Analysis, %				Ultimate Analysis, %						Calorific value, kcal/kg	
	Moisture	Volatile matter	Fixed carbon	Ash	C	H_2	O_2	N_2	S	Ash	As reported	Dry basis
Peat	57	26	11	6	23	10	59	1.5	0.5	6	2000	4600
Lignite	35	28	31	6	42	7	43	1	1	6	4000	6000
Bituminous coal:												
- Low volatile	3.5	16	72	8.5	79.5	4.5	4.5	1.5	1	9	7600	7900
- Mediumvolatile	3	24	62	11	77	5	5	1.5	0.5	11	7500	7700
- High volatile	8	36	49	7	68.5	5.5	16.5	1.5	1	7	6750	7400
Sub-bituminous	19	31	46	4	59	6	29.5	1	0.5	4	5600	7000
Semi-anthracite	3	8.5	79	9.5	80	3.5	4.5	1.5	0.5	10	7500	7700
Anthracite	2.5	3	87.5	7	86.5	2.5	3	0.5	0.5	7	7500	7700
Pine wood	—	—	—	—	52.5	7	40	—	—	0.5	—	5330
Oak wood, dry,	—	—	—	—	50	6	43.5	—	—	0.5	—	4600
Coke (by- productoven)	1	1	87	11	85	1	1	1	1	11	7050	7100
Coke breeze (0—15 mm)		4	66	20	67	2.5	11	1	0.5	18	5650	6450
Charcoal	12	2	83	3	85	2.5	10.5	—	—	2	6300	7150

Characteristics of Different Coals are described below:

PEAT

It is the first stage product in the formation of coal from wood under the action of temperature, pressure and bacteria. Freshly dug peat contains large amount of water (upto 90%), hence it is sun dried before using as a fuel. Its calorific value is (around 4500 kcal/kg) slightly higher than that of the wood and it is mainly used as a domestic fuel as well as for power generation. Near the surface of the deposit, peat is light brown in colour and highly fibrous in nature. With the increase in the depth, the colour becomes darker and finally black, when vegetable structure is not so obvious. A part of the water content of freshly dug peat is drained off while larger part is removed by drying in air for 40-50 days.

The composition and properties of peat vary widely from place to place, depending on the nature of the original plant material and the agencies and extent of decay. The lower layers of peat have usually higher ash than the upper layers.

Peat is extracted either by hand-cutting or by the mechanized excavation method. Its main use is as domestic fuel. It is easy to ignite and burns freely with long flame.

Average composition of Peat is given in table 4.4.

Table 4.4: Composition of Peat

Proximate Analysis (%)

% Air-dried basis				% d.m.m.f. basis
Moisture	Ash	Volatile matter	Fixed carbon	Volatile matter
15-25	3-10	50-55	25-30	65-70

Ultimate Analysis (% d.m.f.)

Carbon	Hydrogen	Oxygen	Nitrogen	Sulphur
55-60	6-6.5	30-35	1.5-2	0.6-1
Calorific value, (d.m.m.f.) = 4,500-5,000 kcal/kg				

Peat is largely used in steam boilers, power stations and gas producers. The low temperature carbonization of peat is also practised for getting peat coke and by-products. Peat coke is a valuable fuel for some metallurgical processes. Peat is also used as a fertilizer or for making fertilizer. However most of peat is used in heat generation.

Lignite

It is the second stage product in the formation of coal from wood. It is friable and occurs in thick seams (upto 30 metres thickness) near the earth's surface. Its moisture content is upto 60% and calorific value around 5000 kcal/kg (on 10% moisture basis).

On exposure to the atmosphere, the brown colour of lignite darkens and the moisture content reduces to an equilibrium value of 10 to 20%. On drying, lignite shrinks and breaks up in an irregular manner. Hence it cannot be moved far from the mine. It is likely to ignite spontaneously as it adsorbs oxygen readily and must not be stored in the open without care.

Its extensive deposits occur in the U.S.S.R., U.S.A., Germany, Australia, Canada and India. The major *deposit of true lignite* in India is in *Neyveli* in Tamil Nadu (2000 million tons). Nichahom (Kashmir), Kerala coasts, Palana and Khari (Rajasthan) and Umarsar (Gujarat) have smaller deposits. The lignite deposits in many areas are relatively near the earth's surface and are quite thick. The deposit of Neyveli occurs about *60 m below the ground level and is 15-20 m thick.*

Compositions and properties of lignite varies widely. The carbon content is 70-75% and the oxygen content is 21-26%. The volatile matter is often over 50% (d.m.m.f.) and in a large number of cases the ratios of volatile matter to fixed carbon are 1: 1. The high oxygen content is responsible for relatively low calorific value (6,500-7,000 kcal/kg) even on d.m.m.f. basis. On air-dried basis, the value may be as low as 3,800 kcal/kg due to high moisture content. The ash of lignites is generally low. However Nichahom lignites in Kashmir have very high ash (upto 50%).

Composition and Properties of Lignite found at Neyveli is given in table 4.5.

Table 4.5. Composition and Properties of Lignite of Neyveli (India)

Proximate Analysis:

% air dried				% d.m.m.f.	
Moisture	Ash	Volatile matter	Fixed carbon	Volatile matter	Fixed carbon
10-30	3.5-7.5	40-45	30-35	52-60	40-48

Ultimate Analysis (% d.m.m.f.):

Carbon	Hydrogen	Oxygen	Nitrogen	Sulphur
70-73	4.6–5.5	22–26	0.6-1.0	0.6-1.5

Calorific Value, kcal/kg:

Location of coal field	Air-dried	d.m.m.f.
Neyveli	4,000-5,500	6,500-6,600

Montan wax present in lignites is 1 to 10%. Neyveli lignite yields up to 5% montan wax. The quality of montan wax of Indian lignites is poor.

Raw lignite is an inferior fuel due to high moisture, low calorific value, small size and bad weathering properties. Lignite is of economic importance in those places where it is available and other fuels do not occur in abundance. In Neyveli, lignite is used in the generation of electricity in thermal power stations and carbonized briquettes are used as smokeless fuel. Kashmir lignite is also briquetted and used as fuel. Other uses of lignite are in gas production and metallurgical furnaces. The tar obtained by the low temperature carbonization of lignite is a valuable raw material for the production of synthetic petrol and other liquid fuels. Lignite is extensively used in the manufacture of producer gas. It is also gasified into synthesis gas for ammonia production.

Sub-Bituminous Coal

It is a black, homogeneous and smooth mass having high moisture and volatile matter content which breaks into smaller pieces on exposure to air. Its carbon content is around 70-80% and oxygen content is 10–20%. It is a non-coking coal having a calorific value about 7000 kcal/kg.

It is a variety of mature lignite resembling true coal in colour and appearance. It is black in colour with a dull, waxy lustre. It is denser and harder than lignite and has a lower moisture content (12–25%). Most sub-bituminous coals appear banded like bituminous coal. Like lignite, sub-bituminous coal disintegrates on exposure to atmosphere and is therefore difficult to transport.

Sub-bituminous coal occurs in the U.S.A., U.S.S.R, Germany, Canada, Australia, India etc. Some of the tertiary coals in Assam, Kashmir and Rajasthan are of the sub-bituminous type.

The sub-bituminous coal has 70–78% carbon, 4.5–5.5% hydrogen and about 20% oxygen (on d.m.m.f. basis). The air-dried moisture is 10–20%. The volatile matter is 40% (d.m.m.f.) and above. The calorific value is 6,800-7,600 kcal/kg d.m.m.f. It ignites easily and is used in raising steam and or manufacturing gaseous fuels also, if low in sulphur.

Bituminous Coal

It is the most common variety of coal known as *"Koela"* in Hindi. It is black & brittle which burns & ignites readily with yellow smoky flame. It has low moisture content (< 10%) and the carbon content varies from 75–90% whereas the volatile matter content is 20–45%. Depending upon the volatile matter content, it is termed as low volatile, medium volatile and high volatile coal. Its calorific value on mineral free basis goes upto 9000 kcal/kg. Most of the coking coals are essentially bituminous coal. It is used for power generation, coke making, gasification, domestic heating etc. Non-coking bituminous coals are generally used for purposes other than coke making which requires coking coal.

The name bituminous is perhaps due to the fact that it burns with a smoky yellow flame similar to that of bitumen and that the pitch obtained from coal tar is of a bituminous nature. The term 'coal' alone ordinarily refers to bituminous coal.

Bituminous coal is black and usually banded, the bands being parallel to the bedding plane. The coal breaks along vertical joints (cleats) into rectangular, columnar or cubical pieces. Sometimes the fracture is conchoidal. The lustre varies from bright to dull. Bituminous coal is denser and harder than lignite and sub-bituminous coal and does not disintegrate into slacks on exposure to the atmosphere.

Bituminous coals vary widely in composition and properties. The air-dried moisture varies from 14 to 0.5%. On the dry and mineral-matter-free basis, the volatile matter varies from about 45% down to about 20%, carbon from under 80% to about 92%, hydrogen from 6 to 4%, oxygen from 15 to 0.5% and the calorific value from 8,900 to 7,500 kcal/kg. *The figures given above are for the bituminous coals of India and may differ slightly in coals of other countries.* Unlike the coals of Europe and America, Indian coals have high percentage of ash (which range between *12 and over 40%, the usual range being 15–30%).* In the tertiary coals of Assam, the ash is often low (3 to 8%) and only in particular areas it is high (24 to 26%).

The caking and coking coals form sub-groups of bituminous coal. A demerit of the Indian coals is that 80% of the total resources is non-caking. Again, out of the 20% of the caking coals, only 4% is of the coking variety. The coals of the upper seam of Jharia coalfield and the Upper and Lower Karharbari seams of the Giridih coalfields are almost the sole deposits of truly coking coals in India. There are some coals *(e.g.* Dishergarh seam in Raniganj, Mahuda in Jharia and Makum in Assam), which are highly caking but do not produce strong coke.

Bituminous coal is used for combustion in domestic ovens, industrial furnaces and boilers, railway locomotives and thermal power stations. Two other important uses are carbonization and gasification, whereby coal is converted into solid fuels *(e.g.,* coke and semicoke) gaseous fuels *(e.g.,* producer gas, water gas and coal gas) and liquid fuels *(e.g.* coal tar fuels). It is also a source of a wide range of coal chemicals, fertilizers and synthetic liquid fuels.

Typical composition & properties of some Indian Bituminous coals are given in table 4.6.

Semi-Anthracite Coal

Its properties lie between that of bituminous and anthracite coal.

It is harder than the most mature bituminous coal, and ignites more easily than anthracite to give a short flame changing from yellow to blue. Some of the properties of semi-anthracite are :

Moisture, % air dried	= 12
Volatile Matter , % d.m.m.f	= 10-15

Calorific Value, (d.m.m.f.) = 8,500 to 8,800 Kcal/kg
Semi-anthracites do not occur India. It is non-coking.

Table 4.6: Typical Composition and Properties of some Indian Bituminous Coals

Coalfield	Basis-Air-dried, %			Basis : d.m.m.f., %				Calorific value, kcal/kg (d.m.m.f)	Caking properties
	Moisture	Ash	Volatile matter	Carbon	Hydrogen	Nitrogen	Oxygen		
Raniganj, Dishergarh	2.5-3.5	15-20	39-44	83-85	5.3-5.8	2.0-2.3	8.0-9.0	8100-8500	Caking
Jharia, Mahuda	1.5-2.2	20-25	36-40	85-87	5.4-5.8	1.7-2.2	3.9-7.5	8400-8600	Caking
Giridih, Karharbari	0.6-1.3	12-22	27-33	89-92	4.7-5.2	1.6-1.9	0.6-2.4	8700-8900	Caking
East Bokaro, Jarangdih	0.8-2.4	13-21	28-36	85-90	4.5-5.4	1.8-2.0	4.0-8.0	8300-8700	Caking
West Bokaro, Kuju	4.2-4.7	14-19	34-37	84-86	4.9-5.1	1.9-2.2	6.0-8.0	8200-8400	Caking
Singrauli	7-9	16-24	37-45	78-81	4.4-5.3	1.5-1.8	12-14	7600-7800	Noncaking
Korba	6-9	12-38	28-41	81-85	4.5-5.3	1.8-2.0	8-12	7800-8200	Noncaking
Wardha Valley, Ballarpur	6-10	12-30	35-42	80-83	4.5-5.0	1.6-2.0	12-14	7300-7900	Noncaking
Talcher	7.0	18-28	41-44	79-82	4.9-5.3	1.6-1.8	10-12	7800-7900	Noncaking

Anthracite Coal

This is the most matured coal; hence is of the highest rank. Thus high carbon content (85 - 95%) and low volatile matter (< 10%) coal is hard, non-coking and burns without smoke with a short non-luminous flame thereby imparting intense localised heating. It ignites with difficulty due to low volatile matter content. The calorific value may be up to 8000 - 8500 kcal/kg which is slightly lower than that of bituminous coal due to its lower hydrogen content.

It has sub-metallic lustre, sometimes even a graphitic appearance. It does not soil the hand. Although there is banded structure this is not always obvious. Anthracites are characterized by low volatile matter (3-10% d.m.m.f.) and high carbon content (over 92% d.m.m.f.). The air-dried moisture is 2-4%. The hydrogen content is 2.8-3.9% and the calorific value 8,400 to 8,700 kcal/kg, both on d.m.m.f. basis. Anthracites are non-caking.

The most notable deposits of anthracites are in South Wales (U.K.) and Pennsylvania (U.S.A.). True anthracites do not occur in India. The tertiary coals in Jammu and the Lower Gondwana coals in Darjeeling have properties approaching those of anthracites and are known as anthracitic coals. There are about 100 million tons of such coals in each of the above two deposits. Composition & Properties of Indian anthracite coal is given in table 4.7.

Table 4.7: Composition and Properties of Anthracitic Coals of India

Location of coalfield	% Air-dried		% d.m.m.f.				Calorific value, Kcal/Kg (d.m.m.f.)
	Moisture	Ash	Volatile matter	Carbon	Hydrogen	Oxygen	
Jammu	0.5-2.0	10-35	13-17	91-93	3.9-4.2	0.4-3	8,400-8,700
Darjeeling	1.0-5.0	19-29	14-18	-	-	-	8,500-8,700

The chief uses of anthracites are in boilers, domestic ovens and metallurgical furnaces. It is also used in small quantities as a component of coke oven charges. On calcining it gives thermo-anthracite which is a raw material for the production of carbon electrodes.

CANNEL COAL AND BOGHEAD COAL

Cannel coal and boghead coal are examples of *sapropelic coals.* By their nature they cannot be fitted into the series : peat-lignite-bituminous-anthracite which constitute the *humic coals.* Both *humic coals and sapropelic coals*are of vegetable origin, the former being derived from higher plants and the latter from smaller plant organisms like *algae.*

Cannel coal and boghead coal have dull lustre and yield conchoidal fracture. They are quite massive, tough and unbanded. A splinter of these coals can be ignited with a match. Cannel coal owes its name to the word "candle" because it burns with a long and steady flame. Cannel is dull black in colour while boghead is dull brownish.

These coals have higher volatile matter and higher hydrogen than the corresponding *normal (humic) coals.* The high hydrogen content is responsible for high calorific value and also high yields (45-65% d.m.m.f.) of tarry products on low temperature carbonization. Typical composition & properties of Cannel Coal and boghead coal is given in table 4.8.

Table 4.8: Composition and Properties of Cannel Coal and Boghead Coal

Type of Coal	% Air-dried.		% d.m.m.f.				Calorific value, Kcal/Kg (d.m.m.f.)
	Moisture	Ash	Volatile matter	Carbon	Hydrogen	Oxygen	
Boghead	4-8	5-38	65-90	72-79	6-10	8-15	9,000-9,500
Cannel	2-5	2-40	45-56	75-84	6-10	5-12	9,000-9,500

Cannel coal and boghead coals usually occur in isolated bands in rocks of normal coal seams. When they occur in a separate seam, this is of limited extent and thickness. These coals do not occur in India.

Earlier cannel and boghead were used in the U.K. for the manufacture of town gas, yielding a gas of high illuminating power. The advent of incandescent mantles has restricted the use of these coals for this purpose.

Significance of the Various Constituents of Coal:

Moisture. In general, high moisture content in coal is undesirable, because:

- it reduces the calorific value of the fuel.
- it increases the consumption of coal for heating purposes.
- it lengthens the time of heating.
- we pay for it, when the coal is purchased and transported. But there are certain specific cases where moisture is deliberately added or is desirable to some extent *e.g.* dust nuissance is checked while loading, unloading, transporting and charging coal into the oven, when it is wet.
- moisture undergoes endothermic decomposition which reduces the temperature of the hot coal bed (either when burning it on grate or during the manufacture of producer gas) therer by eliminating its fusion and clinker formation which otherwise would have blocked the passage of air.
- Sometimes, water is sprayed on the coal dust charged in the coke oven. This apart from checking dust nuissance during coal charging helps in protecting the methane and other hydrocarbons present in the coke oven gas from cracking in presence of hot coke and hot oven walls because, it undergoes endothermic decomposition to reduce the temperature of the coke oven gas as well as that of the coke mass thereby eliminating its fusion (called graphitisation).

Coal/coke absorbs moisture due to rain, washing (of coal) and quenching (of coke).

Owing to its nature, origin and occurrence, coal is always associated with moisture. When a wet coal is exposed to atmosphere the external moisture evaporates but the apparently dry coal still contains some moisture which can be removed only on heating above 100°C. External moisture is also called *accidental* or *free* moisture while inherent moisture is called *equilibrium* or *air-dried* or *hygroscopic* moisture. The quantity of external moisture depends mainly on the mode of occurrence and handling of coal but the air- dried moisture is related to the inherent hygroscopic nature of coal. Variation of inherent moisture with carbon percentage in coal is shown in Fig. 4.1.

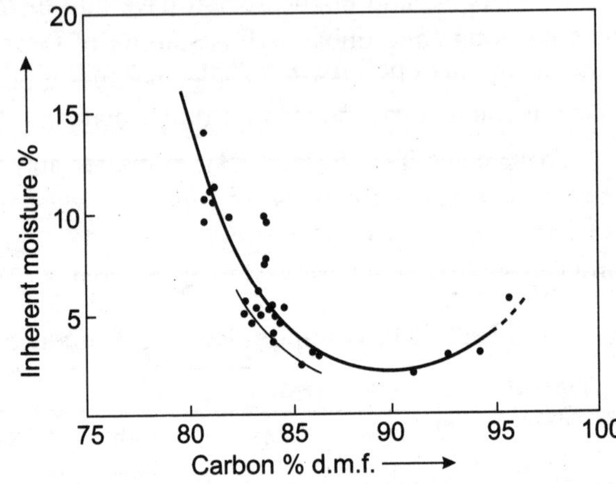

Fig.4.1. Inherent moisture and percent carbon in coal.

Air-dried moisture is determined by observing the loss in weight of the coal sample on heating to 105°C. In the case of peat and lignite which start decomposing at low temperature (below 105°C), this method is not applicable and the *Dean and Stark method* is followed in which coal is heated with excess of toluene when moisture distils over as an azeotrope with toluene at 84°C. The vapours are condensed and separated into two layers. The amount of water is then read off in the graduated receiver.

Air-dried moisture content of coal decreases with increasing rank from a value of 25% for lignite to a minimum value of 0.5% for a low volatile (15% VM) bituminous coal. Beyond the minimum, however the value increases to about 3% for anthracites . For coals with a moisture content of above 3%, it is a useful measure of the rank and hence of the other qualities indicated by the rank.

If coal is exposed to an atmosphere of increasing humidity at a fixed temperature, it absorbs more and more moisture. The quantity of moisture held by coal at an approximately saturated atmosphere (96-99% relative humidity) is called *near-saturation moisture* or *capacity moisture* or *bed moisture.* This is a better measure than the air-dried moisture for determination of rank of high volatile bituminous, sub-bituminous and lignitic coals and serves as a parameter in some coal classification systems.

Volatile matter. Volatile matter is not a constituent of coal. It represents the volatile products of thermal-decomposition of coal under specified conditions. Volatile matter is the loss of weight of coal when heated in absence of air for a standard time at standard temperature.

The coal with higher volatile matter content:

- ignites easily *i.e.* it has lower ignition temperature.
- burns with long smoky yellow flame.
- has lower calorific value.
- will give more quantity of coke oven gas when it is heated in absence of air.
- will require larger furnace volume for its combustion.
- has a higher tendency of catching fire (due to low temperature exothermic oxidation) when stored in open space.

Volatile matter does not include the moisture of coal, but it contains water that is formed from the hydrogen and oxygen of coal during decomposition. When volatile ma'tter is reported on air-dried and d.a.f. basis, it also includes the part of the mineral matter which escapes into the gaseous and vaporous state. On d.m.m.f. basis however it represents only the volatile products from the organic mass.

The volatile matter and fixed carbon, both expressed as per cent on d.m.m.f. basis, add to 100 parts of pure coal (organic mass). The higher the volatile matter, the lower the fixed carbon.

With the increase in the maturity or rank of coal, its volatile matter decreases. However, above a critical value of volatile matter (32-33% d.m.m.f.), this relationship is not valid.

Different micro constituents of the same coal yield different amounts of volatile matter. The highest yields are from the exinites and the lowest from the fusinite. Vitrinite occupies an intermediate position.

The physical nature of the residue left after the volatile matter test is noted as an approximate measure of the caking properties of coal.

Mineral matter and Ash, Ash is the combustion product of mineral matters presents in the coal. It comprises mainly of silica (SiO_2), alumina (Al_2O_3) & ferric oxide with varying amounts of other oxides such as CaO, MgO, Na_2O etc. High ash content in coal is undesirable in general, because a coal with *high ash* content:

- is harder and stronger

- has lower calorific value
- produces more slag (impurities) in the blast furnace when coke made out of it is used in B.F.

Ash content of the coal is reduced by its washing.

Coal contains inorganic mineral substances which are converted into ash by chemical reactions during the combustion of coal. Ash and mineral matter of coal are therefore not same. Mineral matter present in coal consists of silicates of alumina together with basic oxides like lime (CaO) magnesia (MgO) & iron (Fe_2O_3) along with traces of carbonates, sulphates & phosphates.

The mineral matter in coals is either *inherent* or *extraneous.* The inorganic material of the original vegetable substances is responsible for the inherent mineral matter. The extraneous mineral matter is due to the substances which got associated with the decaying vegetable material during its conversion into coal, and also due to the rocks and dirt getting mixed up during mining and handling of coal. The former type of extraneous mineral matter is in a fine state and intimately associated with the organic mass of coal. Hence difficulties are experienced in removing this from coal by mechanical methods. The second type of extraneous mineral matter is more amenable to coal cleaning methods. The inherent mineral matter cannot be removed by any mechanical means. The inherent mineral matter content is low (about 2 to 3% or less) in all coals. In Indian coals, the mineral matter content is not only high but is of intimately associated type due to the drift origin.

The bulk of the mineral matter of coal is due to clay or shale consisting of alumino-silicates of different composition. Other major constituents may be calcite (calcium carbonate) and pyrites. When coal burns, shale and other hydrated minerals lose water of hydration while sulphides, sulphates and carbonates usually decompose, or oxidize, resulting in a loss of weight hence the ash of coal is always less than the mineral matter content.

The coal ash consists mainly of silica, alumina, iron oxide and lime. When heated, coal ash does not melt sharply at any temperature, but begins to soften at a much lower temperature than that at which it becomes molten and flows. Higher alumina content makes the ash refractory while lime, magnesia, iron oxides and alkali oxides have fluxing effect and lower the ash fusion temperatures. The ash of Indian coals is generally refractory in nature and the fusion range is mostly above 1400°C.

The quantity of mineral matter or ash has no relation with the rank or maturity of coal. Among the macro constituents of coal, the vitrain is of low ash (1 to 2%), the clarain higher (3 to 5%), and the durain still higher (5 to 8%). Fusain has very high ash (9-12%).

The mineral matter creates many difficulties in the efficient utilization of coal. In a furnace grate, the ash may seriously restrict the passage of the air and lower the rate of combustion. High ash leads to large heat losses, carbon losses, boiler deposits and clinkering. 1% rise in coal ash is equivalent to 0.3-0.4% fall in boiler efficiency. High ash is therefore undesirable and harmful. Clinkering troubles are due to the lower fusion temperatures. On the other hand coals of refractory ash *e.g.* Indian coals create a problem in their utilization in slagging furnaces and slagging gas generators where ash is removed in the molten state. A non-clinkering ash contains more than 45% silica and 35% alumina and less than 15% ferric oxide and 10% calcium oxide.

The coal ash may be used for extracting valuable rare elements like germanium etc.

Fly-ash produced from coal-based thermal power stations is useful in the production of cement, - concrete bricks filling material and light weight aggregates. A plant at Ennore in Tamil Nadu produces bricks from cement, lime and fly-ash.

The ash of coal is directly determined in the laboratory by completely burning the coal. The content of the mineral matter in coal is determined by the method of *Radmacher and Mohrhauer* which consists of dissolving out the mineral substances by treatment with hydrofluoric acid and hydrochloric acid and finding the weight loss.

Empirical formulae are available for calculating the mineral matter content from the ash values. The *Parr formula* is :

$$MM = 1.08\ A + 0.55\ S$$

Indian coals in general are low in sulphur. A *simplified version* of the *Parr formula* is used for these coals:

$$MM = 1.1\ A$$

The formula assumes that the difference in the weights of mineral matter and ash is due to the water of hydration which is taken to be 10% of the ash value (0.1 A).

Fixed Carbon. It is the pure carbon present in the coal. Higher the fixed carbon content of the coal, higher will be its calorific value.

In anthracites, the volatile matter is very small and the values of fixed carbon and total carbon are almost equal. In other coals, fixed carbon is less than total carbon. The carbon content is determined by Liebig's *method* by completely burning the coal in a current of pure oxygen and finding the amount of resultant carbon dioxide. The carbon content of coal increases with increase in the rank. Among the microconstituents of the same coal, the exinites and vitrinite have almost similar carbon content while fusinite has higher values.

Total Carbon. It means the fixed carbon plus the carbon present in the volatile matters *e.g.* in CO, CO_2, CH_4, hydrocarbons etc. Total carbon is always more than fixed carbon in any coal. High total carbon containing coal will have higher calorific value.

Hydrogen. It increases the calorific value of the coal. It is associated with the volatile matter of the coal. Also, the percentage of NH_3 in the coke oven gas is more if the coal from which it has come out contains more H_2. This NH_3 is recovered as $(NH_4)_2\ SO_4$ (by reacting NH_3 with H_2SO_4), which is a valuable fertiliser.

Part of the hydrogen of coal remains as such in the coke oven gas thereby increasing its calorific value. Also, if hydrogen is to be removed from this coke oven gas for production of NH_3 (to produce nitrogeneous fertiliser subsequently as was done in Rourkela's fertiliser plant of SAIL), its availability will be more if the coal with high H_2 content has been used for the purpose.

The content of hydrogen in coals from the peat to the bituminous stage varies between 4.5 to 6.5% and is not related to the rank. Beyond the bituminous stage, the hydrogen content sharply decreases to a value of 1-2% in anthracites. Among the microconstituents, the exinites are characterised by high and the fusinites by low hydrogen contents. Cannel coal and boghead coals also have high hydrogen contents.

Nitrogen. The nitrogen in coal is present upto 1-2.5% and comes from the proteinous matter present in vegetable matter. Presence of inert nitrogen decreases the calorific value of the coal. However, when coal is carbonised *(i.e.* heated in absence of air), its N_2 and H_2 combine thereby producing NH_3, which is recovered as $(NH_4)_2\ SO_4$, a valuable fertiliser. The proportion of extractable nitrogen decreases with maturity of coal.

The nitrogen content does not bear any relation to the rank of coal. In most coals, it is between 1 and 2%. Lower values down to 0.5 and upper values up to 2.5 are also found. The nitrogen in coal is useful in carbonization industries which recover a part of it as valuable by-products like ammonia, ammonium sulphate and pyridine bases. When coal is carbonised about 15% of N_2 present in it is converted into ammonia. More than 50% of N_2 remains in the coke. Rest is present as N_2 compounds in tar.

Sulphur. Though its presence (mainly as sulphides) increases the calorific value of the coal, but, it has several undesirable effects. The oxidation product of sulphur *(e.g.* SO_2, SO_3 etc.) especially in presence of moisture cause corrosion (damage due to chemical and electrochemical action) of the equipment and cause atmospheric pollution. Sulphur is highly undesirable in metallurgical coal used in iron and steel making as it badly affects the properties of iron and steel. It causes *hot shortness of steel, i.e.* cracking of steel surface during hot rolling.

Sulphur is commonly present in coal in three forms, *viz.* pyritic (FeS_2), organic and sulphate but occurrence of sulphides and elementary sulphur in coal have also been reported. Only organic sulphur should be included in the ultimate analysis. When coal is carbonised, the coke produced contains about 80% of the original sulphur present in coal.

Total sulphur is determined by *Eschka method* or *Bomb method.* In the former method the entire content of sulphur is converted into soluble sulphate by heating the coal with an oxidizing mixture of magnesium oxide and sodium carbonate and then the sulphate is estimated. In the *Bomb method* the total sulphur is also converted into the sulphate form during the determination of calorific value in Bomb calorimeter. The pyritic and sulphatesulphur in coal are determined by the usual methods of analytical chemistry and then the organic sulphur is calculated by subtracting the inorganic sulphur from the total sulphur.

Indian coals are usually low in sulphur (< 0.7%), but the majority of the coals in the *Assam coalfields* have high sulphur (around 4%). The coals of most countries of the world contain between 1 and 3% sulphur. Higher and lower values are also reported. The high sulphur content is generally due to pyrites but Assam coals form an exception, where much of the sulphur is in the organic form.

The sulphur content of coal has no relation to its rank or composition. Although elementary, sulphide, pyritic and organic forms of sulphur contribute to the calorific value of coal, it is an undesirable constituent.

Oxygen. The less the oxygen content, the better is the coal as it reduces its calorific value. It decreases from lignite to anthracite as the maturity of the coal increases. As the oxygen content of the coal increases, its moisture holding capacity increases and the caking power decreases.

An increase of 1% in oxygen content of coal, reduces the C.V. of bituminons coal by 1.7%. High oxygen coals are non-coking and hold over 10% of moisture even when air dried. However, low oxygen bituminous coals are strongly caking and hold only 1-2% of moisture, when air dried.

The oxygen content of coal has a close relation to the rank. The former decreases with the increase in the latter. Among the micro-components of the same coal, vitrinite has the highest oxygen, fusinite the lowest, and exinites has the intermediate value.

Phosphorus. It is undesirable in metallurgical coal as it badly affects the properties of the steel and causes *cold shortness of steel i.e.* cracking of steel surface during cold rolling.

Phosphorus occurs in small quantities in coal. Indians coking coals have normally less than 0.15% phosphorus which offers no problem for common use in iron and steel production. The coking coals of Giridih are still lower in phosphorus. (0.01-0.04%) and suitable for producing coke for making ferro-alloys. There are certain areas in the coalfields of M.P. and Assam, where again coals contain < 0.01% of phosphorus.

Chlorine. It is present mainly as chlorides in the coal. It reduces the fusion point of the ash of the coal. It is substantially removed during washing of the coal.

Problem on Coal Composition

Example 1. A coal has following proximate analysis on air dried basis :

M = Moisture = 1.5%, A = Ash = 15.5%

VM = Volatile matter = 28%, FC = fixed carbon = 55%

Calculate its ash % on dry basis and volatile matter on d.a.f. (dry ash free) and d.m.m.f. (dry mineral matter free) basis.

Solution. Ash %, (dry basis) $= \dfrac{A}{(100-M)} \times 100 = \dfrac{15}{(100-1.5)} \times 100 = 15.7\%$

Volatile matter %, (d.a.f. basis) $= \dfrac{VM}{(100-M-A)} \times 100 = \dfrac{28}{(100-1.5-15.5)} \times 100 = 33.7\%$

Volatile matter %, (d.m.m.f.) $= \dfrac{(VM-0.1A)}{(100-M-1.1A)} \times 100 = \dfrac{(28-0.1\times 15.5)}{(100-1.5-1.1\times 15.5)} \times 100$

$$= \frac{2645}{81.45} = \mathbf{32.7\%}$$

Note. The mineral matter in coal contributes to the volatile matter of the coal to the extent of 10% of ash (since, mineral matter in coal =1.1 times ash). Hence, correction has been made in the *VM* of the coal (as *VM* - 0.1 A) while calculating composition on d.m.m.f. basis. Fixed carbon does not need this correction.

Alternate method of solution

The V.M. can be calculated through fixed carbon also, since these two together comprise the whole coal on d.a.f. and d.m.m.f. basis.

Hence, *FC%*, (d.a.f. basis) $= \dfrac{FC}{(100-M-A)} \times 100 = \dfrac{55}{(100-1.5-15.5)} \times 100 = \mathbf{66.3\%}$

∴ *VM%*, (d.a.f. basis) = 100 - 66.3 = **33.7%**

and, *FC%*, (d.m.m.f. basis) $= \dfrac{FC}{(100-M-11A)} \times 100 = \dfrac{55}{(100-1.5-1.1\times 15.5)} \times 100$

= **67.6%**

VM (d.m.m.f. basis) = 100 - 67.6 = **32.4%**

PETROGRAPHY OF COAL

Coal is a non-homogeneous sedimentaiy rock containing many types of organic and inorganic compounds distributed in layers/bands parallel to the bedding plane of the seam.

Non-homogenity of coal seam is recognisable macroscopically. Petrography is the study of the individual mineral components of a mass of rock or coal by visual means.

All efforts to isolate completely and identify the organic compounds of coal have failed. Coal composition is therefore studied by indirect methods.

Coal appears heterogeneous even to the naked eye. With most coals, differences in texture are apparent, some bands having a bright and others a dull appearance. These differences are most marked in bituminous coals but can be traced in sub-bituminous coals, in some lignites and, to a less extent, in anthracitic coals. The heterogeneity of coal is much more marked under a powerful microscope. An apparently homogeneous part of coal is found to consist of microscopically identifiable distinct components.

Coal petrography is the study of coal components by visual methods, with or without the help of microscopes. The components identified with the naked eye are called macroscopic components. Similarly microscopic studies identify microscopic components.

There are four types of macro-components in coal *viz.* vitrain, clarain, durain and fussain. This terminology is due to *Stopes.* According to the terminology of *Thiessen,* these components correspond to *anthraxylon, translucent attritus, opaque attritus* and *fussain* respectively.

Petrographic constituents of coal *(e.g.*vitrain, clarain, durain and fussain) which are in the form of layers and are visually observable (with naked eye) are described below :

Vitrain. Bright, black and brittle thin band of the homogenous part of coal having a massive texture is vitrain or anthraxylon. It fractures conchoidally. It is derived from wood plant, tissues and is surrounded by a dull ground mass. It is coking in nature and is responsible for lustrous black band of bituminous coal.

Clarain. It is a silky, shinning, minutely striated layer within coal with a banded structure. It contains more plant remains than vitrain and is the commonest of the four types of coal substance. It is not so bright (semi-bright) and fractures irregularly. When fractured at right angle to the bedding plane, the faces have a pronounced gloss/shine.

Durain. It is duller, greyer and very hard component of coal which cracks irregularly. It is highly charged with durable plant remains and is supposed to be formed from silts or muds of small particles of vegetable matter. It is heterogenous and has a firm granular texture.

Fussain. It is soft powdery form occurring in thin seams between the bands of other types. It is a friable, charcoal like substance which dirties the hand when coal is touched. It is non- coking but when fines are present in small percentage in coal charge, they help in increasing the strength of the coke produced therefrom. Fixed carbon content is higher and volatile matter content is lower in fussain than in other banded ingredients. Fussain occurs as soft patches of fibrous material which can be powdered by fingers. It is about 2 to 5% of coal and is responsible for the dusty & dirty character of an ordinary coal.

In a typical coal seam, 50% of the seam may be clarain, 15-30% durain, 10-15% vitrain and 1-2% fussain.

Petrographic components of coal have got different physical & chemical properties and they can be separated by chemical & mechanical methods.

Typical analysis and properties for a medium volatile bituminous coal is given in the Table 5.1.

Table 5.1: Petrographic Analysis of a Typical Coal

Analysis. %	Vitrain	Clarain	Durain	Fussain
Proximate .				
Moisture (air-dried)	1.7	1.4	1.2	0.9
Volatile matter	34.6	37.6	32.2	19.1
Fixed carbon	63.7	57.5	62.0	70.4
Ash	0.6	3.5	4.6	9.6
Ultimate (d.a.f.) :				
Carbon	84.4	82.2	85.8	88.7
Hydrogen	5.4	5.7	5.3	4.0
Sulphur	1.0	2.3	0.9	1.0
Nitrogen	1.5	1.9	1.4	0.7
Oxygen	7.7	7.9	6.6	5.6
C.V., kcal/kg	8216.6	8216.6	8388.8	8244.4
Caking power (crucible)	Swollen coke	Swollen coke	Non-swollen coke	Non-coking

On microscopic study, it is found that the banded components of the coal is further composed of identifiable plant entities called phyterals. Vitrain, clarain, durain and fussain are made up of numerous components or macerals. Durain, for example, includes several macerals like *vitrinite, semi-fussinite, micrinite, cutinite etc.* which are distinguished by the different optical properties.

Clarain and durain are mixture of macerals called exinite (which is rich in hydrogen), *fussinite* (of relatively low hydrogen content) and vitrinite (of intermediate hydrogen content).

The Indian coals are generally dull in appearance, although they are often interbanded with bright and dull coals. Fussain is present in small amounts (< 5%). In regard to micro components, the coals are high (> 50%) in vitrinite content in spite of their predominantly dull appearance and apparently durainous nature.

Usual range of chemical components in banded coal is given in the Table 5.2.

Table 5.2: Chemical Components in Banded Coal

Maceral (Chemical Components)	Vitrain, %	Clarain, %	Durain, %	Fussain, %
Vitrinite (Ulmins)	90-97	70-95	20-80	33-66
Micrinite (Opaque matter)	Nil	0-5	5-10	Nil
Exinite (spore exines and cuticles)	Nil	2-25	5-35	Nil
Resinite (resins, waxes, hydrocarbons)	3-9	3-10	3-15	0.2
Fusinite (carbonisedfibre)	Nil	0-2	0-10	66-33

The petrographic components of coal have different chemical & physical properties and they can be partly separated by mechanical & chemical methods and is gainfully used in the petrographic preparation of coal for coke production.

The *'rational analysis'* of coal determines a number of components on the basis of the differences in their resistance to nitric acid and the solubility in some organic solvents. The components studied by the rational analysis correspond to the micro components of coal to some extent. The approximate equivalence is given below:

Ulmins	Vitrinite
Carbonized Fibres	Fusinite
Opaque Matter	Micrinite
Spore Exins and Cuticles	Exinite } Liptinite (Exinite)
Waxes, Resins and Hydrocarbons	Resinite } Liptinite (Exinite)

The above components are fully developed in bituminous coals and gradually lose their identity and coverage into a homogeneous mass in anthracites. These components have been produced from the original plant material through the peat and lignite stages.

Peat consists of waxes, resins, simple carbohydrates, pectin, hemi-cellulose, cellulose, lignin and humic acids.

Lignite contains waxes, resins, humic acids and a 'residual mass' which is composed of the components corresponding to those of bituminous coal. Humic acids constitute the major part of lignite.

Petrographic study of coal is helpful in ***selective preparation and crushing*** of coking coals for *proper blending* to produce good quality coke and also for fundamental studies. In conventional crushing of coking coal in hammer mills, the softer (easily crushable vitrain is softer than durain) & fusible constituents become smaller in size, while the harder & non-fusible materials become larger in size. On carbonisation the coke produced from coals crushed in this manner are poor in strength due to non-uniform texture and the formation of fissures in the coke mass. A selective crushing of coal will enable to crush the harder components to a smaller size and softer components to a comparatively larger size, so that when carbonised, these coals will produce coke of uniform texture & high mechanical strength. A study shows that vitrain of 3 mm average size, if blended with durain of 2 mm average size produces good metallurgical coke.

6

WASHING OF COAL

Most of the coals when mined contain impurities associated with it which must be removed before the coal is used. Impurities are removed from coal by washing.

Coal contains two types of impurities :

— **Fixed or inherent impurities.** It is derived from coal forming plant and can not be separated from coal by washing. A new process developed at *Central Fuel Research Institute (CERI), Dhanbad called oil agglomeration technique* which consists mainly of leaching of coal impurities by petroleum oil can to some extent remove these impurities. Inherent mineral matter in coal is $<3\%$.

— **Free impurities.** These are the impurities adhering to the surface of the coal and comprise mainly of dirt band and rock particles which can be removed by washing of coal.

Nature of impurities present in coal are of the following eight types :

- residual inorganic matter of coal forming plants
- mineral matter washed or blown into the coal forming mass during the period of its formation
- pyrites (FeS_2) formed by the reaction of iron sulphate with coal forming matter
- sedimentary deposits during the coal forming periods
- deposits formed through deposition on bedding planes
- saline deposits
- slate, shale, clay etc. from the underlying and overlying strata
- moisture.

Objectives of coal washing. Washing of coal :

- reduces its ash content
- reduces its sulphur & phosphorous contents which are detrimental particularly to metallurgical coals
- increases its heating value
- improves its coking properties
- increases the fusion point of its ash by removing alkali chlorides (which is responsible for lowering the fusion point)
- reduces its clinkering tendency
- increases its efficiency in use.

Principle of coal washing. The specific gravity of pure coal varies from 1.2 to 1.7 and that of free impurities from 1.7 to 4.9 as given in Table 6.1.

Table 6.1: Specific Gravity of Mineral Matters

Mineral matters (impurities)	Specific gravity
Coaly shale	1.7-2
Gypsum	2.3
Calcite	2.7
Shale, clay and sandstone	2-2.6
Pyrites	4-4.9

If the average specific gravity of pure coal is 1.3 and it is suspended in a liquid (called washing medium) of specific gravity 1.5 then the impurities being heavier will sink in it, whereas the pure coal will float. Thereby, the pure coal and the impurities can be separated by utilising the difference in specific gravity of the two.

Coals are assayed for cleaning in the laboratory based on density difference using mixed liquid (typically a mixture of toluene & carbon tetrachloride or solutions of calcium chloride). Laboratory float and sink test result of a coal containing about 15% ash is tabulated in table 6.2.

Table 6.2: Float and sink test result of the coal (ash = 15%)

Specific gravity of separating liquid	Floating material (Clean coal)		Sinking material (Dirt)	
	Yield, weight %	% of ash	Yield, weight %	% of ash
1.25	5	1	95	16
1.30	65	2.5	35	39.5
1.35	75	3	25	51.5
1.40	80	4	20	58.5
1.50	82	4.5	18	64.0
1.60	85	5.5	15	70.0

It follows from the above result that, if a clean coal of 1.5% ash is required, the cut off specific gravity of separating medium will be 1.5 which would correspond to a yield of 82% clean coal having 4.5% ash.

The above test result has been shown in Figure 6.1 in what is known as washability curves.

The increment curve shows the ash content of the dirtiest particle which is included in the clean coal which is about 31%. If the cut were made at about 75% clean coal (nearer to the elbow of increment curve), then the ash content would be about 12% only.

Apparently, to get a clean coal, the cut should be made at the elbow of clean coal curve, *i.e.*, about 5% ash and a yield of over 80% of clean coal.

Washing medium used in industrial washeries is a slurry of sand & water or that of water & iron ore (magnetite or haemetite).

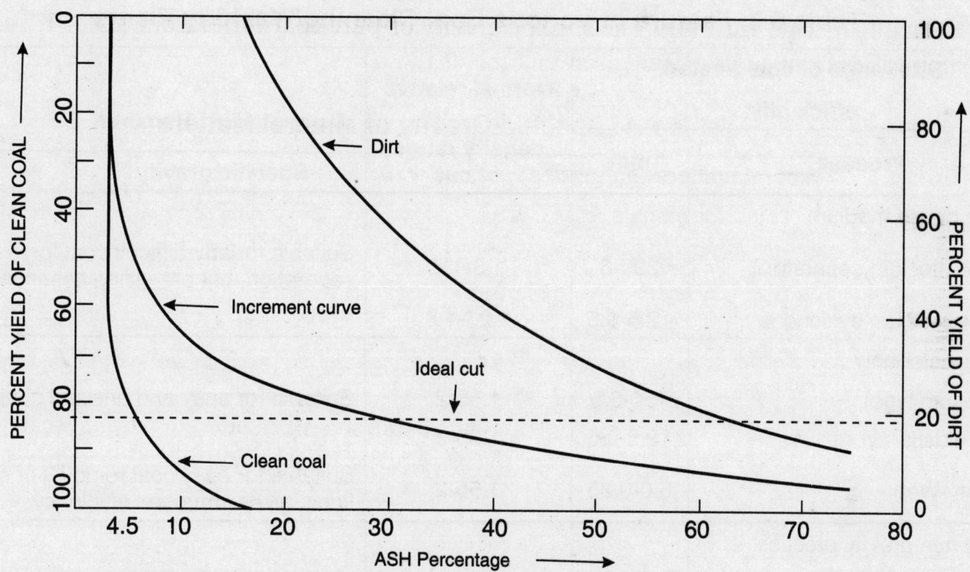

***Figure 6.1** Washability curves.*

ADVANTAGES OF CLEAN (WASHED) COAL:

- higher efficiency of utilisation
- economy of transport and storage
- less clinkering trouble in furnaces and gasifier
- less of ash disposal problems
- lesser content of S, P, Cl and other harmful elements
- improved coking properties.

TYPES OF COAL WASHING PROCESSES

They are of following three types :

(i) Gravity Separation method :

Wet processes — üüüüüüü / Jigs / Rising current classifier / Wilfley tables

Dry processes — Jigs / Tables

(ii) Float and sink method

Fine solids and water floatation — Sand / Magnetite and other materials

(iii) Froth floatation method

Most of the coal is cleaned by wet processes. Features of various coal cleaning washing plants is given in table 6.3.

Table 6.3: Feature of Various Coal Cleaning/Washing Plants

Size range of coal treated efficiently		Normal relative density range of cut	Remarks
Process	mm		
Using dense medium			
Dense medium separator	200-6	1.3-1.8	Suitable for any difficult coal for separation, but pre-sizing essential.
Dense medium cyclone	12.5-0.5	1.25-1.8	"
Using water only			
Jig (Baum type)	150-0.5	1.55-2	Suitable for easy and unsized coal.
Jig (Feldspar type)	12.5-0.5	1.55-2	"
Tables (Wet)	6.0-0.25	1.55-2	Suitable for easy coal removal of coal fines, gives improved efficiency.
Using non-gravity process			
Froth floatation	0.5-0	-	Suitable for slurry and middlings upgrade.
Oil agglomeration technique			Suitable for de-ashing high-inherent dirt containing coal, not yet successfully commercialised. Developed by CFRI, Dhanbad.

Commercial Coal Washing Methods

Based on physical properties of pure coal and mineral matter, maximum processes use difference in their specific gravity & wettability.

The specific gravity at which coal is to be cleaned is determined from the washability data and economic considerations. The ease of washing at this separation gravity is judged from the amount of near gravity material present in the coal, which is defined as the percentage of the coal that will float in a range within 0.10 specific gravity of the separation value. The larger the amount of near gravity material, the more difficult the cleaning operation, and *vice versa.*

Coefficient of friction, wettability and resilience are used in a few commercial washing processes. Some methods of coal preparation have been suggested also on the basis of optical and electrical properties. Size and shape of the solid particles influence all coal cleaning processes to a greater or lesser extent.

Processes are termed dry and wet depending upon the medium used for coal cleaning.

DESCRIPTION OF DRY COAL WASHING PROCESSES

Pneumatic Table. It is a perforated and riffled sloping deck which is rapidly reciprocated. The coal is fed from the upper corner of the table which is transversely inclined. The table is given a reciprocating movement and air is blown through the deck. (Since air is used for separation, it is called pneumatic process.) The coal spreads out over the deck and is segregated into clean coal and refuse under the combined action of air flow, reciprocating movement and frictional resistance of table.

The heavy refuse is trapped in the riffles and transported to the end of the table. The clean coal passes over the riffles and travels towards the lower corner of the table. The *refuse, middlings and clean coal* are collected in separate receivers.

Air-sand Process. A suspension of sand in air of a particular suitable specific gravity is used as medium for coal washing in which clean coal floats and the dirts sink.

Berrisford **Process.** It is based upon difference in resilience of clean coal and dirt. The feed coal is dropped onto an inclined plane containing a gap of choosen width. The clean coal bounces and falls on a receiver while the dirt falls through the gap.

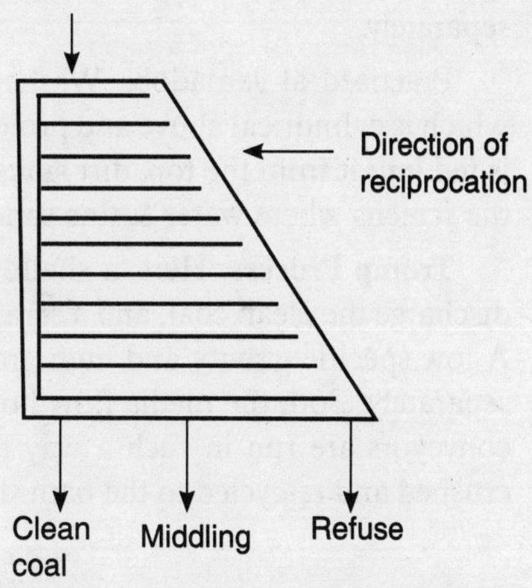

***Figure 6.2:** Pneumatic Table for dry coal washing*

Advantages of Dry Processes :

— Elimination of cost of drying wet coals

— No problem of disposal of slurry of water and fine dust.

Disadvantages of dry processes :

— Feed should be very closely sized lest the separation efficiency will be very poor.

— Coals of difficult washability cannot be cleaned satisfactorily. No dry cleaning process is working in India. Features of dry coal washing processes is given in table 6.4.

Table 6.4: Basis of Separation of Various Dry Coal Washing Processes

Dry Processes	Basis of Separation
Pneumatic table	Density and friction
Air-sand process	Density and friction
Berrisford process	Resilience and density
Wet Processes	***Basis of Separation***
Dense/Heavy medium separator	Density, shape and size
Jig washer	— density, shape and size
Trough washer or launder	— density, shape and size
Concentration table (wilfley table)	— density and friction
Door thickener/radial settling tank	— density, shape and size
Froth floatation process	Wettability and density

Description of Wet Coal Washing Processes

(a) **Dense or heavy medium separator.** The various dense medium processes are the large scale application of float and sink test. Mainly there are two processes under this heading.

(i) Chance process employing deep bath.

(ii) Tromp process employing shallow bath.

Chance Process. Here the washing medium is a mixture of sand and water in such proportion to have an intermediate specific gravity between clean coal and ash/dirt. A deep bath of sand and water mixture is maintained in a trough where the clean coal floats and the dirt sinks & removed separately.

Practised at Jamadoba Washery Dhanbad, it is a dense medium washer employing a tank which is cylindrical above and prolongs downward into an inverted truncated cone. As the raw coal is fed into it from the top, dirt sinks to the bottom while clean coal, water & sand over flows on to the screens where water & fine sand drains out and the clean coal is separated.

Tromp Process. Here, a shallow bath of mild steel is provided with a push plate conveyor to discharge the clean coal, and a scraper conveyor to remove separately the middlings and the refuse. A low specific gravity and high gravity washing medium (say water and sand or hametite) is fed separately. Both the media flow horizontally over the whole width and depth of washer. The two conveyors are run in such a way that the swirl of the media is avoided. The middlings may be crushed and re-cycled to the bath. It is shown in Figure 6.3.

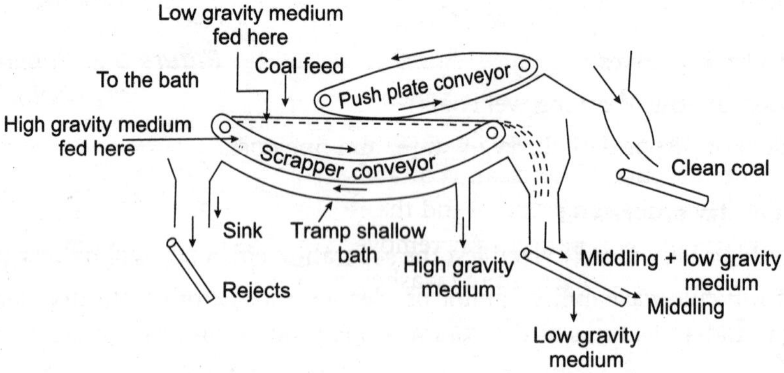

Figure 6.3. *Tromp Process of coal washing*

Advantages of shallow bath process over deep bath process :

- reduction in the building height
- requirement of much lesser volume of fluid
- increase in the capacity for extracting middlings.

Advantages of dense medium process :

- Coals containing high proportions of near gravity material can be efficiently washed (efficiency = 99%).
- It efficiently separates at as low as sp. gr. as 1.28 and at any high practical gravity level.
- Minimum and maximum capacity is similar to jigs.
- Most widely used process in washing of coal.
- **Disadvantages** of dense medium process :
- It is inefficient in cleaning small size coal (<6 mm) as the viscosity of media effects these sizes and reduces the capacity of the plant.
- Quality of solid media removed with clean products is much greater.

Jig Washer. It is very widely used in coal washing plant. In a jig washer, a coal bed is maintained on a perforated plate and subjected to action of periodic upward & downward current of water to separate the coal bed in different layers depending on the specific gravity. The clean coal being lighter is concentrated in the upper strata and the dirt accumulates in lower strata. Pulsating movement of water is created either by reciprocating plunger, diaphragm or alternately admitting and releasing compressed air as in a Baum Jig. The most common type is Baum Jig shown in Figure 6.4.

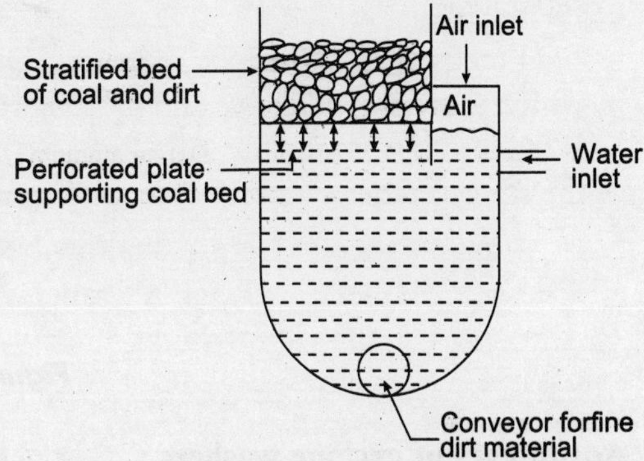

Figure 6.4. *Baug Jig*

Baum Jig is a U-shaped box divided longitudinally by a vertical partition into washing and air compartments. Washing section is divided into two parts each having a sieve. Water pulsation is created in air compartment and relayed to the adjoining washing compartment.

The coal is fed at one end of washing compartment. The larger pieces of refuse collect at the bottom of the first section. The clean coal and finer dirt is carried forward over the weir to the second section. Here, the refuse sinks and the clean coal is carried by the flowing water over the discharge weir. The refuse is periodically removed through valves from both sections. Fine dirt falls through the sieve into the bottom of the washer and are removed by a conveyor.

Advantages of jig washer :

- It gives better separation efficiency for finer (25-200 mm) sizes of coal than dense medium separator.
- High efficiency (98%) for coals with 10% near gravity material.
- High optimum capacity (200-400 tons/hr.); Maximum capacity - 600 tons/hr.

Disadvantages of jig washer :

— It can not wash fines below 0.5 mm.

Cyclone Washer. It is a settling chamber in which gravitational settling force is replaced by centrifugal force. A pulp of deslimed raw coal and heavy medium magnetite is fed tengentially to the cyclone. The overflow from the cyclone contains the cleans and the underflow contains the sinks. These are depulped and the products are rinsed with water before final disposal. The recovered slurry of heavy medium is reused after suitable treatment.

Capacity of cyclone washer is normally 70-150 tons/hr. It is shown in Figure 6.5.

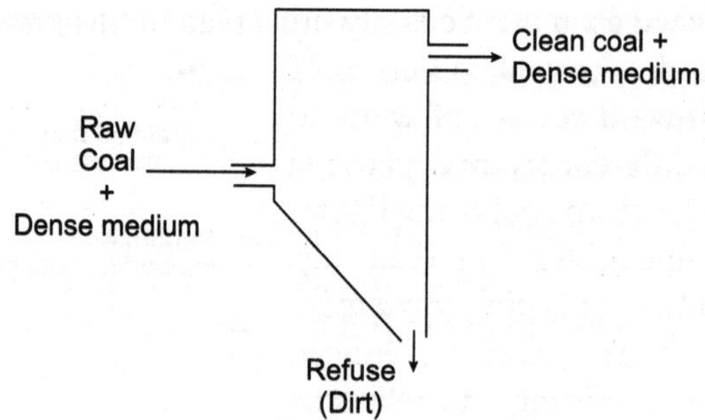

Figure 6.5. *Cyclone washer*

Advantages of cyclone washers :

— efficient for small sized coal.

— high flexibility of operation with regard to percentage of near gravity material, coal Raw size, feed rate and separation gravity.

Disadvantages of cyclone washers :

— Power consumption is more.

— Maintenance cost is high

Trough washers or launders. In a trough washer or launder shown in figure 6.6, the coal feed and the water enters at the top of the inclined trough where the heavier dirt settles at the lower layer of the bed and the lighter clean coal remains in the upper strata & overflows at the lower end of the trough. The flowing refuse is continuously removed through pockets (Rheo box) located along the length of the trough. The refuse may be rewashed into another trough for recovering further amount of clean coal.

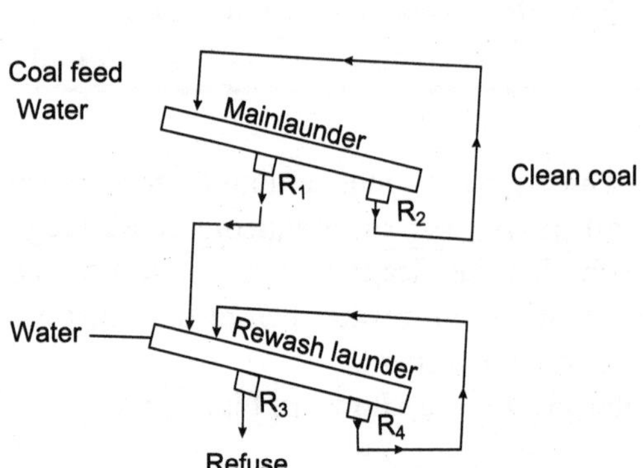

Figure 6.6. *Rheolaveur Launder*

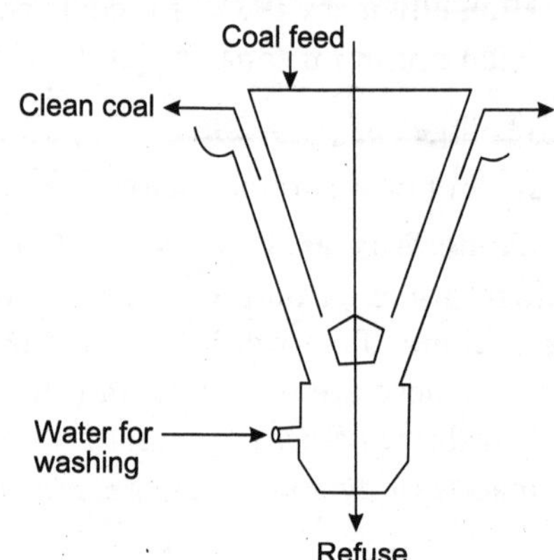

Figure 6.7. *Double cone classifier*

Upward current classifier. A typical system is double cone classifier shown in Figure 6.7. Here, the dirty coal is fed at the top of an inverted conical vessel and a continuous upward current of water enters through the base. Separation is done by adjusting the velocity of water between the terminal velocity of fall of the clean coal particles and that of dirts.

Advantage, *(i)* Large feed sizes can be easily processed in this system.

Concentration tables {*e.g.* **Wilfley table).** Its working principle is same as that of pneumatic table with the only difference that water instead of air is used for washing. Washing efficiency is high with unsized coal of easy washability.

The use of concentration tables, trough washers, upward current classifiers is limited to coals of close size ranges and good washability characteristics. Efficiency of these washers is low in the case of difficult coals. They are not in use in India.

Froth floatation process. In this process, froth is made in a floatation cell by bubbling air through water in presence of frothing agents like cresol, pine oil or alcohols. When the dirty coal powder is added to the cell, the pure coal particles adhere to the bubbles of the froth while the dirts are wetted by the water and sink to the bottom of the cell. The clean coal particles adhere to the froth which floats and is separated. This separation is improved by adding collecting agents like spindle oil, creosote oil, xanthates and soap to the system. The cleaned coal is recovered by vacuum filtration of the froth.

Normally, a froth floatation plant consists of a battery of number of floatation cells with a capacity of 5-10 tons/hr each.

Advantages of froth floatation process :

- It is suitable for fine coals (< 0.5 mm) *e.g.* mine dust, slack coal, washing rejects and slurries.
- It is especially used for producing ultra pure coal for electrode carbon manufacture.

Disadvantages:

- It has a high capital and running cost. —It can't clean lumpy coals unless powdered.

Oil agglomeration process for coal cleaning. Developed by CFRI, Dhanbad, coal fines (<200 mesh) are agitated with water & middle distillate petroleum (5-9% of dry coal) in a floatation cell. Dirt particles (hydrophilic) are removed from liquid suspension by selective wetting and agglomeration with a second immiscible liquid. Coal particles (being hydrophobic) is agglomerated with many oils in an agitated aqueous slurry under high shear. Hydrophilic mineral particles are not affected and remain suspended in water. Coal agglomerates of 0.7 to 2 mm size being bigger than mineral particles can be readily removed. Besides agglomeration of coal fines & deashing, this process does dewatering also as the collecting oil is adsorbed on the surface of the particles displacing moisture.

COAL WASHING IN INDIA

Indian coals have high ash and difficult washability, hence our washeries generally produce coal middlings (having 30-45% ash used in thermal power plant) in addition to clean product and reject. Ash content in Indian metallurgical coal is as high as 20-24% as compared to the upper limit of 10% abroad. Moreover, thermal power plants in India have to tolerate 30 to 45% ash in coal middling used in the boiler (which is a co-product of coal washeries) at the cost of its efficiency. Washing of Indian coals posses many problems hence it is practised only when essential. Better Indian coals are being exhausted at a fast rate with the increase in the coal production rate. More and more dirtier coals are being raised. Modern mechanised mining practice also increases the rock contents in the coal. Some coals improve on their coking properties on washing due to decrease in ash per cent and also due to enrichment in vitrain. The status of coal washeries in India is given in Table 6.5.

Table 6.5: The Status of Coal Washeries in India

Location	Capacity, tons / hr	Washing Process used
For prime coking coal		
West Bokaro	135	Heavy medium-deep bath (chance process).
Lodna	45	Feldspar Jig (Oil Agglomeration Test Unit)
Jamadoba	300	Heavy medium-deep bath (chance process).
Kargali	650	Heavy medium-shallow bath and Baum Jig.
Durgapur Steel plant	360	Heavy medium-shallow bath and Baum Jig.
Sudamdih	500	HM cyclone and floatation
Chasnala	550	Baum Jig and Heavy medium washer.
Bhojudih	500	Heavy medium (shallow bath), HM cyclone and Baum Jig.
Patherdih	500	Baum jig, heavy medium-shallow bath and HM cyclone.
Barora I	57	Baum jig, floatation and HM cyclone.
Munidih	500	Baum Jig, HM cyclone and hydrocyclone.
Dugda (no. 1 and no. 2)	600/600	HM-shallow bath, H.M. cyclones/H.M. cyclone and hydrocyclone.
Non-coking coal/Medium coking coal		
West Bokaro	65	H.M. washer (Chance Process)
Kathara	800	H.M.-shallow bath, HM cyclone and floatation cell.
Gidi	800	H.M.-shallow bath and Baum Jig.
Sawang	115	HM cyclone and Hydrocyclone
Nandan	115	Jig and Floatation
Ramgarh	345	H.M. cyclone and Batac Jig

Typical Indian coal washeries. At Dugda washery, the run of mine coal is screened and the coal above 7.5 cms in size is crushed to less than 7.5 cms size. Crushed coal is fed to a Baum Jig Washer having a slurry of sand and water of specific gravity 1.7 to 1.8. The purer coal floats and the impurities settle and are rejected. The clean coal is separated into three different sizes *viz.* 2.5-7.5 cms, 0.6-2.5 cms and below 0.6 cm. Coal below 0.6 cm size is not washed further as it has lower ash content. The other two sizes are further washed in a slurry of specific gravity 1.5 to yield the products namely clean coal (17% ash), coal middlings (17— 35% ash) and rejects (> 50% ash).

Coal washing is an integral part of coal production. Raw coal coming from mines is washed to remove the ash contents to make them fit for feeding into boilers, particularly those of steel plants. Barring a few instances, a coal washery does not form part of a coal mine in India.

Total installed capacity of washeries in the country is around 131.24 Million tonne per year (MTY) as on 31.3.2015 . As on 31.03.15, a total of 52 washeries, both PSUs and Private, were operating in the country considering both Coking (29.69 MTY) and Non-Coking Coal (101.55 MTY).

Table 6.5(B): List of Coal Washeries to be Setup by 2016-17 in public & private sector

	COKING COAL		Capacity(MTY) as on 31.03.2015
1	Dudga-II, CIL	Jharkhand	2.00
2	Bhojudih, CIL	West Bengal	1.70
3	Patherdih, CIL	Jharkhand	1.60
4	Moonidih, CIL	Jharkhand	1.60
5	Sudamdih, CIL	Jharkhand	1.60
6	Mahuda, CIL	Jharkhand	0.63
7	Kathara, CIL	Jharkhand	3.00
8	Swang, CIL	Jharkhand	0.75
9	Rajrappa, CIL	Jharkhand	3.00
10	Kedla, CIL	Jharkhand	2.60
11	Nandan, CIL	Madhya Pradesh	1.20
(B) Private			**19.68**
12	Durgapur, SAIL	West Bengal	1.50
13	DCOP, DPL	West Bengal	1.35
14	Chasnala, IISCO	Jharkhand	1.50
15	Jamadoba, TISCO	Jharkhand	0.90
16	West Bokaro-II, TISCO	Jharkhand	1.80
17	West Boakaro-III,TISCO	Jharkhand	2.10
18	Bhelatand	Jharkhand	0.86
(B) PSU& Private			**10.01**
TOTAL (A + B) NON-COKING COAL			**29.69**
1	Dugda-I,CIL	Jharkhand	2.50
2	Madhuban,CIL	Jharkhand	2.50
3	Gidi,CIL	Jharkhand	2.50
4	Piparwar,CIL	Jharkhand	6.50
5	Kargali,CIL	Jharkhand	2.72
6	Bina,CIL	Uttar Pradesh	4.50
(A) CIL			**21.22**
7	Dipka, Aryan coal beneficiation pvt. Ltd.	Chattisgarh	12.00
8	Gevra, Aryan coal beneficiation pvt. Ltd.	Chattisgarh	5.00
9	Panderpauni,Aryan coal beneficiation pvt. Ltd. Maharashtra		3.00
10	Chakabuwa, Aryan Energy private ltd.	Chattisgarh	4.00
11	Indaram, Aryan Coal Benefication Pvt.Ltd.	Andhra Pradesh	-
12	Talcher, Aryan Energy Pvt. Ltd.	Odisha	2.00
13	Wani, Kartikay Coal washeries pvt. ltd.(Aryan)	Maharashtra	2.50

14	Korba, ST-CLI Coal washeries ltd.	Chattisgarh	5.20
15	Ramagundam, Gupta coalfield & washeries ltd.	Andhra Pradesh	2.40
16	Sasti, Gupta coalfield & washeries ltd.	Maharashtra	2.40
17	Wani, Gupta coalfield & washeries ltd.	Maharashtra	1.92
18	Umrer, Gupta coalfield & washeries ltd.	Maharashtra	0.75
19	Bhandara, Gupta coalfield & washeries ltd.	Maharashtra	0.75
20	Gondegaon, Gupta coalfield & washeries ltd.	Maharashtra	2.40
21	Majri, Gupta coalfield & washeries ltd.	Maharashtra	2.40
22	Bilaspur, Gupta coalfield & washeries ltd.	Chattisgarh	3.50
23	Ghugus, Gupta coalfield & washeries ltd.	Maharashtra	2.40
24	Talcher, Global coal Mining (P) Ltd.	Odisha	2.50
25	Ib Valley, Global coal Mining (P) Ltd.	Odisha	3.25
26	Ramagundam, Global coal Mining (P) Ltd.	Andhra Pradesh	1.00
27	Wani, Bhatia International Ltd.	Maharashtra	3.73
28	Ghugus, Bhatia International Ltd.	Maharashtra	4.00
29	Jharsuguda, Bhatia International Ltd.	Odisha	1.50
30	Tamnar, Jindal Steel & Power Ltd.	Chattisgarh	6.00
31	Wani, Indo Unique Flame Ltd.	Maharashtra	2.40
32	Nagpur, Indo Unique Flame Ltd.	Maharashtra	0.60
33	Punwat, Indo Unique Flame Ltd.	Maharashtra	2.40
34	Dharamsthal, BLA Industries	Madhya Pradesh	0.33
(B) Private			**80.33**
TOTAL (A+B)			**101.55**
Gross Total (Coking+Non-Coking)			**131.24**

CIL (Coal India Limited) is going to set up 19 nos. of coal washeries on Build, Operate & Maintain basis in the first phase. These washeries are expected to be commissioned by 2016-17. These washeries will be set up in different subsidiaries of CIL except NCL. Details of, subsidiary wise, no. of washeries to be set up and their capacities are listed below in table 6.5 (C).

Table 6.5 (C): List of Coal Washeries to be Setup by 2016-17 in Subsidiary Companies of CIL

Subsidiary of CIL		Nos. of washery	Capacity, Million tons/year (MTY)
ECL	Chitra (2.5 MTY), SonpurBazari (5.0 MTY).	02	7.50
BCCL	Madhuban (5.0 MTY), Patherdih (5.0 MTY), Patherdig (2.5 MTY), Bhojudih (2.0 MTY), Dugda (2.5 MTY), Dahibari (1.6 MTY),	06	18.60
CCL	Ashoka (10.0 MTY), Piparwar (3.5 MTY), Karo (2.5 MTY), Konar (3.5 MTY)	04	19.5

MCL	Basundhara (10.0 MTY), Hingula (10.0 MTY), Jagannath (10.0 MTY), Samleshwari (10.0 MTY).	04	40.0
SECL	Kusumunda (5.0 MTY), Baroud (5.0 MTY).	02	10.0
WCL	Ghughus/Wani (5.0 MTY)	01	5.0
	Total :	19	100.6

ABBREVIATIONS

BCCL - Bharat Coking Coal Limited, CCL - Central Coalfields Limited
WCL - Western Coalfields Limited, NCL - Northerm Coalfields Limited
ECL - Eastern Coalfields Limited, SECL - South Eastern Coalfields Limited
WCL - Western Coalfields Limited

Post washing treatment of coal. In wet washing process, the washed coal is associated with considerable moisture. The coarser fractions of coal are dewatered in oscillating screen, drainage bunker or centrifuge. The finer fractions which form a slurry are dewatered by thickeners and filters. In slurry dewatering, coagulating agents are used.

In heavy medium plants, a circuit for heavy medium recovery and purification is included.

WASHABILITY OF COAL

Cleaning processes generally depend upon differences in density between clean coal and its impurities. They suitably remove the free dirt but not the inherent dirt. The extent of removal of free dirt or the amenability of a coal to improvement in quality is more commonly known as the 'washability' of coal and is indicated by 'float-and-sink' analysis of the coal. These washability investigations are conducted before any proposal for installation of a coal washery is considered. The float-and-sink test is done on sized fractions of the raw coal by separating them into fractions of increasing density by means of a series of liquids with relative densities increasing in steps of 0.05 from 1.30 to 1.80. Each fraction is weighed and the ash is determined and the standard washability curves are plotted. A coal which is easy to clean gives a high yield of low ash fraction ('cleans') of relative density below 1.50 and low yield of 'middlings' of relative density between 1.50 and 1.80. The ease of separation of coal depends on the 'near gravity' fraction *i.e.* percentage of coal with ±0.1 of the relative density selected for cleaning. The fraction sinking at relative density 1.80 is called 'rejects'. Easily washable coals yield high ash in the rejects. A typical coal washability curve is given in Figure 6.8. In washing a coal of 17.5% ash content to cleans containing 8.0% ash, the following data may be noted from the Figure 6.8 :

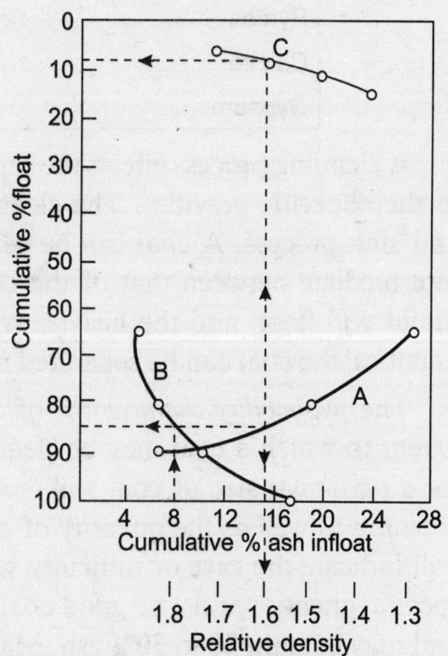

Figure 6.8. Typical washability curves of coal

Relative density-Yield, (B) Yield-Ash, (C) Distribution of near gravity material.

Yield of clean coal : 85% (curve B)

Relative density of the heavy medium to be used : 1.6 :(curve A)

Near gravity material in the clean coal : 8.0% (curve C)

For the purpose of density separation in the laboratory, mixtures of carbon tetrachloride, bromoform and benzene, or solutions of zinc chloride or sulphuric acid are used. In industrial practice, suspensions of fine grains of sand, baryte, magnetite, shale etc. in water are used due to prohibitive cost of those chemicals.

The specific gravity of coal depends on the kind of coal and the moisture & mineral contents in it.

Specific gravities of several kinds of coal and minerals matters which are normally associated with coal is given in the Table 6.6.

Table 6.6. Specific Gravity of Coal and Mineral Matters

Coals and Mineral Matters	Specific Gravity
Pure bituminous coal	1.28-1.37
Bone coal (coal containing layers of shale and state)	1.40-1.60
Shale, clay and sand-stone,	2.0-2.64
Pyrites	4.0-4.90
Calcite	2.7
Gypsum	2.3

A cleaning process effects the separation between coal and impurities by utilizing the difference in their specific gravities. The cleaning quality of a coal is studied by what is known as a float and sink process. A coal can be separated by immersing it in a solution with a specific gravity intermediate between that of the coal and the principal impurity. The material lighter than the liquid will float, and the heavier will sink. By using a series of solutions with different specific gravities, the coal can be separated into a number of fractions.

The *washability characteristic* of a coal is the most important tool available to determine the extent to which a coal may be cleaned. Examination of washability data for a particular coal or for a particular size of coal will reveal the quality of coal which may be obtained by mechanical cleaning as well as the quantity of coal of the particular quality. Further examination of the data will indicate the ease or difficulty with which the coal may be cleaned. The fractions below 1.40 specific gravity consist of good coal with minimum ash. The 1.40 to 1.60 fractions are middlings and may contain 25 to 30% ash. Material within this range is not appreciably affected by ordinary washing processes. The fractions beyond 1.60 are high in ash and discarded as refuse. An ideal coal should have no fractions between the first and last. In India however, coal middling having ash upto 45% are used in power plant.

The *washability data* is obtained by the float and sink test of a coal sample. Liquids of different specific gravities are prepared either by mixing organic liquids in different proportions or by making inorganic solutions of different concentrations. The liquids are taken in tall jars and arranged in the order of increasing specific gravities. The sample is first placed in the lowest specific gravity liquid. The fraction lighter than the liquid floats and heavier one sinks. The portion which floats on a particular specific gravity, is known as float fraction at that specific gravity and the portion which sinks is known as sink fraction. The sink is then placed in next higher specific gravity and the float and sink fractions separated. In this way the float and sink fractions of different specific gravities are collected and weighed. Proximate analysis mainly ash determinations are done for all float and sink fractions and the data'thus tabulated are plotted to give the washability curves.

The washability curve (of yield *vs.* ash type) indicates the quantity and quality of coal to be obtained by cleaning at certain specific gravities. To obtain the specific gravity to be used in order to meet ash requirement an estimate had to be made between the specific gravity levels.

The data necessary to plot washability curves are obtained in the laboratory by examining the float-and-sink characteristics of a particular size fraction of the coal. To determine these characteristics, it is necessary to have a representative sample of the coal crushed to a specific top size. Generally, the top size is determined by market specifications, by the properties of the coal, by the method of mining and by the capacity of the test equipment.

A typical result of a float and sink test of coal is tabulated in Table 6.7 and recorded graphically as "washability curves" following calculations of yield and ash contents.

The washability curves are shown in Figure 6.9.

The ease of separation of mineral matters from coal depends upon its nature which may be grouped into two classes. One class consists of the substances derived from the coal forming plants, and the other consists of the constituents added to coal from external sources known as

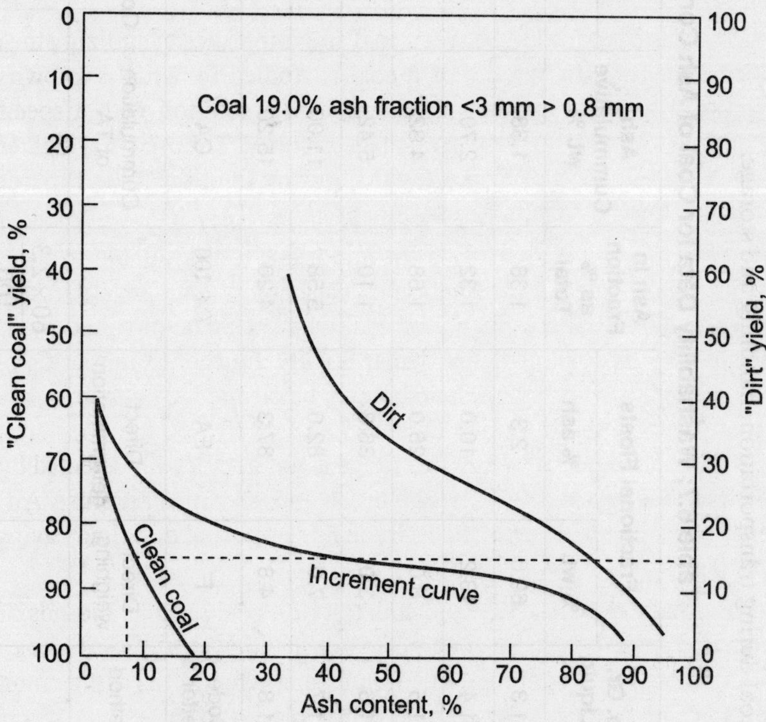

Figure 6.9: Typical washability curves

'inherent' and 'extraneous' mineral matter respectively. A part of the extraneous mineral matter is intimately associated with the organic mass of coal. This is due to their addition to the coal during its early formation stage. The inherent mineral matter cannot be removed from coal by any mechanical method. Chemical treatment like acid leaching are to be used for this purpose. The content of inherent mineral matter in coal is small (< 5%). The extraneous mineral matter is removable by mechanical means. However, the intimately associated variety can be separated only with difficulty. A coal is easy to wash when the bulk of its mineral matter is due to dirt bands and rock particles which get mixed with the coal during mining or due to the dirts settled on coal during transportation, handling and storage.

Table 6.7: Washability Data for Coal of Ash Content 19.0% (A) for Fractions < 3 mm > 0.8 mm

Sp. Gr. of Liquid	Fractional Floats		Ash in Fraction as % Total	Ash Cummulative wt. %	Cummulative Floats		Fractional Sinks % ash	Cummulative Sinks	
	% wt.	% ash			% wt.	% ash		% wt.	% ash
1.3	60.0	2.3	1.38	1.38	60.0	2.3	17.62	40.0	44.0
1.4	13.2	10.0	1.32	2.70	73.2	3.7	16.30	26.8	60.7
1.5	8.0	26.0	1.68	4.32	81.2	5.3	14.68	18.8	79.0
1.6	3.0	36.8	1.10	5.42	84.2	6.8	14.58	15.8	84.0
1.7	7.0	82.0	5.58	11.00	91.0	12.0	8.0	8.8	91.5
1.8	4.8	87.3	4.20	15.20	96.0	15.8	8.8	3.0	95.0
Code letter	F	FA	TA 100	CA	CF	CFA	FSA	CS	SA
Method	Direct weighing	Direct determination		Cummulation of TA	Commulation of F	$\frac{CA}{CF} \times 100$	A-CA	100-CF	$\frac{FSA}{CS} \times 100$
Example			$\frac{60 \times 2.3}{100}$ = 1.38	1.38+1.32 = 2.70	60+13.2 = 73.2	$\frac{2.7}{73.2} \times 100$ = 3.7	19-2.7 = 16.3	100 - 73.2 = 26.8	$\frac{16.3}{26.8}$ = 60.7
Use		For increment curve			For clean coal curve			For dirt curve	

The washability characteristics of coal is evaluated by the float and sink test which utilises the fact that the mineral substances have much higher specific gravity than the pure coal. The specific gravities of pure bituminous coal, shales and pyrites are of the range : 1.28-1.37, 2.0—2.64 and 2.4-4.95 respectively. By using liquids of different specific gravities, the coal sample is divided into a number of fractions in the increasing order of specific gravity and hence of ash value. Usually three curves are constructed from the results of float and sink test *viz.* Total Floats-Ash Curve, Total Sinks- Ash Curve and Washability characteristic Curve or Instantaneous Ash Curve.

Table 6.8: A Typical Set of Float and Sink Test Results of Coal

Specific Gravity of separation	Yield of each fraction (%)	Ash of each fraction (%)	Cumulative Yield of total Floats (%)	Ash of total Floats (%)	Cumulative Yield of total sinks (%)	Ash of total sinks (%)	Cumulative yield upto middle of fraction (CMF)(%)
(1)	(2)	(3)	(4)	(5)	(6)	(7)	(8)
1.40	30.0	9.0	30.0	9.0	70.0	36.1	15.0
1.50	20.0	12.0	50.0	10.2	50.0	45.7	40.0
1.60	26.0	23.0	76.0	14.6	24.0	70.2	63.0
1.70	6.0	47.0	82.0	16.9	18.0	78.0	79.0
> 1.70	18.0	78.0	100.0	27.9	-	-	91.0

From the first two curves, we can find the yields & ash of the clean product (float) and the heavy waste (sink) at a certain specific gravity of washing. Similarly, if it is intended to know at which specific gravity the cleaning should be done with a view to producing the maximum yield of clean coals within an upper limit of ash value, the Total Floats-Ash Curve is consulted. The yield of sink and its ash are read from the Total Sinks-Ash Curve at the same specific gravity. The third curve gives the washability characteristics of coal. It is so constructed (Data of column 8 *vs.* data of column 3 of Table 6.7) as to indicate the variation of ash from particle to particle in the increasing order of higher specific gravity. Hence this curve clearly shows how far it is possible to separate the dirts from the clean coal by mechanical methods. For an easily washable coal, this curve should have a sharp bend. On the other hand, it tends to a straight line joining the cleanest and the dirtiest particles in the case of a very difficult coal as shown in Fig. 6.10.

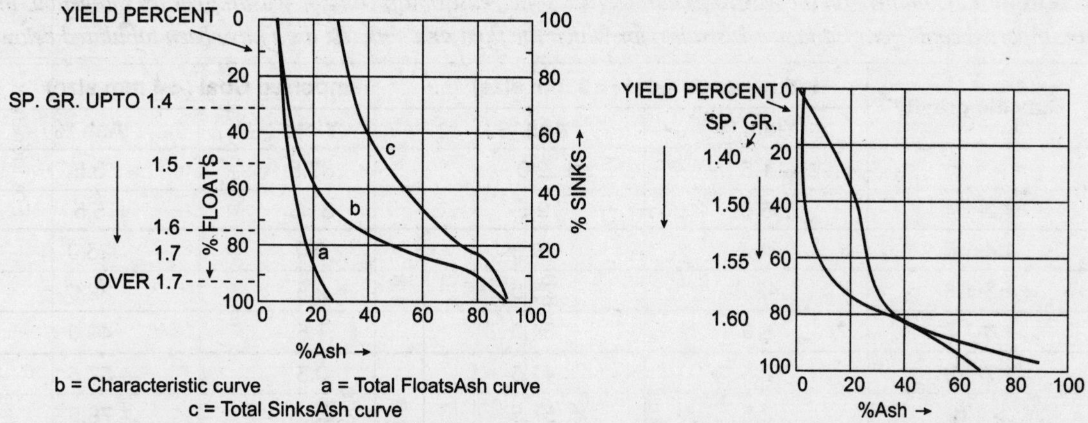

Figure 6.9 (A). *The washability curves*

Fig. 6.10 *A. Characteristic for Easily washable coal, B. Characteristic fo difficult coal*

Preliminary information about the washability of a coal is obtained from the numerical data of the float and sink test. If the yields of fractions between 1.40 and 1.60 specific gravities are very small, it indicates that the mineral particles are not intimately associated with the organic mass and the coal is easily washable. If the yields of these fractions are considerable, it means that much of the mineral matter is intimately associated with the pure coal and hence difficult to separate.

INFORMATION OBTAINED FROM TABLE 6.7 AND CURVES OF FIG. 6.9 :

If a "cut" is made at specific gravity 1.6, then :

Yield of "clean" coal = 84.2% } From clean coal curve
= 6.8% }

Ash in dirt = 84.0% – from dirt curve

Ash content of dirtiest particle included in clean coal = 36.8% – from increment curve.

— If clean coal is required to contain 5.3% ash, the cut should be made at 1.5 specific gravity, and the yield of clean coal would be 81.2%.

— The clean coal curve represents the maximum possible cleaning efficiency by gravity or jig separation for this size grading of coal.

— Variation in ash content of different sized fractions of coal is the practical limitation. In any sized fraction of coal, only dirt that appears as separate particles (free from coal) can be separated by the aid of gravity.

— Instantaneous ash curve (increment curve) shows the rate of change of ash content for different yields. If a coal gives a L-shaped curve {*e.g.* ash content increases rapidly for a small increase in yield in the middle portion of the curve) the coal is easy to clean. If the curve is more nearly straight, the coal is hard to clean and a high proportion of middlings is obtained (coal of medium to high ash content).

— Specific gravity distribution curve (middlings curve) is sometimes drawn by plotting the difference in yields at two chosen densities against the mean of these densities. This gives the weight of coal falling within say 0.1 of a given density. If this weight is less than 10%, the coal is easy to clean.

Example 1. *Comment on the washing characteristics (i.e., ease or difficulty of washing) of an indigenous and an imported coal (from Canada / Australia) for which the float and sink test data have been tabulated below.*

Specific gravity	Indigeneous Coal (< 3 mm size)		Imported Coal (< 4 mm size)	
	Yield %	Ash %	Yield	Ash %
< 1.3	1.5	2.0	37.5	3.5
1.3-1.4	6.0	9.5	31.5	5.5
1.4-1.5	38.5	19.5	3.0	25.0
1.5-1.6	43.0	25.0	2.5	32.5
1.6-1.7	6.0	34.0	1.5	44.0
1.7-1.8	2.5	41.5	0.5	57.5
> 1.8	2.5	57.5	23.5	78.5
Total	100.0	23.5	100.0	23.5

Solution. Yield of fraction between 1.4 to 1.6 specific gravity is 38.5 + 43.0 = 81.5%, for *indigeneous coal* and, 3 + 2.5 = 5.5%, for *imported coal*

This shows that *indigenous (Indian) coal* even after crushing to < 3 mm size contains large quantities of particles where organic mass and mineral matter are intimately associated while the imported coal has a very small portion of such particles. Hence, minerals can be separated with great difficulty in case of indigenous coal while it can be very easily separated from *imported coal.*

This inference is further confirmed by high yield of 37.5 + 37.1 = 69% with ash below 5.5% obtainable from the imported coal at a specific gravity of 1.4 while the corresponding yield is only 1.5 + 6 = 7.5% from indigenous coal. The high ash value (between 19.5 and 25%) of the 1.4-1.6 specific gravity fraction with a high yield of 38.5 + 43 = 81.5% also speaks for the intimately associated nature of the mineral matter in the indigenous coal.

If 1.5 is taken as the specific gravity of washing, the fraction 1.4-1.6 forms the near gravity material. The data in the table shows that coal washing problem will be 'formidable' for indigenous coal while it will be 'simple' for the imported coal.

If we take 1.7 as the specific gravity of washing for indigenous coal by tolerating high ash in the clean 'product,' near gravity material will be 6 + 2.5 = 8.5%. Hence, the coal washing problem will become "moderately difficult".

And, the ash% of the product at 1.7 specific gravity in case of indigenous coal is

$$= \frac{1.5 \times 0.02 + 6 \times 0.095 + 38.5 \times 0.195 + 43 \times 0.25 + 6 \times 0.34}{1.5 + 6 + 38.5 + 43 + 6} \times 100 = 22\%$$

WASHING EFFICIENCY

The usefulness of a coal cleaning process is on the efficiency with which it separates the clean product from the dirty part. Ideally, the separation should be complete and there should not be any wrong migration of particles *i.e.* heavier particles to the product and lighter particles to the tailings. Three formulae for calculating the washer efficiency are given below :

Fraser and Yancey formula :

$$\text{Efficiency\%} = \frac{\text{recovery of clean coal}}{\text{theoretical recovery at the given specific gravity}} \times \frac{\text{\% ash in feed coal} - \text{\% ash in clean coal}}{\text{\% ash in feed coal} - \text{\% ash in theoretical clean coal}} \times 100$$

Yancey and Geer formula :

$$\text{Effeciency \%} = \frac{\text{recovery of clean coal x 100}}{\text{theoretical recovery at the same ash level of the clean coal}} \times 100$$

Anderson formula :

Efficiency % = 100 - (sink in clean coal + float in refuse).

The recovery of clean coal obtained in the washer is calculated from the ash balance as given Below :

$$\text{Recovery \%} = \frac{\text{ash in refuse} - \text{\% ash in feed}}{\text{\% ash in refuse} - \text{\% ash in clean coal}} \times 100$$

The *Fraser and Yancey formula* is quite widely used but it does not allow for any size degradation taking place in the actual washer operation which releases more dirt from the pure coal and alters the washability characteristics of the feed. Moreover a high recovery of a low grade product may result in an efficiency greater than 100%. Similarly an exceptionally clean coal with a low recovery will give misleading results.

The *Yancey and Geer formula* is an improvement upon the Fraser and Yancey formula. It is more simple and not likely to give misleading results, as it is based upon the theoretical recovery at the actual ash level and not at the originally fixed specific gravity. In actual operation, the gravity of separation may differ from the set value. This formula, however, does not guard against the inaccuracy resulting from the size degradation of coal in the washer operation.

The demerits of the above two formulae are absent in the *Anderson formula.* The separation gravity and the quantity of wrongly migrated particles (sink in clean coal plus float in refuse) are determined from the float and sink tests of the clean coal and the refuse. Any change in the set value of the specific gravity and the washability characteristics of the feed during washer operation is therefore taken into consideration. The *Anderson formula* however does not indicate the quantity of the non-combustible matter carried by the wrongly migrated particles. The total amount of migration cannot satisfactorily measure the quality of products.

Example 2. *Following data is available for a coal washery washing an Indian coal:*
Ash in the raw coal to be washed = 21%, Ash in the 'clean product' = 18.5%, Ash in the 'sinks' = 41%
If the theoretical recovery is 91.5% for a product of the same ash level, calculate the performance/efficiency of the washery.

Solution. From the ash balance, we get,

Plant yield of 'clean product' $= \dfrac{41-21}{41-18.5} \times 100 = 88.9\%$

Hence, Yancey and Geer efficiency $= \dfrac{88.9}{91.5} \times 100 = 97\%$

THE STORAGE OF COAL-OXIDATION & SPONTANEOUS COMBUSTION

Freshly mined high volatile coal when stored in bulk undergoes low temperature atmospheric oxidation due to the presence of methane and other volatile matter on the surface. This exothermic oxidation causes the rise in temperature of the coal and if the heat is not removed, a stage comes when coal begins to burn. This is called spontaneous combustion which leads to outbreak of fire in the stored coal.

If the temperature rise due to oxidation does not exceed a critical value (say 50°C for lignite and 80°C for bituminous coals), spontaneous combustion does not take place but the quality of coal is affected depending on the degree of oxidation.

Low temperature oxidation of coal *causes :*

- decrease in its caking power and calorific value
- decrease in carbon and hydrogen content
- decrease in its size grading (physical breakdown and increase in friability)
- increase in its oxygen content
- fire, if the temperature rise exceeds the critical value, *i.e.* 70°C
- decrease in gas and tar yields on its carbonisation

Oxidation of high volatile matter containing coal (*i.e.* low rank coal) is more compared to that of low volatile coals (high rank). That means the chances of catching fire in anthracite is less compared to bituminous coal. As the maturity of the coal increases, its tendency to catch fire during storage decreases. It is worth noting that the ***rate of oxidation doubles for a rise in temperature of 10°C*** and increases proportionally to the cube root of the surface area per unit weight of coal.

Low temperature oxidation of coal is accelerated by:

- temperature rise
- access to diffused air
- presence of small particles
- presence of porous or friable particles
- storage in large heaps with small surface to volume ratio.

On the other hand, low temperature oxidation is retarded by :

- cooling either by ventilation or water spray
- reducing access to air, *i.e.* by storage in compressed piles or storage under water
- reducing the size of the storage piles so that heat losses by conduction and convection to the atmosphere is high thus preventing temperature rise of the stored coal. Size and nature of storage piles is determined by the rank and type of coal. The lower the rank of the coal, the more shallow and smaller the pile and smaller the proportion of fine coal should be for avoiding fire risk. Anthracite is stored with less risk than bituminous coal.

Favourable conditions for safe storage of coal have been summarised in the Table 7.1.

Spontaneous combustion usually occurs within one to four months of mining. Coals already stored longer than six months with exposure to air are usually not liable to spontaneous combustion.

Spontaneous combustion may be avoided by decreasing the exposed surface of coal lumps and suppressing the ventilation. The rate of oxidation increases proportionally to cube root of the surface area per unit weight of the coal. The exposed surface area is reduced by avoiding segregation and by packing the coal tightly and uniformly. Blanketing the pile with fine coal also helps in decreasing the ventilation. Another useful measure is to surround the coal pile by a wall. All these measures minimize the extent of oxidation. Another approach is to enhance the ventilation in such a way that the liberated heat is fast dissipated away, but this method is impractical.

Table 7.1: Conditions for Safe Storage of Coal

Factors	Conditions favourable to safe storage	Conditions not favourable to safe storage
Size grading	Exclusion of fine coal	run of mine coals containing natural fines, coals having finely divided pyrites (sulphides)
Size of pile	< 200 tons for bituminous and higher rank coals	> 200 tons for bituminous coal.
	< 50 tons for lignites	> 50 tons for lignites
Height of pile	< 3 metres for bituminous coals	> 3 metres for bituminous coals
	< 1 metre for lignites	> 1 metre for lignites
Storage period after mining	> 6 months	upto 4 months for bituminous, upto 1 month for lignites
History of storage	Stored after initial heating and cooling	freshly mined
Coal type	Cannel coal, anthracite, Dry steam coal	brights, lignite, dirty coal containing pyrites
Site of store	Open positions on clean firm ground	in contact with hot pipes walls or surfaces, enclosed stores, contact with oily waste.

Whenever coal is stored in bulk for a long time, its temperature should be measured at various points of the pile from time to time. Above 70°C, the risk of spontaneous inflammation is more and hence the coal should be consumed before this critical temperature is reached. Water spraying helps only in the initial stage but not after crossing the danger point. Coal should not be stored near a source of heat.

The problems connected with the storage of coal also apply to the coal in the ship's bunkers and cargoes. Here the conditions of ventilation and proximity of heat source are such that the spontaneous inflammation may occur in piles of smaller quantities *i.e.* less than 200 tons. Therefore greater care must be taken to prevent any rise of coal temperature beyond the critical value.

Freshly mined coals are most susceptible to spontaneous combustion. This hazard usually occurs within 4 to 5 weeks of mining. If coal has been safely stored for more than six months, it is not usually liable to spontaneous combustion.

Though keeping sufficient stock of coal has many disadvantages like blockage of large amount of money, blockage of useful factory space, labour cost in complex type of handling/ stacking, carpet loss, deterioration in quality and fire hazard, still it is necessary to store, coal in sufficient quantities to counter the uncertainties in availability & transportation. There are certain inherent losses with respect to quality and quantity of coal in stock in addition to it being a potential fire hazard. These *losses* enlisted below should be controlled to a minimum and adequate safety measures should be taken to avoid the spontaneous combustion. *Losses* liable to occur *during coal storage are :*

- The oxidation loss (weathering of coal)
- Wind loss
- Carpet loss
- Loss of coal by spontaneous combustion.

Oxidation loss of coal. The moment coal is mined out, it starts absorbing oxygen from the atmospheric air, even at normal temperature. With the increase in temperature of coal, oxygen absorption accelerates.

Absorption of oxygen by coal results in deterioration of coal quality. When oxygen absorption reaches its saturation point, further deterioration retards and ceases.

Experiments on the effect of oxygen on coal, have shown that 1% of oxygen in coal has the same effect as 1% of ash in coal. Oxygen and ash have the same anti-calorific effect. Fig. 7.1 gives a rough idea of oxygen absorption by coal with time. Fig. 7.2 shows the effect of time of storage on the C.V of coal. Fig. 7.3 shows the relation between the heating value of coal and the ratio $\frac{\text{Carbon}}{\text{Oxygen} + \text{ash}}$ Ash remaining constant, higher the oxygen in coal (absorbed oxygen), the lower will be the C. V. of coal. However, if the temperature of coal is not allowed to rise above 60°C, bituminous coal suffers relatively small depreciation in the heating value. Keeping the above in mind, long time storage of coal should not be allowed.

In case of bituminous coal, absorption of oxygen becomes significant at 27°C while at about 120°C, its slow combustion starts with the evolution of carbon dioxide and water vapor.

The rate accelerates at 140 – 160°C and the rate of increase of temperature of coal becomes much greater between 200 to 275°C. Active combustion of coal occurs, when it reaches its ignition point which may be between 300 to 400°C.

- All coals are liable to atmospheric oxidation except anthracite. The liability increases with decreasing rank. Some lignites are storable with safety only under water.
- The changes occurring in bituminous coal due to its slow oxidation is recognised by increased solubility of the coal in caustic soda solution.Moisture in coal does not have significant effect on the rate of oxidation of coal.

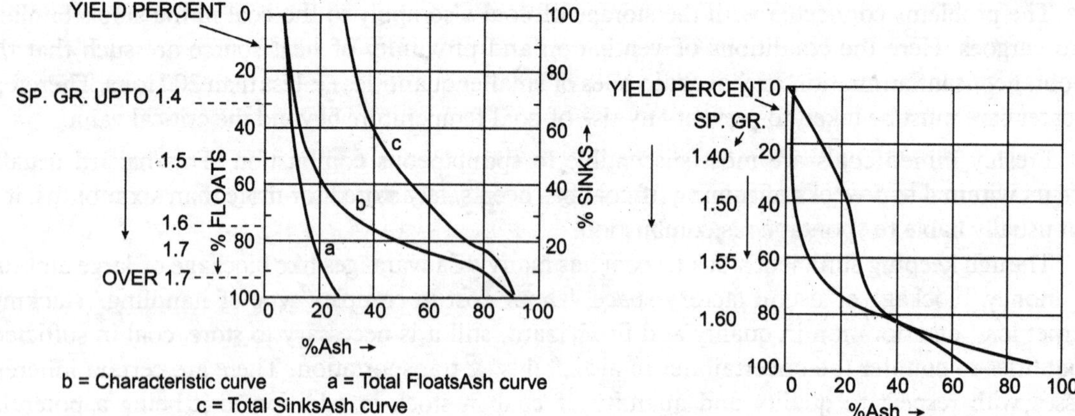

Fig. 7.1 Rate of absorption of oxygen by a coal.

Fig. 7.2 Decrease in heating value of a coal on weathering.

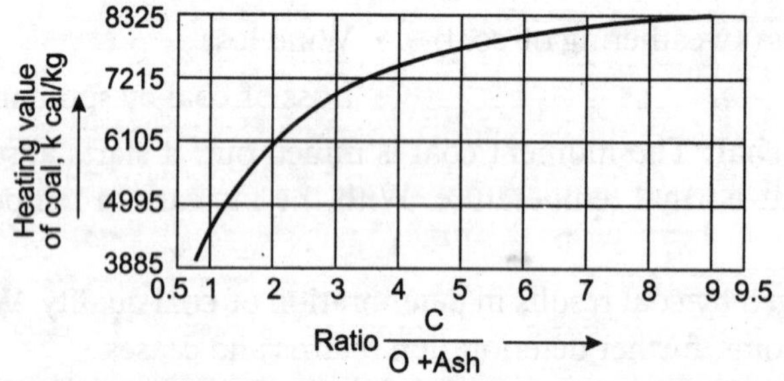

Fig. 7.3 Relation between the heating value of coals and the ratio $\frac{C}{O + Ash}$

- Rate of adsorption of oxygen by coal increases with increasing exposed surface area of coal and increasing proportion of fines in coal.
- Rate of oxidation of coal varies as the square root of the concentration of oxygen.

Wind loss of coal. Due to blowing of wind, it has been estimated that **0.5 to 1.0%** of coal is lost. Providing scientific bays/silos/enclosures and modifying methods of handling according to the situation helps.

Carpet loss of coal. Generally, coal is unloaded from wagons and stored on the ground which is not prepared for the purpose. This results in mixing up of ground clay and soil with the bottom most layer of coal. This loss can easily be avoided by paving the storage floor.

Loss of coal due to spontaneous combustion. Due to spontaneous combustion, considerable amount of coal is lost. Higher atmospheric temperature promotes reaction of oxygen with coal in the stock yards. Absorption of oxygen in coal begins as soon as it is broken out in the mine. Oxidation is most rapid in a freshly mined coal at higher temperatures. Some facts in this regard have been established and are given below :

- Oxidation of coal is a continuous process depending upon the time and conditions.
- The rate of oxidation increases with increasing temperature and decreasing size of coal particles.

- For a given coal, there is a critical temperature below which spontaneous combustion cannot take place. The initial attainment of this temperature and continuation thereof, depends on external conditions. That means, it can be prevented if these conditions are withdrawn. But above this critical temperature, oxidation is independent of external conditions.
- The external conditions which cause initial attainment of critical temperatures are :

(i) External source of heat such as sun's rays, nearby steam pipes, furnace flues etc.

(ii) Fineness of size of coal which gives an increased surface area to come in contact with free air (Oxygen).

- Coals with high textural moisture and high oxygen content have relatively low critical temperatures and are liable to ignite spontaneously.

For safe storage of coal, the following guidelines should be adhered to :

Coal dust. As far as possible, dust coal should be separated and used or disposed off as early as possible.

Ventilation. To prevent penetration of air into the coal heap, three sided wall storage with hardened ground should be preferred. Ventilation must be reduced by various means such as compacting the exposed surface etc.

Location. The storage area should be away from any external sources of heat.

Height of the coal heap. Height of the heap should not be more than 3.0 m.

Conical heap. Conical heap of coal should be avoided. It increases the exposed area and promotes entry of fresh air into the heap.

Long period storage. Coal should not be stored for longer period, particularly during summer seasons.

Smaller heaps. The air on one hand supplies oxygen to coal for auto ignition which results in rise in temperature, on the other hand, it also produces cooling effect when it is in excess & flowing and helps in heat dissipation. The balance between rise in temperature and heat dissipation depends on local conditions. It is generally seen that in a particular heap, the coal is ignited inside the heap, and burns without flames turning into ash, but the outside of the heap is cooler, and not burning. The heat dissipation in smaller heaps is faster and hence a number of smaller heaps should be preferred against one big heap.

Flow of stock. Stationary coal should be moved and be consumed to make room for fresh coal. Storage of coal for longer periods should be avoided.

Rod test. At-least once a week, the temperature at different places in the coal stack (at 1-2 meters depth) should be observed by rod tests. Generally, a temperature of 70°C is considered to be an alarm for beginning of spontaneous combustion and we can expect a slow and steady rise in temperature within the heap, leading to flameless combustion within the heap.

Hydrant points should be provided near the stock so that in case of emergency, water is readily available to drench the hot coal.

If there is fire inside the heap, the best step will be to remove the hot coal & use it, and seggregate the hotter stock from the unaffected one. The heated coal should be dug up and scattered thinly on ground. In most cases, this brings down its temperature to normal. In some cases, drenching the burning coal with excess water has been found to be very effective. Sprinkling of small quantity of water aggravates the heat/fire instead of quenching it, possibly due to the formation of CO and H_2, both of which are flammable ($C + H_2O = CO + H_2$).

Coal storage conditions. Ideal condition of storage of coal is in air tight chamber or under water. Former is impracticable, while later causes rapid disintegration of coal due to wetting and drying and is justifiable only in special circumstances. Covered stores, silos and bunkers are expensive.

The flattened floor, where the coal is to be stored should be cleared of vegetation, and covered with impervious layer of ashes or clays. Else, the concrete floor would be ideal. To prevent the access of air to the coal stack, sides should be tightly packed or blanketed with a layer of fine coal or building a wall on at least one side is also helpful. Coal stack depth upto 1.5 metres is completely safe. The practically safe coal stack depth is about 2 metres for mixed sizes, about 5 metres for sized coal, about 6 metres for sized hard coal and about 9 metres for anthracite or really large coal. However, large tonnage of coking coal is stored in tall concrete silos and bunkers. Open cast mines coal which is of small size and highly oxidised already is stored by spreading the coal evenly in thin layers.

During transportation of coal through ships in cargo & bunkers fire occurs at times, because of its external heating from the boiler or engine room and unlagged steam pipes. Other reasons of fire may be loading of fresh coal on the top of small heaps of well dried & aerated fine coal at the bottom of the bunker and in the corners. The fresh coal prevents the escape of heat from this well aerated surface resulting in local heating. All these reasons must be taken care of to avoid fire hazard.

For coal stored in concrete bunkers in land, no coal residues should be left in the bunker unused. It should be so designed that all its walls slope at 60° to the vertical towards the outlet. The inward leakage of air should be avoided by repairing holes/joints/cracks and sealing the airtight outlet valves, if the bunker is left unused for some time.

Site selection and ideal conditions for storage and piling of coal:

- Coal storage area should be free from standing water and artificial draining in the storage area to be provided, if well drained area is not available.
- Storage area should be free from foreign matters such as wood, paper, rags, waste oil or materials having low ignition temperature.
- Storage site should be away from steam line, hot water line, boiler blow
- Offline etc. and storage on hot bright sunny days should be avoided.
- Coal piles should be built up in successive compact layers and it should be dressed to prevent rain water from entering into it.
- Unsized coal should be as compact as possible and sides of piles should not be steep.
- Avoid conical piling and make provisions for fire fighting equipments at the storage site and for checking temperatures at various locations.
- Coal should be piled in such a way that air may circulate freely through it, ensuring effective ventilation to dissipate heat.
- Stoker fired sized coal should be oil treated before storing. The oil treatment on the surface of coal particles slows down the absorption of moisture and oxygen. So compaction is not needed.

PULVERISED FUEL/COAL

The rate of combustion of solid fuels is slow because of the difficulty of contact between the fuel and the oxygen. The combustion rate can be increased by increasing the rate at which air is forced in but this means a supply of excess air resulting in waste of heat. The combustion rate can also be increased by pulverising *(i.e.* powdering) the fuel so that air and fuel come in close contact easily.

Advantages of pulverised fuel:

- It can be intimately and uniformly mixed with air necessary for combustion and hence can be burnt completely.
- It can be handled with ease like a liquid fuel and can be transported through pipes.
- Furnace temperature can be easily controlled by increasing or decreasing the rate of burning through a valve just like liquid/gaseous fuel.
- The atmosphere inside the furnace can be easily maintained as oxidising or reducing which is of great advantage in metallurgical furnaces.
- Wide varieties of coals can be used.
- Low grade coals can be used taking precaution with high ash coals to avoid carrying dust & grit into atmosphere.
- Combustion is completed with low percentage of excess air, hence high flame temperature and high combustion efficiency can be achieved.
- Maximum efficiency is possible because of close regulation of rate of fuel and supply of air by automatic control.
- Labour charges are low and maintenance is largely exterior to the furnace.
- It imparts greater flexibility of control. Besides, banking & stand-by losses are minimum.

Disadvantages of pulverised fuel:

- Cost of drying and grinding to fine size is relatively higher.
- Fine dust is discharged into atmosphere along with the chimney gas.
- There is a tendency for slagging on refractory walls and furnace linings.
- It requires larger combustion space to complete the combustion & heat transfer and to avoid deposition of soot in tubes or flues.

- Furnace stock/work can be contaminated by ash from coal.Its use results in erosion of pressure boiler parts by flyash entrained in flue gases resulting in reduced availability and high maintenance cost of boiler.
- Heavy erosion damage of induced draft (I.D.) fan blades by entrained ash in the flue gas occurs.
- Operation and maintenance cost of pulveriser is high.
- Burning of anthracite poses difficulty.

Pulverised coal combustion mechanism. Pulverised coal undergoes complete combustion in three stages :

- As the volatile matter is evolved, some change in shape and size of particles occurs in the pre-ignition stage.
- Thereafter, the ignition and combustion of volatile matter occurs.
- Finally, the combustion of carbonaceous residue occurs. Pulverised coking coals form hollow, thin walled spheres called *cenospheres.* The large ones are more difficult to burn than small ones. The higher is the temperature of formation, the smaller the size and the more rapid is the ignition.

Cenospheres are not formed during combustion of weakly coking and non-coking coals. Some fragmentation occurs with high volatile matter coal. The solid residue from high volatile matter coal is much more reactive than that from low volatile coals (say anthracite). Combustion of the solid residue requires more time than the volatile matter, hence the combustion rate is determined by the reactivity of the solid residue. At 800-1000°C, the rate depends more upon the rate of diffusion of oxygen to the surface. Higher combustion rates can be obtained by increasing the relative motion between air and particle, *i.e.* by turbulence,

The smaller the combustion chamber and lower the volatiles, the finer and more uniform should be the grinding. High rank coals require finer grinding. Generally, the coal should be pulverised such that 80% of it passes through 200 mesh (74 micron) sieve. Further grinding may prove uneconomical as power required in grinding will be high.

Normally, ball mills are used for pulverising the coal. However, the use of air-swept mills (say Raymond Mill) reduces the necessity for drying before grinding except in the case of very wet coals. Otherwise dryers using hot gas as drying medium should be used for drying the coal.

A series of cyclones near the base of the chimney are installed to remove dust from chimney gas before discharging into atmosphere.

Preheated air (upto a temperature of 110-300°C) may be used for transporting the coal through pipes to the burner. Lower temperature of preheat should be used for low rank high volatile coals which may catch fire due to its low ignition temperature. However, for burning anthracite (low V.M. coal), a high preheat temperature of air can be used.

Thermal power plants and cement rotary kilns fire coal in the powdered/pulverised form. Capacity of pulverised fuel burner is limited by the size of combustion space only. Ball mill or impact mill is used for pulverising the fuel. The size of pulverised fuel depends upon the rank of coal *e.g.*, 80% through 200 mesh (0.074 mm = 74 microns) for sub-bituminous & brown coals and upto 85% for anthracites. Not more than *2%* should be retained on 50 mesh (0.3 mm). Passing preheated air over the coal during its grinding, raises the capacity of the pulveriser by drying the coal. It also transports the coal to the burner and acts as primary air (which is 15-20% of total combustion air). The solid content and temperature of coal-primary air mixture is usually maintained at 0.5 kg/m^3 and 70-80°C respectively. The old central coal storage system, in which coals from different pulveriser were sent to a central bin for further delivery to individual burners suffered from the explosion risk of coal-air mixture and the tendency of stored coal fines to granulate. In modern unit or direct fired system, the pulveriser directly feeds the burner. Pulverised non-coking coal is also injected in B.F. to partly replace coke on 1 : 1 replacement ratio.

9

Usage of Coal

Coal is used for :

- directly burning and getting heat.
- generating steam for producing electricity (nearly 0.5 kg of coal is burnt to generate 1 kWh of electricity).
- manufacturing coke and coal gas (*e.g.* coke oven gas, producer gas, water gas etc.).
- manufacturing synthetic liquid fuels.
- gasification to produce nitrogeneous fertiliser from synthesis gas.

The primary uses of coal are in electricity generation, the creation of liquid fuel, the production of steel and cement manufacturing. There are two primary types of coal, thermal coal and metallurgical coal. Thermal coal goes into fuel and electricity applications, while metallurgical coal is better suited to manufacturing and chemical applications.

As of 2014, 42 percent of the world's electricity comes from coal, making the electric industry the largest consumer of coal worldwide. Coal-fired furnaces heat water into steam, which turns electric turbines to create electrical current. In some cases, processors can also convert thermal coal into a liquid fuel as an alternative to natural gas or other petroleum products. Metallurgical coal is most widely used by the steel industry, with 70 percent of the world's steel output dependent on coal. In addition, coal finds its way into many different products, such as the carbon fiber that strengthens everything from auto bodies to building materials. Another use of coal is in the creation of activated carbon, an important component of air and water filtration systems. Refined coal tar & other coal by-products are also found in a wide variety of chemical and pharmaceutical products.

Uses of peat. It can be carbonised to produce gas and coke. It is also briquetted and used as a domestic fuel.

Uses of lignite

- For manufacture of producer gas.
- For generation of electrical power.
- For gasification to produce nitrogeneous fertiliser.

Uses of bituminous coal

Bituminous coal is the most common type coal used in the world for a variety of purposes. Owing to its relatively high heat value, low moisture content and its availability in abundance, bituminous coal has a wide number of commercial uses among all the types of coals.

Bituminous coals are soft and completely opaque. They are dark gray to black in color, and is banded, or layered. They have moisture content of about 2 to 17%, carbon content of about 45 to 85%, ash content of about 2 to 14% and contains 0.5 to 2 percent nitrogen. Their sulfur is of about 0.5 to 5 % and chlorine of about 340 ± ppm. The sulfur that is present in the coal brings acid rain.

Bituminous coals are highly combustible and releases excessive smoke and soot if not burned properly. The main pollutants from bituminous coal are particulate matter (PM), sulfur oxides, and nitrogen oxides. Some metals like lead and mercury, hydrocarbons like methane, alkanes, alkenes, benzenes and others are also released. Some unburned substances like carbon monoxide and various organic compounds are released even under proper boiler operating conditions.

Bituminous coals have two subtypes; thermal or steam coal and metallurgical coal. Metallurgical or coking coal is carbonised in coke ovens in the absence of oxygen at a temperatures of 1,100 °C to produce coke. Bituminous coal is used to fire power plants that produce steam for the generation of electricity. Bituminous coal is made free from ash, sulfur, and other impurities, and used in coal forges, where metals are heated & shaped.

Non-fuel uses of bituminous and lower rank coals :

- Production of activated carbon used for decoloration of glycerine and petroleum products.
- Treatment with strong sulphuric acid forms a basic exchange agent used in water treatment process.
- Used as pigments for paints.
- Powdered coal is used in dynamite.
- Used as filter for various purposes due to its lightness, chemical inertness and low cost.
- Carbonisation of bituminous coal produces chemicals such as tar, benzol, ammonia and host of other chemicals.
- Used for making carbon brushes.
- Used for burning in domestic oven.

Uses of anthracite coal. It is used

- For recarbonising steel.
- for making carbon electrodes, brushes, battery parts, resistors, carbon refractory, corrosion resisting structural materials.
- as filter and paint pigment.
- for blending with coking coal to check its swelling and improving the coke quality.

Differences between Bituminous Coal and Anthracite Coal

In general, anthracite is harder than bituminous coal. This is because it contains more carbon. In fact, both types of coal started out as the same thick deposits of soaked woody and organic material that collected in swamps about 300 million years ago. As time passed, this material turned into peat deposits that were buried by dirt and silt. Over millions of years this material was buried thousands of feet under the ground.

Some of the peat that was buried will be exposed to high temperatures. These high temperatures will cause some of the easily vaporized compounds in the peat to evaporate, leaving behind a material with a high percentage of carbon. This material is what makes up bituminous coal.

Anthracite coal is formed in a similar manner, but it starts as peat that is buried several more kms. underground, typically under thick sheets of rock. Because it is buried further underground, the peat is exposed to even higher temperatures than bituminous coal. This means that even great amounts of extraneous materials are evaporated from the peat, leaving coal with an even higher percentage of carbon.

In fact, whether coal is considered to be anthracite or bituminous is determined only by its carbon content. By definition, bituminous coal contains less than 86 percent fixed carbon while anthracite contains 86 percent or more fixed carbon.

Because of the difference in their relative levels of carbon, each of these types of coal is used very differently. Bituminous coal is mostly burned in order to generate electricity. It can also be used to produce heat for some industrial purposes. Bituminous coal is used to make coke, a component necessary for the production of steel.

Comparative Study of Solid, Liquid & Gaseous Fuels

Solid, liquid & gaseous fuels available commercially are tabulated in table 10.1.

Table 10.1: Commercially Available Fuels (solid, liquid & gaseous)

Solid fuels	Liquid fuels	Gaseous fuels
coals	gasoline (petrol)	Natural gas
cokes	kerosene	LPG
briquettes	diesel	Blast furnace gas
solid pitch	fuel oils/L.S.H.S,	coke oven gas
coal middlings	coal tar fuels/PCM	producer gas
	tar	coal gas

A number of factors are to be considered when a fuel or type of fuel (*e.g.* solid, liquid & gaseous) is chosen for a particular purpose. The main factors are :

- The type of heating with respect to size or whether a continuous or intermittant operation is involved. Coal, coke, producer gas and water gas are obviously unsuited to intermittant operation.
- The availability and reliability of supply of fuels in question.
- The price of the fuel delivered to the place where it is required as some fuels need storage and feeding equipment *e.g.* expensive steam heated lines in case of heavy fuel oil and some coal tar fuels.
- The efficiency of the heating operation with the selected fuel. An expensive fuel with a high efficiency of utilisation will be cheaper than a cheap fuel operating at a low efficiency.
- The ratio of the cost of heating operation to the cost of the finished product. If this ratio is low then a high grade expensive fuel need be considered.
- The adverse effect of fuel on the quality of finished product.
- Adequate storage space for the fuels should be available.
- The degree of close technical control necessary to ensure a high quality of finished product.
- The cleanliness and hygiene of working conditions.
- The fuel chosen should achieve smokeless combustion.

Ready availability and certainty of delivery are of great importance to the consumer. The final choice may depend upon the location of production & consuming points and the mode of delivery that is usual or possible.

SOLID FUELS

Advantages:

- No special provision need be made for storage in the way of tankers, heaters, lagging or pipelines. Any hard flat surface would suffice.
- A greater proportion of the gross calorific value of solid fuels can be utilised because of its lower hydrogen content.
- Sulphur content of solid fuels is comparatively very less resulting in reduced atmospheric pollution and corrosion of equipments.
- It is cheap, occurs widely and is easy to handle. Besides, no complicated burning mechanism is needed for it.
- It can be pulverised with flexibility of liquid fuels.
- Coal ash is free from vanadium compounds which form corrosive deposits on hot pressure parts of boilers.

Disadvantages :

- There is dust nuissance during its loading and unloading. Conveying and transferring from one container to another causes much breakage resulting in production of unwanted dust & slack.
- Lot of labour is required for moving solid fuel and then for disposing ash etc.
- Enough space is required for its storage.
- Coal deteriorates during storage causing reduction in size and calorific value.
- It is liable to undergo spontaneous combustion and catch fire during storage.
- It has got ash and clinkering trouble. Heat is wasted in ashes.
- Ash causes the following troubles in furnaces:
  (a) Ash reduces the draught on the furnace grate, causes uneven burning and may produce clinker which seriously reduces the output & efficiency of the furnace and damage the grate.
  (b) Fly ash reduces furnace/boiler availability, corrodes & erodes superheaters, economisers and air preheater. It also reduces draught and pollutes the atmosphere.
  (c) Ash handling is an expensive and dirty operation.
- Thermal efficiency is not so high as lot of excess air is required for complete combustion of solid fuels.
- Coal firing causes heavy wear & tear on grates and refractory linings.

LIQUID FUELS

Advantages:

- It can be stored more compactly than solid fuels.
- For equal heat output, it occupies much less space and has much less weight than solid fuels.
- It can be handled easily and readily with little labour.
- It has no ash & clinkering problems.
- The combustion of liquid fuel can be easily regulated.
- Oil burners can be designed to burn efficiently over a wide range of heat output. Sudden fluctuation in heat demand are met more readily with oil than with coal.
- There is no need for maintaining banked boilers or furnaces with liquid fuel firing unlike coal because starting up is easy and rapid.
- Oils can be burnt at higher combustion rates than coal except in pulverised form.
- It requires lower amount of excess air than solid fuels resulting in higher thermal efficiency and flame temperature.
- Delivery of liquid fuel is more uniform in composition than coal deliveries. Besides, there is no dust nuissance.
- It is not liable to spontaneous combustion and deterioration during storage like coal.
- It can be used in I.C. engine.
- Low sulphur oils are next to gases in cleanliness and controllability.

Disadvantages:

- Special provision for storage in the way of tanks, heaters, lagging and pipelines is to be made.
- Because of comparatively higher hydrogen content than coal, a greater proportion of gross calorific value of the liquid fuel can not be utilised.
- Sulphur content in the most of the petroleum oils is high compared to coal, hence atmospheric pollution and corrosion of equipment is a problem.
- Vanadium compounds present in oils form corrosive deposits on hot pressure parts of boilers.
- Risk of fire and explosion during storage/utilization exists.

GASEOUS FUELS

Advantages:

- It has no ash or clinkering trouble. Burning is quite clean without any smoke.
- It requires least amount of excess air for complete combustion hence giving highest thermal efficiency.
- It is easier to maintain oxidising or reducing atmosphere inside the furnace with gaseous fuels.

- It can be easily metered by pipelines at lower pressure drop thus requiring lower pumping power.
- It can be used in I.C. engines.
- Gaseous fuels may be prepared at a central place and distributed over wide area through pipelines.
- Solid fuels unsuitable for direct uses can be converted into gaseous fuels and valuable by-products can be recovered.
- Sulphur compounds from gases can be removed easily and cheaply.

Disadvantages:

- Gaseous fuels occupy large volumes, hence storage is difficult and expensive.
- Some fuel gases are highly poisonous *(e.g.* blast furnace gas, **L.D.** converter gas etc.) and explosive in nature ; (e.g. coke oven gas, acetylene, etc.) hence careful handling in its utilisation is needed.

SELECTION OF COAL FOR VARIOUS USES

There are various factors which are to be taken into account before it is decided to use the coal for a particular purpose.

Coal for metallurgical purposes (*i.e.* **for making coke**). Coking coals must:

- be strong and hard (as judged by shatter index, abrasion index and micum index)
- have low ash, sulphur & phosphorus content
- be consistent, *i.e.* of uniform quality.

As the rank of coking coal decreases, its coking properties also diminishes. Highly swelling coals when used alone produce weak coke whose reactivity is very high hence it must be blended with weakly coking coal, anthracite, steam coal, coke breeze, fussain (about 3%) or non-swelling non-coking coals so as to get moderately reactive coke with optimum hardness and strength.

Ideally, the coking coal should have 8-10% of ash so that coke produced from it will have around 11-14% ash. Indian coking coals having around 20-24% ash produces coke with about 24-28% ash.

Sulphur content in metallurgical coal should be less than 1% and the phosphorus should be about 0.01% in the coke. Volatile matter should be within 22-36% and crucible swelling index should be 4 to 9. Caking index of coal blend should be about 22.

Consistency in quality of coking coal is achieved by cleaning, crushing and blending of coal. The coke produced should be Gray-King coke type G_1 to G_9.

Properties of Coal suitable for manufacture of metallurgical coke used in steel plants :

Differentiation of coking and non-coking coals : From physical appearance it is very difficult to differentiable between non-coking & coking coal.

A non-caking coal is that coal which when heated in the absence of air for *e.g.* as in the determination of VM, leaves a powdery residue. A non caking coal is always a non-coking coal. A non-coking coal may bot be non-caking coal. (Reason : a non-coking coal can also produce coke which is never suitable for steel industry). A non-caking coal is that coal which on heating in absence of air dosen't from a coherent mass of residue.

A caking coal when heated in absence of air leaves a solid coherent residue. The caking coal might give a coke which may not be accepted by the steel plants as it does not posses the requisite physical and chemical properties of the properties of the coke. A coking coal is that coal which on heating in absence of air leaves a solid residue.

A non-coking coal also leaves a solid coherent residue which may not possess the physical and chemical properties of the coke. The non-coking coals are those coals which may form a solid-residue but not be suitable for manufacture of coke. This coal may form a coke which will not meet the physical & chemical properties as laid down by the steel industry, the primary coke consumer.

A Coking coal is that coal which when heated in absence of air leaves a solid coherent residue possessing metallic grayish luster and which possesses all the physical & chemical properties in the coke when manufactured commercially. Coking coal are those coals which after heating in absence of air leaves a resudue coke. This coke when satisfies all the physical & chemical characteristics such as CSR, CRI, Shutter Index, Micum 10, (M_{10}), Micum 40 (M_{40}) & Porosity, are accepted by the steel plants. The coal which produces such coke is termed as coking coal. _

Various properties of coals which gives rise to the manufacture of coke suitable for steel plant are as follows :

(A) Chemical properties:

- **Carbon.** The elemental carbon expressed on dry mineral matter free (DMMF) basis should vary between 85-88%.
- **Hydrogen.** The hydrogen on DMMF basis should vary between 3.7 to 4.5%.
- **Sulphur.**Sulhur (both organic and inorganic) should not exceed 0.75% on DMMF basis.
- **Phosphorous.** The Phosphorous present in the coking coal should not exceed 0.15-0.25%
- **Other elements.** Such as Nitrogen, Iron and Other rarer elements should be present in traces, only.

(B) Mineral Matter (MM) :

The mineral matter in the coking coal normally should not exceed 20% since the mineral matter constitute a portion of inert matter. There are 2 kinds of MM in the coal, (i) Inherent MM and (ii) Extraneous MM. It has not yet been possible to reduces the extraneous MM Completely in coal washeries.

(C) Physical Properties :

The development of coking property is an intrinsic property. The reason for development of coking property has not yet been clearly established. By physical tests it can be determined whether the coal is coking or non-coking. The following destructive physical test is used for finding the coking propensity of coal.

Geisler Plastometer : Equipment should give a reading between 500 to 2000 dial divisions per minute. Higher the fluidity of coal mass, better will be the dials division per minute.

Caking Index: In this test certain amount of powered coal is thoroughly mixed with graded and sized sand. The total quantity of sand and coal should not exceed 25 grams. This mixture when heated at 925 °C in the absence of air, after cooling down should give a coherent residue which should be able to withstand a weight of 500 grams without generating more than 5% of powder out of the residue. If the solid residue generates more than 5% power, then proportion of coal in the mixture of sand and coal should be increased and vice versa. Normally caking index for coking coal blend should vary from 20 to 24. The minimum caking index of coal should be 13 and maximum 24.

Swelling Index : During determination of volatile matter in coking coal a solid reside is left, comprising of fixed carbon and mineral matter. The solid residue that is the coke bead is viewed horizontally at the same level of the eye. It will be observed that the top surface of the bead has developed some amount of swelling. This swelling is compared with standard chart indicating the quantum of swelling and a number indicating swelling index. Coal swelling index of 5 is ideal for coke manufacturing. The high swelling coal is not charged for coke making as it would create unnecessary pressure on the side wall of the oven and will also produce a highly porous coke structure, (reduced coke strength).

Volatile Matter. The volatile matter of coking coal should vary between 19 to 26% on DMMF basis. A coal with less VM than 19% will give rise to a coke which will not have proper physical property. A coal with VM higher than 26% will give rise to coke with more porosity and less physical strength (poor CSR and CRI values). The VM in coal consist of various gaseous products which is produced when the coal is heated in absence of air at temperature more than 900 °C. The gases, consists of combination of carbon and hydro-aromatic compounds. The gases mainly contain Methane, Hydrogen, Carbon monoxide, carbon dioxide, nitrogen and hydrogen-aromatic compounds such as phenol and benzole. Some amount of tar in gaseous from is also produced.

Petrographic Analysis and Reflectance. Another most important non-destructive test of coal which is used for determining properties of coking coal is the petrographic analysis & reflectance. The coking coal should have a minimum of 60% virtrain (active constituents) and maximum of 40% Inertinite (non-reactive constituent). For finding out the physical strength of the coke, reflectance studies on coking coal are also done. In this study the ideal value of reflectance will be within 1.3-1.5. Generally the coking property develops in coal, if it reflectance is between 0.9-1.3.

Coal for producing gas. Gas coal should have :

- high volatile matter content, that gas yield is more
- less ash content (around 8-10%)
- less sulphur content (< 1%)
- weak coking properties so that coal bed does not fuse during gasification.
- high ash fusion temperature so that clinker formation can be avoided.

Coal for railway locomotives. It should be :

- large, well screened and free from smalls.
- low in ash content.
- weakly coking.
- free from clinkering trouble, *i.e.* its ash fusion temperature should be high.

Coal for domestic ovens. The preferred coal for domestic ovens are the larger graded sizes of the higher volatile free burning ranks. It should be low in ash. Ash fusibility is not that important because the grate of the domestic oven seldom reach a temperature high enough to cause clinkering difficulties.

Coal for steam raising. Coals used in boilers for burning and generating steam is called *steam coal.* Almost all other types of coal which are not suitable for carbonisation or gasification can be used for steam raising. Even coal middlings (produced in coal washeries) containing ash upto 45% can be pulverised and used in the boiler. The design of plant and the method of operation during combustion of steam coals should be satisfactory.

Coal for Combustion Practices : The desirable qualities in a coal for combustion practices are :

- Calorific value should be high.
- Ash and moisture content should be low.
- Uniform size is desirable ; fine dust and large lumps are objectionable.
- Successive deliveries of coal should be as far as possible of the same quality.

In addition to the above, the ash should be refractory and infusible in the furnace. Coals having ash which is easily fusible will form a considerable amount of clinker which may spread over the grates if the temperature is high enough.

Fieldner and Selrig have classified coals on the basis of the fusion/softening point temperatures of their ashes which is as follows :

Class I	: Coals whose ash does not fuse at 1427°C.
Class II	: Coals whose ash fusion temperatures are between 1427°C to 1204°C.
Class III	: Coals whose ash fusion temperatures are below 1204°C.

Class III coals are considered to be the *worst* from the point of view of *clinker formation* and should never be used for boiler firing. The clinkering properties of Class II coals depend upon the nature of mineral matter in the coal and the kind of stoker used. *Class I* coals are the *best* from the point of view of *clinker formation* and can be used for boiler firing without any trouble.

A coal which gives a hard and non porous coke should be avoided as it is not sufficiently free burning to permit sudden increases in the load on the boiler to be rapidly taken up. Moreover, a large excess of air is necessary to avoid excessive carbon loss in the solid refuse. Hence strongly caking coals should be avoided.

Hand fired boilers of low furnace volume, such as Cornish, Lancashire, Locomotive and Scotch Marine boilers are likely to give smoke with accompanying heavy loss of combustible gases and vapours when high volatile coals are used. Therefore, for boilers of these classes, coals with a low percentage of volatiles would be most suitable. Mechanically fired water tube boilers, on the other hand, can handle with high efficiency coals of more widely varying characteristics.

While low ash is desirable, if the coal is to be used on a travelling grate stoker, the ash should not be below 7%. If the ash content falls appreciably below this minimum, the links of the grate burnout rapidly, being unprotected from the full heat of the fire by a sufficient layer of ash. As far as Indian Coals are concerned, this problem does not arise except in a few cases (*e.g.* Low Sulphur Assam Coals).

The coal should be free from large lumps (over 7.5 cms cube) and should contain little dust (below 4 mesh), particularly, if intended for use in hand fired boilers.

The sulphur in coal should be low. The mixture of sulphurous and sulphuric acids formed by the combustion of sulphur may set up serious corrosion in economisers & air heaters.

MINERAL MATTERS IN COAL-ASH AND CLINKER FORMATION

Mineral matters present in the coal are shown below in Table 12.1.

Table 12.1. Mineral Matters in Coal

Normal state of combination	% of total mineral matter
Silica and silicates of aluminium	50-90
Pyrites (FeS_2), ferrous carbonate, ferrous & ferric sulphate and silicates	0-20. It is usually < 10% if the resulting ash has high fusion point.
Carbonate, sulphate and silicates of calcium	0-20
Carbonate and silicate of magnesium	0-8
Chloride, carbonate andsilicate of sodium and potassium	0-4
Titanium oxide Carbonate and silicate of manganese	0-2
Sulphur in combination with the coal substance as	Traces
sulphides and sulphates	0.5-10
Oxides of phosphorus, gallium, germanium, nickel, beryllium etc.	< 1

Most of the mineral matters in the coal is in the form of shale mixed with silica, silicates and other compounds of iron, calcium, magnesium, titanium & the alkali materials.

On combustion of coal, original compounds decompose and residual materials may recombine or interact with other derivatives of the coal ash *e.g.* the sulphides, sulphates and carbonates normally decompose or oxidise leaving their basic radicals free to combine with excess of silica whereas the shales loose water of hydration.

Hence, the ash left on combustion of mineral matters is not a true measure of the nature or amount of mineral matters present originally in the coal.

Ash produced after combustion of coal differs from the mineral matter present in the coal and from which it is formed. It is less than the original mineral matter by the amount of combined moisture and carbon dioxide expelled from clays and other minerals during burning of coal; and also by the difference in weight between mineral matter originally present *(e.g.* iron pyrites) and the substance produced from it during combustion *(e.g.* ferric oxide). Roughly mineral matter in coal is 10% more than the actual ash formed on combustion of coal.

Composition of ash. The chemical composition of ash varies with the type of coal. It normally depends upon the nature of mineral matter content of the original cellulosic plant debris from which coal has been formed and also upon nature of the mineral content of mud & water which percolates through peat bog.

Ash normally consists of silica, alumina, iron oxide and small quantities of lime, magnesia & other alkalies.

Two typical compositions of ash are given in Table 12.2.

Table 12.2. Typical Composition of Coal Ash

Ash constituents	% of ash of two typical coals	
Silica	40-60	25-40
Alumina	20-35	20 - 40
Ferric oxide	5-25	0-30
Calcium oxide	1-15	1 - 10
Magnesium oxide	0.5 - 4	0.5 - 5
Titanium oxide	0.3 - 3	0 - 3
Alkalis	1 - 4	1 - 6
Sulphur trioxide	-	1-12

CALCULATION OF MINERAL MATTERS

Mineral matters in coal is calculated from the following formulae :

***(i)* Parr's formula.** $M = 1.08\,A + 0.55\,S$

where, M = percentage of mineral impurities in coal

A = percentage ash in coal (found after combustion)

S = percentage of total sulphur in coal.

This formula is based on the incorrect assumption that all the sulphur is present in the coal as pyrites and also it does not account for CO_2 present in coal.

***(ii)* King, Maries and Crossley (KMC) formula**

$M = 1.09\,A + 0.5\,S_{pyr}, + 0.84\,CO_2 - 1.1\,SO_{3ash} + SO_{3coal} + 0.5\,CI$

The extra symbols in the KMC formula represent the percentage found by analysis of the constituents of coal or ash *viz.*.

S_{pyr} = Pyritic sulphur in the coal, CO_2 = Carbon dioxide as carbonates in the coal, SO_{3ash} = Sulphate in the ash, SO_{3coal} = Sulphate in the coal, CI = Chlorine in the coal.

***(iii)* KMC formula revised by National Coal Board.** According to this formula

$M = 1.13\,A + 0.5S_{pyr} + 0.8\,CO_2 - 2.8\,S_{ash} + 2.8\,S_{sulphur} + 0.5\,CI$

where, A = Ash obtained on combustion, S_{pyr} = Pyritic sulphur in the coal,

CO_2 = Carbonate CO_2 in the coal, S_{ash} = Total sulphur in the ash,

$S_{sulphur}$ = Sulphur present as sulphate in the coal, CI = Total chlorine in the coal.

***(iv)* Formula of British Coal Utilisation Research Association (B.C.U.R.A.).** According to this formula :

$M = 1.1\ A + 0.53\ S + 0.74\ CO_2 - 0.32$

where, A = ash, S = total sulphur, CO_2 = total carbon dioxide.

***(v)* For rough estimation,** $M = 1.1$ A.

FUSION OF COAL ASH AND CLINKER FORMATION

On heating the coal ash, it first softens and then finally fuses & melts. This temperature is called the fusion point of ash. The composition of ash affects the fusion temperature. The higher the fusion point of ash, the better is the quality of coal. Fusion point of ash is increased with increase in its alumina & silica content. However, presence of Na_2O, K_2O, lime, magnesia, ferric oxide, ferrous oxide, FeS_2, $FeCO_3$, $CaCO_3$ decrease the fusion point of ash. Coals producing dark red ash and having high proportion of sulphur have a low fusion point of its ash.

Clinker is a hard mass of refractory particles of ash, fine coal and fused ash. If ash of the coal has low *fusion point*, then it fuses when coal is burnt on grates. On fusion of this ash, clinker is formed.

Clinker formation is increased by the following factors :

- Low fusion point of ash. Under normal burning conditions minimum safe fusion point of ash is 1300°C. Ashes with fusion point less than 1100°C are liable to form clinker. Fusion point of coal ash varies from 1000° to 1700°C.
- Coals having high calorific value and high fixed carbon content are more conducive to clinker formation. Low volatile steam coals and anthracite require high forced draught and fuel bed temperatures. These factors accelerate clinkering troubles.
- Thick fire beds are more dangerous than thin beds, because reducing conditions prevail in thick beds which form ferrous sulphides and oxide of low fusion point. Oxidising conditions (in this beds) convert iron sulphides to ferric oxide of relatively high fusion point.
- Exposure of masses of ash and residual coal to furnace heat for longer time (called soaking) is undesirable.
- Molten or softened ash sticks more readily to hot surfaces than cold ones. Water cooling prevents trouble of ash and clinker sticking to furnace side walls.
- Small, dirty coals containing high amount of iron and sulphur are conducive to clinker formation in furnace.
- Preheated primary air accelerates clinkering problems on grates.

Adverse effects of clinker formation :

- Clinker clogs the furnace grate and reduces the flow of air because spaces in between coal particles on grates get blocked resulting in uneven combustion of coal on grates thereby producing less heat and uneven distribution of heat.
- Clinker embeds in itself some of the unburnt coal and ash thus causing loss of fuel.
- Clinker removal from grate is very difficult.

Briquetting of Solid Fuels/Coal

Briquetting consists of applying pressure to a mass of particles with or without the addition of a binder and converting it into an agglomerate. The products are called either briquettes containing binders or binderless briquettes, as the case may be. Briquetting converts a low grade solid fuel into one of higher quality. Saw dust, peat, lignite, coal fines etc. may all be briquetted into lumps of suitable size, shape & strength and used as convenient domestic & industrial fuels.

Briquetting of Peat. In Ireland, peat briquettes are a common type of solid fuel, largely replacing sods of raw peat as a domestic fuel. These briquettes consist of shredded peat, compressed to form a virtually smokeless, slow-burning, easily stored and transported fuel. Although often used as the sole fuel for a fire, they are also used to quickly and easily light a coal fire.

Briquetting of Lignites. Briquetting is most widely used in the case of lignites. Due to the low structural quality of lignite, considerable amounts are recovered as fines during its mining. Moreover the coarser pieces disintegrate into fines after air drying. Briquetting is therefore essential prerequisite for burning lignite in domestic ovens and stoker furnaces. However pulverized fuel combustion as in thermal power stations does not need this preparative step.

No binders are required in the briquetting of lignite and comparatively low pressures suffice in achieving the compaction of its particles.

The briquetting of lignite involves its grinding, drying and pressing. The formation of a good briquette requires that the grains are of certain uniform size & optimum moisture content so that the particles can approach very close to each other and allow the cohesive forces to come into play when an external pressure is applied.

In the briquetting plant of Neyveli Lignite Corporation, raw lignite from the mine is crushed to below 250 mm size in a blade crusher and stored in a bunker. This is further ground to below 50 mm in a roller crusher and finally to below 8 mm in a hammer mill. The ground and wet, lignite (moisture 50%) is dried in a steam-tube rotary drier to a moisture content of 10%. The discharge from the rotary drier has a grain size of below 6 mm. Before undergoing final grinding, it needs pretreatment which involves screening and redrying of coarse fraction. The pretreated material is ground to below 1 mm in a hammer mill and then converted into briquettes by passing through quadruple extrusion presses at a pressure of about 1000 kg/cm^2. No binder is used and the optimum temperature for briquetting is 70°C. Cooling of the dried lignite occurs during its transport from the drier to the press by a conveyor. The temperature rises during the pressing operation. It is necessary to allow for cooling before the briquettes are disposed off.

The lignite briquettes at Neyveli are carbonized at 650-700°C. After cooling, the carbonized briquettes are sieved over 10 mm screens. The oversize is marketed under the trade name 'Leco'. The undersize char fines are also used as industrial fuel.

Briquetting of Bituminous coal. Fine coals are produced in large quantities during mining and processing. Before the advent of pulverized fuel combustion these fines could be burnt in ovens and furnaces only after briquetting. The process of briquetting of coal fines is used *because generation of mechanical mining results in fines and the technique of pulverized fuel firing is not suitable for domestic ovens & small industrial furnaces.*

Briquetting bituminous coal without binder is difficult and much higher pressures & much finer grain sizes are required than in the case of lignites. Binder used may be coal tar, pitch or petroleum bitumen. The whole process consists in the operations of drying & grinding the coal, mixing it with the binder, heating the mixture, pressing it into moulds and cooling the briquettes before final disposal. The moisture content of the dried coal is maintained at 2-4%. Pitch and dried coal are often ground together. The grain size of the ground coal is about 80% less than 0.5 mm and 93% less than 0.88 mm. The pitch content is about 7-8% of the coal. The temperature of the mixture is kept at a few degrees within the softening point of the binder before it is charged into the press. Briquetting pressure is normally about 150 kg/cm^2. The quality of the briquettes depends upon the nature of coal, properties of binder and the precise way in which the various operations are carried out.

Many binderless briquetting process have been developed. Economic considerations stand in the way of their wide commercial exploitation. Hot briquetting technique consists in preheating a coal blend containing one or more caking coal to a temperature in the plastic range and then compacting the mass. The "softened" coal acts as the binder and the required briquetting pressure is much lower than in the other binderless briquetting processes for bituminous coals.

Briquetting of charcoal. Charcoal fines cannot be burned by the usual simple burning methods. Briquetting - turning fines into lumps of charcoal - seems an obvious answer. Though it is technically possible to briquette charcoal fines, the economics are not usually favourable, except where the price of lump charcoal is very high and the fines are available at a very low cost.

Briquetting requires a binder to be mixed with the charcoal fines, a press to form the mixture into a cake or briquette which is then passed through a drying oven to cure or set it by drying out the water so that the briquette is strong enough to be used in the same burning apparatus as normal lump charcoal.

Charcoal is a material totally lacking plasticity and hence needs addition of a sticking or agglomerating material to enable a briquette to be formed. The binder should preferably be combustible, though a non-combustible binder effective at low concentrations can be suitable. Starch is preferred as a binder though it is usually expensive. Highly plastic clays are suitable providing not more than about 15% is used. Tar and pitch from coal distillation or from charcoal retorts have been used for special purpose briquettes but they must be carbonised again before use to form a properly bonded briquette. They are of good quality but costly to produce.

The press for briquetting must be capable of agglomerating the mixture of charcoal and binder sufficiently for it to be handled through the curing or drying process. Charcoal is quite abrasive so that equipment for screening fines, grinding, mixing them with binder, briquetting and so on must be abrasive-resistant & well designed.

The most common effective binder is starch. About 4-8% of starch made into paste with hot water is adequate. First, the fines are dried and screened. Undersized fines are rejected and oversized hammer milled. This powder is blended with the starch paste and fed to the briquetting press. The briquettes are dried in a continuous oven at about 80°C. The starch sets through loss of water, binding the charcoal into a briquette which can be handled and burned like ordinary lump charcoal in domestic ovens. Generally briquettes are not suitable for use as industrial charcoal in blast furnaces and foundry cupolas, since the bond disintegrates on slight heating. For this briquettes bonded with tar or pitch and subsequently carbonised in charcoal furnaces to produce a metallurgical charcoal briquette of adequate crushing strength are needed. The cost is too high for the process to find industrial application in most countries.

It is possible to add material to aid combustion of briquettes such as waxes, sodium nitrate, and so on, during manufacture to give a more acceptable product. Also clay as a binder, silica, and so on, can be mixed with the fines to reduce the cost of the briquette .

Successful briquette operations are found mostly in developed countries. An example is the industry based on carbonization of sawdust and bark in the southern U.S.A. using rotary multiple hearth furnaces which produce perhaps 25 to 50 tons of fine charcoal per day. When briquetted, this charcoal, intended for barbecues, can be sold in retail outlets. The furnace gases are burned to produce steam of electric power, thus transforming the waste sawdust & bark into two useful products, electric power and charcoal briquettes. Air pollution and waste disposal problems are minimised at the same time.

Briquetting of Biomass. Biomass briquettes are made from agricultural waste and are a replacement for fossil fuels such as oil or coal, and can be used to heat boilers in manufacturing plants, & also have applications in developing countries. Biomass briquettes are a renewable source of energy and avoid adding fossil carbon to the atmosphere.

A number of companies in India have switched from furnace oil to biomass briquettes to save costs on boiler fuels. The use of biomass briquettes is predominant in the some parts of India, where coal and furnace oil are being replaced by biomass briquettes. A number of units in Maharashtra are also using biomass briquettes as boiler fuel. Use of biomass briquettes can earn Carbon Credits for reducing emissions in the atmosphere Biomass briquettes also provide more calorific value/kg and save around 30-40 percent of boiler fuel costs.

A popular biomass briquette emerging in developed countries takes a waste produce such as sawdust, compresses it and then extrudes it to make a reconstituted log that can replace firewood. It is a similar process to forming a wood pellet but on a larger scale. There are no binders involved in this process. The natural lignin in the wood binds the particles of wood together to form a solid. Burning a wood briquette is far more efficient than burning firewood. Moisture content of a briquette can be as low as 4%, whereas green firewood may be as high as 65%.

Parameters of fuel briquettes made by extrusion from sawdust in Ukraine are as follows :

Parameter	Value
Briquette density, t/m³	1.0-1.2
Heat content, MJ/kg	19.3-20.5
Ash content, %	0.5-1.5

(MJ = Megajoules., 3.6 MJ equals 1 kWh.)

The extrusion production technology of briquettes is the process of extrusion screw wastes (straw, sunflower husks, buckwheat, etc.) or finely shredded wood waste (sawdust) under high pressure when heated from 160 to 350 °C .

Sawdust briquettes have developed over time with two distinct types: those with holes through the centre, and those that are solid. Both types are classified as briquettes but are formed using different techniques. A solid briquette is manufactured using a piston press that compresses sandwiched layers of sawdust together. Briquettes with a hole are produced with a screw press. The hole is from the screw thread passing through the centre, but it also increases the surface area of the log and aids efficient combustion.

Formed Coke/Carbonized Briquettes. Briquettes of lignites & bituminous coals are carbonized for producing smokeless lumpy fuel and substitutes for metallurgical coke. The manufacture of carbonized briquettes requires an additional stage over and above that required for the conventional coking practice. However, the process enables the production of coke-like material (from lignites, sub-bituminous coals and non-coking bituminous coals) called "formed coke".

There are many processes for the production of carbonized briquettes. In Lurgi-Spiilgas, process, binderless lignite briquettes are first produced and then carbonized in a three-chamber unit for low temperature carbonization. The carbonization temperature varies between 500 and 850°C depending upon the use of the products as the domestic or metallurgical fuel. The higher temperature reduces the volatile matter and considerably increases the strength of the briquette. Briquettes meant for metallurgical fuel production are made at high pressures of the order of 2000 kg/cm^2 in ring-roll presses.

Briquettes are susceptible to deformation, fissuration and crumbling into pieces during carbonization. The rate of heating during carbonization is very important. In the case of coal briquettes, the quantity of pitch must not be excessive and the highly coking coals need pretreatment like oxidation or partial carbonization before briquettes are made from them.

The technique of briquetting has also been employed for the production of agglomerates of fuel with one or both of iron ore and lime. The products - 'Ferrocoke' and 'Fluxed coke' - are meant for iron production.

PROPERTIES AND TESTING OF COAL

Proximate Analysis of Coal

Determination of moisture, volatile matter, ash and fixed carbon in coal comprises its proximate analysis.

Determination of moisture content in coal. Loss in weight of coal caused by heating of weighed quantity of coal sample for one hour at 105°C is the moisture content in coal.

A known amount of finely powdered coal sample is kept in a silica crucible and heated in a muffle furnace at 105-110°C for one hour. Thereafter the crucible is taken out, cooled in a dessicator and weighed. The process of heating, cooling and weighing is repeated a number of times till the constant weight of coal (anhydrous) is achieved. The percentage of moisture is given by

$$\% \text{ Moisture in coal} = \frac{\text{Loss in weight of coal}}{\text{Weight of coal initially taken}} \times 100$$

Determination of volatile matter in coal. It is the loss in weight of moisture free powdered coal when heated in a crucible fitted with cover in a muffle furnace at 950°C for 7 minutes.

$$\% \text{ Volatile matter} = \frac{\text{Loss in weight of moisture free coal}}{\text{Weight of moisture free coal}} \times 100$$

Determination of ash in coal. It is the weight of residue obtained after burning a weighed quantity of coal in an open crucible *(i.e.* in presence of air) at 750°C in a muffle furnace till a constant weight is achieved.

$$\% \text{ Ash in coal} = \frac{\text{Weight of residue ash formed}}{\text{Weight of coal initially taken}} \times 100$$

Determination of fixed carbon. It is determined indirectly by deducting the sum total of moisture, volatile matter and ash percentage from 100.

% fixed carbon in coal = 100 - (% moisture + % volatile matter + % ash).

Ultimate Analysis of Coal

Determination of total carbon, hydrogen, nitrogen, oxygen and sulphur percentages in coal comprises its ultimate analysis.

Determination of carbon and hydrogen in coal. A known amount of coal is burnt in a current of dry oxygen thereby converting C and H of coal into CO2 (C+O2 = CO2) and H2O (H2 + 1/2 O_2 = H2O) respectively. The products of combustion (CO2 and H2O) are passed over weighed tubes of anhydrous calcium chloride and potassium hydroxide which absorb H20 and CO2 respectively. The increase in the weight of CaCl2 tube represents the weight of water (H2O) formed while increase in the weight of KOH tube represents the weight of CO2 formed.

% of H and C in coal can be calculated as below :

Let, x = weight of coal sample taken, y = increase in the weight of $CaCl_2$ tube , z = increase in the weight of KOH tube

$\therefore$ Amount of carbon in the coal sample = $\frac{12}{44} \times z.$

$$\left[\begin{array}{l}\text{Since, 44 gms of } C0_2 \text{ is produced from 12 gms} \\ \text{of carbon and 32 gms of oxygen} \longrightarrow \quad + \; 0_2 = C0_2 \\ \qquad\qquad\qquad\qquad\qquad\qquad\qquad 12 \quad 32 \quad 44\end{array}\right]$$

$$\therefore \% \text{ carbon in coal } \left(\frac{12 \times z}{44 \times y}\right) \times 100$$

Similarly, amount of hydrogen in coal sample = $\frac{1}{18} \times y$

$$\left[\begin{array}{l}\text{Since, 18 gms of water is formed by 2 gms} \\ \text{of hydrogen and 16 gms oxygen} \rightleftharpoons H_2 + 1/2 \text{ O2} = CO_2 \\ \qquad\qquad\qquad\qquad\qquad\qquad\qquad 2 \quad 16 \quad 18\end{array}\right]$$

$$\therefore \% \text{ hydrogen in coal} = \left[\frac{2 \times y}{18 \times x}\right] \times 100$$

Determination of nitrogen in coal. Nitrogen estimation in coal is done by Kjeldahal's method. A known weight of powdered coal is heated with concentrated sulphuric acid in presence of potassium sulphate and copper sulphate in a long necked flask (called Kjeldahal's flask) thereby converting nitrogen of coal to ammonium sulphate. When clear solution is obtained (*i.e.* when whole nitrogen is converted into ammonium sulphate) it is treated with 50% NaOH solution. The ammonia thus formed is distilled over and absorbed in a known quantity of standard sulphuric acid solution. The volume of unused sulphuric acid is then determined by titrating against standard NaOH solution. Thus, the amount of acid neutralised by liberated ammonia (from coal) is determined.

$$\% \text{ of nitrogen in coal} = \frac{\text{Volume of acid used x Normality}}{\text{Weight of coal taken}} \times 1.4$$

Since, $(NH_4)_2SO_4 + 2NaOH = Na_2SO_4 + 2NH_4OH$

One litre of N/10 H_2SO_4 consumed is equivalent to 0.1 gm mole of ammonia or 1.4 gm of nitrogen.

Volume of acid used = $V_1 - V_2$

where, V_1 = Volume of H_2SO_4 neutralised in blank, c.c.

V_2 = Volume of H_2SO_4 neutralised in determination, c.c.

Determination of sulphur in coal. A known quantity of coal is burnt completely in bomb calorimeter in a current of oxygen. Ash, thus obtained contains sulphur of the coal as sulphate which is extracted with dilute hydrochloric acid and the acid extract is treated with barium chloride solution to precipitate the sulphate as barium sulphate

$$\underset{32\text{ gms}}{S} \xrightarrow{O_2} SO_4 = S \xrightarrow{Bacl_2} \underset{233\text{ gm}}{BaSO_4}$$

The precipitate of $BaSO_4$ is filtered, washed, dried and weighed from which the sulphur in coal can be computed as follows :

32 gms sulphur in the coal will give 233 gms $BaSO_4$.

If, x = weight of coal sample taken and, y = weight of $BaSO_4$ precipitate formed

then, amount of sulphur in coal sample = $\frac{32y}{233} = 0.1374$

$$\text{So, \% of sulphur in coal} = \frac{0.1374y}{23x} \times 100$$

Alternative method for sulphur determination. A known quantity of coal is heated with Eschka mixture (which consists of 2 parts of MgO and 1 part of anhydrous Na_2CO_3) at 800°C. The sulphate formed is precipitated as $BaSO_4$ (by treating with $BaCl_2$) and it is weighed as done in previous method of sulphur determination.

$$\text{\% of sulphur in coal} = \frac{0.1374y}{23x} \times 100$$

where x = weight of coal sample taken

y = weight of $BaSO_4$ precipitate formed.

Determination of oxygen in coal. It is deduced indirectly as follows.

% oxygen in coal = 100 - (% of C + H + N + S + ash)

Calorific Value

It is defined as the quantity of heat liberated by the combustion of unit quantity of fuel.

Principle of C.V. determination. When a weighed quantity of fuel is burnt in the calorimeter, the heat liberated is used up in heating the calorimeter and the water in the calorimeter. By equating the heat given out by the fuel with heat taken up by the calorimeter and water, the calorific value (also called heating value) is determined.

It is of two types *viz.* higher (gross) calorific value and lower (net) calorific value.

Higher and Lower Calorific Values :

Higher or gross calorific value. Hydrogen is found to be present in almost all fuels and when the calorific value of such fuel is determined hydrogen is converted into steam. Now, when the products of combustion are condensed to room temperature (25°C), the latent heat of condensation of steam also gets included in the measured heat. The total value so calculated is known as *higher or gross calorific value and thus may be defined as the total amount of heat liberated when one unit of the fuel is burnt completely and the combustion products are cooled to room temperature.*

Lower or Net calorific value. In actual practice, the water vapours escape as such along with hot combustion gases and thus are not condensed with the result a lesser amount of heat is available. The amount of heat, so available, is called as *lower or net calorific value and may be defined as the amount of heat liberated when one unit of the fuel is burnt completely and the combustion products are allowed to escape.* Thus,

Net C.V. = Gross C.V. - Latent heat of water vapours formed.

Now, since 1 part by weight of hydrogen gives 9 parts by weight of water, the above equation becomes

Net C.V. = Gross C.V. - Weight of $H_2 \times 9 \times$ Latent heat of steam.

$$\text{G.C. V - N.C. V} = \%\text{ of } H_2 \times \frac{9 \times 587}{100} = 53 \times \%\text{ of } H_2\text{, kcal/kg}$$

Units of Calorific Value :

(i) For solid and liquid fuels. Since solids and liquids are measured in terms of their masses, their calorific values are expressed as below.

Calories/gm (In C.G.S. system) , Calories/kg (In M.K.S. system)
B.T.U./lb (In F.P.S. or British system) , 1 kcal/kg = 1.8 B.T.U./lb

(ii) For gaseous fuels. Since gases are measured in terms of their volumes, their calorific values are expressed as below.

kcal/cubic metre (kcal/m) , kcal/m^3 = 0.1077 B.T.U./ft^3
B.T.U./cubic feet (B.T.U./fr) , 1 B.T.U./ft^3 = 9.3 kcal/m^3

Determination of Calorific Value by Bomb Calorimeter

Bomb calorimeter is used for determining the calorific value of solid and liquid fuels.

Bomb calorimeter consists of a strong stainless steel vessel, called bomb, capable of withstanding high pressures. The bomb is provided with a lid which can be screwed firmly on the bomb. The lid in turn is provided with two electrodes and an oxygen inlet valve. One of the electrodes is provided with a ring to accommodate the silica crucible.

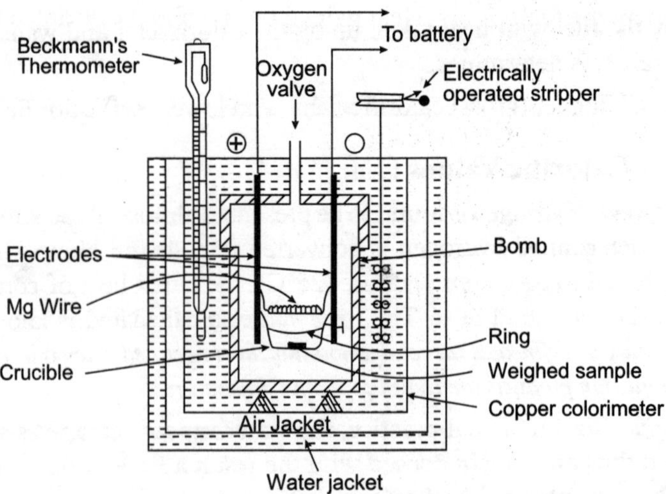

Fig. 13.1 Bomb colorimeter

The bomb is placed in a copper calorimeter having a known weight of water. The copper calorimeter, in turn, is surrounded by an air-jacket and water jacket to prevent loss of heat due to radiation. The calorimeter is provided with an electrical stirrer for stirring water and a *Beckmann thermometer.*

Working. A weighed amount of the fuel is placed in the silica crucible. The crucible is supported over the ring. A fine magnesium wire touching the fuel sample is stretched across the electrodes. Oxygen supply is forced into the bomb till a pressure of 25-30 atmospheres is reached. Initial temperature of the water in the calorimeter is noted after thorough stirring. The current is switched on the fuel in the crucible burns with the evolution of heat. The heat produced by burning of the fuel is transferred to water which is stirred throughout the experiment by the electric stirrer. Maximum temperature shown by thermometer is recorded. The calorific value of the fuel can now be calculated as below.

Observations:

Weight of the fuel taken in crucible = x kg, Weight of water in the calorimeter =y kg

Water equivalent[1] of the calorimeter, stirrer, thermometer and bomb = z kg

Initial temperature of water in calorimeter = $t_1{}^0C$,

Final temperature of water in calorimeter = t_2°C

Let, the higher (gross) calorific value of the fuel = C kcal/kg

Calculations:

Heat gained by water $= y \times (t_2 - t_1)$ kcal

Heat gained by calorimeter $= z \times (t_2 - t_1)$ kcal

Total heat gained $= y\ (t_2 - t_1) + z\ (t_2 - t_1)$ kcal $= (y + z)\ (t_2 - t_1)$ kcal

Heat liberated by fuel $= x \times C$ kcal

1. *Water equivalent of apparatus = weight of apparatus* specified heat

Now, heat liberated by the fuel = Heat gained by water and calorimeter

i.e. $x \times C = (y + z)\ (t_2 - t_1)$

$$\therefore \text{Gross calorific value, } C = \frac{(y+z)(t_2 - t_1)}{x} \text{ kcal/kg}$$

Calculation of lower (net) calorific value :

Let the percentage of hydrogen in fuel = H

Weight of water produced from 1 gm of fuel = $\frac{9H}{100}$ gm = 0.09H gm

∴ Heat taken by water in forming steam = 0.09 $H \times 587$ kcal

(because Latent heat of steam = 587 kcal/kg)

Hence, Lower (net) calorific value of fuel

= Higher calorific value - Latent heat of water vapor formed

= (C - 0.09 $H \times 587$) kcal/kg.

Corrections. In addition to the hydrogen correction, following other corrections must be considered while calculating the calorific value of a fuel.

(i) Cooling correction. Rate and time taken for cooling the water in calorimeter from maximum temperature to room temperature must be considered. From the rate of cooling $dt°/$ minute and the actual time taken for cooling (t minutes), the cooling correction $(dt \times t)$ must be added to the rise in temperature $(t_2 - t_1)$.

(ii) Acid correction. During ignition, sulphur and nitrogen (if present) in the fuel are oxidised to the corresponding acids along with the evolution of heat.

$$S + 2H + 2O_2 \longrightarrow H_2SO_4 + \text{Heat}$$

$$2N + 2H + 3O_2 \longrightarrow 2HNO_3 + \text{Heat}$$

So this heat is also included in the measured heat and hence must be substracted. The amount of these acids are analysed from washings of bomb by titration, while sulphuric acid alone is determined by precipitation as $BaSO_4$. Correction for 1 mg of S is 2.25 cal, while for 1 ml of N/10 HNO_3 formed, it is 1.43 cal.

(iii) Fuse wire correction. The measured heat also includes the heat given by ignition of the fuse wire and hence must be subtracted from the total value.

Thus, higher (gross) calorific value *'C'* is given by

$$C = \frac{(y + z)\ (t_2 - t_1 + \text{cooling correction}) - (\text{acid} + \text{fuse corrections})}{x} \text{ kcal/kg}$$

Modified Dulong formula for C.V.: It is expressed as

$$CG = \text{Gross C.V.} = \frac{1}{100}\left[8084\ C + 34500\ \frac{(0 + N - 1)}{8} + 2220\ S\right] \text{kcal/kg}$$

Assumption in Dulong formulae are :

(a) Gross C.V. of C, H and S are 8080, 34500 and 2220 kcal/kg respectively.

(b) Oxygen is combined with hydrogen (as in water) so that surplus hydrogen available for combustion is $\left(H - \frac{0}{8}\right)$

(c) N_2 in the coal has been neglected. If accounted, then replace $\left[H - \frac{0}{8}\right]$ by $(O+N-1)/8$.

(d) Heat of formation of coal is zero.

(e) C.V. of a fuel is the sum of C.V. of its constituent elements.

Goutal Formula for CV Calculation :

$$C_G = 82\,F + a.\,V$$

where, C_G = gross calorific value, kcal/kg, F = fixed carbon, percent

V = volatile matter, percent

and. a = a constant depending upon the value of volatile matter expressed as percent d.a.f. (V') as shown below :

V'	a	V'	a	V'	a
5	145	20	109	35	94
10	130	25	103	38	85
15	117	30	98	40	80

This formula assumes that the coal consists of volatile matter and fixed carbon, each contributing to the calorific value of the coal. The fixed-carbon of different coals is assumed of a fixed composition and hence of fixed calorific value. The composition and calorific value of the volatile matter differ from coal to coal and are assumed to depend upon the nature of coal as indicated by the volatile matter on d.a.f. basis. These assumptions limit the utility of the Goutal formula.

The following formulae have been developed by the Central Fuel Research Institute, Dhanbad (CFRI), for the calculation of calorific value of Indian coals from their proximate analysis :

Central Fuel Research Institute (CFRI) formulae for CV calculation :

For *low moisture coals* ($M \leq 2\%$), $C_G = 91.7\,F + 75.6\,(V - 0.1\,A) - 60\,M$

For *high moisture coals* ($M > 2\%$), $C_G = 85.6\,[100 - (1.1\,A + M)] - 60\,M$

where *M, A, V* and *F* denote moisture, ash, volatile matter and fixed carbon (all in per cent air-dried), respectively.

The calorific value of coal bears a relationship with the rank. The higher the rank, the higher the calorific value. Anthracites, however, form an exception to this regularity and have lower calorific values than many bituminous coals. This is due to the sharp fall of hydrogen content in anthracites, the hydrogen having more than four times the calorific value of carbon. High calorific value of cannel coal, boghead coal and exinites of humic coals are due to their high hydrogen content.

Problem on Calorific Value Calculation

Example 1. The proximate analysis (% air dried basis), the ultimate analysis (% d.m.m.f. basis) and experimental gross calorific value (kcal/kg on air dried basis) is as follows :

Proximate Analysis %				Ultimate Analysis %					C.V., Kcal/Kg
Moisture	Volatile matter	Ash	Fixed carbon	C	H	O	N	S	(experimental value)
2	30	16	52	86	6	5	2	1	6900

Calculate its gross calorific value (C_G, Kcal/Kg) using

(i) Modified Dulong formula (ii) Goutal formula (iii) CFRI formula

(iv) Difference of gross and net C.V. of coal on (a) air dried basis (b) complete dry basis

Solution, *(i)* G.C.V. by modified Dulong Formula is given by

$$C_G = \frac{1}{100}\left[8080\text{ C} + 34500\left(H - \frac{O}{8}\right) + 2200\text{ S}\right]\text{kcal/kg}$$

or, $$CG = 80.8 \times 86 + 345\left(6 - \frac{5}{8}\right) + 22.2 \times 1$$

or, $$C_G = 8825\text{ kcal/kg}$$

This C.V. which has been calculated by the ultimate analysis of coal is on d.m.m.f. basis. This is to be converted to air dried basis for comparing with the experimental value of gross C.V. *i.e.,* 6900 kcal/kg.

Hence, C.V. on air dried basis is

$$= 8825\ (100 - \text{Moisture} - 1.1\text{ Ash}) \times \frac{1}{100}$$

[Since, mineral matter in coal = 1.1 × ash of coal]

$$= 8825\ (100 - 2 - 1.1 \times 16) \times \frac{1}{100} = 7095\text{ kcal/kg}$$

This figure of C.V. is higher than experimental value of C.V. by 7095 - 6900 = 195 kcal/kg.

(ii) Gross C.V. by *Goutal formula* is given by, $= 82\,F + a\,V$

To find the value of constant *'a'* we first find the V.M. on % d.a.f. basis

$$V = \text{V.M. on \% d.a.f. basis} = \frac{30}{100 - 2 - 16} \times 100 = 36.6\%$$

The value of 'a' from the table is found as follows :

V'	a
35	94
38	85

$a = 94 - \frac{94-85}{38-35} \times (36.6 - 35)$ or, $a = 94 - 3 \times 1.6 = 89.2$

$\therefore$ C.V. on air dried basis is $C_G = 82 \times 52 + 89.2 \times 30 = 6940$ kcal/kg

This C.V. is higher than experimental value by 6940 - 6900 = 40 kcal/kg

(iii) Gross C.V. by CFRI formula (for moisture, $M < 2\%$), on air dried basis is

$$C_G = 91.7\,F + 75.6\,(V - 0.1\,A) - 60\,M$$

$$= 91.7 \times 52 + 75.6\,(30 - 0.1 \times 16) - 60 \times 2 = 6795 \text{ kcal/kg}$$

This C.V. is less than the experimental value by 6900 - 6795 = 105 kcal/kg

(iv) Difference of gross C.V. (C_G) and net C.V.(C_N) is given by

$$C_G - C_N = \text{weight of } H_2 \times 9 \times \text{latent heat of steam}$$

Since, 1 part by weight of hydrogen gives 9 parts by weight of water.

Latent heat of steam = 587 kcal/kg

$\therefore$ $C_G - C_n = \%\text{ of hydrogen} \times \frac{9 \times 587}{100}$ or, $C_G - C_N = 53 \times \%$ of H_2 in coal

(a) $C_G - C_N$ on air dried basis is found as follows :

Moisture = 2%, Ash = 16%, from proximate analysis

Total H = 6%, from ultimate analysis

First, % of hydrogen in coal in organic form is found by deducting the hydrogen present in moisture and ash.

$$\%\text{ hydrogen in moisture} = \frac{6 \times (100 - 2 - 16)}{100} = 4.9\%$$

$$\%\text{ hydrogen in moisture} = \frac{2}{9}\%$$

$$\%\text{ hydrogen in mineral matter} = \frac{0.1\text{ Ash}}{9} = \frac{0.1 \times 16}{9} = \frac{1.6}{9}\%$$

$$\therefore \text{Total \% of hydrogen in coal on air dried basis} = 4.9 + \frac{2}{9} + \frac{1.6}{9} = 4.9 + \frac{3.6}{9} = 5.3\%$$

$\therefore$ $C_G - C_N = 53 \times \%$ total hydrogen $= 53 \times 5.3 = 281$ kcal/kg

Hence, gross C.V. of coal is higher than the net C.V. by 281 kcal/kg on air dried basis

(b) $C_G - C_N$ on complete dry basis is found as follows :

$$\%\text{ ash on dry basis} = 16 \times \frac{100}{100 - 2} = 16.3\%$$

$$\%\text{ hydrogen (organic)} = 6 - \frac{100 - 2 - 1.1 \times 16}{100 - 2} = 5.1\%$$

$$\% \text{ of total hydrogen} = 5.1 + \frac{0.1 \times 16.3}{9} = 5.2\%$$

$$C_G\text{-}C_N = 53 \times 5.2 = 275 \text{ kcal/kg}$$

Hence, the gross C.V. is higher than net C.V. by 275 kcal/kg, on complete dry basis.

Weathering or Slacking Index of Coal

The weathering or slacking index of a coal is an indication of its size stability when stored/exposed to the whether. It denotes the tendency of coal to break on exposure to weather or alternate wet and dry periods.

Coal consuming industries store large quantities of coal at the plant as a reserve. During storage, it undergoes a series of changes by exposure to the atmosphere. This is known as the weathering of coal. Weathering is more pronounced with lower rank coals and greater exposure to air. By storing coal under water and in trenches, the weathering may be minimized.

Weathering reduces the coal size and increases its friability. The formation of slacks or small sizes of coal is particularly pronounced with sub-bituminous coal, lignite and peat which imposes a serious problem in their handling, transport × storage. Anthracite and mature bituminous coals do not slack.

The caking capacity is very sensitive to aerial oxidation and decreases on weathering. A slight decrease in the caking capacity improves the coke size in some cases but generally the coking properties deteriorate during the storage of coal. Other parameters which fall by weathering are carbon content, hydrogen content, calorific value and tar yield. The oxygen content and yields of gas & liquor increase by atmospheric oxidation.

Weathering of coal is a result of mild oxidation and hence an exothermic process. If the heat liberated is not completely dissipated, the temperature of the coal rises. Since the rate of a chemical reaction is approximately doubled by a rise of 10°C, the extent of atmospheric oxidation goes on increasing at an accelerated rate with a greater quantity of heat liberated. Ultimately, the coal may reach its ignition point and burst into flame. This phenomenon is known as the spontaneous inflammation of coal and is associated with the storage of bituminous coal in bulk (say, 200 tons for bituminous coal and 50 tons for lignite).

To determine this property, a known quantity of coal lump (2.5-4.0 cms size) is air dried and then immersed in water for one hour. Water is then drained and coal lump is air dried for 24 hours. The amount of disintegration is determined by weighing the coal which passes through a seive having 0.263 inch square openings. The weathering index is the percentage passing through the seive.

Hardgrove Grindability Index (HGI) of Coal

It gives an idea about the relative ease of grinding coals or the power required for grinding coals in a pulveriser. A standard test called Hardgrove method is employed to determine the *Hardgrove Grindability Index.*

50 gms of air dried coal of -16 + 30 mesh is subjected to 60 revolutions in a small mill (miniature pulveriser). After grinding, the coal is screened through a 200 mesh sieve. Hardgrove grindability index (HGI) is then calculated as,

G = Hardgrove grindability index = $6.93\ W + 13$

where, W = weight of sample passing through the seive, gm

Coals which are easy to grind have grindability index near to 100.

There are two methods of determining the ease of grinding coal to fine sizes, *viz.* Ball mill method and Hardgrove method. The former measures the amount of work done in grinding a pre-sized material to a given fineness and the later method measures the increase of surface produced by the application of a standard amount of work. A high value of G indicates a soft and easily grindable coal. Hardgrove grindability index of coal initially increases with the rank, reaches a maximum of about 105 for bright coals of 89-90 % carbon, and then falls sharply to about 35 for anthracites. The average Hardgrove index for Indian coals is 55-65 but coking coals of Jharia and Giridih show value up to 75.

Figure13.2 gives below the variation of Hardgrove grindability index of coal with the percentage of corbon in coal *(d.a.f.)*.

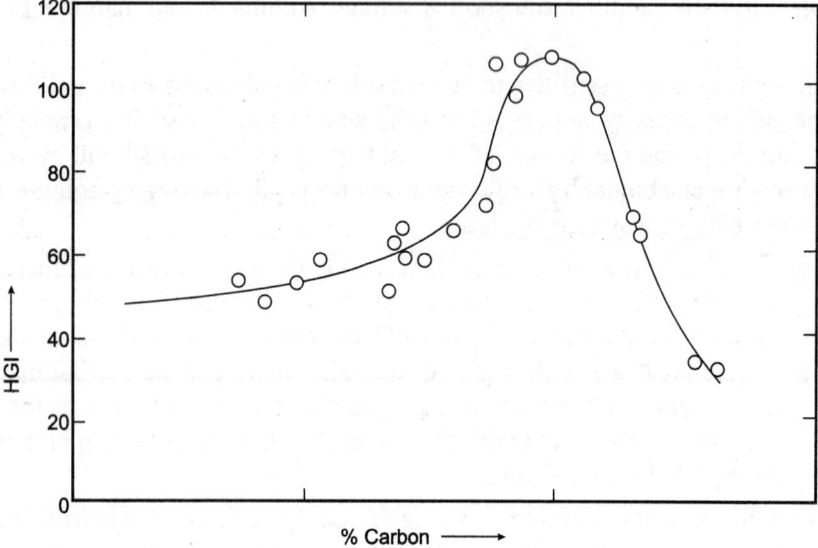

Fig. 13.2 *Hardgrove grindability index (HGI) Vs Carbon percent of coal (d.a.f.)*

Caking Properties of Coal

Caking coals form a part of the bituminous group. Peat, lignite, sub-bituminous coal, semi-anthracite and anthracite are all non-caking. So also are some of the bituminous coals. A measure of the caking properties is necessary for coal classification and for selection of coal for different uses. Caking properties influence the production of coke and the performance of the coal in combustion & gasification.

Caking, swelling, agglutinating and plastic properties are all interrelated. Numerous laboratory test have been proposed for measuring the propensity of these properties. The standard tests in vogue in different countries are: ***Free swelling index, Gray-King (Low Temperature) Coke Type, Caking (Agglutinating) Index, Roga Index*** and ***Thickness of Plastic Layer.***

The appearance of the residue (coke-button) in the volatile matter test may give an idea of the caking capacity of coal. The coke-button may indicate five types - pulverulent, sintered, weakly caked, caked and strongly caked.

Caking index of coal. It is the measure of binding or agglutinating property of coking coal. Numerically, it is the maximum number ratio of sand to coal in a 25 gm mixture of the two (sand + coal) which on heating under standard conditions produces a residue coke capable of supporting a weight of 500 gms without producing more than 5% of loose grains of coke.

The higher the agglutinating propensity of coal (*i.e.* the caking index), the higher the amount of sand (inert) that can be bound by coal and hence higher the caking index.

The caking index of coal blend charge for coke ovens (for producing metallurgical coke) is about 21 to 22.

This is one of three standard tests in India and is judged the best suited for the Indian coals. If the coal has higher than 17% ash, then the sample has to be washed before testing.

Swelling index/Number of coal. It denotes the caking capacity of coal. In this test 1 gm of air dried coal (of-72 mesh size is heated in a crucible with a lid on until volatile matter ceases to evolve. Rate of heating is maintained such that the temperature of the inner surface of the base of the crucible reaches 800°C within 1.5 minutes and 820°C within 2.5 minutes of lighting the burner. After the flame from burning volatile matter has died out or after 2.5 minutes whichever is the greater period of time, the crucible is cooled and the coke button is removed. The shape and size of the coke button is compared with the standard coke numbered from 1 to 9 at an interval of 0.5 (*e.g.* 1, 1.5, 2, 2.5, 3............... 8.5, 9).

This is called swelling number or swelling index of coal which is the number of standard profile most nearly corresponding to the coke button obtained under the test taking the average of five determination.

Standard profile of coke button numbered from 1 to 9 (at an interval of 0.5 unit) is approximately shown in Fig. 13.3.

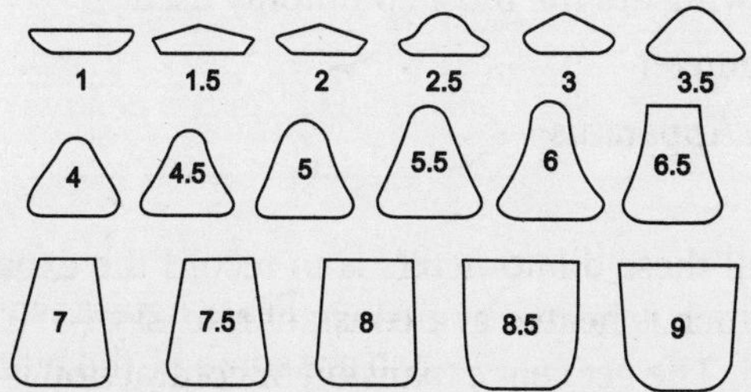

***Fig. 13.3** Standard profiles for swelling index number.*

Free swelling index is a simple and very useful test. It is one of the three tests standardized in India. The higher the swelling number, the better the caking and swelling properties. The limitation is imposed by the maximum possible number 9 and the very highly swelling coals cannot be differentiated from one another. It is a modified Gray-Campredon test and responds well to high ash Indian coals. The Bureau of Indian Standards has adopted it as one of the three tests for caking properties. The demerit of the test lies in its trial-and-error approach.

Roga Index. Roga index is a variation of the agglutinating index. A mixture of coal and specified anthracite in a fixed proportion is carbonized under standard conditions. The strength of the residue is expressed as a numerical figure called Roga Index. A mixture of 1 gm of (0.211 mm) coal and 5 g of anthracite (0.3-0.4 mm) is compacted in a porcelain crucible under a load of 6 kg applied for 30 sec. After being brought to a temperature of 850°C in 15 mins, the coke button is weighed (Q) and screened through 1 mm seive, the weight of the +1 mm fraction being designated as *'a'*. This material is placed in a rotating drum of diameter 20 cm and is rotated for three periods of 5 mins each. At the end of each period the -1 mm fraction is removed by screening. If *'b'*, *'c'* and *'d'* are the weights of the +1 mm residues at the end of first, second and third periods respectively, the Roga Index is given by

$$\text{Roga Index} = \frac{(a+\mathrm{d})/2+b+c}{3} \times \frac{100}{Q} \quad \text{where, Q=weight of coke button, gm.}$$

This ranges from 0 to 70 in practice.

Gray-king Assay (G-K index). Gray King Assay test is also carried out to observe the coking property of coal. In this test, 20 gms of fine coal (-72 mesh size) is heated in a silica tube to 600°C. After cooling, the coke formed is compared with standard profiles numbers *A, B, C,* .... G_1 G_2, G_3 ....G_{10}. If the residue remained is a powder, it is classified as A. Stage *B* is classified as *non-caking coal, C & D* are *weakly caking coal* while *E, F* & G are classified as medium caking coal. If the residue remains a hard coherent mass with the same volume as initially of coal, it corresponds to *G.*

Strongly caking coals swell and they are designated as G1,G2,G3.............G10. The suffix 1, 2,.--------10 indicates the number of grams of inert carbon which must be added to 20 gms coal charge to give zero swelling or a standard G type coke.

Roga index is related with *Gray-king Assay* as index,

0 - 5 = A - *B,* 5 - 10 = *B* - *D,* 20 = G, 45 = G_4

Dilatometric properties. The dilatometric test is done for the knowledge of the swelling and contraction of a coal during heating under load without the access of air. In this test the temperature range of plasticity of a coal can also be determined. There are a number of dilatometers used in different countries. The following are the most commonly used.

(1) Audibert-Arnu Dilatometer

(2) Sheffield Coking Test Apparatus

(3) Giesler Plastometer.

The basic principles of all these dilatometers is to record the expansion or contraction or the fluidity of the coal sample which is heated at a constant rate, say 1-3°C/min to a temperature until the plastic mass is resolidified. The per cent expansion or contraction is plotted against temperature for the first two apparatus, whereas in the third one the number of dial divisions per minute is plotted against temperature. The nature of curve gives the range of plasticity of the coal sample. Sheffield coking test set up is shown in Fig. 13.4 *(a).* Two typical dilatometric curves for coking coals are shown for Sheffield Coking Test Apparatus in Fig. 13.4 (b).

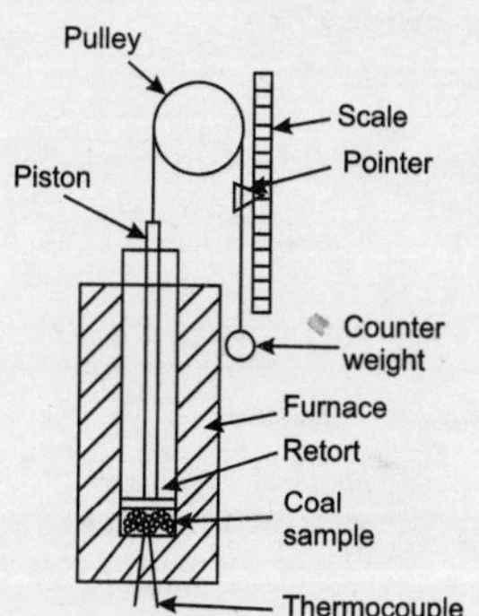

Fig. 13.4 (a) Sheffield coking test set-up.

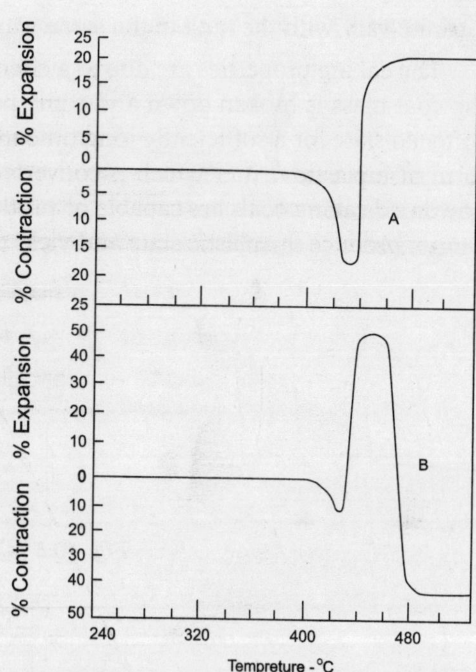

Fig. 13.4 (b) Typical dilatometric curves A and B

Thickness of plastic layer. In Russian practice, the carbon and plastic properties of coals are determined by the thickness of plastic layer. The apparatus used is known as *Sapozhnikov Plastometer* shown in Fig. 13.5 (a). The apparatus consists of a metallic cup with perforated bottom where 100 gm of coal sample (- 2 mm) is charged and a pressure of 1 kg/cm^2 is exerted on the coal by a weight suspended on a horizontal layer. The temperature is measured by a thermocouple placed at the bottom of the sample. Heating is done by an electrical heating element from the bottom. The sample is quickly heated to 250°C in 30 minutes and then the rate is controlled at *3°C/minute* upto 730°C.

The volumetric curve of the coal is plotted on the graph paper as shown in Fig. 13.5 (b) fixed on a rotating drum as the drum rotates at the rate of 1 mm/minute by a clock work mechanism. The maximum contraction after the heating is ***'x'*** (mm). During the heating period, by the help of a needle, fixed on a metallic scale, the formation of the plastic layer is measured in two levels. The readings of the upper level and the lower level of the plastic layer are plotted on the same graph paper and the maximum distance between these two curves give the thickness of the plastic layer, ***'y'*** (mm) which is the index for caking property. The nature of volumetric curve gives an idea of the coking nature of the coal sample.

The very young ***(i.e.*** peat, lignite and sub-bituminous) and the very mature ***(i.e.*** semi- anthracite and anthracite) coals are completely devoid of cokability. A typical plastometric curve for a weakly caking coal is shown in Fig. 13.5 (c). In the case of bituminous coals, the caking capacity first increases with the rank, attains a maximum and then decreases with the further rise in the maturity. The coals of the *highest caking capacity* are called **fat coals.** Among the microcomponents of the same coal, the fusinite is non-caking and the exinites are good caking. The caking properties of vitrinite vary with the rank in the same way as those of the whole coal.

The caking properties are due to a chemical process in which the macromolecular structure of the coal mass is broken down and some products of relatively low molecular weight remain in a softened state for a sufficiently long time in the reaction zone. As a result the entire mass takes the form of a plastic matter which is converted into a solid lump on further heating. In this way, the powdered caking coals are capable of producing lumpy semi-coke and coke. The non-caking coals cannot produce the plastic state and yield on carbonization only a sintered mass in place of coke.

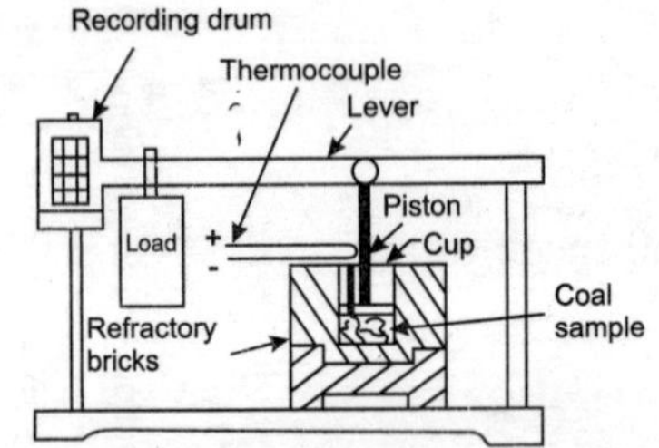

Fig. 13.5 (a) Sapozhnikov Plastometer

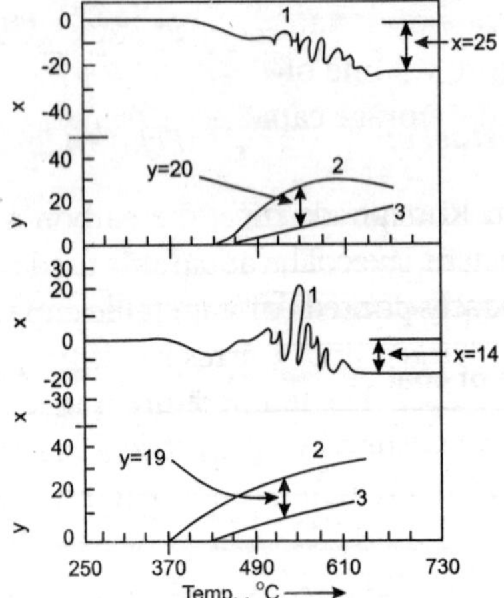

1-Volumetric 2-Curves for upper 3-Curves for lower level

Fig. 13.5 (b) Typical plastometric curves for 2 coking coals – A and B

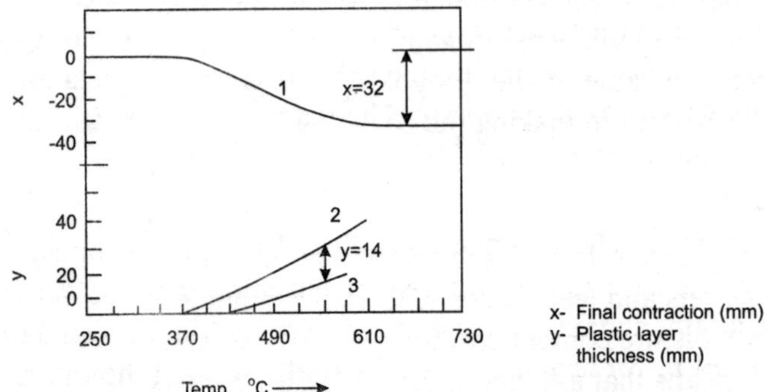

Fig. 13.5 (c) Typical plastometric curve for a weakly caking coal – C.

Specific Gravity

The proportion and nature of both the organic mass & mineral matter influence the specific gravity of coal. For the same type of coal, the higher ash coals have higher specific gravity. The true specific gravity of bituminous coals varies between 1.27 and 1.45. The following formula is valid for many coals of India.

$$S = k + \frac{A}{100}$$

Where, s = specific gravity,

k = a constant having an average value of **1.25**,

and, A = percentage of ash.

This formula assumes an overall specific gravity of 2.25 for the mineral matter of coal.

Among the coals of different types ; the specific gravity increases with maturity *e.g.* peat 1.15 to 1.25, lignite 1.25 to 1.3, bituminous coal 1.27 to 1.45, and anthracite 1.4 to 1.7.

Bulk Density

The bulk density is the overall density of a material kept in large quantities and depends upon the true density of the material, the pore volume of the particles/lumps and the voidage between the particles lumps. It is a measure of the storage capacity or the amount which can be stored in a given volume.

The bulk density of packed coal depends on the size analysis of the coal, the size of the container, the total moisture content of the coal, the shape of the particles/lumps and the method of packing. The bulk density of loosely packed dry coal may vary as (true density = 1300 kg/m^3) :

Forms of coal	Bulk Density, Kg/m^3
Lumpy coal (sized or free from fines)	600-670
Lumpy coal (with fines)	690-830
Coal dust	480-580
Pulverized coal	400-500

The bulk density of small coal decreases with increase of free-moisture content and then reaches a minimum value which may be 16% lower than for the air-dried coal and occurs at 4-6% moisture content. After the minimum, the bulk density again increases with the rise in free-moisture content but never attains other original value of the dry coal. The variation of bulk density with the free moisture content is very important in making the coal charge for coke ovens.

Angle of Repose

It is the angle that a heap of a coal forms with the horizontal and is of importance in its storage and in its flow in the conveyers and feed hoppers. It may vary by 2-3 degrees depending upon the roughness of the surface, moisture content etc. In general the higher the sizes, of coal, the higher the angle of repose. This means that a larger bulk of bigger sizes coal may be stored in a heap of given height. Approximate value of angle of repose for different sizes of dry coal in bulk are :

Size of coal, mm	18-30	12-18	6-12	0-6
Angle with horizontal, Degree	41	40	38	32

Porosity, Surface Area and Heat of Wetting

Coal is a porous substance and when brought in contact with a organic liquid (*eg.* methanol), it enters the very fine pores and the surface of coal is wetted. The process is exothermic and the heat of wetting is released. Heat of wetting is a measure of surface area of coal.

Porosity (per cent pore volume), surface area (m^2/gm coal) and heat of wetting (kcal/kg coal) vary with the rank of coal in the same manner as the moisture varies with the rank. There is initially a fall in the values till a minimum is obtained for the coal of 89-90% carbon. The values again rise in the anthracitic region. Some average data are given below :

Parameters	I	II	III
Carbon % (d.m.m.f.)	80	89	93
Heat of wetting, kcal/kg (d.m.m.f.)	18-20	2-3	6-8
Surface area, m^2/gm (d.m.m.f.)	180-200	20-30	60-80
Porosity, % (d.m.m.f.)	18-20	2-3	6-8

Approximately, 1 kcal/kg heat of wetting = 10 m^2/g surface area = 1% porosity.

Refractive Index and Reflectance. Both these optical properties refractive index and reflectance increase with increase in coal rank. For vitrain-rich samples, the following values are reported :

Carbon % (d.m.m.f.)	75	80	84	91	95
Refractive Index	-	1.6	—	1.9	—
Reflectance, %incident light	0.51	0.67	0.92	2.23	4.17

Coals of higher rank become anisotropic showing a higher refractive index in a direction perpendicular to the bedding plane of the coal in the seam due to the closer packing of atoms in this direction varying from 1.6 at 78% carbon to 1.85 at 90% carbon (using Whattman filter 70 at 7000 ^{0}A).

Among the microcomponents of the same coal, fusinite & micrinite have very high reflectance and exinite has very low reflectance ; the vitrinite having intermediate values.

SPECIFIC HEAT

The specific heat of coal increases with increase in the volatile matter and decrease in the C/H ratio. Typical values on d.a.f. basis at normal temperature are :

	Specific heat, Kcal / Kg °C
Bituminous coal	0.24-0.26
Fussain	0.21-0.22
Anthracite	0.22-0.23
Coal ash	0.16-0.17

The relationship between specific heat & moisture content and between specific heat & ash percentage is linear. The specific heat of a coal of given moisture and ash values can thus be calculated.

Solubility of Coal

Peat is slightly soluble in water. Lignite and other coals are completely insoluble in water, either cold or hot. Alkali solubility is shown by peat and lignite. These fuels contain humic acids which dissolve as humates in alkali solutions and get precipitated on acidification of the solutions. Higher rank coals normally do not contain humic acids but produce them on mild oxidation. Oxidized bituminous coal is therefore partly soluble in alkali solutions. Dilute hydrochloric acid dissolves the hemi-cellujose present in peat. The concentrated acid dissolves most of the mineral matter of coals. The demineralized or ultrapure coal is prepared by treated coal with hydrochloric acid. Nitric acid has an oxidizing effect on the organic mass of coal and produces various aliphatic & aromatic carboxylic acids. Concentrated sulphuric acid decomposes the cellulose present in peat.

Benzene, mixture of benzene & alcohol, chloroform etc. dissolve the waxes & resins which constitute 5-20% of peat & lignite and, generally, 1% or less of higher rank coals. Anthracene, phenanthrene, anthracene oil and tetralin, pyridine, quinoline, monoethanolamine & ethylene diamine can extract considerable portions of coal. The organic mass of a medium rank bituminous coal may be dissolved in tetralin or anthracene oil to the extent of 85-95% on treatment at, 350-400°C. The coal extracts obtained in such large yields are used in the production of liquid fuels, carbon electrodes etc. The solubility of anthracite in the above mentioned solvents is negligible.

Action of Heat on Coal

When a sample of powdered coal is heated out of contact with air, it loses occluded gases consisting of methane ethane, nitrogen, carbon dioxide etc. at temperature below 100°C. Moisture is evolved between 100° and 150°C. The organic mass of coal then starts decomposing with the evolution of gaseous and vaporous products. The decomposition is called 'active' when oil appears in the vaporous products. Peat and young lignites start decomposing at 100°C or less. The initial temperature of decomposition of bituminous coals is 200-300°C while active decomposition starts at 300-375°C for these coals.

Pyrogenic water, primary tar and gases evolve during the primary devolatilization (at 350-550°C), while gases (mainly hydrogen) are evolved during the secondary devolatilization at around 700°C. Peat and lignite show pronounced primary devolatilization & insignificant secondary devolatilization. Anthracite is characterized by the absence of primary devolatilization and occurrence of secondary devolatilization only. All the three processes *viz.* drying, primary devolatilization & secondary devolatilization are endothermic in nature.

While coal undergoes decomposition on heating, the residue becomes richer in carbon content. In the case of caking coals, the residue passes through a plastic state in the range 300- 350°C to 500-550°C. The fluidity of the plastic mass initially increases, attains a maximum and then decreases to zero. The last point corresponds to the resolidification of the plastic mass into semi-coke. This is also the upper limit of the primary devolatilization process. On further heating, secondary devolatilization takes place and semi-coke is converted into true coke. This conversion is marked by a sudden rise in the electrical conductivity of the solid material. Therefore, one method of ensuring proper coking is to check the electrical conductivity of the carbonaceous residue. If coke is heated further, significant changes take place around 2000°C when graphite like material results. This process is known as graphitization and is used in the production of graphite articles. Non-caking coals are not amenable to *graphitization.*

The porosity of coal decreases on heating and attains a minimum in the plastic state. After the resolidification, porosity again rises considerably. Because of the simultaneous formation of the plastic state and volatile products of thermal decomposition, the carbonaceous residue exhibits an initial contraction and fall in porosity followed by swelling, dilatation & rise in porosity. This phenomenon forms the basis of many tests for the caking properties of coal.

CLASSIFICATION OF COAL

Coals are classified into various categories to assess the useful value of a coal for a given purpose based on a few tests.

Coals are classified based on various parameters like proximate analysis, calorific value, maturity (rank), ultimate analysis, caking properties etc.

There is no universal classification system of coal because of the reason that:

- Coal is composed of diverse elements *i.e.* it is a heterogeneous mixture of organic mass, mineral matters and moisture in uncertain proportions.
- The organic mass itself being heterogeneous comprises of many microcomponents which means that the coals of same ultimate analysis do not necessarily have the same properties while coals of different compositions may in some respect behave alike.
- Both the amount and composition of mineral matters in coal varies widely which greatly influence the properties of the whole coal.
- The quality and quantity of the microcomponents vary from coal to coal. The quality of microcomponents is determined by the initial plant material and the coalification processes both of which vary from place to place & time to time.

Because of the above reasons, large number of coal classification systems exist. Grade of coal as used in India means the purity of the coal as assessed from ash and moisture content.

The nature of the plant debris forms the primary basis of coal classification according to which there are *three types* of coal namely :

- *Humic coals* or humites derived from the higher plant kingdom *e.g.* peat, lignite, bituminous and anthracite.
- *Sapropelic coals* or sapropelites derived from small plant organisms *e.g.* cannel coal and boghead coal.
- *Liptobiolites* derived from some selected parts of higher plants *e.g.* paper coal of Moscow basin which originated from a concentration of cuticles in the Palaezoic age superficially resembling accumulations of sheets of brown paper.

Rank of coal (which is the most important variable in the nature of coal) is determined by volatile matter, calorific value and carbon content of coal, singly or in combination. This classifies the coal broadly into groups like, lignite, bituminous and anthracite which is satisfactory for practical purpose. Bituminous coals are subdivided into sub-classes based on their caking capacity.

Since only *humic coals* are of industrial importance, their classification is only described in the following pages :

CLASSIFICATION OF HUMIC COALS

Regnault's classification. This was the first satisfactory classification given by Regnault in 1837 which is based on ultimate analysis of coal.

Regnault Gruner's classification. Regnault's classification was modified in 1874 by Gruner in which numerical limits were given for carbon, hydrogen and volatile matter to the higher rank coals defined by Regnault.

Regnault-Gruner classification is for bituminous coal only which have been divided into five classes as given in Table 14.1.

Table 14.1: Regnault-Gruner Classification of Bituminous Coals

No.of class	Characteristic of coal	Carbon %	Hydrogen %	Oxygen %	O/H ratio	Volatile matter	Nature of coke
I	Dry, long flame, non-coking type	75-80	4.5-5.5	15-19.5	4-3	40-50	Powdery or slightly coherent
II	Fat, long flame	80-85	5-5.8	10-14.2	3-2	32-40	Caked but friable
III	Fat	84-89	5-5.5	5.5-11	2-1	26-32	Caked, mode rately compact
IV	Fat, short flame	88-91	4.5-5.5	4.5-6.5	1	18-26	Caked, very compact, lustrous
V	Lean coals —anthracite	90-93	4-4.5	3-5.5	1	10-18	Powdery or slightly coherent

Brame's classification. It is the modified and extended form of **Gruner**'s classification which includes all the coals and has been tabulated in Table 14.2.

Table 14.2: Brame's Classification (*i.e.* Modified Gruner's Classification) of Coals

No. of class	Name of Class		Carbon	Hydrogen	Oxygen	Volatile matter
I	Lignitous		75-80	4.8-5.5	12-20	35-47
II	Ligno-bituminous		78-84	4.5-6	8-13.5	35-45
		Long flame non-caking				
III	Bituminus	(steam coal)	82-86	5-6	6-12	30-40
		Long flame, partly caking	82-86	4.5-5.5	5-9	30-40
V	Coals	(gas coal)	85-89	4.5-5.5	4-7.5	20-30
		Short flame (coking coal)				
VI	Semi-bituminous		89-92	4-5	2-4.5	13-20
VII	Semi-anthracite		91-93	3-4.5	3-5	8-13
VIII	Anthrace		> 92.5	< 4	< 3	< 8

Being not very suitable for the British coals, this classification system was further modified by Brame in 1925 to include other types of coals also. Among low rank coals, there are dry coals with high oxygen content between the lignites and usual steam 'non-caking' long flame coals which are termed as lignitous & ligno-bituminous.

Frazer's classification. Depending on the ratio of **fixed carbon to volatile matter,** Frazer classified the coals as follows :

Type of Coal	Fixed carbon / Volatile matter ratio
Anthracite	10-12
Semi-anthracite	12-8
Semi-bituminous	8-5
Bituminous	5-1

This coal classification system failed because of appreciable overlapping of classes.

Gruner-Bousquet's classification. This is a modification of the earlier classification by Gruner. In this system, coals above the rank of lignite have been divided into six classes based on their carbon contents. A table was drawn in which the main features of the six classes of coal were defined in terms of hydrogen, fixed carbon, volatile matter, calorific value, the nature of residue on carbonisation and behaviour of coals during combustion. Names were given to the six classes of coal based upon combustion characteristics as "dry, long flame", "fat, short flame" coal etc. The features of this classification is shown in the Table 14.3.

Table 14.3. Features of Gruner — Bousquets Classification

No. of class	Character of coal	C%	H%	F.C.%	V.M.%	C.V., Kcal / Kg	Nature of coke
I	Dry, long flame, non-caking	75-80	4.5-5.5	60-55	40-45	8000-8500	Powdery or slightly coherent
II	Fat, long flame	80-85	5-5.8	68-60	32-40	8500-8800	Caked but friable
III	Fat, "properly named"	84-89	5-5.5	74-68	26-32	8800-9300	Caked, moderately compact
IV	Fat, short flame	88-91	4.5-5.5	82-74	18-26	9300-9600	Caked, very compact, lustrous
V	Lean coals, semi anthracites	90-93	4-5.5	90-82	10-18	9200-9600	Powdery or slightly coherent
VI	Anthracites	93-95	4-2	92-90	8-10	9110-9200	Non-coherent, powdery
N.B. Above data are on ash free dry basis.							

Grout and Ralston classification. Grout in 1907 plotted the ultimate analysis of a large number of coals calculated on ash free dry basis on a triaxial diagram and found that humic coals lie on a narrow band while cannel coals which are rich in hydrogen lie well above the band while cannel coals which are rich in hydrogen lie well above the band.

In 1915, Ralston confirmed the observations made by grout. He found that coals of equal volatile matter and equal colorific value fell on approximately straight lines inclined at an angle to the C, H & O axes.

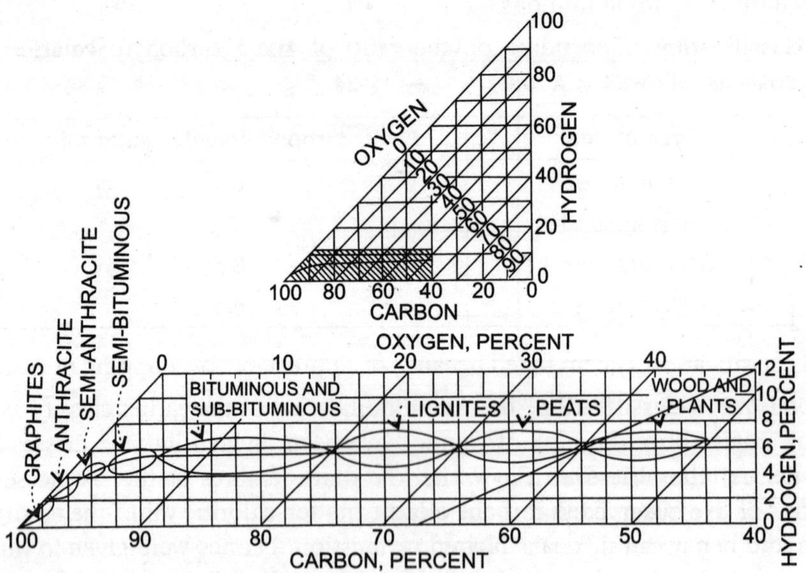

Fig. 14.1 Ralston's classification of coal

Ralston classification of coal was the most comprehensive system which included the percentages of oxygen, hydrogen and carbon in a triangular diagram. The percentages of the three elements were plotted on a trilinear co-ordinates as shown in the Fig. 14.1.

The above diagram distinctly shows the variation in oxygen content from wood to anthracite.

Parr's Classification. Parr classified coal **in** terms of heat of combustion (C.V.) of the coals plotted against the percentages of volatile matters.

The *calorific value* of coal is given by :

$$\frac{\text{Indicated B.T.U.} - 5000\text{ S}}{1-(1.08\text{A} + 22/40\text{S})}$$

and the *volatile matter* is given by :

$$\frac{\text{determined volatile matter} - (0.08 + 0.4\text{S})}{1-(1.08\text{A} + 22/40\text{S})}$$

where, A = ash and, *S* = combustible sulphur expressed as fractions.

The limits of volatile matter (%) and calorific value (BTU/lb) for various types of coals as classified by Parr are tabulated in Table 14.4.

Table 14.4: Parr's Classification of Coal

Type of Coal		Volatile matter, %	Calorific value, BTU/lb
Anthracite		0-8	15000-16500
Semi-anthracite		8-12	15000-16500
Bituminous	A	12-24	15000-16500
	B	24-50	15000-16500
	C	30-55	14000-15000
	D	35-60	12500-14000
Lignite		35-60	11000-12500
Peat		55-80	9000-11000
Cannels		60-80	15000-16500

Mott later devised a chart having volatile matter and heat of combustion as the principal rectangular co-ordinates.

Seyler's Coal Classification. Seyler in 1900 gave **a** classification based on the elementary analysis of coal. He observed that if the elementary constituents **of** the coal, carbon & hydrogen (on the basis of dry mineral free) is plotted on a chart, then there is a close relation between their composition and properties.

The coals are classified into four main groups or species (vertical columns) according to carbon content and each of these species is sub divided into varieties or 'genera' (horizontal columns) according to hydrogen content. Seyler's coal chart gives the technological properties of coal related to its elementary composition.

Seyler's classification permits greater differentiation between various coal types. Seyler's chart showing the Seyler's classification of coal is shown in Fig. 14.2.

Isovalant lines in the above figures show the regular variation of different properties such as caking power, capacity moisture and calorific value. In the above chart, the common types of coal fall within the rectangular areas to the left and together form a curved band which is termed as the coal band. Coals of higher hydrogen content (perhydrous) are bogheads & cannels of less common occurrence. Sub-hydrous coals are fussain and certain durains. The lines of equal calorific values (isocals) slope upward from left to right while those of equal volatile matter (isovols) cross them at right angles. The former are parallel, equally spaced lines following Dulong's law but the latter are less regular outside the region of 10 to 40% volatile matter. These lines correlate calorific value and volatile matter. The isocals are also lines of equal air requirements for the combustion of coal and have been correlated with flue gas volume, calorific intensity & flame temperature under conditions of adiabatic combustion. A number of scales show the combustion characteristics appropriate to the isocals lines. The top & bottom scales are the metric & British units respectively.

A recent & simplified form of Seyler's chart shows C/H axes and the coal band only as given in Fig. 14.3.

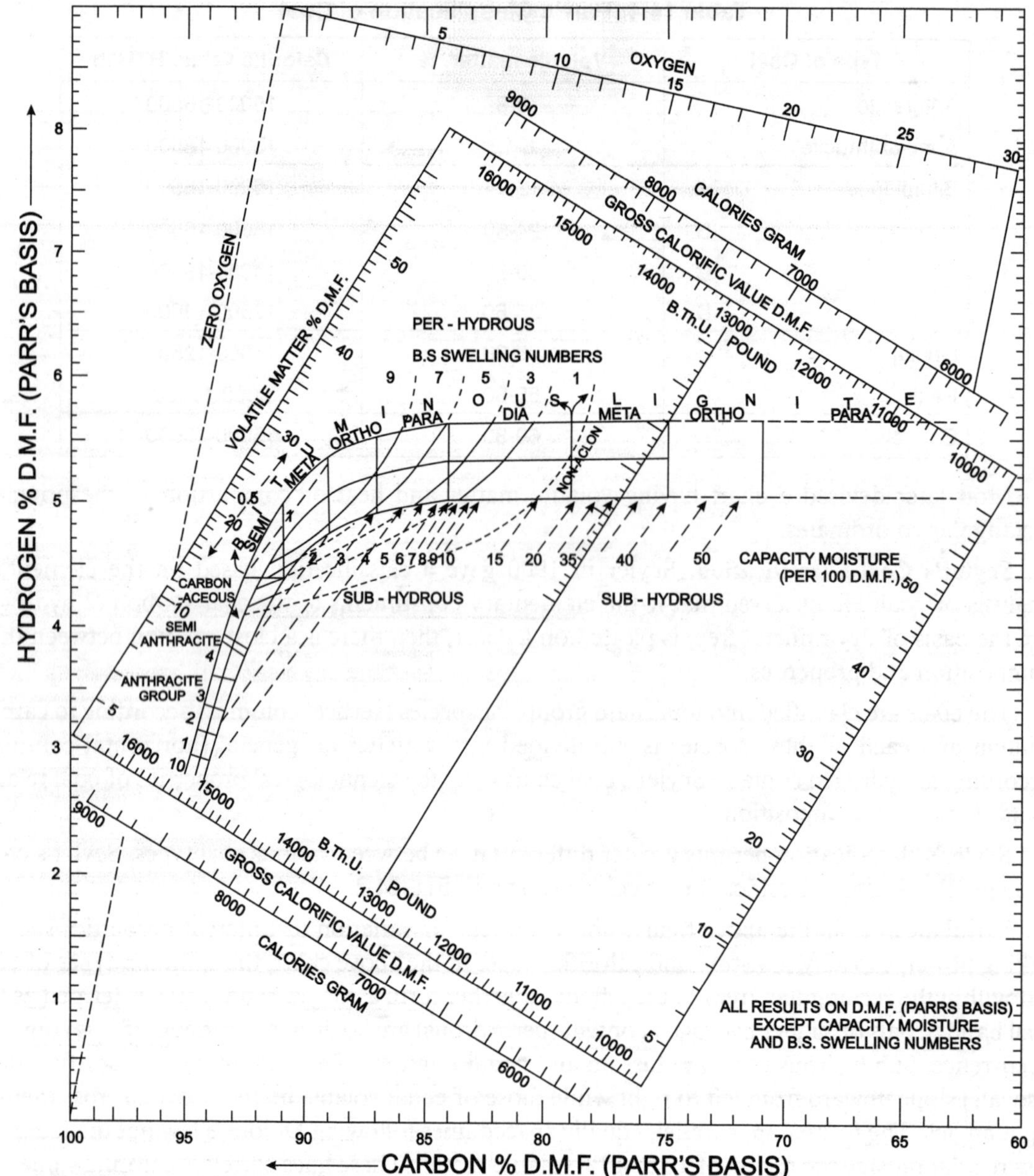

***Fig. 14.2** Seyler's coal classification (simplified)*

On rectangular co-ordinates representing % C and % H on pure coal basis (C + H + 0 = 100), he found that all normal high rank coals are placed in a narrow band between C = 75 & 95% and H from 3 to 5.7% within a maximum width equivalent to 0.7% hydrogen. Any coal falling above this band was considered to be abnormally rich in hydrogen (and so is called per-hydrous) and any coal falling below this band is called sub-hydrous. He also later added lines of equal caking characteristics to his chart.

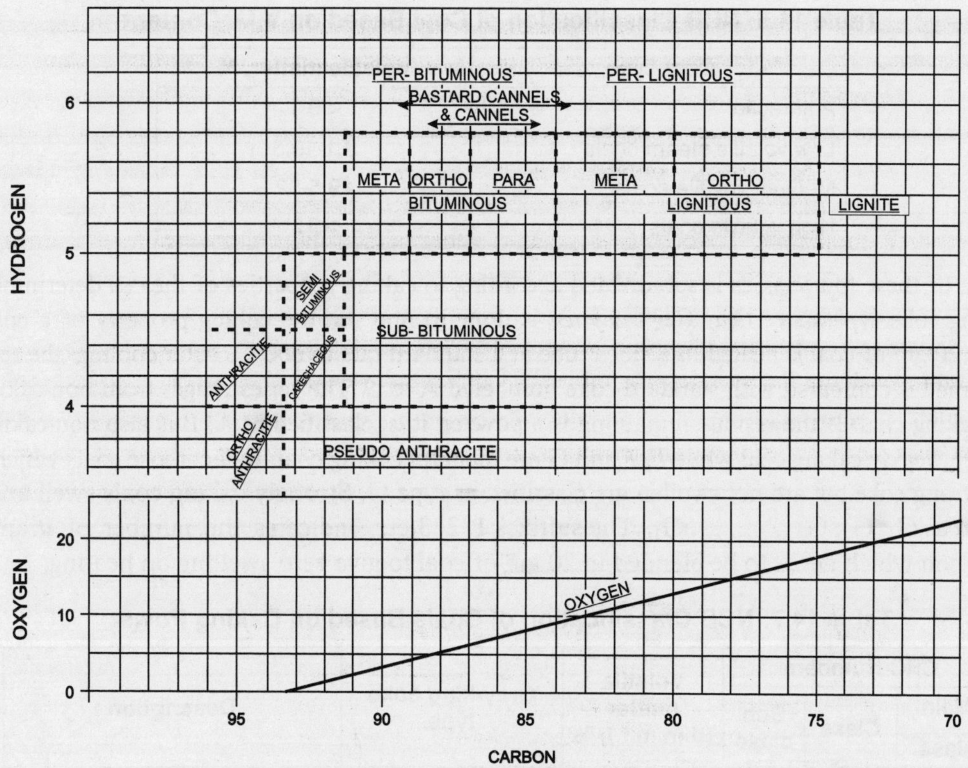

Fig. 14.3 Simplified Seyler chart

Table 14.5. Seyler's Coal Classification-simplified

Coal rank	Carbon %	Hydrogen %	Volatile matter %	Calorific value kcal/kg	B.S. Swelling index
Anthracite	>93.3	3-3.9	5-10	8555	1
Semi-anthracite	93.3-91.2	3.8-4.4	10-14	9722	1
Semi-bituminous	-	4.4-5	14-20	8877	3.5
Bituminous :					
Meta	91.2-89	4.4-5.4	20-28	8722	9
Ortho	89-87	4.7-5.6	28-31	8611	9
Para	87-84.4	4.9-5.7	31-36	8333	6
Lignite	< 75%	5-5.7	49-59	6500	1

Important feature of Seyler's classification is given in Table 14.5.

National Coal Board (NCB) Classification. NCB classification is based on two parameters: *(a)* the percentage of volatile matter on the dry mineral matter free basis and *(b)* the caking power of clean coal (containing < 10% ash) as indicated by Gray-king coke type.

Based on the volatile matter content of the coal, they are classified in four main classes as given in table 14.6.

Table 14.6: NCB Classification of Coal Based on V.M. Content

Coal Type	Volatile matter, %
Anthracite	< 9.1
Low volatile steam coals	9.1-19.5
Medium volatile coals	19.5-32
High volatile coals	> 32

Each of these four classes is sub-divided according to caking properties of the Coal determined by Gray-king coke type assay. *Gray-king assay test* is done to find out the coking property of a coal. In this test, 20 gm of coal is heated in a silica tube at a standard rate to 600°C. After cooling, the residue coke formed is compared with standard coke numbered A to G. The series ranges from non-caking to highly caking coal. If the residue remaining is a powder, it is; classified as A. B is also non-caking. C and D are weakly caking coal while *E, F* and G are medium coking coal. In fact those coals which give a hard strong coke but are not swollen are classified as type G. Strongly caking coals swell and are classified as G_1; G_2, G_3 .... G_{10}.The suffixes 1, 2, 3 etc. indicates the number of grams of inert carbon which needs to be blended in 20 gm of coal to give zero swelling on heating.

Table 14.7: NCB Classification of Coals Based on Caking Power

CRC Numbers			Volatile matter (d.m.m.f.), %	Gray-king coke type	Description
Main class	Class	Sub class			
100	101 102	- - -	< 9.1 < 6.1 6.1-9	A A A	Anthracite
200	- 201 202 203 204	- 201a 201b	9.1-13.5 9.1-19.5 9.1-11.5 11.6-13.5 13.6-15 15.1-17 17.1-19.5	A to G_8 A to C A to B A to C B to G E to G_4 G_1 to G_8	Low volatile steam coal Dry steam coal coking steam coal

300	301		19.6-32	A to G_9	Medium volatile coals
			19.6-32		
		-	$19.6-32$	$>G_4$	Prime caking coals
		-	$19.6-27.5$		
		301a			Medium caking coals
		301b	$27.6-32$		
	302		19.6-32	G to G_3	Weakly caking to non-caking
	303		19.6-32	A to F	coal
400	400 401 402	- - -	>32 $32.1-36$ >36	A to G_9 $\leq.\ G_9$	High volatile coals Very strongly caking coals
500	501 502	-	>32 $32.1-36$ >36	G_5 to G_4	Strongly caking coals
600	601 602	-	>32 $32.1-36$ >36	$G_1$to G_4	Medium caking coals
700	701 702	-	>32 $32.1-36$ >36	E to G	Weakly caking coals
800	801 802	-	>32 $32.1-36$ >36	*C* to *D*	Very weakly caking coals
900	901 902	-	>32 $32.1-36$ >36	A to B	Non caking coals

Table 14.8: INDIAN CLASSIFICATION OF COALS (IS : 770 - 1977)

Class	Type	Symbol	Nature	Basic Parameters			Other Properties			Code number	Uses
				C.V., Kcal/kg, (dmmf)		G-K coke type	Moisture, % *(60% R H)*	Carbon, % (dmmf)	Hydrogen, % (dmmf)		
Lignite	Consolidated	L	Non-caking	6150-7300	>50	A	>20	67-73	4.5-5.5	100, 200	Combustion, hydrogenation, gasification, briquetting
Sub-bituminous	High Volatile	SB	Non-caking	6950-7500	35-50	A	10-20	76-79.5	4.5-5.1	210, 220, 230	Gasification, Combustion
	High Volatile (N.E. coals)	B1	Non-caking to strongly caking	7500-8500	>45	A-G_3 +	2-9	75-82.5	5.3-6.3	300, 400, 500, 501-3, 601-3, 701-4, 310, 410, 510, 511-15, 611-15, 711-15	Combustion, blending Hydrogenation
Bituminous	Medium volatile to high volatile	B2	Non-caking	7500-8250	27-43	A-B	7-11	79.5-83.0	4.7-5.2	320, 420, 520, 330, 430, 530, 340, 440, 540, 450	Combustion, hydrogenation, gasification
	High volatile	B3	Weakly caking	8250-8400	33-43	C- D	5-7	82.5-83.5	5-5.4	521, 531, 621, 631, 721, 731	Blending, hydrogenation Low temp, carbonization
	High volatile, strongly caking	B4	Medium to strongly caking	8250-8500	33-43	E-G3+	2-5	83.5-87.5	5-5.8	622-5, 722-5, 632, 732-5, 832-5	Blending, Hydrogenation

	Medium volatile	B5	Weakly to medium caking	8500-8700	22-33	C-F	<2	86.5-88.0	4.7-5.0	741-2,841-2,851-2	Blending
	Medium volatile	B6	Strongly Caking (Prime coking coal)	8500-8900	22-33	G-G3+	<2	88-90.5	4.8-5.2	843-5, 943-5, 853-5, 953-5	Metallurgical coke making
	Low volatile	B7	Weakly to medium caking	8500-8900	18-22	C-G	<2	90.5-91.5	4.5-4.9	861-2, 961-3	Blending Combustion
	Low volatile	B8	Non-caking to weakly caking	8250-8700	15-18	A-D	<2	91.5-92.0	4.2-4.5	670, 770, 870-1	Combustion, domestic use
Anthracite	Semi-anthracite	SA	Non-caking	8250-8700	10-15	A	<2	92-93	3.7-4.2	670, 770, 870-1, 680, 780, 880	Combustion, Carbon artefacts, domestic use
	Anthracite	A	Non-caking	8500-8700	<10	A	-	>93	0.7	890	-do-

In this system of classification, coal is assigned a Coal Rank Code Number (CRC) which runs from 100 to 900 + with intermediate values as shown in the table. Coals with lower CRC numbers are ranked high and those with higher numbers are low rank coals. As the rank of coal increases, the volatile matter decreases. Anthracite is of the highest rank.

INDIAN CLASSIFICATION OF COAL

Current classification (IS 770-1977) of Indian coals divides them in four classes (namely lignite, sub-bituminous, bituminous and anthracite) & twelve types and also mentions their usage. Coals found in north eastern India (N.E. Coal) which shows very widely different coal characteristics is grouped as B_l constituting a special formation.

Basic parameters used for classification of Indian Coals are *calorific value, volatile matter* and *Gray-King (G-K) coke type* while other properties included are moisture, carbon & hydrogen percentages in coal. Table 14.8 gives the Indian classification of coal as per IS 770- 1977.

Code classification system of Indian Coal. This system of code classification of Indian coal is shown in table 14.9, in which each variety of coal is assigned a three digit basic code number. The group number of coal increases from bottom to top (first digit, calorific value) or from left to right (second digit, volatile matter.) The third digit is based on Gray-King (G-K) coke type. A fourth supplementary digit provided in the code number of coal indicates the *maximum thickness of plastic layer (MTPL)* in case of caking coals or moisture content in case of non- caking coals. Division of basic type of coal into sub group (using supplementary parameters) is shown on the right side of table 14.9.

The decoded property found from Table 14.9 of a caking coal having integrated code number (3 digit basic code number + 1 digit supplementary parameter) 8456 and a non-caking coal having code number 3203 is given below.

Parameters	Coal code number 8456	Coal code number 3203
Calorific value, Kcal/kg (dmmf)	8505-8700	7505-7800
V.M. (dmmf), %	33-28.1	43-37.1
G-K (LT) coke type	G_3 +	A
MTPL, mrn	26	—
Moisture (dmmf), %	—	15-9.1
Nature of coal	Prime coking coal	High moisture, high volatile, non-caking coal

All high volatile, non-caking coals giving G-K coke types A&B have zero as the third digit in the three digit basic code number. But all these non-caking coals do not have same/equivalent technological properties, that is why fourth supplementary parameter sub-divides such coals based on the moisture content ranges.

Increasing group number of coals as given in table 14.9 indicates increasing desirable properties-level depending on their uses *i.e.,* lower moisture content, higher C.V., lower V.M content, higher

MTPL and G-K type approaching towards G. Blank spaces in the table 14.9 indicates that coal of such type do not exist presently.

Table 14.9: Classification of Indian Coals based on code numbers (IS:770:1977)

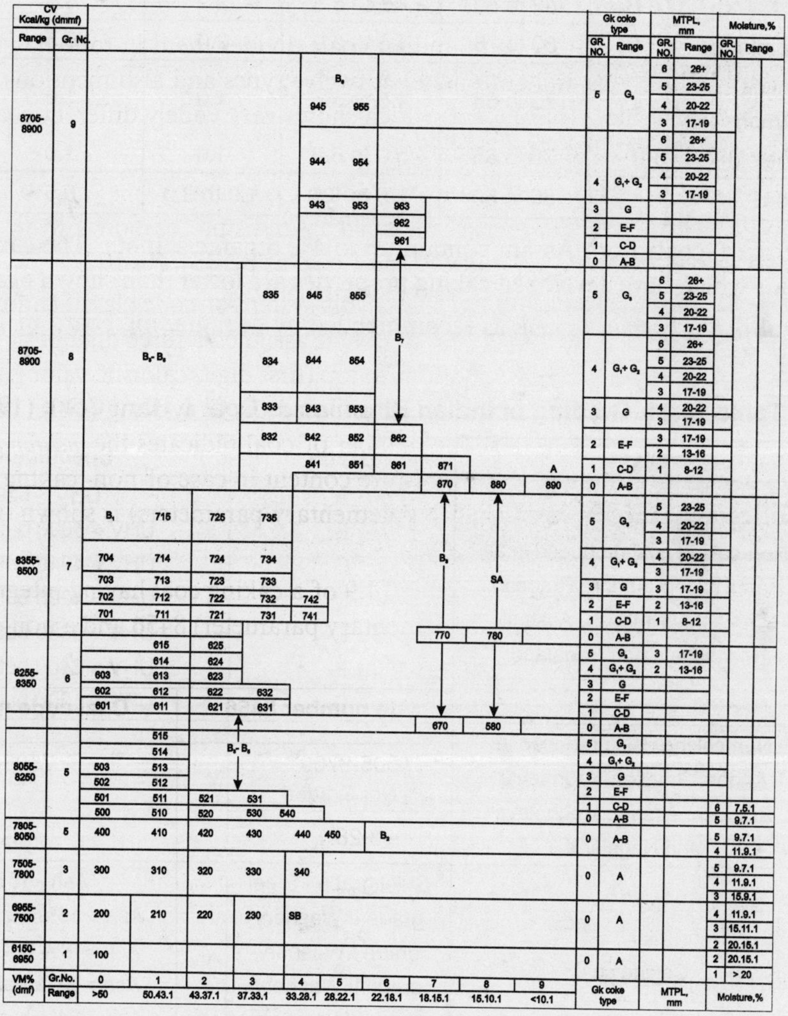

CV Kcal/kg (dmmf) Range	Gr. No.											Gk coke type GR. NO.	Range	MTPL, mm GR. NO.	Range	Moisture,% GR. NO.	Range
8705-8900	9					B_9						5	G_3	6	26+		
														5	23-25		
						945	955							4	20-22		
														3	17-19		
												4	G_1+G_2	6	26+		
						944	954							5	23-25		
														4	20-22		
														3	17-19		
						943	953	963				3	G				
								962				2	E-F				
								961				1	C-D				
												0	A-B				
8705-8900	8											5	G_3	6	26+		
					835	845	855							5	23-25		
														4	20-22		
														3	17-19		
								B_7				4	G_1+G_2	6	26+		
			B_5- B_6		834	844	854							5	23-25		
														4	20-22		
														3	17-19		
					833	843	853					3	G	4	20-22		
														3	17-19		
					832	842	852	862				2	E-F	3	17-19		
														2	13-16		
						841	851	861	871		A	1	C-D	1	8-12		
									870	880	890	0	A-B				
8355-8500	7											5	G_3	5	23-25		
		B_1	715	725	735									4	20-22		
														3	17-19		
		704	714	724	734				B_8			4	G_1+G_2	4	20-22		
														3	17-19		
		703	713	723	733					SA		3	G	3	17-19		
		702	712	722	732	742						2	E-F	2	13-16		
		701	711	721	731	741						1	C-D	1	8-12		
									770	780		0	A-B				
8255-8350	6		615	625								5	G_3	3	17-19		
			614	624								4	G_1+G_2	2	13-16		
		603	613	623								3	G				
		602	612	622	632							2	E-F				
		601	611	621	631							1	C-D				
									670	580		0	A-B				
8055-8250	5		515									5	G_3				
			514		B_5- B_6							4	G_1+G_2				
		503	513									3	G				
		502	512									2	E-F				
		501	511	521	531							1	C-D			6	7.5.1
		500	510	520	530	540						0	A-B			5	9.7.1
7805-8050	5	400	410	420	430	440	450	B_2				0	A-B			5	9.7.1
																4	11.9.1
7505-7800	3	300	310	320	330	340						0	A			5	9.7.1
																4	11.9.1
																3	15.9.1
6955-7500	2	200	210	220	230	SB						0	A			4	11.9.1
																3	15.11.1
6150-6950	1	100										0	A			2	20.15.1
																2	20.15.1
																1	> 20
VM% (dmf)	Gr.No.	0	1	2	3	4	5	6	7	8	9	Gk coke type		MTPL, mm		Moisture,%	
	Range	>50	50.43.1	43.37.1	37.33.1	33.28.1	28.22.1	22.18.1	18.15.1	15.10.1	<10.1						

The ultimate analysis based classification of Indian coals is tabulated in Table 14.10.

Table 14.10. Ultimate Analysis of Indian Coals (I.S.: 770-1964)

Type		Percentage in coal (d.m.m.f. basis)				
		Carbon	Hydrogen	Nitrogen	Sulphur (organic)	Oxygen
Anthracite	A_1	over 93	3.0 to 4.0	1.0	0.5	1 to 2
	A_2	93 to 92	3.0 to 4.0	1.0	0.5	1 to 2

	A_3	92 to 91	4.4 to 4.6	1.5	0.5	1 to 2
	A4	91 to 87	4.5 to 5.3	1.5	0.5	2 to 6
Bituminous	B_3	87 to 84	5.0 to 5.8	2.0	0.5	5 to 8
	B_4	84 to 82	5.0 to 5.5	2.5	0.5	8 to 11
	B_5	82 to 80	4.5 to 5.5	2.5	0.5	10 to 15
Sub-bituminous	B_6	80 to 76	4.5 to 5.0	2.0	0.5	15 to 17
Lignites (Normal)	L_1	75 to 65	4.5 to 5.5	1.0	1.0	20 to 30
(Canneloid)	L_2	75 to 65	5.0 to 6.0	1.0	1.0	20 to 30
Peat	—	60 to 50	5.5 to 6.5	1.0 to 3.0	0.5	30 to 40

Some (tertiary) coals from Assam contain up to 4% organic sulphur. The carbon percentages of Assam coals which have developed caking properties are lower than shown against the groups.

Grading of *Indian bituminous coals* as recommended by coal grading board in 1981 is tabulated in table 14.11.

Table 14.11. Grading of Indian Bituminous Coal & Hard Coke (1981)

S.N.	Class	Grade	Specification
1.	Non-coking coal produced in all states except Andhra Pradesh, Arunachal Pradesh, Assam, Meghalaya and Nagaland	A B C D E F G	UHV > 6200 UHV > 5600 but < 6200 UHV > 4940 but < 5600 UHV > 4200 but < 4940 UHV > 3360 but < 4200 UHV > 2400 but < 3360 UHV > 1300 but < 2400
2.	Non-coking coal produced in Andhra Pradesh, Arunachal Pradesh, Assam, Meghalaya and Nagaland	Not Graded	—
3.	Coking coal	Steel grade I Steel grade II Washery grade I Washery grade II Washery grade III Washery grade IV	Ash <15% Ash > 15% but < 18% Ash >18% but < 21% Ash > 21% but< 24% Ash > 24% but < 28% Ash > 28% but < 35%
4.	Semi-coking and weakly coking coal	Semi-coking I Semi coking II	(Ash + Moisture) < 19% (Ash + Moisture) > 19% but < 24%
5.	Hard coke	By-product Premium By product Ordinary Beehive Premium Beehive superior Beehive ordinary	Ash < 25% Ash > 25% but < 30% Ash <27% Ash > 27% but < 31% Ash >31% but < 36%

ASTM Classification. Coals containing less than **31%** volatile matter **(V.M.)** on the mineral matter free basis are classified only on the basis of fixed carbon *i.e.* **100-V.M.** %. They are divided into five groups *i.e.* > 98% fixed carbon FC, 98-92% (FC), 92-86% FC, 86-78%> FC and 78-69% FC.

The first three of these groups are called anthracites and the last two are called bituminous coal. The remaining bituminous coals, sub-bituminous coals and lignites are then classified into groups determined by the calorific value of the coal containing their natural bed moisture *i.e.* the coals as mined but free from any moisture on the surface of the lump. The classification includes three groups of bituminous coals with moist C.V. from > 14000 to > 13000 BTU/lb (1 kcal/kg=1.8 BTU/lb); three groups of sub-bituminous coals with moist C.V. < 13000 to < 8300 BTU/lb to 1 BTU/lb and two groups of lignitic coals with moist CV < 8300 BTU/lb. This classification also differentiates between consolidated and unconsolidated lignites and between the weathering characteristics of sub-bituminous and lignitic coals.

E.C.E. Classification. This classification system devised by coals committee of the *Economic Committee for Europe* (ECE) is based partly on the NCB code system and partly upon the ASTM classification.

The parameters used are the caking and coking properties for the coals containing < 33% volatile matter (V.M.) and C.V. on the moist, ash free basis (30°C and 96% humidity) for coals containing > 33% V.M.

Indices of *caking* properties are crucible swelling number or the Roga index (measure of agglutinating power).

Alternative indices for *coking* properties are the Gray-king coke type or the Audibert-Arnu dilatometer number.

The distinction between *caking* and *coking* properties is in the rate of heating. High rate of heating applies to the determination of caking and slow rate of heating applies to the determination of coking properties.

In this system of classification, three digits code numbers for groups and classes is used in which the first digit is the group number, the second the caking index and the third the coking index.

Thus a coal containing 10-14% **V.M.** with a crucible swelling number of 1-2 and a Gray king coke type *E-G* would be given a code number 212. The code numbers resemble those used in the N.C.B. code ranging from 100 for anthracite to 900 for high volatile, non-caking coals (of low C.V.), but the parameters used for sub-groups and class types are more appropriate. Hence this classification system is more scientific.

Classification of coking coals. Coking coals are classified as prime coking coal, medium coking coals and semi (or weakly) coking coal depending upon various parameters differently in different countries.

However as a general classification, coals having caking index > 24 are called prime coking coal whereas coals having caking index between 12-23 are termed as medium coking coals having caking index < 12 are called semi/weakly coking coals.

COKING COAL CLASSIFICATION SYSTEM IN U.S.A.

Prime coking coal	Sulphur = 1%, Ash = 6%, Free swelling index (FSI) = 7
Medium coking coal	S = 1.25%, Ash = 7%, FSI = 5.7
Semi-coking coal	S = 1.75%, Ash = 10%, FSI = 4.5

COKING COAL CLASSIFICATION SYSTEM IN AUSTRALIA

Prime Coking Coal :	
V.M. (% d.m.m.f. basis)	20.1 to 24
crucible swelling No.	6.5 to 9
G-K index	G_9
Ash (dry basis)	4.1 to 8%
Vitrinite reflectance	1.2 to 1.3
Medium Coking Coal :	
V.M. (% d.m.m.f. basis)	24.1 to 28
Crucible swelling No.	4.5 to 6
G-K Index	G_1 to G_4
Ash (dry basis)	8.1 to 12%
Vitrinite reflectance	1.
Semi-coking Coal :	
V.M. (% d.m.m.f. basis)	>33
Crucible swelling No.	6.5 to 9
G-K coke type	G_1 to G_4
Ash (dry basis)	8.1 to 12%
Vitrinite reflectance	0.72.

COKING COAL CLASSIFICATION SYSTEM IN JAPAN

Prime Coking Coal:	
Carbon (% d.a.f. basis)	85 to 90
Dilatometer index	< 50
Free swelling index	> 4
G-K index	>G_5
Roga index	> 45
Medium Coking Coal: :	
Carbon (% d.a.f.)	80 to 83
Dilatometer index	10-15
Free swelling index	2.5 to 4
G-K index	G_1 to G_5
Roga index	20 to 45
Semi Coking Coal :	
Carbon content (% d.a.f. basis)	80-83
Dilatometer index	> 10
Free swelling index	1 to 2
G-K index\	G_1
Roga index	5 to 20

Carbonisation of Coal-Coke Making & By-Products Recovery

Carbonisation. Heating of coal in absence of air to produce coke is called its carbonisation or destructive distillation. Its main purpose is to produce coke whereas the co-product is called coke oven gas from which various liquid products like tar, benzol, naphthalene, phenol, anthracene etc. are separated.

Types of Carbonisation. There are mainly two types of carbonisation depending upon the temperature upto which the coal is heated in absence of air namely - Low temperature carbonisation (LTC) and High temperature carbonisation (HTC). Their salient features are tabulated in Table 15.1.

Table 15.1: Salient Feature of LTC and HTC of coal

Low Temperature Carbonisation (LTC)	High Temperature Carbonisation (HTC)
(a) It is carried out at 700°C.	*(a)* It is carried out at 1100°C.
(b) It produces semi-coke which is used as a smokeless domestic fuel. It can sometimes be used in boiler also to avoid smoke.	*(b)* It produces metallurgical coke for use in blast furnace and cupolas in foundry etc.
(c) Yield of coke oven gas is less in LTC. It is about 150-160 Nm^3 gas/ton dry coal. Less gas yield is due to less devolatalisation of coal and less cracking of hydrocarbons at lower temperature of carbonisation.	*(c)* Yield of coke oven gas is more in HTC due to more cracking of hydrocarbons (mainly methane in coke oven gas) at higher temperature. Yield is about 290-300 Nm^3/ton of dry coal.
(d) Yield of tar is high in low temp, carbonisation. It is about 10% of dry coal.	*(d)* Tar yield is less here. It is about 3% of dry coal charged.
(e) Ammonia yield is low.	*(e)* Ammonia yield is more (10-15 gm/Nm^3 of coke oven gas).
(f) Calorific value of coke oven gas produced in LTC is more due to higher percentage of methane and unsaturated hydrocarbons in it. C.V. is about 6000-6500 kcal/Nm^3.	*(f)* C.V. of coke oven gas produced in H.T.C. is less. It is about 4200-4400 kcal/Nm^3 only due to lesser percentage of hydrocarbons resulting from its cracking at higher temperature of carbonisation.
(g) The tar produced is aliphatic in nature. It contains less quantity of aromatic ring compounds like benzene, toluene, naphthalene, phenol, anthracene etc. However, tar acid content is higher. Tar produced is called primary tar and has a lower market value than high temperature.	*(g)* Tar produced has more of aromatic ring compounds, (due to cyclisation reaction of straight chain compounds being favoured at higher temperature).

Low Temperature Carbonisation (LTC)	High Temperature Carbonisation (HTC)
(*h*) After carbonisation, discharging of coke is difficult as it swells a lot but does not shrink much finally at the end of coking due to lower temperature of carbonisation.	(*h*) Discharging of coke from ovens is easier as it shrinks finally to a more extent comparatively due to higher temperature of carbonisation.
(*i*) Free carbon in tar (which results from the cracking of hydrocarbons) is less. It is about 5-10% of tar. Since cracking (HC $\longrightarrow$ H + C) is less severe at lower temperature.	(*i*) Free carbon in tar is more (due to more intense cracking of hydrocarbon at higher temperature. It is about 15-20% of tar. Higher carbon in tar chokes the hydraulic main and other tar flow pipelines).
(*j*) Coke produced is weaker (due to less shrinkage), bigger in size and more reactive (due to higher porosity). Coke produced is termed as soft coke.	(*j*) Coke produced is stronger (i.e. shatter index, micum index, abrasion index are more), smaller in size and less reactive (due to low porosity) due to higher amount of shrinkage of coke at higher temperature. Coke produced is called hard coke
(*k*) Volatile matter content in coke is more (5-7%) hence it is easier to ignite it because of lower ignition temperature of high volatile matter containing coke. Ignition temperature of LTC coke is about 425°C.	(*k*) V.M. in coke is less (1-2%), hence its ignition temperature is more. Ignition temperature of high temperature, coke is about 600°C.
(*l*) Hydrogen content in coke oven gas is less (35-40%). Hence, difference in gross and net calorific value is less.	(*l*) H_2 content in coke oven gas is more (55-60%). It is beneficial for an adjoining nitrogeneous fertiliser plant attached to steel plant (as in the case of Rourkela Steel Plant) which gets hydrogen (for ammonia making) from coke oven gas by its cryogenic cooling.
(*m*) Coke yield is more. It is about 80% of dry coal.	(*m*) Coke yield is less (about 75% of dry coal).
(*n*) Density of coke oven gas produced is more, i.e. about 0.65 kg/Nm^3.	(*n*) Density of coke oven gas produced is less i.e., about 0.45 kg/Nm^3.

Physico-Chemical Changes During Coal Carbonisation. When coal is subjected to carbonisation, various physico-chemical changes occur at different temperatures.

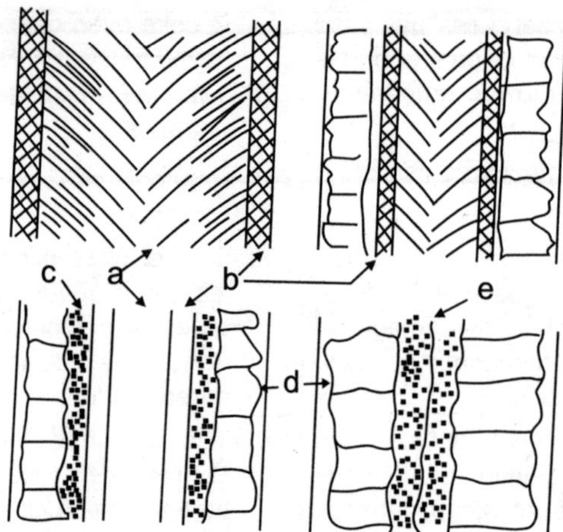

Fig. 15.1. *Stages of carbonization of coal in a coke oven.*

(a) Raw coal (b) Plastic layer
(c) Semi-cock (d) High temp. Coke
(e) Median Crack in Coke.

Raw coal at higher temperature (400^0C) gets converted into plastic layer. Then it is converted to semi-coke (from 450-600^0C). At higher temperature, coke is formed which develop crack in the middle of coke mass when further heated. Stages of carbonisation of coal in a coke oven is shown in Fig. 15.1.

Progressive events during carbonisation of a medium volatile coking coal is given in the Table 15.2.

Table 15.2: Physico-Chemical Changes During Carbonisation of coal

Temperature, °C	Effect	Nature of products		
		Carbonaceous residue (solid)	Tar and Oil	Gas
300	Initial decomposition temperature.	Coal	Nil	Some CO + CO_2, H_20.
320	Slight appearance of oil.	Coal	Thin, light coloured oil.	Above plus some CH_4 and unsaturated hydrocarbons.
360	Marked evolution of thick coal, oils and hydrocarbon gases, coal residue begins to soften.	Partially softened	Darker, red or brown oil.	More CH_4 and higher paraffins, some H_2-
430	Evolution of viscous oil and tar. Coal residue becomes softer and swells. Pronounced gas evolution causes bubble formation, rapid decomposition of coal.	Soft Carbonaceous mass of maximum volume and bubble structure.	Darker brown more viscous oil.	Maximum evolution of paraffin and un saturated hydro carbons with some CO, H_2 and H_20.
460	Oil and tar yield reduces, plastic expanded mass solidifies to semi-coke.	Solid semicoke with maximum bubble structure and weak cell walls.	Nearly viscous oil or tar.,	
600	Oil and tar cease, hard semi-coke starts to shrinks.	Semi-coke is harder and shrunken, colouris still black.	Non-evolved.	Diminishing yield of hydrocarbons water, some in CO and H_2O.

900-1050	Continued shrinkage hardens the coke, structure changes with commencement of formation of graphite lattice.	Hard, silvery Grey appearance.	Non-evolved.	Gasses namely CO, H_2 and CH_4.

Hence mechanism of coal carbonisation may be summarised as occurring in three stages.

Stage-I. Primary breakdown of coal at 700°C yields decomposition products like H_2O, CO, CO_2, H_2S, aromatics paraffins, olefins, phenolics, nitrogen containing compounds etc.

Stage-II. Secondary thermal reaction among these liberated primary products as they pass through hot coke, along hot oven walls and through highly heated free space in the oven involve both synthesis and degradation. Evolution of H_2 and formation of aromatics & methane occur at 700°C. Decomposition of complex nitrogen containing compounds produces NH_3, HCN, N_2 and pyridine bases.

Stage-III. Progressive removal of hydrogen from residue in the oven produces hard coke.

Commercial Production of Low Temperature Coke is done by:

- Carbonisation in thin layers in externally heated, vertical metal retorts-coalite or parker process..
- Internal heating by heat from the products of combustion of part of the gas yield-Rexco Process.
- Carbonising in continuous, vertical gas retorts.

LOW TEMPERATURE CARBONISATION (LTC) OF COAL

It is done mainly for manufacturing domestic smokeless fuel. Open stack burning of coal to produce soft coke is still practised in India which is of course an undesirable method of LTC of coal as :

- it pollutes the air.
- it produces inconsistent quality of soft coke which is partly burnt.
- liquid and gaseous by products can not be utilised.
- heat liberated by combustion is not used.

Heat transmission through the coal charge is very poor in LTC which is overcome in modern commercial LTC plant by using either externally heated metal or refractory retorts of narrow width or internally heated rotary or stationary retorts or a combination of the two.

Coal and semicoke are poor conductors of heat. In a static bed of externally heated coal, a high temperature gradient is needed to achieve the carbonisation in a reasonable time which may affect the product by overcarbonizing the surface layers. Therefore many other methods have been developed with a view to overcoming this difficulty. The heat-transfer rate is enhanced in these methods by using one or more of the following steps : (1) direct heating of the charge by hot gases, (2) movement of the charge by vertical fall, (3) rotating the carbonizer, and (4) using fluidized bed technique.

A pilot plant at the Central Fuel Research Institute, Dhanbad, uses the second step. It has two narrow continuous vertical retorts made of bricks and heated externally. Each retort has a capacity of 10 t/day. Lurgi-spill gas process, uses the first two steps. A pilot plant based on this process is in operation at the Regional Research Laboratory, Hyderabad, and another commercial unit produces carbonized lignite briquettes at Neyveli.

The various products of low temperature carbonization of coal are :Semicoke, low temperature tar, liquor, crude low temperature spirit and gas. The quality & yield of these products depend upon the process and the coal used. L.T.C. plants normally use low rank coals and lignite which lead to excessive smoke formation during burning but the semicoke produced is a clean fuel. Some typical yields (dry basis) from Gray-King assay at 600°C are :Semicoke 75-80%, Tar 9-12%, Liquor 5-7%, and Gas 100-120 Nm^3 /ton. In commercial plants, the yields of semicoke and tar are less, but that of gas is higher. The crude spirit recovered is 0.5-1%.

Semicoke has volatile matter of 6-12%, is highly reactive, can be easily ignited into a smokeless flame and is an ideal domestic solid fuel. The char produced by L.T.C. in fluidized bed is converted into suitable briquettes for domestic uses. Other uses of semicoke or char are: production of iron in low shaft furnaces, gasification into fuel gas or synthesis gas for fertilizers & chemicals, preparation of coke oven blends and as an industrial fuel.

Characteristics of solid smokeless domestic fuels produced by low temperature carbonisation (LTC) is given in Table 15.3.

Table 15.3: Characteristics of Solid Smokeless Domestic Fuel produced by LTC (I.S. 4286-1967)

Characteristics	Grade 1	Grade 2	Grade 3
Volatile Matter, %			
Minimum	6	6	6
Maximum	12	12	12
Moisture, %			
Maximum	6	6	6
Total inerts (ash + moisture), %	upto 30	30.1-40.1	40.1-50
Calorific value, kcal/kg	> 5,510	4,510-5,500	3,500-4,500
In the case of processed fuel made from lignite, VM, not exceeding 15% and moisture not exceeding 15% may be permitted.			

Low temperature tar is called 'primary tar' since the alteration by secondary cracking reactions is small due to relatively low temperature conditions. It is a very complex mixture of higher phenols, substituted aromatic hydrocarbons, various nitrogenous & oxygenated compounds of carbon and hydrogen etc. Many disinfectant preparations are made from this tar. Chemicals and liquid fuels may be produced from it.

The crude spirit corresponds to the crude benzol of H.T.C. but contains larger proportions of paraffins, naphthenes and olefins. Motor fuel may be produced by processing the crude spirit.

The typical properties of the coke oven gas obtained by low temperature carbonization of coal are : specific gravity (referred to air) 0.6, calorific value 6,500 kcal/Nm^3, composition (in volume %) -CO_2 = 4.0, C_mH_n = 4.0, CO = 7.0, H_2 = 33.0, CH_4 = 45 and N_2 = 7.0. It is richer in hydrocarbons and poorer in hydrogen than the gas produced by H.T.C. Its yield is lower but calorific value is higher. The L.T. gas is used up in the plant. The gas obtained from direct heated L.T.C. plant is a mixture of carbonization & combustion gases and hence has a low calorific value of about 1,500 kcal/Nm^3.

Barring one commercial plant each at Neyveli & Dankuni and two pilot plants in research establishments, there is no other L.T.C. plant in the country.

The products available from the LTC plant at Neyveli are carbonized lignite briquettes, char fines, tar, neutral oil (solvent/fuel), carbolic acid, ortho cresol, m - & p - cresol mixture etc.

CFRI process for LTC. In India, Central Fuel Research Institute (CFRI), Dhanbad has developed a process of low temperature carbonisation of coal which has a vertical retort with gas recirculation facility.

Dimension of CFRI retorts are :

Length = 1.98 metres, Height = 6.4 metres, Width = 178 mm at top and 279 mm at bottom.

The yield of products of coal carbonisation at 500-700°C are :

Soft coke	: 70% of coal
Gas	: 85-140 Nm^3/ton of coal with a C.V. of 6200 kcal/Nm^3
Tar	: 45-90 litre/ton of coal
Benzol or light oil	:5-10 litre/ton of coal

The gas can be used for various purposes and the manufacture of chemicals, mainly phenols.

METALLURGICAL COKE PRODUCTION (BY HIGH TEMPERATURE CARBONISATION)

Mainly two types of ovens are used for metallurgical coke production. They are :

(a) Beehive coke oven

(b) By-product coke ovens :

(i) Waste heat ovens

(ii) Regenerative ovens.

High Temperature Carbonisation (HTC). It is used for making metallurgical coke. There are two types of coke ovens for HTC namely Beehive oven and by-product oven.

Beehive Coke Oven. It is made of firebricks with an arched roof having a shape of beehive. There is a circular opening at the top for coal charging and subsequently for exhausting flue gas out of the oven. There is a side door through which the coke is discharged after coking.

Typical dimensions of a beehive oven are ; height = 2 metres, base diameter = 3.5 metres, capacity = 5.7 tons of coal.

Heat for carbonisation is supplied by burning of volatile matter of coal and partly by coal itself. Evolved gases burn with the air sucked through an opening at the top of the closed door.

Carbonisation proceeds from top to the bottom of coal charge. At the end of carbonisation, the door is opened and the coke is quenched with water and then discharged manually or by machine.

Demerits and merits of beehive oven practice of coke making are given below.

Demerits of Beehive Coke Oven :

- No by-products recovery, is possible.
- Lower coke yield due to partial combustion of coal.
- Blends of inferior coal can't produce good coke.
- Lack of flexibility of operation.
- Coke yield is only 60% as against about 77% of dry coal in by-products oven.
- Large coking time i.e. 2-3 days is required for carbonisation, (reduced coke production rate)

Merits of Beehive Coke Oven :

- Strong, blocky coke having good reactivity are produced.
- Low capital and running cost.
- Coke production is not connected with the availability of market for by-products.
- It can be shut down without any damage to refractory unlike by-product ovens.
- Produces hard coke for foundry.
- Simplicity of construction.
- No extra fuel is required for heating as in by-product oven.

This method of coke making has become obsolete though it is being used in India to some extent.

By- product coke ovens. It consists of a rectangular refractory chamber of length 15 metres, height 5 metres and width 0.4 metres with removable doors at both ends to enable the hot coke to be pushed out with a ram. The width of the coke discharge side is slightly more (3-4 cm more than the pusher side). A large number of ovens with heating flues in between, is arranged in a battery with regenerators for heat exchange between hot flue gases and combustion air, placed below.

The coking coal blend is usually introduced by several (typically three) top charging holes. Each pair of ovens which is separated by a system of vertical flues can be heated by producer gas, coke oven gas, or blast furnace gas or mixed gas. Air for combustion is preheated in regenerators. Blast furnace gas, mixed gas or producer gas, if used for heating can also be preheated but coke oven gas is not preheated as hydrocarbons (mainly methane) present in it crack $(HC \xrightarrow{\text{cracking}} C + H)$ resulting in deposition of carbon soots on regenerator walls thereby

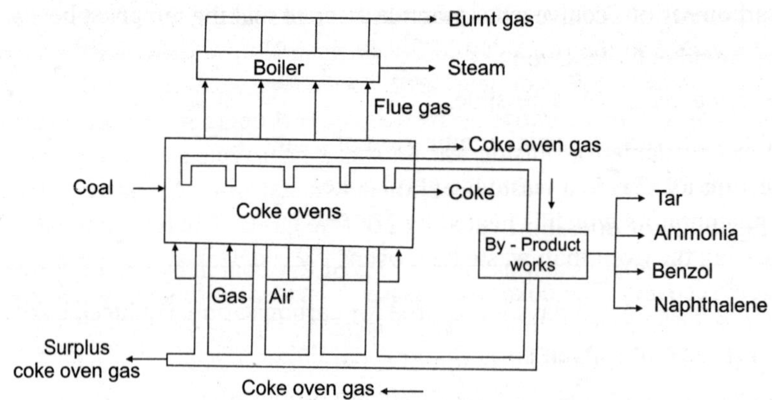

Fig. 15.2. Waste heat type coke oven.

reducing its heat transfer capacity. Carbon soot may also choke the smaller diameter burners in the heating chambers.

Flue temperature in the heating chamber is maintained at about 1200-1300°C. Carbonisation time for a 22 tons wet coal charge varies from 16 to 18 hours depending upon the width of the oven.

The oven walls are made of silica bricks, the high temperature strength of which ensures a very long life and enables the walls to be built thin for a high heat transfer rate through them.

Ovens are charged and discharged in a special order to maintain the uniform temperature over all the oven walls to avoid thermal damage to the battery. Uniform temperature is also maintained by reversing the flow of combustion gases through the flues and regenerators every 30 minutes. The ends of the oven are sealed with heavy brick-lined doors.

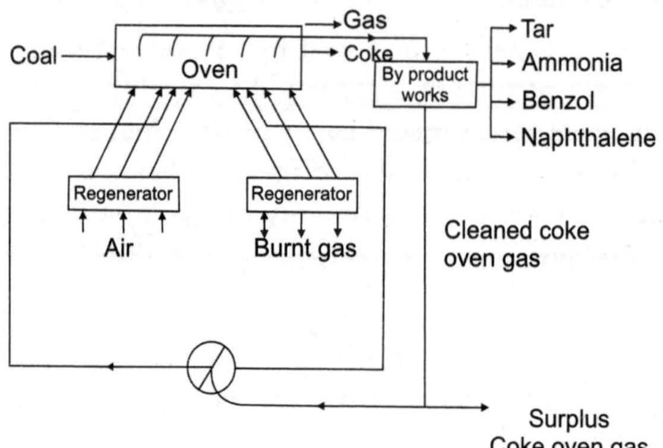

***Fig.15.3.** Regenerative Coke Oven.*

During coking, gas and tar are taken off via an ascension pipe and passed into the gas collecting main called hydraulic main. Coke when fully carbonized is pushed by a ram out of the oven into a quenching car and quenched with water. An alternative dry-quenching uses inert circulating gases (N_2) which transfers the heat to waste heat boilers hence recovering the potential heat but corrosion problem is acute.

During carbonisation, coal swells at first but shrinks as the temperature rises. This puts some pressure on the walls and the oven should not be pushed until the coke has shrunk clear of walls. Hence blending of coal charge is done to limit its swelling.

Waste heat ovens and regenerative ovens are similar in operation and differ only in the proportion of coke oven gas available for outside use. In waste heat type, steam is generated by the hot flue gas by transferring its heat to a waste heat boiler whereas in regenerative ovens, combustion air is preheated in regenerators which is heated by hot flue gases. The regenerative ovens, require much less gas for heating the oven than waste heat ovens. Waste heat ovens have become obsolete now.

By-products recovery The coke oven gas is subjected to following treatments to recover the by-products (chemicals) present in it, before it is used as a fuel gas :

- Cooling of the gas for condensation of tar and ammoniacal liquors.
- Electrostatic separation of remaining tar.
- Recovery of ammonia as ammonium sulphate by reaction with H_2S0_4.
- Gas is oil washed to recover naphthalene and benzol.
- Desulphurisation of gas with moist ferric hydroxide (bog iron).
- Distribution bf fuel gas to consumer furnaces.

Typical yields in high temperature carbonisation (HTC) of coking coals in a byproduct oven is given in Table 15.4.

Table 15.4: Products Yield in HTC of coal

Products	Yield, Weight%
Coke	75
Gas	15
Tar	3
Benzol	1
Ammonia	1
Water Vapour	5
Total	100

Industrial coke making practice in coke ovens attached to steel plant. The modern By-product Coke Ovens are designed to produce coke as well as to recover the products of carbonization. The ovens are narrow and rectangular in cross section. These are heated from both the sides through vertical flues. The walls are made of silica bricks. The necessary heat is produced by burning of gases inside the narrow flue chambers in between two walls. The ovens are arranged in a row to form a battery of coke ovens.

The usual dimensions of coke ovens :

(a) Length : 13-15 metres.

(b) Height : 4.5-5 metres

(7 metre tall ovens are also in use in TISCO,Rourkela, Visakhapatnam & Bhilai steel plant .The tallest coke oven in the world measuring 8.2 metres is operating in Germany;but not without many operational problems)

(c) Width :

(i) Average — 400 mm. (600 mm average width ovens are also in use)

(ii) Coke side — 425 mm.

(iii) Pusher Ram side — 375 mm.

The coal blend is charged into the coke oven chambers from the top by the charging car and continuously levelled. The charging holes are closed after completion of charging operation. The carbonization proceeds for the average coking period of 15-18 hours. At the end of coking, the doors on both sides of coke ovens are opened and the pushing ram pushes the red hot coke into the quenching car. The wall of coke oven is tapered from the pushers side. The side of the coke oven where the pusher machine works is known as ram side or pusher side and the other side is known as coke side. The quenching car receives the red hot coke from this side for quenching. The red hot coke is then quenched with large amount of water and ready for use after sizing in crushers and screening.

The gaseous fuels generally used for heating coke ovens are coke ovens gas, blast furnace gas, mixed gas and producer gas. Only last two gases are preheated along with air. Coke oven gas is not preheated but burnt in preheated air. Preheating is done by regenerators.

In some coke ovens, mixed gas (a mixture of blast furnace gas and coke oven gas) is used for heating the coke ovens which gives higher flame length and better control of heating rate.

Heating chamber (located on both the sides of an oven) is divided into several sections by partition walls. Burners are located at the bottom and are (called underfiring burners) in pairs. In the first cycle the gas burns in one and the flue gas passes through other set of burners. After *20* or *30* minutes the gas flow is reversed i.e. the gas burns in other set of burners and the flue gas is taken out through the burners which were initially fired. This reversal of flow of gas/air and the waste gas (flue gas) is done continuously at fixed interval (generally 20 or 30 minutes). This type of burner is called twin vertical flues or hair pin flue. In some design, there is a provision of recycling of a part of the flue gas at the bottom in the heating chamber. Such heating designs are termed as twin vertical flues with recirculation. Regenerators which are normally made of fireclay bricks are used for waste heat, recovery from the flue gas and in the next cycle

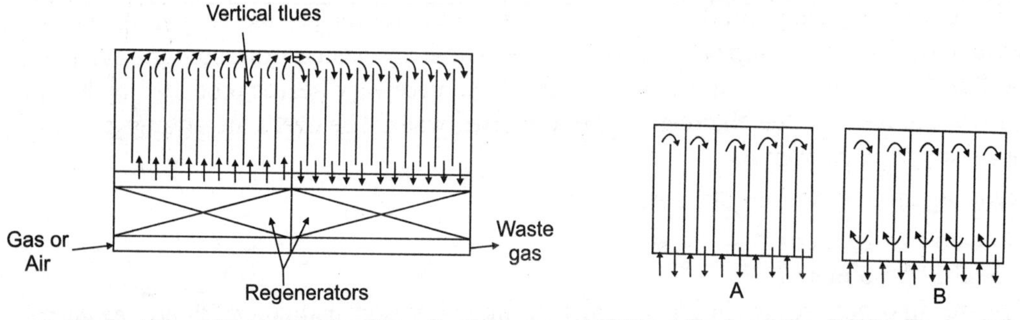

Vertical Flues Divided in 2 or More Sections — A Without Recirculation B-With Recirculation

Fig. 15.4. *Twin vertical flues or hair-pin flues.*

air or gas are preheated in the hot regenerator before entering the burners.

Each heating chamber is provided with two regenerators. While the waste gas (flue gas) passes through the one set (thereby heating its bricks) the fuel gas (B.F. gas mixed gas, producer gas etc.) and air is heated in the other set before firing in the burner.

In yet another heating systems of coke ovens called cross over flue systems, the flue gas passes from one heating chamber to another through the cross over flue at the top of the oven.

The fuel gas & combustion air and flue gas movement in heating chamber is shown in the simplified figures no. 15.4 and 15.5 respectively.

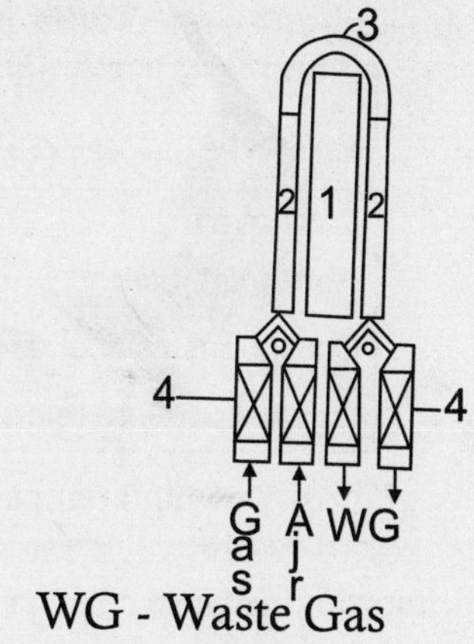

1. Coke oven
2. Vertical flue
3. Cross over-flue

Fig. 15.5. Cross overflue heating system of coke ovens.

PHYSICAL AND CHEMICAL PROPERTIES OF METALLURGICAL COKE

Ash and moisture content in coke. Low moisture & ash content area. Regenerator (2 pairs) desirable in metallurgical coke.

Low moisture in coke can be achieved by suitable control of the water heating system of cokesupply during quenching. That is why dry quenching using inert gas has been suggested but dry coke is dusty and can create handling nuissance. Metallurgical coke should contain < 1.5% moisture. Each additional 1% H_2O in blast furnace coke increases fuel consumption by 1.2%.

Maximum desirable ash in metallurgical coke is 8-10%. Each additional 1% ash in blast furnace coke increases fuel consumption by 2%. In addition, high ash content becomes very undesirable, if it causes trouble in fusion. Also, volume of slag produced in blast furnace becomes more with the use of high ash coke resulting in reduction of the useful volume of the blast furnace and hence production of pig iron from blast furnace is reduced. High ash coke has high resistance to abrasion and has high strength.

Density & Porosity. The more completely the coal is devolatalised, the more closely the density of coke approach that of the graphite (i.e. 2.3). Coking follows formation of a porous structure which increases in density by shrinkage during further heating at a high temperature. Since all cokes contain traces of hydrogen and mineral matter of the coal, its absolute density never reaches that of graphite. The higher the rank of the coal, the higher is the density of coke made (Varies from 1.7 to 1.95 for C% in coal from 81 to 89%).With increase in time and temperature of carbonisation (with faster heating rate) the density of the coke increases while the porosity decreases. Denser coke (i.e. having low porosity) are stronger and harder. An important property of coke for its storage and transport is its bulk density whose values for different types of coke are given in table 15.5.

Table 15.5: Bulk Density of Coke from Various Sources

Sources of coke	Bulk density, Kg/m^3	
	Sized (18-40 mm)	Unsized
Coke oven	417-480	465-513
Horizontal retort	368-417	400-448
Continuous vertical retort	337-353	368
Low temperature retort	305	320

Porosity of high temperature coke varies from 40 to 50%. High porosity is desirable in furnace cokes to obtain high rates of combustion. In blast furnace, high strength is considered more important because coke has to be dropped in the blast furnace from a great height and it should not break before reaching the hearth for better performance and reducibility in the furnace.

Hardness and Strength. It is defined as the resistance power of material to breakage by impact. Shatter test for metallurgical coke gives the measurement of resistance to shatter by impact. Surface hardness or resistance to abrasion is measured by Cochrane test. Foundry coke requires higher resistance to shatter. Low temperature coke shows surprisingly a high shatter resistance despite its higher porosity and abradability due to the quality of toughness which softens shock.

Strength of coke is measured by 'shatter index' while its hardness is measured by 'Cochrane abrasion index'. Both hardness and strength of coke are measured by a parameter 'Micum index'. In Russia, 'Sundgren drum test' is used to determine the coke strength and hardness.

Shatter index of coke. It measures the resistance of coke to breakage by impact i.e. of its strength. 50 lb. of 2 inches size coke is placed in a rectangular box of dimension 28" × 18" × 15" placed 6 ft above a steel base plate 1/2 inch thick. The hinged base of the box is released suddenly,

when the content drop onto the base plate. Box is dropped several times and coke is then screened through a series of seives made of square stamped sheet. The percentage of coke retained on 2", 1.5" and 0.5" seives are recorded and called shatter index. Desirable values of shatter index for blast furnace coke are 80% on 2" screen, 90% on 1.5" screen and 97% on 0.5" screen.

Cochrane Abrasion index of coke. It is measure of hardness of the coke and gives the resistance of coke to breakage by attrition, rubbing or abrasive action. To find out abrasion index, 28 lb. of 3" coke is rotated in a welded steel drum (18" dia. fitted with 2.5" angle iron lifting plates at 17° to axis) at 18 r.p.m. for 1000 revolutions. The coke is then taken out and seived through a screen of 1/8" size. The percentage coke remaining on the seive is called the abrasion index. A high abrasion index means a high resistance to abrasion.

Abrasion index for B.F. coke should be about 80. It is increased by the ash content or by the use of coal of low swelling power. Abrasion index increases roughly by 1% for increase of 0.2% in the 0.5" shatter index.

Micum index of coke. This is a measure of both hardness and strength of the coke. 50 kg of coke of 50 mm size is rotated in the micum drum for 4 minutes at the rate of 25 rotation per minute (rpm). Micum drum is a cylinderical steel drum (whose length and diameter both are 1 metre each) fitted lengthwise with four angle irons (100 × 50 × 10 mm) 90° apart inside the drum. After rotating the coke, it is taken out and screened through 60 mm, 40 mm, 20 mm and 10 mm round hole screen.

The percentage of coke retained on a 40 mm screen is called M_{40} index (Micum forty index) whereas, the percentage of coke that passes through a 10 mm screen is called M_{10} index (Micum ten index). M_{40} gives the resistance of the coke to breakage by impact i.e. it is a measure of the strength of the coke. M_{10} gives the resistance of the coke to breakage by abrasion (i.e. rubbing) and it is a measure of hardness of the coke.

High M_{40} and low M_{10} values are desirable for metallurgical coke. As per BIS minimum M_{40} index should be 75% and maximum M_{10} index should be 14% for metallurgical coke.

For use in 2000 m^3 volume blast furnace, coke M_{40} should be more than 78 and M_{10} should be less .than 10.

Sundgren drum index of coke. Sundgren drum test is used in USSR to determine the strength and hardness of coke. Sundgren drum is made from two discs of 2 metres diameter connected by rods placed 25 mm apart and width 0.8 metre. The coke sample (410 kg coke of 25 mm size) is rotated for 150 revolutions (at 10 rpm for 15 minutes) and then the coke inside the drum is weighed. The amount (kg) of coke left inside the drum gives directly the index of hardness of coke. The drum index should not be less than 340 for good coke.

For metallurgical coke to be used in the blast furnace both hardness and strength are important. If the coke is not strong enough, then it may break in the intermediate height position (say stack/bosh etc.) in the blast furnace while descending. The smaller coke has greater combustibility and reactivity, thereby the concentration of carbon monoxide formed will be more in the upper intermediate position and the time of contact between the descending burden and ascending reducing gases (CO) will be less resulting in lower reduction efficiency of the blast furnace. That is why coke should be so strong that it doesn't break until it reaches the hearth (near the tuyeres through which hot air is injected in the blast furnace) so that carbon monoxide concentration (resulting from the reaction $2C + O_2 \rightarrow 2CO$) is maximum at the hearth thereby facilitating maximum time of contact between the burden and the reducing CO gas. Hardness of coke is also important because of the reason that while falling down along the wall of the blast furnace, it may be broken due to rubbing (attrition) action. The coke should be hard enough to resist the size degradation of coke due to furnace wall rubbing while descending otherwise this also may result in greater concentration of carbon monoxide in the upper region of the blast furnace as well as along the periphery.

Reactivity of coke. It is defined as the ability of coke to react with O_2, CO_2 or steam (H_2O). For industrial fuel, reactivity to oxygen is most important. For domestic fuels, ease of ignition and high reactivity are important. Formation of CO on hearth of blast furnace is promoted by high reactivity of coke. In cupola, high temperature oxidation zone is large so unreactive coke in large pieces is preferred. In producer gas manufacture, high reactivity is favoured to increase CO in the reduction zone. In water gas manufacture, high reactivity is desirable in "make" but undesirable in "blow" period—on the whole unreactive coke is preferred.

More reactive cokes have higher thermal values of their volatile matter. Coke of high reactivity ignites easily and gives rapid pick up of fuel bed temperature. However, low reactivity coke gives a higher fuel bed temperature than a highly reactive coke. Low reactivity coke is therefore desired in coke for cupola operation to produce high hearth temperature and reduce 'solution loss' (i.e. loss of heat as carbon monoxide).

Reactivity is inversely proportional to the absolute density. It is affected by the presence of easily reducible iron compounds in ash.

Coke of high reactivity are obtained from weakly caking coals or blends. Strongly coking, high rank coals produce coke with low reactivity.

For blast furnace coke, size and hardness are more important than reactivity. Satisfactory hearth temperature is obtained with unreactive coke containing little breeze.

Reactivity of coke is measured by Critical Air Blast method and is reported as Critical air blast (CAB) value of coke. The CAB value of coke is the minimum rate of flow of air in ft^3/minute necessary to maintain combustion in a column of closely graded material (14 to 25 B.S.) which is 25 mm deep and 40 mm in diameter. The typical CAB value for oven coke is 0.065 ft^3/minute. More reactive coke has got lower CAB value.

Another modern and current method of expressing the reactivity and strength of coke is Coke Reactivity Index (CRI) and Coke Strength After Reaction (CSR) which is being followed in Indian steel plants.

Coke Reactivity Index (CRI). To determine CRI, 200 gm of coke sample (size + 20-25 mm) is taken in a stainless steel tube and heated in electric furnace to 1100°C. CO_2 gas at 5 kg/cm^2 pressure is passed through the coke bed for two hours. CO formed (by reaction $C + CO_2 \rightarrow 2CO$) is burnt in a burner and is exhausted out. Carbon of coke reacts with CO_2 (depending upon the reactivity level of the coke) and there is a loss of weight of coke depending upon its reactivity. More is the loss in weight of the coke, reactivity is more. % loss in weight of coke is reported as coke reactivity index (CRI). Ideal CRI value of a good blast furnace coke should be about 20%. Typically CRI of Indian blast furnace coke is about 25%.

Coke Strength After Reaction (CSR). The left out coke from the CRI determination test is rotated for 60 rotation in a micum drum. And the % of coke retained on a 10 mm size screen is reported as coke strength after reaction (CSR). Stronger the coke, more is its CSR value. Ideal value of CSR for blast furnace coke is a minimum of about 55%. Typically CSR of Indian blast furnace coke is about 60-65.

Phosphorous &sulphur content. Coke used for making pig iron by acid process (i.e. furnace with acidic brick) should have phosphorous < 0.012% as phosphorous can not be removed by the lining. However, when making basic steel (e.g. in furnaces with basic brick linings), phosphorous can be removed from pig iron by the lining, hence the phosphorous content of the coke need not be strictly limited. Phosphorous present in coke causes cold shortness in the steel produced whereas sulphur causes hot shortness of steel during rolling.

Sulphur in coke becomes part of the metallurgical load, entering metal and slag. In iron making, 90% of sulphur comes from coke and most of it should be removed as calcium sulphide renecessitating an addition of limestone and corresponding amount of coke to heat & melt it. Sulphur content in blast furnace coke should be less than 1% as it finds its way into the metal with injurious result causing brittleness of steel and surface cracks during rolling especially of alloy steels.

Typical Specification for Indian metallurgical coke. Ash = 20-25% Minimum, S = < 0.7%, P = < 0.2%, moisture = 4.5%, volatile matter = 2%, Porosity = 35-45%, Bulk density = 420-480 kg/m^3, True density = 1750-2000 kg/m^3, Abrasion index = 77-82%, Micumin dex : M_{40}> 78% and M_{10}< 10%, Shatter index: 1.5" index = 85% and 0.5" index = 97%, caking index of coal blend used for carbonisation = 21-22, CRI = 20, CSR = 65.

Requirements of hard coke used for gasification, in water gas generators, gas producer, boilers etc. are much less stringent than that of metallurgical coke. However, at present, there is no separate standard for hard coke used for different purposes.

The ISI specifications for hard coke for blast furnaces and foundaries in India are tabulated in Table 15.6 and 15.7 respectively.

Table 15.6: Specifications for Hard Coke for Indian Blast Furnaces(IS : 439-1965)

Characteristics	Requirements for Grade		
	Special	No. 1	No. 2
Proximate analysis, per cent, dry			
(a) Ash, max	20	24	28
(b) Volatile matter, max	2	2	2
Sulphur, per cent, dry, max	0.70	-0.70	0.70
Phosphorus, per cent, dry, max	0.30	0.30	0.30
Shatter index :			
(a) Over 38 mm, per cent, min	85	85	-
(b) Over 13 mm, per cent, min	97	97	-
Micumindex :			
(a) Total on 40 mm, per cent, min	75	75	-
(b) Through 10 mm, per cent, max	14	14	-
Haven index :			
Total on 25 mm, per cent, min	40	40	-
Porosity, per cent, min	35 to 48	35 to 48	-

Table 15.7: Specification for Foundry Coke in India (IS : 4836-1968)

Characteristics	Type 1		Type 2			Type 3		
	Grade 1	Grade 2	Grade 1	Grade 2	Grade 3	Grade 1	Grade 2	Grade 3
Porosity, %	upto 35	35-45	35-45	35-45	35-45	35-45	35-45	35-45
Volatile Matter, % max	0.5	1.0	1.5	2.0	2.0	1.5	2.0	2.0
Ash %	upto 20	20-24	upto 20	20-24	24-28	upto 20	20-24	24-28
Sulphur,/% max	0.5	0.5	0.7	0.7	0.7	1.0	1.0	1.0
Phosphorus, % max	0.1	0.1	0.15	0.15	0.15	0.3	0.3	0.3
Shatter Index for material :								
(a) Over 75 mm, % min	35	30	30	25	25	25	25	25
(b) Over 50 mm, % min	80	75	70	70	70	70	70	70

EFFECT OF COAL AND PROCESS VARIABLES ON COKE PROPERTIES

Coal rank, type & blending. A coking coal should have a volatile matter content of 22- 30% and crucible swelling index of 4 to 9. If a coal falls outside these limits but still possesses coking properties, it may be used for lower grade metallurgical cokes or coke for furnace use.

The most highly swelling coals within good coking range may produce weak cokes because of excessive after shrinkage, with the formation of cracks. It may create sufficient pressure during the swelling process to damage coke oven walls. Such adverse effects are overcome by blending with non-swelling coals, anthracite, coke breeze or fussion. Fussion is soft powdery form of coal occurring in thin seams between the bands of other types.

Alternately, coke breeze or anthracite dust can also be blended to improve the density and hardness of coke produced from high swelling.

Coal Storage. The swelling index of coal falls rapidly with low temperature oxidation i.e. storing finely ground coal. The quality of coke produced from weakly coking coal is greatly reduced by storing for a few days after preparation. High swelling coals are not greatly affected by a few days storage, but longer storage decreases the swelling index. This may be advantageous with some dangerously high swelling coals as an alternative to blending with non- swelling coal.

Rate of heating of coal charge. High rate of heating causes maximum swelling excessive in the case of strongly coking coal, but improving the coke formed from low swelling coals with modern narrow ovens (400 mm wide) and flue temperatures 1350°C, rate of carbonisation upto 2.5 cms/ hr. are obtained. At such rates, strongly coking require blending with non-swelling, finely divided materials such as fussain, coke breeze or non-coking high rank coal. This in turn requires more uniformly mixed & graded charges and produces harder, less reactive & more uniform cokes as required for blast furnace. Faster heating rate produces smaller coke.

Coke Quenching. The hot coke discharged from the ovens is quenched with controlled amount of phenolic water so as to leave practically free moisture in the range of 3-4%.

Inert gas quenching using N_2 is being practised in some plants abroad and is used at Vizag Steel Plant Rourkela steel plant , Burnpur steel plant and Tata Iron & Steel Company (TISCO),Jamshedpur in India. This method extracts the potential heat of coke to generate steam electricity, but corrosion problems of equipments is acute plus the dust nuissiance also exists.

PARAMETERS OF A TYPICAL INDIAN COKE OVEN BATTERY WORKING IN A STEEL PLANT

- Washed coal received in box wagon (also through aerial rope way from coal washery) and unloaded in rotary wagon tippler. Unwashed coal received by trucks and wagons.
- No. of Silos - 46, arranged in two rows of 23 each, Capacity of each Silo-2500 tonnes, Dimension of Silo-38 × 12 m (L × D of 23 each).
- **Blend of coal used.** Normally 50% prime coking coal + 50% medium coking coal. Sometimes, Assam coal, also used upto 5%. Imported coal is about 50% in the coal blend for carbonisation.
- **Coal size** = -3.2 mm size is 80% in the blend.

- **Capacity of each coal tower** - 3200 tons for storing blended crushed coal.
- **Types of coal crushers used.** Roll crusher and hammer crusher.
- **Coke Oven Battery :**

No. of ovens per battery	=69
Total No. of batteries	=8
Height of the oven	=15m (4.7 m is useful height)
Length of the oven	=15 m
Thickness of oven wall	= 110 mm
Width of oven on pusher side.	= 39cm
Width of oven on coke side	= 43 cms.
Volume of the oven	=27.3 M^3
Capacity of the oven	=20.7 tons dry coal.
Height of regenerators	=2.52 m.

Type of oven. Regenerative, recirculative, twin flue, under jet, compound ovens. No. of burners = 30 (16 pusher side + 14 coke side) - odd and even burning. Temp, on coke side is 40°C more than that on pusher side.

- **No. of coal towers.** 1 for 2 batteries, each coal tower contains 4 pockets having 3 rows in each pocket.
- No. of charging holes in each oven = 3
- No. of ascension (stand) pipes/oven = 2

Average composition of coal charge. Ash = 20%, moisture = 7%, Volatile matter = 25%, Sulphur = 0.56%, Phosphorous = 0.1%

- Automatic water jet (at 500 kg/cm^2) oven door cleaner is provided.
- Sequence of charging coal through the charging hole : Coke side - pusher side - Middle charging hole.
- **Type of charging.** Gravity charging (not stamped charging).

Systems of charging:

(a) **Off main charging.** Both stand pipes caps are kept opened during coal charging in the oven hence smoke is ejected outside.

(b) **On main charging.** Both stand pipes caps are kept closed and steam is injected on both sides to create suction of gas. This is a smokeless charging method.

Normally, mixed system of charging is used. 2-9 series of oven pushing/charging is used.

Temperature, of coke oven gas coming out of the oven = 700-800°C.

Analysis of coke produced :

Ash	= 20%	Moisture	= 4%
Sulphur	= 0.36%	Volatile matter	= 0.4%
Phosphorous	= 0.1%	Fixed carbon	= 75%

Micum index of coke produced. M_{40} = 80 (Minimum), M_{10} = 9 (Maximum)

Specific heat consumption = 0.6 Giga calorie/ton of dry coal carbonized

Specific energy consumption = 1.3 Giga calorie/ton of gross coke produced

Specific power consumption = 28 KWh/ton of gross coke produced

Yields of Carbonisation:

Gross coke yield = 77% of dry coal charged in coke ovens

B.F. grade coke (25-80 mm size) yield= 87% of gross coke = 67% of dry coal carbonized

Coke oven gas yield = 290 Nm^3/ton of dry coal

Ammonia = 3 kg/ton of dry coal

Tar = 30 kg/ton dry coal (3% of dry coal)

Benzol = 7 litres/ton dry coal

Typical Composition of coke oven gas :

CO_2 = 2.5%,	O_2 = 0.3%
C_mH_n = 2.5%,	CO = 8%
H_2 = 58%,	CH_4 = 25%
N_2 = 3.7%,	C.V. = 4200 kcal/Nm^3

A simplified figure of Otto oven is given in Fig. 15.6.

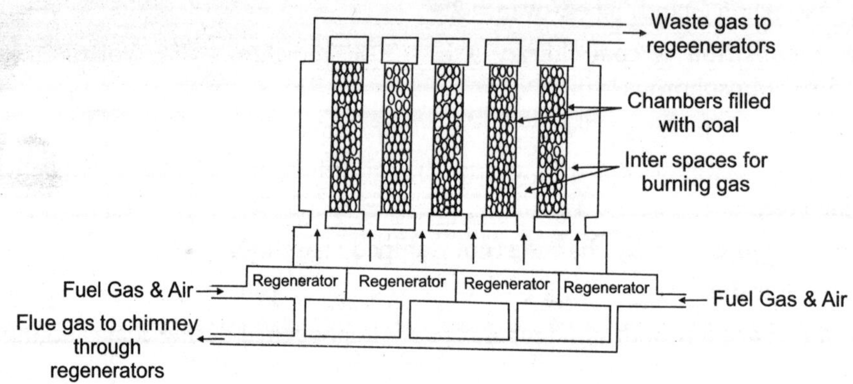

Fig. 15.6. *Otto Hoffmann by-product oven.*

Final sizes of the coke produced :

B.F. grade coke : 25-80 mm size - used in blast furnace

Pearl or nut coke : 15-25 mm size - sold outside for use in foundary

Coke breeze: 0-15 mm size - used in Sintering Plant and soaking pits bottom

Types of screens used in coke sorting plant. Grizzly and vibrating screen. Undercharging of coal in coke ovens is not done.

Table 15.8 gives the list of major coke ovens in India and their manufacturers.

Table 15.8: Major Coke Ovens in India and their manufacturers

Sindri Fertilizers (F.C.I.)	Cart Still, W. Germany
Bhilai Steel Plant	Giprokoks, U.S.S.R.
Rourkela Steel Plant	Otto, West Germany
Durgapur Steel Plant	Simon Carves, England
TISCO, Jamshedpur	Simon Carves, England
IISCO, Burnpur	Simon Carves, England
Durgapur Projects Ltd., Durgapur	Cart Still, England
Bokaro Steel Plant	U.S.S.R.
Vizag Steel Plant	U.S.S.R.

MODERN TRENDS IN COKE MAKING

Bulk consumer of the main product (i.e. coke) of carbonisation industry is the blast furnace in steel plant which is a highly energy intensive industry. Modern trends in coke making aims at reducing specific heat consumption in coke making by way of fuel economy techniques and high productivity of the coke oven.

Our limited coking coal reserve is inferior in quality containing around 20% ash. Around 1.35 ton of coal is required for producing 1 ton hot metal (pig iron from blast furnace). Carbonisation of 1 ton coal requires 0.6×10^6 kcal of energy out of which 50% is lost in the way of sensible heat of coke. Energy conservation in coke making can be done by the following techniques.

Some of the methods of economising fuel consumption in coke making are discussed below :

Washing of coal. Ash content in the coke can be reduced by suitably washing the coking coal. Most of the coking coals the world over are having an ash content of 10-20% resulting in coke ash of 15-28%. Each 1% ash increase in coke gives rise to an increase in coke rate of between 2-3%. Therefore, the coking coal is always washed so as to get a coal having 16-17% ash. This level of ash content in our coking coal is about the optimum to ensure that not too much of coal is rejected as middlings and at the same time the coke rate in the blast furnace is also kept at satisfactory level. The ash in our coke has been averaging at about 24-27% . If the ash content in the coke could be brought down to 22-23% from 26-27%, the coke rate in the blast furnace would be reduced by 75 to 100 kg/tonne hot metal primarily because there will be less slag to melt in addition to the increased carbon content. A very useful experiment has been conducted by Central Fuel Research Institute (CFRI), Dhanbad, for de-ashing coke by the oil agglomeration process by which ash in the coke can be reduced to 10%. If this process is stabilised and commercial plants are established, we can substantially reduce the coke rate in the blast furnace.

Besides, low ash content reduces the load of transportation of coal. Decrease in coke ash content increases the productivity of blast furnace as slag produced will be less thereby reducing the fuel consumption.

Also, high ash content in the coking coal increases the time of carbonisation and hence the fuel consumption in coking as it offers resistance to heat transfer.

Blending of coal. Highly coking coals are always blended with non-coking coal, anthracite, coke breeze, medium coking coal, fussain etc. to check its excessive swelling during carbonisation as it might damage the coke oven walls. Blending of coal improves the yield of uniform coke and its shatter index hence the yield of coke breeze will be low and that of the blast furnace grade will be high.

Blending of coal with non-coking coal increases the porosity and hence the reactivity of coke.

Preheating of coal charge. Coal charge of the coke even can be preheated by the hot flue gas coming out of coke oven regenerators. Preheating of coal increases its bulk density resulting in increased productivity of the oven besides lowering the time of carbonisation and heat required for carbonisation.

Residual heat in the waste flue gas can be recovered to some extent by preheating the coal charge. Preheating of coal upto 250°C in absence of air causes increase in its fixed carbon content and heating value without affecting the agglutinating value.

Preheating of coal having high volatile matter and oxygen content results in the production of coke of improved quality, uniformity & stability. Besides, wider range and variation of coal can be used.

Use of coal briquettes. Use of 30-50% of coal briquettes made with the help of a binder increases the charge density by 7–9% and improves the coking properties of the charge by expansion of briquettes during carbonisation. This allows the use of non-coking coals upto 10-20% in the coke oven. This technique has been widely used in Japan and is suitable for India too.

Addition of coking agents. Low cokability of some weakly coking coal can be improved by the addition of coking agent like asphalt, tar, pitch etc. but the presence of sulphur in the coking agent is detrimental. It permits the use of non-coking coals upto 5% and is suitable for existing ovens.

Formed coke. Formed coke is made from low rank, highly volatile non-coking coal with low swelling number. It is carbonised at a low temperature of 500°C to produce char either in a fluidised bed or in some retort where rapid heating of finely crushed charge is possible. The hot char produced is blended with either about 20% of coking coal or with tar or with a mixture of these and formed by pressing into briquettes of suitable shape & size. These briquettes are then heated to carbonise the binder (tar) and to consolidate the char to a suitable strength & reactivity. The final product is called formed coke.

Since, it has got standard shape & size; better control over its behaviour in the blast furnace can be achieved as the size & size distribution determines the burning characteristics of coke and influences the flow of materials through blast furnace and hence the productivity & efficiency of iron making process. Formed coke made from low rank non-coking coal can be used as a metallurgical coke with suitable control of quality, thus conserving the reserve of coking coal.

Dry-Quenching of coke. Coke is discharged from the coke oven at over 900°C. About 30-40% of energy consumed in coke making is wasted as sensible heat of waste coke while it is quenched with phenolic water as the steam generated can not be used because of its highly corrosive nature.

Dry quenching of coke aims at recovering the sensible heat of the hot coke. In this process, inert gas (N_2) is circulated in a closed circuit which takes up the heat of the coke and generates steam in a waste heat boiler. The coke temperature is thus brought down to 200°C from 900°C and finally quenched by water spray cooling.

About 450 kg of steam per tonne of coke can be generated out of which 50 kg may be required for driving the inert gas circulation fan and the remaining can be used as process steam or for driving a turbogenerator to produce electricity.

According to another estimate, steam can be generated at 55–70 tons/hr at 20 kg/cm^2 to 40 kg/cm^2 at 280-490°C utilising 0.33×10^6 kcal/tonne potential heat of hot coke giving an efficiency of 80% heat recovery. This implies a saving of around 5% in overall energy consumption.

But the equipments used in this heat recovery system is very costly and the corrosion problem of the equipments is very acute.

Inert quenching is being used in many coke ovens abroad, particularly in Russia and Japan. It is used for the first time in India at Vizag Steel Plant followed by TISCO, Rourkela and Burnpur Steel Plant.

Stamped charging of coal. In gravity charging of coal in the coke ovens, the coal particles are loosely packed. In stamped charging, the coal is first pressed mechanically (stamping) and then charged in the coke ovens.

Stamp charging is a means of increasing the bulk density of coal. Stamping enables the utilisation of larger proportion of substandard coking coal.

Coking coal of relatively lesser agglomerating properties when carbonised alone or in blends by the conventional top charging process produce coke of substandard quality.

Stamping of high rank, low fluid coals produces maximum improvement in coke quality. With decrease in rank of coal, the influence of stamping on coke quality decreases.

20 to 30% non-coking coal can be used by stamping process where even 15% non coking coal fail to produce satisfactory coke by top charging process (i.e. gravity charging).

Standard metallurgical coke can be produced by stamped charging process from substandard coal.

Use of small quantities of straw in layers while preparing the coke for stamped charging imparts better stability to the coke and produces blockier coke.

The coke made by stamped charging is more uniform with respect to coherence, shape, texture and lustre. The coke is also denser and larger in size.

Since about 12-16% moisture is required for binding the coal during stamping (pressing), hence the coking time is more by 15-20% as compared to gravity charged ovens due to higher charge quantity and moisture content.

A few coke ovens with stamped charging facility is working at Tata Iron and Steel Company (TISCO) since 1989.

A comparison of the parameters in top gravity charging vis-a-vis stamped charging is given in tabular form in Table 15.9.

Table 15.9: Comparison of Stamp Charging Vs. Top Charging of Coal in Coke Ovens

Parameters	Top charging	Stamped charging	Remarks
Crushing fineness of coal charge	80% thro 3 mm screen	90% thro 3 mm screen	finer size coal required
Extra material - input	-	Blend moisture to be adjusted to 10%	for better binding during stamping
Oven charge	100	130-135	increase of 30-35%
Carbonisation time	100	115-120	increase of 15-20%
Net increase in coke oven productivity	-	10-12%	hence specific heat consumption is less
Increase in B.F. coke yield	-	4-5%	substantial increase
Coke strength, M_{10} index	13-14	9-10	Considerable improvement
Possibility of use of non coking coal	5%	20-25% (with improvement in coke quality)	very good under Indian condition of coking coal reserve
Capital cost	Rs. 100 crores for one battery in 2002	Rs. 180 crores for one battery in 2002	though high capital cost, but payback period is not very high
Bulk density of coal charge	0.75 ton/m^3	1 ton/m^3	on dry coal basis

Use of tall coke ovens. Most of the coke ovens world-over are 4.5-5 metres tall. Use of taller ovens increases the throughput of the oven and hence the production rate. At many places abroad, 7 metres tall coke oven have been built and operated. In India too, 7 metres tall coke ovens are operational at Vishakhapatnam (Vizag) Steel Plant,Bhilai,Rourkela and Burnpur Steel plant. Taller ovens in some designs have got horizontal burners (instead of vertical under firing burners in 5 metres tall ovens) to ensure uniform heating along the height of the oven.

BY-PRODUCTS RECOVERY (BRIEF OUTLINE OF A TYPICAL DIRECT PROCESS)

The gas which comes out of the coke oven (which is the volatile matter of the coal) is called coke oven gas. It contains many valuable chemicals which are the by-products of carbonisation. Coke oven gas is treated in the by-product plant to recover some important chemicals from it and then the coke oven gas is used as a gaseous fuel in the furnaces of steel plant.

Coke oven gas comes out at about 800°C from the coke oven. It is cooled to 80°C with ammoniacal liquor (NH_4OH) by spraying it at the goose neck. Here, the tar from the coke oven gas is condensed, which is separated in the separator. The gas is further cooled to 30-35°C in primary gas coolers by water (indirect cooling) where some more tar gets condensed from coke oven gas.

Cooled coke oven gas is brought to electrostatic tar precipitator, where the last traces of tar vapor/fog present in the C.O. gas is removed. Electrostatic tar precipitator consists of a number of cylindrical tubes inside which a thin wire hangs centrally. The tubes are called collecting (positive) electrode and the wire is called discharge (negative) electrode. A high potential of 60,000 volt DC (secondary current = 150 mA) is applied across these two electrodes. When the coke oven gas passes through these tubes, the tar present in it gets negatively charged and moves towards the pot lively charged collecting electrodes and thereby it is separated from the coke oven gas. Tar contains phenol, benzol, naphthalene, pitch etc., which can be recovered by its distillation.

C.O. gas from the electrostatic tar precipitator goes to the exhauster, which is a turboblower sucking gas from the coke oven and discharging at high pressure. The gas discharged by the exhauster gets heated on compression to 60-65°C and is bubbled through sulphuric acid in the saturator. Here in the saturator, the ammonia present in the coke oven gas (10-15 gm/Nm^3) reacts with sulphuric acid and forms ammonium sulphate which is a fertiliser. The reaction in the saturator is $2NH_3 + H_2SO_4 = (NH_4)_2SO_4$. The gas from the saturator goes to final cooler, where it is cooled to 25-30°C by water. The cooled gas is treated with wash oil in the scrubber where the naphthalene present in the coke oven gas along with absorption benzol is removed by counter current.

Now this clean coke oven gas is separated into three streams. Part of the coke oven gas is passed through a bog-iron bed (mixture of iron dust and saw dust) which removes the sulphur compounds (H_2S, RSH and SO_2) present in the gas. This desulphurised gas is used for hot scarfing of steel slab in the slabbing mill to remove oxidised surface of steel slabs.

Another stream of coke oven gas is further treated in another set of electrostatic precipitator called Return Electro-Static Tar Precipitator (RETP). This gas is used for heating of coke oven batteries which require very pure gas to avoid the choking of smaller diameter (12 mm) burners of coke ovens heating chambers.

Rest of the coke oven gas is used as a fuel gas in steel plant furnaces like reheating furnace soaking pits, thermal power plant boilers, blast furnace stoves, foundary ovens, laddie drier etc. In most of its usage coke oven gas is mixed with Blast furnace gas i.e. it is used as a mixed gas. Extra coke oven gas can be stored in a gas holder in absence of which it is bled into atmosphere. Flow sheet of the above process is shown in Fig. 15.7.

Depending upon the method of recovery of ammonia from the coke oven gas, the process of by-products recovery is termed as direct, indirect or semi-direct methods. In the indirect process, ammonia from coke oven gas is recovered fully as an aqueous ammonia solution (Fosam Process) which may be subsequently converted into ammonium sulphate (a fertiliser). In the direct process, ammonia is fully recovered as ammonium sulphate by bubbling coke oven gas through sulphuric acid bath in the saturator (invariably followed in by-products plant of coke ovens attached to integrated steel plants). In the semi-direct process, ammonia is recovered partly as the aqueous solution and partly as ammonium sulphate. Operating parameters of a by-products plant attached to coke ovens in an integrated steel plant employing direct process for ammonia recovery from coke oven gas is given below :

Typical Parameters of a by-Products Plant (BPP)

- Number of coke ovens pushed / day = 500 (each oven capacity = 20.6 tons of dry coal)
- Coke oven (C.O.) gas produced = 135 x 10^3 Nm^3/hr

- C.O. gas pressure in the hydraulic main = 12 mm water column (mmwc)
- Temperature of C.O. gas coming out of the oven = 800°C
- Temperature of C.O. gas at goose neck after ammonia liquor flushing = 80°C (70% of total tar present in raw C.O. gas is condensed here & subsequently separated)

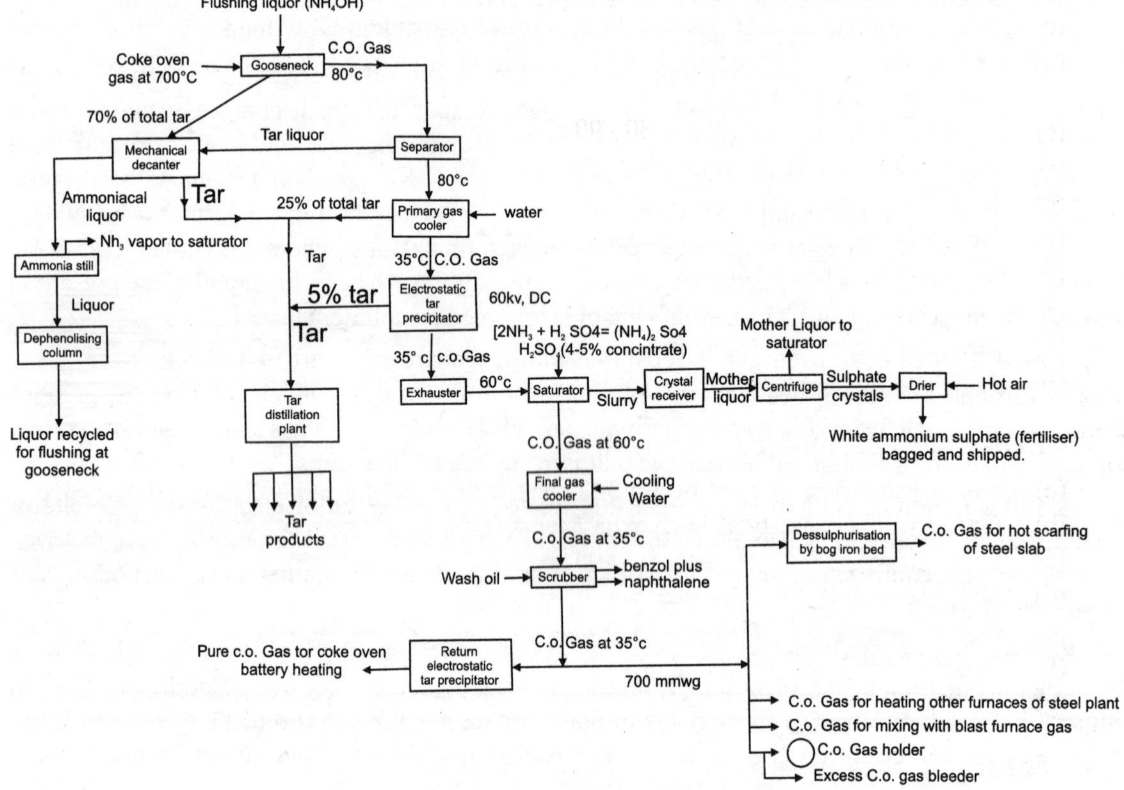

Fig.15.7. *Flow sheet of by-products recovery process.*

- ***Primary gas cooler (PGC) parameters :***

(a) No. of PGC - 8 (indirectly cooled shell & tube type vertical cooler)
(b) C.O. Gas temp, before PGC = 80°C
(c) Gas temp, after PGC = 30 - 35°C (28% of total tar present in raw C.O. gas is condensed & removed in PGC)
(d) Suction pressure of PGC = - 240 MMWC
(e) Pressure drop of PGC = 100 MMWC
(f) Gas handling capacity of each PGC = 17.5 × 103 Nm^3/hr
(g) Cooling water temperature before PGC = 25 - 30°C
(h) Cooling water temperature after PGC = 40 - 45°C
(i) Cooling water pressure = 4.5 kg/cm^2
(j) Water flow rate to one PGC = 400 m^3/hr
(k) Cooling system for recirculating water for PGC - 4 nos forced draft cooling toers
(l) Tar temp, for PGC flushing = 80 - 90°C

(m) PGC dimensions - cooler height = 25 metres, No. of tube = 5500, tube length = 3 metres, cooling surface area = 3000 m^2

- **Parameters of Electrostatic tar precipitator (ETP)**

(a) No. of ETP in operation = 4
(b) C.O. gas handling capacity of each ETP = 35×10^3 Nm^3/hr
(c) Oxygen content in C.O. gas = <1% (to avoid explosion in ETP due to electrical sparks)
(d) Secondary voltage & current in ETP for corona discharge = 60000 Volt D.C. & 150 mA (for corona discharge)
(e) Insulator box temp, of ETP = 80 - 90°C
(f) No. of tube & wire electrodes in each ETP = 15
(g) Height & diameter of ETP = 10 metres & 4 metres respectively
(h) Height of tar fog precipitating tube = 3.5 metres
(i) Diameter of tubes = 250 mm
(j) Pressure drop in ETP = 100 mmwc-
(k) Rest 2% of total tar present in C.O. gas is removed in ETP
(l) Tar content in C.O. gas after ETP = 2-4 mg/Nm^3 of clean C.O. gas

- **Exhauster's parameters**

(a) No. of exhausters in operation = 3 [2 electric driven & 1 steam (35 atm) driven]
(b) Gas handling capacity of each exhauster = 60000 Nm^3/hr
(c) Suction pressure of exhauster = - 400 mmwc
(d) Delivery pressure of exhauster = + 2000 mmwg
(e) Gas temp, before exhauster = 35°C Gas temp after exhauster = 60 - 65°C
(f) Speed of exhauster = 3900 rpm
(g) 35 atm. pressure steam consumption in the steam exhauster = 16 tons/hr

- **Saturator parameters :**

(a) No. of saturators in operation = 3 (.Direct process of ammonia recovery from C.O gas is used)
(b) Ammonia content in C.O. gas before & after saturator = 10 & 0.03 gm/Nm^3 respectively
(c) C.O. gas temp, before saturator = 60-65°C
(d) Concentration of H_2SO_4 in the saturator = 5%
(e) Pressure drop of saturator = 450 mmwg (/) ,
(f) Pressure drop of acid catcher = 50 mmwg
(g) Acidity of ammonium sulphate (white crystals) = 0.04%
(h) Maximum moisture content in ammonium sulphate before & after dryer = 3 & 0.5% respectively
(i) Temp, of water used for washing of ammonium sulphate in the centrifuge = 75°C
(j) Temp of hot air used in fluidised bed dryer for ammonium sulphate = 125C°

- **Final Gas Cooler (FGC) parameters :**

(a) No. of FGC in operation = 3

(b) C.O. gas temp, after FGC = 25-30°C
(c) Temp, of water supplied to FGC = 25-30°C
(d) Cooling water consumption in FGC = 400 m^3/hr (6.5 m^3 water per 1000 Nm^3 of C.O. gas)
(e) C.O. gas pressure drop in FGC = 100 mmwg

- **Naphthalene Scrubber parameters :**

(a) Total naphthalene content in raw C.O. gas coming out from coke oven = 10-15 gm/ Nm^3 (Tar dissolves naphthalene)
(b) Naphthalene content in C.O. gas entering the naphthalene scrubber (after FGC) = 0.7 -0.9 gm/Nm^3
(c) Naphthalene content in cleaned C.O. gas (after naphthalene scrubber) = 0.200 gm/ Nm^3

- **Benzol Scrubber parameters :**

(a) Temp, of coal tar wash oil used for scrubbing benzol from C.O. gas = 3 - 5°C more than gas temp (to avoid water vapor condensation from gas)
(b) C.O. gas pressure drop in benzol scrubber = 300 mmwg
(c) Temp, of C.O. gas entering the benzol scrubber = 25 - 30°C
(d) Benzol recovery in benzol scrubber = 6 - 8 litres benzol/ton of dry coal carbonised
(e) Regeneration of benzolised wash oil (in which benzol is dissolved) heated to 165°C is done by live steam.

- **Bog Iron Bed (for H_2S removal from coke oven gas) :**

(a) H_2S content in C.O. gas before purifying box (containing bog iron bed) = 1 gm/Nm^3
(b) H_2S content in C.O. gas after purifying box = < 0.02 gm/Nm^3
(c) Pressure of clean C.O. gas in pipeline = 700 mmwg

- **Sulphuric acid plant attached to BPP :**

(a) Capacity of DCDA (double contact double absorption) H_2SO_4 plant = 120 tons/day
(b) Molten sulphur pit temperature = 135°C (8 atm. pressure steam is used in heating coils in sulphur pit)
(c) Sulphur burning (with excess air) furnace temperature = 900°C (having SO_2< 8%)
(d) Waste heat steam (at 8 atm pressure) production by cooling of SO_2 gas (from 900 to 450°C) = 6 tons/hr
(e) Steam consumption in the H_2SO_4 plant for sulphur melting = 6 tons/hr.
(f) SO_2 gas entry temp to converter (4 stage) = 450°C
(g) SO_3 gas inlet temp to absorption tower = 160 - 180°C
(h) Absorbing sulphuric acid concentration in the absorption tower = 98.4 to 98.7% (at 50 -55°C)
(i) Concentration of sulphuric acid used in air drying absorption tower = 93 to 95% (at 40 - 45°C)
(j) Conversion efficiency of the converter (from (SO_2 to SO_3) = 98%
(k) Stack gas content = 11 to 13% O_2 & 0.13 to 0.19% SO_2.

COAL TAR DISTILLATION

Coal tar is a black to brown oily and viscous fluid of characteristic odour produced during high or low temperature carbonisation of coal during coke manufacture.

Tar produced during low temperature carbonisation of coal (at 700 °C) is called low temperature tar whereas that produced during high temperature carbonisation (at 1100°C) is called high temperature tar.

Most of the discussion in the following pages is about high temperature tar unless mentioned specifically.

Low temperature tar. Its yield is about 10% of dry coal. It is a brown colour oily liquid having very low viscosity, hence flows readily. It is paraffinic in nature. Its typical ultimate analysis is

C = 84% S = 0.5%	H = 8%	N= 1% O = 6.5%

It contains very less free carbon (about 3%).

High temperature tar. It is a thick, high viscosity, black coloured liquid with an aromatic nature. It contains more amount of useful aromatics like phenol, naphthalene, cresote, anthracene, benzol, pyridine etc. Its free carbon content is high (10%) and the yield is lower (3% of dry coal) as compared to low temperature tar. Typical analysis of H.T. tar is

C = 90.5% S = 0.8%	H = 5.5% O = 2.3%	N = 0.9%

Recovery of tar. During the carbonisation of coal in by-product coke oven, the tar vapors present in the coke oven gas are separated by cooling it by ammoniacal liquor resulting in its condensation. Remaining tar is separated in primary gas cooler and electrostatic tar precipitator thereby recovering the bulk of the tar present in the coke oven gas.

Distillation of tar. Tar contains around 300 identified chemicals. Hence, to recover economically some of these chemicals and to produce better fuel from tar, it is distilled. The main fraction of tar (which are recovered by distillation) their constituents, yield and boiling ranges are given in Table 15.10.

Table 15.10: Products of Tar Distillation

Fractions	Boiling range , °C	Yield, weight%	Main constituents of the fractions
Light oil	< 170	2-4	Benzol, naphtha and phenol
Middle (or carbolic) oil	170-230	5-7	Phenol, tar acids, heavy naphtha, naphthalene and pyridine
Heavy (or creosote) oil	230-270	15-25	Impure naphthalene and creosote oil
Anthracene oil	270-350	14-17	Anthracene and wash oil
Pitch	Residue	60-70	—

Typical Tar Composition. Pitch-64.8%, Creosote oil-28.3%, Naphthalene–4.3%, Phenol–0.3%, Anthracene-0.3%, Benzol-1%, Other light oils-0.7%.

A flow diagram of recovery of various chemicals and fractions from coal tar is given in Fig. 15.8.

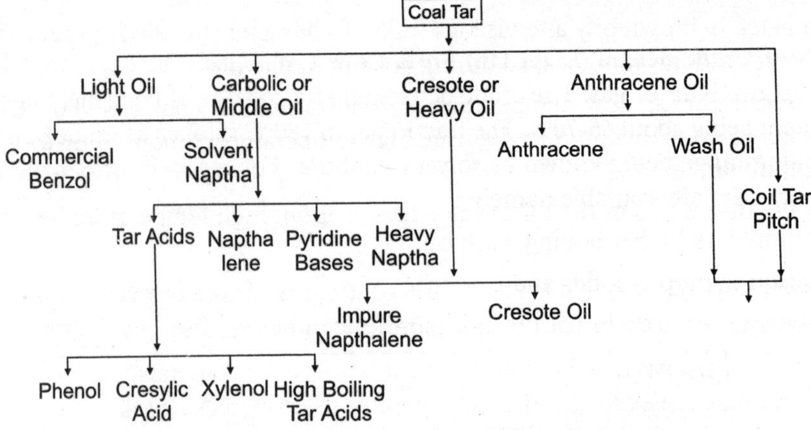

Fig. 15.8. Various fractions and chemicals from tar.

Tar Distillation Plant. The crude tar prior to distillation are separated from the ammoniacal liquor fraction. The corrosion in tar stills is attributed to the ammonium chloride present in the crude tar acting in conjuction with the compounds readily soluble in aqueous caustic soda. To reduce the corrosion effect, 8% caustic soda solution is added to tar, before it is fed to the pipe still heater. Tar from storage tank is heated upto 120-140°C in the first stage of pipe still. After dehydrating this tar in an evaporator (1st stage) it is brought back to 2nd stage pipe still for heating/dehydrating upto 380-410°C (depending upon the quantity of pitch to be produced). From second stage of pipe still heater, dehydrated tar is fed to another evaporator thereby generating vapour mixture of fractions. Superheated steam is fed at the bottom of the 2nd stage evaporator. Hot pitch is obtained from the bottom of the 2nd stage evaporator and Anthracene II is drawn form the side. The top temperature is maintained by means of Anthracene I Reflux. The vapours of the fractions from the top of the 2nd stage evaporator are fed to the fractionating column. A supply of saturated Steam is given at the bottom of the column. Anthracene is withdrawn from the bottom of the column. The other fractions viz. wash oil, phenol and naphthalene are taken out through side tappings. The fractions taken out from the fractionating column flow to respective coolers where they are cooled to 60-80°C. Except for Anthracene I, rest all the fractions are taken to their storage tanks.

A flow diagram of a typical tar distillation plant is given in Fig. 15.9.

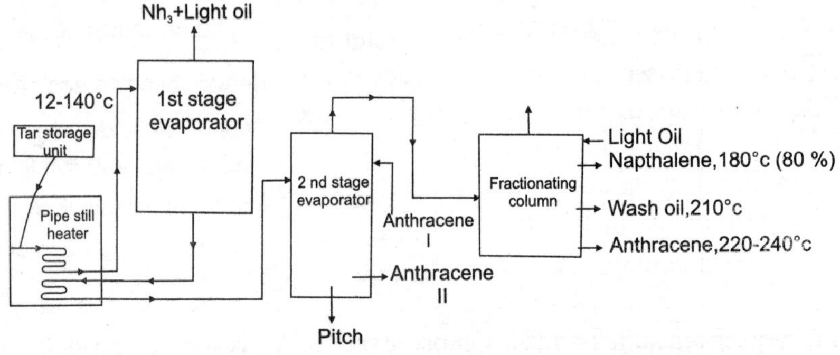

Fig. 15.9. Tar Distillation Plant.

Different techniques of recovery of light oil (crude benzol) from coke oven gas is described below.

Recovery of Crude Benzol (Light Oil) From Coke Oven Gas. The content of crude benzol in gas is about 30 gm/Nm^3 of gas. The crude benzol chiefly consists of benzene, toluene, xylene etc., benzene content being about **65-70%.** The fraction **upto 180°C** is taken as crude benzol, the fraction at higher boiling range being known as solvent naphtha. For recovery of benzol from CO gas, a number of processes are available namely :

- absorption with higher boiling wash oil,
- adsorption in porous solids and
- Compression and cooling of the gas under high pressure.

The process of recovering the crude benzol by absorbing in wash oil or solar oil either in atmospheric or high pressure absorber is generally used in most of the coking industries. The process is based on the absorption of benzol vapour by wash oil, and then regeneration of wash oil by heating. Either wash oil (which is a product of coal tar distillation or solar oil (a petroleum fraction) are used as the absorbing agent. A good absorbing oil must completely absorb benzolvapour present in small concentration in gas and easily evolve it on heating. The boiling point of the oil must be higher than that of the crude benzol to prevent the oil from escaping into crude benzol.

The following should be the minimum requirements of the absorbing oils.

Parameters	Wash oil	Solar oil
Specific gravity	1.04-1.07	Not more than 0.890
Water content	Not more than 0.5%	Not more than 0.2%
Boiling point, °C	230-300 or littlemore	265-350 or little more
Viscosity	Not more than 3°E (Engler) at 25°C.	Not more than 1.5°E (Engler) at 50°C

Sometimes high pressure absorption is done in which equilibrium concentration is increased from 1.8% to 5% in the liquid. The yields of benzol is increased and steam consumption is reduced. The pressure is about 5-8 atmospheres. This process is costly because of compression of gas and is used for special circumstances, i.e. for long distance transportation.

In a typical process used, the coke oven gas (from which benzol is to be recovered) after the removal of tar, ammonia etc. is passed through the benzol scrubber where the benzol vapours are scrubbed by the coal tar based wash oil flowing counter current to the gas. Benzolized wash oil is then pumped to the recovery section where the crude benzol, absorbed in the wash oil is stripped off by steam. The steam-vapour mixture, coming from the top of the stripping column is then condensed and liquified. Benzol and water is separated in two layers. Water is drained off and the crude benzol is stored for further distillation. The debenzolized oil coming out from the bottom is cooled and stored for supply to the benzol scrubber.

Benzol thus recovered contains the following valuable chemicals which are recovered by its distillation.

Benzene = 60-70%, Toluene = 15-20%, Xylene = 5-10%, solvent naphtha, phenol, naphthalene etc. = 1-5%.

To recover these chemicals, benzol is subjected to distillation in benzol distillation plant. A small percentage of benzol is also present in tar from which it is recovered by its distillation.

Crude benzol contains impurities like sulphur compounds (carbon disulphide and thiopene), phenol, pyridine, naphthalene, traces of wash oil etc.

A typical composition of benzol is benzene = 70%, toluene = 15% and xylene = 10%, solvent naphtha & phenol = 5%.

Benzol is mainly subjected to distillation to recover the valuable products like benzene, toluene and xylene.

The net calorific value of motor benzol is 8% lower than that of gasoline (petrol) of average composition, but its density is 19% higher. Consequently, when purchased by volume (the usual method) gasoline is 11% dearer than benzol on a thermal value basis.

USES OF COAL CHEMICALS

Coal chemical recovered from coke oven gas in by-product plant are subjected to various uses. Important uses of some of the chemicals are described below.

Clean coke oven gas. It is a fuel for steel plant furnaces mainly. Besides it is a source of H_2 for ammonia production in an adjoining fertiliser plant as in Rourkela Steel Plant. H_2 is recovered from C.O gas by cryogenic cooling.

Benzol. It is used as a motor fuel blend i.e. blended with petrol upto 15-20% to give better anti-knock rating of the fuel. Also used for the manufacture of photographic material as solvent, for rubber manufacture, for manufacture of varnish and lacquors, for lighting purposes, for manufacture of lamp black and for manufacture of safety explosives. Benzol is also a source of benzene, toluene & xylene which are obtained by its distillation and are valuable chemicals.

Tar. It is used:

- for road making purposes
- as a fuel in furnaces
- for recovering various chemicals by its distillation. Pitch creosote mixture (PCM) thus produced is a fuel which is used in the rotary kiln for calcination of limestone and dolomite. Various grades of coal tar fuels (CTF) are produced from it.

Pitch. It is used:

- as a fuel in furnaces. This semi-solid pitch is mixed with creosote to impart it fluidity whereas the hard pitch (softening point > 100°C) is pulverised and fired in the furnace
- for making carbon electrodes
- for road making purposes
- as moisture proof coating on roof, fibres, underground pipelines etc.

Creosote. It is used:

- as a fuel after mixing with pitch. The fuel is then called pitch-creosote mixture (PCM) or CTF 200.
- for moisture proof pressure impregnation of wood e.g. poles, piling, railroad ties etc.

- in the manufacture of plywood and sunmica to impart it moisture resistance properties.
- as timber preservative also (by way of coating it) against fungi, ants etc.

Phenol. It is used:

- for making the plastic, phenol-formaldehyde which is also called bakelite and is used as an electrical insulating material.
- for the manufacture of perfumes, photographic chemicals, plasticisers, detergents, leather tanning agents etc.
- for making medicines like aspirin, methyl salicylate (e.g.,Amrutanjan).
- for making many chemicals e.g. salicyclic acid which is used for making drugs, dyes and picric acid which is used for making explosives.
- for making adipic acid which is used for making nylon.
- for making phenolic resins which are used as an adhesive and protective coatings.
- for making insecticides e.g. PHENEOL (trade name).

Naphthalene. It is used:

- for making insecticides e.g. Moth balls.
- for manufacture of Beta-naphthol which is used for making dyes, pharmaceuticals and perfumes.
- for making phthalic anhydride which is used for making unsaturated polyster resins of which helmets, reinforced plastics, protective coating, adhesives, electrical insulation material etc. are made. Besides, phthalic anhydride is also used for making plasticiser for plastics and mosquito repellent.

Naphthalene is converted into Beta-naphthol by sulfonating and causticising it. (3-naphthol is also used as an antioxidant for synthetic rubber and for preparation of leather tanning agents.

Pyridine. It is used for separation of various chemical products (as it forms azeotropes with water), for the manufacture of colouring substances and also for medicinal purposes.

Anthracene. It is mainly used in the production of dyes.

Benzene. It is used for:

- the production of styrene, which when polymerised produces polystyrene which is a versatile plastic.
- making styrene-butadiene rubber (SBR), which is the most common synthetic rubber.
- for making phenol whose uses have been listed already.
- for making insecticides like DDT and BHC (benzene-hexachloride).
- for making detergents. Most of the detergents are chemically alkyl benzene sulfonate (ABS). To be specific, 'SURF' is chemically keryl benzene sulfonate, whereas most other detergents are do-decyl benzene sulfonate which are all made from benzene.
- for making nylons.

Toluene. It is used:

- in the manufacture of foam mattresses e.g.polyurathanes.

— for making dyes, plasticiser, pharmaceuticals, phenol, benzoic acid, di-isocynate (used in polyurathane manufacture), benzene etc.

— as a solvent and diluent in coatings.

— for making the explosive 'Trinitrotoluene' (TNT).

Xylene. It is used as a solvent for paints, pharmaceuticals etc. It can be converted to toluene and benzene for subsequent uses.

Characteristics and Distribution of Indian Coal

Indian coals are generally dull in appearance and very often interbanded. The caking coals usually have moisture less than 2%. The moisture content for weakly caking and non-caking coals lie between 3 and 12%. Due to their drift origin, Indian coals have high percentage of mineral matter, most of which is finely disseminated and intimately mixed with the coal substance unlike the coals in Europe & America. The ash usually lies between 15 and 30%. The coals of Assam, however, are low in ash, normally between 4 and 10%, though in particular regions the values may be as high as 24-26%. Sulphur in Indian coals (except in Assam coal which have 3-8% S) is below 1%. Phosphorus is generally less than 0.15% in the coal. Giridih coals and some Madhya Pradesh and Assam coals have phosphorus less than 0.01%. The ash of Indian coals is generally refractory in nature. The heat of combustion (on unit coal basis r the best coking coals range between 8450-8950 kcal/kg; medium to semi-caking coals have he2: of combustion between 8100 and 8900 kcal/kg ; the value for weakly-caking coals lie between 7900 and 8400 kcal/kg and for non-caking coals they range between 7300 and 8000 kcal/kg.

The coals of Jharia coalfield, Bermo and Kargali seams of East Bokaro coalfields, and the Upper and Lower Karharbaree seams of Giridih coalfield are the only deposits of prime coking coals in India (28-32% volatile matter on unit coal basis). Some coals in Raniganj Jharia, Central India and Assam give weak cokes due to their high volatile matter content

The reserve of good coking coal is very poor in India. Out of a total reserve of about 1- Gt (Giga ton) of coal & lignite in India, coking coal is 29 Gt (18% of total coal reserve) and non coking coal is 131 Gt (82%). Prime coking coal reserve is only 5.6 Gt out of the total coking coal available (1 Gt = 1 billion ton = 109 tons).

The major coalfields in India are in the states of Bihar, West Bengal, Madhya Pradesh Maharashtra, Assam, Andhra Pradesh, Orissa, Tamilnadu (Neyveli where lignite is found) and Kashmir.

Good quality coking coals are concentrated in Bihar-West Bengal area. The Jharia coalfields in Bihar is the most important as they are only of coking type. The calorific value of high volatile coal in Jharia ranges from 8260 to 8680 kcal/kg and for low volatile coal it is 831 to 8780 kcal/kg. The reserve is about 5000 million tons (i.e. 5 Gt).

The coals occurring in Bengal is of caking type and the calorific value ranges from 781 to 8200 kcal/kg with a moisture content of 3-12%. Estimated reserve is about 10 Gt.

Madhya Pradesh coals are mostly of weakly coking type with a C.V. of 4200-6200 Kcal/kg. Maharashtra coals are mostly non-coking type (reserve is 2 Gt). Lignite is found mainly at Neyveli in Tamilnadu. Minor coalfields exist in Andhra Pradesh, Kashmir and Assam. ASSAM coals have very high sulphur content (3-8%). Kashmir coals are artificial anthracite converted from lignite deposits.

The coal deposits in seams more than 1.2 m thick and occurring down to depth of 600 m are estimated to be about 80,000 million tonnes. Out of this only about 4,000 million are Tertiary coals and the remainder are Gondwana coals. The coking coals are in short supply being only about 5,000 million tonnes. Deposits of inferior variety coals are very vast. The quantity of coking coal is insufficient for the requirements of Indian metallurgical industries, therefore the inferior coals are being washed to remove mineral matters and coking coal is being suitably blended with weakly coking coal.

Most of the Indian coals is what is known as *Gondwana coal.* 'Gondwana' is a geological term which refers to a certain rock system which is about 200 million years old. Gondwana coal is a laminated bituminous coal with alternate dull and bright layers. In India, anthracite coal is found to a negligible extent. Coal deposits of the tertiary era (60 million years old) are found in Assam, Rajasthan and Jammu.

Coalfield of Bihar & Bengal produce about 80% whereas the coalfields of Madhya Pradesh 10%, Orissa 5% and the rest of the country produce about 5% of the coal. This shows the uneven geographical distribution of coal and the consequent distribution problems. The important coal-fields of Bihar and Bengal are Barjora, Bokaro, Giridih, Karanpura, Jharia & Raniganj. The last two named are the biggest and the best coalfields of the country.

The coalfields have seams or layers of coal beds from which coal is mined. Most of the coalfields have less than half a dozen seams but Raniganj, Jharia and Karanpura fields have more than twenty seams. Most of the coal seams are less than 10 metres thick. Some of the thickest coal seams of India are the Upper Seam No. 1 (45 m) of the Talchar coalfield, the Argada Seam (47.5 m) of the south Karanpura and the Jhingurda seam (75 m) of Singrauli.

Most of the Indian coals have a high ash content. This is a serious disadvantage. Coal suitable for metallurgical industry is not at all sufficient for the needs of the country and is mostly available from some of the seams of Jharia, Raniganj and Bokaro coalfields. Hence coking coal is imported in huge quantity from Canada,Australia ,Newzealand and U.S.A.

The *Gondwana coal,* which contributes about 98% of our coal, has been classed into the following two broad groups :

Raniganj Coal has greater moisture content (about 5%) and more volatile matter (30- 38%) and therefore burns easily with a long flame. It has lower calorific value, the ash does not form clinkers, and it is unsuitable for making coke but is good for making gas.

Jharia or Bokaro Coal has less than 2% moisture and volatile matter content is about 28%; consequently it burns with a shorter flame, and is more suitable for steam raising locomotive boilers (because less combustion space is needed). Its calorific value is higher and ash clinkers at higher temperature. It is suitable for making coke.

The tertiary coal is much younger than Gondwana coal and is found in comparatively small amounts in widely separated regions like Assam, Jammu and Kashmir, Rajasthan and Tamil Nadu.

The seams are quite thin, being 0.5 to 3 m thick. The coal being younger in age is mostly of lower rank like the lignite. Due to the Geological upheavals some coalfields of Jammu and Kashmir & Assam are of higher ranks. These have low moisture (< 3%) and ash (2 to 10%) contents. The tertiary coal has higher sulphur content-Assam coals have up to 7% sulphur- making them unsuitable for metallurgical purposes.

One of the thickest coal seam (131.5 metres) of India has been discovered at Singrauli coalfield while one of the deepest coal mine (440 metres shaft) is at Sudamdih in Jharia coalfield.

Major *coalfields in India* and their approximate coal reserves are tabulated in Table 16.1.

Coal mines in India are worked by *Board and Pillar method.* Initially, tunnels are driven by blasting along and across the extension of the coal seam. These tunnels (3 metres wide) are usually driven 12 metres apart and are known as galleries. The galleries divide the coal seam into many blocks known as coal pillars. In the first stage, about 40% of the coal is removed by driving galleries and rest 60% is left in the pillars. In the second stage (depillaring), another

Table 16.1 Major Indian Coalfields and their Approximate Reserves

age	States	Coalfield location	Approximate Reserves, million tons
Gondwana	West Bengal Bihar Orissa Madhya Pradesh Maharashtra Andhra Pradesh	Raniganj Jharia Giridih Bokaro-Ramgarh-Karanpura Talcher Korea-Hingir-Rampur Singrauli Korba Kanham Valley Pench Valley Wardha Valley (Ghugus-Chanda-Ballarpur) Kamptee-Umrer Singareni-Kottagudem	20,000 13,000 40 15,000 30,000 2,000 29,000 250 2,000 4,000 8,000
Tertiary	Assam Madras Rajasthan Jammu and Kashmir Kerala Gujarat (Kutch	Makum 1 Bituminous and Khasi-Jaintia Hills ? sub-bituminous) Garo Hills J Neyveli (Lignite) Palana (Lignite) Jammu (Anthracite) Nichahom (Lignite) Travancore Malabar Coast (Lignite) Umarsar (Lignite)	1,00 2,000 20 100 100 100 15

40% is extracted from the pillars leaving behind 20% of the coal in the mines. However, use of sand stowing techniques (in which empty spaces in mines created by the extraction of coal is filled up by sand bit by bit) reduces this loss to 10% or even less.

The coal in the mine is brought to the surface either by vertical lifts in a pit or by haulage along an incline. Coal extraction is also done by open cast method in quarries. Quarrying is more economical, when coal occurs at a small depth and the overburden is soft. The open cast method is also applicable to the coal seams near their outcrops. Coal extracted by open cast method is generally of inferior quality.

Categorisation of Resources. The coal resources of India are available in sedimentary rocks of older Gondwana Formations of peninsular India and younger Tertiary formations of north-eastern/northern hilly region Based on the results of regional/Promotional Exploration, where the boreholes are normally placed 1-2 Km apart, the resources are classified into Indicated or Inferred category. Subsequent Detailed Exploration in selected blocks, where boreholes are less than 400 meters apart, upgrades the resources into more reliable 'Proved' categoiy. The Formation-wise and Category-wise coal resources of India as on 1.4.2009 are given below: in Table 16.2.

Table 16.2. Formationwise and Categorywise Coal Reserves of India (as on 1.4.2009)

(.ill "Million Tonnes)

Formation wise	Proved	Indicated	Inferred	In Total
Gondwana Coals	105343	123380	37414	266137
Tertiary Coals	477	90	506*	1073
Total	**105820**	**123470**	**37920***	**267210**

* Includes 456 Mt of Inferred resources established through mapping in NE region. The Type and Category-wise coal resources **of** India as on 1.4.2009 are given in table 16.3.

Table 16.3. Typewise and Categorywise Coal Reserve of India (as on 1.4.2009)

(in Million Tonnes)

Types of Coal I	Proved	Indicated	Inferred	Total
(A) Coking :				
- Prime Coking Coal	4614	699	0	5313
- Medium Coking Coal	12449	12064	1880	26393
- Semi-Coking Coal	482	1003	222	1707
Sub-Total Coking Coal	**17545**	**13766**	**2102**	**33413**
(B) Non-Coking:	87798	109614	35312	232724
(C) Tertiary Coal	477	90	506	1073
Grand Total	**105820**	**123470**	**37920**	267210

Coal Grading: Following table no. 16.4 depicts coal grading with both the ash percent (ash and moisture in case of high volatile coals) as also the corresponding useful heat value (UHV) in Kcal/Kg which is calculated by the emperical formula :

UHV = 8900—138 (A + M), where, A = Ash%. & M = Moisture % in coal

Table 16.4. Ash Percentage and Calorific Value (UHV) of Different Grades of Coal

Grades of Coal	Equivalent ash Percent	Useful heat value (UHV) of coal, Kcal/Kg
A	18.12	6400
	20.29	6100
B	21.01	6000
	25.36	5400
C	26.09	5300
	28.26	5000
D	29.71	4800
	30.43	4700
	33.33	4300
	35.51	4000
E	36.23	3900
	39.86	3400
	42.03	3100
F	42.75	3000
	46.38	2500
	50.00	2000
G	50.72	1900
	54.35	1400
	57.25	1000

Latest informations on indian coal & lignite reserve, production & consumption and import & export of coal is given below.

Reserve of Coal and Lignite

- Coal deposits are mainly confined to eastern and south central parts of the country. The states of Jharkhand, Odisha, Chhattisgarh, West Bengal, Madhya Pradesh, Telangana and Maharashtra account for 99.08% of the total coal reserves in the country. The State of Jharkhand had the maximum share (26.44%) in the overall reserves of coal in the country as on 31st March 2015 followed by the State of Odisha (24.72%) .
- As on 31.03.15, the estimated reserve of coal was 306.60 billion tonnes, an addition of 5.04 billion over the last year . There has been an increase of 1.67% in the estimated coal reserves during the year 2014-15 with Chattisgarh accounting for maximum increase of 4.53%.
- The estimated total reserve of lignite as on 31.03.15 was 43.25 billion Tonne which is equivalent to the total reserve as on 31.03.14.

Statewise Estimated Reserves of Coal in India as on 31.03.2014 and 31.03.2015 is given below.

(in Billion Tonne)

States/ UTs	Proved		Indicated		Inferred		Total		Distribution (%)	
	31.03. 2014	31.03. 2015	31.03. 2014	31.03. 2015	31.03. 2014	31.03. 2015	31.03. 2014	31.03. 2015	31.03. 2014	31.03. 2015
Arunachal Pradesh	0.47	0.47	0.05	0.05	0.00	0.00	0.52	0.52	0.17	0.17
Assam	0.03	0.03	0.04	0.04	0.02	0.02	0.09	0.09	0.03	0.03
Bihar	0.00	0.00	0.00	0.00	0.16	0.16	0.16	0.16	0.05	0.05
Chhattisgarh	16.05	18.24	33.25	34.39	3.23	2.29	52.53	54.91	17.42	17.91
Jharkhand	41.38	41.46	32.78	33.03	6.56	6.56	80.72	81.05	26.77	26.44
Madhya Pradesh	10.41	10.41	12.38	12.78	2.88	3.34	25.67	26.54	8.51	8.66
Maharashtra	5.67	5.95	3.19	3.19	2.11	2.11	10.96	11.25	3.64	3.67
Meghalaya	0.09	0.09	0.02	0.02	0.47	0.47	0.58	0.58	0.19	0.19
Nagaland	0.01	0.01	0.00	0.00	0.31	0.31	0.32	0.32	0.10	0.10
Odisha	27.79	30.75	37.87	36.55	9.41	8.51	75.07	75.80	24.89	24.72
Sikkim	0.00	0.00	0.06	0.06	0.04	0.04	0.10	0.10	0.03	0.03
Uttar Pradesh	0.88	0.88	0.18	0.18	0.00	0.00	1.06	1.06	0.35	0.35
West Bengal	13.40	13.52	13.02	13.01	4.89	4.91	31.32	31.44	10.39	10.25
Telangana	9.73	9.81	9.67	9.96	3.07	3.03	22.47	22.79	7.45	7.43
All India Total	125.91	131.61	142.50	143.24	33.15	31.74	301.56	306.60	100.00	100.00
Distribution (%)	41.07	42.93	46.48	46.72	10.81	10.35	100.00	100.00		

Statewise Estimated Reserves of Lignite in India as on 31.03.2014 and 31.03.2015

(in Billion Tonne)

States/ UTs	Proved		Indicated		Inferred		Total		Distribution (%)	
	31.03. 2014	31.03. 2015	31.03. 2014	31.03. 2015	31.03. 2014	31.03. 2015	31.03. 2014	31.03. 2015	31.03. 2014	31.03 2015
Gujarat	1.28	1.28	0.28	0.28	1.16	1.16	2.72	2.72	6.29	6.29
Jammu & Kashmi	0.00	0.00	0.02	0.02	0.01	0.01	0.03	0.03	0.06	0.06
Kerala	0.00	0.00	0.00	0.00	0.01	0.01	0.01	0.01	0.02	0.02
Pondicherry	0.00	0.00	0.41	0.41	0.01	0.01	0.42	0.42	0.96	0.96
Rajasthan	1.17	1.17	2.67	2.67	1.88	1.89	5.72	5.72	13.23	13.23
TamilNadu	3.74	3.74	22.90	22.90	7.71	8.57	34.35	34.35	79.42	79.42
West Bengal	0.00	0.00	0.00	0.00	0.00	0.00	0.00	0.00	0.01	0.01
All India	6.18	6.18	26.28	26.28	10.78	11.65	43.25	43.25	100.00	100.00
Distribution (%)	14.29	14.29	60.77	60.77	24.93	26.94	100.00	100.00		

Production of Coal and lignite

- Coal production in the country during the year 2014-15 was 612.44 million tonne (MTs) as compared to 556.77 MTs during 2013-14, registering a growth of 8.25% .
- The Lignite production during 2014-15 was 48.26 MTs, which is 9% higher than the production during 2013-14 (44.27 MTs).
- Coal production in India was about 407.04 MTs during 2005-06, which increased to 612.44 MTs during 2014-15 .
- Lignite production increased from 30.23 MTs in 2005-06 to 48.26 MTs in 2014-15.

Trends in Production of Coal and Lignite in India

(in million tonnes per year)

Year	Coal			Lignite	Grand Total
	Coking	Non-coking	Total		
1	2	3	4=(2)+(3)	5	6=(4)+(5)
2013-14	56.82	508.95	565.77	44.27	610.00
2014-15(p)	57.45	554.98	612.44	48.25	660.69
Growth rate of 2014-15 over 2013-14(%)	1.11	9.04	8.25	8.99	8.31

Consumption of Coal and Lignite

- The estimated total consumption of raw coal by industry has increased from 433.27 MT (million tons) during 2005-06 to 827.57 MT during 2014-15 with a CAGR(compound annual growth rate) of 6.69% . The annual growth rate from 2013-14 to 2014-15 is 14.28%.
- Consumption of Lignite increased from 30.24 MT in 2005-06 to 49.57 MT in 2014-15 registering a compound growth of 5.07%. Consumption of Lignite is highest in Electricity Generation sector, accounting for about 84.09% of the total lignite consumption.
- Electricity generation is the biggest consumer of coal, followed by steel industries. Industry wise estimates of consumption of coal show that during 2014-15 electricity generating units consumed 527.10 MT of coal, followed by steel & washery industries (66.37 MT), cement industries(37.95 MT) and paper industries (1.54 MT).

Import and export of coal

- The average quality of the Indian coal is not very high and this necessitates the import of high quality coal to meet the requirement of steel plants. There has been an increasing trend in the import of coal.
- Import of coal has steadily increased from 38.59 MTs during 2005-06 to 212.10 MTs during 2014-15 . During this period, the quantum of coal exported increased from 1.99 MTs during

2005-06 to 2.44 MT during 2012-13 and then decreased to 1.24 MTs during 2014-15.

- There is growth rate of 27.12% of gross import and 28.05% in net imports of coal in 2014-15 over the previous year. However, there was decrease of 43.42% in export of coal during the same period.

Origin, Composition & Classification Of Petroleum-Indian Crude Oil

Petroleum (crude oil or rock oil) is a naturally occurring brown to black oil comprising mainly of hydrocarbons found under the crust of the earth on shore or off shore. It is obtained from the ground either by natural seepage or by drilling wells to various depths. Either petroleum oil flows out itself due to underground gas pressure or these are mechanically pumped out.

ORIGIN OF PETROLEUM

Although several theories were put forward to explain the formation of petroleum, only three widely accepted ones are given below.

Carbide theory. According to this theory, hydrocarbons present in petroleum are formed by the action of water on inorganic carbides. Inorganic carbides, in turn, are formed by the reaction of metal and carbon under high temperature & pressure conditions inside the earth.

$$Ca+2C \xrightarrow[\text{pressure}]{\text{High temp. and}} \underset{\text{üüü=üüü}}{CaC_2}$$

$$4AI+3C \xrightarrow[\text{pressure}]{\text{High temp. and}} \underset{\text{Aluminium carbide}}{AI_4C_3}$$

$$CaC_2 + 2H_2O \longrightarrow Ca(OH)_2 + \underset{\text{acetylene}}{C_2H_2}$$

$$AI_4C_3 + 12H_2O \longrightarrow 4AI(OH)_3 +$$

These lower hydrocarbons then undergo hydrogenation and polymerization to give various types of hydrocarbons (paraffins, aromatic and cycloparaffms).

$$C_2H_2 + H_2 \text{---->} C_2H_4 + H_2 \text{---->} C_2H_6$$

acetylene ethane ethylene

$$\underset{\text{actylene}}{3C_2H_2} \xrightarrow{\text{polymerisation}} \underset{\text{benzene}}{C_6H_6}$$

$$3C_2H_4 \xrightarrow{\text{polymerisation}} \underset{\text{cyclohexane}}{C_6H_{12}}$$

$$\underset{\text{ethylene}}{3C_2H_4} \longrightarrow \underset{\text{butane-2}}{CH_3CH{=}CH.CH_3} \xrightarrow{H_2} \underset{\text{butane}}{CH_3.CH_2.CH_2CH_3}$$

However, this theory fails to explain the following facts :

- Presence of nitrogen and sulphur compounds.
- Presence of chlorophyll and haemin.
- Presence of optically active compounds.

Engler theory. According to Engler, petroleum is of animal origin. Engler (1900) suggested that petroleum is formed by the decay and decomposition of marine animals under high pressure & temperature. The sulphur dioxide gas given out by the volcanoes beside the sea-side kills the fish and other sea animals which go on piling beside the volcano. After hundreds of years, these animals start decomposing under the influence of high pressure and temperature to form petroleum.

Engler's theory is supported by the following facts.

- Experimental destructive decomposition of fish oil & other animal fats under high pressure and temperature gives a product similar to natural petroleum.
- Presence of brine or sea water together with petroleum.
- Presence of nitrogen and sulphur compounds.
- Presence of optically active compounds.
- Presence of fossils (the remains of animals or plants which died long ago) in the petroleum area.

However, this theory fails to account for the presence of chlorophyll (green-colouring matter of plants) in petroleum. Moreover, it does not explain the presence of coal deposits in the vicinity of the oil fields.

Modern theory. According to modern views, petroleum is believed to be formed by the decay and decomposition of marine animals as well as that of vegetable organism of the pre-historic forests, i.e. it is of animal as well as of plant origin. It is thought that due to some upheavals or earthquakes, these prehistoric forests, and sea animals got buried under the crust of earth. Due to action of prolonged action of high temperature and pressure in the interior of earth for ages, the biological matter decomposed into petroleum.

This modern theory explains the presence of brine and coal in the vicinity of petroleum. Presence of brine is explained on the basis of animal origin while the presence of coal is explained by plant origin. It also explains the presence of N & S compounds, chlorophyll and optically active compounds.

Difference Between Origins of Petroleum Oil and Coal:

- Petroleum oil was formed mainly from sea plants and animals decaying under strongly reducing conditions while the coal was formed mainly from land plants decaying under mildly reducing conditions.
- Coal seams remained static where formed/deposited while oil can migrate under the effects of pressure and temperature so that the location of the existing deposits may not be the location of the initial accumulation of oil forming debris. The source rock is sedimentary and mainly or entirely of marine origin.

Composition of Petroleum. A typical composition (by weight %) of petroleum is given below:

Carbon = 84-87%	Hydrogen = 11-15%	Sulphur = 0.1-3%
Nitrogen = 0.1-1.5%	Oxygen = 0.3-1.8%	

The principal components of petroleum are hydrocarbons (paraffins, naphthenes, aromatics, olefins etc.), small amounts of sulphur, nitrogen & oxygen compounds as impurities and some inorganic compounds & metals (vanadium & platinum in traces).

Sulphur compounds found in petroleum are hydrogen sulphide, thiophenes, mercaptans. Oxygen occurs in combined form in alcohols, phenols, resins and organic acids present in petroleum. Nitrogen compound include pyridines, quinolines, pyroles etc. Inorganic compounds present in petroleum are salt, clay and sand etc.

Typical properties of crude oil. Specific gravity = 0.80 - 0.95, viscosity at 37.8°C = 2.3 cst , carbon/hydrogen ratio = 6-8, pour point = 18-30 °C, wax content = 6 - 12%, flash point = 15-27°C, water content = 0.1 - 1.5% Salt content = 1.5 - 8%, Ash content = 0.004 - 0.006%.

NATURE AND CHARACTERISTICS OF CRUDE OILS

It is a complex mixture containing many different hydrocarbons compounds that vary in appearance and composition from one oil field to another. Crude oils range in consistency from black to tar like liquids and in color from clear to black.

- An "Average" crude oil contains about 84% carbon, 14% hydrogen, 1–3% sulfur, and less than 1% each of nitrogen, oxygen, metals, & salts. Crude oils are generally classified as paraffinic, naphthenic, or aromatic, based on the predominant proportion of similar hydrocarbon molecules. Mixed-base crudes have varying amounts of each type of hydrocarbon. Refinery crude base stocks usually consist of mixtures of two or more different crude oils.
- Relatively simple oil assays are used to classify crude oils as paraffinic, naphthenic, aromatic, or mixed. One assay method (United States Bureau of Mines) is based on distillation, and another method (UOP "K" factor) is based on gravity boiling points. More comprehensive crude assays determine the value of the crude (i.e., its yield & quality of useful products) and processing parameters. Crude oils are usually grouped according to yield structure.
- Crude oils are also defined in terms of API (American Petroleum Institute) gravity. The higher the API gravity, the lighter the crude. For example, light crude oils have high

API gravities and low specific gravities. Crude oils with low carbon, high hydrogen, and high API gravity are usually rich in paraffins and tend to yield greater proportions of gasoline & light petroleum products; those with high carbon, low hydrogen, and low API gravities are usually rich in aromatics.

- Crude oils that contain appreciable quantities of hydrogen sulfide or other reactive sulfur compounds are called "sour," Those with less sulfur are called "sweet."
- Crude oil is a mixture of hydrocarbon molecules, that contains from 1 to 60 carbon atoms. The properties of hydrocarbons depend on the number and arrangement of the carbon & hydrogen atoms in the molecules.
- Hydrocarbons containing up to four carbon atoms are usually gases, those with 5 to 19 carbon atoms are usually liquids, and those with 20 or more are solids. The crude oil refining process uses chemicals, catalysts, heat, & pressure to separate and combine the basic types of hydrocarbon molecules naturally found in crude oil into groups of similar molecules. The refining process also rearranges their structures and bonding patterns into different hydrocarbon molecules and compounds.
- Sulfur may be present in crude oil as hydrogen sulfide (H_2S), as compounds (e.g. mercaptans, sulfides, disulfides, thiophenes, *etc.*) or as elemental sulfur. Each crude oil has different amounts and types of sulfur compounds, but as a rule the proportion, stability, & complexity of the compounds are greater in heavier crude oil fractions. Hydrogen sulfide is a primary contributor to corrosion in refinery processing units. Other corrosive substances are elemental sulfur and mercaptans. Moreover, the corrosive sulfur compounds have an obnoxious odor.
- Pyrophoric iron sulfide from the corrosive action of sulfur compounds on the iron and steel used in refinery process equipment, piping, & tanks. The combustion of petroleum products containing sulfur compounds produces undesirables such as sulfuric acid and sulfur dioxide. Catalytic hydrotreating processes such as hydrodesulfurization remove sulfur compounds from refinery product streams. Sweetening processes either remove the obnoxious sulfur compounds or convert them to odorless disulfides, as in the case of mercaptans.
- Oxygen compounds such as phenols, ketones, and carboxylic acids occur in crude oils in varying amounts.
- Nitrogen is found in lighter fractions of crude oil as basic compounds, & in heavier fractions of crude oil as nonbasic compounds that may also include trace such as copper, vanadium, and/or nickel. Nitrogen oxides can form in process furnaces. The decomposition of nitrogen compounds in catalytic cracking & hydrocracking processes forms ammonia and cyanides that can cause corrosion.
- Metals, including nickel, iron, & vanadium are often found in crude oils in small quantities and are removed during the refining process. Burning heavy fuel oils in refinery furnaces and boilers can leave deposits of vanadium oxide & nickel oxide in furnace boxes, ducts, & tubes. It is also desirable to remove trace amounts of arsenic, vanadium, and nickel prior to processing as they can poison certain catalysts.

- Crude oils often contain inorganic salts such as sodium chloride, magnesium chloride, and calcium chloride in suspension or dissolved in entrained water (brine). These salts must be removed or neutralized before processing to prevent catalyst poisoning, equipment corrosion, and fouling. Salt corrosion is caused by the hydrolysis of some metal chlorides to hydrogen chloride (HCl) and the subsequent formation of hydrochloric acid when crude is heated. Hydrogen chloride may also combine with ammonia to form ammonium chloride (NH_4Cl), which causes fouling and corrosion.
- Some crude oils contain naphthenic (organic) acids, which may become corrosive at temperatures above 230°C, when the acid value of the crude is above a certain level.

MAJOR OIL REFINERY PRODUCTS AND THEIR CHARACTERISTICS

- **Gasoline.** The most important refinery product is motor gasoline, a blend of hydrocarbons with boiling ranges from ambient temperatures to about 205°C The important qualities for gasoline are octane number, volatility, and vapor pressure. Additives are often used to enhance performance and provide protection against oxidation & rust formation.
- **Kerosene.** Kerosene is a refined middle-distillate petroleum product that finds considerable use as a jet fuel and in cooking & space heating. When used as a jet fuel, some of the critical qualities are freeze point, flash point, and smoke point. Commercial jet fuel has a boiling range of about 190-275°C and military jet fuel 55-288°C. Kerosene, with less critical specifications, is used for lighting, heating, solvents, and blending into diesel fuel.
- **Liquified Petroleum Gas (LPG).** LPG, which consists principally of propane & butane, is produced for use as fuel and is an intermediate material in the manufacture of petrochemicals. The important specifications for proper performance include vapor pressure and control of contaminants.
- **Distillate Fuels.** Diesel fuels and domestic heating oils have boiling ranges of about 205-370°C. The desirable qualities required for distillate fuels include controlled flash and pour points, clean burning, no deposit formation in storage tanks, and a proper diesel fuel cetane rating for good starting and combustion.
- **Residual Fuels.** Many marine vessels, power plants, commercial building & industrial facilities use residual fuels or combinations of residual & distillate fuels for heating and processing. The two most critical specifications of residual fuels are viscosity and low sulfur content for environmental control.
- **Coke and Asphalt.** Coke is almost pure carbon with a variety of uses from electrodes to charcoal briquets. Asphalt, used for roads & roofing materials, must be inert to most chemicals and weather conditions.
- **Solvents.** A variety or products, whose boiling points and hydrocarbon composition are closely controlled, are produced for use as solvents. These include benzene, toluene, and xylene.
- **Petrochemicals.** Many products derived from crude oil refining, such as ethylene, propylene, butylene, & isobutylene, are primarily intended for use as petrochemical feedstock in the production of plastics, synthetic fibres, synthetic rubbers, and other products.

- **Lubricants.** Special refining processes produce lubricating oil base stocks. Additives such as demulsifiers, antioxidants, and viscosity improvers are blended into the base stocks to provide the characteristics required for motor oils, industrial greases, lubricants, & cutting oils. The most critical quality for lubricating oil base stock is a high viscosity index, which provides for greater consistency under varying temperatures.

DETECTION OF PETROLEUM DEPOSIT

Petroleum deposits are detected by visual, geophysical, geological and drilling methods. **Visual** method includes oil seepages at the surface of the earth.

Geophysical method includes the measurement of density, elasticity, magnetic and electrical properties of the rock in the crust of the earth from which the inference are drawn whether the petroleum deposit occurs there or not.

Geological method includes the measurement of age and nature of rocks inside the earth's crust which gives an idea of oil deposit.

Drilling in the crust of earth is the final test when the petroleum deposits have been proved by either of the earlier methods.

PRODUCTION OF PETROLEUM

Petroleum almost always occurs along with gas called natural gas. After drilling in the crust of earth both natural gas and oil flow up through pipe under pressure initially. When the oil pressure decreases, then the residual oil is either sucked by a pump or pressure is created by injecting compressed gas or high pressure water through a pipe bored by the side of oil delivery pipe so that it can flow out.

When the oil well contains both oil & gas it is called wet well and if it contains only gas then it is called a dry well.

A sketch of typical oil pool formation inside the crust of earth is given in Fig. 17.1. Hard cap rock and hard rock at the bottom, both are impermeable to oil & gas. Geological studies indicates that petroleum was not formed in the pools where it is found today. Deposits of crude petroleum were formed near sea shores, but the action of the surrounding water gradually shifted the location of oil pool. After a period of many centuries, the oil was forced through layers of porous rock strata until it became trapped under a dome capped by hard rock. This makes it impossible for the gas or oil to escape and the water keeps the pool of water under pressure.

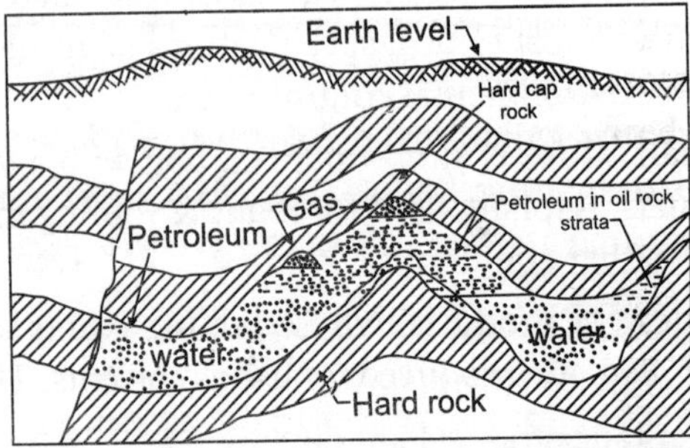

Fig.17.1. Typical oil-pool formations. In the centre, pool water has forced the petroleum and gas up into a dome, or anticline, in the porous oil rock layer. Hard cap rock prevents the escape of the oil. A fault, at the left, or a stratigraphic trap, as shown at the right, similarly traps oil pools between water under high pressure and hard rock strata.

For digging an oil well, the rotary drilling system, utilizes the principle of torsional grinding or cutting. The drilling rig consists of four main parts, viz. (a) derrick—a tall steel structure equipped with hoisting gear for raising and lowering the drilling equipment, (b) drill column or stem - this consists of a main section of steel pipes (drill pipe) with at its lower end some extra heavy pipe (drill collars), to which the bit is attached ; (c) rotating mechanism at the upper end of the drill cloumn, i.e. at the surface, is a hollow bar of square or hexagonal cross-section (thekelly or grief stem), and this slides through a square or hexagonal hole in the machine (rotary table) which provides the rotary movement; (d) mud circulation system — at the top of the kelly is the so-called swivel through which a thin mud (drilling fluid) passes from a flexible hose ; the bottom half of the swivel revolves with the kelly while the top does not, and to it the flexible hose is attached ; the top part of the swivel is attached also to the hoisting mechanism.

The hoisting system is controlled from a working platform built within the derrick above the ground level. The mud is pumped down the drill pipe through the holes of the bit and returned to the pump section via the annular space between the drill pipe and the walls of the bore hole.

An efficient mud circulation is required to drill an oil well rapidly. The drilling fluid serves four principal functions, viz. (a) cooling the bit, (b) flushing out the cuttings, (c) supporting the bore wall by hydraulic pressure, and (d) penetrating porous rock to form a seal of mud cake and prevent the escape of oil or gas through pores & cracks. Drilling muds are colloidal clay suspensions with additions to control viscosity.

The main boring is of 20 to 30 cm diameter and its depth may vary from 1.5 to 4.5 km. Wells of more than 7.5 km have also been drilled while quite shallow wells also exist. The entire well is provided with a casing of steel pipes right upto the productive work layer. This prevents the collapse of the bore walls and ensures that other extraneous fluids do not enter the well. A steel tubing is suspended from the well-head up to oil-bearing rock and is provided with control valves and outlet pipe at the wall head.

Oil is normally under pressure in the reservoir. The source of the pressure may be a body of compressed gas above the oil or a mass of water underlying the oil. These pressures cause the oil to flow towards the foot of the well and then drive it up the tubing to the surface (natural flow). When oil pressure is low or it falls on production, the lifting is done by sending a high pressure gas through the annulus between the casing and the tubing or by providing a suitable pumping device at the lower end of the tubing. Another method of secondary recovery of oil is by water flooding ;water is injected into the reservoir rock through separate wells on the periphery of the oil field.

The total production from each well is piped to the gathering station. The gas is separated by directing the fluid against baffles in cylindrical pressure vessels. The water and solids are separated either by settling out in tanks or by centrifuging. The gas is called wet natural gas. It is processed for the recovery of natural gasoline and the crude is sent to the refinery.

PETROLEUM REFINING OPERATIONS IN OIL REFINERIES

Petroleum refining begins with the distillation, or fractionation of crude oils into separate hydrocarbon groups. The resultant products are directly related to the characteristics of the crude processed. Most distillation products are further converted into more usable products by changing the size & structure of the hydrocarbon molecules through cracking, reforming, and other conversion processes. These converted products are then subjected to various treatment & separation processes such as extraction, hydrotreating, & sweetening to remove undesirable constituents and improve product quality. Integrated refineries incorporate fractionation, conversion, treatment, & blending operations and may also include petrochemical processing.

Petroleum refining processes and operations can be separated into five basic areas:

— **Fractionation** (distillation) is the separation of crude oil in atmospheric and vacuum distillation towers into groups of hydrocarbon compounds of differing boiling point ranges called "fractions" or "cuts."

— **Conversion** processes change the size and/or structure of hydrocarbon molecules. These processes include:
  - Decomposition (dividing) by thermal & catalytic cracking;
  - Unification (combining) through alkylation & polymerization; and
  - Alteration (rearranging) with isomerisation & catalytic reforming.

— **Treatment** processes are intended to prepare hydrocarbon streams for additional processing and to prepare finished products. Treatment may include the removal or separation of aromatics and naphthenes as well as impurities & undesirable contaminants. Treatment may involve chemical or physical separation such as dissolving, absorption, or precipitation using a variety and combination of processes including desalting, drying, hydrodesulfurzing, solvent refining, sweetening, solvent extraction, & solvent dewaxing.

— **Formulating and Blending** is the process of mixing & combining hydrocarbon fractions, additives, and other components to produce finished products with specific performance properties.

— **Other Refining Operations** include: light ends recovery; sour water stripping; solid waste & wastewater treatment; process-water treatment & cooling; storage & handling; product movement; hydrogen production; acid & tail-gas treatment; and sulfur recovery.

PRETREATMENT OF OIL AT OIL FIELD BEFORE REFINING

Oil and gas when they come out of oil field are separated. The natural gas is compressed to liquify it which is used for heating of domestic and industrial ovens. Petroleum oil is made free of water, sediments and salts present in it. It is then made free of some dissolved gases into it by the process called 'stabilisation'. It is then sent to oil refineries for separation into various petroleum products by distillation mainly and auxiliary operations.

Crude oil as it comes out of well may contain upto 25% water, salts ($MgCl_2$, $CaCl_2$, NaCl etc.) upto 2000-5000 mg/litre and sediments upto 1-1.5%. For refining crude oil, the salt content in it should be < 50 mg/litre and water < 0.3%. Excessive water in crude requires extra heat for its distillation, increases its cost of transportation, forms emulsion which absorb materials like resin (hence emulsion breakers are to be used). Salt in crude oil causes scaling, corrosion and reduces heat transfer co-efficient during its processing. Sediments present in crude causes erosion and scaling.

In mechanical method of separation of impurities from crude oil, it is subjected to centrifuging, filtration and settling after heating it to 120-160°C at 6-8 atm. pressure.

In physico-chemical method, emulsion breakers are added. But they are costly and cause corrosion & sludge formation.

Dehydration Of Crude Oil. In electrical dehydration method of crude oil shown in Fig. 17.2 (widely practised in Russia and America), 10% water is added to crude and it is heated before passing it through an electrical dehydrator which has two concentric metal plates as electrodes in which 30000 volts is supplied thereby separating water from crude.

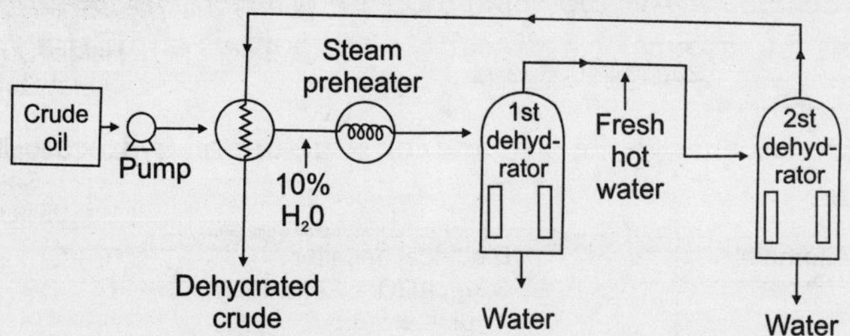

Fig. 17.2. A scheme for electrical dehydration of crude oil.

Details of dehydrators are given below :

First dehydrator □T = 90-95°C, P - 6.8 atm.

Second dehydrator → T = 80-90°C, P - 5.5-6 atm.

In Russian practice, however, spherical dehydrators are used whose separation efficiency is more.

Electrical Desalting of Crude Oil. All crude oil contain salt water together with sediment. Salt content in crude oil is reported as pounds of salt (measured as sodium chloride) per thousand barrel of oil (PTB) and it varies from 10 to 200 PTB. Besides being present as brine during its production in the oil well, crude oil gets contaminated with salt water when shipped in tankers as ballast water mixes into the crude oil during shipment and may be emulsified with crude oil while pumping it to shore installation.

Desalting of crude oil is done electrically in which, first an emulsion of crude oil & water is formed and then the emulsion is broken by means of an electrical field. Crude oil is heated to 120-130°C to reduce its viscosity (<40SUS) to produce good emulsion. Water (3-8% of crude oil) is then added in the crude oil and the mixture is then passed through a mixing valve across which a pressure drop of about 1.5 kg/cm^2 is maintained for proper emulsification of crude oil and water. The emulsified crude oil is fed to the electrical desalter, where it is subjected to high voltage (20000-24000 Volts) electrical field which results in coalescence of water droplets and its separation from the emulsion together with the salt from crude oil. Desalted crude overflows while water along with salt and crude sludge is removed from the bottom of the desalter.

Salt content of crude oil can be brought down to a level of 3 PTB or to the extent of 95% removal. Proper pressure drop (usually 1.5 kg/cm^2) through the mixing valve should be maintained to create a good emulsion. However, maximum pressure drop (usually 4 kg/cm^2) should not be exceeded otherwise an unbreakable emulsion may be formed. Sometimes, chemical demulsifier is injected to maintain satisfactory desalter operation. Quantity of water added should not be <3%, otherwise desalting efficiency is reduced due to inability to contact all the droplets of high concentration brine. If too much water (>8%) is added, then current flow in the desalter will rise making its operation expensive. Pressure inside the desalter is kept at 2-3 kg/cm^2 to prevent vaporisation within the desalter. pH value in the desalter is maintained at 6.5 to 8.5 by caustic addition as the conductivity of fluid flowing between electrodes becomes high at either low or high value of pH and emulsion becomes hard to break due to formation of sodium naphthenates (formed by reaction of caustic and naphthenic acid in crude) which acts as an emulsion stabiliser.

A flow sheet of electrical desalting process of crude oil is given in Fig. 17.3.

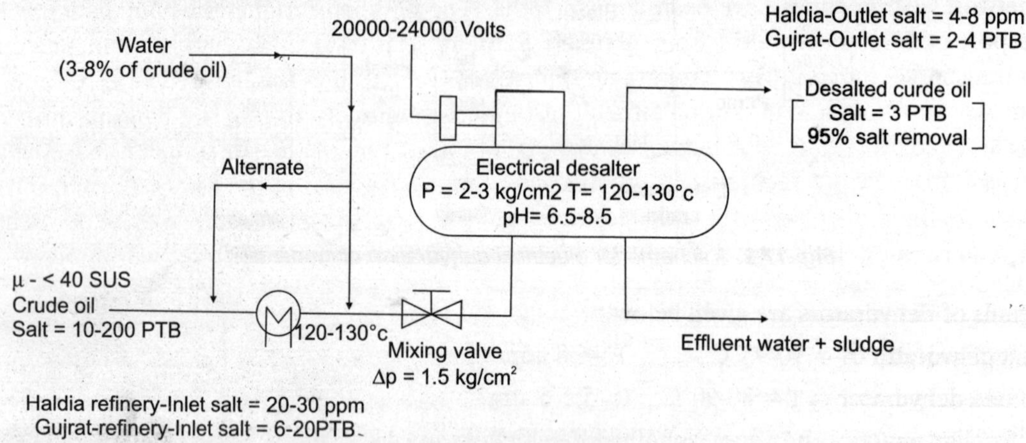

Fig.17.3: Electrical desalting process of crude oil

STABILISATION OF CRUDE OIL

Removal of dissolved gases from crude oil by heating it is called its stabilisation. Gas accompanied with crude oil must be removed to avoid breathing loss (loss of gasoline, if gas is not removed from oil during pre refining). Breathing loss occurs due to changes of humidity of ambient air and day and night temperature change also during emptying & filling of crude tanker. Breathing loss is about 0.4-0.75 kg/m^3 of tank volume per month.

A scheme for stabilisation of crude oil is shown in the flow sheet in Fig. 17.4.

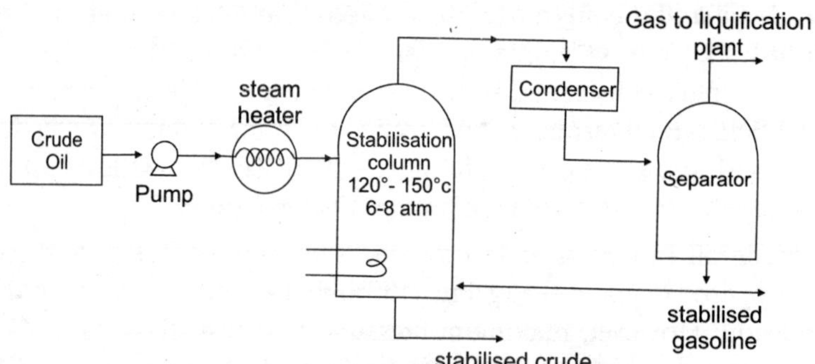

Fig.17.4: Scheme for stabilisation of crude oil

TRANSPORTATION OF CRUDE OIL

Most of the crude oil is transported from oil fields to refineries by pipelines. However, it is also transported by rail road (tank cars), tankers, tank truck and barges (small tankers). Underwater super tankers (in marine ships) are used for inter-continental transport of crude oil.

Trans-Arabian underground pipeline for crude transportation has a diameter of 75 cms equipped with 6000 hp turbine pumping 340000 bbl/day at a pressure of about 35-60 kg/cm^2.

The cheapest mode of transporting oil and gas is by taking them through pipelines. Several large networks of pipelines have been built in India. The Gauhati product pipeline, is designed to transport petroleum products from Gauhati Refinery. Rudrasagar-Gauhati-Barauni pipeline system transports crude oil from Rudrasagar oilfields to Gauhati and Barauni refineries. It brings oil from Assam through 1200 km of difficult terrain crossing many rivers. The Barauni refinery pipeline system is very long. It is in two sections, Barauni-Kanpur and Barauni-Haldia. Barauni to Kanpur is a products line, *i.e.* finished products are pumped from Barauni (Bihar) to Kanpur. At Haldia port, crude oil brought by tankers from abroad are pumped to Barauni. Gujarat pipeline system handles gas, crude oil and refined products. Cambay- Dhuvaran, Ankleshwar-Baroda and Ankleshwar-Uttaran lines are for gas ;Ankleshwar-Koyali line is for crude oil and Koyali-Ahmedabad line is a product line.

Transportation pipeline for crude oil ,natural gas & petroleum products in India. Underground pipeline for the transportation of crude oil and its products are laid in various parts of India. Middle east crude oil having pour point below 0°C and wax content upto 7% pose no problem in its transportation through pipeline at low temperature. Crude oil pipelines are buried 1-2 metres below the ground level, so that low night temperature in winter does not restrict the transportation of crude oil at economic pressure drop due to gel formation in waxy crude oil. Pour point depressants like alkyl acrylate polymers and alkylated polystyrene are added in the crude oil (300-600 ppm) during transportation to reduce its viscosity & pour point. Details of, crude oil and natural gas transportation pipelines in India, are given below in Table 17.1. .

Table 17.1: Crude Oil, Natural Gas and Petroleum Product Pipelines in India

Type of pipeline / Owner	**Length (in Kms.) as on 31st March**							
	2008	**2009**	**2010**	**2011**	**2012**	**2013**	**2014**	**2015**
(1)	**(2)**	**(3)**	**(4)**	**(5)**	**(6)**	**(7)**	**(8)**	**(9)**
Crude Pipelines (Onshore)								
ONGC	566	569	647	790	835	820	820	820
OIL	1193	1193	1193	1193	1193	1193	1193	1193
IOC	3984	4332	4332	4332	4342	4448	4448	4448
BORL (JV with BPCL)	-	-	-	-	937	937	937	937
HMEL (JV with HPCL)	-	-	-	-	1017	1017	1017	1017
CAIRN @			N.A.				670	670
Subtotal	5742	6094	6172	6315	8323	8414	9084	9084
Crude Pipelines (Offshore)								
ONGC	488	488	488	488	488	488	488	488
Subtotal	488	488	488	488	488	488	488	488
Total Crude Oil Pipelines	6230	6582	6660	6803	8811	8902	9572	9572

Gas Pipelines (Onshore)								
ONGC	-	-	-	-	-	53	53	53
IOC	-	-	-	132	132	134	134	140
GAIL	7001	7190	7803	8197	9454	10702	10909	10977
GSPL	N.A.	N.A.	N.A.	N.A.	N.A.	2163	2174	2192
RIL2	-	1480	1480	1480	1480	1480	1480	1480
AGC,DNPL	N.A.	N.A.	N.A.	N.A.	N.A.	N.A.	1000	1000
CAIRN @	N.A.	N.A.	N.A.	N.A.	N.A.	N.A.	592	592
Subtotal	7001	8670	9283	9809	11066	14532	16342	16434
Gas Pipelines (Offshore)								
ONGC	963	963	963	963	963	987	987	987
Subtotal	963	963	963	963	963	987	987	987
Total Gas Pipelines	7964	9633	10246	10772	12029	15519	17329	17421
Product Pipelines (Onshore)								
IOC	5286	5963	6175	6401	6401	6632	6632	6633
BPCL	1389	1389	1389	1389	1648	1648	1697	1935
HPCL	2134	2134	2134	2134	2164	2446	2446	2553
OIL	-	654	654	654	654	654	654	654
PCCK	293	293	293	293	293	293	293	293
PHMB	364	364	364	364	364	364	364	364
GAIL	1927	1964	1978	1978	2038	2038	2032	2032
Subtotal	11393	12761	12987	13213	13562	14074	14117	14463
Product Pipelines (Offshore)	-	-	-	-	-	-	-	-
Total Product Pipeline	11393	12761	12987	13213	13562	14074	14117	14463
Grand Total	25587	28975	29893	30788	34402	38495	41019	41456

1.: RIL's Gas Pipeline (Onshore) comprises 1375 KMs Trunk line + approx. 105 KMs spur line.

2: Total Product Pipeline includes 2334 KMs LPG Pipeline which 2032 KMs from Gail, 274 KMs from IOCL and 28KMs from BPCL.

@: Reported by DGH, Crude Pipeline (Onshore) 590 KMs Mainline + approx. 79 KMs spur line etc. & Crude Pipeline (offshore) 73 KMs mainline + 132 kms others and Gas Pipeline (onshore) 775 KMs others & 348 KMs mainline + 156 KMs others

Hazira, Bijapur, Jagdishpur (HBJ) natural gas pipeline passing through Gujarat, Madhya Pradesh, Rajasthan and Uttar Pradesh for transportation of natural gas from Gujarat (Hazira) oil field, feeds natural gas to some fertiliser plants and some thermal power plants enroute.This pipeline will be further extended to Haldia (West Bengal) .

CLASSIFICATION OF PETROLEUM

Depending on the ***nature of hydrocarbons*** present in it, crude petroleum oil is classified as :

- *paraffinic crude oil* (residue left on distillation is paraffin wax)
- *naphthenic crude oil* (residue left on distillation contains asphalt plus little of paraffin wax)
- *asphaltic (aromatic) crude oil* (residue on distillation contains asphaltic materials)
- *mixed crude oil* (containing all paraffinic, naphthenic, and aromatic constituents)

Indian crudes are mostly mixed crude oil. Naharkatia (Assam) crude contains 40% paraffins, 25% naphthenes and 35% aromatics.

Paraffins having general formula C_nH_{2n+2} are saturated hydrocarbons (where n = 1 to 35) *e.g.* methane (CH_4), propane (C_3H_3), pentane (C_5H_{12}) etc.

When n = 1 to 5, then constituents are gases *e.g.* C_4H_{10} (butane) when n = 5 to 15, then constituents are liquids, *e.g.* $C_{10}H_{22}$ (decane) & when n = 16 and above then constituents are solids/semisolids *e.g.* $C_{17}H_{36}$ (wax).

Naphthenes having general formula C_nH_{2n} are saturated ring compounds (where n = 1 to 9) *e.g.* C_4H_g (cyclo-butane), C_6H_{12} (Cyclohexane) etc.

CH2 CH2 CH2 CH2

CH2 CH2 CH2 CH2

CH2 CH2 Ch2 CH2

Cyclo-butane Cyclo-hexane

Another hydrocarbon group having the same general formula C_nH_{2n} but straight chain (instead of ring structure as in case of naphthenes) is called *olefin e.g.* ethylene (C_2H_4), propylene (C_3H_6) butylene (C_4H_g) etc.

Aromatics having general formula $C_6H_{2n\ 6}$ (n > 6) contain 6 carbon atoms in the form of a hexagon shaped ring and are unsaturated ring compounds *e.g.* benzene (C_6H_6), toluene ($C_6H_5CH_3$), naphthalene ($C_{10}H_g$), anthracene ($C_{14}H_{10}$).

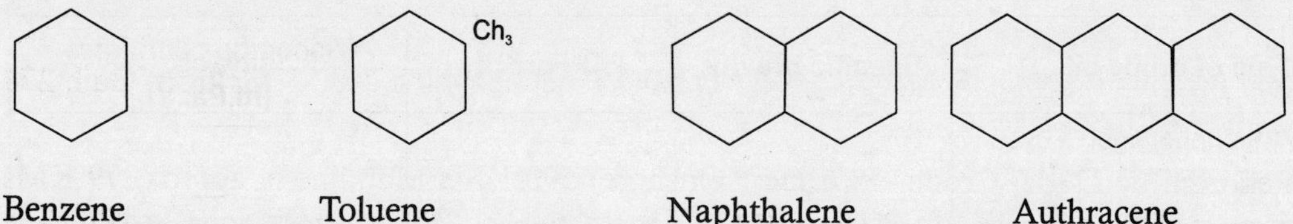

Benzene Toluene Naphthalene Authracene

Lower hydrocarbons in these groups are generally gases, intermediate ones are liquids and higher ones are solids/semi-solids.

The above ***group crude classification*** is based on the principle that the nature of the residue determines the nature of the whole crude which is of limited validity. Moreover the grouping of the bulk of the crudes under mixed base also gives an incomplete picture. These considerations have led to the development of ***Lane and Garton classification*** which is the most widely used in which two arbitrary fractions are recovered from the crudes on the basis of boiling range ; (a) 250-275°C at 1 atmosphere pressure and (6) 275-300°C at a pressure of 40 mm of mercury. The former is termed Key Fraction 1 and the latter Key Fraction 2. Each fraction is classified as paraffin, intermediate or naphthene according to specific gravity, and the crude is classified on the nature of the key fractions as given in Table 17.2. Following this system, a crude oil is called *wax free* or *wax bearing* according as the cloud point of the key fraction 2 is below or above (- 15°C) respectively. Lane and Garton have found that 85% of the crudes examined fell into the three simple classes *viz.* paraffin, intermediate or naphthene. Paraffin-naphthene or intermediate-naphthene crudes were not found to exist.

Lane and Garton classification is an improvement upon the four group classification in as much as it recognised that one type of hydrocarbon might predominate in one fraction and another might predominate in another fraction. However it characterizes the nature of only two fractions by the method of using specific gravity data. More detailed studies are necessary for the complete characterization of crudes.

Table 17.2: Lane and Garton Classification of Petroleum

Class	Key Fraction 1 Sp.gr. (15.5° /15.5° C)	Key Fraction 2 Sp.gr (15.5° / 15.5°C)
Paraffin	≤ 0.8251	≤ 0.8762
Paraffin-intermediate	-do-	0.934-0.8762
Paraffin-nephthene	-do-	≤ 0.934
Intermediate	0.8251-0.8602	0.934-0.8762
Intermediate-paraffin	-do-	≤ 0.8762
Intermediate-naphthene	-do-	≥ 0.934
Naphthene	≥ 0.8602	≥ 0.934
Naphthene-paraffin	-do-	≤ 0.8762
Naphthene-intermediate	-do-	0.934-0.8762

Another crude oil classification based on its specific gravity, API gravity and viscosity is as follows :

Type of crude oil	Specific gravity	°API gravity	Viscosity, centipoise (m.Pa. S)
Conventional or light crude oil	< 0.934	>20	—
Heavy crude oil	0.934-1.000	10 to < 20	10000 (maximum)
Extra heavy crude oil	> 1.0	< 10	10000 (maximum)
Tar sand bitumen or natural asphalt	> 1.0	< 10	> 10000

Crude oil classification using specific gravity, API gravity and viscosity is shown below diagramatically.

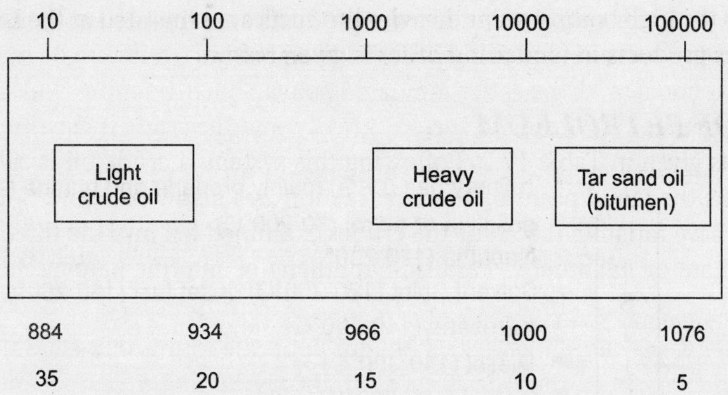

General classification of crude oil by viscosity,density and specific gravity

Classification of crude oil based on its correlation index (CI). Developed by U.S. Bureau of Mines, CI = 0 to 15 for paraffinic crude, CI = 15 to 50 for naphthenic or mixed crude and CI = > 50 for asphaltic (aromatic) crude. CI is defined as $CI = (48460/T_B) + 473.7\,G - 456.8$,

where, T_B - average boiling point, °R and G = specific gravity at 15.5°C.

Characteristics of Paraffinic and asphaltic crude oil. These two types of crudes yield products with different specification, properties and yields. The properties of naphthenic crudes are intermediate between that of the paraffinic and asphaltic crude. Table 17.3 gives the salient features of these two types of crudes.

Table 17.3: Characteristics of Paraffinic and Asphaltic Crude Oil

Properties and yields	Paraffinic crude	Asphaltic crude
Density	low	high
Yield of gasoline (petrol)	high	low
Octane No. (quality) of petrol	low	high
Sulphur content	low	high
Hydrogen/carbon ratio	high	low
Smoke point (quality) of Kerosene	high (better)	low
Cetane No. (quality) of diesel	high	low
Freezing point of fuel oil	high	low (better)
Yield of lubricants	high	low
Viscosity index of lubricating oil	high (better)	low
Yield of wax	high	low
Yield of bitumen/asphalt	low	high
Odour	sweet	pungent
Colour	light	dark

Constituents of Petroleum, Following commercial constituents of petroleum are recovered by its refining which has got different boiling ranges, indicated in parenthesis. The products which are lighter i.e. having low boiling point and molecular weight are separated from top of the distillation column whereas the high boiling point, heavier products are separated at the bottom. Boiling point ranges of various products in increasing order is given below.

REFINING OF PETROLEUM

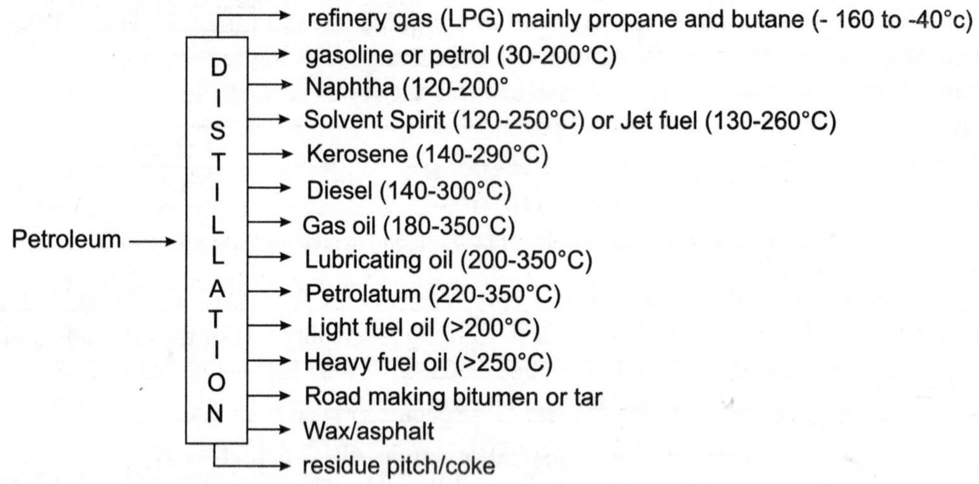

Major End uses of Petroleum Products are given below:

S.No.	Petroleum Products	Major Enduses
1.	LPG (Liquefied Petroleum Gas)	Domestic fuel. Also for Industrial application where technically essential. Now permitted as auto fuel, used for cutting/welding work in industries.
2.	NAPTHA/NGL	Feedstock/fuel for Fertilizer Units, feedstock for petro-chemical sector and fuel for Power Plants, used substitute fuel in steel plants for coke over gas.
3.	Petrol (Gasoline)	Fuel for passenger cars, taxies, two & three wheelers
4.	ATF (Aviation Turbine Fuel)	Fuel for aircrafts.
5.	SKO (Superior kerosene oil)	Fuel for cooking & lighting.
6.	HSD (High speed diesel)	Fuel for transport sector (Railways Road), Agriculture (tractors pump sets, threshers, etc.) and Captive Power Generation.
7.	LDO (Light diesel oil)	Fuel for agrucultural pump sets, small industrial units, start up fuel for power generation
8.	FO/LSHS (Furnace oil/Low Sulfur Heavy stock)	Secondary fuel for Thermal Power Plants, Fuel/ feedstock for fertilizer plants, industrial units.
9.	BITUMEN	Surfacing of roads.
10.	LUBES (Lubricating oils)	Lubrication for automotive and industrial applications.
11.	Other Minor Products (Benzen, Toluene, Paraffin Wax etc.	Feedstock for value added products.

Pre-treated crude oil is heated in pipes in a furnace and its vapors are passed through a tall cylinderical fractionating column which has a number of plates inside it. As the petroleum vapor moves upward, the higher boiling point fractions condense & fall back and only lower boiling point vapors move to higher plates where further condensation takes place and some more liquid falls back. As the freshly condensed liquid comes to the lower plates it gets heated & vaporised and again condenses on higher plates. This process of condensation and vaporisation takes place many times causing separation of the constituents of petroleum according to their boiling points. Thus, the higher boiling point fractions condense towards the lower part of the column and the lower boiling point fractions towards the upper part.

Normally, all the crude is not distilled at atmospheric pressure, otherwise the higher boiling hydrocarbons will thermally decompose (crack). The unvaporised crude oil removed from the bottom is distilled under vacuum at lower temperature to get higher boiling point products.

Heavier hydrocarbons fractions can be further converted into more useful lighter products by processes such as cracking, reforming, visbreaking etc.

General Scheme of Petroleum Distillation in a refinery is shown below :

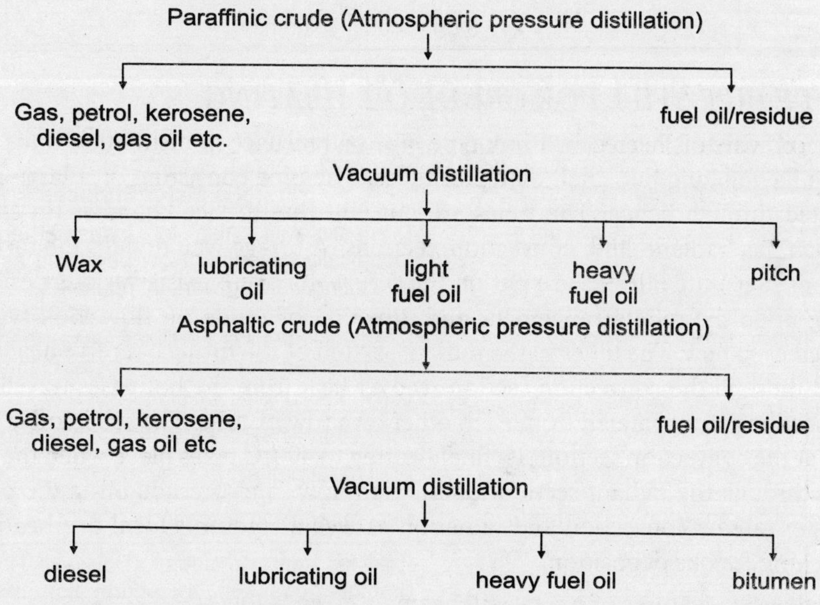

Primary distillation of petroleum (called *topping* of crude oil) is carried out first under *atmospheric pressure* and the residue (called reduced crude) is then distilled under vacuum. Maximum temperature upto which crude can be heated without cracking in primary distillation is 400°C. Steam is used for vaporisation of high boiling constituents. Heavier fractions boiling upto 600°C is recovered in vacuum distillation units.

Primary distillation recovers fractions upto gas oils while vacuum distillation recovers vacuum gas oil, raw lubricating oil distillates and asphalt or bitumen. The actual distillation practice varies from plant to plant depending upon type of crude oil and products demand. Different crudes yield fractions/products of different quality and quantity & their (product's) boiling range overlap each other. Primary distillation involves both high pressure (3-4 kgf/cm^2) and atmospheric pressure distillation. High pressure distillation produces gases and light distillates boiling upto 90°C and its bottom product is heated to 400°C and then sent to atmospheric pressure distillation until where naphtha, petrol, kerosene & gas oil are recovered. The bottom reduced crude (from atmospheric distillation column) is reheated to 400°C and sent to the vacuum (30-80 mm Hg) distillation column where heavy gas oil, raw lubricating oil, heavy fuel oil, bitumen, asphalt and wax are produced. Paraffin wax is made from vacuum gas oil while micro crystalline wax is produced/recovered from heavy lubricating oil distillates or residue from paraffinic crude oil. Lighter component impurities present in the side stream products are removed by steam-stripping.

Typical dimensions of a distillation column :

Diameter	= 3 metres	Height = 30—35 metres
Plates spacing	= 0.55 metre	
No. of plates between two neighbouring products stream :		
	= 3 to 5 (in atmospheric columns)	
	= 2 to 3 (in vacuum columns)	

BOX TYPE PIPE STILL FOR CRUDE OIL HEATING

Crude oil which is heated in a box type pipe still furnace in distillation units is a special type of furnace having long but small diameter tubes. The pipe consisting of a large number of tubes is connected through bends. The tubes are put into the furnace chamber (in single or multiple rows) which has 'radiant' and 'convection' sections. A bridge wall divides the furnace chamber in two sections. Separate burners are put on the side wall. There exists various designs of pipe stills depending upon the tube arrangement and direction of crude oil flow. Box type pipe stills are widely used even now. The tubes in the radiant section of the furnace receive heat directly from the flame and those in the convection section recover heat from the hot gases travelling to the stack. The two sections are separated from each other by a row of tube. The crude oil is pumped into the pipe inlet at the convection section. Its final heating to 400°C takes place when the preheated crude oil passes through the radiant section tubes. High flow rate of crude oil in the pipe ensures high heat transfer rate by convection and radiation as well as prevents local overheating which might cause cracking & coke deposition.

Typical design features of a pipe still heater is given below :

Input capacity	= 5 to 20 ×10^6 kcal/hr
Heat release rate	= 20000 to 45000 kcal/hr/m^3 of furnace volume
Heat flux	= 15000 to 40000 kcal/hr/m^2 of circumferential tube area
Proportion of heat transfer by radiation = 55 to 60%	

A box type pipe still is shown in Fig. 17.5.

Liquefied Petroleum Gas (LPG) - Domestic and industrial fuel. It is also called refinery gas. It is liquefied before use. It is mainly a mixture of propane and butane.

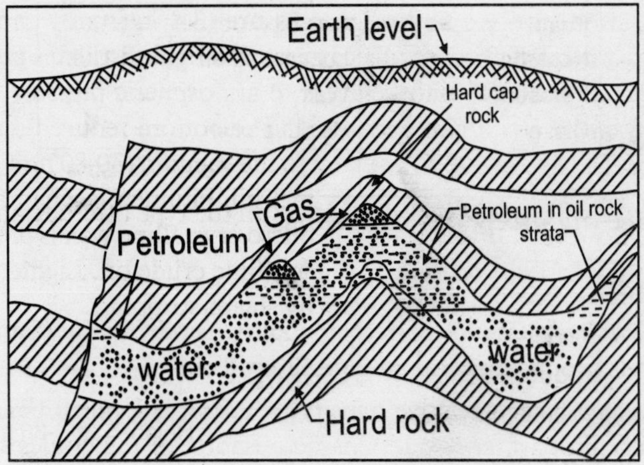

Fig. 17.5. Box type pipe-still.

USES OF PETROLEUM PRODUCTS

Gasoline (Petrol) - Fuel for spark ignition internal combustion engine (motor/aviation), dry cleaning of clothes.

Naphtha - Used as solvent, paint thinner, blending of motor fuel. Its most important use is in the production of H_2 (used in making nitrogeneous fertiliser) by its steam reforming. It is used as a fuel gas (naphtha vapor) in steel plants.

Jet Fuel - Fuel for jet plane, aviation turbine engines.

Kerosene — Domestic illuminant and tractor fuel.

Diesel - Fuel for diesel engines, (compression ignition engine)

Gas oil - It is gasified for fuel gas production. It is a fuel for industrial furnace. It is also used as blend for heavy fuel oil.

Lubricating oils - Used as a lubricant in machines and engines.

Petrolatum - It is also a lubricant used as a base material for grease manufacture.

Light fuel oil - Used as a fuel in industrial furnaces.

Heavy fuel oil - Used as furnace fuel after blending with light fuel oil or naphtha to reduce its viscosity and improve fluidity & atomisation characteristics. It is also cracked to produce gasoline, diesel, gas, light fuel oil etc.

Bitumen or Tar - Used for road making as binder and for moisture proof coating.

Wax - It is used for making candles, waxed papers and cartons, match stick coating, rust prevention etc.

Residue Pitch - For making road and as industrial fuel and electrodes.

Typical Products Yields in Crude oil Distillation. The yield of products of a typical crude oil distillation is given below (volume percent) :

Petrol	= 18%	Gas oil (Fuel oil)	= 12%
Naphtha	= 6%	Lubricating oil	= 12%
Kerosene	= 10%	Wax	= 9%
Diesel	= 8%	Residue pitch	= 25%
		Total	= 100%

This yield percentage will widely vary depending on the type of crude and refining operations.

INDIAN CRUDE OIL

The major *oil fields in India* are in Assam (Naharkatiya) and Gujarat (Cambay and Ankleswar). Besides, Bombay high is producing good amount of crude oil.

The *Naharkatiya crudes* have a relative density of 0.857, sulphur 0.17% & high wax content (7.8%) and contain very little water & salt. Characterisation of the crudes shows that they are of mixed nature.

The oil fields of India yield mixed base crudes. The crudes from the *Assam fields* has high aromatic content (20-25%) of gasoline, a common feature for Burma and other South- East Asian crudes. The yield of straight run gasoline is about 17-18% of the crude. The Assam crudes are also highly paraffinic containing high melting waxes but are deficient in good quality kerosene.

The *Cambay crude* has been found to be associated with substantial amounts of water, has a relative density of 0.863 sulphur content 0.16% and gives somewhat low yields of lighter distillates. The dehydrated oil has high wax content (10.6%). The oil is predominantly paraffinic in nature.

The *Ankleswar crude* is a light oil containing high amounts of lighter & middle boiling fractions and only negligible amount of water. In these respects, it differs very much from the Cambay oil though both are paraffinic base crudes. The Ankleswar crude has relative density of 0.803, sulphur 0.14% and wax 7%.

The crudes from the *Gujarat oil fields* give gasoline with a much smaller content of aromatic hydrocarbons (3-5%). These are very highly paraffinic and wax bearing. Straight run gasoline content from Cambay crude is low (9-10%). The yield of total distillates from this crude is much lower than from Naharkatiya crude. Gujarat crudes are nearer to the Middle East crudes than to the Assam crudes. The sulphur content of Indian crudes is low (0.15%).

Exploration for oil in all prospective sedimentary basins of India is continuing. The areas which are still under survey are : Kutch regions of Gujarat, Brahmaputra valley, plains of Punjab, Uttar Pradesh & Bihar and the adjoining Himalayan foot hills, Rajasthan, the Cauveri, Mahanadi & Godavari deltas, the Andaman and Nicobar Islands and offshore areas of Cambay, Kutch & Madras coast. Of the total requirement of crude in India, only about one-third is produced within the country and the rest is imported from the Middle Eastern countries (Iran, Iraq, Saudi Arabia, Kuwait, etc.) & from Indonesia in the East. The Government of India has a share in Rostam crude from its joint venture in the Persian Gulf. USSR also exports crude oil to India.

The crudes from the eastern oil fields in Assam resemble those from the Far Eastern countries (like Indonesia, Borneo, Burma) in aromatic contents. The Ankleswar crudes, on the other hand, are similar to those of Darius and Aghajari crudes of Iran and other light Arabian crudes. The Middle Eastern crudes vary in relative density from 0.825 to 0.953 and the sulphur contents may be as high as 4.7%. The Kuwait crude has a relative density of 0.871 and a sulphur content of 2.5%.

The total estimated reserve of crude in India is about 750 million tons spread over an area of 1.04×10^6 km^2. Exploration of oil in India is done mainly by Oil and Natural Gas Corporation Ltd. (ONGC) and Oil India Limited (OIL). Most of the refineries in Public sector are under Hindustan Petroleum, Indian Oil Corporation and Bharat Petroleum. Refineries are located at Gauhati (Assam), Barauni (Bihar), Baroda (Gujarat), Cochin, Haldia, Mathura, Bongaigaon (Assam), Madras, Mangalore ,Bina, Karnal (Haryana) Paradeep etc.

Some of the properties of three Indian crudes are tabulated in Table 17.4.

Table 17.4: Properties of Indian Crudes

Characteristics	Cambay (Gujarat)	Naharkatiya (Assam)	Bombay High off shore
Specific gravity (15.5°C/15.5°C)	0.8638	0.8574	0.8324
Water content % volume	2.60	0.06	—
Salt content, % weight	0.014	nil	0.05
Sulphur content, % weight	0.13	0.017	5.10
Kinematic viscosity at 37.8°C, CSt	35.6	4.91	—
Calorific Value, kcal/kg	10,800	–	27
Pour Point °C	35	30-32	
Distillation characteristics :			
I.B.P.,"C			
Upto 150°C, % volume	62.5	41.0	38
150—300°C, % volume	9.2	23.0	22
300-465°C, % volume	26.5	32.5	33
	35.25	32.2	–
Total upto 465°C, % volume	70.95	87.7	–
Fraction, upto 150°C :			
Sp. gr. (15.5715.5°C)	0.7920	0.8516	–
Fraction, 150-300°C :			
Sp. gr., (15.5715.5°C)	0.7195	0.7471	–
Fraction 300-450°C :			
Sp. gr. (15.5715.5°C)	0.8698	0.9273	–
Sulphur, % weight	0.05	0.24	–
Wax content, % weight			–
Short residue (7465°C) :	32.2	24.4	–
		1	
Sp. gr. (15.5715.5°C)	0.9554	0050	
Penetration number	50	16	–
Softening Point (R & B), °C	55	67	–
Wax (300-465°C) on the basis of crude, % weight	10.5	7.8	12

India's requirements of crude oil is more than its refining capacity and production hence, the balance is to be imported. India has gone for collaboration with Iran in the exploration & production of crude oil there leading to the production of oil in Rustam oil fields and discovery of a new oil field at Rakhash in the off shore areas of Persian Gulf. Crude oils are imported from middle east countries, Russia, Indonesia etc. Existing oil refineries can meet the present requirements of all products except kerosene and lubricating oils. Lubricating oils are made in the refineries at Digboi, Barauni, Madras, Haldia etc.

CRUDE OIL REFINNING IN INDIA

To meet the growing demand of petroleum products, the refining capacity in the country has gradually increased over the years by setting up of new refineries in the country as well as by expanding the refining capacity of the existing refineries. The country is not only self sufficient in refining capacity for its domestic consumption but also exports petroleum products substantially .

STATUS ON INDIAN PETROLEUM & NATURAL GAS SECTOR DURING 2014-15

India has total reserves of 763.476 MMT(Million Metric Ton) of Crude Oil and 1488.73 BCM(Billion Cubic Metre) of Natural Gas as on 01.04.2015.

Crude Oil production during 2014-15 is 37.461 MMT which is 0.87% lower than 37.788 MMT produced during 2013-14.

Gross Production of Natural Gas during 2014-15 is 33.656 BCM which is 4.94% lower than the production of 35.407 BCM during 2013-14 .

- The flaring of Natural Gas in 2014-15 is 2.59% of gross production.
- The refining capacity stood at 215.066 MMTPA(Million Metric Ton Per Anum) as on 01.04.2015. By the end of 12th Five Year Plan, refinery capacity is expected to reach 307.366 MMTPA.
- The Crude processed during 2014-15 at 223.242 MMT is 0.33% higher than 222.497 MMT(Million Metric Ton) crude processed in 2013-14 and the pro-rata capacity utilization in 2014-15 was 103.80% as compared to 103.46% in 2013-14 .
- The production of petroleum products during 2014-15 was 221.059 MMT registering an increase of 0.17% over last year's production at 220.756 MMT.
- The country exported 63.928 MMT of petroleum products against the imports of 20.423 MMT during 2014-15.
- The sales/consumption of petroleum products during 2014-48 was 164.987 MMT (including sales through private imports) which is 4.15% higher than the sales of 158.407 MMT during 2013-14.
- The total number of retail outlets of Oil Marketing Companies as on 31.03.2015 has gone up to 53419 from 51868 on 31.03.2013.

The total number of LPG consumers of Public Sector Oil Marketing Companies as on 31.03.2015 were 18,19,02,266 against 16,62,58,759 as on 31.03.2014.

Table below gives the trend in Availability of Crude Oil, Petroleum Products and Natural Gas in India

Year	Crude Oil*			Petroleum Products*			Natural Gas**		
	Production	Net Imports	Gross Availability	Production @	Net Imports	Gross Availability	Production	Net Imports	Gross Availability
1	2	3	4=2+3	5	6	7=5+6	8	9	10=8+9
2005-06	32.19	99.41	131.60	124.41	-10.02	114.39	31.33	5.06	36.39
2006-07	33.99	111.50	145.49	139.75	-15.96	123.78	30.79	6.81	37.60
2007-08	34.12	121.67	155.79	149.47	-18.38	131.10	31.48	8.32	39.80
2008-09	33.51	132.78	166.28	155.15	-20.38	134.77	32.84	8.06	40.90
2009-10	33.69	159.26	192.95	184.61	-36.31	148.30	47.50	9.15	56.65
2010-11	37.68	163.60	201.28	194.82	-42.26	152.56	52.21	9.93	62.75
2011-12	38.09	171.73	209.82	203.20	-44.99	158.21	47.56	13.21	60.77
2012-13	37.86	184.80	222.66	217.74	-47.63	170.10	40.68	13.14	53.82
2013-14	37.79	189.24	227.03	220.76	-51.15	169.61	35.41	13.03	48.44
2014-15(p)	37.46	189.43	226.89	221.14	-43.51	177.63	33.66	15.47	49.13
Growth rate of 2014-15 over 2013 14(%)	-0.87	0.10	-0.06	0.17	-14.95	4.73	-4.93	18.73	1.42

P-provisional

*-Million tons

**-Billion cubic metre

@-Excludes LPG production from natural gas

Table below gives the petroleum refining capacity in India in MMTPA.

(Note : Paradip refinery has also been commissioned in September 2016 with 15.0 MMTPA capacity.)

Sl. No.	Refinery Location & Year of Commissioning	Capacity as on 1st April (MMTPA)						
		2009	2010	2011	2012	2013	2014	2015
(1)	(2)	(3)	(4)	(5)	(6)	(7)	(8)	(9)
A	Public Sector	105.456	112.886	116.886	120.066	120.066	120.066	120.066
	Indian Oil Corporation Ltd.	49.700	51.200	54.200	54.200	54.200	54.200	54.200
1	Digboi, Assam - 1901	0.650	0.650	0.650	0.650	0.650	0.650	0.650

2	Guwahati, Assam - 1962	1.000	1.000	1.000	1.000	1.000	1.000	1.000
3	Barauni, Bihar - 1964	6.000	6.000	6.000	6.000	6.000	6.000	6.000
4	Koyali, Gujarat - 1965	13.700	13.700	13.700	13.700	13.700	13.700	13.700
5	Haldia, West Bengal - 1975	6.000	7.500	7.500	7.500	7.500	7.500	7.500
6	Mathura, Uttar Pradesh - 1982	8.000	8.000	8.000	8.000	8.000	8.000	8.000
7	Panipat, Haryana - 1998	12.000	12.000	15.000	15.000	15.000	15.000	15.000
8	Bongaigaon, Assam - 1974	2.350	2.350	2.350	2.350	2.350	2.350	2.350
	Bharat Petroleum Corporation Ltd.	19.500	21.500	21.500	21.500	21.500	21.500	21.500
9	Mumbai, Maharashtra - 1955	12.000	12.000	12.000	12.000	12.000	12.000	12.000
10	Kochi, Kerala - 1963	7.500	9.500	9.500	9.500	9.500	9.500	9.500
	Hindustan Petroleum Corporation Ltd.	13.000	13.800	14.800	14.800	14.800	14.800	14.800
11	Mumbai, Maharashtra - 1954	5.500	5.500	6.500	6.500	6.500	6.500	6.500
12	Visakh, Andhra Pradesh - 1957	7.500	8.300	8.300	8.300	8.300	8.300	8.300
	Chennai Petroleum Corporation Ltd.	10.500	11.500	11.500	11.500	11.500	11.500	11.500
13	Manali, Tamil Nadu - 1965	9.500	10.500	10.500	10.500	10.500	10.500	10.500
14	Narimanam, Tamil Nadu - 1993	1.000	1.000	1.000	1.000	1.000	1.000	1.000
	Numaligarh Refinery Ltd.	3.000	3.000	3.000	3.000	3.000	3.000	3.000
15	Numaligarh, Assam	3.000	3.000	3.000	3.000	3.000	3.000	3.000
	Oil & Natural Gas Corporation Ltd.	0.066	0.066	0.066	0.066	0.066	0.066	0.066
16	Tatipaka, Andhra Pradesh - 2001	0.066	0.066	0.066	0.066	0.066	0.066	0.066
	Mangalore Refinery & Petrochemical Ltd.	9.690	11.820	11.820	15.000	15.000	15.000	15.000
17	MRPL, Mangalore- 1996	9.690	11.820	11.820	15.000	15.000	15.000	15.000
B	Private Sector	70.500	70.500	70.500	78.000	80.000	80.000	80.000
	Reliance Industries Ltd.	60.000	60.000	60.000	60.000	60.000	60.000	60.000
18	Jamnagar, Gujarat - 1999	33.000	33.000	33.000	33.000	33.000	33.000	33.000
19	Jamnagar (SEZ), Gujarat - 2008	27.000	27.000	27.000	27.000	27.000	27.000	27.000
	Essar Oil Ltd.	10.500	10.500	10.500	18.000	20.000	20.000	20.000
20	Vadinar, Gujarat - 2006	10.500	10.500	10.500	18.000	20.000	20.000	20.000
C	Joint Venture	-	-	-	15.000	15.000	15.000	15.000

	Bharat Oman Refinery Ltd.	-	-	-	6.000	6.000	6.000	6.000
21	Bina, Madhya Pradesh - 2011	-	-	-	6.000	6.000	6.000	6.000
	HPCL Mittal Energy Ltd.	-	-	-	9.000	9.000	9.000	9.000
22	Bathinda, Punjab - 2012	-	-	-	9.000	9.000	9.000	9.000
	Total (A+B+C)	175.956	183.386	187.386	213.066	215.066	215.066	215.066

Import of Crude Oil:

Import of Crude Oil during 2014-15 was 189.432 MMT valued at Rs. 6,87,350 crore which marked a decrease of 0.10% in quantity terms and 20.53% decrease in value terms over the same period of last year. The decline in value terms is related to reduced crude oil prices by around 50% from June'14 to Dec'14. During the year 2013-14 the import of crude oil was 189.238 MMT valued at Rs. 8,64,875 crore.

Table : Import of Crude Oil and average Crude Oil Prices

Year	Imports of Crude Oil (MMT)	% Growth in Imports of Crude Oil	Average Crude oil Prices (US$/bbl.)	% Growth in Average Crude oil Prices
2009-10	159.259	19.95	69.760	-16.53
2010-11	163.595	2.72	85.090	21.98
2011-12	171.729	4.97	111.890	31.50
2012-13	184.795	7.61	107.970	-3.50
2013-14	189.238	2.40	105.520	-2.27
2014-15*	189.432	0.10	84.200	-20.20

: Imports value is Provisional and Crude price is on Average basis in $/bbl.

Imports & Exports of Petroleum Products:

During 2014-15 imports of petroleum products were 20.423 MMT valued at Rs. 72,778 crore which shows an increase of 22.16% in quantity terms and 2.45% decrease in value terms against the same period of previous year. The quantity of petroleum products imported during 2013-14 was 16.718 MMT valued at Rs. 74,605 crore.

During April-March, 2014-15 the exports of petroleum products were 63.928 MMT valued at Rs. 2,88,563 crore which shows a decrease of 5.80% in quantity terms and 21.65% decrease in value terms against the exports of 67.864 MMT valued at Rs. 3,68,279 crore for the same period of last year(2013-14).

During April-March, 2014-15 the import of LNG is 15.470 MMT valued at Rs. 63,110 crore which marked an increase of 18.71% in quantity terms and 18.39% increase in value terms against imports of 13.032 MMT of LNG, valued at Rs. 53,307 crore was imported for the same period of previous year

The trends in quantity of petroleum products and LNG imports & exports are depicted below in Table.

Table : Imports & Exports of Petroleum Products

Year	Imports of Petroleum Products (MMT)	% Growth in Imports of Petroleum Products	Exports of Petroleum Products (MMT)	% Growth in Export of Petroleum Products	Imports of LNG (MMT)	% Growth in Imports of LNG
2009-10	14.665	-21.10	51.155	31.35	9.148	13.50
2010-11	17.379	18.51	59.077	15.49	9.931	8.56
2011-12	15.849	-8.80	60.837	2.98	13.214	33.06
2012-13	15.774	-0.48	63.408	4.23	13.136	-0.60
2013-14	16.718	5.99	67.864	7.03	13.032	-0.79
2014-15*	20.423	22.16	63.928	-5.80	15.470	18.71

*: Provisional

Production and Consumption of Petroleum Products

The production of petroleum products is targeted at 220.964 MMT in year 2015-16 as against 221.136 MMT achieved in 2014-15. During April-December, 2015, production of petroleum products was at 170.345 MMT i.e. an increase of 2.98% over 165.412 MMT of production achieved during the corresponding period last year.

During the year 2015-16, the consumption of petroleum products in India is estimated at 176.972 MMT with a growth of 6.92% as compared to consumption of 165.520 MMT during 2014-15. The consumption of petroleum product during April-December, 2015 is 134.310 MMT i.e. an increase of 9.50% over 122.659 MMT in April-December, 2014.

Year-wise production and consumption of petroleum products since 2010-11 to 2015-16 (April-December) are tabulated below .

Table-: Production and Consumption of Petroleum Products

Year	Production of Petro-Products (MMT)	% Growth in Production of Petro-Products	Consumption of Petro-Products (MMT)	% Growth in Consumption of Petro-Products
2010-11	194.821	5.53	141.040	2.35
2011-12	203.202	4.30	148.132	5.03
2012-13	217.736	7.15	157.057	6.02
2013-14	220.756	1.39	158.197	0.73
2014-15	221.136	0.17	165.520	4.63
2015-16**	220.964	-0.08	176.972	6.92
2014-15 (Apr-Dec)	165.412	-	122.659	-
2015-16 (Apr-Dec)*	170.345	2.98	134.310	9.50

*: Provisional **: Targeted Production

Notes: 1. Production of petroleum products includes Production of Petroleum Products from fractionators.

2. Consumption of Petroleum Products excludes refinery fuels but includes imports.

Table :Oil refining capacity of Major countries(MMTPA)

Country	Qty (MMT)*
US	885.91
Canada	97.87
Mexico	75.78
Argentina	30.97
Brazil	111.28
Netherlands	15.93
Antilles	64.88
Venezuela	79.12
Other S. & Cent.	38.64
America	68.46
Belgium	102.60
France	24.80
Germany	98.79
Greece	63.46
Italy	15.74
Netherlands	315.60
Norway	77.00
Russian Federation	21.71
Spain	30.52
Sweden	68.10
Turkey	
United Kingdom	
Other Europe & Eurasia	255.94
Iran	98.84
Iraq	54.48
Kuwai	46.61
Saudi Arabia	40.52
United Arab Emirates	56.92
Other Middle East	72.10
Africa	176.92
Australia	26.70
China	702.00
India	230.07
Indonesia	54.53
Japan	186.70
Singapore	75.39
South Korea	143.76
Taiwan	59.61
Thailand	61.82
Other Asia Pacific	90.85
* for 2014	

Note: MMTPA-Miliion Metric Tons Per Annum

Table : International Petroleum Statistics

Item Unit	2008	2009	2010	2011	2012	2013	2014*
(1) (2)	(3)	(4)	(5)	(6)	(7)	(8)	(9)
1. World Reserves :							
Oil & Gas at year end							
(i) Crude Oil Tonne	170.8	181.7	221.3	234.3	235.8	238.2	239.8
of which OPEC"	129.8	145.3	159.2	168.4	169.9	170.2	170.5
(ii)Natural Gas Tr.Cu.ft.	6534.0	6621.0	6925.2	7360.9	6614.1	6557.8	6606.4
of which OPEC"	3295.0	3335.0	3312.5	3333.3	3360.1	3372.174	3375.549
2. World Prod. of Oil & Gas							
(i) Crude Oil Mn. Tonne	1991.6	3885.8	1975.4	4008.1	4116.4	4126.6	4220.6
of which OPEC"	1746.0	1622.6	1667.0	1705.3	1779.0	1734.4	1729.6
(ii)Natural Gas Net Bn.Cu. Mtr.	3073.4	2989.4	3202.6	3315.7	3380.2	3408.8	3460.6
of which OPEC"	572.3	580.7	580.4	618.6	657.0	0.0	0.0
3. World Oil Refinery							
(i) Refinery capacity Mn. Tonne/	4423.9	4496.7	4543.0	4565.4	4623.6	4725.579	4790.955
at year end Year							
(ii)Crude throughput"	3683.2	3597.6	3700.7	3715.4	3751.0	3761.818	3813.99
4. World Oil Consumption							
(i) World Total Mn. Tonne	3999.0	3922.9	4041.8	4085.4	4133.2	4179.1	4211.1
(ii) Main consuming area"	2296.4	2244.3	2348.3	2380.2	2429.0	2458.1	2489.1
(iii) Consumption in main							
areas as % of world %	57.4	57.2	58.1	58.3	58.8	58.8	59.1
5. World Imports/							
(i) Crude Oil Mn.Tonne	1970.0	1892.5	1875.8	1894.7	1927.3	1878.5	1876.4
(ii) Pol. Products"	728.0	714.0	757.7	790.7	801.8	897.3	911.5
Total"	2698.0	2606.5	2633.5	2685.4	2729.1	2775.8	2787.9
6. Export of Crude Oil/							
Pol. from OPEC							
(i) Crude Oil"	1196.4	1110.8	1150.7	1168.0	1258.7	1177.2	1127.4
(ii) Pol. Products"	209.6	201.3	181.3	211.6	217.5	230.9	223.9
Total"	1406.1	1312.1	1331.9	1379.6	1476.2	1408.0	1351.3
7. World Primary Energy							
Consumption Mn. Tonne							
(i) Oil Oil Eqv.	3999.0	3922.9	4041.8	4085.4	4133.2	4185.1	4211
(ii) Natural Gas"	2717.7	2655.7	2864.1	2914.2	2986.3	3020.4	3065.5

(iii)Coal"	3256.3	3238.7	3464.0	3628.8	3723.7	3826.7	3881.8
(iv) Hydro-Electric "	727.6	737.7	782.1	794.7	833.6	855.8	878.9
(v) Nuclear Energy"	619.0	614.1	626.4	600.4	559.9	563.2	574
(vi) Renewables"	123.2	142.0	168.6	205.6	240.8	279.3	316.8
Total"	11442.8	11311.1	11947.0	12229.0	12477.5	12730.5	12928.0

@ : Angola & Ecuador is not included.

Note: ! : Includes USA, Canada, Brazil, Germany, Japan, China, Russia, Saudi Arabia, India and South Korea.

Source : (i) BP Statistical Review of World Energy , June-2015 (ii) OPEC Annual Statistical Bulletin.

18

CRUDE OIL DISTILLATION

The process of distillation is the most common method adopted for separating the constituents of crude oil into compounds or groups of compounds having industrial uses.

Crude oil consists of a complex mixture of hydrocarbons widely differing in boiling points. Distillation is done to separate the crude oil into the basic fractions like motor gasoline, kerosene, gas oil and fuel oil. The order in which these components emerge as crude oil is distilled is given in Table 18.1.

Table 18.1: Components of Crude Oil Distillation

Components	Approximate Boiling Range, °C	Uses
Gases	-162 to -42	Domestic cooking, Heating furnaces in refinery plants.
Gasoline/light naphtha	-1 to 150	Motor Fuel (straight run gasoline)
Heavy naphtha	150-205	Catalytic reformer feed, jet fuel, solvent
Kerosene	205-250	Illuminant and fuel.
Light Gas Oil	250-315	High speed diesel oil, fuel oil, absorbent
Heavy Gas Oil	315-430	Catalytic cracking feed.
Vacuum gas oil	430-540	Lubricating oil, catalytic cracking feed
FuelOil/residuum	> 595	Fuel for furnaces, asphalt, bitumen, coke making feed

The proportions of these fractions vary considerably with the source of crude oil but each fraction is identified by its boiling range which overlap considerably and hence nomenclature of fraction also changes accordingly.

The crude oil is pumped from tankers to the crude oil tanks where it is allowed to settle and the water is drained off. Water associated with crude is highly objectionable as it gives rise to sudden pressure changes in the plant, deposits salts on equipment where it vaporizes off, dissolves harmful acids and causes deterioration of equipment by way of corrosion.

The crude oil is pumped from the tanks to the primary column of the distillation unit by centrifugal pumps. The pressure at the discharge of these pumps is maintained high enough to force the crude oil through preheat exchangers and a preheat furnace before it enters the column.

All oil streams from the columns like kerosene, gas oils and residue are at high temperatures and their heat is transferred to the incoming crude oil in heat exchangers. This effects a great economy in the process as this heat would otherwise have to be obtained by burning fuel oil or gas.

Bulk of the heat to the crude oil is given by oil or refinery gas burnt in tube-still heaters. Each heater has a convection section in which the oil is gradually heated by the fuel gas and a radiant section where the tubes are exposed to radiant heat and the oil is brought upto the required temperature. The oil flows through straight tubes joined at their ends by expanding them into box headers.

The air for combustion is provided by a blower. A tall stack sucks the hot gases and discharges them into the atmosphere. The heater outlet temperature is maintained at the required level by varying the quantity of gas or oil burnt as fuel. The air to the furnace is regulated by dampers. The furnaces are operated at a temperature which would not lead to deterioration of brick work or excessive coke formation in the tubes. Emergency valves are provided to drop out the contents of the furnace tubes quickly, if the flow of oils fails for any reason. Stagnant oil in the tubes will be converted into coke, if the heating medium is not cut out promptly.

The crude oil, partly vaporized, enters the primary tower which is a conventional distillation column with a number of plates. Gasoline and gas rise to the top, are condensed in the overhead condenser and flow into an accumulator. The heavier components such as gas oil and residue fall to the bottom of the column. In order to provide sufficient liquid flow down the column, to effect fractionation, a part of the condensed liquid from the overhead accumulator is pumped back to the column, as reflux. Within certain limits, greater the reflux, better is the quality of the overhead product. Sea water (salty water) is used to condense the overhead gasoline vapours which is in abundance, where the refineries are situated on the sea shore. To minimise corrosion by sea water, aluminium brass tubes are used. In places where water is fresh, direct contact condensers are used. This type is, however unsuitable with salty water, because the latter does not separate very well from oils. The primary tower is operated under pressure to retain and dissolve substances like butane which improve the antiknock value of gasoline.

The straight run gasoline from the primary tower contains certain very low boiling compounds which render it unsuitable for storage under atmospheric conditions. The stabilisation of gasoline is effected by preheating and fractionating it under pressure to get a product with the desired vapour pressure. A portion of the kerosene is drawn off as a side stream from the primary tower. The maximum temperature upto which crude oil can be heated without cracking is 400°C, which sets the limit of the maximum amount of distillate that can be collected by atmospheric distillation.

The reduced crude at the bottom of the primary tower consists of kerosene, gas oils and residue. It is pumped through a furnace into the secondary tower. The components of reduced crude are all high boiling materials and would require very high temperature to vaporise them. High temperature will destroy the quality of oil by cracking it and will prove very corrosive to the material of the tower itself. The boiling points are therefore lowered by applying a vacuum to the tower by the use of steam jet ejectors. Operation at a pressure of 300 mm. Hg absolute lowers boiling points well enough to vaporize all the oils with a heater outlet temperature of approximately 320°C.

Kerosene and light & heavy gas oils are drawn off as side streams. The last two liquids still contain some low boiling material which lower the flash point and render them unsafe for storage.

They are led into columns called strippers where steam or oil heat is applied to drive away the light material. Steam reduces the partial pressure of oil vapour and causes the low boiling components to be driven off.

The residue from the secondary tower is pumped away to storage or to the feed preparation unit where it is split into asphalt for bitumen manufacture and waxy distillate for catalytic cracker feed.

A typical distillation operation carried out in an Indian oil refinery is shown in the Fig. 19.1.

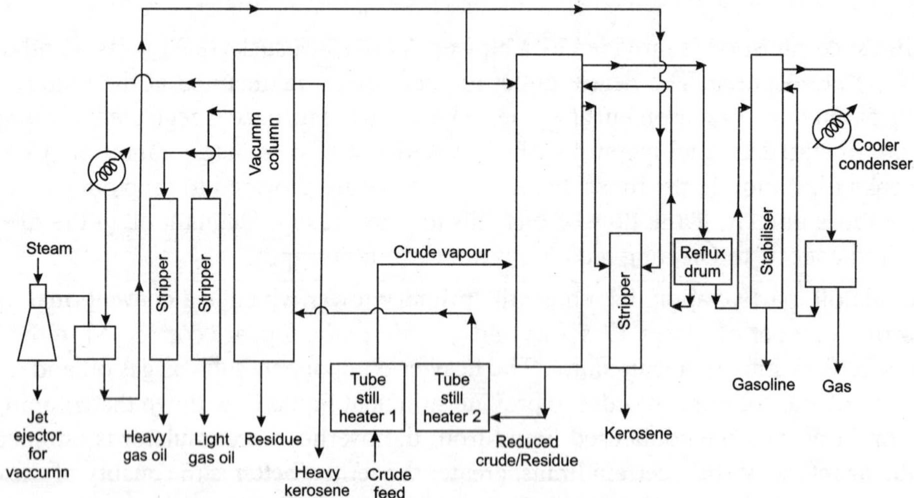

Fig. 18.1. Flow-sheet of a Typical Indian Oil Refinery for Crude Distillation.

Various crude oil distillation systems. There are four ways in which the crude oil is distilled *i.e.*

- *Shell still distillation* (Obsolete)
- *Single stage distillation unit* (Atmospheric pressure)
- *Two-stage distillation unit* (Atmospheric and high pressure units)
- *Three-stage distillation unit* (High pressure, atmospheric Pressure and vacuum units)

Shell still distillation of crude oil. It is the oldest and obsolete method of crude distillation in which distillation is accomplished in a battery of shell stills as shown in the Fig. 18.2.

Fig. 18.2. Battery of shell stills.

Preheated crude is charged continuously to horizontal, cylinderical stills which are externally fired. The lightest product (having lowest boiling point and molecular weight) is taken off at the first still *(i.e.* light naphtha) and the remaining liquid passes on to the following stills where the process is repeated, successively higher boiling point products being obtained in each still which is equipped with bubble cap columns to improve separation of products with overlapping boiling points.

Single stage atmospheric distillation unit. A single-stage crude distillation is shown in Fig. 19.3. The crude feed is preheated by the outgoing streams and then enters a direct fired furnace-type heater (pipe still). The materials separate in the distillation column according to their boiling points, the lowest boiling fraction leaving at the top of the column. Desired products may be withdrawn as side streams at appropriate points on the column. The side streams are further fractionated in small columns called strippers. Here, steam is used to free the cut from its more volatile components so that the initial boiling point of the products can be adjusted to its desired value. It is also called straight run distillation of crude oil.

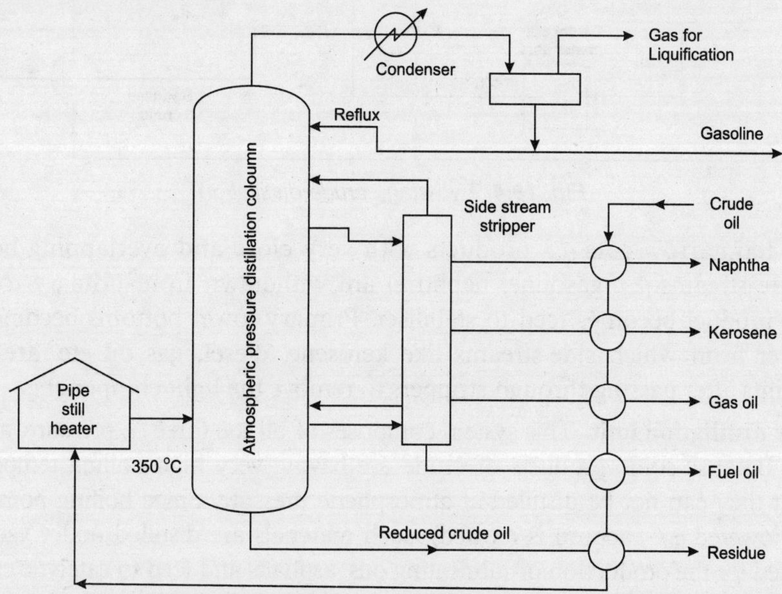

Fig. 18.3. Single-stage crude-refining process.

Typical yield (weight %) from atmospheric distillation of crude oil is (gas + LPG + losses) = 2 to 3%, (gasoline + naphtha) C_5- 145°C = 8 to 15%, Kerosene (145 - 250°C) = 15 to 20%, High speed diesel oil (270 - 370°C) = 20 to 35%, reduced crude (> 370°C) = 30 to 45%.

However capacity of single stage distillation unit for processing crude is limited which calls for addition of pressure units. Moreover, high boiling point products can not be separated as operating it at higher temperature may result in cracking of products. Therefore vacuum units are added to get the high boiling point products vaporised and separated at lower temperature.

This is also called crude *topping* or atmospheric distillation of crude as the pressure in the distillation column is slightly above atmospheric.

Two stage distillation unit. This scheme of distillation includes a *primary tower* which operates at about *3-4 atmospheric pressure* and a *secondary tower* operating at *atmospheric pressure* together with stabiliser. This scheme is used when crude is to be separated

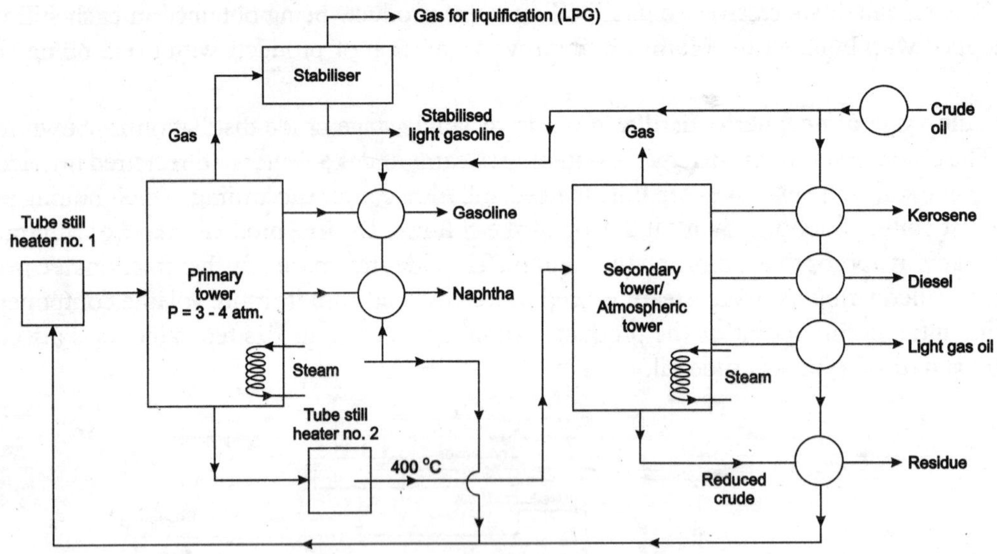

Fig. 18.4. Two-stage crude distillation

into six to ten narrow cuts (*i.e.* products with very close and overlapping boiling points). Two/ three side streams (*e.g.* gasoline, naphtha) are withdrawn from primary tower while the overhead (gas) product becomes feed to stabiliser. Primary tower bottoms becomes the feed for secondary tower from which side-streams like kerosene, diesel, gas oil etc. are withdrawn at appropriate points after passing through strippers to remove the light components.

Three stage distillation unit. This system comprises of all the three *i.e.* pressure, atmosphere and vacuum tower. Because some products of crude are having very high boiling temperature or heat sensitive so that they can not be distilled at atmospheric pressure. Since boiling point is reduced as the pressure is lowered (*i.e.* vacuum is created), such materials are distilled under vacuum. Vacuum distillation is used for the production of lubricating oils, asphalts and feed to catalytic cracking unit (*i.e.* heavy gas oil). Hence, a vacuum tower is added to a two stage distillation unit for the production of above products thus making it a three stage unit. Light vacuum gas oil (LVGO) is a blending component for light diesel oil (LDO) and a feedstock for catalytic cracker/hydrocracker. Heavy vacuum gas oil (HVGO) is a feedstock for visbreaker and catalytic cracker/hydrocracker. Vacuum residue is used for bitumen production as well as is a feed component for visbreaker.

The residue from the atmospheric tower is fed at about 425°C to the vacuum tower, where pressure is maintained at 80-110 mm Hg absolute. The overhead gas oil fraction is removed and the side streams are withdrawn through strippers. The bottom product is tar or asphalt.

Vacuum distillation units are similar to atmospheric crude distillation unit except that their diameter is larger due to larger crude vapor volumes at lower pressure (vacuum). 10 metre diameter vacuum columns are in operation, steam jet ejector or barometric condensers are used for maintaining desired vacuum in the vacuum tower.

A flow sheet of three stage crude distillation is shown in Fig. 18.5.

Pressurised units can handle larger throughput of crude though the pumping cost will be higher and cost of material of construction will be more.

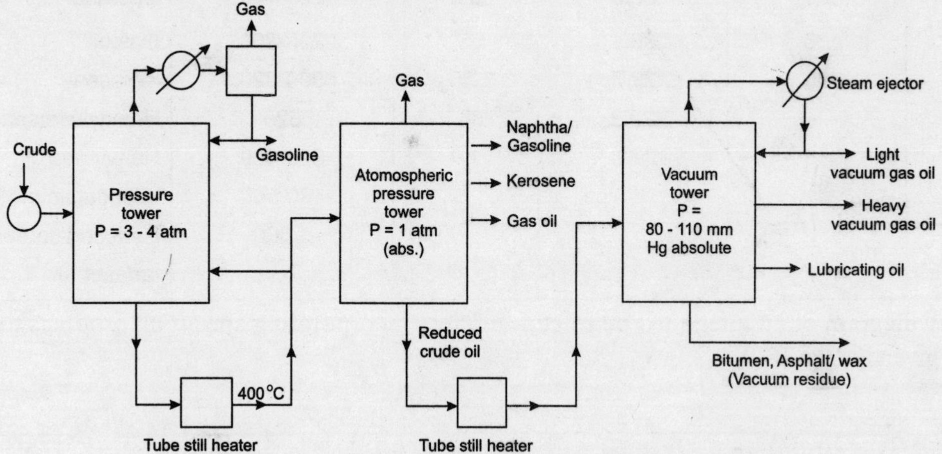

Fig. 18.5. *Three stage processing of crude-General scheme.*

A detailed typical layout scheme for a 2 million tons per year crude processing unit is shown in the Fig. 19.6 which includes all the Three stages (pressure, atmospheric and vacuum) of distillation. Tower details have been given and the products are designated by their boiling point ranges.

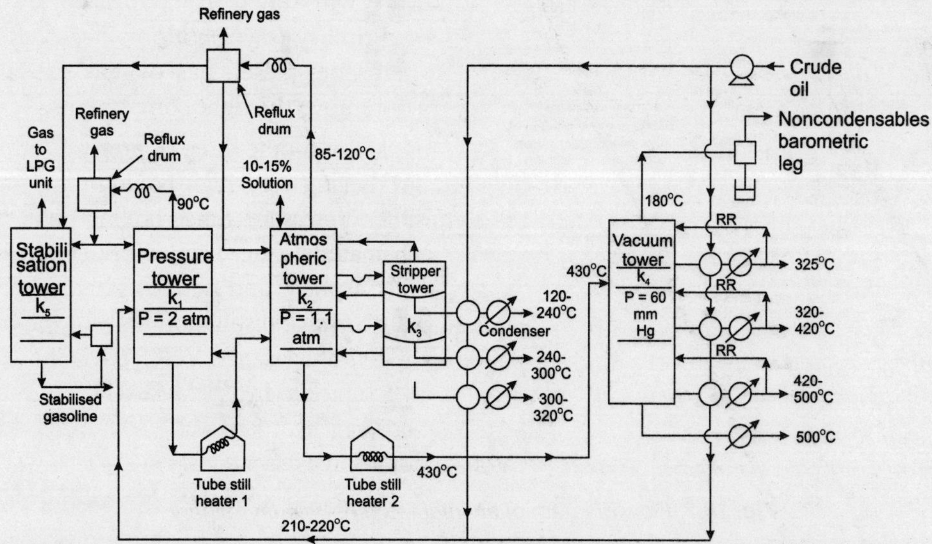

Fig. 18.6. *Detailed layout scheme for a typical three stage crude processing unit.*

Distillation Tower details. (Refer Fig. 18.6)

Tower	No. of Plates	Height, metre	Diameter, metre	Boiling point, °C	Products
K_1	26	30.3	3.8	120-140	Naphtha
K_2	38	30.1	5	240-300	Gas oil
K_3	20	22.5	1.6	300-320	Kerosene
K_4	-	22.8	8	320	Heavy Kerosene
K_5	35	27.8	1.4	320-420	Heavy gas oil
				420-500	Lubricating oil
				> 500	Reduced crude/ tar/asphalt

Flow diagram of an integrated petroleum refinery incorporating almost all products treatment units is given in Fig. 18.7.

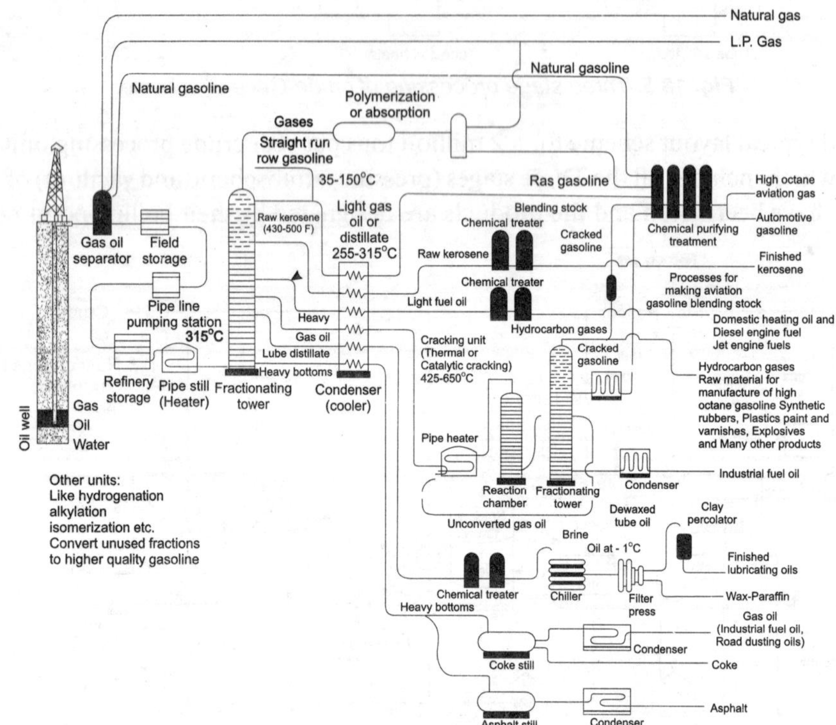

Fig. 18.7. Flow diagram of an integrated petroleum refinery.

A schematic simplified flowsheet of various refinning processes generally followed in Petroleum refineries is given in Fig. 18.8. Flowsheet of yet another refinery incorporating various unit operations is given in Fig. 18.9. Figure 19.10 gives the process flow chart of a modern integrated crude oil refinery incorporating all units.

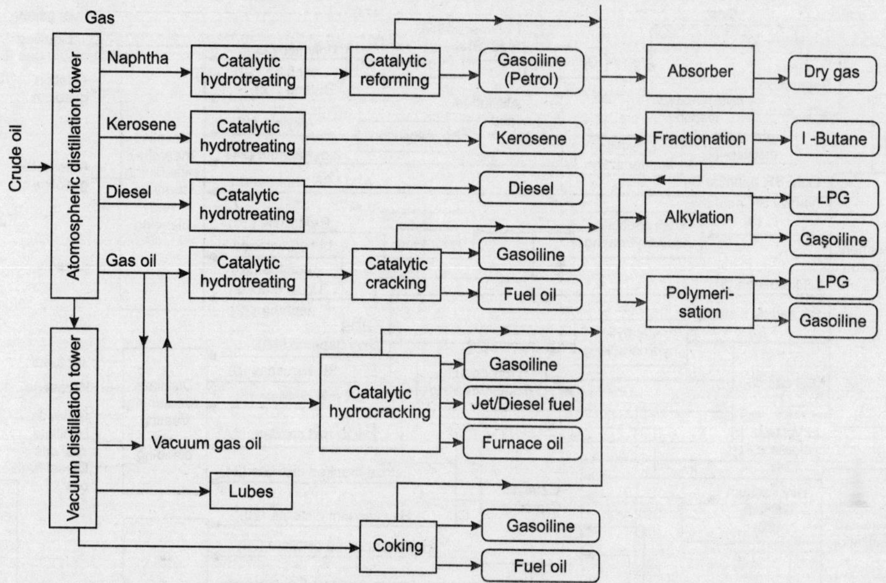

Fig. 18.8. *Flow sheet of a Various Petroleum Refinning Processes followed in an oil refinery.*

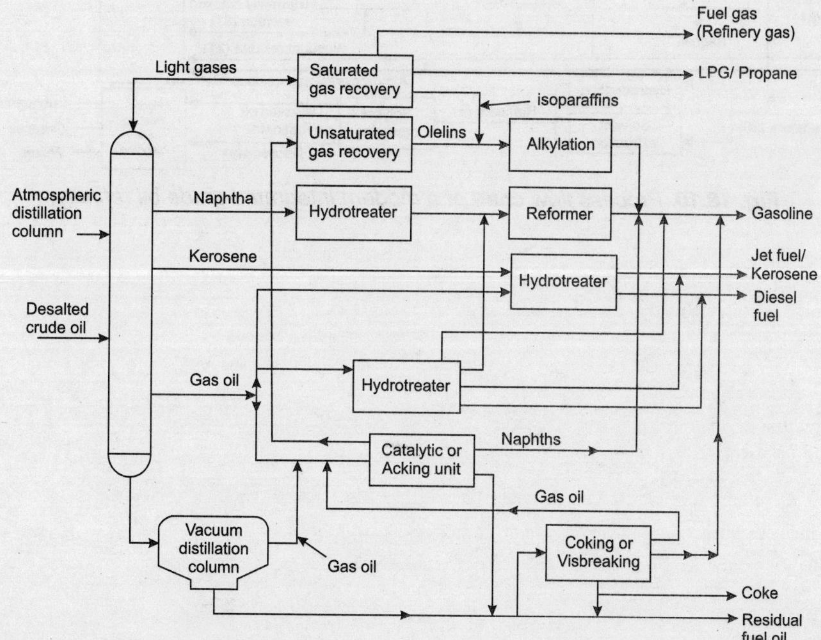

Fig. 18.9. *A modern integrated petroleum refinery incorporating various unit operations.*

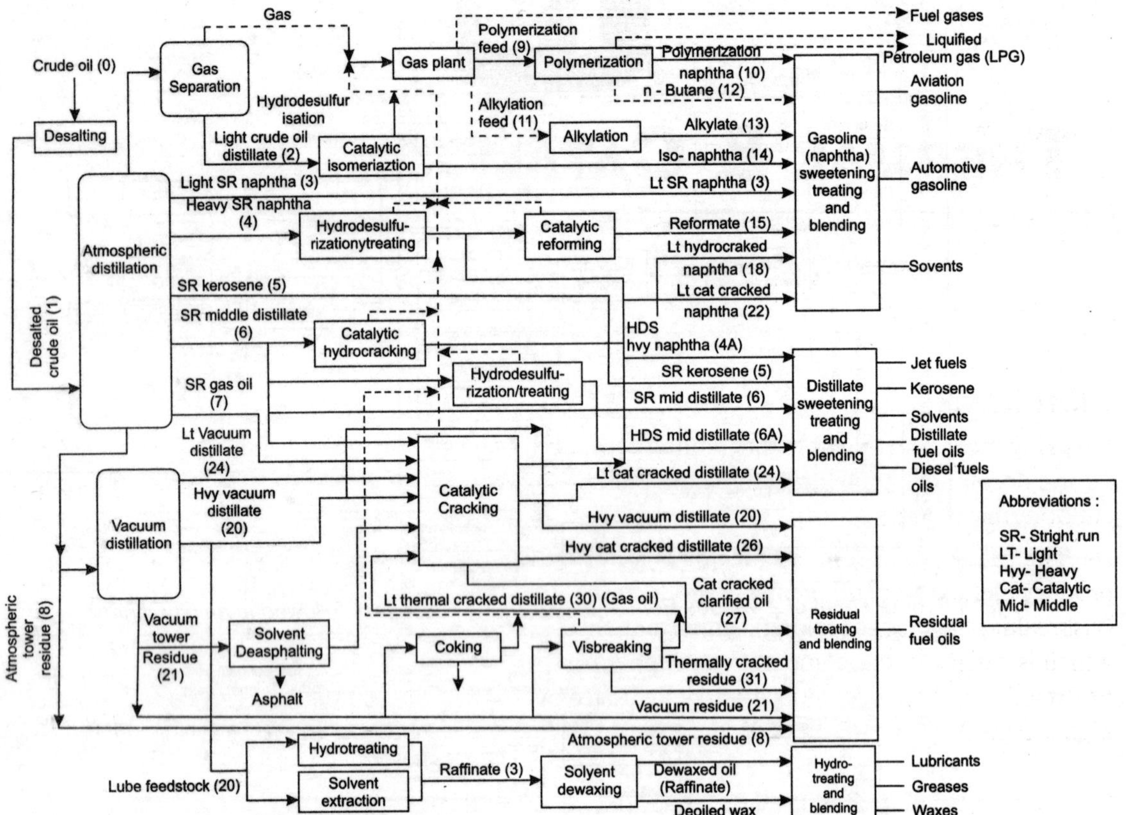

***Fig. 18.10.** Process flow chart of a modern integrated crude oil refinery,*

19

THERMAL & CATALYTIC CRACKING

CRACKING

Cracking means heating of higher boiling petroleum fractions like heavy fuel oil at high temperature (above decomposition temperature) and pressure to produce lower boiling lighter fractions. It is an endothermic reaction.

The friain application of cracking is for production of gasoline from gas oil. It is also done to produce olefins (for petrochemicals production) from gas oil and naphtha, lower viscosity furnace oil (visbreaking) and coke (coking). Important chemical reactions in thermal cracking (non- catalytic) which is carried out at comparatively higher pressure (1-70 atm.) and temperature (450-750°C) by heating the feed are decomposition, dehydrogenation, isomerisation and polymerisation. First, the higher paraffin (say n-decane) decomposes to a lower paraffin and an olefin.

e.g.,

$$\underset{\substack{\text{n-decane}\\\text{(gas oil component)}}}{CH_3(CH_2)_8CH_3} \xrightarrow[\text{T and p}]{\text{cracking}} \underset{\text{n-hexane}}{CH_3(CH_2)_4CH_3} + \underset{\text{butane}}{CH_3.CH_2CH{=}CH_2}$$

(gasoline component)

Correct these equations

The olefins then undergo isomerisation, dehydrogenation and polymerisation reaction

e.g.,

$$\underset{\text{butene}}{2CH_3.CH_2-CH{=}CH_2} \xrightarrow{\text{Polymerisation}} \underset{\substack{\text{iso-octene}\\\text{(gasoline component)}}}{CH_3-C-CH_3-CH_2-CH_3-\overset{CH_3}{\overset{|}{C}}{=}CH_2}$$

gases, solids (coke) are also produced during cracking apart from liquid products. Catalytic cracking compared to the thermal (non-catalytic) cracking produces less coke, less gas but more liquid products. Thermal cracking is now mainly used for production of olefins, coking and visbreaking. Due to presence of higher quantity of iso-paraffins and aromatics gasoline produced by catalytic cracking is far more superior.

The catalyst used in cracking are in the form of powder, beads or pellets and include silica, alumina, natural clay, zeolites etc. Use of catalyst minimises the formation of coke during cracking.

Catalyst need to be regenerated by burning the deposited coke particles in air as it deactivates the catalyst. Pellets and beads of catalyst are used as fixed bed in a cyclic process (*e.g.* Houdry process) or as a moving bed in a continuous process (*e.g.* TCC process). The powder catalyst (5-500 m size) is used either in suspension process or as a fluid bed with the vaporised feed as the fluidising medium (*e.g.* FCC process). Heat requirement of endothermic cracking reaction is met by burning of deposited coke in the catalyst regenerator. In fixed bed process, catalyst is regenerated in the reactor itself. Catalyst movement from reactor to regenerator is more easily achieved in the fluidised cracking process. The demerits of fixed bed processes include variation of temperature and activity of the catalyst along the bed height and with the progress of the cycle causing non-uniformity of the yield products quality. While the merits of fluidised bed processes include high heat transfer rate, isothermal reactor conditions and high reaction rate (due to high mass transfer rates between the catalyst surface and gas phase).

Necessity for cracking. An average crude oil on distillation yields 20-25% petrol, 30-45% intermediate oils (kerosene, naphtha, diesel, lube oil etc.) and 25-50% residual fuel oils. Increasing demand for high quality (octane number) petrol exceed the availability of petrol by distillation alone. Hence, cracking of heavier fuel oils (obtained by distillation of crude) is done to produce mainly petrol. Cracking is also done to reduce the viscosity of fuel oil (by visbreak- ing), to produce coke (by coking) and to produce olefins which is feed for petrochemical industry.

Types of Cracking. There are two types of cracking—thermal cracking and catalytic cracking.

When cracking (*i.e.* breaking larger molecules into smaller ones by heating) is done without any catalyst (which increases the rate of reaction), then it is called thermal cracking.

Cracking in presence of a catalyst is called catalytic cracking.

Thermal Cracking. Here in this process, the fuel which may be gas oil, fuel oil (heavy or light), atmospheric or vacuum residue is heated upto 450-750°C at pressure ranging from 1-70 atmosphere to produce gas, petrol, diesel cracked residue (coke) etc. Octane number of petrol obtained may be upto 75. The yields and quality of products will depend upon the type of feed, temperature and pressure.

Catalytic Cracking. Here the feed (which may be kerosene, fuel oil, gas oil, lubricating oil; atmospheric, vacuum and cracked residue etc.) is heated in presence of a catalyst (like, platinum, nickel, iron, chromium, silica-alumina, aluminium chloride, sulphuric acid, phosphoric acid etc.) at 350-650°C and 1-15 atm to produce gas, petrol, diesel and residue (coke).

Higher temperature gives higher yield of gasoline, whereas higher pressure reduces the octane number of gasoline.

A typical *yield* pattern in cracking is given below :

Feed-residue of vacuum distillation of crude
Yields - petrol = 55%, diesel = 20%, fuel oil = 15%, coke = 10%
Octane number of petrol = 92

Advantages of Catalytic Cracking over Thermal Cracking

- Pressure and temperature required for catalytic cracking is lower.
- Yield and octane number of petrol produced is higher.
- Petrol produced has less quantity of gum and gum forming material.
- Sulphur content of the products is low.
- susceptibility of petrol produced is more.

Thermal cracking processes. Depending upon the pressure and temperature employed for the cracking and the characteristics of feed, there are various thermal cracking processes in which the product yields and characteristics are different. Salient features of some of them are given below :

Low temperature and high pressure thermal cracking process. The feed is residue from atmospheric distillation unit. Products are mainly gas, fuel oil and gasoline. Temperature and pressure employed are 500°C and 20 atm. respectively. This process is also called visbreak- ing when heavy fuel oil is thermally cracked to reduce its viscosity so that it can be properly atomised through the burners. This process is employed for heavy oil cracking.

High temperature and high pressure process :

Feed Gas oil, *Products* - gas and gasoline mainly, *T and P are* 530°C, and 570 atm.

This is also called light oil thermal cracking process. When feed is naphtha, it is called *thermal reforming.*

Low pressure and high temperature process :

Feed – residue from atmospheric distillation unit,
T and P – 550°C and 25 atm.
Products – gosoline and gas rich in unsaturated hydrocarbons (a feed for petrochemicals).

High temperature and low pressure process. This is also called pyrolysis. Pressure and temperature employed is 1 atm. and 700°C. Product is mainly gas rich in aromatic and unsaturated hydrocarbons.

Coking. Feed is vacuum residue. Pressure and temperature is 2 atm. and 400-450°C respectively.

The products are gas, coker gasoline, coke for electrode manufacture (main product) and coke distillate.

THERMAL CRACKING REACTIONS

Few important thermal cracking reactions are :

- Decomposition and destructive condensation.
- Hydrogenation and dehydrogenation.
- Polymerisation.
- Cyclisation.

Octane number increases in the order paraffin Olefin→Naphthene ® iso-paraffin ® aromatic. That means aromatics have the highest octane number and paraffins have the lowest. Hence, cracking reactions should aim for producing hydrocarbons with higher octane number, if production of gasoline is the main product.

Decomposition and destructive condensation of olefins to produce high octane number aromatics is shown below :

$$\underset{\substack{\text{ethylene} \\ \text{(olefin)}}}{2C_2H_4} \xrightarrow{\text{decompostion}} \underset{\substack{\text{ethylene} \\ \text{(olefin)}}}{C_4H_6} + H_6$$

$$C_4H_6 + C_2H_6 \xrightarrow{\text{Condensation}} \underset{\text{Benesene(aromatic)}}{C_6H_6} + H_2$$

Hydrogenation *(i.e.* hydrocracking) of higher boiling paraffin to lower boiling paraffin is shown below :

$$CH_3(CH_2)_6CH_3 + H_2 \xrightarrow{\text{hydrogenation}} \underbrace{\underset{\text{Propane}}{CH_3CH_2CH_3} + \underset{\text{Pentane}}{CH_3(CH_2)_3CH_3}}_{\text{Lower boiling paraffins}}$$

Since, lower boiling paraffins have higher octane number, hydrogenation *(i.e.* hydrocracking) improves octane rating *i.e.,* quality of gasoline (petrol).

Dehydrogenation of naphthenes to aromatics also helps in increasing the octane number of cracking products.

```
      CH2                                CH
      / \                                / \
   CH2   CH2   dehydrogenation        CH   CH
    |     |   ----------------->      |     |    + 3H2
   CH2   CH2                          CH   CH
      \ /                                \ /
      CH2                                CH
  cyclo-hexane                         benzene
  (naphthene)                         (aromatic)
```

Polymerisation. This reaction Converts C_3 and C_4 Olefins to higher diolefins *e.g.*

Cyclisation. This reaction converts n-paraffins to aromatics which have higher octane number. This reaction is also called dehydrocyclisation because hydrogen is a co-product.

$$2\ (CH_3)_2C{=}CH_2 \xrightarrow{\text{Polymerisation}} CH_3{-}C(CH_3)_2{-}CH_2{-}C(CH_3){=}CH_2$$

iso–butylene → di–isobutylene

$$CH_3(CH_2)_6CH_3 + H_2 \xrightarrow{\text{hydrogenation}} \underset{\text{Propane}}{CH_3CH_2CH_3} + \underset{\text{Pentane}}{CH_3(CH_2)_3CH_3}$$

Lower boiling paraffins

Gas-Oil Cracking. For the cracking of gas oil, the reaction pattern is shown below. Paraffins crack to olefins and smaller paraffins in the gas, gasoline, and gas-oil boiling ranges. Side chains of alkyl naphthenes and alkyl aromatics crack to yield olefins plus naphthenes and aromatics

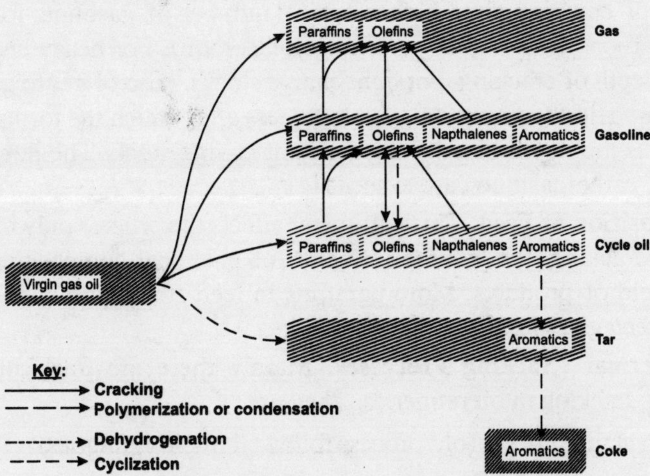

Fig. 19.1. Reactions in thermal cracking of gas oil.

with shorter side chains. Aromatics condense to large molecules. All the primary products except the aromatics crack further as the reaction proceeds. Also, olefins polymerize to larger molecules and cyclize to naphthenes, including condesed-ring naphthenes. Naphthenes dehydrogenate to aromatics, and condensed-ring aromatics condense to larger molecules and eventually coke. The *overall effects* may be *summarized* as follows :

- Cracking of paraffins, alkyl naphthenes, and alkyl aromatics to gas, gasoline, and cracked gas oil.
- Cracking of gasoline to gas.
- Reaction of olefins to produce larger molecules, including aromatics heavy ends ("tar") and eventually coke.
- Further cracking of cracked gas oil to produce gasoline and gas.

INFLUENCE OF VARIOUS PARAMETERS IN CRACKING

Effect of Temperature. Higher temperature decreases the residence time of charge **stock.** Small change in temperature can change the depth of cracking to a considerable extent, **hence,**

it is necessary to obtain the temperature very accurately. Between 450°C to 550°C there is no change in product quality at the same depth of cracking.

Effect of Residence Time. Residence time has a great effect on quality of products as increased time leads to more severity in cracking.

Effect of Pressure. The effect of pressure is limited only to secondary reactions. Lower pressure favours the production of gas whereas higher pressure favours condensation and tar production increases. That is, at low pressure yield of light hydrocarbons are more. Rise in pressure decreases the amount of unsaturated hydrocarbons in gas and gasoline. It has been found that increase of pressure from 10 to 40 atm. decreased the amount of unsaturated hydrocarbons in gasoline by 50%. According to some investigator, increase of pressure increases the gasoline yield *i.e.* to say increase of pressure increases residence time of charge stock and hence the depth of cracking increases.

Effect of Depth of Cracking. Depth of cracking depends upon temperature and residence time. Greater depth of cracking gives higher octane number of gasoline, formation of gas and carboids {*i.e.* hydrocarbons with higher percentage of carbon). For heavy charge stocks, carboid formation limits the depth of cracking. For light charge stocks, ratio of gas to gasoline is same upto about 60% of the completion (conversion) of the process after which the formation of gas is more. This critical yield fixes the depth of cracking for light charge stock. The depth of cracking is so fixed that the yield of carboids in no case exceeds 0.1%.

Chemical Composition of Feed. This parameter effects the process only in the beginning. The greater the depth of cracking, the less the influence of chemical composition. But composition of feed effects the yield of products. More paraffins in feed stock gives higher yield of gasoline whereas naphthenic feed gives less quantity of gasoline.

Commercial Thermal Cracking Processes. Mainly there are four commercial processes employed for thermal cracking in oil refineries. They are :

— Dubbs thermal cracking process (old process followed in old refineries).

— Pyrolysis — Visbreaking — Coking.

Dubbs Thermal Cracking Process. It is an old thermal cracking process used in oil refineries upto about 1939 on gas oil and having fuel oil stocks.

Heavy oil or topped crude (residue of atmospheric distillation of crude oil) is pumped into a bubble cap fractionator where it is mixed with heavy cracked oil from the process. The mixed feed at pressure upto 70 kg/cm^2 is passed through a tube furnace where it is heated maximum upto 540°C before passing to the soaking chamber. This permits a time dwell during which a

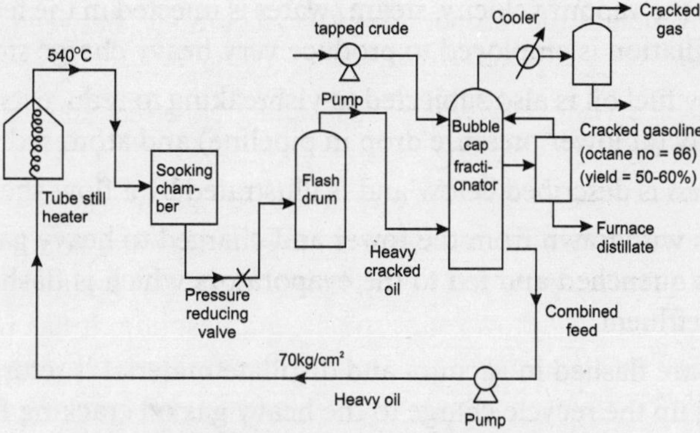

Fig. 19.2. Dubbs Thermal cracking process.

controlled amount of cracking can take place. Pressure is reduced before passing into a flash chamber where heavy cracked fuel oil separates and then to the bubble cap fractionator (distillation column) from which the cracked gas and gasoline passes out at the top and furnace distillate from side. Its flow-sheet is given in Fig. 20.2.

Pyrolysis. Pyrolysis (mild thermal cracking) is done mainly for the production of lighter products mainly unsaturates like olefin (ethylene) and naphthene polymers, diolefins, benzenes and toluenes etc. It is carried out at high temperature (650-700°C) and low pressure. Initially this process was used for the production of toluene. The feed to pyrolysis is raw crude oil, kerosene and natural gasoline. A higher yield of gas, benzene and toluene are obtained by this process.

The feed is heated in special furnaces. The products of the pyrolysis are cooled and delivered to the fractionating column. The pyrolysis gas is drawn from the top of the fractionator. The aromatics (benzene and toluene) are taken from middle section of the column. The residue (aromatic tar, carbon etc.) which is resinous in nature is discharged from its bottom.

The pyrolysis process is used mainly for producing ethylene (by pyrolysis of propane- propylene and low octane gasoline fractions), benzene and toluene. Optimum temperature in pyrolysis is 650°C to 700°C respectively for benzene and toluene to be the principal product.

Visbreaking. It literally means viscosity breaking (i.e. reduction). It is done to reduce the viscosity and pour point of heavy fuel oil without coke formation. Feed is residue from atmospheric distillation unit *i.e.* topped crude charge (having gravity ÅPI = 20.3 and Conradson carbon from 10.5 to 13%). The main product is fuel oil of lower viscosity though gasoline and gas oil are also produced depending upon the intensity of operation.

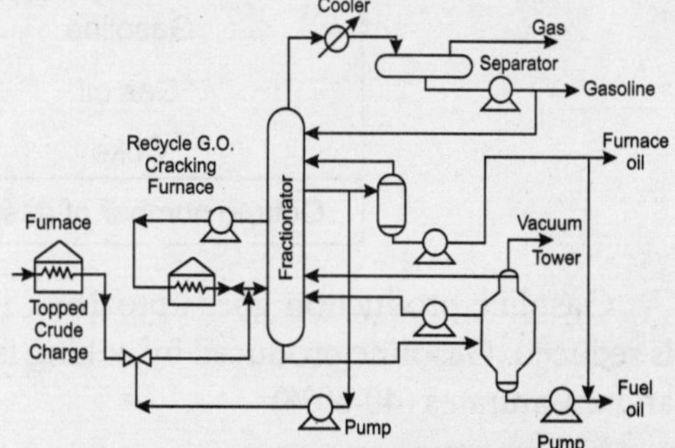

Fig. 19.3. Visbreaking of heavy fuel oil

Lower temperature is employed in visbreaking than in thermal cracking because residual stocks have greater coking tendency than gas oils. Typical reaction condition in visbreaking is 510°C and 17 kg/cm^2. Typical *product yield* is (volume basis): gasoline = 7%, gas oil = 12.4% and fuel oil = 80.6%.

To minimise coking during visbreaking, rapid and even heating of feed and high vapour velocity is desirable. To increase the vapour velocity, steam/water is injected in the feed. Coking tendency is reduced if vacuum distillation is employed to produce very heavy charge stock.

High viscosity heavy fuel oil is also subjected to visbreaking to reduce its viscosity so that it can be pumped economically (at lower pressure drop in pipeline) and atomised smoothly in burners.

The *visbreaking process* is described below and is illustrated by a flow-sheet in Fig. 20.3.

The heavy gas oil is withdrawn from the tower and charged to heavy gas oil cracking furnace. The effluent material is quenched and fed to the evaporators which is flashed with the admixture of visbreaking furnace effluent.

The tower bottoms are flashed in vacuum and distillate material is returned to the fractionator where it aids in making up the recycle charge to the heavy gas oil cracking furnace.

The vacuum oil is blended with light gas oil to meet the desired specifications, having a net production of furnace oil.

The topped crude charge is pumped to a visbreaking furnace where it is heated and thermally cracked slightly. The effluent product from visbreaking furnace is quenched with light gas oil. Now the quenched product is fed to the evaporator section of fractionator where it is flashed. In the base of the tower the tar accumulates with the result of flashing. The vapours are fractionated into gas, gasoline, light gas oil and heavy distillate in the upper part of the fractionator.

Coking. In coking more severe reaction conditions are used than visbreaking so that the feed is completely converted to lighter products resulting from cracking (i.e. gas, gasoline are gas oil and heavy product resulting from condensation reaction *(i.e.* coke). The feed to coking are thermal cracking residue, pitch of pyrolysis plant and atmospheric column residue (topped crude charge).

When primary product in coking process is coke, then gas and gasoline are secondary products. But when gasoline is primary product, then coke is a secondary product. High quality *petroleum coke* (containing ash > 0.5% and volatile matter 5-7%) can be produced by this process.

A typical product yield in coking when feed is thermal cracking residue is given below :

Products	Yield, weight %
Gas	13
Gasoline	7
Gas oil	49
Coke	31
Octane number of gasoline	65

Gasoline production goes upto 16%, if feed is atmospheric column residue and the gas yield is reduced. Gasoline produced by coking is called coker gasoline and is rich in aromatics (20-30%) and unsaturates (40-60%).

Quality of product from a coking unit depends upon the feed. The yield of gas and coker gasoline increases with increasing temperature but the yield of gas oil decreases. The yield of gas oil increases by coking at lower pressure. At higher pressure more gas and coke are produced and the liquid product contains more gasoline. The Octane number of coker gasoline increases with temperature. However coker gasoline produced at higher temperature are unstable and require finishing operations such as clay treating or mild hydrogenation. The gases produced at higher temperature (say 525°C) are more olefinic (upto 50% olefins).

There are batch, semi-continuous (also called delayed coking process) and continuous processes for coking. Coke produced is called petroleum coke which is used in the manufacture of carbon electrodes (used mainly in ahiminium industry), motor brushes, dry cells etc.

Delayed Coking Process. In this process coke is produced from the feed by increasing the time of liquid phase cracking. The furnace type coking unit is called delayed coker because it heats up the reduced crude feed to coking temperature (500°C) so rapidly that little reaction occurs while the charge is in the furnace. Effluent from the furnace is discharged at about 500°C into a large coke drum operating at 2-4 atmosphere pressure where it remains until it cracks and passes off as a vapour or condenses into coke. When the drum is filled up, it is taken out of service and steamed to remove volatile matter in the coke. The coke is then cut out with water jet at a pressure of about 80 atmosphere and removed from the drum. The cleaned drum is returned to service. Two or more drums are used in parallel and is typically on stream from 24 hours and out for cleaning for 24 hours.

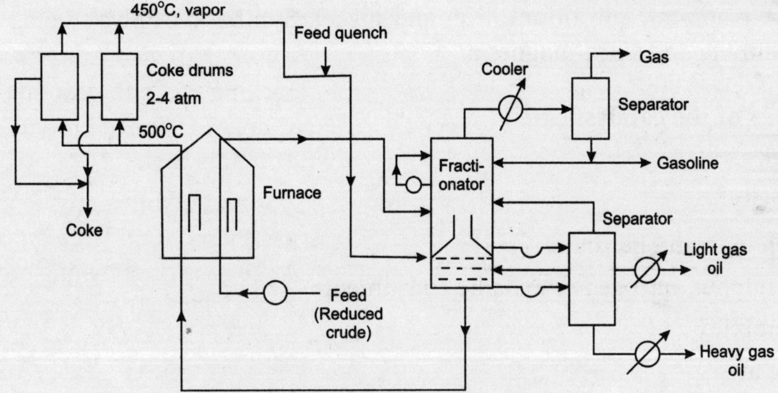

***Fig. 19.4.** Delayed coking process.*

A flow-sheet of delayed coking process is given in Fig. 19.4.

Catalytic Cracking. Cracking of feedstock in presence of a catalyst is called catalytic cracking which gives greater yield and higher octane number of gasoline than is possible by thermal cracking or reforming. Moreover, it employs reduced severity of operation as the temperature and pressure is low *i.e.* 450-550°C and 0.5-1.5 atm. (gage).

Feed for Catalytic Cracking. Feed ranges from light gas oil to reduced crude. For producing aviation gasoline, the boiling range of feed should be 350-420°C. Whereas for motor gasoline the same should be about 350-500°C. *Feed stocks* are generally :

(a) Crude oil/primary products of crude oil.
(b) Kerosene and gas oil from atmospheric distillation unit.
(c) Coker distillate - This promotes coke deposition on catalyst.
(d) Light lubricating oil (from lube oil reprocessing plant) - Its cracking produces aromatics which decompose to give deposits on catalyst.

(e) Deasphalted crude - It may contain deleterious metal component.

However, *straight run distillates are the best feed for catalytic cracking.* Performance of feed depends upon the boiling range, distribution of hydrocarbon types, the concentration of sulphur and nitrogen, the metallic constituents and Conradson carbon.

At constant conversion, gasoline and coke yield increases with boiling range of feed whereas the yields of dry gas, butane and butenes decrease. The octane number of catalytically cracked gasoline is lowered if initial boiling point (ASTM distillation) of the feed drops below 250°C.

The chemical composition of the feed affects the yields from catalytic cracking. The more naphthenic the feed, the higher is the yield of gasoline.

High boiling point heavy feed (particularly residues) contain appreciable quantities of Conradson carbon and metals which gets deposited on the catalyst. Carbon deposit reduces the capacity of those units whose regeneration capacity is limited. Metal deposit on catalyst changes its selectivity in the direction of less gasoline and more coke. To counteract this effect, a portion of the circulating catalyst is discarded and replaced by fresh catalyst.

Catalysts used in Catalytic Cracking. Primary catalysts used were natural or artificial clays *e.g.* bentonite clay and montmorillonite activated by sulphuric acid. Metallic catalysts being used present days are platinum, chromium, iron and nickel. Synthetic catalysts being used presently include silica-alumina and silica-magnesia.

Desirable properties **of the catalysts are :**

- high reactivity
- good selectivity
- sufficient strength and hardness
- absence of sulphur, nitrogen & metallic constituents
- easy regenerability
- high surface area
- high porosity (about 50%).

Size of the fresh catalyst in moving bed catalytic cracking process is 3-5 mm and that of regenerated catalyst is 3-4 mm. In fluidised bed catalytic cracking process the size of the catalyst is about 20-40 microns.

In fluidised bed catalytic cracking, Silica-magnesia catalyst gives the highest gasoline yield whereas silica-alumina gives the highest octane number of gasoline. Bentonite clays are intermediate in both respects.

Metallic contamination in feed lowers the gasoline yield and increases the coke production. Salt such as sodium chloride if present in feed deactivates the catalyst. Nitrogen compounds present in feed are adsorbed on catalyst and deactivate it. However, burning of nitrogen during regeneration of catalyst restores its activity completely. Catalysts also gets deactivated when exposed to steam at higher temperatures.

Catalyst consumption is 0.1-0.4% of charge stock including losses in cyclone separator.

Catalytic cracking reactions. Cracking reactions are endothermic whereas the regeneration of catalysts is exothermic. The reaction occurring during thermal cracking are accelerated in catalytic cracking.

- Paraffin crack faster to lower paraffins and olefins, n-paraffins crack 5-60 times faster than corresponding thermal process.
- i-paraffins crack faster than n-paraffins.
- Dehydrocyclisation reaction takes place resulting in the formation of cyclics (*e.g.* aromatics) from paraffins.

e.g.,

$$\underset{\text{Paraffins}}{-C-C-C-C-C-C} \xrightarrow{\text{dehydrocyclisation}} \bigcirc$$

- Olefins (most reactive) crack 1000-10000 times faster than thermal process.
- Olefins undergoes isomerisation to give iso-olefins and coke (with high C/H ratio).
- Olefins undergo polymerisation.
- Cracking of naphthenes or alkenes.
- Aromatics are not responsive to cracking catalytic conditions as it is stable under the conditions
- unlike thermal cracking. Under severe conditions, it gives carbon deposits.
- However, alkyl aromatics (*e.g.* alkyl benzene) with side chain of carbon atom greater than 2 gives aromatic and olefins on catalytic cracking.

$$\text{alkyl benzene} \rightarrow \text{benzene + olefin}$$

The ease of cracking of alkyl aromatics increases with the length of the side chain at least upto a length of four carbon atoms.

Catalytic Cracking Processes. Depending upon the physical condition of the catalyst bed there are three main processes for catalytic cracking namely :

- Hourdry's fixed bed catalytic cracking process.
- T.C.C. (Thermofer catalytic cracking) moving bed process.
- Fluidised bed catalytic cracking (F.C.C.) Process.

Houdry's fixed bed process. This catalytic cracking process was first used in 1936 has become obsolete now by the development of moving bed and fluidised bed process.

In this process, catalyst (silica-alumina) in the form of compressed pellets is contained in reactors (catalyst cases) arranged normally in a group of three. The reactors are tubular heat exchangers which supply heat necessary to break down the heavier molecules of feed and to remove the heat evolved when burning of the carbon on the catalyst during the regeneration cycle.

Operation of the process is cyclic occurring in four stages and consuming a total time of 30 minutes *e.g.*

- reaction time = 10 minutes at about 2 kg/cm^2 and 480°C.
- purging time = 5 minutes.
- regeneration of catalyst =10 minutes at 510°C.
- purging = 5 minutes.

The temperature of the catalyst bed is controlled by circulating a molten mixture of alkali metal salts through vertical tubes within the reactor core. Life of catalyst is about 18 months but the yield of gasoline falls from 52 to 40% during this period. Octane number of gasoline produced is about 73.

Several reactors (generally in a multiple of 3 *i.e.* 3, 6, 9 or 12) are provided in commercial unit so that feed may be switched cyclically from a deactivated reactor to an active one. Provision is made to heat the catalyst during cracking reaction (which is endothermic) and cool the catalyst during regeneration (which is exothermic).

A flow-sheet of the process is given in Fig. 19.5.

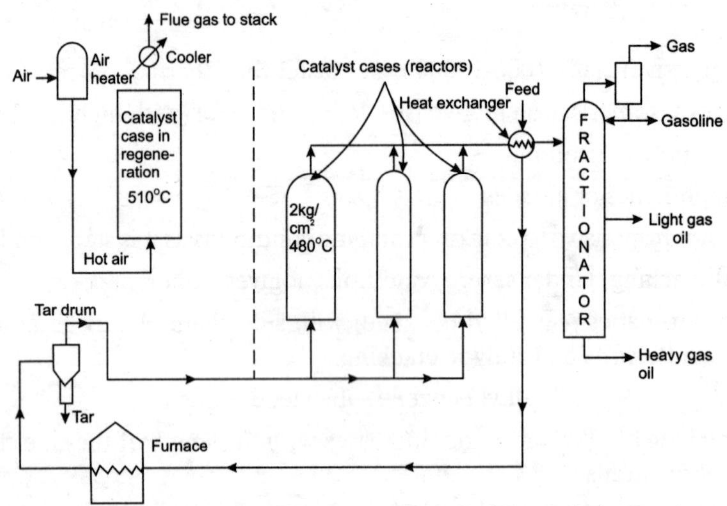

***Fig. 19.5.** Houdry's fixed bed catalytic cracking process.*

Typical material balance for a catalytic cracker. For a charge stock having

(a) Boiling range = 320-500°C, *(c)* Sulphur = 1.9%

(b) Specific gravity = 0.9128, *(d)* Pour point = 30°C.

When subjected to catalytic cracking the above charge stock at 450°C and 0.5 atm. (gage) gave the following product distribution :

Gas (including C_4 fractions) = 13% by weight
C_5 Gasoline (upto final boiling point of 200°C) = 24%
Heavy gas oil (H.G.O.) = 26.9% Light gas oil (L.G.O.) (200-320°C) = 24%
Coke = 5%, Losses = 2.3%.

Effect of reaction conditions in catalytic cracking. What happens in a moving bed or fluidised bed catalytic cracking reactor depends upon the temperature, pressure, space velocity, ratio of catalyst to oil passing through the reactor, concentration of coke on the incoming & outgoing catalyst and the catalyst activity. The normal range of each variable is given in Table 19.1.

Table 19.1: Parameters of Catalytic Crackers

Parameters	In fluidised bed catalytic cracker	In moving bed catalytic cracker
Reactor temperature, °C	475-540	455-500
Reactor pressure, kg/cm2	0.7-1.1	0.4-1
Space velocity, kg oil/hr/kg catalyst	0.5-3	1-4
Catalyst to oil weight ratio	5-20	1.5-7
Carbon on regenerated catalyst, weight %	0.3-0.8	0.1
Carbon on spent catalyst, weight %	0.5-2	—
Maximum regenerator temperature, °C	595	680

Effect of temperature. When temperature is increased the rate of reaction increases. For every 40°C rise, the decomposition rate doubles. For a given conversion, gasoline yield is reduced at higher temperature. When temperature is more and gasoline yield is maintained then the coke formation rate decreases when residence time is less. This is achieved in two stage operation. Higher temperature results in gasoline of slightly higher sulphur content.

Effect of pressure. When reactor pressure is more, then the residence time is also more which results in more unsaturates in the product and less octane number of gasoline. At a given conversion, increasing the pressure increases the production of coke.

Effect of catalyst to oil ratio. Conversion increases with increase in catalyst to oil ratio as it shortens the time required for the catalyst to pass through the reactor and thereby reduces the extent of its deactivation from coke production. At a given conversion and temperature, coke yield is higher at higher catalyst to oil ratio.

Effect of coke concentration on catalyst. Lesser is the concentration of coke on catalyst, better is the effective activity of the catalyst. Lower coke concentration is obtained by increasing catalyst to oil ratio. On silica-alumina catalyst, higher coke concentration catalyses the production of more coke.

Effect of catalyst activity. With decrease in catalyst activity, the conversion declines. Product yield is poorer at low activity hence catalyst is sometimes deliberately discarded so that more fresh catalyst can be added to increase activity and improve yields.

Product characteristics. Gas produced comprises mainly of propane and propylene (used in polymerisation).

- Gasoline has more aromatics and iso-paraffin hence has a high octane number. One third of total gasoline fraction is olefins. About one fourth of C_4 is associated with gasoline fraction. Gasoline has a final boiling point of 200°C. 40% of it evaporates within 100°C.
- Light gas oil (having boiling range 200-320°C) produced has a cetane number of 25-35, if the feedstock is rich in naphthenes and aromatics. However, if feed is of paraffinic base oil then cetane number is 45-55. It contains about 30-55% aromatics and 6% olefins.
- Heavy gas oil is the last fraction from cracking unit. Its viscosity and pour point is similar to the feedstock. Its specific gravity is about 0.89-0.96. This H.G.O. is used as feed to the coking or conventional thermal cracking unit or as a fuel oil.

Thermofer Catalytic Cracking (TCC) Moving Bed Process

In this process, the regenerated catalyst beads (similar to that in Houdry's process) goes down from an elevated hopper to the reactor where it comes in contact with partially vapourised feed. Catalyst and oil can flow concurrently into a section where the cracked products are separated and removed. Additional oil is often obtained by steam stripping of catalyst (*i.e.* by steaming the catalyst at the base of the reactor). The spent catalyst (hot, carbon coated) then descends by gravity into the regenerator where a counter current low oxygen content flue gas is used to burn off the deposited coke. The catalyst containing 2-4% by weight coke at a temperature of about 450°C is regenerated to a coke per cent of 0.5 and leaves at 550°C.

A cooler at the bottom of the regenerator reduces catalyst temperature to the level desired for entry in lift line where a low pressure of 0.15-0.2 kg/cm^2 is used for the lifting of catalyst to the catalyst hopper. A high velocity of 25-30 metres/second is employed in catalyst lift line.

A flowsheet of T.C.C. moving bed process is shown in Fig. 19.6.

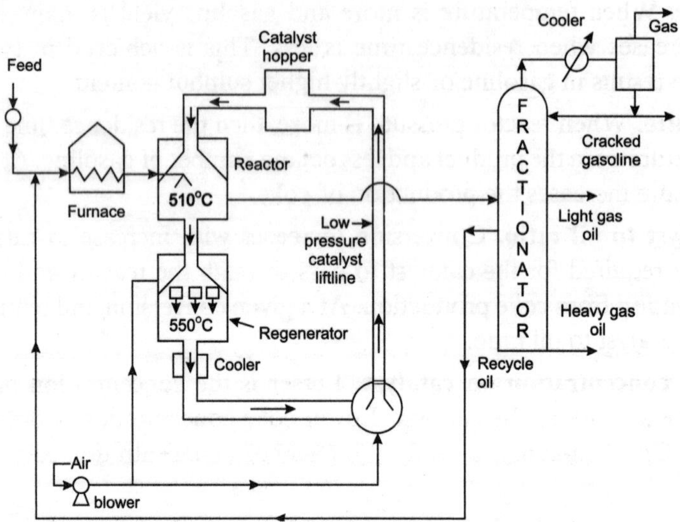

Fig. 19.6. Flowsheet of T.C.C. moving bed process.

A catalyst/oil ratio = 2/5 and recycle ratio = 0-2 is maintained in T.C.C. moving bed process. A typical result shows :

Gasoline yield	43-56%
Light gas oil	15-27%
Heavy gas oil	17-19%
Coke	5%
Octane number of gasoline	95
Catalyst consumption	0.2 lb/barrel oil = 0.57 kg/m3 oil.

FLUIDISED BED CATALYTIC CRACKING (F.C.C.) PROCESS

Fluidised bed catalytic cracking is the most widely used process in oil refineries. The catalyst is in powdered form and is recirculated between two vertical vessels operating as reactor and regenerator.

Fluidised bed of powder is obtained by forming the vapour through columns of powder until the minimum fluidised velocity is reached at which the particles of catalyst are arranged as a loose packing in the gas stream so that they will flow as a fluid. A slight increase in gas velocity causes vapour to flow through a zone of low solids concentration in which high mass transfer rate between catalyst surface and gas enables high reaction rates and high heat transfer rate to be maintained.

A catalyst stand pipe with control valve projects from the bottom of each vessel, the outlet of which is connected by pipeline to the base of the other vessel.

The principle of operation of a single stage fluidised catalytic cracking process is shown in Fig. 19.7.

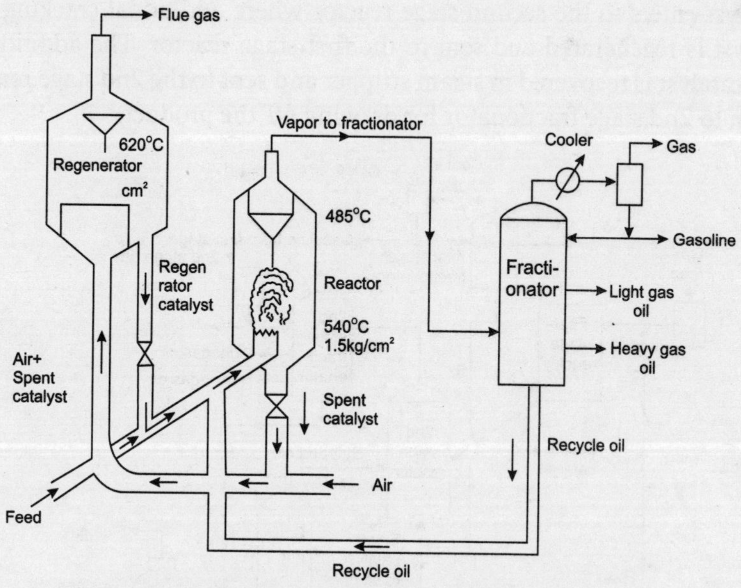

Fig. 19.7. Fluidised bed catalytic cracking (single stage.)

The sequence of events in the fluidised catalytic cracking process as evident from the above figure is as follows :

(a) The feed is heated and vapourised by a stream of hot catalyst passing into the reactor at about 540°C. The catalyst separates from the vapour and sinks to the base of reactor from which it flows, with air into the regenerator.

(b) In the regenerator, the temperature rises to about 680°C due to combustion of carbon on the catalyst. The regenerated catalyst flows down the stand pipe into the fresh oil feed.

(c) The cracked vapour pass from top of the reactor to the fractionating column from which the heavier fractions are recycled through the reactor and the products are drawn off.

The catalyst to oil ratio is 5-15 and the reactor space velocity is 1-3 kg oil/hr/kg catalyst. A typical result from a single stage F.C.C. process shows :

Gasoline yield	= 53-59%	Heavy gas oil = 2-3%
Light gas oil	= 18-25%	Coke = 5-8%
Octane number of gasoline	= 92-94 (unleaded)	

TWO STAGE FLUIDISED CATALYTIC CRACKING (F.C.C.) PROCESS

Here, the catalytic cracking reaction is carried out in two stages and so is the product removal from fractionator. In the two stage process, considerable flexibility is available for changing the yield structure from a given feedstock *e.g.* gasoline yield can be increased further by incorporating recycle streams into cracking stages.

In this process, the feed along with the hot regenerated catalyst travel concurrently up through a riser to the first stage reactor. The material from this reactor is separated into gas and gasoline in the first stage fractionator. The unconverted and insufficiently converted feed along with partially deactivated catalyst enters to the second stage reactor where additional cracking can be effected. This spent catalyst is regenerated and sent to the first stage reactor. The additional unconverted charge from the catalyst is recovered in steam stripper and sent to the 2nd stage reactor where from the vapour is sent to 2nd stage fractionator for drawing off the product.

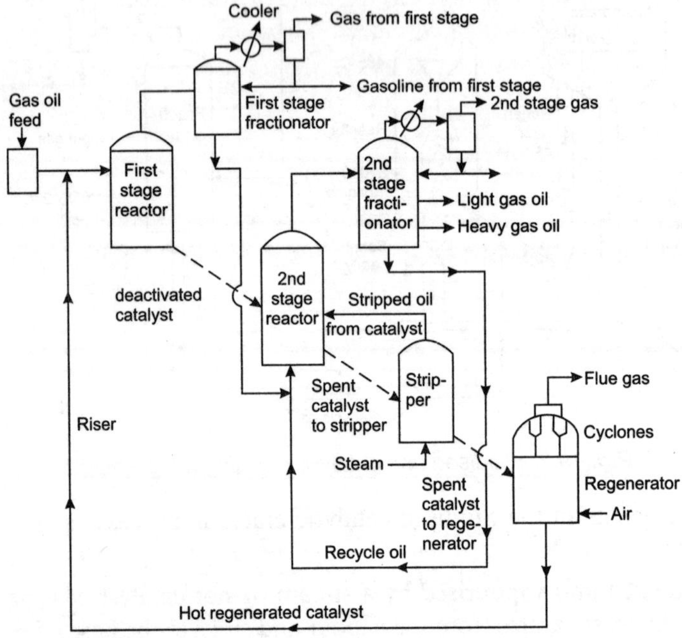

Fig.19.8: Two Stage Fluidised Catalytic Cracking (F.C.C.) Process

Reaction conditions -450-520°C and 8-15 kg/cm^2.

Regeneration condition - 550-600°C and 1.3-1.6 kg/cm^2.

Catalyst /oil ratio = 5-15

Average consumption of catalyst = 0.6 lb/bbl of feed = 1.7 kg/m^3 feed.

Octane number of gasoline produced from this process is about 97.

HYDROCRACKING

Cracking in presence of hydrogen is called hydrocracking. It is also called destructive hydrogenation and leads to the formation of saturated compounds of lower molecular weight by the cracking of heavy fractions and residues under pressure of hydrogen. Hydrogen suppresses the formation of tar & coke and also raises the crackability of polycyclic aromatic compounds.

Hydrocracking achieves high conversion into gasoline and produces high quality diesel fuels. Mild hydrocracking of petroleum residues supply feed for other refinery processes.

Important catalyst used are tungsten sulphide for vapour phase hydrocracking and an iron catalyst or bauxite for liquid phase hydrocracking. Hydrocracking of residues and heavy fractions boiling over 330°C is done in the liquid phase. Reaction conditions in hydrocracking are more severe *viz.* 400-550°C and 70-700 kg/cm^2.

Hydrocracking and other hydrogenation processes in refinery has been possible mainly because of the availability of surplus hydrogen rich gases from the catalytic reforming of naphthas. Hydrocracking produces desired products with a lower molecular weight than the feedstock and any fraction from naphtha to asphalt can be processed. It completely removes sulphur, nitrogen & oxygen besides saturating the olefins thereby giving stable products comprising of paraffins, naphthenes & aromatics. Production of high quality lubricating oil base stock by selective hydrocracking of vacuum gas oils & propane deasphalted oils is its important application. Besides, it increases the yield of lube oil. Further, avoids range of stable distillate products like gasoline or jet fuel or diesel can be obtained in the hydrocracker. However, compared to fluid catalytic cracker, hydrocracking unit is 1.6 times costlier and has a higher operating cost too.

Application of hydrocracking by using different feed stock for getting the desired products is given in table 19.2.

Table 19.2: Hydrocracking applications to get desired products

Feedstock charged to hydrocracker	Desired Products obtained
Naphtha	Butane & Propane (LPG)
Condensate from natural gas	Naphtha
Kerosene	Naphtha
Atmospheric gas oils	Naphtha, jet fuel and/or distillates
Vacuum gas oils	Naphtha, jet fuel, distillates, lube oils
Deasphalted oils	Naphtha, jet fuels, distillate, lube oils
Reduced crude oil	Distillate & LSHS fuel oil
Vacuum residue	Naphtha, distillate, Vacuum gas oil & LSHS fuel oil
Light & heavy cycle oils (from FCC)	Naphtha and/or distillates
Coker distillates & heavy gas oils	Naphtha and/or distillates

Hydrotreating. It is a comparatively mild operation carried out at 300 - 400°C & 15-17 Kg/cm^2, aimed at removal of impurities like sulphur, oxygen, nitrogen, halides & trace metals from products of all boiling ranges from naphtha to lubricating oils. Hydrotreating is also used for stabilisation of petroleum products by saturating olefins present in them. Catalyst used is cobaltmolybdena. Hydrogenation processes like hydrocracking, hydrotreating and aromatic saturation has been possible due to availability of surplus hydrogen rich gases from catalytic reforming of naphtha.

THERMAL & CATALYTIC REFORMING

Reforming means rearrangement of molecules without much affecting the average molecular weight of feed which is generally naphtha of gasoline boiling range. Reforming is carried out to produce high quality (octane number) gasoline by heating with or without catalyst the naphtha. Low octane number thermally cracked gasoline (is also a feed for reforming) is also subjected to reforming to improve its quality (octane number).

Reforming can be thermal or catalytic as in the case of cracking. Catalyst apart from accelerating the process also enhances the yield and quality (octane number) of gasoline. The gasoline produced by reforming is called reformed gasoline or reformate. Thermal reforming has been almost completely replaced by catalytic reforming.

Feed for reforming. The feed is generally naphtha produced during straight run distillation of crude oil, catalytic cracking and coking process. The best result is obtained with naphtha (produced during straight run distillation) having high naphthene content.

Reforming products. Besides the main product i.e. reformate or reformed gasoline, reforming also produces lighter hydrocarbons (gases), hydrogen and traces of very high boiling materials. The hydrogen rich gaseous product is recycled with the naphtha feed. The reformed gasoline has a wider boiling range than the feed and its volatility is increased.

Chemical Reactions in catalytic reforming. Since, aromatics have the highest octane number and iso-paraffins are next to them, most of the desirable reactions during reforming tend to produce high octane number hydrocarbons from lower ones. Octane number increases in the order paraffin □ olefin □ naphthene □ iso-paraffin □ aromatic.

The main reaction in the reforming process is dehydrogenation of naphthene to produce aromatics. Other reactions occurring during reforming are dehydrocyclisation of paraffins, isomerisation, hydrocracking, dehydroisomerisation, dehydrogenation etc.

The various reforming reactions are illustrated below.

Dehydrogenation of naphthene to aromatic :

CH$_2$ / H$_2$C CH$_2$ / H$_2$C CH$_2$ / CH$_2$ — CH3 —dehydrogentaion→ Toluene (aromatic) + CH$_2$

methyl cyclohexane
(naphthene)

Dehydroisomerisation. In a typical reaction methyl cyclopentane isomerises to cyclohexane which dehydrogenates to yield benzene

CH$_2$ / CH$_2$ CH — CH$_3$ / CH$_2$ — CH$_2$ —isomerisation→ CH$_2$ / H$_2$C CH$_2$ / H$_2$C CH$_2$ / CH$_2$ —dehydrogentaion→ Benzene (aromatic) + 3H$_2$

methyl cyclohexane
(naphthene)

cyclohexane

Dehydrocyclisation of paraffins to aromatic :

C_7H_{16} —cyclisation→ CH$_2$ / H$_2$C CH$_2$ — CH$_3$ / H$_2$C CH$_2$ / CH$_2$ —dehydrogentaion→ CH$_3$ + 3H$_2$

n - pantane
(paraffin)

methyl cyclohexane

Toluene
(aromatic)

Hydrocracking of higher paraffins to lower paraffins :

$$C_8H_{18} + H_2 \xrightarrow{\text{Hydrocracking}} C_3H_6 + C_5H_{12}$$

octane (higher bolling point paraffin) → Propane Pentane (lower bolling point paraffin)

$$\underset{\text{Decane}}{C_{10}H_{22}} + H_2 \xrightarrow{\text{Hydrocracking}} \underset{\text{n-pentane}}{n - C_5H_{12}} + \underset{\text{iso-pentane}}{i - C_5H_{12}}$$

Lower bolling point paraffins have higher octane number.

Isomerisation of paraffins: Here n–paraffins are converted into iso–paraffin which have higher octane number.

$$\underset{\text{n– paraffins}}{-C-C-C-C-C-C-} \longrightarrow \underset{\text{i– paraffin}}{-\overset{\overset{C}{|}}{C}-C-C-C-C-}$$

Dehydrogenation of n–paraffins to olefins:

$$\underset{\text{n– paraffin}}{CnH_2n+2} \xrightarrow{\text{dehydrogenation}} \underset{\text{olefin}}{CnH_2n} + H_2$$

Though olefins have higher octane number than corresponding n-paraffins, still this reaction is undesirable because lower olefins polymerise to produce gummy materials which is unwanted.

Thermal reforming. This is carried out in absence of catalyst and is similar to high temperature, low pressure thermal cracking. Feed is usually naphtha and the products are mainly gas (13-14%) and gasoline (75-80%) of octane number 81-86 ; rest being polymer (unwanted gum) and losses. A flowsheet for thermal reforming is given in Fig. 20.1.

Catalytic Reforming. Reforming in the presence of a catalyst is called catalytic reforming. Two types of catalyst are used *viz.* non-precious metal oxide type (*e.g.*molybdena or chromia supported on alumina base) and precious metal oxide type (*e.g.* platinum on a silica-alumina or alumina base.) Platinum is more active and selective catalyst, however, it is easily poisoned by compounds containing sulphur, nitrogen, arsenic etc. and needs pretreatment of the feed for the removal of these impurities. Catalyst may be in the form of powder, bead or pellet.

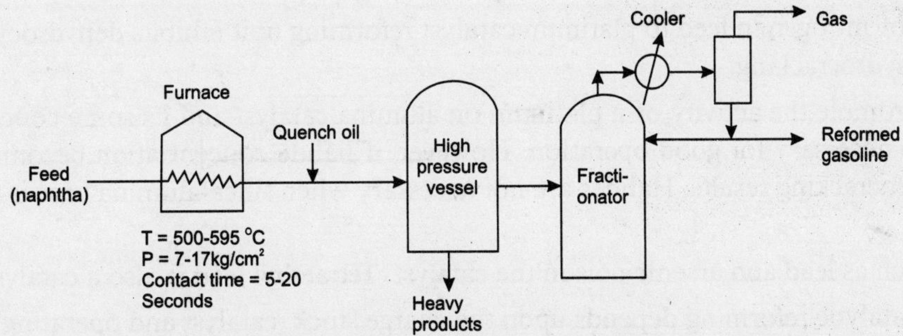

***Fig.20.1:** Thermal reforming of naphtha*

Fixed bed, moving bed and fluidised bed processes are in use. Various commercial catalytic reforming processes in use in oil refineries are Platforming, Catforming, Hydroforming, Fluidised Hydroforming and Thermofer Catalytic Reforming (TCR).

Reforming is normally a once through process. The reactor conditions are normally in the range of T = 450-550°C and P = 15-45 kg/cm2. Fluidised bed and moving bed processes are run in high temperature ranges at about 15 kg/cm2. High pressure upto 45 kg/cm2 are for the fixed bed process.

Effect of variables in catalytic reforming. Catalytic reforming is carried out over a wide range of reaction conditions, the most important of which is pressure. Other variables are recycle gas rate, temperature, space velocity and the catalyst particle size.

Effect of Temperature. Temperature in catalytic reforming is in the range of 450-550°C. If temperature is less than 450°C then catalyst is inactive, whereas if temperature exceeds 550°C then hydrocracking predominates. Cyclisation reaction occurs at 470-490°C whereas isomerisation reaction occurs at 40-50°C above that for cyclisation. Dehydrogenation reaction occurs upto 500-550°C. At 700-800°C, olefin formation starts.

Effect of Pressure. Dehydrogenation of naphthene which is desirable in reforming reaction and is favoured by low pressure and high temperature; however, this condition also favours the formation of coke which deactivates the catalyst. Deposition of coke can be reduced by operating in an atmosphere of hydrogen; the higher the pressure of hydrogen, the slower is the coke deposition. Increasing the pressure of hydrogen does not severely retards the dehydrogenation of naphthenes which produce high octane number aromatics. Hence, catalytic reforming is carried out under substantial pressure with the hydrogen product recycled to the reaction zone. As pressure and hydrogen recycle ratio is increased, gasoline yield (at a given octane level) reduces. Lower pressure are permissible with lower boiling feed which has a lower tendency to form coke.

Effect of catalyst particle size. With very active catalysts, the interior portion is ineffective because reaction occurs before the reactants can penetrate to the core. In this situation, the activity of the catalyst increases as the particle size, is reduced. With platinum catalyst the effect is substantial. In deciding the particle size, catalyst activity is to be balanced against pressure drop through the catalyst bed.

Effect of trace materials in feed. High sulphur in feed inhibits dehydrogenation and promotes hydrocracking over platinum catalyst thereby reducing the yield of liquid products. High sulphur concentration also accelerates deactivation of catalyst especially at low hydrogen pressure.

Presence of nitrogen in feed to platinum catalyst reforming unit inhibits dehydrocyclisation of paraffin and hydrocracking.

Halides promote the activity of a platinum on alumina catalyst and its some concentration in the catalyst is necessary for good operation. However, if halide concentration becomes too high, excessive hydrocracking results. Halides are not necessary when silica-alumina base is used for the catalyst.

Metals such as lead and arsenic poison the catalyst. Tetraethyl lead is also a catalyst poison.

Yield in catalytic reforming depends upon the charge stock, catalyst and operating conditions. If feed has 15-25% naphthenes then the reformate will have 11-18% benzene, 18-20% toluene and 22-24% xylene. With increase in severity of operation, yield increases at the cost of quality (*i.e.* octane number reduces).

In a typical catalytic reforming of naphtha of boiling range 85-180°C, the yield is as follow :

gasoline component	= 84%	unsaturates	= 4%
hydrogen	= 11%	losses	= 1%

Commercial catalytic reforming processes

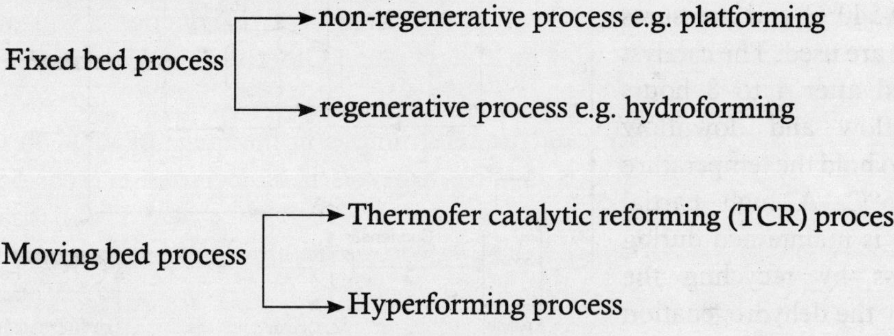

Fluidised bed process -Fluidised hydroforming process

Some of the important commercial processes are described below.

Non-Regenerative Fixed Bed Platforming Process

One of the most widely used catalytic reforming process is the non-regenerative platforming process which employs normally a three to five reactor system. The catalyst is platinum and halogen (normally platinic chloride) on alumina base. The catalyst is not regenerated in situ but is replaced when operating temperature can not be increased further to maintain the octane number of gasoline produced. A catalyst life of 200 barrels *(i.e.* 7000 imperial gallon) of feed per pound of catalyst has been reported.

Preheated naphtha with an end point of 190°C is mixed with hydrogen rich recycle gas and partially preheated by indirect heat. Final preheat is obtained in a preheat furnace and the feed to reactor is at a temperature between 455 to 530°C. Pressures of 33-47 kg/cm2 is used. Reactor space velocity varies from 1.5 to 3 volume of feed per hour per volume of catalyst. Hydrogen rich recycle gas is dehydrated and freed from hydrogen sulphide by scrubbing with monoethanolamine (MEA).

Product after passing through the water cooled condensers enters a high pressure separator where the unstabilised reformate is recovered. The major portion of the gas is recycled while the tail gas is sent to the absorber system for removal of C4 and higher hydrocarbons. The two most important reactions are aromatisation (endothermic) which takes place mainly in the first reactor, and the hydrocracking (exothermic) which takes place in the later reactors.

Gasoline yield and its octane number varies with the reforming severity *(i.e.* temperature and pressure mainly). Reforming of a sample of naphtha gave the following result.

Low severity — gasoline yield = 91.5% and octane number = 85 (unleaded)
High severity — gasoline yield = 85.2% and octane number = 95 (unleaded)

'Catforming' is a similar reforming process which permits occasional regeneration of catalyst in situ ; intervals between regeneration varying from several months to one year.

A flowsheet of platforming process is given in Fig. 20.2.

Regenerative Fixed Bed Hydroforming Process

This process uses a catalyst containing 8-10% of molybdena on an alumina support. Temperatures of 480-540°C and pressures from 14 to 20 kg/cm2 are used. The catalyst beads are regenerated after 4 to 8 hours operation. Both upflow and downflow regeneration is done to hold the temperature between 565 to 595°C. A high partial pressure of hydrogen is maintained during hydroforming process by recycling the hydrogen produced by the dehydrogenation of naphthene and cyclisation of paraffins. Desulphurisation also takes place, most of the sulphur present being removed as hydrogen sulphide. Gasoline yield is more than 75% with an octane number of 80-85 (unleaded).

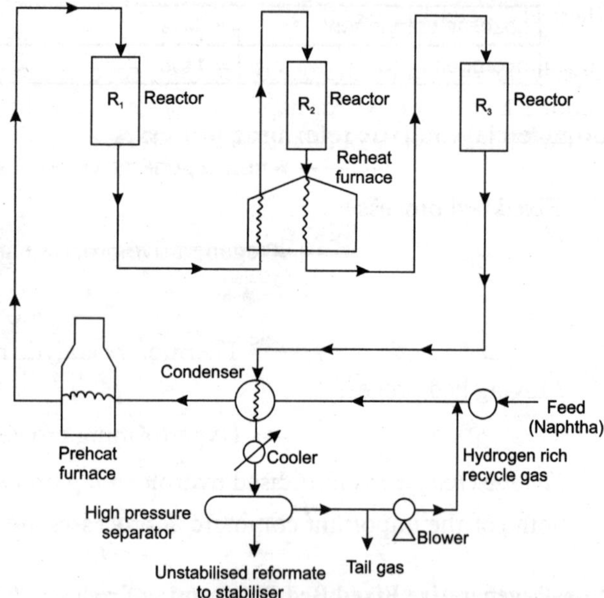

Fig. 20.2. Flowsheet of Platforming

Thermofer Catalytic Reforming (TCR) Moving Bed Process

TCR process uses a single reactor operating at 12 kg/cm2 and 510-540°C. A catalyst regenerating kiln operating at atmospheric pressure and 540-570°C is used to remove coke from the deactivated catalyst. Catalyst is used in the form of beads and contains about 32% chromia on alumina base. Since, it is fairly resistant to poisons in feedstock, no pretreating of feed is necessary.

After being heated separately, the feedstock & recycle stream are combined and enter the reactor. Reactor space velocity is maintained at 0.7 volume oil per hour per volume catalyst. The catalyst to oil ratio (volume) is 0.3. The products are cooled and then sent to fractionator where recycle gas stream and liquid reformate are separated. Catalyst is injected from top hopper and after moving down the reactor goes to a depressurising leg & then to the regenerator. About 25 tons/hr. of catalyst circulates in a reformer designed for 15000 barrels of naphtha/day. Normal catalyst attrition rate is about 0.2 lb/barrel of feed.

A flowsheet of TCR moving bed process is given in Fig. 20.3.

Fluidised Bed Hydroforming Process

In this process, a powdered catalyst containing 10% of molybdena and 90% alumina is circulated between a reactor and a regenerator by the fluidised technique. Fluid hydroformers operate at 480°C and 14-20 kg/cm2 (same as fixed bed hydroforming unit). To maintain this reaction temperature while supplying the endothermic heat of reaction, the recycle gas is heated to 650°C. The feed temperature is kept below 540°C to minimise thermal cracking. Fluidisation enables the temperature to be controlled in the reactor very closely preventing the excessive reforming with production of gases that take place in fixed bed units. Similarly, regeneration temperature is more closely controlled by the use of cooling coils in the regeneration unit. Catalyst circulation rate is about 1 kg catalyst per kg naphtha charge. A gasoline yield of about 80% with octane number of 95 (unleaded) can be obtained by this process.

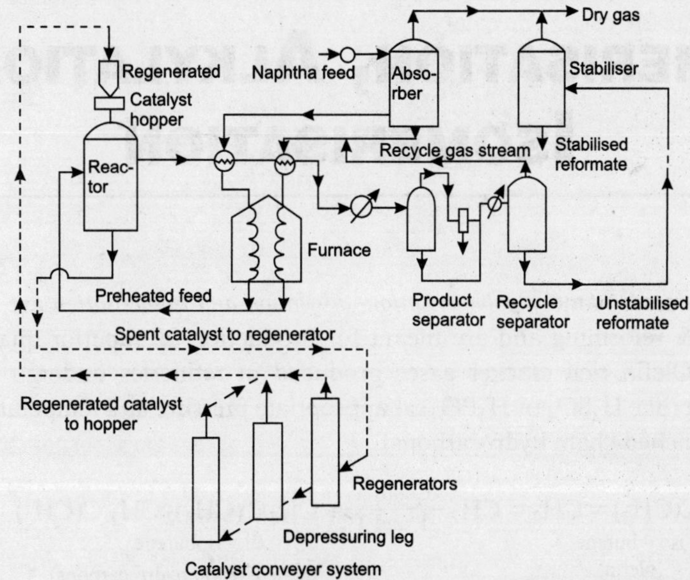

Fig. 20.3. Flowsheet of TCR moving Bed Process.

A flowsheet of fluidised bed hydroforming unit is given in Fig. 20.4.

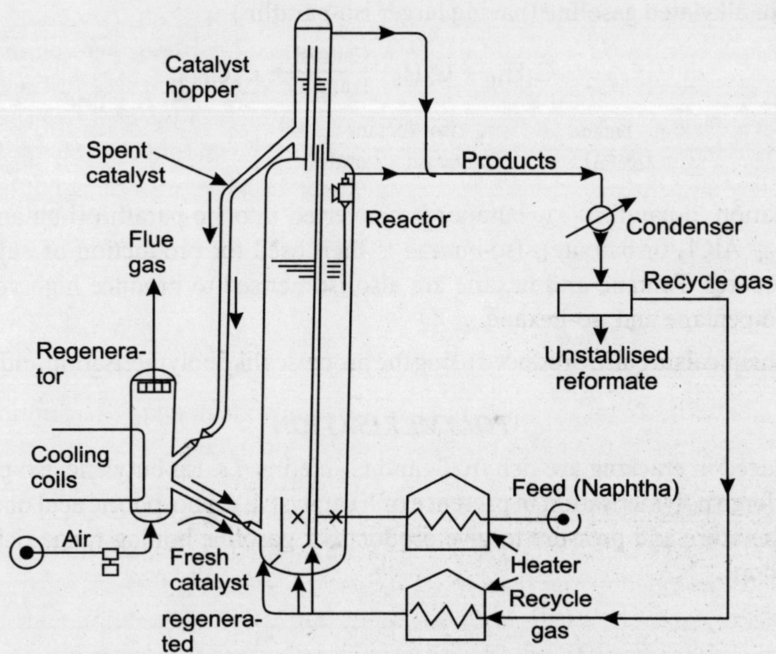

Fig. 20.4. Flowsheet of Fluidsed Bed Hydroforming Unit.

POLYMERISATION, ALKYLATION AND ISOMERISATION

All these three processes namely *polymerisation, alkylation and isomerisation* are complimentary to catalytic cracking & reforming and are meant for production of superior quality higher octane number gasoline. Olefin rich cracker gases produced in refineries undergo polymerisation in presence of catalyst (like H_2SO_4 or H_3PO_4) at appropriate pressure and temperature to give *polymer gasoline* (rich in branched-chain hydrocarbons).

e.g.,

$$\underset{\text{(olefin)}}{\underset{\text{iso – butene}}{2CH_3C(CH_3) = CH_2 = CH_2 = CH_2}} \xrightarrow[\text{P and T}]{\text{Catalyst}} \underset{\text{(branched–chain hydro carbons)}}{\underset{\text{di – isobutene}}{CH_3C(CH_3)_2CH_2.C(CH_3)}}$$

In alkylation, a gaseous iso-paraffin *(e.g.* iso-butane) is reacted with an olefin *(e.g.* propene or butene) in presence of a catalyst *(e.g.* H_2SO_4 or H_3PO_4) at suitable pressure and temperature to produce superior alkylated gasoline (having larger iso-paraffin)

$$C_4H_{10} + C_4H_8 \xrightarrow[\text{P and T}]{\text{Catalyst}} C_8C_{18}$$

iso – butene (iso – paraffin)   n – butene (olefin)   iso – octane (larger iso – paraffin)

In isomerisation, n-paraffin *(e.g.* butane) is converted into iso-paraffin (i-butane) in presence of a catalyst *(e.g.* $AlCl_3$ or bauxite). Iso-butane is then used for production of superior gasoline by alkylation process. Pentane and hexane are also isomerised to produce high volatile gasoline components, iso-pentane and iso-hexane.

Pure petrochemicals are also produced using the processes like polymerisation and isomerisation.

POLYMERISATION

The gases produced in cracking are rich in C_2 and C_4 olefins *(i.e.* iso-butylene, iso-propylene etc.). These gases undergo polymerisation in presence of a catalyst *(e.g.* phosphoric acid or sulphuric acid) at suitable temperature and pressure to give products of gasoline boiling range rich in branched-chain hydrocarbons.

$$\text{e.g., } 2 \underset{\text{isobutylene}}{\overset{CH_3}{\underset{CH_3}{>}}C=CH_2} \xrightarrow{\text{Polymerisation}} \underset{\text{di-isobutylene}}{CH_3-\overset{\overset{CH_3}{|}}{\underset{\underset{CH_3}{|}}{C}}-CH_2-\overset{\overset{CH_3}{|}}{C}=CH_2}$$

The higher liquid hydrocarbons produced (by polymerisation of lower gaseous olefinic hydrocarbon) in the gasoline boiling range is called polymer gasoline. Hence, polymerisation process is mainly for the production of superior gasoline and is complimentary to catalytic cracking & reforming.

Polymerisation is of two types namely thermal polymerisation and catalytic polymerisation. In thermal polymerisation (which is carried out without a catalyst) category, there are two processes namely Unitary thermal polymerisation and Multiple coil thermal polymerisation

Thermal Polymerisation

In Unitary thermal polymerisation, the gases (containing mainly propane, propylene, butane and butylene) from cracker unit are subjected to polymerisation at 510-595°C and 70-310 kg/cm^2. The liquid products produced are polymer gasoline and gas oil. In multiple coil thermal polymerisation, the feed (olefinic gases + paraffinic gases) are subjected to polymerisation at 480-705°C and 3-55 kg/cm^2. This polymerisation takes place in three stages. In the first stage, the olefinic hydrocarbons polymerise. In the second stage, the polymerised olefinic hydrocarbons and the residue gas containing paraffinic gas after polymerisation are cracked. In the third stage, the olefinic hydrocarbons obtained from cracking in the second stage are polymerised. The products obtained are gasoline, fuel oil and tar.

Though polymer gasoline can be produced from any hydrocarbon (whether olefinic or paraffinic), very severe conditions are required for paraffinic gases lighter than butane. It is not necessary to convert all butane and butene produced in refinery because they can be blended into gasoline upto 10%.

Catalytic Polymerisation

The first catalytic polymerisation process employed sulphuric acid as a catalyst. There are two processes under this heading namely cold acid process and hot acid process. The other process uses phosphoric acid as catalyst.

Cold Sulphuric Acid Polymerisation Process. In this process, i-butene is polymerised selectively to di-isobutylene and iri-isobutylene. The cold acid process consists of two steps i.e. extraction of i-butene followed by polymerisation. The extraction of i-butene is done by 65% sulphuric acid at 21 to 38°C in two stages. The extract phase is separated and heated to about 100°C for about 1 minute to effect polymerisation. It is then cooled to 38°C and the product is separated from acid catalyst by settling. Conversion of i-butene is about 90%. The product is generally 75% dimer *(i.e.* di-isobutylene) plus 25% trimer (i.e. tri-isobutylene).

A flowsheet of cold acid process is given in Fig. 21.1.

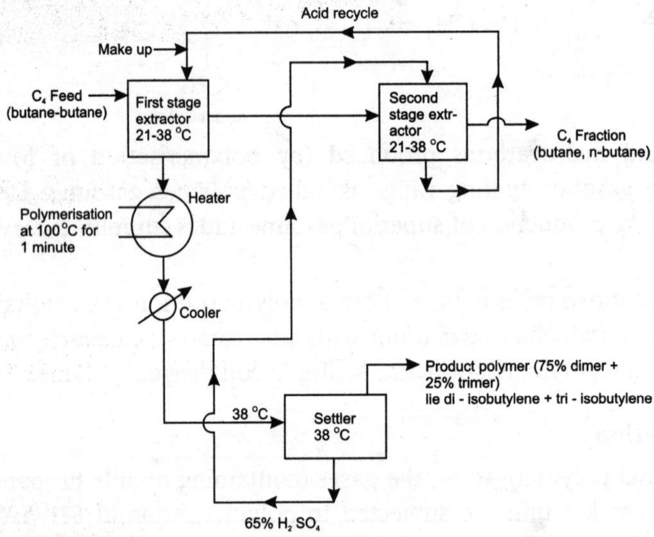

Fig. 21.1 Cold Acid Polymerisation Process.

Cold acid process is also important in preparing feedstock for the manufacture of butadiene. *Iso*-butene which interferes with the production of butadiene is removed from butanes-butenes*via*. cold acid route.

Hot Sulpharic Acid Polymerisation Process

In hot acid process, all the butenes is polymerised (unlike cold acid process which polymerises only *i*-butene). In this process, extraction and polymerisation are carried out in single step at about 100-140°C. A hydrocarbon-acid emulsion is circulated through the reactor to reduce the concentration of iso-butene in the feed. Circulation promotes copolymerisation of more reactive *i*-butene with the less reactive n-butenes and thereby improves yields. Reactor effluent is settled to separate product from the catalyst, which is recycled. The product is fractionated to eliminate butenes and then hydrolysed with caustic to breakdown any acidic substances. At contact time of 10-15 minutes, all the i-butene and an equal amount of n-butenes are converted. The product contains 90-95% octenes.

A flow sheet of hot acid polymerisation process is given in Fig. 21.2.

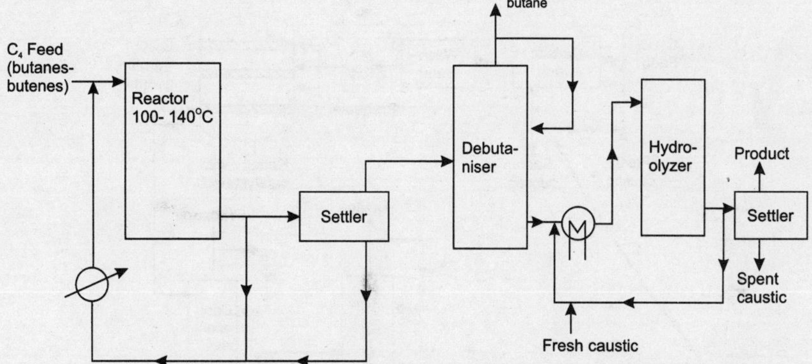

Fig. 21.2. *Hot Acid (H_2SO_4) Polymerisation Process.*

Earlier products from sulphuric acid polymerisation units were fractionated to separate the dimers which were then hydrogenated to produce aviation gasoline components. Except for those units which prepared feedstocks for butadiene production, cold acid plants were later converted to hot acid plants due to higher production rate of the latter.

Phosphoric Acid (H_3PO_4) Polymerisation Process

In this process of catalytic polymerisation, H_3PO_4 catalyst used are of four types—three of them use solid catalyst and the fourth employs liquid phosphoric acid. Commercial solid catalysts include (a) phosphoric acid on kieselguhr or on silica (b) copper pyrophosphate on charcoal. However, the latest is the use of liquid phosphoric acid as polymerisation catalyst. Liquid phosphoric acid is always used with 2% steam to prevent meta and para phosphoric acid formation which are inactive.

Feed is pretreated to remove hydrogen sulphide and mercaptans which would otherwise enter the product and reduce its octane rating. Feed is washed with caustic to remove H_2S &mercaptans and is then washed with water to remove organic bases and any carry over of caustic which poisons the catalyst. Both the feedstock and wash water should be free of dissolved oxygen as it promotes the deposition of tarry material on the catalyst.

Since polymerisation units are operated at high pressure, coke deposition on beaded solid catalyst is greatly reduced, hence catalyst regeneration facility is not provided. Most units employ solid catalyst.

After pretreatment of feed, it is passed over a bed of catalyst at a temperature of 150-225°C and a pressure of 10-80 atm. Olefin conversion to gasoline are 85% or more. To remove the large

exothermic heat of reaction, olefin lean stream (*i.e.* C_3 - C_4 paraffin quench) is recycled to act as heat absorber to control the reaction temperature. When the liquid H_3P0_4 is used as catalyst, only a reaction tank will substitute the packed bed reactor. In this case, temperature is controlled by circulating the liquid catalyst through a cooler.

Typical process conditions in the reactor having beaded solid catalyst for polymerisation are 230°C and 60-70 atm. for C_3 rich feed and 200°C and 40-45 atm. for C_4 rich feed. Quench stream is C_3-C_4 paraffins for cooling and control of olefin concentration in reactor to 25-30%. Beyond this concentration, polymerisation is vigorous. Product has an octane number (motor) of 80-81. A flow sheet of phosphoric acid polymerisation unit is given in Fig. 21.3.

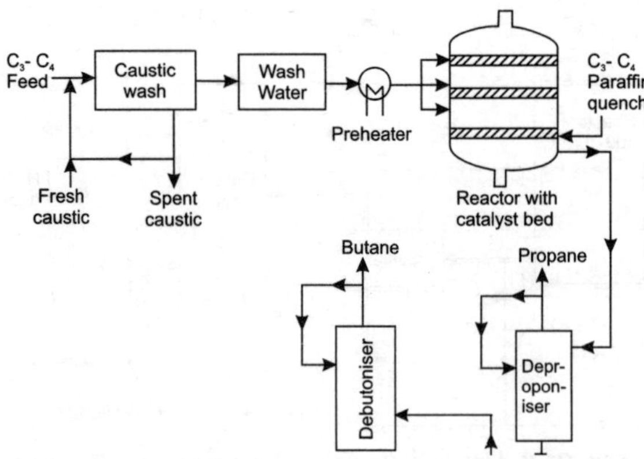

Fig. 21.3. Phosphoric Acid Polymerisation Unit.

Effect of reaction conditions in polymerisation. Conversion of olefins to polymers depend upon temperature, pressure, reaction time, catalyst activity and the feed composition. Reaction rates increase with temperature, however higher temperature also increases rate of accumulation of tarry deposits on catalyst thereby shortening its life.

Increasing the pressure increases conversion by increasing contact time. Higher pressure permits the use of lower temperatures and thus reduce the formation of heavy polymers & coke.

Reaction time must be long enough to give satisfactory conversions, but not so long as to cause the production of polymer chains too long to be included in gasoline.

Reaction rate differs for the several olefins, /so-butene is the most reactive which accelerates the polymerisation of n-butene which in turn accelerates the polymerisation of propene.

Polymerisation product quality. Polymer gasolines are chemically stable, clean burning fuel of high octane quality. They are usually mixture of polymers, mainly dimers and trimers. With a propene feed, trimers predominate while with a butene feed, dimers are major constituents.

Catalyst life. Catalyst life ranges from 100 to 200 gallons of polymer per pound for phosphoric acid on kieselguhr and between 100 to150 gallon per pound for copper pyrophosphate. Catalyst life in a liquid phosphoric acid operation exceeds 45 gallons of polymer per pound of acid.

ALKYLATION

Alkylation is a reaction in which an alkyl group is added to a compound. In petroleum refining, it normally means reaction of an olefin with an i-paraffin to produce a larger i-paraffin having higher octane number.

In typical alkylation process, a gaseous *i*-paraffin *(e.g.*iso-butane) is reacted with an olefin *(e.g.* propene or butene) in presence of sulphuric acid or hydrofluoric acid as catalyst and the result is a larger iso-paraffin (called alkylate) boiling in the gasoline range

$$\underset{\text{i-butene}}{C_4H_{10}} + \underset{\text{n-butene}}{C_4H_8} \xrightarrow{\text{alkylation}} \underset{\text{i- octane}}{C_8C_{18}}$$

Another typical alkylation reaction is given below

$$i\text{-butane+propene} \xrightarrow{\text{alkylation}} i\text{-hetane}$$

Alkylation reactions. The reactions occurring during alkylation are :

(a) *iso* - paraffin + olefin □higher iso - paraffin. This is the main reaction.

(b) Olefins $\xrightarrow{\text{Polymerisation}}$ dimer +*iso* - butane → iso - paraffin

(c) Hydrogen transfer : $2i - C_4H_{10} + nC_4H_8 \rightarrow C_8H_{18} + C_4H_{10}$

iso-butane n-butane iso-octane n-butane

Alkylation processes. There are two processes namely thermal and catalytic alkylation. In thermal alkylation, iso-paraffins are alkylated without catalyst but higher temperature (550°C) & pressure (250-350 atm) are required. As for example, neohexane (2-2 dimethyl butane) having an octane number of 94 can be produced by thermal alkylation by the reaction as shown below :

iso - butane + C_2 olefin→neohexane

Normally, reactivity of iso-butane decreases with increase in olefin carbon atom.

Catalytic alkylation process. In this process, catalyst like sulphuric acid, hydrogen fluoride and aluminium chloride are used for alkylation.

Sulphuric acid is used as an alkylation catalyst with propene and higher boiling olefins. The acid concentration is maintained between 90 to 98%. Less than 90% acid concentration promotes polymerisation whereas more than 98% acid concentration promotes cracking reaction. Acid catalyst deactivation rate depends upon the type of feed and the i-butane charge rate. Acid consumption ranges from 0.3 to 2 pounds per gallon of alkylated product.

Hydrogen fluoride having an acidity *(i.e.* concentration) of 85-92% is also used as catalyst for alkylation of i-butane with propene and higher boiling olefins. High acidity is used to maintain high conversion and high antiknock quality of alkylated gasoline. Hydrogen fluoride consumption ranges from 0.2 to 0.8 pound per barrel of product.

Aluminium chloride (with HCl as promoter), is used as a catalyst to produce di-isopropyl (for aviation gasoline) from *i*-butane and ethylene by alkylation. The optimum aluminium chloride concentration in the catalyst is 63-84% and its consumption is 1.4 to 3 pounds per barrel of product.

Alkylation process variables. Important process variables in alkylation are i-butane/ olefin charge ratio, i-butane content in reactor effluent, temperature, olefin space velocity and contact time. Their effects are discussed below :

***i*-butane/olefin ratio,.** iso-butane/olefin ratio is maintained at 5 to 10 in commercial alkylation plants. Higher ratio produces product with higher octane number'and having lower final boiling point besides reducing catalyst consumption rate.

***i*-butane in effluent.** Increasing the *i*-butane concentration in the reactor effluent has the same effect as increasing the i-butane/olefin ratio and it can be done by reducing the amount of n-butane charged to the reactors in the fresh feed and in the i-butane recycle stream.

Temperature. Low temperatures are used in all alkylation processes.

Operating temperature in various processes are:

(a) H_2SO_4 alkylation process — 5-10°C (This requires refrigeration).

(b) HF alkylation process — 25-38°C

(c) $AlCl_3$ alkylation process — 45-60°C.

If temperature is less, then the octane number of final alkylate is more and yield of polymer is low. If temperature is too low then alkylate yield is reduced whereas at very high temperature, acid attacks olefins.

Space velocity of olefins. It means volume of olefins charged per hour per unit volume of acid in the reactor. Lower olefin space velocity improves the octane number of product, reduces the production of heavy materials and lowers the consumption of acid catalyst.

Contact time. It is defined as the residence time of the fresh feed and externally recycled i-butane in the reactor. Contact time for H_2SO_4 alkylation process is 5-45 minutes and for HF alkylation process it is 5-25 minutes. Long contact time in the absence of i-butane is undesirable because olefins undergo side reaction thereby lowering the product quality.

Feed stocks for alkylation. Any olefin containing hydrocarbon can be used to alkylate i-butane. The chief sources of olefins are catalytic cracking and dehydrogenation of paraffins (mainly butane). Butane is also isomerised to i-butane which is cracked thermally to produce an alkylation feed. Iso-butane is obtained from crude oils, from cracking operations, from catalytic reformers and from natural gas.

Alkylation feed should be free from hydrogen sulphide and mercaptans which makes the product corrosive, lowers leaded octane rating of products & increases the catalyst consumption rate.

Sulphuric acid alkylation process. Feed is given caustic wash to remove sulphur compound and then contacted with recycle i-butane. It is cooled to the required low temperature and contacted with the acid catalyst. Catalyst acid is settled from the hydrocarbon product and recycled. Caustic wash is then applied to remove traces of acid from the product before fractionating recycle i-butane stream from it. To prevent propane from building up in the recycle i-butane, a depropaniser is also provided. Also, a debutaniser is usually included to separate n-butane from the product and a rerun tower is required to remove heavy ends if the product is to be blended into aviation gasoline.

A flow sheet of the process is given in Fig. 21.4.

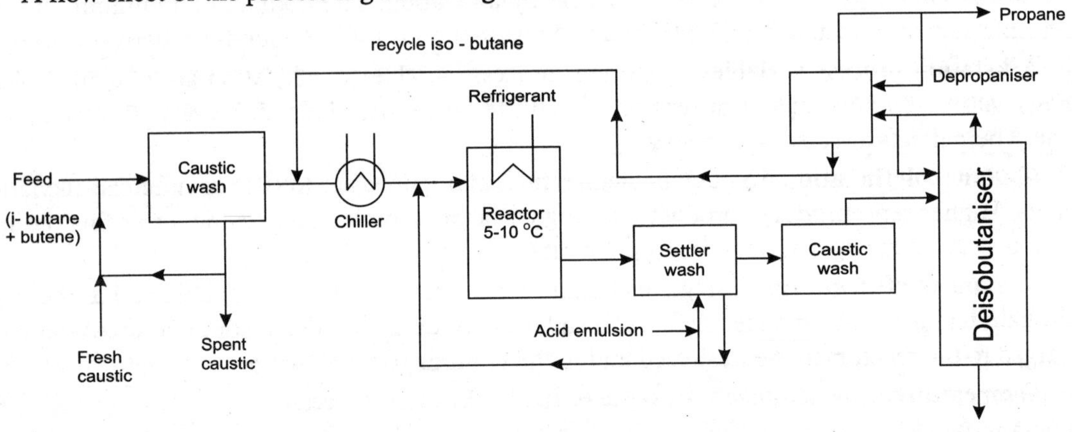

Fig. 21.4: Sulphuric Acid Alkylation Process.

Sulphuric acid alkylation unit needs a refrigeration system to cool the reactants to 5-10°C.

Hydrogen Fluoride Alkylation Process. The process is almost same as H_2SO_4 alkylation on process. This process does not require refrigeration equipment as the reactor temperature is 25-38°C. However, it requires, additional equipment to dry the feed, to strip HF from product, to rerun a slip stream of the circulating catalyst to reject water & tar and to remove traces of flourinated compounds from the product. HF is corrosive at working concentration, hence costly alloy monel is used in the equipment.

Acid/hydrocarbon ratio varies from 0.2 to 1 and optimum*i* C_4/olefin ratio is about 7 in this process. A flow sheet of the process is given in Fig. 21.5.

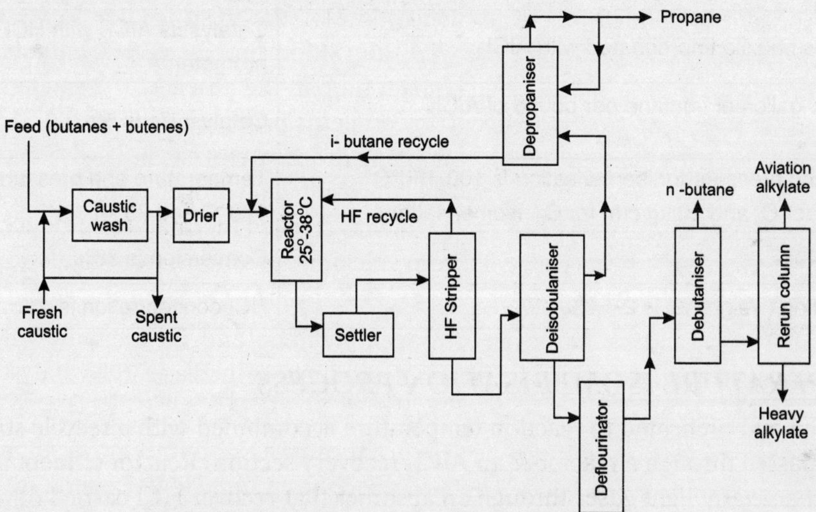

Fig. 21.5: Hydrogen Flouride Alkylation Unit.

Since, hydrogen fluoride is slightly soluble (upto 1%) in the hydrocarbon product, the reactor effluent is stripped in a column packed with aluminium rings.

Table 21.1: Comparison of Alkylation Processes

H_2SO_4 alkylation process	HF alkylation process
Octane number (research) of product is 95.	Octane number of product is 93.
Plant running cost is more as refrigeration unit is there.	Initial investment is more as costly monel clad equipments needed.
Acid consumption is more.	Acid consumption is less.
Spent H_2SO_4 disposal is a problem.	HF poses safety hazard, hence protective clothing required for handling.

ISOMERISATION

Isomerisation process converts *n*-paraffins to *iso*-paraffins (which has a higher octane number) in presence of $AlCl_3$ as catalyst. Commercially, it is mostly practised to obtain iso-butane from *n*-butane which is then used in the production of good quality gasoline by alkylation process.

Pentane and hexane are also isomerised to produce *i*-pentane and *i*-hexane which are high volatile gasoline components.

Isomerisation processes. There are two processes for isomerisation of paraffins namely low temperature process and high temperature process, the former being the more common process. Features of the low temperature process are given in the Table 22.2.

Table 21.2: Features of Low Temperature Process (for butane isomerisation)

Vapor phase batch process	Liquid phase continuous process
This process has been developed by shell oil company.	This process has been developed by standard oil company.
Catalyst used is bauxite impregnated with $AlCl_3$.	Catalyst is $AlCl_3$ with HCl as promoter.
Catalyst life *i.e.* gallon of i-butane per pound of $AlCl_3$ used is 200.	Catalyst life is 50-100.
Temperature and pressure for isomerisation is 100-150°C and 14 kg/cm^2 for C_4 and 20 kg/cm^2 for C_3 isomerisation.	Temperature and pressure is 100°C and 25 kg/cm^2.
Conversion is 40%.	Conversion is 38%.
HCl concentration (weight %) is 2-14%.	HCl concentration is 4%.

LOW TEMPERATURE ISOMERISATION PROCESS

Feedstock, dried and preheated to reaction temperature is combined with a recycle stream, mixed with HCl and passed through a reactor & an $AlCl_3$ recovery section. Reactor effluent is cooled and flashed to discharge any light gases through an absorber that recover HCl carried off in the gases. Liquid from the flash drum is stripped to recover HCl and is caustic washed to remove the traces of acid. Recycle paraffin stream is fractionated from the product isomer and then recycled back to the reactor.

Flow sheet of liquid phase, low temperature continuous isomerisation is given in Fig. 21.6.

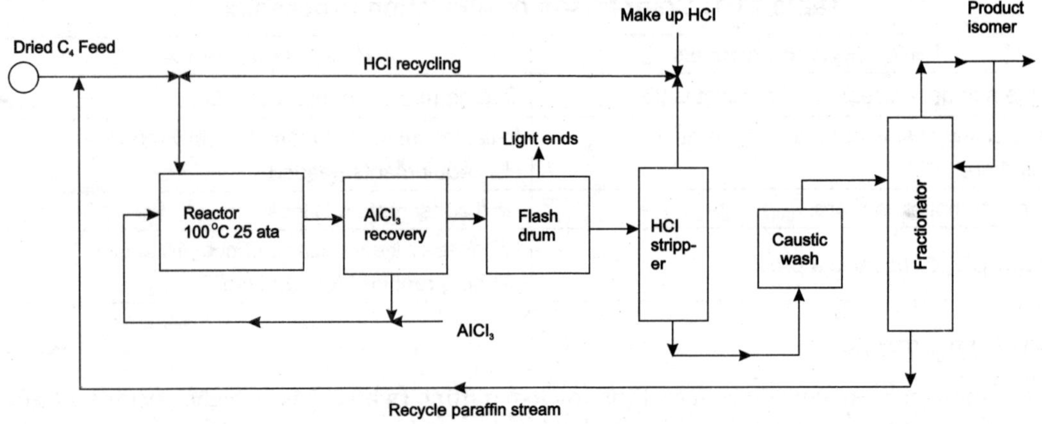

Fig. 21.6: *Low Temperature Liquid Phase Butane Isomerisation Process.*

The HCl stripping column is operated at high enough pressure so that the stripped HCl can be returned directly to the reactor.

In the vapour phase butane isomerisation process, $AlCl_3$ is adsorbed on refined bauxite to prepare the catalyst. In vapour phase batch process, the plant has to be shutdown periodically to change the catalyst.

For pentane and hexane isomerisation, the process conditions are same as that for butane except that inhibitor *(e.g.* hydrogen at high pressure) are used to suppress the side reactions.

high temperature isomerisation process. In this process, catalyst used in the reactor is platinum on silica-alumina and the reaction conditions are 370-480°C and 20-45 kg/cm^2. Typically, a pentane-hexane feed combined with pentane recycle stream is fractionated to take i-pentane overhead. The fractionator bottoms are mixed with hydrogen, preheated and charged to the reactor. Reactor effluent is cooled and flashed to separate recycle hydrogen from the product which is stabilised and depentanised. The pentanes are recycled to the deisopentaniser. *Iso*-pentane and the iso-hexane bottoms from the depentaniser constitute the products.

Hexanes isomerise more readily than pentanes. However, with a nickel catalyst, equal rates are obtained when hexanes are processed at a temperature about 20°C lower than pentanes. To minimise cracking reactions hexanes are processed at lower temperature than pentanes. Desulphurisation of feed and removal of nitrogen is done to improve product quality and life of catalyst.

Product yield varies from 94 to 98% (by volume) and the octane number ranges from 66 to72.

A flowsheet of high temperature isomerisation is given in Fig. 21.7

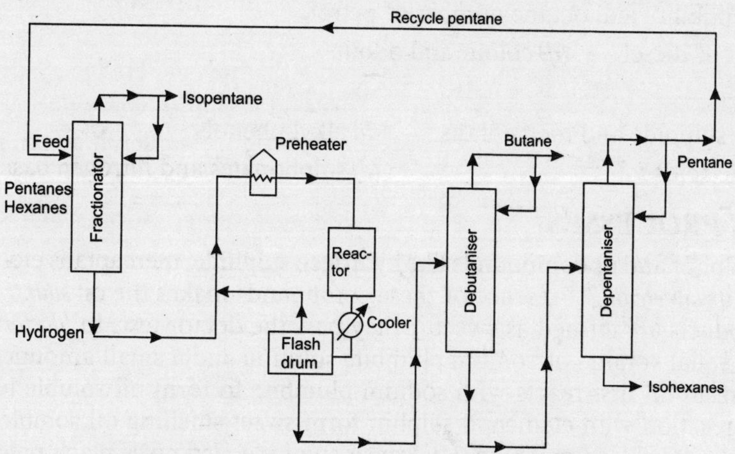

Fig.21.7: High temperature isomerisation process

Low temperature isomerisation processes are more common for butane isomerisation. High temperature isomerisation process, however, has the advantage of avoiding the handling of $AlCl_3$ and HCl. Balanced against this advantage are the less favourable paraffin equilibriums at higher operating temperature.

PURIFICATION OF PETROLEUM PRODUCTS

Impurities and undesirable components in petroleum products come from the original crude or may develop during its processing.

Purification processes are aimed at

- minimising
  - *(a)* acid and corrosion
  - *(b)* carbon residue
  - *(c)* catalyst deactivation
  - *(d)* gum formation.
- improving
  - *(a)* burning characteristics of kerosene, diesel fuels etc.
  - *(b)* lead susceptibility and octane number of petrol.
  - *(c)* pour point of diesel.
  - *(d)* colour and odour.
- *recovering*
  - *(a)* hydrogen sulphide and mercaptans
  - *(b)* alkyl phenols
  - *(c)* naphthenic acids
  - *(d)* sulphonates and nitrogen bases.

SWEETENING PROCESSES

Removal of sulphur and its compounds like hydrogen sulphide, mercaptans etc. from petroleum products is called its *sweetening.* Presence of these compounds makes the oil *sour.*

Petroleum products are termed as sweet if they pass the doctor test. In *'doctor test',* oil sample is shaken with an equal volume of sodium plumbite solution and a small amount of sulphur. The mercaptans present in oil first reacts with sodium plumbite to form oil soluble lead mercaptides which on further reaction with elemental sulphur form sweet smelling oil soluble disulphide and give a black precipitate of lead sulphide. Hydrogen sulphide also gives black precipitate with the plumbite solution. If this test does not produce dark colour at the interface, the oil sample is sweet.

Sour stocks have bad odour. The processes by which the foul smelling sour petroleum products are converted into sweet smelling or odourless products are called sweetening processes.

Important sweetening processes are :

- Doctor's sweetening process
- Copper chloride sweetening process
- Solutizer process
- Catalytic desulphurisation process
- Hydrofining process.

Doctor's Sweetening Process. In this process, mercaptans present in the petroleum products are converted into disulphide by treatment with doctor's solution (sodium plumbite solution) and sulphur. Lead sulphide is converted back (regeneration) to sodium plumbite by air blowing the solution.

Reactions : $$\underset{\text{mercaptans}}{2RSH} + NaPbO_2 \rightarrow Pb(RS)_2 + 2NaOH$$

$$Pb(RS)_2 + S \rightarrow PbS + RSSR$$

Regeneration : $$PbS + 2O_2 \rightarrow PbSO_4$$

$$PbSO_4 + 4NaOH \rightarrow Na_2PbO_2 + Na_2SO_4 + 2H_2O$$

Because the presence of sodium sulphate reduces the solubility of lead sulphate, a portion of the sodium is discarded periodically to control the build up of sodium sulphate.

Discarded doctor solution may be regenerated by settling to separate an emulsion of caustic, oil & lead sulphide from a top oil phase and a bottom sodium plumbite phase ; breaking the emulsion by heating it to 65°C & separating a slurry of lead sulphide and air blowing it at 80°C for about 43 hours at atmospheric pressure.

Doctor's sweetening process is carried out at *65°C.* 20 parts of doctors solution are circulated per 100 parts of lead. Feed and doctor solution are contacted for 30 seconds before sulphur solution is added and the total mixture is then stirred for 5-15 minutes. Sulphur addition should be controlled as excess sulphur produces polysulphides [$Pb(RS)_2$+2S ->PbS+RSSR] which reduces octane rating of gasoline and lower the burning qualities of distillate fuels.

Regeneration of doctor's solution is carried out at 80-110°C and 1-7 atmosphere pressure.

Flowsheet of the process is given in Fig. 22.1.

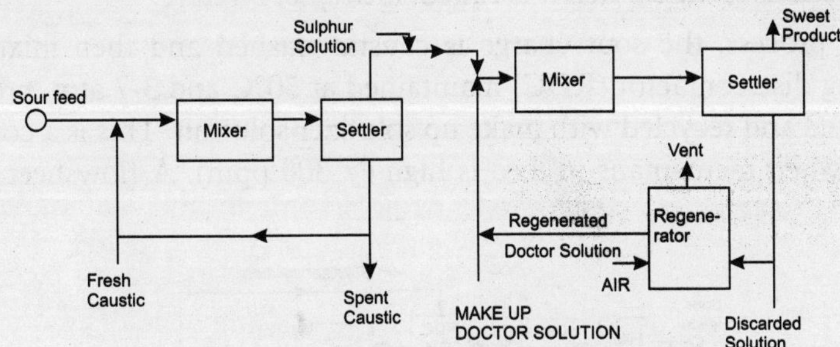

Figure:22.1. Doctor's sweetening process

Copper Chloride Sweetening Process. In this process, no extra sulphur is added. Catalyst is a slurry of $CuCl_2$.

Reactions : $$4RSH + CuCl_2 \rightarrow RSSR + HCI + Cu_2Cl_2$$

$$Cu_2Cl_2 + 4HC1 + O_2 \rightarrow 4CuC1_2 + 2H_2O$$

This process is used for the sweetening butane, gasoline, fuel oils and kerosene. After caustic washing & warming, the feed is mixed with air or oxygen and then with a slurry of copper chloride (one part) & clay (four parts). The mixture is sent to a reactor-separator from which catalyst is recycled to the feed. Treated product is water washed before withdrawing from the system. Reaction temperature is between 25 to 50°C. At lower temperature, the reactions are slow whereas at higher temperature there are undesirable side reactions.

A flowsheet of the process is given in Fig. 22.2.

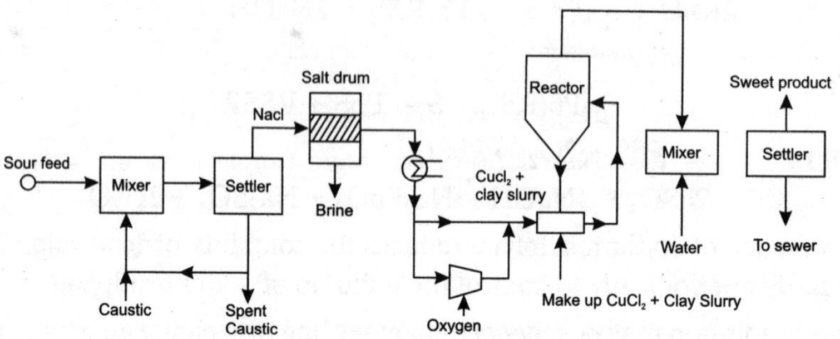

Fig. 22.2. $CuCl_2$ Sweetening-Slurry Process.

Solutizer Sweetening Process

This process removes the mercaptans by dissolving out. Caustic soda treatment to sour stock removes only lower mercaptans. The higher molecular weight mercaptans have only limited solubility in this reagent. However, some solutizer like fatty acids, cresols and naphthenic acid enhance the solubility of higher mercaptans in caustic potash. This forms the underlying principle of solutizer process.

When methanol is used as solutizer solution then the process is known as *Unisol Process.* When naphthenic acid is used as *solutizer,* then it is called *'Mercapsol Process'.*

In the solutizer process, the sour charge is caustic washed and then mixed with solutizer solution in a rotating disc contactor (RDC) maintained at 30°C and 3-7 atm. pressure. The spent solution is regenerated and recycled with make up solutizer solution. This is a costlier process and hence is used only when mercaptans in feed is high (> 300 ppm). A flowsheet of the process is given in Fig. 22.3.

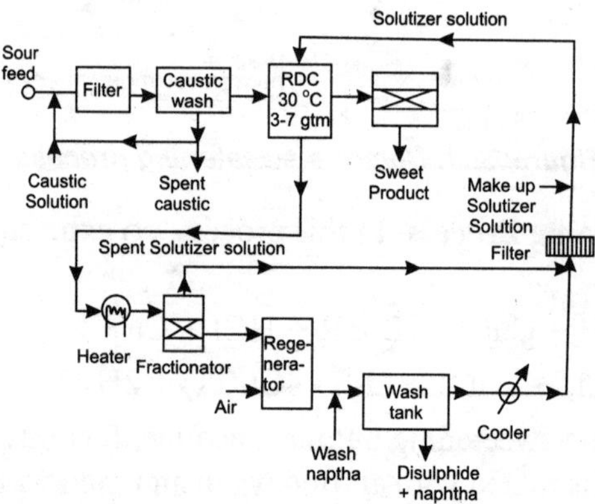

Fig. 22.3. Solutizer sweetening process.

Catalytic Desulphurisation Process

Catalysts like bauxite, fuller's earth, activated clay etc. catalyse the conversion of organic sulphides and mercaptans to hydrogen sulphide& hydrocarbons at 350-400°C.

eg; $C_9H_{19}SH = C_9H_8 + H_2S$

mercaptans

$$2C_9H_{19}SH = C_9H_{19}SC_9H_{19} + H_2S$$

$$C_9H_{19}SC_9H_{19} = C_9H_{19}SH + C_9H_{18}$$

It is used for sweetening (desulphurisation) of straight run gasoline and kerosene. Removal of sulphur from 0.425 to 0.011% and octane number improvement from 45 to 48 can be achieved.

Hydrofining Desulphurisation Process

This process is the most recent and the most effective desulphurisation process. In this process, catalytic desulphurisation is carried out in presence of hydrogen recirculation. Hydrogen enhances the process of desulphurisation and minimises coke formation. Reaction conditions are 350-450°C and 20-50 atmosphere depending upon the nature of feedstocks. Operating conditions for various feedstocks are given in Table 22.1.

Table 22.1: Operating Parameters for Various Feedstocks

Parameters	Unit	Straight run gasoline	Jet fuel	Catalytically cracked light gas oil (LGO)
Temperature	°C	390	350	395
Pressure in reactors	atm.	36	30	40
Temperature in the separator	°C	200	170	190
Yield	%	96.5	98	97.5

Catalyst like molybdenum oxide, cobalt oxide and a mixture of the two are used for desulphurisation in presence of H_2. The exothermic heat of reaction (30 kcal/kg mole) is removed by cold H_2 circulation at the rate of 300-600 Nm^3/m^3 of charge. Concentration of hydrogen in recycle gas stream is maintained at 70-90%. A flowsheet of the process is given in Fig. 22.4.

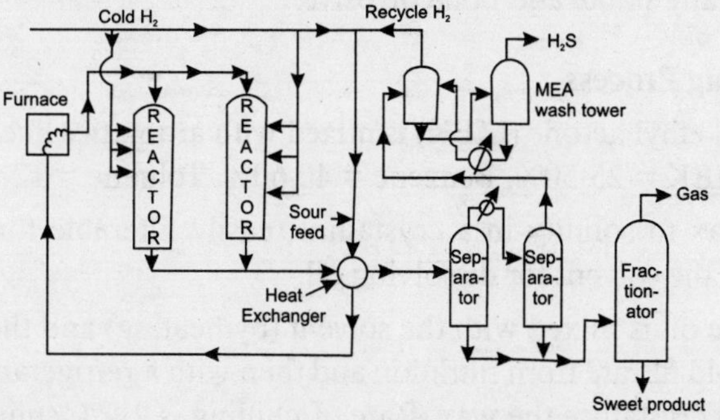

Fig. 22.4. Hydrofining desulphurisation/sweetening process.

DEWAXING OF PETROLEUM PRODUCTS

Removal of wax from petroleum product (mainly the lubricating oil) is called dewaxing. Petroleum waxes are solid hydrocarbons (melting point 35-95°C) and are soluble in crude oils & their products. Waxes can't be separated by distillation as their boiling points overlap with the petroleum products. At lower temperatures, waxes crystallise & precipitate out from diesel fuels and lubricating oils. Hence, to maintain their fluidity they must be dewaxed. There are three methods of dewaxing :

(a) Chilling and Pressing Process *(b) Solvent Dewaxing Process .* — MEK Dewaxing / Propane Dewaxing

(c) Urea dewaxing Process.

Dewaxing raises the viscosity of the oil (*i.e.* the viscosity index is lowered) because the viscosity of dissolved wax is lower than that of oil with which it is associated.

Chilling and Pressing Dewaxing Process

This is the oldest method in which wax bearing oil is chilled by refrigeration in double pipe heat exchangers to crystallise the waxes which are then separated by filtration, centrifuging or settling. Since the viscosity of oil increases after chilling, filtration becomes difficult, time consuming and costly. Hence solvent dewaxing method has been developed.

Solvent Dewaxing Process

In this process, wax bearing oil is first dissolved in the solvent, and the solution is chilled. The crystallised wax is then more conveniently separated from the solution in vacuum filters. Methyl Ethyl Ketone (MEK) and propane are the two solvents widely used in the process.

Requisites of a good dewaxing solvent:

- It should have complete miscibility with oil at high temperature.
- It should have negligible solubility of wax.
- Presence of wax in solvent should be in filterable form.
- Its boiling temperature should be low so as to facilitate easy separation for easy recovery without much heat expense.
- It should have low solvent to feed oil ratio.
- It should be thermally stable and non-corrosive.

MEK (Solvent) Dewaxing Process

Normally, solvent methyl ethyl ketone (MEK) is mixed with aromatics like benzene and toluene in following proportion : MEK = 25-50%, Benzene = 40-60%, Toluene = 12-15%.

The MEK causes wax to solidify in a crystalline, easily filterable form while the aromatics increases the capacity of the solvent for dissolving oil.

The wax bearing lube oil is mixed with the solvent (by heating) and the mixture is chilled first by heat exchange with cold filtrate from filtration and then with a refrigerant (ammonia or propane refrigeration to -20°C) to crystallise the wax. Rate of chilling is 2-7°C/minute. The chilled feed is filtered continuously on rotary drum vacuum filter to separate the crystallised wax from the oil-solvent solution. Wax cake is washed with chilled solvent to displace retained oil. Filtered solution is heated to flash off the solvent which is condensed and recycled. Oil from the flashing operation still contains traces of solvent which are stripped out by steam.

The wax cake contains the original water content as ice and also some solvent. The cake is melted and sent to a settling drum where melted wax & water separate, solvent distributing between them. Dilute solvent from oil stripper is also charged to the settling drum, some of its solvent passing to the melted wax. The aqueous phase from the drum is fractionated to reject water as bottoms, the overhead being recycled to the drum to recover its solvent content by transfer to the melted wax. Melted wax from the drum is sent to an evaporator and a steam stripper. Solvent from the evaporator is returned to the warm solvent stream. Dilute solvent from the wax stripper is sent back to the wax settling drum. A flowsheet of the MEK dewaxing process is given in Fig. 22.5.

Solvent to feed oil ratio in MEK Dewaxing process generally ranges from 1 : 1 to 4 : 1. Heavier feedstocks require larger ratio. Yield of dewaxed oil is 89 to 92% while that of wax is 8 to 11%.

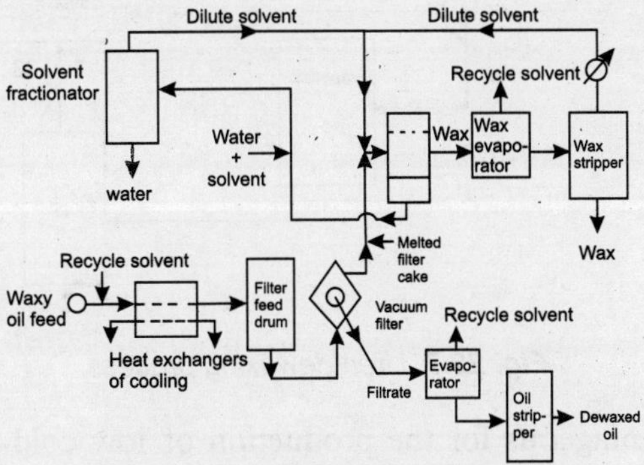

Fig. 22.5. MEK dewaxing process.

Propane Dewaxing Process

Propane dewaxing unit is almost similar to MEK dewaxing plant. However, compared to MEK dewaxing process, it is advantageous in the sense that a part of solvent propane itself vaporises and provides refrigeration effect. The comparative disadvantages are that a high pressure is maintained in the propane evaporation system. A flowsheet of the propane Dewaxing process is given in Fig. 22.6.

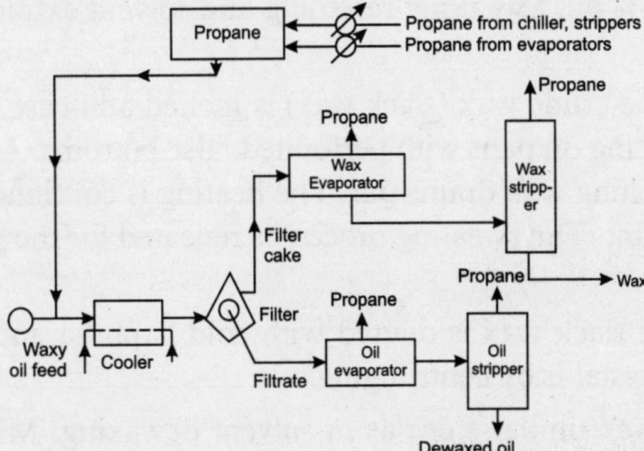

Fig. 22.6. Propane Dewaxing Process.

Solvent to feed ratio is generally higher (between 2 to 3). If it is less than 1.6, then gel formation occurs. Propane dewaxing requires a slightly lower temperature than MEK dewaxing.

Urea Dewaxing Process

This process is based on the fact that urea forms solid, filterable complexes (adducts) with large paraffins (constituents of wax) having little or no branching. Urea dewaxing is very effective on higher stocks. Activator like methyl alcohol is used for adduction to be completed in a reasonable time. Decomposition of adduct is done by dissolving the urea in a suitable solvent or by heating it to about 90°C. A flowsheet of the process is given in Fig. 22.7.

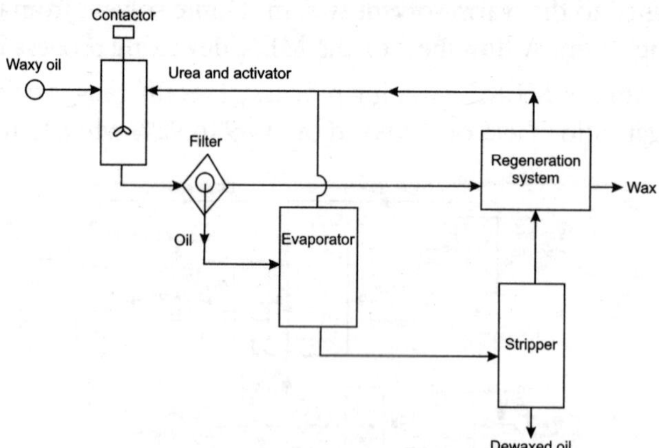

Fig. 22.7. *Urea dewaxing process.*

Urea dewaxing is advantageous for the production of low cold-test oils from light oils that contain little amount of wax *e.g.* from solvent dewaxed oils. Because, the wax in lower viscosity oils contains a higher percentage of straight chain hydrocarbons, urea dewaxing is more effective on lighter stocks.

Deoiling of Wax

Deoiling of wax is done to remove the oil left in the wax produced in dewaxing process. The oil content in wax may be upto 50%, if chilling is done without solvent. Finished wax should not contain more than 0.5% oil. Sweating, resettling and solvent extraction are the three deoiling processes.

In sweating process, the crude wax (slack wax) is melted and cast into thick sheets which are then subjected to slow heating on pans with perforated false bottoms. As the temperature rises, the mixture of oil and low melting wax drains out. The heating is continued until the remaining wax has the desired melting point. The sweating process is repeated for the production of good quality wax.

In settling process, the slack wax is diluted with cold naphtha and recrystallised by chilling. Final separation of wax crystal is by centrifuging.

Solvent deoiling involves similar steps as in solvent dewaxing. MEK or propane is used for deoiling wax.

ACID, ALKALI AND CLAY TREATMENT OF PETROLEUM PRODUCTS

Acid Treatment. Petroleum products of all boiling ranges (naphtha to lubricating oils) are given finishing treatment with sulphuric acid. A mild acid treatment removes the unstable gum forming & colour forming materials (mainly diolefins) and severer treatment removes sulphur as well from petroleum products. In nutshell, *sulphuric acid treatment* of petroleum products :

- removes nitrogen bases, asphaltic substances, diolefins, olefins, acetylenes, aromatics (by higher concentration acid), sulphur compounds (like thiophenes, mercaptans, sulphides).
- Improves colour, stability and odour.
- reduces corrosiveness and gum formation.
- does not remove elemental sulphur, carbon disulphide and some high molecular weight sulphur compounds.

Drawback of sulphuric acid treatment of petroleum products are :

- large refining losses due to formation of acid sludge.
- problem of disposal of acid sludge.

Acid treatment comprises of bringing together acid and raw stock in desired proportion at a suitable temperature and then separating the phases by settling. The oily phase is neutralised and the spent acid is discarded. The product is then fractionated to remove high boiling material produced during acid treatment.

Generally, 93% strength (66°Be) H_2SO_4 is used. If the aim is to only improve the colour of light distillates, then a more dilute acid is used. Lube oil refining needs upto 98% strength acid. Lower contacting temperature yields better colour products and favours the removal of sulphur compounds. Gasolines are treated at 20-30°C. Higher temperatures (30-50°C) are used for treating kerosenes because removal of aromatics, unsaturated hydrocarbons and asphaltic material is favoured at those temperatures. High viscosity lubricating oil need still higher temperature (40-80°C) to ensure intimate contact with the acid. Higher temperature enhances refining losses. Heavier products need a longer time of contact.

Use of acetic anhydride and hydrofluoric acid to improve colour stability & odor of distillates have found limited use.

Alkali (caustic soda) Treatment. After processing with acid, the refined petroleum products are given a caustic soda wash to neutralise the acidity. Alkali wash treatment also removes naturally occurring acidic substances *viz.*, hydrogen sulphide, carbonyl sulphide, lower aliphatic mercaptans, phenol, aryl mercaptans, fatty acids and naphthenic acids from petroleum products.

Clay Treatment. Clay is used as an adsorbant of undesirable substances from petroleum products. Clay treatment is done to

- decolorise and stabilise cracked gasoline.
- desulphurise straight run gasolines and kerosenes.
- decolorise heavy oil products.
- neutralise acid treated heavy oils that can not be caustic washed because of emulsification difficulty.

- remove (adsorb) arsenic from feedstock to catalytic reforming.
- remove asphaltic resinous and acidic materials from heavy oils and waxes.
- improve oxidation resistance of products.

In the treatment of gasolines & kerosene, the clay acts as a catalyst whereas for reformer feedstocks & heavy oil it is used as an adsorbant.

High asphalt oils are deasphalted by acid treatment or solvent extraction and then given a clay treatment. Conventional acid treatment of petroleum products are followed by clay treatment for neutralisation of the charge by the absorption of sludge and acidic materials.

Clays (*e.g.*, fuller's earth and bauxite) catalyse the polymerisation of color and gum forming materials in cracked gasolines. It also catalyses the conversion of organic sulphides & mercaptans to hydrogen sulphide & hydrocarbons at 300-350°C.

The liquid phase clay treatment is carried out by either percolation or by contact filtration. The former consists in percolating the liquid charge through a fixed or moving bed of clay. In filtration, liquid is mixed with the absorbant and filtered.

In vapour phase clay treatment, vapors are passed over a fixed bed of clays. Under desulphurisation conditions, coke deposits deactivate the catalyst which can be regenerated by burning off the coke.

Spent clay disposal is a big problem, as it has limited demand in the manufacture of cinder bricks. Spent clays from the processes other than desulphurisation are usually not regenerated.

DEASPHALTING OF PETROLEUM PRODUCTS

Asphalts are high molecular weight compounds present in petroleum products in concentration from 0 to 65% and have high boiling points in the range of lube oil & heavy gas oils (HGO). Asphalt readily oxidise & form carbonaceous sludges and hence must be removed from lubricating oils. Asphalts are undesirable in catalytic cracking feedstocks (*i.e.* H.G.O.) because they form coke. The processes by which asphalts are removed from heavy stocks is called 'deasphalting'. Deasphalting is done by distillation (atmospheric and vacuum), acid treating and solvent (propane) extraction.

Acid treatment for deasphalting is obsolete due to heavy losses of hydrocarbons. Bulk of deasphalting is done by atmospheric and vacuum distillation. However, lot of desirable materials for lubricating oils is still left in the residue from vacuum distillation. Solvent deasphalting is the best process from this point of view.

Propane (solvent) Deasphalting Process. Propane dissolves hydrocarbons and rejects asphaltic materials near its critical temperature. It also dissolves smaller molecules in preference to larger ones. In the range of conditions used, raising the temperature of propane reduces its dissolving capacity and improves its selectivity. Increasing the solvent (propane) to oil ratio increases the sharpness of separation.

Reduced crude (feed) is fed near the top and liquid propane near the bottom of a liquid-liquid contacting column (Deasphalting Tower). The propane passes up the column, dissolving the oil and carrying it overhead. Asphaltic materials pass out to the bottom. The top of bottom of the column is steam heated to reduce the oil carrying capacity of propane. As a result, some oil refluxes down the column to improve the sharpness of separation. Some propane is dissolved in the reject asphalt which is removed by evaporation and steam stripping. Similarly propane from deasphalted oil is also removed. These are condensed and recycled.

Typical propane deasphalting process parameters are given below.

Deasphalting tower details

Temp, at the bottom	= 40-55°C
Temp, at the top	= 65-85°C
Pressure	= 37.5 kg/cm^2
Height	= 1 2 metres
No. of perforated trays	= 16

Propane to oil ratio is 4-12 (by volume) and the softening point of asphalt is about 60°C.

At a higher deasphalting tower top temperature (say 90°C), the heaviest lubricating stock is rejected with the asphalt at the bottom. The bottoms then are charged to a second column operating at the usual conditions to separate heaviest lube oil stock from asphalt. This is then called propane fractionation because such an operation separates the potential lubricating oil into two fractions.

A flowsheet of the propane deasphalting process is given in Fig. 22.8.

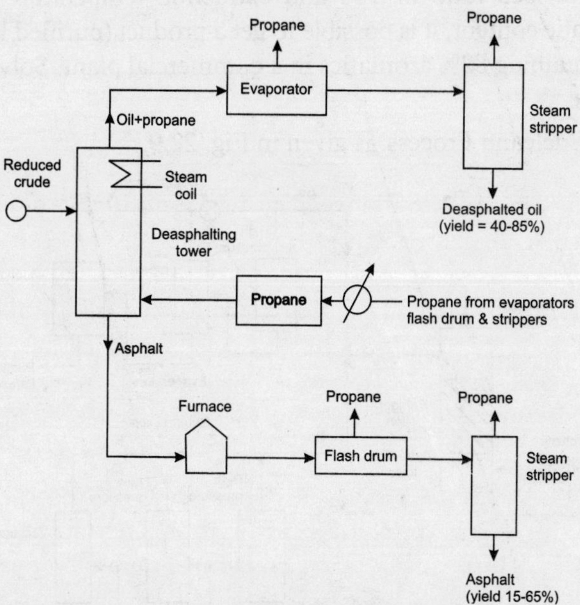

***Fig. 22.8.** Propane Deasphalting Process.*

Asphalt thus recovered is further purified by air blowing in presence of catalyst like ferric chloride and phosphorous pentoxide thereby raising its softening point.

DEAROMATISATION (PURIFICATION) OF KEROSENE (EDELEANU PROCESS)

Removal of aromatic compounds from kerosene is called dearomatisation of kerosene. Presence of aromatics in kerosene makes it smoky (*i.e* reduces its smoke point) hence, they are removed from it by liquid sulphur dioxide solvent extraction in a process called *Edeleanu Process.* At its boiling point (-10°C), liquid sulphur dioxide completely dissolves aromatics, unsaturates, mercaptans and many other sulphur compounds but the paraffins & naphthenes remain insoluble in it. Demerit of this solvent (liquid SO_2) is that moisture can not be tolerated in the system because the solvent then becomes highly corrosive.

In the Edeleanu Process of kerosene purification (ie,dearomatisation) to improve its smoke point, raw kerosene is dried, cooled and then charged to a packed extraction tower near the bottom. The recycled solvent (liquid SO_2) is refrigerated and fed at the top. Both the overhead raffinate and the bottom extract streams are fed to triple effect evaporators for complete recovery of sulphur dioxide. The first stage in each of these two evaporators is operated at high pressure (6 atm.) which is sufficient to condense solvent vapour without compression. The other two stages are operated at low pressures (25 mm Hg in the last stage) and need compression for liquifying the solvent vapours. There is a small drying tower (in the SO_2 stream for the continuous removal of water) made of corrosion resistant material.

By using a solvent to feed ratio of 100 and extraction temperature of -18°C with a feed (kerosene) of 14% aromatic content, it is possible to get a product (purified kerosene) with only 2% aromatic and extract containing 82% aromatics in a commercial plant. Solvent loss is about 0.4 kg per kilolitre of the charge.

A flowsheet of the 'Edeleanu Process' is given in Fig. 22.9.

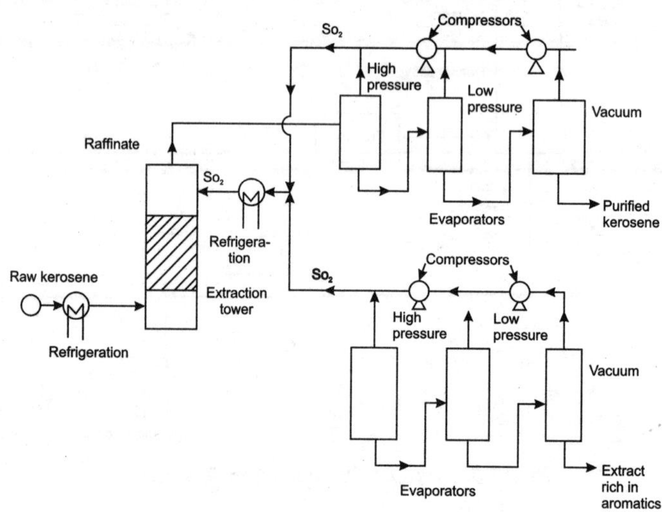

Fig. 22.9. *Edeleanu Process (for Purification/dearomatisation of Kerosene).*

Sulphur dioxide extraction can be used for recovering aromatics from gas oil, diesel oil and naphtha also.

Specification for Kerosene is tabulated in Table 22.2.

Table 23.2: Specification for Kerosene (IS: 1459-1968, Ammendment July, 1972)

Characteristic	Superior Grade	Inferior Grade
Acidity, inorganic	Nil	Nil
Burning Quality		
(a) Char value, mg/kg of oil consumed, Maximum	20	-
(b) Bloom on glass chimney	Not darker than grey	-
Colour (ASTM), not deeper than	-	5.0
Viscosity (Saybolt), Minimum	21	-
Copper-strip corrosion for 3 hours at 50°C	Not worse than No. 1	Not worse than No. 2
Distillation :		
(a) Per cent recovered below 200°C, Min	20	-
(b) Per cent recovered below 300°C, Min	-	60
(c) Final boiling point, °C, Max	300	-
Flash point (Abel), °C, Min	35	35
Smoke point, mm, Min	20	12 or less
Sulphur, total, per cent by weight, Max	0.25	0.3
Uses for superior kerosene :	wick-fed chimney lamps, stoves, cooking ranges, blow lamps, refrigerators, space heating.	
Uses for inferior kerosene :	wick-fed lamps not provided with chimney or air regulator, cleaning fluid, solvent in antimosquito work.	

Other Solvent Extraction Processes

Differences in solubility in suitable solvents is the basis of resolving a stock into two or more products or purify a product from harmful impurities. *Solvent extraction* is used in modern refineries for:

(a) *upgradation of lubricating oils* by removing aromatic materials and sulphur compounds,
(b) *separation of reduced crude and residuums* into lubricating oils & asphalts (deasphalting),
(c) *upgradation of kerosene, heating oils and diesel fuels* by removing aromatics, olefins & sulphur compounds,
(d) *improving cracker feedstocks* by removing asphaltic and other coke-producing materials (decarbonization),
(e) *separation of wax bearing oils* into wax and wax-free oil (dewaxing),
(f) recovery of benzene, toluene and xylenes from suitable naphthas,
(g) *separation of i-butane* from butane-butene stream, etc.

Particularly for improving the quality of lubricating oils, other solvent extraction processes like furfural extraction, phenol extraction, Duo-sol extraction processes are used. Polycyclic materials are undesirable in lubricating oils because of their low viscosity index. An oil rich in polycyclic naphthenes is unsuitable for the production of motor oils. Polycyclic aromatics can be removed by furfural extraction, phenol extraction, and Duo-sol extraction processes.

Furfural Extraction Process

Furfural extraction is being increasingly used in the *upgrading of lube oils.* Aromatics are highly soluble in it. Low moisture in the system does not pose a problem and in fact a small amount of it is purposefully left in the solvent for increasing its selectivity. Moreover, it is readily available, relatively cheap and non-toxic.

In the furfural extraction process, feed stock (lube oil) is charged to the middle of an extraction column and passes up counter current to the solvent (furfural) which is introduced at the top. Solvent extracts polycyclic aromatics together with some desirable stock. The bottom furfural-extract stream (about 90% furfural) is heated in a furnace & flashed to drive off furfural vapors and final traces are stripped by steam. The raffinate stream also contains 10% furfural which is recovered by flashing off. They are recycled after condensation.

Furfural is stable, nontoxic, relatively cheap, readily available and is selective at higher temperatures than are other solvents. Consequently, it has two advantages. Liquids are less viscous at higher temperatures and contacting of the feed & solvent is facilitated. Also, waxy stocks may be processed without difficulty. Because the selectivity of furfural is destroyed by water, the circulating solvent must be dehydrated continuously.

Wet furfural from steam stripping is condensed, and the two liquid phases that result - one, principally furfural, the other, principally water - are charged to a settling drum . The furfural layer that separates is stripped of water by countercurrent contact in a tower with dry furfural vapors. Wet furfural taken overhead is condensed and returned to the settling drum. The water layer from the drum is stripped of its furfural content by countercurrent contact with steam. The wet furfural taken overhead from this tower is also condensed and returned to the settling drum. Thus, the furfural-water condensates are separated into dry furfural and furfural-free water. Dry furfural is recycled to extraction tower and the water is discarded. A flowsheet of furfural extraction process is given in Fig. 22.10.

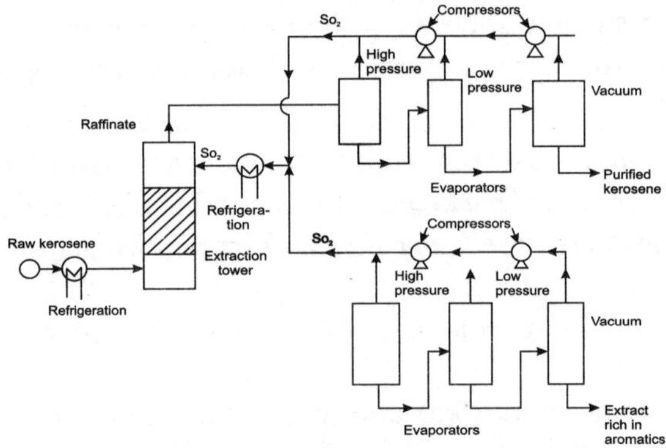

***Fig.22.10:** Furfural extraction process.*

Modern furfural extraction columns contain about 10-15 metres of Raschig rings, with redistribution equipment at about 1.5 metres intervals. Charge rates range from 1.5 to 7.5 m^3 of oil per hr per m^2 of column cross section. The temperature of the entering hydrocarbon ranges from 45 to 105°C and the temperature of the entering solvent from 95 to 145°C. Typical temperatures at the top of the extractor range between 110 to 145°C and bottom temperatures are 0.5 to 25°C lower. Two to five volumes of solvent are circulated per volume of oil charged. Furfural loss is about 0.3% of the amount circulated.

Phenol Extraction Process

But for the fact that phenol is toxic and solidifies at atmospheric temperature, its solvency properties are similar to that of furfural. However, solvent recovery is easier and cheaper. It is extensively used in the upgrading of lubricating oils.

In phenol extraction process, the feed (lube oil) is first contacted with waste vapor (mainly water vapor) to recover phenol and is then charged to the lower part of a refluxed extraction column through which it rises counter-current to the solvent phenol. Reflux is introduced by the introduction of water at the bottom of the column. The rich extract is fractionated to drive off its water content, the water-phenol vapor taken overhead being sent to the phenol absorber. Phenol is recovered from extract & raffinate phases by heating & flashing off and is recycled to extraction tower. Residual phenol is stripped from both streams by steam in vacuum towers. The phenol extraction system operates at slightly lower temperature than furfural extraction.

A schematic flow diagram for a phenol-extraction plant is shown in Fig. 22.11.

With certain feed stocks, aqueous phenol (< 12% water) are used to permit adequate separations of the phases.

Most phenol extraction columns contain 14 to 20 perforated trays, and oil charge rates range from about 1.5 to about 4.5 m^3 per hr per m^2 of column cross section. Column-top temperatures range between 70 and 125°C, bottom temperatures are upto 25°C lower. The amount of solvent circulated ranges from 1.4 to 8 times the oil charged. Phenol losses are 0.02 to 0.03% of the amount circulated.

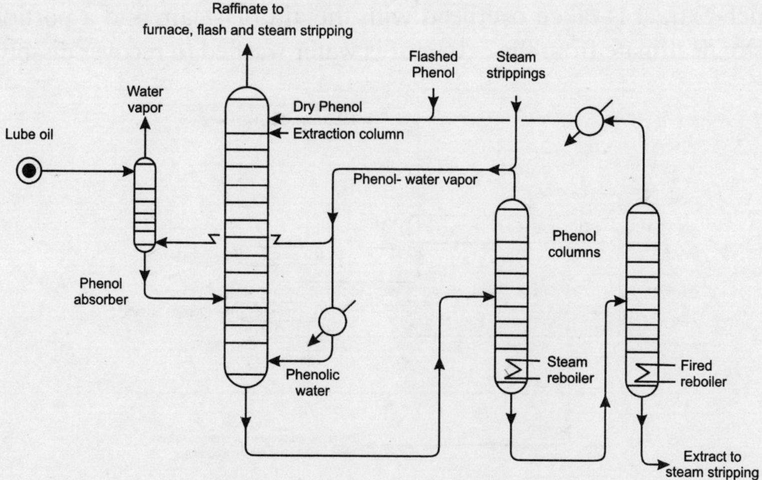

Fig. 22.11. Phenol extraction process.

Duo-Sol Extraction Process

A mixture of propane and phenol-cresol mixture (called selecto) is used as a solvent in the double solvent (Duo-Sol) extraction process for upgrading lube oils. Selecto (phenol-cresol mixture) dissolves asphalt, polycyclic aromatics, naphthenic hydrocarbons, colour bodies whereas other solvent (liquid propane) dissolves paraffinic hydrocarbons and precipitates asphalts. Thus the Duo-sol process has a dual function of removing asphalt as well as undesirable hydrocarbons removed in other extraction processes.

The extraction equipment consists of 7-9 mixer-settler stages operating in series. Propane is charged to one end stage, a mixture of phenol and cresol to the other. The two solvents pass through the extractor counter currently. Feed (lube oil) is charged to an intermediate stage where it distributes itself between the two solvents. The raffinate rich propane stream passes counter-current to the entering phenol-cresol stream, which purifies the raffinate by extracting residual asphalt and polycyclics. The extract rich-phenol cresol streams passes counter-current to the entering propane stream which strips raffinate type material from the extract and thus improves recovery. The raffinate & extract are stripped of their propane & phenol-cresol contents by several stages of evaporation and stream stripping.

Duo-Sol process operates at 15-65°C. The upper temperature is limited by the vapor pressure of propane and the lower temperature by crystallisation of wax or phenol. Solvent to oil ratio is about 4. Propane and phenol-cresol mixture solvent losses are 0.2 and 0.03% respectively of the circulation. The water content of the phenol-cresol mixture is held below 0.15%.

A schematic flow diagram for a Duo-Sol unit is shown in Fig. 22.12.

Udex (Glycol) Extraction Process

Aqueous solution (10% water) of diethylene glycol are used for the extraction of light aromatics (*e.g.* benzene, toluene and xylene) from catalyst reformate or cracked naphtha by Udex process. The product is used as a high quality gasoline or further processed for recovery of pure aromatics.

In Udex process, the feed is fed to a refluxed extractor. The rich solvent passes to a solvent stripper from which extract is taken overhead with the aid of steam and a portion of extract is refluxed to extractor. Raffinate from the extractor is water washed to recover dissolved glycol.

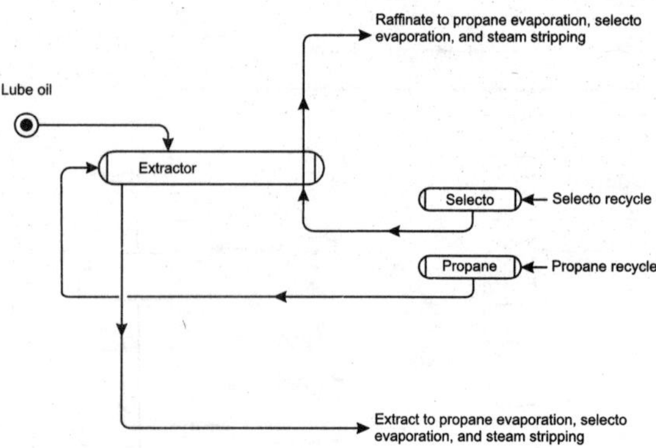

***Fig. 22.12.** Duo-Sol extraction process.*

The washings are combined with the solvent from the bottom of the solvent stripper and the mixed stream is recycled to the extractor.-

Since, most of the aromatics products from Udex process serves as a feed to petrochemical industry, hence they must be very pure. For that reason, refluxed system is used.

A schematic flow diagram is shown in Fig. 22.13. The feed stock is usually a catalytic reformate, although cracked naphthas may also be charged. Rich solvent from the extractor passes to a solvent stripper, from which extract is taken overhead with the aid of steam.

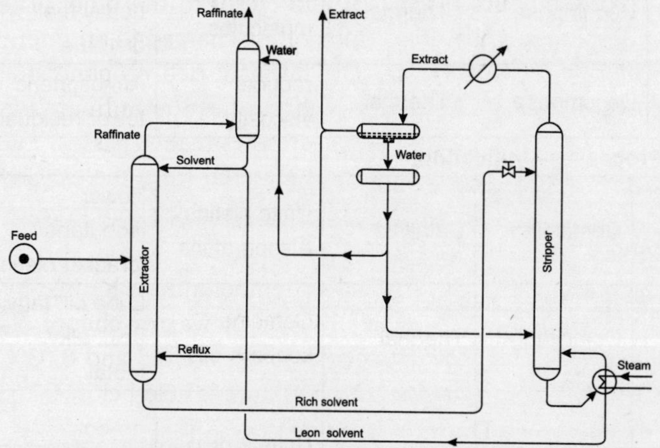

Fig. 22.13. Udex extraction process.

Overview of various petroleum refining operations and processes have been summarised below in table 22.3.

Table 22.3: Summary of Petroleum Refining Processes

S.N.	Process name	Action	Method	Purpose	Feedstock (s)	Product (s)
1.	**Fractionation Processes:**					
	Atmospheric distillation	Separation	Thermal	Separate fractions	Desalted crude oil	Gas, gas oil, distillate, residual
	Vacuum distillation	Separation	Thermal	Separate w/o cracking	Atmospheric tower residual	Gas oil, lube stock, residual
2.	**Conversion Processes-Decomposition:**					
	Catalytic cracking	Alteration	Catalytic	Upgrade gasoline	Gas oil, coke distillate	Gasoline, Petrochemical feedstock
	Coking	Polymerize	Thermal	Convert vacuum residuals	Gas oil, coke distillate	Gasoline, Petrochemical feedstock

	Hydro-cracking	Hydrogenate	Catalytic	Convert lighter HC's	Gas oil, cracked oil, residual	Lighter, higher-quality products
	*Hydrogen steam reforming	Decompose	Thermal/ catalytic	Produce hydrogen	Desulfurized gas, 02, steam	Hydrogen, CO, C02
	*Steam cracking	Decompose	Thermal	Crack large molecules	Atm tower heavy fuel/ distillate	Cracked naphtha, coke, residual
	Visbreaking	Decompose	Thermal	reduce viscosity	Atmospheric tower residual	Distillage, tar
3.	**Conversion Processes-Unification:**					
	Alkylation	Combining	Catalytic	Unite olefins &isoparaffins	Tower isobutane/ /cracker olefin	Iso-octance (alkylate)
	Grease compounding	Combining	Thermal	Combine soaps & oils	Lube oil, fatty acid, alky metal	Lubricating grease
	Polymerizing	Polymerize	Catalytic	Unite 2 or more olefins	Cracker olefins	High-octane naphtah, petrochemical stocks
4.	**Conversion Processes-Alteration or Rearrangement:**					
	Catalytic reforming	Alteration/ dehydration	Catalytic	Upgrade low-octane naptha	Coker/hydro-cracker naphtha	High octance Reformate/ aromatic
	Isomerization	Rearrange	Catalytic	Convert straight chain branch	Butane, pentane, hexane	Isobutane/ pentane/hexane
5.	**Treatment Processes:**					
	*Amine treating	Treatment	Absorption	Remove acidic contaminants	Sour gas, HCs w/CO_2& H_2S	Acid free gases & liquid Hydrocarbons (HCs)
	Desalting	Dehydration	Absorption	Remove contaminants	Crude oil	Desalted crude oil
	Drying & sweetening	Treatment	Absortion/ thermal	Remove H_20 & sulfur cmpds	LiqHCS, LPG, alky feedstock	Sweet & dry hydrocarbons

	*Furfural	Solvent	Absorption	Upgrade mid	Cycle oils & lube	High quality
	extraction	extraction		distillate & lubes	feed-stocks	diesel & lube oil
	Hydrodesulfu-	Treatment	Catalytic	Remove sulfur,	High-sulfur	Desulfurized
	rization			contaminants	residual/gas oil	olefins
	Hydrotreating	Hydrogen- ation	Catalytic	Remove impurities,	Residuals, cracked HC's	Cracker feed, distillate,
				saturate HC's		lube
	*Phenol extr	Solvent ext	Absorpiton/	Improve viscosity	Lube oil base	High quality
	action	raction	thermal	index, color	stocks	lube oils
	Solvent deas	Treatment	Absorption	Remove	Vacuum tower	Heavy lube oil,
	phalting			asphalt	residual, propane	asphalt
	Solvent	Treatment	Cool/ filter	Remove wax	Vacuum tower	Dewaxed lube
	dewaxing			lube stocks	lube Oils	basestock
	Solvent	Solvent	Absorption/	Separate	Gas oil, reform-	High-octane
	extraction	extraction	precipitation	unsaturated oils	ate, distillate	gasoline
	Sweetening	Treatment	Catalytic	Remove H_2S, convert mercaptan	Unteated distill ate/gasoline	High-quality dis tillate/ gasoline

History of oil refinning processes have been given below in Table 22.4.

Table 22.4: History of Oil Refining Processes

Year	Process name	Purpose of refinning processes	By-products produced
1862	Atmospheric distillation	Produce kerosene	Naphtha, tar, etc.
1870	Vacuum distillation	Lubricants (original) Cracking	Asphalt, residual coker
		feedstocks (1930's)	feedstocks
1913	Thermal cracking	Increase gasoline yield	Residual, bunker fuel
1916	Sweetening	reduce sulfur & odor	Sulfur
1930	Thermal reforming	Improve octane number	Residual
1932	Hydrogenation	Remove sulfur	Sulfur
1932	Coking	Produce gasoline basestocks	Coke
1933	Solvent extraction	Improve lubricant viscosity index	Aromatics

1935	Solvent dewaxing	Improve pour point	Waxes
1935	Catalytic	Improve gasoline yield	Petrochemical
	polymerization	&octance number	feedstocks
1937	Catalytic cracking	Higher octance gasoline	Petrochemical feedstocks
1939	Visbreaking	reduce viscosity	Increased distillate, tar
1940	Alkylation	Increase gasoline octance& yield	High-octane aviation gasoline
1940	Isomerization	Produce alkylation feedstock	Naphtha
1942	Fluid catalytic cracking	Increase gasoline yield & octane number	Petrochemical feedstocks
1950	Deasphalting	Increase cracking feedstock	Asphalt
1952	Catalytifc reforming	Convert low-quality naphtha	Aromatics
1954	Hydrodesulfurization	Remove sulfur	Sulfur
1956	Inhibitor sweetening	Remove mercaptan	Disulfides
1957	Catalytic isomerization	Convert to molecules with high octane number	Alkylation feedstocks
1960	Hydrocraking	Improve quality and reduce sulfur	Alkylation feedstocks
1974	Catalytic dewaxing	Improve pour point	Wax
1975	Residual hydrocracking	Increase gasoline yield from residual	Heavy residuals

Properties of Petroleum Products

The important properties of liquid fuels/petroleum/petroleum products that make them suitable for specific uses differ from case to case. No single test or property can evaluate them. Some of the important properties of petroleum products are discussed below.

SPECIFIC GRAVITY

It determines the maximum power per unit of weight/volume. Hydrocarbons of low specific gravity (paraffins) possess the maximum thermal energy per unit volume. Hydrocarbons of high specific gravity (aromatics) possess the maximum thermal energy per unit weight. However, high calorific value of normal paraffins can not be fully utilised because of their low anti-knock ratings. Aromatics produce more carbon deposits than paraffins. These considerations together favour the use of iso- paraffins for aviation gasolines. The lighter the crude oil (the higher ÅPI), the larger is the content of the lighter constituents like gasoline and kerosene. Aromatics have higher Sp. gr. than paraffins.

Hydrometer is used for determination of specific gravity upto 0.001. For higher values, specific gravity bottle is used. For a semi-solid mass like tar, an indirect method is used. Tar is mixed with equal volume of kerosene to make a thick solution. Specific gravity of tar is given as

$$\rho_T = 2\rho_m - \rho_K$$

where, p_T = specific gravity of Tar

p_M = specific gravity of mixture of tar and kerosene.

P_k = specific gravity of kerosene.

°API gravity and gravity on Baume scale is given as (for liquids lighter than water)

$$°Be = \frac{140}{\rho} - 130 \text{ and, } °API = \frac{141.5}{\rho} - 131.5$$

Specific gravity can be used to find the gross calorific value (GCV) of petroleum products using following formula of U.S. Bureau of Mines.

$$G.C.V. = 12400 - 2100\rho^2$$

where, ρ = specific gravity of oil at 15.5°C.

Typical specific gravity of crude and its products are given below :

crude	0.78-0.96
Naphtha	0.70-0.72
gasoline	0.7-0.76
diesel	0.82-0.86
lube oil	<0.9
reduced crude	0.92-0.96

Higher specific gravity means higher carbon to hydrogen ratio. Hence, heavier oils have lower gross calorific value on weight basis but higher gross C.V. on volume basis.

Aromatics have higher specific gravity than paraffins. Hence, a knowledge of specific gravity can predict the quality of a product.

Increase in specific gravity means decrease in paraffin content. An increase in specific gravity increases the amount of heat per unit volume.

Density of Petroleum Products Versus Temperature. Following figure 23.1 shows the approximate relationship of density with respect to temperature. It can be used to convert density of fuel oil or any other Petroleum product at 15°C to density at any other desired test temperature. t°C.

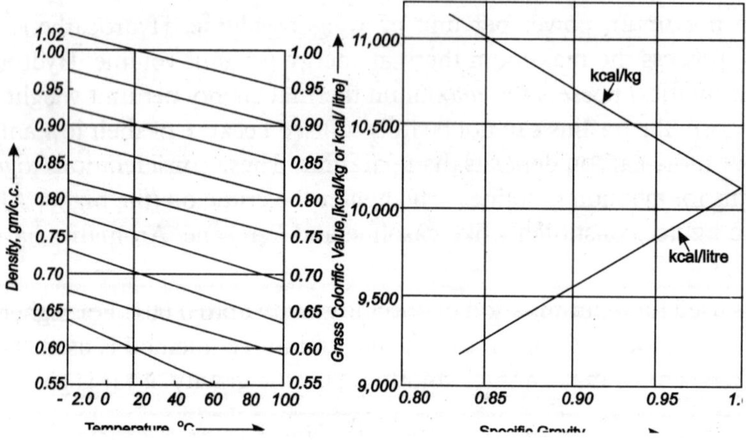

Figure. 23.1. Temperature-density relationship for petroleum products

Figure. 23.2. Specific gravity Vs. calorific value of fuel oils

To determine the density of fuel oil (having density 0.95 at 15°C) at 40°C, draw a vertical line at 40°C to the point where it meets 0.95 Density-Temperature line. From the point of intersection, draw a horizontal line. This line, meets the Density t°C line at 0.935. So the density of the fuel oil at 40°C is 0.935. The inverse approximation can be made similarly.

The Fig. 23.2 shows gross calorific value in kcal/kg and kcal/litre of fuel oils of different specific gravities at 15°C.

MOLECULAR WEIGHT

It is given by (for $\rho < 0.92$),

Molecular weight $= \dfrac{44.29 \times \rho}{1.03 - \rho}$ where, ρ = sp. gr. at 15.5°C.

It is also given by *Eigenson's equation* as :

Molecular weight, $M = a + bT + CT^2$', where, T = average molal boiling point, °R

a, b, and c are constant depending upon the characterisation factor, *K,* which is given by

$$K = \frac{T^{1/3}}{S}$$

where, S = specific gravity and T = average molal boiling point, °R

Molecular weights of crude oil and some of its products are given below :

Crude oil	= 230–250	bitumen	= 400 –600
gasoline	=100–130	diesel	= 200 –280

VAPOR PRESSURE

Vapor pressure of petroleum products is important, because :

- With decreased atmospheric pressure at high altitudes, the vapor comes out of the liquid fuel causing vapor-locking and difficulty in starting of spark ignition engines.
- If the vapor pressure of fuel is too low, starting of engine and lubricating oil dilution becomes difficult.
- If the vapor pressure is high, vapor occurs in fuel tanks & pipelines causing pumping and metering difficulties.
- If the vapor pressure is high that means flash point is low resulting in danger of fire and explosion. Besides, breathing loss is high, *i.e.,* evaporation losses during storage of petrofuels is high.
- Lubricants having high vapor pressure will be lost quickly particularly in case of thin film boundary lubrication.

Vapor pressure of a liquid fuel is reported as *Reid vapor pressure* (RVP) in psi or mm. Hg at 100°F (37.8°C). This apparatus consists of a 'Reid bomb' i.e. a metallic cylinder consisting of an upper expansion air chamber & a lower liquid fuel chamber and fitted with a dial pressure gauge or a mercury monometer. The volume of air expansion chamber is four times that of fuel chamber. It is shown in Fig. 23.3.

The liquid oil is cooled and filled up in the lower chamber. The temperature of air in the upper chamber is taken and the two chambers connected together in a gas tight manner. The bomb is immersed upright in a water bath at 100°F and shaken repeatedly until a constant pressure reading is obtained. This is corrected for initial pressure and temperature of air to get the Reid vapor pressure. True vapor pressure is about 5-9% higher than Reid vapor pressure.

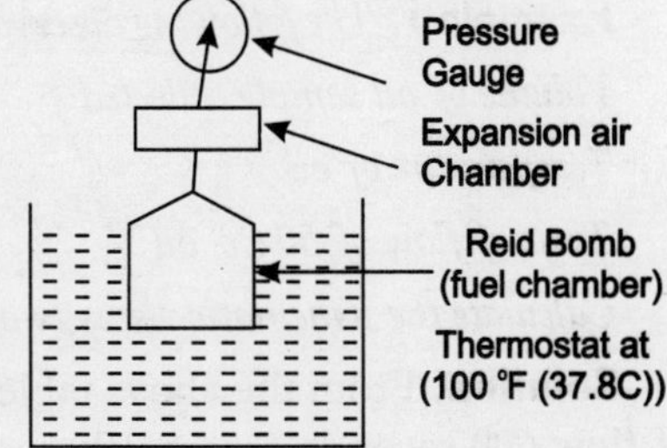

Fig. 23.3. *Reid Bomb.*

VISCOSITY

Viscosity of oil/petrofuels is an important property, because :

- high viscosity fuel oil can not be properly atomised resulting in loss of fuel by way of incomplete combustion.

- highly viscous oil has to be preheated to reduce the viscosity to minimise the pressure drop in pumping to different refineries from the oil field.
- high viscosity lubricating oil reduces its fluidity besides causing undue friction.
- in case of lubricants, it determines the bearing friction, heat generation and rate of flow under particular conditions of load, speed & design.
- it affects the amount of fuel that can be drawn through a wick to a flame by capillaiy action.
- it affects the flow through pipelines, injector nozzles & orifices.
- it decides the pump clearance; which if out of range, results in pump seizer.

Viscosity is determined by three makes of commercial viscometers. The *Redwood viscometer* (used in commonwealth countries), *Saybolt viscometer* (used in U.S.A.) and *Engler viscometer* (used in Europe). In case of viscometers, a fixed volume of liquid at a fixed temperature is allowed to flow through a standard capillary tube and the time of flow is noted. The result is sometimes expressed in terms of time taken by oil to flow through a particular instrument *e.g.* viscosity = 100 Redwood seconds at 20°C.

Kinematic viscosity measured by this method is expressed in stokes or centistokes. Kinematic viscosity (η) liquid fuels is given by

$$\eta = AT - \frac{B}{T}$$

where, T = Time of flow of a fixed amount of oil (at a fixed temperature) through the viscometer. A and B are viscometer constants and depend upon the dimensions of the viscometer capillary through which oil flows. For Redwood Viscometer No. I, T is the time of flow of 50 c.c. of oil and values of constants B and A for various values of' T are given in Table 24.1.

Table 24.1: Values of Redwood Viscometer Constants 'B' & 'A'

Value of T	Value of B	Value of A
34-100 seconds	1.78	0.0026
> 100 seconds	0.50	0.00247

Approximately, n= AT, for fluids having n > 10 cst. (centistoke).

Example 1. *The following observations were made with a sample of oil using Redwopd viscometer No. I.*

Volume of oil sample collected = 50 *c.c.*

Temperature of oil = 20°*C*

Time of flow of 50 c.c. oil = 600 *seconds*

Calculate the Kinematic viscosity of the oil.

Solution. From the above table, the values of viscometer constants corresponding to the time of flow 600 seconds (*i.e.* > 100 seconds) are B = 0.50 and A = 0.00247.

Hence, Kinematic viscosity of oil, $\eta = AT - (B/T) = 0.00247 \times 600 - \frac{0.50}{600}$ stokes

or, $\eta = (1.4820 - 0.0008) = 1.474$ st = 147.4 cst (1 stoke= 100 cst)

Redwood viscometer I is used for low viscosity oils whereas Redwood viscometer II is used for highly viscous oil (its oil flow port area is larger). The value of constants A and B are different for the two viscometers as their flow port diameters are different. These values are supplied with the viscometers. No. of seconds required for flow of fixed quantity of oil (usually 50 c.c. at a fixed temperature) is a measure of viscosity. High viscosity oil takes longer time to flow out.

The commonly used methods measure the viscosity in *Arbitrary Units.* The time required for a standard volume of oil to flow through a standard orifice is measured and the viscosity is stated as a number of seconds at a certain temperature as measured by a named apparatus.

In British practice, the standard instrument for measuring viscosity is Redwood Viscometer which is shown in Fig. 23.4. It consists of a metal cup with an axially-placed orifice in the base. The hole can be closed by a metal ball or a rod. The metal cup can be heated and the oil stirred to ensure uniform temperature conditions. When the ball is removed, a thin stream of oil runs into a small graduated glass flask and the time to fill the flask is recorded which represents the viscosity of oil. Two types of Redwood Viscometer are used. No. 1 is used for less viscous oils and No. 2 is used for more viscous oils. The time required for a standard volume of oil to flow

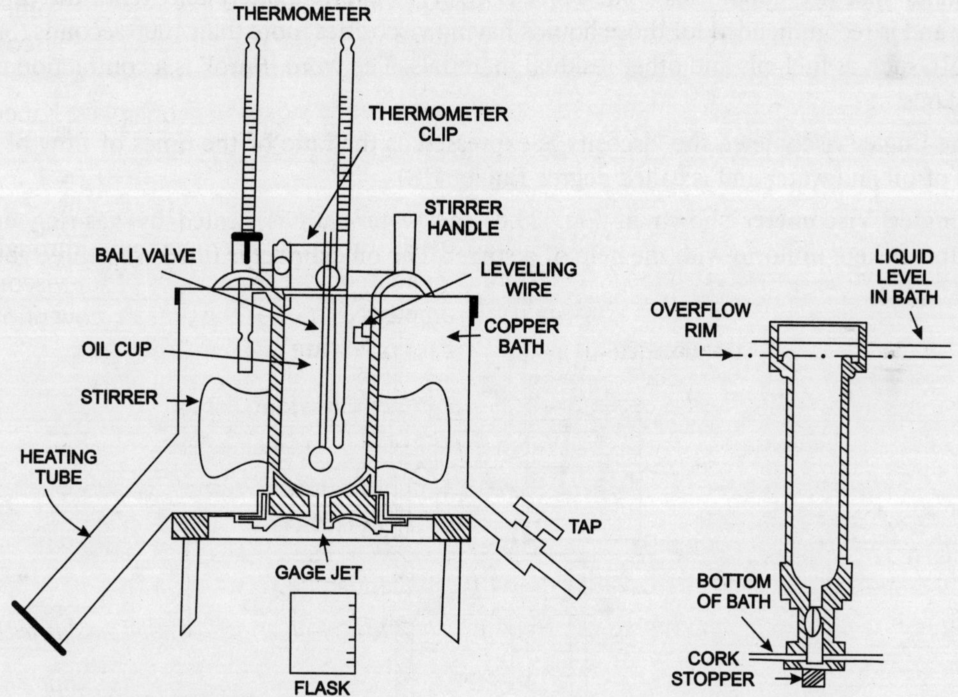

Fig. 23.4. *Redwood Viscometer.*

Fig. 23.5. *Saybolt Viscometer.*

from Redwood No. 2 viscometer is 1/10 of that required for an equal volume of same oil at the same temperature to flow from the Redwood No. 1 viscometer. Viscosities measured by the Redwood viscometer are reported as *"n* seconds Redwood No. (1 or 2) at a given temperature.

Kinematic viscosity (expressed in stokes/centistokes) is related to *absolute / dynamic viscosity* (expressed as poise/centipoise) as;

$$\eta = \frac{\mu}{\rho}$$

where, ρ = density of oil, gm/cm^3

μ = absolute viscosity of oil, poise = gm/cm sec.

η = Kinematic viscosity of oil, stoke

$$\begin{bmatrix} 1 \text{ poise } 100 \text{ centipoise} \\ 1 \text{ stoke } 100 \text{ centistokes} \end{bmatrix}$$

Saybolt universal and Saybolt Furol are similarly different viscometers used for different viscosity oils.

In American practice, *Saybolt viscometer* (Fig. 23.5) is used. Similar to Redwood viscometer No. 1 and No. 2, Saybolt also uses two standards, *i.e. Saybolt Universal Viscosity* (time taken in seconds for 60 ml of sample flowing through a calibrated universal orifice under specified conditions) and *Saybolt Furol Viscosity* (time taken in seconds for 60 ml of sample through a calibrated Furol orifice under specified conditions). The *Furol* viscosity is approximately one- tenth the universal viscosity and is recommended for those liquids having viscosities more than 1000 seconds (Saybolt Universal), such as fuel oils and other residual materials. The word 'Furol' is a contraction of fuel and road oils.

In the Engler viscometer, the viscosity is expressed as the ratio of the times of flow of equal volumes of oil and water and is called degree Engler (°E).

In **Engler Viscometer** shown in Fig. 23.6, the water-bath is heated by gas-ring and its temperature is kept uniform with the help of a stirrer. The oil cylinder is fitted with three gauge

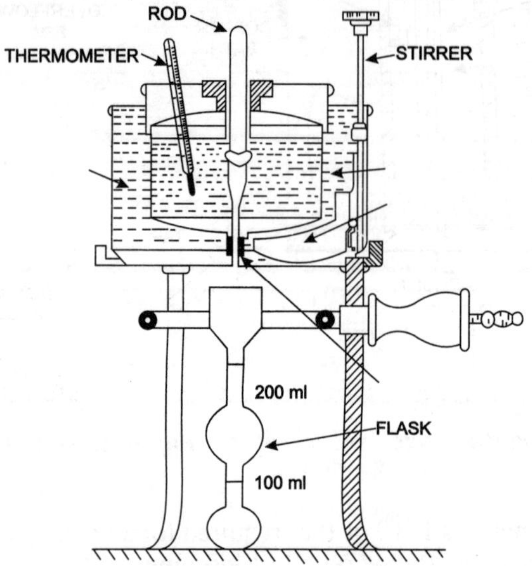

Fig. 23.6. *Engler's Viscometer.*

points, which indicates the amount of oil to be poured in it and also helps in levelling the instrument. The loosely fitted oil cylinder cover carries a thermometer which can be rotated to stir the oil. The jet fitted at the bottom of oil cylinder is slightly tapered. The valve pin which seats itself on the jet, is lifted to start the test and the time for flow of 200 c.c. of oil is noted. The time of outflow of 200 c.c. of water at 20°C is taken as 52 seconds. The viscosity in degree Engler (°E) is determined by dividing the time of outflow of 200 c.c. of oil by 52 seconds (which is the time of outflow of corresponding volume of 200 c.c. of water).

Viscosity data of one standard is converted to another by following formulae.

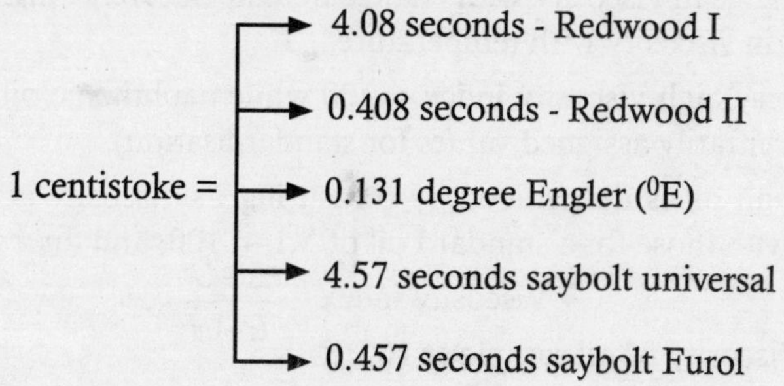

Table 23.2: Kinematic viscosity (η) conversion formula

Viscosity scale	Range of flow time, T, seconds	Kinematic viscosity, stokes
Saybolt Universal	T = 32 to 100	0.00226 T - 1.95/T
	T = > 100	0.00220 T - 1.35/T
Saybolt Furol	T = 25 to 40	0.0224 T - 1.84/T
	T = >40	0.0216 T - 0.60/T
Redwood number I	T = 34 to 100	0.00260 T - 1.79/T
	T = > 100	0.00247 T - 0.50/T
Redwood number (II)	T = 32 to 90	0.02458 T - 1/T
	T = > 90	0.0244 T - 0/T
Engler	°E = 1.35 to 3.2	0.08 T - 0.0864/T
	0 E = > 3.2	0.076 T - 0.04/T

Viscosity of fuel oils influences the ease of handling, transport and nature of storage. Oil viscosity affects its atomisation and combustion efficiency. Preheating of fuel oils is done to reduce its viscosity (above 120°C viscosity is almost constant) to achieve better atomisation (finer oil droplets for better combustion). Maximum viscosity for easy atomisation in conventional burner is 25 centistokes (100 seconds Redwood I). For easy and economic pumping the maximum viscosity should be 1200 centistokes (486 Seconds Redwood II). At higher preheating temperature of oil, the decreased viscosity facilitates easy flow and better atomisation in the burner but the decreased density results in the delivery of lesser weight of oil to the burner. Hence, beyond an optimum oil preheating temperature, the burner capacity is reduced.

In case of diesel fuels, too low viscosity causes excessive leakage at the injection pistons while too high viscosity produces coarse oil droplets which forms engine deposits due to incomplete combustion.

Viscosity of oil blends is experimentally determined (as viscosity is not an additive property or calculated with the help of emperical blending charts).

At high pressures, the viscosity of oil increases very rapidly. At 70 atm. pressure, viscosity increases by 15%. Most oils begin to solidify at 3500 kg/cm^2 pressure.

Viscosity Index (V.I.). The viscosity of a liquid decreases with increase in temperature. The change in viscosity with change in temperature is expressed by viscosity index. It is an emperical number indicating the rate of change of viscosity of an oil from 100°F to 210°F. Low viscosity index means a large change in viscosity with change in temperature. A high viscosity index shows relatively small change in viscosity with temperature.

Paraffinic oil has very high viscosity index = 100 while naphthenic oil has very low viscosity index = 0 (These are arbitrarily assigned values for standardisation).

Viscosity index of an oil is determined by measuring its viscosity at two temperatures and comparing the results with those for a standard oil of V.I. = 100, and for a standard oil of V.I.= 0.

$$\text{Viscosity index} = \frac{L-U}{L-H}$$

where, U = Viscosity of oil sample at 100°F

L = Viscosity of a standard oil of V.I. = 0, at 210° F

H = Viscosity of a standard oil of V.I. = 100 at 210°F.

Values of L and H are obtained from tables after the viscosity of sample has been determined at 100°F.

Viscosity index measures the paraffinity of an oil. Lubricating oil should in general have high viscosity index. Viscosity index improvers (*e.g.* polybutene) are used for improving it. High V.I. oils are used where there is a wide temperature variation. High V.I. lubricating oils are called multigrade oils.

Viscosity temperature relationship of various petrofuels is shown in the Fig. 23.7.

Viscosity Gravity Constant (V.G.C.), It is given by the formula :

$$\text{VGC} = \frac{10\text{G}-1.0752 \log (\text{V}-38)}{10-\log (\text{V}-38)}$$

where, G = specific gravity at 15.5°C, V= Saybolt viscosity at 37.8°C.

This correlation of specific gravity and viscosity gives the measure of paraffinity in lubricating oil as the paraffins are less dense than naphthenic hydrocarbons. V.G.C. is related to other physical properties of the oil also.

Viscosity Index Extension. For those oils which have a viscosity index more than 100, another parameter called viscosity index extension (VI_E) is used which is defined as,

$$VI_E = 100 + \frac{(\text{anti log} N)-1}{0.0075} \qquad \text{where, } N = \frac{\log H - \log U}{KVI_{200}} \quad (H>U)$$

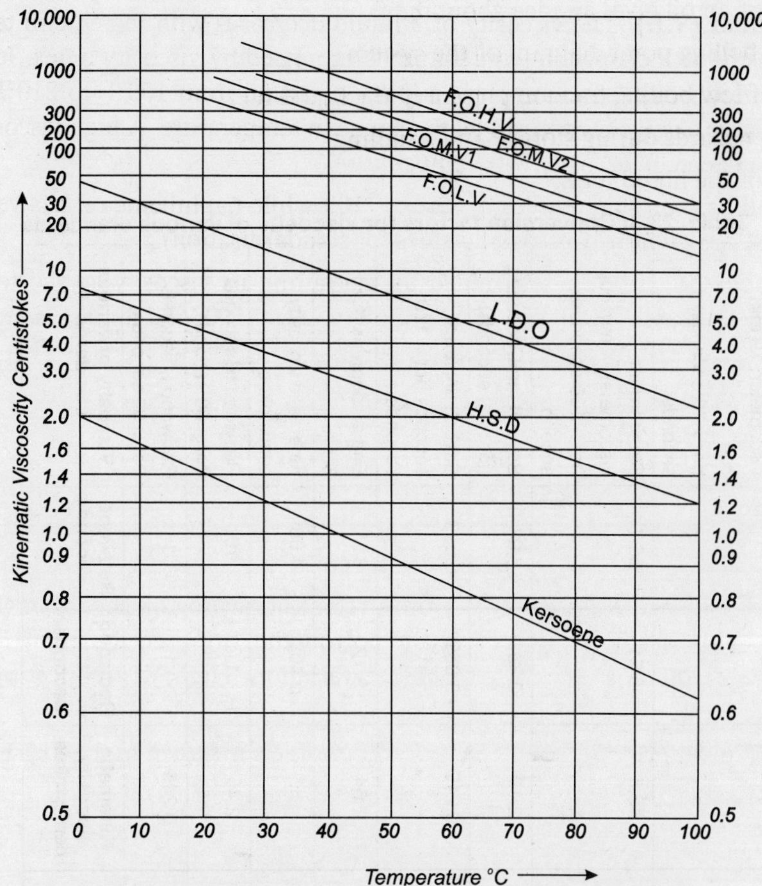

Figure.23.7. Viscosity-temperature relationship of petrofuels

Legends:

H.S.D — high speed diesel fuel F.O.M.V1 — Furnace oil, Medium viscosity grade I

L.D.O. — Light Diesel oil F.O.M.V2 — Furnace oil, Medium viscosity grade II

F.O.L.V. — Furnace oil, Low viscosity grade F.O.H.V —Furnace oil, High viscosity grade

where, H = kinematic viscosity of an oil having a VI of 100 at 100°F using the standard method and having the same viscosity as the unknown oil at 210°F

U = kinematic viscosity at 100°F of unknown oil

KVI_{200} = kinematic viscosity of oil sample at 210°F.

A comprehensive table for conversion of viscosity from one standard to another is given in Table 23.3.

FLASH POINT

It is the minimum temperature at which an oil/petrofuel gives out sufficient vapour to form an inflammable mixture with air and catches-fire momentarily *i.e.* flashes, when flame is applied.

Flash point of an oil gives an idea about the :

- nature of boiling point diagram of the system.
- amount of low boiling fraction present in the liquid fuel.
- explosion hazards during storage and handling.
- volatility of the liquid fuels.

Table. 23.3: Conversion factors for viscosity of various standards

Viscosity of various standards	**Standard tar viscometer, 10 mm cup, seconds**	**Standard tar viscometer, 4 mm cup, seconds**	**Saybolt furol, seconds**	**Saybolt universal, seconds**	**Engler, degree**	**Kinematic viscosity, centistokes**	**Redwood II, seconds**	**Redwood I, seconds**
Standard tar viscometer, 10 mm cup (for > 10 seconds), seconds	1	29.6	183	1830	52.4	400	163	1630
Standard tar viscometer, 4 mm cup (for > 100 seconds), seconds	0.034	1	6.16	61.6	1.77	13.5	5.51	55.1
Saybolt furol, seconds (for > 40 seconds)	0.162	0.00547	1	10	0.287	2.19	0.893	8.93
Saybolt universal, seconds (for > 120 seconds)	0.000547	0.0162	0.1	1	0.0287	0.219	0.089	0.893
Engler, degrees (for > 6°)	0.0191	0.565	3.49	34.9	1	7.64	3.12	31.2
Kinematic viscosity, centistokes	0.0025	0.074	0.457	4.57	0.131	1	0.408	4.08
Redwood II, seconds (for > 30 seconds)	0.00612	0.181	1.12	11.2	0.0321	2.45	1	10
Redwood I, seconds (for > 90 seconds)	0.000612	0.0181	0.112	1.12	0.0321	0.245	0.1	1

Determination of flash point. It consists of heating a given volume of petroleum product at a standard rate of temperature rise until the vapor produced gives a flammable mixture with air in an enclosed space *(i.e.,* closed flash point temperature) or with air in an open cup *(i.e.,* the higher open flash point temperature) ignition as a result of application of flame. At fire point, not only the vapor air mixture flashes but the liquid will continue to burn for 5 seconds.

Flash points of different liquid fuels are as follows :

Crude oil = -10 to 0°C	Gasoline = - 40 to 30°C
Kerosene = 28-55°C	Diesel = > 55°C
Lube oil = 130-300°C	Fuel oil = > 66°C

Flash point less than 23°C is dangerous and highly inflammable. The safe value is > 60-66°C.

Pensky-Marten's apparatus is used for the determination of flash point above 50°C *e.g.,* for fuel oils, lubricating oils, bitumens etc. Abel closed cup apparatus is used for more volatile oils with flash point below 50°C. It is determined by Cleveland *open cup* apparatus also.

In *Cleveland apparatus* used for the determination of flash point & fire point of petroleum products (except fuel oils) having open cup flash point below 79°C, an open brass cup is filled with oil and heated at a certain constant rate. Naked flame is repeatedly applied to the oil vapor at definite interval of temperature. When the vapor catches fire momentarily, the temperature of oil is noted as flash point.

In *Pensky-Marten closed cup apparatus* dried oil is filled up in the cup with lid and heated by a burner. Oil is exposed to flame at every 2°F rise in temperature of the oil. The temperature at which the distinct flame first appears inside the cup by applying test flame through the opening meant for the purpose is the flash point.

Pensky-Marten's apparatus shown in Fig. 23.8 comprises of

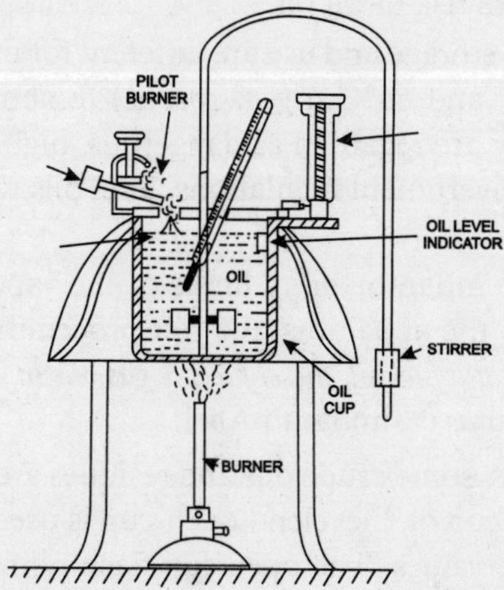

Fig. 23.8. Pensky-Marten's flash-point apparatus.

- An *oil cup* of 5 cms diameter and 5.5 cms deep in which oil is filled up to the level marked inside. Four openings of standard sizes are provided in the lid of the cup. Through one of these passes a thermometer and the second opening is used for introducing test flame.

Through third opening passes the stirrer carrying two brass blades. Fourth opening is meant for admission of air.

- *Shutter* is a lever mechanism, provided at the top of the cup. By moving the shutter, opening in the lid opens and flame (carried by a flame exposure device) is dipped in to this opening & brings the flame over the oil surface.
- *Flame exposure device* is a small flame and this is connected to the shutter by a lever mechanism.
- *Air bath.* Oil cup is supported by its flange over an air bath, which is heated by a gas burner.
- *Pilot burner.* As the test flame is introduced in the opening, it gets extinguished, but when the test flame is returned to its original position it is automatically lighted by the pilot burner.

Working. Oil under test is filled up to the mark in the oil cup and it is heated by heating the air bath by a burner. Stirrer is worked between tests at a rate of about 1 to 2 revolutions per second. *Heat is applied so as to raise the oil temperature by about 5°C per minute.* At every 1°C rise of temperature, flame is introduced for a moment by working the shutter. The temperature at which a distinct flash appears inside the cup is recorded as the flash point. The heating is continued thereafter and the test flame is applied as before. When the oil ignites and continues to burn for at least 5 seconds, the temperature reading is recorded as the fire point of the oil.

FIRE POINT

It is the lowest temperature at which vapors given off by oil ignites and continues to burn for at least 5 seconds. In most cases, fire point is 5–40°C higher than flash point and is determined in the same apparatus as for flash point determination. It gives an idea of fire hazards during the storage and use of oil.

It is customary to determine both the temperatures, although the lower temperature (flash point) is itself a measure of the fire risk of an oil.

The permissible conditions of storage and use are different for products with flash points below 23°C (*e.g.* gasoline), between 23° and 66°C (*e.g.* kerosene) & above 66°C (*e.g.* fuel oils and gas oils). The oils of the first category are regarded as dangerous, highly inflammable for transport & storage, and are subject to strict government regulations. Fuel oils with flash points above 66°C are regarded as 'safe'.

Besides oil's volatility and the inflammability limits of the vapour-air mixture, the flash point also depends upon the design of the apparatus, the test procedure and the barometric pressure. *Kerosene is tested in Abel, apparatus and fuel oil, diesel fuels & gas oils in Pensky-Martens apparatus (closed-cup type). The American 'Tag' apparatus* is similar to Abel.

Flash points of lubricating oils, some crude oils and residues are often determined by the open-cup tests. Either Pensky-Martens cup or Clevelend open-cup is used. The open-cup test is liable to error and gives only approximate values. The open-cup flash point is higher than the closed-cup value for the same sample.

The flash point & the fire point give indications of the range and nature of the boiling point curve. The flash point of lubricants helps to indicate the relative amount of low boiling material present in it. Flash point or fire point should not be confused with completely unrelated spontaneous ignition temperature.

CLOUD POINT, POUR POINT AND FREEZING POINT

These properties are related to the flow conditions of crude oil and its products at low temperature.

Cloud Point. When an oil is cooled at a specified rate, the temperature at which it becomes cloudy or hazy is called the *cloud point of oil.* This haziness is due to the separation of crystals of wax or increase of viscosity at low temperature. It is determined by cold filter plugging point test to find out the extent to which diesel fuel or gas oil will flow even though the temperature is below which the wax crystal will appear *i.e.,* cloud point. It is normally determined for those petroleum products that are transparent in a 40 mm thick layer and have cloud point below 49°C.

Cloud point is important for fuel oils which have to pass through unheated filters of fine mesh *e.g.* a jet plane may be exposed to -60°C and if solid wax separates from fuel oil, the carburettor may be blocked up.

Pour Point. The temperature at which the oil just ceases to flow (or pour) is called the pour point. It determines the temperature below which, an oil can not be used as a lubricant as well as the waxy nature of oil.

To determine the pour point, the oil is cooled in the test jar and with every 3°C fall in temperature tilted to observe the flow of oil. The temperature at which it does not flow even when kept horizontal for 5 seconds is taken as the pour point. Naphthenes and aromatics have lower pour point than paraffins but they are undesirable in fuel oils and diesel fuels. Asphalts act as pour point depressants as they inhibit the growth of wax crystals. Increasing lighter hydrocarbons reduces pour point.

Determination of pour point is carried out with the help of pour-point apparatus shown in Fig. 24.9. It consists essentially of flat-bottomed tube (about 3 cm in diameter and 12 cm high) enclosed in an air-jacket. The jacket is surrounded by freezing mixture (ice+$CaCl_2$) contained in a jar. The tube is half-filled with oil. A thermometer is introduced in the oil. As the cooling proceeds slowly *via* air-jacket, the temperature falls continuously. With every degree fall of temperature in the oil, the tube is withdrawn from the air-jacket for a moment (about 2-3 seconds) and examined. It is then replaced immediately. The temperature at which cloudiness is noticed is recorded as the cloud-point. After this, cooling is continued and the test-tube is withdrawn after every 3°C fall of temperature and tilted to observe the flow or pour of oil. The temperature at which oil does not flow in the test-tube, even when kept horizontal for 5 seconds, is recorded as the pour-point.

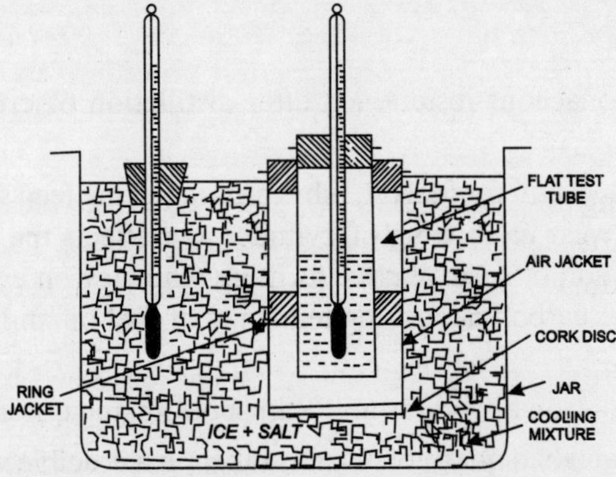

Fig. 23.9. *Pour Point Apparatus.*

Cloud point and pour-point indicate the suitability of lubricants in cold conditions. Lubricants used in a machine working at low temperatures should possess low pour-points ; otherwise solidification of lubricant will cause jamming of the machine. Presence of waxes in the lubricating oil raise the pour-point.

The difference of pour-point and cloud point is 4-6°F usually.

Freezing point. It is the temperature at which the fuel oil freezes completely and can not flow at all. This is important in case of aviation gasoline because at high altitudes where low temperatures are encountered, the fuel supply from fuel tank to engine may be impeded due to choking of pipeline if the freezing point of the fuel oil is not sufficiently low.

Paraffins possess higher freezing points than naphthenes and aromatics. Aviation gasoline should have freezing point below 60°C to avoid trouble due to crystal formation in feed lines and filters.

SMOKE POINT AND CHAR VALUE

These properties of kerosene determine the suitability of kerosene as a fuel and as an illuminant.

Smoke point. This is the maximum height of flame in mm without smoke formation when the kerosene is burned in a standard lamp under closely controlled conditions. Smoking of kerosene is due mainly to the presence of aromatic hydrocarbons.

To determine, smoke point, kerosene sample is burned in a standard lamp with a specified wick for five minutes and the height of flame is read when it leaves no smoky tail. A good quality kerosene used for domestic use should have high smoke point (about 25 mm).

Paraffins are desirable in Kerosene as it has got the highest smoke point whereas aromatics are undesirable as it has got the lowest smoke point. Standard smoke point for Kerosene is 20-30 mm. Aromatics are, hence removed from Kerosene to improve smoke point.

Char value. It is the amount of charred oil deposition on the wick obtained from Kerosene after burning in a standard wick lamp at a standard rate for 24 hours. The weight of char is estimated from that present in the upper portion of the wick after washing with light petroleum spirit and drying at 100-110°C.

Char value should be less than 20 mg/kg of kerosene for a good quality kerosene.

CARBON RESIDUE

It is the percentage of carbonaceous residue left after distillation of crude oil or its products in absence of air.

This determination is important for fuel oil, lubricants, cracking feed stocks etc. Carbon deposit fouls the surface, results in wear & choking of cylinder and affects the regeneration of cracking catalyst as it gives an indication of relative coke forming tendency on evaporation & pyrolysis of an oil. It is also a measure of carbon depositing tendency of fuel oil and engine deposits in diesel fuel fired engine.

It is determined by Conradson apparatus and Ramsbottom apparatus.

In ***Conradson method,*** a weighed quantity of oil is taken in a crucible which is placed in an iron crucible having a cover with a small opening. These are, then placed in a third crucible and covered with a hood and strongly heated at a specified rate and the residue is cooled & weighed. The result (Conradson carbon) is expressed as percentage of the residue on the original weight taken.

In ***Ramsbottom method,*** a weighed quantity of oil is heated in a hemispherical glass bulb of standard dimension for 20 minutes at 550°C. The residual carbon is weighed and recorded as pet cent carbon residue (Ramsbottom carbon).

Conradson carbon residue in high speed diesel fuels should be < 0.1% (on a 10% distillation residue). Atomising burners are insensitive to carbon residue of fuels but gas oil used in vaporising burners should have < 0.05% carbon residue.

Conradson method is normally used for fuel oil and Ramsbottom method for lubricating oil.

Conradson carbon and Ramsbottom carbon are related by *ASTM* relation as

$$3.8\ C = R^2 + 4.25\ R - 0.35$$

where, C = Carbon residue by Conradson method

R = Carbon residue by Ramsbottom method.

ANILINE POINT

It is the lowest temperature at which an oil is completely miscible with an equal volume of aniline. It is the measure of the aromatic content of the oil, hence is a characteristic property of diesel.

Since aromatics dissolve aniline (itself an aromatic substance) more readily than paraffins or *iso*-paraffins, the higher the aniline point, lower the aromatics and higher the paraffin content with very high cetane number making the oil suitable for use in diesel engine.

Aniline point of an oil also gives an indication of the possible deterioration of rubber sealing, packing etc. in contact with the oil. The aromatics have a tendency to dissolve natural rubber and certain types of synthetic rubber. Therefore in such cases, a low aromatic content in the lubricant (*i.e.* with high aniline point) is desirable. Aniline point of cetane is 95°C and for hexyl benzene is -12°C. High aniline point indicates highly paraffinic diesel fuel with high diesel index and good ignition quality. For aromatics, aniline point is low and the ignition quality is poor.

Hence, higher the aniline point, better the diesel fuel and the lubricant.

DIESEL INDEX

An alternative method of expressing the quality of diesel oil is by use of the "Diesel index". It is given by

$$\text{Diesel index} = \frac{\text{Aniline point, °F} \times \text{API gravity}}{100}$$

Diesel index is roughly related with cetane number as :

$$\text{Diesel index} = \text{Cetane number} + 3$$

The use of diesel index does not necessitate the use of a test engine for determining the cetane number and hence the quality of diesel.

OCTANE NUMBER

This is a property of gasoline which is used in a spark ignition engine and expresses its knocking characteristics.

When a gasoline engine is made to operate at high load and low speed, a sort of rattling noise may develop from the combustion chamber resulting from unsteady and uncontrolled combustion. This noise is called knocking or detonation. Knocking is harmful to engine as its parts are set to vibration there by reducing the life of engine. Knocking depends upon the quality of fuel also. Octane number requirement of petrol engines depends on their compression ratio.

It has been found that n-heptane knocks very badly so its anti-knock property has been arbitrarily taken as zero and *iso*- octane (2, 2, 4 trimethyl pentane) gives very little knocking so its anti-knock value has been taken as 100. Octane number of the fuel is defined as "the percentage by volume of iso-octane in a mixture of n-heptane and iso-octane with the same knocking tendency as fuel". Thus if a sample of gasoline gives as much knocking as a mixture of 80 parts of iso-octane and 20 parts of n-heptane, its octane number is 80.

Higher the octane number, better is the fuel as maximum permissible power increases with the octane number besides causing minimum knocking.

Octane number of hydrocarbon fuels increases in the order n-paraffins — olefins — naphthenes — iso-paraffins — aromatics.

To determine the octane number of a fuel, it is burnt in a standard engine for the purpose and the knock is measured by a knockmeter. This is compared with the standard knockmeter readings of fuels of different octane numbers to determine the octane number.

In India, motor gasoline of octane number 87 is sold whereas the same of aviation gasoline is more than 100. In U.S.A. octane number of standard & premier gasoline is 94 & 100 respectively.

Octane number of premium gasoline sold in India is 94. Octane number of gasoline is improved by the addition of *tetra ethyl lead (TEL),* the quantity being 3.6 c.c. TEL/gallon of motor gasoline and 5.5 c.c. TEL/gallon of aviation gasoline. TEL is a very good anti-knocking agent.

Constituents of gasoline respond differently to the addition of TEL. This is called lead response and decreases in the order *paraffins - naphthenes - olefins - aromatics.* Alcohols respond negatively as do sulphur compounds. Use of tetra-ethyl-lead(TEL) for improving the octane number of motor gasoline has been discontinued in most countries including India due to environmental pollution reasons. However its use is permitted in aviation gasoline.

PERFORMANCE NUMBER

The addition of small quantity of TEL to gasoline increases its octane number very rapidly, the maximum quantity used being 6 c.c./gallon. It is possible to increase the octane number to a value greater than 100 *i.e.* in excess of that of iso-octane. Values above 100 octane number are known as "Performance number". Aviation gasoline invariably have octane number greater than 100 hence their anti-knock characteristics are expressed by performance numbers. On performance number scale, 100 octane number equals 100 performance number.

The best aviation gasoline has performance number of 160.

For octane number above 100, the following relationship exists between performance number (P.N) and octane number (O.N):

$$O.N = 100 + \frac{P.N - 100}{3}$$

CETANE NUMBER

It is a characteristic property of diesel and is used to indicate its quality and performance in compression ignition engine.

In case of diesel engine, the fuel should ignite as soon as it is injected into the cylinder. If it does not ignite instantaneously, (the time lag between fuel injection and fuel ignition is called *ignition delay period)* abnormal combustion takes place resulting in shock waves, due to fluctuating pressure rise in the cylinder. With diesel of long ignition delay period, much of the charge is injected into the cylinder before ignition is initiated, causing violent combustion, sudden increase in pressure and rough & bumpy running. This is called *diesel knock.*

Cetane ($C_{16}H_{34}$) has a very small ignition delay period hence it is given a cetane number rating of 100. α-methyl naphthalene has a very high ignition delay period, hence it is given a rating of zero. Cetane number of diesel oil is the percentage by volume of cetane in a cetane/α- methyl naphthalene mixture that has the same ignition delay period and performance in a standard compression ignition engine as that of the fuel.

Diesel oils for high speed (> 1500 rpm), medium speed & low speed engine should have cetane number not less than 50, 35 & 20 respectively.

Cetane number increases in the order aromatics - iso-paraffins - naphthene-olefins - *n*- paraffins, whereas the octane number decreases in the same order. The result is that a oil of high cetane number has a low octane number and *vice-versa.* A petroleum crude which gives diesel oil of high cetane number, gives gasoline of low octane number. A good spark ignition fuel makes a poor compression ignition fuel and *vice-versa.*

1-1.5% of additives called "dopes" like ethyl nitrate, acetone, *iso*-amyl nitrate etc. are added to improve the cetane number. Cetane number is determined by a standard test compression—ignition engine by measuring the ignition delay.

High cetane number diesel fuel will facilitate easy starting of C.I. engines particularly in cold weather, faster engine warm up, increased engine efficiency & power output and reduced exhaust smoke, odour & combustion noise.

EMULSIFICATION

Some oils mix intimately with water and the mixture is called emulsion which has the tendency to collect dirt, grit, particles & other foreign matters.

The emulsification tendency is undesirable in most applications of the oil. Formation of emulsion in a circulating oiling system is undesirable, because sticky, emulsion and sludges interfere with oil circulation. A good lubricating oil should not form emulsion under lubrication condition and even if it forms an emulsion, it should break off very rapidly. Additives are added to check tendency of emulsification. In case of steam turbines, the oil should separate quickly from the emulsion so that it can be recirculated for use. Formation of stable emulsion may be desirable when the oil is to be used in presence of water as in the case of certain steam cylinder oils and water soluble cutting lubricants.

In emulsion test, steam is passed into the oil under specified conditions and the time taken for the separation of oil from the emulsion in minutes is recorded as steam emulsion number which depends upon the refinement and viscosity of oil. Good refinement and lower viscosity are conducive to quicker separation of the oil from the emulsion.

OXIDATION STABILITY

Oils tend to oxidise in contact with air rapidly at higher temperature. Oxidation products are insoluble, gummy and sticky materials which may clog oil holes, filters, pipelines & other parts of the oil circulation system. Soluble oxidation products are corrosive. Hence resistance to oxidation is a desirable property of oil. Oxidation inhibitors are added to improve their resistance to oxidation. The inhibitors get oxidised first and the oil is protected.

A common test for it is the ***Sligh oxidation test*** in which 10 gm sample of oil is heated in a flask containing pure oxygen at 200°C for 2.5 hours. The flask is cooled, the contents are diluted with naphtha and allowed to stand for an hour. The precipitate is filtered, washed, dried and weighed. The weight is expressed as percentage of the weight of the sample and multiplied by 100 to give ***Sligh oxidation number.***

ACID VALUE/NUMBER

Acid value or number determines the acidity of an oil. It is defined as the number of milligrams of the KOH required to neutralise the free acid present in 1 gm of oil.

New or used petroleum products contain acidic constituents present as additive or as degradation products formed during service. Total acidity is a measure of the combined organic and inorganic acidity.

Generally, mineral acids are not present in the lubricants unless refined in a faulty manner. Organic acids are not so corrosive. Acid value of petroleum products should be less than 0.1. Increase in acid number indicates oxidation of the oil which will lead to gum & sludge formation besides corrosion.

COLOUR AND FLUORESCENCE

Colour of petroleum products indicates the degree of refinement. Distillates are lighter in colour than the residual oils. A pale colour does not indicate lower viscosity.

The mineral oils show fluorescence in reflected light whereas fixed oils do not show fluorescence. Hence fluorescence helps to detect adulteration.

DISTILLATION RANGE

This enters into the specification of distillates like gasoline, kerosene etc. In the distillation test, a measured volume of the oil is distilled at the specified rate which is suitably increased as the distillation proceeds. The *initial boiling point* (I.B.P.) is the temperature when the first drop of condensed liquid drops from the condenser and the *final boiling point* (F.B.P.) is the maximum temperature recorded at the end of distillation. *Mid-boiling point* is the temperature at which 50% of the oil distils off. Distillation is stopped at 360°C as further heating causes cracking of the residue. If necessary, distillation is done under vacuum also. Distillation range helps in evaluation of petroleum products. Mid-boiling point is often used for correlating with other parameters.

Distillation data are used for making important curves like *true boiling point* (T.B.P.) curve, equilibrium/flash vaporisation curve. T.B.P. curve gives the variation of boiling point with the yield of distillates. Each point in the flash vaporisation curves gives the total amount of vapor in equilibrium with the liquid at that particular temperature. These curves are used in the design of continuous distillation equipments in refineries.

Distillation characteristics is a measure of volatility of petrofuels and determines the system of fuel metering required (wick feed, carburation or atomisation) and indicates the vapor losses during fuel storage & vapor locking tendency as well as uniformity of distribution to cylinders 10% distillation for motor spirit indicates the ease of starting the engine while too high a FBP will cause crankcase oil dilution.

SULPHUR CONTENT

Even after purification, petroleum products do contain some or other quantity of sulphur. Sulphur content increases with the increase in the boiling range of the products. Some typical sulphur content values are, Kerosene = 0.05 to 0.2%, gas oil/diesel oil = 0.3 to 1.5% and fuel oil = 2 to 4% (LSHS = 1%).

Sulphur content in oil is determined accurately by *bomb calorimeter* and *roughly by lamp method.* In lamp method, a known weight of oil is burned in an air current and the combustion products are passed through an alkaline solution, the sulphur being finally estimated as H_2SO_4 in the solution. Sulphur is largely present in oils in the form of corrosive organic compounds. On combustion, it gives off foul gases corrosive in nature.

MOISTURE

Water solubility in oil is very low (0.005 to 0.05%). However, it is present in fuel oil upto 1% in dispersed state. Undesirable water is separated from oil by preheating. *Dean and Stark method* is used for its determination in which oil sample is heated with xylene which forms an azeotrope with water. The distilled vapor on condensation and settling gives water as a separate layer which is measured.

The presence of water in fuel oil is undesirable because, among other harmful effects, arrival of water globules at the outlet of an oil burner may lead to *'flame failure"* (the flame is extinguished) either by interruption of the oil stream or by the *'quenching'* (extinguishing) effect of a suddenly evolved mass of steam. The amount of water present in oil is determined by dissolving the oil in xylene (it acts as a carrier for water) and distilling off all the water & some xylene. In a common form of apparatus the distillation flask is fitted with a vertical, but offset, reflux condenser which discharges into a vertical, graduated tube. Water collects at the bottom of the tube and is measured, while excess xylene is returned to the boiling flask or a metal still continuously. The standard apparatus of this type is that of *Dean and Stark* as shown in Fig. 24.10. This method is suitable for variety of bituminous materials, but it is especially applicable to crude petroleums and products derived from them, such as fuel oils, road oils, creosotes, road tars, asphalts & tars.

A suitable quantity of sample is transferred to the flask and 100 ml of solvent (xylene) is added to that. When large samples are analysed, a solvent volume in excess of 100 ml may be necessary.

The components of the apparatus are assembled as shown in Fig. 23.10, making all connections vapour and liquid tight. The condenser tube and trap must be chemically clean to ensure free drainage of water into the bottom of the trap. A loose cotton plug is inserted in the top of the condenser to prevent condensation of atmospheric moisture inside it. Cold water is circulated through the jacket of the condenser.

The heat is applied to the flask/still adjusting the rate of boiling so that condensed distillate discharges from the condenser at the rate of 2-5 drops per second. The distillation is continued until no water is visible in any part of the apparatus except in trap. If there is a persistent ring of water in the condenser denser tube, the rate of distillation is increased or condenser water is cut off for few minutes.

When the evolution of water is completed, the trap and contents are allowed to cool to room temperature. Any water drops adhering to the sides of the trap are dislodged with a glass or polytetrafluoroethylene (PTFE) fluoroethylene or other suitable means and transferred them to water layer. The volume of the water in the trap is read to the nearest scale division and the water in the sample is reported as per cent by volume.

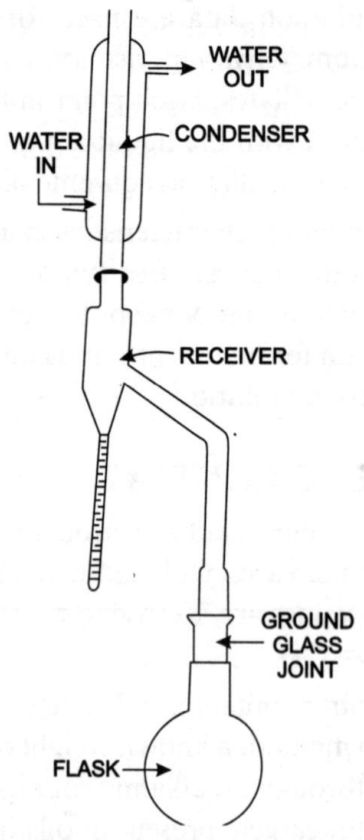

Fig. 23.10. *Dean and Stark Apparatus for Determination of Moisture in Liquid Fules.*

$$\% \text{ Water} = \frac{\text{Volume of water in trap}}{\text{Volume of sample}}$$

Volatile water soluble material present may also be measured as water in this method.

SEDIMENT IN LIQUID FUELS

Sediments comprising of pipe & storage tank scale and a few fibers is defined as material insoluble in toluene. It should be < 0.25% (by weight) in fuel oil. Sludge and sediment formation determine the possible life of oil during storage. Sediment cause clogging of filter & nozzles and corrodes storage tanks & pipelines.

The insolubles in 90's benzole is reported as sediment content. The determination is carried out by an extraction method. The apparatus employed for the purpose is shown in Fig. 23.11. This consists of a conical extraction flask of 1 litre capacity fitted with a block of tin coil condenser. A fused alumina (porous grade) thimble (cylindrical tube 2.5 cm dia and 7 cm in height) is suspended from the condenser coil by means of a platinum wire so that it hangs approximately 4.5 cm above the extracting solvent and 4.5 cm below the condenser coil. An electric hot plate or steam bath may be used for heating the flask.

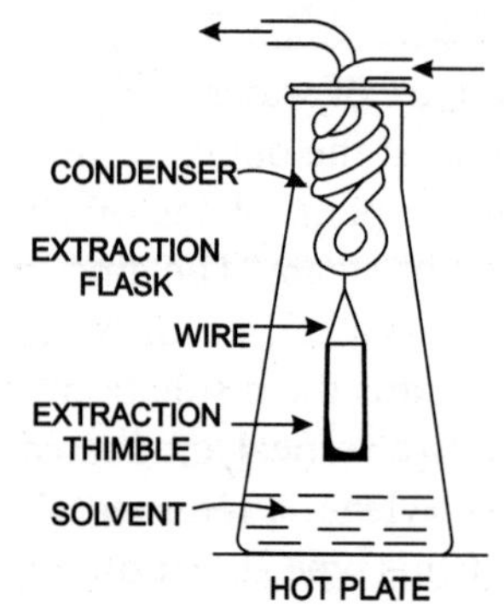

Fig. 23.11. *Sediment Determination Apparatus.*

A preliminary extraction with 90's benzol is given to thimble before its use and the solvent is allowed to drip for 1 hour at a temperature of 100-110°C. The thimble is cooled to room temperature and weighed to the nearest 0.1 mg. The extraction is repeated until the weights of thimble after two successive extractions do not differ by more than 0.2 mg. Now 10 gm of the sample is placed in the extraction apparatus and the sample is extracted with 90's benzol until the solvent dripping from the thimble is colourless. The rate of extraction is so maintained that the surface of the mixture of oil and benzole in the thimble does not rise more than 2 cm from the top. After the completion of extraction, the thimble is dried for 1 hour at 100-110°C, cooled and weighed. The extraction is repeated for 60-75 minutes and again followed by drying, cooling & weighing. The extraction is repeated for 1 hour, until the weights of the dried thimble after two successive extractions do not differ by more than 0.2 mg. The sediment content is represented as per cent of original sample.

$$\%\ \text{Sediment} = \frac{\text{Weight of sediment}}{\text{Weight of sample}} \times 100$$

CALORIFIC VALUE (C.V.)

Gross C.V. of petroleum fuels are in the range of 10000 to 11300 kcal/kg , the higher figure being for gasolines and lower for heavy fuel oils. Petroleum products are high in hydrogen content (11.8-14.5%) and hence their net calorific value is less than the gross by 620-760 kcal/kg. C.V. of oil is determined in a *bomb calorimeter.*

Following emperical formulae are also used for its determination.

(a) *Mendeleev formula* $Q_G = 81C + 300H + 26\,(S - O)$

(b) *Also,* $Q_N = Q_G - 6\,(9H + W)$ *where,* $H = 2.6 - 1.5\,p$

(c) *Also,* $Q_G = 12400 - 2100 \times p^2$

where, Q_G = Gross calorific value, kcal/kg, Q_N = Net calorific value, kcal/kg

C, H, S, O & W are weight percentages of carbon, hydrogen, sulphur, oxygen & water in oil. p = specific gravity at 15.5°C

Gross calorific value Q_G of most hydrocarbon liquid fuels is also given by *Cragoe formula,* which is as follows :

$$Q_G = 51860 - 8790 - S^2\,(1 - x - y - z) + 9410\,.\,z \text{ KJ/kg}$$

$$= 12390 - 2100S^2\,(1 - x - y - z) + 2248\,z\text{Kcal/kg}$$

where *s, x, y, z* are the specific gravity, moisture, ash & sulphur respectively in fuel expressed as fractions.

ASH IN PETROLEUM PRODUCTS

Ash in petroleum products can result from oil, water soluble metallic compounds or extraneous solids such as dirt & rust. An idea of the ash forming material present in oil can give information about its suitability for use in a given application.

This method describes a procedure for determining the ash content in distillate, residual fuel oils, crude oils, lubricating oils, waxes, and other petroleum products. However, this process is undesirable where ash forming additives are used and the compounds of lead & zinc are present because these volatalise appreciably during incineration. The ash content of an oil is given as the percentage by weight of the inorganic residue left after ignition of the oil. The amount of ash in petroleum oil is normally very low. It increases with the viscosity of the fuel oil, since most of the original inorganic impurities allowed to remain in the crude are concentrated in the primary residues. Ash in oil is determined by gradually burning a weighed quantity of oil in a silica crucible, finally igniting the crucible and its charred contents at 800°C.

For estimation of ash, 50 gm sample (nearest to 0.1 gm) is taken in a weighed dish of platinum, porcelain or silica which is already heated at 700°C for 10 minutes and cooled. The dish and sample are heated until the contents can be ignited with a flame. This dish is maintained at such a temperature that the sample continues to burn at a uniform & moderate rate, leaving only ash and carbon when the burning ceases. Then the residue is heated in a furnace at 800°C until all carbonaceous material has disappeared. The dish is cooled to room temperature in a desiccator and then weighed to the nearest 0.1 mg. The dish is reheated for 10-20 minutes at 800°C, cooled in desiccator and reweighed. Heating and weighing are repeated until consecutive weighings differ by not more than 0.5 mg. The ash content in oil is given by:

$$\% \text{ Ash} = \frac{w}{W} \times 100$$

where w = weight of ash, grams and W = weight of sample, grams.

If the fuel oil contains very low ash *i.e.* 0.01-0.02%, more quantity of oil may be taken in a beaker & transferred to the dish in successive portions and igniting them until a suitable quantity of oil has been used.

If the sample contains sufficient moisture to cause foaming and loss of material, the sample is discarded, and a fresh sample is taken to which 1-2 ml of isopropyl alcohol (99%) is added before heating. If this is not satisfactory, 10 ml of an equivolume mixture of benzene & isopropyl alcohol is added and mixed thoroughly. Several strips of ashless filter paper are placed in the mixture and heated, when the paper begins to burn, most of the water would have been removed.

MECHANICAL STABILITY OF LUBE OILS

To judge the stability of a lubricant under conditions of very high pressure, different mechanical tests are carried out. One of such tests is "four ball extreme-pressure lubricant test" shown in Fig. 23.12 in which the lubricant is poured in a machine containing four balls in which lower three balls are stationary ; while the upper ball is rotated. Load is gradually increased and the balls withdrawn and examined at specific intervals for scars formation, on them. If the lubricant is satisfactory under the given load, the ball bearings after the test comes out clean. However, when the load is progressively increased, liberated heat welds the balls, together

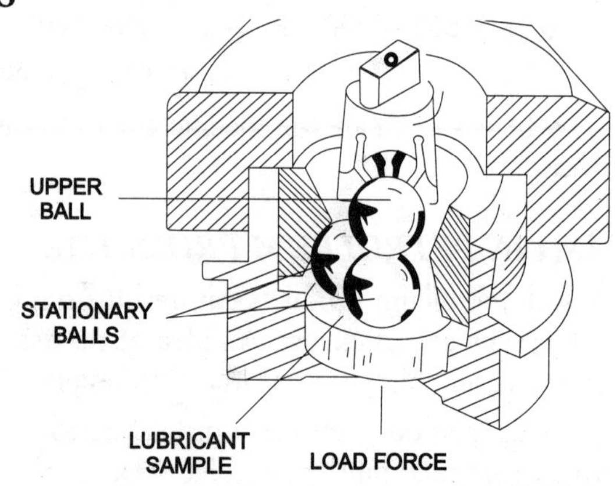

Fig. 23.12. Four ball extreme-pressure tester.

and the lubricant is said to fail completely. Thus, this test enables to determine the load that can be carried safely by a lubricant.

PROPERTIES OF GREASES

While selecting greases for practical jobs the two properties of main consideration are consistency and drop point.

Consistency or yield value of grease. It is expressed in terms of penetration, which is defined as *"the distance in tenth of millimetre that a standard cone penetrates vertically into the sample under the standard conditions of load, temperature and time"*. The value of load, temperature and time are taken respectively as 150 g, 25°C, and 5 seconds. Consistency of greases depends on the structure & interaction of the gelling elements in it and to some extent on the viscosity of oil used. The consistency is determined by using 'Penetrometer" shown in Fig. 23.13. The apparatus consists of:

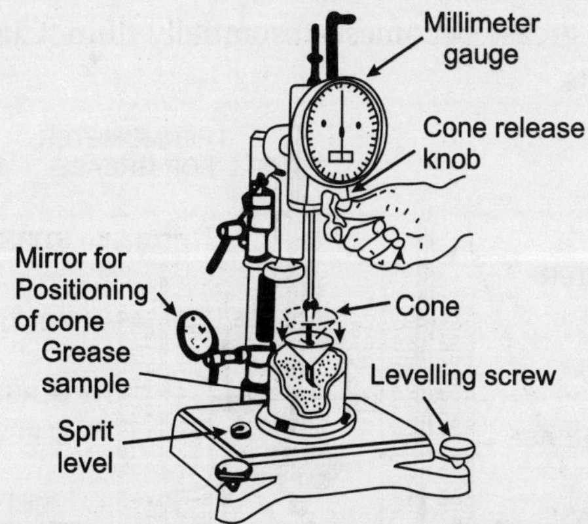

Fig. 23.13. *Penetrometer used for determining consistency of grease.*

- *Heavy base* (of cast iron alloy) which is provided with a spirit level, levelling screws and a plain table over which a box containing the grease under test is placed.
- *Vertical support* is an iron rod fitted to the base. On this are slotted marks, around which a holder can be moved up and down. The holder has a screw which can be tightened in any of the slots.
- *Circular dial.* The holder carries a circular dial gauge which is graduated in millimeters.
- *Moving dial rod* is arranged behind the dial by a mechanical mechanism. The rod is provided with clutch arrangement for disconnecting or connecting it to the circular dial.
- *Standard cone.* The moving dial rod is fitted at its lower end with a standard cone (weighing 150 g).
- *A mirror.* Vertical rod is provided with an adjustable mirror for removing parallax while positioning the cone in contact with grease sample surface.

The apparatus is levelled, the cone is cleaned and the sample of grease under test in a box is placed below the cone. The height of the cone is so adjusted, with the help of mirror, that tip of the cone just touches the sample. Initial dial reading is noted. The cone is then released for exact

5 seconds, by pressing a button provided for this. After 5 seconds, the press button is released and final dial reading noted. The difference of the two dial readings gives the penetration.

Drop-point of grease. It is "the temperature at which it passes from the semi-solid to the liquid state". This temperature determines the upper temperature limit of the applicability of the grease. The apparatus used for determining drop-point is shown in Fig. 23.14.

The sample is taken in a metal cup, which has an opening of standard size in the bottom. It is then enclosed in a glass case, having a tight lid. A thermometer is also inserted in the cup, so that the bulb of the thermometer is just above the surface of grease sample. The combination is then placed in a glass beaker containing water and provided with a stirrer. The beaker is heated slowly at a rate of 1°C/minute. As the temperature is raised, the grease sample passes from a semi-solid to a fluid state. The temperature at which its first drop falls from the opening is recorded as drop-point.

Drop point helps in assessing whether a grease has been properly made or to obtain an idea of the type of thickener used in a grease. It indicates the temperature at which the thickener is so soluble in the base oil that the grease becomes substantially fluid. Clay and dye thickened greases have no measurable drop points.

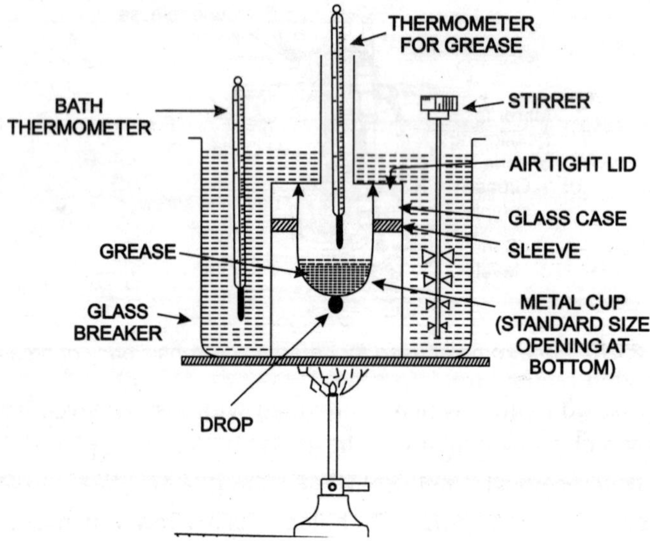

Fig. 23.14. Drop- point apparatus.

PROPERTIES OF BITUMENS

Suitability of several grades of bitumens commercially available are assessed by penetration, penetration index, softening point and ductility for various applications.

Penetration of bitumen. It determines the type and grade of bitumens used for a particular application. The bitumen sample is maintained at 25°C in a container kept in water bath. The weighted needle is brought to the surface and the penetration of the needle into the bitumen at the end of 5 seconds interval in units of 0.1 mm is termed as "*penetration*" of the bitumen. The penetration of a bitumen at the softening point is about 800.

Penetration index of bitumen. Penetration at 25°C and the softening point or penetration at two different temperatures (say at 25°C & 10°C) defines the extent to which the consistency of bitumen changes with temperature defined by a factor called "Penetration Index". It is defined as follows :

$$\frac{\log(800) - \log(\text{PEN}, 25°)}{T_{RB} - 25} = \left[\frac{\text{C } 20 - \text{PI}}{10 - \text{PI}}\right]\left[\frac{1}{50}\right]$$

where, PEN, 25° = Penetration of bitumen at 25°C

PI = Penetration index of bitumen

T_{RB} = Ring and ball softening point of bitumen, °C

A positive value of penetration index (PI) shows that the bitumen is less effected by temperature changes while a negative value of PI indicates large effect in the consistency of bitumen with temperature changes.

Softening point of bitumen. Since there is no sharp temperature at which a bituminous material changes from solid to liquid state (as it gradually becomes softer and less viscous as temperature rises), determination of its softening point is done by a fixed arbitrary & closely defined method called ***"ring and ball test"***.

In this test, the temperature (°C) is measured at which a standard disc of bitumen confined in a metal ring softens to such an extent, when heated at a rate of 5°C/minute, that it deforms under the loading imposed by a small steel ball (of, specified weight), which falls a distance of 2.54 cms.

RING AND BALL TEST

Softening point is used for designating the oxidised bitumen and hard bitumen.

Ductility of bitumen. Bituminous surfaces crack depending on its ductility due to traffic stress on road resulting in varying temperature exposure causing their expansion & contraction. The *ductility* of bitumen is expressed as the distance in cm to which a standard briquette of bitumen can be elongated before the thread formed breaks under the specified conditions. Bitumen having high ductility may be more temperature susceptible.

To determine the ductility of bitumen, a molten bitumen sample is poured into a standard mould, allowed to cool to room temperature and then placed in a thermostatically controlled water bath to bring the briquette of bitumen to the test temperature before mounting in the ductility testing machine which consists of a moving carriage moving over a lead screw. Smooth constant speed and continuous operation is ensured by an electric motor driven reduction gear unit.

SETTING POINT AND MELTING POINT OF WAX

Setting point of wax is the constant temperature observed within a range of 0.1°C for one minute, when the molten wax is allowed to cool in a specified apparatus and the temperature is recorded at frequent interval. This method is not suitable for micro crystalline wax or intermediate wax or blends of paraffin waxes with these or any additives.

Congealing point of all petroleum waxes or petrolatum is determined by applying a drop of molten wax to the thermometer bulb and noting the temperature at which it congeals, when the thermometer is rotated under standardised cooling conditions.

Drop melting point of petroleum waxes or petrolatum is the temperature at which a drop of the waxes sample falls from the bulb of a thermometer, when heated under standard conditions.

GUM CONTENT OF PETROFUELS

Gums present in petrofuels cause piston ring sticking & deposits on engines and indicate the amount of olefins present which have poor storage stability. There are two types of gums in petrofuels namely *existent gum* and *potential gum.*

Existent gum is that gum which is already formed & present in the fuel and can be deposited as the petrofuel evaporates. It is the amount of non volatile heptane insoluble residue left when the sample of oil is evaporated in a jet of hot air at 160°C. For jet fuels, evaporation of oil sample is done in a jet of superheated steam at 232°C.

Potential gum is the amount of gum that may be formed under long storage condition by polymerisation & oxidation of unstable hydrocarbons like olefins present in the oil. It is the amount of gum formed after the sample is aged in an oxidation stability bath and evaporated under specified conditions.

COPPER STRIP CORROSION TEST OF PETROFUELS

Petrofuels contain varying amount of sulphur and sulphur compounds which are corrosive in nature. Copper strip corrosion test serves as a measure of difficulties encountered with copper, brass or bronze parts of the petrofuels handling and utilisation systems.

In this test, a smoothly polished and cleaned copper strip is immersed in the oil sample which is then maintained at the specified temperature for the specified length of time. The strip is then removed, washed with aromatic & sulphur free petroleum spirit and examined for etching, pitting or discoloration. It is then matched with ASTM (American Society for testing materials) copper strip corrosion standard color code number 1, 2, 3 and 4 designating slight tarnish, moderate tarnish, dark tarnish and corrosion respectively. Subscripts '*a*' to '*e*' describes a standard color reproduction in the standard chart. The classification code '1*a*' for example indicates slight tarnish with a light orange color.

Since some of the parts of fuel pump in aircrafts are made of silver, the corrosive tendency of the *aviation turbine fuel* (ATF) is determined by *silver strip corrosion test* in which a polished silver strip is immersed completely in ATF at 45 ± 1°C for 16 hours. At the end of 16 hours, the silvers strip is removed, washed and evaluated for corrosion against the set of standard.

REFRACTIVE INDEX OF PETROFUELS (HYDROCARBON LIQUIDS)

Refractive index (defined as ratio of velocity of light in air to that in liquid under examination) gives an indication of aromatic content and molecular weight of petrofuels. The refractive index of liquids (also defined as the ratio of sine of angle of incidence to the sine of the angle of refraction) varies inversely with both wavelengths and temperature. Naphthene content in naphtha can be calculated by knowing its refractive index and density of the saturates fraction as determined by the refractive intercept method. Refractive intercept of liquid hydrocarbons is given by $\eta - 0.5p$, where η= refractive index and ρ = density (gm/cc) both at 20°C.

WATER SEPAROMETER INDEX (MODIFIED) (WSIM)

WSIM, which is a measure of fuel cleanliness (relative'to its freedom from surfactant materials) measures the water separation characteristics of fuels. In water separometer, a prepared emulsion of water & petrofuel is passed through a cell containing a coalescer. Entrained water is found in the effluent from cell by light transmission. The ease with which the fuel releases the emulsified water is rated on a numerical scale 1–100.

THERMAL STABILITY OF JET FUELS

Jet fuel thermal oxidation tester (JTFOT) measures the high temperature stability of jet fuel (gas turbine fuel) to assess the deposit level resulting from the fuel contacting a heated surface.

The jet fuel under test is pumped at a fixed rate (600 c.c in 2.5 hours) through an aluminium heater followed by a stainless steel filter (17 microns) meant for trapping the fuel degradation products (*i.e.,* deposits). Amount of deposit on heater tube and the rate of plugging of filter gives the indication of thermal stability of jet fuels.

SMOKE VOLATILITY INDEX (SVI) OF ATF

SVI is a measurement parameter for combustion deposit by wide cut gasoline type of aviation turbine fuel (ATF). It is a function of its smoke point and volatility defined as :

SVI = smoke point + 0.42 (% of ATF boiling below 204.4°C)

KAURI-BUTANOL (KB) NUMBER OF HYDROCARBON SOLVENT i.e., SOLVENT POWER

KB number is a measure of solvent power (or solvency) of a hydrocarbon solvent *i.e.,* its ability to dissolve a given product. **KB** number is a point at which a solvent added as a thinner would cloud a solution of kauri resin dissolved in *n*- butanol. For most polymers & resins, solvent power decreases in the order aromatics > naphthene > *n*-paraffins >*i*-paraffins. Higher **KB** number implies higher solvent power. Standards are toluene (**KB** number = 105) and 25% toluene + 75% *n*-heptane (**KB** number = 40). It measures the relative solvent power. **KB** number ranges from 25 to 105 for most commercial solvents.

COAL TAR FUELS (C.T.F.)

COAL TAR

Coal tar is a black to brown oily and viscous fluid of characteristic odour produced during high or low temperature carbonisation of coal during coke manufacture.

Tar produced during low temperature carbonisation of coal (700°C) is called low temperature tar whereas, that produced during high temperature carbonisation (at 1100°C) is called high temperature tar. Tar usually means high temperature tar. Specific heat of tar is 0.3—0.4 kcal/kg °C at 40°C.

Low temperature tar. Its yield is about 10% of dry coal. It is a brown colour oily liquid having very *low viscosity,* hence flows readily. It is paraffinic in nature. Its typical analysis is C = 84%, H = 8%, S = 0.5%, N = 1%, O = 6.5%.

It contains very less free carbon (3%). It is free from naphthalene.

High temperature tar. It is a thick, high viscosity, black coloured liquid with an aromatic nature. It contains more amount of useful aromatics like phenol, naphthalene, creosote, anthracene, benzol, pyridine etc. Its free carbon content is high (5—8%) and the yield is lower (3% of dry coal). Specific gravity is more (1.18) compared to low temperature tar (1.06).

Difference between low temperature tar and high temperature tar :

The salient points of difference in the tar produced by low temperature carbonisation of coal and high temperature carbonisation of coal has been tabulated below in table 24.1.

Table 24.1: Difference between low temperature tar and high temperature tar

	Low temperature tar (L. T. tar)		High temperature tar (H.T. tar)
(a)	It is produced during low temperature carbonisation of coal at 700°C and is also called *primary tar.*	*(a)*	It is produced during high temperature carbonisation of coal at 1100°C and is also called *secondary reformed tar.*
(b)	Its free carbon content is practically nil (0 to 1%). Tar yield is higher ; about 10% of dry coal carbonised.	*(b)*	Its free carbon content is high (5-8%). Tar yield is lower; about 3% of dry coal carbonised.
(c)	Its specific gravity is lower (1.02 to 1.06).	*(c)*	Its specific gravity is higher (1.10 to 1.15).

(d)	Its viscosity is less,, hence it has a better fluidity.	*(d)*	Its viscosity is comparatively higher, hence it has poorer fluidity.
(e)	It is mostly paraffinic in nature having little aromatics.	*(e)*	It is mostly aromatic in nature.
(f)	It contains very little of naphthalene and anthracene (<1%).	*(f)*	It contains higher amount of naphthalene & anthracene (about *5%).*
(g)	It has a higher content of tar acids (upto 35%) having higher molecular weight than the tar acids of high temperature tar.	*(g)*	It has a very low content of tar acids (3%).
(h)	Pitch yield on distillation is lower (35%).	*(h)*	Pitch yield is higher (65%).
(i)	It is a raw material for the production of motor spirit and diesel fuel via catalytic hydrogenation. It has a low market value.	*(i)*	It has a high market value at it is a source of production of valuable aromatic chemicals like benzol, phenol, naphthalene, anthracene, creosote oil etc.
(j)	Typical composition (weight%) in various distillation range is as follows : 0 - 170°C = 6%, 170 - 230°C = 16%, 230 - 270°C = 12%, 270°C Pitch = 31%, hard Pitch = 35%	*(j)*	Typical composition is as follows: 0 – 170°C = 2%, 170 - 230°C = 7%, 230 – 270°C = 12%, 270°C pitch = 14%, hard Pitch = 65%.

A comparative feature of typical L.T. tar and H.T. tar is given below :

Parameters	Unit	L.T. tar	H.T. tar
Yield of tar	litre/ton coal	95.5	33.6
Specific gravity	No unit	1.029	1.169
Water content	weight %	2.2	4.9
Carbon content	weight %	84	90.3
Hydrogen content	weight %	8.3	5.5
Nitrogen content	weight %	1.08	0.95
Sulphur content	weight %	0.74	0.84
Ash content	weight %	0.10	0.24
Toluene insoluble (TI) components	weight %	1.2	6.7

Recovery of tar. During the carbonisation of coal in by-product coke oven, the tar vapors present in the coke oven gas are separated by cooling it by ammoniacal liquor resulting in its condensation. Remaining tar is separated in primary cooler and electrostatic tar precipitator thereby recovering the bulk of the tar present in the coke oven gas.

Distillation of tar. Tar contains around 300 identified chemicals. Hence to recover some of these chemicals economically and to produce better fuel, tar is distilled. The main fractions of tar (which are recovered by distillation), their constituents, yield and boiling ranges are given in the Table 24.2.

Table 24.2: Main Fractions of Tar and Their Yields (weight %)

Fractions	Boiling range, °C	Yield, weight %	Main constituents of the fractions
Light oil	upto 170	2-4	Benzol, naphtha and phenol
Middle (or carbolic) oil	170-230	5-7	Phenol, naphthalene and pyridine
Heavy (or creosote) oil	230-270	15-25	Naphthalene and creosote oil
Anthracene oil	270-350	14-17	Anthracene and wash oil
Pitch	Residue	60-70	—

Uses of tar. Tar is used as a furnace fuel. Its calorific value may be 8500 kcal/kg. Use of crude tar as a fuel possesses a number of disadvantages *viz.*

- the presence of volatile component which are lost on storage and which increase the fire hazard.
- the presence of water in tar which reduces the calorific value, causes corrosion and its irregular combustion.

the presence of solid impurities such as coke dust which may choke the burners and pipelines.

Besides, the fact that tar is a 'storehouse of chemicals', its direct use as a fuel is discouraged. Coal tar is also used as a binder for coal briquettes and for road making purposes.

Coal Tar Fuels (C.T.F.)

Coal tar fuels (C.T.F.) are the liquid fuels obtained by mixing the different tar distillation products (fractions). Though the different products of tar distillation contain valuable chemicals, the demand for them is limited, hence large parts of tar is prepared into a series of coal tar fuels.

Six types of C.T.F. are prepared and designated as :

(a) CTF 50 (b) CTF 100 *(c)* CTF 200
(d) CTF 250 *(e)* CTF 300 *(f)* CTF 400

The number indicates the temperature in °F to which the fuel (CTF) should be preheated to reduce its viscosity (upto maximum 25 centistokes) so as to make it suitable for atomisation through burner. For example, CTF 200 is the fuel which should be preheated at 200°F to decrease its viscosity for smooth atomisation. CTF 50 does not need any preheating.

CTF 50 and CTF 100 are the mixture of carbolic oil, creosote oil and anthracene oil fractions of tar. CTF 200, CTF 250 and CTF 300 are the mixture of creosote oil, anthracene oil and residual pitch.

CTF 400 is nothing but residual medium soft pitch having a softening point of 75-80°C. This pitch can be pulverised and used as a pulverised fuel in the furnace. Pitch is also used for coating of roof and pipeline for guarding against corrosion. Soft pitch having a softening temperature of 30°C can be used as liquid fuel after preheating. Properties of CTF is given in Table 24.3.

Table 24.3: Properties of Coal Tar Fuels (CTF)

Properties	Unit	CTF 50	CTF100	CTF200	CTF250	CTF 300	CTF 400
Calorific value	kcal/kg	9150 to 9750	9150 to 9750	9000 to 9450	9000 to 9300	8900 to 9200	8750 to 9050
Specific gravity	–	–	1.01	1.13	1.16	1.19	1.23
Water content ,Max	volume %	1	1	1	1	0.5	0.5
Ash content, Max	vyeight %	0.05	0.05	0.25	0.30	0.30	0.75
Softening point	°C	–	–	–	–	33.38	75.80
Viscosity, Redwood No. 1	seconds	60	100	1000 to 1500	–	–	–
Flash point	°C	> 65	> 65	> 65	> 65	> 65	> 65
Sulphur content	weight %	< 1	< 1	< 1	< 1	< 1	< 1

Ultimate analysis of coal tar fuels (CTF) is given below in table 24.4

Table 24.4: Ultimate analysis of coal tar fuels (CTF)

S. No.	Coal tar fuels (CTF)	Ultimate analysis,% by weight				
		Carbon	Hydrogen	Sulfur	Nitrogen	Oxygen (by difference)
1.	CTF 100	86.50-89.30	8.0-6.9	0.13-0.25	0.84	4.41-2.83
2.	CTF 200	89.86	6.05	0.39	1.11	2.59
3.	CTF 250	89.97	5.90	0.44	1.16	2.53
4.	CTF 300	90.08	5.73	0.49	1.22	2.48
5.	CTF 400	90.42	5.23	0.65	1.38	2.32
6.	Pulverisable CTF	90.66	4.90	0.86	1.42	2.16

Normally, carbon content increases and hydrogen & oxygen content decreases in the series CTF 50 to 400. C/H ratio increases from 11 to 16.5 in this series.

Uses of CTF. Because of low sulphur content and high flame emissivity, it is one of the ideal fuels for metallurgical furnaces. Besides, it is used for power generation, rotary kilns in calcination & cement units and glass melting furnaces.

Pitch Creosote Mixture (PCM). This liquid fuel is mainly a mixture of a soft pitch & creosote and is a fuel like fuel oil. As a matter of fact this PCM is nothing but CTF 200. PCM or CTF 200 is the most popular coal tar fuel.

A typical composition and properties of PCM are given in table 24.5.

Table 24.5: Composition and Properties of PCM (CTF-200)

Carbon	90%
Hydrogen	6%
Nitrogen	1%
Sulphur	0.5%
Oxygen	2.5%

Specific gravity	1.1
Preheating temperature	200°F.
Viscosity at 20°C, Redwood I seconds	6000 seconds
Specific heat at 40°C	0.3 to 0.38 kcal/kg °C
Flash point	65°C
Net calorific value	9000 kcal/kg.

PCM is most suitable for metallurgical furnaces where low sulphur and high emissivity is desired.

Combustion Properties of Coal Tar Fuels (C.T.F.) :

- Calorific value (kcal/kg) of all the C.T.F. are lower than corresponding petroleum fuel oils.
- Average carbon/hydrogen ratio for C.T.F. is 14 : 1 as compared to 6 : 1 for petroleum fuel oils. As a result of which, C.T.F. burns with a luminous flame which has a high emissivity factor (0.6-0.9).
- Higher emissivity factor of C.T.F. flame results in higher heat transfer rate by radiation given by

$$Q = \sigma \varepsilon (T_1^4 - T_2^4)$$

where, Q= radiation heat transfer rate kcal/hour. m^2

σ = Stefan-Boltzman constant = 4.97 x 10^8kcal/hr.m^2k^4 = 5.67 x 10^8 W/m^2.k^4

ε = Emissivity factor varies from 0.6 to 0.9

T_1 = Flame temperature of C.T.F., °K

T_2 = Temperature of the substance being heated, °K.

- Higher emissivity factor implies higher heat transfer rate resulting in higher thermal efficiency of the furnace. Higher heat transfer rate means lesser time for heating a substance *i.e.* higher production rate of the furnace.
- C.T.F. (produced from high temperature carbonisation tar) is aromatic in nature as compared to paraffinic structure of fuel oils. Since the emissivity factor of aromatic hydrocarbons are 25% higher than paraffins, the former has higher emissivity factor and hence higher radiation heat transfer rate. Hence consumption (in kg) of C.T.F. is 10% less for the same heat load in a furnace, even though its calorific value is 10% less than petroleum fuel oil.
- Sulphur content of C.T.F. is much lower (< 0.5%) than the corresponding petroleum fuel oils (upto 4%). This makes it most suitable for heating the metallurgical furnaces like open hearth furnace, soaking pit, reheating furnace, annealing furnace etc. As the high sulphur petroleum fuel oil causes the brittleness of steel and surface crack during rolling (particularly cold rolling of alloy steel) when the steel slabs are contaminated with sulphur.
- Lighter C.T.F. can be atomised similar to petroleum fuel oil. Heavier C.T.F. are difficult to atomise. Steam or compressed air are used for atomisation of fuel. A mixture of compressed air and steam are most effective giving short, luminous flame with a high heat release rate.
- Because of low percentage of hydrogen (upto 6%) in C.T.F. (as compared to upto 12% in fuel oil), the difference between gross & net calorific value is less (upto 325 kcal/kg) for C.T.F.

as compared to 625 kcal/kg for fuel oil. Because of less difference, the heat losses due to hydrogen in fuel is much less in case of C.T.F. as compared to fuel oil.

- Theoretical air requirement for combustion of C.T.F. is bit less than that for fuel oil, hence the volume of flue gas generated on combustion is less.
- The dew point of flue gas generated by combustion of C.T.F. is lower compared to that by fuel oil. Hence more waste heat recovery from flue gas is possible in case of C.T.F. than fuel oil.
- Specific heat of C.T.F. is 0.3-0.4 kcal/kg°C. Higher specific heat improves fuel cooling capacity (by high heat transfer rate) and reduces the heat transfer area.
- Calorific value of C.T.F. (9000 kcal/kg) is less than that of corresponding fuel oil (10000 kcal/kg).
- The change of viscosity of C.T.F. with temperature (i.e., logarithmic reduction) is greater than fuel oils but the slope of the temperature viscosity curve is much the same for all coal tar fuels.

Requisites of Good Quality Gasoline, Diesel, Fuel Oil & Other Petroleum Products

KNOCKING

When the fuel-air mixture undergoes very fast or intermittant combustion inside the combustion chamber of an internal combustion engine, then a sort of rattling noise develops from the combustion chamber (due to uneven, unsteady and uncontrollable combustion) which is called knocking or detonation.

Reasons for knocking. Knocking in petrol (gasoline) and diesel engines are due to different reasons.

In case of petrol engines, the carburettor (which mixes petrol and combustion air) supplies the petrol air mixture to the combustion chamber which is ignited by the spark from the spark plug. If the first portion of the fuel burns in a normal manner and the last portion burns instantaneously, then a large momentary pressure imbalance is caused which sets up a pressure or shock wave in the combustion chamber resulting in knocking. This is a characteristic of petrol constituents.

Another reason is the preignition of petrol before the regular spark occurs. It occurs at high compression ratio (which is the ratio of cylinder volume at the end of suction stroke to the cylinder volume at the end of compression stroke) which increases the temperature of fuel air mixture in the combustion chamber above ignition point even before spark occurs. Hence, the fuel ignites before the spark occurs. This pre-ignition of petrol causes uneven and unsteady combustion resulting in knocking of engine.

In case of diesel engine (compression ignition engine), there is no spark plug and the combustion is initiated by increasing the temperature of air by compression inside the combustion chamber by piston till the ignition temperature is achieved.

In diesel engines, as soon as the fuel is injected into the combustion chamber, it should instantaneously ignite. If it does not ignite instantaneously *i.e.* if there is an ignition delay, then more and more quantity of unburnt diesel gets accumulated in the combustion chamber which burns at once causing knocking of the engine. This is called *diesel knock.* Hence for a good diesel, its spontaneous combustion should take place as soon as it is injected into the cylinder.

Knocking depends upon the characteristics of fuel, design of engine and the operation of engine. If the engine is operated at high load and low speed, knocking results. If the petrol and diesel are adulterated then also knocking result.

Effects of knocking. Knocking in an engine results in wear and tear of cylinder, piston & other engine parts. Besides, it lowers the efficiency of the engine. Knocking reduces the life of the engine.

OCTANE NUMBER

Octane number expresses the anti-knocking characteristics of petrol. Normal heptane (a constituent of petrol) when burnt in a petrol engine knocks it very badly *i.e.* the anti-knocking value of n-heptane is very poor and it is assigned an octane number equal to zero. On the other hand, *iso-octane* (also a constituent of petrol has got a very good anti-knock characteristics, hence it is assigned an octane number of 100.

Definition of octane number. Octane number of petrol is the percentage by volume of iso-octane in a mixture of iso-octane and normal heptane with the same knocking tendency as the petrol in question.

If the octane number of a petrol is 80, then it means that this petrol gives as much knocking as a mixture of 80% iso-octane and 20% normal heptane.

Determination of octane number (O.N.). The petrol whose octane number is to be determined is burnt in a standard spark ignition internal combustion test engine and its knocking tendency is measured by an instrument called knockmeter. Now this reading of knockmeter is matched with the standard chart in which the value of octane number corresponding to the knockmeter reading is given. This gives the octane number of the petrol.

Three methods are used for O.N. determinations namely :

(a) *Motor method* - for low O.N. motor fuels
(b) *Research method* - for high O.N. motor fuels
(c) *Aviation method* - for very high O.N. aviation fuels.

Parameters of the above methods are as follows :

Parameters in various methods	Motor method	Research method	Aviation method
rpm	900	600	1200
spark advance	variable	13° btdc	—
mixture heating	Yes	No	—

Octane number of Indian petrol. Regular petrol having octane number 87 and premium petrol having an octane number 94 are sold in India for use in motor vehicles. In U.S.A. the octane number of premier gasoline is 94. The octane number of petrol used in air-crafts are more than 100 and they are designated by performance number.

Significance of octane number. Higher octane number of petrol means higher efficiency and higher power output by the engine. Hence, higher the octane number, better the petrol.

Limitation of octane number as a guide. Tendency of a petrol to knock depends not only on its octane number but it depends also upon the design and operation of the engine. So, the octane number of a gasoline is simply an emperical figure and not related to the actual knocking in an automobile engine.

Means to improve octane number. Octane number of petrol is increased by adding some additive compounds in it. The most important additive is Tetra Ethyl Lead (TEL). Quantity added is 3.6 cc/gallon of motor petrol and, 5.6 cc/gallon of aviation petrol.

TEL is extremely poisonous, hence petrol containing TEL is always coloured for identification and safety. However,use of leaded petrol has been discontinued in most countries including India due to environmental reasons.

The use of petrol having higher octane number than necessary to avoid knocking is not at all useful as it is more susceptible to pre-ignition and gum formation on storage.

Advantages of high octane number of petrol. As per an estimate, as the octane number increases from 87 to 100, the cruising speed (which is very important for airplanes) increases by **25** km/hour. There is increase in speed of climb by 80% and the maximum attainable height increases from 8000 metres to 11000 metres. The required take-off distance decreases from 800 metres to 600 metres and pay load also increases by **2** tons. These all are very important for aeroplanes.

Anti-knocking value. Anti-knock value of the hydrocarbons present in petrol increases with increase in the compactness of the molecules, cyclic structure and double bond. Thus butane has an octane number of 90 whereas heptane has 0.

Anti-knock value *i.e.* octane number increases in the order paraffins→ olefins → naphthenes → iso-paraffins → aromatics. It means that aromatics have highest anti-knock value whereas paraffins have lowest anti-knock value (hence octane number). So, presence of maximum quantity of aromatic and minimum quantity of paraffins is desirable in petrol.

In case of paraffins and olefins, anti-knock value *(i.e.* octane number) decreases with increasing chain length.

In case of naphthenes, the larger the size of the ring and the length of the side chain, the lower is the anti-knock value.

In case of aromatics, side chain addition upto 3 carbon atoms length improves the octane number, but afterwards, the octane number decreases in proportion to the length of the longest chain.

Octane Number of Blends. When two paraffins are blended, then the octane number of the resultant blend is directly proportional to the percentage of each paraffin in the blend.

With blends of paraffins and aromatics, the octane number of the resultant blend is lower than expected from a linear relationship.

With blends of paraffins and olefins, the octane number of resultant blend is higher than expected from a linear relationship.

Lead Response. The response of gasoline to the addition of tetra ethyl lead [$(C_2H_5)_4$ Pb] for increasing its octane number is different for different categories of hydrocarbons. The order of decreasing effect of TEL addition (called lead response) is paraffins ----> naphthenes ----> olefins □ aromatics. Alcohols, sulphur compounds and lubricating oils respond negatively.

CETANE NUMBER

Cetane number is the measure of knocking in a diesel engine. Cetane ($C_{16}H_{34}$) has a very short ignition delay hence its cetane number is taken as 100 whereas alpha methyl naphthalene ($C_{11}H_{10}$) has a very large ignition delay hence its cetane number is taken as 0.

The cetane number of a diesel is defined as the percentage of cetane (by volume) in a mixture of cetane and alpha methyl naphthalene which has the same ignition delay as the diesel under test.

Values of cetane number of diesel. The cetane number of low speed, medium speed and high speed diesels are 25, 35 and 45 respectively.

Engine speed, rpm	> 1500	800-1500	400-800	100-400	< 100
Suitable cetane number range	50-60	45-55	35-50	30-40	15-30

Cetane number of diesel can be increased by adding additives like ethyl nitrate and acetone peroxide.

Cetane number *increases* in the order → aromatics → iso-paraffins → naphthenes → olefins → paraffins (AINOP) whereas the octane number *decreases* in this order. Therefore, an oil of high octane number has a low cetane number and *vice versa.* Consequently, a petroleum crude which gives petrol of high octane number gives diesel of low cetane number.

Higher cetane number diesel fuel results in easier cold weather engine starting, faster engine warm up, reduced exhaust smoke & odor, increased engine efficiency & power output, less combustion noise and improved combustion control.

Two important motor fuels namely gasoline and diesel are produced from crude oil. *Petrol is the commercial name of gasoline.*

GASOLINE (PETROL)

Production of petrol (gasoline) from crude oil. Petrol is produced from crude oil by the following processes :

- distillation of crude oil produces *straight run gasoline.*
- cracking of fuel oil produced from crude oil to produce *cracked gasoline.*
- alkylation and polymerisation of refinery gases produced during distillation of crude oil.
- reforming of heavy gasoline and naphtha to get *reformed gasoline.*
- condensation of natural gas to get *natural gasoline* called *"casing head gasoline".*

The gasoline produced by the distillation of crude oil is called *straight run gasoline* which is about 20% of the crude and has a low octane number of 55-60. Gasoline comprises of C_5 to C_{10} hydrocarbons fraction boiling in the range of 30 to 180°C. It is also called motor spirit or motor fuel. On prolonged storage, it undergoes slow but progressive oxidation deterioration resulting in formation of non-volatile gummy residue & darkening of its color. Gummy residues cause deposit formation in carburettor parts, venturi, manifold of distribution system and in the combustion chamber.

Gasoline produced by catalytic cracking of heavy fuel oil at high temperature & pressure is called *cracked gasoline* and has a high octane number of 90.

Refinery gases produced during distillation of crude oil are subjected to some auxiliary processes like polymerisation (gasoline produced in this case is called *polymer gasoline),* alkylation and isomerisation to produce high octane number (90-98) gasoline.

Reforming process of heavy gasoline and straight run naphtha produces *reformed gasoline* of high octane number (90-98) by increasing the proportion of high octane iso-paraffins and aromatics.

There are still other methods called hydro-cracking and hydro-reforming of heavier fractions of crude oil to produce superior gasoline with high octane number.

Requisites of good gasoline. - It should have high octane number to develop more power at higher compression ratio without knocking.

- It should have controlled volatility and vapor pressure. Very low vapor pressure of petrol causes starting trouble in the engine whereas very high vapor pressure of petrol causes vapor locking of fuel pipelines.
- Its calorific value should be high for higher efficiency. C.V. of petrol is around 10000- 10500 kcal/kg.
- Its freezing point should be low to avoid the choking of fuel lines of air-crafts flying at high altitude where the temperature is low and for engines operated during cold season.
- It should not contain abrasive, corrosive & gum forming substance otherwise it will cause abrasion, corrosion & choking respectively of fuel lines, carburettor jets etc.
- It should have optimum flash point for easy starting of engine and avoidance of explosion & fire hazards during storage. Indian specification of motor gasoline is given in table 25.1.

Table 25.1: Specification for Motor Gasoline

(IS : 1585—1960, Amendments in 1967 and 1972 IS: 2796—1971, First Revision)

Characteristics	87 Octane no. Gasoline	83 Octane Gasoline	93 Octane Gasoline
Colour, visual	Orange	Orange	Red
Octane number (Research Method), Minimum	87	83	93
Lead Content *(a)* as lead, g/1. Maximum	0.42	0.56	0.80
Distillation characteristics : (a) Initial Boiling Point (IBP)	—	Not limited, but to be reported	
(b) Recovery upto 70°C, % by volume, Minimum	10	10	10
(c) Recovery upto 125°C, % by volume, Minimum	50	50	50
(d) Recovery upto 180°C, % by volume, Minimum	90	90	90
(e) Final boiling point (FBP), °C, Maximum	215	215	215
(f) Residue, percent by volume, Maximum	2	2	2
Reid vapour pressure at 38°C, kg/cwm^2(g), Maximum	0.70	0.70	0.70
Sulphur, total, percent by weight, Maximum	0.25	0.25	0.20
Residue on evaporation, mg/100 ml, Maximum	4	4	4
Oxidation stability, minutes, Minimum	360	360	360
Copper-strip corrosion for 3 hours at 50°C	Not worse than No. 1		Not worse than No. 1
Density at 15°C	Not limited, but to be reported		

Refining of petrol. The crude petrol is refined by treating with sulphuric acid to remove sulphur compounds present in it (the process is called *'sweetening'* of petrol). It is then passed through a bed of fuller's earth to remove its colour, which is due to the presence of gum in it. Finally, the dissolved gases in it are driven off by a process called *stabilisation.* Tetraethyl lead or other additives are then added to increase the octane number (if the need be) or to inhibit the gum formation before the petrol is ready for use.

Additives in petrol (gasoline)

Commercial grade motor gasoline contains the following additives :

(i) *'Anti-knocking agents', e.g.* tetra ethyl lead (TEL). Currently unleaded petrol (without TEL) is being sold/used in India, as toxic lead in petrol causes atmospheric air pollution.

(ii) *'Antioxidants'* to prevent gum formation and improve storage stability *e.g.,* 2, 6-ditertiary butyl-4-methyl phenol.

(iii) *'Metal deactivators'* to prevent gum formation due to trace metals like iron and copper.

(iv) *'Combustion chamber deposit modifier' e.g.,* tricresyl phosphate (TCP) to minimise misfire due to lead fouling of spark plugs & to reduce ignition and preignition.

(v) *'Anti-icing agents'* to minimise engine stalling & power loss due to ice formation in carburettor and fue lines *e.g.,* ethanol and hexylene glycol.

(vi) 'Anti rust agents'.

(vii) *'Cleansing agents'* to prevent deposits in the fuel supply system.

(viii) *'Dying agents'* to give distinctive colors to different brands and names.

AVIATION GASOLINE

Petrol used in the aeroplanes has comparatively higher volatility and higher octane number (> 100) than motor gasoline. Though higher proportion of TEL (4.6 c.c/U.S. gallon) is added in aviation gasoline, other additives are lesser in number and quantity. The main components of aviation gasoline are isoparaffin alkylates and aromatic reformates.

A typical specification of aviation gasoline is tabulated in Table 25.2.

Table 25.2: Typical Specification of Aviation Gasoline

S.No.	Parameters	Unit	Value
1.	Reid vapor pressure at 37.8°C (100°F)	kg/cm^2	0.385-0.490
2.	Freezing point, maximum	°C	60
3.	Performance number	—	130-145
4.	Aromatic content, minimum	% by volume	5
5.	Diesel index, minimum	—	75-90
6.	Net calorific value, minimum	kcal/kg	10390-10500
7.	Total sulphur, maximum	% by weight	0.05
8.	Tetra ethyl lead	mg/litre	nil
9.	Oxidation stability (16 hours ageing)		
	(a) Gum, maximum	mg/100 ml	6
	(b) Precipitate, maximum	mg/100 ml	2
10.	I.B.P/F.B.P.	°C	<75/170

DIESEL

It is a straight run product of crude petroleum oil distillation obtained in between kerosene and lubricating oil. Diesel obtained from paraffinic crude are most suitable as higher amount of paraffin present in it increases its cetane number. Diesel is also produced by cracking of heavy fuel oil, vacuum distillation of atmospheric distillation residue and coking process to supplement the limited yield of straight run diesel to meet its demand (six times of petrol in India). *High speed diesel* (cetane no. > 42) are used in diesel engines for automative purpose *e.g.*, bus, truck, locomotive, etc. Light Diesel Oil (LDO) are used in engines in marine ships and for electricity generation. Middle distillates of crude oil boiling in the range of 150 - 400°C are suitable as diesel fuels. Iso-octyl nitrate is added to improve the cetane number by 3 to 6 units. In cetane number increase requires 100 pounds per thousand barrels (Ptb) cetane improver. Other additives used in diesel fuel include *detergents* (act as cleaners to remove carbonaceous & gummy deposits) upto 50 Ptb, *Stabiliser* (to retard oxidation during storage), *anti-corrosion & rust inhibitor* (3-5 Ptb), *cold weather flow improvers* (65-125 Ptb), *Anti-wax settling additive* (40-60 Ptb), *Lubricity additive* (3-8 Ptb), *metal deactivator* (2-3 Ptb) etc.

Though the most desired components of diesel fuels are the straight run distillates from paraffinic and mixed base crude oils, these are often used as feedstock for gasoline production via cracking. Since, there is some overlap in the boiling ranges of kerosene and diesel fuel hence diesel is prepared from heavy distillates obtained from catalytic cracking unit. Since, the distillates are rich in aromatics and isoparaffins (having low cetane number), additives are added to improve the cetane number of diesel fuel. The suitable volatility is obtained by blending with light fractions. Residual oils are also used in significant proportions in meeting large demand of diesel fuel.

Requisites of good diesel fuel:

- It should have a high cetane number.
- It should contain very low amount of sulphur (<0.5%), ash & carbon residue so that fouling corrosion and wear of engine parts would be minimum.
- Its calorific value should be high.
- Its flash point should not be less than 65°C to avoid fire and explosion hazard.
- It should have a very low freezing point so that it can be successfully used as a jet fuel after blending with kerosene or naphtha.

Indian standard specification for diesel oils is tabulated in Table 25.3.

Table 25.3. Specification for Diesel Oils

(IS : 1460—1968/1974)

Characteristic	Unit	Grade Special	Grade A	Grade B (LDO) Light Diesal Oil
Flash point, (Pensky-Martens, closed)	°C, Minimum	38	38	66
Kinematic viscosity	centistokes at 37.8°C	2.0 to 7.5	2.0 to 7.5	2.5 to 15.7

Carbon residue (Ramsbottom)	percent by weight, Maximum	0.20	0.20	1.50
Cetane number, Minimum	—	42	40	—
Diesel Index, Minimum	—	45	43	—
Distillation	per cent recovery, at 366 ^{0}C, Minimum	90	90	—
Copper-strip corrosion for 3 hours at 100°C	—	Not worse than No. 1	Not worse than No. 1	Not worse than No. 2
Water content	per cent by volume, Maximum	0.05	0.05	0.25
Sediment	per cent by weight, Maximum	0.05	0.05	0.10
Ash	per cent by Weight, Maximum	0.01	0.01	0.02
Acidity, inorganic	—	Nil	Nil	Nil
Acidity	total mg of KOH/gm, Maximum	0.50	0.50	—
Sulphur	total per cent by weight, Maximum	1.0	1.0	1.8
Pour point	°C, Maximum	6	6	12 for winter 18 for summer

Winter — November to February

Summer — March to October

Special grade and grade A cover distillates of low volatility.

Grade B covers the class of more viscous distillates and blends of these distillates with residuum oil.
Diesel fuel for naval application shall have a flash point of 66°C, Min and cetane number of 45, Minimum.

FUEL OIL (FURNACE OIL)

Fuel oils are the fractions distilled between lubricating oil and bitumen. The heavy fuel oil is thermally cracked/decomposed at high pressure and temperature (process is called Visbreaking) to reduce its viscosity before it is used as fuel oil. At times, heavy fuel oil is blended with naphtha, gas oil and light fuel oil to improve its atomising capability through burners. Fuel oil is composed of higher molecular weight compound known as asphaltic bitumens.

On one extreme is the light fuel oils produced from cycle gas oil of cracking units and on the other extreme is heavy fuel oils produced from residues of crude distillation units, thermal catalytic cracking units, & coker (coking unit). Intermediate grade/viscosity of fuel oils are made by blending of distillates and residual stocks. 'Visbreaking' of residual stocks is a very important process in the production of fuel oils. Cycle gas is filtered through clay and hydrofined. Tricresyl phosphate and sludge dispersing agent are added to improve combustion in burners. Magnesium bearing additives are used in case of residual fuel oils to eliminate fouling and corrosion.

Properties of fuel oils

Desirable properties of fuel oil are given in Table 25.4.

Table 25.4: Desirable properties of fuel oil

Properties	Remarks
Viscosity Pour point and freezing point	It should be low (25 cst) for its proper atomisation through burner and low pressure drop during pumping. But viscosity of fuel oil below 18 cst causes leak back through the pump and wear on the moving parts which are lubricated by the fuel oil will increase.
Sulphur content	It should be low to permit easy flow in cold weather. Desirable pour point is 18°C.
Sludge, gummy material and ash	Should be low (< 4%) to avoid corrosion, pungent odor and increase in dew point of flue gases.
	Should be low to avoid chocking of pipelines, erosion of furnaces/ engines and contamination of products. Ash comprising mainly of vanadium and sodium should be < 0.05%.
Calorific value	Should be high. It varies between 9000-9400 kcal/kg.
Specific gravity	Should be low as the pressure drop on pumping it will be low.
Flash point	Should be high for avoiding evaporation loss and fire hazards.
Water content	Should be very low as it decreases the calorific value of oil.

Analysis of fuel oil. *A* typical analysis *of* fuel oil is given below :

Carbon = 86%, Hydrogen = 12%, Sulphur = 1%, Water = 0.6%, Sediments = 0.4%.

Typical specification of fuel oil:

Calorific value	= 9200 kcal/kg
Specific gravity at 25°C	= 0.9
Flash point	= 75°C
Viscosity at room temperature	= 85 centistokes
Theoretical flame temperature	= 1900°C
Maximum freezing point	= 15°C
Storage temperature	= minimum 20°C
Atomisation temperature	= 125°C
Emissivity factor	= 0.5-0.7

Furnace oil produced in India has a maximum viscosity of 1500 seconds (Redwood no. I) at 50°C. However, *Low sulphur heavy stock (LSHS)* furnace oil (S<1%) is semi-solid at ambient temperature and requires heating to 60°C for pumping. For proper atomisation through burners, fuel oil needs to be heated to 100°C for achieving a viscosity of about 25 Cst.

Indian standard specification of fuel oil is given in table 25.5.

Table 25.5: Indian Standard Specification of Fuel Oils (furnace oils)
(IS : 1593-1960)

Characteristic	Unit	Grades of furnace oil		
		Low Viscosity	Medium Viscosity	High Viscosity
Flash point (Pensky-Martens) (closed)	°C, Minimum,	66	66	66
Kinematic viscosity	centistokes at 50°C Maximum	80	125	370
Water content	percent by volume, Maximum	1.0	1.0	1.0
Ash	percent by wt., Maximum	0.1	0.1	0.1
Sulphur	total percent by wt., Maximum	3.5	4.0	4.5
Sediment	percent by wt., Maximum	0.25	0.25	0.25
Acidity, inorganic	—	nil	nil	nil

A typical specification of low sulphur heavy stock (LSHS) fuel oil produced at *Barauni oil refinery* is given in Table 25.6. LSHS is a class-C petroleum product.

Table 25.6: Typical specifications of LSHS produced at Barauni Oil Refinery

Characteristics	Unit	Value
Kinematic viscosity (at 100°C)	centistokes	26
Inorganic acidity	—	NIL
Specific gravity	—	0.93
Kinematic viscosity, maximum	Cst at 50°C	500
Gross calorific value	kcal/kg	10000-10500
Sediment, maximum	weight %	0.25
Pour point	°C, maximum	60
Flash point (Pensky-Marten, closed cup)	°C, minimum,	66-72
Sulphur content	weight %, maximum	1
Ash content	weight %, maximum	0.1
Moisture	weight %, maximum	1

Combustion characteristics of fuel oil:

- Combustion of fuel oil takes place initially in vapor state, hence atomisation must be sufficient to initiate combustion.
- In case of heavy fuel oil, cracking of hydrocarbons takes place at the surface of droplets

of oil resulting in liberation of carbon particles which burn with a luminous flame having higher emissivity factor, hence higher radiative heat transfer capacity.

- Higher the aromatic content of fuel oil *i.e.* higher the carbon/hydrogen ratio, the more luminous is the flame & higher is the emissivity factor but C.V. decreases and specific gravity increases with increase in C/H ratio.
- About 15 kg of air per kg of fuel oil is required for complete combustion with low excess air to the tune of 10% only.
- The shape and size of flame affects the thermal efficiency. Optimum flame can be produced with fuel oil by suitable burner and combustion air control.
- With 10% excess air (200°C), the flame temperature obtained from fuel oil is around 1700°C which will radiate energy at the rate of 1000000 kcal/m^3 of combustion volume (e = 0.5).
- High amount of sulphur (upto 4%) present in fuel oil leads to high dew point (upto 160°C) of flue gas. Hence the flue gas going out of chimney cannot be cooled below this temperature otherwise SO_2 & SO_3 present in it will condense and corrode the boiler parts, induced draught fan blades & chimney checker bricks etc.
- Sodium and vanadium in the ash of fuel oil are corrosive and get deposited on the hot pressure parts of boilers.
- Pre heating of oil in storage tank is needed to make it less viscous and easily pumpable. Heating an entire oil tank or outflow heater may be needed to heat oil which is to be pumped away. Outflow heaters with steam or electricity are also used to ensure smooth pumping.
- For pre-heating of oil, in-line heaters are to be used to raise oil temperature from pumping to combustion temperature. In-line heaters are of either electrical or steam tracing type. The pre-heat temperatures of oil are determined by the viscosity of oil supplied at room temperature and the recommended viscosity at the burner tip.
- Fuel oil viscosity and pumping temperature are tabulated below :

Viscosity (Centistokes) of oil at room temperature	Pumping Temperature, °C
50	60
230	104
900	121

Parameters for obtaining good efficiency in fuel oil combustion is given below :

- Maintain viscosity of oil of 100 Redwood sec I at burner tip by proper setting of pre-heat temperature.
- Provide atomizing air at 1-3 kg/cm^2 (about 2% of total combustion air requirement)
- To control excess air, flue gases should be continuously analyzed for $C0_2$ or 0_2.
- Acid formation occurs at around 160°C, and cold end corrosin should be avoided by maintaining stack temperature above 160°C.
- Slightest damage to the burner tip may increase fuel consumption by 10-15%, and hence any worn out burner tips should be replaced immediately.

- Oil pressure at the burner tip for pressure jet burners should be 17-20 kg/cm^2.
- Correct flame is normally short. Flame impingement on walls and tubes is to be avoided since it causes carbon/soot deposits.

NAPHTHA

Naphtha is a light hydrocarbon boiling in the gasoline range having boiling point 60 to 200°C. When produced by atmospheric distillation of crude oil, it is called *straight run naphtha* (octane no. = 50 - 70), while when produced by conversion processes like visbreaking, fluidised bed catalytic cracking, hydrocracking & coking, it is termed as *cracked naphtha* (octane no. = 92 - 94). Depending upon its boiling point (bp), naphtha is also classified as *light naphtha (bp = < 100°C)*, *intermediate naphtha (bp* = 100 - 150°C) and *heavy naphtha (bp* = > 150°C). It contains C_5 to C_{13} hydrocarbons comprising typically of 55 - 65% paraffins, 20 - 30% naphthenes and 10 - 15% aromatics depending upon type of crude oil processed.

Typical specifications of naphtha which is a class-A product aş per petroleum act 1934 is given below in table 25.6 (A).

Table 25.6 (A): Typical specifications of naphtha

S.No.	Characteristics	Unit	Value
1.	Specific gravity	dimensionless	0.70-0.75 (0.7165 at 15°C)
2.	Flash point (Pensky Martens closed cup)	°C	< 23
3.	Gross calorific value	Kcal/kg	10,200 (7200 Kcal/litre)
4.	Kinematic Viscosity at 37.8 °C	Centistokes	< 0.5
5.	Carbon/hydrogen (C/H) ratio	dimensionless	5.5
6.	Reid vapor pressure (RVP) at 37.8°C	Kg/cm^2	0.42
7.	Residue on evaporation (maximum)	mg/ml	5
8.	Total sulfur content	ppm	100
9.	Lead content	ppm	3
10.	Initial boiling point (IBP)	°C	45
11.	Final boiling point	°C	155
12.	Aromatics content	% by volume	11.5
13.	olefins content	% by volume	NIL

Main uses of naphtha is as follows :

- Straight run naphtha is subjected to catalytic reforming to produce gasoline of high octane number in the range of 93 - 94, which can be blended with straight run naphtha/gasoline (octane no. = 50 - 70) to produce 87 octane number motor gasoline presently used in India.
- Catalytic steam reforming of naphtha is done to produce synthesis gas for ammonia production in nitrogenous fertiliser plants.

- Naphtha (C_5 - 140°C) is used as feedstock for steam crackers producing hydrogen, methane, ethylene, propylene, butene, butadiene, benzene and isoprene in petrochemical plant. Dearomatised naphtha (4 - 6% aromatics) is used in both fertiliser plant and petrochemical plants.
- Intermediate naphtha is used as motor & aviation gasoline whereas heavy naphtha is used as kerosene and jet fuels.
- Heavy naphtha is also used as solvent (for dry cleaning of clothes) as well as thinner for paints, varnishes & lacquers.
- It is vaporised and mixed with blast furnace gas to improve its calorific value thereby replacing high C.V. coke oven gas used for mixing in B.F. gas in steel plant. Mixed gas of desired C.V. is prepared during shortage of coke oven gas by naphtha vapor injection in steel plant.

KEROSENE

Straight run kerosene produced by distillation of crude oil contains C_{10} - C_{14} hydrocarbons boiling in the range of 150 - 250°C. It burns with smoky flame because of high aromatics content in it. *Smoking tendency decreases in the order aromatics → naphthenes → iso-paraffins → paraffins.* That means, aromatics having same carbon atoms have the highest smoking tendency, while paraffins are the least smoking constituent of kerosene.

- Tendency of aromatics to smoke decreases with increase in molecular weight.
- Tendency of naphthenes to smoke decreases with increasing molecular weight and increases on addition of double bond.
- Tendency of paraffins to smoke increases with increasing molecular weight. Branching and olefin linkage in paraffins also increases, its smoking tendency.

A good quality kerosene should have high proportion of paraffinic hydrocarbons and less of aromatics. Aromatics are removed from kerosene by liquid sulphur dioxide extraction *(Edeleanu Process)* thereby improving its smoke point *i.e.* reducing its smoking tendency. Dearomatisation of kerosene is also done by *hydrotreating*, in which aromatics are converted into naphthenes. Kerosene is also subjected to a process called *hydrofining* which improves its sulphur content, color, odour & burning quality (as measured by wick char value).

Kerosene is used for the following purposes :

- It is a domestic fuel for cooking as well as an illuminant for lighting.
- It is used in the manufacture of insecticides, fungicides & herbicides used for controlling pests, fungi & weeds respectively.
- It is used in the manufacture of detergents (keryl benzene sulphonate)
- Extracted aromatics from kerosene are used for manufacture of aromatic solvents & oil.

AVIATION TURBINE FUEL (ATF)

Aviation turbine fuel (ATF) is the fraction of crude oil boiling in the range of 150 - 270°C, comprising mainly of paraffins and naphthenes with <20% aromatics & <5% olefins. Kerosene type ATF and wide cut gasoline type ATF are in general use, the former being a modified development of the illuminant kerosene originally used in gas turbine engines while, the later is a wider boiling range fuel including some gasoline fraction.

Since, straight run ATF alone can not meet its large demand, it is also produced by hydrocracking of residual stocks (Isomax process) and hydrotreating of petroleum distillates boiling in the range of 60 - 350°C (Arofing process) thereby catalytically converting aromatics into naphthenes.

Important desirable properties of good ATF are :

- High smoke point (i.e., low aromatics content), n-paraffins are the best from smoke point & luminousity point of view.
- High heat content per unit volume in high speed aircraft and weight is more important in low speed aircraft. C.V. on weight basis (i.e. Kcal/kg) is the highest for paraffins, lowest for aromatics and intermediate for naphthenes. However, for C.V. on volume basis (i.e., Kcal/m^3), the order is reversed.
- Denser molecular structure as it permits more energy to be packed in a given tank space. A high density fuel is normally desirable as C.V. on volume basis (Kcal/m^3) increases with increasing density and it decreases on weight basis *(i.e.,* Kcal/kg).
- High luminometer number *(i.e.*, high n-paraffin content) to minimise the temperature rise in combustion zone from flame radiation.
- Low olefins content (<5%) as they are unstable, chemically reactive and have gum forming tendency.
- High thermal stability, which is achieved by the use of an anti-oxidant and a metal deactivator.
- Low freezing point ((- 50°C maximum for kerosene type ATF and - 60°C maximum for gasoline type ATF) as air crafts fly at high altitude with very low outside air temperature. Freezing point parameter affects the yield of ATF from a particular crude oil, because it controls the final boiling point which decreases with decrease in freezing point of ATF.

Controlled low volatility to minimise fuel loss from unpressurised air craft tanks due to boiling off at high altitudes. Desirable value for kerosene type ATF is 20 volume percent recovery upto 200°C, which controls front end volatility and ensures that ATF will readily vaporise for cold starting of engine as well as limit evaporation losses at high altitudes.

Additives added in ATF include.

- **Antioxidants** which prevent gum formation and retard peroxide formation which attacks nitrile rubber hoses used in fuel supply system in the air-crafts.
- **Metal deactivators** for removing any dissolved copper (by chelating) that catalyses the gum formation in ATF which may block burner jets & foul up heat exchangers.
- **Anti-icing additives** (0.5 - 2 volume % of ethanol/isopropanol or 0.05 - 0.2 volume % of hexylene glycol/dipropylene glycol) to avoid freezing of dissolved water in ATF, which may clog the fuel filter.
- **Fuel soluble corrosion inhibitor** (like alkyl aminophosphate) to protect the fuel distribution pipeline. It also increases fuel lubricity thereby helping in the operation of high pressure fuel pump, which is entirely fuel lubricated.

- **Anti-smoke additive** which helps in reducing the amount of smoke emitted from jet engines while taking off.
- **Fuel soluble biocides,** which inhibit fungoidal and microbial growth in fuel supply system of aircraft.
- **Static charge dissipator additive** which increases the electrical conductivity of ATF & thus reduces the possibility of an electrostatic discharge causing explosion during high speed mechanised refuelling through a micronic filter or in refinery ATF storage tanks.

Some important requirements/specification of kerosene type ATF (K-50 grade) as per IS : 1571-1976 is tabulated in Table 25.7.

Table 25.7: Important Specifications of kerosene type ATF

S.No.	Parameters	Unit	Specifications
1.	Aromatics & olefins content	Volume % (maximum)	20 & 5 respectively
2.	Net calorific value	Kcal/kg (minimum)	10225
3.	Copper content	micro gm./Kg ATF (maximum)	150
4.	Freezing point	°C (maximum)	50
5.	Smoke point	mm (minimum)	20
6.	Total sulphur	weight % (maximum,'	0.20
7.	Kinematic viscosity	Centistokes (maximum)	6 at - 17.8°C , at - 34.4°C
8.	Existent gum	mg/100 cc (maximum;	7
9.	Flash point (Abel)	°C	38
10.	Specific gravity	—	0.775 - 0.830
11.	Antioxidants	M/100 c.c,	8.6-24.0
12.	Water separometer index modified	—	70-85
	(WSIM)		
13.	Total acidity	mg KOH/gm (maximum)	0.012
14.	Copper strip corrosion for 2 hr at 100°C	—	Not worse than no. 1
15.	Silver strip corrosion for 16 hrs at 45 ± 1°C	—	Not exceeding (0)
16.	Initial boiling point	°C	Not limited but to be reported
17.	Final boiling point	°C (maximum)	288

Table 25.8: Additives used in the formulation of lubricants

S.No.	Additives used for various purposes	Specific examples of additives
1.	Anti-foaming agents	Polyorgano siloxanes
2.	Anti-wear agents	Tricresylphosphate
3.	Antioxidants	2, 6 ditertiary butyl paracresol, phenyl naphthylamine
4.	Bacteriocides	Phenols
5.	Corrosion inhibitors	Lead napthenate, sodium sulfonate
6.	Detergents	Metal sulfonates, metal phenates
7.	Emulsifiers	Fatty oils, Fatty acids
8.	Extreme pressure agents	Sulphurised fatty oils
9.	Viscosity index improver	Tricresylphosphate

LIQUID FUELS FROM COAL BY ITS HYDROGENATION AND LIQUEFACTION

Synthetic liquid fuels like gasoline, diesel, fuel oil can be produced by destructive *hydrogenation* of coal *(Bergius process)* and liquefaction of coal *(Fischer-Tropsch Process).*

Coal is a solid with hydrogen to carbon (H/C) molar ratio of 0.8. This H/C ratio is between 1.3-1.9 for crude oil and around 2 for gasoline & diesel fuel. To convert coal into liquid fuels, it is necessary to introduce the lacking hydrogen into its structure, which can be done either by pyrolysis or – more efficiently – by direct or indirect liquefaction. There are several different methods of coal liquefaction. The applied technology and the type of coal used have a considerable impact on yield, characteristics & mixture of products. During the process pollutants of coal such as sulphur & nitrogen are largely eliminated and produced hydrocarbons are refined in a way similar to distillation of crude oil. Final products are clean liquid fuels (gasoline, diesel, jet fuel) and other chemical compounds (naphtha, lubricants, sulphur, ammonia).

Indirect Liquefaction is a commercially proven technology, currently used by sasol. During the first stage of processing, coal is gasified with steam to produce syngas (a mixture of hydrogen and carbon monoxide). After removing sulphur compounds and particles the syngas.as is reacted over a catalyst at a relatively low pressure & temperature. Final products depend on the type of reaction and on the catalyst. Indirect liquefaction is characterised by ultra clean fuels and is well suited for ccs, though the process is less efficient (around 40% of overall energy efficiency) and produces less calorific value fuel.

Direct Liquefaction is a potentially more effective way (overall energy efficiency is in the range of 60-70%) . There are a couple of different variations of technologies, yet the main features of the process are the dissolution of coal in a hydrogen-rich solvent at high temperature &pressure and further hydrocracking with a catalyst. Apart from higher efficiency, this method allows adjusting for the type of product necessary and produces fuels of high energy density. Obtained oil products require refining before going to the market. Drawbacks of this method are higher operating costs and higher CO_2 emissions.

A Hybrid Concept combines the advantages and limits the disadvantages of the above mentioned methods, yet with higher costs. In principle, synthetic fuels can also be produced together with electricity in advanced IGCC plants (integrated gasification combined cycle). This process is known as polygeneration. The liquefaction process is characterised by high process water demand and higher CO_2 intensity compared to refinery oil processing.

BERGIUS DIRECT LIQUEFACTION

In the early 20th century, German researcher Friedrich Bergius developed a process to directly liquefy coal under high temperature and pressure (coal begins to dissolve above 250°C) and then "crack" the coal molecules into smaller molecules using hydrogen. Bergius termed the process "coal hydrogenation," which was later referred to as "direct liquefaction." Coal also served as the source of hydrogen. (In modern refining, hydrogen is manufactured from methane gas (CH_4) decomposed by a process termed "steam reforming." Modern refineries rely extensively on hydrogen for hydrocracking and hydrotreating.)

FISCHER-TROPSCH SYNTHESIS

As Bergius was perfecting direct liquefaction, German scientists Franz Fischer and Hans Tropsch were developing a means of indirectly converting coal into a liquid fuel. In 1926, Fischer and Tropsch reported a process to synthesize hydrocarbons using an iron or cobalt catalyst to react hydrogen (H_2) with carbon monoxide (CO) under lower temperatures and pressures than Bergius' process.

Essentially, Fischer-Tropsch (F-T) synthesizes straight molecular chains of carbon and hydrogen, whereas Bergius breaks heavier-weight hydrocarbons into lighter-weight, shorter-length molecules. Both processes involve hydrogen. Fischer-Tropsch synthesis, however, relies on carbon monoxide's potential for exchanging oxygen with hydrogen in the presence of a catalyst. As in the manufacture of water gas, coal is burned to produce the carbon monoxide and steam reacting with hot coal disassociates to produce hydrogen, as shown in the following "water gas shift" equations.

$$C + H_2O \rightarrow CO + H_2 \text{ and } CO + H_2O \rightarrow CO_2 + H_2$$

The CO_2 byproduct of these reactions can be scrubbed from the "syngas" stream before it is introduced to the synthesis reactor.

In the following simplification, Fischer-Tropsch synthesis occurs through two simultaneous reactions promoted by the contact of CO and H_2 with a catalyst:

$$2H_2 + CO \rightarrow -CH_2- + H_2O \quad \text{and } CO + H_2O \rightarrow CO_2 + H_2$$

which can be simplified as:

$$2CO + H_2 \rightarrow CH_2- + CO_2.$$

BERGIUS PROCESS

This method was developed by Bergius in Germany during the first World War. The method uses coal as the raw material which is a mixture of high molecular complex organic compounds deficient in hydrogen. In practice, the finely powdered coal is made into a paste with *heavy fuel oil* or *coal tar* and then a *catalyst composed of organic compound of tin* is added. The coal paste along with the catalyst is prepared and then pumped to the converter where it is heated to *400-500°C under 200-250 atmospheric pressure* in the presence of hydrogen.

$$\underset{\text{(Suspended in heavy oil)}}{\text{Coal dust}} + H_2 \xrightarrow[450^\circ C,\ 200\text{ atm.}]{\text{catalyst}} \text{Mixture of hydrocarbons} \xrightarrow[\text{(ii)hydrogenation}]{\text{(i) cracking}} \text{Crude oil}$$

Hydrogenation takes place to form higher saturated compounds which undergo cracking and hydrogenation processes to yield mixture of alkanes. Thus the vapours leaving the converter upon condensation gives crude oil or synthetic petroleum. Crude oil is fractionally distilled to give *(i)* petrol, *(ii)* middle oil, and *(iii)* heavy oil. The middle oil is again hydrogenated in the vapour phase in presence of a solid catalyst to give more gasoline. Actually, the processing of middle oil gives four times the gasoline obtained from the primary hydrogenation of coal. Heavy oil is again used for making a paste with fresh coal dust. *One ton of soft coal yields 140 gallons of gasoline on above treatment.*

The immediate product from the reactor must be stabilized by passing it over a conventional hydrotreating catalyst. The product stream is high in naphthenes & aromatics, low in paraffins and very low in olefins. The different fractions can be passed to further processing (cracking, reforming) to output synthetic fuel of desirable quality. If passed through a process such as Platforming, most of the naphthenes are converted to aromatics and the recovered hydrogen recycled to the process. The liquid product from Platforming will contain over 75% aromatics and has a Research Octane Number (RON) of over 105.

Overall, about 97% of input carbon fed directly to the process can be converted into synthetic fuel. However, any carbon used in generating hydrogen will be lost as carbon dioxide, so reducing the overall carbon efficiency of the process.

There is a residue of unreactive tarry compounds mixed with ash from the coal and catalyst. To minimise the loss of carbon in the residue stream, it is necessary to have a low-ash feed. Typically the coal should be <10% ash by weight. The hydrogen required for the process can be also produced from coal or the residue by steam reforming. A typical hydrogen demand is ~8 kg hydrogen per ton of dry, ash-free coal. Generally, this process is similar to hydrogenation. The output is at three levels: heavy oil, middle oil, gasoline. The middle oil is hydrogenated in order to get more gasoline and the heavy oil is mixed with the coal again and the process restarts. In this way, heavy oil and middle oil fractions are also reused in this process.

The Bergius Process is a method of production of liquid hydrocarbons for use as synthetic fuel by hydrogenation of high-volatile bituminous coal at high temperature and pressure. It was first developed by Friedrich Bergius in 1913, in 1931 Bergius was awarded the Nobel Prize in Chemistry for his development of high pressure chemistry.

FISCHER TROPSCH PROCESS

This method was devised in Germany in 1925. The raw material is the hard coke which is converted into a mixture of carbon monoxide and hydrogen by passing steam over red hot coke (water gas reaction).

$$C + H_2O \xrightarrow{1200°C} CO + H_2 \text{ (water gas)}$$

From a part of the water gas, carbon monoxide is removed by converting it into carbon dioxide which is obtained by passing the water gas with extra steam over a promoted reduced iron oxide catalyst at 450°C.

$$CO + H_2O \text{ Fe—>à } CO_2 + H_2$$

The residual carbon dioxide is again removed by absorption in ammonical cuprous formate solution. Hydrogen so obtained is mixed with the rest of the water gas in a ratio of synthesis gas *i.e.* hydrogen and carbon monoxide in a ratio of 2 : 1. The synthesis gas is passed at atmospheric pressure over a cobalt-thorium oxide catalyst on kieselguhr at 180-200°C. The reaction products are predominantly straight chain paraffins and olefins, the formation of which may be represented by the following equations.

$$(2n + 1)\,H_2 + n\,CO \rightarrow \underset{\text{n-paraffins}}{C_nH_{2n+2}} + n\,H_20$$

$$\underset{\text{synthesis gas}}{2nH_2 + nCO} \rightarrow \underset{\text{olefins}}{C_nH_{2n}}$$

The gases and vapours leaving the reaction vessel are passed through water cooled condensers, where the vapours are condensed out. From the residual gaseous mixture, propane and butanes are obtained by absorbing in activated carbon (regenerated by steam). The liquid on fractional distillation produces low grade motor spirit, high grade diesel oil, heavy oil and waxes.

The main impurities in synthesis gas (required in *Fischer-Tropsch process)* are hydrogen sulphide traces or organic sulphides and carbon dioxide. These are removed prior to the conversion process. A typical method is washing with mono or diethanolamine. At present the major portion of the operational costs, about 80% is taken up by the manufacture and purification of the synthesis gas. One of the best methods of coal gasification for this purpose is the Lurgi process which operates at 20 to 30 atmospheres and thus provides a gas at the required pressure for synthesis.

The original catalysts for Fischer-Tropsch process was cobalt. Now cheaper iron-oxide catalysts are used. All the reactions are highly exothermic, hence control of reactor temperature is critical.

The main operating variables in the *Fischer-Tropsch synthesis* are reaction temperatures & pressure, recycle ratio *i.e.* ratio of product gases recycled to gas feed, composition of the feed gas and type of catalyst. Typical operating conditions of the Kellogg and Arge synthesis units are given below in the Table 26.1.

Table 26.1: Operating Conditions of Fischer-Tropsch Reactors (High pressure)

Process	Temperature, °C	Pressure, atm	Catalyst
Arge fixed bed	Up to 270	25-30	Precipitate Iron
Kellogg fluidized bed	300-340	25-30	Mill scale or similar iron oxide

By the selection of suitable temperature, pressure and catalysts, the yields of gasoline and diesel fuel are maximised. Because of straight chain hydrocarbons, gasoline is of low octane number (but diesel produced has a high cetane number exceeding 100) and needs catalytic reforming before use.

A conceptual Fischer-Tropsch plant is shown below in Figure .

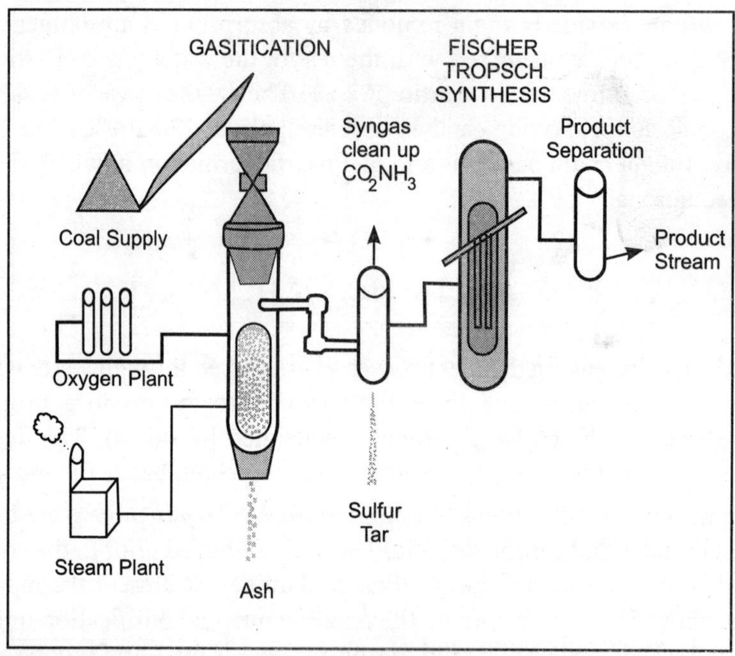

The only large commercial *Fischer-Tropsch plant* in the world for manufacture of liquid fuels from coal is located in *South Africa* (SASOL plant). SASOL plant has been on stream since 1955 and supplies 20% of South Africa's total liquid fuel requirements. The economics of the plant is due to cheap coal and lack of crude oil resources in South Africa.

In *SASOL plant,* the sub-bituminous coal is gasified and the synthesis gas is purified from sulphur compounds and CO_2 by washing with methanol at 60°C at high pressure (by Rectisol process). The purified gas contains 56% H_2, 27% CO, 14% CH_4 and small quantity of carbon dioxide & nitrogen. It is either passed through an *Arge fixed bed reactor* or reformed with oxygen & steam (nickel catalyst, 20 atm, 1100°C) to convert methane into additional quantities of hydrogen and carbon monoxide before passing into a *Kellogg fluidised bed reactor.* The higher temperature in Kellogg unit gives increased yield of gasoline and comparatively low yield of heavier oil & waxes. Arge unit uses lower temperature and produces small yield of gasoline but larger yield of heavier products. The gasoline from the *Kellogg unit* is of much higher octane number.

Besides the fuels, SASOL plant also produces chemicals like methanol, ethanol, methyl ethyl ketone and acetone.

OTHER LIQUID FUELS—BENZOL, SHALE OIL, ALCOHOL AND COLLOIDAL FUELS

BENZOL

Benzol (a product of coal carbonisation) is a mixture of benzene, toluene and xylene. Crude benzol is water white in color and is classified as 90's benzol, 75's benzol etc. depending upon the percentages distilling below 100°C. Crude benzol is associated with impurities like acids and bases. A wash with alkali removes the acids and another wash with 40% sulphuric acid removes the bases. The neutral benzol on rectification gives the first fraction containing carbon disulphide and some low boiling unsaturated hydrocarbons. The next fraction (80-150°C) represents refined benzol and the heaviest fraction (> 150°C) consists mainly of high boiling unsaturated hydrocarbons.

Rectification of distillation refined benzol gives roughly 60-70% benzene, 10-15% toluene, 5-10% xylene and < 5% solvent naphtha, phenol & naphthalene. Crude benzol is recovered by tar distillation (present in light oil) or absorption from coke oven gas by wash oil.

Typical Specifications of benzol:

Specific gravity (15°C)	= 0.87 - 0.89
Distillation range	= 0-150°C
Crystallisation point	= < 5°C
Sulphur	= < 0.5%
Calorific value (Gross/Net), kcal/kg	= 10030/9600
Inflammability limit in air	= 2.65-6.5%
Initial boiling point	= 80°C
Octane number	= 115
Freezing point	= - 50°C

Because of high octane number, benzol can be blended with motor gasoline upto 5-30% in the blend. By blending 50% motor benzol, the octane number of gasoline can be raised from 70 to 80. Benzol is not used as such as a motor fuel owing to its high initial boiling point, high freezing point and large carbon deposits. Benzol can also be used as a ternery blend with gasoline and alcohols ('gasohol') as motor fuel.

SHALE OIL

Shale oil is obtained by destructive distillation of oil shale which is a dark brown sedimentary rock (resembling slate) found over the surface of the earth mainly in USA, USSR, France, Sweden, U.K. and South Africa. Oil shales are a mixture of an organic mass (*kerogen*) and mineral matter (30-70%). Thermal decomposition of oil shale produces about 10-12% shale oil. Analysis of *kerogen of oil shale* is as follows :

C = 76.5-77.5%	Cl = 0.5-0.7%
H = 9.1-9.3%	N = 0.2-0.5%
S = 1.7-2.2%	0_2 = 10.5-12%

Shale mining is similar to coal mining. The run of mine shale is crushed, screened and then charged into retorts (externally heated) maintained at 480°C (in the top) to 750°C (in the base). The gases & vapors are condensed and the recovered products are crude oil, ammonia liquor, crude naphtha (absorbed from the gas after recovery of oil) and gas. The crude oil is dark green in color (sp. gr. 0.86-0.89). Compared to crude petroleum, it contains larger amount of unsaturates, nitrogen bases and phenolic compounds.

Refining of crude shale oil is similar to petroleum refining. Motor gasoline and diesel oil are main products. Wax is also recovered from wax bearing oils. The disposal of spent shale is a problem as it requires a large idle dumping area. Brick-making is one of its uses.

The present shale oil production is very less owing to availability of petroleum. It is an important standby resource of liquid fuels.

ALCOHOL

Alcohols (*methyl alcohol / ethyl alcohol*) are used in the motor gasoline blends upto 10-20% and the mixture is called *'Gasohol.* The octane number of methyl and ethyl alcohols are very high (98 and 99 respectively) but their calorific values are low (5335 and 7120 kcal/kg respectively). Their flash points are -1.1°C and 12.8°C respectively. Cost considerations prohibit the use of alcohols in burners directly.

Alcohols are mainly produced from carbohydrates which are synthesised in nature. Modern technology produces almost anhydrous (absolute) alcohol at a reasonable cost.

Power alcohol. *Alcohol used for the generation of power is called power alcohol.* Absolute alcohol (100% ethyl alcohol) and rectified spirit (93-95% ethyl alcohol) are not used as prime fuels but they are only used as additives to petrol. Generally, power alcohol has petrol and alcohol in the ratio of 4 : 1. It is important to note that rectified spirit mixes well with petrol only in presence of benzene, tetralin or ether. Hence whenever it is used as power alcohol, either of the solvent must be used.

Manufacture of alcohol from Molasses. In India, ethyl alcohol is mainly manufactured from molasses. Molasses is the dark brown coloured mother liquor left after the crystallisation of sucrose from concentrated sugarcane juice. Molasses contain about 50% uncrystallisable cane — sugar, glucose and fructose. Since the recovery of these sugars from molasses is not an economical process, it forms an excellent and cheap raw material for the manufacture of alcohol in India. Molasses is converted into ethyl alcohol by the following steps.

Dilution. Since the high concentration of sugars is undesirable for the fermentation reaction, molasses is first of all diluted with water to get about 8-10% solution of sugars.

Addition of dilute sulphuric acid. The diluted molasses is acidified with small amount of dilute sulphuric acid. Dilute sulphuric acid checks the growth of undesirable bacteria which would otherwise form lactic acid and acetic acid.

Addition of ammonium salts. To the above solution a small quantity of ammonium salts (ammonium sulphate and ammonium phosphate) is added. Ammonium salts act as a food for the ferment and hence these are added only when molasses contain insufficient yeast food.

Addition of yeast **(alcoholic fermentation).** The above solution of molasses is taken in a fermentation tank and yeast is now added to it (approximately 5% of the liquid). The mixture is maintained at about 30°C and kept for 2-3 days. During this period, the enzymes *invtrtase* and *zymase,* present in yeast decompose sugars to ethyl alcohol with the evolution of carbon dioxide (alcoholic fermentation).

$$\underset{\text{sucrose}}{C_{12}H_{22}O_{11}} + H_2O \longrightarrow \underset{\text{glucose}}{C_6H_{12}O_6} + \underset{\text{fructose}}{C_6H_{12}O_6}$$

$$\underset{\text{glucose or fructose}}{C_6H_{12}O_6} \xrightarrow{\text{zymase}} \underset{\text{ethyl alcohol}}{2C_2H_5OH}$$

The process of fermentation is completed in 2-3 days. When the evolution of carbon dioxide ceases, the yeast is filtered off. The filtrate, commonly known as *wash* or *wort* contains 6-10% ethyl alcohol, 3-5% glycerol, some acetaldehyde, higher alcohols etc. It is distilled for obtaining ethyl alcohol.

Distillation. Wash is distilled in a specially designed continuous plant. It consists of two tall fractionating columns called the *analyser* and the *rectifier* provided with a number of perforated plates with valves opening upwards. It works on the principle of counter current, since steam and alcohol travel in opposite directions.

Preheated wash is admitted near the top of the rectifier while a current of steam is introduced at the bottom end of the analyser. As the steam comes in contact with wash, it vaporises the alcohol present in the wash and carries the alcohol vapours to the rectifier, condensation of steam (b.p 100°C) takes place while the uncondensed alcohol vapours (b.p. 78.3°C) leave the rectifier at its top end and are condensed separately with the help of a condenser. The alcohol thus obtained is 90% pure and is subjected to fractional distillation. The process being known as *rectification.*

Rectification. Rectification of 90% alcohol gives following three fractions :

(a) *First fraction.* This fraction consists mainly of low boiling liquids like acetaldehyde

(b) *Middle fraction.* It contains 93-95% ethyl alcohol. It is sold under the name of **rectified spirit** or **commercial alcohol.**

(c) *Last fraction (b.p. 125-140°C) or fusel oil.* It contains water and fusel oil. Fusel oil is a mixture of n-propyl, n-butyl, n-amyl, iso-amyl and optically active amyl alcohols.

In modern plants, both distillation and rectification are done in one operation.

Advantages of Pure Alcohol (i.e., ethyl alcohol) as Fuel:

- It offers the advantages of great safety due to its low degree of volatility and higher flash point (about 17°C).

- Its vapor is half as heavy as those of petrol, so that it does not creep & accumulate in dangerous quantity in low levels and higher proportion is needed to form an explosive mixture.
- Burning alcohol can be extinguished with water as it is miscible with water in all proportions.
- Storage, handling and transport of alcohol is relatively much safer as compared to that of petrol and benzol.
- Inspite of its much lower calorific value compared to benzol and petrol, it shows good hermal efficiency and hence its actual consumption for a given power is not very much higher than these fuels.

It has a very high anti-knocking capacity.

Disadvantages of pure Alcohol as a Fuel:

- Its calorific value is low (about 66% of that of petrol).
- Modification of carburettor is needed to use it as a fuel because of the fact that it has a higher specific gravity and viscosity than petrol.
- Because of higher surface tension, it is comparitively less easy to atomise.
- It does not mix with oil and thus causes lubrication difficulty in the upper parts of the cylinder.
- Due to its high solvent power which affects joints and wash previously deposited gum from tanks and transfer them to engine.
- The corrosion trouble is not very serious and so is that of vaporisation, once the engine become hot, but the difficulty of starting from cold is inherent due to its low vapor pressure.

Advantages of Alcohol Blended Petrol as a Fuel:

- Since alcohol has high octane number (98) than the straight run gasoline/petrol (60-70), its presence in petrol increases the octane number. Hence alcohol blended petrol has better anti-knock properties.
- There is no decrease in power output or increase in specific fuel consumption.
- There is no starting difficulty.
- Alcohol is capable of absorbing any traces of moisture which may otherwise enter into the petrol system.

Disadvantages of Alcohol Blended Petrol as a Fuel:

- The calorific value of alcohol is low, hence alcohol blended petrol has low calorific value.
- Alcohol is easily oxidised to acids. Hence it is liable to cause corrosion.
- The ratio of flow of blend is to be increased to 1.56 times to that of pure petrol by altering the size of carburettor jet.
- Since the alcohol blended petrol requires less air than the pure gasoline, the air entering the cylinders is to be regulated either by modifying air inlet or by operating choke.
- Due to higher specific gravity of the blend, the float is to be weighed down to the correct level.

- Below 70°C, starting problem arises due to low vapor pressure of alcohol necessitating carburettor heating.

Composition and properties of gasoline and alcohols are tabulated below in Table 27.1."

Table 27.1. Composition & Properties of alcohols and gasoline

Composition and Properties	Ethyl Alcohol	Methyl Alcohol	Petrol / Gasoline
% Composition : C	52.2	37.5	—
H	13.0	12.5	—
O	34.8	50.0	—
Properties :			
Specific gravity	0.7946	0.8102	0.70-0.76
Boiling point, °C	77.8	64.7	215, maximum FBP
Specific heat	0.53	0.55	0.53
Ignition temperature in oxygen/air, °C	392/425	464	320
Viscosity at 20°C, est	1.19	0.59	0.45
Flash point, °C	12	-6.5	-40 to 30
Explosion limit in air, % vapor in vapor-air mixture	4-13	7—30	2-6
Reid vapor pressure maximum, kgf/cm^2	0.170	—	0.70
Latent heat, kcal/kg	203	263	75
Octane number (motor/Research method)	99/121	93/122	60 (straight run)
Calorific value, kcal/kg	7120	5335	10000
Oxygen content, weight %	34.7	50.0	—
Heat of vaporisation, KJ/Kg	839	1104	—
Water solubility at 20°C	00	00	—

Compared to a typical petrol, methyl alcohol requires about 45% as much air for combustion, produces 50% as much energy and requires 3.7 times as much heat for vaporisation. The corresponding values for ethyl alcohol are about 60% as much air, 65% as much energy and 2.6 times the heat of vaporisation. For a given energy output, methyl alcohol requires about 7.5 times and ethyl alcohol about 4 times as much heat to vaporise the fuel. In contrast to petrol, both methyl alcohol & ethyl alcohol have a single sharp boiling point in the lower end of the gasoline range and their vapor pressure & limit of inflammability are such that saturated vapor is flammable at common ambient temperature, but is too lean to ignite below 10 - 15 °C. These features along with high heats of vaporisation lead to difficulty in starting an engine in cold weather on pure alcohol. Both alcohols are highly water soluble, hence its spill into water poses serious hazards to aquatic life. Methyl alcohol being highly toxic can be absorbed through the skin.

COLLOIDAL FUELS

'Colloidal fuels' are essentially suspensions of finely ground coal in fuel oil. By increasing the fineness of the grain size (less than 70 μ), it is possible to make the suspension tolerably stable. Colloidal fuels (coal : oil = 40 : 60) can be successfully burnt in oil furnaces but economic considerations prevent their commercial utilization. Recently permanent suspensions of coal in oil have been made in the laboratory by disintegrating coal grains to particle sizes around 1 μ by irradiation. The suspension containing up to 5% coal has been successfully used even as diesel fuel. Better quality colloidal fuels are prepared by using coal extracts instead of the whole coal. If this process can be made more economical, the colloidal fuels may be widely used. Solvent extraction offers greater promise than milling. The advantages of colloidal fuels are their greater efficiency, the ease of combustion compared to coal, and the higher density compared to oil. Basically, they offer the means of converting coal into fluid fuel.

28

Storage & Handling of Liquid Fuels/Fuel Oils

FUEL STORAGE TANK

Low flash point liquid fuels are stored in tanks fitted with roofs which float on the surface of the liquids while those having flash points well above room temperature *(e.g.* kerosene, diesel, fuel oils etc.) are stored in tanks provided with fixed conical shaped roofs. Fuel oil is normally stored in the storage tank erected above the ground when the capacity exceeds 70000 litres. For lower capacities, underground storage tanks are economical.

Storage tanks are normally made of welded mild steel plates with a height to diameter ratio of about 1.5. The overhead tank is mounted on concrete blocks which provide a slight gradient of 1 : 50 towards the drain valves. All fuel oils contain water which settles to the bottom of the tank and should be removed through a drain cock. In case of coal tar fuels *(e.g.* PCM or raw tar), the water being lighter floats on the fuel and is removed by a draw-off arrangement. The outlet or draw-off pipe should be located at the raised tank end (not less than 15 cms from the bottom). The elimination of any air space above the fuels reduces fire hazards and evaporation losses. The vent pipe (gooseneck type) is located at the highest point of the storage tank to facilitate filling by venting air and oil vapor during liquid fuels filling operation. It should be about 5 cms in diameter and provided with a strainer at the outlet. The oil filling pipe can be located either at the top or at the bottom of the tank. When located at the top, the pipe must descend to the bottom of the tank to avoid free fall of oil from the top on the oil surface during tank filling causing frictional/impact heat generation. The drain pipe located at the lower end bottom is used periodically to remove the accumulated water.

Vertical cylindrical tanks upto 2.5 metre diameter may be erected direct on to a reinforced concrete foundation with the base having bitumen seal to prevent entry of water between tank bottom and concrete. One manhole at the top should be provided for tank upto 3.5 metre height. An additional manhole at the bottom should be provided if the tank height is more than 3.5 metres. Brass dip stick should be used to measure the tank level. Alternatively, float or remote reading gauges can be employed. Glass level indicator should not be used for level measurement in the tank. Oil draw off level should be above the oil heater mat and always > 7.5 cms above the bottom. Water accumulated at the bottom of the tank should be drained regularly, otherwise it carried to the burner with oil will cause sputtering of flame and possible flame failure. Water should be drained from main storage tank once every 15 days and from service tank once every 7 days as fuel oil contains upto *1%* water which accelerates the sludge formation.

The liquid fuels are transported from the main installations to the distribution depots and to the consumers either by pipeline or in road tankers or other containers. The storage tanks in the filling stations are sited underground. They are then delivered to the consumers by road tanks or rail tanks and then stored in tanks of bulk storage capacity. A minimum safe distance from the surrounding structures need to be maintained while erecting a storage tank above the ground.

The permissible condition of safe storage and use are different for petrofuels with flash point < 23°C (*e.g.* gasoline which is dangerous and inflammable), between 23 to 66°C (*e.g.* kerosene) and above 66°C (*e.g.* fuel oils and gas oils) which are safe.

STORAGE TANK HEATING

Heating of liquid fuels in storage tank is done usually by low pressure steam coils (called mat heater) located as low as possible in the tank preferably below the level of the offtake. Hot water heater or electrical heaters are also used which are economical for small tanks. Normally, lagging/ insulation of these tanks is not necessary. Hot oil pipelines are lagged with electrical or steam tracer lines within the lagging to balance the heat loss from the pipes through the lagging. Temperature of the liquid fuels in storage tank should be so maintained that it will have a viscosity of about 1000 seconds (Redwood No. I). Recommended preheat temperatures for easy pumping of some liquid fuels/oils is tabulated in Table 28.1.

Table 28.1: Oil Preheat Temperatures for Liquid Fuels for easy pumping.

Oil Grade	Preheating Temperature Required
Diesel/LDO (Light diesel oil)	no preheating
80 CS at 50°C (600 RWI at 100°F)	10°C
125 CS at 50C (1000 RWI at 100°F)	20°C
176 CS at 50°C (1500 RWI at 100°F)	25°C
370 CS at 50°C (3500 RWI at 100°F)	35°C
PCM/CTF-200	200°F
LSHS (Low sulphur heavy stock) furnace oil	70°C

Pumps used for oil pumping. The pumps used for pumping oil are normally of following three types:

- bucket type
- reciprocating pumps
- gear or screw type rotary pumps.

If driven by electric motors, it should be ensured that all electrical insulations are sound and no loose connections are made. The safe current carrying capacity for each pump motor should be noted. High viscosity of oil means more load on motor and high pressure drop of oil. Hence, correct pumping temperature of the oil is very important.

In case pumping capacity is much more than burner oil output, then recirculation of more hot oil (at 100°C) in return line may cause cavitation in pump due to higher vapor pressure. Hence, it should be ensured that minimum quantity of hot fuel oil is returned from the burner.

Liquid Fuels Preheating. The preheating of heavier liquid fuels before pumping & before atomisation should be only upto the temperature required by the viscosity consideration for easy pumping and without excessive pressure drop. The optimum degree of preheat is decided by the plant trials. For reasons of heat economy, excessive heating of oil should be avoided. A temperature-viscosity chart for the oil may be used to get the correct preheat temperature. Normally, a low air pressure oil burner specifies a viscosity of 100 seconds Redwood No. 1 at the burner tip. Preheater capacity depends upon the oil rate and the required rise in temperature. The oil can be preheated either electrically or by steam, though the latter is preferred. Heating coils are placed at the bottom of the tank above the suction level of oil and at least 15 cms from the bottom to permit sludge cleaning. Thermostatic control is kept at a higher level (say 20 cms above the bottom) to one side of heater .

Liquid Fuels Supply System. There are two systems namely gravity feed system and ring *main system* for supply of liquid fuels to furnaces/burners. In gravity feed system (which is used when firing through only one or two burners are involved), either an overhead feed tank or a ground level feed tank with a pump in delivery lines supplies the fuel to the burner through a single feed line.

The ring main system is used for medium and heavy fuel oils burnt in a number of burners supplying fuel to one or more furnaces. This ensures the minimum temperature and pressure differences between the burners. The oil pumps are located nearer to the consuming points. Oil transfer to the consuming end is through preheated pipes. Steam tracers are usually provided in these pipes. These are small diameter steam lines which are laid along sides of the oil line and then lagged in the same insulation bundle. Thus radiation losses are minimised and oil preheat temperatures are maintained at optimum level right upto the burner.

In the ring main oil supply system, the oil is normally circulated through a ring main round the plant at a rate of 2-3 times the actual consumption of the burners. Branch lines feed the individual burners. Both suction and discharge sides of the oil delivery pump are provided with oil filters for protecting the pump as well as the burner nozzles from dirt. There is an oil preheater in the ring main and no separate heating is required in the branch lines as they are having steam tracers. The return pipe of the ring main is connected to the suction end of the pump (not to the storage tank) to increase the heat economy. Standby pump, preheaters and oil filters are provided to ensure continuous plant operation. The oil in the ring main system is kept at the atomisation temperature by the steam tracer lines. Alternately, no heating is done in the ring main and the oil is kept at a pumpable temperature only. The branch line to the furnaces are in that case provided with small heaters to feed the burners with oil at required atomisation temperature. However, highly viscous oils, the fuel handling system comprises of concentrically steam jacketted pipelines, heated pumps and heated filters.

Flash point = 66°C, Viscosity at 70°C, Min = 40 centistokes, Pour point = 50°C,

Total sulphur = 1% by weight, maximum, Water content = 1.0% by volume, maximum

Steam availability at wagons is at 10 kg/cm^2 and 220°C, LSHS pumpable temp. = 85°C

Procedures for correctly quantifying the quantity of oil in the road tankers & railway wagons and statutory rules for handling petroleum oils are described below :

Procedure to be adopted before assessing the quantity of oil in the Road Tankers :

- Ensure that the tank lorry is parked on firm level led ground.
- The dipstick should be checked to ensure that it is the one calibrated by the Weights and Measures Department for the compartment concerned.
- The dip readings of all the compartments should be checked and shortfall, if any one can be determined by topping up of the compartments with measured quantities of furnace oil from the main storage tank.
- If the tank of truck has been filled hot, the product in the compartment is likely to shrink to the extent of 0.7 litre/KL per degree centigrade fall in temperature. If the shortfall is higher, it may be due to pilferage.
- Stationery storage tanks should also be calibrated. Though dip readings in storage tanks are reliable, a major deviation may provide a clue to malpractices. However, the geometry of stationery tanks may not be true during construction stage, for it gets disturbed due to external damage or build-up of the tank in use.

Procedure to be adopted for handling Railway Tank Wagons Carraying furnace oil:

- Dip of the tank wagon is to be taken.
- The valves are to be examined whether these are operative or not before connecting the unloading hose pipe.
- Seal also is to be examined.
- After decanting, again dip of the wagon is to be taken to make sure complete decantation.
- If there is accumulation of the oil in the sumps, that also should be pumped to the main storage tank while decanting the tank wagons by using sump pumps.
- Release memo is to be given only after the hose pipes are disconnected.
- Split oil is irretrievable. A little care while oil is decanted from the tank wagons to the storage tank will eliminate spillage. All oil leaks from joints, flanges and pipelines must be eliminated.

LEAKAGE LOSS OF EVEN ONE DROP OF OIL EVERY SECOND CAN COST/ WASTE OVER 4000 LITRES A YEAR .

All petroleum *oil storage installations* should conform to the following *Petroleum Rules* which is statutory/safety requirement.

- The storage tank shall be electrically connected with the earth in an efficient manner by means of not less than two separate and distinct connections placed at opposite extremities of the tank.
- Every oil installation should be surrounded by a wall atleast 2 metres in height.
- Storage capacity and flash point are to be written on the body of the tank.
- Provisions of fire fighting equipment and adequate supply of dry sand in buckets with necessary implements for its convenient application in extinguishing petroleum fires should be made.

- The ground in the interior of the installation & protected area surrounding the storage tank should be kept clean and free from all inflammable materials & rubbish.
- Care should be taken to prevent any oil escaping into drain, sewer or public road.
- Petroleum having flash point 65°C and above but below 93°C comes under petroleum class 'C' *i.e.* Furnace oil and L.D.O.
- For storage capacity above 45 K.L., licence from the *Chief Inspector of Explosives, Nagpur* is to be obtained.
- Every tank or receptacle for storage of petroleum is bulk (other than a well-head tank shall be constructed of iron or steel in accordance with the relevant Indian standards or specifications approved by chief controller as per petroleum rules 1976. However underground petroleum storage tanks may be made of Fibre Reinforced Plastics (FRP) also.
- When two or more tanks are installed in one enclosure, the total capacity of the tanks in the enclosure shall not exceed 60,000 KL in case of fixed roof tanks and 1,20,000 KL in case of floating roof tanks. A combination of fixed and floating head tanks can also be located in the same enclosure/premises but the total capacity should not exceed 60,000 KL.
- Height of storage tank should not exceed 1.5 times its diameter or 20 m whichever is less.
- An air space of not less then 5% of the total storage capacity shall be considered while designing oil storage tank.
- All petroleum storage tanks installed above or below the ground shall be protected against corrosion by the use of protective coating and cathodic protection.
- Cost iron valves are not permitted in oil installations.
- Tanks after erection or repair will be tested with clean water.
- All storage and handling facility shall maintain the specified distance from any other facility nearby.
- No fire water/foam ring main shall pass through dyked enclosure.
- Grouping of petroleum products for storage purpose shall be based on the product classification class - A and /or class - B petroleum products can be stored in the same dyked enclosure. Class - C should preferably be stored in separate enclosure. However, where class - C product is stored in the same or common dyke along with class - A and/or class - B products, all safety stipulations applicable for class A or B products should apply.
- Tanks shall be arranged in maximum two rows so that each tank is approachable from the road/area surrounding the enclosure. However, tank having capacity of 50000 m^3 and above shall be laid in a single row.
- Piping from/to any tank located in a single dyked enclosure should not pass through any other dyked enclosure. Piping within the dyked enclosure should be minimised.
- Any tank having a diameter of more than 30 metres should be separated with firewalls from other tanks.
- The minimum distance between a tank shell and inside of the dyke wall shall not be less than half of the tank diameter.

- Firewalls should be constructed by limiting the aggregate capacity of each group of tanks within 20000 m^3.

Explosives Regulations on the Storage of Petroleum Products

(a) Petroleum products are classified as under as per the Explosive Regulations.

Flash Point	Classification	Products
Below 23°C	A	Petrol, Naphtha
23°C—less than 65°C	B	Kerosene, Mineral Turpentine High
65°C—less than 93°C	C	Light Diesel Oil (LDO) High Speed Diesel Oil
93°C and above	Excluded	(LDO), Furnace Oil LSHS, Lubricatiing Oils & Greases

(b) If any two classes of petroleum are stored together, the storage rules applicable for more dangerous product between the two shall be applicable *e.g.* if furnace oil and LSHS are stored adjacent to each other, even the storage of LSHS which is normally excluded from the purview of the Explosive Regulations should conform to those applicable to furnace oil. The whole installation in such case will be deemed to be that of furnace oil alone.

(c) If LSHS alone is being stored in the premises, any amount of it could be stored without either the approval or a license from the Dept. of Explosives since it falls under exempted category.;

29

METHANE, WOOD GAS, GOBAR GAS, SEWAGE GAS AND GAS FROM UNDERGROUND GASIFICATION OF COAL

INTRODUCTION

Gaseous fuels are those which are burnt in gaseous state in air or oxygen to give heat for utilisation in domestic/commercial sector. A remarkable feature of gaseous fuels is the absence of mineral impurities, consistency in quality and convenience & efficiency in use. On industrial scale, a gas handling system is the least costly of all the fuels. Some *important gaseous fuels* are as mentioned below :

- *(a)* *Methane* from coal mines *i.e.* CBM (coal bed methane)
- *(b)* *Wood Gas*
- *(c)* *Gobar Gas*
- *(d)* *Sewage Gas*
- *(e)* *Gas from underground gasification of coal*
- *(f)* *Natural Gas*
- *(g)* *Liquified Petroleum Gas* (LPG)
- *(h)* *Refinery Gas*
- *(i)* *Producer Gas*
- *(j)* *Water gas and Carburetted Water Gas*
- *(k)* *Blast Furnace Gas*
- *(l)* *Coke Oven Gas*
- *(m)* *L.D. Converter (or BOF) Gas*
- *(n)* *Coal Gas from Coal Gasification Processes*
- *(o)* *Oil Gas from Oil Gasification Processes.*

Two more important gaseous fuels namely hydrogen and acetylene find wide use in metal cutting, welding etc. Gaseous fuels are termed as *rich gas* (if calorific value is > 4000 kcal/Nm^3) and *lean gas* (if C.V. is < 1500 kcal/Nm^3). Hydrocarbon gases & coal gas are example of rich gas, while blast furnace gas & producer gas are example of lean gas.

Calorific value is not a satisfactory parameter for the technological *classification of gaseous fuels.* Thermal output of a gaseous fuel burner depends upon the *'Wobbe Index'* which is defined as C.V./ $\sqrt{\text{sp.gr}}$. Wobbe index is a dimensional group and its unit depends upon the unit of C.V.

The rapidity of combustion of gaseous fuel depends upon its flame speed. Gases rich in hydrocarbons have Wobbe Index above 8500 (*e.g.* Natural gas = 9000-13500, LPG = 18000-22000).

Wobbe index of some fuel gases are tabulated below :

The Wobbe index indicates the interchange ability of fuel gases and is the best indicator of the similarity between natural gas and a specific propane-air mixture.

S.No.	Fuel Gas	Upper Wobbe Index (kcal/ Nm^3)	Lower Wobbe Index (kcal/ Nm^3)
1.	Hydrogen	11.528	9,714
2.	Methane	12,735	11,452
3.	Ethane	16,298	14,931
4.	Ethylene	15,253	14,344
5.	Natural Gas	12,837	11,597
6.	Propane	19,376	17,814
7.	Propylene	18,413	17,180
8.	n-Butane	22,066	20,336
9.	Iso-Butane	21,980	20,247
10.	Butylene-1	21,142	19,728
11.	LPG	20,755	19,106
12.	Acetylene	14,655	14,141
13.	Carbon Monoxide	3,060	3,060

Since the Wobbe index relates heating characteristics of blended fuel gasses it can be used to obtain constant heat flows from gases of varying compositions.

- The Wobbe Index does not relate to flame temperatures, heat transfer coefficients or temperature gradients.
- Upper and lower Wobbe index corrosponds to upper heating value (Gross calorific value) and lower heating value (Net calorific value) respectively.

A *gas classification system* based on Wobbe Index (followed in the U.K.) is given in Table 29.1.

Table 29.1: Wobbe Index Schedule of Classification of Fuel Gas Quality

Gas Group	Wobbe Index (C.V. in kcal/Nm^3)
1	—
2	—
3	7,650 ± 380
4	6,970 ± 285
5	6,400 ± 285
6	5,870 ± 240
7	5,350 ± 285

Each group covers a range of gas qualities which can be interchanged for a given heating appliance with the minimum deviation from the designed thermal rating of the appliance.

Rich fuel gases (group 1 and 2) are classified on the basis of flame speed ('*Weaver flame speed factor)* which is expressed on a arbitrary scale in which hydrogen is assumed to have a factor of 100 arbitrarily.

The three groups of gases based on *Weaver flame speed factor* are :

- *High flame speed gases :*

  C.V. = 4000-5000 kcal/Nm^3

  Weaver flame speed factor = 32-45

- *Low flame speed gases :*

  C.V. = 7,500-10,000 kcal/Nm^3

  Weaver flame speed factor = 13-25

- *Intermediate type gases :*

  C.V. = 5000-7,500 kcal/Nm^3

  Weaver flame speed factor = 25-32.

Inert gases (N_2 and CO_2) and carbon monoxide reduce the flame speed and Wobbe while presence of hydrogen in the fuel gas does the reverse. Hydrocarbons reduce the speed but raise the Wobbe Index.

Example 1. *Calculate the Wobbe index of natural gas comprising of 89% CH_4, 8% 2% C_3H_8 and 1% C_4H_{10} by volume. The calorific values (kcal/Nm^3) of the constituents are as given below :*

$$CH_4 = 9500,\ C_,H_6 = 16644,\ C_3H_8 = 23688,\ C_4H_{10} = 30714.$$

Solution. Wobbe index = Calorific value/ $\sqrt{\text{Specific gravity}}$

The gross calorific value of natural gas

$$= 0.89 \times 9500 + 0.08 \times 16644 + 0.02 \times 23688 + 0.01 \times 30714 = 10571 \text{ kcal/Nm}^3$$

Now, Average molecular weight of natural gas

$$= 0.89 \times 16 + 0.08 \times 30 + 0.02 \times 44 + 0.01 \times 58 = 18$$

Average molecular weight of air = $0.79 \times 28 + 0.21 \times 32 = 29$

Hence, specific gravity of natural gas (when Air = 1) $\frac{18}{29} = 0.62$

so, the Wobbe index of natural gas $= \frac{\text{Calorific value}}{\sqrt{\text{Sq. gravity}}} = \frac{10571}{\sqrt{0.62}} = 13350 \text{ kcal/Nm}^3$ **. Ans.**

Composition and properties of some manufactured gaseous fuels is given in table 29.2.

METHANE, WOOD GAS, GOBAR GAS AND SEWAGE GAS

These gases are produced in relatively small quantities and used locally. They are described below in detail.

METHANE FROM COAL BED/MINES (CBM)

Methane is associated with coal in mines as it (coal) adsorbs the methane formed during coalification process. At the time of coal mining, the gas is gradually desorbed and it must be removed by proper ventilation otherwise it forms explosive gas-air mixture. It is a source of dangerous explosion in gassy mines and the gas is often called as 'firedamp'.

Table 29.2: Composition and Properties of Some Manufactured Gaseous Fuels.

Composition (volume %) And properties	Producer Gas				Water Gas		Lurgi gas from Bituminous coal		Coke Oven Gas	Blast Fur-nace Gas	Segas / Onia-Gegi Gas Using		
	Air Blast with coal	Mixed (Air + Steam) Blast with			Blue	Carbu-retted	Crude	Purified			Fuel Oil	Gas Oil	Naphtha
		Coal	Coke	Anthra-cite									
CO_2	1.0	4.0	5.0	6.0	4.7	5.6	28.0	0.8	2.0	11.0	9.2	5.8	3.0
C_mH_n (unsaturated)	-	0.4	-	-	-	7.0	0.2	0.2	2.6	-	6.6	4.2	5.1
O_2	-	-	-	-	-	0.4	-	-	0.4	-	0.5	0.4	0.8
CO	33.5	29.0	29.0	26.0	41.0	30.5	22.4	31.5	2.0	27.0	14.8	8.3	15.1
H_2	1.5	12.0	11.0	17.0	49.0	37.0	38.0	53.3	54.0	2.0	49.1	50.3	57.6
CH_4	-	2.6	0.5	1.2	0.8	14.0	10.9	13.4	28.0	-	15.6	16.2	14.7
N_2	64.0	52.0	54.5	49.8	4.5	5.5	0.5	0.8	5.6	60.0	4.2	4.8	3.7
Gross C.V, kcal/Nm3 (dry)	1060	1550	1260	1420	2800	4770	-	3900	5020	520	4780	4700	4680
Sp. Gr. (Air = 1)	0.98	0.87	0.90	0.85	0.55	0.63	-	0.48	0.38	1.01	0.56	0.52	0.40
Theoretical Air requirement, Nm^3/Nm^3		0.87	1.00		2.22	4.27			4.57	0.69			

The methane rich gas from coal mines may be recovered through a system of boreholes and may be used as a fuel. Methane rich mine gas have been recovered in U.K., France, Germany but the recovery process has not been developed on a wide scale due to availability of natural gas on a large scale.

The composition of methane gas from coal mines is variable but it is essentially a high concentration methane gas as shown below.

$$CH_4 = 93\text{-}99\%, \; C_2H_6 = 0\text{-}3\%$$

$$CO_2 = 0\text{-}4\%, \; N_2 = 2\text{-}6\%$$

Hydrogen and rare gases are present in traces. In India, this gas is not being recovered in any coal mine for use as a fuel and this is a potential source of fire in mines under the crust of earth. Recently, exploration of *coal béd methane (CBM)* has been started at *Telgharia near Bokaro Steel city* in Jharkhand.

WOOD GAS

Wood gas is obtained either by carbonisation of wood in metal retorts or by gasification of wood. It is a medium calorific value gas not of much commercial interest. Typical yield and composition of wood gas produced by the carbonisation of wood is given in Table 29.3.

Table 29.3: Composition and Properties of Wood Gas

Temperature of carbonisation' ℃	< 400°C (low temperature carbonisation)	> 1000°C (High temp, carbonisation)
Yield of gas, Nm^3/ton dry wood	125	550
c.v. of wood gas, kcal/Nm^3	3320	3280
Specific gravity of gas (Air = 1)	0.68	0.50
Composition (% dry)		
CO_2	30	20
CmH	4	2
CO	25	25
CH_4	14	12
H_2	20	35
N_2	7	6

Wood gas is also produced by gasification of logs, chips etc. in many countries similar to the principle of producer gas generation from coal. Gas yield in typical wood gasification process is about 194 Nm^3/ton dry wood. Crude methyl alcohol is also recovered (0.5% of dry wood) in gasification. A typical wood gas composition (produced by gasification) is :

$$CO_2 = 6\%, CO = 29\%, CH_4 = 2.5\%$$

$$C_mH_n = 0.7\%, H_2 = 14.5\%, O_2 = 0.3\%, N_2 = 47\%$$

$$\text{C.V.} = 1660 \text{ kcal/Nm}^3, \text{ Sp. gravity} = 0.85 \text{ (Air} = 1)$$

About 50% of the wood gas is consumed in gas producer while rest 50% is available for use elsewhere *e.g.* in thermal power plant. India does not have commercial generation facility for wood gas.

GOBAR GAS

Gobar gas is obtained by the fermentation of gobar (dung) in absence of air. *Gobar gas* (cattle dung) is derived from cattle dung (called 'gobar' in Hindi). In a gobar gas plant, fuel value of the dung is extracted while the refuse can still be used as a fertiliser.

Operating Principle. Gobar gas is produced by the fermentation of dung and other organic materials in absence of air. It consists mainly of CH_4 and CO_2. The extraction of CH_4 does not affect the fertiliser value of the dung. Rather, after the dung is processed in Gobar gas plant, the mannure becomes rich in nutrients and odourless. In addition, it becomes free from flies and other sources of infection. Gobar gas burns with a *blue flame* without smell at an efficiency *five times greater than* that of burning *cowdung cakes.* Hence, this plant provides improved fuel and organic mannure from the dung.

The plant comprises of a digestor to ferment cowdung and any other finally chopped left-over fodder or vegetable efficiently. The digestor is covered by a gas holder which collects and diverts the gas produced into the gas pipe at correct pressure.

Cattle dung & urine is collected and dumped into the mixing tank, where it is mixed with water and fed to the inlet pipe. This slurry, when enters the digestor will discharge an equal quantity of ripe slurry through the outlet pipe which is let into the compost pit and covered with the farm sweepings and other organic wastes from the household which is finally used as a fertiliser. Gobar gas plant is shown in Fig. 29.1.

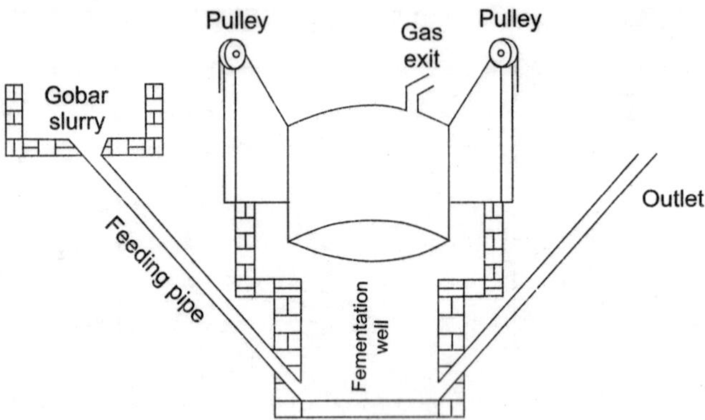

Fig. 29.1. *Gobar gas plant.*

Digestor is a masonary well constructed below the ground level and is 3—6 metres deep having diameter 1—6 metres depending upon the size of the plant. It is divided into two compartments by a wall in the middle. Two slanting cement pipes one on each side reach the bottom of the well and have their openings on the surface of the ground by the side of the top of the well. One pipe serves as the inlet and other as outlet.

The cattle dung is mixed with water in the proportion 4 : 5 . This . fixture is introduced through the inlet pipe and as the well fills up, an equal quantity of dung slurry is discharged out through the outlet pipe; the outlet opening being lower than the inlet. The partition wall stops short of the top of the well and thus remains submerged as the dung slurry. The well is so designed as to hold 50 days material. Initially, it is filled up so that whenever any material is put up from one side, an equal quantity goes out from the other.

The temperature in the digestor is maintained at 30 - 40°C. In the initial acid forming stage, the pH is 6, when much CO_2 is given off. pH, value increases to 8 in 2 - 3 weeks time as the volatile acids & nitrogen compounds are digested and methane is formed. Slight mixing improves the gas yield, but violent agitation retards the process.

Gas holder comprises of a mild steel drum which fits like a cap on the mouth of the well. It dips into the slurry and rests on a ledge built inside the well for this purpose. A central guide pipe guides the up and down movement of the drum. The drum collects the gas which bubbles out from the dung slurry in the digestor. As the gas collects on the top of the slurry, the drum rises and the accumulated gas flows out through the gas pipe. The pressure of the gas depends upon the weight of the drum which is normally 75 to 150 mm wg sufficient to provide pressure for kitchen stoves/

gas lamps. The gas may serve the kitchen or can be used for gas lamps within a distance of 30 metres.

Utilisation of Gobar Gas Plant Products. Two main products of gobar gas plant are *fuel gas and mannure.*

Fuel Gas. Presence of substantial quantity of methane (55—60%) makes it a very suitable fuel gas. Khadi and Village Industry Commission (KVIC) has finalised the designs of appliances suitable for this gas. Even diesel oil engine can be run on gobar gas. An engine developing 1 hp will need 425 litres of gas per hour. An engine developing 5 hp running an *electric generator* or a *water pump,* would need 18 m^3 of gas for 8 hours, that would require 45 to 50 animals.

Yield of gobar gas varies from 0.036 m^3/kg of gobar in winter to 0.09 m^3/kg in summer. Warm water may be used in winter (to stimulate the gas production) for diluting and mixing the cow dung whose solid content is about 20 percent.

Mannure. The outlet slurry from the plant which is rich in nitrogen (> 2%) and humus can be applied directly to the land by mixing with irrigation water.

If the slurry can't be used immediately, it is used for rapid fermentation of compost. The outlet slurry is flown into the pits dug near the gobar gas plant which is dumped with vegetable refuse like corn cobs, leaves and other waste materials. It is again covered with outlet slurry. The nutrient material in the slurry and bacteria accelerates the process of composting.

The residual slurry which is used as mannure contains 1.8 - 2.4% N_2,1 - 1.2% P_2O_5 and 0.6 - 0.8% K_2O.

In India, Khadi and Village Industry Commission (KVIC) scrutinizes the proposals, surveys the sites and supervises the construction of plants. It gives grant upto 25% of the cost and arranges 75% loan from banks.

Pre-requisites for Setting up a Gobar Gas Plant. A farmer with 5 or more cattle can instal a gobar gas plant with financial and technical assistance from the KVIC. Even for the smallest plant, 45 kg of fresh dung (gobar) is required everyday which will facilitate installation of a 2 m^3 (60 ft^3) gas plant. One kg of wet dung gives about 0.027 m^3 of gobar gas and 0.75 kg of useful mannure. Gobar gas being more efficient, it enables recovery of 20% more heat than by burning an equivalent amount of dung directly. Besides, it yields mannure which has more enriching value than that of directly dung turned mannure. The daily average dung droppling from a medium size cow is about 10—12 kg. The gas plant should be located near cowshed and also close to the place where the gas is to be used (< 20 metres). It can be made near the residential quarters since there is no danger of explosion, problem of foul odour and fly-breeding. It should not be constructed within 15 metres of a well, as sometimes there is a chance of the slurry percolating into the well.

In a typical gobar gas plant, the dung in the form of slurry is introduced into a brick lined well (3.7 m deep, 1.7 m wide), called *fermentation well.* An inverted drum of about 1.5 m diameter and 1.2 m height is placed air-tightly on the fermentation well. It acts as a gas holder and can be made up & down with the help of pulleys.

Formation of gas starts in a week. As the gas starts collecting in the drum, the latter begins to rise and float. The gas is taken out from the exit provided at the top of the drum. It is devoid of its condensed moisture and used as a domestic fuel. The spent slurry overflows from the top of the well. It is taken out and used as valuable manure.

An average typical volumetric composition and properties of the gobar gas is given in table 29.4.

Table 29.4: Composition and Properties of Gobar Gas

Methane	= 60%	H_2S and 02 = trace
Carbon dioxide	= 30%	N_2 = 0.5%
Hydrogen	= 9.5%	Calorific value = 5400 kcal/Nm^3
Ignition temperature.	= 650°C	Explosion limit in air = 5-15% gas in gas-air mixture
Octane rating	= 110	Air/gas ratio for complete combustion = 10 : 1.

It is used as a domestic cooking fuel especially in villages where dung is available in plenty. About 45 kg of dung per day is required to produce 2.8 m^3 of the gas which is sufficient for one family. It is also used for domestic lighting, small water pump running, small diesel & petrol engine running etc.

Khadi and Village Industries Commission (KVIC) has set up thousands of gobar gas plants in India and provides subsidies for its setting up as well.

SEWAGE GAS

Sewage gas is produced during sewage disposal by way of aneorobic decomposition of organic wastes (by suitable bacteria). Sewage gas is a methane rich gas which can be used as a town gas component if available in commercial quantity. Composition and properties of sewage gas is given in table 29.5.

Table 29.5: Composition and Properties of Sewage Gas

CH_4 = 58.5-76.5%, CO_2 = 30-41%
N_2 = 3-8%, O_2 = 0-1.8%, H_2S = 0.1%
H_2 = 0-4.5%, H_2S = 0.1%
C.V = 5700-7400 kcal/Nm^3(dry), Sq. gravity = 0.7-0.85,(Air = 1)
Gas yield = 8-30 liters per capita of population per day

Higher H_2S in sewage gas makes its use difficult in gas engines or boilers without purification.

Gases similar to sewage gas are produced by fermentation of agricultural wastes (carbonaceous) of plants of animal origin (*e.g.* straw, fruit peeling etc.) under aneorobic conditions. Residue left after the fermentation of sewage and agricultural waste is a good manure (fertiliser).

COAL GAS FROM UNDERGROUND GASIFICATION OF COAL

Coal in the underground mines can be gasified in situ even without mining it out of the crust of the earth. This is termed as underground gasification of coal and the gas thus produced is called coal gas from underground gasification of coal. Typical composition and properties of the coal gas produced by this process is given in table 29.6.

Table 29.6: Composition and Properties of coal gas produced by under ground gasification of coal

$CO = 10.7\%$, $H_2 = 8.4\%$, $CH_4 = 1.8\%$, $N_2 = 67.4\%$, $CO_2 = 10.5\%$, CmHn = 0.3%, $O_2 = 0.9\%$
C.V. = 820 kcal/Nm^3 (dry), Sp. gravity = 0.95 (Air = 1).

Some commercial plants in USSR produce gas by this method for electricity generation. Bituminous, sub-bituminous and mature brown coal deposits are suitable for underground gasification using air & oxygen as gasifying medium.

Deep vertical channels (boreholes) are made side by side in the ground (where coal deposits are there) both for supplying the gasifying medium (air/oxygen) and for taking out product fuel gas. Link is made under ground between inlet of gasifying medium and outlet for product gas. In shaft system, parallel galleries are made within the coal bed and then connected by making horizontal boreholes through the confined coal mass. Boreholes are ignited in succession or in small groups. Air is passed through one borehole and the product gas is taken out from the other.

Underground gasification of coal eliminates its mining and facilitates utilisation of unworkable coal seams & left over coal in mines. The gas produced is used as a fuel and also in the manufacture of synthesis gas for ammonia, methanol & liquid fuels.

However, underground gasification of coal is not widespread presently but subsequently may become competitive with increasing cost of mining and gasification of coal by other routes. *India does not have any such commercial gasification system.*

The principle of underground gasification of coal has been shown in the Fig. 29.2. Borehole producer for underground gasification of coal has also been shown in Fig. 29.3.

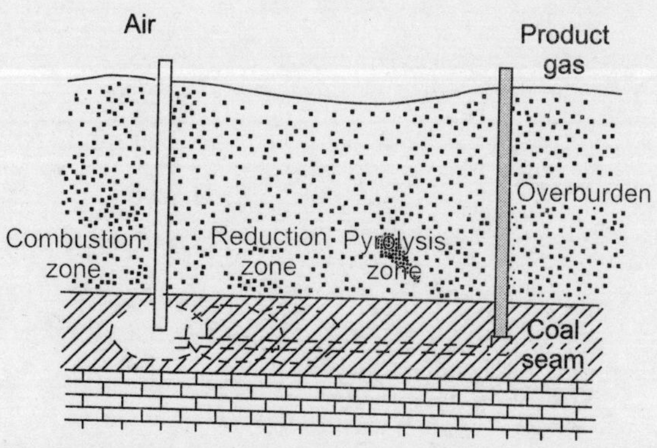

Fig. 29.2. *Priciple of underground coal gasification.*

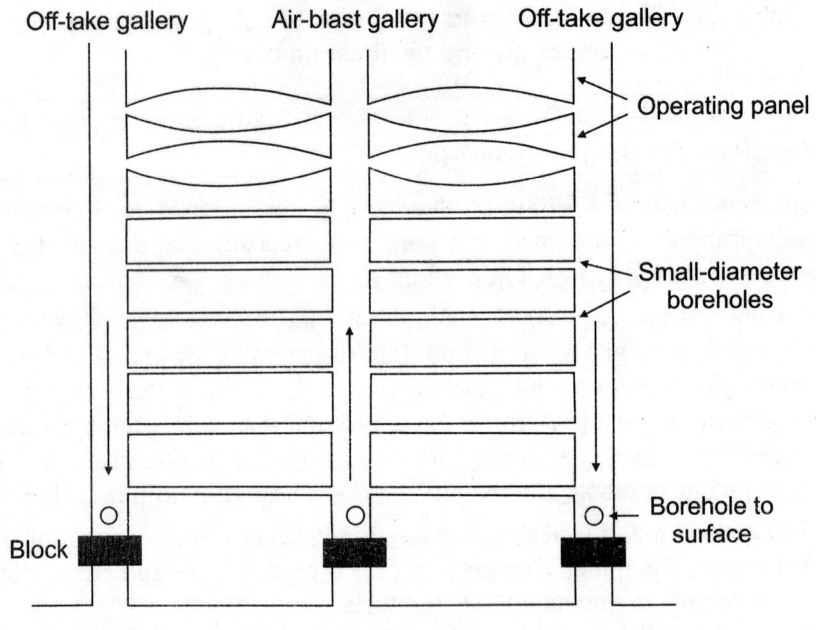

Fig. 29.3. *Early 'borehole producer' method for underground gasification of flat-lying coal.*

Natural Gas, Liquefied Petroleum Gas (LPG) and Refinery Gases

NATURAL GAS

It is a mixture of paraffinic hydrocarbons, in which methane is the main constituent. It occurs in the gas field (under the crust of earth) and also in association with crude petroleum in oil fields. After delivery from wells, natural gas is processed to remove the solids. It is then treated for recovery of gasoline and *liquified natural gas (LNG).* When natural gas contains very less recoverable condensate (< *15* gm/m^3), it is known as *dry natural gas* while it is termed as wet natural gas if recoverable condensate is more (> *50* gm/m^3). This condensate is recovered and is known as *natural gasoline* which is used as a blending agent for motor fuel. The condensate {*natural gasoline)* recovery from gas is done by compressing & cooling or adsorption on a porous solid.

Typical composition and properties of natural gas are given in Table 30.1.

Table 30.1: Composition and Properties of Natural Gas

Parameters	Dry and sweet gas	Wet and sweet gas
Composition (volume %) :		
CH_4	96.91	77.22
C_2H_6	1.33	11.18
C_3H_6	0.19	5.83
C_4H_{10}	0.05	2.34
C_5H_{12}	0.02	1.18
CO_2	0.82	0.80
N_2	0.68	1.39
Gross c.v., kcal/Nm^3 (dry)	9000	11200
Sp. gravity (Air = 1)	0.574	0.741

Natural gas of some localities contains large quantity of N_2 and CO plus recoverable amount of helium. Sometimes H_2S are also present in the gas which is recovered as elemental sulphur. H_2S free gas is called a *sweet gas.*

Natural gas is used as a fuel for cooking, domestic & industrial heating, process furnace and boilers. It is also used for fertiliser production and as a source of carbon & hydrogen in chemical industries.

USES OF NATURAL GAS. Natural gas is used for :

- Enriching the gaseous fuels of low calorific value *e.g.* Lurgi gas.
- Internal combustion (I.C.) engine as its high anti-knocking qualities permits use of high compression ratios possible resulting in improved engine efficiency.
- Gas turbines also use natural gas as a fuel.
- Producing carbon black by burning in a limited quantity of air in the presence of cooled metal surfaces. Carbon black is a pigment as well as a reinforcing agent for rubber.
- Making water gas as its methane content can be reacted with steam for this purpose.
- Making synthesis gas (by reaction with steam) which is further used for the production of methyl alcohol or gasoline (by Fischer-Tropsch process) or hydrogen for ammonia synthesis.
- Partially converting into hydrogen, methane, benzene and other aromatic compounds by pyrolysis which are all valuable chemicals.
- The production of methyl alcohol and formaldehyde by controlled oxidation of methane present in it.
- Producing methyl chloride by the chlorination of methane present in it.

Natural gas is normally delivered to the consumer by a network of pipelines (dia 0.5-0.6 metres) from the oil field (after treatment) at 15-30 kg/cm^2 pressure. In cold climate the gas must be dehydrated to avoid choking of pipelines owing to formation of crystalline hydrates of methane, ethane, propane and i-butane which are stable solids below 19°C.

In USA, natural gas is widely used in domestic heating & cooking and a wide network gas pipeline has made it available to almost each and every home. In India natural gas is available from Gujarat, Assam and Bombay high oil wells. A 1700 km long pipeline (HBJ pipeline) running from Hazira to Jagdishpur *via* Bijapur supplies natural gas to fertiliser plants and thermal power stations en route. This pipeline is being further extended to Haldia in West Bengal.

Liquefied natural gas (LNG) is shipped also. Oil and Natural Gas Corporation Ltd. in India has set up a bottling plant at Uran (Near Bombay High) for filling LNG in cylinders.

COMPRESSED NATURAL GAS (CNG).

Natural gas has a very low energy density ; hence to use it as an alternative fuel for a spark ignition engine on board the vehicle/car, it has to be either *compressed* to a high pressure of 200 atm or cooled to - 160°C to liquefy it for storage. Hence, it is used as compressed natural gas (CNG).

Advantages of CNG over petrol as a S.I. engine fuel.

- Octane number of CNG is 130, hence it does not need anti knock additive like TEL, which causes pollution.

- Its combustion is more regular and it rapidly disperses into atmosphere at normal pressure & temperature.
- It is a light weight gas with a high ignition temperature and is no more dangerous than petrol.
- Emission of carbon monoxide and unburnt hydrocarbons is comparatively less in the exhaust.
- CNG is insoluble in engine oil, hence lubricating oil retains its properties for longer duration resulting in longer interval between oil changes.
- CNG fired engines have longer life, as it does not carbonise spark plugs or wash out lubricating oil films.
- It has a comparatively better cold starting characteristics, because unlike petrol, it does not first have to be vaporised and it does not condense on cold engine parts.

Disadvantages of CNG

- CNG storage tanks add weight because heavy cylinders are required for the storage of natural gas at high pressures.
- CNG storage tanks take up five times higher trunk space for comparable energy storage.
- Refuelling station cost of CNG is very high.
- Compared to petrol driven vehicles, CNG driven vehicles have shorter operating range.

Use of *catalytic converter* to reduce exhaust emissions like carbon monoxide, unburnt hydrocarbons and nitrogen oxides helps in controlling the pollution from vehicles. Catalytic converter fitted in the exhaust system of vehicle comprises of honeycomb ceramic or metallic disc coated with platinum, palladium & rhodium, which catalyses the conversion of carbon monoxide into carbon dioxide, hydrocarbons into carbon dioxide & water and nitrogen oxides. There are two types of catalytic converter namely *(i) two way system* employing platinum & palladium which remove only carbon monoxide and unburnt hydrocarbons leaving nitrogen oxides unchanged, and *three way system* employing platinum, palladium & rhodium, which remove all the three pollutants simultaneously.

Presence of lead in the petrol poisons the catalysts coated on the honeycomb disc of the catalytic converter thereby reducing its potency.

LIQUIFIED PETROLEUM GAS (LPG)

It is mainly a mixture of propane and butane used mainly as cooking gas in domestic ovens as well as for industrial heating in furnaces {*e.g.* heating of annealing furnaces). C_3 (propane etc.) and C_4 (butane etc.) hydrocarbons are easily liquified at room temperature with the application of very low pressure. Hence, (4-10 atm.) propane & butane is liquified, stored and transported in light cylinders. Generally, a mixture of about 80% butane and 20% propane is used for filling in LPG cylinders (also called bottled gas). LPG is prepared from wet natural gas and refinery gases. Even pure butane & propane are also liquified and used as fuels.

LPG is a highly volatile liquid, which expands 247 times its volume as vapor (1 litre LPG liquid = 247 litres of vapor). It is non-toxic, does not support life and is heavier than air.

LPG have high calorific value, high specific gravity, high air requirement and low flame speeds. Since, LPG is odorless, hence odorants like mercaptans (50 ppm) or sulphides are added to detect its leakage from cylinders by smell as mercaptans have very pungent smell. The lighter grade LPG commercially available contains 90% propane and 10% butane.

Properties of main components of LPG *i.e.* butane and propane are given in table 30.2.

Table 30.2: Properties of main components of LPG i.e., butane and propane

	Properties	Unit	Butane	Propane
1.	Gross calorific value	Kcal/kg	30680	23670
2.	Specific gravity of gas (Air = 1)	—	2	1.52
3.	Specific gravity of liquid	—	0.5844	0.5079
4.	Theoretical air requirement	Nm^3 air/Nm^3 of LPG	30.94	23.80
5.	Theoretical amount of flue gas generated :	Nm^3/Nm^3 of LPG		
	CO_2	-do-	3.9	3.1
	H_2O	-do-	4.9	4.1
	N_2	-do-	24.4	18.8
	Total	-do-	33.2	26.0
6.	Sulphur content	%	0.02	0.02
7.	Liquefaction pressure at room temperature	Kg/cm^2	3	10
8.	Limit of inflammability in air	% LPG vapor in vapor-air mixture	1.9-8.5	2.2-9.5
9.	Ignition temperature	°C	500	480
10.	Maximum flame temperature in air	°C	1925	1995
11.	Flash point	°C	-60	- 105
12.	Boiling point	°C	-0.5	-42
13.	Critical temperature	°C	-96.8	- 152

LPG is a popular domestic fuel because of the fact that a large concentration of thermal energy is contained in a small portable cylinder. LPG is quite popular fuel for industrial heating also where alternative fuel is not available. There are more than 150 LPG bottling plant in the country. It is also used as a fuel in I.C. engine as well as for welding & cutting.

Propane and butane are also used for enriching the low C.V. gases in a town gas supply system abroad upto limits of permissible specific gravity. Butane is also diluted with air to get a suitable fuel with a controlled heat release rate. A mixture of 20% butane & 80% air has a C.V. of 6136 kcal/Nm^3 and is outside the limit of inflammability of butane in air. In fact, the LPG used in domestic heating abroad is diluted by adding inert nitrogen or air to reduce its high calorific value so as to obtain a controlled heat release rate in domestic ovens which do not require very high temperature. The amount of air/inert gas added is such that it is well outside the limit of inflammability (explosion limit) of LPG which is about 1.8-8.5% of LPG in LPG/air mixture.

Density of liquid LPG, which is non-corrosive to steel, copper, copper alloys and aluminium is, half that of water. LPG is slightly toxic & anaesthetic, if inhaled in very high concentration. It has no lubricating property and is colorless in either liquid or vapour phase. It is used also for welding, cutting, soldering, brazing and portable blow lambs. Yield (weight %) of LPG is 1 - 2% in straight run distillation of crude oil; 5 - 10% in catalytic reforming of naphtha ; 12% in catalytic cracking of vacuum gas oil; 3 - 14% in hydrocracking of vacuum gas oil and 5 - 10% in coking of residuum.

Propane and butane are also used in the production of petrochemicals. LPG being heavier than air has a tendency to settle down while leaking from cylinder. At times methane rich natural gas is also liquified and sold as bottled gas in cylinders. In that case, it is called Liquified Natural Gas (LNG).

For industrial grade LPG (80% butane + 20% propane):

- C.V. = 29275 kcal/Nm^3 (11200 Kcal/Kg)
- Sp. gravity =1.9 (Air = 1), Theoretical combustion air requirement = 29 $Nm^3/Nm^3$of LPG
  And for domestic grade LPG (70% propane + 30% butane):
- C.V. = 25775 kcal/Nm^3
- Sp. gravity =1.65 (Air = 1)
- Theoretical Combustion air requirement = 26 Nm^3/Nm^3 of LPG

Use of LPG as a S.I. engine fuel. Use of LPG in a spark ignition engine offers following ***advantages :***

- LPG is readily miscible with air and can burn at leaner air/fuel ratios.
- Being resistant to preignition, it leaves less combustion chamber deposits.
- It has an excellent cold starting & warm up performance and gives lower emission of pollutants like unburnt hydrocarbons & carbon monoxide.
- Its anti-knocking properties are excellent provided large amount of unsaturates are not present in it.

However, LPG suffers from the following *disadvantages:*

- LPG has to be stored & transported in bulky tanks capable of withstanding very high pressure.
- Some of the unsaturated compounds present in LPG have a comparatively poor octane number and are prone to causing high speed knock.

The performance of LPG as a S.I. engine fuel with respect to exhaust emissions, power developed & economy depends upon its composition (C_3/C_4 ratio & the level of unsaturates).

REFINERY GASES

Gases obtained during distillation, cracking and other processing of petroleum & its fractions which contain paraffins (e.g. methane, ethane, propane and butane), olefins (e.g. ethylene, propene and butane), hydrogen sulphide and hydrogen are called refinery gases. A possible range of volumetric composition and properties of refinery gases is given in Table 30.3.

Table 30.3: Composition and Properties of Refinery Gas

Propene + Propane	40-55%
Butene + Butane	upto 30%
Ethylene + Ethane	8-12%
Methane	8-10%
Hydrogen	6-8%
Hydrogen sulphide	6-8%
Calorific value, kcal/Nm^3 (dry)	20500
Specific gravity (Air = 1)	1.35

Normally, C_3 and C_4 hydrocarbons {i.e. propane, propene, butane, butene etc.) are separated from refinery gases by liquifaction and sold as LPG/bottled gas. Similarly, sulphur is recovered from hydrogen sulphide. Hydrogen and higher hydrocarbons are utilised in the production of chemicals, fertiliser & plastics. These may be used to enrich town gas also. Surplus refinery gas (after using as refinery fuel and for above uses) is burnt off in atmosphere.

The yield of refinery gases (in the form of LPG) during various petroleum refinery conversion processes is tabulated in Table 30.4.

Table 30.4: Yield of refinery gases in the form of LPG in various refinery processes

Refinery process	Feedstock	Main products	Yield of LPG (weight % on feed)
Visbreaking/coking	Residuum, heavy gas oil	Coke	5—10
Catalytic reforming	Light virgin naphtha	Aromatics, gasoline blend stocks	5—10
Alkylation/ Palymerisaton	Butane-unsaturated gases	Gasoline	10-15
Catalytic cracking	Gas oil	Ethylene, propylene (for petrochemicals)	15—20

PRODUCER GAS AND WATER GAS

PRODUCER GAS

It comprises mainly of CO & N_2 and is produced (in a furnace called 'producer') by blowing air or a mixture of air & steam through hot bed of solid fuels (coal/coke).

In the simple gas producer, dry air is blown through a bed of coal/coke of sufficient depth and at a sufficiently high temperature. Gas obtained consists of CO & N_2 in the ratio of 1 : 2 by volume, with traces of CO_2 & H_2. If coal is used under similar conditions, the gas contains additional hydrogen and methane derived from the volatile matter of the coal. Above method of producing gas was modified as it excessively raised the temperature of fuel bed and much of the potential heat of the fuel leaves the producer as sensible heat of the gas. Another draw-back is that due to very high temperature, ash may get fused and clinker formation may take place.

Above method was modified by addition of steam to the blast to increase the thermal efficiency and moderate the temperature of the fuel bed by generating additional combustible gas by the reaction of steam with carbon, (*e.g.* $C + H_2O = CO + H_2$) which takes place with absorption of heat.

Producer gas made from either coke or coal, under normal operating condition has a composition (volume %) within the range given below :

CO_2	CO	H_2	CH_4	N_2
4-6	20-30	11-20	0-3	46-55

Calorific value (C.V) = 1250-1550 kcal/Nm^3
Sp. Gravity (Air = 1) = 0.85-0.90
Combustion Air requirement = 1-1.3 Nm^3/Nm^3 of producer gas

The producer gas may also contain steam, tarry vapors, sulphur compound, NH_3 in minor quantity.

The commonest types of gas producer is a vertical cylindrical chamber into which the fuel is fed continuously from the top. Air is blown through a grate at the base, and the gas exit is at the top. Ash and clinker are discharged at the grate. The chamber may be lined with fire brick or may have an annular water jacket in which steam is raised for addition to the blast.

A schematic diagram of producer gas plant is given in Fig. 31.1.

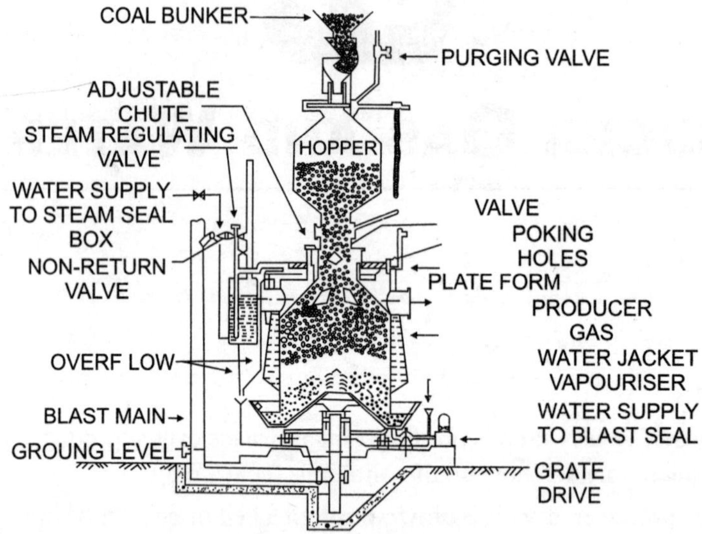

Fig. 31.1. Producer gas plant.

REACTIONS IN GAS PRODUCER

Principal chemical reactions in producer gas plant using coal/coke as fuel and moist blast (air + steam) for gasification are :

Air-carbon reaction :

$C + O_2 \square CO_2$, $\Delta H = - 97000$ kcal/kmole (exothermic reaction)

$C + CO_2 \rightleftharpoons 2CO$, $\Delta H = + 38270$ kcal/kmole (Boudouard reaction-endothermic)

The net overall reaction is :

$2C + O_2 \rightleftharpoons 2CO$, $\Delta H = - 58730$ kcal/kmole. (endothermic reaction)

Steam — carbon reactions :

$C + H_2O \rightleftharpoons CO + H_2O$, $\Delta H = + 28440$ kcal/kmole (endothermic)

$C + 2H_2O \rightleftharpoons CO_2 + 2H_2$, $\Delta H = +18600$ kcal/kmole (endothermic)

$CO + H_2O \rightleftharpoons CO_2 + H_2$, $\Delta H = - 9840$ kcal/kmole. *(shift conversion-reaction exothermic)*

Methanation reaction :

$C + 2H_2 \rightleftharpoons CH_4$, $\Delta H = - 20840$ kcal/kmole. (exothermic)

Producer gas process is favoured by the high temperature of the fuel bed, adequate time of contact between blast & fuel *(i.e.* thick fuel bed) and reactive fuel. Dry air blown producer gas is a mixture of one third CO and two third nitrogen. Its C.V. and cold gas efficiency is low. All these demerits are removed by using a wet blast *i.e.* by adding steam to the blast of air. The advantages of using wet blast are :

- The total content of combustibles ($CO + H_2 + CH_4$) is raised and the inerts ($N_2 + CO_2$) content is lowered thereby increasing C.V.
- A part of the sensible heat liberated by the combustion of carbon is converted into the potential heat of hydrogen and carbon monoxide. The cold gas efficiency of wet blast producer is higher than that of only air blown producer.
- Endothermic reaction between carbon and steam prevents the clinker formation.

FACTORS AFFECTING THE COMPOSITION OF PRODUCER GAS

Nature of fuel. High volatile bituminous coal gives a richer gas, containing small Proportion of methane. Tar vapor also enriches the gas when it is used hot. Coke gives a gas free of tar vapor.

Operating temperature. Low temperature favours high production of CO_2. High temperature favours high production of CO.

Effect of Steam. Water in the coal feed or steam in the air blast increases the proportion of H_2 and CO in the gas, thus raising its calorific value. If excess steam is added, the temperature of gasification is reduced ; more CO_2 is formed and the calorific value of the gas is decreased. If steam is not added, there are chances of clinker formation.

IMPURITIES IN RAW PRODUCER GAS AND THEIR EFFECTS

Steam. It lowers the flame temperature of the producer gas.

Tar. It increases the calorific value of gas, the flame temperature and the radiative heat transfer rate.

Dust. It must be removed by dust catchers, cyclones, electrostatic precipitator etc., otherwise would foul & choke the gas mains.

Sulphur. It should be removed to prevent corrosion and atmospheric pollution .

Ammonia. More NH_3 is present in producer gas than in the coal gas, because thermal cracking is prevented by the lower temperature and by steaming which causes endothermic water gas reaction.

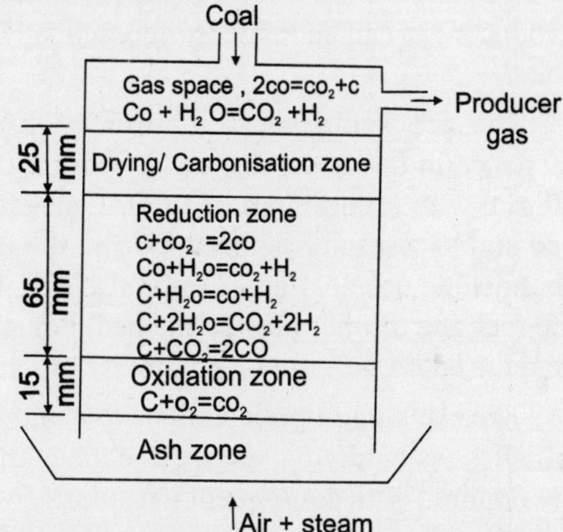

Fig. 31.2. Reaction zones in the fuel bed of a gas producer.

REACTION ZONES IN A GAS PRODUCER

The fuel bed in a producer rests on a metallic grate which is divided into a number of reaction zones *viz.* ash zone, oxidation zone, reduction zone, drying/ carbonisation zone with the depth of the fuel bed as shown in Fig. 31.2.

Ash zone. The air-steam blast is preheated by ash zone which also uniformly distributes blast and protects the grate from intense heat. Ash zone may be considered from the grate to the level at which the descending fuel still contains burning carbon. The thickness of the zone can be varied by controlling the rate of removal of ash and must be sufficient to prevent the overheating of grate by fire.

Combustion (or oxidation) zone. In this zone, oxygen in the air blast reacts with the carbon in the fuel. The oxygen in the blast is completely consumed after passage through a thin layer of fuel (about 75 mm). Gas samples taken from this zone show that the main product of combustion is $C0_2$. Due to high heat of combustion of carbon to carbon di-oxcide the temperature reached by the fuel in this zone is highest in the fuel bed. All the oxygen is consumed within this zone. CO_2 concentration reaches the maximum at the top of oxidation zone when CO just begins to appear.

Reduction Zone. This extends from the level at which no significant amount of free oxygen remains in the gas upto that at which no further reaction of fixed carbon with carbon-dioxide and steam occurs. The gas entering this zone from below has a high content of CO_2 & water vapor and is at a high temperature. At the temperature of 1100 to 1200°C to which the fuel at the base of this zone is heated, the reaction like, $H_2O + C = H_2 + CO$ and $CO_2 + C = CO$ are very rapid so that the content of combustible gases (Carbon monoxide and Hydrogen) rises rapidly and the CO_2 & steam content falls, as the gas passes through reduction zone. Since these reactions absorb heat (endothermic reactions) the temperature falls with increasing height above grate and a level it goes down to 800°C, where no significant gasification of fixed carbon occurs by above reactions.

Drying/carbonisation Zone. Fresh fuel fed to the top of the bed comes into contact with fuel that has already become heated and with hot gases leaving the reduction zone. It is successfully dried, preheated, and (if the fuel is coal) carbonized, adding water vapor, tarry vapors &carbonization gases to the gas leaving the producer.

C.V. of the producer gas is slightly decreased in gas space due to the reactions $CO + H_2O \rightleftharpoons CO_2 + H_2, \Delta H = -9840$ kcal/kmole (exothermic) and $\underset{\text{(Neumann reversal reaction)}}{2CO \downarrow} \rightleftharpoons CO_2 + C, \Delta H = -38270$ kcal/kmole (exothermic).

SOLID FUELS USED FOR PRODUCER GAS MANUFACTURE

Producer gas can be made from practically any solid fuel ; wood-waste, peat, coals of all ranks and coke.

The size of the fuel is important ; close & even sizing results in a more uniform fuel bed and therefore in better conditions of working. For bituminous coals, the best size is from 30 mm to 20 mm with a minimum of fines. Excessive dust causes areas of local high resistance in the fuel-bed and as a result, the blast follows the lines of least resistance where the bed is most open and channelling occurs. Large lumps also cause lack of uniformity in the bed. Within limits, the smaller the fuel, the thinner may be the bed, but at the same time, the permissible rate of gasification may be reduced on account of possibility of carrying fine material forward with the gas.

The ash should preferably be low, but the quality is more important than the quantity. As ash which is easily fused gives rise to clinker and, as a result, extra poking & disturbance of the fuel-bed are required with consequent irregularity of working and reduction in the quality of gas. Where the use of coal of long storing, clinkering propensities cannot be avoided, the saturation temperature of the blast must be raised to cool the bottom of the fire ; this may lead to a decrease in the quality of the gas & the efficiency.

For certain purposes, the content of sulphur may be important. Approximately three- quarters of the sulphur in the fuel is gasified, mainly as hydrogen sulphide, but partly also as organic compounds. Where, there is danger that sulphur in the gas may contaminate the material being heated, either a low-sulphur fuel must be used or the gas must be purified.

High moisture is a disadvantage ; if it exceeds (say) 10%, the thickness of the bed of fuel should be increased so that the moisture may be evaporated at the expense of part of the sensible heat in the gas before the fuel reaches the active zones of the fire.

The higher the moisture in the fuel, the lower will be the temperature of the gas leaving the producer. The quantity of heat in the gas will, however, be the same, although part is latent in the steam. Therefore, if the gas is to be cooled before use, the total quantity of heat to be removed, remains unaltered.

For use in a gas-producer, coal should be noncaking or, at most, weakly caking. Strongly caking coals should be avoided ; it gives rise to unevenness in the fuel-bed owing to the forma tion of masses of coke which cause bridging and irregular settlement of the fuel. (Extra poking is necessary to break up the coke ; it is difficult to prevent channelling, and the rate of gasification falls). The best coals are those of groups 800 and 900, which yield Gray-King residues from *A* to *D.* High volatiles are advantageous, because the gas is enriched with the products of thermal decomposition of the coal.

Anthracite and semi-anthracite are often used where the gas is to be used in gas-engines. These fuels yield less tar than bituminous coke and the gas therefore requires less cleaning. In addition, formation of clinker is less likely.

Coke is a satisfactory fuel, but the gas is of lower calorific value, because the content of methane is low (0.5% compared with the usual 2.5 to 3% from coal).

Composition & properties of producer gas produced by using different fuels is given in Table 31.1

Table 31.1: Composition & Properties of Producer Gas

Composition and properties of producer gas	Dry air blast with coke	Wet blast (air + steam) with the fuel as		
		Coal	Coke	Anthracite
Composition (volume %) :				
CO_2	1	4.0	5.0	6.0
C_mH_n	—	0.4	—	—
O_2	—	—	—	—
CO	33.5	29.0	29.0	26.0
H_2	1.5	12.0	11.0	17.0
CH_4	—	2.6	0.5	1.2
N_2	64	52.0	54.5	49.8
Properties :				
Gross c.v., kcal/Nm^3 (dry)	1060.	1550	1260	1420
Specific gravity (Air =1)	0.98	0.87	0.90	0.85
Theoretical air requirement for combustion, Nm^3/Nm^3 gas	—	1.28	1.0	—

USES OF PRODUCER GAS

It is used for firing in furnaces (glass melting and steel making in open hearth furnaces), coke oven heating and in internal combustion engines.

Producer gas for firing in glass melting and open hearth furnaces. It is an established fact that the gas which is most suitable for firing open hearth furnaces and glass tank should contain as high a percentage of carbon monoxide as possible and not more than 12-14% of hydrogen. For melting steel, the most suitable gas is one which contains 10-12% of hydrogen. As the hydrogen in the gas increases, the carbon monoxide decreases accordingly; but while this entails a reduction in the net calorific value of the gas, its suitability for melting steel falls more rapidly than the calorific value. The deterioration in the quality of the gas cannot be therefore ascribed solely to the fall in calorific value but must be sought for elsewhere.

Apart from the calorific value, two properties are important in determining the value of a gaseous fuel, namely, the radiation from the flame, and the rate of propagation of flame in the burning mixture.

The disadvantage of high percentages of hydrogen seems to be more closely connected with the increased rate of propagation of flame which results from a rise in the percentage of hydrogen. The higher the content of hydrogen, therefore, the shorter the flame, and it is more difficult to maintain the furnace chamber full of burning gases. The higher the rate of propagation of flame, the greater is the tendency of the flame to burn back in the ports, with consequent danger of damage to the refractory materials. Further more, hydrogen shows a greater disposition to burn in contact with surfaces, and although this is advantageous in certain circumstances, it is liable to be objectionable in an ordinary furnace, where accelerated combustion on the linings causes the development of local high temperatures and causes the linings to fail. Experience shows that the destruction of brickwork in steel-melting furnaces is more rapid when the producer gas contain a high percentage of hydrogen.

So long as the content of hydrogen in the gas does not exceed 12 to 13%, the efficiency in a steel-making furnace is independent of the percentage hydrogen, but, as the hydrogen increases above 13%, the efficiency falls off.

The volumetric composition of a producer gas suitable for combustion in a furnace may be :

CO_2 = 2.5%, CO = 30.0% , H_2 = 12.0%, CH_4 = 3.0% , N_2 = 52.5% and C.V. = 1480 kcal/Nm^3.

A gas of this composition would be generated by gasifying a suitable bituminous coal in a mixed air/steam blast with a steam-saturation temperature of 50°C. The gas should enter the regenerators saturated with water vapor at or about 20°C, and it would then enter the furnace at between 1,000°C and 1,200°C with the composition unaltered.

Producer Gas for Use in Internal Combustion Engines. The primary consideration is **that** the calorific value of the gas should be high ; the composition matters comparatively little, and provided that the content of combustibles is high. The saturation temperature of the blast is immaterial.

Gas which is to be used in internal combustion engines must be completely cleaned from tar and dust and it also be cooled to atmospheric temperature before entering the engine.

THERMAL EFFICIENCY OF GAS PRODUCERS

There are three ways of expressing thermal efficiency of gas producers as follows :

Cold gas efficiency (η_c). It is defined as

$$\eta_c = \frac{\text{Potential heat in cold producer gas}}{\text{Total heat in gasified fuel}} \times 100$$

Or,

$$\eta_c = \frac{C_g \times V_g}{C_f + C_{pf} + \Delta t_f} \times 100$$

Hot gas efficiency (η_h). It is defined as

$$\eta_h = \frac{\text{Totalheat (sensible + potential) in hot producer gas}}{\text{Total heat in gasified fuel}} \times 100$$

Or

$$\eta_h = \frac{(C_g + C_{p,g} \times \Delta t_g) V_g}{C_f + C_{p,f} \times \Delta t_f} \times 100$$

Comprehensive efficiency (η_{com}). It is defined as

$$\eta_{com} = \frac{\text{Total heat of producer gas + Heat in surplus steam}}{\text{(Total heat of gasified fuel + total heat of make up steam + thermal equivalent of other energy consumed)}} \times 100$$

Nomenclatures are as follows :

C_g = Calorific value (either gross or net) of producer gas, kcal/Nm^3

C_f = Calorific value (gross/net, but same as in C_g) of fuel gasified, kcal/kg

V_g = Producer gas yield, Nm^3 gas/kg of fuel gasified

$C_{p,g}$ = Avg. sp. heat of producer gas, kcal/Nm^3 °C

$C_{p,f}$ = Avg. sp. heat of gasified fuel, kcal/kg °C

Δt_g = Producer gas. temperature above ambient temperature, °C

Δt_f = Gasified fuel temperature above ambient temperature, °C

The gasification efficiency is also expressed as 'carbon efficiency' defined as :

$$\text{Carbon efficiency} = \frac{\text{Total carbon in producer gas}}{\text{Total carbon in gasified fuel}} \times 100$$

Since, the carbon of the fuel gasified to CO_2 does not contribute to the heating value of the producer gas, hence 'carbon efficiency' is of limited significance.

The most versatile of all the above is the cold gas thermal efficiency.

The above expression of the various efficiencies are also applicable to the calculation of the efficiencies of other coal/oil gasifiers using solid and liquid fuel as the raw material for gasification.

Problem 1. *Find the cold gas thermal efficiency of a producer gas plant from the following data :*

C.V. of producer gas = *1150 kcal/Nm*^{3}

Yield of producer gas = *2800 Nm*^{3}*/ton of coke* = *2.8 Nm*^{3}*/kg coke*

C.V. of the coke gasified = *6000 kcal/kg*

Solution.

$$\eta_c = \frac{\text{Potential heat in cold producer gas}}{\text{Total heat in coke gasified}} \times 100 = \frac{2.8 \times 1150}{6000} \times 100 = 53.7\% \qquad \text{Answer}$$

Example 2. *Following data is available for an operating producer gas plant using coal as a gasifying fuel:*

(a) Analysis of coal gasified :

Proximate analysis, % as charged				Ultimate analysis, % d.m.m.f. basis				
Moisture	Fixed carbon	Volatile matter	Ash	Carbon (C)	Oxygen (O)	Hydrogen (H)	Nitrogen (N)	Sulphur (S)
3.0	47.0	35.5	14.5	85.0	6.0	5.5	3.0	0.5

(b) Composition (Volume %) of producer gas produced (dry basis-orsat analysis) :

N_2	CO	H_2	CO_2	CH_4	C_2H_4	O_2	Total
56.0	20.5	12.5	7.0	3.0	0.5	0.5	100.0

(c) Calorific value (C.V.) of combustibles :

Constituents	CO	H_2	C_2H_4	CH_4
C.V., Kcal/Kgmole	67636	68317	337234	212798

(d) Clinker analysis (% dry basis) : Carbon = 15%, ash = 85%

(e) Loss of carbon with tar = 10%

(f) (i) *Blast saturation temperature (B.S.T.) - 60 °C*

(ii) Vapor pressure of water at 60°C = 150 mm Hg

(iii) 30% of N_2 in coal is converted into ammonia.

From the above data, calculate :

(a) Cold gas thermal efficiency (η_c) of the gas producer using dry blast

(b) Specific blast (air) con- sumtion Nm^3*/ kg of coal charged.*

(c) Specified steam consumption, kg / kg of coal charged.

Solution. Basis of calculation = 100 kg coal (as charged)

Cold gas thermal efficiency (η_c) :

Or,
$$\eta_c = \frac{\text{Potential heat in cold producer gas}}{\text{Total heat in the coal gasified}} \times 100$$

$$\eta_c = \frac{\text{Yield of producer gas (dry basis)} \times \text{C.V. of producer gas (dry basis)}}{\text{C.V. of coal}}$$

(i) Calculation of gas yield (on dry basis). Dry gas yield is found from carbon balance. Carbon is distributed over the gas, clinker and tar.

First the ultimate analysis of coal is converted into "as charged basis" as follows :

$$C = 85 \times \frac{100 - 3 - 1.1 \times 14.5}{100} \qquad \text{(Since, min eral matter} = 1.1 \times \text{Ash)}$$

or, $C = 85 \times 0.81 = 68.9$, $O = 6 \times 0.81 = 4.9$, $H = 5.5 \times 0.81 = 4.4$

$N = 3 \times 0{:}81 = 2.4$, $S = 0.5 \times 0.81 = 0.4$

Moisture = 3.0, Ash = 14.5

Water of hydration of minerals

= 0.1 × Ash = 0.1 × 14.5 = 1.45 ~ 1.5 (Approx.)

Total = 100.0

By ash balance, we get,

Loss of carbon in the clinker = $\frac{15}{85} \times 14.5 = 2.6$ kg

Loss of carbon in tar = 0.1 × 68.9 = 6.9 kg

∴ Carbon gasified = 68.9 - (2.6 + 6.9) = 59.4 kg = $\frac{59.4}{12}$ = 4.95 kg mole

Carbon present in dry producer gas = 7 + 20.5 + 2×0.5 + 3

= 31.3 kg mole/100 kg mole dry gas

∴ Yield of dry producer gas = $\frac{100}{31.3} \times 4.95$ = 15.8 kg mole/100 kg of coal charged

= 0.158 kg mole/kg of coal charged

Now, C.V. of dry producer gas = 0.205 × 67636 + 0.125 × 68317 + 0.005 × 337234 + 0.03×2127

= 30330 kcal/k mole

C.V. of coal is calculated from *modified Dulong formula* as:

$$\text{C.V.} = \left[80.8 \times C + 345\left(H - \frac{O}{8}\right) + 22.2 \times S\right] \text{kcal / kg}$$

Or, $$\text{C.V.} = 80.8 \times 68.9 + 345\left(4.4 - \frac{4.9}{8}\right) + 22.2 \times 0.4 = 6883 \text{kcal / kg}$$

Hence, □c = $\frac{0.158 \times 30330}{6883}$ = ×100 = **70% Answer**

(b) Specific air (blast) consumption :

It is calculated from N_2 balance. Both air and coal supply nitrogen which are received by the producer gas and the ammonia liquour. Out of the total nitrogen in the coal, the producer gas receives 70%, while the ammonia receives 30%.

N_2 in the producer gas = 56 kg mole/100 kg mole dry producer gas

$$= \frac{56}{100} \times 15.8 \text{ kg mole/kg coal charged}$$

= 8.84 kg mole/100 kg coal charged

N_2 supplied from coal to the producer gas

$$= 2.4 \times 0.7 \times \frac{1}{28} \qquad (\text{mol. wt. of } N_2 = 28)$$

= 0.06 kg mole/100 kg coal charged

N_2 supplied by air = 8.84 - 0.06 = 8.78 kg mole/100 kg coal charged

= 0.0878 kg mole/kg coal charged

Now, 1 kmol of any gas occupies 22.4 Nm^3

∴ N_2 supplied by air = 0.0878 × 22.4 Nm^3/kg coal charged

= 1.967 Nm^3/kg coal charged

Since, air contains 79% by volume of N_2, hence air supplied

= 1.967×(100/79)=2.5 = 2.5 Nm^3/kg coal charged. **Answer**

Specific steam consumption. Assuming blast pressure to be 760 mm Hg, we get on application of Dalton's law of partial pressure,

$$\text{Steam content of the blast} = \frac{150}{760-150} = 0.246 \text{ kg mole/kg mole of dry air}$$

= 0.246 × 18 kg steam/kg mole of dry air

= 4.428 kg steam/kg mole of dry air

∴ Steam supplied in 2.5 Nm_3 of blast/kg coal charged

$$= 4.428 \times \frac{2.5}{22.4} = 0.5$$

∴ Specific steam consumption in the blast

= 0.5 kg/kg of coal charged. Answer

Example 3. *Producer gas is manufactured in a producer gas plant using coal which analyses (on dry basis in weight percentage) as follows :*

C = 78%, H = 5.4%, N - 1.4%, S = 1%, O = 10%, ash = 4.2%

Moisture content of coal as charged in the plant is 5%.

Volumetric analysis of dry producer gas produced is as follows :

$CO = 28\%$, $CO_2 = 4.5\%$, $H_2 = 15.5\%$, $CH_4 = 3\%$, $N_2 = 49\%$

Other relavant data are :

Specific heat of steam in the required range = 1.93 kJ/kg.°C

Blast saturation temperature = 50°C

Latent heat of condensation and cooling of steam from 100 to 15°C = 2454 kJ/kg

Temperature of hot producer gas = 205°C

Gross C.V. of CO, H_2 and CH_4 = 12.7, 12.8 and 39.7 MJ/Nm^3 respectively

Mean specific heat, C_p, of producer gas = 1.341 kJ/Nm^3. °C

% of carbon charged to the producer appearing as tar fog = 10%

% of nitrogen charged to the producer appearing as ammonia = 40%

From the above informations and data, *find out* for each kg of dry coal, the

(a) Volume (Nm^3) of producer gas produced.

(b) Volume (Nm^3) of air supplied to gasify it.

(c) Amount (kg) of steam used in the blast.

(d) Percentage of steam decomposed in the blast.

(e) Heat content (kJ) of producer gas produced.

(f) Potential heat in the producer gas produced.

(g) Volume (Nm^3) of producer gas produced if clinker analysis on dry basis is, C = 15% and ash = 85%

Solution, *(a)* Taking 100 kg of coal,

$$\text{C in coal} = \frac{78 \times 0.95}{100 \times 12} = 6.18 \text{ kg atom C (moisture = 5\% = 0.05 .. dry coal = 0.95)}$$

$$\therefore \quad \text{C appearing in the producer gas} = \frac{6.18 \times 90}{100} = 5.56 \text{ kg. atom}$$

(since 10% carbon is lost, hence available carbon for gasification is only 90%)

In, 100 kg mole of producer gas, carbon content = 28 + 4.5 + 3 = 35.5 kg. atom

$$\% \text{ Carbon content in producer gas} = \frac{35.5}{100} = 35.5\%$$

$$\text{for the producer gas, \% Carbon} = \frac{\text{amount of carbon appearing in gas (kg. mole)}}{\text{Total producer gas volume, kg mole}} \times 100$$

or,

$$35.5 = \frac{5.56}{\text{Total producer gas volume}} \times 100$$

$$\text{or, Total producer gas volume} = \frac{5.56 \times 100}{35.5} = 15.6 \text{ kg mole}$$

$\therefore$ Producer gas made per kg of coal charged to the producer

$$= \frac{15.6 \times 22.4}{100} = 3.5 \text{ Nm}^3 \quad \textbf{Ans.} \qquad \text{(since, 1 kg mole = 22.4 Nm}^3\text{)}$$

(b) For finding, the air supplied for gasification, nitrogen balance based on unit weight of 100 kg mole of producer gas is done.

N_2 present in producer gas $= \dfrac{15.6 \times 49}{100} = 7.65$ kg mole

N_2 from coal $= \dfrac{1.4 \times (1-0.4) \times 0.95}{28} = 0.03$ kg mole

(Since, 40% nitrogen in coal is lost as NH_3 in producer gas and coal contains 5% moisture 28 ; = molecular weight of N_2)

hence, N_2 supplied for gasification as air = 7.65 - 0.03 = 7.62 kg mole

Air required for gasification $= 7.62 \times \dfrac{100}{79} = 9.65$ kg mole

$= 9.65 \times 22.4 = 216.1$ Nm^3/100 kg coal charged

hence, air supplied for gasification/kg of coal charged $= \dfrac{216.1}{100} = 2.16$ Nm^3 **Ans.**

(c) Steam used in the blast is calculated by H_2 balance.

For n mole of an ideal gas law, $pV = nRT$, for gas mixtures $p_V = (n_1 + n2 + \ldots) RT$

The partial pressure of each gas is given by $p V = n_2RT$ and $p_1V = n_2RT$

hence, $\dfrac{p_1}{p_2} = \dfrac{n_1}{n_2}$

Since the blast/air saturation temperature is 50°C, then the vapor pressure of water at 50°C is = 92.55 mm Hg (from tables)

If the total pressure = 760 mm Hg

then, the partial pressure of dry air = 760 - 92.55 = 667.45 mm Hg

$\dfrac{p_1}{p_2} = \dfrac{n_1}{n_2}$ *i.e.*, water vapor/mole of dry air $= \dfrac{92.55}{667.45} = 0.133$ kg mole

now, air required for gasification of 100 kg of coal = 9.65 kg mole (calculated earlier in *'b'*)

total steam consumption per kg of coal $= \dfrac{9.65 \times 0.113 \times 18}{100}$ (18 = molecular weight of steam)

= 0.22 kg **Ans.**

(d) When steam is mixed with blast (air) under adiabatic condition to producer a saturated mixture, some condensation results because the increase in enthalpy of air is usually greater than the drop in the enthalpy of steam. Hence to have a dry saturated mixture of air and steam, either the steam is superheated or the air is preheated. In absence of these conditions, a mixture of saturated air and droplets of liquid water is obtained.

Blast is normally preheated before introducing into producer. There may also be indeterminate quantity of moisture introduced via any water seal.

Percentage of steam decomposed in the blast is calculated by hydrogen balance (per 100 kg of coal).

H_2 supplied to producer in coal $= \frac{5.4 \times 0.95}{2} = 2.57$ kg mole

[2 = molecular weight of H_2 and coal contains 5% moisture]

H_2 supplied to producer as moisture $= \frac{5}{18} = 0.28$ kg mole

H_2 in steam in blast $= 0.133 \times 9.65 = 1.28$ kg mole

$$\begin{bmatrix} \text{here, } H_2O \text{ per kg mole of dry air } = 0.133 \text{ kg mole} \\ \text{and, air required } = 9.65 \text{ kg mole} \end{bmatrix}$$

Total H_2 supplied $= 2.57 + 0.28 + 1.28 = 4.13$ kg mole

H_2 appearing in 100 kg mole of producer gas, as combustibles

$= 15.5\%$ as H_2 + 6% as CH_4 $(2H_2)$ = 21.5 kg mole/100 kg mole producer gas

Actual H_2 in producer gas as combustible = Total gas volume × percentage combustible

$$= 15.6 \times \frac{21.5}{100} = 3.34 \text{ kg mole}$$

$$H_2 \text{ converted to } NH_3 = \frac{1.4 \times 0.95 \times 0.4 \times 3}{28} = 0.06 \text{ kg mole}$$

$$\begin{bmatrix} N_2 + 3H_2 = 2\,NH_3, \text{1 mole of N combines with 3 moles of } H_2,\ 28 = \text{molecular weight of } N_2\ 40\% \text{ of} \\ N_2 \text{ charged through coal appears as } NH_3 \text{ in the gas, which is lost in ammonia liquor.} \end{bmatrix}$$

H_2 leaving as steam = 4.13 - (3.34 + 0.06) = **0.73 kg mole**

$$\text{Steam equivalent of oxygen in coal} = \frac{10 \times 0.95 \times 2}{32} = 0.62 \text{ kg mole}$$

$$\begin{bmatrix} 2H_2 + O_2 = 2\,H_2O \text{ i.e., 1 mole of } O_2 \text{ produces 2 moles of } H_2O \\ 32 = \text{molecular weight of oxygen} \end{bmatrix}$$

Steam as moisture in 100 kg coal $= \frac{5}{18} = 0.28$ kg mole

[5% moisture in coal, 18 = molecular weight of steam (H_2O)]

Steam remaining undecomposed = 0.73 - (0.62 + 0.28) = - 0.17 = - ve value or zero. It means that all the steam in the blast has been decomposed. **Ans.**

(e) Sensible heat in hot gases = sensible heat in steam + sensible heat in hot producer gas

$$\text{Sensible heat in steam} = \frac{0.73 \times 18 \times 2454}{100} + \frac{0.73 \times 18 \times 1.93 \times (205 - 100)}{100}$$

$$= 322.5 + 27.5 = 350 \text{ kJ/kg coal}$$

here, H_2 leaving as steam = 0.73 kg mole, coal taken = 100 kg,
18 = molecular weight of steam, temperature of hot producer gas = 205°C,
temperature of steam = 100°C, H_2 leaving as steam = 0.73 kg mole
Latent heat of condensation and cooling of steam = 2454 kg/kg,
specific heat of steam = 1.93 kJ/kg. °C
Now, sensible heat in dry hot producer gas above 15°C = 3.5 x 1.341 x (205 - 15) = 892 kJ
Total heat content in producer gas obtained from gasification of 1 kg coal

= 350 + 892 = 1242 kJ/kg **Ans.**

(f) Potential heat in the producer gas produced by gasification of 1 kg of coal which will be realised on its burning

$$= 3.5 \times 12.7 \times 0.28 + 3.5 \times 12.8 \times 0.155 + 3.5 \times 39.7 \times 0.03$$
$$= 12.5 + 6.95 + 4.15 = 23.6 \text{ MJ/kg of coal}$$ **Ans.**

here, 3.5 Nm^3 = producer gas produced by gasification of 1 kg of coal
12.7, 12.8 and 39.7 = Gross CV (MJ/Nm^3) of CO, H_2 and CH_4 respectively
0.28, 0.155 and 0.03 = fraction (i.e. %) of CO, H_2 and CH_4 respectively in producer gas.

(g) Carbon content of 100 kg of dry coal $= \dfrac{78 \times 0.95 \times 0.90}{12} = 5.55$ kg atom

[Moisture in coal = 5% and 10% carbon is lost as tar fog]

100 kg of dry coal when gasified gives rise to $\dfrac{4.2}{0.95}$ kg ash but the clinker produced contains 15% carbon. Ash produced $= \dfrac{4.2}{0.95} = 4.42$ kg atom

$\therefore$ 4.42 kg ash $= \dfrac{4.92 \times 100}{(100-15)} = 5.2\text{kg}$ clinker

Carbon lost in clinker per 100 kg coal charged :

$$15\% = \frac{5.2 - 4.42}{5.2} \times 100$$

Actual amount of carbon = 5.20 - 4.42 = 0.78 kg $= \dfrac{0.78}{12} = 0.065$ kg atom C lost in clinker
Actual carbon gasified = 5.55 - 0.065 = 5.48 kg atom
The carbon content of 100 kg mole of producer gas is calculated as follows :
C in kg atom in producer gas = 4.4 (CO_2) + 28.1 (CO) + 3 (CH_4)

= 35.5 kg atom/100 kg mole of producer gas

$$\%\text{ C in producer gas} = 35.5 = \frac{5.48}{\text{producer gas volume (kg mole)}} \times 100$$

$$\text{or, Producer gas volume} = \frac{5.48 \times 100}{35.5} = 15.4 \text{ kg mole/100 kg dry coal}$$

∴ Producer gas volume produced/kg dry coal

$$= \frac{15.4}{100} \times 22.4 = 3.45 \text{ Nm}^3 \text{ producer gas/kg of dry coal} \quad \textbf{Ans.} \quad [1 \therefore 1 \text{ kg mole} = 22.4 \text{ Nm}^3]$$

Example 4. *A coke is gasified in a producer gas plant to produce the producer gas of following composition (% by volume) :*

$$CO_2 = 5.0, CO = 25, H_2 = 12.5, CH_4 = 0.5, N_2 = 57$$

The coke as used for gasification contains 4% ash & 10% moisture containing 94% C on d.a.f. basis. Calculate the

(a) Volume (Nm^3) of producer gas produced per kg of coke gasified.

(b) Volume (Nm^3) of air required for gasification of one kg of coke (Assume coke contains no nitrogen)

(c) Theoretical volume (Nm^3) of air required for combustion of producer gas produced by the gasification of one kg of coke.

***Solution,** (a)* Carbon charged in the gasifier per 100 kg of coke (containing 4% ash & 10% moisture) used

$$= \frac{94 \times (100 - 4 - 10)}{100} \text{Kg} = \frac{94 \times 86}{12 \times 100} \text{kg mole} = 6.736 \text{ Kg moles}$$

% carbon present in 100 kg moles of producer gas

$$= 5.0 \text{ (in } CO_2) + 25 \text{ (in CO)} + 0.5 \text{ (in } CH_4) = 30.5\%$$

By carbon balance, we get

Total carbon in producer gas = Total carbon present in coke

$$\text{or,} \quad \frac{30.5}{100} \times \text{Total producer gas volume} = 6.736 \text{ Kg moles}$$

$$\text{Total producer gas volume} = \frac{6.736}{30.5} \times 100 = \mathbf{22 \text{ kg moles}} = 22 \times 22.4 \text{ Nm}^3 = 492.8 \text{ Nm}^3$$

$$\text{Volume of producer gas produced per kg of coke gasified} = \frac{492.8}{100} = 4.921 \text{ Nm}^3 \quad \textbf{Ans.}$$

(b) Nitrogen present in 100 kg mole of producer gas = 57 kg moles

Nitrogen present in the producer gas generated by gasification of 1 kg of coke

$$= \frac{22 \times 57}{100 \times 100} \text{ kg moles}$$

$$= 0.1254 \text{ kg mole} = 0.1254 \times 22.4 \text{ Nm}^3 = 2.8 \text{ Nm}^3$$

This nitrogen has come from the air (blast) supplied for gasification only as coke does not contain any nitrogen.

Hence, volume of air used for gasification of one kg of coke $= \frac{2.80}{0.79} = 3.55\ Nm^3$ **Ans.**

[∴ Air contains 79% N_2 by volume]

(c) Now, producer gas produced by gasification of 1 kg of coke = 4.[illegible] Nm^3 (as found in 'a')

∴ This Producer gas contains :

$$CO_2 = \frac{5}{100} \times 4.92 = 0.246\ Nm^3$$

$$CO = \frac{25}{100} \times 4.92 = 1.230\ Nm^3$$

$$H_2 = \frac{12.5}{100} \times 4.92 = 0.615\ Nm^3$$

$$CH_4 = \frac{05}{100} \times 4.92 = 0.025\ Nm^3$$

$$N_2 = \frac{57}{100} \times 4.92 = 2.804\ Nm^3$$

Total = 4.92 Nm^3

Combustion reactions for combustible components of producer gas are :

$$2CO + O_2 = 2CO_2;\ 2H_2 + O_2 = 2H_2O\ CH_4 + 2O_2 = CO_2 + 2H_2O$$

Theoretical volume of *oxygen* required for combustion of producer gas produced by gasification of one kg of coke

$$= \frac{1.230}{2} \text{ (for combustion of CO)} + \frac{0.615}{2} \text{ (for } H_2\text{)} + 0.025 \times 2 \text{ (for } CH_4\text{)}$$

$$= 0.615 + 0.3075 + 0.05 = 0.9725\ Nm^3$$

∴ Theoretical air required $= 0.9725 \times \frac{100}{21} = 4.63\ Nm^3$ **Ans.**

WATER GAS

It is a medium calorific value gas (about 2800 kcal/Nm^3) comprising mainly of CO and H_2. It is prepared by action of superheated steam on a bed of hot coke/coal at about 1000°C as per the reaction.

$$C + H_2O = CO + H_2,\ \Delta H = +\ 28 \text{ kcal/kmol (endothermic reaction)}$$

Since the above reaction is endothermic, the coal cools down after a few minutes and the reaction proceeds in a different way to form carbon dioxide and hydrogen instead of water gas ($CO + H_2$)

$$C + 2H_2O = CO_2 + 2H_2,\ \Delta H = +\ 19000 \text{ kcal/kmol (endothermic reaction)}$$

In order to avoid the above undesirable reaction, the current of steam is intermittently replaced by a blast of air. The following reactions now occur.

$$C + O_2 = CO_2, \Delta H = -97000 \text{ kcal/kmol}$$
$$2C + O_2 = 2CO, \Delta H = -29000 \text{ kcal/kmol}$$

Thus due to exothermic reaction the temperature of carbon again rises and when the temperature increases to 1000°C, air entry is stopped and steam is again passed. Thus in modern gas plants steam and air are blown alternatively. The period of **steam blow (cold blow)** is usually 4 minutes while the period of **air blow (hot blow),** is very short (about 1-2 minutes). The duration of these periods are adjusted in such a way that the maximum yield of water gas is obtained.

A water gas generator is a cylindrical vessel made of steel. It is about 4-5 m in height and 2-3 m in diameter. At the top, it is provided with a hopper for adding coke. Water gas outlet is provided near the top. It has two separate inlets for blowing steam and air at will. At the bottom, it is provided with an arrangement for taking out ash formed. It is shown in Fig. 31.3.

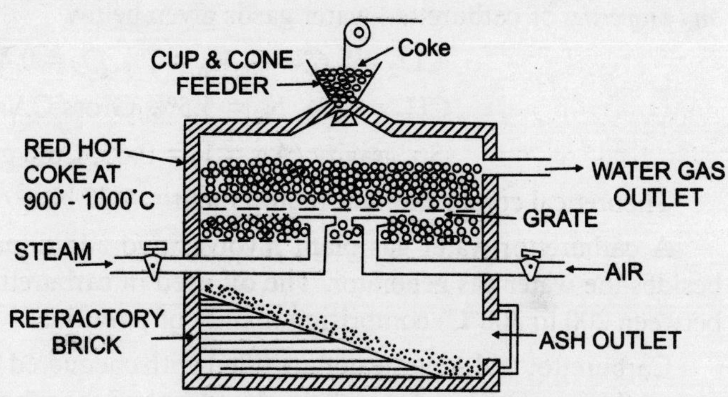

Fig. 31.3. *Manufacture of water gas*

During the steam blow, water gas is produced. This is led out through the water gas outlet. On the other hand, nitrogen, carbon dioxide and carbon monoxide are formed during the air blow. These are allowed to escape in the atmosphere. Thus the manufacture of water gas is intermittent.

The best fuel for water gas manufacture is hard coke and anthracite. A typical composition of water gas is

$$H_2 = 48\text{-}51\%, CO = 40\text{-}42\%, CH_4 = 0.1\text{-}0.5\%, CO_2 = 3\text{-}5\%, N_2 = 3\text{-}6\%$$

Theoretical combustion air requirement = 2-2.5 Nm^3/Nm^3 gas

C.V. = 2500-2800 kcal/Nm^3, sp. gravity (Air = 1) = 0.5-0.55

Water gas is also called *blue water gas* or simply *blue gas* as it burns with a bluish flame due to the presence of high quantity of carbon monoxide.

Uses of water gas. It is used as a fuel in furnaces. Normally, it is enriched by adding hydrocarbon gas (oil gas) and the mixture is called carburetted water gas which has a high C.V.

It is also used as a source of hydrogen for ammonia synthesis in fertiliser plant.

When water gas is specifically produced for ammonia synthesis, a proportion of air is added to the steam so as to get a mixture of H_2, CO, CO_2 and N_2. This is called **'semi water gas'.** With suitable treatment, a mixture of nitrogen and hydrogen in desired proportion is ultimately made from this gas.

CARBURETTED WATER GAS

C.V. of water gas is too low for being an effective gaseous fuel in town gas distribution system. However, it can be enriched to make 'carburetted water gas' by mixing with hydrocarbon oil vapor (having a high C.V. of 10000-13000 kcal/Nm^3). Process of making carburetted water gas involves both gasification of coal as well as oil and then mixing them (oil gas + coal gas). That means *carburetted water gas is a mixture of water gas and oil gas.*

Carburetted water gas is made by passing water gas through a hot chamber (called carburettor) into which hydrocarbon oil is sprayed. The oil cracks and produces methane, ethane, ethylene, propane and other unsaturated hydrocarbons which have high calorific values. The resultant gas mixture is the carburetted water gas with a high C.V. upto 4300 kcal/Nm^3. A typical *composition and properties* of carburetted water gas is given below.

CO_2 = 5.6%, C_mH_n = 7%, O_2 = 0.4%, CO = 30.5%, H_2 = 37.0%

CH_4 = 14%, N_2 = 5.5%, Gross C.V. = 4770 kcal/Nm^3 dry

Sp. gravity (Air = 1) = 0.63, Cold gas efficiency = 65% (with light oil)

Theoretical combustion air requirement = 4.27 Nm^3/Nm^3 gas

A carburettor water gas plant involves two additional units *viz.* carburettor and superheater besides the water gas generator. The oil used in carburetting is a light petroleum fraction (boiling between 200 to 300°C) comprising mainly of paraffins.

Carburettor and superheater are filled with chequered bricks. The gas produced during the blow period (blow gas) is burned in the carburettor and then the combustion is completed in superheater. The flue gases are routed through a waste heat boiler (to generate steam) and finally leave through the chimney.

During the run period, oil is sprayed into the chequor of carburettor and the water gas is passed through the chamber. Endothermic oil cracking process is completed in the superheater which results in the fall in the brickwork temperature which are again heated up during the blow period.

The cyclic process of carburetting water gas is synchronised with the parent water gas generation process. And, the resultant gas is passed out of the plant through a wash box.

Part of the heat produced by blow gas in carburettor and superheater is fed back to the fuel bed by carrying out back run. Steam is blown through superheater and carburettor into the top of the fuel bed. Oil is sprayed into the steam which cracks on passage through the fuel bed. This back-run increases the thermal efficiency of the plant.

Production of 1 Nm^3 of carburetted water gas typically requires : steam -0.65 kg, coke -0.45 kg, oil -0.3 kg, air -1.5 Nm^3.

A schematic diagram of carburetted water gas plant is given in Fig. 31.4.

Carburetted water gas is mainly used as an ingradient of town gas and as substitute of coal gas to meet peak loads of gas supply.

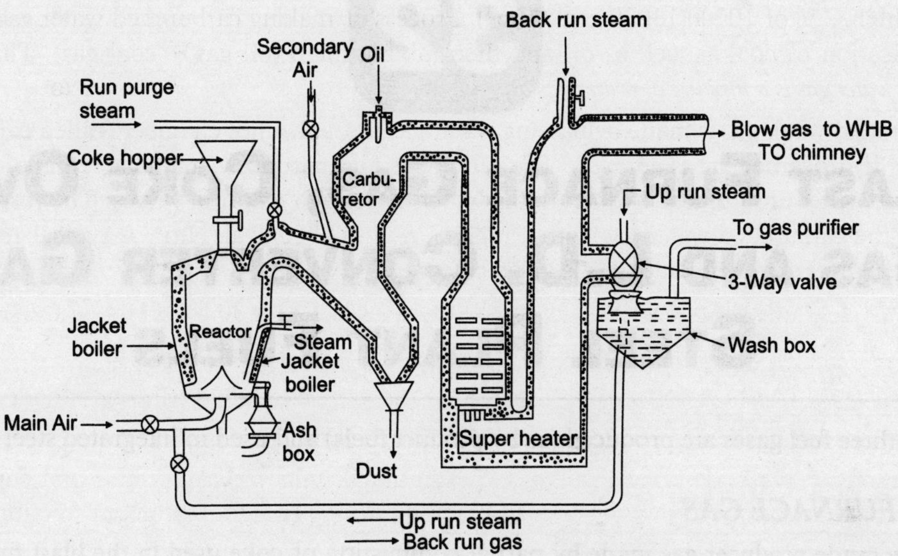

Fig. 31.4. Carburetted Water Gas Plant.

BLAST FURNACE GAS, COKE OVEN GAS AND L.D. CONVERTER GAS-STEEL PLANT FUELS

All these three fuel gases are produced (as by product fuels) and used in integrated steel plants.

BLAST FURNACE GAS

It is a low grade producer gas made by partial combustion of coke used in the blast furnace and modified by the partial reduction of iron ore. It is a by-product of the iron blast furnace. During the manufacture of pig iron, when the air enters the tuyers, its oxygen reacts with the coke. The resulting gas passes up through the shaft of the furnace which has been charged with coke, iron ore, manganese, quartz, limestone, dolomite, sinter etc., and after a number of chemical reactions, issues as a heated dust laden, lean, combustible gas.

Yield of B.F. gas. About 2.5-3 tons of blast furnace gas is generated for one ton of pig iron produced, the coke rate in the furnace being 550—700 kg/ton pig iron. Volumetric yield of gas is about 1800—2000 Nm^3/ton pig iron or about 3000—3200 Nm^3/ton of coke charged in B.F.

Composition of B.F. Gas. A typical range of composition is given below :

Constituents	CO	H_2	CO_2	N_2	O_2
Volume %	21—23	4—5	18—20	53—55	0.2—0.5

Blast furnace may be regarded as gas producer in which a few other reactions take place in addition to the usual producer reactions. The resultant gas has a higher carbon dioxide content and lower hydrogen content than producer gas.

Though the C.V. of B.F. gas is very low (800-850 kcal/Nm^3), still because of its very large quantity produced in steel plant, it is used in most of the furnaces in steel plants. In a modern integrated steel plant about 30—35% of total B.F. gas produced is used in the blast furnace stoves itself for heating the blast (air) and rest is available for other steel plant furnaces. Surplus gas is normally used in captive thermal power plant boilers. Because of its low calorific value, it is usually pre-heated or burnt with preheated air to get high flame temperature.

The pressure of B.F. gas in burners should not be allowed to go below 50 mm water gauge otherwise there is a risk of backfiring and explosion in furnaces.

Normally, floating head type wet gas holders are used for storing B.F. gas. Gas holder also arrests the fluctuation in gas pressure at the time of sudden reduction of wind rate in the blast furnaces.

B.F. gas leakage in atmospheric air is detected by carbon monoxide (CO) detecting tubes, the chemical yellow coating (of palladium sulphide) inside which turns green to blue on exposure to carbon monoxide depending upon the extent of CO concentration (in ppm) in air.

Properties of B.F. Gas. It is a very poisonous gas due to the presence of high amount of carbon monoxide in the gas. B.F. gas pipes and vessels should be leakproof. Under some conditions, mixture of gas and air forms an explosive mixture which explodes when coming in contact with any source of fire. The air-gas mixture can explode near gas pipes in burners in case of rapid gas pressure decrease or in case of getting the flame inside the pipe. That is why a specific gas pressure should be strictly maintained in the gas carrying pipe lines. Properties of B.F. gas is tabulated in Table 32.1.

Table 32.1: Properties of B.F. Gas

Properties	Unit	Values	Remarks
Density	kg/Nm^3	1.3—1.4	Slightly heavier than air (1.28 kg/Nm^3)
Calorific value	$kcal/Nm^3$	800—850	Contains more CO, less H_2 and has a much lower calorific value than coke oven gas and other fuel gases.
Maximum Permissible dust content in the clean gas	mg/Nm^3	10	Higher dust content chokes the burner, pipelines and fouls the heat transfer surfaces in furnaces.
Ignition temperature	°C	680	Higher than many fuel gases.
Theoretical Flame temperature	°C	1450	Lower compared to other fuel gases.
Explosion limit	% of gas in gas air mixture	Lower-37] Higher-71]	Wider explosion limit, hence danger of explosion is more.
Normal gas pressure in the pipe lines,	mm water gauge	900—1000	To prevent backfiring in the burner and to move the gas to different furnaces located farther as pressure drop will be more due to higher density.
Permissible CO content in air	ppm	50	Greater amount in air would cause headache, giddiness etc. 10000 ppm (about 1%) is sufficient to kill a man.
Color, odour and taste	—	—	Colourless, odourless and tasteless in nature.

B.F. gas Processing and Cleaning. Blast Furnace gas leaves the furnace at a temperature of 200 to 250°C and a pressure of 0.2 to 2.2 kg/cm^2 carrying with it water vapour and maximum 10—25 gms of dust/Nm^3 gas. Diameter of dust particle ranges from 0.1 micron to 5 mm (1 mm = 1000 microns).

The dust content in the gas varies with the top pressure of the blast furnace. Dust content in the low top pressure blast furnace is more (because of higher entrainment of dust by high velocity gas) compared to high top pressure furnace.

In early days of blast furnace operation, the raw hot gas was used as it came from the furnace without cleaning causing great deal of trouble with flues, combustion chambers and stoves due to clogging. A heat exchanger installed at the top of the furnace used to recover the sensible heat of the raw gas but not without many problems leading to frequent shutdown of the furnace thereby reducing its availability. Presently the gas is universally cleaned before usage. Gravity settling chambers are used for removing coarser dust particles, cyclones & wet scrubber for coarse particles and electrostatic precipitator for sub-micronic finest particles. The clean (10 mg dust/Nm^3 gas) and cool gas at 30—40°C is transported through pipelines to different furnaces in iron and steel works. Drain pots are provided in the gas pipelines to remove the condensate from the gas continuously. The excess blast furnace gas which can not be utilised in the furnace is bled (ejected into the atmosphere) after burning it as it contains toxic CO which must be burnt to CO_2 before bleeding.

Characteristics of B.F. Gas

- It has a very low calorific value of 800 to 850 kcal/Nm^3, depending upon blast furnace coke consumption rate.
- It has got a low theoretical flame temperature.
- It has a low rate of flame propagation — relatively lower than any other common gaseous fuels.
- It has a high specific gravity (highest of all common gaseous fuels.)
- It burns with a non-luminous blue flame.

Poisonous effect of B.F. Gas. When a man inhales blast furnace gas, the carbon monoxide present in it reacts with haemoglobin of the blood forming carboxyhaemoglobin and replacing oxygen out of it. The affinity of CO for haemoglobin is 300 times more than that of 0_2. Due to 0_2 starvation in the haemoglobin of blood, the tissues in the blood which depends upon 0_2 can not function properly resulting in the effects like dizziness, headache and shivering or weakness in moderate quantity of inhalation of gas. Prolonged exposure in excess of 0.2% of the gas will kill the man exposed.

Determination of Low Carbon Monoxide Concentrations in Atmosphere

Palladosulphite tubes — A small glass tube is packed with a short length (7-8 cms) of plain white silica gel followed by a length (2 cms) of gel impregnated with yellow potassium palladosulphite. Each end of the glass tube is drawn to a fine point and sealed. In use, both the ends of the tubes are broken off and the tube is inserted into gas sampling system. Gas is drawn through the tube at a stated rate for two minutes. Carbon monoxide causes the colour of the yellow portion to change to dark brown to blue. Data on length of the stain and colour change corresponding to various carbon monoxide concentrations is supplied with each box of the tubes.

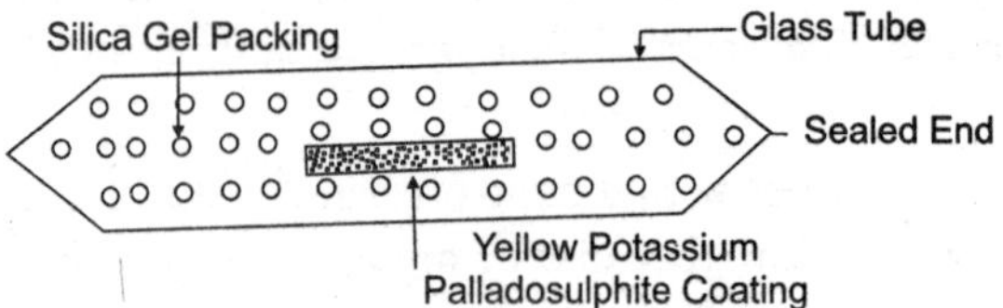

Fig. 32.1. Carbon monoxide detecting tube.

Hence a rough idea of carbon monoxide concentration (in ppm) is found out by the change in colour of the yellow portion in the tube. If the colour changes to dark blue, it corresponds to a concentration of about 1000 ppm of carbon monoxide in atmosphere. The cost of each carbon monoxide detecting tube is about Rs. 50. Now a days , portable CO detecting digital meters based on electro chemical cells are available in the market which can be clipped in the shirt's pocket.

Uses of B.F. Gas. Though it has a very low calorific value, yet because of the large quantity of the gas produced, it is one of the most important fuels in an integrated iron and steel plant. It is used as such or after preheating in :

- Hot dip galvanising line (HDGL) furnace
- Blast Furnace stoves
- Foundary ovens
- Soaking pits
- Annealing furnaces
- Reheating furnace
- Boilers
- Billet heating
- Coke oven underfiring

Great care is taken in its utilisation, because of its toxic & poisonous nature and the surplus gas may be stored in gas holders or bled out in the atmosphere after burning it, as the case may be. Mostly B.F. gas is mixed with coke oven gas and is used in various furnaces of steel plants.

COKE OVEN GAS

It is produced during high temperature carbonisation of coking coal. It is the most important fuel in an integrated steel plant. *Yield of coke oven gas* is 290-300 Nm^3 of gas/ton of dry coal carbonised. A typical volumetric composition of clean coke oven gas is given below.

Constituents	CO	H_2	CH_4	N_2	CO_2	O_2	CmHn
Volume, %	7-8	58-60	23-25	2-3	2-4	0.5-0.8	2-3

About 20% of the sulphur in coal is evolved with the distillation products. Much of this remains in the coke oven gas as H_2S. Gases high in sulphur content are undesirable for metallurgical purposes as they cause brittleness of surface while rolling the steel after hot scarfing the slabs of steel.

FACTORS AFFECTING THE COMPOSITION OF COKE OVEN GAS

Composition of coke oven gas varies with the temperature of carbonisation, time of carbonisation and rank & type of coal.

Effect of temperature. With increase in the temperature of carbonisation :

- hydrogen content of the coke oven gas increases because of cracking of hydrocarbons.
- methane and other hydrocarbon content of the gas reduces due to their cracking.
- carbon monoxide content of the gas increases because of the reaction, $CO_2 + C = 2CO$ (carbon is produced from cracking of hydrocarbons).
- yield of the gas increases due to volatalisation and cracking of liquid hydrocarbons (mainly tar).

- calorific value of the gas decreases because of substantially large reduction in hydrocarbon content which is not compensated by the increase in CO and H_2 content of the gas.

Effect of time. With increase in the time of carbonisation at a given temperature, the hydrogen and the hydrocarbon content decreases resulting in the greatly reduced calorific value of the gas.

Effect of rank of coal. With increasing rank, the proportion of hydrocarbon decreases and the proportion of hydrogen increases *e.g.* the composition of the gas obtained by carbonising anthracite resembles that obtained in the final stage of carbonisation of bituminous coal and is rich in hydrogen. Other way round, a high volatile bituminous coal gives a gas relatively rich in hydrocarbons and poor in hydrogen.

Effect of type of coal. Black durains and cannel coals yield gases containing high proportion of hydrocarbons (particularly unsaturates) of high illuminating value.

Properties of coke oven gas are tabulated in Table 32.2.

Table 32.2: Properties of Coke Oven Gas

Properties	Unit	Value	Remarks
Density	kg/Nm^3	0.43—0.45	It is much lighter than air (air=1.28 kg/Nm^3)
Calorific value	kcal/Nm^3	4200	High calorific value. C.V. may become even 6000 kcal/Nm^3 after removal of H_2 from CO gas.
Dust content	mg/Nm^3	nil	Dust is not there in the gas.
Ignition temperature	°C	560	Lower than blast furnace gas (680°C).
Explosion limit	% of gas in gas-air mixture	lower-5% higher -32%.	Narrower range than blast furnace gas. More explosive than B.F. gas (37-71%).
Theoretical flame temperature Normal gas pressure in the pipeline	°C mm water gage	2000 700	Much higher than blast furnace gas (1400°C), Less than B.F. gas as its sp. gravity is less hence pressure drop in its transmission will be comparatively less.

Though coke oven gas is not as poisonous as B.F. gas, it is highly inflammable and explosive. It can not be preheated because the methane present in it would crack in the regenerator chambers and deposit soot causing fouling and choking of regenerators. Though it is mildly preheated to 60 0C by steam to keep the naphthalene present in it in vapour state to avoid the clogging of burners.

Coke oven gas is deprived of its valuable chemicals in the by-product recovery plant before use. It is saturated with water vapor. Hence in the gas distribution system, means are provided (e.g. drainpots) to drain off the condensate formed due to changes in temperature continuously.

Characteristics of Coke Oven Gas :

- Unlike B.F. gas, surplus coke oven gas can be bled into atmosphere without burning as it is not that toxic.
- It burns with a luminous to semi luminous flame depending upon the degree of mixing of gas and air.

- Its rate of flame propagation is considerably higher than that of natural gas, producer gas and blast furnace gas.
- It has a low specific gravity-lowest of all the pure gaseous fuels used in steel industry.
- It has a high theoretical flame temperature—little higher than that of natural gas even.
- Its explosive range is about twice of natural gas.

Uses of Coke Oven Gas : Coke oven gas has a more extended use than B.F. gas, because of :

- relatively low distribution cost due to its low specific gravity, high calorific value and cleanliness.
- its ability to develop extremely high temperature by combustion.
- high heat release rate thereby eliminating excessively large combustion chamber furnaces. Sulphur in coke oven gas is disadvantageous when used in making low sulphur heats in open hearth furnace and in heating certain grade of alloy steel for rolling. Its presence requires the use of materials resistant to sulphur attack in pipelines, valves and burners. Coke oven gas is used mostly in mixture with B.F. gas in the iron and steel works itself for the following usage :
- Coke Oven heating
- Blast Furnace Stoves heating
- Boilers of captive thermal power plant
- Sinter Machine furnace
- Soaking pits, reheating furnaces, annealing furnaces etc.
- Forge furnaces and foundry ovens
- Blast furnace runner and laddie drying
- Calcination kilns etc.

Coke oven gas produced in a low temperature coal carbonisation plant has a higher C.V. (6000—6500 kcal/Nm^3) and is used as an ingredient of town gas supply system. It is also suitable for synthesis gas. In Rourkela Steel Plant, earlier, hydrogen was taken out from coke oven gas to produce ammonia (using excess N_2 from oxygen plant) for fertiliser manufacture. This is now stooped due to closure of fertiliser plant. A low temperature carbonisation plant at Dankuni in West Bengal will supplies coke oven gas to Calcutta for domestic heating ovens.

MIXED GAS

It is a mixture of blast furnace gas and coke oven gas in different ratios to have a gaseous fuel of a desired calorific value and combustion characteristics.

There are a number of applications where neither B.F. gas nor coke oven gas, when used alone, develop the desired flame characteristics or temperature level for optimum results. By mixing the two fuels of great variance in characteristics, a more ideal fuel can be obtained for specific applications with specific properties.

The calorific value and the *speed of combustion* is very high for coke oven gas and very low for blast furnace gas. The desired speed & calorific value can be attained by the proper proportioning and mixing of the two fuels.

In most of the gaseous fuel applications in an integrated iron and steel plant, mixed gas is used. There may be a common gas mixing station for the mixed gas of a fixed calorific value to be supplied to the entire furnaces of the steel plant or the individual gas mixing stations to mix the two gases in different proportions suitable for different applications. The mixed gas pressure may be boosted when required by the consuming furnace depending on the burner design.

L.D. CONVERTER GAS (BOF GAS)

It is mixture of mainly CO & CO_2 and is produced in L.D. converter during the process of steel making. When oxygen is blown to remove carbon from pig iron in L.D. converter, the converter gas is produced ($C + O_2 = CO_2$, $C + 1/2O_2 = CO$). The converter gas having a very high concentration of carbon monoxide (upto 80% during peak period) in earlier L.D. converter used to be burnt at the mouth of the converter itself thereby eliminating its toxicity by conversion of CO into CO_2 (*viz.* $CO + 1/2O_2 = CO_2$). However, in modern steel making process, the converter gas is recovered as such by suppressing its combustion at the mouth of the converter. It is cooled, cleaned of dust and stored in gas holder for use as a fuel gas.

When oxygen blow is started in the converter (to decarburise the pig iron) the skirt and hood (which covers the mouth of converter) is kept in raised position so that initially the gas burns (for about 2.5 minutes in oxygen blow cycle of 20 minutes) and bled to the atmosphere through a flare stack. The moment carbon monoxide content in converter gas reaches more than 35% and O_2 is less than 2% the changeover valve is operated to divert the gas from bleeder to the gas storage holder.

A schematic diagram of converter gas recovery system is given in Fig. 32.2.

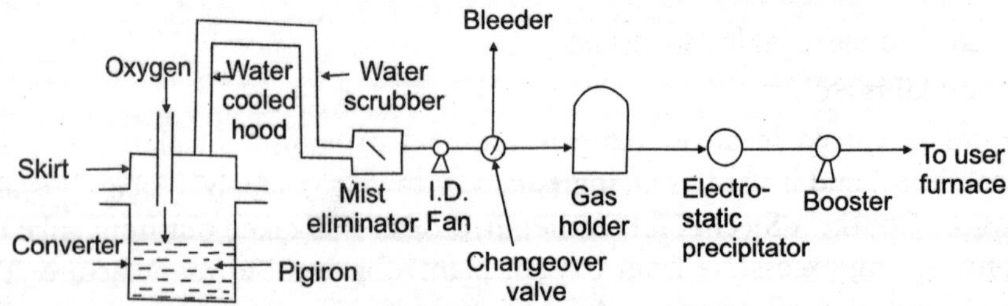

Fig. 32.2. Schematic diagram of converter gas recovery (By IRSID CA-FL method).

Operational Parameters for Converter Gas Recovery and Storage. During suppressed combustion period of oxygen blowing cycle, the converter gas has high proportion of carbon monoxide and can be stored as a fuel. The gases leaving the I.D. fan are continuously sampled and analysed for chemical composition. When CO exceeds 35% and O_2 is below 2%, the change of gas route from bleeder to the gas holder by operating the changeover valve is done.

Interlocks in gas collection system. The time at which converter gas storage is initiated is normally governed by four interlocks :

- Time from start to blow of oxygen.
- Carbon monoxide content in the gas.
- Oxygen content in the gas.

- Gas holder capacity/volume.

Recent practice is to remove the time interlock and allow gas collection to commence immediately when carbon monoxide content in the gas reaches upto 35% and oxygen less than 2% if the gas storage capacity in the gas holder is available. In case the gas storage stops due to > 2% oxygen in the gas (which is sensed by an oxygen analyser in the gas main to holder) after the storage has been started, a bleeder on this gas line automatically opens and purges back the line by back flow of the gas from holder till oxygen percentage in the gas comes below 2%.

Sizing of gas holder is important for maximum recovery of the gas.

Factors Affecting the Quantity of Converter Gas Recovered :

Quantity of converter gas recovered depends upon the following factors:

- Amount of decarburisation in the converter.
- Hot metal/scrap ratio in the converter charge.
- Degree of suppressed combustion (*i.e.* combustion air factor).
- Switch on and switch off level for changeover valve operation.
- Recovery period.
- Gas holder size.

Higher the degree of decarburisation, higher will be the gas quantity generated/recovered during steel making.

Higher degree of suppressed combustion can be achieved by operating the converter with minimum skirt gap and the hood pressure should be controlled to prevent the ingress of air. These are difficult to be maintained due to the problem of skulling around the converter mouth and the difficulty in controlling the hood pressure owing to uneven evolution of gas. With bottom blown converter and newer gas bubbling processes, the tendency to build up nose skull on the mouth of converter is minimised. Hence, skirt gap can be narrower and since gas evolution is more even during blow, the hood pressure can be accurately controlled always on the positive side.

By setting the switch on and switch off level of change over valve to a lower value of CO in the gas (*i.e.* practically say 35% CO), maximisation of gas recovery can be achieved.

Gas recovery period can further be increased by maintaining all the operating parameters within norms so that blow is completed in one stroke without any disruption/interruption by any interlock in the system. Ideally, in a oxygen blow cycle of 20 minutes, gas recovery should be achieved for about 15 minutes (barring 2.5 minutes each at the beginning and end of the oxygen blow period).

Finally gas holder should have enough capacity to store all the gas recovered in one blow. In India, maximum gas recovery has been upto 65 Nm^3/ton steel whereas Japanese Steel Plants have recovered upto 110 Nm^3/ton steel.

Composition of converter gas :

Converter gas during storage period will comprise mainly of CO & CO_2 with little bit of nitrogen and traces of oxygen. CO content in the gas will vary with flow rate of oxygen and degree of suppressed combustion. A typical composition during peak period is

$$CO = 65\%, CO_2 = 15\%, O_2 = 0.2\%, N_2 = 19.8\%, CV = 2000 \text{ Kcal/Nm}^3.$$

Permissible dust content. Normally, the dust content in clean gas should be < 10 mg/Nm^3. However in event of non-functioning of electrostatic precipitator, 80 mg/Nm^3 dust level can be tolerated in all other heating applications of the gas converter except the heating of B.F. stoves, its heating in Coke Oven regenerators prior to underfiring and wherever small diameter burners are used.

Calorific Valve of the Gas

It varies normally from 1600 to 1800 kcal/Nm^3 depending upon the CO percent in the gas. Average C.V. of the gas *can* be increased upto 2000 Kcal/Nm^3 by :

- Maintaining lower skirt gap at the converter mouth.
- Maintaining positive hood pressure constantly.
- Completing the oxygen blow in one stroke without interruptions.
- Bottom/side/combined blowing of converter.

Explosion Limit of the Gas:

For the converter gas, the explosion limit is about 20—70% gas in gas-air mixture. Upper explosion limit rises markedly with increasing temperature of the gas.

Flame Temperature :

Maximum adiabatic flame temperature of the converter gas in air would be about 1800— 1900°C.

IGNITION TEMPERATURE. Spontaneous ignition temperature of the converter gas is about 610°C in air and 590°C in oxygen. Ignition temperature decreases with increase of gas pressure.

FLAME VELOCITY. It is in the range of 0.3—0.4 metre/second (fuel-air system). However, if oxygen is substituted for air, then the flame speed rises to 5—10 times this value. Flame speed rises, if the initial temperature of combustible mixture is increased by preheating.

Combustion Characteristics of Converter Gas

The dust laden gas will burn with an yellowish flame, however, the dust free gas would burn with a bluish flame. If quantity of primary air supply is less, then it will burn with a long and lazy flame giving less heat release rate. If burnt with excess air, it will give short, intense flame with large heat release rate.

Toxicity. Because of high CO concentration, the converter gas is a very poisonous gas. Permissible concentration limit for CO in air is 50 ppm only.

Uses of Converter gas. It can be used in most of the steel plant furnaces like captive thermal power plant boilers, rotary kilns for calcining limestone & dolomite, soaking pits, blast furnace stoves, coke ovens underfiring. In some of the Indian steel plant, the converter gas is being mixed with coke oven gas and is being used for heating furnaces like reheating furnaces, soaking pits, annealing furnaces etc. It is mixed with B.F. gas also in some steel plants *(e.g.* TISCO).

Converter gas recovery system has been installed at Bhilai, TISCO, Rourkela, Durgapur, Bokaro and Vizag Steel Plant in Visakhapatnam. Converter gas is also called *BOF (Basic Oxygen Furnace) gas, L.D. Gas* or, *OG gas.*

Coal Gas from Coal Gasification Processes

Producer gas and water gas manufacture involves incomplete gasification of coal/coke. Also when coke is used as the raw material, gasification involves two stages *viz.* first carbonisation of coal to produce coke and then its partial gasification (at low gasification efficiency).

The complete/total coal gasification in single stage (using processes like Lurgi process, Kopper-Totzek process & Winkler Process etc.) maximises the conversion of coal into gaseous fuel and minimises the tar formation. The gasification efficiency of these processes are very high and they are suitable for a wide range of feed material. The gases produced by the gasifier are mainly used as a fuel and a source of synthesis[1] gas (hydrogen rich) for ammonia making in fertilizer plants.

Gasifiers are *classified* as :

- Early and current gasifiers
- Second generation gasifiers.

Current gasifiers are further classified as :

Fixed bed gasifier	— Lurgi process
Fluidised bed gasifier	— Winkler process
Entrained bed gasifier	— Koppers-Totzek process.

The above three are the most important gasifiers currently in use :

Second generation gasifiers are mainly for synthesis gas production and divided into three groups namely.':

Group I gasifiers :

- Bi-Gas Gasifier
- Synthane Gasifier
- CO_2 acceptor gasification process
- Union Carbide UCC Process.

1 **Synthesis gas** is so called because it is required for the production of chemicals and fertilisers by synthesis. It consists of N_2 and H_2 for ammonia synthesis, CO and H_2 for methanol (methyl alcohol) synthesis & CO and H_2 for manufacture of liquid fuels. The H_2 is mainly derived from steam & partly from the fuel, CO by reaction of Oxygen with carbon of the fuel & N_2 from the atmospheric air during separation of O_2.

Group II gasifiers. (Hydrogasification) :

- HYGAS Gasification Process —
- HYDRANE Gasification Process

Group III gasifiers. It employs coal gasification in liquid medium.

Gasification Reactions:

At > 815°C, following reactions occurs :

$$C + O_2 \rightleftharpoons CO_2 \text{ (Combustion reaction-exothermic)}$$

$$2C + O_2 \rightleftharpoons 2CO \text{ (Boudouard reaction-endothermic)}$$

$$C + H_2O \rightleftharpoons CO + H_2 \text{ (Carbon-steam reaction-endothermic)}$$

$$CO + H_2O \rightleftharpoons CO_2 + H_2 \text{ (Shift reaction-exothermic)}$$

At < 1150°C and elevated pressure, following reactions occurs :

$$C + 2H_2 \rightleftharpoons CH_4 \text{ (Methanation reaction-exothermic)}$$

$$C_mH_n \rightleftharpoons \frac{n}{4} CH_4 + \frac{m-n}{4} C \text{ (Cracking reaction-endothermic).}$$

Table 33:1 Difference between direct coal combustion and coal gasification

S.No.	Parameters	Direct coal combustion	Coal gosification
1	Operating pressure	Usually atmospheric pressure	Often high pressure
2	Operating temperature	Lower	Higher
3	Ash condition	often dry	Often slagging
4	Feed gases	Air	Oxygen, steam
5	Product gases	CO_2, H_2O	CO, H_2, CH_4
6	Gas clean up	Post-scrubbing	Intermediate scrubbing
7	Pollutants generated	SO_x,NO_x	H_2S, HCN, NH_3
8	Char reaction rate	Fast (with O_2)	slow (with CO_2, H_2O)
9	oxidiser	In excess	Deficient
10	Tar production	None	Sometimes
11	Purpose	High temperature gas	Fuel rich gas

The various gasification processes are briefly described below :

LURGI GASIFICATION PROCESS

This is a single stage fixed bed high pressure (20 atm) gasification process. Lurgi generator comprises of :

- coal charging box (at the top) from which coal is intermittantly charged through lock and hopper system.
- vertical water cooled pressure shell in which gasifying medium (steam + oxygen/air) is injected through rotary bottom grate.
- ash received (at the bottom) from which clinker is continuously withdrawn into a lock-hopper under generator.
- *Recent improvements* in Lurgi generator are ;
- rotary coal distributor
- steam cooled grate
- optional mechanical stirrer for weakly caking/ non-caking coal.

Lurgi gasification unit is shown in Fig. 33.1.

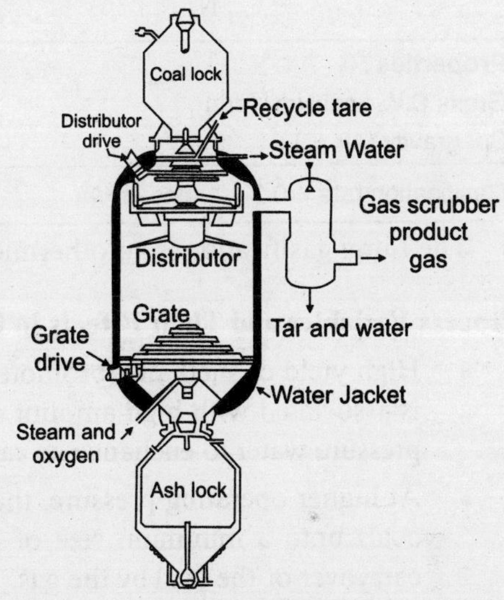

***Fig. 33.1** Schematic diagram of a modern Lurgi coal gasifier.*

Lurgi process is suitable for whole range of coal (3—25 mm size) of < 20% ash excepting medium and strongly caking coals. The gasifying medium is a mixture of superheated steam (at 450-500°C) & oxygen in the ratio of 10 : 1. The high steam/oxygen ratio is used to control the clinkering trouble and H_2CO ratio in the gas. The production of methane increases with increase in pressure and decreases with increase in temperature. The coal-oxygen exothermic reaction supplies the heat thereby rendering the process autothermic and continuous. The hydrogenation of volatile matter of coal under pressure also forms methane. Methanation reaction being exothermic, reduces the oxygen requirement of the gasification process. The presence of large quantity of CO_2 in the gas is due to the use of high amount of steam.

Lurgi gas produced at high pressure facilitates the purification of gas, its distribution through pipes and its use in the synthesis of ammonia, methanol and liquid fuel. Lurgi gas can be used as town gas after CO_2 removal and adjustment of CO/H_2 ratio through shift conversion &methanation. For production of synthesis gas or hydrogen, the processing steps include initial CO_2 removal after steam reforming of CH_4, shift conversion (CO to CO_2) and finally removal of CO_2 by scrubbing.

Typical Composition of Lurgi Gas. Cleaned & CO_2 free gas from oxygen and steam blown Lurgi generator is of following composition (by volume).

$$H_2 = 50\%, CO = 35\%, CH_4 = 15\%$$

Another typical composition and properties of Lurgi gas using bituminous coal (both crude as well as purified by CO_2 removal) is given in Table 33.2.

Table 33.2: Composition and Properties of Lurgi Gas

Composition and properties	Crude gas	Purified gas
Composition (volume %) :		
CO_2	28.0	0.8
C_mH_n	0.2	0.2
O_2	—	—
CO	22.4	31.5
H_2	38.0	53.3
CH_4	10.9	13.4
N_2	0.5	0.8
Properties :		
Gross **C.V.**, kcal/Nm^3 (dry)	2630	3900
Sp. gravity (Air = 1)	0.73	0.48
Gasification rate = 0.2-1.1 tons /m^2/hr		

The Lurgi gasification is autothermic as the fuel-oxygen reaction supplies the heat.

Process Variables and Their Effects in Lurgi gasification :

- High yield of methane (promoted by higher pressure and lower gasification temperature) is associated with high amount of CO_2 in the Lurgi gas which is scrubbed out with high pressure water to enchance its calorific value.
- At higher operating pressure, the gas velocity in the Lurgi gasifier is less hence even fine coals upto a minimum size of 2 mm can also be used for gasification without undue carryover of the coal by the gas.
- Phenomenon like distillation of tar in the gasifier reduces the gas yield as it is dependent on the amount of carbon reacting.
- To avoid clinker formation and to use a revolving grate necessitates the use of high steam/ oxygen ratio (*i.e.,* 8 : 1).
- Because of low steam decomposition, its consumption is more resulting in large production of effluents.
- Volume of the gasifier is reduced due to its operation under higher pressure as compared to conventional gasifiers. Moreover, scrubbing out of H_2S, CO_2 & benzol from the lurgi gas is more efficient when operating under high pressure.
- Because of bituminous coal being of lower reactivity as compared to lignites, the gasification rate of coal as well as methane content of the Lurgi gas decreases while steam & oxygen consumption increases.
- At higher operating pressure of the Lurgi gasifier, some bituminous coals which are non-coking at atmospheric pressure may form weak coke at high pressure which of course can be broken by fitting stirring arms to the revolving grate.

Effect of high gasification pressure. High pressure (20 atm.) in the Lurgi gasifier favours the methanation reaction thereby increasing the CH_4 content of the gas. The hydrogenation of the volatile matter of the coal under pressure also forms methane. Since methane formation is an exothermic reaction, the oxygen requirement of the gasification process is reduced. High pressure Lurgi gas has low velocity thereby reducing the dust carry over from the coal bed surface and permits the use of finer coal for gasification.

The high pressure gas is purified and distributed through pipe for its use in synthesis of ammonia, methanol & liquid fuel manufacture.

Cold gas efficiency of Lurgi plant is 75-80% of heat in the solid fuel which can be further increased by recycling recovered tar (during gas purification) in the gasifier. For 1 Nm^3 gas production, 0.85 kg coal, 1.4 kg steam and 0.21 Nm^3 oxygen are required.

Excepting medium and strongly caking coal, all other solid fuels (> 3 mm size) can be used.

Lurgi gasifiers are the most widely used gasifier all over the world (mainly Germany and U.K.) using mainly lignites & non-caking bituminous coal for supply of town gas (for domestic cooking) and synthesis gas for ammonia manufacture for fertiliser industry. In South Africa, Lurgi gasifiers have been used for the production of synthesis gas for Fischer-Tropsch synthesis to produce liquid fuels and petrochemicals.

A pilot plant of capacity 0.8 tons coal/hr has been operating in Central Fuel Research Institute (CFRI), Dhanbad (India) since 1962.

WINKLER GASIFICATION PROCESS

It has a *fluidised bed type atmospheric pressure gasifier.* The gasifier is a refractory brick lined cylindrical shaft. Coal is continuously fed by a screw feeder from a pressurised coal hopper and fluidised by primary blast just above the bottom grate. A secondary blast just above the fluidised bed gasifies unreacted char in product gas stream. Ash not leaving with this stream (on removed in an external cyclone) is withdrawn through the base of the reactor by means of a rotating scrapper.

Gasification temperature is about 800-950°C at which tar and gaseous hydrocarbons is reformed to CO & H_2. The product gas contains traces of methane. The hot product gas is passed through a waste heat boiler and then in a cyclone & scrubber for dust removal and cooling (dust carry over is high upto 20% due to low pressure in the gasifier).

A typical composition and properties of the gas produced is given below :

CO = 44.4%, H_2 = 36%, CO_2 = 15.7%, N_2 = 2.3%, CH_4 = 1.6%

Gross C.V. = 2600 kcal/Nm^3 (dry), Sp. gravity (Air = 1) = 0.70

Gasification efficiency (carbon conversion) = 80%

Cold gas efficiency = 58%

1 Nm^3 of gas production requires 1.1 kg lignite, 0.31 kg steam and 0.36 Nm^3 oxygen.

Due to construction of the Winkler gasifier, maximum temperature must not exceed 980°C, so fuel for Winkler gasifier is limited to lignite and sub-bituminous coal. High rank coals which are very reactive at such low témperature can not be used.

Commercial Winkler gasifiers are operating in various countries for producing synthesis gas, water gas and also heating gas. Gas produced by Winkler process is more suitable for the manufacture of chemicals, fertilisers and synthetic liquid fuels. Its use as gaseous fuel is restricted due to low calorific value and high specific gravity.

In India, this gasifier is operating in fertiliser plant at Neyveli (Tamilnadu). Ash fusion poses lot of problem in its operation.

A simplified diagram of Winkler gasifier is given in Fig. 33.2.

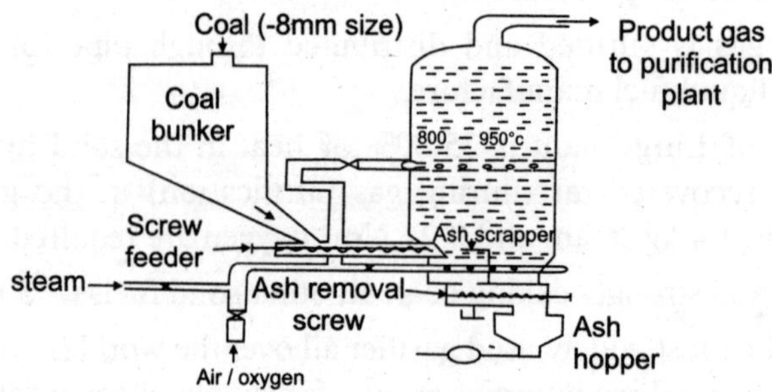

Fig. 33.2. Simplified Schematic Diagram of Modern Winkler Coal Gasifier.

The demerit of this process is high dust entrainment, which reduces the gasification efficiency.

KOPPER TOTZEK GASIFICATION PROCESS

This process gasifies the dry pulverised coal (< 0.1 mm) in suspension (entrained bed) in oxygen (at atmospheric pressure) in the presence of steam. The gasifier is a horizontal, cylindrical (refractory lined) vessel with a conical end. A suspension of pulverised coal in oxygen is injected tangentially through two burner heads. Superheated steam is introduced around these heads to shroud reaction zone and avoid overheating of vessel walls. Gasification occurs at 1600°C very rapidly. The product gas (at 1000-1300°C) is taken out through collector pipe at the top while ash leaves as molten slag from the bottom.

The hot product gas is passed through waste heat boiler and the dust catcher (to remove half the ash and one tenth of the coal carried with the gas).

At the high temperature of reaction zone, the volatile matter of the coal is converted into CO, CO_2 and H_2. The methane is in traces. A typical composition and properties of the gas produced is as follows :

$$CO = 53.3\%,\ H_2 = 31.1\%,\ CO_2 + H_2S = 12.2\%,\ N_2 = 1.3\%,\ CH_4 = 0.1\%,$$

Gross C.V. = 2670 kcal/Nm^3 (dry),

Sp. gravity (Air = 1) = 0.76

Due to high specific gravity and low C.V. the gas is not suitable as a fuel but is used as synthesis gas and for production of methyl alcohol.

Cold gas efficiency of the gasifier is 67%, carbon conversion 97% and overall efficiency (including waste heat recovery is) 75.5%.

A diagram of Kopper-Totzek gasifier is given in Fig. 33.3.

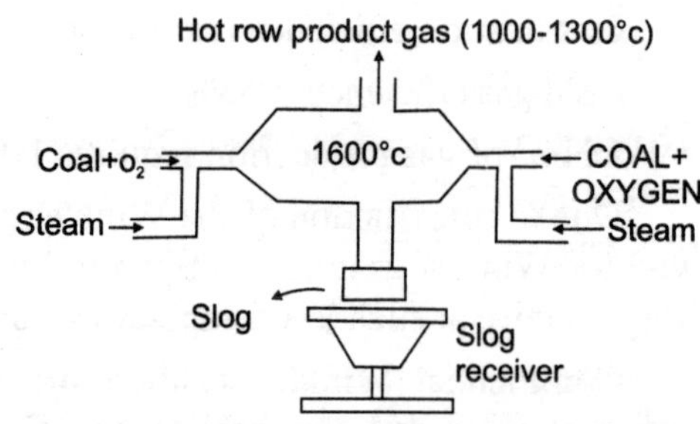

Fig. 33.3. Kopper-Totzek Coal Gassifier.

Operating data from a commercial plant (Kopper-Totzek gasifier)

Gas Yield = 1530 Nm^3/ton coal (ash = 35%)

Consumption/Nm^3 of gas—>4 Coal = 0.65 kg, Steam = 0.36 kg, 0_2 = 0.36 Nm^3.

Advantages of K.T. gasifier :

- It accepts all types of coal
- Ash can also be removed as flyash with raw gas separated in cyclone and also tapped off as slag.
- High operating temperature allows very fast gasification.
- Absence of tar in the gas.
- Adaptability to the gasification of heavy oils.
- Simple design of the gasifier.

Disadvantages of K.T. gasifier :

- Sensible heat of the raw gas is lost during cleaning.
- The gas has to be compressed for downstream chemical synthesis.

A few coal based fertiliser plants *(e.g.* Talcher in Orissa) have this gasifier in operation in India.

The Koppers gas after suitable scrubbing, shift conversion & purification is suitable for ammonia synthesis and is being used in coal-based fertiliser plants in India.

CFRI, Dhanbad has been operating a Kopper-Totzek pilot plant (of capacity 100 kg coal/hr) having facilities for coal preparation (80-90% < 200 mesh size), ash removal, final cooling and cleaning of the synthesis gas.

SECOND GENERATION GASIFIERS

Synthesis gas production using very high pressure for gasification (upto 70 atm) has following advantages:

- Enhances CH_4% in gas therby increasing its calorific value.
- Obviates final compression of SNG (substitute to natural gas) prior to injection in pipe line distribution system.
- Lower cost of production.
- Large scale gasification.

Three processes (For Second Generation Gasifiers):

Group I. Gasify fluidised or gas entrained coal to produce synthetic gas having substantially high CH_4 percentage but still requires "shifting" and "methanation" for using as "high C.V. pipeline gas".

Group II. *Hydrogasification* - gasify fluidised or entrained bed of coal with H_2 and steam to yield CH_4 rich gas needing no "shifting".

Group III. Produces same syngas as in process I but gasifies the coal in liquid medium that may also catalyse the gasification reaction.

GROUP I GASIFIERS — BI-GAS GASIFIER

It employs two stage reactor operating at 75 atm. Pulverised coal and steam (or a coal slurry) is injected in upper part (dense-phase-second-stage) of the reactor and there contacted with hot syngas moving up from the lower section to form additional CO, H_2, CH_4 at 925°C. Residual unreacted char is then carried out through top with product gas, separated in cyclone and reintroduced near the bottom (first stage) of the reactor where it is almost completely gasified in a more dilute phase with O_2 & steam at 1480°C. Ash is withdrawn as a molten slag into a water reservoir below the reactor and residual clinker particles are periodically withdrawn through a lock-hopper device. CH_4 in crude gas is around 20%. It is shown in Fig. 33.4.

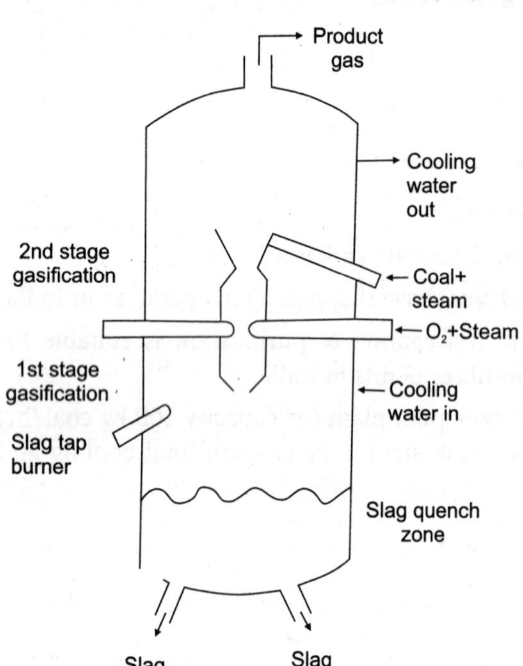

***Fig. 33.4.** Schematic of Bi-Gas Gasifier.*

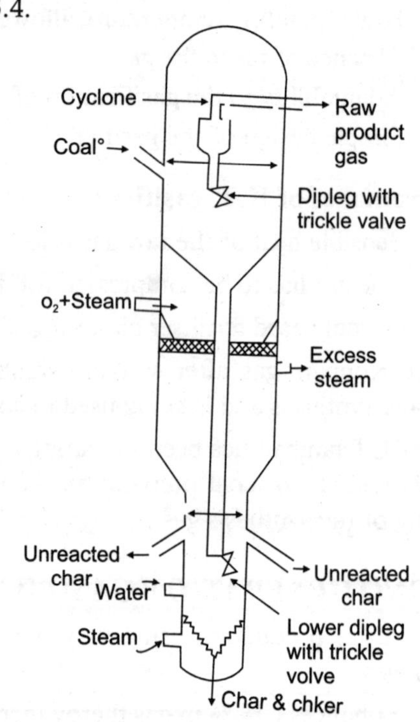

***Fig. 33.5.** Schematic of the Synthane Gasifier. (of height 30.5 m and dia 1.5m).*

SYNTHANE GASIFIER

It has been developed in a 75 ton/day pilot plant at Bruceton, Pennsylvania. Crushed (-20 mesh) dry cool fed from a pressurised lock-hopper into a pretreatment unit where it is fluidised with O_2 and steam at 400°C/70 atm in order to devolatalise it and destroy caking properties. The resultant char with free V.M. and unreacted steam then moves to the generator, where it is partially gasified in a dense fluidised state at 590-790°C/70 atm. and finally reacted with additional oxygen and steam in a more dilute phase (in the constricted lower section of the gasifier) at 950-1000°C/70 atm. Unreacted char and some ash are withdrawn from the bottom of the generator while raw syngas (with 22-23% CH_4) leaves at the top for cyclone separator & a water scrubber which free the gas of entrained dust & tarry matter. It is shown in Fig. 33.5.

CO_2 ACCEPTOR GASIFICATION PROCESS

It is patented by "Consolidation Coal Company". Gasification feedstock is coal and calcined dolomite (MgO. CaO). Dolomite removes CO_2 by carbonate formation and generates some exothermic heat which is utilised to sustain the endothermic carbon-steam reaction. Crushed coal and calcined dolomite (from the regenerator) are fed to the top of the devolataliser and fluidised with steam (and gas from the gasifier) at 815°C & 10-20 atm. The net reaction taking here are :

$$2C + H_2 + H_2O \rightarrow CH_4 + CO$$

$$CO + H_2O \rightarrow CO_2 + H_2$$

$$MgO . CaO + CO_2 \rightarrow MgO . CaCO_3$$

As these reactions proceed, much of spent dolomite, being heavier than the partly gasified char, moves down towards the bottom of the fluidised bed.

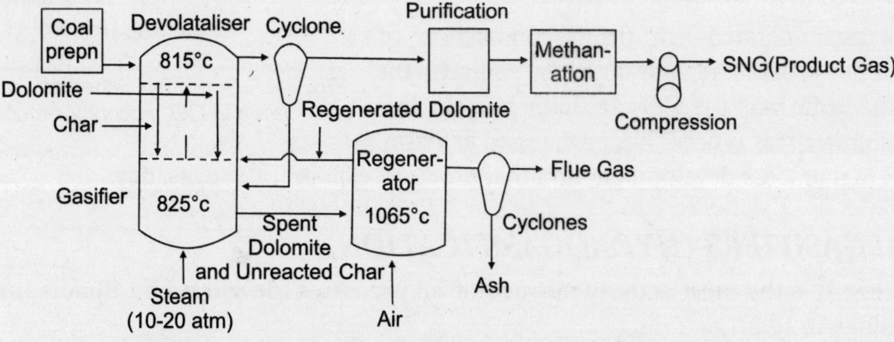

Fig. 33.6. *Simplified flowsheet of the CO_2 acceptor process.*

From the upper part of the bed, unconsumed char is then sent to the gasifier where it is contacted with steam and more calcined dolomite to allow the following reaction to take place.

$$C + H_2O \rightarrow CO + H_2$$

$$CO + H_2O \rightarrow CO_2 + H_2$$

$$MgO . CaO + CO_2 \rightarrow MgO . CaCO_2$$

The resultant product gases are returned to the devolataliser from which they finally exit as raw syngas. Unreacted char from gasifier, together with spent dolomite from devolataliser gasifier is sent to regenerator where the char is burnt in air at 1065°C and heat is used to calcine the dolomite before recycling it. Flowsheet of CO_2 acceptor process is given in Fig. 33.6 while the gasifier used in the process is shown in Fig. 33.7.

Advantages:

- Requires no offsite oxygen or hydrogen and yields a substantially CO_2 free syngas. It uses cheaper air instead of oxygen.
- The product gas is of high heating value and may be used as *substitute to natural gas (SNG).*
- This process can use reactive coal like lignite.

Disadvantages:

- Use of dolo in devolataliser and gasifier necessitates operation of unit below 870°C and since steam gasification of bituminous coal is slow at this temperature, hence the process like Winkler gasification is restricted to lignite and sub-bituminous coal.
- Some difficulty arises from the need to balance material flows between three independent fluidised beds.
- This process is not suitable for higher rank coals.

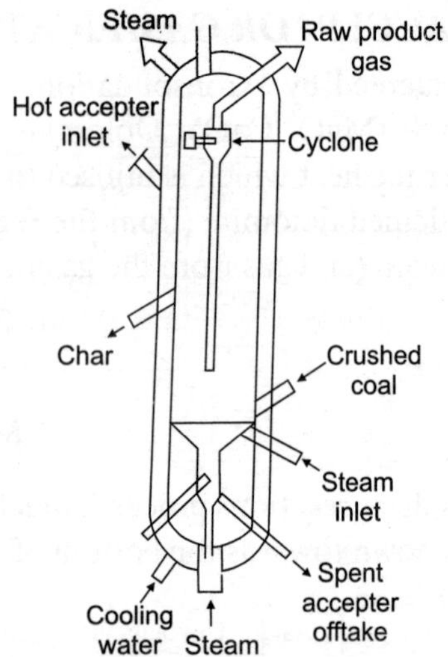

Fig. 33.7. Simplified Schematic of Gasifier used in CO_2 acceptor process.

UNION CARBIDE UCC PROCESS

It uses hot agglomerated ash (from combustion of residual char in a separate reactor) as a heat source in the gasifier in the same way the CO_2 acceptor process uses calcined dolomite. This process has been tested at 7 atm. and 1025°C in a process development unit that produces 800 m^3 of syngas/day.

GROUP II GASIFIERS-(HYDROGASIFICATION)

Hygas Process. It is the most actively pursued of all processes (developed by Illinois Institute of Gas Technology in 1940).

Here, direct hydrogenation of coal with steam & H_2 at 70-100 atm. yields raw gas of composition CH_4=37.5%, H_2=45%, CO=13%, C.V. = 4770 kcal/Nm^3 and is therefore sent to purification and final methanation without prior shifting.

Finely crushed dry coal is pretreated with air at 425°C to destroy caking properties and thereafter slurried with oil. The slurry is then pumped to the dryer section of gasifier at 70-100 atm. where oil vaporises & is carried out with product gas and the coal moves downward through a rising stream of H_2 & steam, reacting in the 1st stage at 650°C and in the 2nd stage at 980°C. Unconsumed charge (which is 50% of original fuel charge) withdrawn from bottom of the gasifier and passed to a hydrogen generator where it is fluidised with steam to produce CO+H_2 that is returned to the gasifier. Flowsheet of Hygas process and the Hygas gasifier is shown in Fig. 34.8 and 34.9 respectively.

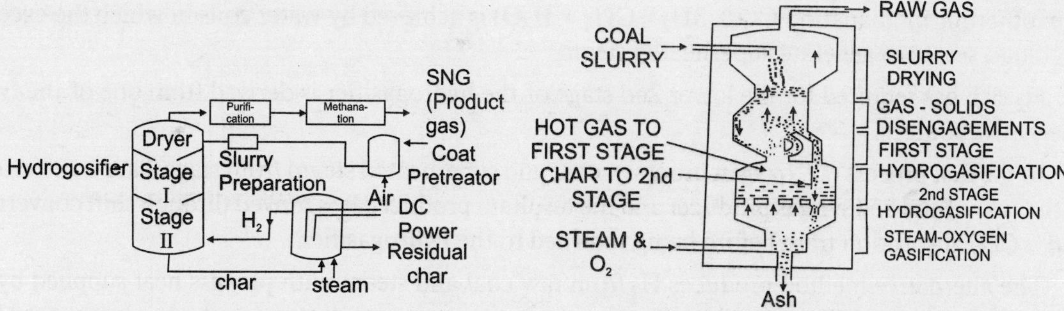

***Fig. 33.8.** Simplified Flowsheet of Hygas Process*

***Fig. 33.9.** Schematic of the Hygas Gasifier (height = 39.5), (max dia intrnal =1.7 m).*

Earlier hydrogen generator employed an electrothermal process, with heat for the reaction, $C + H_2O \rightarrow CO + H_2$, supplied by passing an electric current through the fluidised char. Since, power requirement is more so later work centred on producing $(CO+H_2)$ by steam-iron process or by conventional steam-oxygen gasification of char.

Unconsumed char from the gasifier is fluidised with steam and O_2 in a separate reactor to produce CO, H_2 & CO_2. The crude syngas is then shifted to enrich it in H_2 and after removal of CO_2 returned to the 2nd stage of main gasifier.

It is a well established technology. Hot char from gasifier is transferred to a gas producer and fluidised with air & steam at 1650°C/70-100 atm. The resultant gas is taken out through a reducer where it reacts with magnetite at 800°C/100 atm. ($3Fe + 4H_2O = Fe_3O_4 + 4H_2$). Spent gas is vented and reduced (FeO + Fe) are sent to an oxidiser where they are contacted with steam at 800°C/100 atm. Fe_3O_4 regenerated thereby is returned to the reducer while steam+hydrogen are sent to the gasifier. Flowsheets of Hygas steam-oxygen process & steam- iron process respectively for H_2 generation is shown in Fig. 33.10 & 33.11.

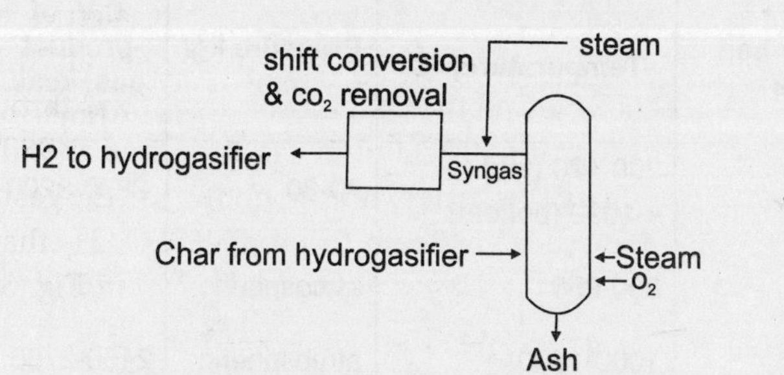

***Fig. 33.10.** Flowsheet of Hygas steam-oxygen process for H_2 generation.*

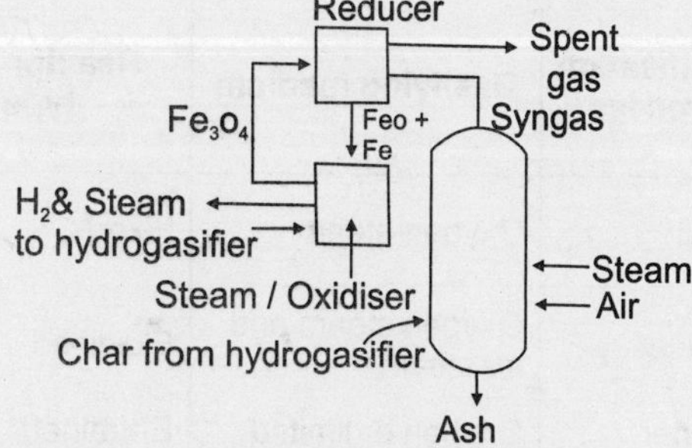

***Fig. 33.11.** Flowsheet of Hygas, steam-iron process for H_2 generation.*

HYDRANE GASIFICATION PROCESS

It is a two stage hydrogasification process resembling HYGAS process but can accept caking as well as noncaking coals without prior treatment. In the hydrogasifier, a dilute phase suspension of coal in a CH_4 & H_2 rich hot gas first devolatalises at 900°C/100 atm. and then flows concurrently with the gas into stage II where it is partially gasified by H_2. Temperature control in this stage (due to exothermic methanation $CO + 3H_2 = CH_4 + H_2O$) is achieved by water coils in which the excess heat load serves to generate superheated steam.

H_2 rich gas required for the lower 2nd stage of the hydrogasifier is derived from one of the two alternative sources.

In the first process (*i.e.* oxygen process), char and superheated steam from stage 2 are transferred with O_2 to a fluid bed syngas producer and the resultant product gas is moved through shift converter and a CO_2 absorption train before being returned to the hydrogasifier.

The alternative method produces H_2 from raw coal and steam with process heat supplied by a solid carrier or 'grog'. The 'grog" itself comes from *a* separate combustor in which it is prepared by adding 5% of carbon to it and burning it in air.

GROUP III GASIFIRES

It produces same synthetic gas as group I gasifiers but gasifies the coal in liquid medium that also catalyses the gasification reaction.

Lurgi, Koppers-Totzek and Winkler process yield high heating value gases. Solid-gas contact in various gasification processes is secured by fixed, fluidised or entrained beds. Fixed bed processes require sized non-coking or weakly coking coal and preserve methane & other distillation products of the coal. Entrained processes use all sizes and types of coal. The operating characteristics of the fluid bed processes are between those of fixed and entrained bed processes. The temperature of the product gases leaving are 260-425°C in fixed bed processes, 815-980°C in fluid bed processes and over 1090°C in entrained bed processes.

The main features of various gasification processes for production of high heating value gases are given in Table 33.2.

Table 33.2: Features of Various Coal Gasification Processes

Gasification process	Gasifying medium	Reactor bed type	Temperature, °C	Pressure kg/ cm^2	C.V. of product gas, kcal/ Nm^3
Lurgi	Oxygen-steam	Fixed	260-420 (top) < 1090 (bottom)	20-30	2850-3200
Winkler	Oxygen steam and air steam	Fluidised	800-950	atmospheric	2600
Kopper-Totzek	Oxygen or limited oxygen-steam	Entrained	1000-1300	atmospheric	2450-2700
Bigas	Oxygen-steam	Entrained	1480 (first stage) 925 (second stage)	75	—
Synthane	Oxygen-steam	Fluidised	950-1000	70	—
CO_2 acceptor	Air in regenerator steam in gasifier	Fluidised	860	10-20	—
Hygas	Hydrogen in hydrogasifier	Fluidised	650 (1st stage) 925-980 (2nd stage)	70-100	4770

Table 33.3 gives the comparative features of various coal gasification processes *viz.* fixed bed, fluidised bed and entrained bed processes.

Table 33.3: Comparison of Coal Gasification Process Types (Air and Oxygen Blown)

Parameters	Fixed bed	Fluidised bed	Entrained bed
Residence time	1-3 hours	0.3-2.5 hours	0.4-12 seconds
Coal type used	most types, no fines	non-caking	all types
Coal size	6-50 mm	500 -2400 μm	10-150 μm
Steam/Coal ratio	0.28-3.09	0.11-1.93	0.1-1.20
O_2/Coal ratio	0.14-0.81	0.25-0.97	0.28-1.17
Temperature range, °K	1150-1300	600-1470	1150-2500
Pressure range, MPa (atm)	0.1-2 (1-20)	0.1-10 (1-100)	0.1-30 (1-300)
Product gases,volume %			
$CO + H_2$	39-66	2-80	35-91
CH_4	2-15	3-68	0.1-17
GCV, kJ/Sm^3	9000-12000	11000-30000	4000-20000
Commercial operation	Gasifier Extensively used is Lurgi Gasifier	Gasifier Slightly used is Winkler Gasifier	Gasifier Moderately used is Kopper-Totzek Gasifier
Main advantages	High turn down ratio Low thermal losses Mature technology	Lower temperature Reduced thermal loss Variety of coal sizes Moderate residence time	Smaller, simple design All coal types Highest capacity/volume

OIL GAS FROM OIL GASIFICATION PROCESSES

Oil gas is produced by gasification of oils (petroleum products) and is used as a gaseous fuel or as a source of synthesis gas for ammonia production in a nitrogenous fertiliser plant. Feedstock for gasification are light & middle distillates and heavy oils (with low C/H ratio). Oil gasification involves cracking, addition of hydrogen or oxygen and removal of carbon (in the form of tar, coke and oxides of carbon). C_5 to C_{15} hydrocarbons are liquid oil at ambient temperature.

Gasification medium is (oxygen + steam) which reacts with oil as follows :

$$C_XH_y + \frac{X}{2}O_2 = XCO + \frac{Y}{2}H_2 \qquad \text{(Partial oxidation reaction-exothermic)}$$

$$C_xH_y + XH_2O = XCO + \left(X + \frac{Y}{2}\right)H_2 \qquad \text{(Steam reforming reaction-endothermic)}$$

$$C_nH_{2n+2} + nH_2O = nCO + (2n + 1)\,H_2 \qquad \text{(Cracking of hydrocarbons-endothermic)}$$

The *heat requirement* of endothermic reaction is supplied by one of the following three methods.

- cyclic method of dividing the process into blow and run period .
- continuous method of burning a part of feed in the gasifier using oxygen & steam in place of air and steam.
- external heating by enclosing the gasifier in a furnace.
- The oil gasification processes are *classified* as follows :
- Non-catalytic cyclic process *e.g. Semet-Solvay Process.*
- Catalytic cyclic processes e.g. Segas Process and Onia-Gegi Process.
- Continuous catalytic steam reforming (of naphtha) processes.
- Partial combustion process e.g. Koppers-Totzek, Shell and Texaco.
- Hydrogenation processes.

Out of the above processes, *Semet-Solvay Process, Segas Process, Onia-Gegi Process and Steam Reforming (of naphtha) processes are important.*

SEMET-SOLVAY PROCESS

This cyclic and non-catalytic process produces rich fuel gas by gasification of heavy fuel oils/gas oil/light distillates.

Two symmetrical vessels (having an upper vaporising chamber and lower chequor filled cracking chamber) connected at the top comprises the gasifier. During the run period, process steam is passed through the hot chequor of one vessel and on preheating mixes with the vapor obtained by oil sprayed at the top of the same vessel. The mixture passes downward through the second vessel where the desired gasification is achieved. The carbon formed during the process is deposited on the walls of the vaporising chamber & chequors of the second vessel. Tarry matter escapes in the gas. The product gas is passed through a wash box and condensers and then sent to the gas holder.

During blow (regeneration), the supplies of oil & steam are cut off and air blast is made counter flow to the gas direction in the preceding run. The burning of the deposited carbon restores the temperature of the system (650-900°C). Steam purging is used between blow & run and also between run & blow. Complete cycle takes about 4 minutes.

A typical composition of oil gas produced by this process is as follows :

CH_4	= 40%	C_mH_n	= 30%
H_2	=18%	N_2	= 6.5%
CO	= 3%	CO2	= 2.5%
Gross C.V.		= 10000 kcal/Nm^3 (dry)	
Sp. Gravity (Air = 1)		= 0.72	
Steam requirement		= 0.6 kg/kg oil.	
Cold gas efficiency		= 55% of total oil	

The equipment arrangement in Semet-Solvay oil gasification plant is shown in Fig. 34.1.

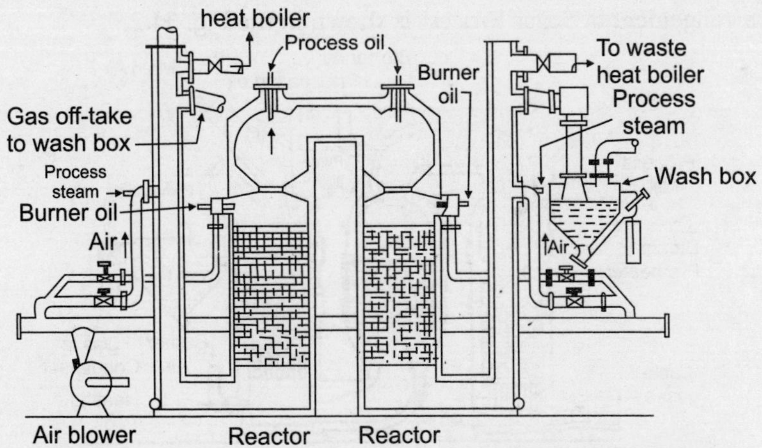

Fig. 34.1. *Equipment arrangement in Semet-Solvay Oil-gas plant.*

SEGAS PROCESS

It is a catalytic, cyclic and regenerative process. The gasifier unit consists of three vessels. The first vessel has top oil vaporising chamber and a bottom chequor filled steam preheater. The middle vessel contains catalyst (magnesia with 1 % lime as the active agent) and the last vessel is the chequor filled air preheater.

During the run, steam is admitted to the bottom of the steam preheater and the oil is sprayed in the vaporising chamber. The mixture of highly superheated steam and oil vapors pass downwards through the catalyst chamber where gasification occurs. The hot gas is sent to the last vessel to preheat the air during blow. Finally, the gas enters the wash box and the condenser.

During blow the air is sent counter current to the gas direction during run period. The carbon deposits on the catalyst and the walls of the vaporising chamber are burnt off. Some heating oil is supplied above the catalyst chamber to maintain 1000°C. Waste gas is routed through steam preheater and a waste heat boiler.

Cycle time is about 4 minutes and all fuel oils can be used as feedstock to produce oil gas with a C.V. of 4500 kcal/Nm^3 (similar to coke oven gas). Typical composition of oil gas produced by this process is given below.

CO_2	= 5.8%	C_mH_n	= 4.2%
O_2	= 0.4%	CO	= 8.3%
H_2	= 50.3%	CH_4	= 16.2%
N_2	= 4.8%		
Gross C.V.		= 4700 kcal/Nm^3 (dry).	
Sp. Gravity (Air = 1)		= 0.52	

This gas can be substituted for coal gas in a town gas supply system in countries with traditional supplies of coal gas in pipes (for domestic ovens/heating).

The vessel arrangement in Segas Process is shown in the Fig. 34.2

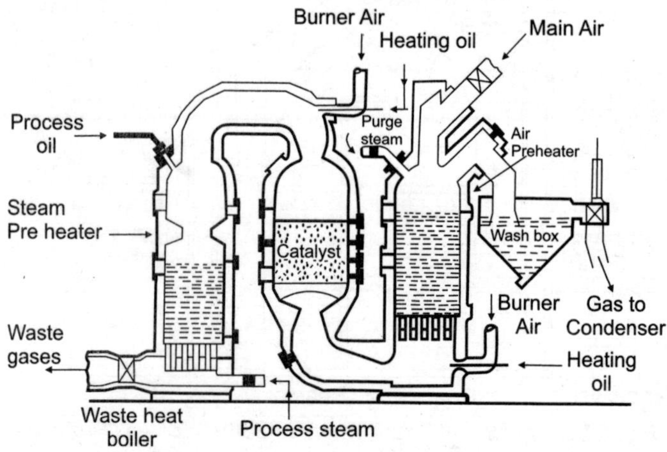

***Fig. 34.2.** Segas Process—vessel arangement.*

A flow diagram of Segas process during heating phase and gasification phase is shown in the Fig. 34.3.

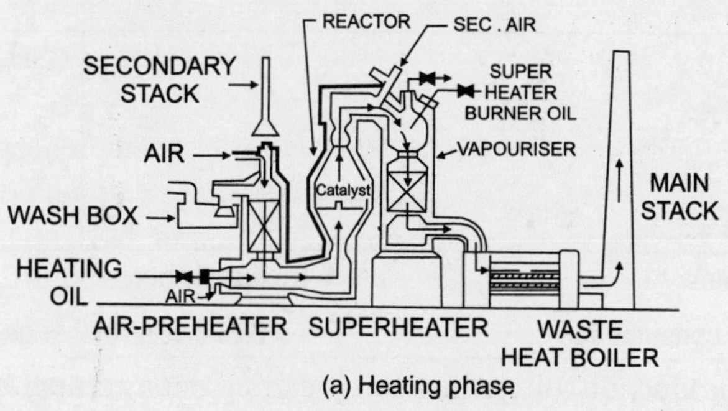

(a) Heating phase

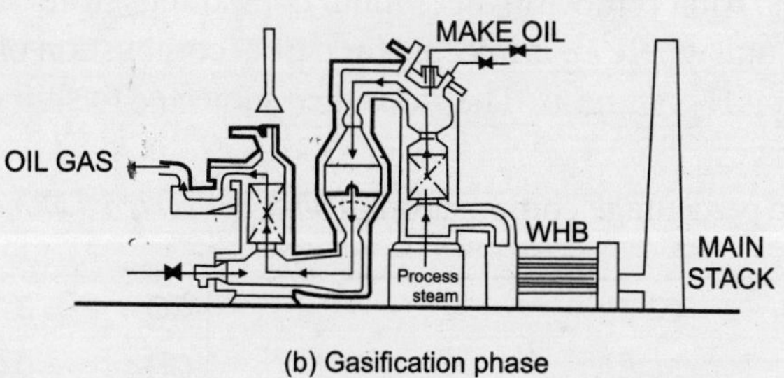

(b) Gasification phase

Fig. 34.3. Flow diagram of Segas process.

ONIA-GEGI PROCESS

This is also a catalytic and cyclic process similar to the Segas process. It uses a nickel catalyst and produces oil gas of C.V. about 4500 kcal/Nm^3 and Sp. gravity of 0.5. This can be substituted for coal gas in town gas supply system.

Feed for Onia-Gegi-process (developed in France) for town gas manufacture ranges from light hydrocarbons to heavy oil. Oil gas is produced using steam as gasifying agent following the reaction

$C_mH_n + \underline{n}H_2O = nCO + \underline{(1/2\ m+n)}\ H_2$. In addition, carbon formation and shift conversion reactions also occur.

Since the reaction between steam and oil is endothermic, a heating phase follows the gas making phase when a small quantity of oil is burnt with air to provide heat as well as to burn away sulphur and carbon deposited on catalyst bed.

The main units in Onia-Gegi plant are nickel catalyst filled main vessel, an air regenerator and a steam regenerator. In the heating phase, air admitted through the top of the regenerator moves down & mixes with the products of combustion of fuel (oil/tar) and air admitted to the bottom section of the regenerator. The mixture of hot air & the gas moves up the catalyst bed through steam regenerator and waste heat boiler & is vented through stack. The heating phase (which is half of total cycle time) raises the catalyst bed temperature to 800-900°C. The process oil is sprayed downwards from top of steam regenerator. Steam and vaporised process oil move downward through the catalyst bed and the formed gases pass through the air regenerator & wash box.

A typical % composition of *town gas* produced using heavy fuel oil is as follows :

CO_2	= 10.2	C_mH_n	[illegible]
CO	=17.2	H_2	[illegible].8
C_nH_{2n+2}	=13.5	O_2	[illegible]
N_2	=9.4		
Calorific Value	= 4316kcal/Nm^3		
Process oil consumption	= 0.594 litre/Nm^3 of gas		

For *Synthesis gas* production, oil gasification is carried out a comparatively higher temperature than that used for town gas manufacture mainly to bring down the methane ($C_n H_{2n+2}$) content of the reformed gas to 7-8%. After removing the sulphur compounds from the reformed gas, it is led to a catalyst filled cracking unit where air is injected for partial combustion of residual hydrocarbons in gas and providing N_2 for NH_3 synthesis. The gas is then subjected to shift conversion for conversion of CO to CO_2.

A typical volumetric percentage composition of *shift converted synthesis gas* is as follows :

CO_2	= 22.7	CO	= 3
H_2	= 56.5	CH_4	= 0.2
N_2	= 9.4	C_mH_n	= nil
Fuel oil consumption	= 30 gm/Nm^3 of gas		
Calorific value	=1570 kcal/Nm^3		

A schematic diagram of Onia-Gegi oil gas plant is given in Fig. 34.4.

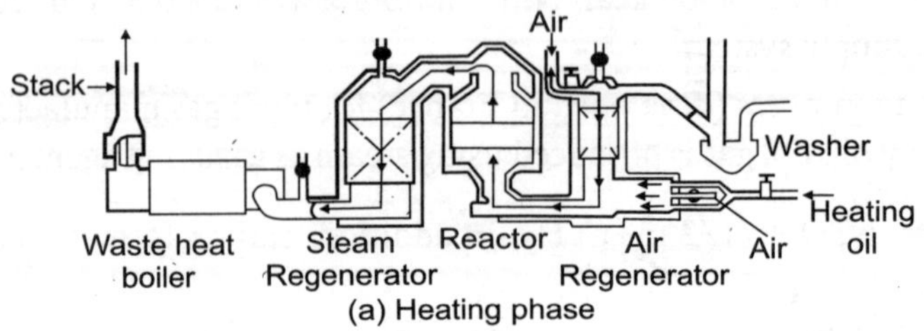

(a) Heating phase

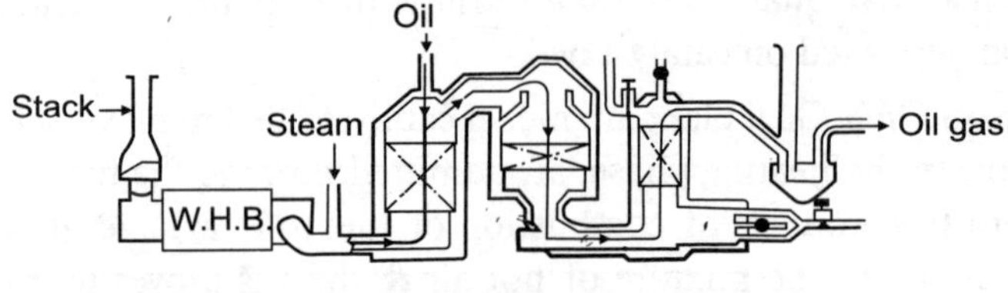

(b) Gasification phase

***Fig. 34.4.** Onia-Gegi oil gas plant.*

STEAM REFORMING OF NAPHTHA

Desulphurised naphtha is gasified/reformed using steam in a tubular reactor packed with nickel catalyst.Since sulphur deactivates nickel catalyst, naphtha needs desulphurisation before steam reforming.

The main reforming/gasification reactor is in the form of three rows of steel tubes charged with the catalyst. They are heated in a furnace at 960°C. The flue gases are used to generate waste heat steam at 20 atm, superheat the reaction steam and preheat the combustion air.

A mixture of desulphurised naphtha vapor, surplus hydrogen and preheated steam enters the reactor from top. The reaction is carried out at a pressure upto 28 atm and temperature upto 840°C.

In a plant for manufacture of town gas, the gas leaves the reactor at 700°C and 19 atm. The sensible heat of the gas is recovered in a waste heat boiler. It is then passed through *shift converter* ($CO+H_2O=CO_2+H_2$) where CO is reduced to *3%*. The sensible heat of the gas is extracted in a waste heat boiler, economiser & naphtha preheater before it is further cooled and washed with a carbonate lye for CO_2 removal. Finally gas is enriched (by natural gas/refinery gas) and supplied to consumers for use as fuel gas. When intention is to produce synthesis gas, then a secondary reformer is used to modify the exit gas from the primary reformer. Air is mixed with primary gas burning a part of combustible and raising the gas to 1200°C at which methane-steam reforming is complete over a special nickel catalyst. The gas leaves the secondary reformer at 900°C. It is then passed through a waste heat boiler before further treatment for ammonia synthesis.

Steam reforming of naphtha is done mostly for fuel gas production and partly for synthesis gas production in fertiliser plants as was earlier done in *Rourkela fertiliser plant* of SAIL.

A simplified flow sheet for steam reforming of naphtha by *ICI process for town gas production* is given in Fig. 34.5.

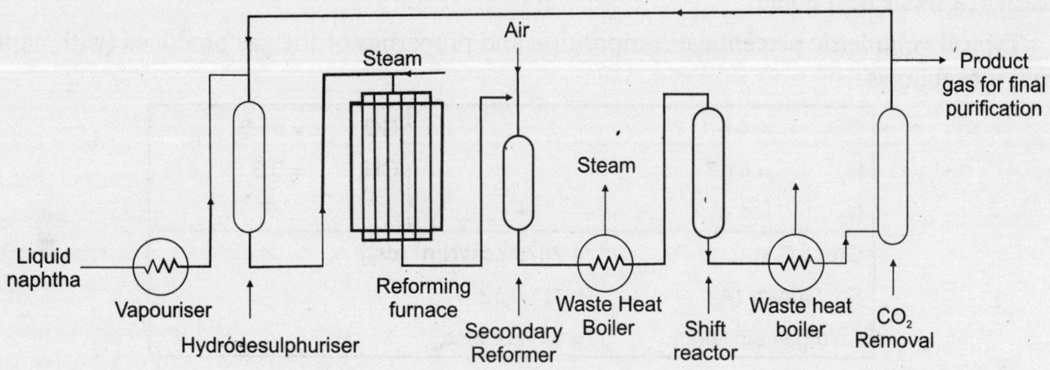

Fig. 34.5. *ICI Process for town gas production by steam reforming of naphtha.*

Another flow sheet for steam reforming of naphtha by *ICI process for ammonia synthesis gas production* is given in Fig. 34.6.

PARTIAL COMBUSTION PROCESS (SHELL GASIFICATION PROCESS)

Mainly used for synthesis gas production and also used for fuel gas production, this is a continuous, autothermic and non-catalytic process.

The feed, oxygen & steam are separately preheated to about 300°C and ·en brought together into the reactor which is essentially a burner where oxygen & steam are mi immediately before partial combustion and a turbulence is set up in the chamber. The brick li cylindrical reactor operates at 1000—1500°C and 10-40 atm. The exit gas is a mixture of

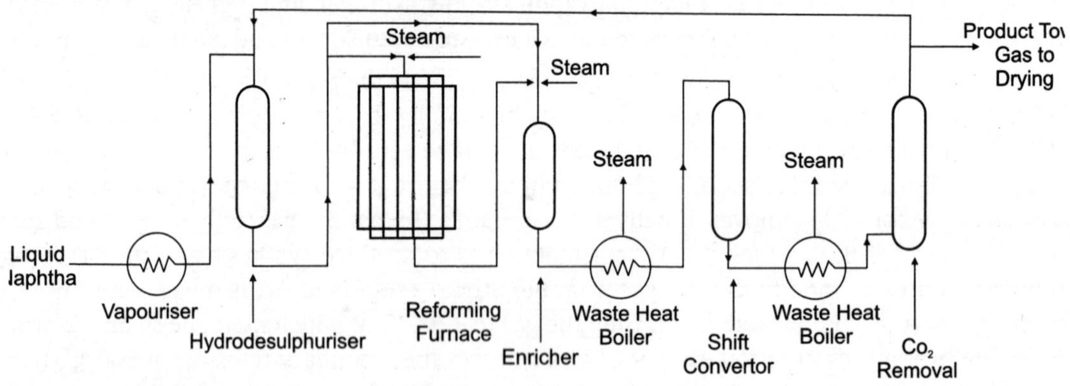

Fig. 34.6. *ICI process for anmonia synthesis gas production by steam reforming of naphtha.*

H_2, CO, CO_2 and small quantity of methane and solid carbon (which is removed by water washing). CO is shift converted and CO_2 is removed. Hot gas leaving at 1000-1200°C produce steam in a waste heat boiler.

Typical volumetric percentage composition and properties of the gas produced (with naphtha feed) is as follows :

CO_2	= 4.8	CO	= 41.8
H_2	= 51.7	CH_4	= 0.3
N_2	= 1.4		
Gross C.V	= 2870 kcal/Nm^3 (dry)		
Sq. Gravity (Air	= 1) =0.52		
Cold gas efficiency	= 84% of feed		

Capital cost of this installation is high.

A flowsheet of shell Gasification Process is given in Fig. 34.7.

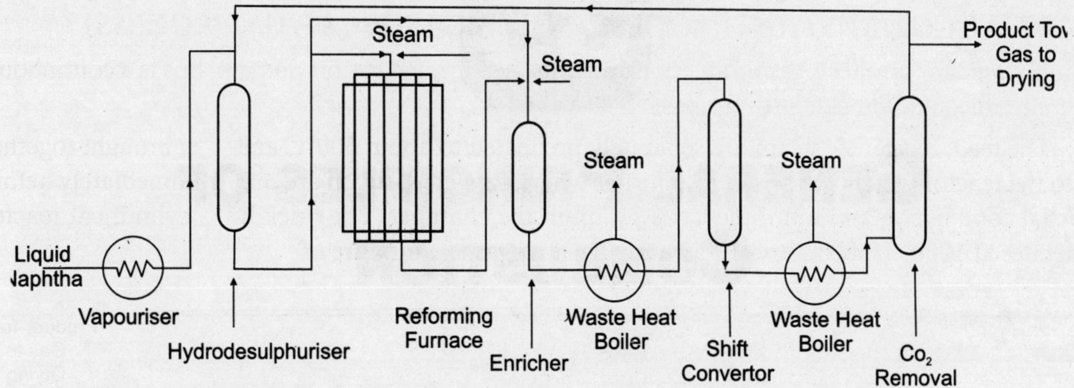

Fig. 34.7. *Shell gasification process.*

HYDROGENATION PROCESS

Hydrocracking of crude oil or heavy fuel oil is used for its gasification. In this process and the oil gas produced can be used as rich fuel gas or further steam reformed into synthesis gas or the lean gas component of town gas.

Preheated feed and hydrogen rich gas are fed into a fluidised coke bed reactor at 700-900°C and upto 50 atm where aliphatics crack and aromatics are dealkylated to give benzene, toluene, naphthalene as byproducts.

Typical gas volumetric percentage composition & properties (produced by hydrogenation of crude oil) are as follows :

CO_2=0.2% CO=2.9% CH_4=45.9%	C_mH_n=0.3% H_2=36.3% C_2H_6=13.5 and N_2=0.9
Gross C.V. = 7800 kcal/Nm^3(dry) Sp. Gravity = 0.44 Cold gas efficiency = 73% of feed.	

Benzene, toluene and naphthalene are recovered as by products.

35

General Principles of Combustion

Principles of Combustion

All conventional fossil fuels (solid, liquid or gaseous) contain basically carbon, hydrogen and sulphur which, when burnt react with oxygen of air to form carbon dioxide, carbon monoxide, water vapour and sulphur dioxide respectively.

The solid fuels are burned in beds in lump (or pellet form) or in pulverised form suspended in the air stream.

The liquid fuels are burned either by vaporising and mixing with air before ignition (when they behave like gaseous fuels) or in the form of fine droplets which get evaporated while mixing with the air stream during burning.

The gaseous fuels are either burned in burners where the fuel & air are premixed or the fuel & air flow separately into a furnace and simultaneously mix together as combustion proceeds. The first type of burning gives a premixed flame whereas the second type is called burning with a diffusion flame.

All the fuels contain basic elements such as carbon, hydrogen, sulphur or its compounds. The combustion of fuel is described with the help of few simple chemical equations as given below :

$$C + O_2 = CO_2 - 97650 \text{ kcal/kmol} \rightarrow \text{exothermic reaction}$$

$$C + \frac{1}{2} O_2 = CO - 29430 \text{ kcal/kmol} \rightarrow \text{ exothermic reaction}$$

$$CO + \frac{1}{2} O_2 = CO_2 - 68220 \text{ kcal/kmol} \rightarrow \text{ exothermic reaction}$$

$$C + CO_2 = 2CO + 38790 \text{ kcal/kmol} \rightarrow \text{ endothermic reaction}$$

$$H_2 + \frac{1}{2} O_2 = H_2O - 57810 \text{ kcal/kmol} \rightarrow \text{exothermic reaction}$$

$$S + O_2 = SO_2 - 69800 \text{ kcal/kmol} \rightarrow \text{exothermic reaction}$$

$$C + H_2O = CO + H_2 - 23380 \text{ kcal/kmol} \rightarrow \text{ endothermic reaction}$$

$$CH_4 + 2O_2 = CO_2 + 2H_2 - 192400 \text{ kcal/kmol} \rightarrow \text{exothermic reaction}$$

$$C_2H_2 + \frac{5}{2}O_2 = 2CO_2 + H_2O - 312400 \text{ kcal/kmol} \rightarrow \text{exothermic reaction}$$

$$H_2S + \frac{3}{2}O_2 = H_2O + SO_2 - 124850 \text{ kcal/kmol} \rightarrow \text{exothermic reaction}$$

$$C_mH_n + (m + n/4)\,O_2 = m\,CO_2 + n\,H_2O - \text{exothermic reaction}$$

$$C_nH_{2n+2} + \frac{3n+1}{2}O_2 = nCO_2 + (n+1)\,H_2O - \text{exothermic reaction}$$

Combustibles present in gaseous fuels are CO, H_2, CH_4, paraffins, unsaturated hydrocarbons etc. Average molecular formula for unsaturated hydrocarbons and paraffins may be assumed as follows for combustion calculations :

Unsaturated H-C (Hydrocarbon):

$C_{2.5}H_5$ - for coke oven gas, oil gas, carburetted water gas & horizontal retort coal gas

C_2H_4 - for producer gas from coal.

C_4H_8 - for vertical retort coal gas and low temperature coal gas.

Paraffins :

$C_{1.25}H_{4.5}$ - for low temperature coal gas.

$C_{1.1}H_{4\,2}$ - for oil gas and carburetted water gas.

Considering the reaction, $CH_4 + 2O_2 = CO_2 + 2H_2O$. We can say that 1 kilomol of methane (CH_4) reacts with 2 kilomols of oxygen (O_2) to produce 1 kilomol of carbon dioxide (CO_2) and 2 kilomols of water (H_2O).

The kilogram mole (called kilomol or kmol) is used in combustion calculation as the unit of weight. The molecular weight of any substance in kilogram (kg) represents 1 kilomol (or kmol). Thus 1 kmol of hydrogen has a mass of 2 kg and 1 kilomol of carbon has a mass of 12 kg. Thus 1 kmol of methane (CH_4) has a mass of 16 kg (12 + 4).

Hence for the above reaction it can be said that 16 kg of methane reacts with 64 kg of Oxygen (since 1 kmole of oxygen weighs 32 kg) to produce 44 kg of carbon monoxide and 36 kg of water (1 kmol of water weighs 18 kg).

Combustion Process :

In the combustion process the rapid chemical combination of oxygen with the combustible portion of the fuel results in heat release. The most common fuels contain carbon and hydrogen, either as elements or as parts of compounds. These combustible elements and compounds react with oxygen to form carbon dioxide (CO_2) and water vapour (H_2O). Carbon monoxide (CO) can also be formed. Sulphur is usually present in some fuels and sulphur's combustion products can cause corrosion and environmental pollution. Combustion of fuel produces flue gases which are also called exit gas, exhaust gas or chimney gas.

The Requirements for Combustion are :

— Fuel — Oxygen

— *The 3 "Ts" (a)* Time *(b)* Temperature *(c)* Turbulence

Fuel. The most commonly used fuels are oil, gas, coal and lignite but other are wood, asphalt, black liquor, peat, domestic waste etc. As the prices of primary fuels continue to increase, more applications using secondary fuels are being encountered. Many industrial plants use their locally produced secondary fuels to reduce their primary energy costs *e.g.* pulp mills, integrated steel works, oil refineries etc.

Oxygen. The normal source of oxygen for combustion is air. Air is a mixture of nitrogen, oxygen and water vapour with very small amounts of inert gases which are usually ignored. Air contains 79% nitrogen by volume and 21% oxygen. Nitrogen does not enter into the combustion process and this means that 79% of the air supplied to the combustion process does not take part in the reaction but has to be heated along with the oxygen to sustain the burning process losing efficiency. A large mass of air must be delivered to the combustion process to get the required amount of oxygen. The dilution effect of the nitrogen causes many problems in terms of lost efficiency.

The 3 "T's" for Combustion :

Time. Sufficient time must be available for complete combustion to occur. This time period is significant because of the dilution effect of the nitrogen in the air.

Temperature. When the fuel reaches its ignition temperature, oxidation is greatly accelerated. The fuel/air mixture must be heated to ignition temperature to promote combustion. The approximate ignition temperatures for the common fuels are :

Oil = 260°C, Wood = 290°C, Bituminous Coal = 400°C, H_2 = 565°C, CH_4 = 705°C, CO = 650°C, S = 245°C

Turbulence. It is vital that the fuel & air are thoroughly mixed otherwise complete combustion cannot take place and turbulent mixing is the best approach to this problem.

Perfect combustion occurs when just enough air has been supplied to burn all the fuel. To ensure that all the fuel has been consumed, additional air is supplied. This extra air is called *excess air.*

Fuel combustion efficiency (or Fuel efficiency). Fuel combustion efficiency depends upon the air-fuel ratio and the exhaust flue gas temperature. It is defined as,

$$\eta = \frac{x}{Q} \times 100,$$ where, η = Fuel efficiency, %

Where, Q = Fuel consumption in furnace, kcal/hr

x = Required heat value in the furnace, kcal/hr

'x' is calculated from the heat in the furnace structure plus heat in the furnace stock (charge) under operating conditions.

Fuel efficiency (η) increases with lowering in the exhaust flue gas temperature and decreases with increase in air-fuel ratio as shown below in Fig. 35.0 for heavy fuel oil. Fuel combustion efficiency is not a fixed quantity for a particular fuel, but varies with the operating conditions like sensible heat of stock, C.V. of fuel, exhaust gas temperature and air-gas ratio .

There is a difference between fuel efficiency and furnace efficiency. Fuel efficiency is concerned with the effective fuel utilisation rate while furnace efficiency is the ratio of heat received in the stock to the total heat input in the furnace.

In fuel combustion system, if there is high loss of combustibles with the flue gas and/or solid refuse, the efficiency of fuel combustion (η) is low which is calculated as below :

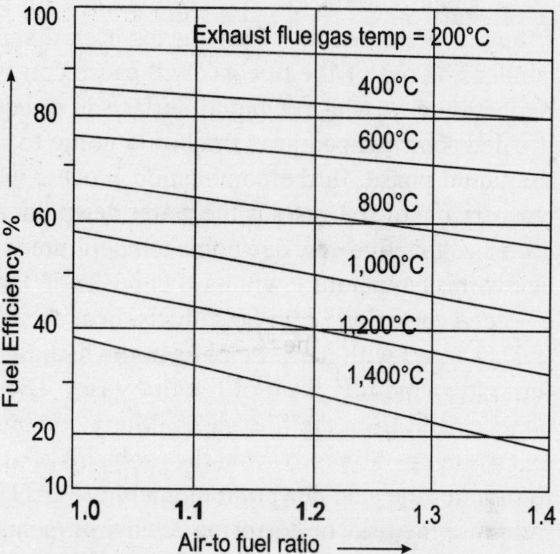

Fig .35.0: Fuel efficiency Vs. Air to Fuel Ratio for fuel oil

$$\eta = 1 - \frac{(\text{Potential heat in flue gas} + \text{Potential heat in refuse})}{\text{Potential heat in fuel}} \times 100$$

where, Potential heat in fuel = heat of combustion (*i.e.* calorific value) of the fuel.

Major fuel Efficiency Losses. The heat carried away by the flue gas is a function of the weight of gas per unit of time, the gas temperature and its specific heat.

Sensible heat loss = $F_g \times S_g \times (T_g - T_a)$

where, F_g = Flue gas weight/hour, S_g = Specific heat of the flue gas

T_g = Flue gas exit temperature, T_a = Combustion air temperature

This expression shows that there are *two possibilities for reducing the heat loss i.e.* by reducing the

- gas exit temperature and
- weight (or mass) of the gas going up the stack.

Losses Due to High Exit Flue Gas Temperature. In combustion systems, high exit flue gas temperature wastes valuable energy and a large increase in efficiency can be obtained by reducing the exit gas temperature. A reduction of approximately 22°C in flue gas temperature results in a 1% increase in efficiency. A heat exchanger either to heat the combustion air or heat water is the first step in improving combustion efficiency. However, when burning fuels containing sulphur the maximum amount of heat that can safely be recovered from the flue gases is a function of the acid dew point temperature of the flue gases. If the flue gases are cooled to below the dewpoint temperature the sulphuric acid will be condensed on the surface of the ducts, fans, chimney etc. resulting in serious corrosion. The reactions are, $S + O_2 = SO_2$, and some of this SO_2 is further oxidised to SO_3. The SO_3 reacts with the water vapour in the flue gas to form sulphuric acid.

$$SO_3 + H_2O = H_2SO_4$$

A small quantity of SO_3 combining with large concentrations of water vapour produces a strong acid condensate. That means even a small amount of SO_3 combined with water vapour condensing out of the flue gas will cause corrosion. The dewpoint temperature is defined as the temperature at which a liquid surface is in equilibrium with the gas immediately above it. At the dewpoint temperature, the liquid going to the gas phase is exactly balanced by gas going to the liquid phase. In the combustion process where fuel contains no sulphur then the dewpoint temperature of the gases is the water dewpoint (*i.e.* around 50 to 80°C). If 20 ppm of SO_3 is now added to this flue gas, dewpoint temperature will increase to about 145°C and any surface at or below this temperature will have deposits of a strong solution of sulphuric acid. This acid causes cold-end corrosion *i.e.* in fans, ducts, heat exchangers and stacks. The acid dewpoint temperature (ADT) varies with the % sulphur in the fuel, the excess air level, the condition of the burners and the water vapour content of the flue gases. The ADT limits the amount of heat which can safely be recovered from the flue gases. Many combustion systems are deliberately operated with high exit gas temperatures to avoid the problems of acid condensation but this is very wasteful. It is easy to monitor the acid dewpoint temperature (ADT) and so maximise the heat recovery benefits and minimise the risks of corrosion & environmental pollution caused by acid smuts. The condensed acid is absorbed by the carbon deposits on surfaces of the ducts, stack etc. and eventually this carbon comes out of the stack damaging property, crops etc. in the vicinity of the plant.

Losses Due to Excess Air. A combustion process with high excess air increases the amount of flue gas going up the stack. When we reduce this excess air, we reduce the volume of flue gas and the hot gases spend more time in contact with the heat transfer surfaces. This reduces the exit flue gas temperature. So to reduce the mass of flue gas we must reduce the excess air to the level so as to put the correct amount of combustion air into the process. If we supply too little air, we get incomplete combustion and the formation of carbon monoxide followed by smoke & solids emission. If we supply too much air we have higher *stack losses* and other problems such as high dewpoint temperatures depending on the fuel being used.

The aim is to supply just enough air to burn all the fuel and then maintain this high combustion efficiency by controlling the fuel/air ratio. Relation between energy losses and exit flue gas temperature is shown in Fig. 35.1.

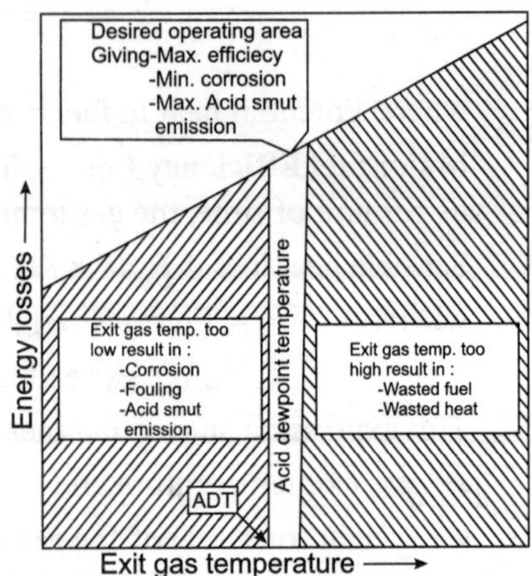

Fig. 35.1. *The relationship between energy losses and exit gas temperature.*

Measurement of Oxygen in the Flue Gas. The most common method of monitoring the quantity of excess air (the fuel/air ratio) is to measure the oxygen content of flue gas and to adjust the fuel/air ratio to obtain an excess air level as low as possible whilst still maintaining complete combustion. The common methods of measuring oxygen in the flue gas use either *(a)* A sampling system where the gas is removed from the duct, cleaned, cooled and dried before being pumped to the gas analyser. There are many maintenance problems with the sampling system due to blockages, corrosion and leaks etc. or (b) Non-sampling system (called an in-situ measurement) using a probe with a zirconia detector. This probe is mounted on the duct with the detector in the flue gas and no sampling, no gas cleaning or cooling systems are required.

The advantage of oxygen measurement, is that it offers the possibility of directly inferring the efficiency of the overall combustion system since the total excess air used is directly related to the excess oxygen.

- **Perfect combustion** is achieved when all the fuel is burned using only the theoretical amount of air, but perfect combustion can rarely be achieved in practice.
- **Good/complete combustion** is achieved when all fuel is burned using the minimal amount of excess air (over and above the theoretical amount of air needed to burn the fuel.). Complete combustion with minimum excess air is always our goal since heat losses due to high excess air in flue gases are unaffordable and unacceptable from the point of view of efficiency.
- **Incomplete combustion** occurs when all the fuel is not completely burned and escapes as CO in flue gases or as unburnts in refuse, both of which result in higher losses & low efficiency.
- **Flue gas analysis** of combustion product is important as it helps to achieve efficient combustion conditions by excess air control and reduction of CO in flue gases.

Using gas analyzers, the excess air quantity can be established from measurement of oxygen or carbon dioxide. Based on oxygen value in flue gas, excess air is given as :

$$\% \text{ of excess air} = 100*\% \text{ of } O_2/(21\text{-}\% \text{ of } O_2).$$

The relation between % O_2 in flue gas and excess air is tabulated below. The advantage of oxygen based analysis is that it is the same for any fuel or fuel combination.

Oxygen content in flue gas and excess air used for combustion

% O_2 in flue gas	% excess air
1	5
2	10.52
3	16.67
4	23.53
5	31.25
6	40
7	50
8	61.7
9	77
10	90.9
11	110

The Disadvantages of using the Oxygen Signal for Fuel/Air Ratio Control. For monitoring combustion efficiency, oxygen measurement is essential. For combustion control however there are a number of disadvantages. This is because the combustion process is affected by a number of variables other than the fuel/air ratio, e.g.

- **Air/Fuel mixing**. Changes in the firing rate involve changes in the flow of air & fuel to the burner and this causes changes in flow pattern in the flame affecting the mixing of fuel & air. Poor mixing will result in part of the combustion air never being involved in the combustion process and the burner would then have to operate at higher excess air levels to ensure complete combustion.

- **Burner Deterioration**. Burners require proper maintenance and they may become blocked or suffer from wear. They are manufactured to fine tolerances and any deterioration will give less efficient combustion.
- **Air and Fuel Distribution**. Multi-burner systems rely on each individual burner receiving the same ratio of fuel and air under all operating conditions. Inevitably some burners will operate with less excess air than others and since to avoid producing smoke the overall combustion air flow must be sufficient to satisfy the most fuel rich burner, the total combustion system operates with higher air levels than the ideal. The result is higher sensible heat losses and lower efficiency.
- **Fuel Oil Atomisation**. In the case of oil firing it is necessary to atomise the oil into a fine spray. Variations in the atomisation efficiency produce changes in the ability of the fuel oil to burn quickly and cleanly. As the atomisation efficiency falls, more air is required to ensure complete combustion.

As the combustion conditions change requiring changes in excess air levels to ensure complete combustion, the point for the excess oxygen controller should be varied accordingly. The exact variations in combustion conditions are not known and to avoid the possibility of making smoke the combustion system operates at an excess oxygen level high enough to maintain complete combustion under the worst possible conditions and the energy is thus wasted because of too high an excess air level.

Another problem is that of air in-leakage into the combustion gases. Most combustion systems operate under negative pressure and this can cause air to leak into the combustion gases. The oxygen (21%) in this "tramp" air can produce serious errors in the oxygen analysis. When determining the system efficiency by measuring the excess oxygen and flue gas temperature this "tramp" air must be measured since process energy is being used to heat this air up to the flue gas temperature.

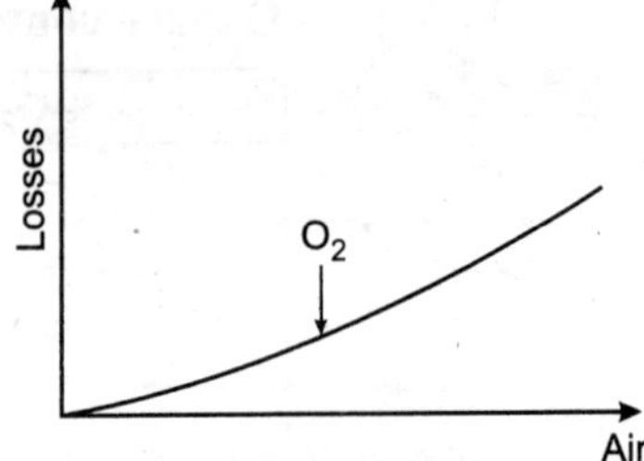

Fig. 35.2. Losses due to excess air

Oxygen is not a product of the combustion and it is assumed that all the oxygen we are measuring has come from the combustion air not consumed in the combustion process which may not be the case always.

To decide the optimum oxygen content of the flue gases, the relationship between excess oxygen and the losses can be seen in Fig. 35.2. The curve is not steep but the losses can be large and there is no reference point that informs us about the correct amount of excess air. More data than is shown in Fig. 35.2 is required to decide the optimum oxygen level.

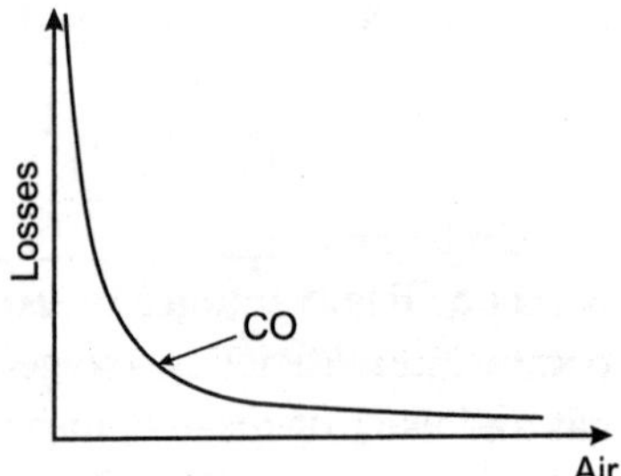

Fig. 35.3. Losses due to CO

Fig. 35.3 shows the curve for carbon monoxide versus air which is a very steep curve and on combining Figs. 35.2 and 35.3 for CO and oxygen we can generate a total stack losses curve as shown in Fig. 35.4, which gives us following informations :

- The excess oxygen does not go to zero before CO is produced. We have CO & excess oxygen at the same time and it is incorrect that all the oxygen goes to zero before the onset of CO. At the minimum of the losses curve (maximum efficiency) or say an oil-fired boiler, we require 1% excess oxygen and somewhere between 50 & 300 ppm of CO.

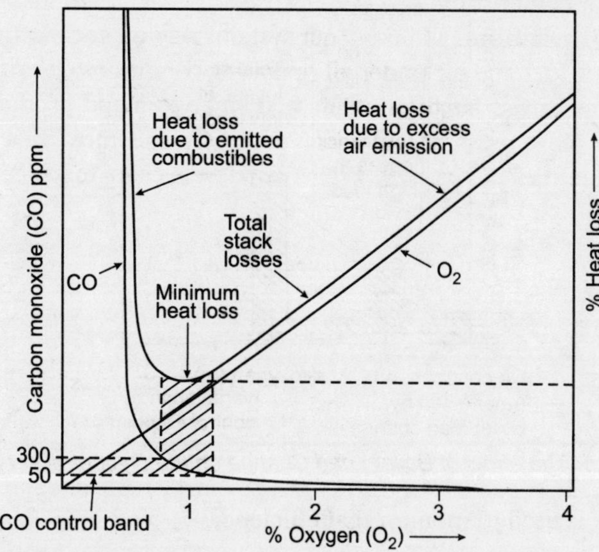

Fig. 35.4. *The relationship between O_2, CO and heat loss*

- The different slopes of the heat loss curve are significant. The curve is much steeper on the high CO side (air deficient or excess fuel) than on the high oxygen side (excess air). The loss in energy when burning carbon to carbon monoxide instead of carbon dioxide is clearly shown. However operating with 50-300 ppm of CO is causing an insignificant reduction in efficiency. If the oxygen & CO curves at a lower load of boiler is plotted and combined them with those of Fig. 35.4 we obtain Fig. 35.5. Here we can see that the minimum stack losses now occur at about 2.4% oxygen (compared to 1% at high load). This is because of the effects of fuel/air mixing etc. discussed earlier. If we are using the oxygen signal to trim the fuel/air ratio, we would require to keep changing the set point of the oxygen controller to ensure minimum losses at all loads. However, if we examine the CO curves in Fig. 35.5, it can be seen that the minimum losses occur between 50-300 ppm of CO for both loads thereby operating the boiler in a safe condition and at maximum efficiency, irrespective of the load or the fuel. (See Fig. 35.5).

Under any set of combustion conditions, the fuel/air ratio required to give maximum efficiency is that at which a small amount of CO is produced, and as CO is a combustion product we can obtain information about the combustion conditions. It is useful to monitor the CO in the flue gas and use this measurement for optimising the fuel/air ratio. The advantage of CO is that it can automatically compensate for changes in combustion conditions, for changes in calorific value of the fuel etc. It is therefore possible to obtain the highest efficiency possible under any set of conditions by controlling the fuel/air ratio to give the correct level of CO in the flue gas.

It is not advisable to operate a combustion system on only the CO measurement or on only oxygen measurement. It is essential to use both oxygen and CO together. The CO measurement tells us about conditions in the combustion chamber and also about the fuel/air ratio. We can use CO signal to trim the fuel/air ratio and hold the CO in the 50-300 ppm range (depending on the fuel being used) and operate at the highest possible efficiency under any set of conditions.

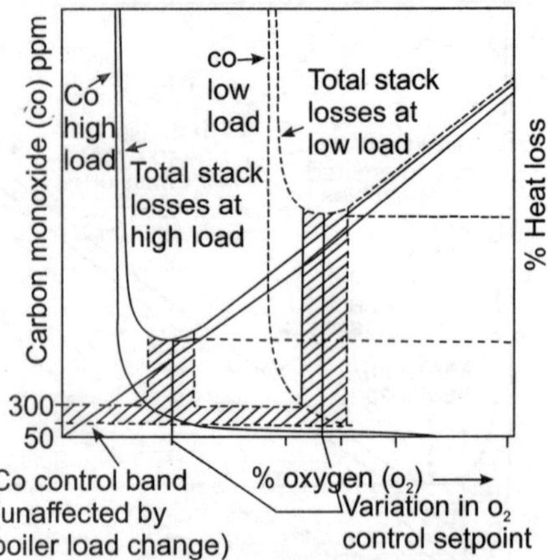

Fig. 35.5. The effect of Boiler Load Change on the O_2 control set point.

The oxygen analyser is used to monitor that efficiency.

By using both CO and oxygen, it is possible to detect any abnormal combustion conditions as shown in Figs. 35.6 and 35.7. A large water-tube boiler is burning gas with low excess air and both the CO and oxygen are being measured. The curves are very different and if we had been using the oxygen signal to trim the fuel/air ratio in Fig. 35.6, we would have about 100 ppm of CO at 0.5% oxygen. If we had the same set-point of 0.5% on Fig. 35.7, we would be operating with 2000 ppm of CO and wasting energy.

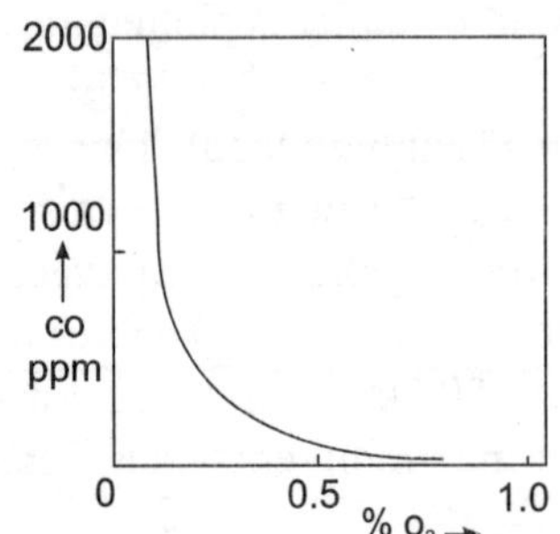

Fig. 35.6. *CO/O_2 Relationship for Good operating conditions.*

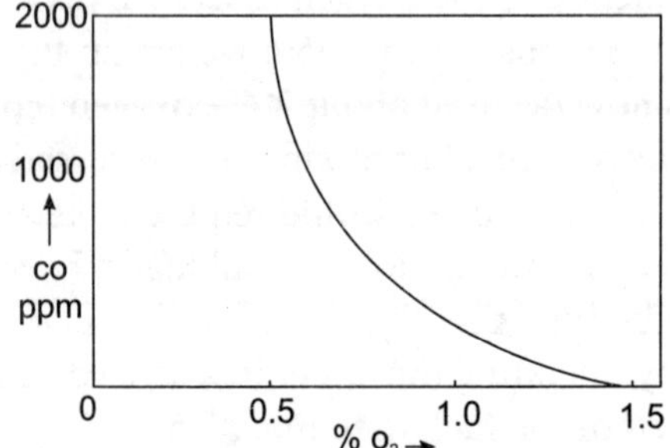

Fig. 35.7. *CO/O_2 Relationship adverse operating conditions (higher levels of CO occurring at low O_2 levels).*

Requirement of Theoretical air and Excess Air for Combustion. For the combustion (i.e. burning of any fuel) the common oxidiser is air. Air is roughly a mixture of 21% oxygen and 79% nitrogen (on volume basis). That means one kmole of oxygen is accompanied by 79/21=3.76 kmole of nitrogen. The chemical equation for combustion of carbon with air is written as :

$$C + O_2 + 3.76\ N_2 = CO_2 + 3.76\ N_2$$

The minimum amount of air required for the complete combustion of a fuel (which is calculated based on the chemical equations involved in the combustion) is called the theoretical air or stoichiometric air. But in practice, it is difficult to achieve the complete combustion with theoretical quantity of air. Therefore fuels are supplied with excess air to burn it completely. Hence, the air supplied in excess of the theoretical air requirement is called the excess air and is expressed as a percentage of the theoretical air. By weight, air comprises of 77% N_2 & 23% O_2, while by volume, air comprises of 79% N_2 & 21% O_2.

Solid fuels require maximum amount of excess air, whereas the gaseous fuels require minimum amount of excess air. Liquid fuels requirement of excess air is in between the solid and gaseous fuels.

Quick Methods of Combustion Calculations.

Formulae derived for theoretical air requirement, flue gas analysis etc. from fundamentals of combustion stoichiometry are given below :

Solid and Liquid Fuels :

Let H_2O, C, H, O, N and S represent the weight per cent of moisture, carbon, hydrogen, oxygen, nitrogen and sulphur in fuel respectively.

Total O_2 required for complete combustion of the fuel

$$= C/12 + H/4 - O/32 + S/32 \text{ kmol/100 kg fuel}$$

$$= 0.224 (C/12 + H/4 - O/32 + S/32) \text{ Nm}^3\text{/kg fuel. (1 kg mole =22.4 Nm3).}$$

Volume of theoretical air required :

$$A_0 = 4.76 (C/12 + H/4 - O/32 + S/32) \text{ kmol/100 kg fuel}$$

$$= \frac{4.76 \times 22.4}{100}\left(\frac{C}{12}+\frac{H}{4}-\frac{O}{32}+\frac{S}{32}\right)$$

$$= 0.089C + 0.267 H + 0.033S - 0.0330 \text{ Nm}^3\text{/kg fuel}$$

Volume of theoretical wet flue gas produced

$$Y_0 = x+y+z$$

where, x = Volume of CO_2 produced, Nm^3/kg fuel

$= 22.4/100 (C/12 + S/32) = 0.01866C + 0.007 S$ Nm^3/kg fuel

y = volume of H_2O produced, Nm^3/kg fuel $= 22.4/100 (H/2 + H_2O/18)$

$= 0.112H + 0.012H_2O$, Nm^3/kg fuel

z = volume of N_2 produced, Nm^3/kg fuel

$$= 0.79 A_0 + \frac{N \times 22.4}{28 \times 100} = 0.79 A_0 + 0.008 N, \text{ Nm}^3\text{/kg fuel}$$

Volume of theoretical dry flue gas (V_0') = x + z Nm^3/kg of fuel

where, z = volume of N_2 produced, Nm^3 / kg of fuel

Analysis of theoretical flue gas (wet), per cent:

$$CO_2 = \frac{100x}{x+y+z};\ H_2O=\frac{100y}{x+y+z};\ N_2=\frac{100z}{x+y+z}$$

Orsat analysis of theoretical flue gas (dry), per cent:

$$CO_2 = \frac{100x}{x+z};\ N_2=\frac{100z}{x+z}$$

Excess air is

calculated from the flue gas analysis :

For *complete combustion*, % excess air = $\frac{100\,O_2}{0.27\,N_2 - O_2}$

For incomplete combustion, % excess air = $\frac{100\,(O_2 - 0.5\,CO - 0.5\,H_2 - 2CH_4)}{0.27\,N_2 - (O_2 - 0.5\,CO - 0.5\,H_2 - 2CH_4)}$

Where O_2, CO, H_2, CH_4 and N_2 are the per cent of the respective components in the flue gas.

Excess air factor (n) = $\frac{A - A_0}{A} \times 100$ per cent excess air

Air supplied : $A = (1 + n)\,A_0$ Nm^3/kg fuel

Flue gas produced : $V = V_0 + nA_0$ Nm^3/kg fuel

Dry flue gas : $V' = V0' + nA_0$ Nm^3/kg fuel

Flue gas analysis (per cent wet) :

$$CO_2=\frac{100x}{V};\ H_2O=\frac{100y}{V};\ N_2=\frac{100\{(z+0.79)(A-A_0)\}}{V};\ O_2=\frac{100n\,A_0}{V}\times 0.21=\frac{21n\,A_0}{V}$$

Gaseous Fuels

Let CO_2, CH_4, $C_n\,H_{2n+2}$, C_nH_m, H_2, CO and N_2 represent the volume or mole per cent of the respective components in the fuel gas as fired.

Volume of theoretical air,

$$A_0=0.0476\left\{2CH_4 + \frac{3n+1}{2}CnH_{2n+2} + \left(n+\frac{m}{4}\right)C_nH_m - O_2 + 0.5\,CO + 0.5\,H_2\right\}$$

Nm^3/Nm^3 fuel gas

Volume of theoretical wet flue gas (V_0) produced :

V_0 = x + y + z Nm^3/Nm^3 fuel gas

where, x = volume of CO_2 produced, Nm^3/Nm^3 wet fuel gas

= 0.01 (CO + CH_4 + nC_nH_m + nC_nH_{2n} + CO)

y = volume of H_2O produced, Nm^3/Nm^3 wet fuel gas

= 0.01 {2 CH_4 + (n + 1) C_nH_{2n+2}+ m/2C_nH_m + H_2}

z = Volume of N_2 from air and fuel, Nm^3/Nm^3 wet fuel gas

= $0.01\ N_2 + 0.79\ A_0$

Volume of theoretical dry flue gas,

$$V_0 = x + z\ Nm^3/Nm^3 \text{ fuel gas}$$

If the gaseous fuel contains significant quantities of nitrogen e.g. producer gas and blast furnace gas, the formula for excess air calculation is based on dry gas analysis.

For complete combustion :

$$\text{Per cent excess air} = \frac{100 \times O_2'}{0.27\ (N_2' - N_2'/V') - O_2'}$$

For incomplete Combustion :

$$\text{Per cent excess air} = \frac{100\ (O_2 - 0.5\ CO + 0.5\ H_2' - 2CH_4')}{0.27\ (N_2' - N_2/V) - (O_2' - 0.5\ CO' - 0.5\ H2' - 2CH_4')}$$

where the symbols with 'prime' denote the per cent composition in dry flue gas.

MAXIMUM CO_2 CONTENT IN DRY FLUE GAS

The content of CO_2 in theoretical dry flue gas is approximately constant for a given class of fuels. This is the maximum CO_2 content of dry flue gas obtained from that class of fuel. The value of $(CO_2)_{max}$ i.e. (CO_2) theoretical may also be calculated from the flue gas orsat analysis.

- **For complete combustion**

$$\%\ (CO_2)_{max} = \frac{100.CO_2}{100 - (O_2 + 3.76\ O_2)} = \frac{100.CO_2}{100 - 4.76\ O_2}$$

- **For incomplete combustion**

$$\%\ (CO_2)_{max} = \frac{100\ (CO_2 + CO + CH_4)}{100 - 4.76\ (O2 - 0.5\ CO - 0.5\ H_2 - 2\ CH_4)}$$

The CO_2 and O_2 contents of flue gas from the complete combustion of a particular class of fuel vary with excess air in a regular manner. This is shown by Ostwald chart which is based on the above two equations for $(CO_2)_{max}$ of fuel has its own Ostwald chart. Maximum CO_2 content of flue gas of a particular fuel along with its actual CO_2 content may be used to determine the excess air ratio, n. This is given by the relation :

$$n = V_0/A_0\{(CO_2)_{theor}/(CO_2)_{actual}\} - 1$$

For solid & liquid fuels, $V_0'/A_0 = 1$, $n = \dfrac{(CO_2)_{theor}}{(CO_2)_{actual}} - 1$

CO_2 content of theoretical dry flue gas is almost constant for a given class of fuel and has been shown in Table 35.1.

Table 35.1. Theoretical CO_2 percent in Flue Gas for Various Fuels

Fuel	(CO_2) max i.e. (CO_2) theoretical, per ce
High temperature coke	20
Bituminous Coal	18.7
Lignite, semi-anthracite,	19.5
Anthracite, semi-coke	19.3
Wood, Charcoal	20.5
Bagasse	20.3
Fuel oil/Coal tar	15.8
Town Gas CN (net calorific value)	
= 4900 kcal/Nm^3	11.0
= 4450 kcal/Nm^3	12.0
= 4000 kcal/Nm^3	13.0
Coke Oven Gas	11.0
Producer gas from coal	19.0
Producer gas from coke	20.5
Blast Furnace gas	24.5
Gasoline	14.9
Kerosene	15.1
Natural gas	12.2
Refinery gas and oil gas	12.8
Propane	13.7
Butane	14.0
Methane	11.8
Water gas	20.5
Carburetted water gas	16.8

However, the above value of $(CO_2)_{max}$ may vary slightly with the composition of fuel.

The two limiting points in the Ostwald chart are the $(CO_2)_{max}$ and the $(O_2)_{max}$ which is 21% (oxygen in air). The flue gas composition containing excess air is shown by a point in the line joining the two limiting points (perfect combustion line). Lines of incomplete combustion in the Ostwald chart is represented by contents of H_2 and CO in the flue gas. Constant air ratio lines are also included in the chart.

$$\text{Air ratio} = \frac{\text{Theoretical air}}{\text{Excess air}} = \frac{1}{1+n} \text{ where, n = excess air factor}$$

If 'n' is known, then volume of actual air supply (A) and actual flue gas produced (V) can be calculated. Ostwald charts are useful in checking flue gas analysis and in combustion calculations. Each flue gas is represented by a single point in the chart meant for the particular fuel.

Ostwald chart for coke oven gas of following typical analysis (by volume %) is shown in Fig. 35.8.

Analysis of coke oven gas :

H_2 = 47.3	CH_4 = 34.2
CO = 6.9	C_2H_4 = 2.6
CO_2 = 2.2	O_2 = 0.3
N_2 = 6	H_2S = 0.5

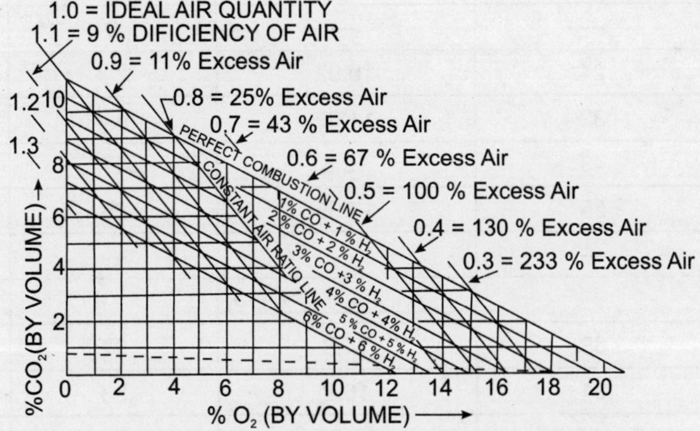

Fig. 35.8. *Ostwald Chart for Coke-Oven Gas.*

Significance of CO_2% in flue gas. The flue gases obtained from any fuel when burned with the theoretical amount of air (without any excess) contains fixed and characteristic percentage of CO_2 (known as theoretical % of CO_2 which is also the maximum % of CO_2 that the gas can have).

When pure dry carbon is burnt with theoretical air, the 0_2 of air is replaced quantitatively by CO_2 so that its theoretical % is 21% by volume. Any air supplied in excess of that theoretically required for any fuel reduces the CO_2% in the flue gas proportionally.

The theoretical CO_2% used in calculations is on dry basis because in practice the CO_2% in flue gas is determined by some form of gas analysis apparatus (say Orsat apparatus) in which the gas is stored and analysed over solutions of reagents in water. Usually, its water vapor content is small (1.3% at 15°C) and is constant during the determination, so that the result obtained is CO_2% by volume on dry basis.

Theoretical CO_2% for fuels containing hydrogen or hydrocarbon gases is lower than that for carbon because air supplied burns hydrogen to form water which condenses, leaving nitrogen to dilute the flue gases formed from carbon alone. SO_2 present in flue gases is analysed and reported as CO_2 (in Orsat apparatus), but this introduces only a small error in low sulphur fuels. Any carbon monoxide produced reduces the proportion of CO_2 formed and must be allowed for when the amount is known.

Excess air for combustion as a function of % CO_2 in flue gas for coal, fuel oil, blast furnace gas and coke oven gas of typical compositions are tabulated in Table 35.2.

Table 35.2. Excess air for Combustion as a Function of % CO_2

% CO_2	Coal		Fuel oil	
	% of excess air	Combustion air, m^3	% excess air	Combustion air
18.7	0	8.81	—	—
18	2.9	9.08	—	—
17.5	5.8	9.32	—	—
17	9.7	9.66		—
15.4	—	—	0	11.01
15	24	10.92	2.4	11.28
14	32	11.62	9.4	12.06
13	43	12.6	17	12.9
12	55	13.65	26	13.9
11.8	—	—	—	—
11.55	—	—	—	—
11	68	14.8	37	15.1
10	84	16.2	53	16.85
9	105	18.05	74	19.2
8	130	20.25	87	20.6
7	162	23.1	112	23.4
6	203	26.7	146	'27.1
5	265	32.3	194	32.4
4	350	39.6	267	40.5

% CO_2	Blast furnace gas		Coke oven gas	
	excess air, %	Combustion air, m^3	excess air, %	Combustion air, m^3
25.1	0	0.687		
25	1	0.693		
23	21	0.825		
21	43	0.98		
19	70	1.17		
17	105	1.4		
15	148	1.7		
13.97	—	—		
13.71	—	—		
13	205	2.09		

12	240	2.4		
11	281	2.62		
10	332	2.97		
9.4	—	—	0	4.1
9	393	3.387	4	4.27
8	470	3.92	16	4.75
7	567	4.58	31	5.37
6	700	5.497	51	6.2
5	883	6.75	81	7.42
4	1160	8.64	126	9.11

Empirical equations for combustion parameters calculation of theoretical air and flue gas volume :

From the knowledge of net calorific value of fuels (which are related to gross calorific values), their theoretical air requirement and volume of theoretical flue gas that they produce can be calculated using the statistical relationships given below.

Conversion of gross calorific value (C_G) into net calorific value (C_N) :

Solid fuels, kcal/kg:

Coal : $C_N = 0.954\ C_N + 110$

Coke : $C_N = C_G - 40$

Liquid fuels, kcal/kg:

$C_N = 0.75\ C_G + 2000$

Gaseous fuels, kcal/Nm³:

Producer gas : CN = $0.95\ C_G$

Water gas : $C_N = 0.915\ C_G$

Blast furnace gas : $C_N = 0.99\ C_G$

Coke oven gas : $C_N = 0.895\ C_G$

Calculation of A_0 (*i.e.* theoretical air required for combustion) and V_0 (*i.e.* theoretical volume of flue gas produced) can be done using the formulae as given in Table 35.3.

Table 35.3. Formulae for Calculation of A_0 and V_0

Kind of fuel	Nm^3/kg (for solid and liquid fuels) and Nm^3/Nm^3 for gaseous fuels	
	Theoretical air required, A_0	Theoretical flue gas volume, V_0
Solid Fuels		
C_N< 5500 kcal/kg	$1.01 \times \frac{C_N}{1000} + 0.5$	$0.89 \times \frac{C_N}{1000} + 1.65$

C_N> 5500 kcal/kg	$1.01\times\frac{C_N}{1000}+0.55$	$\frac{C_N}{1000}+0.9$
Liquid Fuels	$0.85\times\frac{C_N}{1000}+2$	$1.11\times\frac{C_N}{1000}$
Gaseous Fuels **Producer gas**	$0.875\times\frac{C_N}{1000}$	$0.640\times\frac{C_N}{1000}+1.11$
Water gas, carbureted water gas, Coke oven gas	$1.09\times\frac{C_N}{1000}-0.25$	$1.35\times\frac{C_N}{1000}-0.67$
Blast furnace gas	$0.795\times\frac{C_N}{1000}$	$0.634\times\frac{C_N}{1000}+1$
Mixed gas (B.F. gas + C.O. gas)	$1.07\times\frac{C_N}{1000}-0.16$	$1.04\times\frac{C_N}{1000}+0.7$

Primary and Secondary Air

Air admitted through the lump fuel bed and required to burn the fixed carbon of the coal or for effecting the preliminary part of the operation of burning the fuels is called the primary air. In case of pulverised fuel, liquid fuel and gaseous fuel burning, the primary air also acts as a carrier of these fuels to the furnace as it premixes with them. In case of solid lumpy fuels, it passes through the bed of fuel.

The rest of the air necessary for completing the combustion is admitted separately in the furnace and is called secondary air.

Secondary air helps in completing the combustion of the volatile matter (e.g. smoke) expelled from solid fuel during the preliminary stage of combustion. Secondary air is normally admitted over fuel bed.

Complete combustion of solid fuels is thus done by the use of both primary and secondary air. Efficient combustion is effected by correct ratio of the two types of air. Smoke coming from chimney is mainly due to less supply of secondary air.

Dew Point of Flue Gases and Effect of Sulphur & Ash in Fuel

Flue gases flow through ducts and various other parts of the furnace system. It is detrimental to have water deposited on them. Knowledge of dew point of flue gases is therefore desirable.

The dew point of a gaseous mixture is the temperature at which its water vapour starts condensing *i.e.* the temperature at which the saturation pressure for water equals the partial pressure of water vapour in the gaseous mixture.

Most of the fuels contain sulphur in some degree. Removal of sulphur completely from some fuels is technically possible, though not economically feasible. The presence of sulphur in fuels always pose three major problems :

- Low temperature corrosion of economisers, recuperators etc.
- High temperature fouling & corrosion on super heaters, economisers and other equipments & structures in flue path.
- Atmospheric pollution through generation of sulphur-dioxide and trioxides.

The concentration of SO_3 in the flue gas is related to the sulphur content of the fuel varying from about 0.0012% by volume of SO_3 for a fuel containing 1% of sulphur to about 0.004% by volume of SO_3 for a fuel containing 5% sulphur.

During combustion, the sulphur in the fuel is converted largely to SO_2, which is quite corrosive in the presence of moisture and is mainly responsible for low temperature corrosion.

Roughly 1% of the SO_2 is converted to SO_3 under normal conditions of high excess air but it increases slightly more catalytically at high gas temperature when vanadium oxides in ash deposits and iron oxides on metal surfaces act as catalysts, in the presence of sufficient oxygen for conversion of SO_2 to SO_3.

As sulphur content in the fuel increases, the acid dew point also increases which retards the heat recovery from the waste flue gas in order to protect the equipment from corrosion and damage.

Sulphur % in Fuel	Acid Dew Point, °C
1	130
2	133
3	138
4	143
5	150

Flue gas temperature going out of the chimney should always be more than the acid dew point.

The theoretical water dew point is approximately 45°C. The ability to predict acid dew point for given situations would be extremely helpful in anticipating problems and determining corrective actions. Vanadium is found in most of the residual oil and remains in ash during combustion. No economical means for removal of vanadium, from residual fuel oil are available. Vanadium & sodium may cause corrosion and fouling of superheater parts in boilers, gas turbine blades & recuperators where operating temperatures are in excess of 650°C.

Upon combustion, these elements form low melting ash deposits on superheater tubes, supports, hangers, spaces etc. And ash build up depends upon the aerodynamics of the flue path. The vanadium oxide adheres to the colder metallic surface, which insulates the tubes locally, impedes heat transfer making the outer layer hotter. This further traps larger percentage of ash which normally does not adhere to clean tubes, restricting the flue passage and results in fouling of tubes.

Many plants experience the nasty problem of acid stack emission fall out. When flue gas temperature approaches the acid dew point & condensation occurs, ash particles agglomerate and some pass out the stack, while others stick to any surface with which they may be subjected to. Fallout of greenish acidic deposits, small specks of black particles and even large masses of highly corrosive, sticky materials damage automobiles, homes & even agricultural crops.

In case of boilers when sulphur bearing fuel is burned, most of the sulphur is oxidised to form SO_2 and is emitted out through the chimney along with the flue gases. A small amount of SO_2 is further oxidised in the combustion chamber to SO_3 which in the presence of water vapor in the flue gas forms gaseous sulphuric acid (H_2SO_4) which will condense on and corrode any surface (e.g. air preheater, induced draft fans, stacks etc.) which is at a temperature lower than the acid dew point temperature.

A further problem occurs when carbon particles are deposited onto the condensed acid leading to a building up of large acid saturated carbon 'flakes' resulting in "fouling". These flakes may break free and be emitted from the chimney causing pollution/corrosion damage to surrounding buildings & vehicles.

SO_3 in a boiler's flue gas does not pose problems until it combines with water vapor and condenses as sulphuric acid in the flue or stack. Ideally, the flue gas temperature should be maintained slightly above the 'acid dew point temperature' (ADT) to minimise acid condensation whilst giving maximum possible efficiency. However, to avoid internal acid attack, many boiler systems are operated with flue gas temperature far in excess of the acid dew point. Although effective in minimising acid condensation, operating at excess temperature is a waste of heat & fuel and leads to poor efficiency.

The formation of gaseous H_2SO_4 in the combustion process is almost unavoidable, hence monitoring of acid dew point helps in controlling the corrosive effect on 'cold end' components below acid dew point temperature thereby minimising cost and increasing the plant efficiency.

The relationship between desired operating point and flue/exit gas temperature for an operating furnace/boiler has been shown below in Fig. 35.9.

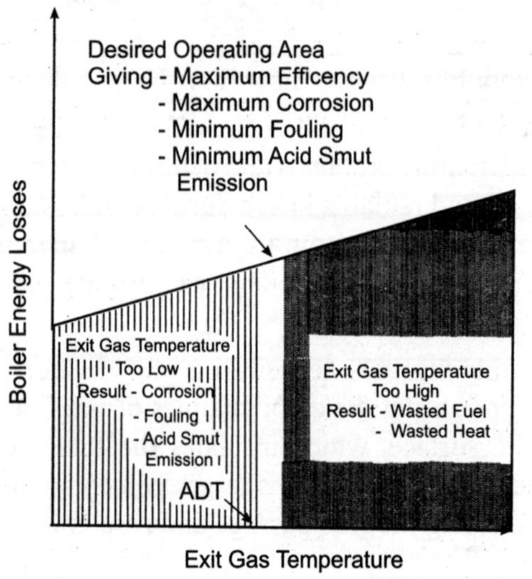

Fig. 35.9. *The relationship between Desired Operating Point and Exit Gas Temperature.*

Fig. 35.10. *The relationship between SO_3 and ADT*

The relationship between SO_3 concentration in flue gas and acid dew point temperature (ADT) is shown in Fig. 35.10.

From the above figure, we find that SO_3 has a significant effect on ADT i.e. just 5 ppm of SO_3 will raise the ADT to 120°C.

Fig. 35.11 gives the theoretical relationship between H_2S04 vapor in flue gas and ADT which shows that ADT increases with an increase of H_2S04 vapor (and therefore $SO_{3)}$ content in the flue gas. It also shows the effect of changes in water vapor content in the flue gas upon the acid dew point temperature (ADT).

Effect of increase in sulphur content in the fuel (fuel oil) and % O_2 in the flue gas (signifying % excess air used for combustion) on ADT is shown in Fig. 35.12.

Pollution control laws enforces to limit ground level concentration of SO_2 forcing either use of low sulphur fuel or installation of pollution control equipment to limit SO_2 emission in the atmosphere.

Fig. 35.13 shows the relationship between the sulphur content in fuel and SO_2 content in the flue gas generated by its combustion.

Fig. 35.14 shows the relationship between the sulphur content in fuel and conversion rate of SO_2 to SO_3. As for example, if the fuel contains 3% sulphur, then 1 to 3% of sulphur out of the total present in fuel is converted to SO_3. If the flue gas temperature is less than the dew point or if thepresent in flue gas contacts the low temperature heat exchanger walls, SO_3 & H_2O present in the flue gas combine producing sulphuric acid causing corrosion.

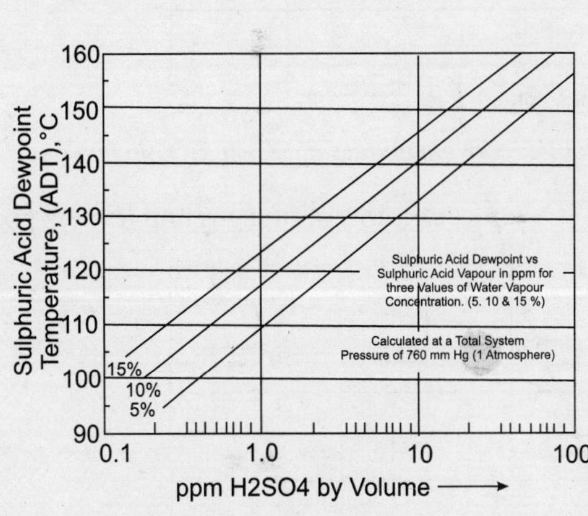

Fig. 35.11. (i) Effect of changes in water vapour content upon the Acid Dewpoint Temperature (ADT) (ii) Relationship between H_2SO_4 vapour and ADT.

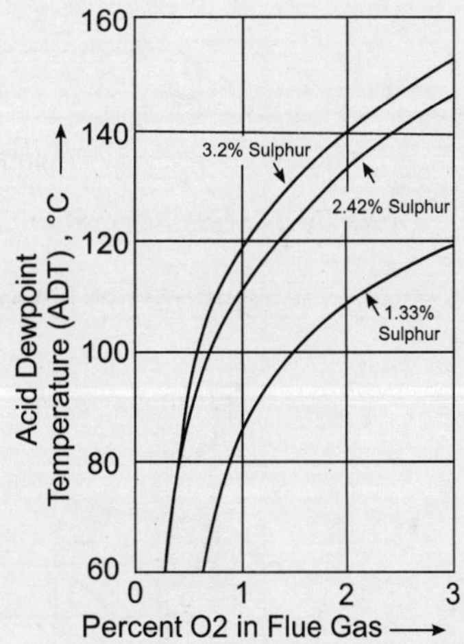

Fig. 35.12. Effect of excess oxygen in flue gas on the acid dewpoint temperature (ADT) when burning fuel oil with different sulphur levels.

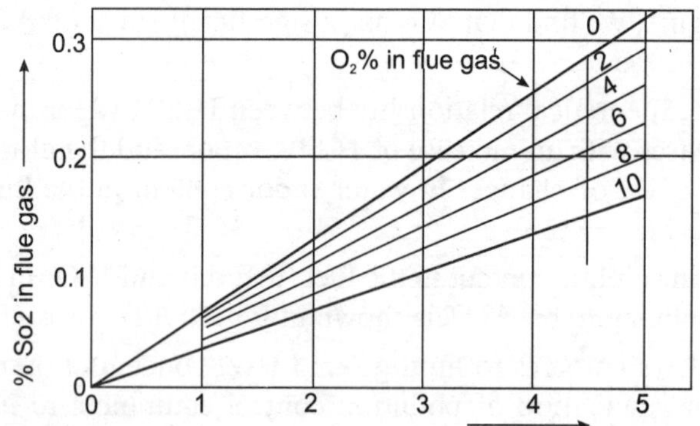

Fig. 35.13. Sulphur content in fuel Vs. SO_2 content in flue gas.

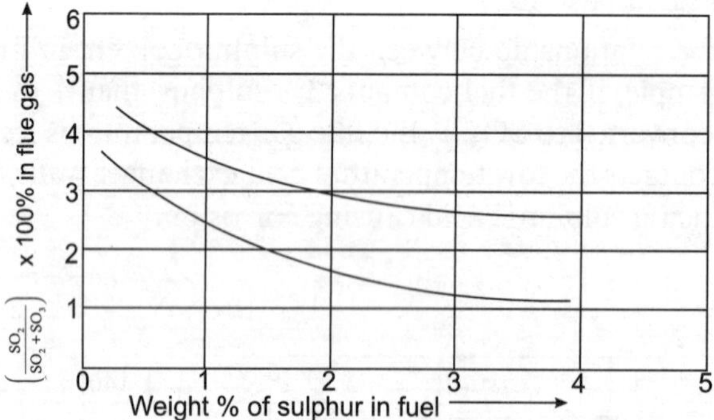

Fig. 35.14: Sulphur content in fuel Vs. conversion rate of SO_2 to SO_3 in flue gas.

Fig. 35.15 shows the relationship between SO_3 concentration in flue gas and its dew point

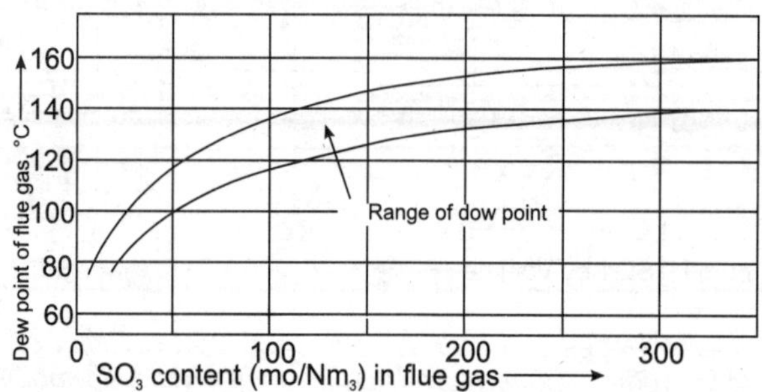

Fig. 35 .15. SO_3 content in flue gas Vs. dew point.

Problem 1. *Find the dew point of a flue gas (at 150°C) being exhausted out of chimney at 760 mm Hg pressure. The composition of the flue gas in volume % is as follows :*

CO_2 = 13%, SO_2 = 0.05%, O_2 = 6.75%, N_2 = 75%, H_2O = 5.2%, Total = 100%.

Solution. Since, the dew point of the flue gas (comprising of gaseous mixture) is the temperature at which its water vapor starts condensing i.e. at this temperature (i.e. at dew point), the saturation pressure for water equals the partial pressure of water vapour in the gaseous mixture (i.e. in flue gases).

Since, water vapor content of flue gas = 5.2%

Hence, partial pressure of water vapor in gaseous mixture

$$= \frac{5.2}{100} \times 760 = 39.5 \text{ mm Hg}$$

Now, the temperature at which water has the saturation pressure (aqueous tension) of 39.5 mm Hg = 33.8°C (found from thermodynamic tables)

Hence, the dew point of flue gas = 33.8°C. **Ans.**

Smoke — Causes, Remedy and Measurement

Smoke is caused by the incomplete combustion of volatile matter and consists mainly of tarry matter, carbon particles and ammonia salts. The following are the causes of smoke formation :

- Smoke is formed when the quantity of secondary air required for complete combustion is not sufficient enough for complete combustion of volatile matter.
- Smoke is also caused when the mixing of air & volatile matter is not proper and thorough.
- Excess of air causes smoke by reducing the temperature of combustible gases below their ignition temperature. Thus no ignition & no combustion of these combustible gases take place and they pass on the chimney without burning and smoke is caused.
- By coming into contact with cold surface, the flames may loose their heat and the combustion may not be complete with the result that smoke is produced.

The following are the methods for avoiding smoke *:*

- Smoke can be avoided by supplying secondary air for complete combustion.
- Thoroughly mixing the combustible gases and air.
- Pre-heating the secondary air.
- Providing sufficient space for combustion.

Smoke Measurement. The darkness of smoke coming out of chimney is judged by the use of 'Reingleman Chart'. There are four shades (no. 1 to no. 4) on this chart increasing in darkness/ blackness from no. 1 to 4. Reingleman chart no. 2 corresponds to 40% black smoke. The colour of the smoke is compared with the various shades of Reingleman chart and then reported as the chart no. on the Reingleman chart. Shade no. 2 or above on the Reingleman chart is defined as black smoke.

Another method of smoke measurement is by smoke density meter in which a photo-electric cell measures the reduction in intensity of a beam of light passing through the chimney or flue duct. This meter can be either indicating or recording type.

Free oxygen in flue gas and air requirement & flue gas production for various fuels of typical compositions have been tabulated in Table 35.4 and 35.5 respectively.

Table 35.4: Free Oxygen Content in Flue Gas for Different Fuels

% Excess Air	Gas Coal	Coal Middlings	LSHS[1]	Coal Tar	Furnace Oil	C.O. Gas	B.F. Gas	Producer Gas	L.D.[2] GAS	Mixed Gas (C.O. Gas + L.D. Gas)					
										L.D. Gas %					
										10	20	30	40	50	60
10	1.97	1.95	2.00	1.95	2.00	2.07	0.95	1.26	1.25	2.02	1.97	1.92	1.86	1.80	1.72
20	3.59	3.57	3.64	3.56	3.65	3.76	1.82	2.38	2.36	3.68	3.61	3.52	3.42	3.31	3.18
30	4.96	4.94	5.02	4.92	5.03	5.12	2.62	3.36	3.35	5.08	4.98	4.87	4.79	4.60	4.43
40	6.12	6.10	6.20	6.08	6.21	6.38	3.35	4.25	4.24	6.27	6.16	6.02	5.88	5.71	5.52
50	7.13	7.10	7.22	7.09	7.22	7.41	4.03	5.04	5.04	7.29	7.17	7.03	6.87	6.69	6.46
60	8.01	7.98	8.11	7.96	8.11	8.30	4.66	5.75	5.77	8.18	8.05	7.90	7.74	7.55	7.32
80	9.46	9.44	9.56	9.42	9.57	9.78	5.80	7.01	7.05	9.65	9.52	9.36	9.19	8.98	8.74

1. Low Sulphur Heavy Stock (furnace oil)
2. L.D. Converter or Basic Oxygen Furnace (BOF) Gas.

Table 35.5: Fuel Vs Air Requirement and Flue Gas Production for Various fuels

Fuels	Air Requirement ‘						Combustion Products					
	Theoretical		Air Volume with different % excess air, Nm^3 / kg				Theoretical		Flue gas volume with varying % of excess air, Nm^3 /kg			
SOLID FUELS	kg/ kg	Nm^3 / kg	10%	20%	30%	40%	kg/ kg	Nm^3 /kg	10%	20%	30%	40%
Coking Coal	9.88	7.35	8.09	9.56	11.03	14.70	10.55	7.62	8.35	9.83	11.30	14.97
Steam Coal	9.31	6.92	7.61	8.99	10.29	13.84	10.00	7.22	7.91	9.29	10.67	14.14
Coal Middlings	7.59	5.64	6.20	7.32	8.44	11.28	8.17	5.88	6.44	7.56	8.68	11.52
Gas Coal	9.91	7.37	8.11	9.59	11.07	14.74	10.64	7.71	8.45	9.93	11.41	15.08
Non-coking Coal	9.77	7.26	7.99	9.45	10.91	14.52	10.48	7.64	8.19	9.65	11.11	14.72
Coke	8.49	6.31	6.94	8.20	9.46	14.62	9.20	6.40	7.03	8.29	9.55	12.71
LIQUID FUELS												
Coal Tar	12.28	9.12	10.03	11.85	13.67	18.24	13.22	9.67	10.18	12.00	13.82	18.39
L.S.H.S.	14.04	10.42	11.46	13.54	15.62	20.84	15.01	11.14	12.18	14.26	16.24	21.56
Furnace Oil	13.70	10.20	11.22	13.26	15.30	20.40	14.63	10.79	11.81	13.85	15.89	20.99

GASEOUS FUELS	Nm³/Nm³						Wet	Dry	Flue Gas Volume (Wet)			
C.O. Gas	—	4.20	4.62	5.46	6.30	8.40	4.90	3.84	5.32	6.16	7.00	9.10
B.F. Gas	—	0.70	0.77	0.91	1.05	1.40	1.55	1.51	1.62	1.76	1.90	2.25
Producer Gas	—	1.15	1.27	1.51	1.75	2^30	1.99	1.78	2.10	2.33	2.56	3.14
L.D. Gas	—	1.27	1.40	1.65	1.91	2.54	2.04	2.01	2.16	2.42	2.67	3.31
Mixed Gas												
(L.D. Gas + C.O. Gas)												
10% L.D. Converter Gas	—	4.03	4.43	5.24	6.05	8.06	4.77	3.79	5.17	5.98	6.78	8.80
20% "	—	3.72	4.09	4.84	5.58	7.44	4.46	3.59	4.84	5.58	6.32	8.19
30% "	—	3.42	3.76	4.45	5.13	6.84	4.16	3.40	4.51	5.19	5.87	7.58
40% "	—	3.11	3.42	4.04	4.67	6.22	3.86	3.20	4.17	4.79	5.41	6.97
50% "	—	2.80	3.08	3.64	4.20	5.60	3.78	3.00	4.06	4.62	5.18	6.58
60% "	—	2.49	2.74	3.24	3.74	4.98	3.25	2.80	3.50	4.00	4.50	5.75

CHIMNEY GASES AND DRAUGHT

Chimney gases which are exhausted to the atmosphere through the chimney normally contains CO, CO_2, N_2, O_2, SO_2, H_2O etc.

The pressure required to supply air to a furnace and to remove the flue gases from the furnace is called the draught. It is measured in mm water gauge which is height of water in U tube gauge equivalent to the pressure in the furnace, flue or chimney. It is of two types namely natural draught and artificial (or mechanical) draught.

Natural draught. It is produced by a chimney. The resultant flow of the chimney gas is controlled by dampers. Natural draught produced by a chimney is proportional to the height of the chimney and to the density of the chimney gas. That means taller chimney and hotter chimney gas produces more draught. Both methods of obtaining more draught are costly due to :

- higher capital cost of taller chimney.
- the fact that in order to create draught we have to use heat of the chimney gas that can be profitably used for raising steam or other heating purposes.

Chimney height required for a given draught and temperature is calculated from the following formula :

$$H = \frac{h}{\left(\frac{7.6}{t_a + 460} - \frac{7.9}{t_g + 460}\right)}$$

where, H = height of the chimney, ft., h = draught required, inches of water gauge

7.6 = weight of 100 ft³ of air at 60°F, 7.9 = weight of 100 ft³ of chimney gas at 60°F

t_a = air temperature, °F, t_g = mean flue gas temperature, °F

If the mean flue gas temperature is not known, it can be calculated approximately by deducting 1°F from the temperature at the base of the chimney for every 3 ft of chimney height.

In fact, chimney is a device in which the draft is set up by the direct expenditure of the part of the heat of combustion of fuel. It depends for its action on the fact that a weight of the column of flue gases within the chimney is less than the weight of an equal column of cold air outside the chimney, the driving force is the difference between the two.

Chimney height can be alternatively calculated as below :

Actual natural draft = Theoretical draft produced by the manometric difference in static head of equal columns of atmospheric air and the flue gases - pressure losses in the flow due to friction.

or, $D = (\rho_g - \rho_a) H - (fv^2/2g)$ where, D = Draft, kg/m^2 = mm water column (MMWC)

ρ_a and ρ_g are flue gas and air densities (gas density is calculated at mean temperature of the chimney)

H = Chimney height, metres

f = friction factor $= 1 + 0.9\ H/d$ (where, d = chimney diameter, metre)

V = exit flue gas velocity, m/sec.

Velocity of the flue gases increases progressively with its mean temperature (in direct proportion to the square root of the difference in temperature) whereas the density decreases. It follows that draft is directly proportional to the height of chimney and increases with the temperature of the flue gas.

- Capacity of the chimney to pass gases varies directly with the cross-sectional area but only as the square root of the height.
- Although the possible draft increases with temperature of the gases, no increase in capacity results after the mean temperature reaches $2t_a + 273$°C (t_a = air temperature). The capacity of a chimney to pass gases is the maximum, when the mean temperature within it is about 315°C.

Artificial (or mechanical) draught. The natural draft of the chimney is limited by practical consideration so that draught for large furnaces is usually supplied by fans which is called artificial or mechanical draft. Natural draught operates entirely by suction, that is the chimney sucks or induces the flow of air or gases through the furnace. Apart from fans, artificial draught is created by steam jet ejector also.

Induced draught. In this system, a suction fan is installed at the base of the chimney to augment the natural draught of the chimney. The draught from the suction fan overcomes the resistance of the fuel bed to the passage of primary air and resistance of the furnace installation to the secondary air drawn over the fire bed. The furnace operates under suction in this case.

Forced draught. In this system, a fan is installed below, or in front of the grate to force primary air through the fire bed and secondary air over the grate. The furnace operates under pressure in this case.

Relative merits and demerits of induced draft & forced draft systems are described below :

The advantages of artificial draught over natural draughts are :

- Cheaper low grade coals/fuels can be used.
- Efficient combustion is more easily attained because of greater ease of air control.
- Furnace control is easier.
- Smaller chimneys can be used.

Artificial draught is of three types :

- Induced draught
- Forced draught
- Balanced draught.

Balanced draught. One fan is installed at the base of the chimney and another at the front end of the furnace system. The pressure and suction effects of forced draught fan & induced draught fan respectively balance in the furnace which operates at or near atmospheric pressure.

Table 35.6 gives the relative merits of induced draught and forced draught systems in a furnace.

Table 35.6: Relative merits of induced draught and forced draught systems

Induced draught	Forced draught
Furnace operates under suction so air leak inwards to the furnace reduces its thermal efficiency.	Furnace operates under pressure, so gas leaks out of the furnace thereby causing little loss of thermal efficiency. Also, there is a danger of blow out through front fire doors.
Fan operates hot so is liable to corrosion and out of balance due to accumulation of deposits.	Fan operates cold so no corrosion or deposits, hence longer life.
Greater horse power fan is required due to larger volume of gases handled.	It gives better control of distribution of air and higher rates of combustion than induced draught.

The best features of both the systems are obtained by balanced draught.

The amount of draught required depends upon :

- The nature of the fuel.
- The depth of fire to be maintained.
- Rate of combustion desired.
- Design of boiler furnace or the flues.
- The combined resistance of the ancillary plant such as economisers, air heaters, regenerators, recuperators, waste heat boilers etc.

FLAME CHARACTERISTICS

Flame is a burning mixture (or product of combustion) of a combustible gas or vapor and air. Solid fuels burn with a glow but with little flame. Main types of flames are :

- *Luminous flame.* Like candle, non-aerated oil, tar, coke oven gas flames etc.
- *Non-luminous flame e.g.* hydrogen, carbon monoxide, blast furnace gas.
- *Long (lazy) flames.* Like partially aerated flames. It is produced when gases are burnt with less quantity of primary air — non turbulent flow.
- *Short (intense)* flames e.g. strongly aerated flames. It is produced when gas is burnt with high amount and current of air as in the case of blow pipe. Air & gas are well mixed and delivered at high speeds with turbulent flow.

Flame Structure. The flame shape and size depends upon its type viz. premixed & diffusion, laminar & turbulent and also on the burner dimensions.

The premixed flame of a simple gas burner is shown in Fig. 35.16.

Here, d = nozzle diameter of the burner

Li = the distance converted by the initial boundary of the flame called the zone of inflammation (height of inner cone)

$Li + \delta + L_s = Lf$ = the total length that indicates the zone of the complete combustion.

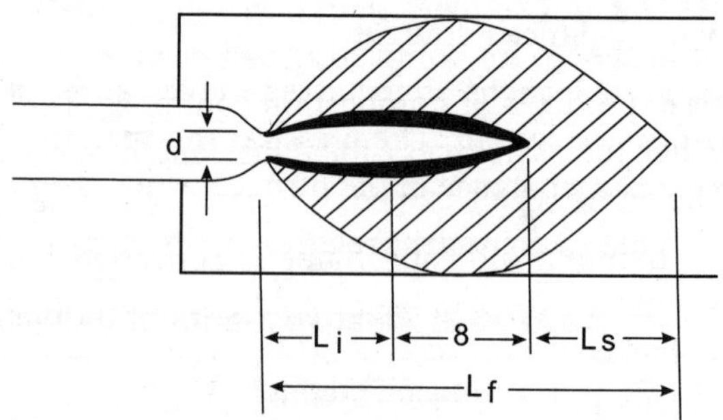

Fig. 35.16. *Premixed flame of a Simple Gas Burner.*

The inflammation zone is sharply pointed (in laminar flame) and has a thin surface (flame front). The flame front is distorted under turbulent conditions and has an appreciable width denoted by 'δ'.

The width of flame front (δ) increases in eddy diffusivity and decreases normal flame speed.

L_s = the distance between the visible front & the flame tip (it represents the boundary of complete combustion and is related mainly to kinetic properties of combustible mixture).

Combustion is slower in the zone, L_s, where diffusion of some air from the atmosphere is involved while it is most intense at the flame front.

'δ' = 0 for laminar premixed flame *i.e.* $L_f = Li + L_s$

L_i increases with increase in average linear gas velocity at the burner nozzle (V_1) and nozzle tube radius (r) and decrease in velocity of flame propagation (V).

$$L_i \approx \frac{V_1 . r}{V}$$

For turbulent flame, the velocity of turbulent flame propagation is approximately equal to the gas velocity (*i.e.* $V = V_1$) and hence $(L_i)_{turbulent} \approx r$

The height of inner cone remains almost constant for a turbulent flame while it increases with increase in gas velocity for a laminar flame.

In a simple laminar diffusion flame (shown in Fig. 35.17) the fuel gas and air flow (at the same velocity) separately through two concentric tubes, the inner tube supplying the fuel jet.

Process of gas-air mixing (achieved by either molecular diffusion or eddy diffusion depending on laminar or turbulent conditions of the flame) determines the structure of diffusion flame (not the velocity of flame propagation).

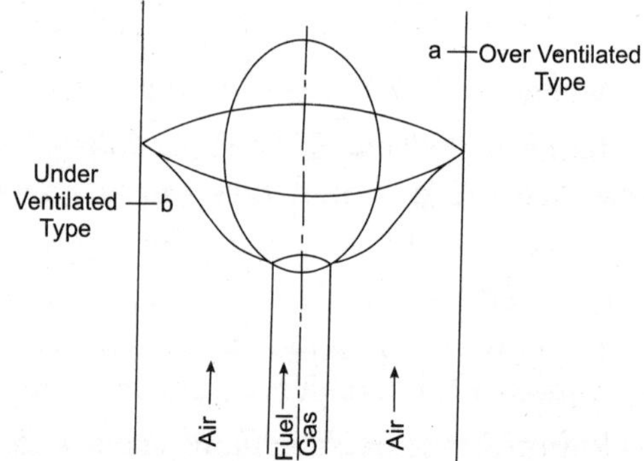

Fig. 35.17. *Simple Laminar Diffusion Flame.*

When air is in excess, over ventilated type *(a)* laminar diffusion flame is formed. With excess fuel gas, the flame forced is of under ventilated type *(b)*. The flame boundary is provided by the surface at which combustion is complete. Fuel and air are assumed to be in stoichiometric ratio at the boundary surface. When air is in excess, the surface closes in towards the axis as the diffusion of gas and air continues tinues. The flame length corresponds to the condition of arrival of sufficient air at the tip for completing combustion. No fuel gas exists beyond this point and the mixing of air & products of combustion continues. The normal flame has the shape of over ventilated type. When fuel is in excess, the flame length corresponds to the arrival of fuel gas at the outer tube wall. The flame is bell shaped.

Laminar diffusion flame is converted into the turbulent type by increasing the gas velocity beyond a critical value of cold Reynolds number (varying widely from 3500 to 10000) depending on fuel composition and quantity of primary air. However, length of a turbulent diffusion flame is directly proportional to the nozzle diameter but is independent of gas velocity. Flame length is a measure of the reciprocal of the combustion intensity (and therefore of heat release) and is useful in the design of combustion chamber.

Flame stability. A free burning flame is said to be stable if, there is no flash-back or blow-off i.e. the normal velocity of flame propagation of the mixture (V_n) is vectorially equal and opposite to the velocity of fuel-air mixture at the flame front (V_{av}).

The normal velocity of flame propagation, V_n depends upon the :

- nature of fuel
- composition and temperature of fuel-air mixture
- burner tube diameter

The temperature of fuel-air mixture at the burner tip depends on the heat transfer from the reaction zone, heat loss to the surroundings and size, shape & material of construction of burner wall.

The average gas velocity V_{av} depends upon the :

- desired flow rate
- burner nozzle diameter

V_{av}, also varies along the cross-sectional area of the nozzle.

If, $V_n = V_{av} \sin \alpha$, then the flame is stable

$V_n > V_{av} \sin \alpha$, then the flame flashes back

$V_n < V_{av} \sin \alpha$, then the flame lifts off the burner tip

where, α = half of the cone angle (degrees) of the flame and is determined from the angle of the flame projected on a screen and photographed or viewed through a telescope. The angle between the tangent to the inner cone surface and the burner tube axis gives α :

The value of α depends on the location of the flame front. When $\alpha > 90°$, the flame will be pushed inside the burner tube and flash back will occur. If $\alpha \rightarrow 0°$ *i.e.* the flame front is developed at a large distance above the burner tip, then there will be undue dilution of combustible mixture by interdiffusion with the surrounding atmosphere thereby causing lowering of flame velocity leading to blow off of the flame. Hence, minimum gas velocity corresponds to the normal velocity of flame propagation.

Increasing the tube diameter enhances the tendency to flash back while decrease in tube diameter tends to cause blow-off. Use of low therm.al conductivity ceramic materials in making the burner tip prevents the blow off tendency of turbulent flames, however, the same augments hold for the flash back tendency of laminar flames.

A stable flame at high gas velocities can be obtained by holding a wire ring at a considerable distance from the burner tip or by inserting a wire partly into the burner tube. The gas velocity is slowed down in the vicinity of the wire in both cases. A conical flame is obtained on the wire ring while the upper part of the straight wire supports a fan-shaped flame.

IGNITION TEMPERATURE OF FUEL

It is the minimum temperature at which the fuel will ignite, if sufficient air/oxygen is present and above which the combustion once initiated will maintain itself and below which active combustion can not occur.

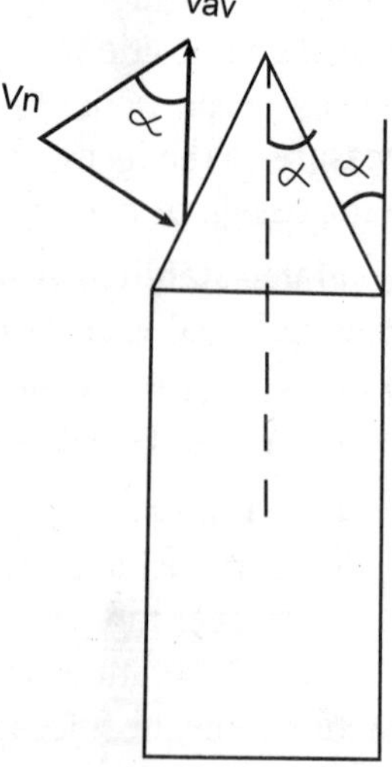

Fig. 35.18. Inner cone of the flame.

Ignition temperature depends upon the inherent property of the system as well as on extraneous factors. If the extraneous factors are kept constant, then the ignition temperature is obtained as a characteristic property of combustibles. It depends on the factors like fuel/oxidant ratio, nature of oxidant, induction period, ignition vessel wall, presence of promoters & inhibitors etc. Ignition temperature of substances also depends upon the method of its determination.

Three methods are normally used for determination of spontaneous ignition temperature (SIT) of vapors and gases *e.g.*

- admission of an explosive mixture of air-vapor/gas to an evacuated and heated vessel.
- heating of fuel (gas/vapor) & oxidant separately and then mixing in a heated vessel.
- adiabatic compression of the premixed system until ignition occurs.

Usually first method is used for gases/vapors.

The ignition temperature of solid fuels e.g. coal and coke is determined in a different way. As for example for coal, about 5 gm of coal powder is thrown into the bottom of the heated crucible. For coke, a slower method is used in which a sample of coke is heated at a constant rate. The temperature at which the rise of coke mass temperature is suddenly accelerated is recorded as its ignition temperature.

The ignition temperature of liquid fuels is determined by dropping oil drops into a heated crucible and observing the minimum temperature at which glow occurs followed by flame or explosion.

Expression for spontaneous ignition temperature of a gas or vapor is derived on the basis of thermal mechanism of combustion. Relation between pressure and initial container temperature is given by the Semenov's equation which is given below :

$$\log_e Q = \frac{P_{cr}}{T_O^2} = \frac{E}{2RT_O} \qquad (a)$$

This equation is valid for successful thermal explosion of combustible mixtures,

where, P_{cr} = Critical pressure, T_0 = Absolute temperature of the vessel wall

E = Energy of activation of the combustible, R = Gas constant

Assuming $T_0 = T_i$ (ignition temperature), we find that, at constant pressure

$$\log_e Q = \frac{E}{RT_i} + \text{Constant} \qquad (b)$$

Where, Q is the heat liberated by combustion

If fuel/oxidant ratio in the system is varied between 0 and 1, Q will vary with the maximum near stoichiometric composition. Ignition temperature (*Ti*) V_s. Pressure (*P*) is plotted in Fig. 35.19.

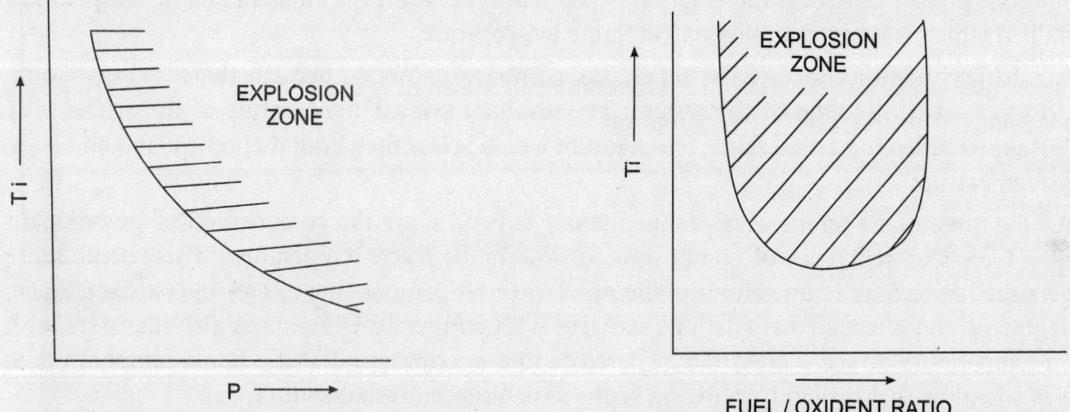

Fig. 35.19. *Ignition Temperature Vs. Pressure ignition (for gases/vapors)*

Fig. 35.20. *Gas-air mixture composition Vs. temperature.*

From equation (b), we get the plot shown in the Fig. 35.20 between fuel/oxidant ratio and the ignition temperature (T_i) which indicates the existence of two limiting compositions of fuel/oxidant mixture for the system to ignite & explode. These are called *limits of inflammability or explosion limits.*

Ignition temperature of a gas varies with its concentration and is reduced by the pressure. If gas is burnt in pure oxygen instead of air, then in some cases there is slight decrease in the ignition temperature.

THEORETICAL FLAME TEMPERATURE OF FUELS

It is the temperature attained by the product of combustion of gas when it is burnt with the theoretical quantity of air and no heat loss is allowed to the surrounding.

The actual flame temperature is however less due to the fact that fuels are burnt with excess air and heat is lost to the surroundings.

Flame temperature of fuels depends upon :

- Calorific value of the gas.
- Volume and specific heat of total gaseous products.
- Heat Losses by radiation.
- Latent heat in steam in the combustion products.
- Endothermic dissociation of gaseous molecules (mainly CO_2 and H_2O present in the combustion products).

Four different types of flame temperatures are recognised viz. theoretical flame temperature, adiabatic flame temperature, actual flame temperature and maximum adiabatic flame temperature (MAFT) or simply maximum flame temperature.

When the combustion of fuel goes to completion and the entire heat of combustion goes to heat the products, the resultant temperature is the theoretical flame temperature. In actual practice, combustion is never complete at high temperature due to dissociation reactions. When the endothermic effect of dissociation reaction is taken into account, the adiabatic flame temperature is obtained which is lower than theoretical flame temperature.

Adiabatic conditions do not exist in practical combustion process because there is always some heat lost to the surroundings of the system. The resultant average temperature of the products of combustion is called the actual flame temperature which is less than both theoretical and adiabatic flame temperature.

All the three flame temperature defined above depend upon the composition of fuel-oxidant mixture. Excess or deficiency of combustion air lowers the flame temperature. Theoretical flame temperature has its maximum volume at the stoichiometric composition of fuel and oxidant. Since, the degree of endothermic dissociation increases with temperature, heat loss also increases with rise in temperature at accelerated rate. Therefore, the maximum adiabatic flame temperature is realised when the fuel is slightly in excess of the stoichiometric composition.

Fuel gases having high inert content have low flame temperatures. Flame temperature of a fuel is much higher in oxygen than in air because of high nitrogen content of air which takes away significant amount of heat.

Maximum adiabatic flame temperature of some fuel gases in air and of a few in oxygen are given in Table 35.7.

From the data, it is inferred that with the exception of acetylene, hydrogen and carbon monoxide, all the gases have their maximum flame temperatures around 1900-1950°C. This is because an increase in the calorific value of a fuel is counterbalanced by the additional air requirement and hence a greater quantity of flue gases. Low flame temperatures of blast furnace gas and producer gas are due to their very high inert (nitrogen) contents.

Flame temperature of fuels is much higher in oxygen than in air. Notably, the depressing effect of dissociation on flame temperature is many times more in oxygen combustion compared to combustion in air. As for example, methane-oxygen flame (stoichiometric ratio 1:2) has an adiabatic flame temperature of 2740°C as against its theoretical flame temperature of 5050°C. The drop in flame temperature due to endothermic dissociation is as high as 2310°C. The corresponding values for stoichiometric composition in air are : theoretical flame temperature 2060°C, adiabatic flame temperature 1870°C, and hence the drop due to endothermic dissociation is 190°C only {i.e. about 1/12th of the drop in oxygen).

Table 35.7: Maximum Adiabatic Flame Temperature (MAFT)

Gas	Formula	In air		In oxygen		Theoretical flame temperature, °C
		% combustible in gas-air mixture giving MAFT	MAFT, °C	% combustibles in gas-0₂ mixture giving MAFT	MAFT, °C	
Hydrogen	H_2	31.6	2045	78	2660	2210
Carbon monoxide	CO	32.0	2100	70	2925	2468
Methane	CH_4	10.0	1875	—	—	2065
Ethane	C_2H_6	5.8	1895	—	—	2104
Propane	C_3H_8	4.15	1925	—	—	2115
Butane	C_4H_{10}	3.2	1895	—	—	2132
Isobutane	C_4H_{10}	3.2	1900	—	—	—
Ethylene	C_2H_4	7.0	1975	—	—	2343
Propene/Propylene	C_3H_6	4.5	1935	—	—	2254
Butene/Butylene	C_4H_8	3.4	1930	—	—	2221
Acetylene	C_2H_2	9.0	2325	44	3137	2632
Natural gas	—	9.0	1930	—	—	2030
Coal gas	—	17.6	1918	—	—	2250
Carburetted water gas	—	18.8	1930	—	—	1680
Producer gas	—	46.7	1663	—	—	1800
Blast Furnace gas	—	—	1460	—	—	—
Benzene	C_6H_6	—	—	—	—	2265
Toluene	C_7H_8	—	—	—	—	2232
Xylene	C_7H_{10}	—	—	—	—	2210
Naphthalene	$C_{10}H_8$	—	—	—	—	2260
Fuel oil	—	—	2125	—	—	
Coal tar	—	—	2230	—	—	
Lignite (dry/wet)	—	—	2068/1890	—	—	
Bituminous coal	—	—	2200	—	—	
Petroleum coke	—	—	2115	—	—	

If combustion products of the system are difficult to dissociate at high temperature then the resultant adiabatic flame temperature should be very high. As for example, a cyanogen-oxygen flame has such products like CO & N_2 and gives a temperature of 4580°C. Similarly, a hydrogen fluorine flame has a maximum adiabatic flame temperature of 4300°C.

Flame temperature can be appreciably increased by preheating the combustion air and/or the fuel. Variation of adiabatic flame temperature with composition of fuel-air mixture for natural gas is shown in Fig. 35.21.

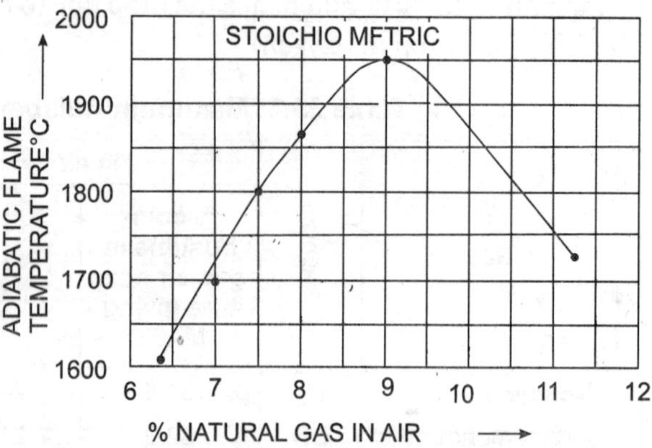

Fig. 35.21. Composition Vs Adiaatic flame temperature.

Adiabatic flame temperature is the maximum temperature that can possibly be achieved under a given set of fuel, oxidant, diluent and temperature conditions. For a satisfactory heating system, flame temperature must be higher than required process temperature. Using a richer, high C.V. fuel, using oxygen in place of air, using near stoichiometric proportions and preheating the reactants are practical ways of increasing the flame temperature. Higher the flame temperature, higher is the efficiency of heat transfer from flame to the heating surface.

CALCULATION OF FLAME TEMPERATURE

Flame temperature is calculated by striking a heat balance between the fuel and air on one hand and the combustion products on the other. For an adiabatic flame, following assumptions are made:

- no heat is lost to the surroundings — combustion is complete
- no thermal dissociation takes place — a datum temperature is used (often 0°C)

Fig. 35.22

Then a heat balance of the flame gives,

$$\text{Flame temperature} = \frac{\text{datum} + \text{net calorific value} + \Sigma\text{preheats}}{\text{volume} \times \text{mean specific heats}}$$

Since the specific heat of combustion products is a function of the unknown flame temperature, the calculation proceeds by trial and error, correcting each cycle for an appropriate specific heat. Above about 2000°C, considerable error arises because of thermal dissociation which absorbs heat and lowers the actual flame temperature obtainable.

It is convenient to base calculations on 1 kg of solid/liquid fuels or 1 Nm^3 of fuel gas. All quantities of oxidant and waste products are then related to unit quantity of fuel. It is then calculated in kcal units. Heat balance in case of actual flame temperature in air gives$C_N + \Delta H_f +$

$$A\text{-}\Delta H_a = V.AH_{fg} + q_d + q_l \quad ...(X)$$

where, C_N = net calorific value of fuel, kcal/Nm³ fuel

ΔH_f = enthalpy of fuel above reference (datum) temperature kcal/Nm³ fuel

A = air supplied, Nm³/Nm³ fuel

ΔH_a = enthalpy of air above reference temperature, kcal/Nm³ fuel

V = flue gases produced, Nm³/Nm³ fuel

ΔH_{fg} = enthalpy of flue gases above reference temperature, kcal/Nm³ flue gas

q_d = heat loss by endothermic dissociation, kcal/Nm³ fuel

q_l = heat loss to surroundings, kcal/Nm³ fuel

Now,
$$\Delta H_{fg} = t_f.C_{pfg\,(0\_t_f)} - t_r.C_{pfg\,(0-t_r)} \quad ...(Y)$$

where, t_r = reference temperature (say 25°C), °C

t_f = flame temperature, °C

$C_{pgf(0\text{-}t_f)}$ = mean specific heat of flue gas between 0°C and t_f

$C_{pfg\,(0-t_r)}$ = mean specific heat of flue gas between 0°C and t_r

If it assumed that there is no thermal dissociation, then $q_d = 0$.

In case of adiabatic flame temperature $q_{loss} = 0$

If the fuel & air are supplied at reference temperature then ΔH_f & ΔH_a both are zero.

In case of theoretical flame temperature, $q_d = 0$ and $q_l = 0$

From equation 'X' and 'Y' we get the generalised formula for calculation of flame temperature which is,

$$t_f = \frac{C_N + H_f + A.H_a - q_d - q_l + V.t_r.C_{pfg(0-t_r)}}{V.C_{pfg(0-tf)}}$$

Appearance and use of t_f on both sides of the formula (in equation Z) complicates the calculation of t_f. Since, the specific heat of flue gases depend upon the unknown t_f, hence a trial and error method is used.

Example 2. Effect of changing conditions on flame temperature :

Calculate the theoretical flame temperature for blast furnace gas under the following conditions:

(a) both theoretical air and blast furnace gas are at 15°C

(b) 50% excess air at 15°C and gas at 15°C

(c) theoretical air at 60°C and gas at 400°C

(d) theoretical oxygen at 15°C and gas at 15°C

Data:

Volumetric Composition of **B.F.** Gas is,

$CO = 22\%$, $CO_2 = 18\%$, $H_2 = 2\%$, $N_2 = 58\%$

Net calorific value of B.F. gas = 719 kcal/Nm³

Mean specific heat of B.F. gas at 600°C = 0.342 kcal/m³°C

Temperature, °C	Mean specific heats, kcal/m^3 °C	
	Air	Theoretical combustion products (flue gas)
600	0.324	—
1000	0.337	0.389
1200	0.343	0.397
1400	0.347	0.404
1600	0.349	0.409
1800	0.355	0.415
2000	0.356	0.419

Assume reference (datum) temperature t_r to be 0°C and that there is no endothermic dissociation of combustion products.

Solution. First of all we find the volume of theoretical O_2, air and flue gas on combustion of B.F. gas. These will be needed in flame temperature calculations.

Basis . 100 m^3 of B.F. gas

Constituents of B.F. gas	Volume, %	Theoretical O_2 required for complete combustion, m^3	Products of combustion		
			CO_2	H_2O	N_2
CO	22	11	22	—	—
CO_2	18	—	18	—	—
H_2	2	1	—	2	—
N_2	58	—	—	—	58
N_2 from air					45.1
Total	100	12	40	2	103.1

Theoretical oxygen required $= 12$ m^3

Theoretical air required $= 12 \times \dfrac{100}{21} = 57.1$ m^3

N_2 present in the theoretical air $= 57.1 \times \dfrac{79}{100} = 45.1$ m^3

which will also remain present in the flue gas.

Total flue gas when using air for combustion (*i.e.* $CO_2 + H_2 + N_2$)

$$= 40 + 2 + 103.1 = 145.1 \text{m}^3$$

Total flue gas when using oxygen for combustion

$$= 40 + 2 + 58 = 100 \text{ m}^3$$

Hence flue gas produced when using oxygen for combustion

$$= \frac{100}{100} = 1 \text{ m}^3/\text{m}^3 \text{ BF gas}$$

Theoretical air required	= 0.571 m^3/m^3 of BF gas
50% excess air	= 0.5 × 0.571 = 0.2805 m^3/m^3 of BF gas
Theoretical O_2 required	= 0.012 m^3/m^3 of BF gas
Theoretical flue gas produced	=1.451 m^3/m^3 of BF gas

Now, writing the generalised equation for flame temperature, we get

$$t_f = \frac{C_N + \Delta H_f + A.\Delta H_a - q_d - q_l + V.t_r.C_{pfg(0-t_r)}}{V.C_{pfg(0-t_f)}}$$

here, ql and $ql = 0$ in case of theoretical flame temperature

t_r = reference temperature = 0°C

hence, $$tf = \frac{C_N + \Delta H_f + A.\Delta H_a}{V.C_{pfg(0-t_f)}}$$

(a) where, C_N = 719 kcal/m^3

ΔH_f = enthalpy of B.F. gas at 15°C = 15 kcal/m^3 B.F. gas

A = air supplied = 0.571 m^3/m^3

ΔH_f = enthalpy of air at 15°C = 15 kcal/m^3 B.F. gas

V = flue gas produced = 1.451 m^3/m^3 B.F. gas

First approximation

Let, C_{pfg} (0 - t_f) = 0.397 kcal/m^3. °C assuming a value of 1200°C for t_f

then, $$t_f = \frac{719+15+0.571\times15}{1.451\times0.397} = 1289°C$$

Since the value of flame temperature assumed was 1200°C and putting the value of specific heat of flue gas at 1200°C i.e. 0.397 we get the flame temperature t_f= 1289°C which is much higher than the assumed value of 1200°C. Hence, a second approximation is made.

Second approximation

Let the flame temperature be 1250°C. Now the value of specific heat C_{pfg} of flue gas at 1250°C is calculated by linear interpolation of C_{pfg}between 1200°C and 1400°C in absence of adequate data.

C_{pfg} at 1200°C = 0.397 kcal/m^3°C, C_{pfg} at 1400°C = 0.404 kcal/m^3°C

$$C_{pfg} \text{at } 1300°C = \frac{1200\times0.397+1400\times0.404}{1200+1400}$$

$$= 0.4007692 \text{ kcal/m}^3{}°C$$

hence, $$C_{pfg} \text{at } 1250°C = \frac{1200\times0.397+1300\times0.4007692}{1200+1300} = 0.399 \text{ kcal/m}^3{}°C$$

hence, $$t_f = \frac{719+15+0.571\times15}{1.451\times0.399} = 1282°C$$

Here also the assumed value of flame temperature i.e. 1250°C is much lower than the calculated flame temperature *i.e.* 1282°C. Hence, we go for the third approximation.

Third approximation :

Let the flame temperature be 1275°C

$$C_{pfg} \text{ at } 1275°C = \frac{1250 \times 0.399 + 1300 \times 0.4007692}{1250+1300} = 0.400 \text{ kcal/m}^3\text{°C}$$

hence, $$t_f = \frac{719 + 15 + 0.571 \times 15}{1.451 \times 0.400} = 1279°C$$

Since calculated value of t_f *i.e.* 1279°C is quite close to the assumed value of flame temperature (for calculation of specific heat) *i.e.* 1275°C, hence, this approximation gives the correct flame temperature *i.e.* 1279°C.

(b) Calculation of flame temperature at 50% excess air at 15°C and gas at 15°C :

As we know that excess air reduces the flame temperature, hence we shall proceed by assuming lower value of flame temperature than in (a).

Now, $$t_f = \frac{C_N + \Delta H_f + A.\Delta H_a}{V.C_{pfg(0-t_f)}}$$

Let the flame temperature be 1100°C.

C_{pfg} at 1100°C is found by interpolating the specific heat of flue gas between 1000 and 1200°C.

$$C_{pfg}' \; 1100°C = \frac{1000 \times 0.389 + 1200 \times 0.397}{1000+1200} = 0.393 \text{ kcal/m}^3 \cdot °C$$

$$C_{p_{air}}' \; 1100°C = (1000 \times 0.337 + 1200 \times 0.343)/(1000 + 1200)$$
$$= 0.340 \text{ kcal/m}^3 .°C$$

$$\rho_{f_g} t_f = \frac{719 + 15 + 1.5 \times 0.571 \times 15}{1.451 \times 0.393 + 0.2805 \times 0.340} = 1122°C$$

(Since 50% excess air has been used)

Since, the assumed value of flame temperature i.e. 1100°C is quite close to the calculated value *i.e.* 1122°C hence, the flame temperature in this case is 1122°C.

(c) Flame temperature when theoretical air is at 600°C and gas is at 400°C:

Here, since both gas and air are preheated, hence appreciable increase in theoretical flame temperature is expected.

$$t_f = \frac{C_N + \Delta H_f + A.\Delta H_a}{V.C_{pf_g}}$$

Here, $\Delta H_f = 400$ kcal/m^3 (since the gas is at 400°C)

and, $\Delta H_a = 600$ kcal/m^3 (since the air is at 600°C)

Let us assume that the flame temperature is 1600°C at which the specific heat of the flue gas is 0.409 kcal/m^3.°C.

Hence, $$t = \frac{719 + 400 + 0.571 \times 600}{1.451 \times 0.409 + 1 \times 0.349} = 1550°C$$

Since the assumed value of 1600°C is quite close to the calculated value of t_f i.e. 1550°C, hence the theoretical flame temperature of the gas in this case (when both gas and air have been preheated) is about 1550°C.

(d) Flame temperature when oxygen is at 15°C (instead of air) and gas is at 15°C :

$$t_f = \frac{C_N + \Delta H_f + A \,.\, \Delta H_a}{V \,.\, C_{pf_g}}$$

Use of oxygen instead of air for combustion will increase the flame temperature further. Assuming a flame temperature of 1800°C, we get

$$t_f = \frac{719 + 15 + 0.012 \times 15}{1 \times 0.415} = 1770°C$$

which is quite close to the assumed value. Hence, the flame temperature in this case is 1770°C.

Example 3. Calculate the theoretical flame temperature of combustion of hydrogen with theoretical amount of air assuming (a) no dissociation of combustion product (*i.e.,* H_2O) and (b) assuming dissociation to the extent of 4% of combustion product i.e., H_2O. Assume that both the hydrogen & air are dry and at 0°C. Net C.V. of H_2 = 10.16 MJ/m^3. Mean value of specific heat, C_p, of product of combustion *i.e.,* H_2O in *kJ*/Nm^3.°C at various temperature is as follows :

Temp, °C	2000	2100	2200
C_p, *kJ*/Nm^3.°C	1.643	1.651	1.657

Solution, (a) $2H_2 + O_2 = 2H_2O$

2 m^3 of H_2 requires 1 m^3 of O_2 for Stoichiometric combustion.

Hence, air required for combustion of 1 m^3 of hydrogen = $\frac{1 \times 100}{2 \times 21}$ = 2.38 m^3

N2 present in 2.38 m^3 air = 2.38 × 0.79 = 1.88 m^3

Flue gas will have, N_2 = 1.88 m^3 and H_2O = 1 m^3, Total flue gas volume = 2.88 m^3/m^3 H_2

Composition of flue gas its $N_2 = \frac{1.88}{1.88 + 1} \times 100 = 65.3\%$ and $H_2O = \frac{1}{1.88 + 1} = 34.7\%$.

(a) Theoretical flame temperature assuming no dissociation of H_2O :

$$T = \frac{10.16 \times 1000}{2.88 \times C_p}$$

Assuming, $T = 2200$,

$$T - 0 = \frac{10.16 \times 1000}{2.88 \times 1.657} = 2125°C$$

Assuming, $T = 2100$

$$T = \frac{10.16 \times 1000}{2.88 \times 1.651} = 2136°C$$

This is quite close to the assumed value of 2100°C.

Hence, theoretical flame temperature without dissociation of H_2O taken in account = 2100°C Ans.

(b) At 2100°C, water vapour/steam is endothermically dissociated to the extent of about 4%, thus the actual heat and temperature developed will be less ($H_2O \rightleftharpoons H_2 + \frac{1}{2} O_2$).

The flame temperature under these conditions will be,

$$T = \frac{10.16 \times 1000 \times (1 - 0.04)}{2.88 \times C_p}$$

Assuming, $T = 2000°C$

$$T = \frac{10.16 \times 1000 \times (1 - 0.04)}{2.88 \times 1.643} = 2061°C$$

This is close to the assumed value of 2000°C.

Hence, the theoretical flame temperature when endothermic dissociation of product of combustion is taken into account is reduced to 2000°C from 2100°C.

However, in practice the temperature is still further reduced by heat loss by radiation from the flame which is generally assumed to be 10%. The maximum flame temperature is achieved when using slightly less than theoretical air.

Example 4. Calculation of flame temperature (MAFT) with theoretical air. Calculate theoretical maximum adiabatic flame temperature (MAFT) of fuel gas of composition 96% CH_4, 0.8% CO_2 and 3.2% N_2 when burnt with theoretical air. Assume fuel and air mixed at 25°C. Consider 1 mole of fuel gas.

Solution : $CH_4 + 2(O_2 + 3.76 N_2) = CO_2 + 2H_2O (g) + 7.52N_2$

POC	Amount (kg mole)
CO_2	0.968
H_2O	1.920
N_2	7.52

[POC-Products of combustion]

Heat balance : Reference Temperature 25°C or 298 K.

Sensible heat in air and fuel + Heat of combustion = Heat in products of combustion ..(1)

Sensible heat of reactants = 0, since they are supplied at 25°C.

$$-\Delta H_{298}^{comb.} = 0.968\,(94.05 \times 10^3) + 1.920\,(57.80)$$

$$\times 10^3\,[0.96 \times 17.89] \times 10^3 =.\ 184 \times 103 \text{ kcal.} \quad ...(2)$$

This heat of combustion raises the temperature of POC to the flame temperature. The heat capacity of POC *i.e.* Cp'.Cp' = $nCO_2Cp\ CO_2 + nH_2OC_p + nN_2C_p\ N_2$

Where nCO_2 nH_2 and nH_2O are moles of CO_2, N_2 H_2O respectively.

$$C_p^1 = 0.968\,(10.55 + 2.16 \times 10^{-3}\,T - \frac{2.04\times10^5}{T^2}) + 1.92\,(7.17 + 2.56 \times 10^{-3}\,T$$

$$+\frac{0.08\times10^5}{T^2}) + 7.25\,(6.66 + 1.02 \times 10 - 3T)$$

$$= 72.27 + 14.41 \times 10 - 3T\frac{2.11\times Ri}{T^2}\frac{k8.24}{kg\ mole^{ok}} \quad ...(3)$$

By 1 and 3, we get

$$184 \times 10^3 = \int_{298}^{T_f} C^1\ P^{dT} \quad ...(4)$$

By 3 and 4, we get

$$184 \times 10^3 = \int_{298}^{T_f} (72.27 + 14.41 \times 10^{-3}T - \frac{2.11\times10^3}{T^2})dT$$

Solution gives T_f = 2300 K.

Consider the use of expression $C_p = a$ + b T and recalculating flame temperature Calculating C_P^1 and making heat balance gives

$$0.0072\ T_f^2 + 72.27\ T_f\text{-}206.\ 157 = 0.$$

This is a quadratic equation whose solution gives

$$T_f = 2319K.$$

Let us calculate flame temperature by using average C_p values of POC

$$184000 = 0.96 \times 12.5\,(T_f\text{-}\ 25) + 1.92 \times 7.73\,(T_f\text{-}\ 25) + 7.12 \times 7.25\,(T_f\text{-}\ 25)$$

Solution gives T = 2643 K.

We note that the accuracy of calculation depends on C_p values. For accurate calculations $C_p = a + b\,T + c/T^2$ must be used. However using $C_p = a + bT$, though simplifies calculation but flame temperature is slightly greater (a difference of 20 Kin this example). Use of average C_p values though simplifies the flame temperature calculation but calculated flame temperature is greater then earlier ones.

Example 5. *Effect of excess air on MAFT (Maximum Adiabatic Flame Temperature). Consider the fuel in example-4. Now it is burnt with (a) 20% excess air and (b) 50% excess air. then calculate MAFT in each case.*

Solution : In the following calculations we will be using

$$C_p = a + bT$$

However, readers may perform calculation using

$$Cp = a + bT + C/T2$$

Take 20% excess air :

Amount of POC : CO_2 = 0.968

H_2O = 1.92 O_2 = 0.40 N_2 = 9.056

We can calculate

$$C_P^T = 87.18 + 16.11\ T\ (kcal/kg\ mol)$$

Heat of combustion = 184 ×103 kcal.

Heat balance

$$184000 = \int_{238}^{TF} (87.18 + 16.11T)\, dT$$

Integration yields

$$184000 = 87.18\ Tf - 259779.6 + 8.05 \times 10^{-3} Tf^2 - 715.3$$

Rearrangement.

$$8.05 \times 10\text{-}3\ Tf + 87.18\ T\text{^-}\ 210695 = 0$$

$$Tf = 2034\ K.$$

(b) Similarly for 50% excess air :

POC : $CO_2 = 0.968$

$H_2O = 1.92$

Heat balance yields.

$$9.18 \times 10^{-3}\ Tf + 102.88\ Tf\text{-}\ 215474 = 0$$

$$Tf^2 = 1819\ K$$

We note that increase in excess air decreases flame temperature. This is due to increase in N_2 and O_2 in the flue gas.

Similar calculation can be done by enriching air with O_2.

FLAME VELOCITY

It is the velocity with which the flame travels through gas-air mixture or it is the velocity of the flame front. When a combustible gas-air mixture is ignited in a glass tube at one end by spark, the flame front travels to the other end until the mixture is burnt off. Flame velocity is also called the velocity of flame propagation. If the linear velocity of gas-air mixture flawing through burner pores is not greater than the flame velocity, then back fire results. However if the gas-air mixture velocity is too high, then flame lifts off the burner tip.

Flame speed depends upon the composition and attains maximum for mixture slightly richer in fuel content than the stoichiometric composition. Flame speed rises if initial temperature of combustible mixture is increased or the pressure of the system is decreased. By increasing the tube/pipe/burner diameter the flame speed increases.

Flame velocity which affects the rate of heat release in combustion is to be considered carefully for solution of many practical problems in combustion. Flame propagation can be demonstrated by explosion of a combustible gas-air mixture in a glass tube by igniting at one end by spark (external heat source). The flame front can be seen to travel to the other end until the whole mixture is burnt off.

For a flame to remain stationary, certain favourable circumstances are required. Considering a system with combustion surface, F (flame front) perpendicular to the axis of a cylindrical tube in which the combustible gas-air mixture flowing from right to the left as shown in Fig 35.23. The flame therefore tends to move from left to the right towards the incoming fresh gas.

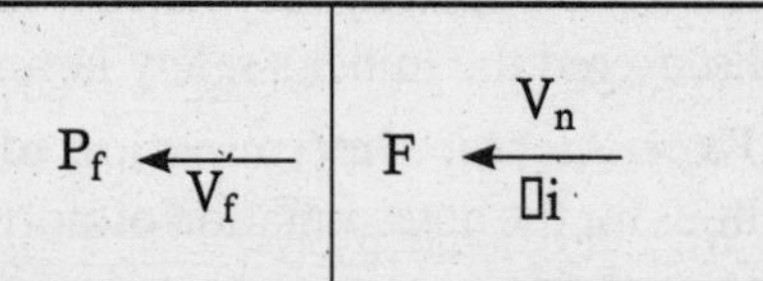

Fig. 35.23: Stationary flame in a tube

[here, ρ_i and ρ_f are density of initial mixture and flue gases respectively and V_n are V_f normal velocity of flame propagation and velocity of the gases respectively.]

If the gas velocity is equal to the flame velocity, the combustion surface will remain stationary. The direction of flame velocity is normal to the surface F. This is called the normal velocity of flame propagation (Vn) for differentiating it from the very much faster propagation of flame in detonation (of the order of several km/sec). Burning velocity (often used instead of velocity of flame propagation) is the velocity relative to the unburnt gas with which a suitably defined flame front travels along the normal to its surface. There is a sharp gradient of gas velocity, gas density, temperature and composition across the flame front.

The flame front in an explosive flame system moves with a velocity which is the vector sum of the normal velocity of flame propagation and the gas velocity. If the mass flow of the gases are prevented in the container, then the actual rate of advance of the flame front into the unburnt gas will be the normal velocity of flame propagation.

Flame propagation phenomenon is explained by a combination of thermal and diffusion mechanisms. The region neighbouring a flame gets heated up by conduction & convection and also chain carriers are diffused from the flame to that region. The pre-flame region as a result becomes the zone of active reaction and converted into the new flame while the completion of reaction and relatively cooling effect extinguish the luminosity of the region where the flame existed earlier. The net effect is the advancement of the flame front.

Normal velocity of flame propagation, Vn (derived from flame propagation by heat conduction alone) is given by

$$V_n = \sqrt{a.Ke^{-E/RT}}$$

where, a = mean thermal diffusivity K = frequency factor, E - energy of activation
R = gas constant, T = flame temperature

Flame velocity is directly proportional to the square root of thermal diffusivity and inversely proportional to the square root of the characteristic time of reaction. The conditions which increase the flame temperature will sharply raise the flame speed.

There is no universally accepted formula for calculating flame speed from the fundamental properties of the system. Surface has a profound effect on flame propagation if the surface to volume ratio of the system is high. The container surface slows down the propagation of flame by exerting a cooling effect and adsorbing chain carriers. As a matter of fact, the combustion can be completely stopped in very narrow tubes which was first observed by Davis on the basis of which he discovered the miner's safety lamp bearing his name.

Experimental Determination of Normal Velocity of Flame Propagation. There are three methods for the determination of normal velocity of flame propagation Bunsen burner or stationary flame method

- Spherical bomb method
- Glass tube method
- Bunsen burner or stationary flame method

All these method with proper care and necessary corrections give comparable & reproducible results of flame velocity.

In Bunsen burner (shown in Fig. 35.24) method, fuel gas & primary air (50-60% of the total air) are mixed in the mixing tube and the mixture burns with a flame anchored at the burner tip.

The flame has an inner cone and an outer envelope. Combustion proceeds at the inner cone until the oxygen of the primary air is exhausted. The outer envelope is the zone of secondary combustion where the remaining combustibles burn in the secondary air derived from the atmosphere. The flame propagates in a direction normal to the surface of the inner cone. As the flame is stationary, the normal velocity of flame propagation (Vn) is vectorially equal and opposite to the velocity of the gas mixture issuing out from the burner tip which can be calculated from the knowledge of gas velocity and cone angle by using the vector triangle shown in Fig. 35.25.

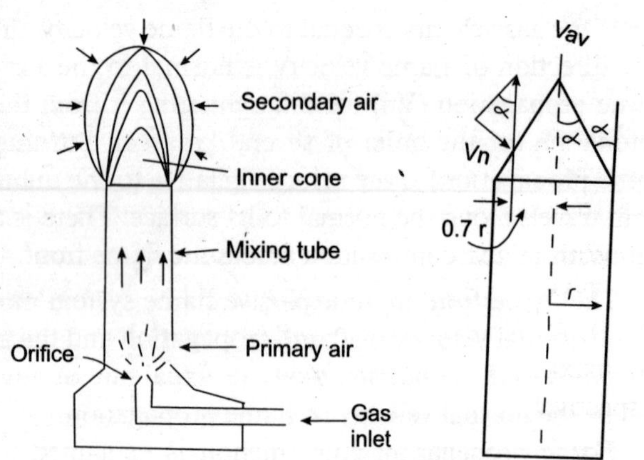

Fig. 35.24. Elements of a Bunsen Burner.

Fig. 35.25. Method for measuring normal velocity of flame propagation (Vn).

The gas velocity varies along the cross-sectional area of the tube being maximum at the tube axis and zero at the wall. The local velocity at a distance of 0.7 times the radius from the axis represents the average velocity. The normal velocity of flame propagation, Vn is given by

$$Vn = V_{av} \sin \alpha$$

where V_{av} = average velocity of gas-air mixture = $\frac{V}{A}$

here, V = cross-sectional area of tube and,

A = flow rate of the mixture ,

2α = cone angle in degrees

The angle α is determined from the angle of the flame projected on a screen, photographed or viewed through a telescope. The angle between the tangent to the inner cone surface and tube axis gives α.

V_n is affected by pressure, temperature, fuel-oxidant mixture, surface etc. Flame speed varies from fuel to fuel depending on the kinetics of com bustion reactions. For a particular fuel, the flame speed varies with the composition and attains maximum for the mixture slightly richer in fuel content than the stoichiometric composition. This relationship is similar to the flame temperature-composition relationship. Flame velocity variation with composition (for various gases) is shown in Fig. 35.26.

Maximum for the mixture slightly richer in fuel content than the stoichiometric composition. This relationship is similar to the flame temperature-composition relationship. Flame velocity variation with composition (for various gases) is shown in Fig. 35.26

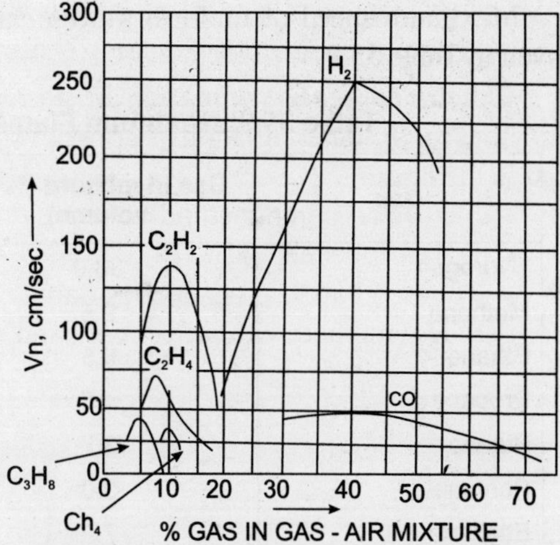

Fig. 35.26. Composition Vs. Flame Velocity

Hydrogen flame is the fastest (V_n = 250 cm/sec) followed by acetylene (Vn ≈150 cm/sec). Carbon monoxide, paraffins, benzene etc. have V_n in the range of 30-40 cm/sec. Propene is slightly above this while ethylene exceeds 60 cm/sec. All these values of V_n (which are for fuel-air systems) rises to 5-10 times these values if, oxygen is substituted for air.

With increase in initial temperature of combustible mixture *(T)*, flame speed rises (following $V_n \propto T^2$) while decrease of pressure *(P)* of the system also raises the flame speed (following V_n Vn $\alpha^{P1/4}$). As for example, by preheating H_2-air mixture to 430°C and CH_4-air mixture to 680°C, the maximum value of Vn can be increased from 2.5 to 13.5 m/sec and from 0.3 to 3 m/sec respectively.

For any gaseous mixture, the flame speed *(s)* is calculated by the following formula.

$$S = \frac{aS_a + bS_b + cS_c + \ldots}{a + b + c + \ldots}$$

where, *a,b,c* = % composition of constituent combustible gas ,
S = flame velocity of the mixture,
Sa, Sb, Sc = flame velocity of constituent gases.

Maximum flame speed of common gases in air in 2.5 cm dia tube is given in Table 35.8.

Table 35.8: Maximum Flame Speed of Gases (tube dia. = 2.5 cm)

Gas	Gas in maximum speed mixture, (% by volume)	Velocity of flame propagation of maximum speed mixture, cm/sec
Hydrogen	38.5	485.0
Carbon monoxide	50.0	125.0
Methane	9.0	76.1
Ethane	6.3	85.5
Ethylene	7.1	143.2
Coke oven gas	17.1	158.8
Producer gas	51.0	73.2

Maximum speed of uniform movement of flame in mixtures with air in horizontal tubes is given in Table 35.9.

Table 35.9: Maximum Flame Speed of Gases in Horizontal Tube

Gas	Gas in mixture (% by volume)	Flame Velocity, cm/sec	
		Tube dia. 2.5 cm	Tube dia. 5.0 cm
Hydrogen	38.0	490	—
Methane	9.5	66	91
Ethane	6.5	86	130
Propane	4.5	80	112
Butane	3.7	83	113
Pentane	3.0	82	115
Ethylene	7.2	142	—
Acetylene	8.9	282	—

EXPLOSIVE LIMITS OR LIMITS OF INFLAMMABILITY

For burning a gaseous fuel, a particular range (minimum and maximum) of concentration of fuel gas is required in gas-air mixture. This concentration range of gas-air mixture is called the limit of inflammability. The minimum value of the concentration range is called the lower limit of inflammability (or lower explosion limit) and the maximum value is called the upper limit of inflammability (or upper explosion limit).

For instance, the limit of inflammability of blast furnace gas is 37 to 71% gas in gas-air mixture. Here the lower limit is 37% and higher limit is 71% of gas in the mixture of gas and air. It means that a gas air mixture in which atleast 37% gas and rest 63% air is there will catch fire & burn if heated to its ignition temperature. On the other side if the gas percentage is upto 71% and air is rest i.e. 29% then also it will burn. The gas-air mixture will not burn if the gas percentage is below 37% or more than 71%. It will burn only in the concentration range of 37 to 71% and explode if burning takes place in a confined space.

Limit of inflammability of coke oven gas is 5-32%, that of hydrogen is 4—75% and for carbon monoxide it is 12.5—74%.

Limit of inflammability is normally widened with increase in temperature and pressure. However, increase of gas pressure narrows down the explosion limit for both CO and H_2 which are exceptions.

A fuel-air/oxygen mixture becomes inflammable when combustion liberates sufficient energy to cause a continued self propagation of flame throughout the unignited regions of the mixture. For a gas-oxidant system to be inflammable, there exists the limits of pressure (*i.e.* explosion limits), the limits of temperature (*i.e.* ignition temperature) and the limits of concentrations (*i.e.* limits of inflammability). In absence of surface effects, the upper inflammability limits refer to the richest mixture and lower inflammability limit to the leanest mixture to propagate a flame.

Existence of two limits of inflammability was earlier indicated by the analysis of the mechanism of the ignition temperature of a fuel-oxidant mixture. Velocity of flame propagation also gives the same indication. The flame speed diminishes on both sides of the optimum fuel- oxidant composition. There is a critical flame speed (10-12 cm/sec) below which flame propagation is not possible due to overwhelming effect of heat loss. The flame should advance at a sufficiently high rate to compensate for the loss of heat. Because the critical flame speed corresponds to the two compositions on the rich and lean sides of the mixture, there are two limits of inflammability. However, a fuel-oxidant mixture below the critical flame speed may be ignited at a point, but the flame will disappear after travelling a short distance through the combustible mixture.

The inflammability limits of gases/vapors in air are significant for precautions against the risks of fire and explosion. Concentrations of the vapors of solvents and other inflammable substances in surrounding air during their use should be kept well outside the limits of inflammability. At times, diluents like N_2, CO_2 etc. are used to bring the atmosphere below the lower limit of inflammability.

Determination of inflammability limits. Limits of inflammability of gases/vapors is determined by confining the gas-air mixture in the tube (a 75 mm dia vertical glass tube of 1.5 metres height) & then igniting it by opening the lower end and applying a spark or a small flame. The two limits of inflammability *i.e.* upper and lower limits (limiting concentrations) are determined at which the flame just fails to propagate upwards, throughout the whole length of the tube.

Factors affecting the inflammability limits. The limits of inflammability are influenced by pressure, temperature, tube/container size, direction of flame propagation, velocity of flame propagation, diluents etc.

Effect of pressure. A reduction in mixture pressure narrows down the two limits of inflammability. It lowers the upper limit and raises the lower limit. However, the upper limit is much more affected than the lower limit. The effect of pressure is explained by the increased diffusivity of chain carriers and other hot products towards the wall at a reduced pressure. The inflammability limits tend to join at a critical pressure below which the gas-oxidant mixture ceases to be inflammable at any concentration.

There are some unexplained effect of pressure on inflamability limits. For example inflammability limit of methane is about 5-15% gas in gas-air mixture at atmospheric pressure. But the upper limit is raised to 46% at 125 kg/cm^2 pressure. Lower limit is not affected very much.

Effect of temperature. Limits of inflammability of the mixture is widened by increasing its initial temperature (i.e. preheating) appreciably. This effect of temperature is explained by the fact that the enhanced flame temperature causes flame propagation in the mixture of unfavourable composition.

For example inflammability limit of methane in air at atmospheric pressure and temperature is 5-14% which becomes 4.8-16.6% when methane-air mixture is heated to 400°C. That means lower limit reduces slightly while the upper limit increases slightly more with increase of temperature i.e., inflammability limit widens.

Effect of tube/container size. The inflammability limits are narrowed down by decreasing the tube diameter below 5 cm due to the result of wall effect. The tube/container size has no effect above 5 cms diameter. The minimum tube sizes below which flame does not propagate at room temperature and pressure depends upon the fuel gases (e.g. H_2 -0.9 mm, CH_4 -4 mm, CO -2.5 mm).

Effect of direction of flame propagation. The variation in limits of inflammability due to change in direction of flame propagation is due to buoyancy effect. Widest limits of inflammability are obtained if the flame is to propagate upwards and the lowest limits result if the flame propagates downwards. Intermediate results are provided by horizontal and inclined tubes.

The conventional inflammability limit for methane-air mixture at atmospheric temperature and pressure determined using a long large diameter tube is as follows :

$$\left.\begin{array}{l}\text{Upward flame propagation} \rightarrow \text{ 5.3 to 15\% methane}\\ \text{Downward flame propagation} \rightarrow \text{6.0to13.3\%methane}\\ \text{Horizontal flame propagation} \rightarrow \text{5.4 to 14\% methane}\end{array}\right\}\text{approximately}$$

Effect of flame velocity. In general, the higher the maximum normal velocity of flame propagation of a combustible gas-oxidant mixture, the wider the limits of inflammability. However, carbon monoxide is an exception. Though, its maximum flame velocity is < 0.5 m/sec, it has a very wide limits of inflammability in air (12.5 to 74% gas in gas air mixture).

Effect of diluents. Presence of diluent like CO_2, N_2, He, Ar etc. in a mixture of gases narrows down the inflammability limits because of diluents influence the flame propagation to a greater or less extent.

H_2 (4-75%), ethylene oxide (3-80%) and acetylene (2.5-80%) have very wide range of inflammability while petrol vapors (1.4-7.6%), LPG (1.8-8.5%) benzene (1.4-7.1%), hexane (1.2-7.5%), acetone (3-11%) etc. have very narrow limits of inflammability.

Effect of Combustion with Oxygen instead of Air. Combustion of gases with pure oxygen instead of air widens the inflammability limits quite considerably. Both the lower as well as upper limits are affected while the effect on lower limit is very less, the upper inflammability limit is affected (increased) quite considerably.

Table 35.10 gives the inflammability limits of some gases at atmospheric pressure in 'air' as well as in 'oxygen' for comparison purposes.

Table 35.10: Inflammability limit of gases in air and oxygen at atmospheric pressure.

Gases	Inflammability limit,% gas in the mixture			
	In Air		In oxygen	
	Lower	Upper	Lower	Upper
Ethylene	3.1	32	3.0	80.0
Methane	5.3	15.0	5.1	61.0
Ethane	3.0	12.5	3.0	66.0
Hydrogen	4.0	75.0	4.0	94.0
Carbon monoxide	12.5	74.0	15.5	94.0

Formula for calculation of limit of inflammability of a mixture of gas. The inflammability limit of a mixture of combustibles in air is given by

$$L = \frac{100}{(x_1/l_1) + (x_2/l_2)(x_3/l_3)}$$

where, L = limit (lower or upper) of inflammability of the mixture in air, percent

l_1, l_2, l_3 = inflammability limit (lower or upper) of pure combustibles in air, percent

$x_1 x_2, x_3$ = volume per cent of combustibles in the mixture.

Inflammability limits of combustibles containing diluents (inerts) is calculated by

$$L_1' = L_1 \times \left(\frac{100}{100 - x_d}\right)$$

and, $$L_2' = L_2 \times \left(\frac{100}{100 - x_d}\right) \text{, if air is the diluent}$$

and, $$L_2' = \left(\frac{1000\ L_2}{1000 x_c + L_2(100 - x_c)}\right) \text{, if there is no oxygen in diluent}$$

$$L_2' = \left(\frac{2100\ L_2}{21 x_c + L_2(100 - x_c)}\right) \text{, if diluent (other than air) contains oxygen}$$

where, L_1 and L_2 = lower and upper limits of inflammability of pure combustibles

L_1' and L_2' = lower and upper limits of inflammability of the combustibles containing diluent

x_c, x_d and x_{o2} = percentage of combustible, diluent and oxygen respectively in the combustible containing diluent.

Lower limit of a mixture of combustibles can be accurately calculated but there is an appreciable divergence between the calculated and experimental values of the upper limit of a mixture. It shows that combustibles interact particularly when they are present in large quantities as in the upper limit of inflammability.

Approximate calculation of the inflammability limits of combustibles containing diluents is based on the fact that the lower limit is actually the criterian of the combustible concentration while the upper limit is that of the combustible to oxygen ratio.

Example 4. *Calculate the limits of inflammability in air of a gas mixture containing (only combustibles) 30% CO, 45%* H_2 *and 25%* CH_4*. The lower* (L_1) *and upper* (L_2) *limits of inflammability of the gases in air are given below:*

Gases	Lower inflammability limit (L_1)	Upper inflammability limit (L_2)
CO	12.5	74
H_2	4	75
CH_4	5.3	14

Solution. Since all the components of the gaseous mixture are combustible here, hence we have,

$$(L_1)_{mixture} = \frac{100}{\frac{30}{12.5}+\frac{45}{4}+\frac{25}{5.3}} = 5.45\%$$

$$(L_2)_{mixture} = \frac{100}{\frac{30}{74}+\frac{45}{75}+\frac{25}{14}} = 35.2\%$$

Hence, the lower & upper limits of inflammability of the gaseous mixture is 5.45% & 35.2% respectively.

Example 5. Calculate the limit of inflammability of coke oven gas whose volumetric composition and limits of inflammability of its combustible constituents are given below :

Constituents	CH_4	H_2	CO	CO_2	N_2	C_mH_n
% by volume	25	55	10	5	5	—
L_l (lower limit)	5.3	4	12.5	diluent	diluent	3
L_2 (upper limit)	14	75	74	—	—	12.5

Solution.

Here, CO_2 and N_2 are diluents

Concentration of diluents (*i.e.* CO_2 and N_2) = 5 + 5 = 10%

Composition of diluent free combustible mixture :

$$CH_4 = 25 \times \frac{100}{100-10} = 27.8\%$$

$$H_2 = 55 \times \frac{100}{100-10} = 61.1\%$$

$$CO = 10 \times \frac{100}{100-10} = 11.1\%$$

Limits of inflammability of diluent free gas mixture is

$$L_1(\text{mixture diluents free}) = \frac{100}{\frac{27.8}{5.3}+\frac{61.1}{4}+\frac{11.1}{12.5}} = 4.67\%$$

$$L_2(\text{mixture diluents free}) = \frac{100}{\frac{27.8}{14}+\frac{61.1}{75}+\frac{11.1}{12.5}} \ 33.9\%$$

Hence, the limit of inflammability of coke oven gas is

$$L_1' = 4.67 \times \frac{100}{100-10} = 5.2\%$$

$$L_2' = 33.96 \times \frac{100}{100-10} = 37.7\%$$

Hence lower and upper limit of inflammability of coke oven gas are 5.2 and 37.7% gas in gas-air mixture.

Example 6. *The composition, limit of inflammability (i.e., explosion limit) and maximum flame velocity of the constituents of a liquefied petroleum gas (LPG) is tabulated below :*

Constituent fuel	Composition	Limit of inflammability in air (*i.e.*, explosion limit), % gas in gas air mixture		Maximum flame velocity of gas air mixture, cm/sec.	
gases of LPG	%by volume	Lower limit (L_1)	Upper limit (L_2)	Volume % of gas in gas-air mixture,	Maximum flame velocity, S
CH_4	1.5	5.3	14	10.0	39
C_2H_6	14.0	3.0	12.5	6.3	46
C_3H_8	53.5	2.2	9.5	4.6	45
C_4H_{10} (iso+ normal)	31.0	1.9	8.5	3.5	44

From the above data, calculate

(a) *Limit of inflammability (Explosion limit) of LPG in air.*

(b) *Composition required for the maximum gas-air mixture for LPG.*

(c) *Maximum flame velocity of the (LPG) gas-air mixture.*

Solution. *(a)* The required limit of inflammability of LPG is calculated as

$$L = \frac{100}{(x_1/l_1) + (x_2/l_2) + (x_3/l_3) + ...} = 37.7\%$$

where, L = limit of inflammability (lower or upper) of the LPG air mixture in%

l_1, l_2, l_3 ... = inflammability limits (lower or upper) of pure combustibles in air in%

x_1, x_2, x_3 ... = volume percent of combustibles in the LPG

hence, lower limit of inflammability of LPG,

$$L_1 = \frac{100}{(1.5/5.3) + (14/3) + (53.5/2.2) + (31/8.5} = \mathbf{2.2\%}$$

Upper limit of inflammability of LPG in air,

$$L_2 = \frac{100}{(1.5/14) + (14/12.5) + (53.5/9.5) + (31/8.5} = \mathbf{9.5\%}$$

Hence, the limit of inflammability of LPG in LPG air mixture is 2.2—9.5%. **Ans.**

(b) The required composition of maximum velocity mixture of LPG in air is given by

$$L \frac{100}{(x_1/m_1)+(x_2/m_2)+(x_3/m_3)....}$$

$$= \frac{100}{(1.5/10)+(14/63)+(53.5/4.6)+(31/3.5} = \mathbf{6.3\%}$$

It means that 6.3% LPG in LPG air mixture will give the maximum flame velocity. **Ans.**

(c) The maximum flame velocity of LPG is calculated from the following formula :

$$S = \frac{aS_a + bS_b + cS_C + ...}{a+b+c+...}$$

where, a, b, c = % composition of constituent combustible gases

S = flame velocity of the mixture

S_a, S_b, S_c ... = flame velocity of constituent gases of LPG

or, $$S = \frac{(1.5\times39)\times(14\times46)+(53.5\times45)+(31.44)}{100}$$

or, $S = \mathbf{44.7\ cm/sec}$

Hence maximum flame velocity of LPG in LPG-air mixture is 44.7 cm/sec corresponding to 6.3% LPG in LPG-air mixture. **Ans.**

Ignition temperature, explosion limit and maximum adiabatic flame temperature (MAFT) of some fuel gases are tabulated in Table 35.11.

Table 35.11: MAFT, SIT & Explosion Limit of Gases

Fuels / Gases	Maximum Adiabatic Flame Temperature, °C		Ignition Temperature, °C		Explosion Limit in air, for upward flame propagation at atm. temp, and pressure % by volume in oxygen				stoichiometric mixture, volume %
	In air	In oxygen	In air	In oxygen	Lower limit	Upper limit	Lower limit	Upper limit	
CO	2100	2925	609	588	12.5	74	15.5	94.0	29.5
H_2	2045	2660	572	560	4	75	—	—	29.5
CH_4	1875	—	632	556	5.3	15	—	—	9.47
C_2H6	1895	—	472	—	3	12.5	—	—	5.64
C_2H_2	2325	3137	305	296	2.5	80	2.8	93.0	7.72
C_6H_6	2290	—	580	566	1.4	7.1	—	—	2.71
C_2H_5OH	—	—	558	—	4.3	19	—	—	6.52
Coal Gas	1918	—	—	—	5.3	32	—	—	—
Water Gas	1930	—	—	—	7	72	—	—	—
Producer Gas	1663	—	—	—	17	70	—	—	—
Gobar Gas			650		5	15	—	—	—

Coke oven gas	2000	—	560	—	5	32	—	—	—
B.F. Gas	1460	—	680	—	37	71	—	—	—
LPG	—	—	—	—	1.8	8.5	—	—	—
Natural Gas	1930	—	—	—	4.8	13.5	—	—	—
Fuel Oil	2125	—	300	290	—	—	—	—	—
Gasoline	—	—	280	—	1.4	7.6	1.4	46.0	—
Diesel	—	—	247	242	—	—	—	—	—
Kerosene	—	—	295	—	0.7	5.0	—	—	—
Coal tar	2230	—	—	—	—	—	—	—	—
Wood charcoal	—	—	200—350	—	—	—	—	—	—
Wood	—	—	300	—	—	—	—	—	—
Lignite (dry/wet)	2060/1890	—	250—350	—	—	—	—	—	—
Bituminous coal	2200	—	380—420	270—290	—	—	—	—	—
Anthracite	—	—	500	340	—	—	—	—	—
Soft coke	—	—	425—500	—	—	—	—	—	—
Hard coke	—	—	500—650	—	—	—			—
Petroleum Coke	2115	—	—	—	—	—	—	—	—

Turn down ratio of the burner. In a premix burner where primary air and gaseous fuel mix before entering in the furnace, if the gas-air mixture velocity is high then the flame lifts off the burner tip whereas if its velocity is very low *i.e.* less than the flame velocity then the flame travels back inside the burner (which is called *backfiring or flashback).* Both the extremes are undesirable.

Flame lift off limits the maximum output of the burner, whereas flame back firing limits the minimum out put of the burner. Hence turn-down ratio of a burner is defined as

$$\text{Turn-down ratio} = \frac{\text{Maximum output of burner}}{\text{Minimum output of burner}}$$

Its value is generally 5 : 1 for most gas fired furnaces.

Character, shape and size of the flame : Fuel gases burnt at very high combustion velocity will produce very little luminousity regardless of the kind of the gas. CO and H_2 burnt with an invisible to a clear flame while the hydrocarbon gases develop highly luminous flame due to thermal cracking into C and H_2, the C particle heated to incandescence thereby giving the flame luminous appearance.

Deficiency of air increases the luminousity and length of the flame. *Excess of air decreases luminousity and flame length.*

Preheating of combustion air reduces luminousity. Use of steam for atomisation also reduces the luminousity. A luminous flame has a greater ability to transfer heat by radiation from a fixed temperature level. The transfer of heat by radiation varies inversely with the square of the distance between the transmitter and receptor of radiant energy.

COMBUSTION OF SOLID FUELS

During combustion of solid fuel/coal in stoker fired furnaces (used for firing coal upto 50 tons/hr), first the volatile matter is driven off and the carbon in the residue is oxidised to produce CO and CO_2.

The volatile matter and CO are burned over the grate by secondary air admitted over the fuel bed. The temperature of the fuel bed and at the top (hottest part of the bed) varies from 1230°C (at low rates) to 1510°C (at high rate). Capacity of stoker fired furnace is determined by the amount of primary air and effective use of secondary air determines the efficiency of combustion. Approx. 20-30% of excess air will permit combustion of gases within 2-2.5 metres above the grate.

Stokers are classified according to the travel of the fuel.

In *Overfired Stoker*— the fuel is fed on the top of the bed.

In *Underfeed or retort stoker*—the fuel is fed at the bottom or the side of the bed.

A travelling or chain grate stoker carries the bed horizontally on the flat upper surface of conveyor.

Modification of these types is spreader stoker which projects the coal into the furnace above the fuel bed and the fuel is burned both in suspension & on the fuel bed.

Fuel bed of a stoker fired furnace is relatively thin (25-75 cms) compared to a gas producer bed but similar reaction zones occur.

In overfired stoker, the ash zone is immediately above the grate, followed by oxidation, reduction and distillation zone.

In underfired / retort stoker, the distillation of volatile matters takes place in oxidising atmosphere and the volatile matters pass through incandescent residue from combustion rather than through green coal, as in the case of overfired stoker.

Normal combustion rate on coal fired stoker is 15—30 kg coal/ft^2 grate area/hr.

COMBUSTION OF LIQUID FUELS

Combustion is obtained by atomisation which breaks up the fuel into fine, mist like globules increasing area for contact between the air and fuel.

The small particles of liquid fuels either vaporise to form gaseous hydrocarbons to burn to CO_2 and H_2O or it cracks to form carbon soots and H_2 which also burn to CO_2 & H_2O when combustion is complete. The first condition predominates with good atomisation and proper mixing with sufficient air. A deficiency of air or poor atomisation causes smoke.

LIQUID FUEL BURNERS

Burners using steam or air for atomisation are classified as :

Inside mixing type. Here fuel and steam/air are mixed inside the burner system. This burner is used in open hearth furnaces and large reheating furnaces. It is also called *emulsion type or nozzle mix type burner.*

Outside mixing type. Here fuel and air/steam meet immediately outside the burner tip.

COMBUSTION OF GASEOUS FUELS

The degree of mixing gas and air is controlled by burner design. Burners have been developed to produce short, intense, non-luminous flame to long, slow burning *luminous flame*. This relation is not always valid as the gas must contain hydrocarbons to develop luminosity.

Burners capable of producing short, intense flame will liberate large amount of heat in small space. Some gases (e.g. CO gas) have high rate of heat release whereas some gas (*i.e.* B.F. Gas) has a very low rate of heat release.

Length of the flame is limited by the ability of the flame to provide enough heat to propagate itself. Carrying an excess or deficiency of air is usually practised to control scale formation and sometime to control flame characteristics.

Excess air shortens the flame and causes the loss of heat. Deficiency of air lengthens the flame and results in loss of fuel by partial incomplete combustion.

Fuels containing H_2 must have high exit stack temp, to avoid condensation of water within the furnace system and subsequent corrosion to furnace parts. Sometimes it is necessary to burn the fuel with excess air to maintain the flue gases above their dew point.

Atomising agent in large furnaces is steam at 4-10 kg/cm^2; consumption of atomising steam is *0.3-0.7 kg / kg fuel*. In smaller furnaces atomisation is by compressed air or by mechanical action (swirl, rotary motion).

Size, shape, luminosity of a liquid fuel flame for a fixed burner design is changed by changing the degree of atomisation controlled by steam pressure/air pressure.

LIQUID FUEL BURNER

Burner using steam or air for atomisation are classified as :

Inside mixing type. Here fuel and steam/air are mixed inside the burner system. This burner is used in open hearth furnaces and large reheating furnace. It is also called *emulsion type or nazzle mix type burner.*

Outside mixing type. Here fuel and air/steam meet immediately outside the burner tip.

Types of Combustion Processes

The potential energy of fuels is released by combustion process which being chemical in nature involves reactions of fuel components with oxygen. Heat of reaction is evolved and the temperature of the combustion products (which are in the gaseous/vaporous form) is raised. It tends to occupy a large volume at the increased temperature attained or generate high pressure if the volume is not allowed to rise.

Combustion processes are characterized by the formation of flame (the hot and light-emitting gaseous zone which separates the combustion products from the fuel). Flame appears just after the fuel is ignited and a sufficiently high temperature is created, and tends to persist slightly beyond the burning zone till a temperature drop renders the gases completely non- luminous. A flame thus covers luminous burning & after burning zones and is confined between the ignition zone & completely non-luminous combustion gases. A non-luminous bunsen flame is actually a flame of low luminosity. The flame front is the surface between the luminous region and the unburnt fuel. A well-defined flame front is obtainable with gaseous fuels. Flame thickness or flame length depends upon many factors *viz.* types of fuel, presence of diluent (nitrogen in air, recirculated flue gases), intensity of mixing of fuel & oxidant, etc. A fast-burning combustible mixture has a very thin flame where ignition, combustion and luminescence occur almost simultaneously. Since a flame separates, fuel from combustion products, it is the zone of very steep concentration and temperature gradients. It is this characteristic which distinguishes flames from chemical reactions in general.

Because of the luminosity of flame, some fuels like kerosene are used as illuminants. Certain devices such as gas mantle are used to augment the flame luminosity.

There are various types of combustion processes depending upon the means of contacting fuel & oxidant and the conditions of the flame. These are briefly described below :

Combustion with Stationary Flame

This is the normal combustion process as practised in ovens/furnaces. The resultant flame front is more or less stationary in space. A stationary flame may be '*premixed*' or *diffusion.* In the premixed type, fuel and oxidant are premixed before they enter the burning zone. When the fuel & air are separately supplied to the burning zone, the flame is called a diffusion flame. In practice, a part of the total air may be premixed with the fuel and the remaining may be directly supplied to the combustion area. The premixed air is known as *primary air* and the rest is called *secondary air.* While solid, liquid & gaseous fuels can all give *premixed flame,* a truly *diffusion flame* is obtained only with *gaseous fuels.*

A stationary flame may be *laminar* or *turbulent* depending upon the velocity of premixed system. In the case of a diffusion flame, both the direction of flow and the velocity of fuel &oxidant determine the nature of the flame. A laminar diffusion flame is favoured by low and equal velocity & parallel flow of gas and air. When they flow at an angle to one another with higher & unequal velocity, a strongly turbulent diffusion flame results.

In general, the diffusion flames are considerably larger than premixed flames for the similar conditions of burning. Laminar flames are longer than turbulent flames. A highly turbulent flame is unsteady and irregular in shape. When the air hole of a Bunsen burner is completely closed, a laminar diffusion flame is obtained which is long, steady, quiet, noiseless, well-shaped and luminous. If the *big air-hole* is fully opened, a turbulent premixed flame results which is *short, unsteady, flickering, noisy, irregular-shaped and of low luminosity.*

In diffusion, unlike the premixed flame both the fuel and air can be preheated before combustion. For an intense combustion in a comparatively short space which is the aim in usual thermal appliances, a truly laminar, diffusion flame is rarely used. However, long flames are desired in coke oven & open hearth furnace where it is desired to have a long flame for setting an extended zone of uniform temperature.

Surface Combustion or Flameless Combustion

All refractory solid surfaces at high temperatures accelerate the rate of combustion of fuel gas and air. Some solids *e.g.* platinum can accelerate the process even at low temperatures. Combustion with a stationary flame is limited by a range of velocity and concentration conditions of the gas & air. Stable combustion is possible even outside this range, if the reaction proceeds in contact with solid surfaces. This is termed as *surface combustion.* Its industrial application is to achieve rapid combustion of a large quantity of fuel in a comparatively small space with the production of high temperature and high heat transfer rate. Tunnel burner is a type of gas burner operating on the surface combustion principle in which normal flames are absent.

Submerged Combustion

It is a special case of application of surface combustion process in which, the burner is partly or fully submerged in a liquid and the hot combustion products bubble through it in an agitated condition. High heat transfer rate results from direct contact of hot gases with the liquid leading to high evaporation rate.. Up to 95% of the potential heat of the fuel may be useful heat for the process. Submerged combustion finds application in the evaporation of severely scaling/corrosive solution/liquids.

Combustion with Explosion Flame

It occurs in a homogeneous mixture of fuel & air and is characterized by the flame front progressing rapidly through the mixture. The process may be either constant pressure *e.g.* mine explosion, or constant volume *e.g.* combustion in a gasoline engine. *Detonation* is a special type of explosion, where the extremely high reaction rate generates high velocity pressure waves (1 to 4 km/sec) and an abnormal rate of pressure rise.

Pulsating Combustion

Pulsating combustion occurs when one end of a long tube is open & the other is closed by a non-return valve, and the air & fuel are introduced at the closed end. On ignition, the pressure in the system rises sharply at near-constant volume and it prevents the flow of the air & fuel momentarily ; when the exhaust gases leave the tube through the open end, a fresh supply of air and fuel arrives at the hot zone and combustion is repeated in the form of pulsations whose frequency corresponds to the resonant frequency of the combustion unit. There is no stationary flame in the system. Pulsating combustion is a specific type of explosion flame. During the operation of a pulsating combustion unit, a standing pressure wave is produced in the pipe with a pressure node (velocity antinode) at the exit end and a pressure antinode (velocity node) at or near the combustion zone. The pulsating combustion system gives higher rates of combustion and heat transfer than a conventional burner. The noise levels of the pulsating combustors can be reduced by simple silencers.

Slow Combustion :

This takes place at sub-flame temperatures (< 400°C) at slow but determinable rates. Slow combustion of higher hydrocarbons is useful in determining the chain-reaction rates. In a premixed system of fuel vapour and air, slow combustion proceeds at a number of points simultaneously in the whole system. No reaction zone or flame-front is visible. This process is called homogeneous combustion which is often characterized by the appearance of *cool flames* in succession which emit small quantities of heat and pale bluish light usually seen only in the dark. Several cool flames (up to four) may follow one another. Cool flames are a feature of gas-phase oxidation of organic substances below 400°C. Slow combustion has no direct industrial application. However, it is indirectly useful in the study of mechanism of combustion.

COMBUSTION OF SOLID FUELS—GRATE FIRING & PULVERISED FUEL FIRING SYSTEM

When coal and air are brought in contact with each other at ignition temperature, then the coal decomposes into fixed carbon & volatile matter causing the fixed carbon to ignite. Combustion then becomes self supporting *i.e.* sufficient heat is evolved to maintain these conditions.

COMBUSTION OF SOLID FUEL LUMPS ON GRATE

The combustion of lumps on grate (which is mounted parallel to one another) in which primary air is passed through the gap between the bars is a class by itself. A glowing fuel bed with an ash layer underneath is maintained on the grate and fresh charge of fuel is fed into it from time to time corresponding to the combustion rate. The insulating ash layer protects the firebars from the intense heat and also helps in air distribution.

The fresh charge on heating undergoes drying & decomposition and the combustible volatiles are burnt in the 'secondary combustion space above the fuel bed by secondary air. This combustion more or less fits in the stationary flame process. The residual solid fuel undergoes true primary combustion on grate. It is a surface process involving heterogeneous reactions between carbon and oxygen. The liberated heat converts the fuel into a glowing mass which radiates heat to the heating surface and also to the fresh charge of fuel. Carbon monoxide is a product of combustion on grate and takes part in the secondary combustion.

The above method of burning solid fuels is known as overfeed firing. In underfeed firing, fresh fuel is fed to the glowing bed from below. In cross-feed firing, the fuel is fed transversely across the front of the grate.

In *down-jet combustion* in which the grate is dispensed with a jet of air is impinged on the surface of the fuel bed. The sides and bottom of the bed are enclosed. The combustion products leave the bed through the upper surface. Fresh fuel is fed from the top while refuse is periodically removed from the bottom. The principle of down-jet combustion is practically not successful due to difficulties in ash removal.

Combustion control: For burning all the combustible matter in the coal, the air supply is to be controlled without using undue excess. If too little air is supplied some combustibles like hydrogen, hydrocarbon, carbon soot, carbon monoxide etc. leave the furnace without burning thus lowering the efficiency of the furnace. If too much excess air is used, the combustion efficiency is reduced because heat is required to raise the excess air to flue gas temperature.

Combustion rate of coal on grates is mainly determined by the rate of air supply which is controlled by the draught available in the furnace. The draught required in the furnace depends upon the suction or pressure available from chimney or fans & the resistance to the flow of air to the furnace and of flue gases from the furnace offered by the fuel on the grates and by the size & the shape of the furnace, heating elements and flues.

Two types of air are supplied to burn coal on a grate efficiently namely:

(a) **Primary air.** It is introduced under the grate to burn the fixed carbon of the coal.

(b) **Secondary air.** It is introduced over the grate and required to burn the volatile matter produced from coal.

Sometimes, at higher altitude over the grate, tertiary air is introduced to burn the left out volatile matter and smoke which could not be burnt completely by secondary air either due to its deficiency or lack of thorough mixing of volatile matter with it.

Ratio of primary air to secondary air increases with increase in rank of coal due to increasing ratio of fixed carbon to volatile matter. Caking characteristics of coal also influences the resistance to the flow of primary air offered by fuel bed.

Sufficient time is required for air to react with solid particles of fixed carbon on the grate and volatile gases above the grate. The former depends upon the thickness of the fuel bed and reactivity of the fixed carbon which is high for low rank coals and low for high rank coals. The later depends on the mixing of air with flue gases and upon the shape & size of combustion chamber. Turbulence in the combustion chamber improves the mixing of air and gas which can be achieved by directing jets of secondary air downwards towards the grate or by supplying this air with a rotary motion.

Excessive cold air supply mainly causes black-smoke or carbon soot to be formed due to cracking of hydrocarbon gases and incomplete combustion. Though the thermal value of the smoke can be neglected but the nuisance value is considerable. High volatile fat coals are liable to produce black smoke whereas low volatile lean coal (or coke) hardly produces smoke. Preheated air supply in controlled amounts, together with good design of furnace prevents the formation of black smoke.

GRATE FIRING SYSTEMS

Lumpy solid fuels are generally burnt on grates. The various systems and overfeed stokers (which includes *sprinkler spreader* and *coking stoker*), *travelling grate or chain grate stoker*, and *underfeed or retort stoker.* Hand firing is resorted when thermal output of the furnace need be low.

Overfeed stokers : In this system fresh or green coal is fed by dropping onto the fuel bed. In sprinkler spreader system, coal is fired by shovel or rotor onto different parts of the grate. The grate may be fixed or rocking. Rocking breaks up the clinker and permits longer run without hand cleaning. But, if the coal contains fines then they are thrown into gas stream and are carried forward into chimney gas causing grit emission nuisance.

In the ram feed or coking stoker which is used for burning coking coal, the coal from the hopper is fed onto the front of the grate where volatile matter is driven off and coke remains. A moving grate then imparts forward motion to the coke. It provides high combustion efficiency but sudden large fluctuation in demand can not be met easily and high draught is necessary to draw air through thick fires.

Majority of industrial grate fired boilers operate on overfeed principles. The fuel bed thickness varies from 10 to 30 cms and is divided into four zones. The topmost zone is distillation zone or green coal zone, the next zone is reduction zone followed by oxidation zone and the ash zone.

The primary air enters from below the fuel bed & passes upward through the grate in the ash zone and gets heated in turn cooling the ash. The oxygen in the primary air reacts with a part of the carbon in the coke bed to form carbon dioxide in oxidation zone. When the oxygen becomes depleted from the rising hot gases through the fuel bed, the coke reduces the carbon dioxide back to carbon monoxide in the reduction zone. The oxidation reaction ($C + O_2 = CO_2$)being exothermic promotes the reaction. The reaction in the reduction zone ($C + CO_2 = 2CO$) being endothermic lowers the temperature of products. In the distillation zone, the combustion gases consisting of CO, CO_2 & H_2 heat up the fresh coal so that the moisture is vaporised and the volatile matter is driven off the coal. The volatile matter (VM) and carbon monoxide then react with the oxygen of secondary air which enters from above to complete the burning. The various reaction zones in an overfed fuel bed have been shown in the Fig. 37.1.

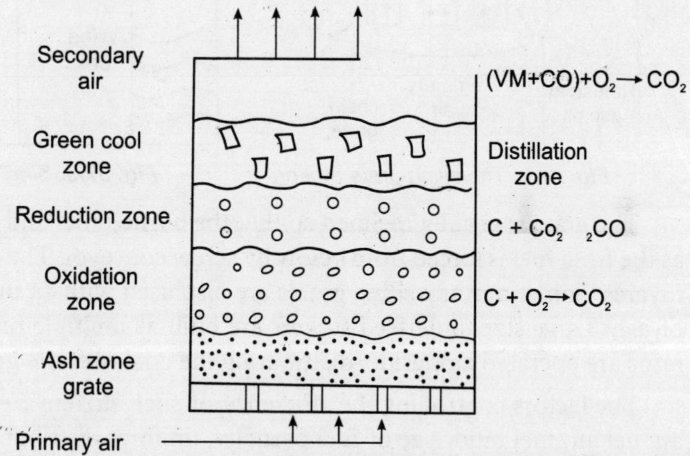

***Fig. 37.1.** Cross-section of overfeed fuel bed.*

Travelling grate or chain grate stoker: This is an endless metal chain conveyor through which primary air passes, which carries fuel into the furnace and ash from the furnace. Coal feed rate is controlled by the thickness of the fuel bed admitted and by the speed of the grate. It can burn any type of small coal effectively. However, it can not burn low ash coal (< 5%) because of the necessity of protecting the grate from heat. It can also not burn slacks containing high proportion of fines (> 30% < 3 mm) which block the air passage in the grate and the coking coals.

Its grate has more closely interlocking bars. The coal enters at one end and by the time it reaches the other end, the combustion is complete & the ash falls into the ash pit.

In the travelling grate stoker, best results are obtained by balanced draught *i.e.* primary & secondary air are supplied under pressure at the front of the furnace and flue gases removed under suction at the back.

Underfeed or retort stoker : In underfeed stokers, the fresh fuel is forced into the fire from below by a screw and moves out to the sides as it burns. The ash that accumulates at the sides is removed by agitating a heap of ash into a hot coke zone and forming a clinker which can be pulled with tongs. The advantage of the stoker is that the volatile matter distilled from the fresh coal passes through the oxidation zone thus ensuring its complete combustion.

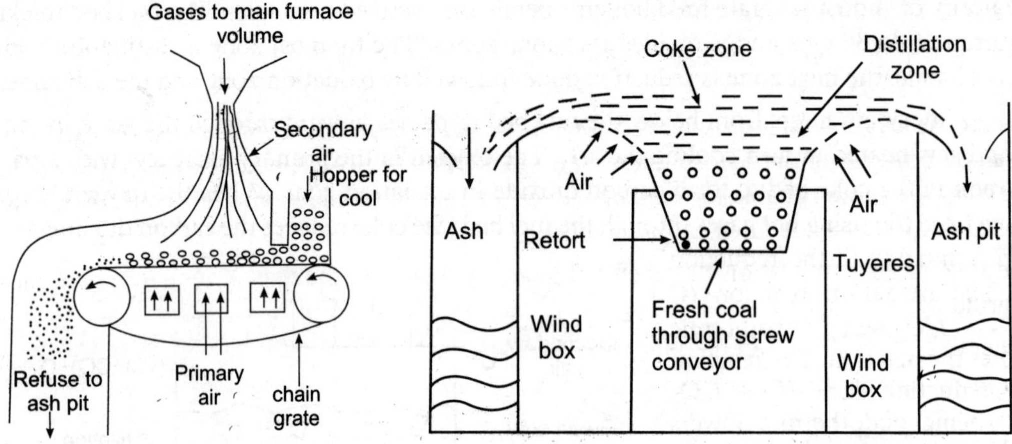

Fig. 37.2. *Travelling grate stoker.*

Fig. 37.3. *Single retort underfeed stoker.*

The grate is usually inclined so that the burned fuel and the ash automatically move outwards as the fresh fuel is forced from below by screw conveyor. The air for combustion is supplied through tuyeres. Sometimes travelling grates are also used with an underfeed mechanism for easy removal of ash. Large size underfeed stokers are built as multiple retort units. Several hoppers and power rams are operated alongside, distributing the coal over the grate surface.

The factors controlling the efficiency of such stokers are proper ratio of primary & secondary air, net air/fuel ratio, size of fuel particles, uniform distribution of coal & air over grate and size & shape of combustion chamber. Some unburnt fuel may pass through the grate while fusion of ash or clinker formation may result in an uneven distribution of air. Such factors reduce heating rate or the efficiency of the stoker.

Advantages of mechanical stokers over hand firing :

- Operation is continuous if stoker is self-cleaning, therefore, no loss of efficiency due to intermittent operation during charging and cleaning.
- Uniform feed rate and combustion operation make automatic control easy.
- Higher CO_2 is possible in flue gases" with consequent higher efficiency.
- Larger grates can be used with higher outputs.
- Lower labour costs.

Characteristics of good mechanical stokers :

- Must burn fuels of varying quality smokelessly and efficiently.
- Must be self-cleaning.
- Maintenance cost must be low and replacement quickly available.
- Observation of fire in furnace must be possible.
- Must be capable of ready adjustment to suit varying loads.
- Must be capable of automatic control.

PULVERISED COAL COMBUSTION

The rate of combustion of solid fuels is slow because of the difficulty of contact between the fuel and air. The combustion rate can be increased by increasing the rate at which air is forced in but this means a supply of excess air resulting in waste of heat. The combustion rate can also be increased by pulverising the fuel so that air and fuel come in close contact easily. Generally high volatile coals are more suitable for making pulverised fuel. The volatile matter is quickly liberated & burns and thus helps the fixed carbon to burn.

The smaller the combustion chamber and lower the volatiles, the finer & more uniform should be the grinding.

Generally, the coal should be pulverised such that 80% of it passes through 200 mesh sieve. Further grinding may prove more uneconomical as power required in grinding will be high.

Advantages of Pulverised Coal :

- It can be intimately and uniformly mixed with air necessary for combustion. Hence it can be burnt completely.
- It can be handled with ease like a liquid fuel and can be transported through pipes.
- Furnace temperature can be easily controlled by increasing or decreasing the rate of burning through a valve just like liquid/gaseous fuel.
- The atmosphere inside the furnace can be easily maintained as oxidising or reducing which is of great advantage in the case of metallurgical furnaces.
- Wide variety of coals can be used.
- Low grade coals can be used provided special precaution are taken with high ash coals to avoid carrying dust and grit into atmosphere.
- Combustion is completed with low percentage of excess air, therefore high flame temperature and high thermal efficiency can be achieved.
- Maximum efficiency is possible because of close regulation of rate of feed and supply of air by automatic control.
- Labour charges are low and maintenance is largely exterior to furnace.

Disadvantages of Pulverised Coal :

- Relatively high cost of drying and grinding to very fine particle size.
- Tendency for slagging on refractory walls and furnace linings.
- Discharge of fine dust and grit into atmosphere.
- Necessity for large combustion space to complete combustion & heat transfer and to avoid deposition of soot in tubes of flues.
- Contamination of product in furnace work by ash from coal.
- Difficulty in burning anthracite.
- Erosion of pressure boiler parts by fly ash entrained in flue gases. Reduced availability and high maintenance cost of boiler.

- Erosion of I.D. fan blades by entrained ash in the flue gas.
- High maintenance and operating cost of pulveriser.

PULVERISED FUEL FIRING SYSTEMS

The object of pulverised fuel burner is to obtain good mixing of air and coal. There are several firing systems for pulverised fuel namely horizontal firing, vertical firing, tangential (or corner) firing and cyclone firing systems.

Coal size in the pulverised coal burner should be 85% passing through 240 mesh screen (*i.e.* < 0.063 mm). These particles are blown into the combustion chamber by hot primary air where it burns like a droplet of liquid fuel. First the volatile matter burns in a spherical diffusion flame surrounding the particle and then the remaining coke undergoes burning at the surface. Higher volatile matter and lower ash content give higher flame speed (maximum being 15 m/sec). For pneumatic transportation of coal particles through horizontal pipe, the air velocity should be more than 20 metres/second.

As coal & air enters the furnace, the speed falls and a stable flame is obtained. The flame should stabilise at distance of 10 to 30 cms from the nozzle.

The size of the combustion chamber depends upon the speed and size of coal particles. The time required for a coal particle to burn depends upon the size of the particle and the temperature of the furnace. Distillation takes place in 0.005 seconds. Half the coal burns in about 0.05 seconds and within 0.1 to 0.3 seconds 95% of coal burns. If combustion is not complete, coal particles are carried away by flue gases. However, carbon monoxide is usually not present in pulverised coal burners.

There are two types of pulverised coal supply system for power plant boilers namely *(a) Centralised Bin and Feeder System* and *(b) Unit feeder type dedicated system*.

Bin and feeder coal supply system shown in Fig. 37.4 comprises of storage bin for raw coal, raw coal feeder, pulveriser (ball mill) with drying facility by hot air (drawn from air preheater of boiler), separator, pulverised coal dust storage bunker and coal dust feeder to the burners of boilers. More than one boilers can be fed coal from this fchis system and coal supply does not get stopped to the boilers in case of breakdown/shutdown of ball mills as stored coal dust in the bin can be supplied during this period.

Unit feeder type coal pulverization & supply dedicated unit system shown in Fig 37.5 supplies coal to only one boiler and in case of any breakdown/shutdown in the ball/bowl mill, the coal supply to the boiler stops as there is no coal dust storage bin provided in this system. The advantage of this system is lower initial investment, simplicity of construction, space saving and better control of amount of coal supplied to each burner. The pulveriser is placed near the boiler & under the raw coal feed hopper and is driven by a shaft that also drives the fan. The coal is dried in the mill by heated primary air (which is 25% of total combustion air) drawn by the mill fan. Hot air carries the coal dust from the bowl mill to the burner. The secondary air for combustion is supplied at openings round the burner and/or into the combustion space.

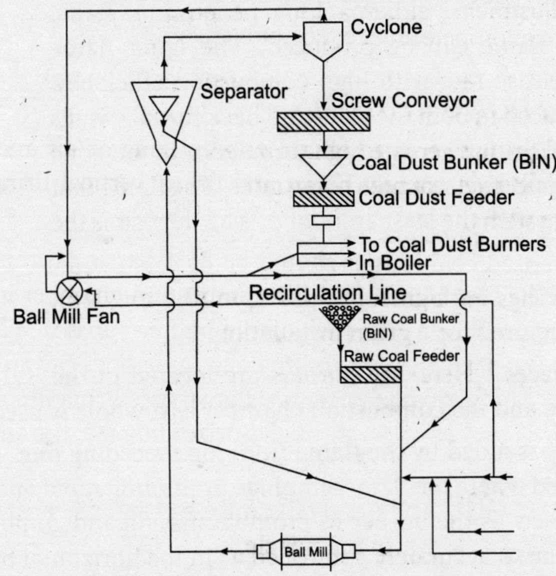

Fig. 37.4. Bin and Feeder System of coal supply.

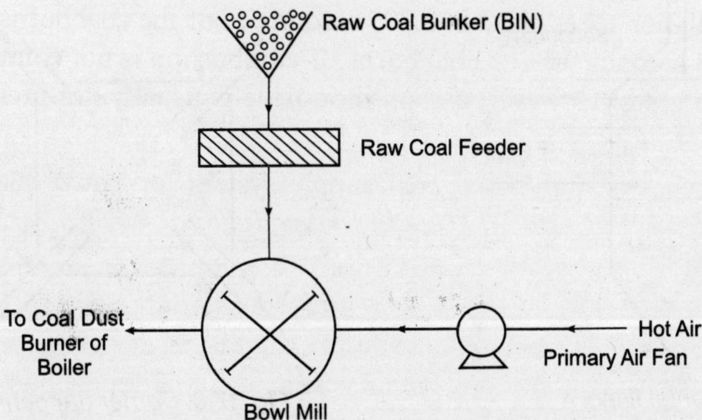

Fig. 37.5. Pulverised coal supply system in boilers (unit feeder type dedicated system)

Vertical fired furnaces : Here, pulverised coal and primary air are fired directly downwards into the combustion space through long narrow rectangular burners. Secondary air is admitted through small parts down the front of furnace walls. By admitting more secondary air high up the furnace wall near burner, the flame produced can be made short and intense. A long diffusion flame may be produced by admitting secondary air progressively down the entire height of the furnace.

Vertical firing provides a uniform temperature distribution throughout the furnace and is especially good for high, narrow furnaces. Low volatile coal which requires a long flame travel to allow time for complete combustion, can be fired vertically better than any other method although this design is equally good for high volatile coals.

Horizontal fired furnaces : Here, burners are placed horizontally or slightly up or down.

By proper burner adjustment, either a long penetrating flame or short sharp turbulent flame can be produced. The latter flame yields a very high heat release rate with high combustion efficiency. Sometimes, burners are placed in both the front and back furnace walls to give opposite horizontal firing to create high turbulence in the zone where flame impinges permitting extra high firing rates. The flame that creates the most turbulence with the least amount of slag deposit is the best.

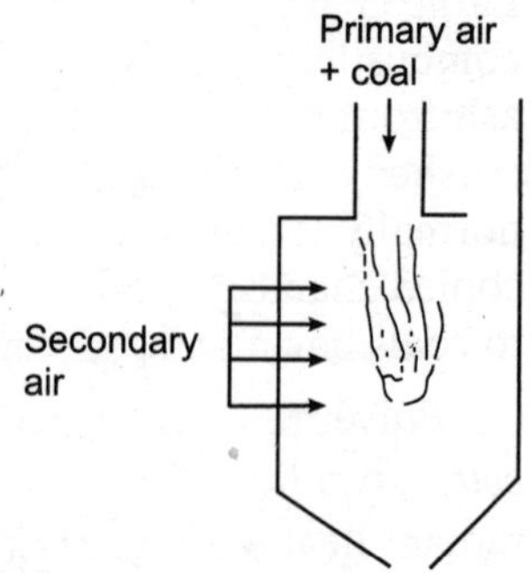

Fig. 37.6. *Vertical firing.*

Horizontal burners are having high capacities upto (0.5 tons/hr), hence fewer burners are required for a given installation.

Tangential fired furnaces : Here, the burners are located in the four corners of the furnace and the combustion chamber as a whole is used to produce swirl.

Ignition at each burner is aided by the flame from the preceding one. An intensive, turbulent, rotative motion is developed where the flames impinge promoting rapid and complete combustion. Since, the whole furnace acts as the burner to promote mixing and combustion, the air and fuel from each burner need to be so accurately controlled as in the horizontal burner.

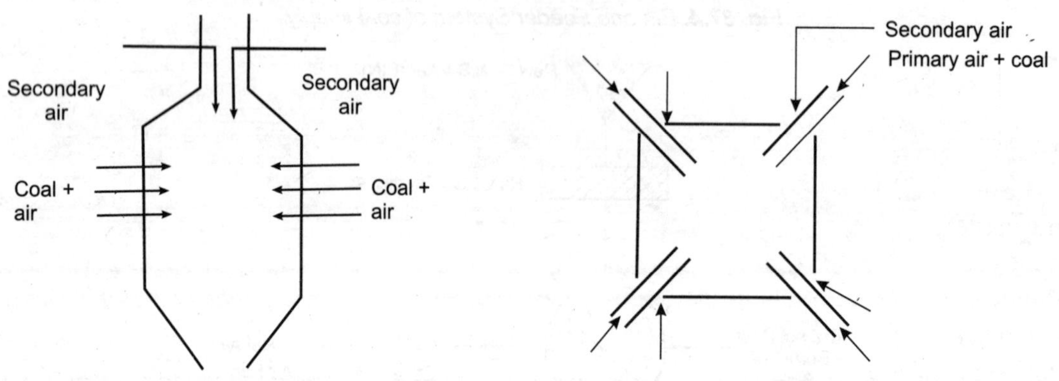

Fig.37.7. *Horizontal firing.*

Fig.37.8. *Corner (tangential) fired furnace.*

Tangential firing permits extra high rates of heat release and gives high combustion efficiency even with little extra air.

Cyclone firing furnaces : Here, the primary air and coal is fired tangentially into the cylindrical combustion chamber.

Here, the fuel swirls forward into the main chamber where it meets the high speed tangential stream of secondary air. Some, tertiary air is supplied at the axis of the chamber to ensure burning of any fine coal particles, suspended near the centre of vortex. The temperature developed inside the furnace is high, hence the ash melts away and ash globules are carried away to the wall by centrifugal force. The furnace is bit inclined to permit the molten ash to flow down.

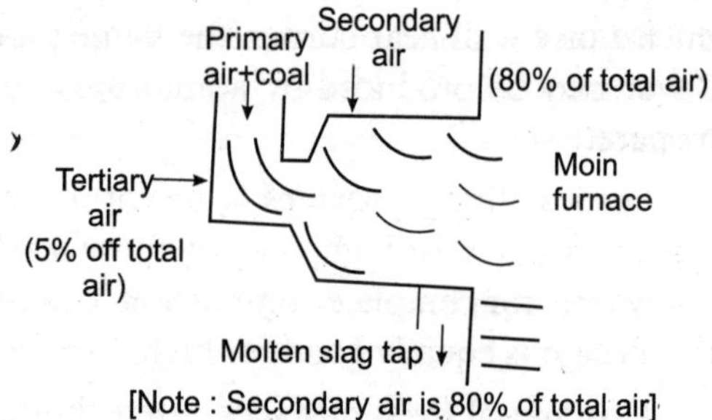

[Note : Secondary air is 80% of total air]

Fig. 37.9. *Cyclone furnace.*

Large coal particles may adhere to the molten ash where they are subsequently burned. A high combustion efficiency is attained with 10-15% excess air. This furnace separates out most of the ash from the flue gases and only about 10% of it is carried through. In cyclone burner, the coal crushed below 6 mm (thus avoiding high cost of grinding especially for lower rank coals which are normally tough) is burnt under conditions of intense turbulence in a comparatively small water cooled chamber. Here, 80-90% of the ash is extracted in the molten state and is therefore applicable to coals having an ash slag viscosity of 250 poise at about 1300°C.

Pulverised fuel furnaces are also classified depending on the method of ash removal as *(i) bottom type furnace (ii) slag tap furnace (H_2) cyclone furnace* etc. In the first, ash particles are cooled by radiant heat exchange with walls ; small particles (called *fly ash)* pass forward with gases (80-85% of ash), larger particles fall to the bottom and are recovered hydraulically (10-20% of ash). Flue gas may be recirculated in convector heat system to cool ash additionally. In the second, ash melts in the furnace and flows down into a tank of water where, it is granulated (40-50% of ash). Low fusion point ashes are preferred in this case.

PULVERISED COAL BURNERS

The types of burners used in firing pulverized coal are the following :

Type of coal used	Burner type
High volatile	Short flame with self induced turbulence
Medium volatile	Short flame with externally induced turbulence
Low volatile	Long flame or U-type.

The principle of short flame burner is that relative motion is induced between the coal particles borne in the primary air stream and the secondary air. Both streams may be given rotary motion so that the coal particles are thrown outwards and dispersed in the fresh supply of oxygen. This rapid & continuous contact of coal with the air is achieved in one burner by applying resonant vibrations to it and in another (the Multi-jet) by dividing the burner area into a number of adjacent streams.

Short flame burners are usually arranged on one furnace wall so that the flame pattern fills die chamber but does not impinge on water cooled walls.

Medium flame burners are in use in many pulverized fuel boilers where coal is of medium volatile type. The nozzles supplying the coal-air mixture are set centrally within rectangular secondary air ports so that a rather long annular jet is obtained surrounded by heated secondary air. The necessary turbulence is achieved by the arrangement of the burners so that an intense swirl is set up. The burners may be arranged in opposite walls or tangentially to an imaginary circle centred on the middle of the combustion space. The walls of the furnace are normally composed of water cooled bare tubes to absorb heat by radiation as well as convection. It is usually necessary to operate such a system on unit mills in order to make sure of uniform distribution.

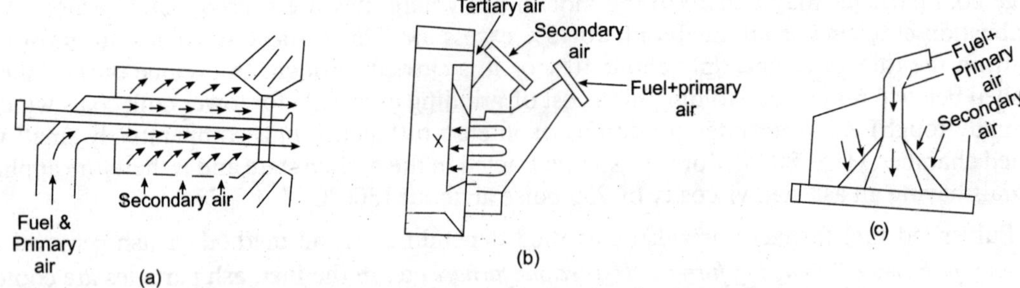

Fig. 37.10. *Pulverised coal burner. (a) Circular burner (b) Multi-jet burner (c) U flame burner*

The long flame burner was first form to be used and was arranged to fire downwards into the combustion chamber. The secondary air, the tertiary air if required, is introduced through the furnace wall at right angles to the flame, the arrangement of the ports is such as to give good control of the flame shape and to achieve complete combustion. The U-shape of the flame has given the name U-flame to these burners. They are most suitable for coal of low-volatile matter since complete combustion can be achieved in the long time of travel. The velocity of flow is however slow and involves the disadvantage of large combustion spaces. They are however, suitable for furnace work in which a number of units are fired from a *bin and feeder system.*

WET BOTTOM FURNACES

The problem of burning very high ash coal or brown coals in boiler furnaces is the removal of ash. Furnaces are therefore designed to remove ash in the liquid form. The temperature of the flame is made very high to enable the ash to melt at the lower part and the molten mass trickles through the furnace wall and is removed. This type of furnaces are called as "Wet Bottom Furnaces". The walls are fitted with heating tubes on all sides and the water is evaporated inside. The molten slag itself is a good heat insulator which protects the lining of furnace wall. The thickness does not exceed a certain value.

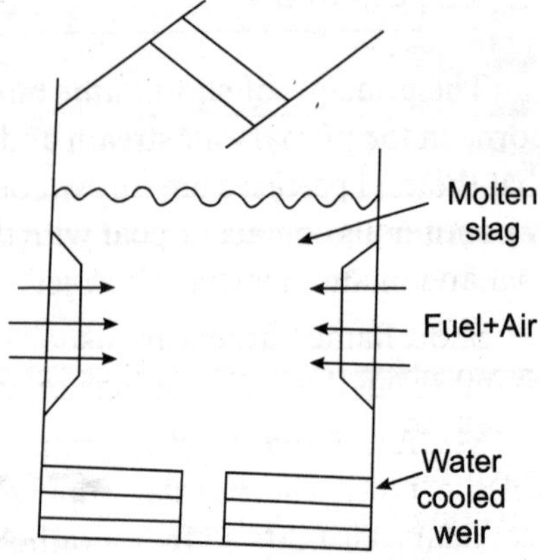

Fig. 37.11. *Wet-bottomed Furnace.*

Droplets coming from first time solidify and a layer is formed of a considerable thickness. The slag does not solidify further. The surface temperature reaches the melting point of the slag as it is a good insulator. The layer is very viscous and it does not move very fast. The rate is about 0.10 cm/sec, which causes a liquid film of slag to form.

These type of furnaces are used in boilers where the fuels obtained are of only inferior quality lignites, having 40-45% moisture and more than 50% ash on dry basis.

Burners for Liquid & Gaseous Fuels Combustion

General Requirement for a Burner/Combustion System

For smooth and efficient combustion, fuel should freely ignite as it enters the burning zone even with load fluctuations within the specified range. Radiation from flame & hot refractory surface, and convection from hot gases should be adequate to ignite the fresh fuel. The composition of the fuel-primary air mixture should be within the limits of inflammability. For obtaining the desired rate of heat release, it is necessary to maintain flame stability throughout the combustion process. Burners should not be operated below or above the stipulated capacity range. The flame should not suddenly come in contact with much cold air surface. All factors causing flame extinction or flash-back are to be avoided. Some flame holders enhance flame stability. High temperature combustion even proceeds with a finite rate hence adequate combustion space should be provided for its completion, otherwise there will be loss of combustibles with the exhaust gases which will also accentuate the smoke nuisance. The brown smoke is due to the unburnt combustible vapours. The black smoke is due to the carbon black produced by the chilling of the flame when it impinges against a cold surface. The flame shape should correspond to the geometry of the furnace and *vice-versa.* The quantity of air supply is important in achieving proper combustion. Excepting some premix systems, excess air is always required for complete combustion. If, however, a reducing atmosphere is desired, excess air is to be avoided. Even otherwise the quantity of air should be the minimum consistent with the requirement of a given system. The method of air supply should be such that there is very intimate contact of the oxygen and the combustibles. This is achieved by creating intense turbulence in the combustion space. A sufficient temperature of the combustion gases should be maintained at all parts of the combustion chamber. This is important in respect of smooth ignition, stable combustion and smoke-free performance of the system. Theoretically the most efficient combustion is that which leads to the maximum temperature possible. However, in practice, keep the furnace temperature below the maximum in many cases where other operation factors are important.

SELECTION OF BURNER

Burner selection for a particular operation depends on the following five characteristics :

- Flame shape
- Combustion volume

- Stability
- Drive
- Turndown ratio

These five characteristics depend primarily on burner design and other factors like increase of primary air pressure, increase/decrease of fuel pressure etc. have very little influence on these characteristics.

Flame Shape :

Burner design, which determines the relative velocities of the fuel and air streams, has the maximum effect upon the flame length & shape. Good mixing, produced by a high degree of turbulence and high velocity, produces a short bushy flame. On the other hand, delayed mixing and low velocities result in long lazy flames.

Combustion Volume :

The space occupied by the fuel and the intermediate products of combustion while burning (flame and invisible combustion) varies considerably with the burner design, the pressures &velocities of the fluid streams, the fuel and the application. Gas Burners can be designed to have a heat release rate as high as 110×10^6 kcal/hr.m^2 of combustion volume. Light oil burners normally operate at the rate of 270000 kcal/hr.m^2 and heavy oil burners at 220000 kcal/hr.m^2.

Stability :

With the improvement in the burner design, the present day burners are much more stable. Burners are now available which are not only the tile stable but also stable in cold and pressurised chambers.

Drive:

Drive relates to the velocity and thrust of the jet stream of hot gases that emanates from a burner. Modern high velocity burners can push the hot gases into a loosely piled load with greater velocities than were possible with most of the older excess air burners.

Turndown Ratio :

A burner's turndown ratio is the heat input rates within which the burner would operate satisfactorily. This is the ratio of the maximum to the minimum heat input rates with which the burner will operate satisfactorily. The maximum input rate is limited by a phenomenon called the flame blow off. This results when the mixture velocity exceeds the flame velocity. The minimum input rate on the other hand is limited by the phenomenon known as flash back.

COMBUSTION SYSTEMS AND BURNER COMPONENTS

Any combustion system must have the following :

- Source of air ;
- Source of fuel;
- A burner ;

- Valves and proportioning devices.

The purposes of a fuel burning system are to :

- position flames at areas of useful heat release
- initiate and maintain ignition
- mix the fuel and air
- volatilize solid and liquid fuels
- proportion the fuel to air and
- supply fuel & air at the proper rates and pressures to facilitate all five previous functions with safety required for the process.

The primary functions of a burner are flame positioning and continuous ignition maintenance without a pilot.

Burner Components :

The heart of any burner is the 'NOSE' which may consist of a single, or multiple nozzles or ports. The nose should be well cooled to protect it from thermal destruction by radiation from the flame & furnace and to prevent flash back in the case of premix burners. Heat resistant cast iron of sufficient cross section assures sufficient cooling by conduction and natural convection in many cases. Other high temperature alloys, refractories and/or artificial cooling. The 'BURNER TILE' is a refractory block with a conical or cylindrical hole through its centre. In addition to serving as the insulating separator between the hot furnace and the cool burner parts, the tile radiates heat into the incoming fuel & air, thus helping to maintain ignition.

The other major elements of a burner are *(a)* The 'BODY' which may incorporate the air/ mixture connection, swirl vanes & the nose *(b)* the 'MOUNTING PLATE' which holds the body, tile, pilot, flame monitoring device & furnace shell in proper alignment *(c)* the 'GAS CONNECTION', gas tube & gas nozzle *(d)* the 'ATOMISER' which includes oil & atomising medium connections, tubes, vanes and nozzles.

The burner 'PILOT' is a small burner used to light up larger main burner. Pilots are generally of three types *viz.,* interrupted pilot, intermittent pilot and continuous pilot. An interrupted pilot burns during the flame establishing period and/or trial for ignition period and is cut off (interrupted) at the end of that period. An intermittent pilot burns during 'light-off & the entire period that the main burner is firing, and is shut-off with the main burner. A continuous pilot burns throughout the entire period that the furnace is in service whether or not the main burner is firing. The interrupted pilot is preferred over continuous or intermittent pilot.

Every burner should have an 'OBSERVATION PORT' for safety's sake, so that the main flame, pilot flame and flame rod (if any) can be seen by the operators.

Burner design should be such as to produce specific flame shape required for a particular furnace design and mode of heat transfer.

In industrial furnaces, the heat transfer techniques used are mainly radiating or convective. Conduction is often a secondary process after the radiation or convection step. Flame shape and velocity are of prime importance in the heat transfer from the combustion processes. The heat transfer technique from the combustion process to the load concerns the heat release pattern which is influenced by flame shape, luminosity & velocity and is a matter of configuration or geometry.

LIQUID FUEL BURNERS

Oils may be burnt in two ways *viz* :

(*i*) it is *vaporised before ignition* so that it burns like a gas (vaporising burner).

(*ii*) it is *broken into fine droplets* which are injected into hot air so that they evaporate while burning (atomising burner).

The first mode is generally not used on industrial scale. In the second mode of liquid fuel's combustion, the *oil is heated to suitably low viscosity and atomised* :

(*a*) *mechanically* by means of a rotating disc or cup or swirler which gives a wide spray of oil and so a wide flame area but with a very uniform droplet size (50 microns).

(*b*) *by high pressure ejection* from a fine orifice which gives a conical spray.

(*c*) *most commonly by entrainment in a blast of air* or steam which gives a narrow jet & a long pointed flame and produces the most variable droplet size.

Though air atomization gives a better theoretical flame temperature compared to steam, still *steam atomisation* is preferred for high viscosity coal tar fuels and residual fuel oils because steam is a source of heat to reduce the viscosity of heavy viscous oils. Besides, steam is cheaper to produce than high pressure air.

For most of the applications *compressed air is the atomising fluid.* The main advantage of atomisation by fluids over mechanical atomisation is the more complete control of output (turn down ratio =10:1) and shape of flame. The most disadvantageous aspect is the cost of supplying fluid under pressure.

When the oil is injected into the hot furnace by preheated air, the droplets immediately evaporate and most of the combustion occurs as vapor phase reaction. The shape of combustion surface determines the rate at which oil jet & air will intermingle and react and hence the temperature attained. Ideally, a burner should be variable to give hot short flames or long cool flames as required. The effect can be attained by controlling the oxygen content of the air, oxygen enrichment of air giving the hotter, shorter flame.

Some typical oil atomising burners are shown in Fig. 38.1.

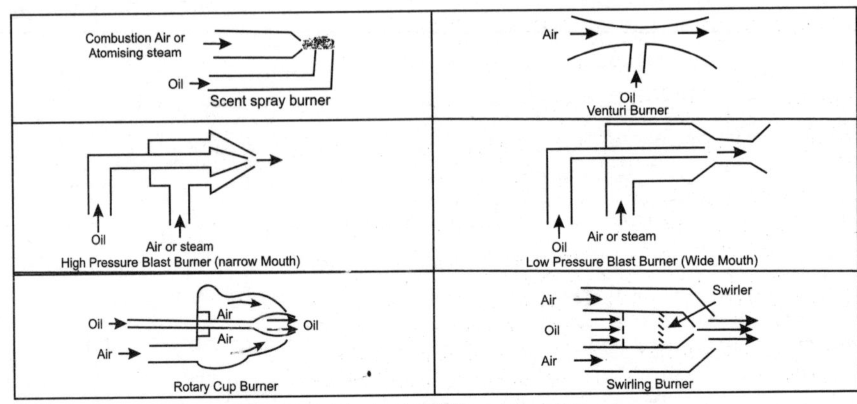

Fig.38.1: *Typical Oil Atomising Burners.*

In some furnaces *(e.g.* compact oil fired boilers for land or marine use) the oil & air are mixed in a special combustion chamber between the burner and furnace chamber itself,. Fuel and air are forced in a tight vertex in this small compartment. A high degree of turbulence is induced and very little heat is lost from the gases before they flow out, fully burned and at maximum temperature into furnace proper.

The vaporizing burners are used for heating units of small size such as portable air heaters, small boilers, cooking stoves etc. & have low cost & quiet operation. The atomizing burners are used mainly in industrial heating.

VAPORISING OIL BURNERS

In vaporizing burners (suitable for naphtha to light fuel oil) like *blow torches* and *portable gasoline stoves* the volatile fuel is passed at a low pressure through a tube adjacent to the flame, where vaporization takes place. The vapour stream issues out of an orifice at a high velocity and entrains primary air. The fuel-air mixture passes through a mixing tube and burns at the burner head.

In *wick-type* vaporizing burners (using superior kerosene), a wick raises the liquid fuel from a trough by capillary action. Radiation from the flame and adjacent hot surfaces vaporizes the kerosene from the wick head. The vapours enter into an annular space between two vertical, perforated metallic cylinders and burn in the air drawn through the perforations by natural draught. The combustion takes place throughout the annular space with a blue flame. These burners are used as cooking stoves. A variation of the wick-type burner (having a capacity upto 5 kg/hr) and is suitable in domestic boilers, central heating etc. Here an asbestos wick receives the fuel by a gravity-feed which may be regulated. In ordinary wick-type burners, the burning rate is controlled by varying the length of the exposed portion of the wick and is less satisfactory.

In a pot-type vaporising burner (shown in Fig. 38.2) there is a fuel reservoir at the bottom, supplied by an oil line. The oil surface is heated by radiation from the flame and the hot walls. Vapours mix with the primary air admitted through a central pillar and perforations in the pot wall. The fuel-air mixture is too rich at the oil surface to support combustion. Further up, it is diluted with air and becomes inflammable. Flame is supported at the top of the pot. These burners work on natural or forced draught. Superior kerosene or light fuel oil is used. Soot formation necessitates the pot burner cleaning periodically. The capacity is upto 10 kg/hr. coarse atomization before the fuel is vaporized), a centrally located centrifugal spinner distributes the oil in the form of a spray of coarse droplets striking the hot stainless steel wall of the burner. The vapours mixed with the air (which is directed towards the wall by a fan) rises upwards and burns on a stabilizing grill with a blue flame. The flame burns around the inner circumference of the furnace only, and from this fact gets its characteristic name. These burners can burn upto 30 kg/hr oil. Power requirement is low and operation is noiseless. Kerosene and light fuel oil are used.

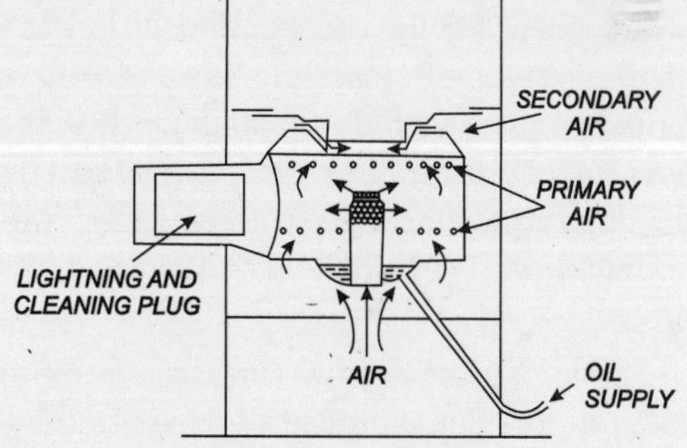

Fig. 38.2. *A Pot-type Burner.*

In *'petromax' stoves* (which burn kerosene), an upward moving fuel jet strikes against a hot metal surface at a high velocity and the vapours produced burn with a short & intense flame over the metal surface. The fuel is stored in a small reservoir below the fuel line and the oil pressure is supplied by compressing air into the reservoir by a hand-driven piston. It has risk of accidents and has noisy operation.

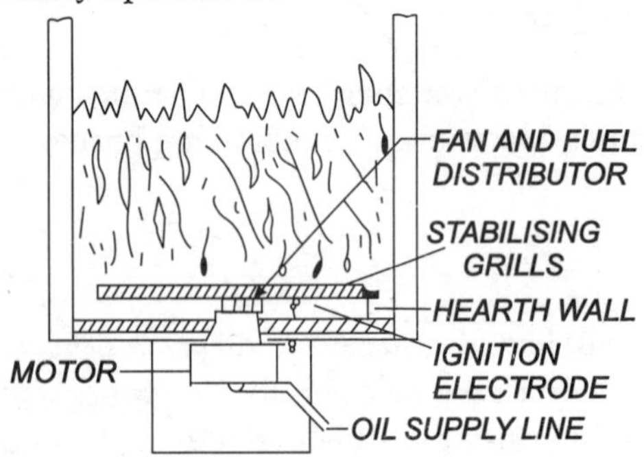

Fig. 38.3. *Wall-wiping Flame Rotary Vaporizing Burner.*

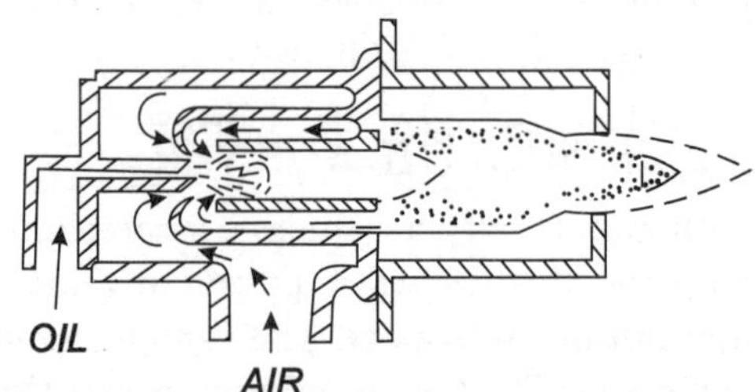

Fig. 38.4. *High-Speed Vaporizing Burner.*

In *high speed vaporizing burner* shown in Fig. 38.4, a high rate of vaporization is achieved by atomization of the volatile liquid fuel and recirculation of hot combustion gases within the burner. The burner bears out the difference between vaporizing and atomizing burners. The combustion chamber receives fuel vapours in the former and fuel droplets in the latter.

ATOMIZING OIL BURNERS

Atomization of oil creates enormous surface area per unit weight of the fuel which helps the heterogeneous combustion of the liquid fuel and the gaseous oxidant. Atomizing burners have an arrangement for the atomization of liquid fuels before the actual combustion takes place. There are three types of atomizing burners (differing from one another on the principle of atomization) namely pressure jet atomizing burner, twin-fluid or blast atomizing burner, and rotary atomizing burner. For atomization, initially a jet or a thin film of liquid is obtained and allowed to emerge into the open atmosphere at a suitable velocity. Both the jet & unsupported film are unstable and break up into drops. Disturbances like swirling of the jet or rotation of the film hastens the disintegration process.

Low viscosity & low surface tension are desirable for ease of dispersion and formation of small drops. The atomization is assisted by the resistance, friction and turbulence of air. The air resistance and friction become effective when any ruffling or distortion of the surface of the jet or film occurs. These forces are significant at furnace conditions due to increased air viscosity at high temperatures. Local turbulence in the air causes deformation and disruption of the drops. The higher the liquid velocity, the greater the disturbance in the liquid & also the resistance, friction and turbulence of air, and hence the greater the extent of atomization. A high-speed fuel jet disintegrates into a number of big droplets and a far greater number of 'secondary' droplets. The atomization of fuel proceeds during the penetration of the spray into the surrounding medium and the resultant spray has a combination of droplets of various sizes. Small size range droplets (< 25 μ) assist in the initial ignition and promote the stability of flame as the small droplets are quickly vaporized/cracked and ignited. The resultant flame supports the combustion of the coarser ones.

Very large droplets do not completely burn in the time available and they then create complications by depositing themselves on the walls of the combustion chamber.

Pressure Jet Atomising Burners. These are of two types namely the plain orifice type (used for fuel injection in diesel & other internal-combustion engines) and *centrifugal swirl type* (used in large industrial boilers).

In the former, *very high pressure of oil upto 350 kg/ cm²* or even higher are used. Much lower oil pressures (*7 to 35 kg /cm²*) are required in the swirl type burners.

A *swirl-type atomizer* is shown in Fig. 38.5. The oil enters the circular swirl/vortex chamber through tangentially spaced ducts/slots (D_1; D_2) and rotates in the chamber around an air core. The rotating mass of liquid issues through an orifice (0), under the influence of the translational force (which moves it forward) and the centrifugal force (which makes it fly outwards tangentially) resulting in the formation of a spray of droplets.

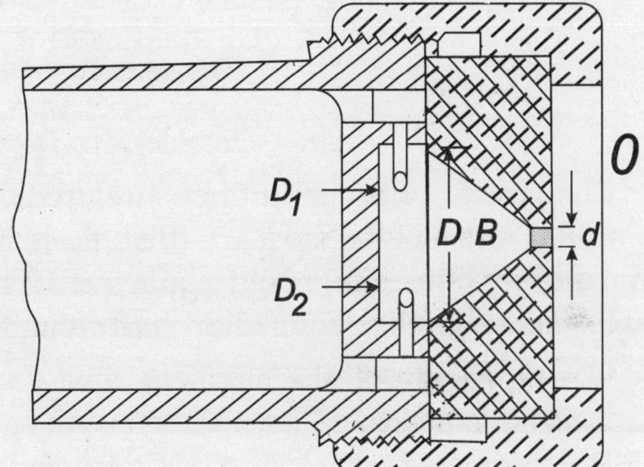

Fig. 38.5. Centrifugal Swirl-type Oil Burner.

The oil viscosity should be about 70 Redwood I seconds for small nozzles and about 100 Redwood I seconds for larger nozzles.

The output of a pressure atomizer *(Q)* increases with the increase in oil pressure *(P)*, (following $Q \alpha \ p^{0.5}$) and decreases with the decrease in oil viscosity. The mean droplet size *(d)* increases with viscosity, varies inversely as $p^{0.35}$ and increases with the orifice diameter (d_0), (following $d \alpha \ d_0^{0\ 5}$). Droplet size ranges between 10 μ and 220 μ.

The simple pressure jet swirl-type atomizers have limited turndown ratio (due to the dependence of output on oil pressure) as a wide variation of oil pressure from 7 to 35 kg/cm² allows a turndown ratio of only 1 : 2.2. This limitation is removed in the two modified designs of pressure jet atomizers in which a turndown ratio of upto 10:1 may be realised. In the spill-type, the rate of discharge is varied by providing a by-pass. In the other design, this is achieved by having a number of tangential slots for the entry of oil into the swirl chamber and varying the number of available slots with a sliding piston.

The pressure jet atomizing burners are very quiet, have low operating cost and power requirement is the lowest of all atomizers. These are the most widely used of all atomizing burners. Such burners with a capacity of upto 4,500 kg/hr of oil are available. The pressure jet atomizers are susceptible to choking by dirt in the oil. The oil should be strained/filtered for this reason.

Twin Fluid or blast atomising burners. These burners use air or steam to atomize the oil and are of three main types : (*i*) *low pressure* (LP) types using air af 35 to 70 cm w.g., *(ii) medium pressure* (MP) types using air at 0.4 to 1 kg/cm² gauge, and *(iii) high pressure* (HP) types using air or steam at pressures exceeding 1 kg/cm² gauge. Each of these types may again be either *'inside-mix'* or *'outside-mix'* type. Some inside-and-outside-mixing systems are shown in Fig. 38.6. The inside-mix types which are more commonly used provide high efficiency at the high rate of discharge, a high turndown ratio and a flexible flame structure.

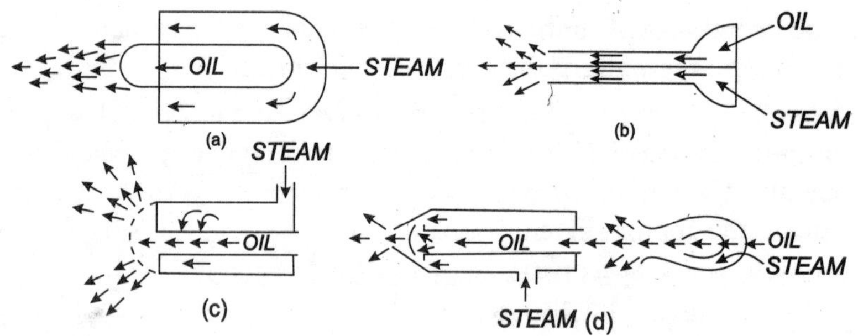

Fig. 38.6. *Twin-Fluid Atomization Burners*

a,b — Outside-Mix Types c,d — Inside-Mix Types.

The general principle in these atomizers is common. Oil flows through a central tube at a controlled rate and the auxiliary fluid meets it obliquely at a predetermined angle, as it emerges from the tube. Since the energy requirement is met by the **auxiliary** fluid hence the oil pressure need not be more than that required to ensure a uniform flow through the nozzle.

Kinetic energy of the auxiliary fluid required for atomization in all types of twin-fluid atomizers is same hence higher the pressure of the auxiliary fluid, the lower the quantity required. HP atomizers need only 2-3% of the total combustion air for atomization while this figure is 3-5% for MP and as high as 20-100% for LP burners. On the weight basis, steam requirement is similar to air requirement (0.3 kg/kg oil) in the HP atomizers.

A high turndown ratio of upto 10 : 1 is obtained with HP & MP burners. The LP units have normally a low turndown ratio of 2 : 1 to 5 : 1. HP and MP burners use most of the combustion air as secondary air and thus permit high waste heat recovery by air preheating. The cost of air compression in HP burners is very high, hence steam is used to atomize heavy residual fuels and coal tar fuels which require preheat for getting the correct viscosity.

LP units are used in domestic and industrial installations. Industrial LP burners (with a capacity of upto 2,000 kg/hr) use lighter grades of fuel oils.

MP burners are used in boilers and furnaces where a great flexibility and heat regeneration or recuperation are required.

When steam is substituted for compressed air, the same quality of atomization can be obtained at considerably lower pressure. HP steam burners are used in large installations where high pressure steam is available. These burners have a capacity upto 3,500 kg/hr. Heavy fuel oils can be burnt by HP steam atomizing burners. Less preheating of the oil is required when steam is used to atomize it. Steam should be at a pressure exceeding the oil pressure. The flame shape is altered by changing the nozzle tips.

Rotary Atomising Burners.

These have a central stationary fuel line which delivers the oil to the inner surface of a rotating hollow tapered cup. The cup is rotated (at 3,600-10,000 rpm) and the friction between the oil and the cup surface causes the oil to rotate with the cup. The centrifugal force, assisted by the tapering of the cup causes oil to flow towards the brim. Finally, the oil is flung off in a thin film which disintegrates into fine droplets. A diagram of rotary atomising burner is given in Fig. 38.7

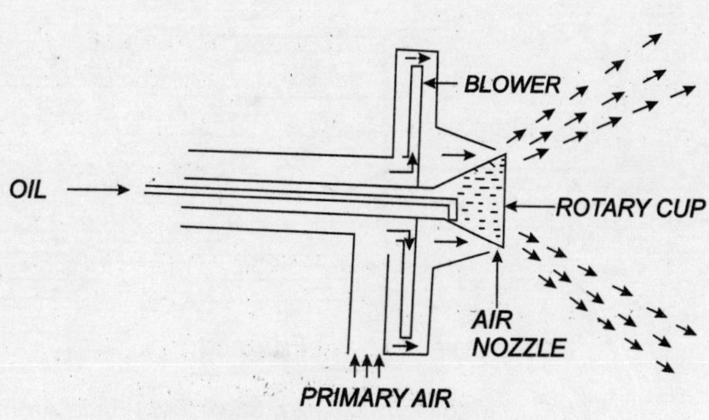

Fig. 38.7. Rotary Atomizing Burner.

About 15% of the combustion air (supplied as primary air by a fan attached to the rotating shaft of the motor driven cup) at a pressure of 10 to 25 cm w.g. gives a forward movement to the spray and also exerts additional atomizing forces on the droplets. Secondary air is supplied to the door underneath or through an opening in a checker work floor, the draught being provided by the stack. A fuel droplet size of 50 micron is achieved.

The oil pressure should be low (5-8 cm w.g.) and the viscosity should be in the range of 100-300 Redwood 1 second. Rotary atomizers can tolerate higher viscosity than the others. They may however lose atomizing efficiency at low viscosities. Such low viscosities as are used in a pressure jet system may cause the oil to 'slip' within the cup. Trouble due to choking by dirt is also less. However there is a greater chance of carbon deposition by cracking caused by the radiation from the surroundings after the burner is turned off.

Rotary atomizing burners, are extensively used in boilers and small installations. A high turndown ratio of 16 : 1 and capacity upto 800 kg/hr is available. Capacity is much lower than the maximum capacity of other atomizer types. Rotary atomizing burners are compact, efficient, comparatively low in first cost and power requirements are relatively low.

The rotary units are usually mounted horizontally in the furnace walls, although a few domestic units have a vertical shaft.

Drip Feed System in Oil Burning. When the combustion chamber is at > 600°C & the feeding system can be conveniently provided in the chamber, a very simple & cheap method of oil burning can be utilized. The fuel oil is gravity-fed through a metal tubing into the hot environment of the furnace. The feed rate is controlled by a valve. The oil droplets fall on a current of combustion air which is drawn from the chambers undergoing cooling. The ignition is quite smooth and the combustion is stable. This system is suitable for some kilns used in silicate industries.

BURNERS USED FOR GASEOUS FUELS FIRING

There are two methods of burning gaseous fuels :

(*i*) The gas and air is *pre-mixed* and then fired in the furnace chamber. This is called premix burner or *inside mixing type* of burner; only a part of the air needed for combustion is mixed with the gas. This is called primary air. The maximum permissible amount of primary air decreases with increasing hydrogen content of the fuel gas. Flame stability range also changes in the same manner. The rest of air which is called secondary air is supplied separately in the furnace. Bunsen burner is an example of this type of burner (also called aerated burner).

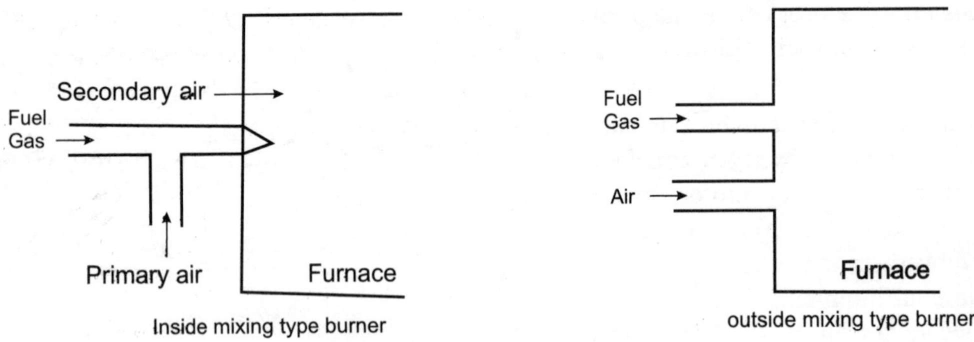

Fig. 38.8. *Gaseous fuel firing burners.*

(ii) The gas and air flow separately and mix together in the furnace as combustion proceeds. This is called *outside mixing type* of burner or diffusion flame burner or nozzle mix type burner, (also-called non-aerated burner). No backfiring is possible in this type of burner, however flame can be blown off the burner tip at high pressure.

The method of premixing *sub-divides* the *premix burners* into the following :

(a) **Atmospheric or 'aerated' gas burners**. *Here the kinetic energy of the gas is used to aspirate air from the atmosphere into a mixing tube which has the burner head at its end.* A bunsen burner is a typical example. *Depending upon the gas pressure, there may be low-pressure gas aspiration* (7.5 to 20 cm w.g.) and *high pressure gas aspiration* (0.25 kg/cm^2 gauge and above). A typical aerated gas burner is shown in Fig. 38.9.

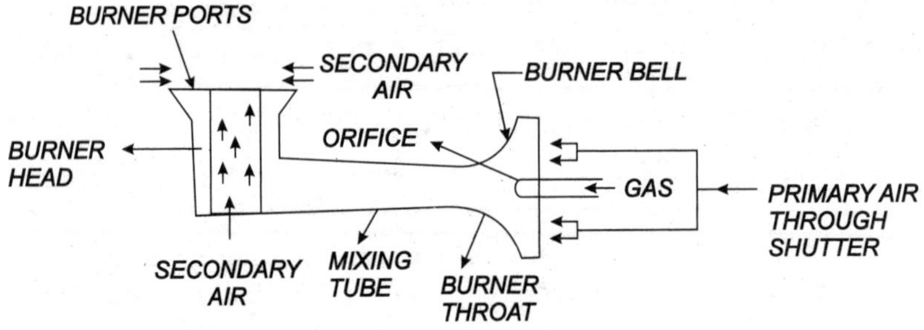

Fig. 38.9. *Atmospheric/Aerated Gas Burner*
(Low Pressure gas Aspiration)

(b) **Air aspiration gas burners.** Here air is supplied at a medium pressure of 0.15 to 0.35 kg/cm^2 gauge or above and its kinetic energy is used to aspirate gas (7.5 to 20 cm w.g.) into an aspirator body for mixing the two and then the mixture enters the combustion chamber. The gas is supplied at a constant pressure controlled by a governor.

(c) **Burners with independent supply of gas and air to mixer.** Here, the fuel & air are mixed in desired proportions and compressed mixtures are then delivered to the combustion chamber under pressure. This system achieves total premix with high efficiency of mixing.

The typical methods of delivering gas and air in the nozzle-mix systems /outside mixing burners are as described below :

(i) **Separate air and gas ports**. Here streams of air and gas are separately led into the combustion chamber through the ports of suitable size & shape. The introduction of almost parallel systems of gas and air is used to get diffusion flame for uniform heating of large spaces *e.g.* coke ovens. By supplying gas and air at an angle and imparting a whirl to air, a short intense flame is obtained. When the gas line is in the form of a long, straight pipe, it is called a 'gun'. This system has been shown in Fig 38.10.

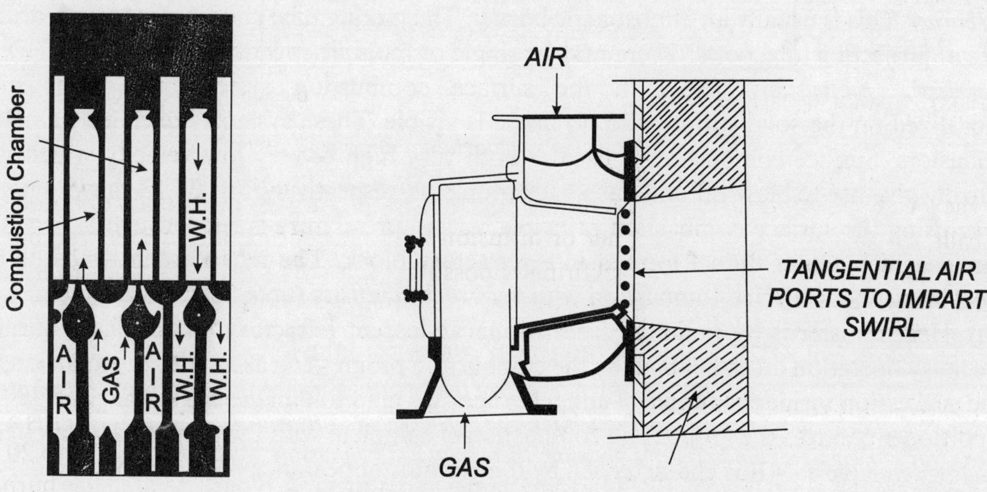

***Fig. 38.10.** Combustion with Separate Gas and Air Ports.*
(a) Long flame (coke oven), (b) Short Flame.

(ii) Gas ports and air register. Here the gas is supplied by a gas gun to a single or multiple port head while air is delivered by a forced-draught "air register. The air register controls its quantity Such units give a short intense and non-luminous flame. A diffuser plate is attached to the burner tip to maintain a low gas velocity in the central zone which helps in steady & prompt ignition and flame stabilization. The diffuser also promotes rapid mixing of air and fuel. These burners are known as blast burners and has been shown in Fig. 38.11.

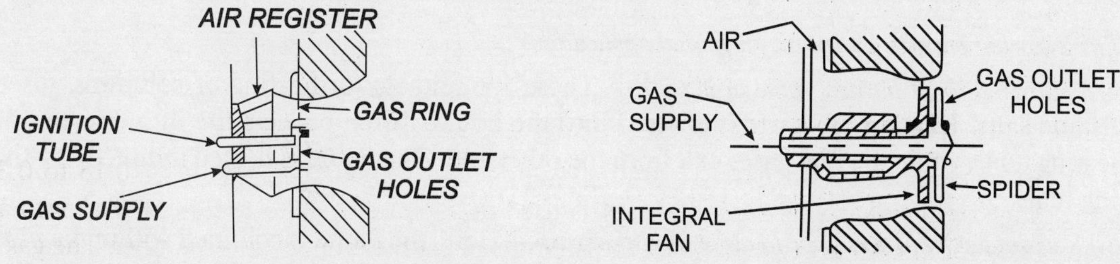

***Fig. 38.11.** A Blast Burner with Air Register and Gas Gun.*

***Fig. 38.12.** Fanmix Burner.*

(iii) **'Fanmix' burner**. High burning rates are obtainable with this nozzle-mix system using as low as 5% excess air. The straight gas line ends in a turbine (spider) having holes drilled on one side of its blades. A fan is attached to the turbine. The emerging gas rotates the turbine and the fan. The fan delivers the combustion air at right angles to the gas streams. High turbulence results and

leads to very thorough mixing of gas and air. The quantity of air is controlled by an air shutter. It is shown in Fig. 38.12.

The structure of the flame classifies gas burners into the following types :

(*a*) *'Torch' burner'*. Here each burner gives one flame visibly projected from the burner tip.

(*b*) *'Ring' burner.* It is usually an atmospheric burner. The burner head is in the form of concentric rings having suitably drilled holes.

(*c*) *Pipe burner.* This is usually an atmospheric burner. The mixing tube ends in a pipe with drilled holes or projecting tube ports. There may be single or multiple-row drilling.

(*d*) *'Flameless' combustion burner.* In the surface combustion system, the combustion is localized on the solid surface and no flame is visible. These systems are called flameless combustors. Surface combustion is used at both very high & very low velocity which will normally give rise to blow-off and flash-back problems, respectively. Fig. 38.13 shows a *tunnel burner* using the surface combustion principle. A gas-air mixture is fed through a port into a circular/rectangular tunnel formed in a refractory block. The incandescent walls of the tunnel bring about surface combustion with very high intensity (upto 15000 kcal/hr cm^2 port area). Heat transfer is by radiation from the incandescent refractory block. These burners have noisy operation brought about by the combustion progressing as a series of explosions in rapid succession inside the tunnel. Tunnel burners are manifold in groups so that one gas-air proportioner controls the group. The combustion is complete with the primary air supplied.

(*e*) *Pulsating combustors* — It is characterised by the periodic appearance of explosion flames.

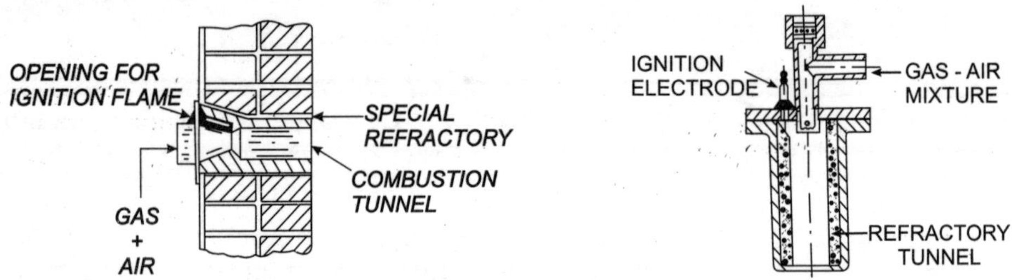

Fig. 38.13. *Surface-Combustion Tunnel Burner.* ***Fig. 38.14.*** *Gas Burner of Submerged Combustion Type.*

The following burners are meant for special applications :

(i) Immersion torch burners (pipe in liquid) — These are suitable for heating of solutions, fused metals and salts. Heating tubes are immersed into the liquid. Torch burners fire into one end of the heating tubes while the flue gases exit from the other end. The liquid is heated indirectly by the heating tubes.

(ii) Radiant torch burners (pipe in air or any other gas) — Here, the flame is confined within the pipe which heats the air or any other gas. Any contamination by the products of combustion is thus avoided. Radiant burners are used in industrial drying.

(iii) Submerged-combustion burner — Since tunnel burners do not need any secondary air for combustion, the refractory tunnel may be submerged in a liquid without disturbing the combustion. The combustion gases will bubble through the liquid which will receive heat from these hot gas bubbles. Fig. 38.14 shows simple burner using the submerged combustion principle. More intense

and more uniform combustion is achieved by providing additional refractory surface inside the tunnel. The main application of the submerged combustion technique is in crystallizing salts for which scaling of the heat-transfer surface is a serious problem. Burners are built with a rated capacity upto 1.3 million kcal/hr. Many such burners may be used in one tank.

Bunsen Burner. The most common example of inside mixing type burner is the Bunsen burner shown in Fig. 38.15.

In the Bunsen burner (or inside mixing type burner), suction is induced by the jet of gas which draws in primary air. Secondary air is supplied from atmosphere to the flame. The primary air gas mixture velocity is kept more than the flame speed of gas-air mixture, so that the flame does not *strike back / flash back / backfire* down the bunsen tube. If rate (velocity) of primary air supply is much greater than the flame speed, the flame can be blown off the tube and the burner get extinguished. With insufficient (less) primary air supply, the flame produced is long, lazy and slightly luminous which gives low heat release rate. With increase in primary air supply, the flame shortens and becomes non-luminous giving high heat release rate.

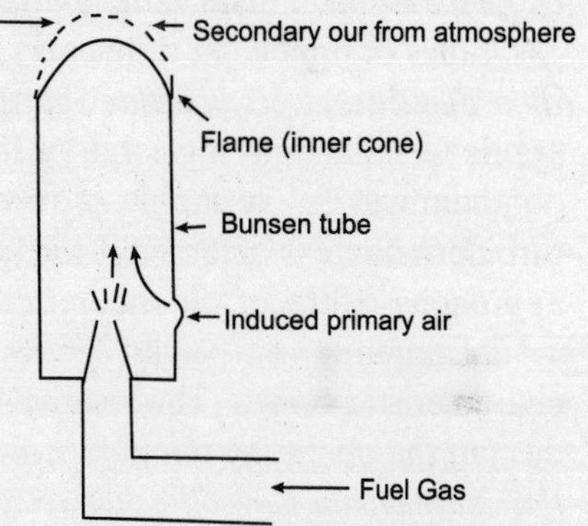

Fig. 38.15. *Bunsen Burner.*

Aerated Atmospheric Gas Burners. These burners are based on the principle of the bunsen burner. The burner shown earlier in Fig. 38.9 consists of a fixed orifice ('spud') for gas inlet, a controllable shutter for air supply, a venturi-shaped mixing tube and a burner 'head' with 'ports' (holes) drilled in it. The narrow zone of the mixing tube is called its 'throat' which diverges into the hind part called its 'bell'. The gas rate is controlled by a gas line valve. When the gas is admitted through the orifice, its kinetic energy is increased owing to high injection velocity and creates a low static pressure at the throat. Primary air is sucked into the bell through the air shutter. During flow through the mixing tube from the bell to the burner head, the gas & air uniformly mix and the static pressure is increased by the slowing down of the fluid stream. The combustible mixture burns in the form of tiny burner flames anchored on the ports of the burner head. The primary air is not enough to complete the combustion. Secondary air is entrained on the flame surface from the surrounding atmosphere by convection currents. The quantity of primary air depends upon the type of flame, burner design, input rate and type of gas.

The operation characteristics of an *aerated gas burner* is shown by a set of curves in Fig. 38.16 whose coordinates are heat input rate and primary air supply. Stable burner operation is possible in the middle region confined by the *curves 'a', 'b', 'c' and 'd'. Curve 'a' represents the conditions of lifting and blow-off* & sets the upper limit of input rate (or gas flow-rate) & the primary air supply. This upper limit varies inversely with the quantity of primary air. A high input rate therefore requires an increased quantity of secondary air. *Curve 'b'* represents the lower limit of primary air. A low proportion of primary air gives a long, wavy & yellow flame in aerated burners and tubes to the formation of soot & carbon monoxide by incomplete combustion. The supply of primary air becomes critical towards the range of high input rate as shown by the *curves 'a' and 'b'* tending to converge on the right-hand side. Burner operation is not satisfactory in this region. The proportion of carbon monoxide in the flue gas sharply increases for operation in the right of *curve 'd'*. The conditions of flash-back prevail on the left of *curve 'c'* which thus sets the lower limit of the input rate. Fig. 38.16 shows that a given burner can operate within a range of gas flow-rates. The maximum gas rate is limited by the blow-off, incomplete combustion or inability to aspirate sufficient primary air. The minimum is limited by the flashback. The ratio of maximum to minimum gas rate obtainable in a burner is called its *turndown ratio.* A value of *5:1* is desired in the most of gas-heated appliances. At a given gas rate, a long luminous flame is obtained with a low supply of primary air while a short turbulent flame is obtained if the quantity of primary air approaches 100% of the theoretical.

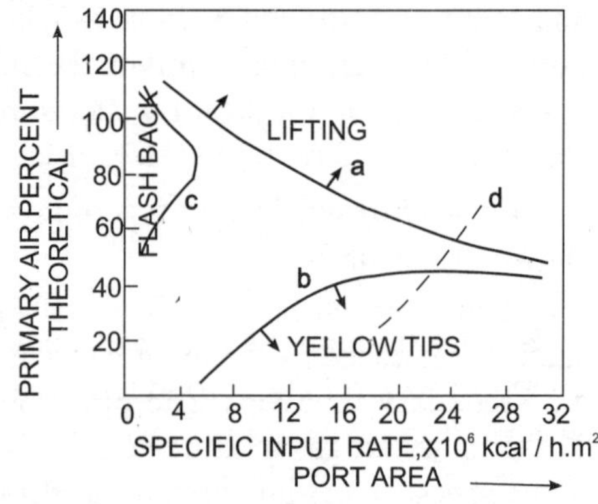

Fig. 38.16. *Operation Characteristics of an Aerated Burner*

All premix or partial premix burners give similar characteristic curves. The relative shape & location of individual curves depend upon the burner design, the operating temperature and the type of gas used. *In outside nozzle-mix burners, there is no flash-back but the blow-off can be related to the corresponding conditions.* In the industrial burners of very high input rate either the flame stabilizer in the form of a physical obstruction *(e.g.* diffuser) or the principle of surface combustion is used.

Design Data for Aerated Atmospheric Gas Burners :

A few equations and design data for a simple aerated burner are as follows:

The gas flow-rate through the orifice is given by :

$$V=83240\ K.A\sqrt{\frac{\triangle P^{p}}{T.S}} \qquad \text{...(A)}$$

where, V = gas flow-rate, Nm^3/hr

K = coefficient of discharge (usually 0.8 to 0.9 for fixed tubular orifice or spud)

A = orifice (port) area, m^2

$\triangle p$ = pressure drop at the orifice, cm H_20

P = initial gas pressure, cm Hg absolute,

T = gas temperature, °K

S = specific gravity of gas (air = 1).

Knowing the input rate and the desired pressure drop at the orifice, it is possible to calculate the orifice diameter from the above equation. The desired pressure drop may be calculated from the primary air requirement with the following empirical equation for the aeration capacity of the burner :

$$P_A=\left(\frac{\sqrt[4]{PS}}{\sqrt{C_N}}\right)\times\left(\frac{5\times10^6\times^4\sqrt{A_mA_p}}{Q}\right)\times\frac{300}{T} \qquad \text{...(B)}$$

where, P_A = primary air, per cent of theoretical air

P = gas pressure of orifice inlet, cm w.g.

C_N = net C.V. of gas, kcal/Nm^3

A_m = average area of mixing tube, m^2 = $\frac{A_1 + A_2}{2}$

A_1 = throat area, m^2

A_2 = area of mixing tube outlet, m^2

A_p = total port area, m^2

Q = heat input rate, kcal/hr.

The first term on the right-hand side of this equation is a function of the nature of the gas and its supply pressure. The second term is a function of the burner dimensions. The third terms shows the effect of mixture temperature.

The nature of application of the burner influences the primary air requirement. When a long flame is permitted in water heaters, furnaces etc., *only 35—40% primary air is required with natural gas & most manufactured gases and at least 55% with LPG. When a short flame is required (as in cooking ranges), the primary air requirement is higher.* Cooking ranges need 55—60% and radiant heaters need at least 65%.

The total port area is computed from the design data of specific heat input rate for the burner (upto 6 million kcal/h.m^2 of port area).

Other design data are :

Throat to total port area ratio	= 0.20 to 1.0
Air shutter area	= 1.25 to 2.25 times total port area
Distance from gas orifice to throat	= 0.5 to 2 times throat diameter
Length of mixing tube	= 6 times throat diameter
Slope of mixing tube	= 3°
Area of burner head	= 1.5 to 2 times total port area.

The aerated burner design is greatly influenced by the nature of the gas. On the other hand the nature and quantity of available gases may widely vary. It is therefore necessary to know the basis of determination of inter-changeability of gases.

The burner output rate is given by :

$$Q = \eta \, . \, V C_N \quad ...(C)$$

where, η = coefficient of combustion efficiency. Combining equation (A) and (C), we get:

$$Q \propto \frac{N}{\sqrt{S}}$$

Furthermore, from equation (B), we have the primary air supply,

$$PA \propto \frac{1}{\left(\frac{C_N}{\sqrt{S}}\right)^{0.5}} \quad ...(D)$$

The ratio $C_N / \sqrt{S}$ is known as *Wobbe Index.* It is an useful parameter in assessing the inter-changeability of gases with respect to aerated burners. It is a dimensional quantity with the dimensions of calorific value of gases.

Wobbe Index is of great importance for gases supplied through pipeline distribution systems because aerated burners are widely used by consumers. This parameter is of limited significance for other types of burners.

An *'aeration test burner' (A.T.B.)* has been specified for evaluating variation in gas quality. It is a Teclu burner of 30.5 cm height and 10.26 mm i.d., and has a nipple of 0.99 mm dia. The air shutter is rectangular and adjustable by a micrometer head. The given gas is burned in it. It is aimed at determining the extent of the opening of the air shutter at which an exact cone of 6.35 cm height is obtained. The result is expressed as A.T.B. Number. A variation of 5 units in the *A.T.B. Number* of gas can seriously affect the performance of an aerated burner.

The maximum heat inputs obtainable with aerated burners of low gas pressure types are given in Table 38.1.

Table 38.1: Capacity of Aerated Gas Burner (Low gas pressure)

Type of fuel Gas	Maximum heat Input Rate x 10^3 kcal/h.m^2 of port area % Primary Air			Wobbe Index, ($C_N/\wedge S$), kcal/Nm^3
	40	55	70	
Coke Oven Gas	20,400	11,800	7,500	8,100
Natural Gas	12,100	7,100	4,700	11,800
L.P.G.	Yellow Flame	8,200	5,100	21,200

The aerated gas burner of high pressure gas aspirator type obtain better mixing and control of gas & air with a two stage aspiration at the orifice (spud) and the burner throat as shown in the Fig. 38.17.

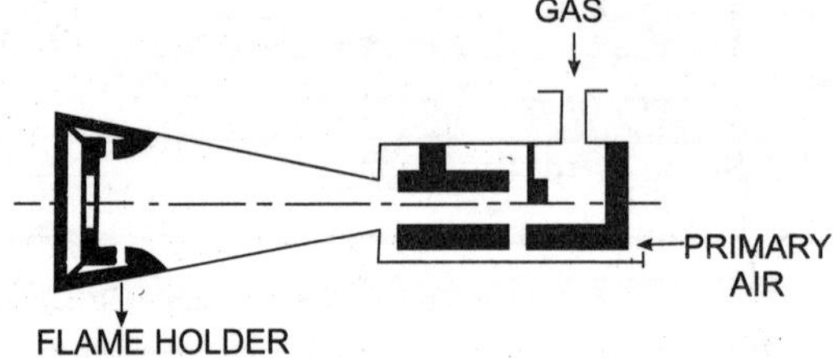

Fig. 38.17. Aerated Gas Burner with two stage aspiration and flame holder.

A flame holder provided at the burner tip stabilises the flame by maintaining a zone of relatively low gas velocity when the gas pressure in the line is high.

In *pressure type aerated burner (blast burner),* both gas and air are supplied at higher pressure. Here, concentric primary air and gas jets under pressure are used instead of single gas jet. Separate jets are used to induce secondary air or they can first be fed to a mixing chamber. Large furnaces are heated by a multiple gas jets set in a common head, each jet being surrounded by compressed air with a concentric orifice for the supply of secondary air. When air is supplied under pressure, control of burner is obtained by adjusting air pressure.

In *premix type of burner,* primary air should not be preheated to a temperature more than the ignition temperature of the gas, otherwise the gas may start burning before entering into the furnace and explosion may take place.

Because of their low flame speeds, gases of low calorific value (*e.g.* B.F. gas) are easily blown from the burner tip by primary air. If this is given a strong rotary motion (swirl), the flame becomes more stable. Similarly, with gases of high calorific value, an air swirl shortens the flame and gives better control.

The flame speed of hydrogen is very high. Hence its percentage in the fuel gas greatly affects the flame speed and flame stability. The maximum permissible amount of primary air decreases with increasing hydrogen content. The flame stability range also changes in the same manner. This range is that between minimum rate of gas supply to prevent back firing of the burner and the maximum rate permissible before the flame is blown off the burner tip. This ratio of maximum output to the minimum output of the burner is called the turndown ratio of burner which is about 5 : 1 for most gas fired furnaces.

Inside mixing type of burners are limited by the size of the burner to which stable flames can be attached & by the fact that preheated air or gas beyond the ignition temperature of the gas can not be used because of pre-ignition and explosion, risk.

In *outside mixing type* burners (diffusion flame) combustion occurs when the gas is allowed to burn at an orifice by surrounding air (*e.g.* luminous flame of a bunsen burner with the air valve closed) or when gas and air stream are passed into the furnace, either parallel or impinging streams.

In jet burners, gas is injected into the furnace space through a specially shaped slit which produces a turbulent jet of gas which in turn entrains air in a flow patterns determined by the front surface of the burner. These burners may be made of ceramics and they produce flat flame (like bat's wings). All the air used here comes from atmosphere outside the burner through adjacent vents.

Where, the volumetric flow rate of gas is high, combustion takes place over a considerable space and some time is required for complete combustion of the fuel (*e.g.* in case of lean gas), the burners may be simply open ends of pipes delivering the gas into the furnace alongside larger ducts carrying preheated air. The flame produced is called diffusion flame because air and gas diffuses into each other before burning. In fact, this diffusion results from the *turbulence* created which can be *increased by the :*

- difference in the velocities of gas and air stream
- difference in directions of the gas and air streams
- difference in densities of the gas and air stream, if the heavier one is at the top.

The greater these differences are, the faster the combustion and hotter the flame can be obtained.

An excess of air accelerates combustion as does a further increases in turbulence by baffles. The use of oxygen enriched air or pure oxygen increases the combustion reaction rate and flame temperature.

Burner design should be such as to produce short hot flames or long cool flames with uniform capacity over a wide area depending upon the specific requirements.

RECENT DEVELOPMENTS IN BURNER DESIGN

Some of the recent development in the burners used in industrial heating are described below:

RADIATION HEATING

Radiation heating is resorted to when heating flat or thin surface of a material or when the material is a good conductor of heat or when it can be easily surrounded by parallel radiant surfaces. Radiation heating is self-spreading, thereby providing a uniform distribution of heat over a surface.

To accomplish the same uniformity with convection heating requires an array of high velocity jets whereas a single spreading type radiation burner can do the equivalent job with much less equipment investment. Although most radiation heating and jet impingement heating jobs require custom made or' wrap around' heat sources, the radiation type is somewhat more versatile in the variety of loads that it can do well.

A spreading type of radiation flame is very much preferred when heating a vessel containing a valuable load or a vessel that in itself is expensive and on which impingement of flame would cause localised hot spots which greatly shorten their lives.

'Holding Heat' for liquid in insulated containers is best applied by radiation because the main heat loss is from the liquid surface and the heating is simply to balance any loss from the surface. From the heat transfer point of view, the surface is then just like a thin flat sheet of solid material.

Radiation Burners

Radiation Burners heat the load primarily by radiation — either by long luminous burner or by convective transfer to a surface that becomes a radiant emitter. These have been specifically developed for Furnaces where radiation is a primary source of heat transfer.

Flat Flame Burners

Where penetration or forward drive by the flame and hot combustion gases is to be avoided or when radiation heat transfer is to be enhanced, flat flame burners are used which actually heats its own refractory tile and the refractory surface of a surrounding furnace wall or roof by convection from the high velocity flue gases thrown sideways from the burner. The hot refractory surfaces then radiate heat to the load. The disc shaped flames produced by these burners have no forward velocity. It has been shown in Fig. 38.18.

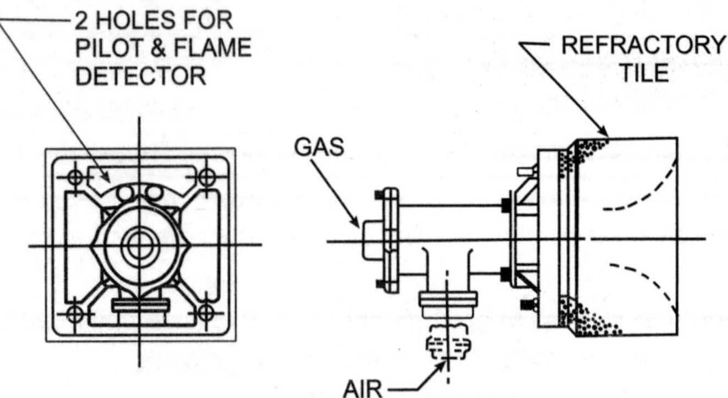

Fig. 38.18. *Flat Flame Gas Burner.*

Fig. 38.19 sketches a variety of commercially available refractory shapes for flat flame burners. Style *'a', 'c', 'f* and *'g'* are *nozzlemix, 'b', 'd'* and 'e' are premix. Style *'e'* utilises a porous refractory membrane. Style *'g'* is designed to focus radiation in an intentional hot spot. All except *'e'* utilises the *Coanda effect.* Coanda effect is the tendency of a fluid issuing from a jet to follow the wall contour even where the wall's curvature is away from the axis of the jet. This *Coanda effect* used in many radiation type burners tends to be counter productive to thorough

Fig. 38.19. Different Styles of Flat Flame Radiation Burner.

atomisation by the usual means. Hence these burners are more suitable for use with gaseous fuels. Dual fuel burners are generally available for gas and light oil only. Fig. 38.20 illustrates applications of flat flame burners.

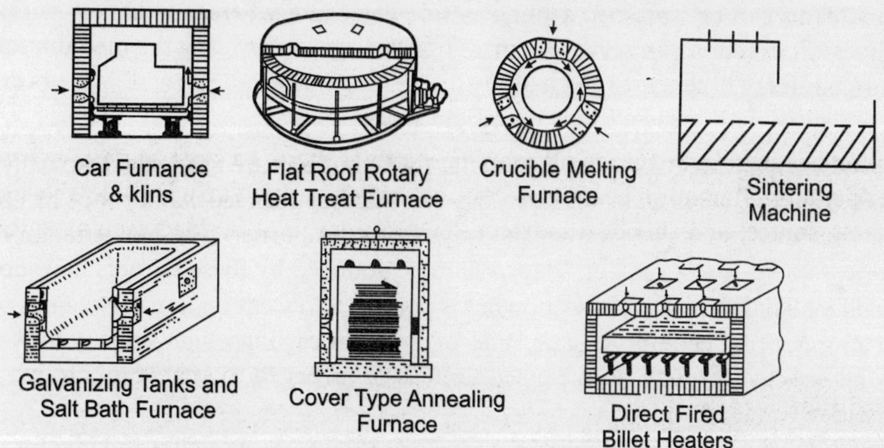

Fig. 38.20. Industrial Applications of Flat Flame Burners.

Radiant Tube Burners

These burners not only crowd the flame into a very confining cross section and need a long strung-out heat release, but they are tight chambers and on start up they are cold chambers. When operating on gas, the radiant tube burners have delayed and partial premix features that provide uniform heat release along the tube length. A unique swirling air entry configuration causes the main air to blanket the inner wall of the tube and protect it from over heating in the region where the flame is the most intense. As the swirl energy dissipates, the air and unburnt fuel vapours diffuse together creating the desired long heat release pattern.

Delayed Mixing Burners

In some operations a direct radiation over a large area is desirable. This is frequently the case in wide or extremely long furnaces where poor heat distribution *i.e.* hot spots and cold spots would be obtained from conventional short clear flames, or where all burners must be located at one end. Luminous flames have a considerable length and they can fill a large volume of combustion space with flames having little temperature variations along their length. This permits more effective use of the hearth area

These burners are designed to have a very low mixing rate of the fuel & air so that the two fluids travel a considerable distance from the burner before completely mixing and burning. The flame is often termed a diffusion flame because the mixing occurs as the parallel laminar air and fluid streams diffuse into one another. Strong heating of the fuel in the absence of air causes thermal cracking of the fuel molecule into light and heavier molecules. These molecules become micron size soot particles. These opaque particles absorb heat from the flame, become luminous & emit radiant energy and hence the name long luminous flame burners.

CONVECTION HEATING

Convection heating with modern velocity forward flame burner is best suited for applications involving heating of loosely piled charges, impingement heating, mass transfer and direct fired air heating. Convection has the advantage of directly penetrating into deep narrow cavities. Convection streams can be 'splashed around' inside enclosure whereas radiation can only travel in straight lines. Convection can serve as a mass transfer agent providing the medium for carrying away vapours, oxides or other gases given off by the heating operation. If these vapours or reactants are not carried away, their presence may retard the processes.

Convection heating can be turned off more promptly if there is a stoppage somewhere else in a continuous operation. Radiation heating, in contrast, may have built up a considerable heat storage in the radiating source, and this heat may continue to pour into the load even after the flame is turned off.

If the total load has a thick cross-section but is open to allow circulation between its parts, such as bricks in a tunnel kiln, ceramic seggars, piles of irregular castings, high velocity forward flames can usually be directed through open topped tunnels or arranged to penetrate through the many pieces of a wide thick load.

The most difficult heating job is when the load has a thick cross-section and is solid so that hot gases cannot circulate through it. Neither radiation nor convection can deliver heat to the interior of the mass. In such cases conduction can be relied on to diffuse the heat to the centre of the load. This process is often called 'soaking'.

High Velocity Burners

High velocity burners ensure very fast & through mixing of the fuel and air & are used with special tiles having constricted opening on the discharge side resulting in the flame & flue gases escaping with very high velocity (380 km/hr). They increase effective heat transfer regardless of the temperature level or loading in the furnace. This effective heat transfer may initially heat the refractory wall as well as the load resulting in considerable radiation heating. Use of High

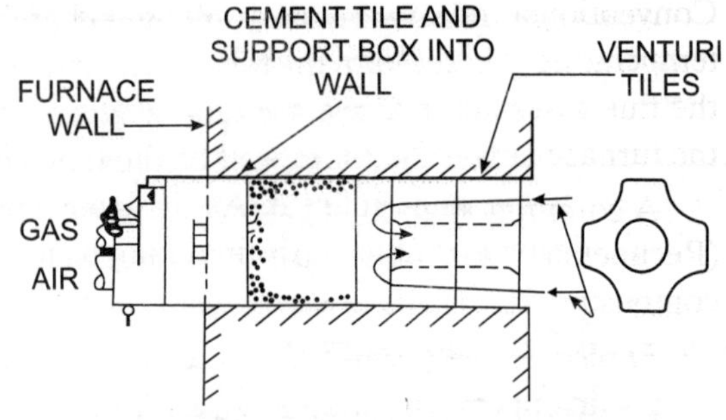

Fig. 38.21. High velocity burner with Ventury Tile showing Re-circulation of Furnace Gases.

Velocity Jets aids in faster heat transfer rates lowering the furnace bring-up time and achieving better temperature uniformity. It has been shown in Fig. 38.21.

High Velocity Burners have been very successfully used in the Heat Treatment Furnaces. The flame jets have so much velocity that they act like Inspirators or Ejectors pulling surrounding gases into their stream, thereby inducing a strong recirculation of furnace gases all around the load. In most Industrial Furnaces, the amount of gas recirculation determines both speed of heating and temperature uniformity.

When fuel is cheap, excess air may be used to aid temperature uniformity by :

- Reducing the Hot-Mix Temperature
- Preventing stratification
- Enhancing effective heat transfer.

With modern High Velocity Burners which induce recirculation of furnace gases, the recirculating gases produce the above three benefits, thereby eliminating use of excess air and therefore wastage of fuel.

Gas recirculation in low temperature ovens and dryers may be provided by a recirculating fan. However, fans in high temperature furnaces & kilns are prohibitively expensive and difficult or impossible to maintain. The high velocity burners can create recirculation right in the furnace itself without the use of a recirculating fan.

In Non-Ferrous Melting Furnaces, the charge is piled high. In such applications high velocity burners successfully 'punch' or 'pierce' through these piles using convection to melt the whole load at once instead of working at the outside pieces only. Additional benefits come from the fast combustion and from the fact that splashing molten metal is likely to plug the small opening in the high velocity burner's refractory tile.

Scrap preheating is another application of the 'punching' or the 'piercing' benefit of high velocity burners, particularly in deep scrap containers.

'PIERCING', 'REACHING', 'STIRRING' AND 'AIR HEATING' are the functions that high velocity burners can do best.

BURNER DESIGN FOR WASTE HEAT RECOVERY

Conventional Industrial Furnaces exhaust combustion products at temperatures above the furnace temperature. These waste gases contain considerable sensible heat. The amount of heat lost through the flue gases and the resultant furnace efficiency vary with the flue gas temperature. The higher the furnace exhaust gas temperature, the higher is the heat loss through the flue.

A common method of salvaging the sensible heat from the flue gases is to use a heat exchanger (Recuperator) for preheating the combustion air which raises fuel efficiency. Use of preheated combustion air results in increased flame temperature and therefore better radiation.

Burners designed to operate with preheated air may require :

- An insulating refractory lining to protect material from excess temperature and to reduce heat loss.
- Special alloy or refractory materials for the internal components to prevent scaling and distortion that can affect performance & capacity.

- Insulated protection for auxiliaries such as pilots, flame monitoring devices, observation ports and valves.
- Proper burner sizing because of the different air density which can be quite complex when all of its ramifications throughout a complete system are considered.

Primary/Atomising air is normally not preheated. As such burners designed for use with preheated combustion air, should have lower percentage of atomising air.

Self Recuperative Burner

As the name suggests, the Self Recuperative Burners have inbuilt heat exchanger for preheating the combustion air. The integral heat exchanger with the burner eliminates many of the disadvantages with external heat recovery device. In the case of external heat exchangers, combustion products have to be routed through them with a resultant heat loss and expensive ducting arrangement. An external heat exchanger with its associated pipe work and ducting can have inertia which is a disadvantage particularly with Batch Type Furnaces.

In the self recuperative burner the combustion air is preheated within the burner itself by the flue gases passing back through the burners as they leave the furnace. A circular tube separates the combustion air from the flue gases. Both combustion air and the flue gas pass through narrow annuli, the air flowing on the inside. The heat exchanger is integral to and directly behind the burner nozzle and tunnel, the front end of the air tube enclosing the refractory burner tunnel. The fuel flows to the burner nozzle through an inner tube concentric with the air tube. The flue gas is removed after the useful heat is extracted by an eductor system.

Twin Bed Heat Reelaimation System

Twin-bed heat reclaimer burners are the latest development in the design of the burners utilising waste heat in the flue gas for preheating of the combustion air. They utilise the regeneration principle to provide fuel saving. The use of conventional external regenerators have been limited because of size and high initial cost.

Twin bed are always set up in pairs. While one is firing, its twin is serving as flue heat extractor and heat storage bed. Each burner contains a bed of heat storing material which extracts heat from the flue gas. Every 20 to 30 seconds flows are reversed. The cold combustion air flows through the heat reclamation bed and is preheated to within 90°C of furnace temperature. The burner that had been firing is now serving as flue heat extractor and storage bed.

Exhaust temperatures are lowered to just above the dewpoint, which not only signifies superior fuel efficiency but also permits use of induced draft fan or venturi eductors of standard low cost materials. Air preheat temperatures produces very high flame temperatures. As the flame radiation varies as the fourth power of the absolute temperature, resultant heat transfer is superior. This burner system is shown in Fig. 38.22.

The very high air temperature causes most fuels to undergo thermal cracking, which turn the flame luminous and further enhances radiation.

Among its many advantages, the heat reclamation system can utilise benefits of excess air without incurring efficiency penalties excess air imposes with cold air burners. With cold air burner operating under excess air condition, extra fuel has to be burnt to heat every m^3 of the excess air to the furnace temperature. The heat reclamation system heats the excess air with the heat stored in the bed and not with fuel.

Twin-bed heat reclamation burners can be used in a variety of Furnaces & Kilns such as Shuttle Kilns, Elevator Kilns, Ladle Heaters, Retort Furnace, Box Forging Furnace, Car Bottom Furnace, Continuous Pusher Furnace, Continuous Annealing/ Normalising Furnace, Slot Forging Furnace, Indirect Fired Metal Melters, Galvanizing Furnace and Full Muffle Furnaces.

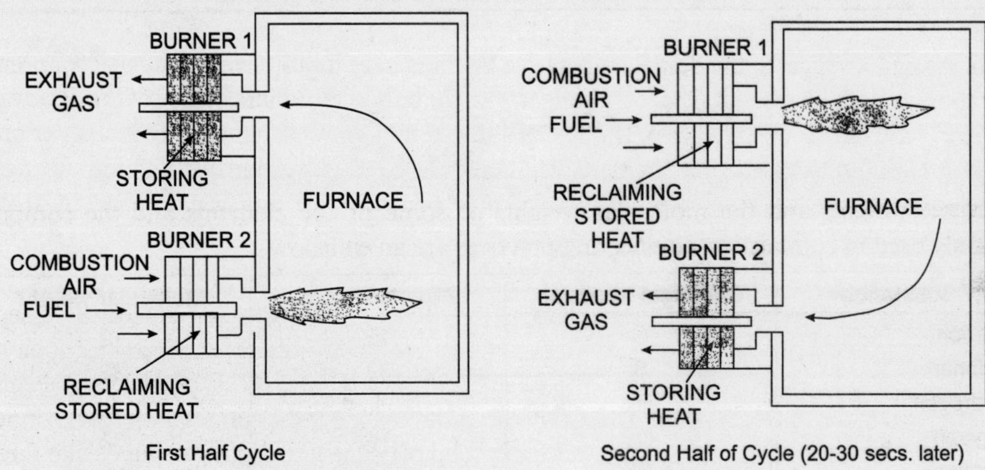

Fig. 38.22. *The Twin Bed Heat Reclamation Principle.*

COMBUSTION CALCULATIONS

The atomic weights and the molecular weights of some of the elements and the compounds commonly used in combustion engineering practices are given below :

Substance	Chemical formulae	Atomic weight	Molecular weight
Carbon	C	12	12
Sulphur	S	32	32
Hydrogen	H_2	1	2
Oxygen	O_2	16	32
Nitrogen	N_2	14	28
Carbon dioxide	CO_2	—	44
Carbon monoxide	CO	—	28
Water vapour	H_2O	—	18
Sulphur dioxide	SO_2	—	64
Methane	CH_4	—	16
Ethylene	C_2H_4	—	28
Air	—	—	29

COMPLETE COMBUSTION

Combustion is a chemical reaction between a fuel and an oxidant which proceeds above some minimum temperature called the spontaneous ignition temperature to which the reactants must be heated.

In combustion of gases & vapours, the proportions of fuel & oxidant must be between the limits of inflammability, and vary with the particular fuel, oxidant, diluent, direction of flame propagation, pressure, temperature, presence of catalyst, etc. In the combustion of all fuels, it is desirable to know the stoichiometric proportions i.e. the amount of oxidant which on completion of combustion with a given amount of fuel would lead to the products like carbondioxide (CO_2), water (H_2O), sulphur dioxide (SO_2) and nitrogen (N_2). This idealized concept is the basis of all combustion calculations.

The majority of hydrocarbon fuels have as active ingredients the elements — carbon (C), hydrogen (H), oxygen (O), nitrogen (N) and sulphur (S). An understanding of the basic chemical reactions in which these elements are involved is fundamental to practical operations in combustion of most solid, liquid and gaseous fuels. Inert ingredients are often observed in fuel in the form of moisture and ash, the latter being the inorganic residue after combustion of the fuel in air or oxygen to constant mass at a temperature of around 800°C. Gaseous fuels may contain water vapour, carbon dioxide, or nitrogen, which are diluents, which generally increase the volume of combustion products and detract from the maximum temperature attainable.

Generally combustion requires reaction with an oxidant or supporter of combustion. Oxygen is the main oxidant, but the halogens (chlorine and fluorine), hydrogen peroxide and also nitric acid, may act as oxidants as in rocket propulsion.

Air is the commonest oxidant because it is cheap and readily available. A typical volumetric percentage composition of dry air is :

N_2 = 78.09, O_2 = 20.95, Ar = 0.93, CO_2 = 0.03, Ne = 0.0018, He = 0.005, CH_4 = 0.002, Kr = 0.0001, NO = 0.000 05, H_2 = 0.000 05, O_3 = 0.000 04, Xe = 0.000 008.

This composition varies with altitude and slightly by industrial pollution. By convention, the following analysis are used in combustion calculations :

Air components	by volume %	by weight %
O_2	21	23
N_2	79	77
Total	100	100

Assuming the fixed proportion of oxygen to 'nitrogen', which includes all the inerts, the stoichiometric oxygen is related to the stoichiometric air requirement.

Combustion calculations are based on a number of assumptions :

(a)Avogadro's hypothesis : Equal volumes of gases measured under the same conditions of temperature and pressure contains equal numbers of molecules.

(b) The ideal gas law generally applies, since most combustion reactions involve high temperature and a total pressure close to one atmosphere, (i.e. PV = nRT) .

(c) Conservation of chemical elements : The total mass of C, H, O, N, S before reaction is the same as that after reaction.

(d) Dalton's law of partial pressures applies to mixture of gases. Each component at the same temperature existing on its own in a vessel would exert a partial pressure p_i, thus

i.e.,
$$P = \frac{1568}{2650} p_i$$

(e) Conservation of energy, for a given mass of fuel, oxidant and diluent.

i.e. Total energy of reactants = Total energy of products.

Enthalpy tables normally use 0°C as the datum temperature for zero enthalpy. Hence sensible heats can be calculated from this datum. Heats of chemical reaction, determined by calorimeter are generally referred to a datum temperature of 25°C and the preheats of reactants & products are corrected to this temperature in calorimetry.

(f) Most combustion work is at constant pressure (in practice near one atmosphere) so constant pressure conditions may be assumed by convention unless stipulated otherwise.

(g) Theoretical air is defined as the minimum quantity of air per unit mass of fuel required which is sufficient on complete combustion to give $C0_2$, H_2O, $S0_2$ and N_2 as products. Theoretical air is a fixed quantity for a given fuel and is calculated by the stoichiometry/ chemical reaction of the various combustible constituents of the fuel. Similarly, theoretical flue gas refers to the flue gas obtained by the complete combustion of fuel using theoretical amount of air. Basis of calculation is generally taken as 100 kg of solid or liquid fuel and 100 Kg moles of gaseous fuels or flue gas as may be the case.

(h) Excess air is the practical amount of air which is supplied in a combustion process usually to ensure that, under the conditions of inadequate mixing of fuel and air, the combustion process is likely to go to completion. Normally 10, 15-20, 20-25 and 50-100% excess air is supplied with gaseous, liquid,pulverised fuel and solid lumpy fuels, respectively.

$$\text{Excess air} = \frac{\text{actual air} - \text{theoretical air}}{\text{theoretical air}} = \frac{\text{actual } O_2 - \text{theoretical } O_2}{\text{theoretical } O_2}$$

$$\text{Excess air factor} = \frac{\text{actual air used}}{\text{theoretical air}}$$; At 40% excess air, the value of excess air factor is 1.4.

(i) Except for the purposes of corrosion, the convention is that sulphur burns to sulphur dioxide. (In practice 1-3% of sulphur in the gases may form SO_3, part of which is subsequent to the combustion process and is completed in the upper atmosphere).

(ii) Oxygen in the fuel is assumed to take part in the combustion reaction. Thus explosives contain nearly stoichiometric oxygen.

(iii) Except for a very small amount of nitrogen which forms NO and NO_2, nitrogen (N) in the fuel is conventionally assumed to form nitrogen gas (N_2) on combustion.

(i) Basic complete combustion reactions are :

C	+	O_2	=	CO_2
1 mole	1 mole			1 mole (molar or volume basis)12 kg
12 kg	32 kg			44 kg(mass basis)
H_2	+	$\frac{1}{2}O_2$	=	H_2O
1 mole		$\frac{1}{2}$ mole		1 mole (molar or volume basis)
2 kg		16 kg		18 kg (mass basis)
S	+	O_2	=	SO_2 1 mole
1 mole		1 mole		(molar or volume basis)
32 kg		32 kg		64 kg (mass basis)

These reactions obey the laws of algebra, as each reaction can be multiplied through by any constant factor, and reactions can subsequently be added & subtracted. Reactions are always shown in a positive sense.

Some common combustion reactions are :

Combustibles	Chemical Reactions
carbon monoxide	$CO + \frac{1}{2} O_2 = CO_2$
hydrogen sulphide	$H_2S + 1\frac{1}{2} O_2 = H_2O + SO_2$
methane	$CH_4 + 2O_2 = CO_2 + 2H_2O$
unsaturated hydrocarbon	$C2_{2\cdot5}H_5 + 15O_2 = 10CO_2 + 10H_2O$
propane	$C_3H_8 + 50_2 = 3CO_2 + 4H_2O$
butane	$C_4H_{10} + 6\ \frac{1}{2}\ O_2 = 4CO_2 + 5H_2O$
ethane	$C_2H_6 + 30_2 = 2CO_2 + 2H_2O$
pentane	$C_5H_{12} + 80_2 = 5CO_2 + 6H_2O$
acetylene	$C_2H_2 + 2\ \frac{1}{2}\ O_2 = 2CO_2 + H_2O$
Ethylene	$C_2H_4 + 3O_2 = 2CO_2 + 2H_2O$
benzene	$C_6H_6 + 7\ \frac{1}{2}\ O_2 = 6CO_2 + 3H_2O$
octane	$C_8H_{10} + 12\ \frac{1}{2}\ O_2 \rightarrow 8CO_2 + 9H_2O$

(j) Analyses of solid and liquid fuels are normally reported on a mass basis, while gaseous fuels are normally analysed on a volume basis. Waste gas analyses are generally reported by volume on the dry basis because condensation takes place from the sample. Moisture is determined separately, if it is measured at all.

(k) Ultimate CO_2 is defined as the volume % CO_2 in waste gas (dry basis) when the fuel is completely burnt with stoichiometric air.

Combustibles in the solid and liquid fuels are expressed as elements in weight percent while for gaseous fuels, they are given in volume percent or mole percent. Usually orsat analysis (on dry basis) for fuel gas & flue gas are used. SO_2 & CO_2 in flue gas are reported together in Orsat analysis as SO_2 is also absorbed in KOH in Orsat apparatus along with CO_2. In case of solid & liquid fuels and some gaseous fuels (having $N_2 < 3\%$), contribution to the N_2 content in respective flue gas is negligible and hence can be neglected. Combustion air is usually assumed to be dry having average molecular weight of 29. Nm^3 refers to volume in m^3 at N.T.P. (*i.e.,* O°C & 760 mm Hg).

INCOMPLETE COMBUSTION

The formation of black smoke, soot, tar, partial decomposition products and unburnt fuel are all symptoms of incomplete combustion. CO, H_2 or CH_4 analysed in the waste gas indicate incomplete combustion. The process may be carried out deliberately as in gasification and sometimes for reducing atmospheres in the heat treatment of metals. More often it is due to insufficient air for complete combustion or bad mixing of fuel and air.

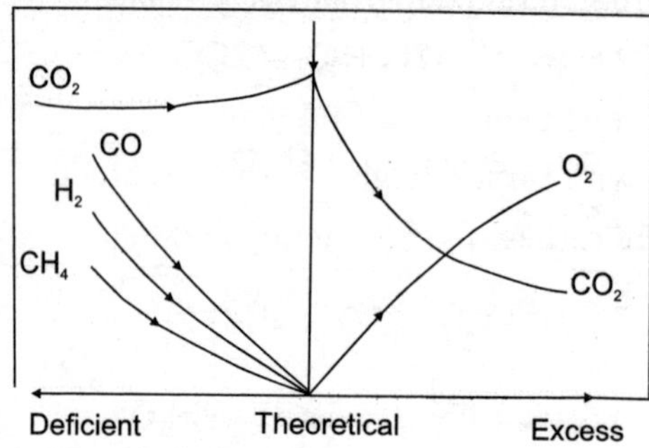

Fig.39.1: Combustion air Vs flue gas components

FLUE GAS COMPOSITION-ULTIMATE CO_2

Fig. 40.1 indicates the approximate variation in composition (dry) of the waste gas for a given fuel for a varying amount of air assuming complete reaction of the oxygen available.

Equations for Stoichiometric or chemically correct mixtures

1. *(a) Carbon burning to CO_2 (Analysis by weight)*

$$C + O_2 = CO_2$$

Inserting the values of the molecular weights,

12 units by weight + 2 ×16 units by weight = 44 units by weight

If we consider the unit of weight as kg.

$$12 \text{ kg of C} + 32 \text{ kg of } O_2 \longrightarrow 44 \text{ kg of } CO_2$$

$$\therefore \quad 1 \text{ kg of C} + \frac{32}{12} \text{kg of } O_2 \longrightarrow \frac{32}{12} \text{ kg of } CO_2$$

$$\text{or,} \quad 1 \text{ kg (C)} + \frac{8}{3} \text{ kg (O)}_2 \longrightarrow \frac{11}{3} \text{ kg (CO)}$$

(b) Carbon burning to CO_2 (Analysis by volume), $C + O_2 \longrightarrow CO_2$

By Avogadro's hypothesis, the molecules of all the gases occupy the same volume under the same conditions of temperature and pressures.

Therefore, negligible volume of solid carbon + 1 volume of O_2 = 1 volume of CO_2 or 1 volume of oxygen combines with carbon to give 1 volume of CO_2.

Thus, there is no molecular expansion or contraction.

2. *(a)* Carbon burning to carbon monoxide *(Analysis by weight)*

$$2C + O_2 = 2CO$$

$$\therefore \quad 2 \times 12 \text{ kg (C)} + 2 \times 16 \text{ kg } (O_2) = 2\,(12 + 6) \text{ kg (CO)}$$

i.e. 24 kg of carbon + 32 kg of oxygen = 56 kg of carbon monoxide

$$\therefore \quad 1 \text{ kg(C)} + \frac{4}{3} \text{ kg}(O_2) = \frac{7}{3} \text{ kg(CO)}$$

(b) Carbon burning to carbon monoxide *(Analysis by volume)*

$$2C + O_2 = 2CO$$

Neglecting the volume of solid carbon, we find that

1 volume of O_2 combines with carbon to give 2 volumes of CO.

There is an increase in volume here.

3. *(a)* Carbon monoxide burning to carbon dioxide *(Analysis by weight).*

$$2CO + O_2 = 2CO_2$$

$$2 \times 28 \text{ kg} + 32 \text{ kg} = 2 \times 44 \text{ kg}$$

$$\therefore \quad 1 \text{ kg of carbon monoxide} + \frac{4}{7} \text{ kg of oxygen} = \frac{11}{7} \text{ kg of carbon dioxide.}$$

(b) Carbon monoxide burning to carbon dioxide (Analysis by volume).

$$2CO + O_2 = 2CO_2$$

$$2 \text{ vol CO} + 1 \text{ vol. } O_2 = 2 \text{ vol. } CO_2$$

There is molecular contraction in volume.

4. *(a) Sulphur burning to sulphur dioxide (Analysis by weight)*

$$S + O_2 = SO_2$$

$$32 \text{ kg} + 32 \text{ kg} = 64 \text{ kg}$$

$$\therefore \quad 1 \text{ kg of S} + 1 \text{ kg of } O_2 \rightarrow 2 \text{ kg of } SO_2$$

(b) Sulphur burning to sulphur dioxide (Analysis by volume)

$$S + O_2 = SO_2$$

Negligible volume of S + 1 vol. of O_2 = 1 vol. of SO_2

There is no molecular contraction or expansion in volume.

5. *(a) Methane, complete combustion (Analysis by weight)*

$$CH_4 + 2O_2 = CO_2 + 2H_2O$$

$$16 \text{ kg} + 64 \text{ kg} = 44 \text{ kg} + 36 \text{ kg}$$

$$\therefore \quad 1 \text{ kg of } CH_4 + 4 \text{ kg of } O_2 = \frac{11}{4} \text{ kg of } CO_2 + \frac{9}{4} \text{ kg of } H_2O$$

(b) Methane, complete combustion (Analysis by volume)

$$CH_4 + 2O_2 = CO_2 + 2H_2O$$

$$1 \text{ vol.} + 2 \text{ vol.} = 1 \text{ vol.} + 2 \text{ vol.}$$

$$\therefore \quad 1 \text{ vol. of } CH_4 + 2 \text{ vol. of } O_2 = 1 \text{ vol. of } C0_2 + 2 \text{ vol. of } H_20 \text{ (steam)}$$

There is no molecular expansion or contraction in volume.

6. *(a) Ethylene, complete combustion (Analysis by weight)*

$$C_2H_4 + 3O_2 = 2CO_2 + 2H_20$$

$$28 \text{ kg} + 96 \text{ kg} = 88 \text{ kg} + 36 \text{ kg}$$

$$\therefore \quad 1 \text{ kg of } C_2H_4 + \frac{24}{7} \text{ kg Of } O_2 = \frac{22}{7} \text{ kg of } CO_2 + \frac{9}{7} \text{ kg of } H_2O$$

(b) Ethylene, complete combustion (Analysis by volume)

$$C_2H_4 + 3O_2 = 2CO_2 + 2H_2O$$

$$1 \text{ volume} + 3 \text{ volume} = 2 \text{ volume} + 2 \text{ volume}$$

∴ 1 volume of C_2H_4 + 3 volume of 0_2 = 2 volume of CO_2 + 2 volume of H_2O (steam)

There is no molecular contraction or expansion in volume.

Mole, Kilomole and Mole Fraction

Mole of a substance is the mass of the substance numerically equal to the molecular weight of the substance. Thus mole can be expressed as

$$\text{Mole} = \frac{\text{Weight of the substance in kg}}{\text{Molecular weight of the substance}}$$

Moles are specially useful in the case of gaseous fuels because the equations of chemical reactions are also molecular equations. Avogadro's law states that under the same conditions of pressure and temperature, a fixed volume of any gas contains the same number of molecules. In other words, the molecular volume of all the gases is the same under the same conditions of pressure and temperature. Thus molecular volume for any gas can be expressed as

$$pV = \text{MRT}$$

where, M = No. of moles of gas and, R = characteristic gas constant

$$V_{mol} = \frac{848 \text{ T}}{p}$$

where, T is expressed in °K, p is expressed in kgf/m^2 and V is expressed in m^3 .

It can also be expressed as $V_{\text{mol}} = \dfrac{0.0848\ T}{p}$ where, p is expressed in kgf/cm^2 .

Kilomole or kg.mole

Molecular weight of a substance is a number proportional to the mass of the molecule. Thus if the mass is expressed in any system of units, the molecule will be correspondingly called. Therefore, 1 kilomole of oxygen will mean 32 kg of oxygen. 1 gram mol. of oxygen will mean 32 grams of oxygen. Similarly 1 kg mole of any gas will occupy a volume of 22.4 Nm^3.

Example 1. The volumetric composition of a gaseous fuel is the following :

$H_2 = 50\%$; $CH_4 = 20\%$; $C_2H_4 = 2\%$; $CO_2 = 5\%$; $CO = 16\%$; $N_2 = 7\%$.

Determine the molecular weight, and the density of the mixture at S.T.P. if the total pressure of the mixture is 1.03 kgf/cm². Determine also the gravimetric composition or composition by weight.

Solution. The universal gas constant 'MR' for any gas is equal to 848. Therefore, we have

Constituent gas	Percentage composition by volume	Molecular fraction	Molecular weight, M
H_2	50	0.5	2
CH_4	20	0.2	16
C_2H_4	2	0.02	28
CO_2	5	0.05	44
CO	16	0.16	28
N_2	7	0.07	28

Therefore, molecular weight of the mixture :

$M_{mixture} = 0.5 \times 2 + 0.2 \times 16 + 0.02 \times 28 + 0.05 \times 44 + 0.16 \times 28 + 0.07 \times 28 = 13.4$

Molecular volume of gas at S.T.P. (*i.e.* at 15°C and 760 mm Hg) is found as follows :

$$T = 15°C = (273 + 15)\ °K$$

and, $P = 1.03\ kg/cm^2 = 1.03 \times 10^4\ kg/m^2$

$$V_{mol} = \frac{848\ T}{p} = \frac{848 \times (273+15)}{1.03 \times 10^4} = \frac{848 \times 288}{1.03 \times 10^4} = 23.67\ m^3$$

Therefore, density of the gas is given by

$$\rho = \frac{\text{Molecular weight expressed in kg}}{\text{Molecular volume expressed in m}^3}$$

$$= \frac{13.4}{23.67} = \mathbf{0.565\ kg/m^3 \text{ at S.T.P}} \text{ (i.e., at 15°C \& 760 mm Hg)}$$

Gravimetric Composition

Constituent gas	Mole fraction x,	Molecular weight M,	Product x.M,	Gravimetric Analysis $\frac{xM}{\sum X.M}$
H_2	0.5	2	1.00	$\frac{1}{13.4}$ = 0.0746 or **7.46%**
CH_4	0.2	16	3.20	$\frac{3.2}{13.4}$ = 0.239 or **23.9%**
C_2H_4	0.02	28	0.56	$\frac{0.56}{13.4}$ = 0.0418 or **4.18%**
CO_2	0.05	44	0.20	$\frac{2.20}{13.4}$ = 0.1642 or **16.42%**
CO	0.16	28	4.48	$\frac{4.48}{13.4}$ = 0.334 or **33.4%**

If the coal supplied to the furnace has 79.5% by weight of carbon, estimate the heat carried away by the exit gases per kg of coal burnt. Assume the average specific heat of the flue gas to be 0.24 Kcal/kg/K per kg per °K.

Solution. The data does not provide the water vapor content in the flue gas. Therefore, the flue gas will be considered dry.

Constituent gas	Analysis by volume % A	Molecular weight B	Proportional weight C=AЧB	Analysis by weight $D=\frac{C}{\Sigma C}$	Carbon per kg constituent gas E	Weight of carbon for kg of flue gas F = DЧE
CO_2	11.6	44	510.0	0.1693	12 44	0.0462
O_2	6.8	32	217.6	0.0722	—	—
CO	0.5	28	14.0	0.00465	12 28	0.00199
N_2	81.1 by difference	28	2.272	0.754	—	—
Total	100		3014			0.04819

Therefore, 0.04819 kg of carbon burns to give 1 kg of flue gas.

Thus, 0.795 kg carbon (in coal) will give = $\frac{0.795}{0.04819}$ kg of dry flue gas (D.F.G.)

= **16.5 kg D.F.G./kg coal.**

And heat carried away by the dry gases per kg of coal burnt

= 0.24 x 16.5 x (275 - 20) = **1002 kcal/Kg of coal burnt**

Conversion of Gravimetric Analysis to Volumetric Analysis and vice-versa

(i) Since, molecular weight of all gases occupy the same volume, proportional volume of constituent gas multiplied by the molecular weight of the constituent gas will give the proportional weight of constituent gas.

(ii) And the proportional weight of the constituent gas divided by its molecular weight will give the proportional weight of the constituent gas.

Example 3. Volumetric analysis of a sample of flue gas is 10.5% CO_2, 0.5% CO, 8% O_2 and 81% N_2 (by difference). Determine the gravimetric composition.

Solution.

Constituent gas	% Volumetric analysis A	Molecular weight B	Proportional weight C =A ЧB	% Gravimetric analysis $D = \frac{c}{\Sigma c}$ l00
CO_2	10.5	44	462	$\frac{462}{3000} \times 100 = 15.4\%$
CO	0.5	28	14	$\frac{14}{3000} \times 100 = 0.4665\%$

O_2	8	32	256	$\frac{256}{3000} \times 100 = 8.533\%$
N_2	81	28	2268	$\frac{2268}{3000}$ × l00= 75 134%
Total	100		3000	100

Example 4. Percentage analysis by weight of flue gas is 16% CO & 0.95% CO, 7.6% O_2 and 75.45% N_2. Determine the percentage volumetric analysis of the sample.

Solution. In tabular form, we have

Constituent gas	% Volumetric analysis A	Molecular weight B	Proportional weight C=A/B	% Volumetric analysis d = $\frac{c}{\sum c}$ ч100
CO_2	16	44	0.362	10.95%
CO	0.95	28	0.00339	1.026%
O_2	7.6	32	0.2375	7.2%
N_2	75.45	28	2.695	80.85%
Total	100		3.29789	100

Minimum Air Required in kg of Solid or Liquid Fuel Burnt:

Let us consider a representative sample of fuel containing C parts by weight of carbon, H_2 parts by weight of hydrogen, O_2 parts by weight of oxygen and S parts by weight of sulphur.

Therefore, we have, C kg of carbon requires C × $\frac{8}{3}$ kg of O_2

H_2 kg of hydrogen requires H × 8 kg of O_2.

S kg of sulphur requires S × 1 kg of S.

And O_2 kg of oxygen is already present in the fuel.

Therefore, oxygen required for complete combustion of 1 kg of fuel is given by :

$$\left(\frac{8}{3}C + 8H_2 + S - O_2\right) = \frac{8}{3}C + 8\left(H_2 - \frac{O_2}{8}\right) + S$$

And atmospheric air contains 23 per cent by weight of oxygen. Therefore,

$$\text{Minimum air/kg of fuel} = \frac{100}{23}\left[\frac{8}{3}c + \left(8H_2 - \frac{O_2}{8}\right) + S\right] = 116C + 34.8 = 11.6\,C + 34.8\left(H_2 - \frac{O_2}{8}\right)$$

+ 4.35 S

The minimum quantity of air required can also be computed in a tabular form which gives an idea of the products of combustion.

Example 5. *A sample of coal has the following percentage composition by weight* $C := 77.7\%$, $H_2 = 6.8\%$, $N_2 = 1.2\%$, $S = 2.2\%$, $O_2 = 8.8\%$ *and, incombustibles 3.3%. Determine the minimum quantity of air required for complete combustion of 1 kg of coal, and the constituent products of combustion in kg per kg of coal.*

Solution.

Constituent	Weight per kg of coal A	O_2 required per kg of constituent B	Oxygen required per kg of fuel C = A×B	Products of combustion in kg per kg of fuel			
				CO_2	H_2O	SO_2	N_2
C	0.777	8/2	2.072	**28.49**	—	—	—
H_2	0.068	8/3	0.544	—	**0.612**	—	—
N_2	0.012	—	—	—	—	—	**0.012**
S	0.022	1	0.022	—	—	**0.022**	—
O_2	0.088	-1	-0.088	—	—	—	—
Ash	0.033	—	—	—	—	—	—
			$\sum C = 2.550$				

Therefore, minimum air required for complete combustion

$$= 2.55 \times \frac{100}{23} = \mathbf{11.1\ kg\ per\ kg\ of\ coal.}$$

Excess Air:

The quantity of air in kg in excess of the theoretical minimum air required for complete combustion of the fuels is called Excess Air. Free O_2 in the products of combustion is an indication of the excess air. Usually, the indication of excess air is provided by the CO_2 recorder. For a fuel containing only carbon, the theoretically correct combustion will yield in the products of combustion only CO_2 and N_2 and they will be in the same proportion by volume as O_2 and N_2 in the atmosphere. Therefore, the CO_2 percentage can never exceed 21%, i.e. the percentage by volume of O_2 in air. Usually CO_2 percentage varies from 10 to 12% by volume in coal fired boilers with natural draught and 10 to 15% by volume in boilers with artificial draught. These lower figures are the result of dilution of the product of combustion by excess air.

Example 6. *The analysis of coal used in a boiler trial is as follows : 82% carbon, 6% hydrogen, 4% oxygen, 2% moisture and 8% ash.*

Determine the theoretical minimum air required for complete combustion of 1 kg of coal. If the actual air supplied is 18 kg per kg of coal the hydrogen is completely burned and 80% of carbon is burned to CO_2*, the remainder to CO, determine the volumetric analysis of the dry products of combustion.*

Solution. For complete combustion, all the carbon in the fuel must burn to CO_2, therefore,

Minimum air required $= 11.6C + 34.8\left(H_2 - \frac{O_2}{8}\right) + 4.35\,S = 11.6 \times 0.82 + 34.8\left(0.06 - \frac{0.04}{8}\right)$

$= 0.5 + 1.912 = 11.412$ kg per kg of fuel.

But actually only 80% of the available carbon is burnt to CO_2. Therefore, CO_2 actually produced

$$= 0.8 \times 0.82 \times \frac{44}{12} = 2.403 \text{ kg of } CO_2$$

Therefore, oxygen actually required by the portion of carbon burning to CO_2

$$= 0.8 \times 0.82 \times \frac{32}{12} = 1.749 \text{ kg of } O_2$$

Similarly 20% of carbon is burnt to CO. It produces CO

$$= 0.2 \times 0.82 \times \frac{28}{12} = 0.383 \text{ kg of CO}$$

The oxygen actually required for 20% carbon burning to CO

$$= 0.8 \times 0.82 \times \frac{16}{12} = 0.219 \text{ kg of } O_2$$

Also oxygen required by hydrogen

$$= 0.06 \times 8 = 0.48 \text{ kg of } O_2$$

and, water produced $= 0.06 \times 9 = 0.54$ kg of H_2O

Air actually supplied is 18 kg. It contains O_2

$$= 18 \times \frac{23}{100} = 4.14 \text{ kg of } O_2$$

Therefore, free O_2 in the flue gas

= 4.14 (from air supplied) + 0.04 (from fuel)

= 1.749 (for CO_2) - 0.219 (for CO) - 0.48 (for H_2O)

= 1.732 kg per kg of coal

and, nitrogen in the flue gas = 18 × 0.77 = 13.87 kg per kg of coal

Constituent gas	kg of constituent gas per kg of coal A	Molecular weight B	Proportional weight C =A/B	% volume of constituent gas $D = \frac{C}{\sum C} \times 100$
CO_2	2.403	44	0.0546	8.84%
CO	0.383	28	0.0137	2.22%
O_2	1.732	32	0.0541	8.74%
N_2	13.87	28	0.4953	80.2%
Total			0.6177	100

Example 7. The analysis of the coal in boiler trial was : C = 81%, H_2 = 4.5%, O_2 = 8%, remainder-incombustible.

The Orsat analysis of the dry flue gas was CO_2 = 8.3%, CO = 1.4%, O_2 = 10%, N_2 = 80.3% (by difference).

Determine :

(a) the weight of air supplied per kg of coal;

(b) the percentage of excess air.

Solution.

Constituent Gas	% Volumetric Composition. A	Molecular weight B	Proportional weight C =AЧB	Analysis by weight $D=\frac{C}{\sum C}$	Carbon per kg of constituent E	Weight of carbon per kg of dry flue gas F = DЧE
CO_2	8.3	44	365.2	0.1228	12 44	0.0335
CO	1.4	28	39.2	0.0132	12 28	0.00566
O_2	10.0	32	320.2	0.1076	—	—
N_2	80.3	28	2249	0.756	—	—
Total	100		2973.4			0.03916

Therefore, for 0.81 kg of carbon in 1 kg of coal the flue gas will be

$$= \frac{0.81}{0.03916} = 20.68 \text{ kg of dry flue gas per kg of coal.}$$

Water formed $= 0.045 \times 9 = 0.405$ kg per kg of coal.

Incombustibles $= 1 - (0.81 \times 0.045 + 0.08) = 0.065$ kg/kg of coal.

Therefore, air supplied per kg of coal

$$= 20.68 \times 0.405 - (1 - 0.065) = 21.085 - 0.935$$

$$= \mathbf{20.15 \text{ kg per kg of coal.}}$$

Minimum air required per kg of coal

$$= 11.6C + 34.8\left(H_2 - \frac{O_2}{8}\right) - 4.35\,S = 11.6 \times 0.81 + 34.8\left(0.045 - \frac{0.08}{8}\right)$$

$$= 9.4 + 1.2175 = 10.6175 \text{ kg air per kg of coal.}$$

Therefore, percentage excess of air

$$= \frac{20.15 - 10.6175}{10.6175} \times 100 = \frac{9.5325}{10.6175} \times 100 = \mathbf{89.8}$$

Example 8. *(a) Calculate the amount of air required for theoretically complete combustion of 100 kg of coal of the following composition.*

C = 82%, H = 6%, O_2 = 4%, ash = 8%, moisture = 2%.

Solution. Minimum O_2 required for combustion of carbon $= 82 \times \frac{32}{12} = 218.7$ kg

Minimum O_2 required for combustion of $H_2 = 6 \times \frac{32}{12} = 48$ kg

O_2 available with the coal = 4 kg

Hence, Net O_2 required for combustion of coal = 218.7 + 48 – 4 = 262.7 kg

Air required for combustion of coal $= 262.7 \times \frac{100}{23} = 1141$ kg.

Example 8. *(b) Calculate amount of air required for theoretically complete combustion of 100 Nm^3 of blast furnace gas of the following composition (by volume %)*

$$CO_2 = 17,\ CO = 22.1,\ H_2 = 4.9,\ N_2 = 55.8,\ O_2 = 0.2.$$

Solution.

O_2 required for combustion of CO $= 1/2 \times 22.1$ m^3 $= 11.05$ m^3

O_2 required for combustion of H_2 $= 1/2 \times 4.9$ m^3 $= 2.45$ m^3

O_2 available with the gas $= 0.2$ m^3

Net O_2 required for complete combustion $= 11.05 + 2.45 - 0.2 = 13.3$ m^3

Air required for complete combustion of gas $= 13.3 \times \frac{100}{21} = 63.4$ m^3.

Air-Fuel Ratio

Air-fuel ratio can be easily found out either by volume for gaseous fuels or by weight for solid/ liquid fuels in the following manner :

In case of example *'8 a'* above, air/fuel ratio (by weight)

$$= \frac{1141}{100} = 11.41 \text{ kg/kg}$$

In case of example *'8 b'* above, air/fuel ratio (by volume)

$$= \frac{63.4}{100} = 0.634 \text{ Nm}^3/\text{Nm}^3$$

Nm^3 means volume of gas in m^3 at NTP (*i.e. at* 0°C and 760 mm Hg).

Example 8. *(c) For calculation of volume of products of combustion, consider 100 Nm^3 of same (B.F. gas) fuel as mentioned in example 8(b).*

Solution.

Constituents	Volume %	Oxygen required for complete combustion, Nm^3	Products of combustion, Nm^3		
			CO_2	H_2O	N_2
CO_2	17	—	17	—	—
O_2	0.2	(-0.2)	—	—	—
CO	22.1	11.05	22.1	—	—
H_2	4.9	2.45	—	4.9	—
N_2	55.8	—	—	—	55.8
Total	100	13.3	39.1	4.9	55.8

Besides nitrogen of the fuel, nitrogen of the air required for combustion will also be present in the products of combustion.

Volume of air required for combustion, as calculated in example *'8b'* = 63.4 Nm^3

Volume of N_2 in the air required for combustion $= 63.4 \times \frac{79}{100} = 50.1$ Nm^3

Now, we have following products of combustion (for 100 Nm^3 of B.F. gas)

Constituents	Volume, Nm^3	% by volume
CO_2	39.1	26.1
H_2O	4.9	3.1
N_2	(55.8 + 50.1) = 105.9	70.8
Total	149.9	100

$$\text{Ratio of products of combustion and the fuel gas (by volume)} = \frac{149.9}{100} = 1.499$$

Thus, 100 Nm^3 of B.F. gas of the above mentioned composition will require 63.4 m^3 of air for theoretically complete combustion and will produce 149.9 m^3 of wet flue gases on combustion.

To find out volume and composition of dry flue gas, we may neglect H_2O from the flue gases as below :

Constituents	Volume, Nm^3	% by volume
CO_2	39.1	27
N_2	105.9	73
Total	145.0	100

Excess Air for Combustion of Fuel:

Complete combustion is not attained in practice unless more than the amount of air theoretically required is used. The need of excess air is due to the difficulty of obtaining intimate contact between the air and the fuel. It is also partly due to the need to complete the combustion within the combustion space.

For maximum efficiency, it is essential that correct percentage of excess air should be used. If too much air is used fuel will be wasted because the extra air will have to be heated and additional heat will thus pass out of the system as sensible heat in the chimney gases. If too little air is used combustion will not be complete and the chimney gases will carry away unused potential heat in the form of unburnt combustible gases, such as carbon monoxide (CO), hydrogen (H_2) & methane (CH_4).

Thus the coefficient of excess air is defined as :

$$\text{Coefficient of excess air 'E'} = \frac{\text{Actual air used for combustion}}{\text{Theoretically required Air for complete combustion}}$$

and percentage of excess air $= (E - 1) \times 100$

Effect of Excess Air on Products of Combustion :

The volume and composition of products of combustion can be calculated, when using excess air for combustion as shown in the following example.

Example 8. *(d) Assuming 10% excess air for combustion of B.F. gas as in example 8 (b) and 8 (c), we find*

Solution. Total volume of air required $= 1.1 \text{ x } 63.4 - 69.7 \text{ Nm}^3$

Volume of excess air $= 69.7 - 63.4 = 6.3 \text{ Nm}^3$

Total volume of products of combustion

$$= \text{volume of products of combustion with theoretical air}$$
$$+ \text{volume of excess air} = 149.9 + 6.3 = 156.2 \text{ Nm}^3$$

O_2 in excess air $= \dfrac{21}{100} \times 6.3 = 1.32 \text{ Nm}^3$

N_2 in excess air $= 6.3 - 1.32 = 4.98 \text{ Nm}^3$

Now with reference to the example *8c'*, we find following products of combustion :

Constituents	Amount (Nm^3) per 100 m^3 of fuel gas	% by volume
CO_2	39.1	25
H_2O	4.9	3.15
N_2	(105.9 + 4.98) = 110.88	71
O_2	1.32	0.85
Total	156.2	100

Conversion of Volumetric Analysis to Analysis by Weight:

If the volumetric analysis of any gas is known it can be converted to a weight analysis by multiplying the volume of each constituent by its own molecular weight. This will give the proportional weights of the constituents, then by adding up these weights and dividing each by the total, the analysis by weight is obtained.

Example 8. *(e) A typical example is shown below by considering the B.F. gas of the same composition as in example 8 (b).*

Solution.

Constituents	% Vol. A	Molecular Weight, B	Proportional weights = AЧB	Weight % analysis
CO_2	17	44	748.0	25.400
O_2	0.2	32	6.4	0.218
CO	22.1	28	618.0	21.000
H_2	4.9	2	9.8	0.332
N_2	55.8	28	1560.0	53.05
Total	100		2942.2	100%

Weight of Flue Gas per kg of Fuel Burned :

The weight of flue gas will be more than that of the fuel on account of the air supplied. The actual weight of the dry flue gases can be obtained by comparing the weight of carbon in flue gases with the weight of carbon in the fuel, as there is no loss of carbon during the process of combustion.

In the case of a steam boiler there may be some unburned carbon in the cinders in the ash pit. This should be subtracted from the carbon in the fuel.

The flue gases leaving the fire grate will contain steam, due to the combination of the hydrogen in the fuel with oxygen, this steam will condense as the gases cool. The flue gas sample analysed is, therefore dry. Thus it will be seen that the flue gases contain the excess air supplied and do not contain the steam formed by combustion. Neither of these condition will effect the weight of carbon in the sample analysed.

Hence, by carbon balance,

Weight of flue gases per kg of fuel burned

$$= \frac{\text{weight of carbon in 1 kg of fuel}}{\text{weight of carbon in 1 kg of flue gas}}$$

Example 8. *(f) Considering the same B.F. gas in Example '8c' and '8e' the weight of carbon in 1 kg of fuel*

$$= \textit{weight of carbon in } CO_2 \textit{ and } CO = \frac{25.4}{100}\times\frac{12}{44}+\frac{12}{28}\times\frac{21}{100} = 0.0694 + 0.09 = 0.1594 \textit{ kg}$$

(Since $CO_2 = 25.4\%$ and $CO = 21\%$ by weight in fuel)

Solution. Now for the weight of carbon in the flue gases, it is necessary to convert the dry flue gas analysis of example *8c* into weight analysis as below :

Constituents	Volume %	Molecular weight	Proportional weight	Weight %
CO_2	27	44	27 × 44 = 1242	37.9%
N_2	73	28	73 × 28 = 2042	62.1%
Total	100		3284	100

$$\text{Weight of carbon in 1 kg of flue gases} = \frac{37.9}{100}\times\frac{12}{44} = 0.1035 = 0.1035 \text{ kg}$$

$$\text{Hence weight of flue gases per kg of fuel burned} = \frac{0.1594}{0.1035} = 1.54 \text{ kg}$$

In other words, ratio of flue gases to fuel (by weight) = 1.54.

Calorific Value of a Mixture of Gases

Calorific value of a mixture of gases can be found out, if we know the calorific values of its individual constituents.

Net calorific value of some individual combustible substances are given below in Kcal/Nm^3.

Carbon monoxide (CO) = 3040	Hydrogen (H_2) = 2590
Methane (CH_4) = 8560	Unsaturated hydrocarbons C_mH_n = 17000

Example 8. *(g) Considering the B.F. gas of the composition as in example '8b', we find*

Solution.

Combustible constituents	Volume %	C.V. of gas, kcal/Nm^3
CO	22.1	3040
H_2	4.9	2590

Hence, C.V. of B.F. gas = 0.221 × 3040 + 0.049 × 2590 = 800 kcal/Nm^3

Density of a Mixture of Gases :

Density of a mixture of gases can also be found out from the density of its fundamental constituents.

Densities of some gases are given below :

Constituents	Volume %	Density, kg/Nm³	Remarks
CO_2	17	$\frac{44}{22.4} = 1.97$	1 Kg mol of any gas at NTP occupies 22.4 Nm³. Hence Density (ρ) is given by,
O_2	0.2	$\frac{32}{22.4} = 1.43$	
CO	22.1	$\frac{28}{22.4} =125$	$\rho = \frac{\text{Molecular weight}}{22.4}$, kg/Nm³
H_2	4.9	$\frac{2}{22.4} = 0.0894$	
N_2	55.8	$\frac{28}{22.4} =125$	

Hence density of B.F. gas of composition in example *(8b)*

$$= \frac{1}{100}(17\times1.97 + 0.2\times1.43 + 22.1\times1.25 + 4.9\times0.0894 + 55.8\times1.25) = 13 \text{ kg/Nm}^3$$

Specific Heat (C_p) of a Mixture of Gases :

Specific heat of a mixture of gases at constant pressure can also be calculated on the basis of specific heat of its constituents. The specific heat can be found out either on volume basis or weight basis and their units are Kcal/m³ °C or Kcal/Kg °C respectively.

Cp, mean specific heat at constant pressure between 0°C to 500°C (on volume basis) for some gases is tabulated below :

Constituents	Cp, Kcal /Nm³ °C
CO_2	0.48
O2	0.334
N_2	0.319
CO	0.321
H_2	0.312

Considering a flue gas of following composition (by volume%)

CO_2	25
O_2	0.85
N_2	71

$$Cp \text{ of dry flue gases} = \frac{1}{100}(25\times0.48 + 0.85\times0.334 + 71\times0.319) = 0.38\ \frac{\text{kcal}}{\text{m}^3\ {}^0\text{C}}$$

Heat Loss in Flue Gases :

Total heat loss in the flue gases is the sum of heat carried away by dry flue gases and the heat carried away by steam in the flue gases caused by the combustion of the hydrogen or due to moisture in the fuel.

Heat carried away by dry flue gases per m^3 of fuel burned = $V.Cp.\ (t_2 - t_1)$

Heat carried away by the steam formed per m^3 of fuel burned = $W.(Hs_2 - Hw_1)$

Hence, total heat loss in flue gases = $V.Cp\ (t_2 - t_1) + W\ (hs_2 - hw_1)$.

where, V = Volume of dry flue gases per kg of fuel , Cp = Mean specific heat of dry flue gases

t_2 = Temperature of flue gases leaving the furnace ,

t_1 = Ambient temperature

W = Weight of H_2O formed per m^3 of fuel burned

Hg_2 = Total heat of super heated steam at temperature t_2

H_{w1} = Sensible heat of water at temperature t_1.

Thus, the heat loss in the flue gases can be calculated by the above formula after substituting the various values, already calculated in the way shown earlier.

Example 9. *A blast furnace gas has the following volumetric analysis : H_2 = 9%, CO = 24%, CH_4 = 2%, CO_2 = 6%, 0_2 = 3% and N_2 = 56%. Determine the ultimate gravimetric analysis.*

Solution. Ultimate analysis is given in terms of the constituent elements and not the constituent chemical compounds. Thus in this case, the ultimate analysis will be expressed in terms of percentage weights of H_2, C, O_2 and N_2 in the mixture.

Constituents gas	Percentage volumetric analysis A	Proportional weight of elements = A ч molecular weight of element			
		C	H_2	O_2	N_2
H_2	9	—	9 × 2 = 18	—	—
CO	24	24 × 12 = 288	—	24 x 16 = 384	—
CH_4	2	2x12 = 24	2 × 4 = 8	—	—
CO_2	6	6 × 12 = 72	—	6 × 32 = 192	—
O_2	3	—	—	3 × 32 = 96	—
N_2	56	—	—	—	56 × 28 =1568
Total	100	384	26	672	1568

Constituent elements	Proportional weight, W	Gravimetric percentage composition D = $\frac{W}{\sum W}$ ч 100
C	384	$\frac{384}{2650}$ × 100 = **14.45**

H_2	26	$\frac{26}{2650} \times 100 =$ **0.98**
O	672	$\frac{672}{2650} \times 100 =$ **25.35**
N_2	1568	$\frac{1568}{2650} \times 100 =$ **58.22**
Total	2650	100

Example 10. *A coal with a calorific value of 7100 K cal/ Kg has a composition by weight % : C = 0.78, H_2 = 0.05, 0_2 = 0.08, S = 0.02 and N = 0.02, remainder is ash. It is burnt in a furnace with 50% excess air. The flue gases enter the chimney at 325°C and the atmospheric temperature is 15°C. Calculate the proportion of heat carried away by the flue gases. Assume perfect combustion. C_p for air = 0.24, C_p for dry products of combustion = 0.25. Heat carried away per kg of moisture in flue gases is 700 K cal.*

Solution. Since complete combustion is assumed, all the C burns to CO_2 and all the H_2 to H_2O and S to SO_2.

Therefore, for 1 kg of coal, we get

$$0.78 \times \frac{44}{12} = 2.86 \text{ kg } CO_2 \text{ (C burning to } CO_2)$$

$$0.05 \times 9 = 0.45 \text{ kg } H_2O \text{ (} H_2 \text{ burning to } H_2O)$$

$$0.02 \times 2 = 0.04 \text{ kg } SO_2 \text{ (S burning to } SO_2)$$

and thus oxygen required is given by

$$0.78 \times \frac{32}{12} = 2.078 \text{ kg } O_2 \text{ (C burning to } CO_2)$$

$$0.05 \times 8 = 0.4 \text{ kg } O_2 \text{ (} H_2 \text{ burning to } H_2O)$$

$$0.02 \times 1 = 0.02 \text{ kg } O_2 \text{ (S burning to } SO_2)$$

$$\text{Total} = 2.498 \text{ kg } O_2$$

Therefore, 2.498 kg oxygen is required per kg coal for complete combustion but coal contains 0.08 kg of O_2.

Therefore, O_2 to be supplied = 2.498 – 0.08 = 2.418 kg per kg of coal, and, with 50% excess air :

O_2 actually supplied = 1.5 × 2.418 = 3.627 kg per kg of coal.

This gives N_2 actually supplied $= 3.627 \times \frac{77}{23}$ = 12.14 kg per kg of coal.

Total free oxygen in flue gas = 3.627 - 2.418 = 1.209 kg per kg coal.

Total free nitrogen in flue gas = 12.14 + 0.02 = 12.16 kg per kg of coal.

Heat carried away by dry products (CO_2 + SO_2)

= (2.86 + 0.04) × 0.25 × (375 – 15) = 252 kcal/kg of coal

Heat carried away by steam = 0.45 × 700 = 315 kcal/kg of coal.

Heat carried away by O_2 and N_2 = (12.16 + 1.209) × 0.24 × (375 - 15) = 1115 kcal/kg of coal.

(assuming Cp ≈ 0.24)

Total quantity of heat carried away per kg of coal

= 252 +.315 + 1115 = 1682 kcal/per kg of coal.

Example 11. *The flue gas from an industrial furnace have the following composition by volume :*

$$CO_2 = 11.73\%, CO = 0.2\%, N_2 = 0.09\%, O_2 = 6.81\% \text{ and } N_2 = 81.17\%$$

Calculate the percentage excess air employed in the combustion, if the loss of carbon in clinker & ash is 1% of the fuel used and the fuel has the following composition by weight:

$$C = 74\%, H_2 = 5\%, O_2 = 5\%, N_2 = 1\%, S = 1\%, H_2O = 9\% \text{ and ash} = 5\%.$$

Solution.

Basis : 100 kg of the fuel charged to the industrial furnace

Reactions : $C + O_2 = CO_2$

$H_2 + 1/2\ O_2 = H_2O$

Oxygen balance:

Oxygen required for complete combustion $= \frac{74}{12} + \frac{5}{2} \times 0.5 + \frac{1}{32}$ = 7.447 kg mole

Oxygen already present in fuel $= \frac{5}{32}$ = 0.157 kg mole

Net oxygen demand from air = 7.447 - 0.157 = 7.29 kg mole.

Carbon balance :

Carbon lost in clinker and ash = 1 kg

Carbon burnt = 74 – 1 = 73 kg = 73/12 = 6.08 kg atom.

Let x kg mols of flue gas are formed.

Therefore, by carbon balance, $(0.1173 + 0.002)\, x = 6.8$ ∴ $x = 50.96$ kg mole.

From flue gas analysis,

N_2 in flue gas = 50.96 × 0.8117 = 41.36 kg mole.

N_2 from fuel = 1 kg = 0.036 kg mole.

N_2 from air = 41.360 – 0.036 = 41.324 kg mole.

Oxygen supplied from air $= 41.324 \times \frac{21}{79}$ = 10.98 kg mole

Excess oxygen = 10.98 - 7.29 = 3.69 kg mole

Percentage excess air used = percentage excess oxygen used

$$= \frac{\text{Excess}}{\text{Theoretical}} \times 100 = \left(\frac{3.69}{7.29}\right) \times 100 = 50.62$$

Example 12. *Octane is burnt with 10% excess air. Calculate :*

(a) Air/fuel ratio by weight.

(b) Air/fuel ratio by volume.

(c) Weight of dry exhaust gas formed per unit of fuel.

(d) Mole of oxygen in the exhaust gas per unit weight of fuel.

(e) Mole of water vapour in exhaust gas per unit weight of fuel.

(f) Volume of exhaust gas at 1 atmosphere and 260°C per unit weight of fuel.

The specific gravity of octane may be taken as 0.7.

Solution.

Basis : 1 kg mole of octane burnt.

Reaction : $C_8H_{18} + 12.5\ O_2 \rightarrow 8CO_2 + 9H_2O$

(a) Theoretical oxygen demand = 12.5 kg mole

Oxygen supplied by 10% excess air = 12.5 x 1.1 = 13.75 kg mols = 13.75 × 32 = 440 kg

Nitrogen supplied by air $= 13.75 \times \frac{79}{21} = 51.73$ kg mole = 1448.4 kg

Amount of air supplied = 13.75 + 51.73 = 65.48 kg mole = 1888.4 kg

Molecular weight of fuel = 114

$$\frac{\textit{weight of air}}{\textit{weight of fuel}} = 1888.4/114 = 16.56$$

(b) Specific gravity of octane = 0.7

Density of octane = 0.7 gm/cc = 700 kg/m^3

∴ Volume of fuel $= \frac{114}{700} = 0.163$ m^3

Assuming ideal gas behaviour, volume of air at N.T.P.

= 65.48 × 22.4 = 1466.75 m^3 = 1466.75 Nm^3

$$\frac{\text{Volume of air}}{\text{Volume of fuel}} = \frac{1466.75}{0.163} = 8998.5$$

(c) Excess O_2 = supplied O_2 – used O_2 = 13.75 – 12.50 = 1.25 kg mole

Dry flue gas analysis :

Constituents	Amount, kg mole	Molecular weight	Amount, kg
CO_2	8.00	44	352.0
O_2	1.25	32	40.0
N_2	51.73	28	1448.4
Total	60.98		1840.4

$$\frac{\text{Weight of dry exhaust gas}}{\text{Weight of fuel}} = \frac{1840.4}{114} = 16.4$$

(d) $$\frac{\text{Mole of } O_2 \text{ in the exhaust gas}}{\text{Weight of fuel}} = \frac{1.25}{114} = \mathbf{0.011}$$

(e) $$\frac{\text{Mole of water vapour in the exhaust gas}}{\text{Weight of fuel}} = \frac{9.0}{114} = \mathbf{0.079}$$

(f) Moles of exhaust gas (wet) = 60.98 + 9.0 = 69.98

Applying ideal gas law, volume at 260°C and 1 atmosphere,

$$V = \frac{nRT}{P} = \frac{69.98 \times 0.08206 \times (260 + 273)}{1.0} = 3060.8 \text{ m}^3$$

$$\frac{\text{Volume of exhaust gas(wet)}}{\text{Weight of fuel}} = \frac{3060.8}{114} = \mathbf{26.85}$$

Example 13. *A producer gas with the composition by volume, CO = 27.3%, CO_2 = 5.4%, O_2 = 0.6%, N_2 = 66.7% is burnt with 20% excess air. If the combustion is 98% complete, calculate the composition by volume of the flue gases.*

Solution.

Basis : 100 kg mole of producer gas burnt.

Oxygen balance :

O_2 required for CO combustion	= 27.3 × 0.5 = 13.65 kg mole
O_2 persent in fuel	= 0.6 kg mole.
Net O_2 required	= 13.65 - 0.60 = 13.05 kg mole
O_2 supplied by 20% excess air	= 13.05 × 1.20 = 15.66 kg mole
O_2 used for 98% combustion of CO	= 27.3 × 0.5 × 0.98 = 13.38 kg mole
O_2 excess	= 15.66 – 13.38 = 2.28 kg mole

Nitrogen balance:

N_2 from air = 15.66 x (79/21)	= 58.91 kg mole
N_2 from producer gas	= 66.7 kg mole
Total N_2 in flue gas	= 66.70 + 58.91 = 125.61 kg mole

Carbon dioxide balance *:*

CO_2 from producer gas	= 5.4 kg mole
CO_2 from the combustion of CO	= 27.3 × 0.98 - 26.75 kg mole
Total CO_2 in flue gas	= 5.4 + 26.75 = 32.15 kg mole

Carbon monoxide balance :

CO burnt = 26.75 kg mole

CO left = 27.30-26.75 = 0.55 kg mole

Flue Gas Analysis:

Constituents	Amount	
	kg mole	Mole%
CO_2	32.15	20.02
CO	0.55	0.34
N_2	125.61	78.22
O_2	2.28	1.42
Total	160.59	100.00

Example 14. *A furnace is fired with a natural gas that consists entirely of hydrocarbons (negligible inert gases and sulphur compounds). The Orsat analysis of the flue gas gives 9.5% CO_2, 2.0% O_2 and 1.8% CO.*

(a) *What is the molar ratio of net hydrogen to carbon in the fuel ?*

(b) *What percentage of excess air js being used ?*

Solution.

Basis : *100 kg mol of dry flue gas.*

From Orsat analysis, Mole of N_2 = 100 - (9.5 + 1.8 + 2.0) = 86.7

Oxygen balance :

O_2 supplied by air $= 86.7 \times \frac{21}{79} = 23.05$ kg mole

O_2 reported in fuel gas (dry) = 9.5 + 0.5 × 1.8 + 2.0 = 12.4 kg mole

O_2 unaccounted = 23.05 – 12.4 = 10.65 kg mole = O_2 reacted with H_2.

(a) Moles of hydrogen reacted = 10.65 × 2 = 21.3,

Amount of carbon =9.5 + 1.8 = 11.3 kg atom.

$$\frac{PV}{nT} = \frac{760 \times 22.4}{1 \times 273} = 62.36 \frac{(\text{mm Hg})(\text{m}^3)}{\text{kg mole °K}} = 1.185$$

(b) Moles of O_2 required for complete combustion

= Mole required for H_2 + Mole required for C = 10.65 + 11.3 = 21.95

Amount of excess O_2 = 23.05 - 21.95 = 1.1 kg mole

% excess air = % excess O_2 $= \frac{1.1}{21.95} \times 100 = 5\%$.

Example 15. The exhaust gas from a hydrocarbon fuel oil fired furnace, show 10.2% CO_2, 7.9% O_2 and 81.9% N_2 by Orsat analysis. Calculate *(i)* % excess air used, and *(ii)* kg of dry air supplied per kg of oil burnt in the engine.

Solution. Basis : 100 kg mole of dry flue gas.

oxygen balance:

N_2 in flue gas = 81.9 kg mole

O_2 supplied from air $= 81.9 \times \frac{21}{79} = 21.77$ kg mole

O_2 reported in flue gas = 10.2 (as CO_2) + 7.9 (as O_2) = 18.1 kg mole

O_2 unaccounted = 21.77 - 18.1 = 3.67 kg mole = O_2 used for H_2.

O_2 used actually = 10.2 + 3.67 = 13.87 kg mole

O_2 excess in flue = 7.9 kg mole

(a) % excess air $= \%$ excess $O_2 = \frac{7.9}{13.87} \times 100 = 56.96\%$

(b) O_2 used for H_2 = 3.67 kg mole

Amount of H_2 in fuel = 3.67 × 2 = 7.34 = 7.34 × 2 = 14.68 kg

Carbon in fuel = 10.2 kg atom = 10.2 × 12 = 122.4 kg

Total weight of hydrocarbon fuel oil

= Weight of H + Weigh of C = 14.68 + 122.4 = 137.08 kg

Weight of air = Mole of O_2 × 32 + Mole of N_2 × 28

= 21.77 × 32 + 81.9 x 28 = 2989.8 kg

$$\frac{\text{Kg dry air}}{\text{kg oil burnt}} = \frac{2989.8}{137.08} = \mathbf{21.81.}$$

Example 16. *Determine the flue gas analysis and air/fuel ratio by weight when a medium fuel oil with 84.9% carbon, 11.4% hydrogen, 3.2% sulphur, 0.4% oxygen and 0.1% ash by weight is burnt with 20% excess air. Assume complete combustion.*

Solution. Basis : 100 kg of medium fuel oil burnt

Oxygen balance : Amount of oxygen used are calculated with the help of combustion equations for C, S and H_2.

O_2 required for $C = 84.9 \times \frac{32}{12} = 226.4$ kg

O_2 required for $S = 3.2 \times \frac{32}{12} = 3.2$ kg

O_2 required for H = 11.4 × 8 = 91.2 kg

Total = 320.8 kg

O_2 present in fuel = 0.4 kg, O_2 required from air = 320.8 - 0.4 = 320.4 kg

O_2 supplied = 320.4 × 1.2 = 384.48 kg, O_2 excess = 384.48 – 320.4 = 64.08 kg

air supplied $= 384.48 \times \frac{100}{23} = 1671.65$ kg ; $\frac{\text{Weight of air}}{\text{Weight of fuel}} = \frac{1671.65}{100} = \mathbf{16.716}$

Flue Gas Analysis (Wet) :

Amount of N_2 $= 1671.65 \times \frac{77}{100} = 1287.17$ kg = 45.970 kg mole

CO_2 produced $= \frac{84.9}{12} \times 44 = 311.3$ kg = 7.075 kg mole

SO_2 produced $= 3.2 \times \frac{64}{32} = 6.4$ kg = 0.10 kg mole

H_2O produced $= \frac{11.4}{2} \times 18 = 102.6$ kg = 5.70 kg mole

O_2 excess = 64.08 kg = 2.003 kg mole.

Flue gas analysis is given below :

Constituents	Amount in kg mole	Mole % = Volume %
CO_2	7.075	11.63
SO_2	0.100	0.16
O_2	2.003	3.29
N_2	45.970	75.55
H_2O	5.700	9.37
Total	60.848	100

Example 17. *A boiler is fired using 200 kg/hr of a pure saturated hydrocarbon gas (C_nH_m)at atmospheric pressure and 20°C. The dry analysis of the flue gas which leaves the boiler at atmospheric pressure and 300°C is CO_2 =12%, O_2 = 3% and N_2 = 85%. Estimate the formula of the fuel and total volumetric flow rate of the gas.*

Solution. Basis : 100 kg mole of dry flue gas.

Oxygen balance:

O_2 in flue gas = 85 kg mole.

O_2 supplied by air $= 85 \times \frac{21}{79} = 22.59$ kg mole

O_2 reported in flue gas = O_2 as CO_2 + O_2 as O_2 = 12 + 3.0 = 15.0 kg mole

O_2 unaccounted = O_2 reacted with H_2 = 22.59 - 15.00 = 7.59 kg mole

H_2 reacted = 7.59 × 2 = 15.18 kg mole = 15.18 × 2 = 30.36 kg mole

Formula of the fuel (hydrocarbon) gas :

Amount of carbon = 12 kg atom

Amount of hydrogen = 30.36 kg atom

$$\frac{\text{Atom of H}}{\text{Atom of C}} = \frac{30.36}{12} = 2.53$$

As the ratio is more than 2.0, hence it is a paraffin of formula C_nH_{2n+2}

Hence,
$$\frac{n}{2n+2} = \frac{1}{2.53}$$

Solving, $n = 3.77 \approx 4.0$

Hence the fuel is *n*-butane (C_4H_{10}).

Volumetric flow rate :

Amount of fuel $= \frac{200}{58} = 3.45$ kg mole/hr

Assuming ideal gas law, volume at 20°C and 1 atmosphere is,

$$V = \frac{nRT}{P} = \frac{3.45 \times 0.08206 \times (273+20)}{1.0} = \mathbf{82.95\ m^3\ /hr.}$$

Example 18. *A furnace is fired by a fuel gas having the volumetric composition :H_2 = 52%, CH_4 = 30%, CO = 8%, C_mH_n = 3.6%, CO_2 = 2%, O_2 = 0.4% and rest N_2. Using a certain quantity of air in excess over stoichiometric, complete combustion of the gas is achieved, giving a dry waste gas of 5 m³ per m³ of fuel burned, estimate :*

(a) Composition by volume of dry waste gas formed.

(b) Per cent excess air used.

(c) Weight of water formed per m³ of gas burned, neglecting water present in the air used. For purpose of calculation, unsaturated hydrocarbon (C_mH_n) can be considered as C_3H_6.

Solution.

Basis 100 kg mole of fuel burnt.

Reactions : $CO + \frac{1}{2} O_2 = CO_2$

$CH_4 + 2O_2 = CO_2 + 2H_2O$

$C_3H_6 + 4.5\ O_2 = 3CO_2 + 3H_2O$

$H_2 + \frac{1}{2} O_2 = H_2O$

Oxygen balance :

O_2 required for CO $= 8 \times \frac{0.158 \times 30330}{6883} = 4$ kg mole

O_2 required for CH_4 $= 30 \times 2 = 60$ kg mole

O_2 required for C_3H_6 $= 3.6 \times 4.5 = 16.2$ kg mole

O_2 required for H_2 $= 52 \times \frac{1}{2} = 26.0$ kg mole

Total O_2 required $= 106.2$ kg mole

O_2 in fuel $= 0.4$ kg mole

Net O_2 required from air $= 106.2 - 0.4 = 105.8$ kg mole

Dry flue gas formed (with theoretical amount of air):

CO_2 formed = 8.0 (from CO) + 30.0 from CH_4) + 10.8 (from C_3H_6) + 2.0 (from fuel)

$= 50.8$ kg mole

N_2 from air $= 105.8 \times \frac{79}{21} = 398.0$ kg mole

Total N_2 $= 398.0 + 4.0 = 402.0$ kg mole

Amount of dry flue gas $= 50.8 + 402.0 = 452.8$ kg mole

Waste gas produced $= 5\ m^3/m^3$ of fuel gas

Let it be assumed that the fuel gas and the flue gas are reported at the same conditions. Hence volume ratio is identical to mole ratio.

Mole of waste gas $= 5 \times$ moles of fuel gas $= 500$ kg mole

Dry flue gas with theoretical amount of air $= 452.8$ kg moles

Hence excess air $= 500 - 452.8 = 47.2$ kg moles

Theoretical air $= 105.8 \times \frac{100}{21} = 503.81$ kg moles

(a) excess air used $= \frac{47.2}{503.81} \times 100 =$ **9.39%**

(b) Composition of waste gas :

Air used $= 503.81 + 47.2 = 551.01$ kg moles

N_2 from air $= 551.01 \times \frac{79}{100} = 435.29$ kg moles

Total N_2 $= 435.29 + 4.0 = 439.29$ kg moles

Excess O_2 $= 47.2 \times \frac{21}{100} = 9.91$ kg moles

Waste gas analysis :

Constituents	Amount, Mole	mole % or volume %
CO_2	50.80	10.16
O_2	9.91	1.98
N_2	439.29	87.86
Total	500	100

(c) Amount of water formed = 52.0 (from H_2) + 60 (from CH_4) + 10.8 (from C_3H_6)

$= 122.8$ kg moles $= 122.8 \times 18 = 2210.4$ kg

Amount of gas burned $= 100$ kg mol $= 2240\ m^3$ at NTP 2240 Nm^3

$$\frac{\text{Weight of water formed}}{m^3 \text{of gas burned}} = \frac{2210.4}{2240} = 0.987.$$

Example 19. *The dry flue gases from an oil fired furnace has a composition of 11.2% CO_2, 5.8% O_2 and 83% N_2 when analysed by an Orsat apparatus. Calculate :*

(a) % excess air, and

(b) Weight of combustion air used per kg of oil fired.

Assume fuel to have 82% C, 12% H, 3% S and balance impurities. Molecular weight of dry flue gas is 30.

Solution.

Basis : 100 kg of oil fired.

Oxygen balance :

O_2 required for 82 kg carbon = 218.67 kg = 6.833 kg mole

O_2 required for 12 kg H = 12 × 8 = 96 kg = 3.0 kg mole

O_2 required for 3 kgS = 3 kg = 0.094 kg mole

Total O_2 required = 6.833 + 3.0 + 0.094 = 9.927 kg mole

Let x kg mole of dry flue gases are formed.

Carbon balance :

Carbon in fuel $= \frac{82}{12} = 6.833$ kg atom,

Carbon in flue gases = $0.112\,x$ (from CO_2) kg atom.

So, $0.112x = 6.833$

or, $x = 61.0$ kg mole

Amount of N_2 in flue gas = 61.0 × 0.83 = 50.63 kg mole

This N_2 has come from combustion air.

Hence, O_2 from air $= 50.63\ x\ \frac{21}{79} = 13.46$ kg molei y

O_2 used = 9.927 kg mole

O_2 excess = 13.46 – 9.927 = 3.533 kg mole

% excess air = % excess O_2 $= \frac{3.533}{9.927} \times 100 =$ **35.60%**

(b) Amount of combustion air $= 50.63 \times \frac{100}{79} = 64.09$ kg mole,

Molecular weight of air = 28.84

Weight of air = 64.09 × 28.84 = 1848.36 kg

$$\frac{\text{Weight of combustion air}}{\text{Weight of fuel}} = \frac{1848}{100} = 18.48.$$

Amount of gas burned = 100 kg mol = 2240 m3 at NTP = 2240 Nm3

Example 20. *A producer gas contains the following (composition in mole %) :*

$CO_2 = 9.2$, $C_2H_4 = 0.4$, $CO = 20.9$, $H_2 = 15.6$, $CH_4 = 1.9$ *and* $N_2 = 52$.

When it is burnt, the products of combustion are found to have the following (% by volume) :

$CO_2 = 10.8$, $CO = 0.4$, $O_2 = 9.2$ *and* $N_2 = 79.6$

Compute :

(a) Litres of air used in the combustion of 1 litre of producer gas, both being at the same temperature and pressure.

(b) The percent excess air used in combustion.

(c) The percent of nitrogen in the products of combustion which came from the producer gas.

Solution.

Basis : 100 kg mole of producer gas.

Constituent	Amount, Kg moles	Atoms of C
CO_2	9.2	9.2
C_2H_4	0.4	0.8
CO	20,9	20.9
H_2	15.6	—
CH_4	1.9	1.9
N_2	52.0	—
Total	100.0	32.8

Basis : 100 kg mole of dry flue gas.

Constituent	kgmole	Atoms of C
CO_2	10.8	10.8
CO	0.4	0.4
O_2	9.2	—
N_2	79.6	—
Total	100.0	11.2

For 100 kg moles of dry combustion gas.

Let, x = kg mole of producer gas, and, y = kg mole of air used.

By carbon balance, 0.328 × = 11.2

or, $x = 34.15$ kg mole

(a) By nitrogen balance, $0.52x + 0.79\,y = 79.6$

or,
$$y = \frac{79.6 - 0.52\,x}{0.79} = 78.28 \text{ kg mole}$$

$$\frac{\text{Mole air}}{\text{Mole producer gas}} = \frac{(\text{litre of air}) \text{ at N.T.P.}}{(\text{litre of producer gas}) \text{ at N.T.P.}} = \frac{78.28}{34.15} = 2.29$$

(b) O_2 supplied by air $= 78.28 \times 0.21 = 16.44$ kg mole

$$O_2 \text{ excess} = O_2 \text{ in combustion gas} - O_2 \text{ required for CO} = 9.2 - 0.4 \times \frac{1}{2} = 9.0 \text{ kg mole}$$

$$\% \text{ excess air} = \% \text{ excess } O_2 = \frac{O_2 \text{ excess}}{O_2 \text{ supplied} - O_2 \text{ excess}} \times 100 \frac{9.0}{16.44 - 9} = \mathbf{121\%}$$

(c) N_2 from producer gas $= 34.15 \times 0.52 = 17.76$ kg mole

% N_2 in the products of combustion which came from the producer gas

$$\frac{17.76}{79.6} \times 100 = 22.31\%.$$

Example 21. *A furnace utilises a mixture of blast furnace gas and coke oven gas as a fuel. The gas mixture is fired along with air for obtaining the necessary temperature. The analysis of the individual gases and the flue gas evolved in the furnace as a result of the combustion of the gas mixture are given on dry basis (in volume %) as under.*

Constituents	Coke oven gas	Blast furnace gas	Air	Flue gas
CO_2	2.4	13.4	—	13.5
O_2	2.0	0.2	21.0	6.9
CO	7.2	24.2	—	0.4
CH_4	30.8	—	—	—
H_2	51.3	3.0	—	—
N_2	6.3	59.2	79.0	79.2

Calculate :

(a) The ratio of blast furnace gas to coke oven gas used.

(b) The per cent excess air used.

Solution. Basis : 100 kg mole of flue gas obtained.

Let, A = kg mole of air used, B = kg mole of blast furnace gas

C = kg mole of coke oven gas used.

Blast furnace gas:

Constituents	kg mole	O_2 required	Atom C
CO_2	13.4	0	13.4
O_2	0.2	0.2	—
CO	24.2	12.1	24.2
H_2	3.0	1.5	—
N_2	59.2	0	—
Total	100.0	13.4	37.6

Coke oven gas :

Constituents	kg mole	O_2 required	Atom of C
CO_2.	2.4	0	2.4
O_2.	2.0	-2.00	—
CO	7.2	3.60	7.2
CH_4	30.8	61.60	30.8
H_2	51.3	25.65	—
N_2	6.3	0.00	—
Total	100.0	88.85	40.4

Making a carbon balance gives, $0.376B + 4.404C = 13.5$ ...(1)

Making a nitrogen balance gives, $0.592B + 0.063C + 0.79A = 79.2$ ...(2)

Making an oxygen balance gives, $0.134B + 0.8885C + 6.9 = 0.21A$ ...(3)

From (1), $B = 36.968 - 1.074C$ ...(4)

From (2), $A = \dfrac{79.2 - 0.592B - 0.063C}{0.79}$ ...(5

From (4), $A = 72.56 + 0.724C$ ...(6)

Putting (4) and (5) in (3) yields, $C = 5.73$ kg mole

From (4), $B = 30.82$ kg mole

From (5), $A = 76.71$ kg mole

(a) $\dfrac{\text{Blast furnace gas}}{\text{Coke oven gas}} = \dfrac{30.82}{5.73} = 5.38$

(b) O_2 required for blast furnace gas $= 0.134 \times 30.82 = 4.13$ kg mole

O_2 required for coke oven gas $= 0.8885 \times 5.73 = 5.09$ kg mole

Total oxygen demand $= 4.13 + 5.09 = 9.22$ kg mole

O_2 supplied from air $= 76.71 \times 0.21 = 16.11$ kg mole

% excess air used = % excess O_2 used $= \dfrac{16.11 - 9.22}{9.22} \times 100 =$ **74.7%**

Example 22. *100 kg per hour of coke are fed to a furnace having a grate efficiency such that 95% carbon present in the coke is burnt. The coke composition is 90% carbon and 10% ash (by weight). 30% excess air is supplied for ensuring complete combustion. If 98% of carbon burnt is oxidised to dioxide and the rest to monoxide,*

(a) Find the composition by volume of the flue gases.

(b) If the flue gases are at a temperature of 300°C and a pressure of 750 mm Hg, calculate their flow rate in m^3/minute.

Solution. Basis : 100 kg of coke feed

Oxygen balance :

O_2 required for complete combustion $= \dfrac{90}{12} \times 32 = 240$ kg

O_2 supplied $= 240 \times 1.3 = 312$ kg

Carbon burnt $= 90 \times .95 = 85.5$ kg

Carbon to CO_2 $= 85.5 \times .98 = 83.79$ kg

Carbon to CO $= 85.5 \times 0.2 = 1.71$ kg

O_2 used for CO $= 83.79 \times \dfrac{32}{12} = 223.44$ kg

O_2 used for CO_2 $= 1.71 \times \dfrac{16}{12} = 2.28$ kg

Total O_2 used = 225.72 kg

Excess O_2 in flue gas= 312 - 225.72 = 86.28 kg

(a) Flue Gas Analysis :

N_2 from air = 312 × = 1044.5 kg

CO_2 formed = 83.79 + 223.44 = 307.23 kg

CO formed = 1.71 + 2.28 = 3.99 kg

Analysis of flue gases is as follows :

Constituents	Amount, kg	Kg Mole	Mole % (= vol.%)
CO_2	307.23	6.98	14.81
CO	3.99	0.14	0.30
O_2	86.28	2.70	5.73
N_2	1044.50	37.30	79.16
Total	1442.00	47.12	100.00

(b) $P = 750$ mm Hg, $T = 300°C = 573°K$

$$R = \frac{PV}{nT} = \frac{760 \times 22.4}{1 \times 273} = 62.36 \frac{(\text{mm Hg})(\text{m}^3)}{\text{kg mole °K}}$$

$$\text{Volume of flue gas} = V = \frac{nRT}{P} = \frac{47.12 \times 62.36 \times 573}{750}$$

$= 2244.94\ \text{m}^3/\text{hr} = 37.42\ \text{m}^3/\text{min}.$

Example 23. *A coal containing C = 67.9%, H = 4.4%, S = 0.8%, N = 16%, O = 7.9%, ash = 4.5% and water = 12.9% is burnt in a furnace. The dry products of combustion (dry basis) analyses, CO_2 = 14.5%, O_2 = 4.7% and N_2 = 80.8% by volume. Calculate the :*

(a) theoretical volume of air used for the complete combustion of 100 kg of coal;

(b) per cent excess air used.

Solution. Basis : 100 kg of coal fired.

Oxygen balance :

O_2 required for carbon $= 67.9 \times \frac{1}{2} = 5.66$ kg mole

O_2 required for H_2 $= \frac{4.4}{2} \times \frac{1}{2} = 1.10$ kg mole

O_2 required for S $= 0.8 \times \frac{1}{32} = 0.025$ kg mole

Total O_2 required = 6.785 kg mole

O_2 present $= \frac{7.9}{32} = 0.247$ kg mole

O_2 to be supplied = 6.785 – 0.247 = 6.538 kg mole

(a) Air required $= 6.538 \times \dfrac{100}{21} = 31.13$ kg mole

$$= \frac{31.13 \times 0.082 \times 273}{1} = 697.4 \text{ m}^3 \text{ at N.T.P.} = 697.4 \text{ Nm}^3$$

(b) Let, Y = mole of flue gas obtained

Carbon in flue gas = 0.145 Y

Carbon in coal = 5.66 kg atom.

So, $0.145\,Y = 5.66$

$\therefore$ $Y = 39.03$ kg mole

N_2 in flue gas = 100 – (14.5 + 4.7) = 80.8%

Amount of N_2 = 39.03 × 0.808 = 31.54 kg mole

N_2 from air = Total $N_2 - N_2$ in Coal

$$= 31.54 - \frac{16}{28} = 30.97 \text{ kg mole (molecular weight of } N_2 = 28)$$

Air supplied $= 30.97 \times \dfrac{100}{28} = 39.2$ kg mole

Percent excess air = Percent excess oxygen $= \dfrac{O_2 \text{ supplied} - O_2 \text{ required}}{O_2 \text{ required}} \times 100$

$$= \frac{(39.2 \times 0.21) - 6.538}{6.538} \times 100 = \mathbf{25.9\%}$$

Example 24. A coal tar fuel containing C = 90%, H = 5%, S = 0.5%, O = 2.5% and N = 2% and of gross calorific value of 38,000 kJ/kg and specific gravity = 1.1 (Air = 1) is burnt with air as the atomizing agent at a rate of 50 kg per hour. Assuming that 20% excess air is used and the combustion is complete, calculate :

(a) The volume at N.T.P. of air supplied to the burner per hour (Nm^3/hr)

(b) The volume of the resulting product of combustion per hour at 700°C.

(c) The volumetric composition of the product of combustion.

Solution. *Basis : 1 hr operation.*

Coal tar fuel used = 50 kg/hr

Oxygen balance :

O_2 required for $C = 50 \times 0.9 \times \dfrac{32}{12} = 120$ kg

O_2 required for $H = 50 \times 0.05 \times \dfrac{16}{2} = 20$ kg

O_2 required for $S = 50 \times 0.005 \times \dfrac{32}{32} = 0.25$ kg

(c) Volumetric composition of the products of combustion is as follows :

Constituents	Amount in mole	Volume (or mole) %
CO_2	3.750	14.70
O_2	0.870	3.41
N_2	19.625	76.96
H_2O	1.250	4.90
SO_2	0.008	0.03
Total	—	100

Example 25. *Blast furnace gas having analysis (volume %) : $CO_2 = 13$, $CO = 25$, $H_2 = 3.5$, $N_2 = 58.5$ is burnt in a furnace. Calculate :*

(a) *Percentage of excess air, when the dry product of combustion contains 3.5% O_2.*

(b) *Percentage of excess air, when the dry flue gases contain 19.5% CO% ,5.8% O_2 and 74.7% N_2.*

Solution. Basis : 100 kg mole of blast furnace gas

Oxygen demand and flue gas formed are computed as under :

Constituents	Amount, kg mole	Oxygen required, kg mole	Flue gas obtained
CO_2	13.0	0	13.0 (CO_2)
CO	25.0	12.50	25.0 (CO_2)
H_2	3.5	1.75	3.5 (H_2O)
N_2	58.5	0	58.5 (N_2)
Total	100.0	14.25	100.0

Let, y_1 = kg mole of dry flue gas obtained, and, z = kg mole of combustion air used.

N_2 from combustion air $= 0.79\,z$

O_2 from combustion air $= 0.021\,z$

O_2 excess in flue gas $= 0.21\,z - 14.25$

Dry flue gas $= (100 - \text{water vapour}) + 0.79z + (0.21z - 14.25)$

$= (100 - 3.5) + z - 14.25 = 82.25 + z = y_1$

$$\frac{\text{Oxygen}}{\text{Dry flue gas}} = \frac{0.21\ z - 14.25}{82.25 + z} = \frac{3.5}{100}$$

Solving, we get, $z = 97.88$ kg mole

∴ Oxygen from air $= 97.88 \times 0.21 = 20.55$ kg mole

Oxygen excess $= 20.55 - 14.25 = 6.3$ kg mole

% excess air $= \%\text{ excess oxygen} = \frac{6.3}{14.25} \times 100 =$ **44.2%**

Let y_2 = kg mole of dry flue gas in this case.

By a *carbon balance* for the blast furnace and the dry flue gases, $0.195\,y_2 = 38$

So, $y_2 = 194.97$ kg mole

Oxygen in flue gas is 5.8%, hence

Amount of oxygen $= 194.87 \times 0.058 = 11.3$ kg mole

% excess air = % excess oxygen $= \frac{\text{Excess oxygen in flue gas}}{\text{Oxygen required}} \times 100 = \frac{11.3}{14.25} \times 100 = \mathbf{79.3\%}$

Example 26. *Calculate the excess air used in a furnace when the flue gas Orsat analysis is $CO_2 = 8.2\%$, $O_2 = 9.9\%$, $CO = 0.1\%$, $H_2 = 0.4\%$ and $N_2 = 81.4\%$ by volume.*

Solution. Basis : 100 kg mole of dry flue gas

Oxygen balance :

Constituents	Amount, kg mole	O_2 reported, kg mole
CO_2	8.2	8.20
O_2	9.9	9.90
CO	0.1	0.05
H_2	0.4	—
N_2	81.4	—
Total	100.0	81.15

Oxygen supplied $= 81.4 \times \frac{21}{79} = 21.64$ kg mole

Oxygen unaccounted $= 21.64 - 18.15 = 3.49$ kg mole (used for H_2 to give H_2O)

Oxygen used $= 8.2$ (for CO_2) + 0.05 (for CO) + 3.49 (for H_2) = 11.74 kg mole

Additional amount of oxygen required for the combustion of CO and H_2 of flue gas

$= 0.05 + 0.2 = 0.25$ kg mole

Total amount of oxygen required for complete combustion $= 11.74 + 0.25 = 11.99$ kg mole

Oxygen excess $= 21.64 - 11.99 = 9.65$ kg mole

% excess air (= % oxygen) used $= \frac{9.65}{11.99} \times 100 = 80.48\%$.

Example 27. *A sample of coke having the following percentage composition viz. C = 80%, H =5.8%, O = 8.0%, N = 1.4%, S = 0.64% and the rest ash is gasified in a generator. The gas produced has the following percentage composition : $CO_2 = 5\%$, CO = 32.0%, $H_2 = 12\%$ and the rest N_2. Calculate :*

(a) m^3 of gas produced per kg of coke gasified.

(b) per m^3 of gas produced, the amount of air used for gasification.

Solution. Basis : 1 kg of coke gasified.

Amount of carbon in coke = 0.80 kg = 0.067 kg atom

Let, y = moles of gas obtained

(CO_2 + CO) in gas = 37%

Making a *carbon balance* for coke and gas, we get, $0.37y = 0.067$

y = 0.18 kg mole = 4.03 m^3 at NTP = 4.03 Nm^3

Amount of N_2 in the gas $= 100 - (32 + 5 + 12) = 51\%$

Mol. of N_2 $= 0.51y = 0.092$

N_2 from coke $= 0.014$ kg = 0.0005 kg mole

N_2 from air $= 0.092 - 0.0005 = 0.0915$ kg mole

Air supplied $= 0.0915 \times \dfrac{100}{79} = 0.116$ kg mole $= 2.60\ m^3$ at NTP $= 2.60\ Nm^3$

(a) $\dfrac{m^3 \text{ gas}}{\text{kg coke}} = \dfrac{4.03}{1.0} = \mathbf{4.03}$ *(b)* $\dfrac{Nm^3 \text{ air}}{Nm^3 \text{ gas}} = \dfrac{2.60}{4.03} = \mathbf{0.645}$

Example 28. *The flue gas from a furnace using a hydrocarbon fuel oil analyses as follows by volume. $CO_2 = 16.5\%$, $CO = 0.2\%$, $O_2 = 1.5\%$, N_2 = rest. Calculate :*

(a) The percentage composition by weight of fuel burnt.

(b) The percentage excess air used.

Solution. Basis : 100 kg mole of flue gas.

Amount of N_2 $= 100 - (16.5 + 0.2 + 1.5) = 81.8$ kg mole

Oxygen balance :

O_2 from air $= 81.8 \times \dfrac{21}{79} = 21.74$ kg mole

O_2 reported in fuel $= 16.5$ (with CO_2) + 0.1 (with CO) + 1.5 (as O_2) = 18.1 kg mole

O_2 unaccounted $= 21.74 - 18.1 = 3.64$ kg mole

O_2 used for H_2 $= 3.64$ kg mole

H_2 reacted $= 3.64 \times 2$ $= 7.28$ kg mole

(a) Carbon in flue gas $= 16.5 + 0.2 = 16.7$ kg atom = Carbon from fuel

Amount of carbon $= 16.7 \times 12 = 200.4$ kg

H_2 in fuel $= 7.28$ kg mole $= 7.28 \times 2 = 14.56$ kg

Composition of fuel oil :

Constituents	Weight, kg	Weight, %
C	200.40	93.23
H	14.56	6.77
Total	214.96	100.00

(b) O_2 required for carbon $= 200.4 \times \dfrac{32}{12} = 534.4$ kg

O_2 required for hydrogen $= 14.56 \times \dfrac{16}{2} = 116.48$ kg

Total O_2 required $= 650.88$ kg $= 20.34$ kg mole

O_2 supplied $= 21.74$ kg mole

O_2 excess $= 21.74 - 20.34 = 1.40$ kg mole

% excess air ($\equiv$ % O_2) used $= \dfrac{1.40}{21.74 - 1.40} \times 100 = \dfrac{1.40}{20.34} = \mathbf{6.88\%}$.

Example 29. *A kiln is fired by producer gas having the volumetric composition : $CO = 25\%$, $H_2 = 11\%$, $O_2 = 0.4\%$, $CO_2 = 5.6\%$, $N_2 = 58\%$. If the combustion is only 98% complete, calculate the composition by*

volume of the dry flue gas formed and the weight of water produced per 100 Nm³ of gas burned.

Solution. Basis : 100 kg mole of producer gas burnt.

Oxygen balance :

O_2 required for CO $= 25 \times \frac{1}{2} = 12.5$ kg mole

O_2 required for H_2 $= 11 \times \frac{1}{2} = 5.5$ kg mole

Total O_2 required = 18.0 kg mole

O_2 present = 0.4 kg mole

Net O_2 required from air = 18.0-0.4 = 17.6 kg mole

O_2 supplied by 20% excess air = 17.6×1.2 = 21.12 kg mole

Combustion is 98% complete.

Oxygen used = 18.0 × 0.98 = 17.64 kg mole

O_2 from air = 17.64 – 0.40 = 17.24 kg mole

O_2 excess = 21.12 – 17.24 = 3.88 kg mole

Flue gas analysis :

N_2 supplied by air $= 21.12 \times \frac{79}{21} = 79.45$ kg mole

N_2 from fuel gas = 58 kg mole

Total N_2 in the flue gas = 137.45 kg mole

CO_2 from 98% combustion of CO = 25.0 × 0.98 = 24.5 kg mole

CO_2 in flue gas = 5.6 kg mole

Total CO_2 in flue gas

H_2 unburnt = 11 × (1.0 - 0.98)= 0.22 kg mole

CO unburnt = 25 × (1.0 - 0.98) = 0.50 kg mole

Excess O_2 in flue gas = 3.88 kg mole

Dry flue gas analysis :

Constituents	kg mole	Volume % (= mole %)
CO_2	30.10	17.49
CO	0.50	0.29
N_2	137.45	79.84
O_2	3.88	2.25
H_2	0.22	0.13
Total	172.15	100.00

Water formed :H_2 reacted = 11 × 0.98 = 10.78 kg mole

Water formed = 10.78 kg mole = 194.04 kg

Fuel gas burned = 100 kg mole = 2240 m³ at N.T.P. = 2240 Nm³

$$\frac{\text{Weight of water produced}}{100 \text{ Nm}^3 \text{of gas burned}} = \frac{194.04}{2240} = 100 = \mathbf{8.66}$$

Example 30. *A fuel gas containing 97% methane and 3% N_2 by volume is burned in a boiler furnace with 200% excess air. 85% of methane goes to CO_2, 10% to CO and 5% remains unburnt. Calculate the composition of stack gas.*

Solution. Basis : 100 kg mole of fuel gas.

Chemical reactions : $CH_4 + 2O_2 \rightarrow CO_2 + 2H_2O$ ...(1)

$2CH_4 + 3O_2 \rightarrow 2CO + 4H_2O$ ...(2)

Oxygen balance :

O_2 required for complete combustion $= 97.0 \times 2 = 194.0$ kg mole

O_2 supplied by 200% excess air $= 194 \times 3.0 = 582$ kg mole

Reaction (1) is 85% and (2) is 10% complete.

Methane converted by (1) $= 97 \times 0.85 = 82.45$ kg mole

Oxygen required $= 164.9$ kg mole

Methane converted by (2) $= 9.7 \times 0.10 = 9.7$ kg mole

Oxygen required $= 9.7 \times \frac{3}{2} = 14.55$ kg mole

Oxygen actually used by the two reactions = 164.9 + 14.55 = 179.45 kg mole

Stack gas formed :

CO_2 formed = 82.45 kg mole

CO formed = 9.70 kg mole

CH_4 unconverted = 5.0 kg mole

N_2 from air $= 582 \times \frac{79}{21} = 2189.43$ kg mole

N_2 from fuel = 3.0 kg mole

Total N_2 in stack = 2192.43 kg mole

H_2O vapour formed $= 82.25 \times 2$ (by 1) $+ 9.7 \times \frac{4}{2}$ (by 2) = 184.3 kg mole

Oxygen excess = 402.55 kg mole

Stack Gas Analysis :

Constituents	kg mole	Mole %
CO_2	82.45	2.87
CO	9.70	0.34
CH_4	4.85	0.17
O_2	402.55	13.99

N_2	2192.43	76.22
H_2O	184.30	6.41
Total	2876.28	100

Example 31. *Wood having an analysis of 46% C, 23% O, 5% ash and rest being moisture and hydrogen is being burnt in a furnace. An Orsat analysis of the flue gas* during a run showed *CO_2 = 14.9%, CO = 16.5%, O_2 = 3.45%, and N_2 = 80%. Calculate :*

(a) The complete analysis of the wood used. *(b) The fuel to air ratio by weight.*

(c) The percentage excess air used. *(d) The composition of stack gas.*

Solution. *Basis : 100 kg of wood burnt.*

Amount of carbon $= 46 \text{ kg} = \dfrac{42}{12} = 3.83$ kg atom

Carbon in flue gas = 14.90 + 1.65 ≐ 16.55 kg atom.

Let, y = kg mole of flue gas obtained. ∴ $0.1655\, y = 3.83$

Hence, $y = 23.16$ kg mole

N_2 in the flue gas = 23.16 × 0.8 = 18.53 kg mole

Air supplied= $18.53 \times \dfrac{100}{79} = 23.45$ kg mole

= 23.45 × 29 = 676.3 kg (Molecular weight of air = 29)

Oxygen balance :

Oxygen supplied $= 18.53 \times \dfrac{21}{79} = 4.93$ kg mole

Oxygen in flue = 23 kg = 0.72 kg mole

Oxygen available = 4.93 + 0.7 = 5.65 kg mole

Oxygen reported in flue gas

$$= \left\{114.9\ (\text{in } CO_2) + \frac{1.65}{2} (\text{in CO}) + 3.45\ (\text{as } O_2)\right\} \times \frac{23.16}{100} = 4.44 \text{ kg mole}$$

Oxygen unaccounted = 5.65 - 4.44 = 1.21 kg mole (used for H_2)

Hydrogen balance :

Oxygen reacting with H_2 = 1.21 kg mole

Moles of H_2 reacted = 1.21×2 = 2.42 kg mole = 2.42 ×2 = 4.84 kg

Moisture balance :

Moisture and Hydrogen in wood = 100 – (46 + 23 + 5) = 26 kg H_2 in wood = 4.84

Moisture in wood = 26 – 4.84 = 21.16 kg = 21.16/18 = 1.18 kg mole

Moisture from combustion of H_2 = 2.42 kg mole

Total moisture (*i.e.* water vapour) in stack gas = 2.42 + 1.18 = 3.6 kg mole.

(a) Analysis of wood :

Constituents	Amount, kg	Weight, %
C	46.00	46.00
O	23.00	23.00
H	4.84	4.84
Ash	5.00	5.00
Moisture	21.16	21.16
Total	100	100

(*b*) Fuel/air ratio $= \frac{100}{676.3} = \mathbf{0.148}$

(*c*) Oxygen required for complete combustion = 3.83 + 1.21 = 5.04 kg mole

Oxygen in fuel = 0.72 kg mole

Oxygen required from air = 5.04 - 0.72 = 4.32 kg mole

Oxygen supplied by air = 4.93 kg mole

Oxygen excess = 0.61 kg mole

% excess air = % excess O_2 $= \frac{0.61}{4.93-0.61} \times 100 = \mathbf{14.12\%}$

Amount of CO_2 in flue gas = 23.16 × 0.149 = 3.45 kg mole

Amount of CO in flue gas = 23.16 × 0.0165 = 0.38 kg mole

Amount of O_2 in flue gas. = 23.16 × 0.0345 = 0.80 kg mole

Amount of N_2 in flue gas = 18.53 kg mole

Amount of water vapour = 3.60 kg mole

Flue gas analysis *:*

Constituents	Amount, kg mole	Mole %
CO_2	3.45	12.89
CO	0.38	1.42
O_2	0.80	2.99
N_2	18.53	69.25
H_2O	3.60	13.45
Total	26.76	100.00

Example 32. *A coal fired furnace emits 52,200 m^3/hr of flue gas at 250°C. The coal used analyses as C = 58.4%, H = 3.3%, N = 1.3%, S = 0.4%, O = 8.6%, H_2O = 8%, and ash = 20%. If the burning rate is 3 tons of coal per hour, calculate :*

(*a*) *Orsat analysis of the flue gas.*

(*b*) *Per cent excess air used.*

Solution. Basis : 100 kg of coal burnt

Oxygen required and flue gas obtained is found as below :

Constituents	Weight, kg	O_2 required, kg	Flue gas obtained, kg mole
C	58.4	155.73	4.867 (CO_2)
H	3.3	26.40	1.650 (H_2O)
N	1.3	—	0.046 (N_2)
S	0.4	0.40	0.013 (SO_2)
O	8.6	–8.60	—
H_2O	8.0	—	0.444 (H_2O)
Ash	20.0	—	—
Total	100.00	173.93	7.020

Net oxygen required $= 173.93 \text{ kg} = 5.435 \text{ kg mole}$

N_2 from air $= 5.435 \times \dfrac{79}{21} = 20.45 \text{ kg mole}$

Wet flue gas obtained with theoretical air for 100 kg of coal

$= 7.02 + 20.45 = 27.47 \text{ kg mole}$

Actual flue gas obtained $= 52{,}200 \text{ m}^3/\text{hr} = \dfrac{52200}{30 \times 100} = 1740 \text{ m}^3/100 \text{ kg coal}$

[Since coal burning rate is 3 tons/hr = 3000 kg/hr.]

Mole of flue gas $= \dfrac{PV}{RT} = \dfrac{1740.0}{0.08204 \times 523} = 40.55 \text{ kg mole m}^3 \text{ coal}$

Assuming complete combustion with theoretical air, excess flue gas is due to excess air used.

Excess flue gas $= 40.55 - 27.47 = 13.08 \text{ kg mole}$

Excess air used $= 13.08 \text{ kg mole}$

Excess O_2 $= 13.08 \times 0.21 = 2.75 \text{ mole}$

N_2 with excess air $= 13.08 \times 0.79 = 10.33 \text{ kg mole}$

Total N_2 in flue gas = N_2 from theoretical air + N_2 from excess air + N_2 from fuel

$= 20.45 + 10.33 + 0.046 = 30.826 \text{ kg mole}$

(a) Orsat analysis of flue gas.

Constituents	Amount, kg mole	Mole %
CO_2	4.867	12.66
N_2	30.826	80.17
SO_2	0.013	0.04
O_2	2.750	7.14
Total	38.456	100.00

(b) Percent excess O_2 (air) = $\frac{2.750}{5.435} \times 100 =$ **50.0%.**

Example 33. *In a boiler furnace, pulverised coal having the following ultimate analysis is fired : C = 68.8%, H = 5%, O = 17.6%, N = 1.5% and ash = 7.1% (all by weight of dry sample). Coal contains 22.6% moisture. Exit gases carry solids which analyse 7.9% carbon and 92.1% ash. Air used is 5% in excess and under these conditions 0.9% of total combined carbon in the exit gases pass as CO, rest being as CO_2.*

Calculate :

(i) What per cent of total carbon of coal is lost in the exit material ?

(ii) What is the exit gas analysis ?

(iii) What weight of water vapour (kg) accompanies each cubic metre of dry gas at 300°C and 1 atmosphere. Air may be considered as dry ?

Solution. Basic : 100 kg of dry pulverised coal fired

Ash in coal = 7.1 kg

Let all the ash comes with the flue gas.

Carbon lost $= 7.1 \times \frac{7.9}{92.1} = 0.61$ kg

Carbon burnt = 68.8 - 0.61 = 68.19 kg

Carbon as CO in the flue gas $= 68.19 \times \frac{0.9}{100} = 0.61$ kg

Carbon as CO_2 = 68.19 – 0.61 = 67.58 kg

Oxygen balance :

O_2 required for Carbon $= 68.8 \times \frac{32}{12} = 183.5$ kg

O_2 required for Hydrogen $= 5 \times \frac{16}{2} = 40.0$ kg

O_2 in fuel = 17.6 kg

Net O_2 required = (183.5 + 40.0) – 17.6 = 205.9 kg

O_2 supplied by 5% excess air = 205.9 × 1.05 = 216.2 kg

O_2 used for combustion of 'C' to CO $= 0.61 \times \frac{16}{12} = 0.82$ kg

O_2 used for combustion of 'C' to $CO_2 = 67.58 \frac{32}{12} = 180.0$ kg

Oxygen used = 180.0 + 0.82 = 180.82 kg

Oxygen available = 216.2+ 17.6 = 233.8 kg

Oxygen excess = 233.80 – 180.82 = 52.98 kg

N_2 from coal = 1.5 kg

N_2 from air $= 216.2 \times \frac{77}{23} = 723.8$ kg

Total N_2 number in fuel gas = 1.5 + 723.8 = 725.3 kg

Flue gas obtained:

CO_2 formed = 180.0 + 67.58 = 247.58 kg

CO formed = 0.61 + 0.82= 1.43 kg

Water formed = 5.0+ 40.0 = 45.0 kg

(a) Per cent of total carbon of coal lost in the exit material

$$= \frac{0.61}{69.8} = \times 100 = 0.89\%$$

(b) Exit gas analysis (dry)

Constituents	Amount		Mole%
	kg	Kg Mole	
CO_2	247.58	5.63	16.94
CO	1.43	0.05	0.15
O_2	52.98	1.66	4.99
N_2	725.30	25.90	77.92
Total	1027.29	33.24	100.00

(c) Coal contains 22.6% moisture.

77.4 kg dry coal contains = 22.6 kg water

100 kg dry coal contains = $\frac{22.6}{77.4} \times 100$ = 29.2 kg water, Water formed = 45.0 kg

Total water vapour with flue gas = 45.0 + 29.2 = 74.2 kg

Volume of 1 kg mole at 300°C and 1 atmosphere = $\frac{22.4 \times (300+273)}{273} = 47.02\ m^3$

Total volume of dry flue gas = 47.02 × 33.24 = 1562.94 m^3

Water vapour per m^3 of dry gas at 300°C and 1 atmosphere = $\frac{74.2}{1562.94}$ = 0 047 kg.

Example 34. *A hydrocarbon fuel gas is burned with 15% excess air yielding a flue gas containing CO_2 = 11.6%, CO = 2%, H_2 = 1.6%, O_2 = 4%, N_2 by difference. Comment on the data reported and the furnace performance.*

Solution. Since percentage of water has not been reported, the data reported is on dry basis. Estimate the water/dry gas ratio by oxygen balance, then check whether oxygen was used 15% in excess or not.

Basis : 100 Kg mole of flue gas

Components	Mole	O_2 Present
CO_2	11.6	11.6
CO	2	1.0
H_2	1.6	—
O_2	4	4
N_2	80.8	—
Total	100	16.6

O_2 accompanying 80.8 kg mole N_2 $= \frac{80.8}{79} \times 21 = 21.47$ kg mole

Unaccounted $O_2 = 21.47 - 16.6 = 4.87$ kg mole which must have combined with hydrogen amounting to $2 \times 4.87 = 9.74$ kg mole to form water.

Theoretical O_2 required for $(11.6 + 2) = 13.6$ gm atom carbon is 13.6 gm mole O_2 and for $(1.6 + 9.74) = 11.34$ gm mole Hydrogen is $= \frac{11.34}{2} = 5.67$ gm mole O_2.

Total theoretical O_2 $= 13.6 + 5.67 = 19.27$ kg mole

Acutal O_2 supplied $= 21.47$ kg mole

Per cent excess O_2 $= \frac{21.47-19.27}{19.27} \times 100 = \frac{22}{19.27} \times 100 = $ **11.4%.**

Remarks

(a) O_2 supplied is 11.4% excess and not 15%

(b) Combustion is incomplete.

Example 35. *The Orsat analysis of the flue gas from an oil fired furnace is $CO_2 = 8\%$, $CO = 3\%$, $O_2 = 4\%$ and $N_2 = 85\%$. An analysis indicates that the oil contains 78% by weight C; the remainder being combustible hydrogen and moisture. Dry air enters at 30°C and 1.013 bar.*

Calculate :

(a) the per cent excess air used

(b) the carbon-hydrogen ratio in the fuel oil

(c) the volume of air used per kg of oil fired

(d) the kg moisture in the flue gas per kg of oil fired.

Solution. Basis : 100 mole flue gas

Components	Mole	C, gm atom	O_2 present	O_2 for formation of CO_2
CO_2	8	8	8	8
CO	3	3	1.5	3
O_2	4	—	4	—
N_2	85	—	—	—
Total	100	11	13.5	11

Air used for 100 mole flue gas from N_2 balance

Nitrogen in flue gas = N_2 from air

$$\therefore \quad 85 = \frac{79}{100} = \text{Air mole}$$

$$\text{or, Air mole} = \frac{85 \times 100}{79} = 107.5 \text{ mole}$$

O_2 associated with 107.5 mole air $= 107.5 \times 0.21 = 22.5$ mole

O_2 accounted for in flue gas = 13.5 mole.

The rest of O_2 = (22.5 – 13.5) = 9 mole must have combined with hydrogen amounting to 9 × 2 = 18 mole to form H_2O.

Therefore theoretical O_2 required (for C + for H_2O) = 11 + 9 = 20 mole

(a) Per cent excess air = % excess $O_2 = \frac{65}{11} = 12.5\%$

(b) In the oil 11 gm atom of carbon are associated with 18 moles of H_2,

$$\therefore \frac{C}{H_2} \text{ weight ratio} = \frac{11 \times 12}{18 \times 2} = 3.66$$

(c) The oil contains 78% of carbon by weight. Therefore 1000 gm of oil would contain 780 gm carbon, $\frac{780}{3.66}$ = 213 gm of H_2 and (1000 - 780 - 213) = 7 gm water, 780 gm carbon = $\frac{780}{12}$ = 65 gm atom of C and would produce $\frac{65}{11} \times 100$ mole dry flue gas carrying $\frac{65}{11} \times 100 \times \frac{18}{100}$ = 106.5 mole

water = 106.5 × 18 = 1920 gm water besides the original 7 gm of water present in oil.

Therefore total moisture present in flue gas per kg of oil burnt = (1920 + 7) gm = 1.927 kg

(d) 1 kg of oil produces $\frac{65}{11} \times 100$ mole flue gas : $\frac{65}{11} \times 100 \times \frac{107.5}{100}$ = 635 mole air

$$= 635 \times \frac{22.4 \times 10^{-3}}{273} (273 + 30) = 15.4 \text{ m}^3 \text{ air/kg of oil.}$$

Example 36. *The steam-raising plant of a refinery uses 500 kg/hr of tar as a fuel. An analysis of the tar indicates that it contains 88% carbon and 0.5% sulphur by weight ; the remainder being combustible hydrogen, moisture and ash. The ash content of this tar is four times the moisture content. For a given test run, the following data were obtained.*

Flue gas Orsat analysis :

CO_2 = 14.5%; CO = 1.5% ; 02 = 2.5% and N_2 = 81.5%

Calculate :

(a) The per cent excess air used

(b) The combustible hydrogen, moisture and ash content of the tar

(c) The kg of water per kg of dry flue gas

(d) The kg of sulfuric acid that could have been produced per hour if all the sulfur in tar were convertedinto sulfuric acid.

Note : *In Orsat analysis CO_2*

Solution. Basis : 100 mole of dry flue gas

Component	Mole	C, gm-atom	O_2 present, gm-mole	Theoretical C for CO_2
CO_2	14.5	14.5	14.5	14.5
CO	1.5	1.5	0.75	1.5
O_2	2.5	—	2.5	—
N_2	81.5	—	—	—
Total	100.0	16.0	17.75	16.0

Air per 100 mole of dry flue gas (by N_2 balance)

$$= \frac{100}{79} \times 81.5 = 103 \text{ mole air which contains } (103 - 81.5) = 21.5 \text{ mole } O_2$$

Mole O_2 which formed H_2O = 21.5 - 17.75 = 3.75 mole O_2. (% H_2O in flue gas is not reported in Orsat analysis).

Water formed = 3.75 × 2 = 7.5 mole

Theoretical O_2 required = 16+ 3.75 = 19.75 mole

(a) % excess O_2 = % excess air $= \frac{21.5 - 19.75}{19.75} \times 100 = 8.85\%$

(b) Analysis of tar (by weight), C = 88 wt%, S = 0.5 wt%

$$\frac{C}{S} \text{ gm-atom ratio} = \frac{88/12}{0.5/32} = \frac{7.33}{0.0156}$$

Now, 100 mole dry flue gas contains 16 gm-atom of (C+S) equivalent and carries 7.5 mole equivalent of H_2.

16 gm-atom of (C + S) would contain $= \frac{0.0156 \times 16}{(7.33 + 0.0156)} = 0.034$ gm atom of sulphur

and (16 – 0.034) = 15.966 gm-atom carbon

Therefore, $\frac{C}{H}$ mole ratio in tar $= \frac{15.966}{7.5}$

$$\frac{C}{H} \text{ weight ratio in tar} = \frac{15.966 \times 12}{7.5 \times 2} = 12.8$$

Therefore 100 gm of tar would contain, C = 88 gm, S = 0.5 gm, $H_2 = \frac{1}{12.8} \times 88 = 6.88$gm

Moisture and ash = (100 – 88 - 0.5 - 6.88) = 4.62 gm.

$$\frac{\text{ash}}{\text{moisture}} \text{ weight ratio} = 4 \text{ (given in the problem)}$$

Hence, Ash $= \frac{4.62}{5} \times 4 = 3.7$ gm, and Mosisture $= \frac{4.62}{5} \times 1 = 0.92$

Tar analysis (by weight) is as follows :

C = 88.00%, S = 0.50%, H_2 = 6.88%, Ash = 3.70%, Moisture = 0.92%

(c) Per 100 mole dry flue gas contains 7.5 mole water besides the moisture originally present in tar.

Component	Mole	Weight, gm
CO_2	14.5	14.5×44 = 638
SO_2	0.034	0.034×64 = 2.18
CO	1.5	1.5×28 = 42
O_2	2.5	2.5×32 = 80
N_2	81.5	81.5×28 = 2280
Total		3042.18

100 mole of dry flue gas containing 15.966 gm atom carbon would be produced from

$$\frac{15.966 \times 12}{0.88} \text{ gm tar containing } \frac{15.966 \times 12}{0.88} \times \frac{0.92}{100} = 2.1 \text{ gm water}$$

Hence total moisture in 3.042 kg dry flue gas = (7.5 × 18 + 2.1) = 135.1 gm = 0.1351

$$\therefore \quad \frac{\text{kg moisture}}{\text{kg dry flue gas}} = \frac{0.1351}{3.042} = 0.0445$$

(d) Total S present in 500 kg tar $= \frac{0.5}{100} = \times 500 = 2.5$ kg atom of S from which

$= \frac{2.5}{32}$ kg mole H_2SO_4 can be produced

$\therefore$ Weight of H_2SO_4 produced $= \frac{2.5 \times 98}{32} = 7.65$ kg H_2SO_4

(molecular weight of H_2SO_4 = 98)

CALORIFIC VALUE BY DULONG FORMULA

The C.V. of fuel can be calculated by knowing the percentages of the constituent elements C, H, O and S. According to Dulong , the calorific value of a fuel is the sum of the calorific values of its constituent elements. The calorific values of C, H and S are found to be 8080, 34500 and 2240 kcal/kg respectively. In case oxygen is also present, it is assumed to be present in the combined form i.e. as H_2O.

Thus the **Dulong's formula** for calculating the higher calorific value (H.C.V.) from the chemical composition of the fuel may be written as below :

$$\text{H.C.V.} = \frac{1}{100}\left[8080C + 34500\left(H - \frac{O}{8}\right) + 2240S\right] \text{ kcal/kg}$$

where C, H, O and S represent the percentage of the corresponding elements. Thus,

Lower calorific value (L.C.V.) is given by

L.C.V. = Higher calorific value – $\frac{9}{100}$ H × 587 kcal/kg = H.C.V $\frac{9}{100} H$ × 587 kcal/kg.

Example 37. A sample of coal contains : C = 60%, O = 33%, H = 6%, S = 0.5%, N = 0.3%, ash = 0.2%. Calculate the higher and lower calorific values of coal.

Solution.

Here, C = 60, 0 = 33, H = 6, S = 0.5

Now gross calorific value

$$= \frac{1}{100}\left[8080C + 34500\left(H - \frac{O}{8}\right) + 2240S\right] = \frac{1}{100}\left[8080C \times 60 + 34500\left(6 - \frac{33}{8}\right) + 2240 \times 0.5\right]$$

$$= \frac{1}{100}[484800 + 34500\,(6 - 4.125)] = \frac{1}{100}[484800 + 64687.5 + 1120]$$

= 4848 + 646.87 + 1120 = **5505.07 kcal/kg**

L.C.V. = H.C.V. – $\frac{9H}{100}$ × 587 kcal/kg = 5505.07 – $\frac{9 \times 6}{100}$ × 587 kcal/kg

= 5505.07 – 318.7 kcal/kg = **5186.37 kcal/kg.**

Example 38. *Find the gross as well as net calorific value of a gas which on ultimate analysis gave the following results.*

C = 70.3%, H = 5.1%, O = 22.4%, S = 0.8% and N = 1.4%.

Solution.

$$\text{Gross C.V.} = \frac{1}{100}\left[8080C + 34500\left(H - \frac{O}{8}\right) + 2240S\right]$$

$$= \frac{1}{100}\left[8080 \times 70.3 + 34500\left(5.1 - \frac{22.4}{8}\right) + 2240 \times 0.8\right]$$

$$= \frac{1}{100}[568024 + 34500\,(5.1 - 2.8) + 1792] + \frac{1}{100}[568024 + 79350 + 1792]$$

= 5680.24 + 793.50 + 17.92 = 6491.66 kcal/kg

Net C.V. = Gross C.V. - $\frac{9H}{100}$ × 587 kcal/kg

= 6491.66 – $\frac{9 \times 5.1}{100}$ × 587 =6491.66– 269.43 = **6222.23 kcal/kg**.

CALCULATIONS OF CHIMNEY HEIGHT AND DRAUGHT

We have, $C + O_2 = CO_2$

$$1 \text{ vol.} + 1 \text{ vol.} = 1 \text{ vol.}$$

$$2H_2 + O_2 = 2H_2O$$

$$2 \text{ vol.} + 1 \text{ vol.} = 2 \text{ vol.}$$

Molecular volumes of all the gases are the same.

From the above equations it is found that one volume of O_2 from air used for combustion of carbon in the fuel, produces one volume of the gaseous product of combustion CO_2 ; but one volume of O_2 from air produces two volumes of steam (H_2O) in the gaseous products. Nitrogen and other constituents in the air do not take part in the combustion. Also, the hydrogen content in solid or liquid fuel is very small and consequently steam formed in the products of combustion is very small as compared to other constituents. Therefore, the assumption that the volume of air required for combustion is approximately the same as the volume of the products of combustion is sufficiently accurate.

Let, w = weight of air used in kg per kg fuel.

T = mean absolute temperature of the chimney gases in °K.

T_1 = absolute temperature of outside air in °K.

Therefore, weight of chimney gas = $(w + 1)$ kg per kg fuel.

Volume per kg of flue gases at 273°K or 0°C = volume per kg of air at 273°K or 0°C

and, $$v = \frac{RT}{p} = \frac{29.27 \times 273}{1.03 \times 10^4} = 0.7734 \text{ m}^3/\text{kg}$$

(Here, $R = 29.27$ kgf. m/kg °K for air)

The pressure difference being very small, of the order of a maximum of 25 mm of water column, the pressure can be treated as constant at furnace and chimney base for the purposes of the calculation of volumes at higher temperatures.

Therefore, volume per kg of air at temperature T_1 °K = $0.7734 \times \dfrac{T_1}{273}$

(i) or, volume of w kg of air at T_1 °K = $\dfrac{0.7734 \times w \times T_1}{273}$

(ii) And volume of $(w+1)$ kg of chimney gases at T °K = $\dfrac{0.7734 \times w \times T_1}{273}$

(iii) Therefore, density of air at T_1°K = $\dfrac{\text{Mass}}{\text{Volume}} = \dfrac{w \times 273}{0.0734 \times w \times T_1} = 1.239\left(\dfrac{273}{T}\right)$ kg/m^3

(iv) and density of chimney gases at T^0 K = $\dfrac{\text{Mass of chimney gases}}{\text{Volume of chimney gas}} = \dfrac{(w+1) \times 273}{0.7734 \times w \times T}$

$$= 1.293\left(\frac{w+1}{w}\right) + \frac{273}{T}$$

Let, H = height of the chimney required in metres measured from the grate level. Pressure exerted per sq. metre at the fire grate level by a column of hot gas 1 metre height.

= weight of one cubic metre = density of gas in kg/m³

Therefore, pressure exerted by a column of hot gas of H metre height = density × H kgf/m²

$$= 1.293\left(\frac{w+1}{w}\right)+\frac{273}{T} \times H \text{ kgf/m}^2$$

Similarly pressure exerted by a column of cool air of H metre height

$$= 1.293 \times \frac{273}{T_1} \times H \text{ kgf/m}^2$$

Since, the draught p in kgf/m² is due to the static pressure difference between the hot gas column in chimney of height H, and cool air column outside of height H.

Therefore, $$p = \left[\left(1.293\times\frac{273}{T_1}\times H\right)-\left\{1.293\left(\frac{w+1}{w}\right)\times\frac{273}{T}\times H\right\}\right]$$

$$= 1.293 \times 273 \times H\left[\frac{1}{T_1}-\left(\frac{w+1}{w}\right)\times\frac{1}{T_1}\right]\text{kgf/m}^2$$

$$= 353\, H\left[\frac{1}{T_1}-\left(\frac{w+1}{w}\right)\times\frac{1}{T_1}\right]\text{ kgf/m}^2$$

We know, 1 kgf/m² = 1 mm of water column.

Therefore, if the draught is h mm of water column, then

$$h = 353\text{H}\left[\frac{1}{T_1}-\left(\frac{w+1}{w}\right)\times\frac{1}{T_1}\right]\text{ mm of } H_2O \quad \text{...(A)}$$

Let the draught pressure p kgf/m² be produced by H' height in metres of hot gas column.

Therefore, $$H' = \frac{p(\text{kg/m}^2)}{\text{density }(\text{kg/m}^3)} \text{ in metres} = \frac{1.293\times273\times H\left\{\frac{1}{T_1}-\frac{W+1}{W+T}\right\}}{1.293\times273\left(\frac{W+1}{W}\right)+\frac{1}{T}}$$

or, $$H' = H\left\{\frac{w}{w+1}\frac{T}{T_1}-1\right\} \quad \text{...(B)}$$

Thus for given temperature of air and flue gas, the draught produced is determined by the height of chimney.

Example 39. *A chimney is 30 metres high and is filled with hot gases at a temperature of 288°C. The temperature of the outside air is 21°C. If the draught is 80% of the theoretical draught, determine the available draught. The boiler uses 18 kg of air per kg of fuel burnt.*

Solution. We have

$$h = 353H\left[\frac{1}{T_1} - \left(\frac{w+1}{w}\right)\frac{1}{T}\right] = 353 \times 30\left[\frac{2}{21+273} - \left(\frac{18+1}{18}\right)\frac{1}{(288+273)}\right]$$

$$= 10590\left[\frac{1}{294} - \frac{19}{18} \times \frac{1}{561}\right] = 16.1 \text{ mm of water column}$$

Therefore, available draught = 0.8 × 16.1 = **12.88 mm of water column.**

Example 40. *Determine the height of chimney to produce a static draught of 20 mm water column. The mean flue gas temperature in the chimney is 270°C and the atmospheric temperature 20°C. Barometer reads 760 mm Hg. The characteristic gas constant R for air is 29.27 kgf m / kg°K and for chimney gas it is 26 kgf. m / kg°K.*

Solution. Density of air at 20°C is given by

$$\rho_{air} = \frac{1}{v} = \frac{p}{RT} = \frac{1.033 \times 10^4}{29.27 \times (273+20)} = \frac{1.033 \times 10^4}{29.27 \times 293} = 1.2 \text{ kg/m}^3$$

Similarly, density of flue gases at 270°C is given by

$$\rho_{flue\ gas} = \frac{1}{v} = \frac{p}{RT} = \frac{1.033 \times 10^4}{26 \times (270+273)} = \frac{1.033 \times 10^4}{26 \times 543} = 0.73 \text{ kg/m}^3$$

Pressure difference is, therefore, given by

$$P = H(P_{air} - P_{flue}) \text{ kg/m}^2$$

Or, $$p = h = 20 = H(1.2 - 0.73)$$

(since h mm of water column = h kg/m²)

Therefore, $$H = \frac{20}{1.2 - 0.73} = \frac{20}{0.47} = \textbf{42.6 metres.}$$

Example 41. *Determine the quantity of air required per kg of coal burnt in a boiler fitted with 32 metre high stack. Draught produced is 18.5 mm of water when the temperature of the flue gases in the chimney is 370°C and that of the boiler house is 30°C. Calculate also the draught produced in terms of height of a column of hot gas in metres.*

Solution. We have $$h = 353H\left[\frac{1}{T_1} - \left(\frac{w+1}{w}\right)\frac{1}{T}\right] \text{ mm of water column}$$

Therefore, $$18.5 = 353 \times 32\left[\frac{1}{(273+30)} - \left(\frac{w+1}{w}\right)\frac{1}{(370+273)}\right]$$

or, $$18.5 = 353 \times 32\left[\frac{1}{303} - \left(\frac{w+1}{w}\right)\frac{1}{643}\right] = 37.3 - 17.55\left(\frac{w+1}{w}\right)$$

Therefore, $\frac{w+1}{w} = \frac{37.3-18.5}{17.55} = \frac{18.8}{17.55} = 1.07$ or, $\frac{1}{w} = 1.07-1 = 0.07$

Therefore, $w = \frac{1}{0.07}$ **= 15.3 kg of air/kg of fuel**

Also the draught in terms of the height of hot column gas is given by

$$H = H\left[\left(\frac{w}{w+1}\right)\frac{T}{T_1} - 1\right] = 32\left[\left(\frac{15.3}{15.3+1}\right)\frac{643}{303} - 1\right] = \mathbf{31.65\ metres.}$$

Example 41 (A). *Find out the available effective draught in a chimney handling flue gases from the following given data :*

Height of the chimney (H) = 33.5 metres, Diameter of chimney (D) = 1.9 metres

Flue gas rate passing through the chimney = 12.2 kg/sec.

Density of flue gas (air = 1) = 0.91

Density of air at N.T.P = 1.29 kg/m³ = 1.29 kg/Nm³

Average temperature of flue gas in chimney = 150°C

Average temperature of atmospheric air = 15.5°C

Friction loss in chimney in terms of one pressure head = 0.05 H/D

g = 9.8 m/sec².

Solution. *Static draught = H (pa - pg)*

Now, density of air at atmospheric temperature, *i.e.,* 15.5°C

Density of flue gas at 150°C, $p = 1.29 \times 0.91 \times \frac{(273+0)}{(273+150)} = 0.76\ \text{kg/m}^3$

∴ Static draught = 33.5 (1.22 – 0.76) = 15.41 kg/m²

This static draught is reduced by friction and kinetic energy to give the effective draught.

Draught loss due to friction $= 0.05\ \frac{H}{D}\ P_g\ \frac{V^2}{2g}$ and Draught loss due to kinetic energy $= p_g\ \frac{V^2}{2g}$

here, $H = 33.5$, $D = 1.9$, $p_g = 0.76\ \text{kg/m}^3$, $g = 9.8\ \text{m/sec}^2$

V = flue gas velocity

∴ Total reduction in draught due to kinetic energy and friction in the chimney

$$= \left(P_g\frac{V^2}{2g} + \frac{0.05H}{D}P_g\frac{V^2}{2g}\right) = \left(1 + 0.05\frac{H}{D}\right)pg\frac{V^2}{2g} = \left(1 + 0.05\times\frac{33.5}{1.9}\right)\times 0.76\times\frac{(5.66)^2}{2\times 9.8}$$

$= 2.34\ \text{kg/m}^2$

hence, the effective draught available

= static draught - reduction in draught due to friction and kinetic energy

= 1.5.41-2.3 = 13.11 kg/m^2 = 13.11 mm water gage1 kg/m^2 = 1 mm wg] **Ans.**

Example 42 *.A fuel gas of composition: CH_4=12%, CS_2=28%, CO_2=11%, H_2=9% and N_2=40% is burnt with 100% excess air.*

Calculate the flue gas analysis on SO_2 free basis (both volume % .and weight %). CS_2 on complete combustion forms CO_2 and SO_2.

Solution.

Basis - 1 Nm^3 of fuel gas which contains :

CH^4 = 0.12 Nm^3, CS_2 = 0.28 Nm^3 CO_2 = 0.11 Nm^3, H_2 = 0.09 Nm^3 and N_2 = 0.4 Nm^3

Constituents	Quantity, Nm^3	Combustion equation	O_2 required for combustion, Nm^3	Products of combustion, Nm^3				
				CO_2	O_2	N_2	H_2O	SO_2
CH_4	0.12	$CH_4 + 2O_2 = H_2O + CO_2$	0.12×2 = 0.24	1×0.12 = 0.12	—	—	2×0.12 = 0.24	—
CS_2	0.28	$CS_2 + 3O_2 = CO_2 + 2SO_2$	0.28 × 3 = 0.84	1×0.28 = 0.28			—	0.28×2 = 0.56
H_2	0.09	$2H_2 + O_2 = 2H_2O$	0.9/2 = 0.045	—	—	—	0.09	
CO_2	0.11	Incombustible	nil	0.11	—	—		
N_2	0.40	Incombustible	nil	—	—	0.40	—	
Total	1.00	—	1.125	0.51	1.125 (from 100% excess air)	8.45 (from air) 8.85	0.33	0.56

Now, total theoretical oxygen required for complete combustion of 1 Nm^3 fuel gas = 1.125 Nm^3.

Hence, theoretical air requirement $= 1.125 \times \frac{100}{21} = 5.35$ Nm^3 air

Since, fuel gas is burnt with 100% excess air, hence actual quantity of air supplied

$= 5.35 \times 2 = 10.7$ Nm^3

N_2 present in actual air used $= 10.7 \times 79 = 8.45$ Nm^3

(since, air contains 79% N_2 by volume)

Volumetric analysis of flue gas is found as follows :

Flue gas constituents	Volume (Nm^3)/Nm^3 flue gas	Volume %	Volume % on SO_2 free basis
CO_2	0.510,	$\frac{0.510}{11.375} \times 100 = 4.5$	$\frac{4.5}{(100-5)} \times 100 = 4\ 7$
O_2	1.125	$\frac{1.125}{11.375} \times$ 100-9.9	$\frac{9.9}{(100-5)} \times 100 = 10\ 4$
N_2	8.850	$\frac{8.850}{11.375} \times 100 = 77.8$	$\frac{77.8}{(100-5)} \times$ l00 = 81-9

H_2O	0.330	$\frac{0.330}{11.375} \times 100 = 2.8$	$\frac{2.8}{(100-5)} \times 100 = 3$
SO_2	0.560	$\frac{0.560}{11.375} \times 100 = 5$	—
Total	11.375	100	100

Now flue gas analysis on weight % basis (SO_2 free basis) is found as follows :

Note. Nm^3 (Normal cubic metre) means volume of gas at Normal Temperature and Pressure (N.T.P.). Since, volume of a given mass of gas varies with pressure (Boyle's law) and temperature (Charle's law), hence both pressure and temperature is specified, when it is expressed in Nm^3.

Flue gas constituents	Volume % or mole % (a)	Molecular weight (b)	Proportional weight = a Ч b	Weight % (on SO_2 free basis)
CO_2	4.7	44	0.047 × 44 = 2.07	$\frac{2.07}{28.27} \times 100 = 7.1$
O_2	10.4	32	0.104×32 = 3.33	$\frac{3.33}{28.87} \times 100 = 11.5$
N_2	81.9	28	0.819×28 = 22.93	$\frac{22.93}{28.87} \times 100 = 79.6$
H_2O	3	18	0.003 × 18 = 0.54	$\frac{0.74}{28.87} \times 100 = 1.8$
Total	100	—	28.87	100

Example 43. *Composition (weight %) of petroleum oil is: C = 84.8%, H_2 = 12.7%, O_2 = 1.4%, S = 1.1%. Calculate the following:*

(a Nm^3 of theoretical air used per kg of petroleum oil burnt.

(b) Nm^3 of flue gas generated per kg of petroleum oil burnt without excess air.

Solution. *Basis : 1 kg of petroleum*

Petroleum contains :

C= 848 gm, H_2 = 127 gm, O_2 = 14 gm, S = 11 gm

Flue gas generated will contain :

(i) CO_2 from combustion of C *(ii)* H_2O from combustion of H_2

(iii) SO_2 from combustion of S *(iv)* N_2 from combustion air

(v) It will not contain any oxygen as no excess air is used

(i) $\underset{12}{C} + \underset{32}{O_2} = \underset{44}{CO_2}$, 12 gm of carbon produces 44 gms of CO_2

848 gm of carbon produces $\frac{48 \times 44}{12}$ = 3109.3 gm CO_2

(ii) $2H_2 + O_2 = 2H_2O$, because 4 gm H_2 produces 36 gm H_2O

$\therefore$ 127 gm H_2 produces $\frac{127 \times 36}{12}$ = 1143 gm H_2O

(iii) $S + O_2 = SO_2$, because 32 gm S produces 64 gms of SO_2

32 32 64

$\therefore$ 11 gm S produces = $\frac{11 \times 64}{32} = 22$ gm SO_2

(a) Oxygen Requirement

Oxygen required for complete combustion of 848 gm of carbon = $\frac{848 \times 32}{12}$ gm = 2261.3 gm

O_2 required for combustion of H_2 = $\frac{127 \times 32}{4} = 1016$ gm

O_2 required for combustion of S = $\frac{11 \times 32}{32}$ =11 gm

$\therefore$ Total oxygen required for combustion = 2261.3 + 1016 + 11 = 2288.3 gm

But oxygen already present in 1 kg petroleum = 14 gm

Net oxygen required from external source = 2288.3 – 14 = 2274.3 gm

because 23 gm oxygen is present in 100 gm air

$\therefore$ 2274.3 gm oxygen is present in $\frac{2274\ 3 \times 100}{23}$ = 9888.2 gm air

$\therefore$ Theoretical air required = 9888.2 gm

because Molecular weight of air = 29 $\therefore$ gm mole of air = $\frac{9888.2}{29}$ =341

1 gm mole of any gas at NTP occupies 22.4 litre = 22400 cm^3

$\therefore$ 341 gm mole air at NTP will occupy 341 × 22400 cm^3 = $\frac{341 \times 22400}{10^6}$ m^3

$\therefore$ Nm^3 of air used/kg oil burnt = $\frac{341 \times 22400}{10^6}$ = **7.64 Nm^3**

(b) Total flue gas generated will be the sum of the volumes of CO_2, H_2O, SO_2 and N_2 from air

CO_2 = 3109.3 gm = $\frac{3109.3}{44}$ = 70.6 gm mole

H_2O = 1143 gm = $\frac{1143}{18}$ = 63.5 gm mole

SO_2 = 22 gm = $\frac{22}{64}$ = 0.35 gm mole

Air required = 9888.2 gm (because air contains 77% N_2 by weight)

$\therefore$ N_2 in the air $= 9888.2 \times \frac{77}{100} = 7614$ gm,

$\therefore$ gm moles of $N_2 = \frac{7614}{28} = 272$ gm moles

$\therefore$ Total flue gas generated $= 70.6 + 63.5 + 0.35 + 272 = 406.45$ gm moles

Since, 1 gm mole of any gas at NTP occupies 22400 cm^3

$\therefore$ 406.45 gm mole of flue gas will occupy 406.45×22400 cm^3

$$= \frac{406.45 \times 22400}{10^6} Nm^3 = 9.1 \ Nm^3$$

$\therefore$ Nm^3 of flue gas generated/kg petroleum burnt without excess air = **9.1 Nm^3** .

Example 44. *Orsat analysis of flue gas obtained on combustion of a solid fuel (which does not contain nitrogen and oxygen) is as follows :*

$$CO_2 = 12.7\%, O_2 = 7.1\%, N_2 = 80.2\%$$

Calculate the percentage of excess combustion air.

Solution.

$$\% \text{ excess air, } = \frac{\text{Excess air}}{\text{Theoretical air}} \times 100 = \frac{\text{Excess air}}{\text{Total air } - \text{ Excess air}} \times 100$$

In this problem, oxygen in the flue gas entirely corresponds to the excess air as there is no combustible in the flue gas. Though oxygen content in the solid fuel is nil, however, the water vapor content in the flue gas is not known (because Orsat analysis is on dry basis) and this restricts the calculation of total air from the oxygen balance of flue gas analysis. Since, the fuel contains no nitrogen which means that the nitrogen of the flue gas is entirely contributed by the combustion air and hence total air can be calculated from nitrogen balance.

Basis : 100 kilomole of flue gas

Constituents	Quantity , Kg mole	Kg mole of Carbon accounted for	Kg mole of oxygen accounted for
CO_2	12.7	12.7	12.7
O_2	7.1	—	7.1
N_2	80.2	—	
Total	100	12.7	19.8

Total oxygen = Total nitrogen/3.76 $= \frac{80.2}{3.76} = 21.35$ kg mole ; Excess oxygen = 7.1 kg mole

hence % excess air = % excess oxygen $= \frac{7.1}{21.35 - 7.1} \times 100 =$ **49.8%** **Ans.**

Example 45. A coal has the following gravimetric composition : Proximate Analysis (% air dried) :

Moisture = 8%, volatile matter = 28.5%, ash = 20%, fixed carbon = 43.5% Ultimate Analysis (% dry ash free) : C = 81%, O = 12%, H = 4.6%, N = 1.8%, S = 0.6%

Assuming complete combustion of the coal using 50% excess air, calculate :

(a) flue gas analysis (b) air/fuel ratio by weight (c) flue gas volume at 250°C and 1 atm.

(d) the air blower capacity required if, the coal is burnt at the rate of 6 tons/hr.

Solution.

(l) The ultimate analysis of coal is converted from dry ash free basis to air dried basis by multiplying by i.e. 0.72 weights are then converted into kilomole by dividing by molecular/atomic weight.

Theoretical oxygen and flue gas quantity is calculated by combustion equations. The air supplied is calculated after substracting the oxygen of the coal from the total oxygen requirement. Composition of flue gas includes excess air. Final flue gas analysis is reported on dry basis after combining together CO_2 and SO_2 (as reported in Orsat analysis). The calculations and results are tabulated below :

Elements	kg/100 kg d.a.f. coal	kg/100 kg air-dried coal	kmol/100 kg air-dried coal	kmol O_2 theoretically required per 100 kg air-dried coal	kmol products/100 kg air dried coal					
					CO_2	H_2O	SO_2	N_2	O_2	Total
C	81.0	58.4	4.87	4.87	4.87	—	—	—	—	4.87
H	4.6	3.3	1.65	0.83	—	1.65	—	—	—	1.65
O	12.0	8.6	_0.27	-0.27	—	—	—	—	—	—
N	1.8	1.3	0.05	—	—	—	—	0.05	—	0.05
S	0.6	0.4	0.01	0.01	—	0.44	0.01	—	—	0.01
Moisture	—	8.0	0.44	—	—	—	—	—	—	0.44
Ash	—	20.0	—	—	—	—	—	—	—	—
Total	100.0	100.0	—	5.44	4.87	2.09	0.01	0.05	0.00	7.02
Excess O_2 at 50 per cent of the theoretical (5.44 × 0.5 = 2.72)					—	—	—	—	2.72	2.72
N_2 from total air supplied (Theoretical 02 × 3.76 × Excess-air factor = 5.44×3.76x 1.5 = 30.7)					—	—	—	30.70	—	30.70
				Grand Total	4.87	2.09	0.01	30.75	2.72	40.44

(a) Flue gas analysis has been calculated and tabulated below :

Components of flue gas	Quantity, kg mole	%volume as produced	% volume on dry basis (Orsat analysis)	Remarks
N_2	30.75	76.0	80.2	CO_2 and SO_2 have been combined in analysis on dry basis. To convert to dry basis the % volume of flue gas as produced, multiplied by
CO_2	4.87	12.06	12.7	
O_2	2.72	6.73	7.1	
H_2O	2.09	5.18	—	
SO_2	0.01	0.03	—	$\frac{100}{100-5.18}=1.055$
Total	40.44	100	100	

(b) Air/fuel ratio by weight

$$= \frac{\text{kmole theoretical } O_2 \times \text{excess air factor} \times \text{molecular weight of air}}{100 \text{ kg air dried coal}}$$

$$= \frac{5.44 \times 4.76 \times 1.5 \times 29}{100} = 11.27$$

(c) Flue gas volume at 250°C and 1 atm.

$$= \text{kmol of flue gas} \times 22.4 \times \frac{273+250}{273} \text{ m}^3/100 \text{ kg air dried coal}$$

$$= 40.44 \times 22.4 \times \frac{523}{273} = 1740 \text{ m}^3/100 \text{ kg air dried coal}$$

(d) Assuming the ambient temperature to be 30°C, the capacity of air blower

$$= \frac{1740}{100} \times 6000 \times \frac{273+30}{273+250} = 6000 \text{ m}^3\text{ /hour.}$$

Example 46. *A producer gas has following composition by volume as reported by Orsat analysis (on dry basis) :*

$N_2 = 52\%$, $CO = 29\%$, $H_2 = 12\%$, $CO_2 = 4\%$, $CH_4 = 2.6\%$ and $C_2H_4 = 0.4\%$.

This gas is burnt in a furnace (at the rate of 7200 Nm³/hr) completely using 10% excess air. The flue gas is discharged at 400°C and 1 atmosphere. Calculate :

(a) flue gas analysis *(b) air flow rate* *(c) flue gas production rate.*

Assume ambient temperature to be 30°C and pressure is 760 mm Hg.

Solution. Basis -100 kg mole of producer gas

Vapor pressure of water at 30°C = 31.89 mm Hg. Assuming that the gas is saturated with water vapor at 30°C and 760 mm Hg.

Hence, water vapor content in the saturated gas = = 4.2 kg mole

The dry composition of the producer gas $\frac{31.89}{760} \times 100$ is converted to the wet (saturated) composition in

volume per cent by multiplying by $\left(\frac{100-4.2}{100}\right)$ *i.e.* 0.958. Using the combustion equations,

Using the combustion equations, theoretical oxygen requirement and flue gas production are calculated and have been tabulated below :

Components	kg mole /100 kg mole dry gas	kg mole /100 kg mole wet gas	kmol O_2 theoretically required	kmol of combustion products				
				CO_2	O_2	N_2	H_2O	Total
CO	4.0	3.8	—	3.8	—	—	—	3.8
C_2H_4	0.4	0.4	1.2	0.8	—	—	0.8	1.6
O_2	—	—	—	—	—	—	—	
CO	29.0	27.8	13.9	27.8	—	—	—	27.8
H_2	12.0	11.5	5.75	—	—	—	11.5	11.5
CH_4	2.6	2.5	5.00	2.5	—	—	5.0	7.5
N_2	52.0	49.8	—	—	—	49.8	—	49.8
HO	—	4.2	—	—	—	—	4.2	4.2
Total	100.0	100.0	25.85	34.9	—	49.8	21.5	106.2
Excess O_2 at 10 per cent of the theoretical (25.85 × 0.1= 2.6). N_2 from total air supplied (25.85 × 3.76 × 1.1 = 100.7)				—	2.6	—	—	2.6
				—	—	100.7	—	100.7
Grand Total				34.9	2.6	150.5	21.5	209.5

(a) Flue gas analysis as produced has been converted to dry basis (Orsat analysis) by multiplying by $\frac{100}{100-10.3}$ *i.e.* 1.115 and tabulated below :

Components	kmole	Volume % as produced	Volume % dry (Orsat analysis)	Remarks
CO_2	34.9	16.7	18.6	Convert volume % as produced to volume % dry by multiplying by $\frac{100}{100-10.3}$ ie 1.115
O_2	2.6	1.2	1.3	
N_2	150.5	71.8	80.1	
H_2O	21.5	10.3	—	
Total	209.5	100.0	100.0	

(b) air flow rate $= 25.85 \times 4.76 \times 1.1 \times \frac{7200}{100}$ Nm³/hr = 9750 Nm³/hr

(c) Flue gases production rate = 209.5 kg mole/100 kg mole producer gas

$$= 209.5 \times \frac{7200}{100} \text{ Nm}^3\text{/hr} = 15{,}100 \text{ Nm}^3\text{/hr}$$

$$= 15{,}100 \times \frac{673}{273} \text{ m}^3\text{/hr at 400°C and 760 mm} = 37{,}200 \text{ m}^3\text{/hr.}$$

Example 47. *The dry refuse from a coal fired furnace has a combustible content of 40% as carbon. The composition of coal used is : C = 68%, N = 1.9%, S = 0.3% and ash = 14.7%. Orsat analysis of flue gas produced is : N_2 = 80.5%, CO_2 = 11.5%, O_2 = 6.5% and CO = 1.5%.*

Assume that flue gas exit temperature is 300°C, ambient temperature 30°C, pressure 760 mm Hg and air is dry. Calculate the following :

(a) *Quantity of dry flue gases* (b) *Volume of air per kg coal*

(c) *Content of surplus hydrogen and water in coal* (d) *Per cent excess air used*

(e) *Volume of flue gases produced per kg of coal*

(f) *Potential heat loss with the solid refuse as % of potential heat (calorific value) in coal*

(g) *Combustion efficiency* (h) *Total heat loss with the flue gas.*

Following data are given :

Components	Enthalpy at 300°C, kcal/Nm³	Specific heat at 30°C, kcal/Nm³ °C
H_2O	110	0.357
CO_2	134.7	0.400
CO	94.5	0.311
N_2	94.2	0.311
O_2	97.2	0.314

Solution.

— Calculate the quantity of dry refuse from ash balance between coal and refuse. Hence determine the carbon lost in the refuse and the carbon burnt to CO_2 and CO.

— Calculate the kmol of dry flue gas from carbon balance, (sulphur in coal is burnt to SO_2 and reported as CO_2 in the flue gas is to be noted).

— Use nitrogen balance for calculating total oxygen supplied. Calculate the oxygen (sum of free and combined) accounted for in the flue gas. The deficit in oxygen is due to the combustion of hydrogen of coal into water.

— The hydrogen calculated from the oxygen deficit is the surplus hydrogen in coal. This is the hydrogen in excess of that required to combine with the oxygen of coal to form water (Vide Dulong's formula for C.V. calculation). The water content of coal is calculated by difference *i.e.* 100 — per cent (C + S + N + ash + surplus H). This water is the total of actual and potential water content of coal. The latter term means the sum of oxygen in coal and the hydrogen required to combine with it to form water.

— Volume of flue gas is calculated from the dry flue gas, the water content and the surplus hydrogen content of coal.

Basis : 100 kg of coal as fired

Components	kg	kg mole
Carbon	68.0	5.67
Nitrogen	1.9	0.07
Sulphur	0.3	0.01
Ash	14.7	—
Total	84.9	
Surplus hydrogen	?	
Water	?	
Grand Total	100.0	

From ash balance :

C in refuse = $14.7\times\frac{40}{60} = 9.8$ kg = 0.82 kg mole

$\therefore$ C burnt = 5.67 – 0.82 = 4.85 kg mole

Total (C + S) burnt = 4.85 + 0.01 = 4.86 kg mole

Dry Flue Gas Analysis

Components	kg mole/100 kg mole dry flue gas	kg mole (C+S) /100 kg mole dry flue gas	kg mole / 37.5 kg mole dry flue gas	Total O_2 accounted for/ 37.5 kg mole dry flue gas
CO_2	11.5	11.5	37.5 × 0.115 = 4.32	4.32
CO	1.5	1.5	37.5 × 0.015 = 0.56	0.28
O_2	6.5	—	37.5 × 0.065 = 2.44	2.44
N_2 (by difference)	80.5	—	37.5 × 0.805 = 30.18	—
Total	100.0	13.0	37.50	7.04

(a) From (C+S) balance,

Quantity of dry flue gas = $\frac{100}{13.0}\times 4.86 = 37.5$ kg mole/100 kg coal.

The kmol of components in 37.5 kmol dry flue gas and the total O_2 accounted for in it are given in the table above.

(b) N_2 from air = total N_2 – N_2 from coal = 30.18 - 0.07 = 30.11 kg mole

Total air supplied = 30.11 × 1.27 kg mole/100 kg coal = $30.11\times 1.27\times 22.4\times\frac{303}{273}\times\frac{1}{100}$

= 9.5 m^3/kg coal at 30°C and 760 mm

(c) Total O_2 from air $= \frac{30.11}{3.76} = 8.02$ kg mole

O_2 deficit = 8.02 – 7.04 = 0.98 kg mole

Hence, surplus H in coal = 0.98 × 2 = 1.96 kg mole/100 kg coal = 3.9 kg/100 kg coal

Water content of coal = 100 – (84.9 + 3.9) = 11.2 kg.

This includes the oxygen of coal and the hydrogen required to combine with it.

(d) Now O_2 required for burning CO in 37.5 kg mole dry flue gas = $0.56\times\frac{1}{2} = 0.28$ kg mole

Excess O_2 in the flue gas = 2.44 – 0.28 = 2.16 kg mole

$\therefore$ excess air $= \frac{2.16}{8.02-2.16}\times 100 = 36.9\%$

(e) Total wet flue gases produced

= dry flue gas + water from coal + water from surplus hydrogen of coal

$= 37.5 + \frac{11.2}{18} + 1.96$ kg mole/100 kg coal = 37.5 + 0.62 + 1.96 kg mole/100 kg

$= \frac{37.50+0.62+1.96}{100}\times 22.4\times\frac{573}{273} = 18.8$ m^3/kg coal at 300°C and 760 mm.

(f) Analysis of coal as fired is : C = 68%, surplus H = 3.9%, S = 0.3%, Ash = 14.7%, Water (including water from oxygen of coal) = 11.2%.

By Dulong's formula :

$$\text{C.V. (gross)} = 80.8\text{ C} + 345\text{ H} + 22.2\text{S} = 80.8 \times 68.0 + 345 \times 3.9 + 22.2 \times 0.3$$

$$= 6.842 \text{ kcal/kg coal fired}$$

$$\text{C.V. (net)} = 6842 - 53\left(\text{net H} + \frac{\text{water}}{18}\right) = 6882 - 53\left(3.9 + \frac{11.2}{18}\right)$$

C in refuse = 0.098 kcal/kg coal.

Potential heat loss with the refuse = 0.098 × 8080 = 792 kcal/kg coal fired

$$= \frac{792}{6593} \times 100 = 12\% \text{ of the potential heat in coal}$$

(g) Total wet flue gas produced = 37.50 + 0.62 + 1.96 = 40.08 kg mole/100 kg coal

$$= \frac{40.08 \times 22.4}{100} \text{ Nm}^3\text{/kg coal} = 8.98 \text{ Nm}^3 \text{ / kg coal}$$

Flue gas analysis (per cent):

$$H_2O = \frac{0.62 + 1.96}{40.08} \times 100 = 6.4\%$$

$$CO_2 = 11.5 \times \frac{100 - 6.4}{100} = 10.8\%$$

$$CO = 1.5 \times \frac{100 - 6.4}{100} = 1.4\%$$

$$N_2 = 80.5 \times \frac{100 - 6.4}{100} = 75.3\%$$

$$O_2 = 6.5 \times \frac{100 - 6.4}{100} = 6.1\%$$

calculated from the Orsat analysis on dry basis.

Enthalpy of flue gas at 300°C

= 0.064 × 110.0 + 0.108 × 134.7 + 0.014 × 94.5 + 0.753 × 94.2 + 0.061 × 97.2

= 99.9 kcal/Nm^3

Let mean sp. heat of flue gas at 30°C (datum temp.)

= 0.064 × 0.357 + 0.108 × 0.400 + 0.014 × 0.311+ 0.753 × 0.311+ 0.061× 0.314

= 0.324 kcal/Nm^3°C

∴ Enthalpy at 30°C = 0.324 × 30 = 9.72 kcal/Nm^3

∴ Sensible heat loss with the flue gas = 99.9 – 9.72 = 90.18 kcal/Nm^3 of flue gas

Again, potential heat loss with the flue gas

$$= (\text{C.V.})_{co} \times \frac{1.4}{100} = 3020 \times 0.014 = 42.28 \text{ kcal/Nm}^3 \text{ of flue gas}$$

$$\therefore \text{ Combustion efficiency } = \left[1 - \frac{792 + 42.28 \times 8.98}{6593}\right] \times 100 = 82.2\%$$

(h) Total heat loss with the flue gas = 90.18 + 42.28 = 132.46 kcal/Nm^3 gas

= 132.46 × 8.98 = 1,190 kcal/kg coal fired.

$$= \frac{1190}{6593} \times 100 = 18.2\% \text{ of the potential heat in coal,}$$

Example 48. *The coal has the following analysis: C = 54.0%, H = 6.5%, O = 3.0%, N = 1.8%, moisture = 17.3% and remaining is ash. This coal, on combustion with excess of air, gave 21.5 kg of dry flue gases per kg of coal burnt. Calculate the percentage of excess air used for combustion.*

Solution. 1 kg of coal contains : C = 0.540 kg ; H = 0.065 kg ; O = 0.030 kg ; N = 0.018 kg. Let us calculate the minimum weight of air required and weight of dry products of combustion per kg of coal burnt.

Constituents	Amount, kg	Minimum weight of oxygen required, kg	Weight of dry products of combustion, kg
C	0.540	0.540 × (32/12) = 1.44	CO_2 = 0.540 × (44/12) = 1.980
H	0.065	0.065 × (16/2) = 0.52	—
O	0.030	= – 0.03	—
N	0.018	—	N_2 = 0.018 + 8.39 (77/100) = 6.478
Total		1.93	8.458
∴ Corresponding weight of air = 1.93 × (100/23) = 8.39 kg			

Since actual weight of dry flue gases per kg of coal burnt is 21.5 kg, so, balance quantity of dry flue gas must have come from excess air. So, excess air supplied per kg of coal burnt

= 21.5 – 8.468 kg = 13.032 kg

$$\text{Hence, excess air used} = \frac{13.032}{8.39} = \mathbf{155.3\%}.$$

Example 49. *A fuel has the following composition by volume : Hydrogen = 24%, methane. = 30%, ethane = 11%, ethylene = 4.5%, butene = 2.5%, carbon monoxide = 6.0%, carbon dioxide = 8%, oxygen = 2% and nitrogen = 12%. Calculate the theoretical amount of air required at 25°C and 750 mm pressure for complete combustion of 1 m^3 of fuel.*

Solution. 1 m^3 of fuel gas contains : H_2 = 0.24 m^3, CH_4 = 0.30 m^3, C_2H_6 = 0.11 m^3, C_2H_4 = 45 m^3, C_4H_8 = 0.025 m^3 combustion equations involved are :

$$\underset{\text{1vol.}}{H_2(g)} + \underset{\text{1/2 vol.}}{\frac{1}{2}O_2(g)} = H_2O(l)$$

$$\underset{\text{1 vol.}}{CH_4(g)} + \underset{\text{2 vol.}}{2O_2(g)} = 2H_2O(l) + CO_2(g)$$

$$\underset{\text{1 vol.}}{C_2H_6(g)} + \underset{\text{3.5 vol.}}{7/2O_2(g)} = 2CO(g) + 3H_2O(l)$$

$$C_2H_4(g) + 3O_2(g) = 2CO_2(g) + 2H_2O(l)$$
1 vol. 3 vol.

$$C_4H_8(g) + 6O_2(g) = 4CO2(g) + 4H_2O(l)$$
1 vol. 6 vol.

$$CO_2(g) + \left(\frac{1}{2}\right) O_2(g) = CO_2(g)$$
1 vol. 0.5 vol.

∴ Volume of oxygen required for combustion

$$= 0.24 \times \left(\frac{1}{2}\right) + 0.3 \times 2 + 0.11 \times 3.5 + 0.045 \times 3 + 0.025 \times 6$$

$$+ 0.06 \times (0.5) - 0.02 \text{ (oxygen of fuel) } m^3$$

$$= 0.120 + 0.600 + 0.385 + 0.135 + 0.150 + 0.030 - 0.02 \, m^3 = 1.400 \, m^3$$

Corresponding volume of oxygen at 25°C and 750 mm Hg pressure can be calculated by using gas equation : $P_1V_1/T_1 = P_2V_2/T_2$

Given data (STP) ; At 25°C and 750 mm Hg :

$P_1 = 760$ mm $P_2 = 750$ mm

$V_1 = 1.4$ m^3 $V_2 = ?$

$T_1 = 273$ °K $T_2 = 298$ °K

$$\therefore \quad \frac{760 \times 1.4 \, m^3}{273} = \frac{750 \times V_2}{298}, \text{ hence } V_2 = \frac{760 \times 1.4 \times 298}{273 \times 750} = \mathbf{1.548 \, m^3}$$

Hence, volume of air required under given conditions

$$= [\text{Volume of oxygen}] \times \left[\frac{100}{23}\right] = 1.548 \, m^3 \times (100/23) = \mathbf{6.733 \, m^3}$$

Example 50. *The percentage composition of a sample of bituminous coal by mass was found to be :C = 76%; H_2 = 5.2; O_2 = 12.8%; S = 1.2%; N_2 = 2.7%, Ash = 2.2%. Calculate the minimum amount of air necessary for complete combustion of 1 kg of coal and percentage composition by weight of dry products of combustion.*

Solution. 1 kg of coal sample contains : C = 760 gm ; H_2 = 52 gm ; O_2 = 128 gm ; S = 12 gm; N_2 = 27 gm. The combustion equations involved are :

$$C(s) + O_2(g) = CO_2(g)$$
12 g 32 g 44 g

$$H_2(g) + \frac{1}{2} O_2(g) = H_2O(l)$$
2g 16 g 18g

$$S(s) + O_2(g) = SO_2(g)$$
32 g 32 g 64 g

$\therefore$ Amount of O_2 required for combustion of C, H and S in 1 kg of coal

$$= 760 \text{ gm} \times (32/12) + 52 \text{ gm} \times (16/2) + 12 \text{ gm} \times (32/32)$$
$$= 2,026.7 \text{ gm} + 416 \text{ gm}` + 12 \text{ gm} = 2,454.7 \text{ gm}$$

But amount of O_2 already present in 1 kg of fuel = 128 gm.

$\therefore$ Amount of O_2 needed from outside = 2,454.7 - 128 = 2,326.7 gm

Then, minimum amount of air required for 1 kg of fuel

$$= 2,326.7 \text{ gm} \times (100/23) = 10,116 \text{ gm} = 10.116 \text{ kg}$$

Calculation of masses of products of combustion :

Amount of CO_2 = Mass of C × (44/12) = 760 gm × (44/12) = 2,786.7 gm

Amount of N_2 = Mass of N_2 in [fuel + air used] = 27 gm + 10,116 gm × (77/100)
= 27 gm + 7,789.3 gm = 7,816.3 gm

Amount of SO_2 = Mass of S × (64/32) = 12 gm × (64/32) = 24 gm

Calculation of mass percentage composition of dry products of combustion.

Constituents	Amount, gm	% Composition by weight
CO_2	2,786.7	2,786.7 x 100/10,627 = 26.223
N_2	7,816.3	7,816.3 x 100/10,627 = 73.551
SO_2	24.0	24 x 100/10,627 = 0.226.
Total	10,627.0	100

Example 51 . *A gaseous fuel has the following composition : CH_4 = 40%; C_2H_6 = 24%; C_2H_4 = 6% ; C_4H_8 = 16% ; CO_2 = 5% ; CO = 5% ; O_2 = 3% and N_2 = 1%. Calculate the air / fuel ratio and the analysis of products of combustion using 50% excess air.*

Solution. 1 m^3 of fuel contains CH_4 = 0.40 m^3; C_2H_6 = 0.24 m^3 ; C_2H_4 = 0.06 m^3 ; C_4H_8 = 0.16 m^3; CO_2 = 0.05 m^3 ; CO = 0.05 m^3 ; O_2 = 0.03 m^3 and N_2 = 0.01 m^3. Their combustion reactions are :

$$CH_4(g) + 2O_2(g) = CO_2(g) + H_2O(l)$$
1 vol. 2 vol. 1 vol.

$$C_2H_6(g) + \frac{7}{2} O_2(g) = 2CO_2(g) + 3H_2O(l)$$
1 vol. 3.5 vol. 2 vol.

$$C_2H_4(g) + 3O_2(g) = 2CO_2(g) + 2H_2O(l)$$
1 vol. 3 vol. 2 vol.

$$C_4H_8(g) + 6O_2(g) = 4CO_2(g) + 4H_2O(l)$$
1 vol. 6 vol. 4 vol.

$$CO(g) + \frac{1}{2} O_2(g) = CO_2(g)$$
1 vol 0.5 vol. 1 vol

$\therefore$ Volume of oxygen needed for complete combustion of 1 m^3 fuel

$$= \text{Vol. of } [CH_4 \times 2 + C_2H_6 \times 3.5 + C_2H_4 \times 3 + C_4H_8 \times 6 + CO \times 0.5 - O_2]$$

$= [0.40 \times 2 + 0.24 \times 3.5 + 0.06 \times 3 + 0.16 \times 6 + 0.05 \times 0.5 - 0.03]\ m^3$

$= 0.80 + 0.84 + 0.18 + 0.96 + 0.025 - 0.03 = 2.775\ m^3$

$\therefore$ Volume of oxygen actually supplied per m^3 of gas = $2.775 \times (150/21)\ m^3$ = **19.821 m^3.** hence air/fuel ratio is **19.821.**

Volume of products of combustion from 1 m^3 of fuel.

CO_2 = Vol. of $[CH_4 + C_2H_6 \times 2 + C_2H_4 \times 2 + C_4H_8 \times 4 + CO + CO_2]$

$= 0.40 + 0.24 \times 2 + 0.06 \times 2 + 0.16 \times 4 + 0.05 + 0.05\ m^3$

$= 0.40 + 0.48 + 0.12 + 0.\times 4 + 0.05 + 0.05\ m^3 = 1.790\ m^3$.

O_2 = From excess air used = $19.821 \times (50/100) \times (21/100)\ m^3 = 2.081\ m^3$

N_2 = from (air + fuel) = $19.821 \times (79/100) + 0.01\ m^3 = 15.658 + 0.01\ m^3 = 15.668\ m^3$

$\therefore$ Total volume of products of combustion = $1.790 + 2.081 - 15.668\ m^3 = 19.539\ m^3$

where, volumetric composition of flue gas is :

$CO_2 = 1.790 \times 100/19.539 =$ **9.16%**

$O_2 = 2.081 \times 100/19.539 =$ **10.65%**

$N_2 = 15.668 \times 100/19.539 =$ **80.19%.**

Example 52. *The composition of a producer gas was found to be; H_2 = 14%, CH_4 = 2%, CO = 22%, CO_2 = 5%, N_2 = 55% and O_2 = 2% by volume. Find the air required for the perfect combustion of 1 m^3 of this gas. If 40% excess air is used, find the volume analysis of the dry products of combustion (Atomic weight: H = 1, O = 16, N = 14, C = 12).*

Solution. 1 m^3 of producer gas contains : H = 0.14 m^3, CH_4 = 0.02 m^3 , CO = 0.22 m^3, CO_2 = 0.05 m^3, N_2 = 0.55 m and O_2 = 0.02 m^3. Combustion equations involved are :

$$\underset{1\ \text{Vol.}}{H_2(g)} + \underset{0.5\ \text{vol.}}{\frac{1}{2}O_2(g)} = H_2O\ (l)$$

$$\underset{1\ \text{vol.}}{CH_{4}(g)} + \underset{2\ \text{vol.}}{2O_2(g)} = \underset{1\ \text{vol.}}{CO_2(g)} + 2H_2O(l)$$

$$\underset{1\text{Vol.}}{CO(g)} + \underset{0.5\ \text{vol}}{\frac{1}{2}O_2(g)} = \underset{1\ \text{Vol.}}{CO_2(g)}$$

$\therefore$ Amount of O_2 required for combustion

$= 0.14 \times 0.5 + 0.02 \times 2 + 0.22 \times 0.5\ m^3$

$= 0.07 + 0.04 + 0.11\ m^3 = 0.22\ m^3$

$\therefore$ Net mass of oxygen required = Oxygen needed - Oxygen in fuel

$= 0.22 - 0.02\ m^3 = 0.20\ m^3$.

Volume air required for perfect combustion of 1 m^3 of gas = $0.20 \times (100/21) = 0.952\ m^3$

Now volume of air actually used (*i.e.,* 40% in excess) for burning 1 m^3 of gas.

$= 0.20 \times 140/21\ m^3 = 1.333\ m^3$

Volumes of products of combustion :

CO_2 = Vol. of [$CH_4 \times 1 + CO \times 1 + CO_2 \times 1$]

$= 0.02 \times 1 + 0.22 \times 1 + 0.05 \times 1 \text{ m}^3$

$= 0.02 + 0.22 + 0.05 \text{ m}^3 = 0.29 \text{ m}^3$

O_2 = From excess air = Minimum oxygen required × (40/100)

$= 0.2 \text{ m}^3 \times (40/100) = 0.08 \text{ m}^3$.

Total N_2 = 79% of air used + N_2 in fuel

$= 1.333 \times (79/100) + 0.55 \text{ m}^3 = 1.053 + 0.55 \text{ m}^3 = 1.603 \text{ m}^3$.

∴ Total volume of dry products of combustion $= 0.29 + 0.08 + 1.603 \text{ m}^3 = 1.973 \text{ m}^3$

Hence, volume analysis of flue gas :

CO_2 = 0.29 × 100/1.973 = **14.70%**

O_2 = 0.08 × 100/1.973 = **4.05%**

N_2 = 1.603 × 100/1.973 = **81.25%**.

Example 53. *The volume analysis of a producer gas is : hydrogen = 14%, methane = 2%, carbon monoxide = 22%, carbon dioxide = 5%, oxygen = 2% and nitrogen = 55%. Find the air required for the perfect combustion of one cubic metre of this gas. If 40% excess air is supplied, find the volume analysis of the dry products.*

Solution. 1 m³ of the producer gas : $H_2 = 0.14 \text{ m}^3$, $CH_4 = 0.02 \text{ m}^3$, $CO = 0.22 \text{ m}^3$. Their combustion equations are :

$$\underset{\text{1 Vol-}}{H_2(g)} + \frac{1}{2}\underset{\text{0.5 vol.}}{O2(g)} = H_2O\,(l)$$

$$\underset{\text{1 vol.}}{CH_4(g)} + \underset{\text{2 vol.}}{2O_2(g)} = \underset{\text{1 vol.}}{CO_2(g)} + 2H_2O(l)$$

$$\underset{\text{1Vol}}{CO(g)} + \underset{\text{0.5 vol}}{\frac{1}{2}O_2(g)} = \underset{\text{1 Vol.}}{CO_2(g)}$$

∴ Volume of oxygen required for their combustion

$$= \text{Volume of}\left[H_2 \times \frac{1}{2} + CH_4 \times 2\,CO \times \frac{1}{4}\right]$$

$$= 0.14 \text{ m}^3 \times \frac{1}{2} + 0.02 \text{ m}^3 \times 2 + 0.22 \text{ m}^3 \times \frac{1}{2} = 0.22 \text{ m}^3$$

But volume of oxygen already present in 1 m³ of producer gas = 0.02 m³

∴ Oxygen needed for complete combustion of 1 m³ of fuel gas = 0.22 – 0.02 m³ = 0.2 m³

∴ Volume of air (containing 21% oxygen by volume) needed for complete combustion of 1 m³ of fuel gas = 0.2 × (100/21) = **0.95 m³**

Calculation of volume of dry products of combustion.

Volume of air required, if 40% in excess = 0.2 × (140/21) = 1.33 m^3

Volume of CO_2 = Volume of CH_4 × 1 + Volume of CO × 1 + Volume in fuel itself
= 0.02 + 0.22 + 0.05 m^3 = 0.29 m^3

Volume of N_2 = Volume of N_2 in [fuel + air used]
= 0.55 m^3 + volume of air used × (79/100)
= 0.55 + 1.33 × (79/100) m^3 = 0.55 + 1.05 = 1.603 m^3.

Volume of O_2 = Excess oxygen left unused = 0.20 (40/100) m^3 = 0.08 m^3.

Total volume of dry products of combustion = 0.29 + 1.60 + 0.08 = 1.97 m^3

∴ *Flue gas analysis is :*

CO_2 = 0.29 × (100/1.97) = **14.72%**

N_2 = 1.60 × (100/1.97) = **81.22%**

O_2 = 0.08 × (100/1.97) = **4.06%.**

Example 54. *The petrol used for petrol engine contains : 84% carbon and 16% hydrogen. The air supplied is 80% of that theoretically required for complete combustion. Assuming that all hydrogen is burnt, and that carbon burns partly to CO and partly to CO_2 without any free carbon, find the volumetric analysis of dry exhaust gas.*

Solution. 1 kg of petrol contains : 0.84 kg C ; 0.16 kg H.

∴ Air required for combustion of 1 kg of fuel

$$= \{0.84 \times (32/12) + 0.16 \times (16/2)\} \times \frac{100}{23} = 15.3 \text{ kg}$$

∴ Actual air supplied = 15.3 × (80/100) = 12.24 kg/kg of fuel

When C burns to CO_2 ($C + O_2 \rightarrow CO_2$), the air required per kg of C = $\frac{32}{12} \times \frac{100}{23}$ kg

and when C burns to CO ($C + \frac{1}{2} O_2 \rightarrow CO$), the air required per kg of C = $\frac{16}{212} \times \frac{100}{23}$ kg

∴ Reduction in quantity of air required for combustion, when C burns to CO instead of CO_2=

$$= \frac{100}{23}\left(\frac{32}{12} - \frac{16}{12}\right) = 5.8 \text{ kg}$$

But actual reduction in air supplied per kg of fuel = 15.3 – 12.24 = 3.06 kg

∴ Weight of C out of 0.84 kg burning to CO = $\frac{3.06}{5.8}$ = 0.527 kg/kg of fuel

∴ C burnt to CO_2 = 0.84 – 0.527 = 0.313 kg/kg of fuel.

∴ The masses of the dry products formed/kg of fuel burnt are :

CO_2 = 0.313 × (44/12) = 1.147 kg

CO = 0.527 × (28/12) = 1.23 kg

N_2 (from air = 12.24 × (77/100) = 9.40 kg

Volumetric analysis of flue gas :

Gas	Mass of gas per kg of fuel (a)	Molecular weight (b)	Proportional volume (a/b)	Percentage by volume
CO_2	1.147	44	0.026	0.026 × 100/0.436 = 5.96
CO	1.23	28	0.044	0.044 × 100/0.436 = 10.09
N_2	9.40	28	0.366	0.366 × 100/0.436 = 83.95
Total	—	—	0.436	100

Example 55. *The percentage composition of sample of anthracite coal was found by mass to be : C = 96.0, H = 3.3, O = 3.0, N = 0.8, S = 0.9 and ash = 2.0. Calculate the minimum amount of air required for the complete combustion of 1 kg of this fuel. If 50% excess air is supplied, find the percentage composition of the dry flue gases by volume and the mass of carbon per kilogram of dry flue gas.*

Solution. 1 kg of coal contains : C = 900 gm ; H = 33 gm ; 0 = 30 gm ; N = 8 gm ; S = 9 gm ; ash = 20 gm. Their combustion equations are :

$$\underset{12\,g}{C\,(s)} + \underset{32\,g}{O_2\,(g)} \,\square\, \underset{44\,g}{CO_2(g)} :$$

$$\underset{2\,g}{H_2(g)} + \underset{16\,g}{\frac{1}{2}\,O_2(g)} \rightarrow H_2O(l)$$

$$\underset{32\,g}{S\,(s)} + \underset{32\,g}{O_2\,(g)} \rightarrow \underset{64\,g}{SO_2(g)}$$

∴ Amount of O_2 required for combustion of C, H and S in 1 kg of coal

= 900 gm × (32/12) + 33 gm × (16/2) + 9 gm × (32/12) = 2,673 gm.

But amount of O_2 already present in 1 kg of fuel = 30 gm

∴ Amount of O_2 needed from outside = 2,673 - 30 = 2,643 gm

∴ Minimum amount of air required for 1 kg of fuel

= 2,643 gm × (100/23) = 11490 gm = **11.490 kg**

Then amount of air, if 50% excess is supplied = 11.490 × 150/100 = 17.235 kg

Calculation of masses of product of combustion :

CO_2 = Mass of C × (44/12) = 900 gm × (44/12) = 3,300 gm = 3.3 kg

N_2 = Mass of N from fuel + Mass of N from air = 0.008 +17.235 × (77/100) = 13.279 kg

Excess oxygen (O_2) = Mass of oxygen used × Excess % = 2.643 × (50/100) = 1.322 kg

SO_2 = Mass of S × (64/32) = 0.009 × (64/32) = 0.018 kg

Calculation of corresponding volumes of products of combustion :

Constituents	Mass of it, kg (a)	Molecular mass (m)	ratio (a/ m)	% composition by volume
CO_2	3.3	44	0.0750	0.0750 × 100/0.5535 = 13.55
N_2	13.279	28	0.4743	0.4743 × 100/0.5536 = 85.58
O_2	1.322	32	0.0041	0.0041 × 100/0.5535 = 0.74
SO_2	0.018	64	0.0002	0.0002 × 100/0.5536=0.04
Total	17.919	—	0.5536	100

Calculation of mass of C/kg of dry flue gas :

Now 17.919 kg dry flue gases contain CO_2 = 3.3 kg.

∴ 17.919 kg dry flue gases contain C $= \frac{3\times3}{44} = 0.9$ kg

∴ 1 kg dry flue gas contains carbon $= \frac{0.9}{17.919} = 0.0502$ kg = **50.2 gm.**

Example 56. *A sample of fuel contains the following : C = 84%, H_2 = 4%, O_2 = 2% and remaining ash. This fuel on combustion gave the following flue gas analysis : CO_2 = 10.5%, CO = 7.5%, O_2 = 4.5% and N_2 = 85.5%. Calculate the percentage excess air or otherwise used for combustion.*

Solution. Conversion of volume to mass analysis of flue gas;

Constituents	Volume per m³ of flue gas (v), m³	Molecular weight (m)	Proportional weight (v 4 m)	Weight per kg of flue gas, kg
CO_2	0.105	44	4.62	4.62/29.98 = 0.1549
CO	0.015	28	0.42	0.42/29.98 = 0.0142
O_2	0.035	32	1.28	1.28/29.98 = 0.0376
N_2	0.845	28	23.66	23.66/29.98 = 0.7934
Total	—	—	29.98	1

Now weight of O_2 required for the combustion of 0.014 kg of CO (present per kg of flue gas) to CO_2 (*i.e.*, CO + $\frac{1}{2}$ O_2 + CO_2) = 0.014 kg × (16/28) = 0.008 kg

∴ Amount of excess of oxygen per kg of flue gas = 0.0376 - 0.008 = 0.0298 kg.

Now amount of C in 1 kg of coal used is 0.84 kg and amount of C present in 1 kg flue gas

= Mass of CO_2 × (12/44) + Mass of CO × (12/28)

= 0.1549 × (12/44) + 0.014 (12/28) = 0.0423 + 0.006 = 0.0483 kg

∴ Amount of flue gas formed from 1 kg of coal burnt

$$= \frac{\text{Amount of C / kg of flue}}{\text{Amount of C / kg of fuel gas}} = \frac{0.84}{0.0483} = 17.39 \text{ kg}$$

∴ Amount Weight of excess oxygen/kg of fuel burnt

$$= \begin{bmatrix}\text{Weight of excess} \\ O_2 / \text{kg flue gas}\end{bmatrix} \times \begin{bmatrix}\text{Weight of flue gas} \\ \text{formed / kg fuel burnt}\end{bmatrix} = 0.0298 \times 17.39 = \mathbf{0.5182\ kg}$$

∴ Amount of excess air/kg of fuel burnt = 0.5182 × (100/23) = 2.253 kg

Also mass of oxygen required for complete combustion of 0.84 kg carbon and 0.04 kg hydrogen = 0.84 × (32/12) + 0.04 × (16/2) = 2.24 + 0.32 = 2.56 kg

Since the amount of oxygen already present in 1 kg of coal = 0.02 kg

 Net oxygen required for complete combustion of 1 kg fuel = 2.56 - 0.02 = 2.54 kg and amount of air for complete combustion of 1 kg of fuel = 2.54 × (100/23) = 11.043 kg

Hence, total amount of air supplied 1 kg of fuel burnt = 11.043 + 2.253 = 13.296 kg.

Example 57. *In a boiler trial, the analysis done on the flue gas shows the result by volume as:* CO_2 *= 10.0%, CO = 1.5%,* O_2 *= 8.0%,* N_2 *= 80.5%. The coal used in the boiler has the following analysis by mass : C = 82%, H = 4.2%, O = 4.8%, ash = rest. Ascertain the amount of excess air per kg of fuel burnt. Also determine the total mass of air supplied per kg of fuel burnt.*

Solution. Conversion of volume to mass analysis;

Constituents	Volume per m³ of flue gas (v), m³	Molecular weight (m)	Proportional weight (v Ч m)	Weight per kg of flue gas, kg
CO_2	0.100	44	4.40	4.4/29.92 = 0.147
CO	0.015	28	0.42	0.42/29.92 = 0.014
O_2	0.080	32	2.56	2.56/29.92 = 0.085
N_2	0.805	28	22.54	22.54/29.92 = 0.754
Total	—	—	29.92	1

Now mass of O_2 required for the combustion of 0.014 kg of CO (per kg of flue gas) to CO_2

$$(i.e.\ CO + \frac{1}{2} O_2 = CO_2) = 0.014 \times (16/28) = 0.008 \text{ kg}$$

∴ Amount of excess oxygen/kg of flue gas = 0.085 – 0.008 = 0.077 kg

Now amount of carbon present in 1 kg of flue gas

$$= \text{weight of } CO_2 \times (12/44) + \text{weight of CO} \times (12/28)$$

$$= 0.147 \times (12/44) + 0.014\ (12/28) = 0.040 + 0.006 = 0.046 \text{ kg}$$

But amount of carbon in 1 kg of coal used = 0.82 kg

∴ Amount of flue gas formed by burning 1 kg of coal

$$= \frac{\text{Amount of C / kg of fuel}}{\text{Amount of C / kg of flue gas}} = \frac{0.82}{0.046} = 17.83 \text{ kg}$$

Amount of excess oxygen/kg of ïuel burnt

$$= \begin{bmatrix}\text{üüüüüüü} \\ O_2 / \text{kg flue gas}\end{bmatrix} \times \begin{bmatrix}\text{Mass of flue gas formed} \\ \text{per kg of fuel burnt}\end{bmatrix} = 0.077 \times 17.83 = 1.373 \text{ kg}$$

∴ Amount of excess air/kg of fuel burnt = 1.373 × (100/23) = 5.969 kg

Also mass of oxygen required for complete combustion of 0.82 kg of C and 0.042 kg of hydrogen

$$= 0.82 \times (32/12) + 0.042 \times (16/2) = 2.187 + 0.336 = 2.523 \text{ kg}$$

Since amount of oxygen already present in 1 kg of fuel = 0.048 kg

Net amount of oxygen required for complete combustion of 1 kg fuel

$$= 2.523 - 0.048 = 2.475 \text{ kg.}$$

∴ Amount of air required for complete combustion of 1 kg of fuel

$$= 2.475 \times (100/23) = 10.761 \text{ kg}$$

Hence, total amount of air supplied per kg of fuel burnt

= Mass of air (for combustion + in flue gases)

= 10.761 + 5.969 = 16.73 kg.

Example 58. *In a boiler trial, the analysis done on the flue gas shows the following results by volume : CO_2 = 10.0%, CO = 1.5%, O_2 = 8.0%, N_2 = 80.5%. The coal used has the percentage composition by mass : C = 82%, H = 4.2%, O = 4.8%. The temperature of the flue gas is 310°C and that of air in the boiler house 16°C. Calculate the heat carried away by the excess air per kg of coal burnt, taking the average specific heat of air as 0.238.*

Solution. Conversion of volume percentage of flue gas to mass percentage.

Constituents	Volume per m³ (v), m³	Mol. weight (m)	Percentage of weight (v Ч m)	Weight of gas per kg flue gas, kg
CO_2	0.100	44	4.40	4.40/29.92 = 0.147
CO	0.015	28	0.42	0.42/29.92 = 0.014
O_2	0.080	32	2.56	2.56/29.92 = 0.085
N_2	0.805	28	22.54	22.54/29.92 = 0.754
Total	—	—	29.92	1

Now amount of C present per kg of flue gas

$$= \text{Weight of } CO_2 \times (12/44) + \text{Weight of CO} \times (12/28)$$

$$= 0.147 \times (12/44) + 0.014 \times (12/28)$$

$$= 0.040 + 0.006 = 0.046 \text{ kg}$$

But the amount of C/kg of coal used = 0.82 kg

∴ Amount of flue gas formed by burning 1 kg of coal

$$= \frac{\text{Amount of C / kg of Coal used}}{\text{Amount of C / kg of flue gas formed}} \; 0.82/0.046 = 17.83 \text{ kg}$$

Now amount of oxygen needed for the combustion of 0.014 kg of CO_2 (present per kg of flue gas) to CO_2 (*i.e.*, $CO + \frac{1}{2} O_2 = CO_2$) = 0.014 × (16/28) = 0.008 kg

∴ Amount of excess oxygen/kg flue gas

$= $ Mass of O_2 [in flue gas - required to convert CO to CO_2]

$= 0.085 - 0.008 = 0.077$ kg

$$= \begin{bmatrix}\text{Weight of excess air} \\ \text{pe kg of coal burnt}\end{bmatrix} \times \begin{bmatrix}\text{Sp. heat} \\ \text{of air}\end{bmatrix} \times \begin{bmatrix}\text{Temp. of outgoing gases} \\ -\text{Temp. of incoming gases}\end{bmatrix}$$

∴ Amount of oxygen/kg of fuel burnt

$$= \begin{bmatrix}\text{weight of excess} \\ O_2 \text{ per kg flue gas}\end{bmatrix} \times \begin{bmatrix}\text{Weight of flue gas formed} \\ \text{per kg fuel burnt}\end{bmatrix} = 0.077 \times 17.83 = 1.373 \text{ kg}$$

hence, excess air/kg of fuel burnt

$= 1.373 \times (100/23)$

Hence, excess of heat carried away by burning 1 kg of coal

$= 5.969 \times 0.238 \times (310 - 16)$ kcal $= 5.969 \times 0.238 \times 294$ kcal $=$ **417.66 kcal.**

Example 59. *A gas mixture containing 87% CH_4, and 13% C_2H_6 by volume is burned with air. Orsat analysis of the flue gas after complete combustion show 5.3% CO_2 by volume but no CO. The air enters at 0°C, 760 mm Hg and contains no water vapour, calculate* :

(a) the per cent excess air used and

(b) the volume in m^3 of the flue gas produced at 300°C and 760 mm Hg per m^3 of the fuel gas measured at NTP.

Solution. Basis : 1 kg mole of fuel gas mixture at NTP.

Reactions:

$$CH_4 + 2O_2 \rightarrow CO_2 + 2H_2O \quad ...(i)$$

$$C_2H_6 + 3.5\, O_2 \rightarrow 2CO_2 + 3H_2O \quad ..(ii)$$

Since the flue gas analysis indicates no CO, complete combustion is assumed,

(a) Oxygen balance :

Theoretical O_2 required for CH_4 $= 0.87 \times 2$ $= 1.74$ kg mole

Theoretical O_2 required for C_2H_6 $= 0.13 \times 3.5$ $= 0.455$ kg mole

Total theoretical O_2 required $= 2.195$ kg mole

CO_2 balance

CO_2 formed by first reaction $=0.87$ kg mole

CO_2 fromed by second reaction $=0.26$ kg mole

Total CO_2 formed $= 0.87 + 0.26 = 1.13$ kg mole

Let the amount of dry flue gas be x kg mole

So, by CO_2 balace, $0.053\, x = 1.13$

or $x = 21.32$ kg mole

Excess O_2 + N_2 from air $= 21.32 - 1.13 = 20.19$ kg mole

Theoretical O_2 demand $= 2.195$ kg mole

Theoretical N_2 supply $= 2.195 \times \frac{79}{21} = 8.26$ kg mole

Excess O_2 + its equivalent N_2 = (excess O_2 + total N_2) - (theoretical N_2)
= 20.19 – 8.26 = 11.93 kg mole

So, excess air = 11.93 kg mole

$$\% \text{ Excess air} = \frac{11.930}{10.455} \times 100 = 114\%$$

$$\left[V = \text{Volume} = \frac{nRT}{P} = \frac{23.45 \times 0.082 \times 573 = 1101.8 \text{ m}^3}{1}\right]$$

(b) Dry flue gas = 21.32 kg mole

Water vapour balance :

H_2O produced from first reaction = 0.87 × 2 = 1.74 kg mole

H_2O produced from second reaction = 0.13 × 3 = 0.39 kg mole

Total H_2O formed = 2.13 kg mole

Total wet flue gas = 21.32 + 2.13 = 23.45 kg mole = 1101.8 m^3

1 kg mole of fuel gas mixture at N.T.P. = 22.4 m^3

So, $$\frac{\text{flue gas at 300°C and 760 mm Hg}}{\text{flue gas at NTP}} = \frac{1101.8}{22.4} = 49.19$$

Example 60. *The weight composition of a fuel oil is 88% carbon and 12% hydrogen. It is burned with air which is known to be enriched but the composition of the air is not known, although it is a mixture of N_2 and O_2 only. The Orsat analysis of the flue gases produced is CO_2 = 13.4%, O_2 = 3.8% and N_2 = 82.8%.*

Determine the per cent excess oxygen, if any.

What is the composition of the air that is being used ?

What would have been the Orsat analysis of the flue gases if air used had its normal composition. The stoichiometric amount been used and the reaction be complete.

Solution. Basis : 1 kg of fuel oil

Reactions: $C + O_2 \rightarrow CO_2$ ...*(i)*

$$H_2 + \frac{1}{2} O_2 \rightarrow H_2O \quad \text{...}(ii)$$

Assumption : Complete combustion is achieved.

Oxygen balance :

O_2 required for first reaction $= 0.88 \times \frac{326}{12} = 2.347$ kg

O_2 required for second reaction $= 0.12 \times \frac{16}{2} = 0.96$ kg

Total O_2 required = 2.347 + 0.96 = 3.307 kg = 0.103 kg mole

Amount of CO_2 formed $= 0.88 \times \frac{44}{12} = 3.227$ kg = 0.073 kg mole

Let the dry gas formed = x kg mole, So, *by CO_2 balance, $0.134x = 0.073$*

or, $x = 0.545$ kg mole

O_2 in flue gas = 0.545 × 0.038 = 0.021 kg mole

N_2 in flue gas = 0.545 × 0.828 = 0.451 kg mole

O_2 required = 0.103 kg mole, O_2 in flue gas = 0.021 kg mole

% excess, O_2 used $= \frac{0.021}{0.103} \times 100 = 20.39\%$.

(b) Composition of enriched air:

O_2 in air = 0.103 + 0.021 = 0.124 kg mole,

N_2 in air = 0.451 kg mole

Constituents	kg Mole	Mole % = volume %
O_2	0.124	21.57
N_2	0.451	78.43
Total	0.575	100.00

(c) Stoichiometric amount of O_2 for complete combustion = 0.103 kg mole

N_2 from air $= 0.103 \times \frac{79}{21} = 0.387$ kg mole

Orsat analysis :

Constituents	kg mole	mole %
CO_2	0.073	15.85
N_2	0.387	84.15
Total	0.460	100.00

Example 61. *A hydrogen fuel containing 16% hydrogen and 84% carbon by weight is burnt with 20% excess air. Find out the*

(a) chemical formula of the fuel

(b) flue gas analysis on dry and wet basis

(c) air/fuel ratio by weight.

Solution. *(a)* The ratio of H/C by weight in the fuel is 16/84. By dividing the weight of each element by its respective atomic weights, we get the atomic ratio of H/C $\frac{16/1}{84/12} = \frac{16 \text{ atoms of H}}{7 \text{ atoms of C}}$

Hence the fuel can be represented by the chemical formula C_7H_{16}

(b) Now, the theoretical/stoichiometric combustion equation of the fuel in oxygen will be

$$C_7H_{16} + 11O_2 = 7CO_2 + 8H_2O$$

Since the fuel is burned in air (which roughly contains 79% N_2 and 21% O_2), the ratio of N_2/O_2 in air is 79/21 = 3.76, which means that 3.76 kmols of N_2 accompany each mole of oxygen when a fuel is burned in air instead of pure oxygen. The combustion equation of C_7H_{16} in air then can be written as :

$$C_7H_{16} + 11O_2 + 3.76 \times 11N_2 \longrightarrow 7CO_2 + 8H_2O + 3.76 \times 11\ N_2$$

or ,

$$C_7H_{11} + 11O_2 + 41.36N_2 \longrightarrow 7CO_2 + 8H_2O + 41.36\ N_2$$

Since, N_2 being inert does not enter into chemical reaction its molal quantity (*i.e.* 3.76 ×11 = 41.36 in this case) remains same on both sides of combustion equation.

Now in this case, since 20% excess air is supplied (*i.e.* excess air factor is 1.2), the co-efficient of oxygen and nitrogen on the left side of the above combustion equation is multiplied by 1.2. N_2 remains unchanged during the reaction, but since more O_2 is supplied (20% excess) then is needed for the chemical reaction, some oxygen (20% of theoretical O_2) appear in the flue gases. Hence, the combustion equation with 20% excess air is then written as follows :

$$C_7H_{16} + (1.2 \times 11)\ O_2 + (1.2 \times 41.36)\ N_2 \longrightarrow 7CO_2 + 8H_2O\ (0.2 \times 11)\ O_2 + (1.2 \times 41.36)\ N_2$$

or ,

$$C_7H_{16} + 13.2O_2 + 49.6\ N_2 \longrightarrow 7CO_2 + 8H_2O + 2.2O_2 + 49.6\ N_2$$

Hence, the analysis of flue gas is as follows :

Components	Wet flue gas analysis		Dry flue gas analysis (i.e. Orsat analysis)		
	K. moles	% Volume (wet basis)	Components	Kmoles	% volume , dry basis
CO_2 H_2O O_2 N_2	7 8 2.2 49.6	(7/66.8) × 100 = 10.5 (8/66.8) × 100 = 12.0 (2.2/66.8) × 100 = 3.3 (49.6/66.8) × 100 = 74.2	CO_2 O_2 N_2	7 4.2 49.6	(7/58.8) ×100 = 11.9 (2.2/58.8) × 100 = 3.7 (49.6/58.8) × 10=84.4
Total	66.8	100	—	58.8	100

(*c*) Air/fuel ratio (by weight) $= \dfrac{(13.2 + 49.2) \times 29 \text{ kg air/K.mol}}{1\,(7 \times 12 + 16\times) \text{ kg fuel/Kmol}} = 18.2$ kg air/ fuel

Example 62. *A coal has the following composition (weight %) :*

C = 71.89%, O_2 = 9.01%, H = 7.66%, N = 2.68%, S = 1.91%, ash = 6.85%.

It is burned with dry combustion air in a furnace entering at 20°C and 740 mm Hg (absolute). The Orsat analysis (i.e on dry basis) of the flue gas generated is CO_2 = 11.5%, O_2 = 5.39% and CO = 13.2% while the dry refuse taken from the ash pit contains 36% unburnt and rest 64% ash. Calculate:

(a) % excess air, *(b) m^3 of air supplied.per 100 kg of coal fired.*

Solution. Basis of calculation = 100 kg coal fired

First of all the quantity of carbon lost in the ash pit is found by equating the ash in the coal fired to the ash in the refuse from the ash pit.

$$\text{kg ash} = 0.64 \times (\text{weight of refuse}) = 6.85 \text{ kg ash/100 kg coal fired}$$

$$\therefore \text{Weight of refuse} = \frac{6.85}{0.64} = 10.7 \text{ kg}$$

$$\therefore \text{Weight of carbon lost in ash pit} = 10.7 \times 0.36 = 3.85 \text{ kg}$$

The weight of carbon lost in the ash pit is to be substracted from the carbon in the coal fired to find the amount burned.

Per 100 kg coal fired

Constituents	Weight, kg	Molecular/ atomic weight	K moles fuel burnt	K moles of O_2 theoretically required
Carbon	71.89			
	–3.85			
	= 68.04	12	5.67	5.67
Hydrogen	7.66	2	3.83	1.91
Oxygen	9.01	32	0.28	– 0.28
Nitrogen	2.68	32	0.28	– 0.28
Sulphur	1.91	28	0.10	—
Ash	6.85	32	0.06	0.06
Total	100	—	—	7.36

Carbon balance :

Carbon in the flue gases (1.32 + 11.50) = 12.82% of dry flue gas, must contain the 5.67 Kmols of carbon burned.

Kmoles Carbon = 0.1282 × (moles of dry flue gas) = 5.67,

$$\text{hence, moles of dry flue gas} = \frac{5.67}{0.1282} = 44.2 \text{ Kmoles.}$$

Since in Orsat analysis SO_2 is reported as CO_2, hence moles of dry flue gases per 100 kg coal fired can be more accurately determined by including the moles of sulphur also with the carbon. And in that case,

$$\text{Kmole of dry flue gas} = \frac{5.67 = 0.06}{0.1282} = 44.7 \text{ Kg moles}$$

Then, Kmoles of O_2 in flue gases $= 0.0539 \times 44.7 = 2.41$ Kg moles

Kmoles of CO in flue gases $= 0.0132 \times 44.7 = 0.59$ Kg moles

$$\text{Kmoles of excess oxygen} = 2.41 - \frac{0.59}{2} = 2.12 \text{ Kg mole} \left(\text{Since, } C + \frac{1}{2}O_2 = CO\right)$$

(a) Hence, % excess air = % excess oxygen $= \dfrac{2.12}{7.36} \times 100 = 28.8\%$.

(b) The volume (m^3) of air supplied/100 kg coal fired

$$= 7.36 \times \frac{100}{21} \times 1.288 \times 22.4 \times \frac{293}{273} \times \frac{760}{273} = 1114.5 \text{ m}^3$$

$$\left[\begin{array}{c} \text{where, } 1.288 = \text{excess air factor, } 1 \text{ kmol} = 22.4 \text{ Nm}^3 (\text{at } 0°\text{C, } 760 \text{ mm Hg}) \\ T_1 = 0°\text{C} = 273 + 0 = 273°\text{K, } T_2 = 273 + 20 = 293°\text{K, } P_1 = 760 \text{ mm Hg, } P_2 = 740 \text{ mm Hg} \end{array}\right]$$

Example 63. *Exhaust gas (flue gas) volumetric composition on dry basis from an internal combustion engine is as follows :*

$CO_2 = 6\%$, $O_2 = 0.2\%$, $CO = 13.4\%$, $H_2 = 8\%$, $CH_4 = 0.5\%$, $N_2 = 71.9\%$

Assuming that a hydrocarbon fuel is burned in the engine, find out the air / fuel ratio by weight and the H / C ratio in the fuel.

Solution. Basis = 100 Kmol of exhaust gas.

Let, x = no. of carbon atoms in the hydrocarbon fuel

and, y = no. of hydrogen atoms in the fuel.

The following is the combustion equation for the hydrocarbon fuel which has been written on the basis that 71.9 Kmol of N_2 were used and 71.9/3.76 Kmoles of O_2 were involved in the combustion reaction :

$$C_x H_y + \frac{71.9}{3.76} O_2 + 71.9 N_2 = 6CO_2 + 0.2O_2 + 8H_2 + 13.4CO + 0.5CH_4 + 71.9N_2 + ZH_2O$$

where, Z = Kmol of H_2O formed in the products of combustion.

Carbon balance. A carbon balance is done to determine the number of carbon atoms appearing in the exhaust products which must also appear as '*x*' *i.e.* the number of carbon atoms in fuel.

Hence, $\underset{\text{in } CO_2}{x = 6} + \underset{\text{in CO}}{13.4} + \underset{\text{in } CH_4}{0.5} = 19.9$ carbon atoms,

Hydrogen balance. A hydrogen balance is done to find the original quantity of hydrogen 'y' from the hydrogen shown in the exhaust gas, noting that much hydrogen is present in H_2O in the exhaust gas which does not appear in the dry exhaust gas analysis.

First of all, water vapour formed by finding where all the 71.9/3.76 = 19.12 Kmoles O_2 have gone, is determined

hence, 19.12 Kmoles $O_2 = 6 + 0.2 + \frac{13.4}{2} + (z/2)$

or, z = 12.44 Kmoles of H_2O formed in the combustion products,

hence, no. of hydrogen atoms in the fuel

$$y = \underset{\text{in } H_2O}{(2 \times 12.44)} + \underset{\text{in } H_2}{(2 \times 8)} + \underset{\text{in } CH_4}{(4 \times 0.5)} = 42.88 \text{ hydrogen atoms}$$

Hence, the ratio of hydrogen atoms to carbon atoms in the fuel is, $H/C = \frac{42.88}{19.9} = 2.16$

And, $\frac{\text{air}}{\text{fuel}} \text{ratio} = \frac{(19.12 + 71.9) \times 29}{19.9 \times 1.2 + 42.88 \times 1} = 94$ kg air / kg fuel

$$\begin{bmatrix} \text{where, } 29 = \text{molecular weight of air} \\ 12 = \text{atomic weight of carbon} \\ 1 = \text{atomic weight of hydrogen} \end{bmatrix}$$

Example 64. *A coal has following composition (weight %) : C = 72%, N = 1.4%, H = 6%, O = 8%, S = 1.6%, ash = 11%. It is completely burnt with 50% dry excess air. Determine*

(a) flue gas analysis on both dry and wet basis

(b) air /fuel ratio by weight

(c) volume of flue gases at 300°C in m^3.

Solution. Basis — 100 kg of coal burnt.

Constituents	kg	Kg mole	Kg mole of O_2 theoretically required	Kg moles of combustion products (flue gases)				
				CO_2	H_2O	SO_2	N_2	O_2
Carbon	72.0	6	6	6	—	—	—	—
hydrogen	6.0	3	1.5	—	3	—	—	—
oxygen	8.0	0.25	–0.25	—	—	—	—	—
nitrogen	1.4	0.05	—	—	—	—	0.5	—
sulphur	1.6	0.05	0.05	—	—	0.5	—	—
ash	11	—	—	—	—	—		—
Total	= 100		Total = 7.30				0.05	0.0 3.
							41.17*	65
							41.22	3.65**

*(7.30 × 1.50) ×3.76 = 41.17 Kgmole N_2 with 50% excess air

**excess oxygen = 0.5 × 7.30 = 3.65 Kg moles

(a)

Constituents	Analysis of wet flue gas, Kg moles	Volume %, (wet)	Analysis of dry flue gas (i.e. Orsat analysis), Kg moles	Volume %, dry
CO_2	6	(6/53.92) × 100= 11.13	$CO_2 + SO_2$ = 6.05	(6.05/50.92) × 100= 11.88
H_2O	3	(3/53.92) × 100 = 5.56	O_2 = 3.65	(3.65/50.92) × 100 = 7.17
SO_2	0.05	(0.05/53.92) × 100 = 0.09	CO = 0.00	0.00
N_2	41.22	(41.22/53.92) × 100 = 76.45	N_2 = 41.22	(41.22/50.92) × 100 =80.95
O_2	3.65	(3.65/53.92) × 100 = 6.77		
Total	53.92	100	50.02	100

(Actually in the dry flue gas analysis as given by Orsat apparatus, the SO_2 is also absorbed along with CO_2 in the first pipette. That is why, % CO_2 appearing in the result of Orsat analysis is actually the sum of the percentages of CO_2 and SO_2 in the flue gas. Hence, Kg moles of SO_2 and CO_2 are added in the table above while reporting the flue gas analysis on dry basis.

(b) $\frac{\text{Air}}{\text{flue}}$ ratio by weight $= \frac{7.31 \times 1.50 \times (100/20) \times 20 \text{ kg air/Kg moles}}{100 \text{ kg coal}}$

= 15.12 kg air/kg coal [where, 1.5 = excess air factor]

(c) Volume of flue gas in m^3 at 300°C

$$= 53.92 \times 22.4 \times \frac{573}{273} = 2.535 \text{ m}^3 \text{ flue gases/100 kg coal}$$

[where, 1 Kg mole = 22.4 Nm^3 *i.e.* volume at 0°C and 760 mm Hg],

Example 66. *A fuel gas has the following composition (volume %) :*

$H_2 = 52.8\%$, $CO = 38.3\%$, $CO_2 = 5.5\%$, $N_2 = 2.9\%$, $CH_4 = 0.4\%$, $O_2 = 0.1\%$

It is burnt with 10% excess air and the resulting flue gas leaves the furnace at 400°C and 760 mm Hg (absolute).

Find out the :

(a) composition (both dry and wet basis) of flue gases

(b) air/ fuel ratio by volume

(c) flue gas volume (in m^3) formed per m^3 of dry fuel gas supplied at 15.5°C and 760 mm Hg (absolute).

Solution. Basis of calculation — 100 Kmole of fuel gas.

Constituents	Kmols	Kmols of theoretical air required	Kmoles of combustion products formed			
			CO_2	O_2	N_2	H_2O
CO_2	5.5	—	5.5	—	—	—
O_2	0.1	–0.1	—	—	—	—
CO	38.3	19.15	38.3	—	—	—
H_2	52.8	26.4	—	—	—	52.8
CH_4	0.4	0.8	0.4	—	—	0.8
N_2	2.9	—	—	—	2.9	—
Total	100	46.25 theoretical + 4.63 (10% excess)	44.2	— 4.63*	2.9 191.3"	53.6
		50.88 supplied			194.2	

*excess oxygen (*i.e.* 4.63 Kmols) resulting from 10% excess air

**N_2 with the air = 50.88 × 3.76 = 191.3 Kmols.

(a) The flue gas analysis (both on wet and dry basis) has been tabulated below :

Constituents	Total flue gas, Kmoles	Volume %, wet basis	Dry flue gases, Kmoles	Volume %, dry basis i.e. Orsat Analysis
CO_2	44.2	(44.2/296.63) × 100 = 14.9	CO_2 = 44.2	(44.2/243.03) × 100 = 18.2
O_2	4.63	(4.63/296.63) × 100 = 1.6	O_2 = 4.63	(4.63/243.03) × 100 = 1.9
N_2	194.2	(194.2/296.63) × 100 = 65.4	N_2 = 194.2	(194.2/243.03) × 100 = 79.9
H_2O	53.6	(53.6/296.63) × 100 = 18.1		
Total	296.63	100	243.03	100

As is evident from the above table, that the dry flue gases are determined by omitting the water vapor from the combustion products.

(b) Air/fuel ratio by volume $= \dfrac{50.88 \times (100/21) \text{ Kmole air}}{\text{Kmole fuel gas}}$

$$= 2.42 \frac{\text{Kmole air}}{\text{Kmole fuel gas}} = 2.4 \frac{\text{m}^3 \text{air}}{\text{m}^3 \text{ fuel gas}}$$

(c) $\dfrac{\text{Volume } (\text{m}^3) \text{ of flue gas at 400°C, 760 mm Hg}}{\text{m}^3 \text{of dry fuel gas at 15.5°C, 760 mm Hg}} = \dfrac{296.6 \times \left(\dfrac{400+273}{273}\right) \times 22.4}{100 \times \left(\dfrac{15+273}{273}\right) \times 22.4}$

$$= 6.9 \frac{\text{m}^3 \text{ flue gas at 400°C}}{\text{m}^3 \text{ fuel gas at 15.5°C}}$$

[here, 1 Kmole = 22.4 Nm^3 i.e.volume in m^3at 0°C and 760 mm Hg
flue gas volume has been determined by $\dfrac{P_1V_1}{T_1} = \dfrac{P_2V_2}{T_2}$]

Example 67. *A water gas has the following composition (volume %) :*
H_2 = 48%, CO = 38%, CO_2 = 6%, N_2 = 5.5% , CH_4 = 2%, O_2 = 0.5%. This water gas at 20°C and 765 mm Hg (abs) is burned with dry air supplied at 20°C and 765 mm Hg (abs.). If the Orsat analysis of flue gas is CO_2 = 15.5%, O_2 = 4.76%, CO = 0.20%, then find out the :

(a) % excess air supplied for combustion

(b) Volume (in m^3) of flue gases at 230°C and 760 mm Hg (abs) formed per 100 m^3 of water gas burnt.

Solution. Basis — 100 Kg mole of water gas.

Carbon balance. Making a carbon balance, we get

Kmols of carbon in the flue gases = Kg moles of carbon in the water gas.

The Orsat analysis shows 15.5% CO_2 and 0.2% CO *i.e.* 15.5 + 0.2 = 15.7% of the dry flue gases contain carbon.

Now, the total number of Kg moles of carbon in 100 Kg moles of water gas is 6 (in CO_2) + 2 (in CH_4) + 38 (in CO) = 46

Hence, *as per carbon balance,*

$$46 = 0.1570 \times (\text{Kg moles of dry flue gas}/100 \text{ Kg moles of water gas})$$

or $$\frac{\text{Kg moles of dry flue gas}}{100 \text{ Kg moles of water gas}} = \frac{46}{0.1570} = 293$$

Hence, $$\frac{\text{Kg moles } O_2 \text{ in the flue gas}}{100 \text{ Kg moles of water gas}} = 0.0476 \times 293 = 13.95.$$

[Since, dry flue gas contains 4.76% oxygen as given by Orsat analysis] All of this oxygen is not excess, because 0.0020 × 293 = 0.586 Kg moles of CO is also present in the flue gases (since flue gas contains 0.2% CO). This CO would have needed 0.586/2 = 0.293

Kg mole of O_2 (Since, CO + + $\frac{1}{2}$ O_2 = CO_2) had it burned completely to CO_2. The Kg mole of oxygen are then 13.95 - 0.293 = 13.66. Only that oxygen which would have appeared in the combustion products had the fuel burned completely to CO_2 and H_2O is considered as excess.

The number of moles of oxygen theoretically required for combustion can now be computed as given below :

Basis — 100 Kg mole of water gas

Constituents	Kg moles	Kg moles of theoretical oxygen required	Kg moles of H_2O formed in the flue gases
CO_2	6.0	—	—
O_2	0.5	–0.5	—
N_2	5.5	—	—
CH_4	2.0	4.0	4
H_2	48.0	24.0	48
CH	38.0	19.0	—
Total	100.0	46.5	52

(a) % excess air = % excess O_2 = $\dfrac{13.66 \text{ Kg moles } O_2\text{, excess}}{46.5 \text{ Kg moles } 0_2\text{, theoretical}} \times 100 = 29.3\%$

(b) Now, water vapor formed by burning 100 Kg moles of water gas = 52 Kg moles (as found in table above)

hence, total wet flue gas

= 293 Kg moles dry flue gas + 52 Kg moles water vapor = 345 Kg moles

$$\text{Hence, } \frac{\text{m}^3 \text{ of flue gases formed}}{\text{m}^3 \text{ of water gas}} = \frac{345 \times 22.4 \times \left(\dfrac{230+273}{273}\right)}{100 \text{kg moles of water gas} \times 22.4 \times \left(\dfrac{230+273}{273}\right) \times \dfrac{760}{765}}$$

= 5.95.

Example 68. *Dry methane gas at 25°C undergoes complete combustion under adiabatic conditions when burnt with dry air at 25°C. The flame temperature developed is 1297°C. Find out the percent excess air used for combustion. Following data are given :*

Heat of reaction = 0.2 × 10^6 kcal/kg. mole

Average specific heat, Cp for various gases in kcal/kg mole. °C is given below :

CO_2 = 12.37, H_2O = 9.6, N_2 = 7.68, air = 7.74.

Solution. Basis — 1 kg mole of methane burnt.

$CH_4 + 2O_2 = CO_2 + 2H_2O$, $\qquad \Delta H_R = 0.2 \times 10^6$ kcal/kg. mole

O_2 required for combustion = 2 kg mole

N_2 with O_2 $= 2 \times \frac{79}{21} = 7.52$ kg mole

Let, x = kg mole of air supplied in excess

Material balance:

Reactants entering : CH_4 = 1 kg mole

$$\text{Air} = 2 + 7.52 + x = 9.52 + x$$

$$\text{Total reactants} = (10.52 + x) \text{ kg moles}$$

Combustion products :

CO_2 formed = 1 kg mole

H_2O formed = 2 kg mole

N_2 left = $(7.52 + 0.79x)$

O_2 unused = $0.21x$

Total products = $(10.52 + x)$ kg mole

Energy balance — Let reference temperature be 25°C, hence, enthalpy *(i.e.* heat content) of reactants = 0 now, enthalpy of products at 1297°C is calculated as below :

Constituents	C_p (Kcal/Kg mole °C)	Enthalpy (heat content), kcal / kg.mole [$=C_p\,dT = C_p\,(1297 - 25)$]
CO_2	12.37	15734.6
H_2O	9.60	12211.2
N_2 (from reacted O_2)	7.68	9769.0
air	7.74	9845.3

Hence, total enthalpy of products

$$= (1 \times 15734.6) + (2 \times 12211.2) + (7.52 \times 9769) + (9845.3)$$

$$= 113619.9 + 9845.3x$$

Since, combustion takes place under adiabatic conditions, and enthalpy of reactants is zero, hence,

$$\text{enthalpy of products} = \text{heat of reaction}$$

or, $$113619.9 + 9845.3x = 0.2 \times 10^6 = 200000$$

or, $$x = \frac{200000 - 113619.9}{98453} = 877 \text{ kg moles}$$

Theoretical air $= 2 + 7.52 = 9.52$ kg moles

Excess air $= 8.77$ kg moles

hence, % excess air used $= \frac{877}{9.5} \times 100 =$ **92.12%.**

Example 69. *Heat of reaction developed when methane gas (CH_4) is burnt with theoretical quantity of dry air (both methane and air being initially at 25°C) is 191.760 kcal / kg mole. Complete combustion of* *methane does not occur with theoretical air. Carbon content of methane which burns goes exclusively to carbon dioxide. Total combustion products (including unburnt CH_4 and unused oxygen) leave at 400°C. Calculate as to what percentage of methane is burnt ?*

The average specific heat (Cp) values in kcal / kg mole. °C over the temperature range is as follows : $O_2 = 7.40$, $N_2 = 7.09$, H_2O vapor = 8.35, $CO_2 = 10.48$, $CH_4 = 10.88$.

Solution. Basis — 1 kg mole of methane

$$CH_4 + 2O_2 = CO_2 + 2H_2O,\ \Delta H = -\ 191,760 \text{ kcal/kg mole}$$

Let, only 'x' kg mole of methane only out of 1 kg mole methane is burnt.

Material balance :

Material entering — CH_4 = 1 kg mole

O_2 required for its combustion = 2 kg moles

$$N_2 \text{ accompanying } O_2 \text{ from air} = 2 \times \frac{79}{21} = 7.52 \text{ kg moles}$$

Total = 10.52 kg moles.

Combustion products leaving, kg moles :

$CH_4 = (1 - x)$, $CO_2 = x$, $H_2O = 2x$, $O_2 = (2 - 2x) = 2(1 - x)$, $N_2 = 7.52$

Total = 10.52

Energy balance — Assuming, the base temperature to be 25°C, the reactants of enthalpy = 0 And, the heat of reaction when only 'x' kg mole methane is burnt = - 191760 x kcal. Combustion products enthalpy is calculated as below :

Constituents	C_p (kcal / kg mole. °C)	Enthalpy, kcal/kg mole [= $C_p\, d_T = C_p\,(400-25)$]
CH_4	10.88	4080
O_2	7.40	2775
N_2	7.09	2658.75
CO_2	10.48	3930
H_2O vapor	8.35	3131.25

hence, total enthalpy of combustion products

$$= 4080\,(1 - x) + 2775 \times 2\,(1 - x) + 2658.75 \times 7.52 + 3930 \times x + 3131.25 \times 2x$$

$$= (29623.8 + 562.5x) \text{ kcal}$$

Assuming combustion to be taking place under adiabatic conditions when enthalpy of reactants is zero, we have,

enthalpy of products = heat of reaction

or, $(29623.8 + 562.5x) = 191760x$

$$\text{or,} \quad x = \frac{29623.8}{(191760 - 562.5)} = 0.155 \text{ kg mole}$$

$$\text{hence, \% methane burnt} = \frac{0.155}{1} \times 100 = \mathbf{15.5\%.}$$

Example 70. *Carbon monoxide (CO) gas at 100°C undergoes complete combustion when burnt in 90% excess than theoretically required dry air at 500°C. Flue gases leave the combustion chamber at 950°C. Calculate the heat evolved in the combustion chamber in kcal / kg mole of CO burned. The heat of combustion of carbon monoxide is 67640 kcal / kg mole.*

The average value of specific heats (Cp) over the temperature range in kcal / kg mole. °C is given below

Gases	Temperature range, °C	Cp, kcal/kg mole.°C
CO	18 — 100	7.00
CO_2	18 — 950	11.80
O_2	18 — 950	7.50
N_2	18 — 950	7.50
air	18 — 500	7.25

Solution. Basis — 1 kg mole of carbon monoxide (CO) burnt

Reaction : $CO + \frac{1}{2} O_2 = CO_2, \Delta H_R = -67640$ kcal/kg mole

Theoretical O_2 required = 0.5 kg mole

Oxygen supplied (by 90% excess air) = 0.5 × 1.9 = 0.95 kg mole

Air supplied $= 0.95 \times \frac{100}{21} = 4.52$ kg moles

N_2 from air $= 4.52 \times \frac{79}{100} = 3.57$ kg moles

Unused O_2 $= 0.95 - 0.50 = 0.45$ kg mole

Assuming the combustion to be occurring under adiabatic conditions, the heat evolved in the combustion chamber (ΔH) is given by

$$\Delta H = \sum H_P - \sum H_R + \Delta H_R$$

where, $\sum H_P$ = Total enthalpy of combustion products,

and, $\sum H_R$ = Total enthalpy of reactants

Let the reference temperature to be 18°C.

Now, total enthalpy of reactants ($\sum H_P$) is calculated as follows :

Reactants	C_P kcal/kg mole. °C	Enthalpy of reactants, $H_R = m.C_P\, d_T$, kcal
CO_2	7.00	1 × 7 × (100 - 18) = 574
air	7.25	4.52 × 7.25 × (500 - 18) = 15795
		$\sum H_p = 16369$

And, total enthalpy of combustion products ($1H_P$) is calculated as follows :

Products	c_p, kcal / kg mole. °C	Enthalpy of reactants, $H_p = m\text{-}C_p\text{-}d_T$, kcal
CO_2	11.80	1 × 11 × (950 - 18) = 10997.6
O_2	7.50	0.45 × 7.50 × (950 - 18) = 3145.5
N_2	7.50	3.57 × 7.50 × (950 - 18) = 24954.3
Total		$\sum H_P$ = 39097.4 kcal

hence, heat evolved in the combustion chamber,

$$\Delta H = \Sigma_{H_P} - \Sigma_{H_R} + \Delta H_R$$
$$= 39097.4 - 16369 + (-67640) = -44911.6 \text{ kcal.}$$

Example 71. *Producer gas is burnt with 50% excess dry air. Both gas and air enter the combustion chamber at 15°C. Flue gases serve as the source of heat for a furnace requiring heat input at ≥ 1000°C. Heat losses from combustion chamber and furnace to the surroundings are negligible and the combustion products are cooled to 1000°C before they leave the furnace. What percentage of the heating value of the fuel is delivered to the furnace ? Following data are given :*

Constituents	CO_2	CO	O_2	CH_4	H_2	N_2	H_2O vapor
Composition, volume % of the producer gas,	4.8	27.0	0.4	1.9	14.1	51.8	—
Heating value, kcal/kg mole	—	67600	—	192000	57800	—	—
Average sp. heat in the required temperature range, Cp, kcal /kg mole. °C	12.1	8.4	8.0	20.0	7.0	7.6	9.4

Solution. Basis — 100 kg mole of producer gas burnt.

Oxygen required for combustion of 100 kg moles of producer gas and the combustion products produced is calculated below in tabular form :

Constituents of producer gas	Volume % = mole %	O_2 required for combustion (theoretical), kg moles	Products of combustion, kg moles			
			CO_2	H_2O	N_2	O_2
CO_2	4.8	—	4.8	—	—	—
CO	27.0	13.5	27.0	—	—	—
O_2	0.4	- 0.4	—	—	—	—
CH_4	1.9	3.8	1.9	3.8	—	—
H_2	14.1	7.05	—	14.1	—	—
N_2	51.8	—	—	—	51.8	—
Total	100%	23.95	33.7	17.9	51.8	—

Now, total theoretical oxygen to be supplied = 23.95 kg mole but O_2 supplied by 50% excess air = 23.95 × 1.5 = 35.93 kg moles hence, excess O_2 = 35.93 - 23.95 = 11.98 kg mole (it will appear in products of combustion).

N_2 supplied from air = $35.93 \times \frac{79}{21}$ = 135.15 kg mole

This N_2 will also remain in the products of combustion along with the N_2 present in producer gas. *i.e.* total N_2 in the combustion products = 135.15 + 51.8 = 186.95 kg mole.

Material balance of the process is summarised below :

Constituents	Reactants entering, kg moles	Combustion products leaving, kg moles
CO_2	4.8	33.7
CO	27.0	—
O_2	35.93 + 0.40 = 36.33	11.98
CH_4	1.90	—
H_2	14.1	—
N_2	186.95	186.95

Energy balance. Reference temperature is assumed to be 0°C.

Total enthalpy of reactants ($\sum_{HR}$) and that of combustion products ($\sum_{HP}$) is tabulated below :

Constituents	kcal Cp, kg mole. °C	Reactants enthalpy, H_R, kcal		Combustion products enthalpy, H_p, kcal	
		amount, m, kg mole	m.Cp.dT – mCp (15 – 0)	amount, m, kg mole	m.Cp.dT = m.Cp (1000 - 0)
CO_2	12.1	4.80	4.8 × 12.1 × (15-0) = 871.5	33.70	33.70 ×12.1 × (1000-0) = 407800
CO	8.4	27.00	3402.5	—	—
O_2	8.0	36.33	4359.0	11.98	95800
CH_4	20.0	1.90	570.0	—	—
H_2	7.0	14.10	1480.5	—	—
N_2	7.6	186.95	21312.0	186.95	1420800
H_2O	9.4	—	21312.0	17.90	168300
		Total	$\sum$ Hr = 31995 kcal		$\sum H_P$ = 2092700 kcal

Now, heat released by the combustion of the combustibles (*i.e.* H_2, CO and CH_4) present in the producer gas

$$= \Delta H_R = 14.1 \times 57800 + 27 \times 67600 + 1.9 + 192000 = 3004980 \text{ kcal}$$

Heat balance gives :

energy input = energy delivered to the furnace + energy output

now, energy input = heat of combustion (ΔH_R) of producer gas + enthalpy of reactants

= 3004980 + 31995 = 3036975 kcal

and, energy output = enthalpy of combustion products = 2092700 kcal

hence, energy delivered to the furnace = 3036975 — 2092700 = 944275 kcal

hence, % heating value of the fuel delivered to the furnace = $\frac{944275}{3004980} \times 100 = 31.4\%$.

Example 72. *A fuel gas containing 20% CO and 80% N_2 is burned with 150% excess air (both air and gas being at 25°C). Calculate the theoretical flame temperature of the gas. Following data is given :*

Constituents	CO_2	O_2	N_2	CO
Average sp. heat C_p, kcal/kg mole. °K	12.10 -	7.90	7.55	—
Heat of formation at 25°C, kcal/kg mole	-94052	—	—	- 26412

Solution. Basis — 100 kg mole of fuel gas entering

$$CO + 1/2\ O_2 = CO_2$$

Theoretical O_2 required $= 20 \times 1/2 = 10$ kg mole

O_2 supplied (is 150% excess) $= 10 \times (1 + 1.5) = 25$ kg mole

N_2 supplied (with 150% excess air) $= 25 \times \frac{79}{21} = 94.05$ kg mole

CO_2 formed in combustion product $= 20$ kg mole.

Material balance

Constituents	Materials entering, kg moles	Materials leaving, kg moles
CO	20	—
N_2	80 + 94.05 = 174.05	174.05
O_2	25	15
CO_2	—	20
Total	219.05	209.05

Energy balance. Let reference temperature be assumed to be 25°C, so that enthalpy of the reactants is zero.

Now, heat of reaction $(\Delta H_R) = \Delta H_{fp} - \Delta H_f = -94052 - (-26412) = -67640$ kcal/kg mole Enthalpy of combustion products $(\sum H_P)$ is calculated fr as below :

Constituents	C_p, kcal / kg,°K	Product enthalpy, kcal/kg, Hp	
		amount, m, kg moles	m.Cp.dT
CO_2	12.10	20	20 × 12.10 × dT = 242 dT
O_2	7.90	15	118.5 dT
N_2	7.55	174.05	1314.08 dT
			$\sum H_p$ = 1674.58 dT kcal

or, $\sum H_P = 1674.58\ (T - 298)$,

where, T = temperature of combustion products, °K

Heat of reaction/combustion of fuel gas

$$= \Delta H_r = -67640 \times 20 = -1352800 \text{ kcal}$$

Assuming adiabatic conditions and no heat losses, temperature of combustion products, T = theoretical flame temperature

now, enthalpy of combustion products = heat of reaction (as the enthalpy of reactants is zero)

or, $1674.58\ (T - 298) = 1352800$

or, $$T = \frac{1352800}{1674.58} + 298 = 1105.8°K = 1105.8 - 273 = 832.8°C.$$

Example 73. *Volumetric composition of a producer gas is as follows :*

$$N_2 = 54\%, H_2 = 12\%, CO = 29\%, CO_2 = 5\%$$

It is burnt with theoretical amount of air. Find out the wet flue gas composition assuming :

(a) no dissociation of products of combustion present.

(b) that 18% of CO_2 and 5% of H_2O in the flue gas is dissociated.

Solution. *(a)* Basis 100 m^3 of producer gas

Combustion reactions are : $2CO + O_2 = 2CO_2$ or, $CO + \frac{1}{2} O_2 = CO_2$

$2H_2 + O_2 = 2H_2O$ or, $H_2 + \frac{1}{2} O_2 = H_2O$

Constituents	Volume, m^3	O_2 required for complete combustion, m^3	Products of combustion, m^3		
			CO_2	H_2O	N_2
CO	29	14.5	29	—	—
H_2	12	6	—	12	—
N_2	54	—	—	—	54
CO_2	5	—	5	—	—
Total	100	20.5	34	12	54

Theoretical air required per 100 m^3 of producer gas = $\frac{20.5 \times 100}{21} = 97.5$ m^3

N_2 present in theoretical air $= 97.5 \times \frac{79}{100} = 77$ m^3

N_2 present in producer gas $= 54$ m^3

Total N_2 $= 77 + 54 = 131$ m^3

Total wet flue gas = $CO_2 + H_2O + N_2 = 34 + 12 + 131 = 177$ m^3

So, for 100 m^3 of producer gas burnt with theoretical air, we have following products of combustion :

Constituents	Volume, m^3	% by volume
CO_2	34	$\frac{34}{177} \times 100 = 19.2\%$
H_2O	12	$\frac{12}{177} \times 100 = 6.8\%$
N_2	131	$\frac{131}{177} \times 100 = 74.0\%$
Total	177	100%

(b) Now when dissociation of CO_2 (18%) and H_2O (5%) present in the flue gas is considered, then the composition of flue gas after dissociation will be as follows : The dissociation reactions are :

$$2CO_2 \rightleftharpoons 2CO + O_2 \text{ and } 2H_2O \rightleftharpoons 2H_2 + O_2$$

Constituents	Volume, m³
CO_2 left after dissociation	34 × (1-0.18) = 27.9
H_2O left after dissociation	12 × (1-0.05) = 11.4
N2	131.0
CO produced from CO_2 dissociation	34×0.18 = 6.1
O_2 produced from CO_2 dissociation	$\frac{34 \times 0.8}{2}$ = 3.0
O_2 produced from H_2O dissociation	$\frac{12 \times 0.5}{2}$ = 0.3
H_2 formed from H_2O dissociation	12×0.05 = 0.6

Hence the final flue gas composition after taking into account the dissociation of H_2O and CO_2 becomes as follows :

Constituents	Volume, m³	% by Volume
CO_2	27.9	$\frac{27.9}{180.3}$ × 100 = 15 5%
H_2O	11.4	$\frac{11.4}{180.3}$ × 100 = 6.3%
N_2	131	$\frac{131}{180.3}$ × l00 = 72.7%
CO	6.1	$\frac{6.1}{180.3}$ × 100= 3.4%
H_2	0.6	$\frac{0.6}{180.3}$ × 100= 0.3%
O2	3 + 0.3 = 3.3	$\frac{3.3}{180.3}$ × 100 = 1.8%
Total	180.3	100%

Example 74. *Steam boiler plant in a petroleum refinery has an overall thermal efficiency on gross C.V. of the fuel of 85% and an output of 90720 kg of steam per hour. The steam is produced at a pressure of 17 kg / cm² and is superheated to 230°C from feed water at 50°C.*

Refinery gas is used to fire the boiler which has the following volumetric composition in percentage.

$$H_2 = 21,\ CH_4 = 15,\ C_2H_6 = 15,\ C_3H_6 = 36$$

$$N_2 = 9,\ H_2S = 2,\ CO_2 = 1,\ CO = 1$$

Find out :

(a) Volumetric flow rate of refinery gas required to fire the boiler plant.

(b) Volumetric capacity of the induced draught fan assuming that the refinery gas is burnt with 20% excess air and the exit flue gas temperature from the boiler plant is 205°C.

Given :

Gases	H_2	CH_4	CO	C_2H_6	C_3H_6	H_2S
Gross C.V, kcal /Nm³	3026	8900	3026	17177	20470	6257

Solution. *(a)* Gross C.V. of refinery gas

$= 0.21 \times 3026 + 0.01 \times 3026 + 0.02 \times 6257 + 0.15 \times 8900 + 0.15 \times 17177 + 0.36 \times 20470$

$= 12072$ kcal/Nm³

Enthalpy of steam at 17 kg/cm² and 230°C = 683 kcal/kg

Heat content of steam produced = $90720 \times 683 = 61961760$ kcal

Heat in feed water = $90720 \times 1 \times (50 - 0) = 4536000$ kcal

∴ Total heat to be supplied by fuel (*i.e.,* refinery gas) for steam generation

= 61961760/4536000 = 57425760 kcal

Heat produced by burning refinery gas = heat gained by steam

or, Heat produced by gas $\times \dfrac{85}{100} = 57425760$ (∵ thermal efficiency of boiler = 85%)

or, Heat produced by gas $= \dfrac{57425760}{0.85} = 67559717$ kcal

Since, C.V. of refinery gas = 12072 kcal/Nm³

∴ Volume of refinery gas required $= \dfrac{67559717}{12072} = 5596$ Nm³/hr. **Ans**

∴ Volumetric fuel *(i.e.* refinery gas) firing rate.

(b) I.D. fan. capacity is computed by calculating the total volume of wet flue gases at the boiler furnace exit temperature.

Total wet flue gas volume per Nm³ of refinery gas on being burnt with 20% excess air is found as below in tabular form :

Constit-uents	Quan-tity, Nm³	Combustion reaction	O_2 required for combustion, Nm³	Products of combustion, nm³				
				CO_2	O_2	N_2	H_2O	SO_2
H_2	0.21	$H_2 + \frac{1}{2} O_2 = H_2O$	0.105	—	—	—	0.21	

CO	0.01	$CO + \frac{1}{2} O_2 = CO_2$	0.005	0.01	—	—	—	
H_2S	0.02	$2H_2S + 2O_2 = 2H_2O + SO_2$	0.020	—	—	—	0.02	0.01
CH_4	0.15	$CH_4 + 2O_2 = CO_2 + 2H_2O$	0.300	0.15	—	—	0.30	—
C_2H_6	0.15	$2C_2H_6 + 7O_2 = 4CO_2 + 6H_2O$	0.525	0.30	—	—	0.45	—
C_3H_6	0.36	$2C_3H_6 + 9O_2 = 6CO_2 + 6H_2O$	1.620	1.08	—	—	1.08	—
N_2	0.09	Incombustible	—	—	—	0.09	—	—
CO_2	0.01	Incombustible	—	0.01	—	—	—	—
Total	1.00	—	2.575	1.55	0.51 FROM excess air	11.62 from total air	2.06	0.01
Grand Total	—		—	1.55	0.51	11.71	2.06	0.01

Theoretical oxygen requirement for burning 1 Nm^3 of refinery gas = 2.575 Nm^3

$\therefore$ Theoretical air requirement = $2.575 \times \frac{100}{21} = 12.26 \ Nm^3$

Actual air used = 1.2 × theoretical air = 1.2 × 12.26 = 14.71 Nm^3

$\therefore$ Excess air = 14.71 - 12.26 = 2.45 Nm^3,

$\therefore$ O_2 present in excess air = 2.45 × 0.21 = 0.51 Nm^3

N_2 present in actual air used = $14.71 \times \frac{79}{100} = 11.62 \ Nm^3$

Hence, total wet flue gas volume produced by combustion of 1 Nm^3 of refinery gas with 20% excess air

= 1.55 + 0.51 + 11.71 + 2.06 + 0.01 = 15.84 Nm^3/Nm^3 of refinery gas

Refinery gas burning rate = 5596 Nm^3/hr

$\therefore$ Total flue gas production rate = 5596 x 15.84 = 88640 Nm^3/hr

$\therefore$ Capacity of I.D. fan required = 88640 Nm^3/hr (at N.T.P. *i.e.,* 0°C and 760 mm Hg)

Since, the flue gas leaves at 205°C, the actual capacity of I.D. fan is calculated using Charle's law :

i.e, $\frac{V_1}{T_1} = \frac{V_2}{T_2}$ or, $\frac{88640}{(273+0)} = \frac{V_2}{(273+205)}$

or, $V_2 = \frac{88640 \times (273+205)}{273}$ = 155201 m^3 at 205°C and 760 mm Hg.

Hence, the I.D. fan capacity required = 155201 Nm^3/hr **Ans.**

Example 75. The gravimetric composition and properties of two different typical fuels i.e., coal and fuel oil is tabulated below :

Parameters	Coal	Fuel Oil
Composition (weight %) :		
C	76	85.0
H	4.5	11.5
S	0.5	3.5
N	1.0	—
O	3.0	—
ash	10.0	—
moisture	5.0	—
Other relevant data :		
Air used for complete combustion	theoretical air	theoretical air
Flue gas temp, leaving the furnace, °C	550	550
Ambient temperature, °C	15	15
Gross C.V., kcal/kg	7200	10100
V.M. on d.a.f. basis	30%	—
Viscosity, Centistokes	—	25
Latent heat of condensing steam, kcal/kg	586	586
B.S. Swelling number	7	—
Mean specific, C_p of flue gas, kcal/kg °C	0.253	0.258

Calculate the efficiency of combustion (i.e., fuel combustion efficiency) of the two fuels i.e., coal & fuel oil and comment as to which fuel is more efficient.

Solution. First we compute the *fuel efficiency of coal* as follows :

Basis-1 kg of coal burnt with theoretical air

Constituents	Quantity, kg	Combustion reaction	Theoretical oxygen required for the combustion, kg	Products of combustion, kg				
				CO_2	O_2	N_2	H_2O	SO_2
C	0.76	$C + O_2 = CO_2$	$0.76 \times \frac{32}{12}$ =2.02	2.78	—	—	—	—
H	0.045	$H_2 + \frac{1}{2} O_2 = 2H_2O$	$0.045 \times = 0.36$	$\left(0.76 \times \frac{44}{12}\right)$	—	—	0.81	—
S	0.005	$S + O_2 = SO_2$	$0.005 \times \frac{32}{32} =$ 0.005	—	—	—	$\left(0.45 \times \frac{36}{2}\right)$	0.01

N	0.01	Incombustible	—	—	—	0.01	—	$\left(0.005\times\frac{64}{32}\right)$
O	0.03	Incombustible	0.03	—	—	—	—	—
ash	0.10	Incombustible	—	—	—	—	—	—
moisture	0.05	Incombustible	—	—	—	—	0.05	—
Total	1.0		2.355	2.78	—	0.01	0.86	0.01

Theoretical oxygen required for burning 1 kg coal = 2.355 kg

$$\therefore \text{ Theoretical air required } = 2.355 \times \frac{100}{23} = 10.24 \text{ kg}$$

$$N_2 \text{ present in theoretical air } = 10.24 \times \frac{77}{100} = 7.88 \text{ kg}$$

(Air contains 23% O_2 & 77% N_2 by weight)

Hence, total weight of wet flue gas/kg of coal

$$= 2.78 + 0.01 + 0.86 + 0.01 + 7.88 = 11.54 \text{ kg/kg of coal}$$

Now, Net C.V. of coal = gross C.V. - (586 × weight of water)

$$= 7200 - [586 \times (0.86 + 0.05)]$$

$$= 7200 - 533 = 6666 \text{ kcal/kg}$$

The heat carried away in flue gas from 1 kg of coal

$$= 11.54 \times 0.253 \times (550 - 15) = 1562 \text{ kcal/kg coal}$$

□ % of the gross potential heat, of coal available in the combustion chamber

$$= \left(\frac{6666 - 1562}{7200}\right) = 100 = 71\%$$

Now, we compute the *fuel efficiency of fuel oil* as follows :

Basis - 1 kg of fuel oil burnt with theoretical air

Constitents	Quantity, kg	Theoretical oxygen required, kg	Products of combustion, kg			
			CO_2	N_2	H_2O	SO_2
C	0.85	2.26	3.12	—	—	—
H	0.115	0.92	—	—	1.035	—
S	0.035	0.035	—	—	—	0.07
Total	1.0	3.215	3.12	—	1.035	0.07

Theoretical oxygen required = 3.215 kg/kg of oil

$$\therefore \text{ Theoretical air required } = 3.215 \times \frac{100}{23} = 14 \text{ kg/kg of oil}$$

N_2 present in theoretical air $= 14 \times \frac{77}{100} = 10.8$ kg/kg of oil

Hence, total weight of wet flue gas produced by combustion of 1 kg of fuel oil with theoretical air

$$= 10.8 + 3.12 + 1.035 + 0.07 = 15 \text{ kg/kg of coal}$$

The heat carried away by flue gases from 1 kg of fuel oil

$$= 15 \times 0.258 \times (550 - 15) = 2070 \text{ kcal/kg oil}$$

Net C.V. $=$ gross c.v. $-$ (586 × weight of water) $= 10100 - [586 \times 1.035]$

$$= 10100 - 606.5 = 9493.5 \text{ kcal/kg}$$

$\therefore$ % of gross potential heat of fuel oil available in the combustion chamber

$$= \frac{9493.5 - 2070}{10100} \times 100 = 73.55\%$$

Since, fuel efficiency of coal is only 71% as against 73.5% for fuel oil; hence it is concluded that fuel oil is more efficient than coal in this case. **Ans.**

Example 76. *A hydrocarbon liquid fuel (containing only C and H_2 is burnt in air which produces a flue gas which analyses (Orsat analysis) on dry basis as $CO_2 = 10.5\%$, $O_2 = 5\%$ and $N_2 = 84.5\%$.*

Find out the :

(a) gravimetric composition (weight %) of the fuel

(b) % excess air used for the combustion.

Solution, *(a)* Combustion reactions are :

$$C + O_2 = CO_2 \text{ and } H_2 + \frac{1}{2} O_2 = H_2O$$

Assuming 100 kg mole of dry flue gases :

O_2 present in flue gas = 10.5 (from CO_2) + 5 (as free oxygen) = 15.5 kg mole of O_2

N_2 present in 100 kg mole of flue gas = 84.5 kg mole

$\therefore$ the actual air used $= 84.5 \times \frac{100}{79} = 107$ kg mole/100 kg mole of flue gas

O_2 present in the air supplied $= 107 \times \frac{21}{100} = 22.5$ kg mole/100 kg mole of flue gas

Hence, the amount of oxygen present in the water (H_2O) produced by the combustion of hydrogen (H_2) $= 22.5 - 15.5 = 7$ kg mole

Hence, the amount of H_2 present $= 7 \times 2 = 14$ kg mole (as $H_2 + \frac{1}{2} O_2 = H_2O$)

$= 14 \times 2 = 28$ kg/kg mole of dry flue gases

And, the amount of carbon present $= 10.5$ kg mole ($C + O_2 = CO_2$),

CO_2 % in flue gas = 10.5%

$= 10.5 \times 12 = 126$ kg/100 kg mole of dry flue gases

Hence, gravimetric composition of fuel (by weight) is C = 126 kg and H = 28 kg

i.e., % of carbon in fuel $= \dfrac{126}{126+28} = 81.5\%$

and, % of hydrogen in fuel $= \dfrac{28}{126+28} = 18.2\%$ **Ans.**

(*b*) In 100 kg mole flue gas, CO_2 = 10.5 kg mole *i.e.* carbon is also 10.5 kg mole ($C + O_2 = CO_2$) and hydrogen is 14 kg mole.

Amount of oxygen required to burn 10.5 kg mole carbon = 10.5 kg mole

And, amount of oxygen required to burn 14 kg mole hydrogen $= \dfrac{14}{2} = 7$ kg mole

∴ Total oxygen required to burn the fuel = 10.5 + 7 = 17.5 kg mole/100 kg mole of dry flue gas but actual oxygen supplied is = 22.5 kg mole/100 kg mole of dry flue gas

∴ % excess O_2 = % excess air $= \dfrac{22.5-17.5}{17.5} = 28.6\%$ **Ans.**

Example 77. *Gravimetric analysis (weight %) of a fuel is as follows :*

C = 82%, H = 10%, S = 3%, O = 2.5%, ash = 2.5%

for an actual air/fuel ratio of 12 : 1, find out the

(a) fuel-air mixture strength as lean or rich

(b) dry flue gas gravimetric (by weight) as well as volumetric analysis

Solution. First we find the theoretical air requirement for burning 1 kg of the fuel

Constituents	Quantity, kg	Combustion reaction	Theoretical O_2 requirement, kg
C	0.82	$C + O_2 = CO_2$	$0.82 \times \dfrac{32}{12} = 2.186$
H	0.10	$H_2 + \dfrac{1}{2} O_2 = H_2O$	$0.1 \times \dfrac{16}{2} = 0.80$
S	0.03	$S + O_2 = SO_2$	$0.03 \times \dfrac{32}{32} = 0.03$
0	0.025	incombustible	0.025
Ash	0.025	incombustible	—
			2.991

∴ Theoretical oxygen required = 2.991 kg/kg of fuel

$\therefore$ Theoretical air required $= 2.991 \times \dfrac{100}{23} = 13$ kg/kg of fuel

hence, theoretical fuel/air ratio $= \dfrac{1}{13}$, But, actual fuel/air ratio $= \dfrac{1}{12}$

$$\therefore \text{Equivalence ratio} = \frac{\text{actual fuel / air ratio}}{\text{theoretial fuel / air ratio}} = \frac{1/2}{1/3} = \frac{13}{2} = 1.083$$

i.e., the fuel-air mixture is 8.3% rich. **Ans.**

(b) Now, deficiency of combustion air = 13 - 12 = 1 kg/kg of fuel

The weight of air saved by burning 1 kg C to CO instead of CO_2

$$= \frac{16}{12} \times \frac{100}{23} = 5.8 \text{ kg}$$

$$\left[\underset{12\text{kg}}{C} + \underset{16\text{ kg}}{\frac{1}{2}O_2} = \underset{28\text{kg}}{CO_2}\right]$$

$\therefore$ 1 kg of air will be saved when $\dfrac{1}{5.8} = 0.1725$ kg carbon burns to CO

$\therefore$ CO formed from 0.1725 kg C $= 0.1725 \times \dfrac{28}{12} = 0.403$ kg/kg of fuel

CO_2 formed from (0.82 - 0.1725) = 0.6475 kg C $= 0.6475 \times \dfrac{44}{12} = 2.38$ kg/kg of fuel

$$\left[\underset{12\text{kg}}{C} + \underset{32\text{ kg}}{O_2} = \underset{44\text{kg}}{CO_2}\right]$$

SO_2 formed from 0.03 kg of S $= 0.03 \times \dfrac{64}{32} = 0.06$ kg/kg of fuel

$$\left[\underset{32\text{kg}}{S} + \underset{32\text{ kg}}{O_2} = \underset{64\text{kg}}{SO_2}\right]$$

N_2 supplied to flue gas from 12 kg air = 12 × 0.77 = 9.25 kg/kg of fuel

The flue gas constituents and its gravimetric & volumetric analysis is tabulated below :

Constituents of flue gases	Quantity, (m) kg / kg fuel	Weight % analysis	Molecular weight, (M)	$\frac{m}{M}$	Volume %, analysis
CO	0.403	3.3%	28	0.0144	$\frac{0.0144}{0.40044} \times 100 = 3.6\%$
CO_2	2.380	19.7%	44	0.0541	$\frac{0.0541}{0.40044} \times 100 = 13\%$

N_2	9.250	76.5%	28	0.3310	$\frac{0.3310}{0.40044} \times 100 = 82.66\%$
SO_2	0.060	0.5%	64	0.00094	$\frac{0.3310}{0.40044} \times 100 = 0.24\%$
Total	12.093	100%	—	0.4004	100%

ANS.

Example 78. *What is the dew point of the flue gas produced by burning of C_8H_{16} with 20% deficiency of air, if the total pressure is 1 atm. ?*

Solution. Combustion reaction for incomplete combustion of C_8H_{16} with 20% deficiency of air is as follows :

$$C_8H_{16} + 0.8 \times 12\,(0_2 + 3.76\,N_2) \rightarrow x\,CO + y\,CO_2 + 8H_2O + 361.1\,N_2$$

$$\text{or, } C_8H_{16} + 9.6\,O_2 + 36.1\,N_2 \quad \rightarrow x\,CO + y\,CO_2 + 8H_2O + 36.1\,N_2$$

The above combustion reaction is based on the assumption that all the hydrogen in the fuel is converted to water vapour and due to 20% deficiency of combustion air, carbon is converted to both CO and CO_2.

from carbon balance, $x+y = 9$ and from balance, $x/2 + y + 4 = 9.6$

Solving these equations, we get, $x = 4.8$ and $y = 3.2$

Hence, the combustion reaction becomes

$$C_8H_{16} + 9.6\,O_2 + 36.1\,N_2 \text{—>} 4.8\,CO + 3.2\,CO_2 + 8H_2O + 36.1\,N_2$$

The total moles of products of combustion = 4.8 + 3.2 + 8 + 36.1 = 52.1 out of which water vapor = 8 moles.

∴ The partial pressure of water vapor (8 moles) present in the flue gas

$$= \frac{8}{52.1} \times 1 \text{ atm} = 0.153 \text{ atm.}$$

The saturation temperature corresponding to 0.153 atm pressure = 55°C (from steam table) hence, the dew point of the flue gas = 55°C. **Ans.**

Example 79. *Ultimate analysis (weight %) of a coal on dry ash free (d.a.f) basis is :*

C = 89.3%, H = 5%, N = 1.5%, O = 3.8%, S = 0.8%. This coal as fired in a furnace contains 7% moisture & 8% ash and has a gross calorific value (G.C.V.) of 36000 kJ / kg. Calculate the:

(a) *Weight (kg) of theoretical air required to burn 100 kg of d.a.f. coal.*

(b) Volume of air required at NTP (Nm^3) to burn 100 kg of d.a.f. coal.

(c) Volume at NTP (Nm^3) of theoretical air required to burn 100 kg of as fired coal (containing 7% moisture & 8% ash).

(d) Volumetric analysis of dry theoretical flue gas at NTP (from combustion of 100 kg of as fired coal).

(e) Volumetric analysis of wet theoretical flue gas at NTP (from combustion of 100 kg of as fired coal)

(f) Net Calorific Value (N.C.V.) of as fired coal (containing 7% moisture & 8% ash).

(g) % of CO_2 & O_2 in flue gas produced by using 20% excess air for combustion.

(h) *% of CO_2 & O_2 in flue gas produced by using 40% excess air for combustion.*

(i) *% of excess air used for combustion of as fired coal, if the dry flue gas contains 12% CO_2.*

(j) *% of excess air used for combustion of 'a fired coal', if the dry flue gas contains 8% O_2.*

Solution. Basis : 100 kg of d.a.f. coal

Combustion reactions are as follows :

$$C + O_2 = CO_2,\ H_2 + \frac{1}{2}O_2 = H_2O,\ S + O_2 = SO_2$$

Theoretical oxygen & air requirement and theoretical flue gas quantity are calculated by combustion equation and tabulated below :

consiti-tuents	Quantity per 100 kg d.a.f coal, kg	kg. atom or kg. mole $= \frac{\text{weight}}{\text{mol. weight}}$	Theoretical O_2 requirement per 100 kg d.a.f coal, kg mole	Theoretical flue gas volume produced by combustion of 100 kg d.a.f coal in kg mole				
				CO_2	H_2O	SO_2	N_2	O_2
C	89.3	$\frac{89.3}{12} = 7.4$	7.4	7.4	—	—	—	—
H_2	5.0	$\frac{5}{2} = 2.5$	$\frac{2.5}{2} = 1.25$	—	1.25	—	—	—
N_2	1.5	$\frac{1.5}{28} = 0.053$	—	—	—	—	0.053	—
O_2	3.4	$\frac{3.4}{32} = 0.106$	–0.106	—	—	—	—	—
S	0.8	$\frac{0.8}{32} = 0.02^5$	0.025	—	—	—	0.025	—
Total	100	—	8.569	7.4	1.25	0.025	0.053	—

Weight of theoretical air required (kg) to burn 100 kg of d.a.f. coal:

Theoretical oxygen requirement = 8.569 kg mole = 8.569 × 32 = 274.2 kg

$\therefore$ Theoretical air requirement $= 274.2 \times \frac{100}{23} = 1196.5$ kg **Ans.**

[$\because$ Air contains 23% O_2 by weight & molecular weight of $O_2 = 32$]

Volume of theoretical air required at NTP (Nm^3) to burn 100 kg of d.a.f. *coal:*

Theoretical oxygen requirement = 8.569 kg mole = 8.569 x 22.4 = 191.945 Nm^3

[$\because$ 1 kg mole = 22.4 Nm^3]

$\therefore$ Theoretical air requirement $= 191.945 \times \frac{100}{21} =$ 914-70. Nm^3 **Ans.**

[$\because$ Air contains 21% 0_2 by **Volume**]

(c) Volume at NTP (Nm^3) of theoretical air required to burn 100 kg of as fired coal (containing 7% moisture & 8% ash) :

Volume of air required to burn 100 kg *of d.a.f.* coal = 914.70 Nm^3 (as found in *'b'*)

∴ Volume of air required to burn 100 kg of *as fired* coal

$$= 914.70 \frac{(100-7-8)}{100} = 777.0 \text{ Nm}^3$$ **Ans.**

[Since, as fired coal contains 7% moisture & 8% ash i.e. out of 100 kg d.a.f. coal, combustibles present in (100 – 7 – 8) = 85 kg will only undergo combustion, as a result, volume of air required for as fired coal (777.5 Nm^3) is less than that for d.a.f. coal (914.70 Nm^3)]

***(d)*Volumetric analysis of dry theoretical flue gas at NTP from combustion of 100 kg of as fired coal:**

From the table in *(a)* we get the theoretical flue gas volume produced by the combustion of 100 kg of d.a.f. in kg mole. The same for 100 kg of as fired coal has been calculated by multiplying the value by 85/100 = 0.85 because as fired coal contains 7% moisture + 8% ash = total 15% more incombustibles. kg mole has been converted into Nm^3 by multiplying by 22.4 since 1 kg mole = 22.4 Nm^3. Results are tabulated below.

Components of flue gas	kg moles present in d.a.f. (100 kg) coal combustion products	kg moles present in 100 kg of as fired coal combustion products	Nm^3 present in 100 kg of as fired coal combustion products	Volumetric % of components on dry basis
CO_2	7.4	7.4 × 0.85 = 6.29	6.29×22.4=140	$\frac{140}{751.5}$×100= 18.7%
S^2	0.025	0.025 × 0.85 = 0.0212	0.0212×22.4 = 0.5	$\frac{0.5}{751.5}$× 100 =0.07%
O_2	—	—	—	—
N_2 from coal	0.053	0.053 × 0.85 = 0.045	0.045×22.4= 1.0	—
N_2 from combustion air	—	—	777.0 × 0.79 = 610	—
Total N_2 in flue gas	—	—	610+1 = 611	$\frac{611}{751.5}$×100= 81.3%
Total	—	—	751.5	≈100%

Since in Orsat analysis, SO_2 is detected & reported as CO_2 hence the theoretical dry flue gas analysis is : CO_2 = 18.7% & N_2 = 81.3% (Total = 100%) **Ans.**

(e) Volumetric analysis of wet theoretical flue gas (from combustion of 100 kg of as 'fired' coal) :

The results as calculated in *'d'* and extended for calculation of water vapor present in flue gas resulting from moisture present in coal & that from combustion of H_2 present in coal, are tabulated below:

Components of flue gas	kg moles present in 100 kg of as fired coal combustion products	Nm³ present in 100 kg of as fired coal combustion products	Volumetric % of components on wet flue gas basis
CO_2	6.29	140	$\frac{140}{808} \times 100 = 17.4\%$
SO_2	0.0212	0.5	$\frac{0.5}{808} \times 100 = 0.06\%$
O_2	—	—	—
Total N_2 in flue gas	—	611	$\frac{611}{808} \times 100 = 75.6\%$
H_2O from 7% moisture present in coal	$\frac{7}{18} = 0.388$	$0.388 \times 22.4 = 8.7$	—
H_2O from combustion of H_2 in coal	$\frac{5}{2} \times 0.85 = 2.125$	$2.125 \times 22.4 = 47.6$	—
Total H_2O in flue gas	—	56.3	$\frac{56.3}{808} \times 100 = 6.94\%$
Total	—	808	100%

Ans.

(f) Net calorific value of as fired coal (containing 7% moisture & 8% ash) :

given, gross calorific volue (GCV) of d.a.f coal = 36000 kJ/kg

Hence, GCV of as fired coal = 36000 × 0.85 = 30700 kJ/kg

[Since 100 kg of d.a.f. coal = 85 kg of as fired coal]

Now, net calorific value (NCV) of $as\ fired\ coal = GCV - 2454$ (weight of water vapor in flue gas)

Now, weight of water vapor from combustion of hydrogen = $\frac{5}{2} \times 0.85 \times 18 = 38$ kg

[18 = Molecular weight of water]

and, weight of water from coal containing 7% moisture = 7 kg

Hence, total weight of water = 38 + 7 = 45 kg/100 kg of as fired coal = 0.45 kg/kg of as fired coal

$$\therefore \qquad NCV = 30700 - (2454 \times 0.45) = 29596 \text{ kJ/kg}$$ **Ans.**

[2454 = Heat of condensation of moisture at 15°C in kJ/kg = 587 kcal/kg]

(g) % of CO_2 & O_2 in flue gas produced by using 20% excess air for combustion of as fired coal:

As calculated earlier in *(c)* theoretical air (*i.e.* 0% excess air) required to burn 100 kg of as fired coal = 777 Nm³

As calculated in *(d)* CO_2% in theoretical flue gas

$$= \frac{\text{Volume of } CO_2}{\text{Total volume of flue gas}} = \frac{140}{751.5} \times 100 = 18.7\% \text{ and } O_2\% \text{ is zero obviously.}$$

With 20% excess air (i.e., excess air factor = 1.2).

Total volume of flue gas = 751.5 + 777 × 0.2 = 906.9 Nm^3

$$\% \, CO_2 \text{ in flue gas} = \frac{\text{Volume of } CO_2}{\text{Total volume of flue gas}} \times 100 = \frac{140}{906.9} \times 15.4\%$$

Volume of oxygen = 1.55 × 21 = 32.6 Nm^3

$$\text{Hence } \% \, O_2 \text{ in flue gas} = \frac{\text{Volume of oxygen} \times 100}{\text{total volume of flue gas}} = \frac{32.6}{906.9} \times 100 = 3.5\%$$

(h) % of CO_2 & O_2 in flue gas produced by using 40% excess air for combustion of as fired coal:

Total volume of flue gas = 751.5 + 777 × 0.4 = 1062.3 Nm^3

$$\% \, CO_2 \text{ in flue gas} = \frac{140}{1062.3} \times 100 = 13.2\%$$

Now, volume of oxygen = 3.1 × 21 = 65 Nm^3

$$\therefore \% \, O_2 \text{ in flue gas} = \frac{65}{1062.3} \times 100 = 6.1\%.$$

(i) % of excess air used for combustion of as fired coal, if dry flue gas contains 12% CO_2:

Dry flue gas from 100 kg of as fired coal contains (as found in *'e'*):

CO_2 = 140 Nm^3, SO_2 = 0.5 Nm^3, N_2 from coal = 1 Nm^3, N_2 from theoretical air = 610 Nm^3

Let, excess air = x Nm^3

∴ Total dry flue gas = (140 + 0.5 + 1 + 610 + x) Nm^3 = (751.5 + x) Nm^3

Since, SO_2 is reported as CO_2 in flue gas in Orsat analysis, then

CO_2 in flue gas = 140 + 0.5 = 140.5 Nm^3

$$\therefore \% \, CO_2 \text{ in flue gas} = 12 = \frac{140.5}{751.5 + x} \times 100$$

or, 12 (751.5 + x) = 140.5 ×100.

$$\therefore \% \text{ excess air} = \frac{\text{excess air quantity}}{\text{theoretical air (as calculated in 'c')}} \times 100$$

$$\% \text{ excess air} = \frac{419}{777} \times 100 = 54\% \text{ Ans.}$$

[N_2 in the fuel can be safely ignored, if it is less than 3% as it will not affect the final result too much]

(j) % excess air used for combustion of as fired coal, if dry flue gas contains 8% O_2:

Total dry flue gas $= (751.5 + x)\ Nm^3$ as found in *(i)*

$$\therefore \%\ O_2 \text{ in flue gas} = 8 = \frac{0.21\,x}{751.5+x} \times 100$$

or, $8\,(751.5 + x) = 21x$ or, $21\,x - 8\,x = 751.5 \times 8 = 6012$

or, $13\,x = 6012$ $\therefore$ $x = 6012/13 = 462\ Nm^3$

$$\therefore \%\ \text{excess air} = \frac{462}{777} \times 100 = 59.5\%$$ **Ans.**

Example 80. *An imported British coal as fired in a furnace analyses : C = 67.9%, H = 4.4%, S = 0.8%, N= 1.6%, O = 7.9%, ash =4.5%, and moisture = 12.9%. When burnt, the flue gas issuing out of the furnace at 545°C analyses on dry basis as : CO_2 = 14.5%, O_2 = 4.7%, N_2 = 80.8%, Other relevant data are :*

Gross C.V. of coal = 28000 Kj/ kg, Ambient temperature = 15°C

Specific heat of dry flue gas, C_p = 1.341 kJ/Nm³ °C

Specific heat of steam =1.92 kJ/kg °C

Heat of condensation of water vapor at 15°C = 2454 kJ/kg

Calculate the:

(a) Volume (Nm³) of actual air used for combustion per kg of as fired coal.

(b) sensible heat in the dry flue gas expressed as % of net C.V. of coal.

(c) heat loss in the combustion of hydrogen and in the moisture in the coal as kJ / kg of coal.

Solution. Basis 1 kg of coal

Constituents of coal	Quantity per kg of coal	kg atom or kg mole	Theoretical oxygen required, kg mole	Theoretical flue gas volume produced, Nm^3			
				CO_2	H_2O	SO_2	N_2
C	0.679	$\frac{0.679}{12} = 0.0565$	0.0565	0.0565×22.4 =1.265	—	—	—
H	0.044	$\frac{0.044}{2} = 0.022$	0.011	—	0.011×22.4 = 0.2464	—	—
S	0.008	$\frac{0.008}{32} =$ 0.00025	0.00025	—	—	0.00025×22.4 = 0.0056	—
N	0.016	$\frac{0.016}{28} = 0{,}0057$	—	—	—	—	0.00057 × 22.4= 0.0127
O	0.079	$\frac{0.079}{32} =$ 0.00246	-0.00246	—	—	—	—
Total	—	—	0.0653	—	—	—	—

$\because$ Total theoretical oxygen requirement = 0.0653 kg mole = 0.0653 × 22.4 = 1.46272 Nm^3

$\therefore$ theoretical air requirement = $\dfrac{1.46273 \times 100}{21}$ = 6.965 Nm^3

Now, dry flue gas from combustion of 1 kg of coal will contain

CO_2 = 1.265 Nm^3, SO_2 = 0.0056 Nm^3, N_2 coal = 0.0127 Nm^3

N_2 from theoretical air = 6.965 × 0.79 = 5.50 Nm^3

Let, excess air = x Nm^3

$\therefore$ Total flue gas volume = 1.265 + 0.0056 + 0.0127 + 5.50 + x = (6.783 + x) Nm^3

Total volume of (CO_2 + SO_2) in flue gas = 1.265 + 0.0056 = 1.27 Nm^3

(Since SO_2 is reported as CO_2 in Orsat-analysis)

By carbon dioxide balance,

$$\% \; CO_2 \text{ in dry flue gas} = \frac{\text{Volume of } CO_2}{\text{Total volume of flue gas}} \times 100$$

$$\text{or,} \quad 14.5 = \frac{1.27}{6.783 + x} \times 100$$

$$\text{or,} \quad 14.5\,(6.783 + x) = 1.27 \times 100 = 127$$

$$\text{or,} \quad 6.783 + x = \frac{1.27}{14.5} = 8.758$$

$$\text{or,} \quad x = 8.758 + 6.783 = 1.975 \text{ Nm}^3$$

(a) Hence, actual air used/kg of as fired coal = theoretical air + excess air

= 6.965 + 1.975 = 8.94 Nm^3 Ans.

(b) Sensible heat loss per kg of coal

= Volume of dry flue gas × specific heat × temperature difference

= (6.783 + 1.975) × 1.341 × (545 - 15) = 8.758 × 1.341 x 530 = 6224.5 kJ

Now, net CV of coal = gross CV - (2454 × total weight of moisture). Weight of water as moisture in 1 kg of coal = 12.9/100 = 0.129 kg

Water produced by combustion of H_2 present in coal = $\dfrac{2.2 \times 18}{100}$ = 0.2958 kg

Total weight of water = 0.129 + 0.2958 = 0.525 kg per kg of as fired coal

$\therefore$ Net C.V. = 28000 - (2454 × 0.525) = 26712 kJ/kg

$\therefore$ Sensible heat loss as % of net C.V. of coal = $\dfrac{6240}{26712}$ × 100 = 23.3%. **Ans.**

Heat loss due to combustion of hydrogen and moisture in coal will be summation of that due to the cooling of steam to 100°C and the latent heat of condensation & cooling of water to 15°C.

$\therefore$ Heat lost per kg of coal = 0.525 × (545 – 100) × 1.92 + 0.525 × 2454

= 347.7 + 1288.3 = 1636 kJ. Ans.

Example 81. *(a) Find theoretical / stoichiometric air / fuel ratio (AFR) for combustion of methane (CH_4).*

(b) If it is burnt in 150% excess of stoichiometric (theoretical air), then write the actual combustion equation and find Air / fuel ratio (AFR).

Solution : *(a)* Stoichiometric equation for combustion of CH_4 is

$$CH_4 + xO_2 + 3.76x.N_2 \rightarrow y\,CO_2 + z\,H_2O + 3.76\,x.N_2 \qquad ...(i)$$

where, x = molecule of oxygen required for theoretical combustion of methane.

y = molecules of CO_2 formed.

z = molecules of H_2O formed.

100 moles of air contains 79 moles of N_2 and 21 moles of air

$\therefore$ For each mole of O_2 in air, 79/21 = 3.76 moles of N_2 is present

$\therefore$ For x mole of O_2 used for combustion, x × 3.76 moles of nitrogen is present both in the reactant as well as product. N_2 being inert remains as such in the combustion product.

Equating the carbon, hydrogen & oxygen atoms respectively on both sides of the equation *(i)* I, we get

By carbon balance: $y = 1$ ...*(ii)*

By hydrogen balance: $2z = 4$ ...*(iii)*

By oxygen balance: $2x = 2y + z$ ...*(iv)*

Hence, $y = 1, z = 2$, and $2x = 2 \times 1 + 2 = 4 \backslash x = 4/2 = 2$

$\therefore$ Stoichiometric equation of combustion of methane is found by putting the value of x, y and z in I.

$$CH_4 + 2O_2 + 7.52\,N_2 = CO_2 + 2\,H_2O + 7.52\,N_2$$

AFR (air/fuel ratio) is the weight of air to the weight of fuel. It is given by

$$AFR = M_a.\,N_a/M_fN_f$$

where, M_a = Molecular weight of air (21% O_2 + 79% N_2) = 0.21 × 32 + 0.79 ×28 = 29

N_a = No. of moles of air = No. of moles of O_2 + No. of moles of N_2

M_f = Molecular weight of fuel.

N_a = No. of moles of fuel.

Molecular weight of methane CH_4 =12 + 1 × 4 = 1 6

$$\therefore \text{ Theoretical or stoichiometric AFR} = \frac{M_a.N_a}{M_f.N_f} = \frac{29\times(2+7.52)}{16\times1} = 17.25 \qquad \textbf{Ans.}$$

(b) If 150% of stoichiometric air is used, then the actual combustion equation will be

$$CH_4 + 2 \times 1.5\,O_2 + 2 \times 3.76 \times 1.5\,N_2 = CO_2 + 2H_2O + 2 \times 3.76 \times 1.5\,N_2$$

or, $$CH_4 + 3O_2 + 11.3\,N_2 = CO_2 + 2H_2O + 11.3\,N_2 \qquad ...(v) \text{ Ans.}$$

And, AFR = 17.25 × 1.5 = 25.875 **Ans.**

Example 82. *Find theoretical air for combustion of methyl mercaptans (CH_3SH).*

Solution : General formula of mercaptans is R-S-H where, R is alkyl group which is methyl group in this case.

Theoretical combustion equation is :

$$CH_4S + XO_2 + 3.76 , x.N_2 \rightarrow y\ CO_2 + zH_2O + k.SO_2 + 3.76.x.N_2 \quad ...(i)$$

Equating carbon, hydrogen sulphur and oxygen atoms respectively on both sides of the equations , we get

$$y = 1,\ 2z = 4 \quad \therefore\ z=2,\ k=1, 2x=2y + z + 2K = 2 \times 1 + 2 + 2 \times 1 = 6$$

or, $$x = 6/2 = 3$$

Hence, on putting $x = 3$, $y = 1$, $z = 2$, $k = 1$ in, *(i)* the stoichiometric combustion equation becomes

$$CH_4S + 3O_2 + 11.3N_2 = CO_2 + 2H_2O + SO_2 + 11.3N_2$$

$$\text{Stoichiometric Air/Fuel ratio (AFR)} = \frac{M_a.N_a}{M_f.N_f} = \frac{29(3+11.3)}{48\times1} = 8.63 \textbf{ Ans.}$$

[Molecular weight of fuel, $CH_4S = 12 + 1 \times 4 + 32 = 48$]

Example 83. *(a) Find AFR (Air/fuel ratio) for combustion of formaldehyde (HCHO), if 100% excess air is used for its combustion.*

(b) Find AFR for combustion of diethyl Ketone (C_2H_5. C=0. C_2H_5,) if 150% excess air is used for its combustion.

Solution : *(a)* Stoichmetric combustion equation is:

$CH_2O + xO_2 + 3.76.x.N_2$ □ $y\ CO_2 + z\ H_2O + 3.76.x.N_2$ ...*(i)*

Balancing carbon, hydrogen and oxygen atoms on both sides of equation *(i)*, we get

$y = 1$, $2z = 2$ ∴ $z = 1$ and $1 + 2x = 2y+z = 2 \times 1 + 1 = 3$ or, $2x = 3 - 1 = 2$ □$x = 1$,

putting, $x = 1$, $y = 1$ and $z =1$ in equating *(i)*, we get stoichiometric combustion equation as

$CH_2O + O_2 + 3.76\ N_2 = CO_2 + H_2O + 3.76\ N_2$

$$\therefore \quad \text{Stoichiometric AFR} = \frac{M_a.N_a}{M_f.N_f} = \frac{29(1+3.76)}{30\times1} = 4.60$$

[Molecular weight of $CH_2O = 30$]

If 100% excess air is used, then the total air used for combustion = 200%

∴ Actual AFR (with 100% excess air) = 4.60 × 2 = 9.20 **Ans.**

(b) Stoichiometric combustion equation is for combustion of diethyl ketone ($C_5H_{10}O$) is

$$C_5H_{10}O + x.O_2 + 3.76.x.N_2 \rightarrow CO_2 + z\ H_2O + 3.76\ N_2 \quad ...(i)$$

Equating the carbon, hydrogen and oxygen atoms, respectively we get

$y = 5$, $2z = 10$ ∴ $z = 5$ and $2x + 1 = 2y + z = 2 \times 5 + 5 = 15$, or, $2x = 15 - 1 = 14$ ∴ $x = 7$

Putting, $x = 7$, $y = 5$, $z = 5$ in *(i)*, we get the stoichiometric combustion equation of diethyl ketone as:

$$C_5H_{10}O + 7O_2 + 26.3\ N_2 = 5CO_2 + 5H_2O + 26.3\ N_2$$

$$\therefore \quad \text{Stoichiometric AFR} = \frac{M_f.N_f}{M_a.N_a} = \frac{29\,(7+26.3)}{86\times 1} = 11.22$$

(Molecular weight of $C_5H_{10}O = 86$)

If 150% excess air is used, the total air used is 100 + 150 = 250%, then

Actual AFR = 11.22 x 2.5 = 28.05 **Ans.**

Example 84. *Anthracite coal analyses, : C = 90%, H_2 = 3%, O_2 = 2.5%, N_2 = 1%, S = 0.5% & ash = 3%. Find dry and wet analysis of flue gas, if 20% excess air is used for its combustion.*

Solution : Basis 1kg of coal

Constituent of coal	Weight, kg	Combustion equation	Theoretical O_z required, kg	Flue gas produced, kg
C	0.90	$C + O_2 = CO_2$	0.9 × 32/12 = 2.4	$CO_2 \rightarrow 0.9 \times 44/12 = 3.3$
H_2	0.030	$H_2 + \frac{1}{2} O_2 = H_2O$	0.03 × 16/2 = 0.24	$H_2O \rightarrow 0.03 \times 18/2 = 0.27$
O_2	0.025	—	– 0.025	—
N_2	0.010	$S + O_2 = SO_2$	—	$N_2 \rightarrow 0.010$
S	0.005	—	0.005 × 32/32 = 0.005	$SO_2 \rightarrow 0.005 \times 104/32 = 0.01$
Ash	0.030	—		
Total	1.000	—	2.620	—

Theoretical O_2 required/kg of coal = 2.620 kg

∴ Theoretical air required/kg of coal = 2.620 × 100/23 = 11.25 kg (Air contains 23% O_2 by weight)

∴ Theoretical AFR (Air/fuel ratio) = 11.25 kg air/kg coal

With 20% excess air, actual AFR = 11.25 x 1.2 = 13.5 kg air/kg coal

Actual O_2 supplied = 2.62 × 1.20 = 3.144 kg/kg coal

Actual N_2 supplied = 0.01 (from coal) + 3.144 × $\frac{77}{23}$ (from air) = 10.36 kg/kg coal

(Air contains 77% N_2 and 23% O_2 by weight)

Hence, total air supplied = 3.144 + 10.36 = 13.5 kg/coal

 Flue gas produced with 20% excess air are:

(a) O_2 = 3.144 - 2.620 = 0.524 kg/kg coal

(b) N_2 = 10.36 kg/kg coal

All the constituents of wet and dry flue gas compositions are tabulated below:

Flue gas components	kg /kg coal	% by weight	Molecular weight	Moles /kg coal	% Volume, wet analysis	% volume, dry analysis
CO_2	3.3	22.80	44	0.075	15.74	16.30
H_2O	0.27	1.87	18	0.015	3.16	–
N_2	10.36	71.67	28	0.370	77.80	80.30
O_2	0.524	3.60	32	0.0162	3.40	3.51
SO_2	0.01	0.07	64	0.0002	0.03	0.03
Total	14.646	100	—	0.4764 $-H_2O = 0.0150$ DRY = 0.4614	100	100

Example 85. *A coal gas analysis is : $H_2 = 49.4\%$, $CO = 18\%$, $CH_4 = 20\%$, $C_4H_8 = 2\%$, $O_2 = 0.4\%$, $N_2 = 6.2\%$ and $CO_2 = 4\%$*

(a) Find Air fuel ratio (AFR) on volume basis if, 20% weak mixture is employed for the combustion of fuel.

(b) Also, find the flue gas analysis on both dry and wet basis.

Solution : Basis of calculation-l kg mole of coal gas

Theoretical O_2 required and theoretical (stoichiometric) flue gas produced is calculated below in the table.

Constituents	kg mole / kg mole of gas	Combustion equation	Theoretical O_2 required, kg moles	Products of combustion (Flue gas) kg moles	
				CO_2	H_2O
H_2	0.494	$H_2 + 1/2O_2 = H_2O$	0.247	—	0.494
CO	0.180	$CO + 1/2O_2 = CO_2$	0.090	0.18	—
CH_4	0.200	$CH_4 + 2O_2 = CO_2 + 2H_2O$	0.400	0.20	0.400
C_4H_8	0.020	$C_2H_8 + 6O_2 = 4CO_2 + 4H_2O$	0.120	0.08	0.080
O_2	0.004	—	– 0.004	—	—
N_2	0.062	—	—	—	—
CO_2	0.040	—	—	0.04	—
Total	1.000	—	0.853	0.50	0.974

∴ Theoretical O_2 required for complete combustion = 0.853 kg mole

∴ Theoretical air required for complete combustion = 0.853/0.21 = 4.06 kg moles/kg mole of coal gas

∴ Theoretical AFR (on volume basis) = 4.06

∴ Theoretical AFR (on weight basis)

$$= 4.06 \times 29/(0.494 \times 2 + 0.18 \times 28 + 0.2 \times 16 + 0.02 \times 56 + 0.004 \times 32 + 0.062 \times 28 + 0.04 \times 44)$$

$$= 8.4$$

Mixture strength $= \dfrac{\text{Theoretical AFR}}{\text{Actual AFR}} = \dfrac{1}{1.2}$ (in this case as 20% weak mixture is employed for combustion)

Mixture means the mixture of fuel gas and combustion air. Higher mixture strength (> 100%) cause incomplete combustion resulting in CO formation in flue gas because actual air supplied is less than the theoretical air. Normally boilers operate with 20% weak mixture ; gas turbines at 300% weak mixture & there is a wide variation in the operation of petrol engines. If excess air is more, nitrogen of air react with excess oxygen forming NO_x emissions. Hence, optimum mixture strength should be provided to minimise emission of air pollutants. Weak mixture means actual combustion air supplied is in excess of theoretical air, whereas rich mixture (> 100%, mixture strength) means actual air supplied is less than the theoretical air requirement.

∴ Actual AFR (on volume basis) = 4.06 × 1.2 = 4.872 kg moles

∴ Excess oxygen = 0.853 × 0.2 = 0.1706 kg mole (20% weak mixture used)

∴ Oxygen present in flue gas = 0.1706 kg mole

N_2 present in flue gas = N_2 present in coal gas + N_2 present in air = 0.062 + 4.872 × 0.79 = 3.911 kg mole.

Hence, dry and wet flue gas analysis is as computed and tabulated below:

Combustion products	kg moles	% by volume, dry	% by volume, weet
CO_2	0.500	10.90	9.00
H_2O	0.974	-	17.50
O_2	0.1706	3.70	3.10
N_2	3.911	85.40	70.40
Total wet	5.557	100	100
H_2O moles	– 0.974		
Total dry	4.583		

Example 86. *Find AFR (Air / fuel ratio) and flue gas analysis (both on dry and wet basis), if (a) 90% and (b) 120% mixture strengths are used for the combustion of ethyl alcohol (C_2H_5OH) in a motor vehicle.*

Solution : Basis of calculation - 1 kg of ethyl alcohol (C_2H_6O)

Stoichiometric/theoretical combustion equation for C_2H_6O is:

$$C_2H_6O + 3O_2 + 3 \times 3.76\ N_2 \rightarrow 2\ CO_2 + 3H_2O + 3 \times 3.76\ N_2$$

or
$$C_2H_60 + 3O_2 + 11.28\ N_2 = 2CO_2 + 3\ H_2O + 11.28\ N_2 \qquad ...(i)$$

AFR (theoretical)

$$\frac{M_a . N_a}{M_f . N_f} = \frac{\text{Molecular weight of air(No. of moles is } O_2 + N_2)}{\text{Molecular weight of fuel } (C_2H_6o) \times \text{No. of moles of fuel (ethyl alcohol)}} =$$

$$\therefore \quad \text{AFR (theoretical)} = \frac{29(3+11.28)}{46 \times 1} = 8.96 \text{ kg air/kg of } C_2H_6O$$

(a) Mixture strength $= \dfrac{\text{Theoretical AFR}}{0.9} = 0.9$ (90% mixture strength means (100/90) × 100=11% weak mixture)

$$\therefore \quad \text{Actual AFR} = \frac{\text{Theoretical AFR}}{0.9} = \frac{8.96}{0.9} = \quad = 9.95$$

Thus, the ethyl alcohol/air mixture is weak and (9.95/8.96) x 100 = 11% excess air has been used.

∴ Actual combustion equation (with 11% excess air than theoretical air) is :

$$C_2H_6O + 1.11\,(3O_2 + 11.28\,N_2) \rightarrow 2CO_2 + 3H_2O + 0.33\,O_2 + 1.11 \times 11.28\,N_2$$

or, $$C_2H_6O + 3.33\,O_2 + 12.52\,N_2 = 2CO_2 + 3H_2O + 0.33\,O_2 + 12.52\,N_2$$

∴ Total no. of kg moles of products of combustion (flue gas) = 2 + 3 + 0.33 + 12.52 = 17.87

Hence, flue gas analysis on wet and dry basis is computed & tabulated below :

Combustion products	No. of kg moles	% wet, analysis	% dry, analysis
CO_2	2	(2/17.87) x 100 = 11.20	(2/14.87) x 100 = 13.40
O_2	0.33	(0.33/17.87) x 100 = 1.80	(0.33/14.87) x 100 = 2.20
H_2O	3	(3/17.87) x 100 = 16.80	—
N_2	12.52	(12.54/17.87) x 100 = 70.20	(12.54/14.87) x 1000 = 84.40
Total wet	17.87	Total = 100%	Total = 100%
Total H_2O	-3.00		
Total dry	14.87		

(b) In this case mixture strength $= \dfrac{\text{Theoretical AFR}}{\text{Actual AFR}} = 120\% = 1.2$

∴ Actual AFR = 8.96/1.20 = 8.96 × 0.834 = 7.47 kg moles

Thus combustion air availability is only 83.4% [(1/1.20) × 100 = 83.4%] of the theoretically required quantity resulting in incomplete combustion of alcohol leading to production of CO also in flue gas.

Theoretical combustion equation is :

$$C_2H_6O + 3O_2 + 11.28N_2 = 2CO_2 + 3H_2O + 11.28N_2$$

But actual combustion equation (with 83.4% of theoretical air) will be (H_2O is the first product to be formed as hydrogen has greater affinity for oxygen than carbon followed by formation of CO and CO_2) :

$$C_2H_6O + 0.834\,(3O_2 + 11.28N_2\,) = CO_2 + x.\,CO + 3H_2O + 0.834 \times 11.28\,N_2 \quad ...(ii)$$

by equating carbon atoms on both sides of the equation *(ii)*, we get, $1 + 0.834 \times 3 \times 2 = 2x + y + 3$

or, $$2x + y = 1 + 5.004 - 3 = 0.004 \quad ...(iii)$$

From, *(iii)* and *(iv)* we get, $x = 1.004$ and $y = 0.996$

Hence, the actual combustion equation becomes (on putting the value of x and y in *(ii)*; $C_2H_6O + 2.502O_2 + 9.41N_2 = 1.004CO_2 + 0.996CO + 3\ H_2O + 9.41N_2$ ...*(iv)*

$\therefore$ Final combustion products are = 1.004 moles CO_2 + 0.996 moles CO + 3 moles H_2O + 9.41 moles N_2 = Total 14.41 moles)

Hence, the flue gas analysis (both wet and dry basis) is calculated and tabulated below :

Combustion products	No. of kg moles	% wet flue gas analysis	% dry, flue gas analysis
CO_2	1.004	(1.004/14.41) × 100 = 6.97	(1.004/11.41) × 100 = 8.8
CO	0.996	(0.996/14.41) ×100 = 6.90	(0.996/11.41) × 100 = 8.7
H_2O	3.00	(3.0/14.41) × 100 = 20.83	—
N_2	9.41	(9.41/14.41) × 100 = 65.30	(9.41/11.41) x 100 = 82.5
Total wet Total H_2O	14.41 –3.00	Total = 100%	Total = 100%
Total dry	11.41		

PROBLEMS ON INCOMPLETE COMBUSTION OF FUELS

Example 87. *Volumetric flue gas analysis (dry basis) from incomplete combustion of methane (CH_4) is :*

$$CO_2 = 10\%,\ O_z = 2.37\%\ O_2,\ CO = 0.53\% \text{ and } N_2 = 87.10\%$$

Find (a) Air fuel ratio (both molal and weight basis) and (b) % theoretical air employed for combustion

Solution : Basis 1kg mole of fuel (CH_4)

(a) Let, the combustion equation be

$x.\ CH_4 + y.\ O_2 + 3.76.\ y.\ N_2 \rightarrow 10\ CO_2 + 0.53\ CO + 2.37\ O_2 + z\ H_2O + 87.1\ N_2$ ...*(i)*

Equating carbon, hydrogen, oxygen and nitrogen atoms on both sides of equation *(i)*, we get $x = 10 + 0.53 = 10.53$, $2y = 10 \times 2 + 0.53 + 2.37 \times 2 + z$ or, $2y - z = 25.27$

$$4x = 2z \text{ or, } z = 2x = 2 \times 10.53 = 21.06$$

$\therefore$ $2y - z = 25.27$ or, $2y - 21.06 = 25.27$ or, $2y = 25.27 + 21.06 = 46.33$

or, $y = 46.33/2 = 23.16$

hence, $x = 10.53$, $y = 23.16$ and $z = 21.06$

Putting values of x, y and z in (i), we get the actual combustion equation as

$$10.53CH_4 + 23.160_2 + 3.76 \times 23.16N_2 = 10CO_2 + 0.53CO + 2.37\ O_2 + 21.06H_2O + 87.1\ N_2$$

or, $$10.53CH_4 + 23.160_2 + 87.1N_2 = 10CO_2 + 0.53CO + 2.37O_2 + 21.06H_2O + 87.1N_2$$

Hence, actual combustion equation per kg mole of fuel (*i.e.*, CH_4) can be written as (by dividing both sides by 10.53)

$$CH_4 + 2.2O_2 + 8.27N_2 = 0.95CO_2 + 0.0500 + 0.0500 + 2H_2O + 0.225O_2 + 8.27N_2 \quad ...(ii)$$

$\therefore$ Air fuel ratio (AFR) on molal basis

$$= \frac{\text{Moles of } O_2 + \text{moles of } N_2}{\text{Moles of fuel}} = \frac{\text{Moles of air}}{\text{Moles of } CH_4} = \frac{2.2 + 8.27}{1} = 10.47$$

i.e., AFR on mole basis = 10.47 kg moles air/kg mole of methane

$$\therefore \text{ AFR on weight basis} = \frac{M_a.M_a}{M_f.M_f} = \frac{29 \times 10.47}{16 \times 1} = 18.97 \text{ kg of air/kg of methane} \quad \textbf{Ans.}$$

[Molecular weight of methane (CH_4) = 12 + 1 × 4 = 16]

(b) The theoretical combustion equation for methane is

$$CH_4 + 2O_2 + 2 \times 3.76N_2 = CO_2 + 2H_2O + 2 \times 3.76N_2$$

or, $$CH_4 + 2O_2 + 7.52N_2 = CO_2 + 2H_2O + 7.52N_2$$

∴ Theoretical AFR (on weight basis)

$$= \frac{M_a.M_a}{M_f.M_f} = \frac{(2+7.52)}{16 \times 1} = 17.23 \text{ kg of air / kg or methane}$$

hence, % by theoretical air employed = $\frac{18.97}{17.23} \times 100 = 110\%$. **Ans.**

EXAMPLE 88. *Volumetric flue gas analysis (dry basis) from the combustion of a fuel of unknown composition is :*

$$CO_2 = 8\%,\ CO = 0.9\%, O_2 = 8.8\%, \text{ and } N_2 = 82.3\%$$

Find (a) Fuel composition on mass basis.

(a) Air fuel ratio (AFR) on weight basis.

(b) % of theoretical air employed for combustion

Solution : *(a)* Since, the combustion products contains CO_2, CO, O_2 and N_2 the actual combustion equation can be written as:

$$C_xH_y + dO_2 + 3.76 \times d \times N_2 \rightarrow 8CO_2 + 0.9CO + 8.80_2 + ZH_2O + 82.3N_2 \quad ...1$$

Equating the nitrogen, oxygen, carbon and hydrogen atoms on both sides of the equation,we get

By nitrogen balance, we have

$$(3.76 \times d) \times 2 = 82.3 \times 2 \text{ or, } d = \frac{82.3 \times 2}{3.76 \times 2} = 21.9$$

By oxygen balance, we have,

$$d \times 2, = 8 \times 2 + 0.9 \times 1 + 8.8 \times 2 + z \times 1 \text{ or, } 2d = 16 + 0.9 + 17.6 + z$$

or, $$2 \times 21.9 = 34.5 + Z \text{ or, } Z = (2 \times 21.9) - 34.5 = 9.3$$

By carbon balance, we have

$$x = 8 + 0.9 = 8.9$$

By hydrogen balance, we have

$$y = 2Z \text{ or, } y = 2 \times 9.3 = 18.6$$

Hence, composition of fuel can thus be written as $C_{8.9}H_{18.6}$. **Ans.**

(b) Theoretical combustion equation of the fuel $C_{8.9}H_{18.6}$ is

$$C_{8.9}H_{18.6} + 13.5O_2 + 13.5 \times 3,76N_2 \rightarrow 8.9CO_2 + 9.3H_2O + 13.5 \times 3.76N_2$$

or, $$C_{8.9}H_{18.6} + 13.5O_2 + 50.8\,N_2 = 8.9\,CO_2 + 9.3\,H_2O + 50.8\,N_2$$

$$\therefore \text{ Theoretical AFR} = \frac{M_a.N_a}{M_f.N_f} = \frac{29\,(13.5+50.8)}{(8.9\times12+18.6\times1)\times1} = 14.9 \text{ kg air/ kg of fuel}$$

However, the actual combustion products contain CO_2, CO, O_2 and N_2. Hence, the actual combustion equation is

$$C_{8.9}H_{18.6} + 21.9O_2 + 21.9 \times 3.76N_2 \rightarrow 8CO_2 + 0.9CO + 8.8O_2 + 9.3\,H_2O + 21.9 \times 3.76N_2$$

or, $$C_{8.9}H_{18.6} + 21.9O_2 + 82.3N_2 = 8CO_2 + 0.9CO + 8.8O_2 + 9.3H_2O + 82.3N_2$$

$$\therefore \quad \text{Actual AFR} = \frac{M_a.N_a}{M_f.N_f} = \frac{29\,(21.9+82.3)}{(8.9\times12+18.6\times1)\times1} = 24.1 \text{ kg air/kg of fuel.} \quad \textbf{Ans.}$$

(c) Hence, % of theoretical air employed for combustion = $\frac{24.1}{14.9} \times 100 = 162\%$. **Ans.**

Example 89. *Volumetric analysis of a fuel gas in as follows :*

$$H_2 = 48\%,\ CH_4 = 26\%,\ CO_2 = 11\%,\ CO = 5\%,\ N_2 = 10\%$$

Its dry flue gas analysis (volumetric) is as follows :

$$CO_2 = 8.8\%,\ O_2 = 5.5\%,\ N_2 = 85.7\%$$

Find out AFR (Air fuel ratio) on volume basis.

Solution: Basis –100 kg moles of fuel gas

Combustion equation for 100 kg moles of fuel gas can be written as (from its analysis)

$$48H_2 + 26CH_4 + 11CO_2 + 5CO + 10N_2 + x\,[O_2 + 3.76N_2] = v.CO_2 + xH_2O + y.\,O_2 + z.N_2 \quad ...(i)$$

[For 1 kg mole of O_2, there are 3.76 kg moles (79/21 = 3.76) of N_2 in air]

Equating the co-efficients of carbon, hydrogen, oxygen and nitrogen on both sides of combustion equation *(i)*, we get

By carbon balance : $V = 26 + 11 + 5 = 42$ ...*(ii)*

By hydrogen balance, $2x = 48 \times 2 + 26 \times 4$, or, $x = 48 + 52 = 100$ ...*(iii)*

By oxygen balance, $2V + x + 2y = 22 + 5 + 10 \times 2 + 2x$

or, $2v + 2y - x = 47$ ...*(iv)*

By nitrogen balance, $10 \times 2 + x \times 3.76 \times 2 = Z$ or, $20 + 100 \times 3.76 \times 2 = 2Z$

or, $2Z = 20 + 752 = 772$ or, $Z = 772/2 = 386$ ...*(v)*

Putting the value of *v* & *x*, we get

$2 \times 42 + 2y - 100 = 47$ or, $2y = 47 + 100 - 84 = 63$ or, $y = 63/2 = 36.5$ ...*(vi)*

Hence on putting $v = 42$, $x = 100$, $y = 36.5$ and $z = 386$ in equation *(i)*, the actual combustion equation is

$$48H_2 + 26CH_4 + 11CO_2 + 5CO + 10N_2 + 100O_2 + 3.76 \times 100N_2$$
$$\rightarrow 42CO_z + 100H_2O + 36.50_2 + 386N_2$$

or,
$$48H_2 + 26CH_4 + 11CO_2 + 5CO + 100O_2 + 386N_2$$
$$= 42CO_2 + 100H_2O + 36.5O_2 + 386N_2 \quad ...(vii$$

$$\text{Actual AFR} = \frac{\text{Total moles of air}}{\text{Total moles of fuel gas}} = \frac{\text{Moles of }(O_2 + N_2)}{100} = \frac{100 + 376}{100} = 4.86$$

Hence, AFR = 4.8. **Ans.**

Example 90. *% Gravimetric analysis (by weight) of a fuel is :*

$C = 82\%$, $H_2 = 10\%$, $S = 3\%$, $O_2 = 2.5\%$, Ash = 2.5%

For an AFR (air fuel ratio by weight) of 12, calculate

(a) fuel-air mixture strength as a percentage rich or lean

(b) dry volumetric flue gas analysis

Solution : basis 100 kg of fuel *(.a)*

Constituents	Weight, kg	Combustion equation	Theoretical O_2 required kg
C	82	$\underset{12}{C} + \underset{32}{O_2} = \underset{44}{CO_2}$	$\frac{18}{12} \times 32 = 218.6$
H_2	10	$H_2 + \frac{1}{2}O_2 = H_2O$	$\frac{10}{2} \times 16 = 80$
S	3	$\underset{32}{S} + \underset{32}{O} = \underset{64}{SO_2}$	$\frac{10}{2} \times 32 = 3$
O_2	2.5	—	–2.5
Ash	2.5	—	—
Total	100	—	299.1

∴ Theoretical oxygen required = 299.1 kg/100 kg fuel

So, theoretical air required = $299.1 \times \frac{100}{23}$ = 1300 kg air/100 kg fuel (Air contains 23% O_2 by weight)

$$\therefore \quad \text{Theoretical AFR (Air fuel ratio)} = \frac{1300}{100} = 13$$

$$\text{But actual ratio} = \frac{\text{Theoretical AFR}}{\text{Actual AFR}} = \frac{13}{12} = 1.083$$

It means the fuel air mixture is 8.3% rich. **Ans.**

(That means actual air supplied is less than the theoretical air and hence, flue gas produced will contain CO also a product of incomplete combustion of carbon along with CO_2)

(b) Basis - 1 kg of fuel

Now, deficiency of combustion air = 13 – 12 = 1 kg air/kg of fuel

Weight of air saved by burning 1 kg carbon to CO instead of CO_2

$$= \frac{16}{12} \times \frac{100}{23} = 5.7971 \text{ kg}$$

Hence, 1 kg of air will be saved when 1/5.791 = 0.1725 kg of carbon burns to CO.

$$\therefore \qquad \text{CO formed} = 0.1725 \times \frac{28}{12} = 0.403 \text{ kg/kg fuel}$$

[Molecular weight of CO = 28, and that of carbon = 12]

$$CO_2 \text{ formed} = (0.82 - 0.1725) \times \frac{44}{12} = 2.38 \text{ kg/kg fuel}$$

[∴ 1 kg fuel contains 0.82 kg carbon out of which 0.1725 kg C burns to CO, hence only (0.82 - 0.1725) kg C burns to CO_2.]

$$\text{Actual } N_2 \text{ supplied from 12 kg of air/kg of fuel} = 12 \times \frac{77}{100} = 9.25 \text{ kg/kg of fuel}$$

[∴ Since, Air contains 77% N_2 by weight]

$$SO_2 \text{ formed} = 0.03 \times \frac{64}{32} = 0.060 \text{ kg/kg of fuel}$$

∴ Dry flue gas volumetric analysis is tabulated below.

Flue gas constituents	Weight, kg/kg fuel, m	Molecular weight, M	% volume $= \frac{m/M}{\sum m/m} \times 100$
CO	0.403	28	3.6
CO_2	2.380	44	13.5
N_2	9.250	28	82.67
SO_2	0.060	64	0.23
		Total	100%

Example 91. *A furnace is heated by burning a gaseous fuel of composition: 29% CO, 9% CO_2, 16% H_2 and 46% N_2 with dry air. The Orsat analysis of products of combustion is 15% CO_2, 7% O_2 and 78%, N_2. Calculate the following on the basis of 1 m^3 of gaseous fu e l:*

(a) Volume of flue gas at NTP, STP and at 1000°C (NTP : 0°C, 1 atm, STP : 25°C, 1atm)

(b) Volume of air at NTP

(c) Percent excess air

(d) % H_2O in flue gas

Solution : *(a)* Volume of flue gas

Consider 1 m^3 of gaseous fuel at NTP (1 atm. and 273 K). Let Y m^3 flue gas at 1 atm, 273 K.

By Carbon balance; we have

Carbon in CO + Carbon in CO_2 = Carbon in flue gas

$$0.29 + 0.09 = 0.15Y,$$

$$\therefore \quad Y = 2.53 \text{ m}^3 \text{ at NTP (1 atm and 273 K)} = 2.53 \text{ Nm}^3$$

$$Y = 2.53 \times 298/273 = 2.765 \text{ m}^3 \text{ at STP (1 atm and 298 K)} = 2.765 \text{ Sm}^3$$

$$Y = 2.53 \times 1273/273 \text{ m}^3 \text{ at (1 atm and 1273 K)}$$

Note the increase in volume of flue gas at 1273 K, which is around 5 times that at 273 K. This knowledge is important in designing combustion chamber for gaseous fuels.

***(b)* Volume of air at NTP**

Let Z m^3 is the volume of air required at NTP. By nitrogen balance, we get

Nitrogen from air + Nitrogen in gaseous fuel = Nitrogen in flue gas

$$\therefore \quad 0.79Z + 0.46 = 0.78 \times 2.53$$

$$\therefore \quad Z = 1.916 \text{ m}^3 \text{ at NTP} = 1.916 \text{ Nm}^3$$

***(c)* Percent excess air**

In order to calculate percent excess air, first we have to calculate theoretical air. Theoretical air is the air required for complete combustion of the combustibles present in gaseous fuel as per following reactions :

$$CO + 0.5O_2 = CO_2 \quad \text{...(1)}$$

$$H_2 + 0.5O_2 = H_2O \quad \text{...(2)}$$

We note that both CO and H_2 require 0.5 mole of oxygen, Hence

Theoretical amount of air

$$= \left(\frac{29 + 0.16}{2}\right) \times \frac{100}{21} = \frac{0.45}{2} \times \frac{100}{21} = 1.071 \text{ m}^3 \text{ at NTP} = 1.071 \text{ Nm}^3$$

Excess air in % = 100 (Actual air - Theoretical air)/Theoretical air = 100(1.916 - 1.071)/ 1.071 = 78.89%

Alternatively excess air can also be calculated from excess oxygen in flue gas, since theoretical oxygen in flue gas will appear only when it is in excess than theoretical in this problem.

Excess air in % = 100(0.07 × 1.916) / (0.21 × 1.071) = 78.74%.

The slight difference in excess air is due to rounding off.

***(d)* % H_2O in flue gas**

From the reaction 2 we get straightway that a mole of hydrogen gives 1 mole of water, hence Percent H_2O in flue gas = (100 × 0.16) / (2.53 + 0.16) = 5.95%

[Since, H_2 = 0.16 Nm^3 in 1 Nm^3 gaseous fuel, volume of flue gas at NTP = 2.53 Nm^3]

Example 92. *A billet heating furnace is heated by burning coal with air. The ultimate analysis of coal is; C = 72.5% , H = 7.5%, O = 6.7%, N = 1.3%, S = 2.5% and ash, = 9.5%. The moisture content of coal, when fired is 3%. The air is moist with the relative humidity (RH) of 45%. The barometer shows 735 mm Hg pressure and 20°C temperature. The water vapour pressure at saturation is 17.54 mm Hg. The Orsat analysis of products of combustion is : 10.3%* CO_2*, 6.2%* O_2*, 2% CO and 81.5%* N_2*. Calculate the following on the basis of 1 kg of coal:*

(i) Percent excess air

(ii) Volume of air at NTP

(iii) Total volume of POC. (Product Of Combustion)

Solution : (i) Percent excess air : Let V is flue gas volume at NTP (1 atm and 273K). Performing carbon balance gives the flue gas volume. Note that coal contains sulphur. Orsat analysis of the flue gas reports both CO_2 and SO_2. Therefore carbon balance here is

Carbon from coal + Sulphur from coal = Carbon flue gas

The flue gas volume is = 11.142 m^3

Now we can calculate actual amount of air that is used in the combustion process by performing nitrogen balance; Let Z kg mole is the amount of nitrogen derived from air.

∴ Nitrogen in coal + Nitrogen in air = Nitrogen in POC. Substituting the values we get, Z = 0.405 kg mole, hence oxygen would be 0.108 kg mole. Now calculate theoretical (stoichiometric) amount of air basis of complete combustion; as per the following combustion reactions :

$$C + \frac{1}{2} O_2 = CO_2$$

$$H_2 + \frac{1}{2} O_2 = H_2O$$

$$S + O_2 = SO_2$$

Theoretical amount of oxygen = 0.07785 kg moles

Percent excess air = 100(Actual oxygen - theoretical oxygen)/ theoretical oxygen. Substituting the values we get Percent excess air = 38.33%

***(ii)* Volume of air at NTP.** Note that the air is moist, so we have to find the composition of moist air by using Dalton's law which is

$$P_{N2} + P_{O2} + P_{H_2O} = 735 \text{ mm Hg} \qquad (P = \text{partial pressure})$$

P_{H_2O} = RH × vapour pressure of water vapour at saturation, we get

$$P_{N2} + P_{O_2} = 574.41$$

Volume of air = 0.405 × 760 / 574.41 = 0.536 kg mole

Volume of air = 13.315 at 20°C and 735 mm Hg or 11.998 at 0°C and 760 mm Hg (11.998 Nm^3)

***(iii)* Total volume of flue gas.** Total volume of flue gas = Dry volume flue gas + H_2O formed by reaction + Moisture of coal + Moisture in air

On substituting the value, we get

Total volume of flue gas = 12.148 m^3 at 0°C and 760 mm Hg and 13.482 m^3 at 20°C and 735 mm Hg.

UNSOLVED PROBLEMS ON COMBUSTION CALCULATIONS (FOR PRACTICE)[SOLID FUEL]

1. Analysis of an air dried coal is as follows :

$C = 60.5\%$, $H_2 = 3\%$, $N_2 = 1.5\%$, $S = 0.5\%$, $S = 0.5\%$, $O_2 = 9.5\%$, $H_2O = 5\%$, ash = 20%

A furnace is fired using this coal at the rate of 200 kg/hr. The dry refuse contains 20% combustibles. Analysis of flue gas (on dry basis) is as follows :

$$CO_2 = 13.5\%, 0_2 = 5.5\%, CO = 1\%, N_2 = 80\%$$

Determine :

(a) hourly heat loss due to unburnt combustibles (kcal/hr)

(b) air blower capacity (m^3/minute)

(c) volumetric flow rate of flue gas at 250°C and 740 mm Hg (m^3/hr)

2. Calculate the Orsat analysis of flue gas (on dry basis) of a coal fired furnace. The coal analysis is as follows :

$C = 58.5\%$, $H_2 = 3.5\%$, $N_2 = 1.5\%$, $S = 0.5\%$, $0_2 = 8.5\%$, Moisture = 8.0%, Ash = 19.5% Coal burning rate = 3 tons/hr and rate of flue gas generation = 52,000 m^3/hr at 250°C.

3. Calculate the complete flue gas orsat analysis and air/coal ratio (by weight), when coal of following analysis : $C = 80\%$, $H_2 = 5\%$ is burnt in air. The flue gas Orsat analysis contains : $CO_2 = 12.5\%$, $O_2 = 3.5\%$, carbon monoxide is suspected in the flue gas.

4. Analysis of a boiler coal is as follows :

$$C = 80\%, H_2 = 4.5\%, O_2 = 5.5\%, \text{Incombustibles} = 10\%$$

Volumetric analysis of flue gas produced by burning it, is as follows :

$$CO_2 = 10\%, O_2 = 5.0\%, CO = 1.5\%, N_2 = 83.5\%$$

Determine :

(a) weight of air supplied per kg of coal *(b)* excess air (%) used.

5. A coal has following analysis (by weight) : $C = 88\%$, $H_2 = 5\%$, $N_2 = 2\%$, $S = 0.5\%$, $O_2 = 4.5\%$, (d.a.f), Moisture = 3%, Ash = 14.5% (as received basis)

It is burnt at the rate of 300 kg/hr in a furnace. Combustion air is supplied at 30°C and 740 mm Hg and is 20% in excess of theoretical. Flue gas leaves at 250°C and 740 mm Hg.

Determine the:

(a) capacity of air blower (m^3/min)

(b) volume of flue gas per minute (m^3/min)

(c) analysis of flue gas (dry basis)

(d) dew point of flue gas

(e) heat loss with the flue gas.

6. The composition of coal (by weight) is as follows :

$C = 72.5\%$, $H_2 = 5.5\%$, $N_2 = 2.0\%$, $O_2 = 8.5\%$, Moisture = 7.0%, $S = 10\%$ and Ash = 4.5% Determine the minimum amount of air required to burn 1 kg of coal on weight and mole basis.

7. An imported coke (C = 88%, ash = 12%) is burnt in a furnace at the rate of 250 kg/hr. Air supplied is 30% in excess of that required for complete combustion of coke. If only 90% of the carbon present in coke is burnt in the furnace, calculate

(a) total combustion air required per hour

(b) flue gas produced per hour

(c) composition of the flue gas.

Assume that 95% of carbon burnt is oxidised to carbon dioxide and the remainder forms carbon monoxide.

8. Pulverised fuel of following analysis (by weight) is fired in a furnace at the rate of 10 tons/hr, C = 65%, H_2 = 5%, O = 12%, N = 1%, H_2O = 5%, ash = 12%.

Orsat analysis of the flue gas produced is as follows :

$$CO_2 = 15\%, O_2 = 5\%, N_2 = 80\%.$$

Determine :

(a) volume of dry air used and % excess air, m^3/hr

(b) flue gas volume leaving the furnace, m^3/hr

(c) % H_2O in flue gas leaving the furnace.

[LIQUID FUELS]

9. A liquid fuel (H to C ratio = 0.175 by weight) is burnt with 25% excess air. Determine :

(a) volume of air at 30°C and 750 mm Hg (m^3/kg)

(b) volume of flue gas at 300°C and 750 mm Hg (m^3/kg)

(c) weight of water formed per kg fuel (kg/kg)

10. The ultimate analysis of coal tar is as follows :

$$C = 90\%, H_2 = 6.0\%, S = 0.5\%, N_2 = 1.0\%, O_2 = 2.5\%$$

It is burnt with 20% excess air and the temp, of outgoing flue gas is 250°C. Determine the heat loss with the flue gases per kg of dry tar.

11. A hydrocarbon liquid fuel after burning exits flue gas of following analysis : CO_2 = 10.5%, O_2 = 5.0%, N_2 = 84.5%.

Determine :

(a) composition of fuel by weight (b) the excess air used.

12. After burning of a liquid hydrocarbon fuel, the flue gas at 200°C and normal pressure having following analysis :

$$CO_2 = 12.0\%, CO = 1.5\%, O_2 = 3.5\%, N_2 = 83.0\% \text{ enters the stack.}$$

Determine :

(a) excess air (%) used *(b)* volume of stack gases (m^3/kg of carbon burnt)

(c) dew point of flue (stack) gases *(d)* atomic ratio of H to C in fuel, (mole/mole)

(e) moles of stack gases/mole of carbon burnt.

13. A liquid fuel containing C = 85% and H = 15% by weight is burnt in a furnace. The volumetric composition of water saturated flue gas produced is CO_2 = 20%, CO = 2.5%, O_2 = 5.5% and N_2 = 72%.

Determine the excess air (or deficiency of air) used for combustion.

14. A fuel has following (%) analysis by weight:

$$C = 85\%, H_2 = 6\%, S = 3\%, O_2 = 2.5\%, Ash = 3.5\%$$

For an air/fuel ratio of 12 : 1, determine

(a) the mixture strength (as % rich or lean) (b) volumetric analysis of the dry flue gas.

15. A furnace oil having following gravimetric analysis is burnt with 50% excess air.

$$C = 85\%, H_2 = 13\%, O_2 = 1\%, S = 1\%$$

Determine *(a)* the weight of air supplied per kg of fuel.

(b) the volumetric analysis of the wet flue gas.

16. A fuel has following ultimate % analysis : C =085%, H = 5.5%, Rest = 9.5% is incombustible. A partial volumetric analysis of the dry products of combustion shows : CO_2 = 15%, O_2 = 3.5% and some amount of CO is suspected.

Determine *(a)* the complete volumetric analysis of the dry exhaust gases.

(b) air/fuel ratio by weight.

Assume that all H and C has been burnt.

17. A furnace oil of sp. gravity 0.90 and having following analysis (by weight) :

$$C = 85.5\%, H_2 = 12.0\%, S = 2.5\%$$

is burnt at the rate of 170 litres/hr in a furnace.

The combustion is supplemented by a gaseous fuel firing of following volumetric composition, : CO_2 = 8.0%, CO = 15.0%, H_2 = 14.5%, CH_4 = 2.0%, N_2 = 60.5%. Oil to gas ratio = 2:1 (by weight) Determine the capacity of fan delivering the air if,

- excess air used for combustion is 20%
- temperature and pressure of combustion air is 30°C and 760 mm Hg respectively
- air humidity = 18.8 gm moisture/kg dry air

[GASEOUS FUELS]

18. A gaseous fuel has following volumetric composition :

$$H_2 = 53\%, CH_4 = 32\%, CO_2 = 2\%, C_2H_4 = 3.5\%, CO = 6\%, O_2 = 0.5\%, N_2 = 3\%$$

If this gas is burnt completely at the rate of 1000 Nm^3/hr with 20% excess air, calculate

(a) flue gas analysis on dry basis (Orsat analysis)

(b) Nm^3 of CO_2 produced per hour.

19. A gaseous fuel (measured at 760 mm Hg and 30°C) has following composition (by volume): H_2 = 30%, CO_2 = 22%, CO = 18%, C_2H_2 = 4%, C_2H_6 = 3%, CH_4 = 15%, O_2 = 0.5%, N_2 = 7.5%

It is burnt with 20% excess air (dry).

Calculate :

(a) The air required to burn 1000 Nm^3/hr of the fuel gas

(b) Volumetric analysis of flue gas (dry basis) at N.T.P.

20. A fuel gas has the following volumetric composition :

$$CH_4 = 30\%, H_2 = 25\%, C_2H_4 = 2\%, C_6H_6 = 0.5\%, CO = 6\%, O_2 = 0.5\%, N_2 = 36\%$$

It is burnt in a furnace with 15% excess air. Determine

(a) combustion air requirement, Nm^3/Nm^3 fuel gas

(b) flue gas analysis (on dry basis).

21. Decane ($C_{10}H_{22}$) is burned using an air/fuel ratio of 12 : 1 by weight. Assuming that all the hydrocarbon is burned to H_2O. Determine the complete volumetric analysis of flue gas.

22. A fuel gas has following volumetric analysis :

$$CH_4 = 20\%, C_2H_6 = 40\%, C_3H_8 = 40\%$$

and the volumetric analysis (dry) of its flue gas is as follows :

$$CO_2 = 10.5\%, O_2 = 5\%, CO = 1\%, N_2 = 83.5\%$$

Determine *(a)* gravimetric analysis of fuel.

(b) the theoretical air/fuel ratio. *(c)* % excess air.

23. The volumetric analysis of a natural gas is as follows :

$$C_2H_6=16\%, CH_4 = 84\%$$

Determine *(a)* the stoichiometric air/fuel ratio.

(b) % by weight of the carbon and hydrogen in the fuel.

(c) the weight of CO_2 and H_2O formed per kg of fuel.

24. A natural gas of following analysis is burnt with 10% excess air : $CO_2 = 0.5\%$, $CH_4 = 92.5\%$, $C_2H_6 = 5.0\%$, $N_2 = 2.0\%$.

Determine *(a)* the volume of combustion air.

(b) the flue gas produced per m^3 of natural gas burnt.

(c) the dew point of flue gas.

(d) composition and average molecular weight of the flue gas.

25. A fuel gas has following analysis :

$$H_2 = 30\%, CO = 25\%, N_2 = 45\%$$

and the Orsat analysis of its combustion product (flue gas) is :

$$CO_2 = 12.0\%, O_2 = 2.5\%, CO = \text{nil}, N_2 = 85.5\%.$$

Determine the amount of excess air (%) used in the combustion.

26. A propane-butane mixture (1 : 4) is burnt in a bunsen burner. If the fuel-air mixture in the mixing tube contains 5% fuel, calculate the primary air in the mixure as a percentage of the total theoretical air.

27. If octane is burnt with 90% theoretical air, determine the CO content of the exhaust gas. Reactivity to oxygen of hydrogen is greater than that of carbon monoxide (assumption).

28. A gaseous fuel has following analysis :

$$CO_2 = 3.5\%, O_2 = 0.5\%, CO = 17.5\%, C_2H_4 = 3.5\%, C_6H_6 = 1.5\%,$$
$$H_2 = 35.5\%, CH_4 = 25.0\%, N_2 = 13.0\%.$$

It is burnt with air in a furnace. The CO_2 content in flue gas is 10%. Determine :

(a) the excess air (%) used. (b) the complete Orsat analysis of flue gases.

29. Burning of CO in air at atmospheric pressure gives a gas in which the molal ratio of CO to CO_2 is unity. The process takes place in an insulated chamber, and the products of combustion exit at 2450°C. CO enters the furnace at 30°C while the air is preheated to the required temperature. Determine :

(a) the composition of exit gas. *(b)* the theoretical air (%) used,

(c) the temperature of the preheated air.

[GENERAL TYPE PROBLEMS]

30. A fuel contains (ultimate analysis weight), : C = 75%, H_2 = 12% incombustibles = rest (*i.e.,* 13%). The combustion air supply is 30% in excess of that required theoretically for complete combustion. Calculate the volumetric analysis of dry flue gases if 0.05 kg of carbon per kg of fuel is burnt to (incomplete combustion) CO and the rest to CO_2 (complete combustion).

31. A coal has following composition (by weight) :

C = 80%, H = 5%, S = 0.5%, O = 6.5%, mineral matter = rest *i.e.* 8%

This coal on gasification produces a fuel gas of following composition (by volume)

$$CO = 30\%, CO_2 = 5\%, H_2 = 10\%, N_2 = 55\%$$

Find out :

(a) volume (m^3) of fuel gas produced per kg of coal.

(b) volume of air (Nm^3) used per kg of coal.

32. A flue gas has following Orsat analysis :

$$CO_2 = 8.0\%, O_2 = 10.0\%, CO = 0.1\%, H_2 = 0.4\%, N_2 = 81.5\%$$

Calculate :

(a) CO_2 content in dry theoretical flue gas

(b) excess air (%) used in combustion.

33. Calculate : *(a)* the theoretical and

(b) the adiabatic flame temperature of pure benzene (C_2H_6) burnt in stoichiometric air.

34. Determine the enthalpy change of the theoretical products of combustion of methane in cooling from 1500°C to 300°C. Express the results in (a) kcal/mole of methane and (b) kcal/Nm^3 of the flue gas.

35. The volumetric analysis of a flue gas is as follows at temperature of 1000°C. It is mixed with the atmospheric air at 30°C in the volume ratio of 1 : 2.

$$CO_2 = 15.0\%, H_2O = 5.5\%, O_2 = 5.0\%, N_2 = 74.5\%$$

Determine the temperature of the mixture.

36. The analysis of a natural gas is as follows :

CO_2 = 0.5%, CH_4 = 94.0%, C_2H_6 = 1.0%, C_3H_8 = 0.5%, C_4H_{10} = 0.5%, C_5H_{12} = 0.5%, N_2 = 3.0%. It is burnt with 20% excess air. The air is preheated to 800°C and its humidity is 18.8 gm moisture/kg of dry air. Determine (a) the theoretical and (b) adiabatic flame temperature.

37. Find the dew point temperature of the high temperature products of combustion of C_8H_{16} with 20% deficiency of air if the total pressure is 1 atm.

38. Gaseous propane is allowed to react with 150% theoretical air at a pressure of 1 atm. If the air has an initial relative humidity of 60% at 27°C, determine the dew point of the combustion products.

39. Ethyl alcohol (C_2H_5OH) is burnt with 20% excess air. Determine the dew point temperature at of the products of combustion if, the total pressure is 1 bar.

40. The net heating value (NHV) for methane at constant pressure and at 25°C is 191,759 kcal/k mole.

Determine : *(a)* gross heating value (GHV) at constant pressure

(b) GHV at constant volume

(c) NHV at constant volume and

(d) GHV at constant pressure at 600°K.

41. 2.5 m^3/sec. of C_2H_6 are supplied along with 120% stoichiometric air (preheated to 400°K) to a furnace at 1.5 atm pressure and 300°K. Heat is transferred from the product of combustion to generate steam. The products leave the boiler at 500°K. Combustion is complete and there is no dissociation of the combustion products. Determine the rate of heat transfer to the steam.

42. Liquid octane (n-C_8H_{18}) and combustion air (in 50% excess), enters a combustion chamber in steady flow at 1 atm and 25°C. Determine the amount of heat which must be removed per mole in order to cool the combustion products to 800°K. The enthalpy of vaporisation of fuel is 9910 kcal/kmole.

43. Determine the flame temperature at the end of constant pressure, adiabatic combustion of CO with stoichiometric air *(i)* with and *(ii)* without dissociation of CO_2. The initial pressure and temperature may be taken as 1 atm and 27°C respectively. Also determine the flame temperature when the combustion takes place at constant volume under adiabatic conditions.

44. Determine the maximum flame temperature when CH_4 is burnt with stoichiometric air at constant pressure of 1 atm, both reactants entering at 25°C. Assume that the combustion with respect to carbon is only 80% complete, i.e. 20% of the carbon in methane is burned only to CO. The dissociation of the combustion products at high temperature may be neglected.

45. Determine the adiabatic flame temperature and explosion pressure for the constant volume combustion of CO with 50% excess air. The initial conditions are 1 atm and 300°K. Assume that no dissociation of the combustion products takes place.

46. Calculate the percentage heat losses due to the heat carried away in flue gases when the following fuels are completely burnt in the theoretical amount of air.

(a) Solid fuel : C = 74%, H = 5%, CV = 27900 kJ/kg, flue gas temperature = 315°C

(b) Gaseous fuel: CH_4 (methane), C.V. = 44.6 MJ/m^3, flue gas temperature = 260°C

Ambient temperature = 15°C

Gases	CO_2	H_2O	N_2	air
Cp at 15°C, kJ/m³.°C	1.516	1.409	1.245	1.316

47. Calculate the static draught of a chimney 100 m high handling flue gases having a mean temperature of 177°C. Find out the loss of draught due to velocity head and friction from the following data :

I.D. of chimney = 1.83 metres

Density of flue gases = 1.312 kg/Nm³

Average velocity of flue gases = 4.675 m/sec.

Frictional co-efficient = 0.026

Ambient temperature = 15°C

Density of air = 1.292 kg/Nm³

48. A furnace oil analysing on weight % basis as : C = 87%, H = 11% and O = 2% is burned in a furnace and the flue gas produced contains 3.3% O_2 (by volume). Furnace operates with a flue gas temperature of 1260°C and the gross C.V. of furnace oil is 45400 kJ/kg. *Cp* of flue gas = 1.474 kJ/Nm³.°C.

Find out the :

(a) theoretical/stoichiometric combustion air requirement in Nm³/kg of furnace oil

(b) % excess air used

(c) sensible heat of the total flue gases expressed as a percentage of the net calorific value of the fuel.

49. A furnace oil analysing (by weight %), : C = 84.5%, H = 10.9%, S = 2.3% and having a C.V. of 44100 kJ/kg is burned in a furance from which the flue gases are exhausted at 350°C; the ambient temperature being 15°C. The dry flue gases contain 12.77% of (CO_2 + SO_2) by volume as analysed by Orsat apparatus. The total heat in the steam at 350°C at the partial pressure of steam in the flue gases is 2590 kJ/kg from 0°C. Mean specific heat of dry flue gases is 1.408 kJ/Nm³. Calculate :

(a) The volume of air in Nm³ required to burn 1 kg of furnace oil

(b) The sensible heat lost (expressed as a % of gross C.V. of furnace oil) in the

(i) wet flue gases

(ii) steam contained in flue gases

40

GAS ANALYSIS & CALORIFIC VALUE DETERMINATION

Gas Analysis includes flue gas analysis as well as fuel gas analysis. Flue gas analysis helps in finding out unburnt fuel (by CO present in flue gas) and excess air (by O_2 present in flue gas) used for combustion as well as air infiltration at different points in combustion systems.

Fuel gas analysis helps in knowing the composition & calorific value and also the toxic *(e.g.* CO) *& corrosive (e.g.* SO_2, SO_3, H_2S etc.) constituents present in the fuel gases.

Calorific value determinations of fuel gases give the values directly which can be checked by its analysis based C.V (Calorific Value).

FLUE GAS ANALYSIS

The apparatus used for the routine analysis of flue gas is the *Orsat apparatus.* The gases usually estimated are carbon dioxide, oxygen and carbon monoxide.

The Orsat apparatus consists of a water-jacketed gas burette, connected by a common line of tube and tapped connections to three absorption pipettes, containing respectively, solutions of caustic potash, alkaline pyrogallol and ammoniacal cuprous chloride for absorption of CO_2, O_2 and CO.

A levelling bottle is connected to the base of the gas burette to enable readings to be taken at constant pressure and for use in transferring gas to & from the absorption pipettes.

The levelling bottle, tubing & gas burette, are filled with acidulated water and connected to the gas flue *via* the aspirator.

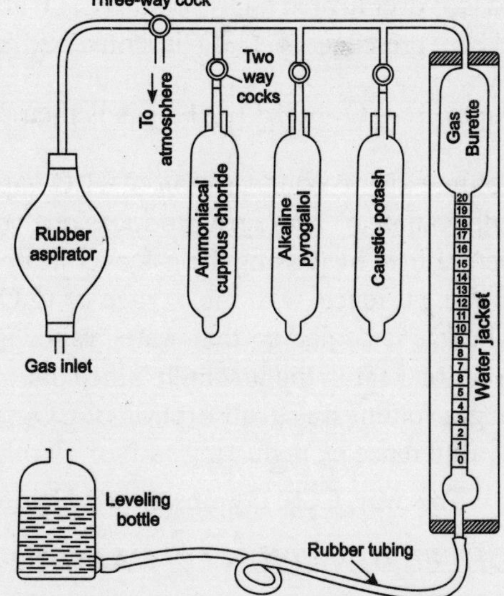

Fig. 40.1. Orsat gas analysis apparatus.

Flue gas is then pumped through these parts (escaping through the levelling bottle, held low) until the gas in the burette is representative of the flue gas.

The gas burette is then closed, and the volume & pressure adjusted so that the volume reads zero when the levels of water in the levelling bottle and & burette are the same, at zero.

The gas is then passed into the potash bulb for 2 minutes and returned for measurement in the gas burette, pressure being adjusted to atmospheric with the levelling bottle.

This procedure is repeated until the volume remains constant. The percentage of CO_2 is the new reading of the gas burette. The percentages of O_2 and CO are measured in turn in the same way, using the alkaline pyroglloI to absorb O_2 and the ammoniacal cuprous chloride to absorb CO.

FUEL GAS ANALYSIS

Fuel gases *(e.g.* blast furnace gas, coke oven gas, producer gas etc.) contain, H_2, CH_4, $C_{2.5}H_5$ (unsaturated hydrocarbons) also apart from CO, CO_2, O_2 and N_2. Complete analysis of fuel gases are done by *Haldane apparatus* or modified Orsat apparatus.

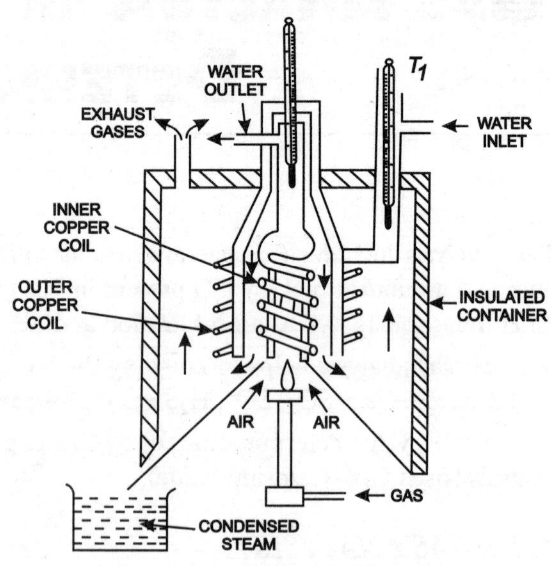

Fig. 40.2. Boy's gas calorimeter.

In the modified Orsat apparatus there are four absorption bulbs for absorption of CO_2, O_2, unsaturated hydrocarbons ($C_{2.5}H_5$) & CO respectively by KOH, pyrogallol, bromine water & ammoniacal cuprous chloride solution. After knowing the percentages of CO_2, O_2, $C_{2.5}H_5$ and CO, the next step is to know the amount of H_2 and CH_4 in flue gas. For this the residual gas is passed over heated cupric oxide (C_uO) at 280°C. H_2 in presence of CuO is converted to H_2O vapor $\left(H_2 \frac{1}{2} O_2 = H_2O\right)$; the gas is then cooled so that H_2O vapor is condensed. Reduction in gas volume due to condensation of H_2O vapor gives the percentage of hydrogen in the gas. (Because one volume of hydrogen produces one volume of H_2O vapor). After that CH_4 content in the gas is determined by passing the gas over heated C_uO kept in the same tube but at 850°C. Here methane in the gas reacts with the oxygen of C_uO and forms CO_2 and H_2O. ($CH_4 + 2O_2 = CO_2 + 2H_2O$). The gas is cooled so that water vapor is condensed and the resulting CO_2 is absorbed in KOH solution kept in the first bulb. Since one volume of CH_4 gives one volume of CO_2, hence reduction in gas volume due to absorption of CO_2 gives the percentage of methane in the fuel gas. N_2 is found by difference by deducting sum of all other costituents from 100 .

This way, CO_2, O_2, unsaturated hydrocarbons, CO, H_2 and CH_4 are found out in fuel gas.

DETERMINATION OF CALORIFIC VALUE

The C.V. of fuel gases are determined directly either by *Boy's calorimeter or by Junker's Calorimeter as described below.*

BOY'S GAS CALORIMETER

This calorimeter is used for determining the calorific values of gaseous fuels and those liquid fuels which are easily vaporized.

It consists of a gas burner in which a known volume of gas can be burnt at a known pressure and at a uniform rate. Around the burner, there is a chimney /combustion chamber having a copper tubing coiled inside as well as outside. Water at a constant rate is passed at the top of the outer coil, it leaves the chamber at the top of the inner coil. Two thermometers T_1 & T_2 are fitted for recording the temperature of the incoming & outgoing water respectively. The whole assembly is closed in an insulated chamber.

Circulation of water and burning gaseous fuel are continued at a constant rate for about 15 minutes for warming up the calorimeter. When the calorimeter gets warmed, the rates of gaseous fuel and water are so adjusted that the outgoing water leaves the apparatus nearly at atmospheric pressure. Heat produced by the burning gaseous fuel is transferred to water in the copper coil and the steam formed (if any) during combustion is condensed back into water which is collected when the steady conditions are reached. Following observations are made :

- Volume of gas (fuel) burnt at a given temperature and pressure in a certain time. The volume of gas burnt is then converted to NTP.
- Quantity of water passed through the coil in the same time.
- Rise in temperature.
- Weight of condensed water.

Calculation of Gross C.V.

Let, volume of gas burnt at NTP in time T $= V$

Weight of water passed through coil in time T $= W$

Temperature of incoming water $= t_1$

Temperature of outgoing water $= t_2$

Weight of condensed water (steam) $= W'$

Let, Higher (gross) calorific value of fuel $= C$

Calculations :

Heat absorbed by circulating water $= W(t_1 - t_2)$

Heat produced by the combustion of fuel $= V \times C$

Heat produced = Heat absorbed *i.e.,* $V \times C = W(t_2 - t_1)$

or, $$C = \frac{W(t_2 - t_1)}{V}$$

Calculation of Lower (net) Calorific Value :

Weight of water condensed from 1 m^3 of gas $= \dfrac{W'}{V}$

Latent heat of steam per m^3 of gas $= \dfrac{W' \times 587}{V}$

(Latent heat of steam = 587 kcal/kg)

Lower calorific value of the fuel

= Higher calorific value - latent heat of steam per m^3 of gas = $C - \frac{W' \times 587}{V}$

Example 1. *The following data were obtained in a Boy's gas calorimeter experiment.*

Volume of gas used = 0.1 cubic meter at NTP = 0.1 Nm^3

Weight of water heated =25 kg Temperature of inlet water = 20°C

Temperature of outlet water = 33°C Weight of steam condensed = 0.025 kg

Calculate the higher & lower calorific values per m^3 at NTP assuming that the heat liberated in condensing water vapour and cooling the condensate is 580 $kcal/kg$.

Solution. Here,

V = 0.1 Nm^3 ; W' = 0.025 kg

W= 25 kg ; $t_2 - t_1 = 33 - 20 = 13\ ^0C$

Latent heat of steam =580 kcal/kg

Now, higher calorific value (C) $= \frac{W(t_1 - t_2}{V} = \frac{25 \times 13}{0.1}$ **= 3250 kcal/Nm^3**

And lower calorific value = C– $\frac{W' \times \text{latent heat of steam}}{V}$

$=3250 - \frac{0.025 \times 580}{0.1}$ =3250-145=3105 kcal/Nm^3.

JUNKER'S CALORIMETER

It consists essentially of an inner combustion chamber in which the gas burns. This chamber is surrounded by a water jacket through which a number of tubes, called flues, pass. The gas on combustion heats the water jacket and then passes down the flues in the water jacket, thus complete transfer of the heat of combustion to water is ensured.

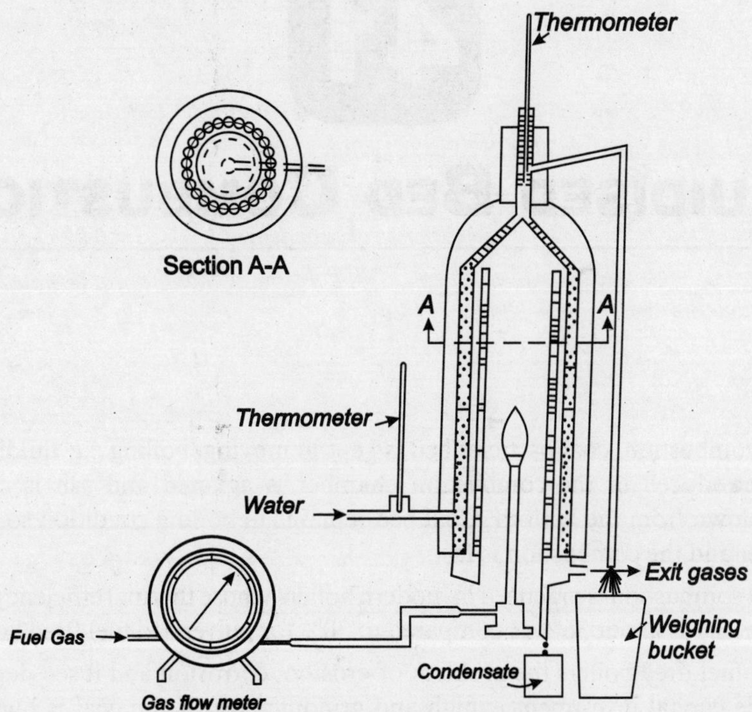

Fig. 40.3. Junker's gas calorimeter.

Water enters the calorimeter through a constant head and its temperature is noted by a thermometer. The gas whose calorific value is to be determined passes through a gas meter, which records its volume, and burns in a burner. The hot products of combustion move upward in the chamber and then downward through the flues provided in the water jacket and escape through the exit. A thermometer at the exit indicates its temperature which should be as near the room temperature as possible so that the transfer of heat to water is complete. Water, moving in a direction opposite to that of the gases in the flues, goes out from the upper portion. Its temperature and volume are recorded. Water formed by the condensation of steam is also collected.

The volume of gas burnt during the experiment is converted to NTP volume (at 0°C and 760 mm Hg pressure). Knowing the weight of the water flowing and the rise in its temperature, the heat given out by the gas is known. The gross calorific value is then obtained by dividing the heat given out by the corrected volume of the gas. For obtaining the net calorific, value the condensed water is weighed and the weight multiplied by the latent heat of vapourization of water. This is substracted from the total heat and the result is then divided by the corrected volume of the gas.

41

FLUIDISED BED COMBUSTION

In fluidised bed combustion, coal particles bed is kept in moving/boiling (*i.e.* fluidised) condition. Coal particles introduced in the combustion chamber is agitated and ash is carried over by combustion air blown from the bottom. Coal bed remains in boiling condition so that contact of air & fuel is better and the combustion is fast.

Fluidised bed combustion is practised in modern boilers whose thermal efficiency for conversion of coal into electricity is about 36% as compared to 30% for pulverised fuel fired boilers.

In pulverised fuel fired boilers the problem of erosion, corrosion and fused deposit formation exists. Besides the capital investment is high and grinding cost of the coal is high. However, in fluidised bed combustion, the heat transfer surface is in direct contact with fluidised bed and thus is able to avoid many disadvantages of pulverised fuel firing and reduces the cost of steam & power generation.

In fluidised bed combustion boilers, the coal/fuel of relatively coarser size (compared to P.F. firing boilers) is fed to the hot fluid bed of ash particles. The bed is maintained at 800-900°C and the heat transfer surfaces are placed direct in contact with the bed. The use of coarser particles eliminates the high cost and abrasion problems involved in pulverisation. The use of lower temperature of combustion eliminates slagging and corrosion problems of high temperature operation. The flyash produced is soft in nature and thus minimises the problem of fly ash erosion in boiler plants.

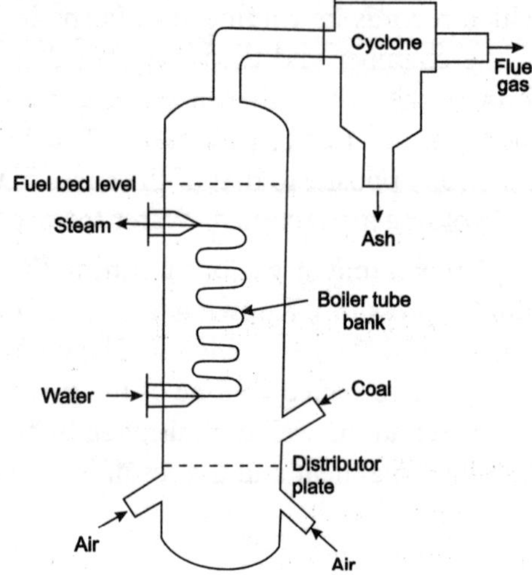

Fig. 41.1. Schematic diagram illustrating the principles of fluidised bed combustion.

Schematic diagram illustrating the principles of fluidised bed combustion is shown in Fig. 41.1.

Salient features of fluidised bed combustion system are as follows :

- Combustion chamber is a vertical cylinder or squared shell; crushed coal (-1.5 mm) is supported on perforated distributor (sometimes water cooled) and fluidised by combustion air pumped through the distributor plate.

- Fuel is continuously admitted through inlet above distributor plate.
- Combustion provides heat for steam generation.
- Boiler tubes can be directly submerged in the fuel bed.
- *Combustion chamber* is refractory lined ; water cooled casing mounted inside a 1.8 mm dia. pressure shell (cyclones are for dust removal).

Advantages of Fluidised Bed Combustion

- Lower capital and operating cost.
- High heat transfer to boiler tubes.
- Reduced NOx formation due to low flame temperature (800 -900°C).
- Pressurised fluidised bed combustion would require much less space.
- Excess air requirement is less (only 5-10%) compared to pulverised fuel (p.f.) fired boilers (which require 15-20% excess air) thus resulting in economy due to less loss of sensible heat along with flue gases.
- Reduction of oxides of sulphur in flue gases even by using high sulphur fuels due to lower operating temperature.
- Low cost fluidised bed boiler is capable of giving at least the same efficiency (around *80%)* attainable in P.F. fired boilers.
- Low grade coals can be burnt efficiently in this system.
- The high rate at which heat can be transferred from the bed for steam raising or other purposes enables smaller boilers and furnaces to be used.
- Combustion is hardly affected by the type of coal, ash or moisture contents as the concentration of coal in the bed is small. The burning coal is surrounded by inert material so that there is no tendency for coal particles to stick together. Low grade coal regarded in the past as unsuitable for combustion can be used in a Fluidised Bed Combustor.
- In contrast to pulverised fuel firing fluidised bed combustors can use commercially available grades of coal without further treatment.
- The temperature in FBC (at a max. 950°C) is well below that in other system *(e.g.* over 1200°C in pulverised fuel fired boiler plant). Emission of deposit forming materials is therefore much less than in other boilers and tube fouling is minimum. The low temperature of combustion eliminates problems associated with fusion of ash.
- Particles velocities are well below that at which steel is eroded by impact. In fact, scouring by the moving particles can be beneficial in preventing building up of deposits on the pipes.
- The evenness of the temperature enables automatic control more precisely than in other coal burning systems.

Disadvantages of Fluidised Bed Combustion

- Because heat transfer rates in the bed are so high, the immersed boiler tubes are easily damaged by excessive bed temperatures. On the other hand, combustion ceases if the bed temperature drops too low. Therefore, the bed must be maintained within rather narrow temperature limits, and this complicates the process of controlling output. It is necessary to control the heat absorption rate as well as the firing rate.

- Relatively little variation can be tolerated in the velocity of the fluidising air. If it is too high, entrainment of particulates is excessive and loss due to unburnt carbon is high. If it is too low, fluidising becomes sluggish, coarse material in the bed segregates, and the heat transfer is impaired. If, on turndown, air flow is held constant while fuel flow is decreased, excess air level becomes unacceptably high. The combustion air must be introduced at a substantial pressure in order to fluidise the bed. For pressurised systems, much higher combustion air pressure are necessary.

CIRCULATING FLUIDIZED BED (CFB) BOILER

A Circulating Fluidised Bed Boiler is a device for generation of steam by burning fossil fuel in a furnace operated under a special hydrodynamic condition, where, fine solids are transported through the furnace at a velocity exceeding the free fall velocity of individual particles. The major fraction of solids entrained out of the vessel, is captured by a separator and is recirculated back to a point near the base of the furnace at a sufficiently high rate to cause a reasonable amount of refluxing of solids in the furnace.

The primary air (sub-stoichiometric in amount) is injected through the base of the furnace and rest of the secondary combustion-air is injected from the sides at a certain height above the base. Fuel particles burn in the furnace generating heat. A part of the heat is extracted by water/steam surfaces located in the furnace but above the secondary air injection level.

The creation of the special hydrodynamic condition of circulating fluidized bed, popularly known as 'Fast Bed' / 'Dilute Phase Refluxing', is key to this process. A certain combination of gas velocity, recirculation rate, solid characteristics, volume of solids and the geometry of the system, gives rise to this special hydrodynamic condition where solid particles are subjected to a upward velocity greater than the terminal velocity of most individual particles. Yet these particles are not entrained immediately as observed in vertical pneumatic transport systems. On the contrary, solids are found to move up and down in groups known as clusters or strands. These long slender solid agglomerates are seen to move vertically, sideways as well as downwards. They are continuously formed, dissolved and reformed again.

- A cross-section of a typical circulating fluidized bed boiler is shown in Fig. 41.2. The boiler furnace contains a mass of granular (100-300 micron) solids known as bed materials which could be :
- sand or gravel (in case of fuels such as wood chips, low ash coal etc.).
- fresh or spent limestone (in case of high sulphur coal requiring sulphur retention).
- ash from coal (in case of high or medium ash coal requiring no sulphur retention).
- Sometimes a combination of two or three of the above materials are also used. Fuel particles constitute a minor fraction (< 1%) of the total bed solids.

The boiler can be divided into two sections. The first pass is a solid circulation loop, in which fuel are burnt and a fraction of the generated heat is absorbed. The solid circulation loop includes the

- Gas-solid separator (cyclone or impact separator)

- Solid recycle device (loop seal, seal pot or L-valve)
- External heat exchanger (optional)

In addition, to the above major items other components attached to the boiler are bed drain and solid classifier.

Walls of the lower section of the furnace are lined up to the level of secondary air with refractory. The furnace above this level is generally cooled by evaporative, superheater or reheater surfaces. The gas-solid separator and the non-mechanical valves for solid recycle are also lined with refractory. These absorb a limited amount of heat. However, if there is a need for heat extraction in addition to that extracted in furnace it is necessary that a part of the hot solid recycling between the cyclone and the furnace is diverted through an external heat exchanger. This heat exchanger is a bubbling fluidized bed with immersed heat transfer surfaces. Very little combustion takes place inside the external heat exchanger, but it absorbs the necessary amount of combustion heat.

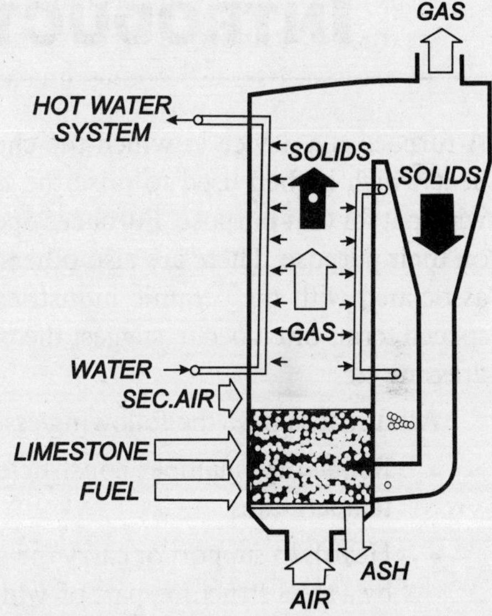

Fig. 41.2. *Circulating Fluidised Bed Boiler Furnace.*

Coal is generally injected into the lower section of the furnace. It is sometimes fed into the loop-seal from where it enters the furnace along with returned hot solids. Limestone is fed into the bed in a similar manner.

A part of the combustion air (primary) enters the furnace through air distributor or grate. The secondary air is injected at a certain height above the grate to complete the combustion. Coal burns when mixed with hot bed solids. Bed solids are well mixed throughout the height of the furnace. So the bed temperature is nearly uniform in the range of 800-900°C though heat is extracted along its height. Coarse unburnt coal or char particles are captured in the gas-solid separator. Ash & spent sorbent generated during combustion & desulphurization leaves the furnace partly through bed drain and partly escaping through the gas-solid separators.

Following are the major *advantages of circulating fluidized bed (CFB) boilers :*

- **Fuel flexibility.** A CFB boiler can burn a wide range of fuels with good efficiency without much change in hardwares of the furnaces.
- **High combustion efficiency.** It can burn fuels particles of wide size ranges with very high efficiency.
- **Smaller size.** For same output a CFB boiler requires smaller plan area than bubbling bed boiler units.
- **Low air pollution.** The NOx emission from the circulating fluidized bed is below acceptable limits. It can reduce the SOx emission by 90% with minimum use of limestone.
- **Good turndown capability.** A circulating fluidized bed boiler can respond to the fluctuating load demand much easily.
- **Smaller feed points.** It requires only a fraction of coal feed points than a bubbling bed boiler of similar size would require.

Introduction to Furnaces

A furnace is a device in which the chemical energy of fuel or electrical energy is converted into heat which is then used to raise the temperature of material, called the burden or *stock,* placed within it for that purpose. Furnaces operating at low temperatures are often called *ovens* depending on their purpose. There are also others used for higher temperatures for different purposes. When associated with the ceramic industries they are called *kilns.* Glass is melted in tanks and other special terms often occur, suggest the purpose, such as sintering machine used in metallurgical ore dressing etc.

All furnaces have the following essential components as shown in Figure 42.1 below :

- Refractory chamber constructed of insulating materials to retain heat at high operating temperatures.
- Hearth to support or carry the stock steel, which consists of refractory materials supported by a steel structure, part of which is water-cooled.
- Burners that use liquid or gaseous fuels to raise and maintain the temperature in the chamber. Coal or electricity are also used in some furnaces.
- Chimney to remove flue gases from the furnace chamber.
- Charging & discharging doors for stock through which the furnace chamber is loaded and unloaded. Loading & unloading equipment include roller tables, conveyors, charging machines and furnace pushers.

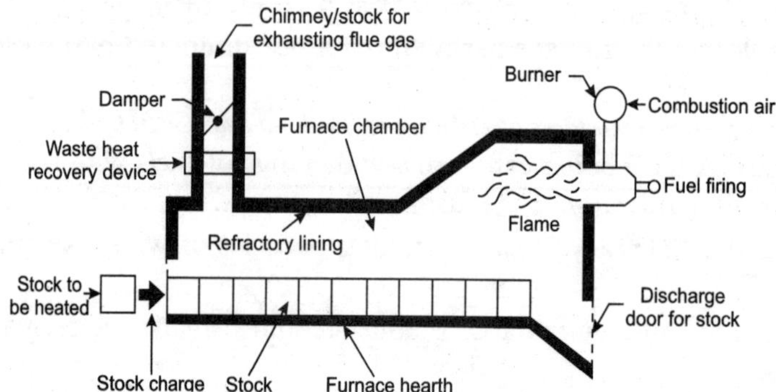

Fig. 42.1. Typical furnace components/parts and accessories

Since flue gases from the fuel come in direct contact with the materials, the type of fuel used is also important. For example, some materials will not tolerate sulphur in the fuel. Solid fuels generate particulate matter, which will interfere with the materials placed inside the furnace. For this reason :

- Most furnaces use liquid fuel, gaseous fuel or electricity as energy input.
- Induction and arc furnaces use electricity to melt steel and cast iron.
- Melting furnaces for nonferrous materials use fuel oil.
- Oil-fired furnaces mostly use furnace oil, especially for reheating and heat treatment of materials.
- Light diesel oil (LDO) and Low sulfur heavy stock (LSHS) oil is used in furnaces where sulphur is undesirable.

Furnace ideally should heat as much of material as possible to a uniform temperature with the least possible fuel and labor. Efficient furnace operation requires complete combustion of fuel with minimum excess air. Furnaces operate with relatively low efficiencies (as low as 7%) compared to other combustion equipment such as the boiler (with efficiencies higher than 90%). This is caused by the high operating temperatures in the furnace. For example, a furnace heating materials to 1200 °C will emit exhaust gases at 1200 °C or more, which results in significant heat losses through the chimney, unless waste heat is recovered by a waste heat recovery device and recycled back in the form of preheated combustion air or fuel gases in the furnace.

Basic diagram of furnace operation incorporating waste heat recovery system is shown in Fig. 42.2.

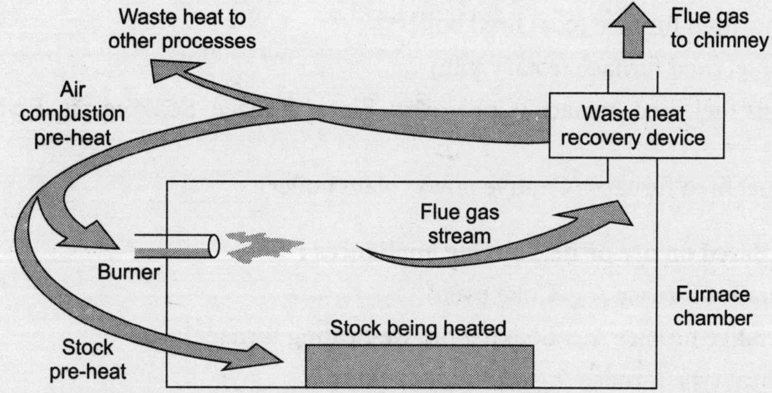

***Fig. 42.2.** Basic diagram of furnace operation incorporating waste heat recovery.*

Classification of Furnaces :

The classification of furnace can be done as :

- batch type
- direct heating type

Different *metallurgical furnaces are :*

- Crucible Furnace
- Shaft Furnace eg, Blast Furnace
- continuous type

- Hearth Furnace eg, Open Hearth furnace
- indirect heating type.
- Converters-Bessemer, side blown or top blown (LD)
- Soaking pits
- Annealing furnaces.
- Muffle Furnace
- Reheating furnaces

Other Types of Furnaces :

- Rotary Kiln-Cement Kiln
- Tunnel Kil
- Boiler Furnace
- Electric furnaces (Three Phase or Single Phase)

(a) Heating by metallic resistors, or

(b) Heating by induction.

Furnaces are further classified based on various factors as below :

Classification Based on type of fuel used :

- Solid fuel fired furnace (coal fired boiler)
- Liquid fuel fired furnace (rotary kiln)
- Gaseous fuel fired furnace (Coke oven, Sinter furnace, Soaking pit, Re-heating furnace etc.)
- Multi fuel fired furnace (Captive power plant boiler)

Classification Based on use of heat saving appliances :

- Regenerative furnace *(e.g.* Coke oven)
- Recuperative furnace *(e.g.* Soaking pit, Re-heating furnace)
- Waste heat type furnace *(e.g.* Rotary lime kiln)

Classification Based on draft control system :

- natural draft furnaces *(e.g.* B.F. stoves, coke ovens etc.)
- forced draft furnaces
- induced draft furnaces
- balanced draft furnaces (most large combustion furnaces come under this category)

Some of the above classified furnaces *e.g.* batch furnace, continuous furnace, Recuperative furnace and regenerative furnace have been shown in Fig. 42.3, 42.4, 42.5 and 42.6 respectively.

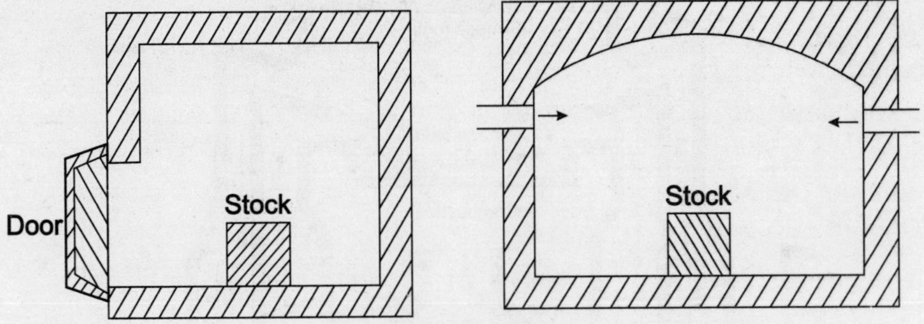

Fig. 42.3. Batch furnace.

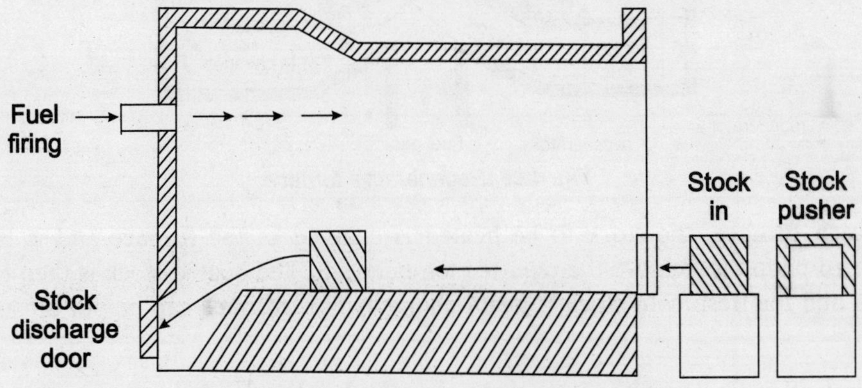

Fig. 42.4. Continuous furnace (End Discharge).

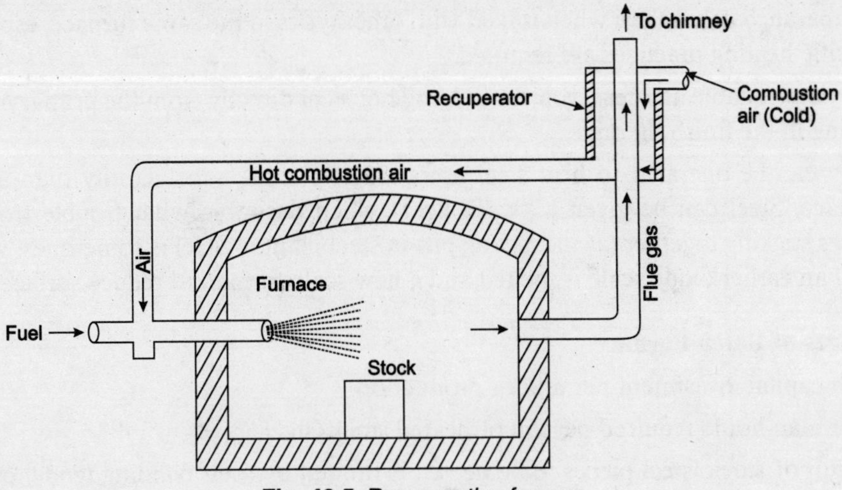

Fig. 42.5. Recuperative furnace.

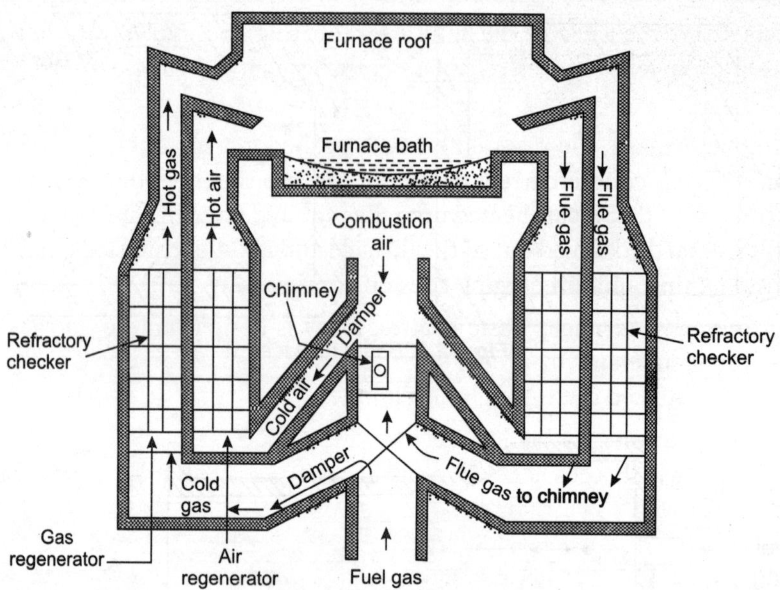

Fig. 42.6. Regenerative furnace.

In a **batch furnace,** the stock to be heated is charged in the furnace and is heated for a predetermined period to achieve the required temperature. The heated stock is then taken out of the furnace and the fresh cold stock is again charged. This furnace can be put off as and when needed.

Advantages of Batch Furnace

- They can heat the stock/steel of various types and sizes which can be heated more properly in separate batches than when mixed with other types in the same furnace, especially when specific heating practices are required.
- They are suitable as a reservoir for holding hot steel directly from the primary mill for later rolling in the finishing mills.
- They can be operated to heat steel upto 1320 °C more satisfactorily than a continuous furnace. Steel can be given a "wash heat" when desirable what trouble from the stock pieces sticking together (as in soaking pits in steel plants). Steel is sometimes 'wash heated' until an earlier oxide scale is melted and a new scale formed to reduce surface defects.

Disadvantages of Batch Furnace

- High capital investment per unit of production.
- High man-hours required per ton of heated stock/steel.
- Length of stock/steel pieces to be heated is limited by their bending tendency when they are removed.
- Low furnace hearh area efficiency because hearth is not utilised fully due to interference of door jambs and clearance required inside the furnace for handling the stock with charging machines.

- It suffers from the lack of flexibility for heating the stock/steel slowly in the lower temperature range. Furnaces must be cooled to charge high carbon steel and some alloy steel.

In a **continuous furnace,** the stock enters or is pushed in from one end (i.e., charging end) of the furnace and it travels countercurrent to the hot flue gases to the other end (i.e. discharge end). Its speed is so controlled that it gets heated upto the required temperature by the time it reaches the discharge end. It is then discharged out of the furnace and another cold stock may be charged. This process of stock charging and discharging goes on continuously.

- In a **recuperative furnace,** waste heat recovery is done from the flue gas in a metallic or ceramic tile recuperator by preheating the combustion air and or fuel gas (if permissible) before it is exhausted out through the chimney.

In a **regenerative furnace,** waste heat recovery from flue gas is done by a matrix of refractory checker work/bricks which gets heated up during the passage of hot flue gas through it for a predetermined time (usually 20-30 minutes). In the next reversal cycle, cold combustion air and/or cold fuel gas (provided its preheating is permissible) is routed through the hot checker work/brickes of the regenerator which preheats them. Again when the hot regenerators bricks loose their heat & become cool, the passage of the hot flue gas & cold air and cold fuel gas is reversed through the reversal damper/valve. That is why two sets of regenerators are provided for each furnace so that while one set of regenerator is being heated by hot flue gas, the other set of already heated hot regenerator is loosing its heat to the cold combustion air and/or cold fuel gas for preheating them. And after a fixed interval (called reversal cycle usually of 20-30 minutes), regenerators duty is interchanged/reversed.

FURNACE AUXILIARIES

The important furnace auxiliaries are :

- Damper
- Fan (forced draft and induced draft)
- Control valve
- Burner

Damper. The function of a damper is to regulate draught and is a prime instrument for furnace economy. Damper is located just after the furnace and before the stack. In modern furnaces, damper is made to work in auto mode and its opening gets automatically adjusted depending upon the draught required in the furnace (positive or negative draft).

Fan. In a furnace, a fan is usually used to provide combustion air and is known as forced draught fan. Sometimes a fan is provided before the stack to remove flue gas from the furnace. This is called induced draft fan. A furnace having both forced draft fan and induced draft fan is said to have a balanced draught system.

Control valve. Control valve regulates the amount of fuel entering the furnace. For each furnace there is an ideal rate of heat consumption or fuel for the particular heating operation. Any departure from the correct setting of valves results in waste of fuel. These control valves can be regulated automatically depending upon the heat load of the furnace. All control valves should be cleaned, checked and adjusted regularly to ensure their good working.

Burner. Burner is a device which produces flame/heat for heating the stock in the furnace. It must mix the fuel and an oxidising agent (air) in proportion that are within the limits of flamability for igniting as well as for steady combustion. Thus, the essential function of a burner is to mix fuel with the combustion air and introduce the mixture into the furnace in such a way that desired flame shape and 7 penetration is obtained. Basically there are two types of burners *i.e.,* premix and direct (outside mix) type. In the premix burner, the fuel & the air oxidiser are mixed prior to ignition and in the direct type, fuel & the combustion air is mixed at the point of ignition.

EFFICIENT ENERGY UTILISATION IN FURNACES

When considering thermal efficiency of high temperature furnaces, there are *three* routes to make energy savings. *Firstly,* the performance of existing equipment can be improved by operating in a way which will reduce heat losses. *Secondly,* some form of waste heat recovery can be practised to recover waste heat from flue gases. *Finally,* consideration can be given to modifying or replacing the furnace, if it has become energy inefficient or obsolete.

Conventional high temperature furnace operate generally with thermal efficiencies between 10 to 50% (excepting steam generating boiler whose thermal efficiency may be upto 85%). The sources of heat loss are radiation losses, losses to furnace structures and flue losses. The major source of heat loss is in the flue gases.

The obvious routes to improved efficiency are to minimise avoidable heat losses from the flue gases by better air/gas ratio control, to avoid openings which permit excessive radiation, and to utilise refractories with better insulating properties.

Minimising flue gas losses :

Heat losses from flue gases fall into two categories :

- Sensible heat loss due to the discharge of flue gases at a temperature exceeding ambient temperature.
- Potential heat loss due to the discharge of products of incomplete combustion in flue gases.

If flue gases are discharged at 1300°C, about 70% of potential energy goes to waste in flue product even after ensuring complete combustion.

The correct air/gas ratio in a furnace can be determined by analysing flue gases and reducing the air/gas ratio from the CO_2 or O_2 concentrations in the flue gas.

Minimising heat losses from the furnace structure :

The amount of heat lost through the furnace structure varies with the amount, thickness & properties of furnace insulation, the size & type of furnace (batch or continuous) and its mode of operation.

Continuously operated furnaces benefit from the use of very well insulated walls, while intermittent furnaces which are frequently heated up and cooled down are better served by using light weight insulation to keep the heat storage effect to a minimum. The additional fuel consumption is again several times the actual heat loss from the wall.

Furnace firing practice :

The temperature in the furnace should be maintained at optimum level. Any undesirable increase in furnace temperature will increase fuel consumption and scale loss in a metal heating furnace. Besides radiation losses would also increase.

Another approach which can be used to realise substantial fuel savings may take the form of changes in the way in which existing burners are used, for example, by altering burner control modes (on/off versus high/low) or adopting different firing patterns for start-up, production and hold up situations. Sometimes improvements may be made by changing the design of burners. These improvements are made based on experience backed up by adequate instrumentation and metering to monitor the subsequent changes. Use of recuperative/ regenerative burner provides an additional advantage from efficient energy utilisation point of view.

Significance of radiation heat losses :

Radiation losses occur mainly through opening in the furnace walls. Its value increases rapidly with furnace temperature because of fourth power dependence of heat losses by thermal radiation on absolute temperature. For a furnace operating at 1000°C, approximately 15 kW (12500 kcal/hr.) of heat will be radiated from 1000 cm^2 of opening in the furnace wall. If furnace temperature is raised to 1400°C, the radiation heat loss will be approximately 42 kW (35000 kcal/hr.) from an opening of 1000 cm^2. Apart from the direct loss, this represents in heat entering the work place from the furnace and this heat loss must be compensated for by turning up the burners. This in turn leads to an increased volume of flue gases and additional flue losses. In fact each unit of energy loss by radiation is compensated by three units of additional input. Poor control of air/gas ratio makes the losses even worse.

FURNACE EFFICIENCY

The furnace thermal efficiency *depends upon :*

- Flame temperature
- Air/gas ratio
- Furnace structure
- Insulation levels of furnace
- Radiation losses from furnace
- Method of furnace operation

Even in a well constructed and operated high temperature furnace, between 50% & 80% of the available heat is wastefully discharged in the flue gases.

The lower this temperature can be made by transferring heat to the combustion air, the load or both ; or the fuel gas (if permissible); higher the furnace efficiency will be.

Heat can be recovered within or outside the process. Waste heat boilers are an example of the latter. They give an overall energy saving but will not affect the furnace efficiency. The *heat* can be *recovered* within the process by *practicing* the following.

Stock/Load/Charge preheating :

It occurs when the flue gases are used to preheat the incoming stock/charge/load. This is generally achieved by passing flue gases in counter flow over the incoming stock in an elongated/tunnel furnace. Examples are pusher or walking beam steel reheating furnaces and rotary cement or lime kilns.

Combustion air/fuel gas preheating :

The use of preheated combustion air increases the flame temperature of the fuel and saves energy. Recuperators or regenerators are used for this purpose. In recuperator, the combustion air & flue gases are brought into close proximity and separated by an interface (ceramic or metallic) across which heat is transferred continuously between the two streams. In a regenerator, thermally dense refractory matrix is used to absorb heat from the flue gases, prior to giving it to the combustion air. This can be done by switching the air and flue gas streams such that their path reverses/alternates, after a fixed time interval, between the two streams. Gaseous fuels can also be preheated in the same way to recover waste heat provided its preheating is otherwise permissible. Fuel gases containing substantial hydrocarbons *(e.g.* coke oven gas, natural gas, refinery gas etc.) are not preheated to high temperature as the carbon produced on its cracking (HC—>H + C) will foul the heat transfer surface as well as may clog the low diameter burners. BF gas, low CV mixed gas (BF gas + C.O gas), converter gas etc. can be preheated.

The Heating Capacity of a Batch Type Furnace :

The heating capacity of a furnace is usually expressed by the weight of metal that can be heated in unit time to a given temperature at the coldest spot of the charge and without over-heating the rest of the charge. Since the cost of a furnace is approximately proportional to its size, heating capacity per unit volume is important. Specific heating capacity is expressed either by :

Weight heated/(hour × furnace volume), or, Weight heated/(hour × hearth area).

The heating capacity depends on several factors, such as rate of heat release, rate of heat transfer to the charge, rate of conduction to the coldest point in the charge and several others.

Effect of rate of heat release : The rate of heat release is expressed in heat units released in unit volume in unit time. Its unit is kcal/m^3. hr. Closely related to rate of heat release is the term "flame length". For open hearth furnaces & glass tanks, flame length means the distance between the burner and the point where the flame is "extinguished" (becomes non-luminous). For industrial furnaces, flame length means the distance between the burner and the point at which combustion is about 95% complete in shorter flames. In the case of gaseous fuel & air if air is mixed ahead of the burner (premixing), combustion occurs more rapidly & less combustion volume is needed than with mixing inside the furnace and if the combustion air is preheated, combustion proceeds even more rapidly.

In addition to the quantity of heat that is released in unit volume and in unit time, the temperature at which it is released affects furnace capacity. Each molecule of the combustion elements burns at the theoretical flame temperature and the heat of reaction is transferred to surrounding objects (gaseous liquid and solid) while combustion proceeds. Only by infinitely rapid combustion or by combustion in a heat tight chamber the theoretical flame temperature can be reached and is not very high with lean fuels. With lean fuels, temperature can be increased only by preheating the air or the fuel or both. Oxygen enrichment in air likewise increases flame temperature.

Effect of rate of absorption of heat by the charge : It is seen that in an industrial furnace when adequate quantity of heat can be released at a considerably high temperature, the capacity of the furnace depends upon the rate of heat transfer to the charge which *depends* on the following factors:

- The temperature of the furnace walls when cold stock is charged.
- Temperature at which the stock is to be heated.
- Temperature of the products of combustion,
- Emissivity of the products of combustion.
- Emissivity of walls.
- Emissivity of the heating stock.
- Ratio of the area of the stock to the area of wall.
- Thickness of the layer of products of combustion.
- Thermal conductivity of the stock (including void).
- Desired uniformity of the temperature in the product.
- Thickness of the heating stock.
- Size of the furnace (hearth area and volume).
- Velocity of gases in the furnace.

Generally, the transfer of heat to the charge determines the heating capacity of thin heating stock, whereas the temperature equalization within the stock determines heating capacity for thick stock having low thermal conductivity.

Effect of arrangement of the stock : The arrangement of stock and their thickness has marked effect while heating pieces. Space must be provided between the pieces for manupulating tongs. Unless the spaces between the pieces are very large, the heating capacity is not noticeably affected, because bare spots of the hearth receive radiation from the bases as well as from the roof and the side walls. Radiation from the hearth (which is not as hot as the roof) increases the transfer of heat to the vertical surfaces.

Effect of thickness of stock : If the volume of a body is small compared to its heat transfer surface, the body can be heated quickly but with a small difference of temperature between the surface and the core. In the heating of steel for rolling or forging, practice pays little attention to the fact that soaking time should increase much more rapidly than the diameter or the thickness of the stock provided the diameter does not increase 30 cm and heat is received uniformly from all sides. In the heating of larger sections of mild steel, the supply of heat is high in the early stages and is reduced when the surface of the stock approaches the desired final temperature.

MUFFLE FURNACE

The principal purpose of muffle is to protect the charge from the effects of the products of combustion. Another object is to smooth the temperature inequalities that exist on the combustion side of the muffle wall. The heating capacity of a muffle that surrounds the charge depends upon the surface and emissivity of the stock and upon properties for the muffle wall (temperature, area, emissivity). For a given set of these variables, the heating capacity is slightly less than that of a direct fired furnace (having the same wall temperature) because of the absence of products of combustion that are hotter than the furnace wall.

The great difference of temperature on each side of a muffle wall that is made of fireclay not only reduces the heating capacity of the muffle but also causes the products of combustion to leave at a very high temperature. For both these reasons, muffle walls are made as thin as practicable with material having both high thermal conductivity and durability against high temperature.

Alloy steel and silicon carbide are the most suitable materials for muffle walls. Muffles are seldom used for heating metals to more than 1000°C because of cost of maintenance. Similar muffle furnaces are used as laboratory furnaces which are heated electrically. A muffle furnace is shown in Fig. 42.7.

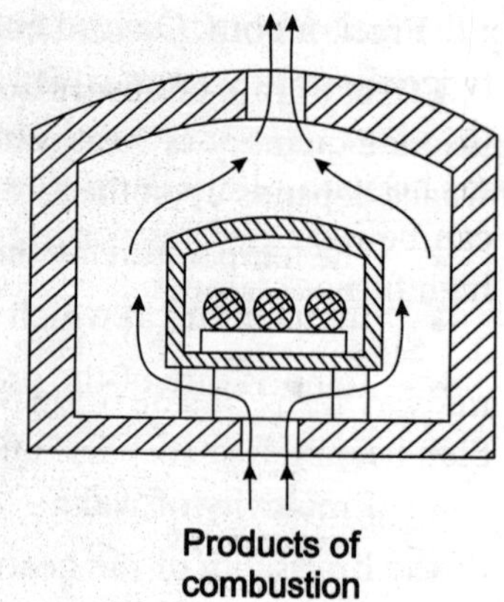

Fig. 42.7. *Muffle furnace.*

LOW TEMPERATURE FURNACES AND OVENS

In the case of low temperature furnaces, heat transfer by radiation is comparatively lower than the high temperature furnaces. At 1230°C, unit area radiates 16 times that of the quantity of heat radiated at 476°C. The rate of heat transfer by convection increases with greater density and with greater velocity of the flowing medium. At low temperatures, the density is greater than that at higher temperature. The velocity of flowing medium is increased either by the jet action of the excess air or in the majority cases by hot fans. A limit to the velocity is set by the balance between the reduction of furnace size (and cost) and increased, temperature uniformity on one side with the cost of power & the condition of the charge on the other, emissivity being one of the factors. The brighter the charge, the higher the optimum velocity. The power delivered to the fan is converted into heat. Convective heat transfer is determined by the coefficient of heat transfer (the value of the coefficient must be ascertained). After coefficient is known, the heating time can be calculated at the place where the hot gases first come in contact with the charge.

Heating Capacity of Continuous Furnaces :

In continuous furnaces, the stock or charge moves continuously intermittently through them. The stock may be pushed or rolled through, or it may rest on a rotation, hearth or be suspended from an overhead conveyor. A continuous furnace may be so heated that the temperature of the products of combustion is practically constant everywhere in the furnace.

The heating capacity of continuous furnaces can exceed that of batch-type furnaces having the same hearth area because the temperature must be held down in a batch-type furnace to prevent overheating. In a continuous furnace, the temperature may be very high if thin stock is removed quickly and for thick stock there is sufficient soaking zone to have equality in temperature at all sections.

Front Fired Continuous Furnaces : The heating capacity of such furnace is limited by the average temperature difference between the products of combustion and the stock which is lower than in a furnace in which the temperature is uniform. Front firing has the disadvantage that the rate of heat transfer is greatest at the place where the surface temperature of the stock is higher.

Front Firing Top and Bottom : The heating capacity of furnaces and bottom heat is less than twice those in the previous case, because a soaking zone must be provided for the elimination of the black spots that were caused by water cooled skid pipes. The maximum rate is about 460 (kg/m^2.hr) with 1290°C furnace temperature and proportionally less for very heavy stock. These values can be increased by altering short-flame and long flame burners. The latter can be increased by high furnace temperature.

Side Firing: The heating capacity of continuous furnaces is greatly increased by adding midway firing or side firing to end firing. By these means capacity can be increased by about 50%. Slab heating furnaces are rated 730 (kg/m^2.hr.) on low carbon steels.

The heating capacity of continuous furnaces varies greatly with the nature and surface condition of the heating stock. High carbon steel is not heated to as high a temperature as the low carbon steel to avoid decarbonisation with the result that the heating capacity is lowered.

Limitation in Hearth Length by Pushing of Stock: Another factor limits the capacity of pusher type continuous furnaces with regard to quantity heated per unit time but not with regard to quantity heated per unit area per unit time. This factor is known as *safe length of hearth.* By 'Safe Length' is meant a length that prevents upward buckling and piling up. The safe length depends upon the flatness of the hearth, thickness of the stock, and shape of the contacting surfaces of the stock.

FUEL ECONOMY OF FURNACES

The 'economy' or 'efficiency' when used in their true sense in connection with industrial furnaces, have reference to heating cost per unit weight of finished saleable product. "Heating Cost" includes not only the cost of fuel but also the cost of firing the furnace, the costs of maintenance & repair, the cost of generating a protective atmosphere and the cost of burned, spoiled or otherwise rejected pieces. It also includes the cost of machining those pieces which, at inspection are found to be defective because of improper heating.

With so many different items entering into the cost of heating, it is quite possible that in some cases, the highest priced fuel or other source of heat energy may be the cheapest in the end so far as the total heating cost is concerned. In practice, in many furnaces the cost of fuel becomes outstanding item of expense and the fuel economy in those cases is worthy of study. Whereas boiler furnace efficiencies range from 60-90% or even slightly higher; furnace efficiencies in metallurgical uses are sometimes as low as 10% in combustion types ; and little higher in electrical types. With favourable conditions, with excellent design and with good operation, furnace efficiencies upto 60% or even slightly higher can be obtained but such high values are the exception rather than the rule. By efficiency, it is meant the fuel efficiency which is the ratio of the heat input into the stock to the potential heat of the fuel. One reason for the difference in the efficiency between boilers and industrial furnace is in the final temperature of the material being heated. Gases can give up heat to the charge only as long as they are hotter than the charge. Consequently, flue gases leave industrial furnace at a very high temperature. This factor is responsible for low furnace efficiency.

The fuel required for combustion is burnt in the combustion zone of the furnace. It would be desired that most of the heat liberated be imparted to the stock but the heat goes elsewhere. Some of it passes into the furnace wall and hearth. Another portion of heat is lost to the surroundings by radiation and convection from the outer surface of the walls or by conduction, into the ground.

Heat, is also radiated through cracks or other openings and furnace gases pass out around the door, frequently burning in the open & carrying off heat. Heat is also lost every time the door is opened. Another form of heat dissipation is though the stock. Finally, the bulk of the heat loss passes out along with the products of combustion, either in the form of sensible heat or as incomplete combustion. Fuel economy demands that the fraction of total heat that passes into the stock be as large as possible and that all the losses be minimised.

FACTORS AFFECTING FUEL ECONOMY IN THE FURNACE

Various factors which affect fuel economy in industrial furnaces are :

- Complete combustion with minimum excess air.
- Proper heat distribution
- Operating at the desired temperature.
- Reducing heat losses from furnace openings.
- Minimising wall losses.
- Control of furnace draught.
- Optimum capacity utilization
- Waste heat recovery from furnace flue gases.
- High emissivity ceramic coating of furnace refractories

Complete combustion with minimum excess air :

To achieve complete combustion of fuel with minimum excess air, a number of factors (such as proper selection and maintenance of control, excess air monitoring, air infiltration, pressure of combustion air) are to be considered. The magnitude of stack losses with different quantities of excess air in the case of a furnace carrying away flue gases at 900°C is tabulated below.

Excess air percentage	Percentage of total heat in the flue carried away by waste gases (flue gas temperature 900°C)
25	48
50	55
75	63
100	71

Besides abnormal increase in stack losses, with the increase in excess air, the ingress of too much excess air lowers the flame temperature and consequently reduces furnace temperature and the heating rate. If too little excess air were used, combustion would be incomplete and chimney gases will carry away unused fuel potential in the form of unburnt combustible gases such as carbon monoxide, hydrogen and unburnt hydrocarbons which should have otherwise been usefully burnt in the combustion chamber.

Proper heat distribution :

Ideally, a furnace should be designed so that in a given time, as much material as possible, be heated to as uniform a temperature as possible with the minimum fuel firing rate. To achieve this, the following points should be considered.

- The flame should not touch the stock and should propagate clear of any solid object. Any obstruction whatsoever, deatomises the fuel particles thus affecting combustion & creating black smoke. If the flame touches the stock, the scale losses increase manifold.
- Refractories are leached, if the flames touch any part of the furnace, as the incomplete combustion products can react with some of the refractory constituents at high flame temperature.
- The flames from various burners in the combustion space should also stay clear of each other. If the flames intersect, inefficient combustion would occur. This can be controlled by staggering the burners of the opposite walls.
- The flame has a tendency to travel freely in the combustion space just above the material. In small reheating furnaces, the burner axis is never placed parallel to the hearth but always at an upward angle. Every precaution should be taken to ensure that the flame never hits the roof.
- The larger burner produces a long flame which may be difficult to contain within the furnace walls. More burners of less capacity give better distribution of heat in the furnace, and also reduce scale losses while increasing furnace life.
- For smaller reheating furnaces, it is advisable to maintain a long flame with a golden yellow colour while firing furnace oil for uniform heating. It should be seen that flame should not be so long that it enters the chimney and comes out at the top or through doors. This occurs when excessive fuel is fired normally to increase the production rate. Such an operational practice in the real sense helps marginally.
- A major portion of the additional fuel fired is carried away into the stack as waste heat. It is therefore, advisable to contain the flame within the furnace.
- It is also desirable to provide the combustion volume adequate to the heat release rate'. As a thumb rule, for direct fired heating furnace, a heat release equivalent to 700-900 kcal/hr. per cubic metre of combustion volume may be used.

Operating furnaces at the desired temperature :

For any given industrial heating or melting operation, there is an optimum temperature for operation of the furnace.

Operating at too high temperature will not only mean unnecessary waste of fuel in terms of heat, but will also cause overheating of the stock, its spoilage or excessive oxidation and decarburization as well as over-stressing the refractories. The furnace are often run without any temperature controls, often with the 'on-off control', which is extremely harmful to the optimum performance of the furnace. In the 'off condition, only the atomising air enters the furnace bringing down its temperature rapidly so that when the fuel firing process is recommended, the amount of fuel supplied to the furnace to bring up the temperature, is much more than would be necessary had the furnace been operated on a 'proportional control'.

Reducing heat losses from furnace openings :

In fuel fired furnaces, substantial heat losses occur through furnace opening. For every large opening, heat loss due to openings may be calculated by computing black body radiation at furnace temperature, and multiplying these value with emissivity (usually 0.8 for furnace brick work), and the factor of radiation through openings.

All unnecessary openings of the furnaces should be sealed and the flame be contained within the furnace.

Doors should be tightly shut and made of light material that is durable & heat proof. Doors that are hinged or swinging type may be provided. To minimise heat losses, the refractory lining should be sufficiently thick and of good insulation quality.

In order to reduce fatigue as well as to ensure ease in closing and opening, vertically lifting doors balanced by counterweights, should be used.

Minimising furnace wall/skin/surface heat losses :

About 30-40% of the fuel input to the furnace generally goes to make up for heat losses in intermittent or continuous furnaces. The appropriate choice of refractory and insulation materials achieves fairly high fuel saving in industrial furnaces.

Radiant heat loss comparison at various temperatures from an open furnace door is tabulated below:

C_1 (0C)	C_2 (°C)	K1 (C_1 + 273)	K2 (C_2 + 273)	$(K_1/K_2)^4$	Relative Radiant heat Transfer
700	20	973	293	122	1.0 (baseline value)
900	20	1173	292	255	2.1
1100	20	1373	293	282	3.96
1300	20	1573	293	830	6.83
1500	20	1773	293	1340	11.02
1700	20	1973	293	2056	16.91

(C_1 is hot body temperature and C_2 is ambient air temperature)

In other words, the radiation heat losses from an open furnace door at 1500°C are 11 times greater than the losses of the same furnace at 700°C.

Determining black body radiation at a particular temperature

If a furnace body has an opening, the heat in the furnace escapes to the outside as radiant heat. Heat loss due to opening can be calculated by computing black body radiation at furnace temperature, and multiplying these value by emissivity (usually 0.8 for furnace brick work), and the factor of radiation through openings. The black body radiation losses can be directly computed from the curves as given in Figure 42.8 A.

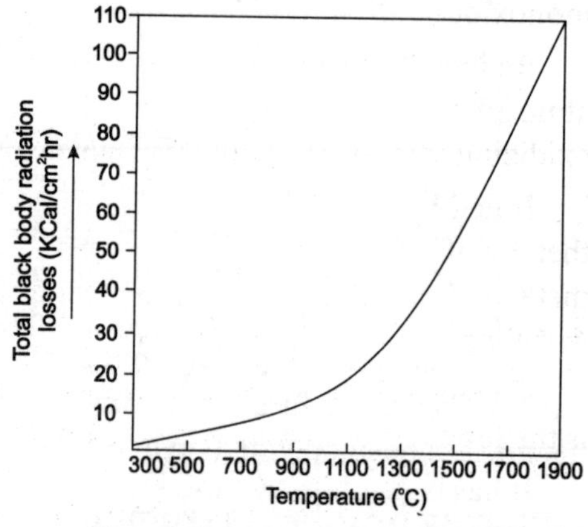

Fig. 42.8 A. Black body radiation

Several material with different combinations of heat insulation and thermal inertia should be considered to minimise heat losses through furnace walls. The use of insulating refractories of appropriate quality and thickness can cut down heat storage capacity of walls, and time required to bring the furnace to operating temperature by as much as 60-70% in intermittent furnaces.

Insulating materials owe their lower thermal conductivity to the pores; while their thermal inertia depends on the weight per unit volume of solids and its specific heat. There are in general, three types of insulating bricks :

- Product made from diatomaceous earth.
- Products derived from vermiculite.
- Refractory (fireclay or silica) based products known as insulating refractories and are used as inner lining in furnace.

There are two methods of insulation normally used for high temperature furnaces. When the furnace is operating continuously under severe condition of atmosphere or high temperature, a lagging of diatomaceous insulation is placed behind a dense refractory base. In other insulations where cleaner conditions or a lower temperature prevail, particularly in short cycle intermittent furnaces and kilns, the insulating material can itself form the refractory lining with no back up refractory bricks.

Materials of the latter type are highly porous refractory-and have fair strength & spalling resistances, and can in general be used up to a temperature as high as 1800°C.

Another class of materials (ceramic fibres) with very low thermal conductivity, and extremely light weight, with the capacity to withstand thermal shocks can also be used in intermittent furnaces.

Control of furnace draught:

In any furnace, the entrance of uncontrolled free air must be prevented. It pays to maintain a slight excess pressure inside the furnaces to avoid air infiltration. Large furnaces, especially those with stack draft, are equipped with pressure sensing devices which adjust damper or the flow of protective gas into the furnace.

At furnace temperatures, oxygen attack the common metals and forms oxides or scale. Carbon monoxides and hydrogen reduce the oxides to metal.

Reducing atmospheres (excess of fuel) form less scale than that which is formed by oxidizing atmospheres (excess of air) but the scale is tight and adheres firmly. More scale is formed in oxidizing atmospheres but this scale is more easily removable.

If negative pressures exists in the furnace, air infiltration occurs through the cracks and openings thereby affecting air-fuel ratio control. Neglecting furnace pressure could mean problems of cold metal and non-uniform metal temperatures, which could affect subsequent operations like forging & rolling and result in increased energy consumption.

The furnace pressure is constant in different parts of the furnace. The zero level in the furnace is the level at which the furnace pressure equalizes the atmospheric pressure.

It has been observed that minimum consumption of fuel occurs at considerably higher furnace pressures, than that required to keep zero level at the hearth. This phenomenon is due to exfiltration. Exfiltration is less serious than infiltration. Some of the associated problems with *exfiltration* are *leaping out of flames, overheating of the furnace refractories* leading to reduced brick life, increased furnace maintenance, *burning out of ducts* & equipments attached to the furnace etc.

Optimum furnace capacity utilisation :

When the initial design of a furnace is considered, the aim should be to obtain the correct furnace for the jobs in hand *i.e.* whether the continuous furnace or a batch type should be used.

If stock can be fed continuously at one end of a furnace and discharged through the other, then the overall efficiency will increase with load recuperation from the waste gas products.

If it is not possible to use a continuous furnace, then careful planning of the loads for batch type furnaces is essential. A furnace should be recharged as soon as possible (within the metallurgical and other physical limitations) to enable any residual furnace heat to be used.

Furnace loading :

One of the most vital factors affecting efficiency is the furnace loading. There is a particular loading at which the furnace will operate at maximum thermal efficiency. If the furnace is under-loaded, a smaller fraction of the available heat in the working chamber will be taken up by the load and therefore the efficiency will be low.

The optimum loading is generally obtained by trial, noting the weight of material put in at each charge, the time it takes to reach a given temperature and the amount of fuel used.

Care should be taken to load a furnace at the rate associated with optimum efficiency, although it must be realised that limitations to achieving this are sometimes imposed by work availability or other factors beyond operational control.

The "no-load" consumption to maintain a simple batch type furnace at 1300°C when empty is about 70% of that required to operate at the optimum loading rate.

The disposition of the load on the furnace hearth should be arranged so that:

- It receives maximum radiation from the hot surfaces of the heating chamber and the flames.
- The hot gases are efficiently circulated around the heat receiving surfaces.
- There should be adequate spacing between the stocks. Overlapping of materials, results in non-uniformity of temperature and should be avoided.

Stock should not be placed in the following positions :

- In the direct path of the burners or where flame impingement is likely to occur.
- In an area which is likely to cause a blockage or restriction of the flue system of the furnace.
- Close to any door openings where cold spots are likely to develop.

For fuel economy and work quality, the materials comprising the load should only remain in the furnace for the minimum stipulated time to obtain the required physical & metallurgical requirements.

When the material attain these properties they should be removed from the furnace to avoid damage and fuel wastage. For heating steels, the general rule is that heat travels 3 mm in 5 minutes. For cylindrical square bars, prescribed heating rate is 20 minutes per 25 mm of diameter of mild steel; and 40 minutes per 25 mm of diameter for alloy steels. For annealing of steel, 60 minutes are needed per 25 mm of diameter or thickness. The above values hold for the furnaces where the stock size is relatively small with respect to size of the furnace.

Furnace should be operated at their optimum rated capacity to achieve best fuel economy by selection of appropriate size & type of the furnace and material handling system. Frequent breakdowns in the production machines make the furnace idle for lengthy periods, thus causing extra expenditure of fuel. Economy could be achieved by collecting an adequate sequence of charges, operating the furnace continuously at the most economical loading and then shutting down entirely for a given period if sufficient work is not available. When shut down, all sources of air leakage should be eliminated.

Waste heat recovery from furnace flue gases

In any industrial furnace the products of combustion leave the furnace at a temperature higher than the stock temperature. Sensible heat losses in the flue gases, while leaving the chimney, carry 35 - 55% of the heat input to the furnace.

The higher the quantum of excess air and flue gas temperature, the higher would be the waste heat availability.

The sensible heat in flue gases can be recovered by :

- Charge preheating
- Preheating of combustion air

Besides, there are other waste heat recovery techniques such as steam generation through waste heat boilers, gas turbine systems, vapour absorption refrigeration system etc.

The direct recovery of energy from waste gases can be achieved in metallurgical furnaces by preheating the incoming raw materials. The effectiveness of charge preheating depends upon the physical and dimensional properties of the material being heated alongwith the handling methods employed.

A continuous furnace of pusher type can be so operated that it utilises the waste heat by preheating the stock. If the rate of heating is low, most of the heat is imparted to the stock at the hot end, and the products of combustion can preheat the slowly moving ingots.

Considerable fuel savings can be done by preheating the combustion air, by recuperator and the regenerator.

In a recuperator, heat exchange takes place between the flue gases and the air through metallic or ceramic walls. On the other hand, in a regenerator, the flue gases and the air to be heated are passed alternatively through a heat storing medium *(e.g.*, chequor bricks), thereby resulting in transfer of heat.

Recuperators

They are of three types namely counter-flow, parallel flow & cross-flow, according to the flow of the hot flue gases and of the cold combustion air.

Counter flow type allows the highest temperature of preheat to be reached whereas the parallel flow type gives the lowest maximum temperature of the recuperator wall.

The disadvantage of the recuperator is that it involves a convection process on each side of the separating wall and conduction through the wall. Consequently, the overall heat transfer co-efficient is low and large surfaces are required for high heat recovery.

Recuperators for relatively low temperature processes are made of steel and cast iron but for high temperature ceramic refractory material *(e.g.* SiC) are used. The high temperature gives a high heat transfer rates but they thick walled and bulky with a lower thermal conductivity than metal.

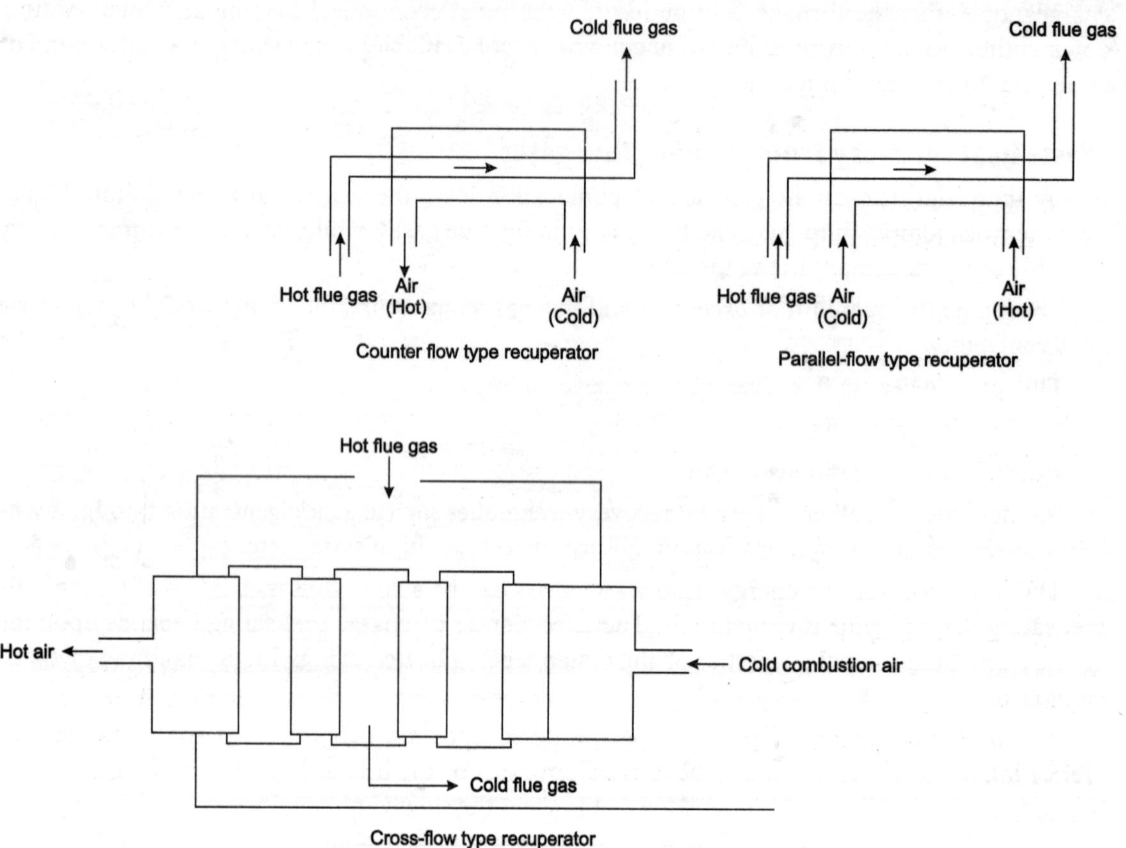

Fig. 42.9. *Various types of recuperators.*

REGENERATOR

This equipment employs a discontinuous process. Two independent masses of chequor bricks are provided. At any moment the hot flue gases are passing through one mass of brickwork in one direction and heating it up while the cold air is passing through the second mass in the opposite direction, being heated in the process. After a fixed interval of time (usually 30 minutes) the process is reversed,

Advantages of Regenerator over Recuperator :

- One convection surface and the separating wall are eliminated
- Brickwork can be used over a wide range of temperatures.

- It is a cheap form of heating surface and can be used in standard brickwork shapes.
- It can store larger quantity of heat because its refractory can be heated upto say 900°C and can be cooled upto say 600°C at the end of the cycle.

Regenerators are commonly used where large quantities of heat are involved *e.g.* in open hearth furnaces, coke ovens, etc.

Disadvantages of Regenerator Compared to the Recuperator :

- Chequer brickwork is heavy, bulky and massive unlike recuperators whose thin metallic walls facilitates rapid heat transfer.
- It cannot be employed usefully on very small furnaces. Moreover it is expensive in initial cost.
- Preheat temperature fluctuates with the cycle which necessitates the use of several chambers for smoothening the fluctuations.
- Changeover valves need automatic control and constant supervision for its smooth operations.

Advantages of hot *i.e,* preheated combustion air are :

- Saving in fuel consumption.
- Increase in flame temperature.
- Improvement in combustion.
- Reduction in initial heating time.
- Reduction in scale losses.

In a recuperator, heat exchange takes place between the flue gases and the air through metallic or ceramic walls. On the other hand, in a regenerator, the flue gases and the air to be heated are passed alternatively through a heat storing medium, thereby resulting in transfer of heat.

DISTRIBUTION OF HEAT IN A FURNACE

Most of the heat is released in the combustion zone at the left and travels from there to the right. The heat thus liberated should be absorbed (1) by the charge, which is most desirable. But heat also goes elsewhere; some of it is passed into the furnace walls and some into the hearth (3) increasing the temperature of those parts. Another portion of the heat is lost to the surroundings by radiation and convection from the outer surface of the walls or by convection into the ground (2). Through cracks or other openings heat radiates away (4), and furnace gases pass out around the door, (5) frequently burning in the open and carrying off heat. Heat is lost everytime a door is opened. Heat is lost along long pieces of the metal from the part which is outside the furnace (6). Finally, heat passes out with the products of combustion and (7) either in the form of sensible heat or as potential heat due to incomplete combustion.

Heat lost from furnace walls or wall loss. The heat loss from the outside of furnace walls is one of the salient features affecting the economy of furnaces. The wall loss during continuous uninterrupted operation of a furnace differs from the wall loss of the same furnace if the latter is operated intermittently.

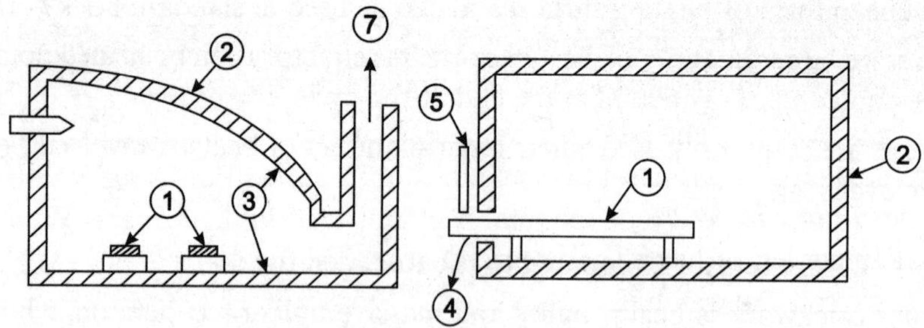

Fig. 42.10. Heat distribution in a furnace.

Wall loss during steady operation. For steady flow of heat through a wall, the conditions as in Fig. 42.11 are used. The temperature of the products of combustion exceeds the temperature of the surface of the wall. The heat loss of walls made of rammed fire clay is lower than the heat loss by dense firebrick because of the thermal conductivity of the plastic clay is (after drying and burning in place) about 60% of the conductivity of dense firebrick. Gas leakage increases the wall loss but it would be wrong to add the wall loss by conduction because leakage increases the temperature of the cold face of the wall and thus reduces the heat flow by conduction. Heat loss by conduction and leakage is greatly reduced by enclosing the furnace in a steel jacket. The loss per unit area is much greater near edges than near the centre. In fact, about 1/3 of the total loss occurs through the outer 5% of the width.

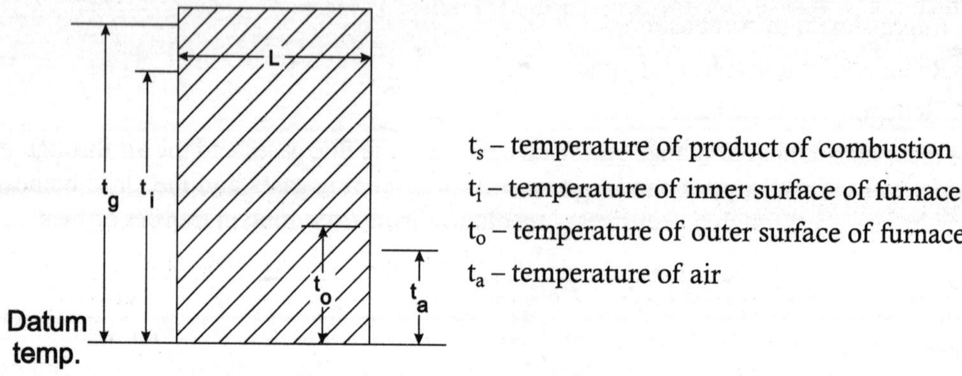

Fig. 42.11. Temperature profile in a furnace.

Wall losses during intermittent operation. In actual practice, operating periods (on) alternate with idle periods (off). During the off-period, the heat stored in refractories during the on-period is gradually dissipated, mainly by radiation and convection from cold face. In addition, some heat is taken out by air flowing through the furnace. Dissipation of stored heat is a loss, because the lost heat is at least in part again imparted to the refractories during the next on period.

For short periods of operations between long shutdowns, these walls on insulating refractories are preferable, because little heat is stored in them. The heat loss by dissipation of heat stored in the refractories is determined by on-time and off-time. If a furnace is operated 24 hours every third day, practically all of the heat stored in the refractories is lost but if the furnace is operated 8 hours per day not all of the heat stored in the refractories is dissipated.

Heat Loss by Radiation through Openings. The inner red bricks radiate considerable heat like black bodies through the more or less wide cracks and door opening.

Heat loss by gas leakage. The loss of heat that is caused by furnace gases out around doors is hard to estimate. Leakage of gas has a distinct effect on heat loss from a wall. With tightly fitting doors, the loss need not exceed 2% of the total heat delivered to the furnace. With loosely fitting doors & with a flame or jets of heated gases projected against the doors it may easily reach 8% and at time may even exceed that amount.

SPECIAL HEAT LOSSES

Heat loss due to part of the stock projecting out of the furnace. Heat is lost due to dissipation of heat from metal portion projecting from the inside of the furnace into the air. Throughout the heating period, heat flows from the part that is in the furnace to the part that is outside, and a portion of this heat is dissipated from the latter part, by radiation & convection.

Heat loss to tongs and charging machines. Cold tongs absorb heat while in the furnace. Tongs are always cold when they enter the furnace because either they are water cooled or else have been in a water trough. Although the flow of heat into tongs is less than the heat that is radiated through the open door, it is by no means inconsiderable.

The heat transmitted to the tongs is the product of (1) exposed surface of tongs, (2) time interval during which the tongs is inside the furnace and (3) radiation factor.

Losses to trays, conveyer chains and rollers. The over looking of this item has caused important furnaces fail to meet the capacities and fuel consumption guarantees. Trays or other containers usually attain the same temperature as the stock and the amount of heat that they take out of the furnace can be ascertained from their weight & the heat content data.

Potential and sensible heat carried out of furnace by products of combustion.

In combustion type furnaces, two additional sources of heat loss occur. These losses are caused by the heat energy that the products of combustion take out of the furnace in the shape of unburnt fuel and in the form of sensible heat due to high temperature. The heat loss is commonly known as stack loss.

THERMAL EFFICIENCY

The attainment of high thermal efficiency is very desirable in all heating operations. This means that the calories available in the fuel must be directed as completely as possible into the stock with the least possible loss to the surroundings and after the heat has had the desired effect on the stock it should be re-used for some other purpose if at all possible.

One way of reducing the rate at which heat is lost to the surroundings is to use insulation or padding on the furnace, between the hot chamber and the outside wall. The use of insulation is well established, where low or moderate temperature are involved in steam lines, hot blast mains and small batch heat-treatment furnaces, but it is also practicable in moderately high temperature units like recuperators, pottery kilns and some heat-treatment & billet heating furnaces, particularly where counterflow heat transfer is being applied.

High Emissivity Ceramic Refractory Coating In Furnaces

Emissivity is the capacity of a material to radiate heat. Emissivity of black body is taken as 1 which is the maximum possible. Emissivity is expressed as fraction or the percentage of the ideal black body radiation of the heat. The closer the emissivity value comes to 1, the more effective is the transfer of heat by radiation. Refractories are used in internal lining of the furnaces which affects its optimum performance and efficiency. High emissivity ceramic coating of the internal refractories of the furnace helps in energy conservation in the furnace operation by way of saving/ reduction in fuel consumption.

Ceramic coatings in the furnace inner chamber promote rapid & efficient transfer of heat, uniform heating and extended life of refractories. The emissivity of conventional refractories decreases with increase in temperature whereas for ceramic coatings it increases slightly. This outstanding property has been exploited by using ceramic coatings in hot face insulation. Ceramic coatings are high emissivity coatings and a have a long life at temperatures up to 1350°C. There are two types of ceramic coatings: those used for coating metal substrates, and those used for coating refractory substrates. The coatings are non-toxic, non-flammable and water based. Applied at room temperatures, they are sprayed and air-dried in less than five minutes. The coatings allow the substrate to maintain its designed metallurgical properties and mechanical strength. Installation is quick and can be completed during furnace shut down. Energy savings of the order of 8-20 percent have been reported depending on the type of furnace and operating conditions.

The heat transfer in furnaces are mainly by radiation at higher temperature, and the higher emissivity of the *refractory* lining of furnace increases the overall net heat transfer. There is a decline in emissivity of conventional refractory materials with increasing furnace operating temperature. But the emissivity of high emissivity ceramic *coating material* increases with rise in temperature as shown in Fig. 42.12. Hence the internal refractory surface of a furnace is coated with high emissivity ceramic coating material. The coating material is zircon based which increases the emissivity of refractory lining at higher temperature thereby enhancing the radiative heat transfer rate to the charge (stock) inside the furnace resulting in uniform heating condition and high fuel efficiency & saving. The emissivity of refractories as well as that of zircon based coating material is about 0.8 at normal temperature but with increase of temperature to say 800°C, the emissivity of various refractories drops to as low as 0.4 to 0.6 while that of the zircon based coating material increases to as high as 0.95. Ceramic coating material for metallic components for furnace are also available and they comprise of silica-alumina or ball clay silica.

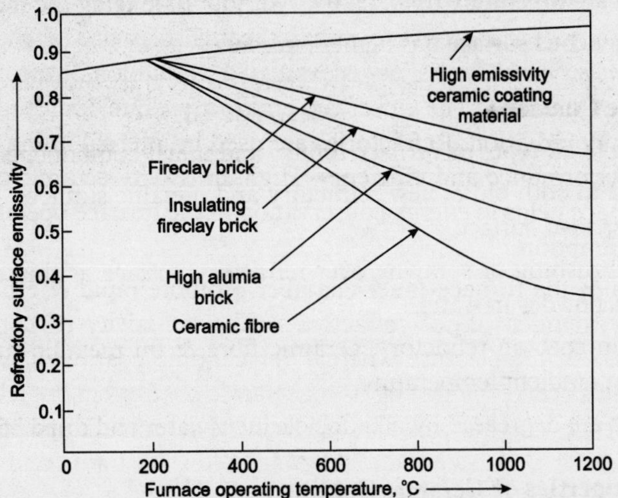

Fig. 42.12. *Emissivity of various refractory materials Vs. furnace operating temperature.*

Composition of ceramic coating material:

- It is a white yellowish zircon powder containing > 50% ZrO_2.
- Solids comprising of zirconium oxide is also used as coating material. Acidic vehicle for zircon and zirconium oxide is a mixture of SiO_2 & Al_2O_g particles that form a colloidal suspension.
- Ceramic coating material for metallic components comprise of silica-alumina and ball clay silica.

Advantages of Ceramic Coating :

- Ensures high emissivity of refractory surface.
- It provides refractory surface protection.
- Results in reduced heat losses and thus energy savings.
- It provides good thermal shock resistance to bricks. There is no crack after repeated heating and cooling at 1350°C several times.
- Provides resistance to chemical attack.
- Results in fuel savings ranging from 15 to 30%.
- It ensures quick heating and uniformity in temperature of refractory bricks
- Ensures longer refractory surface life.
- Increases the radiative power of the refractory surface.
- It keeps the furnace dust free and reduces the maintenance cost.
- Ceramic coating material for metallic components prevents its oxidation upto 1250°C.
- Polished homogeneous surface has higher emissivity.
- Refractoriness of the coated surface is higher than that of the conventional refractories and metals.

- It can withstand a temperature upto 1500°C without damaging the mother structure.
- Polished homogeneous surface has higher emissivity.

Application of Ceramic Coating :

- It can be applied to all types of furnace linings and metal components.
- It can be applied to both old & new refractory and metallic surfaces. It is equally suitable for uneven and spalled surface.
- It is applied by brushing or spraying over refractory surface as thin coat with 25— 40% suspension of its powder in water.
- Applied as a thin coat on refractory, ceramic fibre & on metallic surface, air dried and instantly cured at ambient temperature.
- Metallic surfaces are degreased by alkaline cleaner/water and dried before coating.

Typical Parameters/Properties of Ceramic Coating :

- Coating thickness = 3 mm (maximum)
- Average service temperature = 1350°C (maximum 1500°C)
- Specific consumption of coating material = 6 kg (maximum)/m^2 of surface
- Surface emissivity after coating = 0.95 at > 800°C
- Density after sintering at 1300°C = 3-3.2 gm/cc
- Thermal expansion = < 0.2%
- Water absorption after heating at 1350°C = < 2%
- Wettability to molten metal — nil (high resistance to molten metal like aluminium and steel)
- Nature of coating — It is insert, non-toxic and non-inflammable after curing for 10 minutes at 260°C.
- Scratchability of coated surface — can not be removed with moderate chieseling
- Coated surface look after heating-even, light buff color
- Prevents fluxing and oxidation of refractories at elevated temperature upto 2480°C.
- Presence of black materials like FeO and Cr_2O_3 in refractories increases its emissivity.

Usage/Areas of Applications of Ceramic Coating :

Used in all types of furnace linings and associated metal components (used in both old and new and smooth/uneven and spalled surfaces). Applied mostly in following furnaces:

(a) Blast furnace bosh, stack and throat

(b) Soaking pits

(c) Reheating furnaces

(d) Sintering plant furnace

(e) Coal fired boilers

(f) Primary reformer used for steam reforming of naphtha in fertiliser industry
(g) Continuous annealing furnaces
(h) Forge furnaces
(i) Stress relieving furnaces
(j) Heat treatment furnace
(k) Crude oil heaters
(l) Process heater for distillation
(m) Ceramic kilns
(n) Calcination kilns

Few companies in India have developed the expertise in the technology of ceramic coating and undertake the job on turnkey basis.

WASTE HEAT RECOVERY & FUEL ECONOMY IN FURNACES

HEAT BALANCE IN FURNACES

In case of industrial furnace and boilers, energy is supplied in the form, the heat either to the charge to get heated to the required level or to the water to produce steam of required quality, respectively. The mass and energy (heat) balances are based on the laws of conservation. Thus, the material and energy which are going into the process must come out without any loss either in the same or any other form. For Nuclear fuels, this theory is not true, where mass disintegrates to give energy.

In furnaces operating on combustion of fuels, the fuel becomes an outstanding item of expenditure and its economy is essential. For successful design of a furnace, knowledge of all probable items of consumption and loss is very much essential which ultimately compute the fuel consumption rate.

Thermal efficiency of a furnace which is the ratio of the quantity of heat utilized to the total quantity of heat supplied, varies in wide ranges, from as low as of 10% in some metallurgical furnaces to 60% or even more than 90% in case of a well designed thermal plant. The boiler low efficiency of some metallurgical furnaces is due to small amount of heat that is actually consumed for useful purposes.

The assessment of fuel economy or efficiency can be judged by the heat balance which accounts all the items for the input & output of the process and balancing them. In making a heat balance, all the items should also be expressed in the same unit so that all of them are comparable. Heat balances should be based on either a definite period of time, such as an hour/day or time required for one cycle of operations or on unit quantity of initial/finished products.

For a heat balance the input and output items are separately taken as given below :

Heat Input items :

- Chemicals or potential energy of fuel; sensible heats of fuel, air for combustion and charge; electrical or mechanical energy supplied to auxiliaries such as fans or pumps.
- Heat evolved due to exothermic reaction (if any).

Heat Output items :

- Total heat content of the outgoing material (sensible and/or latent heats).
- Heat absorbed in endothermic reaction or in change of phase or in raising temperature and then change of phase (if any).

- Total heat of the exit gases (sensible heat, and latent heat of steam going out, potential heat of fuel).
- Heat losses through walls, roofs, cracks & fissures by conduction and radiation.
- Miscellaneous and special losses, related to the process involved.

An overall heat balance, comprising of the total energy entering & leaving is relatively simple and gives an idea of the overall performance only, but not the effective contribution of the different sections like fuel bed, heating chamber, recuperator, regenerator, etc. To get the actual performance of each individual parts of the system, a detailed heat balance must be prepared for each item which may be very labourious and not so simple. But unless individual balances are made, design of these units would not be possible. In case of input items, as mentioned earlier, the chemical or potential energy is calculated from product of quantity of fuel & its calorific value, sensible heat from the product of mass of the material, difference in temperature from the datum; and specific heats. The thermal effects, both endothermic & exothermic that may occur in the process may be collected from literature and calculated on an accepted basis. All these can be tabulated inside as input & output items and the balance is made.

HEAT DISTRIBUTION IN A FURNACE

In the operation of a furnace, one must have a clear idea of the distribution of the heat supplied to the furnace. The heat of combustion evolved in the combustion zone is transferred by all three modes of heat transmission, (that is, conduction, convection and radiation) to the stock. This is known as the *actual heat load* of the furnace. Besides this, heat is transferred outside through the walls & roof; through the floor to the ground and through cracks & other openings in the furnace. Heat also passes out through the door when opened. A major quantity of heat goes out of the furnace chamber with the hot flue gases, as potential as well as sensible heat. All these quantities when added together comprise the actual amount of heat to be supplied to the furnace through combustion of fuels so that the furnace operates at its design capacity.

Except the actual heat load, all other items constitute the losses to the furnace, which is termed as waste heat. If proper utilization or recovery of this otherwise wasted heat is made through suitable means or appliances, an equivalent amount of saving in the fuel consumption can be done, improving overall efficiency of the furnace appreciably. The system falling under this is called waste heat recovery system.

SOURCES OF HEAT LOSS FROM A FURNACE

Only a part of the heat input given to a furnace is utilised by stock or the process and a major part goes unused alongwith the flue gases and also get lost through walls & opening of the furnace.

Heat losses from furnace can be classified into three categories :

- Heat loss from furnace walls.
- Heat loss from opening and cracks.
- Heat loss through waste gases.

Heat loss from furnace walls. Heat is lost to the surroundings by radiation & convection from the outer surface of furnace walls and by conduction into the ground. Furnace walls built of insulating refractory bricks and encased in steel shell reduce flow of heat to the surroundings. Reduction of heat loss, by providing better insulation does not necessarily mean saving of money, because the fixed charges or cost of insulation may exceed the cost of fuel saved. Hence, amount of insulation should be carefully decided in view of overall economy.

Heat loss from openings and cracks. There may be various openings on the furnace, for example, peep holes, inspection doors, charging & discharging doors, which cause considerable heat loss. The charging and discharging doors must be opened for the minimum period that is only during charging & discharging of stock. Other openings on the furnace should also normally be kept closed to eleminate heat loss by leakage of hot gases from furnace or by infiltration of cold air into the furnace. Cracks on the furnace walls should be repaired to avoid leakage of gases.

Heat loss through waste gases. It can be reduced by having correct fuel air ratio to ensure complete combustion of fuel with minimum possible air. If combustion is incomplete, unburnt fuel will pass out along with flue gases. On the other hand, higher amount of air would require heat for its being heated to furnace temperature and then same will also pass out with flue gas.

WASTE HEAT RECOVERY

Inspite of using the fuel wisely and not wasting heat by using too much or too little excess air, by having inadequate insulation or by having poor heat transfer characteristics in the appliance, the heat will still be discharged from the equipment because of the temperature at which the process is carried out.

To take an example, if a furnace is heating steel to 1200°C, the primary fuel must be providing a temperature greater than 1200°C; and all the while that there is any load on the furnace, be it the heating of the steel or furnace losses, the flue gas temperature must be above 1200°C. In a batch furnace the gases are discharged from the furnace at this temperature even in the best maintained and set up system. The gases leaving the furnace at this temperature constitute a large proportion of the energy supplied to the furnace in the form of fuel. It is this heat which is being wasted.

This waste heat may be defined as the heat in the products of combustion discharged from an installation when its primary functions are fulfilled. A fundamental consideration is that waste heat is not wastage or surplus on account of any inefficiency of the primary plant but that it is no longer usable for its primary purpose.

The heat content of products of combustion rise with temperature and so, obviously, the hotter the waste gases the greater is the quantity of heat that is available for recovery.

Fig. 43.1 shows the variation of heat content of the stoichiometric combustion products of natural gas with temperature. The heat content axis of the graph is shown in Btu/sft^3 of the combustion products and as the percentage of total heat available from the fuel. From this we can see that if the gases are discharged at 1300^{C}C (a not uncommon occurrence in steel reheating, ceramic firing, or copper melting), 70% of the fuel goes to waste in the flue products. Even at lower temperatures substantial quantities of heat are lost in this way.

Fig. 43.2 shows the maximum possible efficiency of a process over a range of flue gas exit temperatures. This is the converse of Fig. 43.1. The lines on the diagram shows the boundary between useful heat and waste heat at various process temperatures. These diagrams are true for any combustion process including combustion engines and gas turbines.

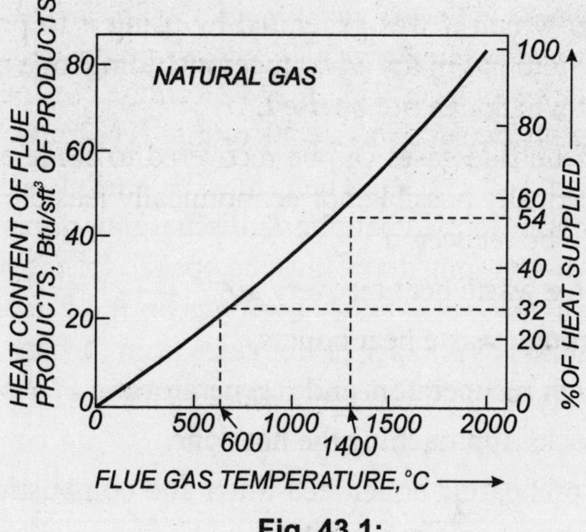

Fig. 43.1:

The quantity of heat available in these waste gases is therefore large and it is of interest to note the relative amounts of heat lost by excess air firing and in waste heat. Fig. 43.3 shows the variation in the quantity of heat in the combustion products produced by a fixed volume of gas at a range of excess air factors. From this diagram it is seen that the flue gas loss increases with increasing excess air levels. However, even at zero excess air the quantity of heat lost at high temperatures is considerable.

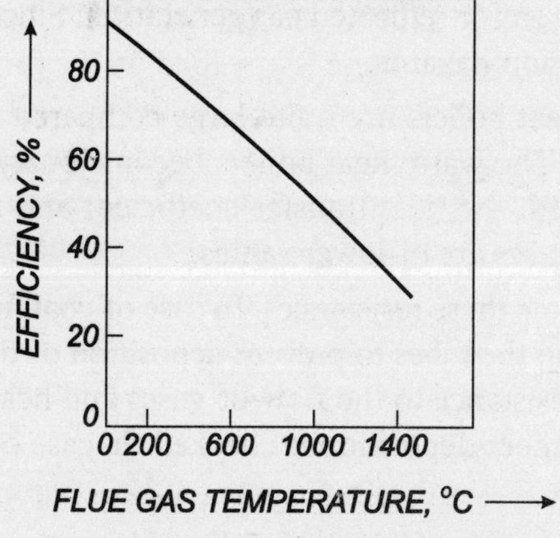

Fig.43.2: Graph of efficiency of heat Transfer from combustion products against flue gas outlet temperature

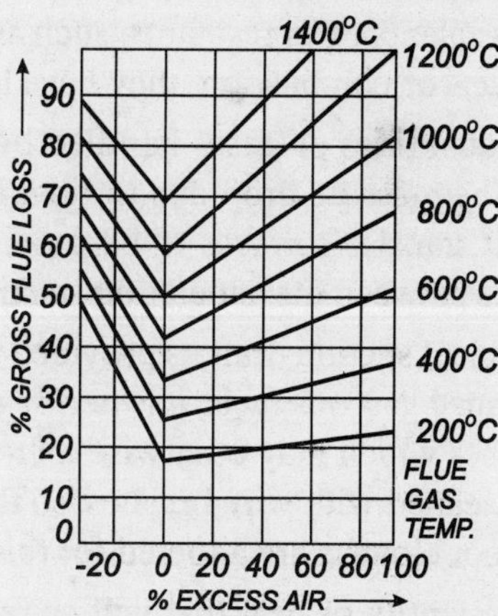

Fig.43.3: Gross Flue Losses –Natural Gas

USE OF WASTE HEAT AS A SECONDARY HEAT SOURCE

In all industrial process heating, valuable heat is wasted in the flue products. This may be regarded as a secondary source of heat and may make a real economic contribution to the overall operation of thermal processes if, it is used effectively/reused.

In some cases, it is possible to use the flue gases directly in a nearby process or for preheating or drying of material to be processed. However, in most cases, the waste heat is most effectively used directly in the process in which it was generated by using it to preheat the air required for combustion. This recycling system has many advantages including; one of which is that the system is never generating waste heat which can not be used.

Only the sensible heat of the flue gases can be recovered to some extent. Complete recovery of waste heat is neither theoretically possible nor economically feasible and hence only optimum quantity of heat is advisable to be recovered.

The principal techniques for waste heat recovery are :

- Recovery as steam through waste heat boilers,
- As sensible heat through recuperators and regenerators.
- Preheating of solid stocks approaching the hot zone.
- Super-heating steam, preheating boiler feed water and combustion air in boilers.

WASTE HEAT BOILERS

These boilers operate on waste heat of the flue gases. They are useful when the flue gas temperature are below 650–750°C. The primary difference between the fuel & fired steam generating boilers and the waste heat boilers is that the heat transmission in the former is largely due to radiation whereas in the latter it is essentially by convection & conduction as the waste gases enter at comparatively lower temperatures. Waste heat boilers are mostly fire tube boilers, where the sensible heat of the gases is only used for converting water to steam. Steam thus produced is used as heating medium in process plants or as reactants, such as in water gas or reformed gas generation. Since it operates on products of combustion, they have less explosion hazards.

The quantities of gases handled by waste heat boilers are quite large compared to fuel fired ones so the pressure drop due to flow is higher. The waste heat boilers become bulky because of high heat transfer surfaces to be provided as both the heat transfer coefficient and temperature difference between the generating steam & flue gases are of lower values.

Sufficient suction draft, is provided to overcome these resistances. In case of coal fired furnaces or uncleaned gaseous fuels, it is necessary to clean the tubes to prevent deposition of dust particles in the tubes which may otherwise increase the resistance to the flow of gases and heat. The tubes may be cleaned with wire brushes, so that permanent deposition is checked. In case of water tube boilers soot blowers are adopted for this purpose.

The quantity of heat that will be recovered in a waste heat boiler depends on the temperature of the gases. Boiler design is made on the basis of total amount of sensible heat in gases which is directly proportional to the amount and the inlet and outlet temperatures. A lower amount of heat will be transferred if the inlet temperature is lowered and hence the steam pressure become lower with the deterioration of the quality of steam.

The temperature of inlet gas may be lowered by infiltration of cold air into the boiler. It not only lowers the steam production rate but also increases the volume of gases to be handled by the fan and power requirements increase. Suitable precaution are taken to eliminate, the air infiltration into the gas line. *Waste heat boilers* have the following limitations:

- When the flue gas temperatures are higher than 650—750°C, the waste heat boiler are generally not considered.

- Since the heat is recovered in the form of steam, it is applicable to those installations where the steam thus raised can be used totally in the immediate vicinity.
- Small and medium furnace installation find it difficult to use this system.
- When the flue gas temperatures are of higher order, these can be used in conjunction with recuperators thus permitting higher overall waste heat recovery.

RECUPERATORS

Heat transfer from a hot fluid stream to a colder stream is done in heat exchangers in which, the two streams are not allowed to mix each other as they are separated by metallic membranes. Most common are the shell and tube heat exchangers where one fluid moves through the tubes & the other through the shell. The fluid passing through the tubes is known as tube-side fluid and other one which passes through the shell is known as shell-side fluid. They are continuous in operation. Recuperators are heat exchangers made from either metallic tubes or refractory channels. Because of limitations of the operating temperatures for the metals, refractory tubes/channels are constructed for operation at higher temperatures. These become bulky and sometimes leaky because of construction difficulties but act as heat reservoir where the fluctuations in temperature between hot & cold fluids are minimized.

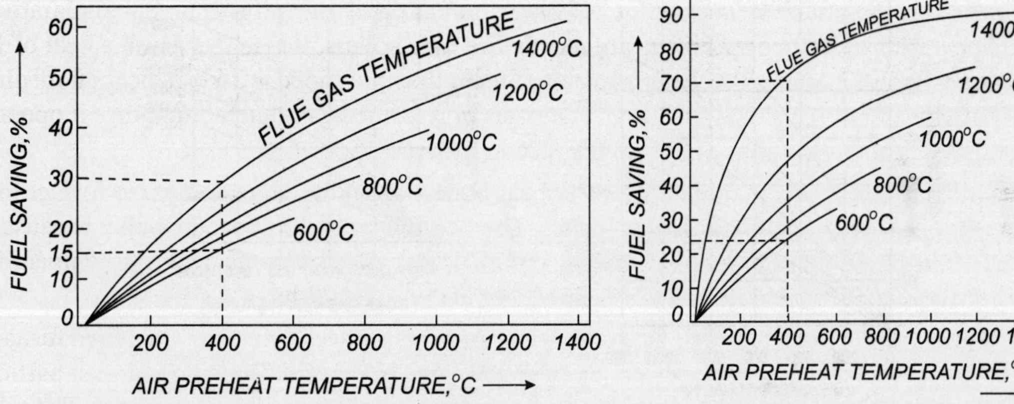

Fig.43.4: Variation of fuel saving against air preheat temperatures for stoichiometric combustion temperatures for 50% excess air (for natural gas)

Fig.43.5: Variation of fuel saving against air preheat temperatures for a number of furnace flue gas outlet (for natural gas)

A recuperator may be considered as a heat exchanger transferring heat across a furnace on one side of which flue gas flows and on the other the combustion air or more rarely fuel gas. The basic difference between this and the regenerator, in operation, is that the recuperator is a continuously operating system while the regenerator works cyclicly (intermittently). In given conditions, the results of recuperation are predictable. Recuperation is normally used on large furnaces due to the high capital cost. The fuel (natural gas) saving that can be achieved by using a recuperator for air preheating can be determined from Figs. 43.4 and 43.5. From this relationship it is clear that the same air preheat represents a greater percentage saving of the total heat input the higher the flue gas temperature. This factor is important when assessing the value of recuperation for various processes. For example (Fig. 43.4), a preheat of 400°C gives approximately a 15% saving with a flue gas temperature of 600°C and a 30% saving with a flue gas temperature of 1400°C. This is because the flue gas loss at 1400°C is 54% of the total heat input whereas at 600°C it is 32%. (Fig. 43.1). Use of excess air for combustion increases the potential of fuel saving as shown in Fig. 43.5. Potential of fuel economy by air preheating by waste heat recovery with the type of fuel is shown in Fig. 43.6.

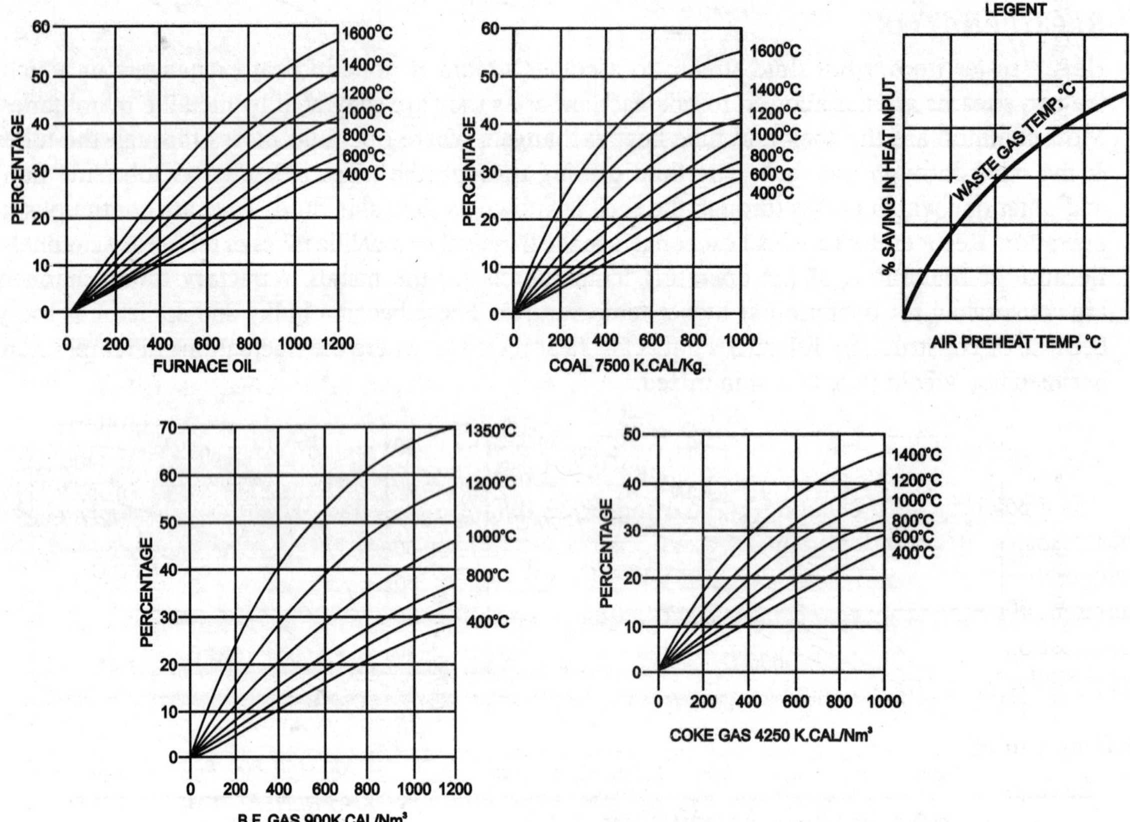

Fig. 43.6: Potential of fuel economy by waste heat recovery for different fuels.

Note: For coke oven gas fuel, preheat air temp. of 4000 ℃ would result in fuel saving of water 20% when the flue gases are leaving at 10000 ℃ frim the furnace.

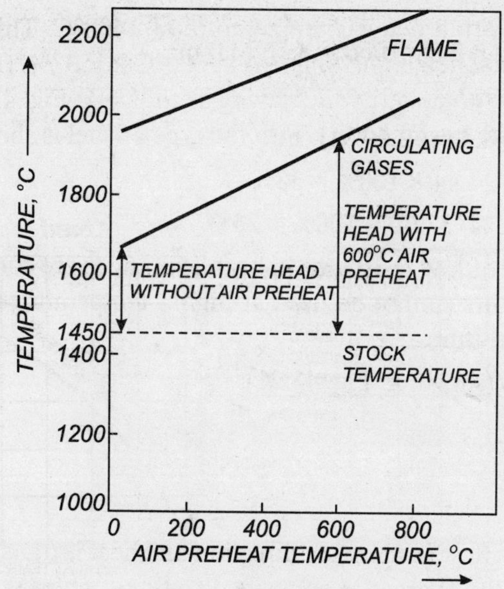

Fig.43.7: Variation of flame temperature with air preheat

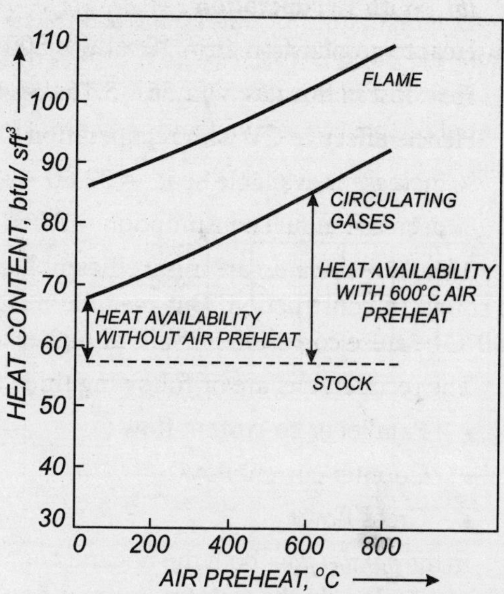

Fig.43.8: Variation of heat content of combustion products at flame temperature with air preheat

As a result of air preheating in the recuperator, flame temperatures are increased and higher heat transfer rates can be obtained. The effect of air preheat on flame temperature and furnace gas temperature is shown in Figs. 43.7 and 43.8 respectively for a process operating at 1450°C. The differential temperature between the cir3culating gases and the stock is doubled by preheating the combustion air to 600°C. The increase thus in heat transfer rate leads to shorter operating cycles, improved efficiency and reduced fuel costs.

Example of energy saving resulting from the use of a recuperator

Example 1. *Consider a gas-fired furnace operating under the following conditions :*

Net CV of fuel: 37 MJ/Nm³ (at S.T.P. i.e. 15°C and 760 mm Hg)

Flue gas temperature : 800°C

Mean specific heat of flue gases 1.60 kJ/m³K

Ratio of combustion products of fuel 10 :1 by volume (same P & T)

Combustion air inlet temperature. : 15°C (1 calorie = 4.18 Joules)

Calculate the fuel saving resulting from the installation of a recuperator designed to cool the flue gases to 200°C if, the recuperator has an efficiency of 85%.

Solution. The solution to this problem involves the concept of Effective Calorific Value given by :

Effective CV = Nett CV – CV of flue gases.

(a) **Without Recuperation** :

Effective CV of fuel = 37 – (10 × 785 × 1.6)/1000 = 37 – 12.56 = 24.44 MJ/m³ *(b)* **With Recuperation** :

Heat to combustion air = 10 ×1.6 × 800 - 200 × 0.85/1000 = 8.16 MJ/m^3

Heat lost in flue gas = 12.56 – 8.16 = 4.4 MJ/m^3

Hence, effective CV with recuperation = 37 – 4.4 = 32.60 MJ/m^3

% increase is available heat = 32.60 – 24.44/24.44 × 100% = **33%**

% decrease in fuel consumption = 32.60 – 24.44/32.60 × 100% = **25%**

Metallic recuperators are more efficient because of low resistance to heat flow and more compact & sound in construction. But metallic recuperators cannot be used at higher temperatures than 700°C because corrosion prevails at higher temperatures.

The recuperators are of following three types (shown in Fig. 43.21).

- Parallel or co-current flow ;
- Counter-current flow ;
- Cross-flow.

In the *parallel flow,* both the hot and cold fluids flow in the same direction and heat is transferred during the flow. In the counter - current flow, the fluids move in opposite directions, whereas in the cross flow, the fluids move at right angles to each other. The temperature of the hot fluid drops and that of cold fluid increases.

The temperature distribution along the length of the recuperator for the parallel, counter- flow and cross flow type is shown in Fig. 43.21. The temperature in and around the tubes vary from top to bottom.

The *counterflow* recuperators are most efficient. The air or fuel gas can be preheated very near to the incoming temperature of the hot flue gases and the temperature difference between the hot & cold fluids is almost constant throughout the length of the passage. In the case of parallel flow the maximum preheat can only be obtained theoretically upto the mean temperature of the two incoming fluids, which is not achieved in practice. The cross-flow type are not very common in use.

Comparative features of parallel and counter-flow recuperators are tabulated in Table 43.1.

Table 43.1: Comparative Features of Parallel and Counter-Flow Recuperators

Parallel Flow Recuperators	Counter-Flow Recuperators
Less heat recovery achieved. Minimum temperature difference between hot and cold ends. Suitable for high temperature applications. Increased life of recuperator. As metal temperatures at hot end is considerably less, cheaper materials can be used.	More heat recovery achieved. Too hot and too cold ends are encountered, hence higher risk of thermal stress failure exists. Suitable for all temperature levels. The materials of construction should be carefully selected for high temperature. Life depends on material of construction. Expensive materials are required to be used at high temperatures.

Recuperator materials — Metallic Vs. Ceramic :

Metals having much higher thermal conductivity, metallic recuperators have certain distinct edges over ceramic ones like higher heat transfer coefficient resulting in smaller volume for the same total heating surface area and less chance of chocking. They are also gas tight, easy to clean and can preheat the combustion air to as high as 500°C. But for high temperature industrial furnaces, metallic recuperators are simply unsuitable for two major limitations *viz.* high waste gas temperature and susceptibility to corrosion by oxygen and other exhaust constituents like sulphur & oxygen under oxidising condition and by carbon under reducing condition due to embrittlement caused by carbide formation. The ceramic recuperators ensure stable operation at waste gas temperatures upto 1200-1400°C and heat up air upto 800-850°C. But poor thermal conductivity of ceramic materials leads to large volume of recuperators and large underground flues.

Desirable properties of ceramic materials to be used as a recuperator material are :

- excellent corrosion resistance against aggressive media (liquid and gas)
- high temperature strength
- high thermal shock resistance
- high refractoriness
- high thermal conductivity
- low linear expansion
- proper mechanical properties

For long time, fireclay materials reigned the recuperator arena but their low thermal conductivity and gas tightness resulted in lower efficiency. Search for substitute yielded high alumina (more than 60% $A1_2O_3$) muliite, cordierite, silicon carbide (SiC) and carbo chamotte (35-39% SiC) materials. SiC is the most ideal as a heat exchanger material due to its high thermal shock resistance, excellent resistance to crack & fatigue and low creep value at high temperature. Different types of SiC and similar materials are available like silicionised silicon carbide, nitride bonded SiC, sintered SiC, Silicon nitride etc.

Depending on the mode of heat transfer, the metallic recuperators are of three types *viz.* radiation, convection and combined radiation & convection type while ceramic recuperators are of combined type normally.

RADIATION RECUPERATOR

In a radiation recuperator, the products of combustion enter the recuperator through an opening in the furnace roof while air flow at higher velocity through a narrow annulus between the outer and inner walls. The outer shell is insulated. Fins are provided on the inner shell to increase the heat transfer area. This type of a simple recuperator is particularly suitable for small forging furnaces. This is shown in Fig. 43.9. A side radiation recuperator is shown in Fig. 43.10.

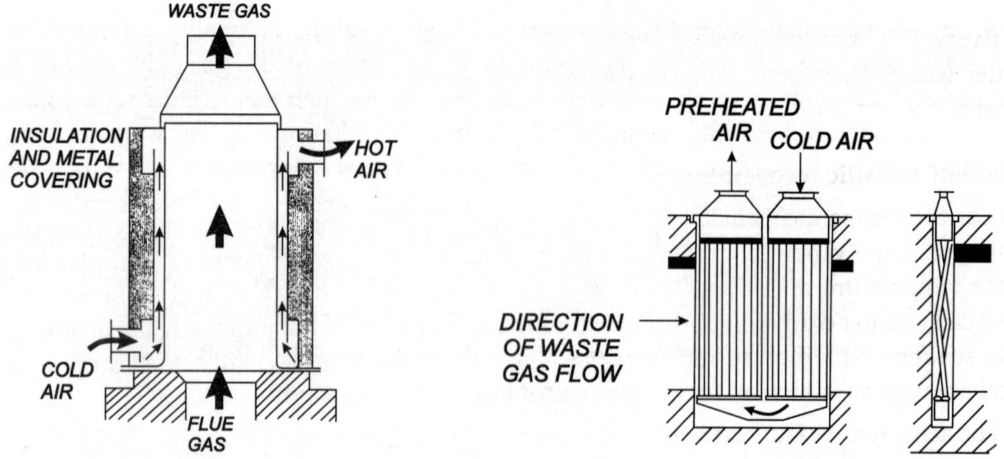

Fig. 43.9: Metallic radiation recuperator.

Fig. 43.10: Side flue radiation recuperator

Radiation recuperators can operate with waste gas temperatures in the range of 1000 to 1500°C to preheat air upto 600°C. The advantage of radiation recuperator over convection recuperator is that the heat transfer is intensive through radiation and also offers possibility of a higher air pre-heat. Other advantages are low pressure drop in flue gas side because of large cross sections, thereby enhancing the possibility of using dust laden waste gases.

Convection recuperator

These types of recuperators can be with a heating system made up of either drawn steel tubes, or cast tubes. Convection recuperators are used with flue gas temperatures from 300 to 1050°C ; the degree of air preheat is limited to 600°C. A tubular convection recuperator is shown in Fig. 43.11.

Combined radiation and convection recuperator :

The advantages of radiation and convection recuperators are combined in the arrangement of heat exchanging surfaces. In this system, a tube bundle is arranged in a ring inside a double shell type recuperator.

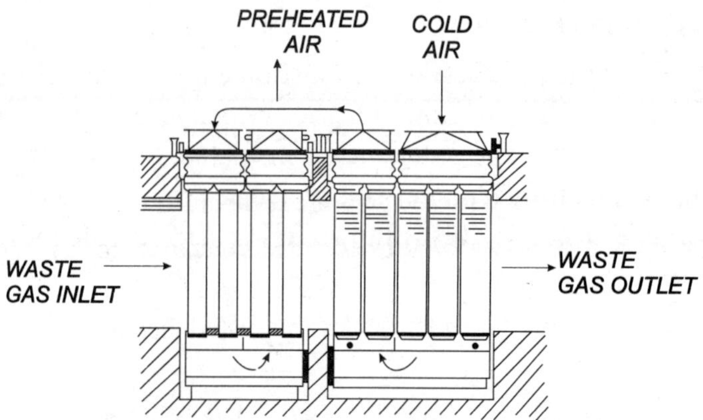

Fig.43.11: Tubular convection recuperator

Every 21°C rise in combustion air temperature results in 1% fuel oil savings. The savings are greater with higher flame temperature, reduced excess air levels and higher productivity of the furnace.

Failure of metallic recuperators :

Parts of the recuperator which reach temperatures of 500°C and above are subjected to oxidation by oxygen, water vapour or carbon dioxide present in waste gases. The intensity of attack depends on the temperature of surfaces as well as the concentration of oxidants in the gases. However, alloys can counter oxidation upto a wall temperature of 1100°C but ultimately the recuperator will suffer scaling and will need replacement. Alloying elements in steel such as chromium, silicon, nickel are used to withstand high temperature services.

The action of sulphur on the waste gases depends on the nature of gases (oxidising or reducing), temperature of gases, wall temperature and on the type of metal used as the heat exchanging surfaces. At the hot end of a recuperator with flue gases of oxidising nature, and the presence of vanadium oxides, sulphur in the waste gases is corrosive. At the cooler end of recuperator, acid corrosion may occur due to the action of sulphur, if metal temperatures drop below the acid dew point of gases. Acid dew point for products of combustion depends mainly upon the sulphur content of the fuel and varies from 130°C with 1% sulphur to about 180°C with 4% sulphur in the fuel. The problem of low temperature corrosion is taken care of at the time of designing the recuperator, by ensuring that the wall temperature in all parts of the recuperator are well above the acid dew point of gases by choosing a suitable gas/air flow pattern and other changes in the constructional features of a recuperator.

High temperature corrosion of recuperators is the corrosion attack on metallic surfaces in a certain temperature range caused through traces of vanadium contained in the fuel ash. The insoluble metal oxides in the fuel ash, viz. vanadium pentoxide and sodium oxide, cause considerable corrosion when the metal temperature is in the range of about 680°C. The hazards of high temperature corrosion can be avoided by maintaining the metal wall temperature of a recuperator below 600°C. This limitation is not there in case of ceramic recuperator.

Cracking and leakage in heat exchanging parts of a recuperator is due to the thermal expansion not being contained fully in different parts of the recuperator. Sometimes, this also occurs in the recuperator which is supposed to expand in a particular direction is restricted due to any shift or changes in the refractories housing the recuperator. This problem can be countered by paying attention to thermal expansion in the metallic recuperator at the design stage.

Comparison of metallic recuperators with the ceramic recuperators :

- Metallic recuperators are 10 to 20 times lighter compared to an equivalent ceramic recuperator.
- Metallic recuperators occupy 15 to 30 times less space and therefore they need lower outlay on civil work.
- Metallic recuperators are practically leakproof and therefore, can handle toxic fuel gases as well, whereas ceramic recuperators leak to the extent of about 50% of the volume of gases and air handled.

- Metallic recuperators are economical when used for preheating air below 650°C. Whereas ceramic recuperators are economical only when air is preheated above 650°C.
- Specially designed metallic recuperators can preheat air upto 650°C.
- Metallic recuperators can have higher pressure differentials between flue gas and air side (1000 mm w.g. compared to 50-70 mm w.g. for ceramic recuperators).
- Metallic recuperators are easier to maintain & install and involve less initial cost.

RECUPERATIVE BURNER

It is a recent development in recuperator technology for gas firing in which the recuperator is integrated with the burner system such that the flue gases are drawn back through the combined unit. The functions of a burner, flue and recuperator are combined in the recuperative burner in a single compact unit which provides an economic means of waste heat recovery in a batch high temperature plant. Here fuel gas and air are supplied to the burner. The flue gases instead of exhausting out through a conventional flue are drawn back through the burner and is used to preheat the combustion air maximum upto 600°C depending on the type of burner and process temperature. Recuperative burner has been shown in Fig. 43.12.

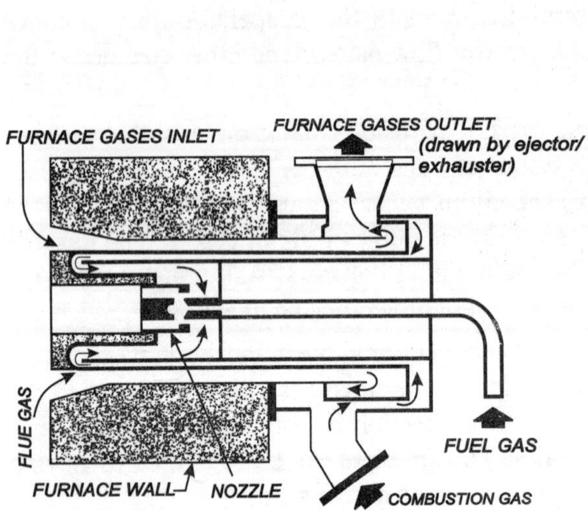

Fig.43.12: Recuperative burner

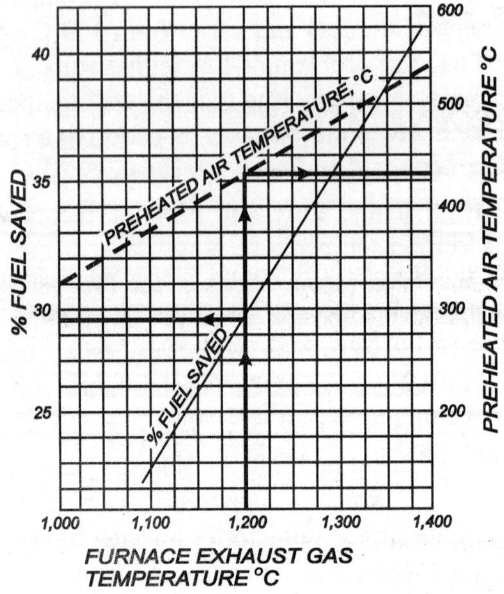

Fig.43.13: Fuel saving by use of recuperative burners

Recuperative burners are used for hot metal/steel laddie heating, firing ceramic ware, heating steel billets/slabs in reheating furnaces etc. Installation of recuperative burners can reduce the fuel consumption in a heating application/furnaces substantially. The saving of energy using recuperator burner depends upon the temperature of flue gases and the degree of heat required in the combustion air. Fig. 43.13 shows the relationship between the amount of preheat given to the combustion air by heat exchange with the furnace exhaust gases and the percentage of fuel which can be saved by use of this type of burner.

Fig. 43.14. Sankey diagram of a non-recuperative batch reheating furnace.

Fig. 43.14 and Fig. 43.15 show the Sankey diagram respectively of a non recuperative and recuperative batch reheating furnace which explains that installation of a recuperator has reduced the flue loss from 70% to 40% and thereby the fuel heat input from a level of 100 to 70.

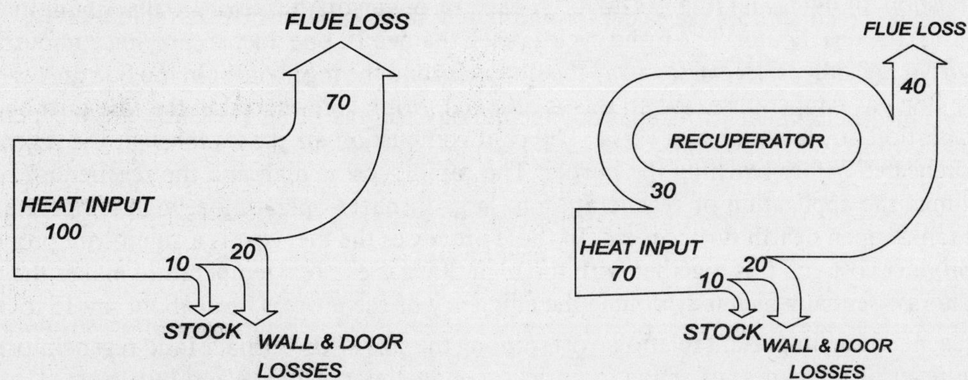

Fig.43.14: Sankey diagram of a non-recuperative furnace

Fig.43.15: Sankey diagram of a batch reheating recuperative batch reheating furnace

REGENERATORS

Regenerators are heat exchangers constructed of refractory bricks and operated in pairs. The bricks are so arranged that gases can flow with low pressure drop through them. The chambers are operated in cycles of alternative heating and cooling. Heating of the chambers are effected by the hot flue gases flowing downwards while leaving the furnace chamber after releasing heat to the charge in the furnace operation. Cooling is then effected by the passage of air and/or fuel gases passing through them in the upward direction in the next cycle.

In the Fig. 43.21, through the left hand chambers air & fuel gas are flowing separately and are being heated through transfer of heat from chequer work. They are on the cooling cycle. The right hand chambers, (through which the hot flue gases are passing) are being heated from top to bottom. They are on the heating cycle. The temperature at the top of the brick layer is higher than the lower end. After some predetermined interval say 20-30 minutes, the flow pattern is reversed by reversal valve & the air and gas are flown through the right hand side chambers & fuel gas are burnt and the flame temperature becomes much higher with a considerable saving of fuel. Regenerator design is generally a compromise between the requirements of mechanical strength due to thermal shocks, minimum resistance to flow of gases, maximum time of contact, freedom from choking due to dust or fumes in the products of combustion and maximum heat transfer surfaces available. To avoid thermal stress in the refractory materials due to drop in temperature between the heating & cooling cycles, reversals are necessary and the time of reversal depends on the allowable drop in temperature for minimum stress. To combine increased heat absorption surface with an economical use of the brick-work, special shapes are sometimes used. The thickness usually is limited to about 50 mm. About 60-85% of the sensible heat is recovered from the waste gases during the passage through a regenerator as preheat to the air and/or fuel gas. Taking, however, the heat returned to the furnace by the regeneration through the fuel gas and air, as percentage of sensible heat in the gases going out, which constitutes the true thermal efficiency of the regenerative systems, the maximum value realisable is of the order of 40%, depending on the degree of preheat which is admissible. Regenerators are used at higher temperatures than the recuperators and beyond 1000°C preheat.

The term regeneration is applied to systems using cyclic firing. Two ducts are used alternately as combustion air ducts and flue gas ducts. A bank of openwork refractory walls is built into each duct. This is the regenerator. The firing cycle is such that whilst one duct is carrying combustion air to the burner, the other duct carries away the flue gases and the regenerator in the flue duct is heated upto the flue gas temperature. When this is attained, firing is reversed i.e. the flue duct becomes the combustion air duct and vice versa. The cold combustion air passes through the regenerator and is preheated before reaching the burner. The capital cost is high and the regenerator is large which limits the application of regeneration to large furnaces operating at high temperature e.g. in glass tanks, open hearth furnaces etc. In these processes the fuel cost is a significant proportion of the product cost and this together with the high flue gas exhaust temperature makes the use of regenerators essential which may double the efficiency of the process from about say 15 to 30.

In regeneration, important relation exists among the size of the furnace (and regenerator), time between reversals, thickness of refractory bricks, conductivity of bricks and heat storage ratios of the bricks. Longer period of time between reversals means higher thermal storage and hence larger regeneration & higher cost. Morever, longer periods of reversal result in lower average temperature of preheat and consequently reduce fuel economy.

Accumulation of dust on the bricks and slagging of the brick surfaces decreases the efficiency of heat transfer in the regenerators as the furnace becomes old. Heat losses from the walls of regenerators and leakage of air inwards during the gas period & leak out during the air period cause an apparent decrease in the heat transfer co-efficient. Considering all these factors, the overall heat transfer co-efficient (for the checkers operating with natural draft) can be taken as 4-6.5 kcal/m^2/hr/°C.

Cawper Stoves are special types of regenerators used in steel industry to preheat the air (blast) for the blast furnaces. In Cawper Stoves, instead of sensible heat, the potential heat of the blast furnace gas is utilized by burning the gas in the chequered work in air and then the blast is preheated while passing through the hot bricks in the following cycle. Temperature of the air rises to more than 1000°C in this manner and blasted through the tuyers into the blast furnace. Regenerators are used in coke ovens, open hearth furnace, batch reheating furnace etc.

Rotary Regenerators (Heat Wheels). The cyclic firing necessary for the application of fixed bed regenerators has been eliminated by the development of rotary regenerators. The original development of such heat wheels (rotary regenerators) was carried out by Ljungstrom in the last century. Commercial supply and installation of Ljunstrom Regenerative Rotary Air Preheater in India is done by Bharat Heavy Electricals Limited (BHEL).

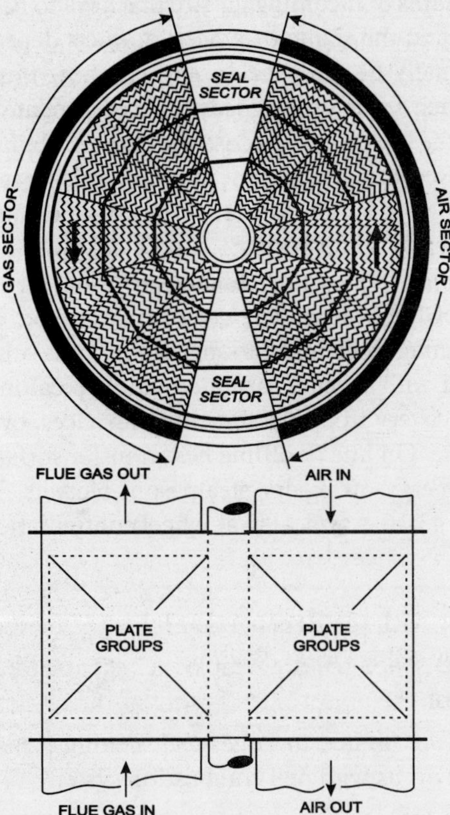

Fig.43.16: Rotary regenerator

Storage & release of thermal energy is the unique feature of fixed bed checker regenerators and industrial regenerative heat exchangers. In the Ljungstrom rotary matrix regenerators, a cylindrical rotor containing the heat transfer surface rotates (at 1-3 rpm) inside a housing divided into two sections. Waste heat is absorbed from the flue gas and transferred to incoming cold air by means of the continuously rotating heat transfer surface (wheel) formed from metal plates (and called the matrix or elements). The matrix is in a cylinder rotating across the waste gas and air streams. As the heat transfer surfaces pass through the normally counter- flowing exhaust gas & incoming air streams, heat is absorbed uniformly and released to combustion air thus maintaining the regenerative cycle. The rotor of the heat wheel, which consists of sectors of either steel mesh or inorganic fibrous materials with a hygroscopic coating of glass ceramic, offers a large surface area to the air or gas flows. This direct method of heat transfer results in uniformly higher metal temperatures compared to recuperators. Lower exit gas temperatures and higher air preheat with minimised corrosion are possible to achieve. Normally, the flue gas & air flows are similar in quantity and the speed of rotation can be (upto 10-20 rpm) varied to give the best efficiency. The main attraction of rotary regenerators is that it can be applied to low temperature exhaust gases and can even transfer latent heat by suitable matrix design. Seals are provided to minimise cross-stream leakage (*i.e.* infiltration of flue gas into air). Also to avoid contamination occurring between the two streams of gas, a purge bleed between the clean & dirty gas streams is incorporated.

Because the flue gas streams & incoming air streams have to be kept separate, ducting (vertical and horizontal) is required and the sizing of wheels & ducts depends upon the flow rates of two streams which should normally be fairly balanced. The heat transfer surfaces are variable and layered within the rotor. Each layer is designed to accommodate the compromising features of maximum heat recovery at the lowest energy cost with the capability of maintaining clean surfaces and minimising corrosion. Recent development in rotary regenerators include deeper rotor matrix to increase heat recovery, cold-end heat transfer surface that can be rotated for extended life and side removed surface. Efficiency upto over 80% can be achieved for this device but it can widely vary depending upon the individual cases. Filters are used for cleaning the dusty flue gas to maintain the performance of the heat wheel. Use of hygroscopic coating allows for the recovery of latent heat *(e.g.*, in swimming bath applications) as such a coating can substantially reduce the latent heat losses associated with venting saturated air. Typical industrial applications of rotary regenerators include heat recovery from metal melting furnaces, ovens, kilns boilers, blast furnace stoves, ventilating systems etc. On line or offline heat transfer surface cleaning is possible by water deluge washing and high energy air & dry steam soot blowers. Working principle of a rotary regenerator is shown in Fig. 43.16, a typical heat wheel configuration in Fig. 43.17 and heat wheel installation in Fig. 43.18.

Comparison of Regenerator with the Recuperator :

- The overall size and space requirement of a regenerator is very high compared to an equivalent recuperator.
- Use of regenerators are limited to large steel melting furnaces *(e.g.* open hearth furnace) and by product coke ovens with fuel combustion systems only.
- The regenerator have higher initial and maintenance cost.
- The regenerators need larger outlay and large & deeper civil works.

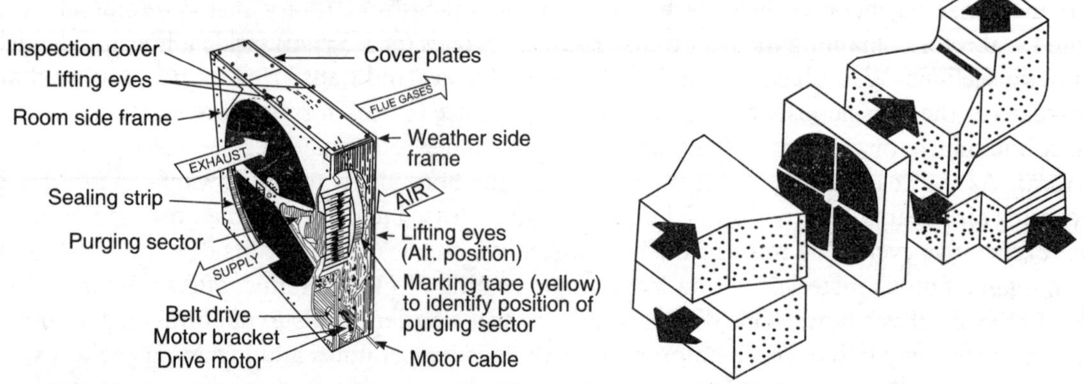

Fig 43.17: A typical heat wheel configuration

Fig 43.18: Heat wheel installation.

PREHEATING OF COMBUSTION AIR AND FUEL GASES

Recuperators and regenerators provide acceptable ways of preheating combustion air and fuel gases. The general methods of preheating are tabulated in Table 43.2.

Table 43.2: General Methods of Preheating

	Type/method of preheating	Preheat temp, achieved, °C	Remarks
I	**Recuperation Types :**		
	(a) Convection :		
	(i) metallic tubes	500	—
	(ii) refactory tubes	750	—
	(b) Radiation :		
	(i) concentric shells	500	Needs high flue gas temperature.
	(ii) metallic tubes-cage	600	
	(c) Rotary :		
	(i) metallic matrix	upto 750	High preheat relative to flue gas temperature.
	(ii) refractory matrix	above 750	
II	**Regeneration Types :**		
	(a) B.F. stoves	upto 1250	
	(b) open hearth	upto 1150	Subject to air in filtration
	(c) continuous	—	under development stage.
III	**Burner Method of Preheating :**		
	(a) Self recuperative	Moderately low	Still relatively small inputs not yet developed to handle B.F. gas. Self regenerative burner does not handle dirty coke oven gas very well.
	(b) Self regenerative	Very low	

In case of fuel gases, top preheat temperature limit is set by the decomposition of methane/ hydrocarbons present in the gas.

The quantitative effects of preheating has been shown below for a low C.V. lean gas *i.e.* B.F. gas and a high C.V. rich gas *i.e.* natural gas. It shows :

- The great advantage of preheating combustion air especially in case of high C.V. fuel gases.
- The great advantage of preheating low C.V. B.F. gas. This is much more effective than the same level of preheating air for combusting B.F. gas.
- The small advantages in preheating high C.V. fuel gases.

QUANTITATIVE EFFECTS OF PREHEATING

(A) For B.F. gas

Case I: Both Gas and Air unpreheated (BASE)

1 Nm^3 B.F. gas (3.12 MJ) (COLD) USEFUL HEAT 1.53 Nm^3 W.G. at 700 ^0C

$\longrightarrow$ LEFT IN FURNACE $\longrightarrow$ and 0.9% O_2

0.66 Nm^3 air 400 ^0C (10% excess) (COLD) Q = 3.12 – 1.686 S.L = 54.0%

$\longrightarrow$ = 1.43 MJ = 1.686 MJ

W.G. — Waste gas 10% excess means 10% excess air, S.L- stack loss

Case II: Air preheated to 400°C and B.F. gas cold

$\therefore$ Q = Heat supplied - Heat in waste gases

$\therefore$ 1.434 = 3.12 x + 0.350 x - 1.686 x

$\therefore$ x = 0.80 Nm^3 B.F. gas

i.e. 20% saving over base (W.G. - Waste gas, S.L. - Stack Loss)

Case Ill-Air preheated to 400°C and B.F. gas to 200°C

y Nm^3 B.F. gas (3.12 y MJ) USEFUL HEAT 1.53 Nm^3 W.G. at 700 ^0C

$\longrightarrow$ $\longrightarrow$ and 0.9% O_2

At 200 °C = 0.281 y MJ

0.66 Nm^3 air 400 ^0C (10% excess) Q = 1.434 MJ $\longrightarrow$

= 0.350 x MJ $\longrightarrow$ S.L = 54.0%

= 1.686 MJ

$\therefore$ Q = Heat supplied - Heat in waste gases

$\therefore$ 1.434 = 3.12 y + 0.350 y + 0.281 y - 1.686 y

$\therefore$ y = 0.69 Nm^3 B.F. gas

i.e. 31% saving over base and 11% further saving over case II

(B) For Natural Gas (N.G.)

Case I: Both gas and air unpreheated (base)

1 Nm^3 N.G.(36.6 MJ) (COLD) USEFUL HEAT 11.519 Nm^3 W.G. at 700 ^0C

$\longrightarrow$ LEFT IN FURNACE $\longrightarrow$ and 0.9% O_2

10.461 Nm^3 air (10% excess) (COLD) Q = 36.6 – 11.8804 S.L = 54.0%

$\longrightarrow$ = 24.7196 MJ = 11.8804 MJ

CASE II: Air Preheated to 400°C and Natural Gas Gold

x Nm^3 B.F. gas (3.12 x MJ) (COLD) USEFUL HEAT 1.53 Nm^3 W.G. at 700 ^0C

$\longrightarrow$ $\longrightarrow$ and 0.9% O_2

0.66 Nm^3 air 400 ^0C (10% excess) S.L = 54.0%

= 0.350x MJ $\longrightarrow$ Q= 1.43 MJ = 1.686 MJ

x Nm^3 N.G.(36.6 y MJ) (COLD) USEFUL HEAT 11.519 Nm^3 W.G. at 700 ^0C

$\longrightarrow$ $\longrightarrow$ and 1.74% O_2

10.461 Nm^3 air (10% excess) at 400 ^0C $Q = 24.7196$ MJ S.L = 32.46%

= 5.5443x MJ $\longrightarrow$ = 11.8804x MJ

$\therefore$ Q = Heat supplied - Heat in waste gases

$\therefore$ $24.7196 = 36.6x + 5.5443\,x - 0.353\,y - 11.8804\,x$

$\therefore$ $x = 0.80$ Nm^3 B.F. gas

i.e. 20% saving over base (W.G. Waste gas, S.L. – Stack Loss)

Case III - Air Preheated to 400°C and Natural gas preheated to 200°C

y Nm^3 N.G.(36.6 y MJ) 200 ^0C USEFUL HEAT 11.519y Nm^3 W.G. at 700 ^0C

$\longrightarrow$ $\longrightarrow$ and 1.74% O_2

10.461 Nm^3 air (10% excess) at 400 ^0C $Q = 24.7196$ MJ S.L = 32.46%

= 5.5443y MJ $\longrightarrow$ = 11.8804y MJ

$\therefore$ Q = Heat supplied - Heat in waste gases

$\therefore$ $24.7196 = 36\% + 5.5443\,y + 0.353\,y - 11.8804\,y$

$\therefore$ $y = 0.81$ Nm^3 Natural gas

i.e. 31% saving over base and 11% further saving over case II

PREHEATING OF SOLID STOCKS FOR HEAT RECOVERY

This technique is also used for recovery of heat from flue gases going out of the systems; examples are Multi-chambered Ring Furnaces used in brick, refractory/ceramic industries, carbon electrode industry, Rotary kilns in cement industry, Tunnel kilns also in ceramic industry. In all these appliances the charges are preheated by the counter current contact of the flue gases which are consequently cooled during this process before it is exhausted through the chimney. The combustion air is preheated by extracting heat from the hot products coming out of furnaces.

In Rotary Kilns, the charged materials flow by gravity due to inclination from the feed end while the combustion products flow in opposite direction. After passing the combustion zone, the hot solids move towards the discharge end and are cooled by the incoming atmospheric air required for combustion. The air thus becomes preheated before it reaches the active combustion zone.

In Tunnel Kilns, the charge travels from one end to the other through some suitable mechanical carriers and passes to the reaction or hot zone on the way. The combustion air is taken in from the discharge end of charge which gets preheated with subsequent cooling of the outgoing charge.

In both the above cases, the charges are actually continuously moved inside the furnace but in King Furnaces, there is no actual movement of the solid materials but the operation is so adjusted through multiple chambers that there is an apparent counter-current movement of the charges and the gases.

Considering only the combustion zone thermal efficiency, it will be a very low figure, but by extraction of waste heat from the flue gases and hot charges, the same may be elevated to a much higher level.

Miscellaneous techniques in heat recovery in thermal power plant boilers i.e. preheating of boiler feed water & combustion air, superheating of steam etc. in boiler furnaces:

Preheating of boiler feed water through *economizers,* wherein the hot flue gases exchange heat with the boiler feed water and sensible heat of the gases is extracted. In some cases even evaporation of water also takes place in economizers.

Superheaters are used to superheat the steam coming out of the boiler drum so that the enthalpy of the steam is increased and condensation during expansion in the turbine is avoided.

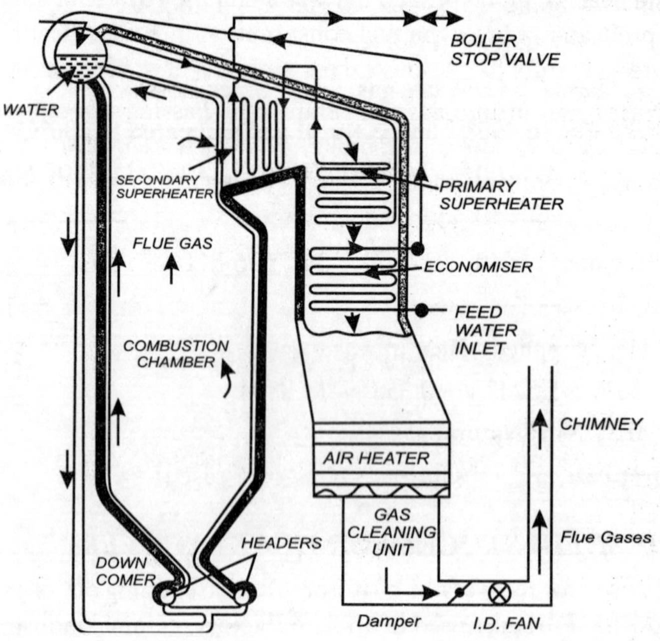

Fig. 43.19: Water tube boiler with waste heat recovery equipments

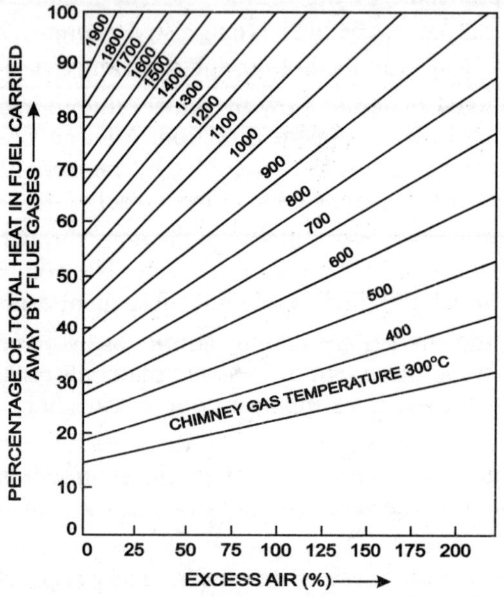

Fig. 43.20: Heat lost in boiler chimney flue gas

Finally the flue gases pass through air heaters to preheat the air required for combustion, specially, in case of pulverised fuel combustion to have a high flame temperature. These are shown in Fig. 43.19.

Thus through the above means of heat recovery systems, the overall thermal efficiencies of boiler furnaces may be brought to a level of 85-90% or even more.

Recovery of waste heat escaping through the boiler chimney by use of superheater, economiser and combustion air preheater reduces the fuel consumption in boiler. Heat lost in boiler chimney flue gases at various temperature for various excess air levels have been shown in Fig. 43.20.

For every 22°C reduction in flue gas temperature by passing through an economiser or air preheater, there is 1% saving of fuel in the boiler. That means, for every 6°C rise in feed water temperature through an economizer or 20°C rise in combustion air temperature through an air preheater, there is 1% saving of fuel. However, chimney, flue gas temperature by these waste heat recovery equipments (*e.g.* superheater, economiser and preheater) should not be reduced below dew point so as to save the chimney brickwork, I.D. fan blades and other metallic ducts & equipments from corrosion damage due to condensation of H_2S0_4, present in flue gases.

Various recuperators and regenerators are shown in Fig. 43.21.

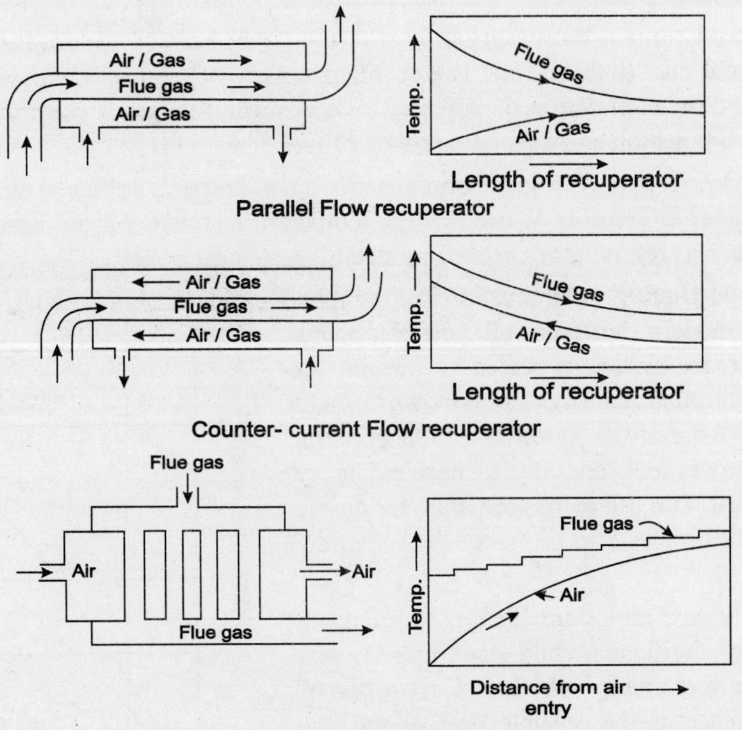

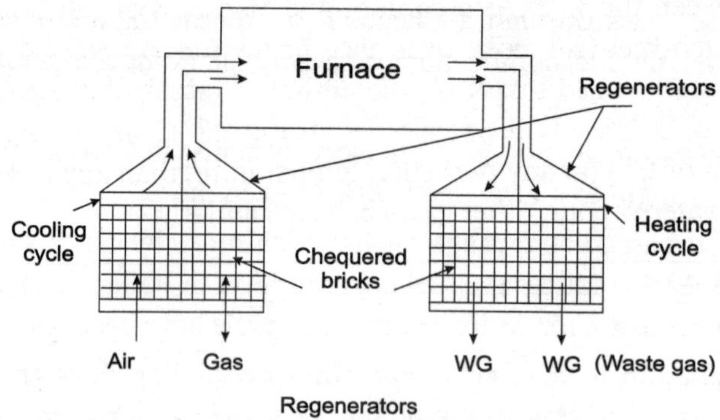

Fig. 43.21: Various recuperators and regenerator

FUEL ECONOMY IN FURNACES

Means employed for heat conservation *i.e.* fuel economy on furnaces ensure its economic operation. Furnace processes having higher temperature levels have the greater thermal loss, hence heat recovery must be made in this case. The largest losses are contained in the waste flue gases and in radiation from furnaces walls.

Waste flue gases contain both sensible heat & the latent heat of vaporisation of water and sometimes potential heat (unburnt fuel gases). The recovery of heat of vaporisation of water is not practicable and the elimination of potential heat in waste flue gas is controlled by providing sufficient air for combustion and adequate mixing of fuel & air.

The total heat loss in waste flue gas is minimised by providing only sufficient air for combustion, and by preventing air infiltration. Waste flue gas temperature is reduced and heat is recovered by heat exchanger such as regenerators, recuperators and waste heat boilers.

Fuel-air proportioning. Combustion air supply should be little above the theoretical requirement. In modern furnaces, all possible points of leakage in furnace system is sealed to prevent air infiltration and controlled quantity of air is blown through the regenerators. Accurate & automatic proportioning of fuel & air improves fuel economy, furnace efficiency and process control. The use of oxygen analyser on the products of combustion leads to improved fuel economy and efficiency.

One of the most important factors influencing the efficient use of the fuels is their stack loss i.e. that proportion of the fuel energy which is carried out of the operating furnace as the sensible heat of the flue gases. The important factors governing this parameter are normally the temperature of the flue gases leaving the furnace and their oxygen content (which is related to the proportion of air used for combustion above strict stoichiometric requirements i.e. % excess air). The

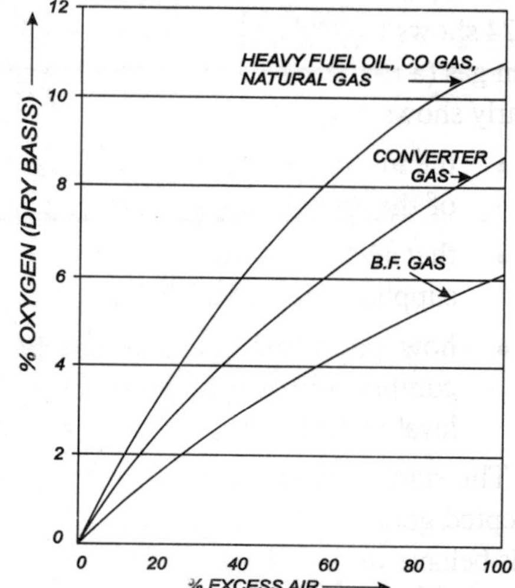

Fig. 43.22: Relationship of % oxygen in waste gases and % excess air

general relationship between the % oxygen content and % excess air for both low C.V. fuel (e.g. blast furnace gas) and rich fuel (e.g. coke oven gas, heavy fuel oil etc). is shown in Fig 43.22. The factors that contribute to stack losses *via* the enthalpy values of the contributing flue gases (comprising normally of CO, CO_2, N_2,O_2) are given in Fig. 43.23 whose significant feature is the much higher heat content of $C0_2$ (at any particular temperature) than any of the other constituents of flue gas. Apart from water vapour, other gases behave similarly.

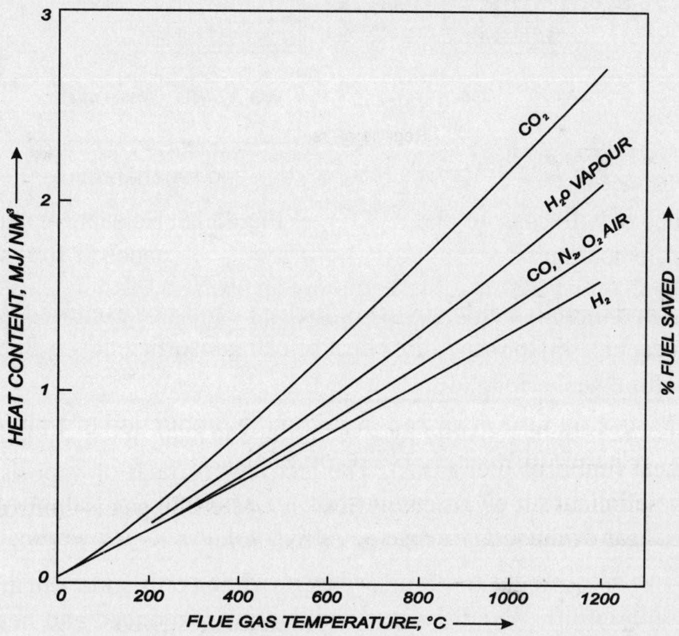

Fig.43.23: Heat contents of common constituents of flue gases.

These data along with the appropriate combustion data are used to calculate stack losses. Fig. 45.24 shows the stack losses for blast furnace gas (the leanest gas *i.e.* low C.V. fuel gas) and for coke oven gas (a rich gas *i.e.* high C.V. fuel gas) over a range of flue gas temperature and % excess air. It clearly shows :

- the influence of exhaust flue gas temperature *e.g.* even with a high C.V. fuel gas, some 50% of the energy supplied ends up in the waste gases at temperature around 1000°C.
- that in the case of a low C.V. fuel gas, even with good combustion control most of the supplied energy leaves *via* the stack at temperatures as low as 600°C.
- how poor fuel can B.F. gas be is demonstrated by the fact that B.F. gas with good combustion control is no better than C.O. gas at 100% excess air *(i.e.* an extremely poor level of combustion control).

The stack loss relationship of some common fuel gases taken at 10% excess air (a normally accepted good level of combustion control) is shown in Fig. 43.25 which reveals that higher C.V. fuels behave very similarly.

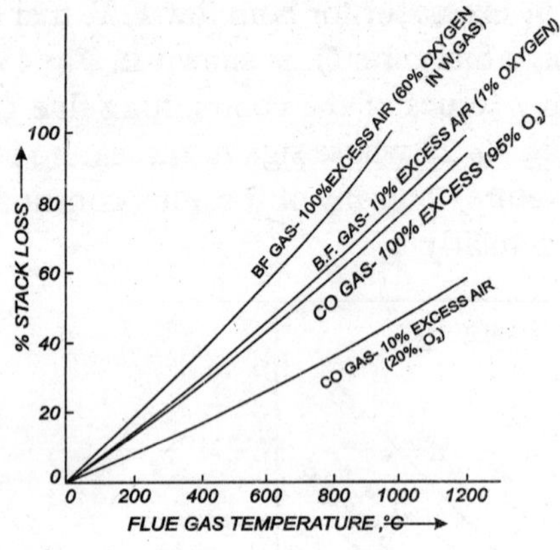

Fig.43.24: Influence of % Excess Air and Exhaust Temperature

Fig.43.25: Relationship for various fuel gases at 10% excess air.

For low stack loss and therefore efficient use of fuel, care must be exercised to ensure that proper combustion takes place. Fig. 43.26 shows the effect of using insufficient (i.e. Substoichiometric) air for combustion of natural gas. It indicates that:

- 10% deficiency in combustion air can .result in the production of unburnts equivalent to a loss of energy release equivalent to 6% of original gas.
- Even where a sufficient supply of combustion air is present, 1% reduction in oxygen content (say for natural gas exhausting flue gas at 700°C) will lead to a lowering of stack loss by 1%.

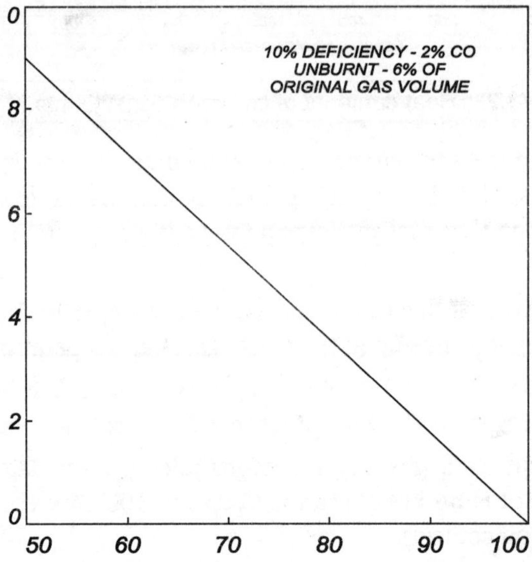

Fig.43.26: Substoichiometric combustion, % unburnt in waste gases for natural gas

Oxygen enrichment of combustion air. Air is mainly composed of 21% (by volume) oxygen and rest (*i.e.,* 79%) inert nitrogen. When combustion takes place, the oxygen combines with the carbon & hydrogen of the fuel and liberates heat. The inert gases of the air absorb heat from the combustion and carry it out of the furnace resulting in loss. It reduces flame temperature by absorbing heat and thus reducing the rate of heat transfer to the stock.

If inert content of air could be reduced, much more efficient combustion would be attained. This can be done by addition of oxygen to combustion air. Increasing the oxygen content lowers the inert content of the air, consequently when a given amount of fuel is burned with oxygen enriched air, the volume of waste gas is less than if ordinary air were used. If the temperature of the waste flue gas is not increased, the sensible heat loss in the flue gas will be decreased due to the smaller heat capacity of the smaller volume. In furnaces operated at higher thermal loads, reduction of inert results in decrease of waste gas temperature. With the same fuel input, enriched air for combustion raises flame temperature of a given fuel thereby increasing heat transfer rate and increasing production rate. The fuel input may be decreased when enriched air is used, to maintain the same production rate as obtained with more fuel using ordinary air.

The use of pure oxygen reduces the stack losses of fuels as shown in Fig. 43.27. The really outstanding point is the transformation of low C.V. B.F. gas to be able to emulate the high C.V. fuels when combusted with cold air - this

without the high flame temperature which limit the application where straight pure oxygen/high C.V fuel can be used.

Use of oxygen enriched combustion air *i.e.* oxygen enrichment of combustion air offers a method of achieving a measure of stack loss reduction when using oxygen to combust the high C.V. fuels with the technical disadvantages associated with the use of pure straight oxygen.

Fig. 43.28 shows the effects of combustion of fuel oil with oxygen enrichment upto 25% oxygen, which indicates that a 1% increase in oxygen enrichment reduces the stack loss at 1000°C by 3%.

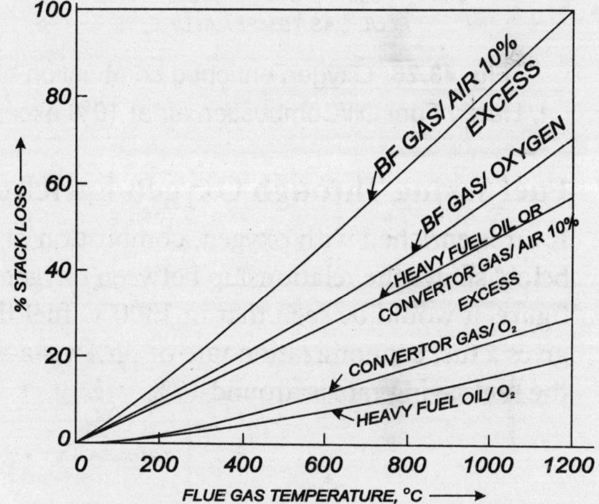

Fig.43.27: fuels with technical disadvantages (Effect on stack loss.)

When steam is used for fuel oil atomising, it influences fuel consumption adversely in two ways:

viz. (i) by the need to heat the steam used from supply temperature to exhaust conditions alongwith *(ii)* the stack loss associated with fuel needed to supply this additional heat.

The adverse effect of using atomising steam, at a rate of 0.5 kg steam/kg oil is shown in Fig. 45.29. This has a small but positive influence on effective stack loss; typically 1% at flue gas temperature of around 1000°C.

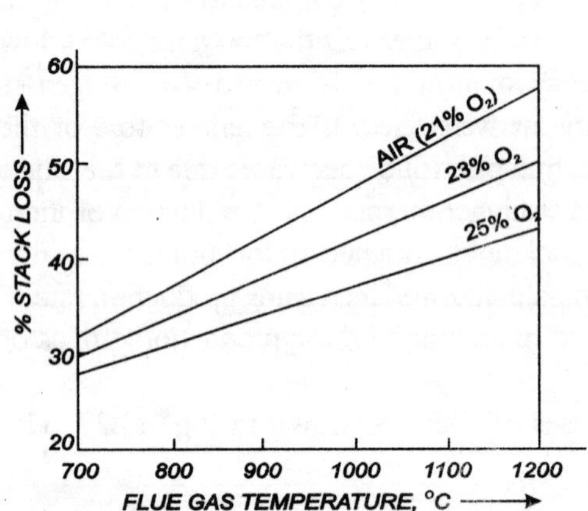

Fig. 43.28: Oxygen enriched combustion air Heavy Fuel Oil/Combustion air at 10% excess

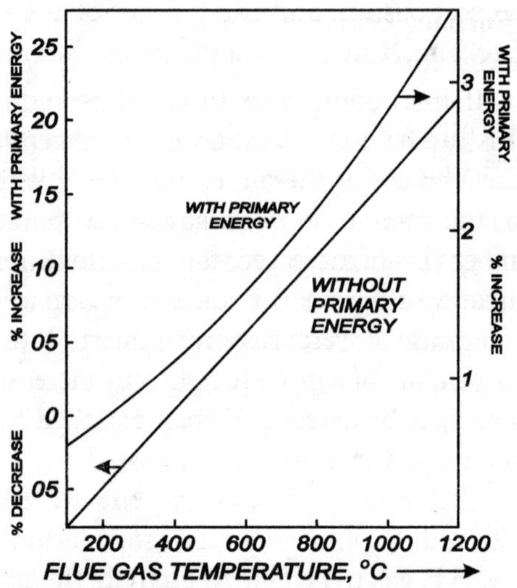

Fig.43.29: Effect of steam atommisation decrease/Increase in energy requirement

Fuel Saving Through Oxygen Enrichment of combustion air :

If air is enriched with oxygen, combustion of the fuel is faster and more complete. The Fig. 45.30 below shows the relationship between oxygen content in the air and the fuel saving rate. From the figure, it would be seen that in 1300°C fuel fired furnace, air of 30 percent oxygen concentration gives a fuel economization rate of 35% vis-a-vis normal air (21% oxygen), while in 900°C furnace, the fuel saving rate is around 17%.

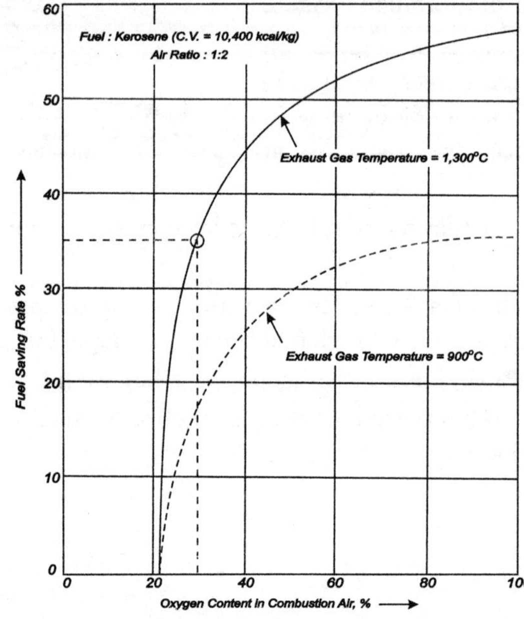

Fig.43.30: Fuel economization rate resulting from oxygen enrichment of the combustion air.

Furnace pressure control. If the pressure of the gases in the heating chamber of a furnace is below atmospheric, cold outside air will be drawn into the furnace through any openings that exist. If the interior pressure is above that of the outside air, the hot gases would be forced out of the furnace through these openings and if too much higher, will in addition, tend to penetrate refractories and overheat the furnace causing damage.

Hence, it is generally desirable to operate a furnace with a slight positive pressure in the heating chamber *(i.e.* furnace pressure slightly higher than atmospheric). Pressure from top to bottom of the heating chamber is not uniform due to the stack effect of hot gases. Hence, control is aimed at maintaining desired pressure at hearth level.

Air drawn into a furnace operating under negative pressure upsets fuel air ratio which is controlled automatically. In such furnaces such as reheating or heat treating ones, this air aggravates the problem of oxidation (scaling) of the stock because of oxygen present.

If the pressure in the furnace is atmospheric or positive, better heating condition is obtained by improving heat transfer and by better control of temperature uniformity. The positive pressure must be controlled to prevent excessive sting out of flame from furnace openings as well as to avoid the build up of excessive back pressure that would interfere with the proper flow of fuel gas and air. Furnace pressure is controlled by adjusting the opening in the stack damper.

Flame sting out is an indication of positive pressure.

Waste gas lost directly from operating furnaces due to flame sting out is usually hotter than that leaving in the normal way, *e.g.,* waste gas lost through soaking pit covers have a stack loss (for high C.V. fuels) of around 75% instead of around 35% *i.e.,* for a 10% loss, a final stack loss 3-4% higher than it should be. In certain slab reheating furnaces, some 20-25% of the furnace flue gases may be lost through the charging doors.

Air infiltration reduces the temperature of waste gases which can reduce the efficiency of heat recovery. Stack losses increases by increase in the oxygen content in flue gases. Air infiltration rates of 50-100% excess rates are not uncommon *(e.g.* in some waste heat boilers). Remedial measures for minimising sting out and air infiltration in a furnace include satisfactory furnace pressure control, closing of furnace doors and good furnace engineering & housekeeping.

Flame sting out and air infiltration are related subjects. Factors influencing flame sting out and air infiltration are discussed below :

Sting out is promoted by :

- high furnace pressures ; especially at the high point of the furnace.
- open charging doors.
- high burner thrusts towards charging doors.
- holes and gaps in the furnace structure-especially the roof.

Air infiltration is promoted by :

- open charging doors of furnace operating with negative (vacuum) pressure.
- high suction associated with waste heat boilers.
- poor furnace pressure control.
- holes and gaps in furnace structure especially at hearth level.
- up discharge slopes of reheating furnaces by burner thrust and buoyancy.

Furnace temperature control. The use of excessive amount of fuel is wasteful and results in high furnace exit gas temperature & damage to refractories. High fuel rates hasten the heat transfer rate to the material and may cause damage to it in some cases. The use of insufficient fuel reduces the rate of heat transfer and prolongs process time. The optimum fuel rate for protection either of the material being heated or the furnace refractories and often for control of production rate or heating, is maintained by automatic temperature measurement.

Insulation. Insulating materials are valuable to conserve fuel, to afford safer & more comfortable working conditions, to protect materials succeptible to damage from heat/thermal strain and to speed up furnace operations.

CONTROL OF FURNACE ATMOSPHERE

During heating of solids in furnaces, chemical reaction between the charge/stock and the surrounding medium (which may be solid, liquid or gaseous) at high furnace temperature does occur. The gaseous medium in contact with charge inside the furnace is called 'furnace atmosphere'. The chemical reaction between the charge & atmosphere depends on the furnace temperature and the composition of the stock & the atmosphere inside the furnace.

Two different kinds of furnace atmosphere are in existence.

- Combustion or natural atmosphere containing the products of complete or incomplete combustion *e.g.*, CO, CO_2, N_2, H_2O, H_2, O_2, SO_2 etc.
- *Prepared /artificial atmosphere* comprising mainly of N_2, CO and H_2.

Heating and heat treatment of metals and alloys require special types of atmosphere to improve the physical properties. Atmosphere inside the furnaces can be of the following types (depending upon the specific duties furnaces are supposed to perform).

- Oxidising atmosphere *(e.g.* in boiler furnace).
- Reducing atmosphere *(e.g.* in blast furnace).
- Carburising/decarburising atmosphere *(e.g.* in L.D. converter).
- Nitriding atmosphere *(e.g.* in metal surface hardening process).
- Neutral/inert atmosphere *(e.g.* in steel annealing furnace).

In many processes in which materials are heated, it is important to maintain a special atmosphere inside the furnace in contact with materials. In a furnace which is heated internally by gas, the combustion of gas can be controlled to give an oxidising or reducing atmosphere.

In some cases, the heat treatment of metals involves chemical treatment of surface layers *(e.g.* carburising or nitriding) which is carried out in carefully controlled atmosphere. In other cases, the furnace atmosphere is controlled to prevent surface attack of metals *(e.g.* oxidation, scaling, decarburisation or sulphur penetration).

At times, atmosphere is produced externally to the furnace *(e.g.* neutral atmosphere for steel annealing is created by nitrogen) and fed to it. For many applications, CO_2 & H_2O vapor are considered to be oxidising agent (irrespective of any free O_2 present) and CO & H_2 are reducing agents. Nitrogen gives inert atmosphere.

OXIDISING ATMOSPHERE

In the low temperature heating furnaces (< 700°C), an oxidising atmosphere does not cause much of scaling of material due to presence of free oxygen in flue gas. If an oxidising atmosphere is desired, then use of as little excess air is desired as will complete combustion and give the desired atmosphere. In a furnace fired with coal gas, it is usual to adjust to 10.5% CO_2 and 1.5% O_2in flue gas which gives mild oxidising atmosphere.

During heating of steel/common metals, flue gas components like O_2, CO_2, SO_2 and H_2O vapor oxidises the metal to form scales at furnace temperature. Oxidation/scaling increases with the time of exposure, but at a diminishing rate and the effect of temperature is very high when the melting temperature of oxides is exceeded (indicated by shiny surface of steel). Scaling by flue gas components is greatly reduced if the charge lies in a quiet motionless part of the furnace atmosphere and is heated almost entirely by radiation heat transfer. At high furnace temperature, oxidation and decarburisation of steel takes place simultaneously in combustion atmosphere. The higher the temperature, the greater is the affinity of carbon for oxygen in comparison with the affinity of iron for oxygen.

REDUCING ATMOSPHERE

Presence of CO & H_2 and absence of O_2 in flue gas from a furnace is the indication of reducing atmosphere in the furnace. When the furnace condition is to be adjusted to reducing conditions (as for steel hardening), the primary and secondary air controls are regulated in conjunction with flue damper to give a faint curtain of flame as the furnace door is opened. A piece of paper thrown into furnace should char but not flame. This practical test indicates the absence of oxygen but gives no indication of amount of CO present. Gas analysis will show about 1-2% CO in flue gases depending on the process.

Reducing atmosphere in the furnace are obtained by adjusting the secondary air quantity such that a small quantity of CO is present in furnace flue gases. Carbonaceous particles such as smoke does not create reducing atmosphere in furnaces.

Reducing atmosphere (i.e. excess of fuel) during heating of steel forms less scale than is formed by oxidising atmosphere *(e.g.* excess of air) but the scale is tight and adheres firmly. During heating of steel at about 1200°C, scale formation does not take place, if CO/CO_2 equals 3 and H_2/H_2O exceeds 1.2 in the flue gas. Since, combustion to carbon monoxide does produce some heat but little and unburnt hydrogen produces no heat, a furnace temperature of 1200°C cannot be attained with these ratios without resorting to combustion with oxygen enriched air or preheating of combustion air. During steel slab heating, manganese is not oxidised, if the furnace atmosphere has H_2, CO and N_2 in the flue gas which is possible in reducing atmosphere.

CARBURISING ATMOSPHERE

Hydrocarbon gases carburise the surface of steels at annealing temperature.

Carburisation which means increase of carbon content, is practiced for the purpose of producing hard surface on mild steel.

DECARBURISING ATMOSPHERE

It involves removal of carbon from iron carbide (Fe_3C) thus affecting the crystalline structure of the steel. $C0_2$ decarburizes steel while CO carburises.

$$Fe_3C + CO_2 \rightleftarrows 3Fe + 2CO$$

Hydrogen also decarburises at high temperature (when methane is formed).

$$Fe_3C + 2H_2 \rightleftarrows 3Fe + CH_4$$

Hence, addition of CH_4 to the atmosphere reduces decarburising by hydrogen.

Nitrogen in presence of steam decarburises high carbon steels. In the absence of steam, it is passive.

Decarburisation is usually considered to be harmful during steel heating because the surface of steel becomes soft.

NITRIDING ATMOSPHERE

Active nitrogen formed by cracking ammonia at the metal surface causes nitride formation with increase in surface hardness.

NEUTRAL ATMOSPHERE

In case of hood annealing of steel coils, a protective neutral/inert gas atmosphere (95% N_2 + 5% H_2) is used, so that, steel coils do not get oxidised on heating in presence of air. It is resorted in those cases, where neither oxidising nor reducing atmosphere can be tolerated.

EXTERNALLY HEATED FURNACES

In bright annealing furnace, a carefully controlled atmosphere is required for achieving desired physical properties of metals by eliminating scaling, decarburisation and other chemical reactions (usually due to sulphur or oxygen which impair the metal surface). This is carried out in externally heated furnaces as necessary control of the furnaces can not be attained if the combustion gases circulate in the furnace. The furnace is heated indirectly and the gases constituting the furnace atmosphere is produced separately.

45

Classification of Furnaces

Furnaces are classified from different points of view in order to have an idea of representative types of various industrial furnaces which comprise mainly of three main portions.

- The fire place where combustion of fuel takes place.
- The working chamber or furnace proper where heat is transferred from the products of combustion to the materials under heating.
- The appliances for removal of flue gases.

Classification of furnaces based on various factors are given below :

Based on the heat source:

(a) **Combustion/flame furnaces** - These are furnaces where heat is developed due to com bustion of fuels.

(b) **Thermo-electric furnace** - In these furnaces, heat is generated by electricity.

Based on type of fuels used:

(a) Solid fuel fired furnaces

(b) Liquid fuel fired furnaces

(c) Gaseous fuel fired furnaces

(d) Mixed/multi fuel fired furnaces - Here combustion of two or more types of fuels are used simultaneously.

Based on methods of heat transfer from fuel to charge under heating:

(a) Furnaces in which the fuel is in direct contact with the material under heating. *Example* -goldsmith's and blacksmiths open fireplaces.

(b) Furnaces in which charge is heated by products of combustion ; an example is heat treatment furnaces.

(c) Furnaces in which material is heated by the way of heat transfer through a solid wall *e.g.* a boiler furnace.

(d) Furnaces in which heat transfer takes place through a liquid medium surrounding the submerged material under heating *e.g. salt bath furnaces or lead bath furnaces.*

Based on use of heat saving appliances:

(a) **Recuperative furnace** - Here, flue gas and air circulate along alternate paths and the heat absorbed by the walls of the passages from the hot flue gas is taken away by cold water.

(b) **Regenerative furnace** - Here, flue gases are allowed to pass through alternate checker-brickwork chambers after a definite interval of time. The cold air or gas passing over the checker brick work takes away a portion of the heat accumulated by the checker brick and the latter again starts absorbing heat from the flue gas.

Based on charging system :

(a) Manual charging furnace *(b)* Mechanical charging furnace

Based on mode of operation :

(a) Batch furnaces *e.g.* Cupola etc.
(b) Periodic furnaces *e.g.* annealing furnaces, coke ovens etc.
(c) Continuous furnaces *e.g.* glass tank furnace, reheating furnace etc.

Based on draught:

(a) **Natural/self draught furnace** - Operates with chimney or with open doors.
(b) **Forced draught furnace** - Here, forced draught fans are used both for supply of combustion air as well as removal of flue gas. This furnace normally operates at positive pressure.
(c) **Induced draught furnace** - Here, induced draft fan **(I.D.** fan) located at the bottom of the chimney sucks out the flue gas from the furnace such that it is at negative (vacuum) pressure.
(d) **Balanced draught furnace** - Here, both forced draft fan for supply of combustion air and induced draught fan for sucking out flue gas are provided. They are so operated that the furnace pressure is almost atmospheric.

Based on shape of furnace:

(a) crucible furnace *(b) shaft furnace e.g.* blast furnace
(c) hearth furnace

Based on the purpose:

(a) melting furnace *(b)* roasting furnace *(c)* reheating furnace

Based on the way in which charge is handled:

(a) shaft furnace *(b)* Tunnel kiln *(c)* Muffle furnace
(d) Pusher type furnace
(e) Bath furnace - fixed, tilting or *rotary*

Based on the industries which use the furnaces:

(a) Steel reheating furnace :
- gas fired continuous billet heating furnace
- pusher type
- walking beam type
- soaking pits

(b) Production of pig iron - blast furnace
(c) Production of iron nodules - rotary kiln
(d) Iron casting manufacture - cupola furnace
(e) *Chemical industry;*
- Radiant pipe still furnace
- Pot furnace

(f) Small iron & steel melts - coke fired crucible furnace
(g) *Steel making furnaces:*
- Open hearth
- Basic oxygen furnace (BOF)/LD converter
- Bessemer converter

(h) Steel heat treatment furnaces :
- Reversing regenerative bogie type furnace
- Steel strip annealing furnace

(i) Refractory bricks and pottery :
- Coal fired beehive kiln for refractory firing
- Continuous brick kiln
- Stoker fired tunnel kiln for refractory firing

(j) *Non-ferrous metals :*
- Oil fired tilting rotary furnace
- Reverberating furnace
- Small rotary brass melting furnace
- Oil fired tilting rotary furnace
- Oil fired tilting crucible furnace

(k) *Cement and lime burning :*
- Pulverised fuel fired counterflow rotary kiln
- Direct fired shaft kiln

(l) *Glass melting :*
- Gas fired regenerative bath
- Pot furnace - oil/producer gas fired recuperative/regenerative muffle

(m) *Steam raising :*
- Stoker fired - fixed grate or moving grate furnace
- Pulverised coal fired furnaces

(i) horizontal firing
(ii) vertical firing
(iii) tangential firing
- Fluidised bed furnaces/circulating fluidised bed (CFB) furnaces
- Cyclone furnaces

(n) *Carbonisation :*

- Gas making - horizontal - gas fired muffle - vertical - gas fired muffle intermittent
- Vertical continuous - gas fired counter flow muffle
- Beehive oven
- By-product oven

(i) regenerative type

(ii) waste heat type

- *Regenerative, recirculating, twin flue, compound ovens.*

No single classification of furnace is suitable because two furnaces may be alike as far as one process is concerned but entirely dissimilar from the viewpoint of another process. For instance, both type of brass furnace is similar to the open hearth furnace as far as the process of heat transfer from the flame to the charge is concerned but entirely dissimilar from the point of view of combustion. Classification of furnaces therefore, does not serve such a useful purpose as classification of the processes taking place in the furnace. A broad classification of furnaces is given in Table 45.1.

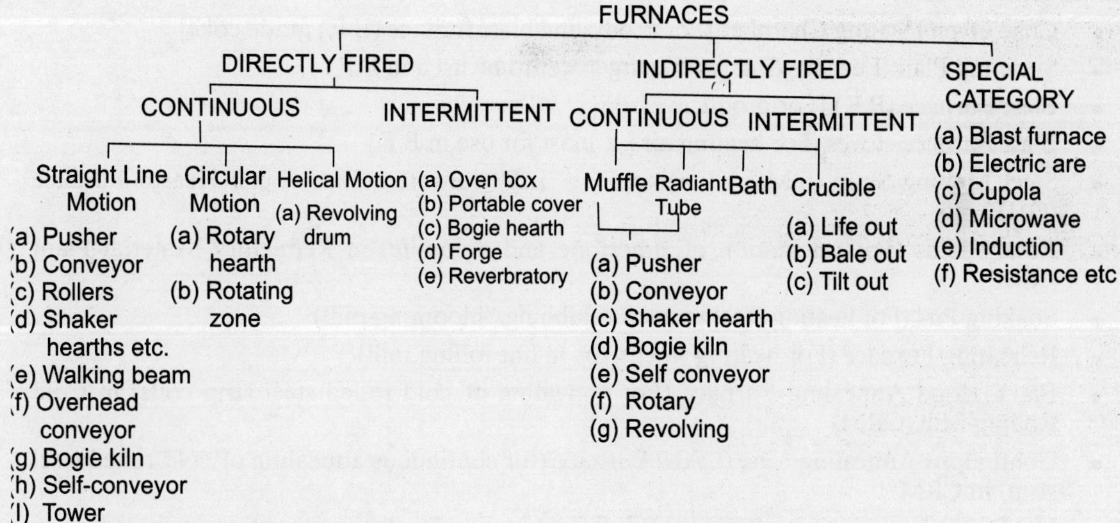

FUELS & FURNACES USED IN STEEL PLANTS

An integrated iron and steel plant employs various types of furnaces which include heating furnaces, metallurgical furnaces, heat treatment furnaces, steam generating boilers etc. The purpose, fuels used, combustion system, end uses, waste heat recovery system etc. vary from furnace to furnace.

Various important/major furnaces installed in various units of a typical Indian integrated steel plant are listed below :

- Coke Oven Heating Chamber [For producing blast furnace (B.F.) grade coke]
- Sintering Plant Furnace (For producing sinter to be used in B.F.)
- Blast Furnace (B.F.) [For producing iron]
- Blast Furnace stoves (For heating air *i.e.* blast for use in B.F.)
- Steel Melting Shop (S.M.S.) Furnaces *e.g.* L.D. converter, Basic open Hearth Furnace (OHF) etc.
- Rotary kilns (For calcination of limestone and dolomite) in Refractory Material Plant (RMP)
- Soaking Pits (For heating steel ingots in slabbing/ blooming mill)
- Reheating Furnace (For heating steel slabs in hot rolling mill)
- Batch Hood Annealing Furnace (For annealing of cold rolled steel strip coils) in Cold Rolling Mill (CRM)
- Continuous Annealing Line (CAL) Furnace (for continuous annealing of cold rolled steel strip) in CRM.
- Hot Dip Galvanising Line (HDGL) Furnace (An annealing furnace used prior to galvanising of cold rolled steel strip) in CRM.
- Captive Thermal Power Plant Boiler Furnace

CLASSIFICATION OF STEEL PLANT FURNACES

Above mentioned furnaces are classified based on various factors as described below :

(i) Based on the purpose of the furnace :

- *Metallurgical Furnaces (e.g.,* Blast Furnace, Sinter Plant Furnace, L.D. Converter etc.)
- *Heating Furnaces (e.g.,* soaking pit, reheating furnaces, coke ovens heating chamber etc.)
- *Heat Treatment Furnaces (e.g.,* annealing furnace, HDGL Furnace etc.)
- Steam Generating Furnaces (e.g., boilers)

(ii) Based on the type of fuel used :

- Solid fuel fired furnaces
- Liquid fuel fired furnaces *(e.g.,* rotary kiln)
- Gaseous fuel fired furnaces *(e.g.,* coke ovens, sinter plant, soaking pits, reheating furnace etc.)
- Multi-fuel fired furnaces *(e.g.,* captive thermal power plant boilers using all solid, liquid & gaseous fuels simultaneously)

(iii) Based on the use of waste heat saving appliances:

- Regenerative furnaces *(e.g.,* coke oven)
- Recuperative furnaces *(e.g.,* soaking pit, reheating furnace etc.)
- Waste heat boiler type furnaces *(e.g.,* rotary lime kilns, L.D. converter)

(iv) Based on draft control system:

- Natural draft furnaces *(e.g.,* B.F. stoves, coke ovens etc.)
- Forced draft furnaces *(e.g.,* soaking pits, etc.)
- Induced draft furnaces
- Balanced draft furnaces (most of the heating/combustion type furnaces used in steel plant fall under this category)

(v) Based on mode of operation of the furnaces:

- Batch/Periodic furnaces *(e.g.,* coke ovens, annealing furnaces etc.)
- Continuous furnaces (rotary lime kilns, reheating furnaces etc.)

WASTE HEAT RECOVERY SYSTEMS INSTALLED IN STEEL PLANT FURNACES

Various types of waste heat recovery equipments typically installed in various steel plant furnaces are listed below :

Coke Ovens and By-product Plant:

(a) Regenerators for preheating combustion air and mixed gas.

(b) Waste heat boiler in sulphuric acid plant

Sintering Plant: Sinter cooler produces hot combustion air.

Blast Furnace :

(a) Combustion air preheater for waste heat recovery from the flue gas of stoves.

(b) Evaporative cooling system (ECS) in hot blast main valve of stoves for its cooling.

(c) B.F. stove itself is a type of regenerative furnace where fuel gas burning instead of waste heat is used for heating of checkers.

Steel Melting Shop :

(a) Waste heat boilers in L.D. converter of open combustion type

(b) Converter gas recovery system in L.D. converter of suppressed combustion type.

Rotary Line Kilns : Waste heat boilers.

Soaking Pits :

(*a*) Ceramic recuperator for combustion air preheating.

(*b*) Metallic recuperator for fuel gas (mixed gas) preheating.

Reheating Furnace :

(*a*) Ceramic recuperator for combustion air preheating.

(b) Evaporative cooling system (ECS) for skid cooling.

Captive Power Plant Boilers :

(*a*) Steam superheater.

(*b*) Air preheater

(*c*) Economiser (Feedwater preheater)

TYPICAL FUELS USED IN STEEL PLANT FURNACES

Steel plant furnaces employ all solid, liquid and gaseous fuels in its various furnaces. Most furnaces are single fuel fired furnace, though dual fuel firing and multifuel firing is also practiced in some furnaces. Most of the furnaces are fired with plant generated fuel gases and the shortfall is met by other purchased fuels from outside.

Typical fuels used in various furnaces of steel plant are listed below in Table 46.1.

Table 46.1: Fuels used in Various Furnaces of Steel Plant

Furnaces	Fuels used in the furnaces
Coke Ovens	Mixed (B.F. + C.O.) gas of C.V. 1000 Kcal/Nm^3 at a pressure of 400 mmwg is used for heating coke ovens. Alternatively pure coke oven (C.O.) gas (CV = 4200 Kcal/Nm^3) is also used during shortage of mixed gas.
Sintering Plant Furnace	In primary (*i.e.*, first) hearth mixed gas (CV = 1500 Kcal/Nm^3) is used. In secondary (*i.e.*, Second) hearth coke oven gas is used. Fuel supply gas pressure in both the hearths is 400 mmwg.
Blast Furnace (B.F.)	B.F. grade coke (25-80 mm size).
B.F. Stoves	Mixed gas (CV = 1200 Kcal/Nm^3) at 400 mmwg is used. Alternatively, pure B.F. gas (CV = 820 Kcal /Nm^3) is also used in case of C.O. gas shortage.
Rotary Lime/Dolomite Kilns	Pitch Creosote Mixture (PCM) or, tar or furnace oil is used. Tar plus coke oven gas (dual firing) is also used if later is in surplus supply.
Soaking pits in Slabbing Mill	Mixed gas (CV - 1600 Kcal/Nm^3) at 700 mmwg pressure is used.
Reheating Furnace in Hot Rolling Mill	Mixed gas (CV = 1800 Kcal/Nm^3) at 2100 mmwg is used for pusher type furnace Mixed gas (CV = 2300 Kcal/Nm^3) at 2100 mmwg is used for walking beam type furnace

Annealing furnace in Cold Rolling Mill (CRM)	Mixed gas (CV = 1800 Kcal /NmNm^3) at 2100 mmwg is used.
Hot Dip Galvanising Line (HDGL) furnace in CRM	(a) In Radiant Tube Fired (RTF) section - Mixed gas (CV = 1800 Kcal/Nm^3)- at 2100 mmwg is used (b) In direct Fuel Fired (DFF) section - Pure C.O. gas at 3000 mmwg. is used.
Captive Power Plant Boilers	Coal middling / steam coal, B.F. gas, C.O. gas, converter gas, furnace oil, tar/PCM all are used simultaneously (multifuel firing)
Mixed gas is a mixture of B.F.gas and C.O. gas.	

Apart from the above, pure coke oven gas is used as fuel in small heat load heating applications namely.

(a) Hot metal (*i.e.* pig iron) runner and ladle drying in B.F.

(b) SMS mixer heating and SMS pit side ladle heating.

(c) Firing smaller furnaces in ingot mould foundry, steel foundry and iron & copper founry.

(d) Firing smaller annealing furnaces in machine shop, forge shop and structural shop.

In case of rolling mill (slabbing mill, HRM, CRM) furnaces, to maintain the required calorific value, equivalent amount of naphtha is vaporised and used during the shortage of coke oven gas (1 KL naphtha = 2000 Nm^3 C.O. gas).

CHARACTERISTICS OF VARIOUS FUELS USED IN STEEL PLANT FURNACES

(i) **Fuel Gases :**

The composition & properties of various fuel gases used in steel plant namely BF Gas, C.O. Gas and Converter gas are tabulated in table 46.2.

Table 46.2: Compositions and Properties of B.F. gas, C.O. gas and converter gas

Parameters	Unit	B.F. Gas	C.O. Gas	Converter Gas
Composition : (% by volume)				
CO		22-24	6-7	60-65
H_2		4-6	58-60	—
CH_4		—	23-25	—
CO_2		17-20	2-3	12-18
N_2		52-54	3^t	20-30
O_2		< 1	< 1	< 2
C_mH_n		—	2-3	—
Properties:				
C.V.,	K.Cal/Nm^3	820-840	4200-4300	1800-2000
Density, kg/Nm^3	mg/Nm^3	1.3-1.35	0.44-0.46	1.33-1.36
Dust content,		< 10.0	nil	<10.0

Ignition temp. Theoretical Flame temp.	°C °C	680 1400	560 2000	610 1900
EXPLOSIVE LIMIT: lower upper	(%gas in gas/air mixture)	37 71	5 32	20 70
Gas pressure, Gas yield	mmwg —	1000 1850-2000 (NM³/THM)	700 290-300 (Nm³/TDC)	1000 50-60 (Nm³/TCS)
Theoretical air/gas ratio	—	0.70	4.20	1.60
Theoretical flue gas/fuel gas volume	—	1.5	4.9	2.3
THM — Ton Hot Metal,	TDC— Ton Dry coal		TCS — Ton Crude Steel Coal,	

B.F. Gas. Though blast furnace gas has a very low C.V., yet because of its large quantity produced, it is one of the most important fuels in an integrated steel plant. Great care is taken in its utilisation because of its toxic and highly poisonous nature. The surplus gas is either stored in a gas holder or may be bled out in the atmosphere (after burning), as the case may be. Mostly, B.F. Gas is mixed with coke oven gas (mixed gas) and is used in various furnaces of the steel plant.

Coke Oven Gas. Produced during high temperature carbonisation of coking coal, it is the most important fuel in a steel plant. C.O. gas is deprived of its valuable chemicals in the byproduct recovery plant before use. Though cleaned C.O. gas is not as poisonous as B.F. gas, it is highly inflammable and explosive. Since the C.O. gas contains sulphur compounds (H_2S, SO_2 etc.) its direct use as fuel in the steel plant's metallurgical furnaces is undesirable as it causes brittleness of steel surface during rolling/hot scarfing of steel slabs. Hence, C.O. gas is normally desulphurised, detarred, debenzolised, denaphthalised and stripped of its ammonia content before use.

Mixed Gas. It is a mixture of B.F. gas and C.O. gas in different ratios to get a gaseous fuel of desired C.V. and combustion characteristic. C.V. and speed of combustion is very high for C.O. gas but it is very low for B.F. gas. The desired C.V. and speed of combustion can be attained by proper mixing of the two fuels in various proportions.

In most of the gaseous fuels applications in the steel plant, mixed gas is used in furnaces. There may be centralised common gas mixing station for supplying mixed gas of a fixed C.V. to all the furnaces of the steel plant. Alternatively individual/separate gas mixing stations are provided to mix the two gases in different proportions to produce mixed gas of different calorific values suitable for different furnaces.

L.D. Converter Gas. In view of its very poisonous nature (carbon monoxide - C.O. is upto 70% and maximum safe working limit for C.O. being only 50 ppm) its use as such in steel plant furnace is avoided. It is mostly mixed with coke oven gas or blast furnace gas. Part of it is used in captive power plant boilers. Though this gas can be used in other steel plant furnaces also, however great care is required in its handling and use due to its deadly poisonous nature.

(ii) **Pitch Creosote Mixture (PCM)**: This in house produced liquid fuel which is mainly a mixture of soft pitch and creosote oil is used as a fuel like furnace oil. Its typical analysis and properties are given below :

PCM supply pressure and temperature are - 25 kg/cm^2 and 90°C respectively.

Composition :	C = 90%, H = 6%, N = 1%, S = 0.4%, O = 2.5% Ash = 0.1%, Moisture = nil, CV = 9000 kcal/kg,
Properties :	
Sp. gravity	= 1.1
Flash point	= 65°C
Pre heating temperature	= 90°C
Redwood I viscosity at 20°C	= 6000 seconds
Viscosity in ^{0}Engler at 80°	= 12.6

In case of non-functioning of tar distillation plant; tar itself is used as a fuel in place of PCM.

(iii) **Furnace Oil.** Mostly low viscosity and medium viscosity furnace oil of following specification is used. High viscosity oil may be used in summer as its unloading & usage during cold winter may pose problems.

Table 46.3: Specification of Furnace Oil

Characteristics	Grades of Furnace oil		
	Low Viscosity (LV)	Medium Viscosity (MV)	High Viscosity (HV)
Flash point (Pensky-Martins) (closed), °C minimum	66	66	66
Kinematic viscosity, centistokes at 50°C, maximum	80	125	370
Density, gm/cc	0.85-0.89	0.9-0.93	0.93-0.99
Water content, % by volume, max.	1.0	1.0	1.0
Ash % by wt. max.	0.1	0.1	0.1
Sulphur, total % by wt. max.	3.5	4.0	4.5
Sediment, % by wt. max.	0.25	0.25	0.25

The desirable characteristics of the furnace oil used in boilers of captive thermal power plant are:

Specific gravity = 0.9, furnace oil supply pressure to burners = 6 to 8 kg/cm^2, atomisation temperature = 90-120°C, oil storage temp. = 20°C (minimum), freezing point = 15°C (maximum), theoretical flame temp. = 1900°C, atomisation viscosity = 20-25 centistokes, emissivity factor = 0.5 to 0.7, combustion air requirement = 15 Kg/kg of oil, boiling range = 200 to 250°C, Residue at 350°C = 50-60%. Flash point = 80-110°C, Calorific Value -10000 Kcal/kg. Typical analysis : C = 86%, H = 12%, S = 2%, N, O, & inorganic acidity = nil.

(iv) **Middling Coal** : A typical composition of a blend of middling coal and ROM (Run of Mine) coal finally used is given below :

Volatile matter	= 20.8%	Ash	= 40.4%	Calorific value	= 4625 kcal/kg
Fixed carbon	= 37.6%	Moisture	= 1.2%		

The calorific value of middling coal supplied from various coal washeries varies from 3000 to

4000 kcal/kg while that of ROM coal is in the range of 4500-5500 kcal/kg. About 20-25% ROM coal is blended with the middling and finally fired in the captive power plant boilers as the main fuel after pulverising in ball mill.

(v) **Naphtha** : Typical properties of Naphtha (used after steam vaporisation) for boosting the CV of mixed gas (meant for rolling mill furnaces *i.e.* soaking pits, reheating furnaces, annealing furnaces, and HDGL furnaces) to the required level during coke oven gas shortage period is given below : Naphtha is a class-A product as per Petroleum act 1934.

Properties of Naphtha:

Boiling point (IBP - FBP)	= 45-156°C
Viscosity at 37.8°C	= < 0.5 Cst
Sulphur content	= 0-0.25% (by weight) (Typically 100 ppm)
Flash point	= < 23°C
Aromatics	= 15-60% (by volume)
Olefins	= 1.5% (by volume)
Maximum residue on evaporation	= 5 mg/100 ml, maximum
Density at 15°C	= 0.70-0.75 gm/cc
RVP at 38°C	= 0.42 kg/cm^2
Lead content,	= 3 ppm maximum
C/H ratio	= 5.5
Gross CV (HHV)	= 10,200 Kcal/kg = 7000 Kcal/Litre.

In a typical integrated steel plant, BF Gas, C.O. Gas and converter gas are produced during iron making, coke making and steel making processes respectively. The average calorific values of BF Gas, C.O. Gas and Converter Gas are 800, 4300 and 1800 Kcal/Nm^3 respectively. BF Gas and C.O. Gas are mixed in different proportions and the mixed gas thus made is supplied to rolling mill furnaces namely soaking pits in Slabbing Mill, reheating furnaces in HSM and annealing furnaces & HDGL furnaces in cold rolling mill.

Various furnaces require mixed gas of different CV and for this purpose gas mixing stations are provided. The pressure of mixed gas is usually boosted (increased) before supplying to the furnaces. Naphtha vapour injection is done to increase the CV of mixed gas for rolling mill furnaces during shortage/non availability of C.O gas due to various reasons. To facilitate this, naphtha storage and injection system is operated. Naphtha is first unloaded from road tankers through unloading pumps into a storage tank. Naphtha from the tank is then pumped, preheated and vapourised (by steam) and then injected into BF gas pipe line. In order to maintain the CV of mixed gas, the available C.O gas is mixed with BF gas and then calculated quantity of naphtha (1 KL naphtha = 2000 Nm^3 of C.O gas) is injected in BF gas line to get the required CV of mixed gas.

The calculated quantity of vapourised naphtha is taken through an insulated pipe line and injected in BF gas line of the Gas Mixing and Boosting Station through a diffuser after the pressure control throttle in BF gas line. And then the available CO gas is mixed to get the required CV of mixed gas. The pressure of this mixed gas is then boosted by centrifugal compressors (700 mmwg for Soaking Pit and 2100 mmwg for Reheating, Annealing & HDGL furnaces) and supplied to the rolling mill furnaces.

TIPS FOR SAFE HANDLING AND UTILISATION OF FUEL GASES IN STEEL-PLANTS

Since the fuel gases used in steel plant *i.e.* BF Gas, CO Gas and Converter gas are poisonous, inflammable and explosive in nature; they need to be handled and utilised carefully. Following precautions and practices are followed for this purpose:

- Regular inspection of gas carrying pipelines, drain pots, and other equipments for leakage (typically about 100 km fuel gas pipeline network exists in steel plant).
- Positive pressure (> 50 mm wg) is to be maintained in gas pipelines. (Automatic bleeders are provided for excess gas pressure release).
- All the gas line valves are blanked for isolating (only closing of valve is not reliable.)
- Isolation/depressurising and charging of gas pipelines is done with steam or Nitrogen purging.
- For cutting and welding job on gas pipelines, O_2> 20% must be ensured inside the pipeline.
- Welding/cutting/inflammable material are not allowed near gas pipelines during its charging/depressurising.
- Fuel Gas pipeline is to be left with full gas or full air. (Air/fuel gas mixture within explosive range must be avoided).
- Pressure/leakage testing of gas pipeline must be done before charging.
- Gas leakages and their concentration are detected/determined using gas *detecting tube*, *canary birds* and *explosimeter.*
- Pressure survey of fuel gas pipeline is done for identifying the area of naphthalene/tar fog deposits inside the pipelines (steaming is done for their removal).
- Gas masks are used while working in gas leakage prone area.
- Artificial breathing apparatus is used for administering oxygen to the lungs of gas affected/poisoned persons.
- Centralised gas control post maintains the optimum fuel gas line pressure (BF Gas-1000 mmwg, C.O. Gas - 700 mmwg, Converter gas - 1000 mmwg). Higher gas pressure may result in gas leakage/pipe bursting while low gas pressure (< 50 mmwg) may cause backfiring and explosion in furnaces. Such emergencies occur during total/partial power failure.

Explosion risk is involved when :

- charging/depressuring a gas pipeline.
- Repairing a gas pipeline/equipment.
- Lighting up/shutting off a furnace.

To prevent the explosion risk :

- Avoid formation of explosive air-fuel gas mixture by proper purging of gas pipeline with steam/nitrogen.
- Naked light/source of fire not to be used near depressurised gas line unless thoroughly purged (oxygen > 20%).

- No source of ignition should be put near gas bleeder.
- Light up and shut off of furnaces should be done by following correct procedure.

PROCEDURE FOR LIGHT UP AND SHUT OFF OF A GAS FIRED FURNACE

For lighting up a furnace following procedure should be followed sequentially:-

- Ensure main and burner gas valve are closed.
- Open damper, all doors and run I.D. fan/F.D. fan to ensure drafting.
- Light fire at the base for natural draft furnace.
- Purge gas line up to burner gas valve.
- Start combustion air fan and adjust air valve for > 4 mmwg furnace pressure.
- Place a lighted torch in front of burner.
- Ensure that gas pressure is > 150 mmwg.
- Open burner valve slowly and see for flame visibility. Open burner valve further and stabilise the flame.
- Do not take B.F. gas in cold furnace.
- If flame extinguishes during lighting, shut off burner valve and ventilate before lighting again.
- Do not turn the gas on to the furnace and then apply light, otherwise explosion will certainly occur.

To *shut off a running furnace,* following procedure should be followed sequentially :

- Close burner gas valve fully till there is no flame.
- Stop combustion air fan.
- Close valve in the branch fuel gas pipeline and open gas bleeder on this branch line.
- Stop I.D./F.D. fan after 20 minutes and close chimney damper.

FEATURES/DESCRIPTION OF VARIOUS STEEL PLANT FURNACES

The steel plant employs varieties of furnaces. The purpose, combustion system, end uses, heat recovery system vary from furnace to furnace. The salient features/description of various steel plant furnaces are given below:

COKE OVEN HEATING CHAMBER

The process of coke making consists of carbonisation/ destructive distillation of coal in absence of air. Each coke oven of the coke oven battery is included in between two heating chambers/walls. The heat generated inside the heating chambers from combustion of gaseous fuels, flow inside the oven through the partition walls and heats the coal. The process of coke making consumes energy about 1.4 to 1.5 GCal/Ton of Crude steel. (TCS). The specific heat consumption in coke making is about 0.620 GCal/Ton of dry coal.

Energy saving in Coke Ovens can be achieved by taking following measures:

- To maintain optimum combustion and thermal regime (i.e., proper fuel gas to air ratio).
- To eliminate door leakages and blazing by proper sealing of doors.
- To maintain proper cleaning and maintenance of door frames.

- To ensure regular cleaning of regenerators.
- To attain uniform rate of oven pushing.
- To avoid under charging.

Other New Energy Saving Technologies in Coke Ovens

- Drying of coal prior to charging in ovens by hot exhaust gases.
- Use of thinner walls between adjacent combustion and carbonisation chambers.
- Use of computer based carbonisation control system in which continuous measurement of flue and C.O. gas temperature determines the carbonisation end point.
- Use of programmed heating which will permit computation and control of the energy input needed at any stage of carbonisation process.
- Stamped charging of coal. — Dry quenching of coke.
- Use of taller (7 meters) Coke Ovens.

Typical Parameters of a Steel Plant Coke Ovens:

- *Blend of coal used* - A typical analysis of coal blend (dry basis) used (having 45% imported coal in the blend) is as follows :
- Ash = 15.4%, Volatile Matter = 23%, Fixed carbon = 61.6%, caking index of blended coal = 21
- Analysis of coke (dry basis) produced is :
- Ash = 19.8%, Volatile Matter = 0.6%, Fixed carbon = 79.0%, S = 0.6%, P = 0.1%,
- CSR = 67% (min), CRI = 19% (max), Porosity = 40% (minimum), M_{40} = 78.5% (min.)
- M_{10} = 10% (max.)
- Coal size : - 3.2 mm size is 80% in the blend of coal
- Carbonisation temperature = 1100°C
- Mixed gas consumption (CV = 1000 Kcal/Nm^3) = 40000 Nm^3/hr/battery at gas supply pressure of 400 mmwg.
- Pure coke oven gas consumption = 9000 Nm^3/hr/battery at gas supply pressure of 400 mmwg. Fuel gas pressure in the burner (for both C.O. gas and mixed gas) is about 100- 120 mmwg.

Total no. of batteries = 8, No. of ovens per battery = 69
Height of the oven = 5 metres (useful height is 4.7 metres upto which coal is charged).
Length of the oven = 15 metres, Thickness of oven wall =110 mm
Width of oven = 430 mm (on coke side) and 390 mm (on pusher side)
No. of coal charging holes and stand pipe per oven = 3 and 2 respectively
Volume of each oven = 27.3 m^3 Capacity of each oven = 20.7 Tons dry coal
Height of regenerators = 2.52 M
Automatic waterjet door cleaner is provided
Type of coal charging = Gravity charging (not stamped charging)
System of coal charging = Both off main and on main (HPLA-High Pressure Liquor Aspiration system)

Types of oven : Regenerative, Recirculative, Twin flue, Underjet, Compound ovens.

No. of burners = 30 (16 on pusher and 14 on coke side)
Temperature on coke side = 40°C more than that on pusher side
Specific power consumption = 30 kWh/T of Gross coke produced
Specific steam consumption = 150 Kg/T of Gross coke produced

Products of carbonisation:

Gross coke yield = 77% of dry coal charged.
BF coke (25-80 mm size) yield = 87% of gross coke produced = 67% of dry coal charged
Coke oven gas yield = 290-300 Nm^3/Ton of dry coal
Tar yield = 3% of dry coal, Ammonia yield = 0.3% of dry coal
Benzol yield = 0.7% of dry coal

Final sizes of coke produced :

BF grade coke (25-80 mm size) - used in BF
Nut coke (15-25 mm size) - sold outside for use in foundry.
Coke breeze - used in Sinter making and soaking pit bottom making.

Coal supply sources :

A blend of 50% indigeneous metallurgical coking coal (from coal washeries) having low ash of 20% plus 50% imported coal containing 10% ash (from Newzealand and Australia) is used for coke making. At times, high sulphur (4-6% S) Assam coal (upto 5% in the blend) is also used. Coal is stored in silos and blended through dozimeters before crushing in hammer crusher and is finally charged in the coke ovens *via* coal storage towers through coal charging car from the oven top coal charging holes.

Typical Analysis of Coke oven gas produced (% by volume) :

O_2 = 0.5% , CmHn (unsaturated hydrocarbons) = 2.5%, CO_2 = 2.5% N_2 = 3.5%, CO = 8%, CH_4 = 2.5%, H_2 = 58%
Net Calorific value = 4250 Kcal/Nm^3, Density = 0.46 Kg/Nm^3

Heating System of Coke Ovens. Each coke oven is located in-between two heating walls on the adjacent sides. Heat produced in the heating chamber by the combustion of either coke oven gas (rich gas) or mixed gas/B.F. gas (lean gas) heats up the coal in the oven through the refractory partition walls. To ensure uniform heating of coal inside the oven, throughout the length and height of the coal charged oven, gaseous fuels are used. A coke oven having facilities to heat it with both rich gas (*i.e.* C.O. gas) as well as lean gas (*i.e.* B.F. gas/mixed gas of C.V. 1000 Kcal/Nm^3) is called a compound oven. In heating chamber provision of more number of vertical flues (30 burners) ensure more uniform heating along the length of the oven. The vertical flues act in pair (*i.e.,* twin flue) and is located at the bottom (hence called underjet type). To ensure uniform heating along the height of the oven, *recirculation of flue gases* through an opening between the pair flue at the bottom is resorted. Flue gas from the heating chamber passes through rgenerator checker bricks (to recover waste heat) before being let off to the atmosphere through the chimney. Hot regenerator preheats the combustion air and the B.F. gas/mixed gas while the cooled regenerator gets heated up by the hot flue gas. As the regenerative system requires alteration of path of flow (*i.e.* reversal) of gases (*i.e.* flue gases and combustion air & mixed gas /B.F. gas), a minimum of two sets of identical paths (*i.e.* regenerative path) is provided for effective heating of each oven.

Such coke ovens are classified as twin flue (pair vertical) underjet type, recirculative, compound, regenerative ovens.

Each coke oven heating chamber has 15 pairs (30 burners) of vertical flues (16 burners on pusher side and 14 burners on coke side). Every two adjacent vertical flues are interconnected at the top by cross over windows and recirculation hole at the bottom but adjacent pairs are not interconnected. Two regenerators are located beneath each oven. Each regenerator section connected with the heating flue (at the upward flow) and the adjacent regenerator section connected with the heating flue (at the downward flow) constitute an individually controlled heating element of the oven. Each regenerator is connected with heating flue (*i.e.* burners) by means of one sloping flue. Air and lean fuel gas (*i.e.* B.F. gas/mixed gas) comes into regenerators *via* gas-air valve and sole flues.

If coke ovens are to be heated by rich gas *e.g.,* coke oven gas, then it is supplied directly to the heating flues (burners) bypassing the regenerator. Coke oven gas is not preheated in the regenerators to a high temperature because hydrocarbons present in it will thermally crack (HC $\rightarrow$ H+C) thereby fouling the heat transfer surface of the regenerator checkers (resulting in reduced heat transfer co-efficient) and chocking the low diameter (12-13 mm) burners/flues of the oven heating chambers. Reversal of path of flue gases and combustion air/mixed gas is done every 20 minutes by the reversal valve.

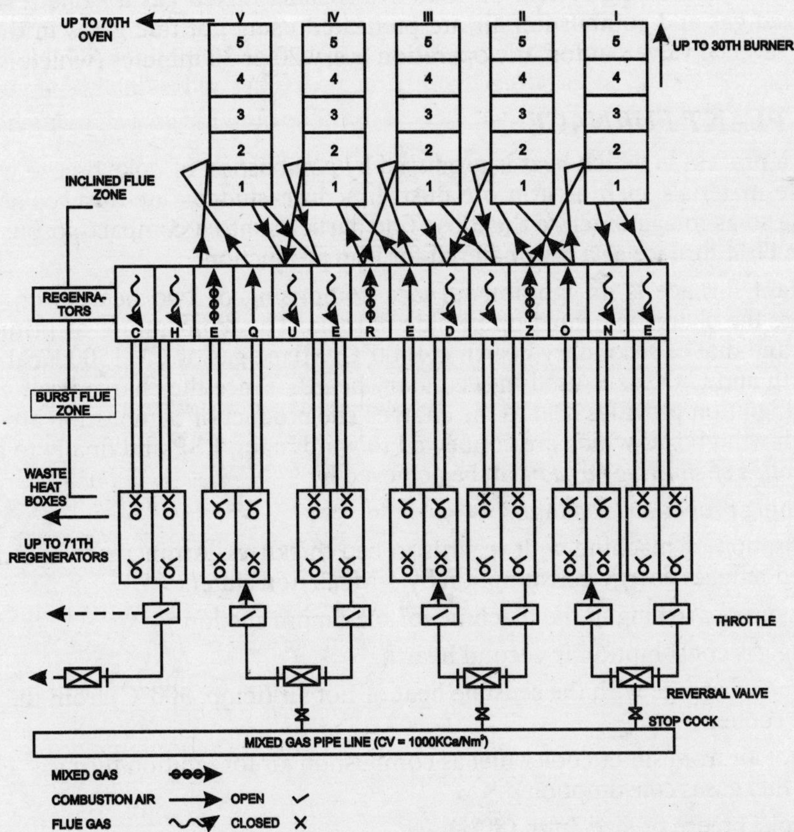

Fig 46.1: Heading system of coke ovens-one reversal route using mixed gas

Coke oven batteries & regenerators are built of refractory mass and supported by structures. Anchorage system comprising of anchor column is provided for :

- holding the brickwork in shape.
- ensuring tightness of the brickwork.
- protecting brickwork from destruction particularly in case of heating walls and regenerator walls.
- absorbing the dynamic load during oven pushing and thermal shock suffered by the brickwork.

Modern coke ovens (including the heating chamber) are built up of silica refractory bricks due to its better strength, thermal conductivity and less post commissioning expansion. Fireclay bricks is used for making regenerators, sole flue (where thermal fluctuation is more), buttresses and oven tops. Insulating bricks are used for heat protection.

The regenerative system requires two sets of identical paths for changing the path of the flow of gases (fuel gas *i.e.* mixed gas and flue gases) for effective heating.

Regenerator system is divided in three zones namely, 'burst flue zone,' 'chequered brick zone' and 'inclined flue zone'. Control of combustion air and mixed gas flow, distribution in the regenerator section and heating flues along the length of the oven is done from the beneath of the ovens by special accessories (called finger plates and bannana bricks).

Fig. 46.1 shows the heating system of coke ovens using mixed gas for one reversal route in which both mixed gas and combustion air are preheated using hot flue gases in the regenerator cyclically using reversal valve's automatic operation every 20 or 30 minutes (which is preset).

SINTERING PLANT FURNACE

Sintering is a process in which heat is produced by combustion of coke breeze within moving beds of fine size materials such as iron ore dust, limedust, sludge, flue dust, coke dust/breeze, rolling mill scale so as to agglomerate the loose fine particles into a compact porous mass (called sinter) for use in blast furnace as a raw material for iron production.

Sintering plant furnace is an ignition furnace comprising of two hearths *i.e.* primary and secondary, where the charge is heated up by side burners. The temperature of primary hearth is around 1200°C and that of secondary hearth is 1000°C. Mixed gas of CV 1500 Kcal/Nm^3 is used in primary hearth and C.O gas is used in secondary hearth. Since the charge itself contains coke breeze which on ignition provides heat to the charge. The product of combustion are exhausted to chimney through wind boxes which are connected to wind main, ESP and finally to an exhauster. The *efficient operation of sintering furnace* can be achieved by :

- Maintaining proper air/gas ratio.
- Full utilisation of machine with regards to bed thickness. An increase in 2 mm depth of sinter bed reduces energy consumption by 1 Mcal/ ton sinter.
- Ensuring proper sealing in the machine for maximum suction.
- Reducing gas consumption in second hearth.
- Preheating of charge with the sensible heat of hot air (temp. 300°C) from the hottest zone of sinter cooler.
- Use of hot air from sinter cooler fuel as combustion air for ignition furnace. This results in reduced fuel gases consumption.
- Use of coke breeze of size 3mm (80%).
- Reduced ignition furnace pressure to the extent possible up to +0.2 mm wg.

- Use of high chrome grate bars on the pallets.
- Computerisation and automatic control of sintering plant furnace and process.

Typical Parameters of a Sintering Plant:

- Capacity of each sinter machine = 300 – 400 tons of sinter/hr.
- No. of vacuum chambers = 26
- Vacuum in vacuum chamber = 800–1000 mmwg
- Maximum depth of sinter bed = 430 mm
- No. of burners in 1st and 2nd hearth = 6 each (side burners)
- No. of flue gas exhausters/machine = 2
- Speed of sinter machine = 3 m/min.
- Specific heat consumption in Sinter furnace = 0.065 Gcal/ton of gross sinter produced
- Specific power consumption in Sinter furnace = 52 kWh/ton of gross sinter produced.
- Specific energy consumption in Sinter furnace = 0.65 Gcal/ton of gross sinter produced.
- Coke breeze consumption = 75 kg/ton of skip sinter *(i.e.* B.F. grade/size sinter)
- Ignition furnace hearth area = 32 m^2 (8x4 m)
- Typical *composition* of the *charge materials* for sinter making :

Iron ore dust = 50%, lime dust = 15%, sinter return fines = 20%
Rolling mill scale + flue dust (from B.F. dust catcher) = 3%
Coke breeze/dust = 5%moisture= 7%

— Skip sinter quantity = 90% of gross sinter sent to B.F. *(i.e.* - 5 mm fraction is 10%).

In sinter making, the various fines raw material (containing iron, carbon and fluxing materials) is fed to the rotating balling drum along with water. Here the charge is mixed and formed into small balls. The charge may be preheated to 60°C with hot air from sinter cooler (rotary or straight line type) and moisture in charge is controlled to about 7%. This is then discharged to the charge feeder hopper from where it is fed to the moving bed sinter machine.

The sinter machine is an endless moving grate consisting of pallets fitted with high chrome grate bars. As the sinter machine with the charge continues to move, it passes through the ignition furnace where the charge is heated up by the burners on both sides of primary and secondary hearths. As the charge is ignited from the top, it is simultaneously under suction from below the sinter machine through vacuum chambers which are connected to the exhauster through wind main & electrostatic precipitator (ESP) and finally waste gases are discharged through the chimney. The moment, charge materials enter the hearth, there is suction from below which continues upto the end of the sinter machine (i.e. upto 26th vacuum chamber). The ignited zone moves slowly from the top surface to the bottom and sintering of the charge material is completed by the time it reaches the last vacuum chamber i.e., at the end of the sinter machine. Sintering process can be compared with the burning of a cigarette while smoking. As the cigarette goes on burning, while applying suction from mouth during smoking, the zone of highest temperature moves from the top to the butt end and the whole cigarette tobacco is converted to ash. While, in case of sintering, the raw fine charge materials (equivalent to tobacco) is converted to sinter (a lumpy mass) instead of ash. However, there is one difference that sinter machine is not stationary (as the cigarette is) and is instead moving continuously facilitating continuous feeding of charge and hence continuous production of sinter.

At the end of the sinter machine, sintering is completed i.e. whole thickness of the charge gets sintered. The sinter is discharged to a single roll crusher, where it is crushed to a size below 150 mm. This hot sinter (- 150 mm) is fed to the 'hot screen' of 6 mm size. The undersize (- 6 mm) is called hot sinter return which is cooled with water and reused in the sinter making. + 6 mm hot sinter falls on the moving sinter cooler which it is cooled to 80°C by forcing air from the bottom. The cooled sinter falls on the 'cold screen' having openings of 6 mm and 12 mm size. + 6-12 mm sinter fraction is called sinter bed material and is reused in the sinter machine for the bottom most portion of the charge. - 6 mm size cold sinter return is also recycled in the sinter making process. + 12 mm size sinter is sent to the blast furnace (B.F.) where it is furtherscreened on a 5 mm size screen. The oversize called skip sinter' is charged in the B.F. through skip. –5 mm sinter return is recycled and reused in sinter making.

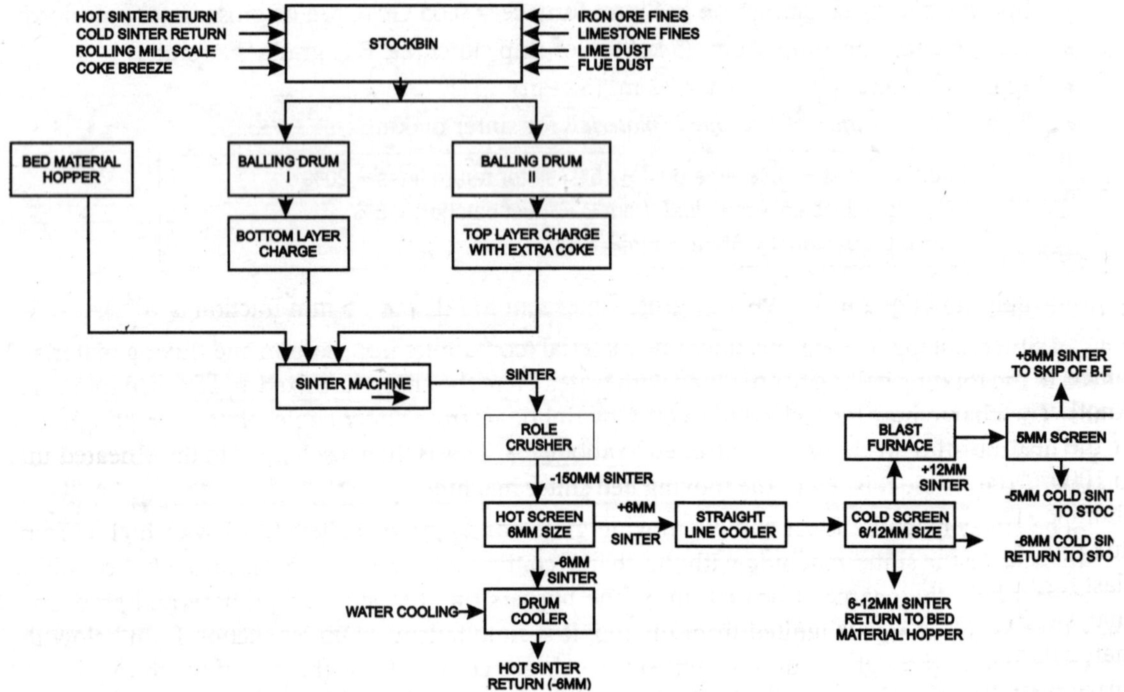

Fig. 46.2: Process Flowchart of a typical sintering plan

The lower most charge material layer on the sinter machine is called the bed material comprising of + 6 –12 mm size cold sinter. Middle/intermediate layer will be other fine raw materials (including sinter returns with less coke (this is called bottom charge layer). The topmost layer will be the sinter charge with extra coke (this is called the top layer).

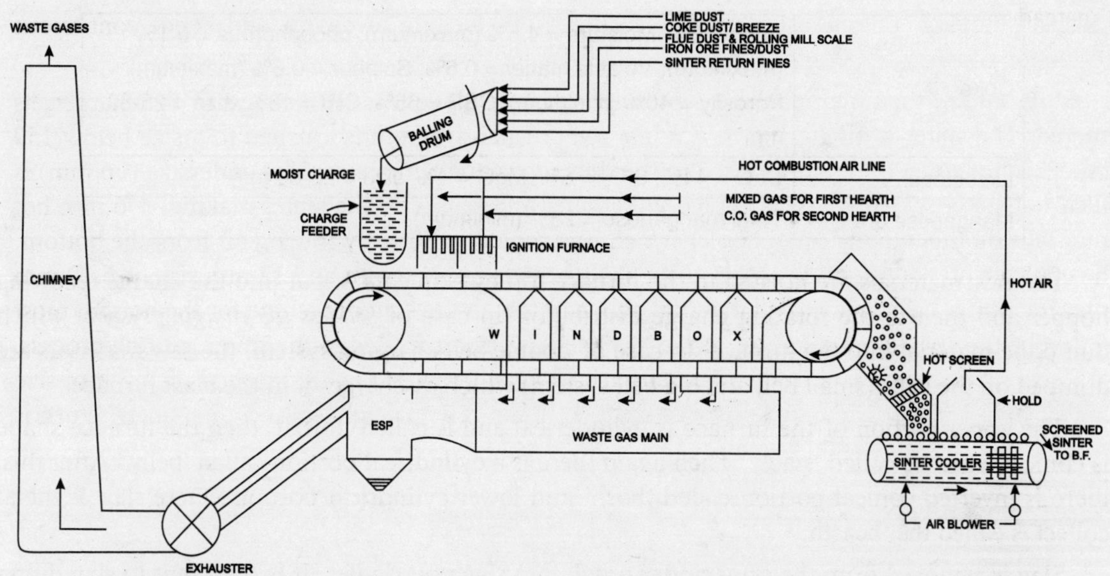

Fig. 46.3: Schematic diagram of sinter plant furnace with waste heat recovery system.

Fig. 46.2 gives the simplified process flow chart of a typical sintering plant. Fig. 46.3 gives the schematic diagram of sinter plant furnace with waste heat recovery system.

BLAST FURNACE (B.F.)

Blast furnace produces pig iron from iron ore and sinter (containing iron bearing as well as fluxing materials). The iron oxide of ore and sinter is reduced to iron with the help of carbon monoxide which is produced inside the furnace by the action of hot air (blast) at about 1000°C with coke. Another purpose of coke is to supply heat for reactions and melting iron inside the furnace. Part of the heat load of the Blast furnace is also provided by the heat of the blast which is heated up to 1000°C (in stoves) before it is introduced into the blast furnace. The gaunge material like silica, alumina etc. of the iron ore, sinter & coke (ash) are removed as slag (impurities) with the help of flux provided by limestone & sinter. Additives like quartzite and manganese ore are charged in the blast furnace in small quantities as corrective materials for proper furnace working and control of analysis. Except air blast which is fed from the lower part of the furnace through tuyers, all other materials are charged from top by means of bell less top charging system. The products of the blast furnace are pig iron (hot metal) & slag from the lower part of the furnace and flue dust & blast furnace gas from the top.

Prepared raw materials (of proper size, composition and specification) for B.F. namely sinter, iron ore lump, coke and additives (like manganese ore, quartzite, limestone etc.) are stocked in the bins. Typical specifications of various raw materials used in an Indian blast furnace is given below :

Iron ore lump	→	Total Fe = 62.5%, Silica (SiO_2) = 1.6% (maximum) Alumina (Al_2O_3) = 3.8% (maximum), ore size = 10—40 mm
Sinter	→	Total Fe = 50%, FeO = 9%, SiO_2 = 7.5% CaO = 12.5%, MgO = 3.5%, basicity (CaO/SiO_2) = 1.7

Coke →	Ash = 20%, Moisture = 4.5% (maximum), phosphorous = 0.1% (maximum), Volatile matter = 0.5%, Sulphur = 0.6% (maximum) Porosity = 40% (minimum), CSR = 65%, CRI = 18%, size = 25-80 mm
Limestone →	CaO = 44%, SiO_2 = 5%, MgO = 4%, size = 25-80 mm
Manganese ore →	Total manganese = 28% (minimum)

The raw materials are hoisted to the furnace top by skip cars & put into the charge receiving hopper and then to the rotating charge distributor (in case of *bell-less top charging system)* which dump the material into the furnace. In case of double bell charging system, the raw materials are dumped on the bells (small bell and big bell system) which discharges it in the blast furnace.

The upper portion of the furnace is cylinderical and is called 'throat' then the furnace shape is conical which is called 'stack.' Then again there is a cylindrical portion called 'belly.' After this, there is inverted conical portion called 'bosh' and lower cylindrical portion where slag & metal collect is called the 'hearth.'

Slag is removed from the slag/cinder notch into slag pots/ladles. It is then sent to slag dump yard or to slag granulation plant (for making granules to be used in cememt making). Hot metal (pig iron) is tapped through the iron notch into ladles and is sent to steel melting shop (if. Si < 1.6%) for steel making. In case Silicon (Si) in hot metal is more than 2%, it is sent to pig casting machine (PCM) for casting it into iron pigs (which is sold outside). If Si in hot metal is in between 1.6% to 2%, it is sent to ingot mould foundry (IMF) for ingot mould and bottom plate making.

Blast furnace gas leaving the top of the furnace is cleaned of its dust content and cooled to room temperature in gas cleaning plant. Part of this clean gas (about 35% of total B.F. gas production) is burnt in stoves to preheat the blast (air) to about 1000°C. Hot blast is taken through the hot blast line to the 'bustle pipe' encircling the blast furnace. Hot blast then enters the furnace through water cooled openings called 'tuyers' located just below the top of the hearth. The 'tuyers' are connected to the 'bustle pipe' by means of goose neck connection *via* 'blowpipe' and 'elbow.' Steam is injected in the cold blast before preheating it in the stoves. Hot blast burns the coke inside the furnace.

The bottom, hearth, tuyer zone, bosh & belly are water cooled by stave cooler and the stack is water cooled by the peripherial cooler. The tuyers are cooled by circulating water along its whole surface. The wearing plates of the top are not cooled. Stave coolers are double walled plate type coolers having arrangement for water circulation. Temperature rise of the cooling water is restricted to a maximum of 5-6°C by controlling the cooling water circulation rate. Modern bell-less top charging system provides almost continuous charging and better distribution of charge materials compared to earlier used double bell charging system which is obsolete now.

Hot metal tap hole (made of cast steel) is closed by forcing anhydrous tar bonded clay mass by mudgun nozzle. For opening of the taphole for casting the metal out, mudgun drilling machine is used.

Typical features and operating data of an Indian blast furnace with bell-less top charging system is given below :

Typical Features and Parameters of an Indian Blast Furnace (B.F.):

Useful volume of the blast furnace = 2000 m^3

Hot metal (pig iron) production capacity = 3000 tons/day
B.F. hearth diameter = 9.75 metres
Hearth height = 3.8 metres
Bosh height = 3 metres

Refractory lining thickness of the various furnace parts :

- Hearth bottom carbon block thickness =1.6 metres
- Hearth bottom alumina block thickness = 2.7 metres
- Hearth wall (carbon block) thickness = 1 metre
- Bosh wall (fireclay bricks) thickness = 0.35 metre
- Stack (upto cooler level) = 0.7 metre
- Stack (above cooler level) = 0.9 metre

Coke consumption rate = 580 kg/ton of hot metal (THM)
Coke burning intensity = 940 kg dry coke/m^3 of furnace volume /day
Blast pressure = 3 kg/cm^2
Blast rate = 3000 Nm^3/minute
Blast humidity = 60 gms/Nm^3 of blast
Hot blast temperature = 1000°C
B.F. gas top pressure = 1.5 kg/cm^2 (gauge pressure)
Raw B.F. gas temperature = 250°C
Hot metal temperature =1400°C
Alumina content in slag = 24%
Silicon content in basic pig iron = 1.2% maximum
Silicon content in foundary grade pig iron = 2% maximum
Slag generation rate = 350 kg/THM
Sinter % in total charge material = 70%
B.F. gas yield = 1900 Nm^3/THM
CO/CO_2 ratio in B.F. Gas = 1.2 (CO_2 = 19%, CO = 23%)
Specific steam consumption = 550 kg/THM
Specific electrical power consumption = 25 kWh/THM
Specific energy consumption = 3.3 Gcal/THM (lGcal = 10^9 calorie)

Energy conservation techniques in Blast Furnace :

- Use of proper quality & size of iron ore and coke.
- Increased use of sinter in the burden.
- Use of high hot blast temperature with steam humidification of blast.
- Auxiliary fuel (coal dust, tar, fuel oil etc.) injection in BF.
- Efficient operation of BF stoves.

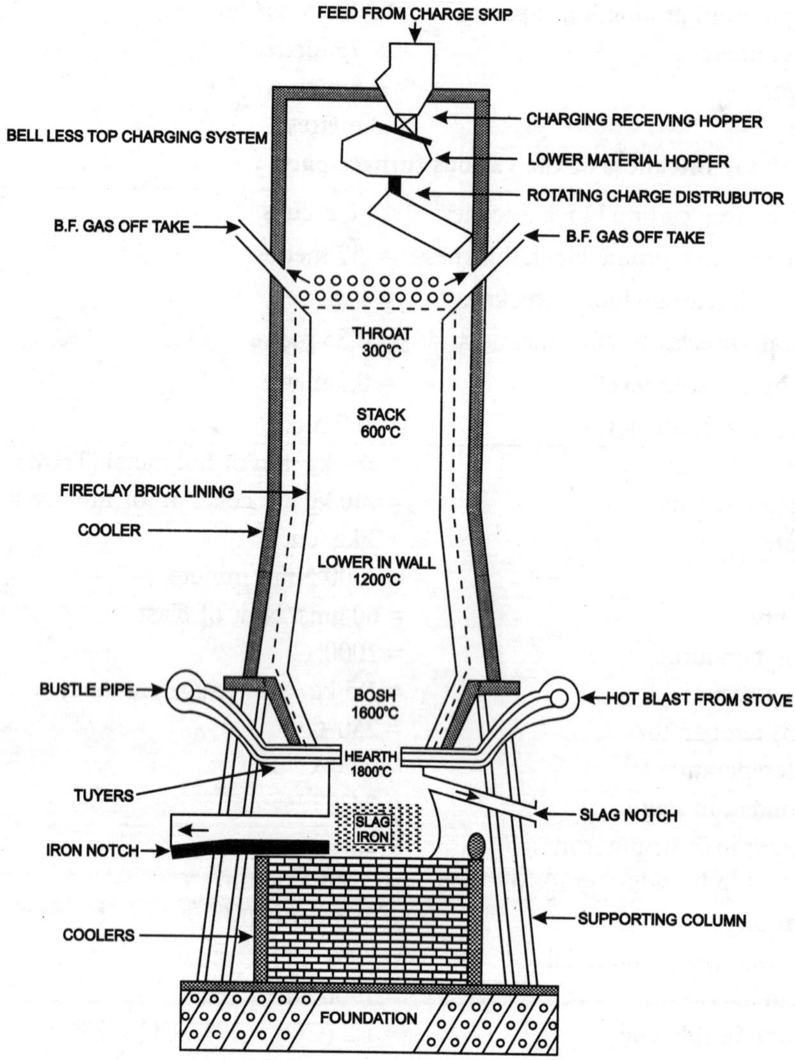

Fig.46.4: Constructional features of a blast furnace with bell-less top charging system. (Double bell charging system has now become obsolete)

- Oxygen enrichment of hot blast.
- High top pressure operation of BF using top pressure recovery turbine (TRT).
- Use of large volume BF with bell-less top (BLT) burden distribution system.
- Reducing hot blast losses from tuyers.
- Use of furnace sensors and other techniques for monitoring & controlling the distribution of charged materials to improve the efficiency in furnace reduction reaction which in turn reduces the coke rate.

Fig. 46.4 shows the constructional features of a blast furnace with bell-less top charging system. Fig. 46.5 shows the material flow diagram for a blast furnace.

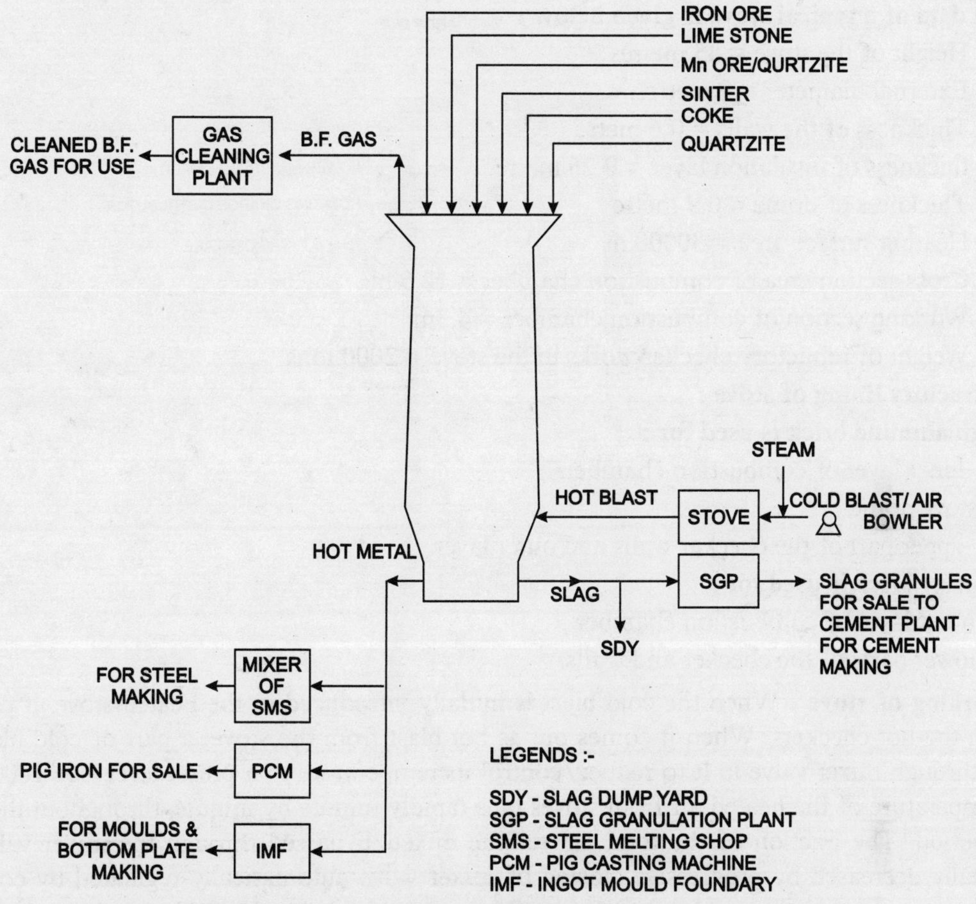

Fig.46.5: Material Flow (In/Out) Daigram for a Blast Furnace.

BLAST FURNACE (B.F.) STOVES/CAWPER STOVES

BF stove is used for heating the air blast to 1000-1200°C before the blast is introduced into the Blast Furnaces. The stoves which is a refractory chamber work on regenerative principle *i.e.,* first the fuel gas (BF gas or mixed gas) is burnt inside the stoves (for about 3 hours) and then as the refractory brickwork (checkers) gets heated up, the gas burning is stopped and air blast is passed through the stoves. The air blast takes out the heat from the checker work and attains the required temperature (1000°C). This period is called on blast cycle which lasts for about 1.5 hours.

A typical blast furnace producing 3000 tons pig iron per day will be blown with 3000 Nm^3/ minute of atmospheric air (cold blast at 140°C) preheated to a temperature of 1000-1100°C. To preheat this blast volume, a set of **four stoves** is normally provided for each blast furnace. Each stove consists of a refractory lined vertical steel cylinder topped with a spherical 'dome.' A side combustion chamber is separated by a bridge wall with a lens shaped horizontal cross- section. The remaining volume is filled with heat absorbing checkerwork.

Design data of a typical stove is given below :

- Height of the stove = 45 metres
- External diameter = 9 metres.
- Thickness of the walls = 0.6 metre
- thickness of insulation layer = 0.25 metre
- Thickness of dome = 0.9 metre
- Heating surface area = 30000 m
- Cross-section area of combustion chamber = 12.5 m^2
- Working section of combustion chamber = 4,3m^2
- Weight of refractory checker works in the stove = 2000 tons

Refractory lining of stove :

High alumina brick is used for :

(a) inner layer of combustion chamber.

(c) stove dome

(b) upper part of the checker walls and outer layer.

Fireclay brick is used for :

(a) outer layer of combustion chamber.,

(b) lower parts of the checker and walls.

Working of stove : When the cold blast is initially introduced in the heated stove, it passes through the hot checkers. When it comes out as hot blast from the stove; a part of cold blast is mixed through mixer valve in it to reduce/control its temperature to a preset level (say 1000°C). The temperature of the heated air from stoves falls rapidly minute by minute, throughout the 'on blast' period. The fraction of the total air volume mixed/bypassed through the mixer valve is continually decreased by progressive closing of mixer valve automatically regulated by control of a thermocouple. At the end of the 'on blast'period (say 1.5 hrs.), the cold blast mixer valve is practically closed i.e., the entire blast passes through the stove checkers. Uniform blast temperature is achieved by automatic control of the mixer valve.

The rapid decrease in temperature exhibited by the air discharging from the stove checkers is a characteristic feature of classical checker work heat transfer. The thickness of refractory flue walls retards the flow of heat by thermal diffusion into the central portion of the brick. Although the heat removed during 'on blast' cycle is < 5% of total sensible heat stored in the stove refractories, the introduction and removal of heat from such large dimensioned refractory elements is sluggish, with the result that the in and out movement of heat is largely a skin effect confined closely to the refractory surface.

Fuel gas/air ratio is very important for proper heating of stoves using either B.F. gas or mixed gas. About 10% excess air gives the right flame appearance of violet colour. With very high excess air, flame becomes light, transparent and blue. With less than required amount of combustion air, the flame becomes yellow and the walls of combustion chamber is hardly visible.

MODE OF OPERATION OF STOVES

The stoves are run in cycles called 'on blast', 'on gas', 'on separation' and 'back draught.' The stove is heated up in 'on gas' cycle and the cold blast (air) is passed through the stoves in 'on blast' cycle. 'On separation' is the period for changing over from one to any other cycle. 'Back draught' is utilised when the blast furnace is connected with the stoves chimney *via* stoves so as to remove B.F. gas inside the furnace before giving the furnace shut down for repair work.

The stoves (normally 4 nos. per B.F.) are operated in following modes :

- Sequential mode
- Parallel mode
- Lap parallel mode
- Staggered parallel mode

Sequential mode of stove operation. This type of operation is carried out with 3 stoves only. The stoves are put 'on blast' one after another. At a particular time, only one stove is 'on blast' and other two are 'on gas' firing.

Parallel mode of stove operation. This is practised with 4 stoves, two stoves are put on blast simultaneously by equally dividing it in these stoves. The other two stoves are on gas firing. Since less blast flows through a stove, higher blast temperature can be achieved in this mode.

Lap parallel mode of stove operation. In this mode of operation two stoves are on equal blast but the commencement of blast period is different. However, the overall blast period for the stoves is constant. As a result blast from one stove is hotter than the other stove. Blasts from both the stoves along with cold blast are mixed in varying proportion to maintain constant blast temperature In this way, higher blast temperature can be maintained for a longer duration.

As lap parallel operation utilises some of the heat of the stove to produce blast below the required blast temperature, it minimises cold blast additions.

Staggered parallel mode of stove operation. This mode of operation is similar to lap parallel operation except that control is achieved by varying the inflow of cold blast through stoves with the help of a valve. Two stove operation is used, with the blast from a 'hot' stove being mixed with blast from a 'COOL' stove to achieve constant blast temperature This eliminates the addition of cold blast to the heated blast after the stoves.

Typical *parameters of a B.F. stove* is given below :

Number of stoves in each blast furnace = 4

Fuel consumption in Nm³ / hr. / stove :

(a) BF gas (CV = 840 Kcal/Nm^3) = 40,000

(b) Mixed gas (CV = 1200 Kcal/Nm^3) = 28,000

Fuel gas / air ratio (theoretical) —

(a) For BF gas = 1 : 0.75 (10% excess air is used in practice)

(b) For mixed gas = 1 : 1.05 (10% excess air is used in practice)

'On gas' period = 3 hours, 'On blast' period = 1.5 hours

Flue gas temperature at the end of:

(a) blast period = 200°C

(b) gas period = 350°C.

Dome temperature at the end of:

(a) blast period = 1150°C, *(b)* gas period = 1300°C.

Hot blast rate = 3000 Nm^3/minute (Max.), Hot blast temperature = 1100°C (max.)
Specific heat consumption in stoves = 0.65 Gcal/THM (ton of hot metal *i.e.*, pig iron)
(1 Gcal = 10^9 calorie)
Cold blast temperature = 140°C,
Blast pressure = 2.5-3 kg/cm^2
Stove thermal efficiency = 80%

Energy efficient operation of B.F. stoves

For achieving high hot blast temperature and increased stove efficiency thereby reduced fuel consumption, the stoves are to be operated efficiently at high dome temperature. An increased maximum dome temperature represents a hotter flame during "gas cycle" which will result in an improved rate of heat transfer. Also because of high top checker (dome) temperature the heat transfer during "blast cycle" will be improved. Both these factors improve stove efficiency.

An increase in stove dome temperature can be achieved by:

- Reducing combustion air flow
- Increasing calorific value of fuel gas
- Enriching combustion air with oxygen
- Preheating the fuel gas and/or air

Type of fuel used is decided on the basis of required dome temperature. BF Gas is used for lower dome temperature while mixed gas (CV 1200 Kcal/Nm^3) should be used for higher dome temperature. Stove efficiency is higher at moderate blast temperature with rich gas (high CV gas) as compared to lean gas (low CV gas). Cycle time should be as short as possible to achieve higher stove efficiency. Increase in CV of fuel gas improves the stove efficiency. Preheating of combustion air/fuel gas is potentially the most cost effective method of increasing the dome temperature and thermal efficiency, if waste heat recovery can be employed. Increasing the cold blast temperature (from 80 to 140°C) by insulating the cold blast main reduces the specific fuel consumption but reduces the stove efficiency to some extent (the saving is more when lean fuel gas is used).

The differences and similarities between stoves fired with rich (e.g., mixed gas) and lean (e.g. pure B.F. gas) gas are summarised below :

- High blast temperatures are not achievable with lean gas.
- At moderate blast temperatures, stove efficiency is much higher with rich gas than with lean gas.
- At low blast temperatures, there is little difference between the efficiency obtained by the use of rich gas and that by the use of lean gas.

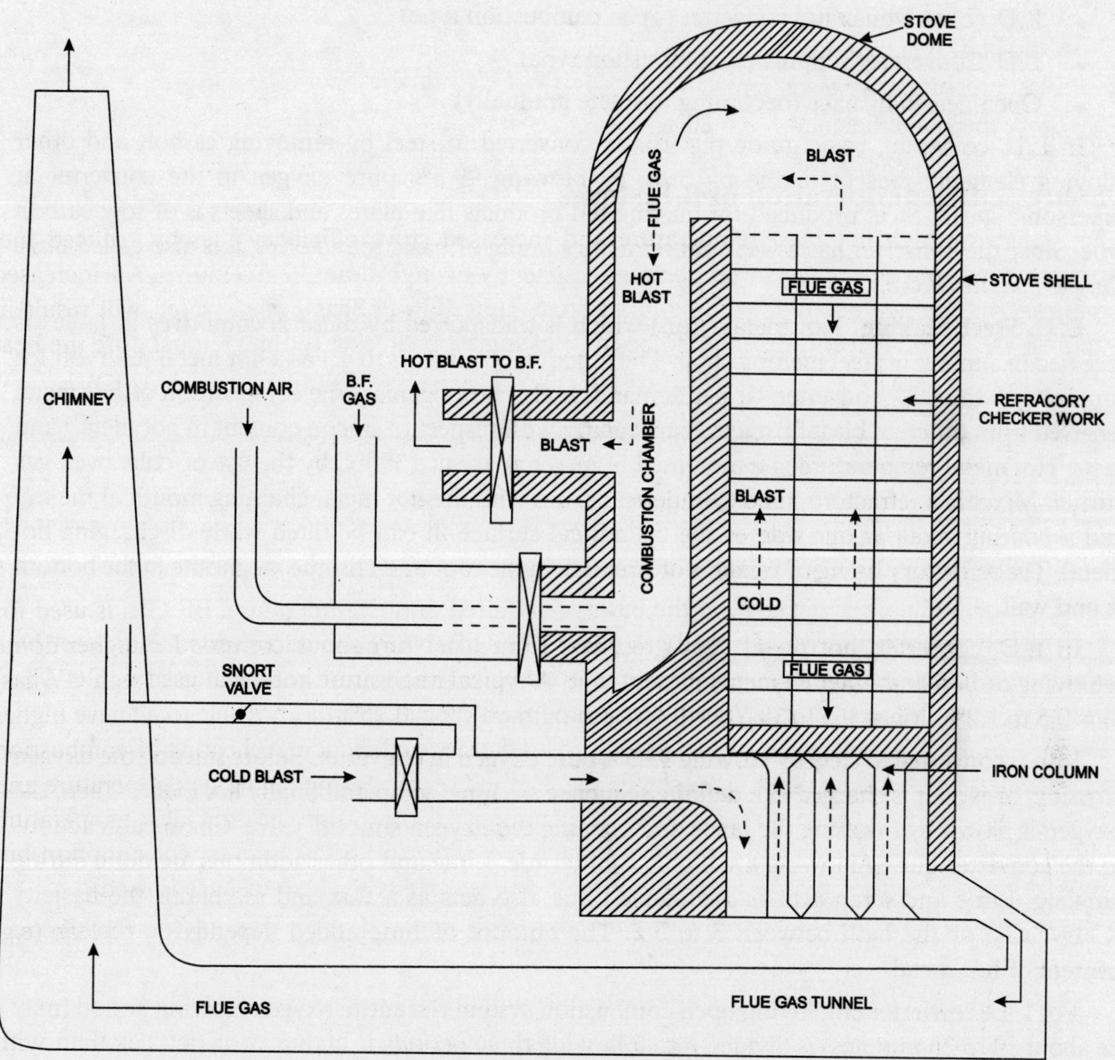

Fig. 46.6: Blast furnace stove

The difference between the dome and blast temperature depends upon the mode of stove operation *viz* sequential, parallel, lap parallel & staggered parallel.

A diagram of blast furnace stove is given in Fig. 46.6.

STEEL MELTING SHOP (S.M.S.) FURNACES

S.M.S. furnaces are decarburising furnaces in which hot metal (iron) is converted into steel by removing the carbon from hot metal by oxygen blowing to the level required in particular quality of steel. Steel making is a refining process carried out on a charge comprising pig iron and steel/iron scrap in which C, Mn & Si are removed by oxidation and P & S by reaction with lime to form slag.

Basically there are three important types of *metallurgical furnaces* for making steel from iron by decarburisation *(i.e.* carbon removal), namely :

- L.D. (Linz Donawitz) converter (open combustion type)
- L.D. Converter (Suppressed combustion type)
- Open hearth furnace (becoming obsolete gradually)

In L.D. converter, basic grade pig iron is converted to steel by removing carbon and other alloying elements present in the pig iron by blowing 99.5% pure oxygen in the converter at supersonic speed. Steel produced for making flat products like plates and sheets is of low carbon type. Since the converter has a wear out refractory lining of basic refractories, it is also called *Basic Oxygen Furnace (B.O.F.).*

L.D. Steel Making. Hot metal (iron) which is transported by diesel locomotives in ladles is received in a mixer in steel melting shop. The function of mixer is to act as a hot metal reservoir for supplying to the L.D. converters as per demand. It also homogenises the composition of hot metal received from different blast furnaces (particularly with respect to silicon content in hot metal) and casts. Hot metal temperature is maintained in mixer at about 1350°C by the use of coke oven gas burner. Mixer is a refractory lined cylinderical drum with the hot metal charging mouth at the top and a pouring spout at one side of the cylindrical surface. It can be tilted while discharging hot metal. The refractory lining of mixer is of firebrick in the roof and chrome megnesite in the bottom & end wall.

In L.D. converter, hot metal (iron) received from blast furnace is converted into steel by removing carbon and other elements present in it. A typical analysis of hot metal used is, C = 4%, Si = 0.5 to 1.2%, Mn = 0.5 to 1%, S = 0.05% (maximum), P = 0.25%.

Iron is converted to steel by blowing 99.5% pure oxygen in converter. Before starting the oxygen blowing, converter is charged in a definite sequence *viz.* lime, scrap and finally hot metal. Then, the oxygen is blown by lowering the lance and opening the oxygen shut off valve. Chemical reactions in the converter during blowing are highly exothermic. Lime and other additions are done during blowing. Lime and scrap acts as a coolant. Lime also acts as a flux and maintains the basicity (CaO/SiO_2) of the bath between 3 to 3.5. The amount of lime added depends on the silicon content of hot metal.

For L.D. converter employing open combustion system the entire oxygen blowing period (may be about 20 to 24 minutes) is divided into following three periods :

(i) Slag formation period. Lasting for about 10 minutes; Fe, Si and Mn in hot metal oxidises during this period. Silicon oxidises more intensively compared to other elements. Lime is continuously added for assimilation of Fe, Si and Mn oxides in the slag. Iron ore/manganese ore is also added at the beginning (at 3rd minute) of this period. The lance is first kept at 2.2 metres for 2-3 minutes and then lowered down and kept between 1.5 to 1 metre.

(ii) Intensive carbon oxidation period. After 10 minutes, brightness of flame starts increasing due to burning of CO and CO_2 at the mouth of the converter which indicates the beginning of this second period. The oxygen lance is brought down to 1.2 to 1 metre level in the last period of blowing.

(iii) End of blow period. End of blow is judged by dropping down of the flame.

Finally when carbon in steel is assessed to have come down to the required level (by flame appearance), oxygen lance is lifted and oxygen blow/flow is stopped. The slag on the top of steel is run off by tilting the converter & steel temperature is measured by dipping a thermocouple and the steel sample in a capsule is pneumatically transported to the laboratory for its chemical analysis.

If steel bath opening temperature is more than the steel tapping temperature (typically 1640°C), lime or scrap is added as a coolant in the converter. While if the temperature is low and carbon content in steel is higher then oxygen is reblown. When carbon content in steel is low, the carbon is picked up by addition of coke breeze in the teeming ladle while tapping the steel.

Overflowing of slag from the converter mouth (called slopping) occurs sometimes during oxygen blowing which is controlled by :

- lime addition to increase the slag viscosity
- lowering the lance with reduced oxygen supply
- lifting the lance and deslagging the converter.

If the heat is found alright, it is tapped in the teeming ladle. During tapping of steel, deoxidisers like ferrosilicon, ferromanganese, aluminium shots etc.) are added in the teeming ladle in required quantity depending upon the steel quality *(i.e.* whether rimming, semi-killed or killed steel is being made). After steel tapping, converter is tilted to the other side and the remaining portion of slag in the converter is dumped in the slag pot. The slag pot is then transferred to the slag dumping yard.

The steel teeming ladle is transferred to the teeming bay and teeming is done in ingot moulds with bottom plates to produce steel ingots on cooling which is subsequently stripped off the mould by stripper crane. Alternatively, the steel may be continuously cast in the form of slabs if continuous casting facility exists.

The L.D. steel making consumes very little energy of about 0.5 Giga calorie/ton of crude steel which includes the energy consumed in the Oxygen plant and lime & dolomite shop. In open combustion system, converter gas is allowed to burn at the mouth of the converter and the resultant hot flue gases are used for generating steam in a waste heat boiler. While in suppressed combustion system, converter gas is not allowed to burn at the mouth of the converter. It is as such recovered and stored in a gas holder after dust cleaning & cooling. Yield of Converter gas is 50-60 Nm^3/ TCS. Fuel economy in BOF can be achieved by reducing the hot metal consumption by increased scrap consumption. This in turn can be achieved by improving the hot metal temperature. Scrap should be used as a coolant to control steel tapping temperature (about 1640°C) instead of lime. Though high silicon in pig iron reduces the refractory lining life, it permits lower hot metal ratio *(i.e.* higher scrap consumption) and hence lower energy consumption in steel making stage though higher energy consumption in Blast furnace. However the net effect is the reduced specific energy consumption.

L.D. converter vessel has been shown in Fig. 46.7.

The permanent refractory lining of L.D. converter is of chrome magnesite brick. While the wear out lining is of either magnesia carbon brick, periclase or tar dolomite bricks.

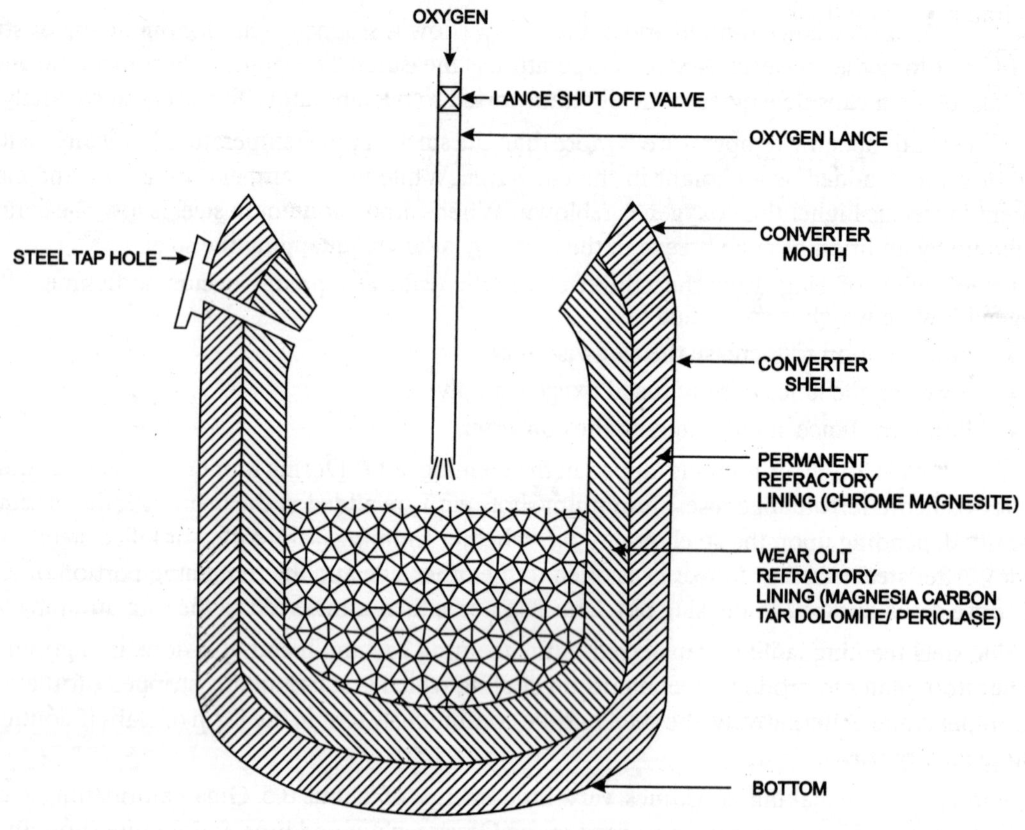

Fig. 46.7: L.D. Converter Vessel.

Typical parameters of a steel melting shop having L.D. converters with **open combustion system** is as follows :

No. of hot metal mixer	= 2
Capacity of each mixer	= 1300 tons of hot metal
Fuel used for heating mixer	= coke oven gas (3 burners, consumption =700 Nm^3/hr)
No. of L.D. converters	= 4
Capacity of each converter	= 100 – 130 tons of steel
No. of water cooled oxygen lances	= 2 per converter
Oxygen blow/flow rate	= 320 Nm^3/minute
Oxygen blow period	= 24-26 minutes
Oxygen consumption per blow	= 7800-8200 Nm^3
Oxygen pressure	= 15 – 20 kg/cm^2
Purity of Oxygen used	= 99.5%
Specific oxygen consumption	= 65-70 Nm^3/ton crude steel (TCS)
Specific hot metal consumption	= 1000 kg/TCS
Hot metal handling losses	= 4%

Refractory lining thickness = (115 + 675)mm, (Permanent + wear out) lining
Converter lining life = 180 – 200 heats (chrome magnesite + tar dolomite)
Waste heat steam generation = 50 tons/hr (1.25 tons/hr/blow)
Scrap consumption = 150 - 200 kg/TCS
Specific power consumption = 35 kWh/TCS
Specific energy consumption in only SMS = 0.35 G.cal/TCS
Tap to Tap time = 60 minutes

Typical parameters of a steel melting shop having L.D. converter employing **suppressed combustion system** is as follows :

No. of hot metal mixer = 2
Capacity of each mixer = 2500 tons of hot metal
No. of L.D. converters = 2
Capacity of each converter = 300 tons of steel
Weight of converter lining = 700 tons
Converter lining thickness = 0.9 metre
Height of converter = 10.5 metres
Length of oxygen lance = 21 metres
Maximum travel of lance = 18 metres
Diameter of lance = 0.430 metre
Cooling water flow to lance = 500 m^3/hr

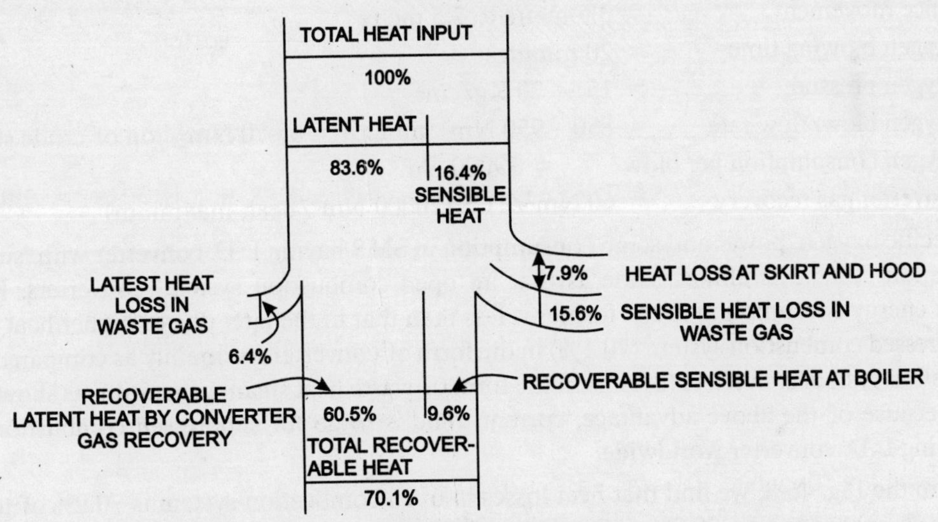

SUPPRESSED COMBUSTION SYSTEM (CONVERTER GAS RECOVERY SYSTEM)

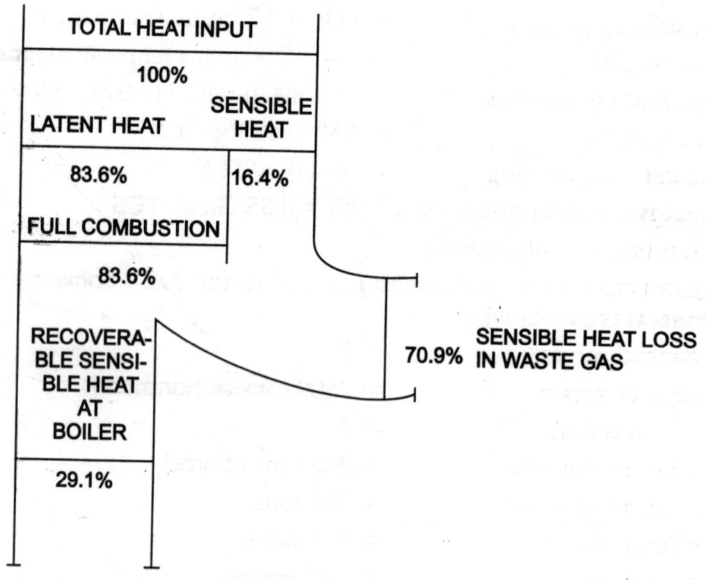

OPEN COMBUSTION SYSTEM (WASTE HEAT BOILER SYSTEM)

Fig.46.8: Heat balance of suppressed combustion system and open combustion system in L.D converter

No. of nozzles per lance	= 5
Lance movement	= 3.5 metre to 2.5 metre
Oxygen blowing time	= 20 minutes
Oxygen pressure	= 15 — 20 Kg/cm^2
Oxygen blow/flow rate	= 850 – 950 Nm^3/minute (= 65-70 Nm^3/ton of crude steel)
Oxygen consumption per blow	= 19000 Nm^3
Converter gas recovery	= 60 Nm^3/TCS (20000 Nm^3/heat, maximum)

Specific oxygen, power and scrap consumption in SMS having L.D. converter with 'suppressed combustion system' is almost same as that in 'open combustion system' converters. However, specific energy consumption in the former is less than that in the later due to higher heat recovery in suppressed combustion system (70.1%) in the form of converter gas mainly as compared to open combustion system converters (29.1%) in the form of waste heat steam generation as shown in Fig. 49.8. Because of the above advantage, current trend is to go for suppressed combustion system employing L.D. converter worldwide.

From the Fig. 46.8, we find that heat losses in open combustion system is 70.9% of total heat input as compared to only 29.9% (7.9 + 15.6 + 6.4) in suppressed combustion system.

Fig. 46.9 gives the schematic diagram of converter gas recovery and utilisation system.

Energy Conservation Measures in SMS are listed below :

- Reduction of track time of hot metal.
- Lower specific hot metal consumption.

- Reduction in mixer loss.
- Improvement in steel yield.
- Maximising scrap consumption
- Non-mixer operation.
- Optimising/minimising oxygen consumption.
- Reduction of heat loss from converter.
- Recovery and utilisation of converter gas.

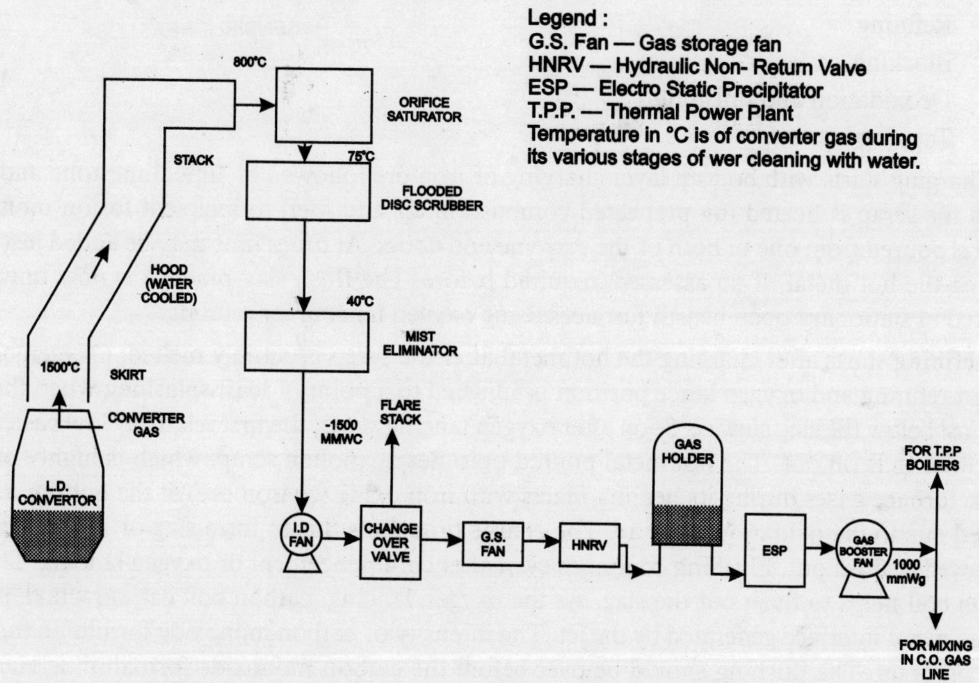

Fig.46.9: Schematic Diagram of converter gas recovery and utllisation system.

- Energy recovery from converter hood-cooling in suppressed combustion system.
- Minimising idle running of equipments (G.S. Fan, I.D. Fan etc.) by installing variable speed control system.
- Recovery of heat from SMS slag.

BASIC OPEN HEARTH FURNACE (OHF)

This furnace used in steel making is getting almost obsolete now gradually. Basic OHF is a shallow basic refractory lined regenerative furnace vessel. It is heated by either liquid fuel (tar/PCM/Fuel oil) or gaseous fuel (mixed gas - B.F. + C.O. gas) or by both the fuels (dual firing) using heat regeneration principle to achieve steel making temperature of about 1650°C. Both the combustion air & fuel gas is preheated in the regenerator and the flow of flue gas & the gaseous fuel & air is reversed every 25-30 minutes.

The charge to the OHF is scrap and molten pig iron. Scrap is initially heated to its softening-point by burning fuel and then molten pig iron is poured on it. **Refining** is then done by blowing oxygen through lances. This makes the refining process autothermic though heating is required initially for softening/almost melting the scrap. The atmosphere in the furnace is always oxidising. The basic slag is prepared by adding lime/limestone and iron ore. A tap to tap time of about 5-9 hours is needed in modern OHF.

Following steps are sequentially involved in steel making using modern basic OHF in each heat after hot refractory repairs (called fettling) using automatic machines have been done :

- Charging and heating of solid materials.
- Charging of hot metal
- Refining.
- Blocking
- Deoxidation and finishing
- Tapping of steel

Charging starts with bottom layer charging of iron ore followed by lime/limestone and scrap. When the scrap is heated (by preheated combustion air and fuel) to incipient fusion molten hot metal is poured from one or both of the extreme end doors. At times lime may be added just before pouring the hot metal, if so assessed/required before. The flush slag practice is now universally adopted in stationary open hearth furnaces using oxygen lancing for refining.

Refining starts after charging the hot metal after the scrap is entirely melted. Oxygen blowing rate for refining and oxygen lance position is adjusted to a point of least splashing when the lance tip is just below the slag surface. Soon after oxygen lancing starts, the first relatively low basicity slag is allowed to flush out. The hot metal poured onto nearly molten scrap, which is highly oxidised due to furnace gases during its heating reacts with iron oxide (or iron ore) at the bottom and that formed due to the oxidation of scrap. The charge time joins this to form slag of low basicity and is allowed to flush out. Flushing continues even after commencement of oxygen lancing. The mild carbon boil helps to flush out the slag. As the oxygen lancing, carbon boil can now take place at the gas-metal interface generated by the jet. The intensity of carbon monoxide formation therefore, soon picks up. The flushing should be over before the carbon monoxide formation is vigrous or else high basicity slag will also be flushed out which is not desirable. After flushing is over, fresh lime is required to make high basicity slag so as to retain phosphorous oxidised from the bath by oxygen lancing. This is a continuous operation in which slag is flushed out from the middle door and fresh slag is made on either side by lime addition. In this way, basicity of slag is slowly raised to the required value and flushing brought to an end. As the generation intensity of carbon monoxide increases, the fuel gas supply may be reduced or at times completely shut off. Oxygen is fed in the OHF through the roof lance to accelerate combustion in the flame.

Blocking of heat is done at the end of refining when oxygen pressure is reduced to decrease the vigour of carbon oxidation and then it is completely stopped if rimming steel is not being made. Lumpy ferro-manganese and/or ferro-silicon are added to the bath for blocking the heat. Fuel is shut off and a steel sample is drawn for chemical analysis.

Deoxidation of heat is carried out further (if the carbon analysis is correct) inside the furnace itself and alloying additions are done. Finishing of heat means obtaining correct chemistry of steel and temperature before the heat is ready for tapping. Still further deoxidation and alloying of bath is usually done in the ladle during tapping.

Tapping of steel is done in the ladle carefully seeing that basic slag containing phosphorous is not allowed to flow along with metal because of danger of phosphorous reversion then exist in the laddle while deoxidising the metal there. Normally about 50% of the refined steel is retained in the furnace for the next heat *i.e.* only 50% of the steel is tapped.

A schematic diagram of regenerative basic open hearth furnace is shown in Fig. 46.10.

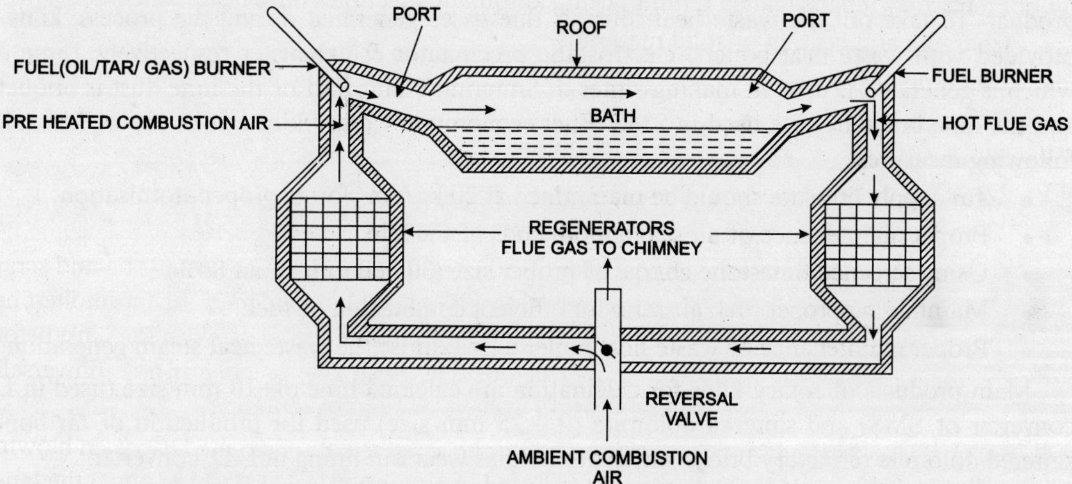

Fig.46.10: schematic diagram of regenerative basic open hearth furnace

A typical parameter of a basic open hearth furnace is given below :

- Capacity of the furnace = 250 tons
- Tap to tap time = 7 hours 30 minutes

(i) fettling from last heat = 30 minutes

(ii) charging time = 1 hr 30 minutes

(iii) charge heating time = 1 hr 30 minutes

(iv) Hot metal pouring = 30 minutes

(v) Melting = 2 hours

(vi) Refining and finishing = 1 hour 30 minutes

Total = 7 hours 30 minutes

- Oxygen lancing rate = 15-30 Nm^3/minute
- Oxygen pressure = 5 kg/cm^2
- Steel tapping temperature = 1650°C
- Hot metal in the charge = 65%
- Iron ore charge = 5% of metallic charge
- Limestone in charge = 5% of metallic charge
- Hot metal temperature = 1250-1300°C
- Capacity of hot metal mixer = 1300 tons
- Scrap charged = = 20% of the total charge
- Campaign life of BOF = 15 weeks with silica bricks roof = 20 weeks with basic bricks roof

ROTARY LIMESTONE/DOLOMITE CALCINATION KILN

Rotary kilns are used for calcination of limestone & dolomite to produce calcined lime & sintered dolomite. Some rotary kilns are used for sintered dolomite production and the other are for production of calcined lime. The kilns use Tar/PCM as fuel. The alternate fuel provided are CO gas and furnace oil/LSHS. The kilns are provided with a water jacketted cooler for cooling its product. To take out the waste heat, dust & flue gases generated during the process, kilns are provided with waste heat boilers, electrostatic precipitator & exhauster respectively. Lime dust which is generated is used in making sinter in Sintering Plant. Part of the lime dust is briquetted and the lime briquettes are used in SMS. Fuel economy in Rotary kiln can be ensured by taking following measures:

- Tar supply pressure should be maintained at 20 kg/cm^2 for its proper atomisation.
- Proper maintenance of air seal at both ends of the kiln.
- Using screened limestone charge of proper size (30 - 60 mm about 80%)
- Maintaining proper fuel/air ratio for efficient combustion of fuel.
- Proper maintenance of waste heat boiler to maximise the waste heat steam generation.

Main products of rotary kilns for calcination are calcined lime of+10 mm size (used in L.D. converter of SMS) and sintered dolomite (0 – 25 mm size) used for production of tar bonded sintered dolomite refractory bricks which are used as wear out lining in L.D. converter.

Secondary products of rotary kilns are lime fines (- 10 mm size), flue dust and waste heat steam which are used for the following purposes.

- (a) as sinter making raw material
- making lime briquettes for using in L.D. converter
- Lime solution preparation for coating the moulds fo hot metal pig casting machines in blast furnace section so that iron pogs do not stick to the mould.
- Water treatment plant for coagulation of suspended impurities in water.

Uses of lime fines (–10mm) and flue dust

- (b) Process steam for heating purpose in furnace oil station (storage tank and furnace oil heating).
- Steam for atomisation of furnace oil.
- Steam for injection in electrostatic precipitator (ESP) to enhance the electrical conductivity of flue dust particles for increasing the cleaning efficiency of ESP.
- Surplus steam is injected into steam grid catering to various other requirements in steel plant.

Waste heat steam usage

Kiln is provided with a burner pipe which discharges tar/PCM/furnace oil in the form of a spray along with primary air encircling it from outside. Steam atomised burner may also be used. Coke oven gas lines/burners are also provided in the kiln to act as an alternative fuel. Even dual firing (C.O. gas + PCM/tar/furnace oil) can also be done. All oil/tar/PCM pipelines are steam jacketted to maintain proper fluidity. Raw materials (limestone/dolomite) are fed into the kiln at its elevated end and the burner pipe is at the other end so that raw materials move in the opposite direction to that of the flue gases.

A schematic diagram of rotary kiln for calcination of limestone and dolomite is shown in Fig. 46.11.

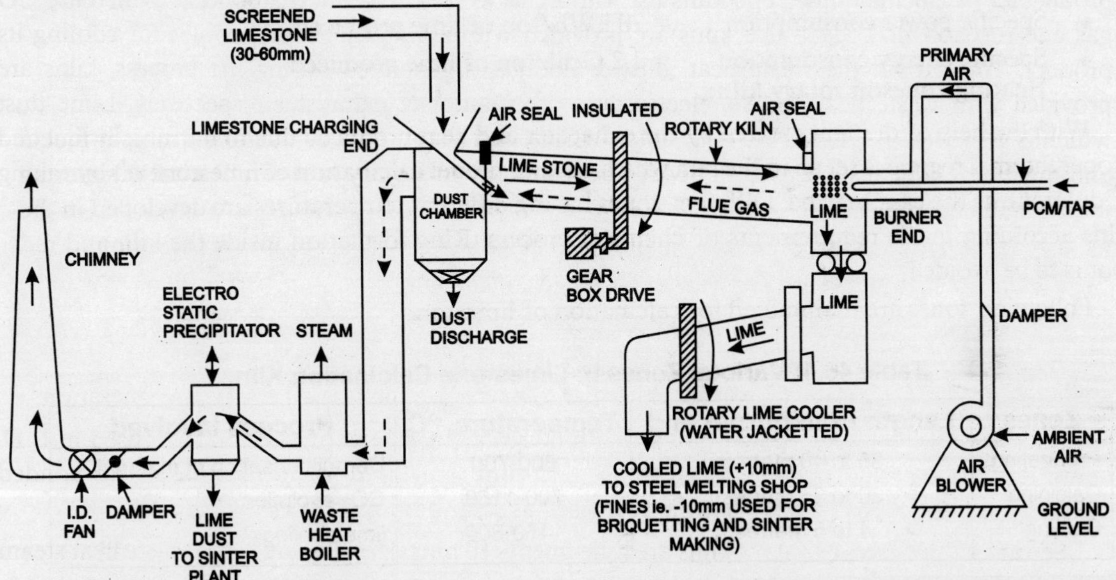

Fig. 46. 11:. Schematic diagram of rotary kiln for calcination of limestone and dolomite.

Typical specifications/description of equipments and raw materials, products & operating parameters of rotary kilns used for calcination of limestone & dolomite are given below :

(a) **Rotary Kilns**

- Diameter & length of kiln = 3.6 & 75 metres respectively
- Normal speed of the kiln = 1 rpm
- Shell material – steel of boiler quality
- Inclination of cooler to the horizontal = 3°
- Limestone charging rate = 20–25 tons/hr.
- PCM/Tar/furnace oil consumption = 3 tons/hr (during normal running period)

  = 1.0 ton/hr. (during idle period)

  (when running on PCM/tar /furnace oil alone)
- Coke oven (C.O.) gas consumption= 6000 Nm^3/hr./kiln (when running on C.O. gas alone)
- C.O. gas supply pressure = 600 mmwg
- Fuel consumption during dual firing

  = 1.5 tons/hr. of PCM/tar/oil + 3000 Nm^3/hr. of C.O. gas.
- PCM/tar/furnace oil supply pressure = 20-25 kgf/cm^2
- Primary air consumption = 10000 Nm^3/hr. (at 400 mmwc)
- Vacuum in the kiln at burner end = 600 mmwc
- Specific PCM/tar consumption in lime kiln = 232 kg/ton of lime

= 20 kg/ton crude steel.

- Specific heat consumption in lime kiln = 2Gcal/ton of lime produced
- Specific heat consumption in dolo kiln = 3.5 Gcal/ton of dolo produced
- Yield of lime from limestone = 48%
- Specific power consumption = 70 kWh/ton of lime produced.
- Specific energy consumption = 2.2 Gcal/ton of lime produced

Heating zones in rotary kiln::

With the help of draught created by the exhauster and heat produced due to burning of fuel, a proper thermal regime is set up in the kiln which ensures proper calcination of limestone & sintering of dolomite as the case may be. Different zones having different temperatures are developed in the kilns according to the requirements of chemical process. Ring formation inside the kiln and red spot is to be avoided.

Following zones are maintained for calcination of limestone :

Table 46.4: Various Zones in Limestone Calcination Kiln

Zones	Length from burner end	Temperature, °C	Process involved
Preheating	35 to 40 metres	600-700	Demoisturisation of material.
Calcining	30 to 35 metres	700-1150	CO_2 evolution
Cooling	4 to 5 metres	1150-600	lime cooling

Following zones are maintained for calcination/sintering of dolomite.

Tale 46.5: Various Zones in Dolomite Calcination Kilns

Zones	Length from burner end	Temperature,°C	Process occuring
Preheating	10 to 15 metres	800-900	Demoisturisation of material
Decomposition	30 to 35 metres	900-1200	Evolution of CO_2
Exothermal	8 to 10 metres	1200-1300	Formation of silicate and calcium aluminoferrite.
Sintering	8 to 10 metres	1300-1700	Partial melting of aluminates and ferrites and material stabilisation and shrinkage.
Cooling	4 to 5 metres	1300-800	Cooling of sintered dolomite.

- Thickness of refractory lining in rotary kiln = 250 mm
- **Refractory lining pattern in rotary kiln (from burner end)**:

(a) For lime Kiln :

(i) 0 to 1.5 metres — High grog fireclay bricks (7-38% $A1_2O_3$)

(ii) 1.5 to 30 metres — Mag-chrome bricks (70% MgO)

(iii) 30 to 75 metres — High grog fireclay bricks.

(b) For dolomite kiln :

(i) 0 to 1.5 metres — High grog fireclay bricks

(ii) 1.5 to 45 metres — Mag-chrome bricks

(*iii)* 45 to 75 metres — High grog fireclay bricks

(c) **Rotary Cooler :**

Diameter of the product cooler = 2.5 metres

Length of the cooler = 40 metres

Inclination of cooler to the horizontal = 3.5%

Speed of the cooler = 3 rpm

cooling system for lime/dolo — water jacket provided in cooler

Linning thickness in cooler = 150 mm of fireclay bricks

(10 metre length from its discharge end is not lined)

(d) **Waste Heat Boiler :**

Capacity = 20 tons steam/hour

Steam pressure = 10 Kg/cm^2

Steam temperature = 250°C (superheated)

Type of boiler — water tube type

(e) **Electro Static Precipitator (ESP) :**

Type — Dry type, Secondary voltage = 80000 volts D.C.

Dust removal done by — pneumatic pump/screw conveyor

(f) **Exhauster:**

Speed = 750 rpm, Pressure rating = 450 mmwc

Flue gas temperature = 230°C (maximum), Capacity = 200000 Nm^3/hr. (maximum)

(g) Composition of Limestone :

General Composition	Typical Composition
CaO = > 50% MgO = < 3.5% SiO_2 = < 3.5% Al_2O_3 = < 1.5% L.O.I. = rest Size = 30 — 60 mm	CaO =51% MgO = 2% SiO_2 = 2.5% Al_2O_3 = 1% FeO = 0.5% L.O.I. = rest (43%) (L.O.I-Loss on ignition).

(h) Composition of Dolomite :

General Composition	Typical Composition
CaO = > 30% MgO = > 20% SiO_2 = < 2.5% Al_2O_3 = < 1% L.O.I. = 45% Size = 5 — 25 mm	CaO = 32% MgO = 21% SiO_2 = 1% Al_2O_3 = 0.5% L.O.I. = rest (45.5%)

(i) Composition of Lime Produced :

General Composition	Typical Composition
CaO + MgO = < 88%	CaO = 74%
MgO = < 6%	MgO = 8%
SiO2 + Al_2O_3 = < 7%	SiO_2 = 5%
L.O.I. = < 5%	Fe_2O_3 = 1%
Size = 10-50 mm (- 10 mm < 10%)	Al_2O_3 = 2%
Slacking time = < 10 minutes	L.O.I. = 10%

(j) Composition of Sintered Dolomite Produced :

MgO = > 36%
SiO_2 = < 4%
R_2O_3 = < 1.5%
L.O.I. = < 1%
Bulk density = > 3 gm/c.c.
Size = 0-25 mm

SOAKING PITS

Soaking Pit is a semi elliptical type pit/furnace used for heating the steel ingots to a suitable rolling temperature *i.e.* 1250-1280°C. It is a gas fired recuperative furnace with side top fired burner. The flame & flue gases in the Soaking Pit flow in "horse shoe" fashion around the steel ingots and are carried to chimney through the channel situated in the bottom part of the end wall on the burner wall (on the burner side). Hot flue gases from the pit are used to preheat the combustion air to 800°C (in ceramic recuperator) and injection primary air & fuel gas (mixed gas) to 300°C (in metallic recuperator). Charging and discharging of steel ingot of the soaking pit is done with the help of a crane. Each pit is provided with a top cover. Coke breeze is used as a bottom making material on which steel ingots are put for heating.

Efficient operation of Soaking Pit can be ensured by :

- Operating optimum/minimum no. of pits.
- Proper scheduling of charging and discharging of steel ingots in pits.
- Maintaining optimum thermal and combustion regime in the pit
- Proper sealing of Soaking Pits from top.
- To charge the pit upto full capacity.
- Installation of supervisory computer for heating control in the pit.

Typical Parameters of Soaking Pits

No. of Soaking Pits = 40

Capacity of each pit = 160 T/charge (Ingots weigh 18 to 28 tons each)
Fuel used = mixed gas (CV = 1600 Kcal/Nm^3)
Maximum mixed gas consumption = 5600 Nm^3/hr/pit.

Pressure inside the pit = + 0.2 to + 0.5 mmwg
Specific heat consumption = 0.45 Gcal/ton ingot charged
Specific power consumption = 25 kWh/ton slab produced
Specific energy consumption = 0.635 Gcal/ton slab produced

Size of soaking pit —

(i) Height = 4.5 metres
(ii) Length = 10 metres
(iii) Width = 3.3 metre at the centre, 2.7 metre at the corner
Hearth area of each pit = 32 m^2
Volume of the pit = 135 m^3
No. of slag holes = 2 per pit
No. of burner = 1 per pit (outside mixing type)
Turn down ratio of burner = 1 : 5
Mixed gas supply pressure = 500 mmwg
Mixed gas pressure at burner = 100 mmwg

Details of mixed gas preheating recuperator :

(i) Type - metallic recuperator (double U-shaped)
(ii) Heating surface area = 140 m^2
(iii) Temperature of preheated mixed gas = 300°C

Details of injection (primary) air preheating recuperator :

(i) Type-metallic recuperator (single U-shaped)
(ii) Temperature of preheated injection air = 300°C
(iii) Maximum consumption of injection air = 1100 Nm^3/hr.
(iv) Injection air pressure = 1.5 kg/cm^2

[90% of total combustion air is drawn from atmosphere (as secondary air) through ceramic recuperator by injection of 10% primary air at 1.5 kg/cm^2.]

Details of atmospheric/Secondary air preheating recuperator :

(i) Type - Ceramic (tile) recuperator of SiC tubes.
(ii) Heating surface area of recuperator = 580 m^2
(iii) Temperature of preheated secondary air = 800°C.
(iv) Secondary air consumption = 9000 Nm^3/hr. (maximum)

Total combustion (primary+secondary) air consumption = 10000 Nm^3/hr. (maximum)

Gas to total combustion air ratio = 1 : 1.5 to 1 : 2

Waste gas temperature = 250 to 450°C (controlled by diluting with air through air dilution valve)

Waste gas volume = 22000 to 130000 Nm^3/hr.

Pressure at stack = 30 mmwg (natural draft stack of 70 m height × 3.5 m diameter)

Average temperature of hot ingots = 700 – 750°C (track time of ingot = 3 hours)

Heating and soaking time of hot ingots = 3 to 4 hours (> 16 hours for cold steel ingots)

Refractory lining of soaking pit:

(i) Pit cover - fireclay bricks

(ii) Upper part of the wall of soaking pit-fireclay bricks

(iii) Hearth and the lower part of the wall-chrome-magnesite bricks

In the slabbing,'/billet/blooming mill of an integrated steel plant, ingots/billets/blooms of various sizes and weights are received from ingot stripper yard of steel melting shops. After heating and soaking them in the soaking pits to the rolling temperature, they are rolled into steel slabs for further rolling in hot strip/rolling mill.

A schematic diagram of soaking pit is shown in Fig. 46.12.

REHEATING FURNACE

Reheating furnaces heat the steel slabs (hot or cold) to the rolling temp, of 1250 - 1280°C. The steel slabs are pushed into the furnace by two pushers from slab charging side. Slabs travel *on water cooled skid pipes* through heating zones and then on a solid hearth in soaking zone. The furnace has five zones out of which four are called heating zones and fifth is known as *soaking zone.* There are in total 59 gas burners. The combustion air is preheated to 500°C in a ceramic recuperator which is sucked through the injection burners (thereby eliminating air blower). Finally the flue gases are exhausted out through stack (7m × 120 m). Furnace pressure in soaking zone is maintained positive (between 2-3 mmwg). Control and measuring instruments required for control of furnace temperature & pressure as well as efficient fuel utilisation are provided. Evaporative cooling system (ECS) is used for skid pipe cooling during which waste heat steam is produced. Flue gases are used for preheating the combustion air before being exhausted to the atmosphere through chimney.

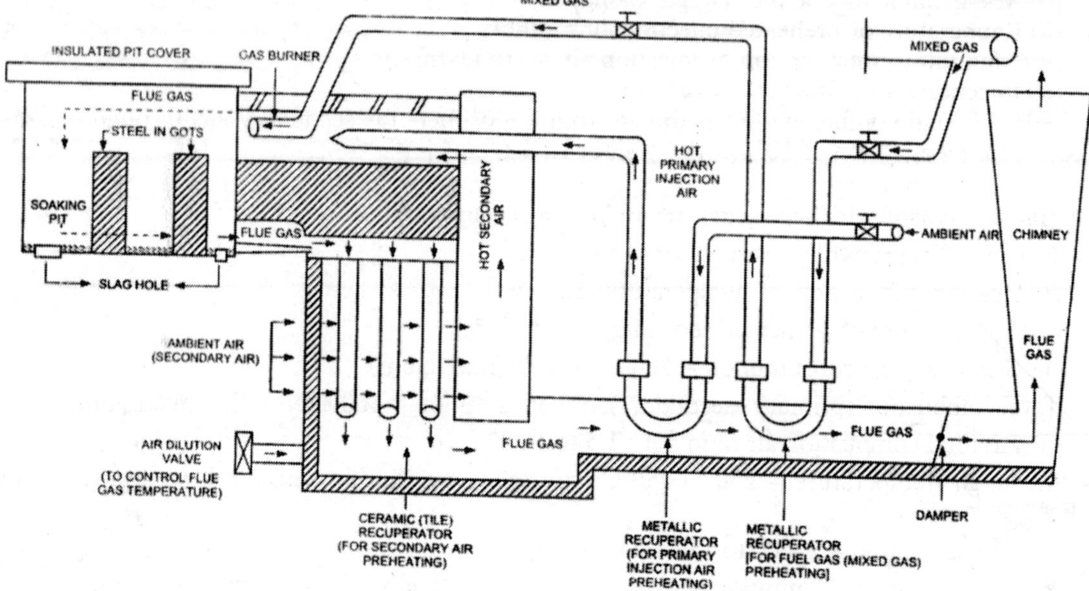

Fig. 46.12: Schematic diagram of soaking pit.

Hot or cold steel slabs are heated in the reheating furnace to the rolling temperature before hot rolling it. Hot rolling of steel refers to the plastic deformation of metal above its recrystallisation temperature by squeezing it between a pair of rollers to produce hot rolled strips in the form of a coil. Slab reheating before its rolling is done to achieve higher plasticity and minimum resistance to deformation.A typical slab and hot rolled steel strip specifications are given below :

Steel slab specifications :

Length of slab = 7.5 to 10 metres, Weight of the slab = 1 to 1.6 metres

Thickness of slab = 0.15 to 0.25 metre, Width of the slab = 20 to 28 tons

Rolled steel strip specifications:

Thickness of steel strip = 2 to 10 mm, Width of steel strip =1.1.6 metres

Steel strip coil diameter (inside/outside) =1/2.2 metres

Steel strip coil weight = 19 to 27 tons

Material flow diagram of a hot rolling mill is shown in Fig. 46.13.

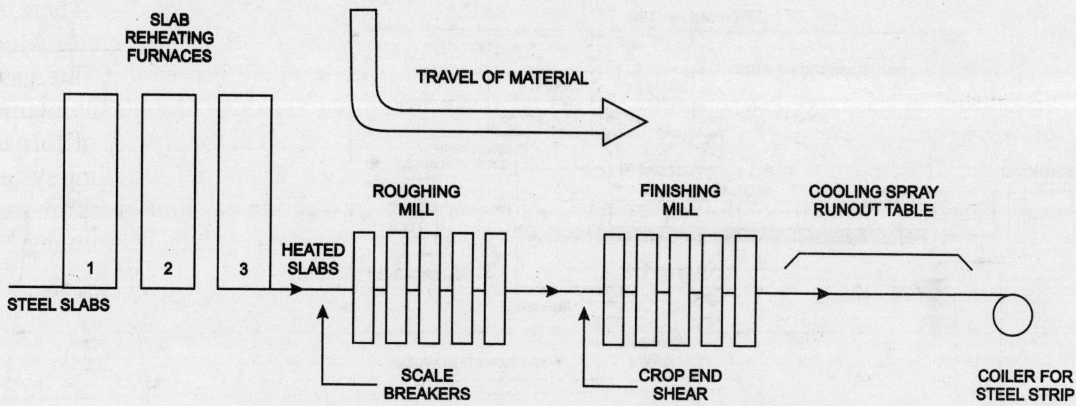

Fig. 46.13: Material flow diagram of hot strip/rolling mill.

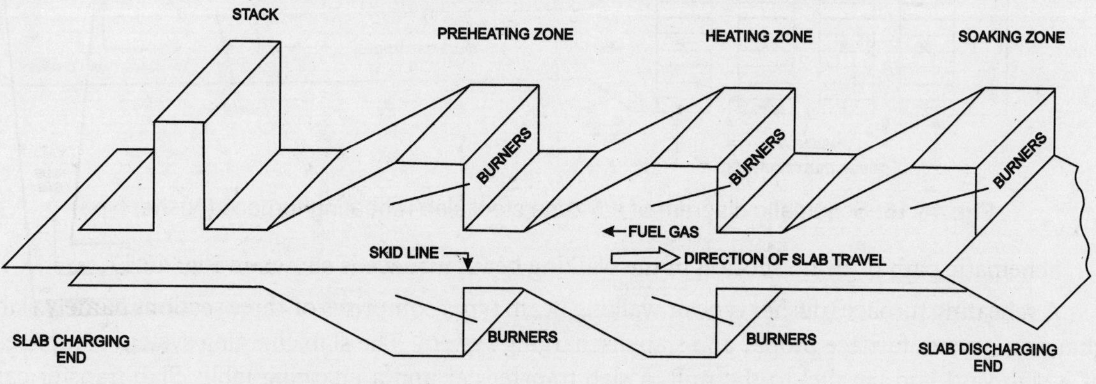

Fig. 46.14: Slab reheating furnace showing direction of slab and flue gas travel.

Burner arrangement and various heating zones in a five zones slab reheating furnace is shown in Fig. 46.15.

A reheating furnace is of two types namely *'pusher type reheating furnace'* and *'Walking beam type reheating furnace'*. In the pusher type furnace, movement of slab is by its pushing by another slab which causes many operational problems. An improved modified version of the reheating furnace is the *'Walking beam type furnace'* which has reduced many inherent problems of the pusher type furnace. For example, skid marks on the slab & mechanical damage caused by abrasion on the skids are minimised and piling up of the slab is obviated. Since, in walking beam furnace, slabs are physically separated, the problem of sticking of one slab to another is eliminated. Walking beam is a device to move the slab in the furnace by lifting it, moving it in forward direction, lowering it down and then again returning to its original position. Walking beam furnaces can be conveniently emptied in either forward or backward direction. Besides specific heat consumption in walking beam furnace is much less compared to a similar pusher type furnace.

Fig. 46.16 shows the schematic diagram of a five zone slab reheating furnace (pusher type) with waste heat recovery system.

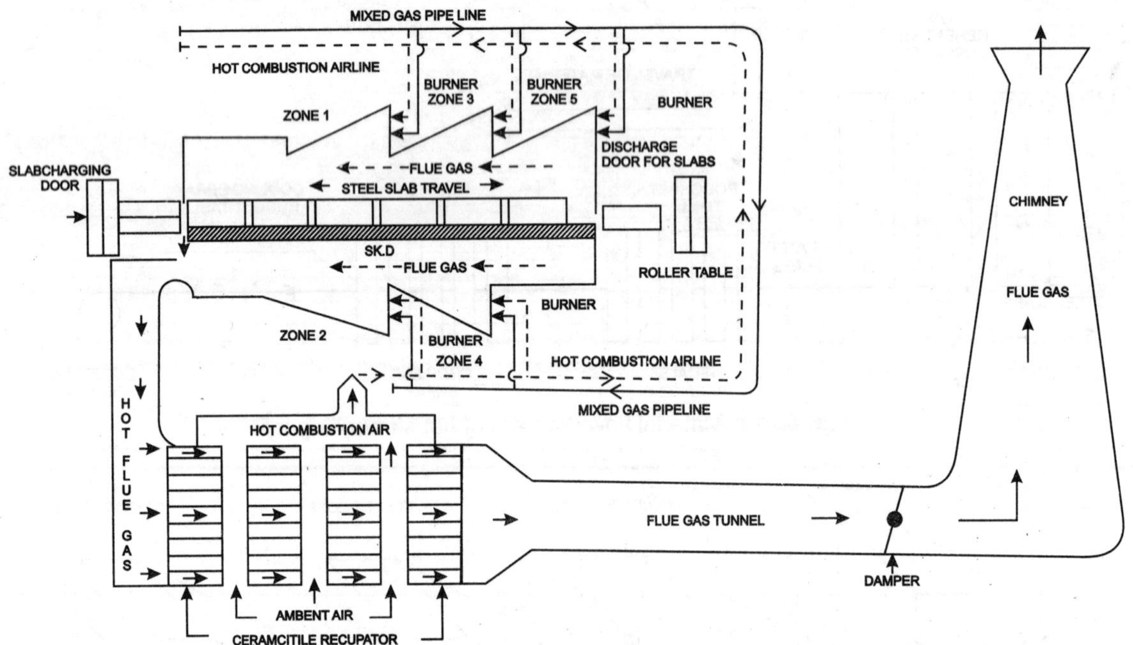

Fig. 46.16: Schematic diagram of a 5 zone steel slab reheating furnace (pusher type).

Schematic path of beam motion in the walking beam furnace is shown in Fig. 46.17.

A reheating furnace (pusher type or walking beam type) comprises of three sections namely slab charging system, furnace proper and slab discharging system. The slab charging system comprises of a slab yard laid parallel to the mill, a slab transfer car and a charging table. Slab transfer car carries a pile of slabs which is depiled on to the charging table by means of a depiler. Slabs may also be directly delivered to the charging roller table from the slabbing mill. After heating the slabs in the furnace, they are discharged (at the delivery side of the furnace) on the slab receivers to avoid damage to roller table due to impact from falling heavy slabs. TV camera is provided to control charging and pushing of slabs in the furnace.

In pusher type reheating furnace, possibility of piling up of slabs is reduced by inclining the hearth downwards in the direction of slab motion so as to reduce the pusher force required. The slabs slide on skid rails thereby avoiding excessive hearth wear and reducing the pusher force required.

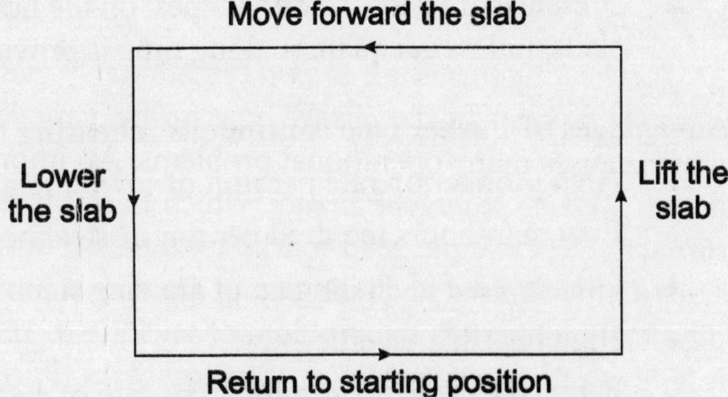

Fig. 46.17. Schematic path of beam motion in the walking-beam furnace.

A walking beam reheating furnace comprises of a slotted stationary hearth with one or usually more moving beams having a lifting & reciprocating movement; lifting, carrying forward and putting the slab again on the hearth. Scale deposition in hearth slots presents some difficulties but walking beam furnaces are widely used for large, heavy stocks/slabs.

Disadvantages of Pusher Type Reheating Furnace :

- Moving slabs/billets rub against one another and against the hearth; this has an adverse effect on the quality of the metal surface.
- Scale breaks off easily from the moving metals and further oxidation of the fresh surface of the metals are likely to occur.
- Scale drops on to the hearth, reacts with the hearth materials which hinder normal progressing of the metals & cleaning.

Special measures are required for emptying the furnace during maintenance.

- The removal of scale (actually donemanually) is difficult with quite wide furnace.
- Size & productivity of this type of furnace is limited due to the fact that different product variation and pushing of great load through the furnace without buckling/pile ups cannot be achieved.
- Lack of flexibility for heating efficiently small lot size of steel of varying thickness necessitating increased heating time to suit the heating cycle of the stock piece requiring lower heat transfer rate which may be detrimental to the heating of the other grades and sizes in the furnace.
- Touble from water cooled skid maintance and from building up of scale on hearth which results in piling of steel in furnace due to climbing of stock/slab pieces over each other, particularly if square contact is lost between adjacent pieces, when pressure is applied by the slab pusher. Face of contacting surface of the slab/stock should be square to prevent piling.
- Difficulty encountered while pushing mixed sizes of the stock through the furnace.
- It is expensive to empty furnace at the end of the schedule.
- Thickness of stock/slab is limited to 300-350 mm, where water cooled skids are used.

- Presence of visibly colder 'stripes' on the hot steel slab caused by contact with water cooled skids, unless adequate soaking time is provided.

Advantages of Pusher type continuous reheating furnaces :

- High production rate per unit of investment.
- Low man-hours required per ton of steel heated.
- Greater ease in charging and drawing steel.
- High hearth area efficiency.
- Better means for controlling the rate of heating at all temperature levels. Gradual rise in temperature levels permits charging all grades of cold steel without cooling furnace.
- Less trouble encountered from temperature inequalities between each succeeding pieces drawn.
- High production per m^2 square foot of ground space occupied.
- It can be built for any reasonable length of stock pieces to be heated resulting in higher hot rolling mill yields, because of fewer crop ends from lengths beyond batch furnace possibilities.

These drawbacks of the pusher furnaces led to the development of the walking beam/hearth furnace. Though walking beam furnaces are costlier than pusher type furnaces, it provides the following advantages.

Disadvantages of Walking beam furnaces :

- Higher cost per ton of capacity.
- Complexity of the walking beam system.
- Rigorous maintenance of hearth seals and hearth refractory required.
- Precaution required to avoid problems from scale that drops off the slab stock being heated.
- There is potential of extension of overall furnace length of improve the utilisation of furnace waste gases and reduce specific fuel consumption, a similar advantage is not available with other furnace types because of limitation on overall furnace length.

Advantages of Walking Beam Type Reheating Furnace :

- Excellent surface condition of the product is achieved.
- To provide a more or less important gap between two slabs/billets so that atleast three side heating is possible.
- Emptying of the furnace in either direction is easier.
- Bottom firing over entire length of the furnace, thereby ensuring a higher specific production rate per m^2 of the hearth.
- Easy control of heating conditions.
- More progressive heating up of the slabs/billets resulting in reduced oxidation and decarburation.

- Reduced maintenance and repairs are involved.
- Minimum time for cleaning the furnace because of the fact that scale does not fall off on the furnace bottom and there is less oxidation.
- No limitation of length of the furnace.
- Slab pile ups and furnace retention line is reduced.
- Slab pieces can be separated from one another on the hearth, eliminating stickers.
- No line contact with water cooled skid occurs, thus eliminating skid marks (also called black mask or stripes) on the heated stock.
- There is no rubbing or friction between the stock and the hearth minimising hearth wear and stock damage.
- Better hearth utilisation can be obtained when charging mixed slab sizes by selecting the proper number of waling beams.

TYPICAL PARAMETERS OF A PUSHER TYPE REHEATING FURNACE

Hearth area of the furnace = 425 m^2 (11.25 m × 38 m)

No. of heating zones = 5 [2 preheating (top and bottom) + 2 heating (top and bottom) + 1 soaking zone]

No. of water cooled longitudinal skid pipes = 8 (in preheating & heating zones)

Cooling system for skid pipes — Evaporative cooling system (E.C.S.)

Cooling system for charging & discharging doors and burners — ordinary water cooling. Heating capacity of the furnace = 260 ton/hr. with cold slab charge

= 350 ton/hr. with hot slab charge

Discharge slab temperature — about 1250°C, Mixed gas supply pressure = 2000 mmwg

Temperature, of preheated combustion air = 500°C (through ceramic recuperator)

CV of mixed gas = 1800 Kcal/Nm^3

Mixed Gas flow (maximum) = 100000 Nm^3/Hr.

Combustion air requirement = 200000 Nm^3/Hr.

Specific heat consumption = 0.68 Gcal/ton of slab rolled

Waste gas volume = 80000 — 2,20000 Nm^3/hr.

Waste gas temperature = 450°C

Specific power consumption =75 kWh/ton of product (i.e. hot rolled steel coil/plate) Specific energy consumption =0.9 GCal/ton of product

Waste heat steam generation (skid cooling) = 40 T/hr.

Table 46.6: Temperature, in Various Zones of Re-heating Furnace

Zone	No. of burners	Temperature. 'C
1st top preheating zone	12	1100-1200
1st bottom preheating zone	12	1100-1200
2nd top heating zone	12	1150-1300
2nd bottom heating zone	10	1150-1250
Soaking zone	13	1200-1270
Total	59 burners	

Furnace pressure = + 2.5 mmwg, Gas/air ratio in the burners = 1:2

Mixed Gas pressure at the inlet of the burner = 500 - 2000 mmwg (combustion air flow is by gas injection)

Refractories used in the reheating furnace :

(i) Furnace roof, chimney and flue channel — fireclay bricks

(ii) Bottom hearth and bottom falls — chrome magnesite bricks

(iii) Monolithic solid hearth in soaking zone—Above fireclay bricks is kept electro cast/fused alumina bricks (called corundum blocks) over which chrome magnesite brick is used as lining.

TYPICAL PARAMETERS OF A WALKING BEAM REHEATING FURNACE

Capacity of the furnace = 260 tons/hr (with mixed gas of CV 1650 Kcal/Nm^3)
= 300 tons/hr (with mixed gas of CV 2300 Kcal/Nm^3)

Dimension of the furnace :

Length = 40 metres
Width = 11.5 metres
Depth below pass line = 2.35 metres
Height above pass line = 1.5 metres (maximum)
No. of walking beams = 4 (hydraulically operated)
No. of fixed beams = 4
No. of slab rows = 1 or 2

Dimension of slab charge :

Thickness = 180 – 250 mm
Width = 750 – 1850 mm
Length = 6500 – 10500 mm (for one row charging)
= 4500 – 5250 mm (for two rows charging)

Dimension of reference slab :

Thickness = 250 mm

Width = 1100 mm
Length = 10250 mm
Maximum slab weight = 32 tons
Gap between slabs = 50 — 70 mm
Slab charging temperature = 30°C for cold slabs , 800°C for hot slabs ,350°C for warm slabs
Maximum furnace operating temperature= 1350°C
Slab discharge temperature = 1250°C
Furnace external roof/wall temperature = 110 — 130°C
Total No. of burners = 153

Total No. of heating zones = 10 comprising of preheating zone (lower and upper), heating zone (lower & upper), heating zone (lower & upper), pre-soaking zone (lower & upper) & soaking zone (lower & upper)

Type of burners:

(i) For lower preheating, heating and soaking zone — single air inlet burner.

(ii) For 2nd lower heating zone — Double air inlet burner.

(iii) For all upper zones — Radiant roof burners.

Turn down ratio of all the burners = 1:10
Mixed gas supply pressure = 2000 mmwg
Mixed gas pressure at burner intake = 300 mmwg
Combustion air pressure = 400 mmwg
Mixed gas C.V.— Designed for both 1650 and 2300 Kcal/Nm^3 C.V.
Mixed gas consumption = 70000 Nm^3/hr. (of CV = 1650 Kcal/Nm^3)
= 52000 Nm^3/hr. (of CV = 2300 Kcal/Nm^3)
Excess air used = 5% above theoretical air

Table 46.7: Mixed gas (B.F. gas + C.O. gas) analysis

Composition, % by volume	Mixed gas C.V., Kcal/Nm^3	
	1650	2300
CO_2	12.2	9.9
CO	22.0	18.6
CH_4	4.8	9.7
CmHn	0.5	0.9
H_2	19.2	28.9
N_2	40.9	31.5
O_2	0.4	0.5
NH_3 mg/Nm^3	12.4	22

$C_{10}H_8$, mg/Nm^3	65	120
H_2S, mg/Nm^3	0.8	1.4
Density, kg/Nm^3	1.11	0.935

Table 46.8: Furnace efficiency

Production rate, tons / hr.	Skid insulation, %	C.V. of mixed gas used, Kcal/Nm^3	Furnace efficiency, %
260	100	1650	64
300	100	2300	65
260	85	1650	54
300	85	2300	56

Table 46.9: Specific Heat Consumption (for cold charge to 1250°C discharge temperature) :

Production rate, tons/hr.	Skid insulation, %	Specific heat consumption, Gcal/ton of slab
260	100	0.315
260	85	0.374
300	100	0.310
300	85	0.362

Table 46.10: Specific heat consumption (for various slab thicknesses)

Slab thickness, mm	Specific heat, Gcal/Ton slab	
	For 260 tons/hr. with mixed gas C.V. = 1650 Kcal/Nm^3	For 300 tons/hr. production with mixed gas C. V. = 2300 Kcal/Nm^3
180	0.303	0.300
200	0.305	0.301
220	0.310	0.307
250	0.313	0.310

Type of recuperator (used for combustion air preheating) — convection type metallic recuperator

Dew point for recuperator = > 200°C

No recuperator is provided for mixed gas preheating.

Table 46.11: Recuperator data (for 100%-85% skid insulation)

Parameters	Unit	For production rate of 300 tons/hr, (100-85%) skid insulation	For production rate of 260 tons/hr, (100-85%) skid insulation
Flue gas flow rate to recuperator inlet	Nm^3/hr	135000—158000	129000-153000
Flue gas temp, to recuperator inlet	°C	940	840
Flue gas temp, behind recuperator	°C	550	510
Combustion air flow rate	Nm^3/hr	101000—118000	84000-100000
Combustion air temp, behind recuperator	°C	650	625
Δp in flue gas in recuperator	mmwg	20-27	17-24
Δp in air in recuperator	mmwg	430-570	300-405
Heat exchange surface area	m^2	2000	2000

Dimension of recuperator — Dia 76 mm ×3 mm ×4000 mm
Chimney—natural draught
Furnace pressure control — by main damper after recuperator
Steam production rate by E.C.S. = 28 tons/hr (at 6 kg/cm^2)

Measures for efficient operation of Reheating Furnace :

- To operate optimum number of the Re-heating furnaces.
- To maintain the optimum thermal and combustion regime in the furnace.
- To minimise heat loss through skid cooling water by proper maintenance of skid pipe insulation.
- To keep the furnace doors and side windows in operating condition.
- Evaporative cooling system (ECS) for skid cooling should be kept operative.
- Flue gas damper should work in auto mode.
- All the burners should be in operative condition.
- Maximising the hot/warm slab charging.
- Maintaining the proper furnace pressure and temperature. Furnace temperature should be reduced during holding/non rolling time.
- Maintaining the recuperator in clean and good condition.

COLD ROLLING MILL FURNACES

Annealing Furnace

During cold rolling of hot rolled coil in Tandem mill of Cold Rolling Mill for further thickness reduction, the grain structure of steel gets elongated and heavy internal stresses are developed. As a result, the cold working properties of the material is lost. To regain the material properties, the material (cold rolled steel coils) is heated in inert atmosphere to 680-720°C and cooled to room temperature. This process of heating & cooling of steel coils is called annealing which is done to release internal stresses in the material and hence improve its mechanical properties.

In modern cold rolling mills, both batch annealing (in Hood Annealing Furnace) and continuous annealing (in Continuous Annealing Line Furnace) facilities exit.

Hood Annealing Furnace (batch type):

Three or four cold rolled steel coils are stacked on the base of the furnace and in between each coil, one convector plate is placed. The stack is then covered with an inner cover and the whole charge is *sealed with sand.* A movable furnace having 16 burners around its circular periphery is put over this inner cover. The air inside the inner cover is taken out and *protective gas (96% N_2 + 4% H_2)* is filled up to create a neutral/inert atmosphere during annealing by heating to avoid the oxidation of steel coils. The burners are lighted up and the annealing temperature is raised upto 680-720°C depending on the weight of charge, quality of steel, its use etc. & furnace is then put off. Then the whole charge is allowed to cool down to 100°C after which the coils are transferred to a forced cooling base where it is cooled till it comes to room temperature. Fig. 46.18 shows a batch hood annealing furnace (for cold rolled steel coils).

Typical parameters of a Batch Hood Annealing Furnace :

Number of batch annealing furnaces	= 40
Number of bases	= 100
Number of fast/forced cooling bases	= 40
Capacity of the furnaces	= 500,000 tons coil/year
Maximum charging height	= 4.7 M
Maximum charging weight	= 120 T/furnace
Annealing cycle time	= 95-135 hours
Furnace weight	= 25T
Maximum outer dia of coil	= 2.2 M
Specific heat consumption	= 0.4 Gcal/T of Annealed coil
Mixed gas C.V.	= 1700 Kcal/Nm^3
Mixed gas supply pressure	= 2000 mmwg
Annealing temperature	= 680-720°C

Annealing process makes the material soft. Steel annealing is divided into two parts namely, *(i)* **Full Annealing.** It involves heating the steel above its critical temperature range and then cooling it slowly. Full annealing results in a structure which is pearlite together with free ferrite or free carbide depending on the carbon-content. Full annealing has an additional effect of grain size refinement which improves the ductility of the softened steel and also the toughness of steel, if it is hardened subsequently. To convert ferrite into austenite, annealing temperature should be maintained just above the critical temperature range. However, in steels without additions for grain size control, the grain starts to coarsen at temperature just above critical temperature and hence annealing temperature should be kept as little as possible above critical temperature. The necessary slow cooling through the critical range is done by usually cooling in the furnace itself. *Bright annealing* is practised using protective/inert gas atmosphere.

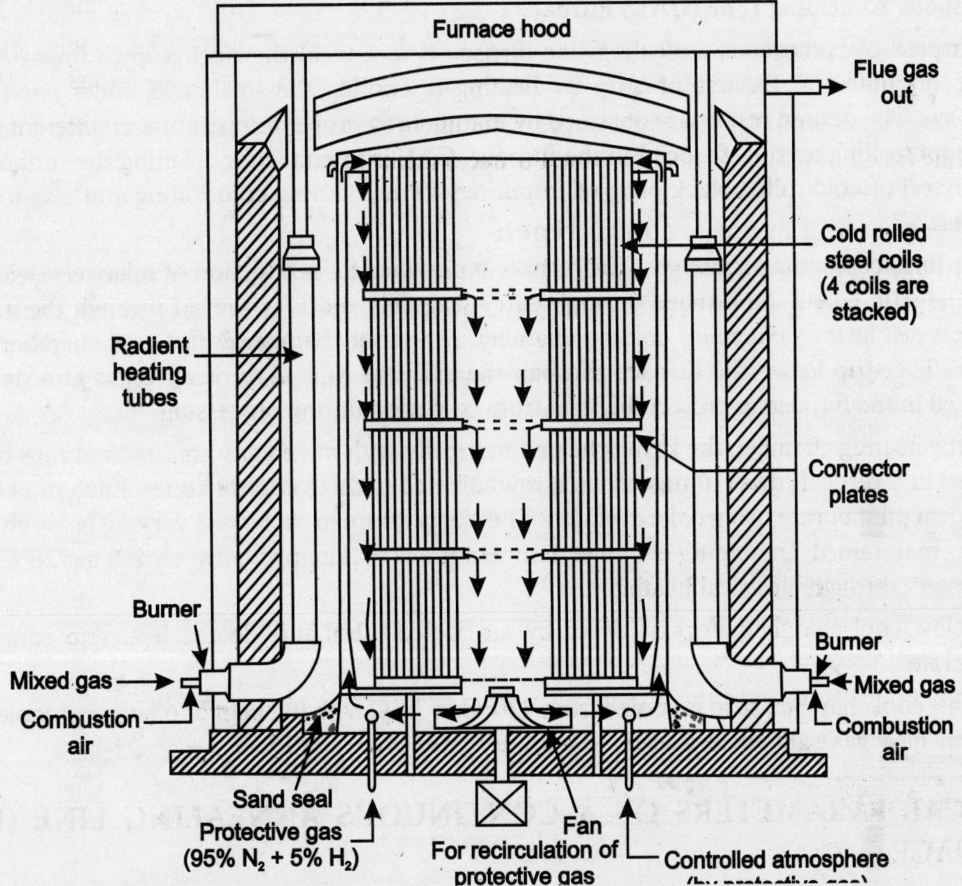

Fig.46.18: Batch hood annealing furnace (for cold rolled steel coils)

(ii) **Sub-critical Annealing.** It involves heating the steel to just below its critical temperature. Its main aim is to remove internal stress usually in low carbon steels (C < 0.25%) which have been cold worked. No major structural changes are promoted but grain distortion is corrected. If steel is not to be further cold rolled, temperature well below the critical range may be employed *e.g.* 550°C. Stress relief is carried out on welded structure to render the welding stress distribution uniform at temperatures in the range of 500 to 650°C. Inert atmosphere is used during annealing.

A typical parameters of another *modern batch hood annealing furnace* is given below :

- No. of bases/furnaces/fast cooling bases : 40/20/30.
- Maximum steel coil charging height & weight = 3.6 m & 50 tons respectively.
- Annealing temperature = 680-720°C
- Total annealing cycle time = 53 hours

(i) heating period = 24 hours *(ii)* cooling period = 24 hours

(iii) ancilliary time = 5 hours

- C.V. of mixed gas = 1700 Kcal/Nm^3
- Composition of protective gas = 96% N_2 + 4% H_2

Continuous Annealing Line (CAL) Furnace :

The purpose of continuous annealing line furnace is to anneal the steel strip to impart desired temper, brightness & flatness of strip by heating & cooling the steel strip under *protective gas atmosphere.* The desired results are obtained by maintaining proper temperature at different regions and by controlling tension & speed in the furnace. CAL is intended for cleaning the surface (with electrolytes) of cold rolled black plate of impurities formed during cold rolling and also for *bright annealing.*

The furnace is *vertical tower type.* The furnace is designed for production of relatively twin gauge strip. The strip enters the furnace through entry seal rolls and is conveyed through the different chambers *i.e.*, heating chamber, holding chamber, slow cool chamber & fast cool chamber of the furnace. The strip leaves the furnace through the exit seal rolls. A protective gas atmosphere is employed in the furnace to protect the strip from oxidation during processing.

In the heating chamber, the strip is heated indirectly with mixed gas fired radiant tube burners arranged in 8 vertical rows (10 nos. in each row) divided in three control zones. Each radiant tube has its own pilot burner using coke oven gas. The temperature in each zone varying between 480 to 720°C is maintained. In holding chamber, temperature is maintained between 480 to 720°C as per requirement through electrical heaters.

In slow cool chamber, strip is cooled by air cooled tubes and electric heater to control the cooling rate.

In fast cool chamber, strip is cooled from 420°C to 70°C with the help of 63 nos. jet coolers. All chambers have gas tight enclosures filled with dry protective gas (96% N_2 + 4% H_2).

TYPICAL PARAMETERS OF A CONTINUOUS ANNEALING LINE (CAL) FURNACE :

Salient features of furnace :

Capacity of CAL furnace - 25 T/Hr. (for 0.26 mm thick × 820 mm wide strip) at 250 metres/ minute strip speed

Furnace pressure = + 8 to 12 mmwg, Cold steel strip thickness processed = < 0.35 mm,

Normal temperature, of strip heating = 680—730°C,

Normal temp, of cooled steel strip = 70°C Protective gas composition = 4 - 6% H_2 + 94 - 96% N_2 (at 1350 - 2000 mmwg)

Heating chamber:

Mode of heating in heating chamber-indirect heating by mixed gas fired radiant tube burner Burner arrangement—8 vertical rows (10 nos. burner in each row), C.O. gas pilot burner for each burner is provided

No. of control zones in heating chamber = 3

Temperature = 480 — 720°C as per metallurgical requirement

No. of vertical passes — 6

Lining material for the chamber—Insulation bricks

Mixed gas consumption — 4400 Nm^3/Hr. for 80 burners.

CV of mixed gas = 1600 – 1800 kcal/Nm^3

C.O gas consumption — 130 Nm^3/hr for 80 pilot burners

Mixed gas pressure — 1700 - 2000 mmwg

C.O gas pressure — 2000 - 3000 mmwg

Material of construction of chamber — thick steel plate lined with insulating refractory.

Table 46.12: Temp(^{0}C); cycle for TMBP

Zone	0.28 mm thick	0.20 mm thick
I	500-550	450-550
II	550-650	500-600
III	650-680	550-650

Holding chamber:

Material of construction—thick steel plate lined with insulating refractory.

Mode of heating — electrical heating in wall brick work with nichrome ribbon (270.90 kw, 415 V)

No. of vertical passes — 6

Temperature, in zones 4 to 6 — 700 - 720°C (for 0.28 mm thick) and 630 - 680°C (for 0.20 mm thick)

Slow cool chamber:

No. of vertical passes-6

Strip cooling temp.—720-420°C

Medium of cooling — exhaust air through cooling tubes

No. of air blowers — 6

Mode of heating in case temperature, drops down — electrical heating (240 kw/415V)

Control of heating & cooling system interlocked.

Similar construction as holding chamber.

Fast/Jet cool chamber:

Similar construction as heating chamber but not insulated.

No. of vertical passes — 8

Strip cooling chamber — from 420 to 70°C (slightly higher temperature, for thicker strip)

Cooling medium-Water through jet coolers (@ 10 gm/min at 2000 mmwg)

No. of jet coolers-63 located in 9 vertical rows (uses cooled protective gas for cooling).

Table 46.13: Steel Strip Speed

Width in mm	Thickness in mm	Speed in m/min
650	0.20	360
800	0.22	307
850	0.26	250
920	0.35	167
1040	0.35	148

Overall Parameters of a Typical Continuous Annealing Line is Given Below :

Electrolytes used in steel strip cleaning tanks — caustic soda and tri-sodium phosphate in 3 : 2 ratio with 3% dilution.

Type —Tower type continuous annealing line (10 metres)

Capacity = 20-25 tons/hr.
I.D. of steel coil = 420 mm
O.D. of steel coil = 700 - 2000 mm
Maximum weight of coil = 18 tons
Furnace chambers = 4 nos.
Steel strip width = 650 - 1050 mm
Steel strip thickness = 0.15 to 0.35 mm

Continuous on line lap welding of strip is done.

Maximum Line Speed (metre/minute) of strip :

(i) entry section = 450 *(ii)* furnace section = 350
(iii) Delivery section = 450

Continuous Annealing Line Furnace is shown in Fig. 46.19.

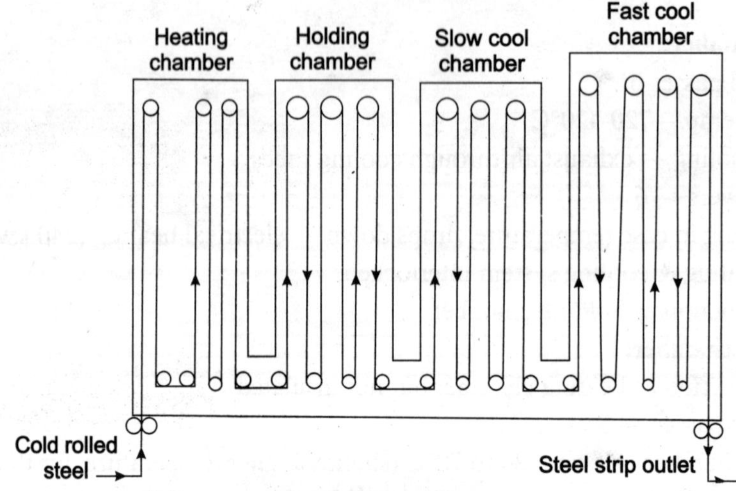

Fig. 46.20: Continous Annealing Line (CAL) furnace in CRM.

HOT DIP GALVANISING LINE (HDGL) FURNACE

Galvanising (zinc coating) of cold rolled steel coil is done to improve its corrosion resistance, appearance and surface finish. Galvanising of cold rolled steel coil is done by dipping in a pot furnace in which molten zinc is maintained at 450°C with the help of electrical heating. Before the steel strip is passed to the zinc pot, it is annealed in a continuous furnace. The continuous annealing furnace (called HDGL furnace) consists of a preheating tunnel maintained at 200°C. The strip from preheating tunnel is passed into a *vertical furnace* known as *direct fired furnace (DFF)* consisting of three zones namely top, middle & bottom zone maintained at 1100°C, 1150°C &

1200°C respectively. Each zone consists of 10 burners fired with coke oven gas. Each burner is supported by a pilot burner which provides support flame. In this furnace, steel strip comes in direct contact with flame and a reducing atmosphere is maintained. The strip from DFF now passes into *radiant tube fired (RTF) furnace* in which there are thirty six burners in 6 zones. Each burner is in the form of U-tube and mixed gas firing takes place inside the tube. There is no direct contact between the steel strip and the product of combustion. An *inert/protective gas atmosphere* is maintained inside the furnace and a temperature, of 730°C is maintained.

The strip is then cooled to 400°C in slow cooling furnace and jet cooling chamber. It is then finally passed through the molten zinc pot (at 450°C) furnace for galvanising (i.e. zinc coating on cold rolled steel strip).

The front end of the new coil and the rear end of the preceeding coil are joined together with the help of an on line continuous welding machine. Steel strip after passing through the annealing section (HDGL furnace) descends down into molten zinc pot. There are air jets pointed at the strip as it comes out of the molten zinc bath for controlling the thickness of zinc coating. After zinc coating, the strip enters the strip cooling unit through a vertical water dip tank. The strip is then given a chemical treatment with chromic acid to prevent zinc rusting. It is then oiled, marked and coiled in tension reels. Modern instrumentation for measuring the thickness of cold rolled steel strip & zinc coating as well as systems for automatic control of parameters of chemical solutions and thermal and gas conditions are provided. Fig. 46.20 gives the line diagram of hot dip galvanising line (HDGL) vertical annealing furnace.

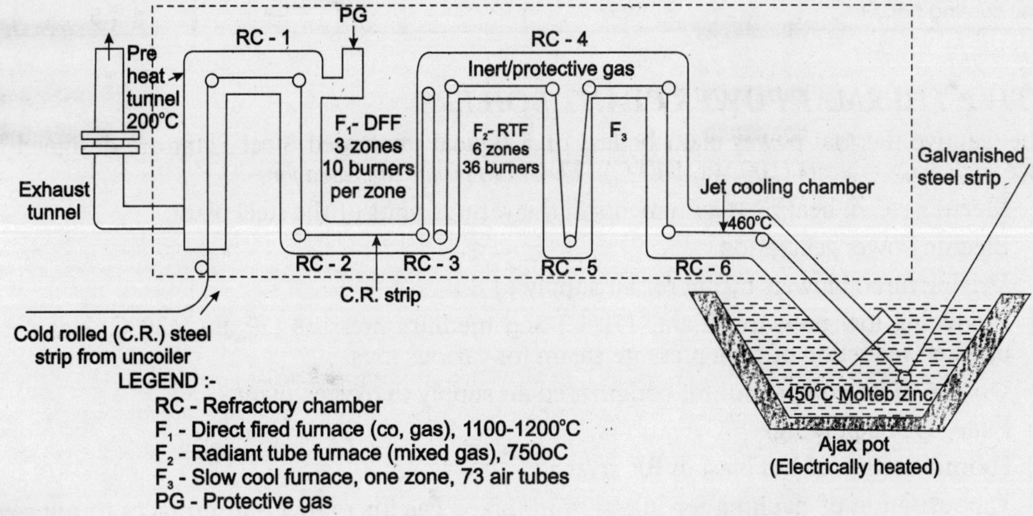

Fig. 46.19: Line diagram of hot dip galvanising line (HDGL) vertical annealing furnace.

Typical *parameters of a hot dip galvanising line* facility provided in the cold rolling mill of an integrated steel plant is given below :

- I.D./O.D. of C.R. Steel coil = 600/2200 (maximum)
- Maximum steel C.R. coil weight = 30 tons
- Width of steel strip = 650-1250 mm

- Thickness of steel strip = 0.3 to 1.6 mm
- Thickness of galvanised strip = 0.4 to 1.6 mm
- Strip annealing temperature = 700-730^{0}C
- Composition of protective gas = 20 to 40% H^2+rest N_2
- Zinc bath temperature = 450°C
- Zinc coating range = 60 to 750 gms/m^2 (on both sides)
- **Steel strip line speed (metres/minute) :**

(i) entry section = 15 to 210
(ii) process section = 20 to 150
(iii) delivery section = 20 to 210

Table 46.14: Zonewise typical temperature in a cycle

Section	Zone temperature, °C	Strip temperature,"C		Residence time, second
		incoming	outgoing	
Preheated unifired convection of heat	650	40	250	65
DFF (Direct fired furnace)	1250	250	595	5
RTF (Radiant Tube furnace)	925	595	730	30
Slow air cool furnace	425	730	595	13
Jet cooling furnace	315	595	450	13
Zinc coating pot	450	—	—	1.5

CAPTIVE THERMAL POWER PLANT BOILERS

The Captive thermal power plant boilers of a typical integrated Steel Plant are designed to produce *high pressure steam* (100 ata, 540°C). *The steam produced is used for :*

- Meeting steam heating requirements in the various units of the steel plant.
- Electric Power generation.
- Driving turbo blowers meant for air supply to B.F.
- Producing low pressure (8 ata, 170°C) and medium pressure (39 ata, 430°C) steam by pressure reduction of high pressure steam for various uses.
- Driving turbo compressor for compressed air supply to oxygen plant.
- Fuel gas line purging.
- Humidification of hot blast in BF. stves.
- Vapourisation of naphtha for injection in mixed gas for rolling mill furnaces to improve C.V.
- To run the coke oven gas exhausters.
- To preheat the boiler feed water.
- To produce chilled water (by vacuum refrigeration) in central refrigeration plant.

Boilers are of *water tube type* having multifuel firing facility. The boilers use middling coal as the main fuel and surplus BF Gas, CO Gas & Converter gas is the auxiliary fuel. Furnace oil is used for lighting and supporting the combustion flame stabilisation (even coal tar firing facility also exists for this purpose) in the boiler furnace. The heat in the flue gas is partly recovered in preheating water through economisers and heating the combustion air through air preheaters. In economisers, water passes through tubes & flue gas is outside while in air preheaters, the flue gas passes through the tubes & air is outside. Electrostatic precipitators are provided for dust removal from flue gases. The bottom of the boiler furnace has a water seal and the bottom coal ash from the slag bath is crushed and flushed off with water. All the fuels are fired in the same burner which have concentric annular spaces for different fuels and preheated combustion air. A water chemical treatment plant provides the demineralised water for high pressure boilers as well as sodium cationated water for various waste heat boilers/ECS throughout the steel plant. A coal unloading, storage and conveying system installed in the coal handling plant supplies coal middling to the pulveriser (ball mill). The pulverised coal is dried and then fed to the boilers. In power plant furnace oil station, facilitates for its unloading, storage, pumping and heating exists. Steam pressure reduction and desuperheating unit (PRDU) reduces the steam pressure to 8 and 39 ata from 100 ata for various needs of the plant. A back pressure turbine (of 12 MW capacity) also reduces the steam pressure from 100 to 39 ata besides generating power.

Fig. 46.21 shows the schematic diagram of a water tube industrial boiler attached to a steel plant.

Typical Parameters for a Thermal Power Plant Boiler Attached to Steel Plant

Total steam generation capacity	= 1880 T/hr.
No. of boilers	= 5 (220 T/hr each) + 3 (260 T/hr each)
Steam temperature and pressure	= 540°C and 100 kg/cm^2
Feed water temperature	= 210°C
Boiler efficiency	= 85%
Flue gas temperature	= 135°C
Boiler furnace chamber volume	= 1220 m^3
Boiler furnace size	= 10 M × 7 M

Maximum fuel gas consumption/boile :

BF gas	= 120000 Nm^3/hr.
C.O Gas	= 15000 Nm^3/hr.
Converter gas	= 10000 Nm^3/hr.

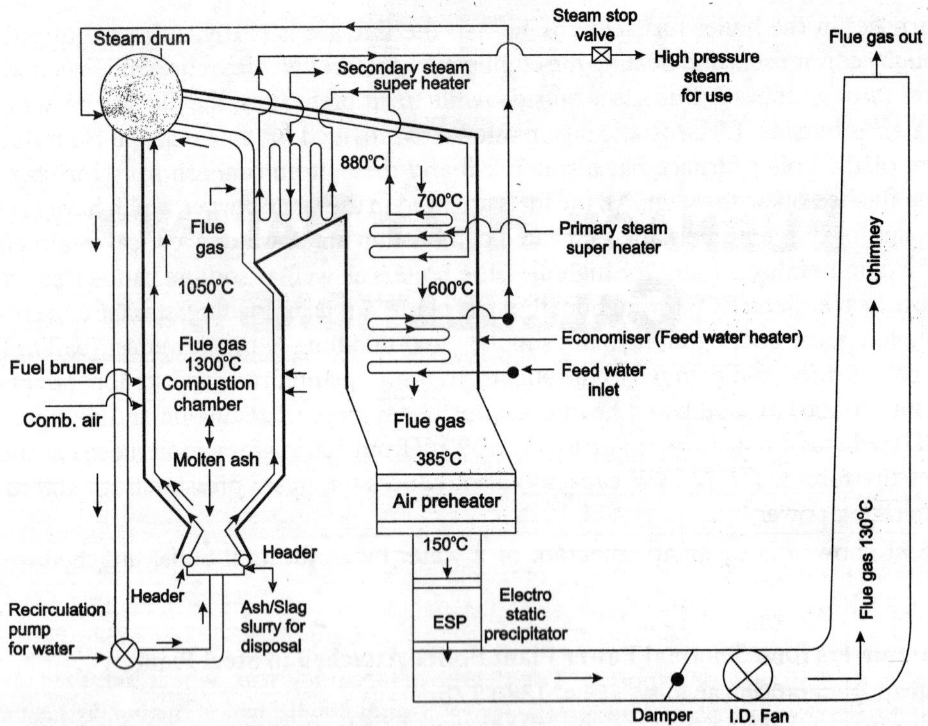

Fig. 46.21: Schematic diagram of a water tube industrial boiler attached to a steel plant.

Maximum steam generation load of the boiler :

= 110 T/hr. when only coal is burnt

= 90 T/hr. when gas is fired

Capacity of each Ball Mill = 20 T/hr

Capacity of induced draft fan = 208000 m^3/hr, 260 mmwg

Capacity of forced draft fan = 112000 m^3/hr. , 408 mmwg

Power generation capacity = (1 ×2) + (2 × 55) + (3 × 60) = 302 MW

Specific heat consumption = 0.75 Gcal/T steam

Specific steam consumption in TG = 4T/hr/MW = 3.6 Gcal/MWh (for power generation)

Specific power consumption = 30 kWh/T steam in power plant auxiliary units

Specification of boiler feed pump (6 Nos.) = 2MW, 270 m^3/hr, 150 kg/cm^2

No. of burners/boiler = 6 (3 horizontal burners on each side)

Temperature of waste gas = 130°C

Excess air for combustion = 20% above theoretical

Heat release rate in boiler furnace = 123 ×10^3 kcal/m^3/hr.

= 10 ×10^3 kcal/m^2/hr.

Furnace Heat Balance Calculations

A heat balance of furnaces is a means of determining the thermal efficiency of the process and comparing the relative values of heat losses. By comparison with a known identical process, areas of inefficiencies can be identified where a change in operational control of equipment could result in an improvement in efficiency.

A heat balance can be carried out on a process, part of a process or over a number of processes in which masses must balance (unless it is a nuclear fission/fusion process) and energy/heat supplied must equate to that used or exported. The degree of accuracy with which the total input items correspond with the total output items reflects the accuracy with which the balance is calculated. The percentage heat unaccounted should be less than 5% of the total in a good heat balance.

A datum temperature, which is usually 0°C or 15°C is required to be fixed for heat balance to which sensible heat calculation can be referred. All input and output items in a heat balance should be expressed in consistent system of heat units. At times, the heat balance is done with reference to unit quantity of :

- raw material
- main product
- time
- standard fuel of fixed calorific value

Heat input items in a heat balance :

It may contain any or all of the following items :

- preheat in combustion air
- potential heat in fuel
- sensible heat/preheat in fuel
- exothermic heat of reaction
- sensible or preheat in raw materials
- total heat of steam *e.g.* steam used for atomisation of fuel oil
- heat equivalent of other forms of energy supplied *e.g.* electrical power for heating, driving pump and fan etc.

Heat output items in a heat balance :

Various heat output items may include any or all of

- sensible + latent *i.e.* total heat in the product
- endothermic heat of chemical reaction
- total heat in flue gas :

(a)potential heat of any combustible gas present

(b) sensible heat of gases

- sensible heat transferred to cooling water
- sensible heat stored in furnace structure if the furnace is started from cold condition
- sensible heat in conveying equipment *e.g.* trays, chains, tongs, carrier plates, bogies etc. used to charge and discharge the furnace
- *total heat in any residue in combustion :*

(a)potential heat in unburnt carbon in ashes, soot, tar etc.

(b) sensible heat in ash and clinker

(c) total heat in molten slag

- *structural loss of heat :*

(a)conduction through the hearth

(b) conduction through the walls & roofs, equivalent to convection and radiation from outer surface of the furnace

- *Unaccounted figure* computed by difference from total input less all output items above. It should be < 5% of the total input items.

Heat items required/used in heat balance

Sensible heat. It is the heat above datum (reference) temperature contained in material supplied to or taken from the process and equals quantity × mean specific heat × temperature rise over datum.

Specific heat. It is the heat stored in unit mass/volume of a solid, liquid or gaseous substance when its temperature is raised by one degree. Over a wide range of temperature, a mean value may be used, since specific heat is a function of temperature.

Latent heat. It is the quantity of heat released or absorbed at constant temperature during a change of phase *e.g.* solid to liquid, liquid to gas or from one crystalline state to another or *vice versa.*

Total heat. It is the sum of sensible and latent heat usually on a unit weight basis. The total heat of steam for example, comprises of :

- sensible heat from 0°C to boiling point
- latent heat at boiling point
- sensible heat of steam above boiling point

Properties of gases change with pressure. Values for steam and other materials are available in chart, equation or tabular form.

Potential heat. It is the heating value of fuel or unburnt material in waste products which could be released on combustion and equals quantity × calorific (gross/net) value.

Heat of reaction. It is the heat of any other chemical reaction (or the net heat if more than one involved in the process. Heats of reaction may be quoted in terms of amounts or reactants which would give the molecular weight on kg/gms/lbs of the main product. A consistent system of units should be used. Standard reactions are quoted at 25°C and may be modified by Hess's law to other conditions.

Heat equivalent. The heat equivalent of steam, electrical power, compressed air or other mechanical source supplied to fans/pumps/motors for circulatory, feeding or withdrawing materials and for other auxiliary equipments should be included in the heat balance.

Period of Heat Balance. For a satisfactory heat balance, the conditions at the end of the period during which observations are made should be same as those at the beginning but not so long as to take into account long term trends *e.g.* due to wear of furnace walls. The normal practice is as follows :

(a) **Cyclic processes.** Observations begin and end at corresponding points in the cycle *e.g.* from one tap to the next tap.

(b) **Batch processes.** Observations cover complete batch or batches.

(c) **Continuous processes.** Observations after conditions have become steady and continue long enough for minor fluctuations to be negligible. Conditions must be kept steady while observations are being made.

Measurements in heat balance :

Direct measurements of mass should be made wherever possible. Net changes in stock should be evaluated. Volumes of oil, steam, air, gases etc. should be measured together with appropriate pressure & temperature of all outgoing & incoming materials and should be recorded during the period of observation.

Physical data are needed to calculate the individual potential, sensible and latent heats. Chemical analysis of fuels, other reacting materials and waste gases are needed together with the calorific value of the fuel. Heats of chemical reactions are available in tables. If possible, calorific value of fuels should be determined directly.

CALCULATION OF FURNACE EFFICIENCY/PERFORMANCE

Furnace efficiency is calculated in two ways similar to that of the boiler *i.e.*, **direct method** and **indirect method.**

Direct Method: The efficiency of a furnace can be determined by measuring the amount of heat absorbed by the (heat stock output) and dividing this by the total amount of fuel consumed (heat input).

$$\therefore \text{Thermal efficiency of furnace} = \frac{\text{Heat in the stock being heated}}{\text{Heat in the fuel consumed for heating the stock}}$$

The quantity of heat transferred to stock,

$$Q = mx\ C_p(t_2\text{-}t_1)$$

where, Q = Quantity of heat in stock in Kcal

m = Weight of stock in kg

C_p = Mean specific heat of the stock in Kcal/kg.°C

t_1 = Final temperature of stock in °C.

t_2 = Initial temperature of stock before entering the furnace in °C.

Indirect method : In this method heat losses are substracted from the heat supplied to the furnace. Different type of heat losses in a furnace are shown diagrammatically below.

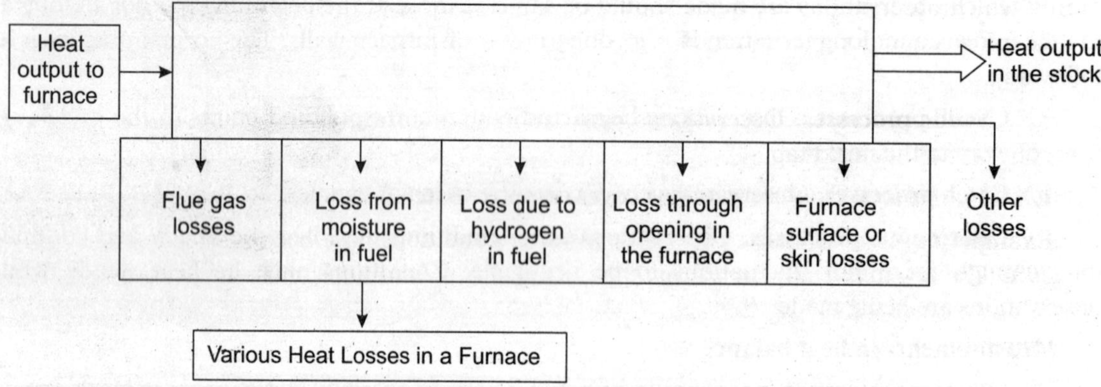

Various Heat Losses in a Furnace

These furnace heat losses are described below in detail:

- **Furnace flue gas losses** : Part of the heat remains in the combustion/flue gases which goes out through the stack of the furnace. This is called waste gas loss or stack loss.
- **Loss from moisture in fuel**: All fuels usually contains some moisture and some of the heat is used to evaporate the moisture inside the furnace.
- **Loss due to hydrogen in the fuel**: It results in the formation of water vapor.
- **Losses due to openings in the furnace** : Radiation heat losses occurs when there are openings in the furnace enclosure. These losses can be significant for furnaces operating at a temperature of more than 540°C as radiation heat transfer is proportional to (absolute temperature)^4. Another loss is due to air infiltraction because the draft of the furnace stacks/ chimneys cause a negative pressure inside the furnace, drawing in air through leaks/ cracks or whenever furnace charging or discharging doors are opened.
- **Furnace surface/skin losses**: This is called **wall losses.** While temperatures inside the furnace is high, heat is conducted through the roof, floor and walls & emitted to the ambient air, once it reaches the furnace skin or surface.
- **Other losses** : There are several other ways in which heat is lost from a furnace, although quantifying these losses is difficult. Some of these losses include :
- **Stored heat losses** : When the furnace is started, the furnace structure and insulation is

also heated and this leave only heats the structure again when the furnace shuts down. Therefore, this type of heat losses increases with the number of times the furnace is turned on and off.

- **Material handling losses** : The equipment used to move the stock through the furnace, such as conveyors, walking beams, bogies etc. also absorb heat. Everytime equipment leaves the furnace, they loose their heat, therefore heat losses increases with the amount of equipment and the frequency by which they enter & leave the furnace.
- **Cooling media losses** : Water and air are used to cool down the equipment, rolls, bearings etc and heat is lost because these cooling media absorb heat.
- **Incomplete combustion losses** : Heat is lost, if combustion of the fuel is incomplete because unburnt fuel or particles have absorbed heat but this heat has not been put to use.
- **Scale losses** : This is the heat loss due to formation of scales.

Furnace efficiency increases when the percentage of heat that is transferred to the stock or the load inside the furnace increases.

HEAT BALANCE CALCULATION OF A REGENERATIVE HEAT EXCHANGER

Example 1. *A regenerative furnace consumes air at a rate of 12750 m³/hr at a steady inlet temperature of 20°C. The air enters a regenerator containing 50 tons of refractory filling of mean specific heat 0.247 kcal/kg °C over the half cycle of 20 minutes and over the other half cycle 14200 m³/hr of waste gas enters at a steady temperature of 1220°C and leaves at a mean temperature of 700 °C.*

Calculate the mean temperature of air leaving the regenerator and also the mean temperature rise of the checker bricks during the heating cycle. Assume that 95% of the heat available reaches the regenerator.

Mean specific heats of air and waste gases are 0.317 and 0.332 kcal/m³°C respectively.

Solution.

Heat available in regenerator = 0.95 × volume × specific heat × temperature drop

= 0.95 × 14200 × 0.332 × (1220 - 20)

= 5374416 kcal/hr

Waste heat rejected = 14200 × 0.332 × (700 - 20)

= 3205792 kcal/hr

Heat stored in regenerator = 5374416 - 3205792

= mass × specific heat × temp, rise

or, 2168624 = 50000 × 0.247 × temperature rise of refractory bricks

hence,

$$\text{temperature rise} = \frac{2168624}{50000 \times 0.247}$$

= 175.6°C

Heat stored in regenerator = 5374416 - 3205792 = 2168624

= Heat taken away by air

or, $12750 \times 0.317 \times (T - 20) = 2168624$

where, T = Temperature of outgoing air from regenerator

$$\therefore \quad T = \frac{2168624}{12750 \times 0.317} + 20$$

$$= 556.5°C$$

Hence, the temperature of outgoing air is 556.5°C while the temperature rise of refractory brick during heating cycle is 175.6°C.

HEAT BALANCE CALCULATION OF A STEEL SLAB REHEATING FURNACE

Example 2. *Make a heat balance, calculate the thermal efficiency and comment on the result obtained for a continuous slab reheating furnace which receives 108 m^3 of coke oven gas (net calorific value = 4180 kcal/m^3) per ton of slab heated. Combustion air is preheated to 410°C and is consumed by the gas at air /gas ratio of 5 and the combustion product/fuel gas volume ratio is 5.60. The waste gases leave the furnace chamber at 1100°C, having entered at ambient 10°C. Water cooling losses amount to 10% of the total heat input and external losses 5%.*

During the heating process 1.5% of the metal is oxidised and the mill scale produced evolves 1593 kcal/kg of metal. Slab discharge temperature is 1180°C.

The mean specific heats (kcal / m^3°C) of air & combustion products are 0.317 and 0.370 respectively & that for steel is 0.163 kcal/I kg °C. Assume complete combustion of fuel.

Solution. Basis — per ton (*i.e.* 1000 kg) of slab

Heat input through :	Heat, kcal
(a) Fuel (coke oven gas) = 108 × 4180	= 451440
(b) Air = 108 × 5 × 0.317 × (410 - 10)	= 68472
(c) Scale = 1000 × 0.015 × 1593	= 23895
Total	= 543807

Heat output by :	Heat, kcal
(a) cooling water = 543807 × 0.10	= 54380
(b) external losses =543807 × 0.05	= 27190.3
(c) Waste gases = 108 × 5.6 × 0.37 × 1100	= 246153.6
(d) Steel = 1000 × 0.985 × 0.163 × 1180	= 189454.9
Enaccounted for (by difference)	= 26627.5
Total	= 543807

$$\text{Thermal efficiency of the furnace} = \frac{\text{Heat of steel}}{\text{Total heat input}} \times \frac{189454.9}{543807} \times 100 = \mathbf{34.8\%}$$

The 'unaccounted for' is $\frac{26627.5}{543807} \times 100 = 4.9\%$ of the total heat input which is acceptable.

Waste gases leaving the furnace chamber at 1100°C would easily preheat the combustion air to 410°C suggesting that some other form of heat recovery such as a waste heat boiler could be used in addition. While this will not directly increase the heating efficiency of steel, it would increase the overall efficiency. The waste gas temperature is not excessive for rapid heating of thick slabs, but this temperature might be excessive for thin strip heating. A longer furnace could achieve more load recuperation.

HEAT BALANCE OF A COMPOSITE STEEL SLAB REHEATING FURNACE

Example 3. *A steel reheating furnace has 10% external heat losses and receives 10000 m³ / hr of mixed gas of net calorific value 2300 kcal/m³. The air/mixed gas ratio by volume is 2.9 and the flue gas/mixed gas ratio is 3.67. The furnace employs a waste heat recovery (W.H.R.) system as shown in the Fig. 47.1.*

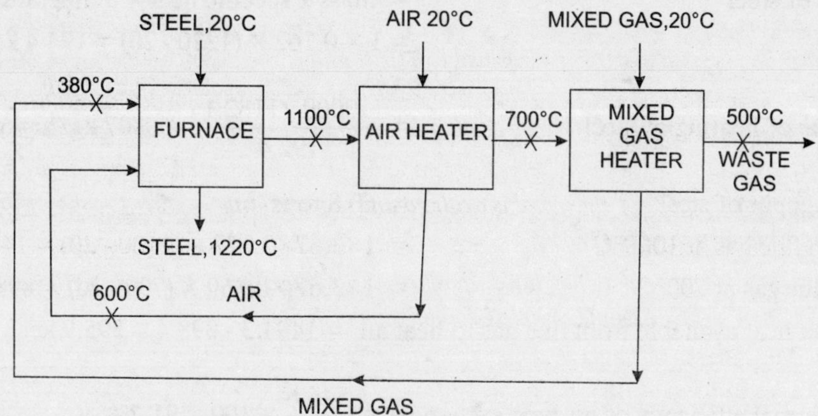

Fig.47.1: W.H.R system in a rehating furnace

The mean specific heats (from 20°C) are as follows :

Gases	Air	Mixed gas	Flue gases		
Temp., °C	600	380	1100	700	500
Sp. heat,kcal/m³.°C	0.325	0.353	0.377	0.360	0.351

The mean specific heat of steel from 20°C is 0.162 kcal/kg.°C. Steel leaves the furnace at a temperature of 1220°C.

Calculate :

(a) the useful heat per m³ of mixed gas

(b) the thermal efficiency (%) of the furnace

(c) the throughput of steel

(d) the thermal efficiency of air heat exchanger

(e) the thermal efficiency of gas heat exchanger.

Solution. Basis —1 m^3 of mixed gas and datum is 20°C

(a) Heat given to air	= 1 × 2.9 × 0.325 × (600 - 20) = 546.6 kcal
Heat given to mixed gas	= 1 × 0.353 × (380 - 20) = 127.1 kcal
Total heat input to the furnace	= net calorific value + preheat of air + preheat of gas
	= 2300 + 546.6 + 127.1 = 2973.7 kcal
Furnace external heat loss = 10%	= 2973.7 × 0.1 = 297.3 kcal
Heat in flue gas	= 1 × 3.67 × 0.377 × (1100 - 20) = 1494.3 kcal
Hence, heat exchanged in furnace	= total heat input - heat loss - heat in flue gas
	= 2973.7 - 297.3 - 1494.3 - 1182.1 kcal

(b) Hence, thermal efficiency of furnace $= \frac{\text{useful heat output}}{\text{total heat input}} \times = \frac{1182.1}{2973.7} \times 100 =$ **39.7%**

(c) Heat of steel	= mass x specific heat × temperature rise
	= 1 × 0.162 × (1220 - 20) = 194.4 kcal/kg

hence, rate of heating of steel $= \frac{10000 \times 1182.1}{192.4}$ =60807 kg/hr=60.80 tons/hr

i.e., throughput of steel	= 60.8 tons/hr
(d) heat in flue gas at 1000°C	= 1×3.67×0.377 × (1100 - 20) = 1494.3 kcal
Heat in flue gas at 700°C	= 1×3.67 ×0.360 × (700 - 20) = 898.4 kcal

Hence, net heat available from flue gas to heat air = 1494.3 - 898.4 = 595.9 kcal

hence, thermal efficiency of air heat exchanger $= \frac{546.6}{595.9} \times 100 =$ **91.7%**

(e) heat in flue gas at 700°C	= 898.4 kcal
heat in flue gas at 500°C	= 1 × 3.67 × 0.351 × (500 - 20) = 618.3 kcal

hence, net heat available from flue gas to heat the mixed gas = 898.4 - 618.3 = 280.1 kcal

So, thermal efficiency of mixed gas heat exchanger $= \frac{127.1}{280.1} \times 100 = 45.4\%$

HEAT BALANCE OF A SLAB REHEATING FURNACE (5 ZONES)

Example 4. *Basic information / data about a steel slab reheating furnace is given below :*

A. *Slab output*	= *125 Tonnes / hr*
— *Slab temperature*	= *1250°C*
— *Base temperature*	= *15°C*
B. *Fuel input (dual fuel firing)*	
— *Heavy fuel oil*	= *7189 kg / hr at 104°C*

— *C.O. Gas* = *1277 Nm³ /hr*

— *Atomising steam usage* = *4333 kg/hr (0.60 kg / kg fuel oil)*

— *Atomising steam conditions* = *2.75 bar (Gage pressure), 200°C*

C. *Skid cooling* = *Evaporative steam cooling (ECS) system*

— Steam make = 8000 kg/hr

— Steam conditions = 29 bar Gage (saturated), 255°C

— Feed water temperature = 70°C

D. *Slab surface oxidation rate* = *75 MJ/Tonne steel*

E. *Fuel analysis*

Heavy fuel oil:

% Carbon = 87

% Hydrogen = 11

% Sulphur = 2

— Calorific value = 40.36 MJ/kg (net)

— Specific heat = 1.947×10^{-3} MJ/kg fuel oil/°C

Coke Oven Gas (C.O. gas) :

% Hydrogen = 48.4 , Methane (CH_4) = 28.0

% Ethane (C_2H_6) = 2.9 , Carbon monoxide = 6.4 %

% Carbon dioxide = 2.2 , Nitrogen =12.1

Calorific value of C.O. gas = 17.88 MJ/Nm³

(F) *Combustion air temperature* = 15°C (No preheat)

(G) *Waste Gas Conditions leaving Furnace :*

— Temperature = 1057°C

— % Oxygen content = 2.5% of dry gases

From the above given informations and data, determine the following :

(i) Hourly heat input to furnace and heat output from furnace

(ii) How this heat is accounted for within the heating process *i.e.* summarise the heat balance

Solution. Heat Input (in GJ) per hour :

	G-J/HR
— Fuel oil = 7189 kg at 40.36 MJ/kg	290.1
— Coke oven gas = 1277 Nm³ at 17.8 MJ/Nm³	22.8
— Sensible heat in fuel oil = $7189 \times 1.947 \times 10^{-3} \times (104 - 15)$	1.3
— Atomising steam = $4333 \times 0.35 \times 10^{-3}$	1.5
— Oxidation = 125×0.075	9.4
Total	= 325.1

[Enthalpy of water vapour (*i.e.* ignoring latent heat) at 200°C relative to zero at 15°C = 0.35 MJ/kg]

Heat Output in GJ-Per hour (GJ-Giga Joule)

— **Stock :** (Assuming same heat in furnace charge before and after balance)	**GJ/hr**
Energy required to heat steel from 15°C to 1250°C = 0.844 GJ/Tonne Total/hour = 125 × 0.844	= 105.5

Skid Cooling:

8000 Kg/hr water at 70°C to steam at 29 bar G (saturated) = 8000 × (2.806 - 0.295) = 20.0

Waste gases + atomising steam (see appendix for calculation)

= (157.0 + 9.8) = 166.8

Furnace losses (by difference) = 32.8

Total = 325.1

[Note: 1 GJ = 10 J
1 MJ = 10^6
1 calorie = 4.2 J]

Heat Balance Summary

Heat Input	GJ/hr
From fuel combustion Sensible heat of fuel oil As atomising steam	290.1 + 22.8
From oxidation of stock	1.5
Total	325.1
Heat Output	
To stock	105.5
For skid cooling	20.0
In combustion waste gases	157.0
In exhaust atomising steam	9.8 (See calculation in appendix)
As furnace losses	32.8
Total	325.1

APPENDIX

Heat in Waste Gases :

(a) **From combustion Products :**

Waste gas volumes :

From Heavy fuel oil combustion :	Nm^3/kg *fuel oil (Stoichiometric)*
CO_2 (Including SO_2)	1.615
H_2O	1.211
N_2	8.360
	11.186

From C.O. gas (Stoichiometric) combustion : Nm^3 / Nm^3 gas

CO_2 :

(a) From Methane — same volume 0.280

From ethane — 2 × volume 0.058

From CO — same volume 0.064

From CO_2 — same volume 0.022

0.424

H_2O: *(a)* From hydrogen — same volume 0.484

(b) From methane — 2 × volume 0.560

(c) From Ethane —13 × volume 0.087

1.131

O_2 : *(a)* For hydrogen — 1/2 volume 0.242

(b) For methane — 2 × volume 0.560

(c) For Ethane — 2.5 × volume 0.073

(d) For carbon monoxide — 1/2 volume 0.032

0.907

∴ Associated air $= \left(\frac{100}{21} \times 0.907\right) = 4.319$

Of which nitrogen $= \frac{79}{21} \times 0.907 = 3.412$

N_2 from C.O. gas itself = 0.121

Total = 3.533

Hourly waste gas volume (Nm^3/hr) — Stoichiometric

Usage/hour	CO_2	H_2O	N_2	O_2
Fuel oil = 7189 kg	11610	8706	60100	—
C.O. gas = 1277 Nm^3	541	1444	4512	—
Total =	12151	10150	64612	—

Dry volume (stoichiometric) = 76763 Nm^3/Hr = (64612 + 12151) Nm^3/hr

But oxygen content is 2.5% on dry basis

∴ Air content is $= \frac{2.5}{0.21} = 11.9\%$

∴ Actual dry volume $= \frac{76763}{1-0.119} = 87132$ Nm^3

∴ O_2 volume = 2178 Nm^3/hr

and associated N_2 volume = 8.93 Nm^3/hr

∴ Actual waste gas volume from combustion in (Nm^3/hr)

CO_2	H_2O	N_2	O_2
12151	10150	64612 + 8194 Total 72806	2178

Heat content in these gases at 1057°C (relative to 15 °C): (ignoring latent heat of steam)

Constituents	Volume, Nm^3/hr	Enthalpy change, MJ/ Nm^3	Heat content in gases at 1057°C, GJ/hr
CO_2	12151	2.340	28.433
H_2O	10150	1.810	18.372
N_2	72806	1.468	106.879
O_2	2178 '	1.551	3.378
			157.062
		Total	= 157.0 GJ/hr

(b) From atomising steam :

Weight of atomising steam =4333kg/hr

∴ Volume =5392 Nm^3/ hr

Heat content at 1057 °C $= \dfrac{5392 \times 1.81}{1000}$ GJ/hr

= **9.760 GJ/hr** = 9.8 *GJ/hr*

Total heat in waste gases*(a+b)* : = 157.062+9.760=166.8 GJ/hr.

HEAT BALANCE OF A SOAKING PIT

Example 5. (a) Using the following data, draw up a heat balance for a steel ingot soaking pit for a twenty hours test period during which 130 tons of cold steel ingots were heated from 20°C to 1220°C.

Data

Mean temperature of:

- air after recuperator = 620°C
- blast furnace gas (fuel) after recuperator = 450°C
- flue gases after soaking pit = 1050°C
- flue gases after air recuperator = 770°C
- flue gases after B.F. gas recuperators = 530 m^3 /hr

Average B.F. gas consumption = $2805\ m^3/hr$

Air / fuel gas ratio = 0.8

Flue gas / fuel gas ratio = 1.65

Net calorific value of B.F. gas = $860\ kcal/m^3$

Mean specific heats (kcal / m^3 °C) are as follow :

air	0.301
B.F. gas	0.310
flue gas	0.358
steel	0.180 (kcal/kg:°C)

(All volumes are measured at 1 atm, 20 °C)

Scale formation = 1.5% (by weight)

Scale formation reaction is

$$3Fe + 2O_2 = Fe_3O_4, \qquad \Delta H_{20^0C} = -266841\ kcal$$

Atomic weight of iron = 56, *Structural loss = 270 KW*

(b) Indicate the ways in which overall thermal efficiency of the soaking pit could be improved. Also, calculate the thermal efficiency of both the air and gas recuperators using the above data.

Solution.

Datum (Reference temperature) = 20°C

(a) Volume of B.F. gas burned in 20 hours = 20 × 2805 = 56100 m^3

Heat input items *:*

Hence, heat of combustion of B.F. gas = volume × calorific value of B.F. gas

= 4333 kg/hr = 5392 Nm^3/hr

Heat picked by air in preheater = 56100 × 0.8 × 0.301 × (620 - 20) = 8105328 kcal

Heat picked by B.F. gas in gas preheater = 56100 × 0.310 × (450 - 20) = 7478130 kcal

Heat produced due to oxidation of steel = $130{,}000 \times 0.015 \times \frac{266841}{3 \times 56}$ = 3097261 kcal

(Since 3Fe *i.e.* 3 × 56 kg iron oxidation produces 266841 kcal of heat)

hence, standard heat of iron oxidation = $\frac{266841}{3 \times 56}$ kcal/kg

Hence, total heat input = 48246000 + 8105328 + 7478130 + 3097261 = 66926719 kcal

Heat output items:

Heat to steel = 130,000 × 0.180 × (1220 - 20) = 28080000 kcal

Heat supplied to air recuperator through flue gas

$= 56100 \times 1.65 \times 0.358 \times (1050 - 770) = 9278715$ kcal

Heat supplied to B.F. gas recuperator through flue gas

$= 56100 \times 1.65 \times 0.358 \times (770 - 530) = 7953185$ kcal

Heat out from chimney through flue gas

$= 56100 \times 1.65 \times 0.358 \times (530 - 20) = 16900517$ kcal

Structural heat loss in 20 hours = $270 \times 20 \times 860 = 4644000$ kcal

(Since 1 kWh = 860 kcal)

hence, unaccounted (by difference) = 70302 kcal

So, total heat output = 66926719 kcal

(b) The thermal efficiency of soaking pit can be improved by :

- limiting the oxidation of steel
- reducing the heating time by hotter ingot charging
- using continuous casting to avoid ingot manufacture
- further recovery of heat from waste gas *e.g.* by a waste heat boiler to generate stea
- reducing the volume of waste gases or by obtaining better heat recovery from the recuperators since,

$$\text{Thermal efficiency of air recuperator} = \frac{\text{Heat picked up by air in recuperator}}{\text{Heat supplied to air recuperator}} \times 100$$

$$= \frac{8105328}{9278715} \times 100 = \mathbf{87.3\%}$$

$$\text{and, thermal efficiency of gas recuperator} = \frac{\text{Heat picked up by gas to preheater}}{\text{Heat supplied to B.F. gas recuperator}} \times 100$$

$$= \frac{7478130}{7953185} \times 100 = \mathbf{94.1\%}.$$

L.D. STEEL MAKING SCRAP COOLANT

Example 6. *A L.D. converter takes a 250 tons charge which normally contains 2.57 % of cold steel scrap. It is proposed to preheat the scrap using 427 m^3 of burnt convertor top gases at a temperature of 1000° C which results from the use of 57.6 m^3 of oxygen for each ton of hot metal blown.*

The enthalpy of liquid steel is 334687 kcal/ton and the excess heat available to preheat scrap is 111084 kcal/ton of hot metal converted. The mean specific heat of top burnt convertor gas is 0.336 kcal/m^3 °C (all volumes referred to 0°C and 1 atm). The mean specific heat of scrap steel is 0.157 kcal/kg.°C and the scrap preheat temperature is 790°C. Calculate the

(a) weight of preheated scrap for thermal balance

(b) thermal efficiency of preheating

(c) oxygen saved per heat assuming that the scrap consumes no oxygen and that there is no change in metal yield.

Solution.

Let, y = tons of scrap in 250 tons total charge

Now, excess heat in converter = heat absorbed by scrap

or, 111084 (250-y) = 334687y

or, $(111084 + 334687)y = 111084 \times 250$

or, $y = \frac{27771000}{445771}$ =62.3 tons (cold scrap is 25%)

(a) When scrap is preheated to 790°C,

The heat supplied $= 10^3 \times y \times 0.157 \times (790 - 0) = 124000$ y

hence, excess heat in converter (when preheated scrap charged) = heat absorbed by scrap

$\therefore$ 111084 (250-y)+ 124000 y = 334687 y

or, y (111084 - 124000 + 334687) = 111084 × 250

or, $y = \frac{27771000}{321771}$ = 86.3 tons preheated scrap

(b) Total waste gas produced = 427(250 - 86.3) = 69900 m^3

Waste heat available = 69900 × 0.336 × (1000 - 0) = 23184000 kcal

Heat absorbed by preheated scrap = 86.3 × 10^3 × 0.157 × (790 - 0) = 10703000 kcal

hence, thermal efficiency of preheating $= \frac{10703000}{23184000} \times 100 = 46.1\%$

(c) Without preheat, oxygen used = 57.6 × (250 - 62.3) = 57.6 × 187.7 m^3

With preheat, oxygen used = 57.6 × (250 - 86.3) = 57.6 × 163.7 m^3

hence, oxygen saved = 57.6 × (187.7 - 163.7) = 1382 m^3 per heat (*i.e.*, charge)

HEAT BALANCE OF A BLAST FURNACE STOVE

Example 7. A blast furnace stove burns B.F. gas at the rate of 760 m^3/minute with 15% excess air. B.F. gas analysis is CO= 26%, CO_2 = 14%, H_2 = 2%, N_2 = 58%,.

Waste gas leaves the stove at 150°C. The net calorific value of the gas is 798 kcal/m^3. Cold air blast at 180°C is preheated to 1000°C in a period equal to the heating time of the stove at a rate of 1980 m^3/ minute. The specific heats of both air and gas may be taken as 0.320.

Assume that the stove is 33.5 metres high with a hemispherical dome 8 metres diameter from the outer surface from which the convection & radiation loss can be calculated and cross-checked with the given value. The metal thickness can be neglected. The stove is lined with 50 mm of insulation and 225 mm of firebrick of thermal conductivity (kcal / hr.m. °C) 0.17 and 1.12 respectively. The outside temperature of the stove is 35°C and the inside temperature is 1050°C. The total heat loss from the stove may be assumed to be 33000 kcal / minute.

Calculate the heat balance and thermal efficiency of the stove using an ambient temperature of 15° C.

Solution.

Volume of flue gas produced for every 100 m^3 of B.F. gas with 15% excess air is first calculated as below :

Constituents	Volume %	O_2 required, m^3	Waste gas produced, m^3
CO_2	14	—	14 (CO_2)
CO	26	13	26 (CO_2)
H_2	2	1	2 (H_2O)
N_2	58	—	58 (N_2)
Total	100	14	100

Total oxygen required for theoretically complete combustion of 100 m^3 of B.F. gas = 14 m^3

hence, air required for complete combustion $= 14 \times \frac{100}{21} = 66.7 \text{ m}^3$

N_2 of air required for combustion also will be present in the waste gas which amounts to

$$66.7 \times \frac{79}{100} = 52.7 \text{ m}^3$$

With 15% excess air, the total air required = 66.7 × 1.15 = 76.7 m^3

hence, volume of excess air = 76.7 - 66.7 = 10 m^3

Hence, total volume of waste gas

= volume of waste gas with theoretical air + volume of excess air

= (100 + 52.7) + 10 = 162.7 m^3

hence, waste gas/B.F. gas ratio $= \frac{162.7 \text{ m}^3}{100 \text{ m}^3} = 1.627$

Now, heat released by burning B.F. gas = 760 × 798 = 606480 kcal/minute

Heat loss through waste gas = 760 × 1.627 × 0.320 × (150 - 15) = 53417 kcal/minute

Heat input through cold blast = 1980 × 0.320 × (180 - 15) = 104544 kcal/minute

Heat taken by hot blast = 1980 × 0.320 × (1000 - 15) = 624096 kcal/minute

Heat Balance

Heat Input (Kcal/minute) in stove :

(a) By burning B.F. gas = 606480

(b) through cold blast = 104544

Total = 711024

Heat output (kcal / minute) from stove

(*a*) through hot blast = 624096

(*b*) through waste gas = 53417

(*c*) heat loss from stove = 33000

(*d*) unaccounted = 511

Total = 711024

$$\text{Thermal efficiency of stove} = \frac{624096 - 104544}{711024} \times 100 = \mathbf{73.1\%.}$$

HEAT BALANCE OF A BOILER INSTALLATION

Example 8. *Compute the heat balance and thermal efficiency of a boiler from the following operating data :*

Weight percentage composition of coal (ash free basis) :

$C = 88.5\%, H = 4.7\%, N_2 = 1.5\%, 0 = 4.6\%$

Weight percentage composition of coal (on as used basis):

Moisture = 13.6%, Ash = 14.9%, Gross calorific value = 6038 kcal / kg

Coal firing rate = 410 kg / hr, Steam production rate = 2610 kg / hr

Steam pressure = $5.5\ kg / cm^2$*, Steam temperature = 163°C*

Temperature of flue gases leaving the boiler = 325°C

Volumetric analysis of flue gas : $CO_2 = 9.82\%$, $O_2 = 9.48\%$

Ash and clinker formation rate = 89.5 kg/hr (combustible = 37.6%)

Grit emission rate = 2.88% of coal firing rate (combustible in grit = 51.2%)

Mean specific heat (C_p) *of flue gases = 0.329 kcal/*Nm^3 *°C .*

Solution. Heat Input Item

Heat of combustion of coal = 410 × 6038 = 2475580 kcal/hr

Heat Output Items

(*a*) **Heat in steam** = 2610 × 657.675 = 1716531 kcal/hr

[here, enthalpy of steam at 5.5 kg/cm^2 and 163°C = 657.675 kcal/kg (from steam table)]

(*b*) **Heat loss as Unburnt Coal in Ash and Clinker**

= 89.5 × 0.376 × 6038 = 203190 kcal/hr

here, Percentage incombustible in ash and clinker = 37.6% = 0.376

Calorific value of coal = 6038 kcal/kg, Ash and clinker formation rate = 89.5 kg/hr

(*c*) **Heat loss as unburnt fuel in grit** = 410 × 0.0288 × 0.512 × 60.38 = 36504 kcal/hr

here, grit emission rate = 2.88% of coal firing rate *i.e.,* 410 kg/hr

and, %combustible in grit = 51.2%, calorific value of coal = 6038 kcal/kg

(*d*) **Heat loss as sensible heat of flue gas**

For computing this item, total volume of wet flue gas hourly is first found as follows :

Coal firing rate = 410 kg/hr

Carbon content of 100 kg of coal = 88.5 kg $= \frac{88.5}{12} = 7.37$ kg.mole (atomic weight of C = 12)

Hydrogen content of 100 kg of coal = 4.7 kg $= \frac{4.7}{2} = 2.35$ kg mole

Carbon used in as fired coal $= \frac{\text{amount} \times (100 - 13.6 - 14.9)}{100} = \text{amount} \times 0.175_n$

Now, %CO_2 in flue gas $= 9.82 = \frac{\text{amount of carbon gasified}}{\text{Total carbon}} \times 100$

Amount of carbon gasified is found as follows :

Carbon available as fired = 7.37 × 0.715 = 5.27 kg mole/100 kg coal fired

Coal firing rate = 410 kg/hr

∴ Carbon available per 410 kg coal $= \frac{410 \times 5.27}{100} = 21.6$ kg mole

Carbon lost in ash and clinker $= \frac{89.5 \times 0.376 \times 0.885}{12} = 2.48$ kg mole

(since C = 88.5% in coal and combustible in ash and clinker = 37.6%)

Carbon lost in grit $= \frac{410 \times 0.0288 \times 0.512 \times 0.885}{12} = 0.4458$ kg mole

(here, percentage combustible in grit = 51.2%, grit emission = 2.88% of coal fired and C% in coal = 88.5%, atomic weight of carbon = 12)

Hence, carbon gasified per hour = 21.6 - 2.48 - 0.4458 = 18.67 kg mole

∴ Volume of dry flue gases formed $= \frac{18.67 \times 100 \times 22.4}{9.82} = 4258$ Nm^3/hr

(since, 1 kg mole = 22.4 Nm^3)

To get the total volume of wet flue gases, the moisture present in coal as such and the moisture produced by combustion of H_2 present in coal is added to the dry flue gas volume.

In 100 kg as fired coal moisture = 2.35 x 0.715 = 1.68 kg mole/hr

Coal firing rate = 410 kg/hr

Hence, moisture volume $= \frac{1.68 \times 410 \times 22.4}{100} = 154.3$ Nm^3 /hr

Moisture present as such per hour = 13.6 kg/100 kg coal $= \frac{13.6}{18} = 0.755$ kg mole/hr

$$= \frac{0.755 \times 410 \times 22.4}{100} = 69.3 \text{ Nm}^3\text{/hr}$$

Hence total volume (dry + wet) of flue gases = 4258 + 154.3 + 69.3 = 4481.6 Nm^3/hr

Hence, the heat loss as sensible heat of flue gas

$$= 4481.6 \times 0.329 \times (325 - 15) = 457078 \text{ kcal/hr}$$

Heat Balance is Summarised Below :

Heat Input Item

Heat input by combustion of coal	= 2475580 kcal/hr
Heat Output Items	
Heat in steam	= 1716531 kcal/hr
Heat loss as unburnt coal in ash and clinker	= 203190 kcal/hr
Heat loss as unburnt coal in grit	= 36504 kcal/hr
Heat loss as sensible heat of flue gas	= 457078 kcal/hr
Total heat output	= 2413303 kcal/hr

hence, unaccounted losses = Total input - Total output = 2475580 - 2413303 = 62277 kcal/hr

$$\text{hence, thermal efficiency of boiler} = \frac{\text{Heat in steam}}{\text{total heat input}} \times 100 = \frac{1716531}{2475580} \times 100 = 69.3\% \textbf{ Ans.}$$

HEAT BALANCE OF STEEL BILLET HEATING CONTINUOUS RECUPERATIVE FURNACE

Example 9. *During performance test carried out on steel heating oil fired continuous recuperative furnace, the following data were reported :*

Furnace oil consumption rate = $850\ kg/hr$

Gross calorific value (C.V.) of furnace oil = $44.04\ MJ/kg$

Net calorific value (C.V.) of furnace oil = $41.24\ MJ/kg$

Specific gravity of furnace oil = 0.95

Price of furnace oil = *Rs. 10 /litre = Rs. 10000/kilo litre*

Sensible heat of flue gas discharged from furnace = 15.145 MJ / kg of oil burned. Sensible heat recovered from flue gas in recuperator by preheating combustion air and returned to the furnace = $3.495\ MJ/\ kg\ of$ *oil burned*

Heating rate of cold steel billet to rolling temperature = $250\ kg/m^2$ *of hearth area/hr. Heat absorbed by the charge (i.e., steel billet)* = $1050\ kJ/kg$. *Effective furnace hearth dimension = 9 m wide × 12 m long From the above data, find out:*

(a) heat loss rate per hour from the furnace chamber

(b) Overall thermal efficiency of the furnace

(c) the increase in furnace oil consumption, if the recuperator is by passed i.e. no waste heat recovery is done

(d) The effect / saving due to recuperator on the cost of oil per ton of steel billet heated.

(e) Compute the heat balance in tabular form.

Solution. Basis of calculation — hourly

(a) Oil consumption rate = 850 kg/hr

Net heat released by oil by its combustion = 850 × 41.24 = 35054 MJ/hr

The heat available in the furnace will be the net C.V. of the furnace oil plus any sensible heat in oil & preheated combustion air minus the heat carried away be flue gases & radiation losses.

hence, the sensible heat loss rate in flue gases from the furnace chamber

$$= (15.145 - 3.495) \times 850 = 9902 \text{ MJ} \qquad \textbf{Ans.}$$

(b) Overall thermal efficiency of the furnace

$$= \frac{\text{heat} - \text{absorbed by the steel billet charge}}{\text{gross potential heat in the furnace oil}} \times 100$$

Effective hearth area of the furnace = 9 × 12 = 108 m^2

Heating rate of billet = 250 kg/m^2 of hearth area/hr

hence, weight of billet heated = 250 ×108 = 27000 kg/hr

Heat absorbed by the charge = 1050 kJ/kg

hence, total heat absorbed by the charge per hour

$$= 27000 \times 1050 = 28350{,}000 \text{ kJ/hr} = 28350 \text{ MJ/hr}$$

Gross potential heat in the furnace oil

= gross heat released by combustion of oil per hour

= 850 x 44.04 = 37434 MJ/hr 28350

$$\therefore \text{ thermal efficiency of furnace} = \frac{28350}{37434} \times 100 = 75.4\% \qquad \textbf{Ans.}$$

(c) Heat returned to the furnace in preheated air by use of recuperator

$$= 850 \times 3.459 = 2970 \text{ MJ/hr}$$

i.e., Heat saved as a result of recuperation = 2970 MJ/hr

Net C.V. of furnace oil = 41.24 MJ/kg

hence, increase in oil consumption, if recuperator is by passed = $\frac{2970}{41.24}$ = 72 kg/hr **Ans.**

(d) here, cost of oil = Rs. 10/litre, saving per kg oil = 10 - 0.95 = 0.5 Rs.

Now, 1 litre = 0.95 kg hence, cost of 0.95 kg oil = Rs. 10

$$\therefore \text{ Cost of 72 kg oil saved by recuperation per hour} = \frac{10}{0.95} \times 72 = 757.9 \text{ Rs./hr}$$

Billet heating rate = 27000 kg/hr = 27 tons/hr

hence, saving due to recuperation= $\frac{757.9}{27}$ = Rs. 28 ton of billet. **Ans.**

(e) Heat balance is tabulated below :

Heat input to the furnace, MJ/ hr	Heat output, MJ/hr
Net heat released by fuel = 35054, Heat as sensible heat in preheated air = 2970,	Heat absorbed by billet = 28350 Heat lost as sensible heat of flue gases = 9902
Total = 38024	Total = 38252
Unaccounted heat = 38252 - 38024 = 228 MJ/hr which may be the sensible heat of furnace oil.	

Example 10. Example on calculation of reheating furnace efficiency (both by direct and indirect method).

Calculate the efficiency of an oil-fired reheating furnace with the direct and indirect method using the data below.

Operating temperature :	*1340°C*
Exit flue gas temperature after preheater :	*750°C*
Ambient temperature :	*40° C*
Preheated air temperature :	*190°C*
Specific gravity of fuel oil:	*0.92*
Average fuel oil consumption:	*400 liters/hr = 400 x 0.92 = 368 kg/hr*
Calorific value of oil	*10000 kCal/kg*
Average 0_2 percentage in flue gas :	*12 percent*
Moisture in 1 kg of fuel oil:	*0.15 kg*
H_2 in 1 kg of fuel oil:	*0.1123 kg*
Theoretical air required to burn 1 kg of oil:	*14 kg*
Weight of stock :	*6000 kg/hr*
Specific heat of billet:	*0.12 kcal/kg/°C*
Furnace wall thickness (D) :	*460 mm*
Billet extraction outlet (X) :	*1 m × 1 m*
Average surface temperature of heating + soaking zone:	*122^0C*
Average surface temperature of area other than heating and soaking zone :	*80°C*
Area of heating + soaking zone :	*70.18 m^2*
Area other than heating and soaking zone :	*12.6 m^2*

Solution : Direct method of calculation of efficiency

The heat input is 400 liters per hour, the specific gravity of fuel is used to convert this into kg.

Therefore : 400 1/hr × 0.92 kg/1 = 368 kg/hr × 10000 kcal/kg = 368 × 10000 kcal/hr

The heat output is calculated as follows :

$$= m \times C_p \times \Delta T = 6000 \text{ kg/hr} \times 0.12 \text{ kcal/kg°C} \times (1340 - 40)\text{°C}$$

$$= 936000 \text{ kcal.hr}$$

The efficiency is :

$$= (\text{heat output/heat input}) \times 100$$

$$= [(936000/((368 \text{ x } 10000)] \times 100 = 25.43\%$$

The approximate heat loss is 100% - 25% = 75%

Indirect method of calculation of efficiency

The different heat losses are calculated below,

***(a)* Heat loss in flue gas**

Excess air (EA)

$$= O_2 \text{ percent}/(21 - O_2 \text{ percent})$$

$$= 21/(21 - 12) = 133\%$$

Mass of air supplied

$$= (1 + EA/100) \times \text{Theoretical air}$$

$$= (1 + 1.13) \times 14 = 32.62 \text{ kg/kg fuel oil}$$

$$\% \text{ Heat loss in flue gas} = \frac{m \times c_p \times \Delta T \times 100}{GCV \text{ of fuel}}$$

Where,

m = weight of flue gas (air + fuel)= 32.62 + 1.0 = 33.62 kg/kg oil

C_p = specific heat

ΔT = temperature difference

$$\% \text{ Heat loss} = \frac{\{(33.62 \times 0.24 \times (750 - 40)\} \times 100}{10000} = 57.29\%$$

(b) Heat loss from moisture in fuel

$$\% \text{ Heat loss from moisture in fuel} = \frac{M \times \{584 \times C_p(T_f - T_{amb})\} \times 100}{\text{GCV of fuel}}$$

Where,

M = kg of moisture in 1 kg of fuel oil

T_{fg} = Flue gas temperature, °C

T_{amb} = Ambient temperature, °C

GCV = Gross Calorific Value of fuel, kcal/kg

$$\%\text{Heat loss} = \frac{0-15 \times (584 + 0.45(750 - 40)\} \times 100}{10000}$$

(c) Loss due to hydrogen in fuel

$$\text{Heat loss due to hydrogen in fuel} = \frac{9 \times H_2 \times (584 + C_p(iy - T_{amb}) \times 100}{\text{CCV of fuel}}$$

Where, H_2 = kg of H_2 in 1 kg of fuel oil (= 0.1123 kg/kg of fuel oil)

$$\%\text{ Heat loss} = \frac{9 \times 0-1123 \times \{584 + 0.45\,(750 - 40)\} \times 100}{10000} = 9.13\%$$

***(d)* Heat loss due to openings in furnace**

$$\%\text{ Heat loss from furnace openings} = \frac{(\text{Black body radiation factor x emissivity} \times \text{factor of radiation} \times \text{area of opening}) \times 100}{\text{Quantity of coil} \times \text{CCV of oil}}$$

The factor of radiation through openings and the black body radiation factor can be obtained from standard graphs as shown in Figure 47.2 and Figure 47.3.

- Factor of radiation (refer Figure 47.2) = 0.71
- Black body radiation at 1340°C (refer Figure 47.3) = 36 kcal/cm^2/hr
- The area of the opening is 100 cm x 100 cm = 10000 cm^2
- Emissivity = 0.8

$$\%\text{ Heat loss from furnace opening} = \frac{36 \times 0.8 \times 0.71 \times 10000 \times 100}{368 \times 10000}$$

$$= 5.56\,\%$$

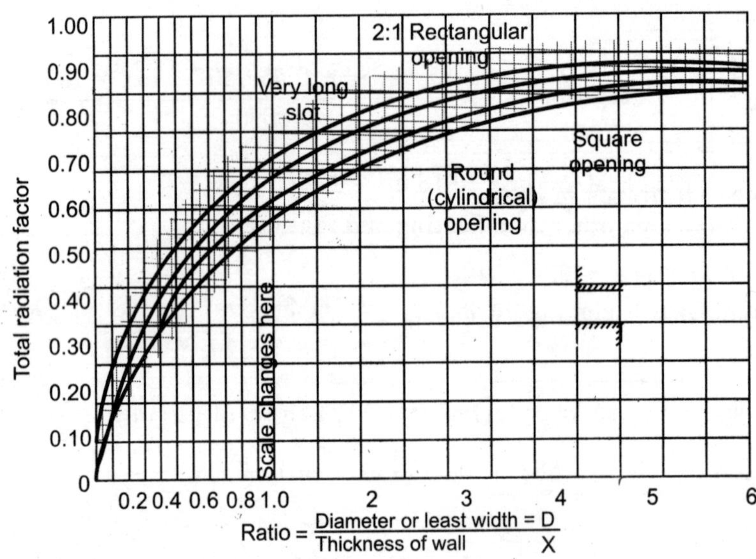

Fig.47.2: Radiation Factor for Heat Release through Opening relative to the Quality of Heat Release from Perfect Black Body.

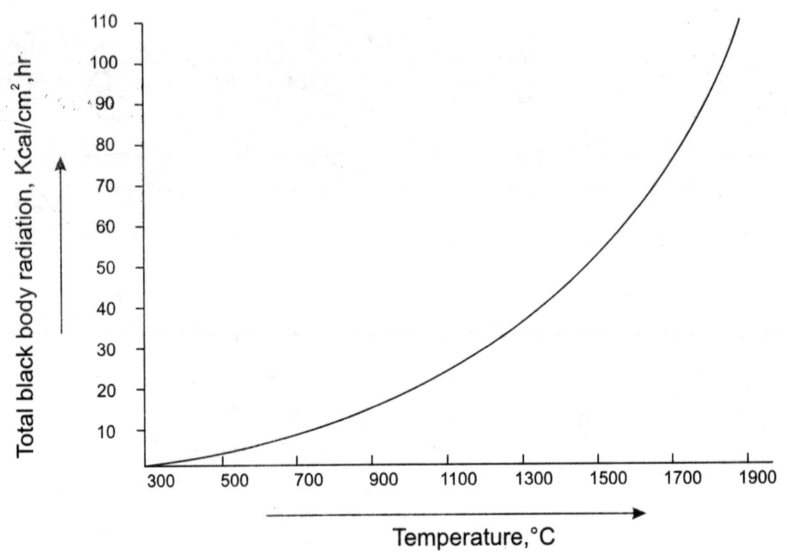

Fig.47.3: Black Body Radiation at Different Temperatures

(e) Heat loss through furnace skin

To determine the heat loss through the furnace skin, first the heat loss through the roof & sidewalls and through other areas must be calculated separately.

(i) Heat loss through roof/ceiling and sidewalls (= heating and soaking zone) :

- Total average surface temperature = 122°C
- Heat loss at 122°C (Refer Figure 50.4) = 1252 kcal/m^2 hr

- Total area of heating + soaking zone = 70.18 m^2

$$\text{Heat loss through furnace roof} = \frac{\text{Heat loss from roof and walls}}{\text{Area of roof and walls}}$$

Total heat loss = 1252 kcal/m^2 hr x 70.18 m^2 = 87865 kcal/hr

(ii) Heat lost from area other than heating and soaking zone :

- Total average surface temperature = 80°C
- Heat loss at 80°C (Refer Figure 48.4) = 740 kcal/m^2 hr
- Total area = 12.6 m^2

$$\text{Heat loss through other areas} = \frac{\text{Heat loss from roof and other areas}}{\text{Area of other areas}}$$

Total heat loss = 740 kcal/m^2 hr × 12.6 m^2 = 9324 kcal/hr

$$\%\ \text{Heat loss through furnace skin} = \frac{(\text{Heat loss(i)} + \text{heat loss(ii)}) \times 100}{\text{GCV of oil} \times \text{Quantity of oil per hour}}$$

$$\%\ \text{Heat loss through furnace skin} = \frac{(87865\ \text{kcal/hr} + 9324\ \text{kcal/hr})}{10000\ \text{kcal/kg} \times 368\ \text{kg/hr}} \times 100 = 2.64\%$$

***(f)* Unacounted losses**

The unaccounted losses cannot be calculated unless the other types of losses are known.

Furnace efficiency

Adding the losses *'a'* to 'f' up gives the total losses :

(a) Flue gas loss = 57.29%

(b) Loss due to moisture in fuel = 1.36%

(c) Loss due to H_2 in fuel = 9.13%

(d) Loss due to opening in furnace = 9.56%

(e) Loss through furnace skin = 2.64%

Total losses = 75.98%

The furnace efficiency calculated through the indirect method = 100 - 75.98 = 24.02%

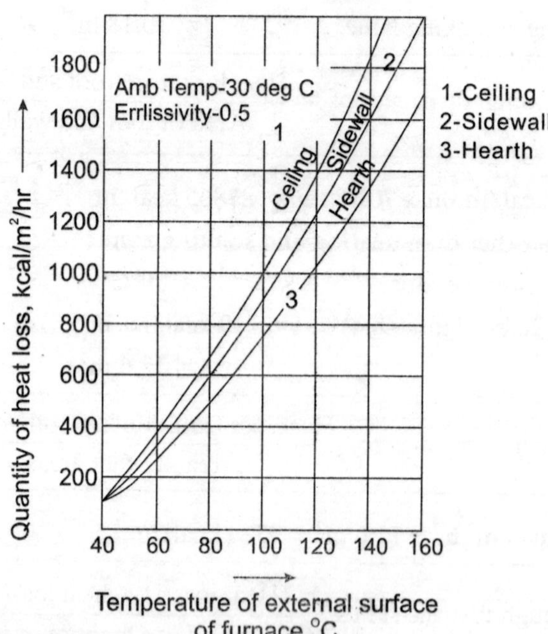

Fig.47.4: Heat Loss from the Ceiling, Sidewall and Hearth of Furnace.

48

Briquetting of Solid Fuels/Coal

Briquetting consists of applying pressure to a mass of particles with or without the addition of a binder and converting it into an agglomerate. The products are called either briquettes containing binders or binderless briquettes, as the case may be. Briquetting converts a low grade solid fuel into one of higher quality. Saw dust, peat, lignite, coal fines etc. may all be briquetted into lumps of suitable size, shape and strength and used as convenient domestic and industrial fuels.

Briquetting of Peat

In Ireland, peat briquettes are a common type of solid fuel, largely replacing sods of raw peat as a domestic fuel. These briquettes consist of shredded peat, compressed to form a virtually smokeless, slow-burning, easily stored and transported fuel. Although often used as the sole fuel for a fire, they are also used to quickly and easily light a coal fire.

Briquetting of Lignites

Briquetting is most widely used in the case of lignites. Due to the low structural quality of lignite, considerable amounts are recovered as fines during its mining. Moreover the coarser pieces disintegrate into fines after air drying. Briquetting is therefore essential prerequisite for burning lignite in domestic ovens and stoker furnaces. However pulverized fuel combustion as in thermal power stations does not need this preparative step.

No binders are required in the briquetting of lignite and comparatively low pressures suffice in achieving the compaction of its particles.

The briquetting of lignite involves its grinding, drying and pressing. The formation of a good briquette requires that the grains are of certain uniform size and optimum moisture content so that the particles can approach very close to each other and allow the cohesive forces to come into play when an external pressure is applied. In the briquetting plant of Neyveli Lignite Corporation, raw lignite from the mine is crushed to below 250 mm size in a blade crusher and stored in a bunker. This is further ground to below 50 mm in a roller crusher and finally to below 8 mm in a hammer mill. The ground and wet, lignite (moisture 50%) is dried in a steam-tube rotary drier to a moisture content of 10%. The discharge from the rotary drier has a grain size of below 6 mm. Before undergoing final grinding, it needs pretreatment which involves screening and redrying of coarse fraction. The pretreated material is ground to below 1 mm in a hammer mill and then converted into briquettes by passing through quadruple extrusion presses at a pressure of about 1000 kg/cm^2. No binder is used and the optimum temperature for briquetting is 70°C. Cooling of the dried lignite occurs during its transport from the drier to the press by a conveyor. The temperature rises during the pressing operation. It is necessary to allow for cooling before the briquettes are disposed off.

The lignite briquettes at Neyveli are carbonized at 650-700°C. After cooling, the carbonized briquettes are sieved over 10 mm screens. The oversize is marketed under the trade name 'Leco'. The undersize char fines are also used as industrial fuel.

Briquetting of Bituminous coal

Fine coals are produced in large quantities during mining and processing. Before the advent of pulverized fuel combustion these fines could be burnt in ovens and furnaces only after briquetting. The process of briquetting of coal fines is used because generation of mechanical mining results in fines and the technique of pulverized fuel firing is not suitable for domestic ovens and small industrial furnaces.

Briquetting bituminous coal without binder is difficult and much higher pressures and much finer grain sizes are required than in the case of lignites. Binder used may be coal tar, pitch or petroleum bitumen. The whole process consists in the operations of drying and grinding the coal, mixing it with the binder, heating the mixture, pressing it into moulds and cooling the briquettes before final disposal. The moisture content of the dried coal is maintained at 2-4%. Pitch and dried coal are often ground together. The grain size of the ground coal is about 80% less than 0.5 mm and 93% less than 0.88 mm. The pitch content is about 7-8% of the coal. The temperature of the mixture is kept at a few degrees within the softening point of the binder before it is charged into the press. Briquetting pressure is normally about 150 kg/cm^2. The quality of the briquettes depends upon the nature of coal, properties of binder and the precise way in which the various operations are carried out.

Many binderless briquetting process have been developed. Economic considerations stand in the way of their wide commercial exploitation. Hot briquetting technique consists in preheating a coal blend containing one or more caking coal to a temperature in the plastic range and then compacting the mass. The "softened" coal acts as the binder and the required briquetting pressure is much lower than in the other binderless briquetting processes for bituminous coals.

Briquetting of charcoal

Charcoal fines cannot be burned by the usual simple burning methods Briquetting - turning fines into lumps of charcoal - seems an obvious answer. Though it is technically possible to briquette charcoal fines, the economics are not usually favourable, except where the price of lump charcoal is very high and the fines are available at a very low cost.

Briquetting requires a binder to be mixed with the charcoal fines, a press to form the mixture into a cake or briquette which is then passed through a drying oven to cure or set it by drying out the water so that the briquette is strong enough to be used in the same burning apparatus as normal lump charcoal.

Charcoal is a material totally lacking plasticity and hence needs addition of a sticking or agglomerating material to enable a briquette to be formed. The binder should preferably be combustible, though a non-combustible binder effective at low concentrations can be suitable. Starch is preferred as a binder though it is usually expensive. Highly plastic clays are suitable providing not more than about 15% is used. Tar and pitch from coal distillation or from charcoal retorts have been used for special purpose briquettes but they must be carbonised again before use to form a properly bonded briquette. They are of good quality but costly to produce.

The press for briquetting must be capable of agglomerating the mixture of charcoal and binder sufficiently for it to be handled through the curing or drying process. Charcoal is quite abrasive so that equipment for screening fines, grinding, mixing them with binder, briquetting and so on must be abrasive-resistant and well designed.

The most common effective binder is starch. About 4-8% of starch made into paste with hot water is adequate. First, the fines are dried and screened. Undersized fines are rejected and oversized hammer milled. This powder is blended with the starch paste and fed to the briquetting press. The briquettes are dried in a continuous oven at about 80°C. The starch sets through loss of water, binding the charcoal into a briquette which can be handled and burned like ordinary lump charcoal in domestic ovens. Generally briquettes are not suitable for use as industrial charcoal in blast furnaces and foundry cupolas, since the bond disintegrates on slight heating. For this briquettes bonded with tar or pitch and subsequently carbonised in charcoal furnaces to produce a metallurgical charcoal briquette of adequate crushing strength are needed. The cost is too high for the process to find industrial application in most countries.

It is possible to add material to aid combustion of briquettes such as waxes, sodium nitrate, and so on, during manufacture to give a more acceptable product. Also clay as a binder, silica, and so on, can be mixed with the fines to reduce the cost of the briquette .

Successful briquette operations are found mostly in developed countries. An example is the industry based on carbonization of sawdust and bark in the southern U.S.A. using rotary multiple hearth furnaces which produce perhaps 25 to 50 tons of fine charcoal per day. When briquetted, this charcoal, intended for barbecues, can be sold in retail outlets. The furnace gases are burned to produce steam of electric power, thus transforming the waste sawdust and bark into two useful products, electric power and charcoal briquettes. Air pollution and waste disposal problems are minimised at the same time.

Briquetting of Biomass

Biomass briquettes are made from agricultural waste and are a replacement for fossil fuels such as oil or coal, and can be used to heat boilers in manufacturing plants, and also have applications in developing countries. Biomass briquettes are a renewable source of energy and avoid adding fossil carbon to the atmosphere.

A number of companies in India have switched from furnace oil to biomass briquettes to save costs on boiler fuels. The use of biomass briquettes is predominant in the some parts of India, where coal and furnace oil are being replaced by biomass briquettes. A number of units in Maharashtra are also using biomass briquettes as boiler fuel. Use of biomass briquettes can earn Carbon Credits for reducing emissions in the atmosphere Biomass briquettes also provide more calorific value/kg and save around 30-40 percent of boiler fuel costs.

A popular biomass briquette emerging in developed countries takes a waste produce such as sawdust, compresses it and then extrudes it to make a reconstituted log that can replace firewood.

It is a similar process to forming a wood pellet but on a larger scale. There are no binders involved in this process. The natural lignin in the wood binds the particles of wood together to form a solid. Burning a wood briquette is far more efficient than burning firewood. Moisture content of a briquette can be as low as 4%, whereas green firewood may be as high as 65%.

Parameters of fuel briquettes made by extrusion from sawdust in Ukraine are as follows :

Parameter	Value
Briquette density, t/m³	1.0-1.2
Heat content, MJ/kg	19.3-20.5
Ash content, %	0.5-1.5

(MJ = Megajoules., 3.6 MJ equals 1 kWh.)

The extrusion production technology of briquettes is the process of extrusion screw wastes (straw, sunflower husks, buckwheat, etc.) or finely shredded wood waste (sawdust) under high pressure when heated from 160 to 350 °C .

Sawdust briquettes have developed over time with two distinct types: those with holes through the centre, and those that are solid. Both types are classified as briquettes but are formed using different techniques. A solid briquette is manufactured using a piston press that compresses sandwiched layers of sawdust together. Briquettes with a hole are produced with a screw press. The hole is from the screw thread passing through the centre, but it also increases the surface area of the log and aids efficient combustion.

Formed Coke/Carbonized Briquettes

Briquettes of lignites and bituminous coals are carbonized for producing smokeless lumpy fuel and substitutes for metallurgical coke. The manufacture of carbonized briquettes requires an additional stage over and above that required for the conventional coking practice. However, the process enables the production of coke-like material (from lignites, sub-bituminous coals and non-coking bituminous coals) called "formed coke".

There are many processes for the production of carbonized briquettes. In Lurgi-Spiilgas, process, binderless lignite briquettes are first produced and then carbonized in a three-chamber unit for low temperature carbonization. The carbonization temperature varies between 500° and 850°C depending upon the use of the products as the domestic or metallurgical fuel. The higher temperature reduces the volatile matter and considerably increases the strength of the briquette. Briquettes meant for metallurgical fuel production are made at high pressures of the order of 2000 kg/cm^2 in ring-roll presses.

Briquettes are susceptible to deformation, fissuration and crumbling into pieces during carbonization. The rate of heating during carbonization is very important. In the case of coal briquettes, the quantity of pitch must not be excessive and the highly coking coals need pretreatment like oxidation or partial carbonization before briquettes are made from them.

The technique of briquetting has also been employed for the production of agglomerates of fuel with one or both of iron ore and lime. The products - 'Ferrocoke' and 'Fluxed coke' - are meant for iron production.

Gas Cleaning and Purification

Gas Cleaning. Gas cleaning is a gas-solid separation phenomenon aimed at removal of solid particles/dust from the gas.

PURPOSES OF GAS CLEANING ARE :

(a) **Air-pollution control,** *e.g.* removal of flyash from power-plant flue gases.

(b) **Equipment maintenance reduction,** *e.g.* filtration of intake air before its entry into I.C. engine cylinder, pyrites furnace gas cleaning before its entry into a chamber sulphuric acid system etc.

(c) **Increase in heat** transfer rate & reduction in power for transportation and chances of transmission pipelines burner chocking e.g. cleaning of blast furnace gas (a fuel gas).

(d) **Safety or health hazard elimination,** *e.g.* collection of metallic dust around grinding & drilling equipment and floor dust from milling or bagging operations.

(e) **Improvement in product quality,** *e.g.* air cleaning in the production of pharmaceutical products and photographic films.

(f) **Recovery of a valuable product,** *e.g.* collection of dusts from dryers and smelters.

(g) **Collection of powdered products,** *e.g.* in pneumatic conveying, spray drying of milk, eggs, detergents etc.

DEFINITION OF TERMINOLOGIES USED IN GAS CLEANING :

Particulates. Solids/droplets which are visible are called particulates.

Dust and Fumes. Particles having diameter 1-7 mp are called dust and those having diameter less than 1 mp are called *fumes.*

Powder. 80% less than 200 mesh size (0.074 mm = 74 microns) is called powder.

$$\begin{bmatrix} 1\mu(\text{micron}) = 10^{-3}\,\text{mm} = 10^{-4}\,\text{cm}, \; 1\,\text{mm} = 1000\,\text{microns} \\ 1\,\text{m}\mu\;(\text{millimicron}) = 10^{-3}\;\mu = 10^{-6}\,\text{mm} \end{bmatrix}$$

Grit. Particles having dia > 75 mμare called grit.

Aggregates (Agglomerates). 149 μ to 4 inches size particles are called aggregates.

Colloid/Aerosole. 5 — 200 mμ size particles are called colloids. There are a variety of gas cleaning equipments which are effective to separate dust of a particular diameter range from the gas.

Table 49.1 gives the principal gas cleaning equipments and the range of particles sizes that they can separate from the gas most effectively.

Table 49.1: Various Parameters of Principal Gas Cleaning Equipments

Equipments	Gravity Settling chamber	Cyclone Separator	Wet Scrubber	Bag filter	Electro-static pre-cipitator
Particle size range, micron	10-1000	0.6-100	0.1-100	0.1-60	0.001-50
Dust concentration in cleaned gas, gm/Nm^3	5	2.5	0.5-0.7	0.03	0.01
Maximum dust size (microns) in clean gas	150	40	20-25	10	5
Pressure drop in cleaner, mmWg	< 15	>50,	< 50 in hollow scrubber, 100-250 in packed scrubber	50-250 (clean-dirty)	< 25

Some of the important gas cleaning equipments are described below :

DUST CATCHER/GRAVITY SETTLING CHAMBER

It is also called the dust catcher and is simplest of dust collection equipments. It consists of a chamber in which the velocity of dust laden gas is reduced to enable the dust particles settle down towards the bottom at a speed equal to their terminal velocities.

If the dusty gas remains in the chamber for a sufficient time, the particles reach the bottom from which they can be periodically removed.Reentrainment of particles from the bottom by the gas stream is prevented by keeping the gas velocity less than about 3 metres/sec. Otherwise the height of the settling chamber has to be increased. Industrial utility of settling chamber is limited to removing particles, larger than 43 μ diameter because for separating smaller particles, larger chamber is required.

For a given gas flow rate, the collection efficiency depends upon the cross-section of the chamber and is independent of the height. The height should be just large enough so that gas velocity in the chamber is not so high as to cause re-entrainment of separated dust. Minimum cross-sectional area of the chamber is dictated by the maximum permissible velocity of the air. The smaller the height, the shorter & wider the chamber and the smaller the volume. If the height is too small, it is difficult to remove the solids from the floor of the chamber.

The size of the chamber is independent of the dust load of the entering gas. Dust catcher is shown in Fig. 49.1.

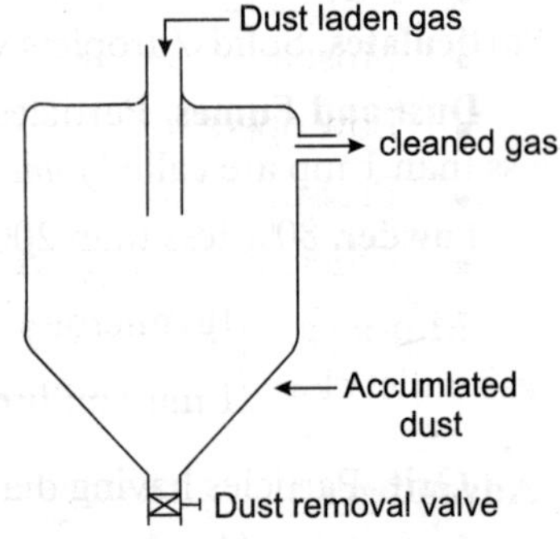

Fig. 49.1: Dust catcher

CYCLONE SEPARATOR

It consists of a vertical cylinder with a conical bottom, a tangential inlet near the top and an outlet for dust at the bottom of the cone. The outlet pipe is extended into the cylinder to prevent short circuiting of gas from inlet to outlet.

The dust laden gas introduced tangentially at the top of the cylindrical portion receives a rotating motion on entrance to the cylinder. The vortex thus forced develops a centrifugal force which throws the particles radially towards the wall which drop and collect to the bottom cone.

It utilises strong centrifugal force for the settling of particles instead of weak gravitational force as in the case of gravity settling chamber. Separation depends both on particle size and its density. Centrifugal force in a cyclone is from 5 times gravity (in large, low velocity units) to 2500 times gravity (in small high pressure units).

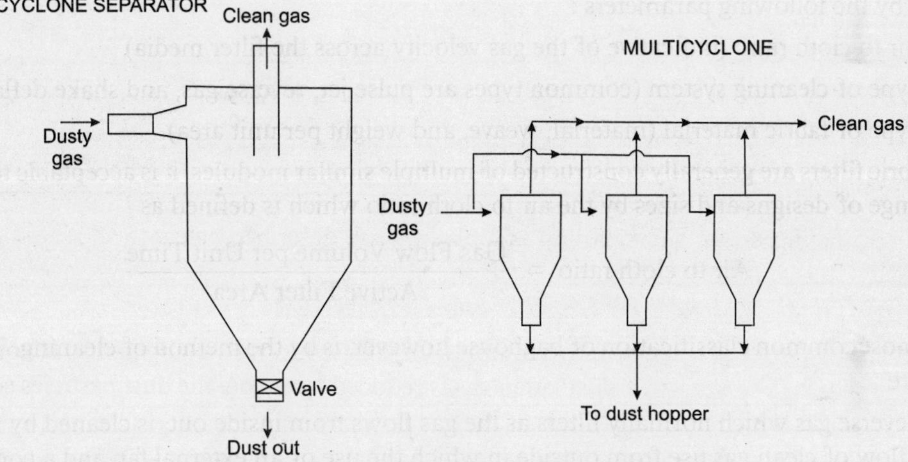

Fig.49.2: Cyclone separator and multicyclones

Cyclones can remove dust particles from liquids and gases and can operate at high pressure (500 atm) and temperature (1000°C). They are most suitable for separating particles larger than 5μ diameter from the gas. Cyclone separator and multicyclone is shown in Fig. 49.2.

- Advantages of cyclone separator:
- Most effective for 5-200 μ dia particles.
- Low operational and construction cost.
- Pressure drop is low
- Efficiency of dust removal is high *i.e.* efficiency = 90% for > 20 μ particles.

Multicyclones. Small cyclones arranged in parallel and delivering to a common dust hopper & gas outlet chamber is called multicyclones. In nutshell, it is a battery of cyclone.

Advantages of multicyclones are :

- Efficient removal of grit at constant loads with low pressure drop.
- High capacity.

- Defective ones can be isolated without affecting others hence easy maintenance and continuous operation of cleaner is possible. In case of single cyclone, the cleaning will be affected if the cyclone is isolated for maintenance.

Bag Filter. Here, dust laden gas is sucked into long cylindrical bags of cotton or wool fabrics. Dust removed remains inside the bag during filtering period. Then the gas flow is cut off at intervals and shake the bag so that dust will be discharged into collecting hopper. Household vacuum cleaner is a simple bag filter. Suction fan keeps the unit under vacuum. For < 5 μ, diameter particles its efficiency $\eta = 90\%$.

Bag fabric filters use a large surface area of filter fabric so that the flue gas can pass from the dirty to clean side at a relatively low velocity. This low velocity is necessary to minimize the pressure differential across the fabric and the layers of collected dust. This collected dust commonly referred to as filter cake actually contributes to the filtering effect of the fabric. Fabric filters are generally classified by the following parameters :

- Air to cloth ratio (indicative of the gas velocity across the filter media)
- Type of cleaning system (common types are pulse jet, reverse gas, and shake deflate)
- Type of fabric material (material, weave, and weight per unit area).

As fabric filters are generally constructed of multiple similar modules, it is acceptable to classify a wide range of designs and sizes by the air to cloth ratio which is defined as

$$\text{Air to cloth ratio} = \frac{\text{Gas Flow Volume per Unit Time}}{\text{Active Filter Area}}$$

The most common classification of baghouse however, is by the method of cleaning. The most applied are :

- Reverse gas which normally filters as the gas flows from inside out, is cleaned by reversing a flow of clean gas use from outside in which the use of an external fan and a complex set of dampers to reverse the flow path to each module as it is cleaned.
- Shake and deflate which utilizes the same logic as reverse gas plus the additional cleaning action of a bag shaking mechanism which operates during the cleaning cycle.
- Pulse jet which filters from the outside in and is constructed with an internal frame to prevent collapse of the bag under the resulting differential pressure. The individual bag or rows of bags within a module are cleaned by pulse or burnt of air down the length of the inside of the bag. The movement of the pulse down the bag provides a vigorous motion (decaying intensity down the length) which cleans the bag quite thoroughly.

The air to cloth ratio achieved at a constant pressure differential improves substantially for a puls jet versus a reverse gas. This is primarily due to the filter cake thickness achieved after many hours of operation with normal cleaning.

The less vigorously cleaned reverse gas builds a substantial amount of filter cake which then contributes to the low emissions but leads to the high pressure differentials and fabric filter areas for a given gas flow. Generally, the reverse gas installation has been chosen when availability and long-term maintenance costs are given strong consideration for larger units. Generally, the pulse jet is applied on smaller industrial or co-generation quality plants where first capital cost is the more important criteria.

The type of fabric material includes material composition, weave, weight, and finish. The choice of each of these parameters is dependant on the :

- Type of cleaning employed
- Flue gas composition and temperature
- Fly ash size and shape
- Availability and long-term cost consideration

Dust collection with fabric filters with bag lives of 1000 hours are typically achieved.

Precoating of the fabric filters with a powder material prior to start up can reduce blinding and the resultant pressure drop. Carefully sized limestone is suitable for this purpose.

Schematic arrangement of bag filters with different cleaning systems have been shown in Fig. 49.3.

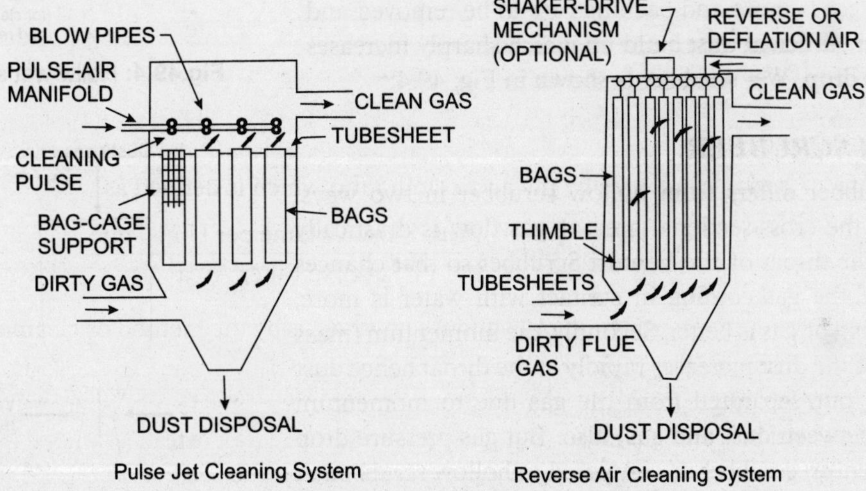

Fig. 49.3: Bag filter cleaning system

Limitation of Bag Filter :

- handles only dry air at low temperature, (< 250°C).
- large space requirement,
- high maintenance cost.

WET SCRUBBER

It cools as well as cleans the gas by direct counter current contact of dust laden gas by water spray. It is mainly of three types namely :

- hollow scrubber
- packed scrubber
- venturi scrubber.,

Hollow scrubber is a tall cylindrical vessel of large diameter in which water is sprayed from the top and the dusty gas ascends up counter currently. The spray of water forms a curtain throughout the diameter of the scrubber so that whole amount of gas does come in contact with water and the dust gets scrubbed (mechanically removed) out. Water is sprayed at 3 to 4 altitudes (heights) in the scrubber so that proper contact of gas and water droplets is ensured. Usually, dust particles of > 5 micron dia are removed by water droplets of dia 100 microns. Small particles need smaller diameter water droplets for its effective removal.

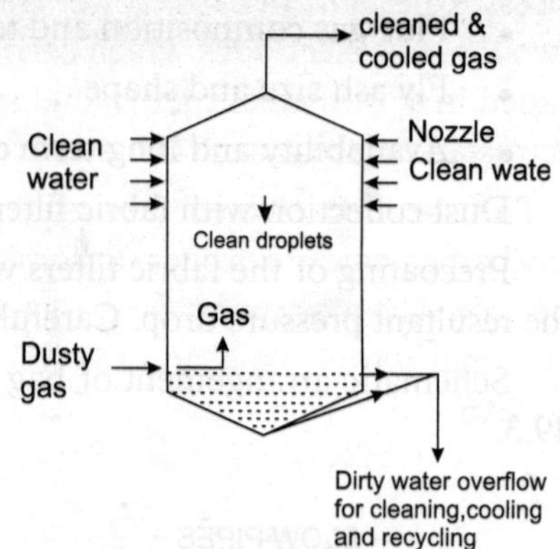

Fig.49.4: Hollw wet scrubber.

Dusty gas Packed scrubber works on the same principle as hollow scrubber. In this, packing are used in the scrubber to increase the contact area and chances between dusty gas and water for better cleaning and efficiency. But the pressure drop of the gas is more and packings are to be removed and cleaned after sufficient dust build up which sharply increases the pressure drop. Wet scrubber is shown in Fig. 49.4.

VENTURI SCRUBBER

Venturi scrubber differs from hollow scrubber in two ways. One is that the cross-sectional area of gas flow is drastically reduced at the throat of the Venturi Scrubber so that chances of whole of the gas coming in contact with water is more, hence cleaning of gas is better. Secondly, the momentum (mass x velocity) of the dust increases rapidly at the throat hence dust gets thrown out/separated from the gas due to momentum difference (between dust and gas) also. But gas pressure drop is more in venturi scrubber as compared to hollow scrubber. A venturi scrubber is shown in Fig. 49.5.

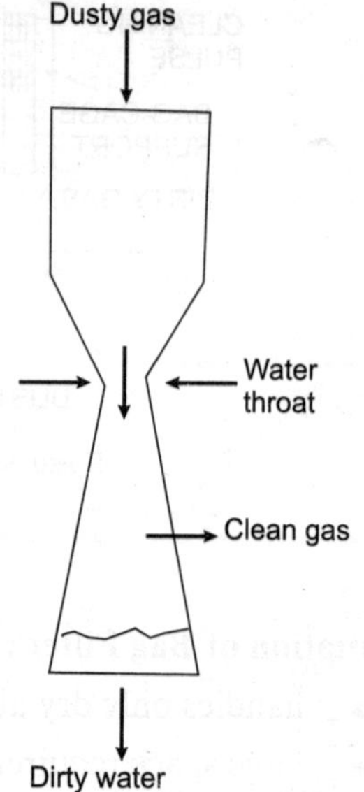

Fig.49.5: Hollow wet scurubber

ELECTROSTATIC PRECIPITATOR

It is very effective for the complete removal of ultrafine dust particles tides present in the gas. It is the final dust cleaning equipment in any Venturi scrubber, gas cleaning plant.

It consists of a hollow cylinder in which many steel tubes are vertically installed and electrically insulated with the body of the cylindrical portion. Inside each tube the steel wires hang centrally which is called wire electrode/negative electrode. The tube acts as a positive electrode as it is earthed. The dusty gas passes up through the tubes. A potential difference of 60—80 kilovolts D.C. is supplied between the tube electrode (*i.e.* positive electrode) and wire electrode (negative electrode) which results in 'corona' discharge (silent and sparkless discharge).

When the dust laden gas passes through the electrically charged tubes, the dust gets ionised *i.e.*, acquires negative charge and hence is attracted towards positively charged tube. On reaching the tube its charge gets neutralised and it sticks to the tube surface. This separated dust sticking to the tube can be removed by dry or wet method. In dry electrostatic precipitator, the tubes are periodically shaken by vibrators so that the dust falls down and can be recovered & reused in the process if needed. In case of wet electrostatic precipitator, a continuous thin film of water flows down the tube electrodes which carries aways the dust and the muddy water thus produced can be cleaned in a 'dorr thickener (radial settling tank) & further chemically treated for recycling in the electrostatic precipitator. The dust free gas goes out from the top.

The cleaning efficiency of electroprecipitator can be further increased by having a twisted wire/barbed wire electrodes. In some precipitators, plate type of +ve electrodes are used instead of tube type. The electroprecipitator is used for removal of tar from Coke Oven gas, flue gas cleaning in thermal power plant and final cleaning of B.F. gas. An electrostatic precipitator is shown in Fig. 49.6.

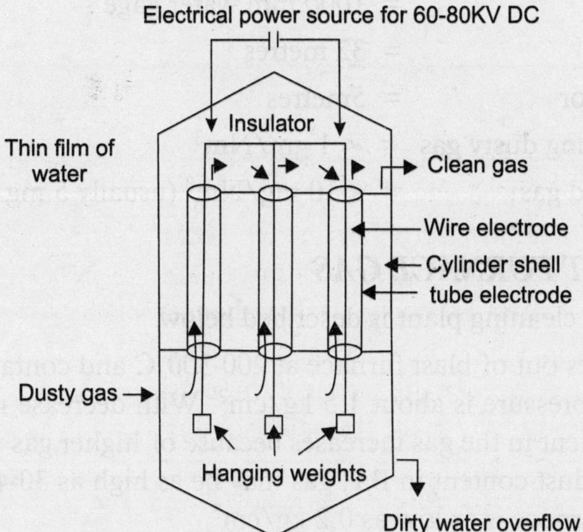

Fig.49.6: Wet electrosatic precipitator

Advantages of electrostatic precipitator:

- High efficiency for < 4 micron dust (η= 90% for < 4 μ and 98% for > 30 μ).
- Low pressure drop
- No moving parts
- Low maintenance cost
- Dust may be removed dry and used/sold/recycled
- Cleaning efficiency is more in wet type than in dry type.

Disadvantages:

- High capital cost of equipment and high space requirement
- Low efficiency during overload due to high velocity of dusty gas
- High electrical power requirement.

Typical Parameters of a Wet Electrostatic Precipitator (used in B F J[1]. Gas Cleaning Plant):

No. of tubes/wire electrodes = 316
Dia of mild steel tube electrodes = 245 mm
Dia of steel wire electrodes = 3 mm
Length of electrodes = 4.5 metres
Voltage applied in the electrodes = 60000 volts B.C.
Current applied to the electrodes = 250 mA
Gas Cleaning capacity = 120000 Nm^3 gas/hr
Gas pressure inside the electrostatic precipitator
= 1000 mm water gage
Height of precipitator = 32 metres
Diameter of precipitator = 5metres
Dust content in incoming dusty gas = < 1 gm/Nm^3
Dust content in cleaned gas = < 10 mg/Nm^3 (usually 5 mg/Nm^3).

CLEANING OF BLAST FURNACE GAS

A typical blast furnace gas cleaning plant is described below.

Blast furnace gas comes out of blast furnace at 200-300°C and contains about 10-15 gm/Nm^3 of dust when the top gas pressure is about 1.5 kg/cm^2. With decrease in top gas pressure of the blast furnace, the dust content in the gas increases because of higher gas velocity and hence higher entrainment of dust. The dust content in B.F, gas may be as high as 30-40 gm/Nm^3 when the top gas pressure of the blast furnace is as low as 0.2 kg/cm^2.

The gas first comes to a dry gravity settling chamber (dust catcher) where coarser particles settle down and get separated. Dust catcher is emptied periodically (once in a day or so depending on dust build up) to avoid high dust accumulation which may affect its dust separation efficiency. Dust concentration in gas going out of dust catcher is about 5 gm/Nm^3.

The gas then comes to a hollow wet scrubber where water is sprayed down counter current to the flow of the gas. Water sprayers cover the whole diameter of the scrubber so that uniform curtain of water droplets are spread in the scrubber and all the gas comes in contact with water spray. Water is sprayed at four different altitude in the scrubber so that no part of the gas remains unscrubbed with water. The dust is removed by contact with water and the gas is cooled to room temperature (say 35°C). The dust content in the gas after the scrubber is about 1 gm/Nm^3. Muddy water over flowing from scrubber is recycled after its treatment and cooling.

The gas is further scrubbed with water in the venturi scrubber whose throat diameter is very less (typically 625 mm). Here the velocity of dust laden gas is increased and the dust being heavier, its momentum (mass x velocity) is more than that of the gas and hence the dust is separated due to momentum difference. Also at narrow throat of venturi scrubber, the chances of all the gas coming in contact with water is more, hence, a part of the dust is removed by scrubbing action also.

The gas then passes through a throttle assembly where its pressure is reduced from 1.5 kg/cm^2 to 0.1 kg/cm^2 (because the clean gas is sent through the pipelines for use as a fuel at lower pressure of 0.1 kg/cm^2). In some modern plants (*e.g.* in Vizag Steel Plant in India) the pressure reduction is done in a top pressure recovery turbine (TRT) where some electrical power (roughly 20 kWh per ton of iron produced in B.F.) is also generated in the process. The reduced pressure B.F. gas (at 0.1 kg/cm^2) is sent to a wet electrostatic precipitator for final cleaning of the gas. The electrostatic precipitator removes the ultrafine particles from B.F. gas and the dust content in the clean gas comes below 10 mg/Nm^3. The cleaned B.F. gas is then sent to the gas holder/distribution pipelines for use in steel plant furnaces as a fuel gas.

The schematic diagram of a typical B.F. gas cleaning plant is given in Fig. 49.7 and its important parameters are given below :

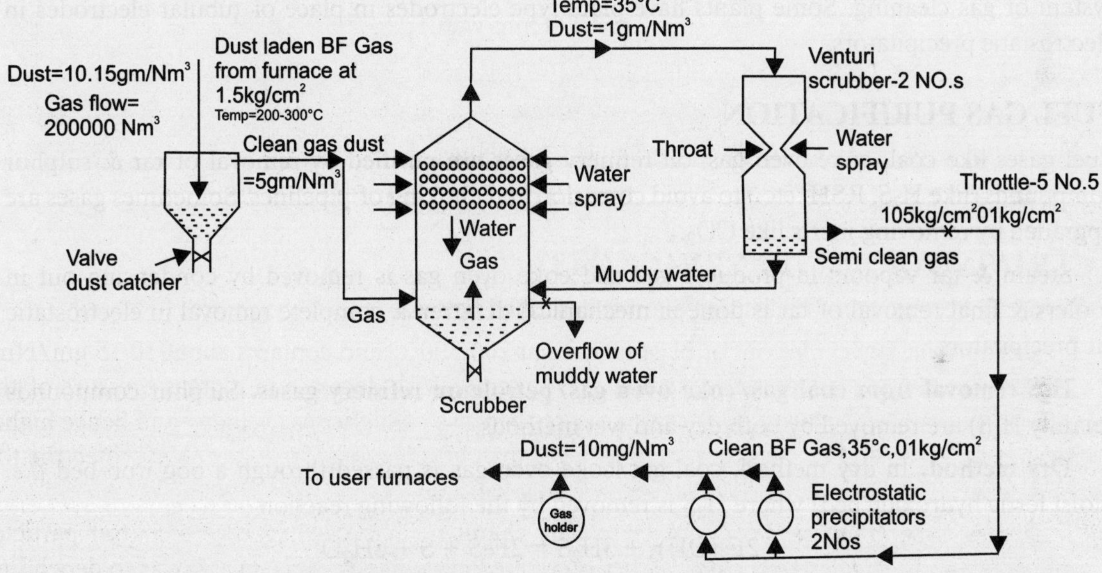

Fig.49.7: Schematic diagram of a typical blast furnace gas cleaning plant.

Typical parameters of a B.F. gas cleaning plant:

Gas cleaning capacity	= 200000 Nm^3/hr
Dia of scrubber	= 8 metres
Number of sprayer in scrubber	= 80 (20 each at four altitudes)
Height of scrubber	= 37 metres
Throat dia of venturi scrubber	= 625 mm,
Number of venturi scrubbers	= 2
Secondary voltage in electrostatic precipitator	= 50000 volts DC.
Height of each electrostatic precipitator	= 32 metres.
Number of tubes in each electrostatic precipitator	= 316

Specific water consumption for gas cleaning, $= 6.5\ m^3/1000\ Nm^3$ gas

Dusty gas parameters : dust content in raw B.F. gas $= 10\text{-}15\ gm/Nm^3$

Temperature of raw B.F. gas $= 200 - 250°C$

Pressure of raw B.F. gas $= 1.5\ kg/cm^2$

Clean gas parameters : dust content in cleaned B.F. gas $= < 10\ mg/Nm^3$

Temperature of clean B.F. gas $= 35°C$

Pressure of clean B.F. gas $= 0.1\ kg/cm^2$

In some of the B.F. gas cleaning plants, packed scrubber (having stacked packing of rasching rings) without venturi scrubber is used. Modern plants use top pressure recovery turbine which generates power as well apart from reducing the gas pressure. This provides a more energy efficient system of gas cleaning. Some plants have plate type electrodes in place of tubular electrodes in electrostatic precipitators.

FUEL GAS PURIFICATION

Fuel gases like coal, coke oven gas, oil refinery gases are purified by removal of tar & sulphur compounds (like H_2S, RSH, etc.) to avoid corrosion and choking of pipelines. Sometimes gases are upgraded by removing inerts like CO_2.

Steam & tar vapours in producer gas and coke oven gas is removed by condensing out in coolers & final removal of tar is done in mechanical detarrer & complete removal in electrostatic tar precipitators.

H_2S removal from coal gas/coke oven gas/petroleum refinery gases. Sulphur compounds (mainly H_2S) are removed by both dry and wet methods.

Dry method. In dry method, coal gas/coke oven gas is passed through a bog iron bed (*i.e.* moist ferric hydroxide bed) where H_2S is removed by the following reaction :

$$2Fe(OH)_3 + 3H_2S = 2FeS + S + 6H_2O$$

The bog iron bed is regenerated by adding small quantity of air to bed as per following reaction.

$$4\ FeS + 3O_2 = 2Fe_2O_3 + 4S$$

Regeneration of iron sulphide proceeds until 30% S is present in the bed. The sulphur in the spent oxide is converted to sulphuric acid *via* sulphur dioxide.

Wet method. In wet method, gas is scrubbed with a suitable reagent which removes H_2S by absorption. The spent reagent is regenerated by oxidation (air blowing) or boiling. Reagents used in some wet H_2S removal processes are given below.

Thylox Process. The gas is washed in scrubbers with a solution of ammonium thioarsenate. The reaction is :

$$(NH_4)_3\ AsO_2S_2 + H_2S \longrightarrow\!\wedge\ (NH_4)_3AsOS_3 + H_2O$$

The solution of sulphur rich thioarsenate is regenerated by blowing air, followed by filtration to remove sulphur as per following reactions :

$$2(NH_4)_3ASOS_3 + O_2 = 2\ (NH_4)_3ASO_2S_2 + S$$

Seaboard Process. The gas is scrubbed by a solution of sodium carbonate. The hydrogen sulphide is dissolved and is subsequently removed, unchanged, by blowing air through the solution.

Ferricyanide Process. Sodium ferricyanide is used as scrubbing medium. This reacts with hydrogen sulphide to form the ferrocyanide, which is then oxidized to reform ferricyanide, precipitating sulphur. The sulphur is recovered by filtration.

Manchester Process. The gas is scrubbed with a suspension of ferric hydroxide in a dilute solution of ammonia or sodium carbonate. The concentration of iron in solution is maintained by addition of ferrous sulphide. Sulphur is recovered by blowing air through the liquor.

Shell Phosphate Process. A solution of tri-potassium phosphate removes H_2S from gases. The hydrogen sulphide is regenerated by boiling, and sulphur is recovered in a kiln.

$$K_3PO_4 + H_2S \longrightarrow\hat{} K_2HP0_4 + KHS$$

The plant consists of a packed tower for absorption, under pressure, and a "bubble-cap" tower for regenerating with steam.

The reagent it stable, non-corrosive to mild steel, and solution loss is negligible.

Girbotol Process. Aqueous solutions of aliphatic amines dissolve H_2S and CO_2 from gases. The dissolved gases may be recovered by boiling. Ethanolamine is used in the Girbotol process. Diethanolamine is the preferred reagent, since monoethanolamine reacts irreversibly with CO_2 and triethanolamine has low absorptive capacity (though H_2S is absorbed preferentially).

$$(CH_2OH\ CH_2)_2\ NH + H_2S \longrightarrow [(CH_2OH\ CH_2)_2NH_2]HS$$

The solution used contains 15 to 30% of diethanolamine. Absorption takes place in packed towers and regeneration is by steam in a bubble-cap tower.

Carbon Disulphide Removal. This can be removed from coal gas or petroleum oil gases by the Carpenter-Evans process. The gas is passed over nickel at 450°C, where it is converted to hydrogen sulphide and is removed by any of the above processes.

Lurgi Coal Gas. The gas is obtained under pressure having H_2S with CO_2, is readily removed by scrubbing with water, or tripotassium phosphate, as in the Shell process.

50

Rocket Fuels

Unlike a conventional piston engine or gas turbine, a rocket motor which is a type of internal combustion engine carries with it both the fuel & oxidant and hence is independent of atmospheric air supply for combustion of fuel. A rocket can operate in any medium and even in a vacuum because it is not dependent on atmospheric air for combustion. During operation of rocket engine, high temperature and high pressure gases are produced due to a set of chemical reactions taking place due to combustion of rocket fuel with the oxidant. And when these gases are ejected through a nozzle, they exert an enormous thrust (upward force) on rockets for its propulsion (flight into space).

Both the fuel and the oxidant are called 'rocket propellant'. A good rocket propellent should have :

- high value of specific impulse
- high value of density impulse
- easy availability
- stability
- high degree of safety in its use
- easier handling characteristics

Depending upon the design of the rocket, the propellants may be of many types like mono or bi-propellants, solid or liquid propellants, self-igniting or non-self-igniting propellants. Gaseous propellants are used after liquefaction owing to the consideration of bulky storage capacity for gases.

Specific impulse of a rocket propellant system means the amount of thrust (in Kgf) per kg of the propellant consumed per second. High value of specific impulse is required because the rocket carries its own propellant which has high weight. The source of the thrust is the volume of gas moving at high velocity through a nozzle.

Specific impulse of the rocket increases with :

- increase in the combustion chamber temperature
- decrease in the average molecular weight of the exhaust gases.

The normal specific impulse range of a liquid propellant system operating at 20 kg/cm^2 is 225-370 kg-seconds/kg. The same for solid propellants are generally lower in the range of 170-230.

High density impulse of the propellant is desirable as more quantity of propellant can be carried in a given storage space. Density impulse is given by the product of specific impulse and the density of unburned propellant system.

Some examples of rocket propellants (Fuel + oxidant) are as follows :

PROPELLANTS

Fuel

(a) *Solid* — Synthetic plastics and rubbers, phenol-furfural, asphalt, phenol formaldehyde etc.

(b) *Liquid* — Kerosene, aniline, methyl alcohol, gasoline ethyl alcohol, nitromethane, boron hydride, hydrazine hydrate.

Oxidant

(a) *Solid* - Nitrates of sodium, potassium & ammonium, potassium perchlorate, ammonium perchlorate

(b) *Liquid* - Hydrogen peroxide, liquid oxygen, red & white fuming nitric acid

Though solid propellants produce lower specific impulse compared to the liquid propellants, it has comparative advantages like :

- ease and rapidity of loading onto launchers and firing
- ease of installation
- lesser sources of failure

When the fuel and the oxidant are kept in a single substance or compound it is called a *monopropellant e.g. hydrogen peroxide,* nitromethane or a homogeneous mixture of several compounds like alcohol mixed with hydrogen peroxide. Monopropellants being stable under atmospheric condition, decompose on heating and pressurisation to give hot combustion gases which when ejected through nozzle provide upward thrust for rocket movement in upward direction. However, most of the successful liquid propellant rocket motors use *bipropellants* which has two separate propellants which are mixed inside the combustion chamber only and not outside. In vary exceptional cases, three or more liquid propellants are used in rockets.

Theoretically speaking, atomic hydrogen (having lowest atomic weight) at > 3000°C should be the best rocket fuel because of its highest specific impulse and density impulse, but it is practically not possible to get appreciable quantity of atomic hydrogen at such a high temperature. The next best rocket fuel is the liquid hydrogen but it has a very low density & very low boiling point hence it won't generate greater amount of specific and density impulses. The more practical rocket fuels are the hydrogen. containing compounds which are less volatile than hydrogen but of higher density and which on combustion produces flue gases of lower molecular weight.

The most powerful liquid propellant in regular practical use is pure ethyl alcohol-liquid oxygen system. Though hydrazine gives higher specific impulse than gasoline and ethyl alcohol but it is expensive. A typical propellant used in a German fighter rocket is hydrazine-(methyl alcohol-hydrazine hydrate-water in 57 : 30 : 13 ratio).

Boron hydride gives very high specific impulse but it is prohibitively costly.

Aniline is a good rocket fuel but with a lesser specific impulse than alcohol and gasoline.

Most widely used liquid oxidant in rocket propulsion is liquid oxygen because :

- it gives highest specific impulse compared to any other oxidant
- it is safe and convenient to store and handle
- it is neither toxic nor corrosive
- it is produced commercially at a low cost
- its low boiling point is more than compensated by many practical advantages.

Nitric acid and hydrogen peroxide are also very common oxidants having high specific impulses; but while nitric acid is toxic, hydrogen peroxide has a poor stability.

51

THERMAL INSULATION

Thermal insulation is defined as a material or combination of materials which retard the flow of heat.

The materials can be adapted to any size, shape or surface. A variety of finishes are used to protect the insulation from mechanical & environmental damage, and to enhance appearance. Thermal insulators have air or gas filled pockets or void spaces that retard the transfer of heat effectively.

For industrial facilities, such as steel plants, power plants, oil refineries, paper mills etc , thermal insulations are installed to control heat gain or heat loss on process piping & equipment, steam & condensate distribution systems, boilers, smoke stacks, bag houses & electro precipitators, and storage tanks.

The most important characteristics of any insulation material include a **low thermal conductivity, low tendency toward absorbing water, and of course the material should be inexpensive.** In the process industry, the most common insulators are various types of **calcium silicate or fiberglass.** Calcium silicate is generally more appropriate for temperatures above 225 °C, while fiberglass is generally used at temperatures below 225 °C.

Principle of Insulating Materials

Insulators are poor conductors of heat and have low conductivity. Insulating materials are porous containing a large number of air cells. The resistance to heat transfer by the insulating materials would depend on the number of dormant air cells packed in the mass. The air cell would be dormant if its diameter is no larger than the mean free path which, for all practical purposes, could be taken as 0.09 microns. This would be possible, if insulating materials could be produced with very low fibre diameters and space between the fibres be so compressed as to be equal or less than the mean free path. Larger air cells are not dormant, and thus convection currents are set up, due to which air temperature rises, this in turn increases the air pressure and consequently the rate of flow. This means that for higher temperatures, high density materials should be used.

Heat is transferred through insulation by anyone or more, of the following modes:

- Solid conduction through matrix or fibres
- Gaseous conduction
- Radiation across gas or air space

- Convection across gas or air space
- Overall convection from the hot face to the cold face

Materials vary widely in heat conducting properties. This is expressed by the term thermal conductivity. Thermal conductivity is the amount of heat transferred in unit time through difference across the face. Thermal conductivity of a given material usually increases with rise in temperature.

Conductivity of an insulating material varies with its density , Porosity , Pore size, (in the case of fibrous materials) , fibre diameter and shot content (i.e. non-fiberized particles of rock or glass wool).

Functions of Thermal Insulation

Insulation is used to perform one or more of the following functions:

- Reduce heat loss or heat gain to achieve energy conservation.
- Protect the environment through the reduction of CO2, NOx and greenhouse gases.
- Control surface temperatures for personnel and equipment protection.
- Control the temperature of commercial & industrial processes.
- Prevent or reduce condensation on surfaces.
- Increase operating efficiency of heating/ventilation/cooling, plumbing, steam, process and power systems.
- Prevent or reduce damage to equipment from exposure to fire or corrosive atmospheres.
- Reduce noise from mechanical systems .

Benefits of thermal insulation:

They are described below .

(a) Energy savings

Substantial quantities of heat energy are wasted daily in industrial plants because of uninsulated under maintained or under insulated heated or cooled surfaces. Properly designed and installed insulation systems will immediately reduce the need for energy. Benefits to industry include cost savings, improved productivity & improved environmental quality.

(b) Process control

By reducing heat loss or gain, thermal insulation can help maintain process temperature to a predetermined value or within a predetermined range. The insulation thickness must be sufficient to limit the heat transfer in a dynamic system or limit the temperature change, with time, in a static system. The need to provide time for owners to take remedial action in emergency situations in the event of loss of electrical power, or heat sources is a major reason for this action in static systems.

(c) Condensation control

Specifying sufficient insulation thickness with an effective vapor retarder system is the most effective means of providing a system for controlling condensation on the membrane surface and within the insulation system on cold piping, ducts, chillers & roof drains. Sufficient insulation thickness is needed to keep the surface temperature of the membance above the highest possible design dew point temperature of the ambient air so condensation does not from on the surface. The effective vapor retarder system is needed to restrict moisture migration into the system through the facing, joints, seams, penetrations, hangers, and supports. By controlling condensation, we control the potential for degrading system service life & performance, corrosion of pipes, valves & fittings caused by water collected & contained within insulation system.

(d) Personnel protection

Thermal insulation is one of the most effective means of protecting workers from burns resulting from skin contact with surfaces of hot piping and equipment operating at high temperatures . Insulation reduces the surface temperature of piping or equipment to a safer level resulting in increased worker safety and the avoidance of worker downtime due to injury.

(e) Fire protection

Used in combination with other materials, insulation helps provide fire protection .

Firestop systems designed to provide an effective barrier against the spread of flame, smoke, gases at penetrations of fire resistance rated assemblies by ducts, pipes & cable .

(f) Sound Attenuation

Insulation materials can be used in the design of an assembly having a high sound transmission loss to be installed between the source and the surrounding area. Sometimes, insulations with high sound absorption characteristics may be used on the source side of an enclosure to help lower the exposure to people to the reduction of the noise level on the other side of the enclosure.

(g) Aesthetics

Most mechanical insulation systems in commercial construction are not generally visible to the occupants of the building. The common exceptions to this are found in mechanical equipment rooms where the heating equipment, cooling equipment, and the associated piping are visible to the personnel who work or otherwise must access these areas. It is common practice to require a finished and neat appearance for insulation surfaces that are visible within the building envelope. These surfaces may also be painted or covered for a more acceptable appearance in the case of hospitals, schools, supermarkets, restaurants and even in industrial facilities in food processing, and computer component manufacturing where visible to the occupants.

(h) Greenhouse Gas Reduction

Thermal insulation for mechanical systems provides immediate reductions in CO2,NOx and other greenhouse gas emissions to the outdoor environment in stack emissions by reducing fuel consumption required at the combustion sites because less heat is gained or lost by the system.

The basic requirement for thermal insulation is to provide a significant resistance path to the flow of heat through the insulation material. To accomplish this, the insulation material must reduce the rate of heat transfer by conduction, convection, radiation, or any combination of these mechanisms.

This provides information on three general types of insulation:

Mass insulation with air or another gas with thermal properties similar to air within the interstices inside the material. Many cellular insulation and all fibrous and granular insulations are of this type.

Mass insulation with low conductivity gas within the interstices inside the material. Some closed cellular insulation are of this type.

Reflective insulation bounding one or both sides of an air space , many insulation facings such as FSK (foil/scrim/kraft) are of this type.

Heat is transferred by radiation, conduction and convecton. Radiation is the primary mode and can occur even in vacuum. The amount of heat transferred for a given area is relative to the temperature differential and emissivity from the radiating to the absorbing surface.

Emissivity is a measure of the ability of a material to radiate energy. It is expressed as a ratio of the radiating ability of a given material to that of a black body. A black body emits radiation at the maximum possible rate at any given temperature, and has an emissivity of 1.0. The emissivity values (k) for various materials can be obtained from heat transfer data tables.

TYPES OF INSULATION

Low Temperature Insulations (<90°C)

This range coves insulating materials for refrigerators, cold & hot water systems, storage tanks, etc. the commonly used insulation materials are Cork, Wood, 85% magnesia, mineral fibers, polyurethane , expanded polystyrene, etc.

Medium Temperature Insulations (90-325°C)

Insulation in this range are used in low temperature, heating & steam raising equipment, steam lines, flue ducts etc. the types of materials used in this temperatures range include 85% magnesia, asbestos, calcium silicate and mineral fibers etc.

High Temperature Insulation (above 325°C)

Typical uses of such insulation materials are super heated steam system, oven dryer and furnaces etc. The most extensively used materials in this range are Asbestos, Calcium Silicate, Mineral Fibre, Mica & vermiculite based insulation, Fireclay or Silica based insulation , Ceramic Fibre etc.

The following table describes the characteristics & applications of various insulating materials

Table 51.1: Types of thermal insulation materials and their applications

S.NO.	TYPE OF INSULATION MATERIALS	APPLICATIONS	REMARKS
1.	**Polystyrene** An organic foam made by polymerizing styrene	Suitable for low temperatures (-167 to 82 °C). mainly used in cool concrete retaining structures	Advantages – Rigid and lightweight Disadvantages – Combustible, has a low melting point, is UV (ultraviolet) degradable, and susceptible to attacks by solvents
2.	**Polyurathane** Made by reacting iscoyanates and alcohols. Made in continuous slab or foamed in situ.	Suitable for low temperature (-178 to 4°C). mainly used in cool rooms, refrigerated transports, deep freezing cabinets, refrigeration piping and floor & foundation insulation	Advantages – closed cell structure, low density and high mechanical strength Disadvantages – combustible, produces toxic vapours and has a tendency to smoulder
3.	**Rockwool (mineral fibre)** Manufactured by melting basalt and coke in a cupola at about 1500°C. phenolic binders used.	Suitable for temperatures up to 820°C , mainly used to insulate industrial ovens, heat exchangers, driers, boilers and high temperature pipes.	Has a wide density range and is available in matts, blankets, loose form or preformed for pipe insulation. It is chemically inert, non- corrosive and maintains mechanical strength during handling .
4.	**Fibreglass** Formed by bonding long glass fibres with a thermo setting resin to form blankets and batts, semi rigid boards, high density rigid boards and preformed pipe sections.	Suitable for temperature up to 540°C. mainly used to insulate industrial ovens, heat exchangers, driers, boilers and pipework	Will not settle or disintegrate with ageing. Fiberglass products are slightly alkaline (pH=9) .It should not promote or accelerate the corrosion of steel, provided it is protected from external contamination
5.	**Calcium silicate** Made from anhydrous calcium silicate material reinforced with a non-asbestos binder. Available in slab form of various sizes.	Suitable for temperature up to 1050°C. mainly used to insulate furnace walls, fire boxes, back-up refractory, flue lining and boilers	Has a minute air cell structure, has a low thermal conductivity and will retain its size and shape in its useable temperature range. It is lightweight, but has good structural strength ; so it can withstand mechanical abrasion. It will not burn or rot, is moisture resistant and non-corrosive
6.	**Ceramic fibre** Made from high purity alumina and silica grains, melted in an electric furnace and blasted by high velocity gases into light fluffy fibres. Made in a variety of forms.	Suitable for temperature up to 1430°C. mainly used to insulate furnace and kiln back-up refractory, fire boxes, glass feeder bowls, furnace repair, induction coil insulation, high temperature gaskets and wrapping material	Suitable for many applications because of the variety of forms. These include cloth, felt, tape, coating cements and variform castable (fire brick)

Definition of terms related to thermal insulation:

Insulations have different properties and limitations depending upon the service, location, and required longevity of the application. These are taken into account when considering the insulation needs of an industrial or commercial application. Various properties of thermal insulation are described below:

(a) Thermal resistance (R) (°c. m^2..hr/cal)

The quantity determined by the temperature difference, at steady state, between two defined surfaces of a material or construction that induces a unit heat flow rate through a unit area. A resistance associated with a material shall be specified as a Material R. A resistance associated with a system or construction shall be specified as a system R.

(b) Apparent Thermal Conductivity (ka) (kcal .m/hr. m^2..°C)

A thermal conductivity assigned to a material that exhibits thermal transmission by several modes of heat transfer resulting in property variation with specimen thickness or surface emittance.

(c) Thermal conductivity (k) (k-cal m./hr.m^2. °C)

The time rate of steady state heat flow through a unit area of a homogenous material induced by a unit temperature gradient in a direction perpendicular to that unit area. Materials with lower k factors are better insulators.

(d) Density (kg/m^3)

This is the weight of a specific volume of material measured in kilograms per cubic meter.

(e) Surface Burning Characteristics

These are comparative measurements of flame spread and smoke development with that of select red oak and inorganic cement board. Result of this test may be used as elements of a fire risk assessment, which takes into, account all of the factors, which are pertinent to an assessment of the fire hazard or fire risk of a particular end use.

(f) Compressive Resistance

This is a measure of the material to resist deformation (reduction in thickness) under a compressive load. It is important when external loads are applied to an insulation installation. Two examples are deforming the insulation on a pipe at a Clevis type hanger due to the combined weight of the pipe and its contents between the hangers and ....resistance of an insulation to compress on an outdoor rectangular duct due to heavy mechanical loads from external source such as wind, snow, or occasional foot traffic.

(g) Thermal Expansion/Contraction and Dimensional Stability

Insulation systems are installed under ambient conditions that may differ from service conditions. When the operating conditions are imposed, metal surfaces may expand or contract differently from the insulation and finish applied. This can create openings and parallel heat flow & moisture flow paths that can degrade system performance.

Long term satisfactory service requires that the insulating materials, closure materials, facings, coating, and accessories withstand the rigors of temperature, vibration, abuse, and ambient conditions without adverse loss of dimensions.

(h) Water Vapour Permeability

This is the time rate of water vapour transmission through unit area of flat material of unit thickness induced by unit vapour pressure difference between two specific surfaces, under specified temperature and humidity conditions. It is important when insulation system will be operating with service temperatures below the ambient air. Materials and systems with low water vapor permeability are needed in this service.

(i) Cleanability

Ability of a material to be washed or otherwise cleaned to maintain its appearance.

(j) Temperature Resistance

Ability of a material to perform its intended function after being subjected to high & low temperatures, which the material might be expected to encounter during normal use.

(k) Weather Resistance

Ability of a material to be exposed for prolonged periods of time to the outdoors without significant loss of mechanical properties.

(l) Abuse Resistance

Ability of a material to be exposed for prolonged periods of time to normal physical abuse without significant deformation or punctures.

(m) Ambient Temperature

The dry bulb temperature of surrounding air when shielded from any sources of incident radiation.

(n) Corrosion Resistance

Ability of a material to be exposed for prolonged periods of time to a corrosive environment without significant onset of corrosion and the consequential loss of mechanical properties.

(o) Fire Resistance/Endurance

Capability of an insulation assembly exposed for a defined period of exposure to heat & flame (fire) with only a limited and measurable loss of mechanical properties. Fire endurance is not a comparative surface burning characteristic for insulation materials.

(p) Fungal Growth Resistance

Ability of a material to be exposed continuously to damp conditions without the growth of mildew or mold.

Insulation Types, Forms & Finishes

(a) Mass Insulation Types

(i) Fibrous insulation

Composed of air finely divided into interstices by small diameter fibers usually chemically or mechanically bonded & formed into boards, blankets, and hollow cylinders.eg;

- Fiber glass or mineral fiber
- Mineral wool or mineral fiber
- Refractory ceramic fiber

(ii) Cellular insulation

Composed of air or some other gas contained within foam of stable small bubbles and formed into boards, blankets, or hollow cylinder eg;

- Cellular glass
- Elastomeric foam
- Phenolic foam
- Polyethylene
- Polyisocyanurates
- Polystyrene
- Polyurethanes
- Polyamides

(iii) Granular insulation

Composed of air or some other gas in the interstices between small granules and formed into blocks, boards, or hollow cylinders eg;
Calcium silicate

- Insulating finishing cements
- Perlite

(B) Forms of insulation

(i) Board

Rigid or semi-rigid self- supporting insulation formed into rectangular or curved shapes eg

- Calcium silicate
- Fiber glass or mineral fiber
- Mineral wool or mineral fiber
- Polyisocyanurates
- Polystyrene

(ii) Block

Rigid insulation formed into rectangular shapes. eg;

- Calcium silicate

- Cellular glass
- Mineral wool or mineral fiber
- Perlite

(iii) Sheet

Semi-rigid insulation formed into rectangular pieces or rolls. eg:

- Fiber glass or mineral fiber
- Elastomeric foam
- Mineral wool or mineral fiber
- Polyurethane

(iv) Flexible Fibrous Blankets

A flexible insulation used to wrap different shapes and forms. eg:

- Fiber glass or mineral fiber
- Mineral wool or mineral fiber
- Refractory ceramic fiber

(v) Pipe & Fitting Insulation Pre formed insulation to fit piping, tubing & fittings eg;

- Calcium silicate
- Cellular glass
- Elastomeric foam
- Fiber glass or mineral fiber
- Mineral wool or mineral fiber
- Perlite
- Phenolic foam
- Polyethylene
- Polyisocyanurates
- Polyurethanes

(vi) Foam

Liquid mixed at the time of application which expands and hardens to insulation irregular areas and voids , eg;

- Polyisocyanurates
- Polyurethane

(vii) Spray Applied Insulation

Liquid binders or water introduced to an insulation while spraying on to flat or irregular surfaces for fire resistance, condensation control, acoustical correction and thermal insulation , eg;

- Mineral wool or mineral fiber

(viii) Loose fill Granular insulation used for pouring expansion joints eg;

- Perlite
- Vermiculite

(ix) Cements (Insulating and Finishing Muds)

Produced with mineral wool and clay insulation, these cements may be hydraulic setting or air drying types eg ;

- Flexible Elastomeric Foam
- Foam sheets and tubing insulation containing vulcanized rubber.

(c) Insulation Finishes

Insulation finishes are important because the ability of an insulation system to perform well is dependent upon protection from moisture, weather, chemical and mechanical damage. Insulation can also be used to enhance system appearance.

(i) weather Barriers

Protect the insulation from rain, snow, ice, sunlight, ultraviolet degradation, ozone and residues of chemical compounds in the atmosphere. eg;

- mastic
- metal
- plastic
- roofing felt

(ii) Vapour Retarders

Retards the passage of moisture vapor from the atmosphere to the interior of the insulation system , eg;

- CPVC
- FRP
- Laminated foil-scrim membranes
- Mastic
- Metal
- Plastic
- PVC
- Reinforced polyester resin

(iii) Mechanical abuse protection:

Rigid racketing provides protection against mechanical abuse from personnel, equipment, machineries etc. eg; metal & plastic.

(iv) Appearance choosen primarily for appearance value in exposed areas eg;

- Fabrics
- Laminate foil/scrim/membranes
- Painted metals and paints
- PVC

For mass insulation types, the most important physical property is thermal conductivity. Materials with low thermal conductivity allow less heat to be transferred per unit time, per unit temperature difference per inch of thickness. All other items being the same, materials with lower thermal conductivities are better insulations. Commercially available mass insulations have thermal conductivities at 75°F mean temperature less than 0.5 Btu in /(hr, S.F.,°F).

For reflective insulation types, the important physical property is low surface emittance. Surface with low emittance have high reflectance. Reflective insulation have emittance (emissive) values in the range of 0.04 to 0.1.

SELECTION OF PROPER THERMAL INSULATION

The owner, engineer, insulation contractor and insulation & accessories manufactures must communicate with each other in order to minimize problems in the stages of design, specification, preparation, construction, operations, and maintenance. Knowledge of the most important insulation criteria as detailed below, will aid the engineer in calculating the insulation thickness required for the intended service.

Insulations have different properties & limitations depending upon the service, location, and required longevity of the application. These are taken into account when considering the insulation needs of an industrial or commercial application.

The insulating material used for steam pipes should possess the following properties:-

It should have high insulating efficiency. The maximum heat loss from insulation pipe should not exceed 1 kcal/hr.m^2 °C.

Insulating efficiency=(bare surface loss-insulated loss)/bare surface loss.

Following points should be considered while selecting a proper thermal insulating material:

- It should have high mechanical strength so that vibrations and knocks will not adversely affect it. It should be able to with stand thermal shock and chemical attack.
- It should not be affected by moisture, heat, weather and chemical attack.
- It should not cause corrosion of pipes if chemically decomposed.
- The material should be easily applied or removed.
- It should be able to withstand the temperature to which it will be subjected.
- It should be stable and resist deterioration over the working life of the pipe.
- It should be easily moulded and applied.
- It should not overload the pipe by its dead weight.
- It should not be too costly. Total cost including the cost of thermal insulation material, its installation & maintenance cost thereafter should be considered.
- Thermal conductivity, diffusivity & permeability of insulation material must be considered in its evolution.
- It should have enough resistance of the material to retard the spread of flame.

INSULATION MATERIALS:

The materials most commonly used for steam pipe insulations are asbestos, magnesia, cork, hairfelt, woolfelt, rockwool and diatomaceous earths. Most commercial insulations are either built up from corrugated asbestos paper or laminated asbestos paper artificially roughened to produce air spaces or are moulded or felted with asbestos.

A very common and effective insulation for temperature up to 400° C is the moulded 85% magnesia (85% carbonate of magnesia and 15% binder). The insulation used for higher temperature should have inner layer of some special high temperature insulation as high temperature .decomposes the inner layer of magnesia. High temperature breaks down the magnesia into asbestos fibre & magnesium oxide and, therefore destroys cohesion between the pipe & insulation.

A layer of glass silk before giving the layer of magnesia is generally used for pipe insulation when the temperature is above 500°C glass silk has an advantage of cleanliness, is non-inflammable and can withstand vibrations and rough handling without losing its form or insulating efficiency. The packing density is varied between 100 to 150 kg/m^3 as per requirements.

Generally , steam pipes are lagged to a thickness of 8.5 cm with plastic magnesia, which is reinforced with galvanised wire netting and covered with 1.5 cm thick hard setting non-conducting material. The lagging finally covered by a wrapping of canvas and two coats of selected paints are given to the surface.

The small steam pipes should have the same insulation thickness as large ones as the heat loss per m^2 on a small pipes is higher than large pipe.

The insulation manufacturers publish 'insulation efficiency' data for different thickness of their different grades of insulation. Such tables generally provide the efficiency data against two variables as pipe size and temperature difference.

The amount of insulation to be applied is an economic problem. The cost of insulation must be weighed against the saving of heat energy obtained due to insulation.

The insulation materials can be classified into five groups according to the temperature range where it is going to be used.

(i) Low temperature insulations (-100 to 100 °C):- This range covers insulating materials for refrigerators, cold & hot water systems, storage tank etc. The commonly used materials are blankets and rigid bond rock wood, cork, foam glass, polyurethane foam, polystyrene foam, fibre glass etc.

(ii) Medium temperature insulators (100 to 500 °C):- Insulators in this range are used in low temperature heat & steam raising equipments, steam lines, fuel ducts etc. The type of materials used in this temperature range includes 85% magnesia, Asbestos, Calcium Silicate and Mineral fibers, Reflective metal sheets etc.

(iii) High Temperature Insulators (500 ° C and above):- Typical use of such materials are super heated steam system, oven , dryer and furnaces etc. The most extensively used materials in this range are Calcium Silicate, mineral Fibre, Mica and Vermiculite based insulation, Fireclay or Silica base insulation and Ceramic Fibre etc.

(iv) Cryogenic Range (-270 to -100 °C):- In this temperature range vaccum wall enclosure, multiple metal reflective foil, opacified powders, low emissivity surfaces (Silver , Aluminium), polyurethane foam etc. are used.

(v) Personal protection:-The insulation is some times required to be bonded on different hot surfaces for personal protection. Generally calcium silicate and mineral wool is used for such insulation.

Features, Choice and Properties of various insulations / lagging materials

(a) Cork, Felt and straw:- These are natural and the cheapest lagging material available. Cork is the best of all materials due to its intricate natural structure. These materials are useful to temporary lagging as they can be conveniently wrapped around pipes with single strands of wireand string. They should not be used for surfaces hotter than 200°C as they are inflammable & get charred and brittle at moderate temperatures. Exposure to the weather is likely to destroy their insulating properties. Earlier, cork was used as an insulation for ice houses. Blocks of ice wear cut from frozen lakes in winter and stored in cork lined ice houses for use in summer. When mechanical refrigeration came in use, steam baked cork board was used to insulate pipes and equipments

(b) 85% Magnesia:- This is light calcined magnesium carbonate with 15% amosite asbestos, moulded into standard pipe sections and other shapes. It is excellent, robust and economical in operation, but requires protection from the weather. It should not be used on surfaces whose temperature is over 300°C . This disadvantage can be overcome by interposing a thin layer of special cement between magnesia and the hot surface. Use of magnesia as insulation has become obsolete these days. This material is similar to calcium silicate used today but has a lower maximum use temperature and contains asbestos fibers as a reinforcing agent.

Table below gives The recommended Thickeness of Insulation (85% Magnesia)

Nominal Pipe Size, (Inches)	Temperature of Pipe °C				
	37 - 92	93 - 148	149 - 203	204 - 259	260 - 315
1½" and under	1	1	1	1½	1½
2"	1	1	1½	1½	2*
2½" and 3"	1	1	1½	2*	2*
4"	1	1½	1½	2*	2½*
5"	1	1½	2	2	2½*
6"	1	1½	2	2½	2½*
8"	1½	1½	2	2½	3
10"	1½	2	2½	2½	3
12" 10"	1½	2	2½	3	3
14"and over	1½	2	2½	3 3	3½

The recommended thickness of 85% magnesia is given above for average conditions. Increased thickness may be required when temperature must be maintained with little fluctuation. Outdoor piping should have additional than is required by indoor piping.

(c) Slag wool and rock wool:- These are made by centrifugal spinning process from molten slag or mineral rocks. They are available in the form of loose and slag wool as filling material where bulk insulation is needed and are commonly applied by hand packing or in the form of stitched mats.

The mats may be faced on one or both side with a confining medium. First users of rock wool were natives of Hawaiian inslands who used rock wool to blanket their huts. The fibers came from volcanic depolosits where escaping steam had broken the molten lava into fluffy fibers.

They are good heat insulation and are comparatively less costlier. They can be applied to hot surfaces with temperature upto 700°C and are available in the form of mattress, blankets, moulded section and insulating boards.

Table 51.2: Below gives the thermal conductivity of slag wool/rock wool

IS: 3677 – 1973				
Mean Temp; (°C)	**Thermal Conductivity (in Kcal /m/h°C)**			
	Apparent density			
	200 kg / m³	150 kg / m³	120 kg / m³	100 kg / m³
50	0.036	0.040	0.041	0.043
100	0.042	0.045	0.046	0.048
150	0.051	0.053	0.056	0.058
200	0.061	0.064	0.067	0.068

The table below gives recommended slag /rock wool insulation thickness in mm, required for different pipe size and different operating / fluid temperatures.

Pipe dia	**Operating/fluid temperature, °C**					
(NB),mm	90	150	200	260	315	375
25	25	40	65	65	75	90
40	25	40	65	65	75	90
50	25	40	65	65	75	90
80	25	40	65	65	75	90
100	25	50	75	75	100	100
150	40	50	75	75	100	115
200	40	50	75	75	100	115
250	40	65	90	90	100	125
300	40	65	90	90	100	125
FLAT	40	65	90	90	100	125

The above insulation thickness is based on following considerations:

- Insulation material assumed is slag wool or rock wool having thermal conductivity 0.042 kcal / hr m°C and average density about 100-120 kg / m³.
- The surface temperature after insulation is 55-60 °C.
- Insulation upto 65 mm should be in single layer, and above 65 mm, multilayer e.g. for 80 mm two layers of 40 mm each; for 90 mm inner layer of 40 mm and outer of 50 mm.

(d) Bonded mineral wool (IS 8183):- This is made from rock slag or glass processed from molten state into fibrous from and shall be bonded with suitable binder.

The slag are normally supplied unfaced. Certain applications require an applied finish of cloth, foil, wire netting on one or both face.

The following guidelines to be followed at the time of insulation:

Group	Bulk density, kg/ m^3	Maximum recommended Hot Face Temperature,°C
Group 1	12 - 50	Up to 250
Group 2	50 - 80	Up to 400
Group 3	80 - 120	Up to 550
Group 4	120 - 160	Up to 750

The thermal conductivity of material is given below:

Mean temp °C	Thermal conductivity, kcal /m/hr °C			
	Group 1	Group 2	Group 3	Group 4
100	0.059	0.044	0.045	0.045
200		0.067	0.063	0.058
300		0.094	0.081	0.077

The table below gives recommended bonded mineral wool insulation thickness in mm. required for different pipe sizes and different operating / fluid temperatures.

Normal dia	OPERATING TEMPERATURE, °C				
(Inches)	Up to 125	201 - 250	301 – 350	401 - 450	501 - 550
0.5	25	50	70	95	115
0.75	25	50	75	100	125
1	25	55	80	105	125
1.5	25	60	85	115	140
2	25	60	90	120	150
3	25	65	100	115	150
4	25	70	105	140	150
6	25	75	110	155	165
8	25	75	115	160	180
10	25	80	120	170	190
12	25	80	125	175	200
14	25	80	125	175	210
16	25	85	130	180	210
18	25	85	130	185	220
20 & above	25	85	130	185	220
Flat surface	25	85	130	185	220

(e) Asbestos Rope:- Short length of pipe within reach from the floor are lagged by spirally winding asbestos rope over it. The rope should be of lagging variety and not of packing type. The lagging also serves the purpose of protecting persons from getting scalded , if they accidentally touch the steam pipe. Recently, a pipe insulation made from corrugated layers of asbestos paper has been developed for hot applications .

The following tables enable to estimate the length of the rope required to lag a given length of steam pipe of different diameter:

Pipe dia, inches	Length in feet of asbestos rope required /foot run of pipe			
	¼" rope	½" rope	1" rope	1¼" rope
½"	11	-	-	-
¾"	12	10	-	-
1"	14	11	10	-
1½"	17	14	12	10

(f) Aluminum paint:- An application of coat of bare aluminum paint helps to reduce heat losses by about 10-15%. However, it is not recommended where temperature exceed 200 °C.

Special Insulation Materials :

(i) Ceramic fibres:- Ceramic fibres are pressed into a blanket for use up to 1250 °C.

(ii) Calcium Silicate:- it is preferred rigid type of insulation having very low thermal conductivity, low density, very high mechanical strength, incombustibility, immunity to water wetting and high' service temperature suitability. The combination of these physio- thermal properties make the product distinctly superior than most of other insulation products for service temperature range of 200 to 850 °C.

Recommended Insulation Thickness in mm for Calcium Silicate Pipe Sections is tabulated below.

Nominal	Temperature, °C				
Pipe Size in mm	200°C	300°C	400°C	500°C	550°C
25	50	75	100	100	100
50	65	75	100	125	125
75	65	75	100	125	125
100	75	100	100	125	125
150	75	100	100	125	150

Important Properties of Different Insulating Materials are tabulated below.

S.No.	Insulation materials	Availability in form	Density	Thermal conductivity	Approximate limiting
			Kg/ m^3	(kcal/m/hr°C)	Temp ,°C
1	Asbestos	Mattress, textile, loose fill	80-250	0.042 at 40 °C	600
2	Glass fiber	Slab, loose fill, mattresses	12-100	0.032 at 40 °C 0.051 at 150 °C	400

3	Rock wool and slag wool	Slab, preformed sections, mattresses loose fill, sprayed	40-250	0.032 at 40 °C 0.036 to 0.045 at 110 °C	700
4	Mineral wool	Slab	50-250	0.044 at 100 °C	750
5	Calcium silicate	Slab, selection, plastic, brick	160-320	0.051 at 40 °C	850
6	Magnesia	Slab, section, plastic loose fill	177-225	0.033 at 40 °C 0.07 at 250 °C 0.081 at 250 °C	300
7	Vermiculate	Plastic, wax fill	50-500	0.093 at 250°C	1100
8	Aluminum	Reflective	10-30	0.0.26 to 0.036 at 20 °C 0.074 at 250 °C	500

High Fiber Content Felt/Paper

Materials made from layers of high fiber content felt or paper, with layers of asphalt saturated felt formerly were used on moderate to cold temperature applications.

Today, materials manufactured from fiberglass, ceramic, mineral wool, calcium silicate, foamed plastic, glass, and other substances are used in many shapes and forms.

The thermal conductivity(W-m^2/m°C or, W m/°C) of an insulation material increases with temperature. So thermal conductivity is always specified at the mean temperature (mean of hot and cold face temperatures) of the insulation material.

Thermal conductivities of typical hot & cold face insulation materials are tabulated below.

Table 51.3: Thermal Conductivity of some hot face insulation materials

Mean Temperature °C	Calcium Silicate	Resin bonded Mineral wool	Ceramic Fiber Blankets
100	-	0.04	-
200	0.07	0.06	0.06
300	0.08	0.08	0.07
400	0.08	0.11	0.09
700	-	-	0.17
1000	-	-	0.26
Specific heat (kJ/kg/°C)	0.96 (at 40°C)	0.921 (at 20°C)	1.07 (at 980°C)
Service temp: (°C).	950	700	1425
Density, kg/m^3	260	48 to144	64 to 128

Table 51.4: Thermal conductivity of some Cold Face Insulation materials

MATERIALS	Thermal Conductivity W/m°C
Mineral Or Glass Fiber Blanket	0.039
Board or Slab	
Cellular Glass	0.058
Cork Board	0.043
Glass Fiber	0.036
Expanded Polystyrene (smooth) - Thermocole	0.029
Expanded Polystyrene (Cut Cell; - Thermocole	0.036
Expanded Polyurethane	0.017
Phenotherm (Trade Name)	0.018
Loose Fill	
3aper or Wood Pulp	0.039
Sawdust 01 Shavings	0.065
Minerals Wool (Rock. Glass, Slag)	0.039
Wood Fiber (Soft)	0.043

Formula for heat loss calculation

Various charts, graphs and references are available for heat loss computation. The surface heat loss can be computed with the help of a simple relation also as given below. This equation can be used up to 2000C surface temperature. Factors like wind velocities, conductivity of insulating material etc has not been considered in the following equation.

$$S = [10+(Ts-Ta)/20] \times (Ts-Ta)$$

Where ,

Total heat loss/hr (Hs) $= S \times A$

Where A is the surface area in m^2

Based on the cost of heat energy, the quantification of heat loss in Rupees can be calculated as under:

Equivalent fuel loss (Hf) (kg/Yr) $= \dfrac{Hs \times \text{yearly hours of operation}}{GCV \times \pi b}$

Annual heat loss in monetary terms (Rs.) $= Hf \times$ Fuel cost (Rs./kg)

Where ,

GCV = Gross Calorific value of fuel , kCal/kg

ηb = Boiler efficiency in %

Example on heat loss and fuel saving calculation by providing insulation :

Steam pipeline 100 mm diameter is not insulated for 100 metre length supplying steam at 10 kg/ cm^2 to the equipment. Find out the fuel savings if it is properly insulated with 65 mm insulating material.

Assumptions:

Boiler efficiency – 80%

Fuel Oil cost - Rs.15000/tonne

Surface temperature without insulation – 170°C Surface temperature after insulation – 65°C Ambient temperature – 25°C

Solution .

Existing Heat Loss:

S = [10+ (Ts-Ta)/20] x (Ts-Ta)

Here,

Ts = 170°C

Ta = 25°C

S = [10+(170-25)/20] x (170-25)= 2500 kCal/hr.m^2

S1 = S =Existing heat loss (2500 kCal/hr-m^2)

Modified System after insulation:

After insulating with 65 mm glass wool with aluminum cladding the hot face temperature will be 65°C .

Here, Ts=65 °C and Ta= 25 °C

Substituting these values

S = [10+(65-25)/20] × (65-25)

= 480 kCal/hr m^2

S2 = S =Existing heat loss (480 kCal/hr-m^2) after providing insulation

Fuel Savings Calculation

Pipe dimension	=	ϕ 100 mm & 100 m length
Surface area existing (A1)	=	$3.14 \times 0.1 \times 100 = 31.4\ m^2$
Surface area after insulation (A2)	=	$3.14 \times 0.23 \times 100 = 72.2\ m^2$
Total heat loss in existing system (S1 × A1)	=	$2500 \times 31.4 = 78500$ kCal/hr
Total heat loss in Modified system (S2 × A2)	=	$480 \times 72.2 = 34656$ kCal/hr
Reduction in heat loss	=	78500 – 34656 = 43844 kCal/hr
No. of hours operation in a year	=	8400
Total heat loss (kCal/yr)	=	$43844 \times 8400 = 368289600$
Calorific value of fuel oil	=	10300 kCal/kg
Boiler efficiency	=	80%
Price of fuel oil	=	Rs.15000/Tonne
Yearly fuel oil savings	=	368289600/10300 x 0.8
		44695 kg//year

The following tables gives the bare surface heat losses in kW/m^2 at different operating temperatures.

PIPE DIA, inches	BARE SURFACE HEAT LOSSES (KW/m²) AT DIFFERENT. OPERATING TEMPERATURES, °C						
	100	200	300	400	500	600	700
0.50	1.15	3.93	8.3	14.83	24.3	37.5	55.6
1.00	1.09	3.74	7.95	14.32	23.6	36.65	54.5
1.50	1.04	3.6	7.7	13.98	23.1	36	53.75
2.00	1.01	3.53	7.46	13.78	22.83	35.6	53.3
3.00	0.97	3.4	7.35	13.45	22.4	35.1	52.6
4.00	0.95	3.3	7.2	13.21	22.1	34.7	52.2
6.00	0.09	3.19	6.95	12.89	21.63	34.1	51.4
8.00	0.89	3.16	6.9	12.75	21.5	33.95	51.25
10.00	0.87	3.1	6.8	12.62	21.4	33.75	51
12.00	0.86	3.06	6.72	12.52	21.18	33.6	50.7
14.00	0.85	3.04	6.68	12.47	21.1	33.45	50.6
Flat surfaces	0.84	3.01	6.63	12.4	21	33.3	50.45

Energy costs would make it economical to insulate even surfaces at temperatures 50 to 60 °C.

Bare pipes carrying high temperature steam lose much heat to the atmosphere. The heat lost through pipe has high potential for work; therefore, practically every hot pipe in the modern plant should be insulated.

By insulating the hot pipes, there is not only conservation of heat energy but it also avoids an uncomfortable overheated atmosphere in the vicinity of pipe.

Heat conducted through an insulating material is transferred to the surroundings air by radiation and convection. The combined effect of radiation and convection heat transfers, from the insulation surface is called surface coefficient and is affected by the velocity of air passing over the surface and by emissivity of the surface which is the rate of heat transfer as compared to that from a black body. Variation of surface coefficient affects a large difference in the insulation's surface temperature, but marginal difference to heat loss.

Table below illustrates the effect of surface coefficient on heat losses and surface temperature for a 150 mm pipe for different temperature values.

Table 51.5: Effect of Surface Coefficient on Heat Loss and Surface Temperature for a 150mm pipe (In ambient Air at 30 °C)

Temperate (°C)	Cloth		Galvanised mild steel		Aluminium	
	Q	ts	Q	ts	Q	ts
100	85.9	41	87.9	43	78.8	46
100	42.0	36	42.0	37	40.0	39
300	205.5	53	203.0	57	196.6	63
300	94.1	42	91.5	44	91.2	48
500	206.5	53	204.9	57	201.3	64
500	127.9	45	126.4	48	124.3	53

$$Q = \text{heat loss} = \frac{\text{kcal}}{\text{hr m2}}$$

ts = surface temperature (°C)

Increased air movement has a greater effect on heat loss from bare piping than from insulated piping.

Table below shows the effects of wind velocity on heat loss from bare and insulated surfaces.

Table: Effect of wind velocity on Heat Loss

	Heat loss (kcal/m2/hr)				
Wind velocity (m/sec)	**Bare surface**	**25 mm Insulation**	**50 mm insulation**	**75 mm insulation**	**100 mm Insulation**
0	3415	507	307	232	191
1	4197	537	319	238	198
5	7086	572	330	247	205
10	10490	576	336	248	206

Note:

(a) 150 mm.pipe at 300 °C

(b) Ambient temperature = 30 °C

(c) Insulation = 0.0515 kcal/m/hr/°C

(d) Finish : Galvanised mild steel

Consider a pipe having R_1 and R_2 as inside and outside radius over which an insulation is provided, whose radius is R_3. To find out the liner heat loss or heat loss per unit length, we need to calculate the overall heat heat transfer coefficient (U) of the pipe including the insulation.

$$U = \frac{1}{\frac{R_3}{R_1 \, hi} + \frac{R_3 \, LN\left(\frac{R_2}{R_1}\right)}{k_{pipe}} + \frac{R_3 \, LN\left(\frac{R_3}{R_2}\right)}{k_{insulation}} + \frac{1}{ho}}$$

Where, U = the overall heat transfer coefficient

hi & ho are the convective heat transfer coefficient.

Generally, the heat transfer coefficient of ambient air is 40 W/ m^2k.

This coefficient can of course increase with velocity , if the pipe is outside. The outdoor air coefficient in warm climates with wind speeds under 25km/hr is around 50 W/m^2k.

The total heat loss per unit length (Q/L) can then be calculated by:

$$\frac{Q}{L} = 2\pi R3 \, U\Delta T$$

Where , $\Delta T = T_{\text{inside pipe}} - T_{\text{ambient}}$

Since heat loss through insulation is a conductive heat transfer, there are instance when adding insulation actually increases heat loss. The thickness at which insulation begins to decrease heat loss is described as the critical thickness. Since the critical thickness is almost always a few millimeters, it is seldom an issue for piping. Critical thickness is a concern however in insulating wires.

CALCULATION OF INSULATION THICKNESS

Insulation economics depends on the cost of insulation, which in turn depends upon the thickness of insulation, which in turn depends upon the thickness of insulation that is to be provided. So it is very important to determine the exact insulation thickness that is required.

Method -1 for economic insulation thickness calculation:

The following formula can be used for flat surfaces:

$$\text{Insulation thickness} = \frac{\text{OperatingTemperature(degC) - SurfaceTemperature(degC)}}{\text{Heat Loss}\left(\frac{\text{kcal}}{\text{h}}\right)} * k$$

The important factor, which affects the thickness, is the emissivity (k). so higher the 'k', higher the insulation thickness required and hence higher the cost. By using the mean value of the insulating material, we get lower 'k' value, and therefore a lower insulation thickness.

To find the 'k' value, find the mean temperature as follows:

$$\text{Mean temperature} = \frac{\text{OperatingTemperature} + \text{SurfaceTemperature}}{2}$$

Based on this mean temperature, 'k' values can be obtained from heat transfer handbooks.

Method -2 for calculation of economic thickness of insulation:

Readymade table are available to calculate the Economic thickness of Insulation.

Example of Economic Thickness Determination:

Using the table (Reference Perry's Chemical Engineer's Handbook) below, assuming a 8.0 inch pipe at 345°C in an indoor setting with an energy cost of $3.792/million kJ energy costs, we see temperature of 343.3°C, 371.1 °C and 565.6°C. Since our temperature does not meet 565°C or 371°C, we use the thickness before it. In this case, for 343.3°C, 50.8 mm (2 inch) thickness of insulation is sufficient. At 371 °C, it can be increased to 63.5 mm (2.5inch) thickness of insulation.

Economic thickness charts from other source will work in much the same way as this example.

Table 51.6: Economic thickness of Indoor Insulation

Addated from parry's chemical Engineer's Handbook		Minimum pipe temperature (°C)							
		Energy costs, $/ million K.J							
Pipe Size (mm)	Insulation Thickness (mm)	0.948	1.896	2.844	3.792	4.72	5.688	6.636	7.584
19.05	38.1	510.0	315.6	287.8	204.4	176.7	148.9	121.1	121.1
	50.8				593.3	537.8	482.2	426.7	398.9
	63.5				954.4	565.6	510.0	454.4	426.7
	76.2								648.9
25.4	38.1	648.9	426.7	315.6	260.0	232.2	204.4	176.7	148.9
	50.8			648.9	537.8	482.2	426.7	371.1	371.1
	63.5					648.9	565.6	537.8	482.2
	76.2						593.3	621.1	510.0
38.1	38.1	593.3	398.9	287.8	232.2	204.4	204.4	176.7	148.9
	50.8			537.8	454.4	371.1	343.3	315.6	260.0
	63.5				565.6	482.2	426.7	398.9	343.3
	76.2						621.1	593.3	537.8
50.8	38.1	565.6	371.1	260.0	232.2	204.4	176.1	148.9	148.9
	50.8			565.6	454.4	398.9	371.1	148.9	315.6
	63.5			593.3	510.0	537.8	398.9	371.1	343.3
	76.2				648.9	565.6	510.0	454.4	426.7
76.2	38.1	510	343.3	260.0	204.4	176.7	148.9	148.9	121.1
	50.8		693.3	482.2	371.1	315.6	287.8	260.0	232.2
	63.5		-17.8	656.6	454.4	398.9	343.3	260.0	260.0
	76.2				565.6	510.0	426.7	398.9	371.1
101.6	38.1	510.0	315.6	260.0	204.4	176.7	148.9	148.9	121.1
	50.8		593.3	454.4	371.1	315.6	315.6	260.0	232.2
	63.5				537.8	454.4	454.4	371.1	343.3
	76.2				565.6	482.2	482.2	398.9	371.1
	88.9							621.1	565.6
152.4	101.6	315.6	176.7	148.9	121.1	121.1	39.3	93.3	93.3
	50.8		593.3	454.4	371.1	315.6	287.8	260.0	260.0
	63.5			482.2	426.7	343.3	315.6	287.8	287.8
	76.2						398.9	371.1	315.6
	88.9			621.1	537.8	454.4	593.3	537.8	482.2
	101.6								648.9

203.2	50.8	-17.8	537.8	426.7	343.3	287.8	260.0	232.2	204.4
	63.5		565.6	454.4	371.1	315.6	287.8	260.0	232.2
	76.2				565.6	482.2	426.7	398.9	371.1
	88.9					648.9	593.3	537.8	482.2
	101.6							621.1	593.3
254	50.8	-17.8	593.3	454.4	371.1	343.3	287.8	260.0	232.2
	63.5		648.9	482.2	398.9	371.1	315.6	260.0	260.0
	76.2			565.6	482.2	398.9	371.1	287.8	287.8
	88.9					648.9	565.6	315.6	482.2
	101.6							510.0	648.9
304.8	50.8	621.1	398.9	315.6	260.0	204.4	204.4	176.7	148.9
	63.5		537.8	426.7	343.3	287.8	260.0	232.2	204.4
	76.2			648.9	537.8	482.2	426.7	371.1	343.3
	88.9					648.9	593.3	537.8	482.2
	101.6						621.1	565.6	510.0
	114.3						648.9	593.3	537.8
355.6	50.8	565.6	343.3	287.8	232.2	204.4	176.7	148.9	148.9
	63.5		537.8	426.7	343.3	287.8	260.0	232.2	204.4
	76.2			593.3	510.0	426.7	371.1	343.3	315.6
	88.9					621.1	537.8	510.0	454.4
	101.6					648.9	565.6	537.8	482.2
	114.3						648.9	565.6	537.8
406.4	50.8	510	343.3	260.0	204.4	176.7	148.9	148.9	148.9
	63.5		537.8.	426.7	371.1	315.6	287.8	260.0	232.2
	76.2		648.9	510.0	426.7	371.1	315.6	287.8	260.0
	88.9					621.1	565.6	510.0	454.4
	101.6					648.9	593.3	537.8	482.2
	114.3						621.1	565.6	510.0
457.2	50.8	537.8	343.3	260.0	204.4	176.7	176.7	148.9	148.9
	63.5		510.0	398.9	315.6	287.8	260.0	232.2	204.4
	76.2		621.1	482.2	396.9	343.3	287.8	260.0	260.0
	88.9					648.9	593.3	537.8	482.2
	101.6						621.1	565.6	510.0
	114.3						648.9	593.3	537.8
508	50.8	565.6	371.1	287.8	232.2	204.4	176.7	176.7	148.9
	63.5		537.8	426.7	315.6	287.8	260.0	232.2	204.4
	76.2		621.1	482.2	398.9	343.3	287.8	10.0	260.0
	88.9						593.3	537.8	510.0
	101.6						621.1	565.6	537.8
	114.3							648.9	593.3

609.6	50.8		15.6	260.0	204.4	176.7	148.9	148.9	121.1
	63.5		621.1	482.2	398.9.	343.3	287.8	260.0	232.2
	76.2					398.9	371.1	315.6	287.8
	88.9		-17.8	565.6	482.2	593.3	537.8	482.2	426.7
	101.6					621.1	565.6	510.0	454.4
	114.3						621.1	565.6	510.0

*** 80 °C Still Ambient Air, Aluminum Jacketed Calcium silicate Insulation

The economic thickness of insulation depends on the first cost (initial cost) & maintenance cost of insulation and annual value of heat loss, which depends on the cost of producing the steam & the thermal conductivity of the lagging. Generally thicker insulation will represent higher owing costs and lower heat loss costs.

Major points of consideration in selection of thermal insulation :

Three major factors play an important role in determining insulation type and thickness. The three key factors are:

- Economics
- Safety
- Process Conditions

Each situation must be studied to determine how to meet each one of these criteria. First, we'll examine each aspect individually, then we'll see how to consider all three for an example.

Economics:

Economic thickness of insulation is a well documented calculation procedure. The calculations typically take in the entire scope of the installation including plant depreciation to wind speed. Data charts for calculating the economic thickness of insulation are widely available.

Safety:

Pipes that are readily accessible by workers are subject to safety constraints. The recommended safe "touch" temperature range is from 55 to 65 °C . Insulation calculations should aim to keep the outside temperature of the insulation around 60 °C. An additional tool employed to help meet this is aluminum foil cladding / covering wrapped around the outside of the insulation. Aluminum's poor thermal conductivity of 0.025 W/m.K makes it a good choice for this application. Typical thickness of aluminum used for this purpose range from 0.2 mm to 0.4 mm. The addition of aluminum adds another resistance term when calculating the total heat loss:

$$U = \frac{1}{\frac{R_3}{R_1\, hi} + \frac{R_3\, LN\left(\frac{R_2}{R_1}\right)}{k_{pipe}} + \frac{R_3\, LN\left(\frac{R_3}{R_2}\right)}{k_{insulation}} + \frac{1}{ho}}$$

And final Equation becomes:

$$\frac{Q}{L} = 2\pi R_4 \, U \, \Delta T$$

Where R4 is the radius of the pipe, insulation & aluminum cover combined.

However, when considering safety, engineers need a quick way to calculate the surface temperature that will come into contact with the works. This can be done with equations or the use of charts. We start by looking at other diagrams given below:

At steady state, the heat transfer rate (Q) will be the same for each layer:

$$Q = \frac{T_1 - T_2}{(R_2 - R_1)(\text{kpipeALMpipe})} = \frac{T_2 - T_3}{(R_3 - R_2)(\text{kinsALMins})} =$$

$$\frac{T_3 - T_4}{(R_4 - R_3)(\text{kinscoverALMinscover})}$$

Where:

$$\text{ALMpipe} = \frac{(2\pi R_2 L) - (2\pi R_1 L)}{LN\left(\frac{2\pi R_2 L}{2\pi R_1 L}\right)}$$

$$\text{ALMins} = \frac{(2\pi R_3 L) - (2\pi R_2 L)}{LN\left(\frac{2\pi R_3 L}{2\pi R_2 L}\right)}$$

$$\text{ALMins cover} = \frac{(2\pi R_4 L) - (2\pi R_3 L)}{LN\left(\frac{2\pi R_4 L}{2\pi R_3 L}\right)}$$

Rearranging the above Equation by solving the three expressions for the temperature difference yields:

$$Q = \frac{T_1 - T_4}{\left(\frac{R_2 - R_1}{\text{kpipeALMpipe}}\right) + \left(\frac{R_3 - R_2}{\text{kinsALMins}}\right) + \left(\frac{R_4 - R_3}{\text{kinscover ALM inscover}}\right)}$$

Each term in the denominator of above Equation is referred to as the "resistance" of each layer. We define this as Rs and rewrite the equation as:

$$Q = \frac{T_1 - T_4}{\text{Rspipe} + \text{Rsins} + \text{Rsinscover}}$$

Since the heat loss is constant for each layer, use Equation to calculate Q from the bare pipe, then solve next Equation for T4 (surface temperature). Use the economic thickness of insulation as a basis for calculation, after all, if the most affordable layer of insulation is safe, that's the one we would want to use. If the economic thickness results in too high a surface temperature, repeat the calculation by increasing the insulation thickness by ½ inch each time until a safe touch temperature (60°C) is reached.

Using heat balance equations is certainly a valid means of estimating surface temperatures, but it may not always be the fastest. Charts are available that utilize a characteristic called "equivalent thickness" to simplify the heat balance equations. This correlation also uses the surface resistance of the outer covering of the pipe.

With the help of above figure & table (or similar data- another material) the relation :

Equivalent Thickness = Kins Rs $\dfrac{T_{pipe} - T_{surfce}}{T_{surfce} - T_{ambient}}$ can be used to easily determine how much insulation will be needed to achieve a specific surface temperature.

Process Conditions:

The temperature of a fluid inside an insulated pipe is an important process variable that must be considered in many situations. Consider the length of pipe connecting two pieces of process equipment shown below:

In order to predict T_2 for a given insulation thickness, we first make the following assumptions:

- Constant fluid heat capacity over the fluid temperature range.
- Constant ambient temperature.
- Constant thermal conductivity for fluid, pipe, and insulation.
- Constant overall heat transfer coefficient.
- Turbulent flow.
- 24km/hr wind

$$Q = 2\pi\, L\, R_3\, U\, \Delta T\, LM$$

Where:

$$U = \frac{1}{\dfrac{R_3}{R_1\, hi} + \dfrac{R_2\, LN\left(\dfrac{R_2}{R_1}\right)}{Kpipe} + \dfrac{R_3\, LN\left(\dfrac{R_3}{R_2}\right)}{Kins} + \dfrac{1}{ho}}$$

$$hi = \frac{0.023\, Cpm.}{Ax\left(\dfrac{\dfrac{Cp}{\mu}\text{fluid}}{\text{Kfluid}}\right)^{\frac{2}{3}}\left(\dfrac{2\, R_1 m.}{Ax\mu\text{fluid}}\right)^{0.2}}$$

K = thermal conductivity

μ = viscosity

Cp = heat capacity of fluid

hi = heat transfer coefficient inside pipe

ho = 40 w/m^{2}K , indoors

ho = 50 w/m^{2}k , outdoors

$$\Delta TLM = \frac{(T_2 - T_{amb}) - (T_1 - T_{amb})}{LN\left(\frac{T_2 - T_{amb}}{T1 - T_{amb}}\right)}$$

Another heat balance equation yields:

$$Q = m.\ Cp\ (T_1 - T_2)$$

Where:

m. = mass flow of fluid

Equating above Equation and solving for T_2 yields:

$$T_2 = (T_1 - Tamb)\ \exp\left(\frac{-2\pi R_3 UL}{m.CP}\right) + T_{amb}$$

Equation above provides another useful tool for analyzing insulation and its impact on a process.

One example may be the importance of designing insulation thickness to prevent condensation on cold lines. Usually, when we hear the word "insulation" we instantly think of hot lines. However, there are times when insulation is used to prevent heat from entering a line. In this situation, the dew point temperature of the ambient air must be considered. which is a function of relative humidity and dry bulb temperatures.

It is crucial that sufficient insulation is added so that the outer temperature of the insulation remains above the dewpoint temperature. At the dewpoint temperature, moisture in the air will condense onto the insulation and essentially ruin it.

The cost of an insulation material relates to its temperature capability. Hence, insulation and refractory systems are designed to include several layers of different materials that offer optimum economic performance. Selecting the most economical system requires consideration of its cost compared to potential savings from reduction of heat losses. In some cases, the most economical thickness may not meet regulatory requirements for the safety of personnel and property. In such cases, appropriate design and materials should be used. In other cases, thickness may be reduced when there is a danger of exceeding the limiting refractory or insulating temperature.

Insulation definitely saves money, but above that it is very vital from safety point of view as well. Pipes that are readily accessible by workers are subject to safety constraints. The recommended safe "touch" temperature range is from 55 to 65 °C.

Economic thickness of an insulation material is a calculated parameter in which the owning costs. i.e.intial & maintenance costs of greater or lesser thicknesses are compared with the relative values of heat energy, which might be saved by such various thicknesses. Basic benefits that we are getting due to insulating pipe or furnaces should be taken into consideration along with the total cost i.e.inital & maintenance cost while selecting insulating material.

PROPERTIES & EFFICIENT UTILISATION OF STEAM

Steam is used for generating power as well as for process heating applications in industries . The following characteristics of steam make it so popular and useful to the industry:

- Highest specific heat and latent heat;
- Highest heat transfer coefficient;
- Easy to control and distribute;
- Cheap and inert.
- Efficient and economic to generate
- Easily transferred to the process
- Steam plant easy to manage

PROPERTIES OF STEAM

Sensible Heat

When water is first heated to the boiling point, it is called sensible heat addition (with change in temperature); liquid enthalpy (h_f) is the "enthalpy" (heat energy) available when it is in the water form, and is represented in kCal/kg.

Latent Heat

Once water reaches its boiling point, the enthalpy of evaporation (hf_g) is the additional heat energy to be added to this hot water in order to change it into steam. There is no change in temperature during evaporation, hence the name latent heat.

$$\boxed{h_g = h_f + h_{fg}}$$

Where , Enthalpy of water (h_f) is the Heat required to raise temperature from 0°C to current temperature .

Enthalpy of evaporation (h_{fg}) is the Heat required to change water into steam at boiling point .

Enthalpy of saturated steam (h_g) is the Total energy in saturated steam

Boiling Point

The temperature at which water boils is also called the boiling point or **saturation temperature**. This value depends on the steam pressure and increases as the steam pressure increases. As the steam pressure increases, the useful latent heat energy in the steam decreases.

The total heat carried by dry saturated steam or enthalpy of saturated steam is given by the sum of these two enthalpies ($h_f + h_{fg}$).

If the heat energy added is less than required latent heat, **wet steam** is produced, which contains partial water content, and the total heat carried by such wet steam will be lower than that of **dry saturated steam**. It is detrimental to use wet steam for process and steam distribution.

Dryness fraction is the property which tells us how dry the steam is. **Dryness fraction (x) is 0.95 if water content of steam = 5%**

Actual enthalpy of evaporation = dryness fraction X specific enthalpy h_{fg}

If saturated steam is heated further to a still higher temperature, it starts behaving like a hot gas, and this steam is called **superheated steam**. Superheating is the addition of heat to dry saturated steam without an increase in pressure. The temperature of superheated steam, expressed as degrees above saturation corresponding to the pressure, is known as the degree of superheat.

THE TEMPERATURE–ENTHALPY RELATIONSHIP

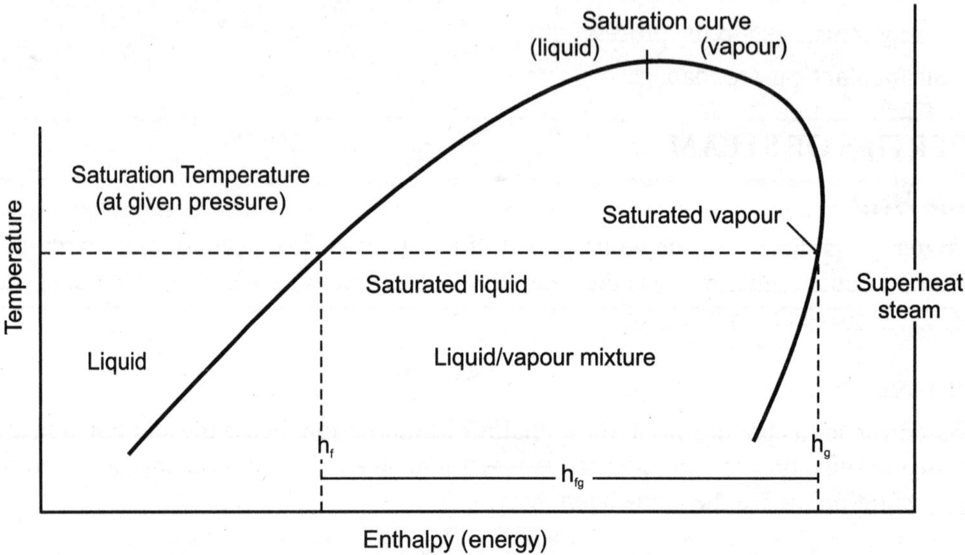

Fig.52.1: Temperature–Enthalpy Relationship of Steam

The properties of steam provided in the steam tables can also be expressed in a graphic form. Figure below shows the relationship between the enthalpy and the temperature at different pressures, and is known as a **phase diagram**. This is also called the Mollier diagram, and it helps us to understand the conditions and transition zones of water to steam and to superheated steam.

Another version of phase diagram is shown below.

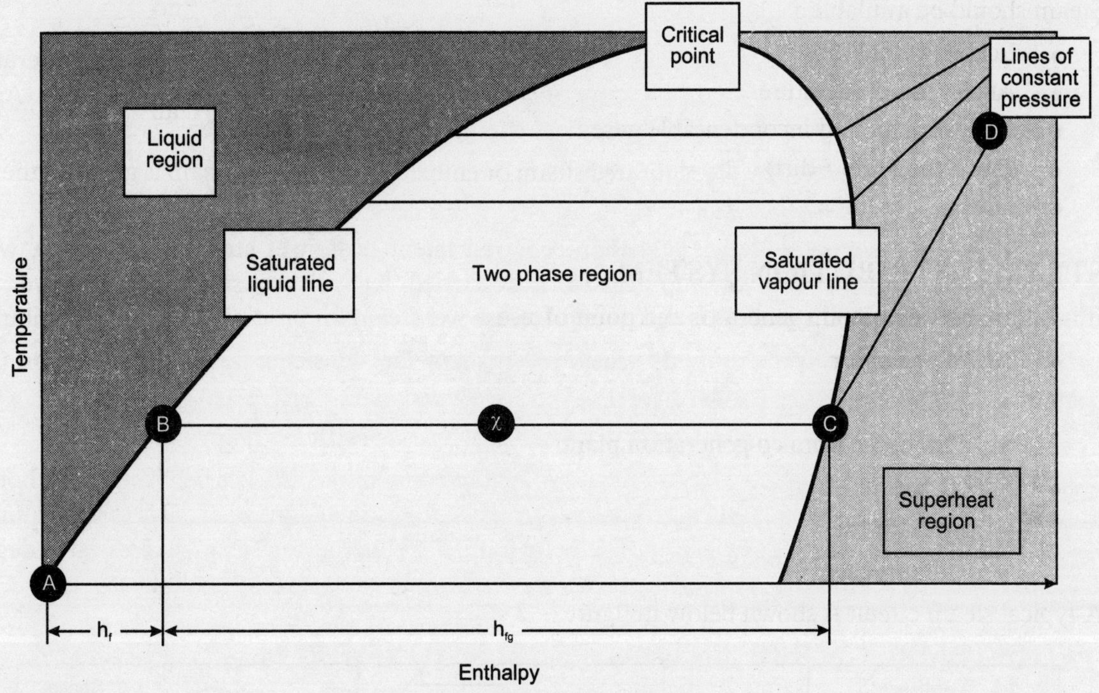

Fig.52.2: Phase diagram

STEAM TABLES

The steam tables shown below present various steam parameters at different pressure & temperature state points and are a helpful tool for design and analysis of steam systems.

Table. 52.1: Extract from the steam table

Pressure (kg/cm^2)	Temperature ºC	Enthalpy in Kcal/kg			Specific Volume (m^3/kg)
		Water (hf)	Evaporation (hfg)	Steam (hg)	
1	100	100.09	539.06	639.15	1.673
2	120	119.92	526.26	646.18	0.901
3	133	133.42	517.15	650.57	0.616
4	143	143.70	509.96	653.66	0.470
5	151	152.13	503.90	656.03	0.381
6	158	159.33	498.59	657.92	0.321
7	164	165.67	493.82	659.49	0.277
8	170	171.35	489.46	660.81	0.244

Steam should be available :

- In correct quantity
- At correct temperature
- Free from air and incondensable gases
- Clean (no scale / dirt)
- Dry

STEAM DISTRIBUTION SYSTEM

It is a Link between steam generator and point of use .

- Steam generator
  - Boiler
  - Discharge from co-generation plant
- Boilers use
  - primary fuel
  - exhaust gases

A typical steam circuit is shown below in figure .

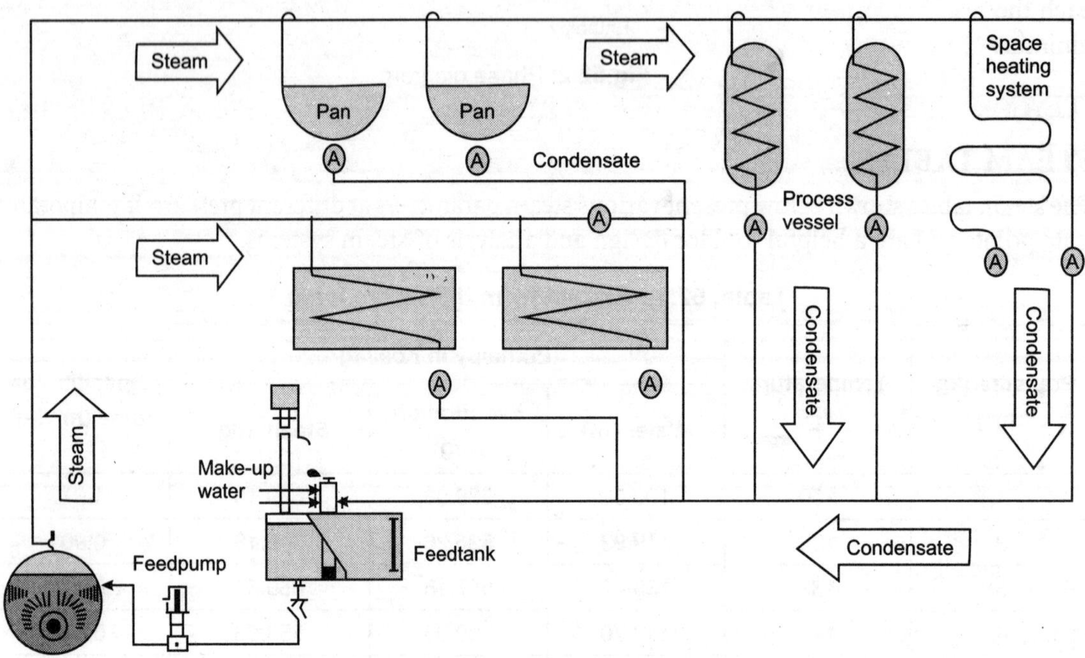

Fig.52.3: A typical steam circuit

Pressure and steam

- Steam pressure is influenced by many factors .
- Steam loses pressure in distribution pipework .

- Advantages of high pressure steam are :
  - Increased thermal storage capacity of boiler .
  - Smaller bore steam mains required .
  - Less insulation of smaller bore steam mains .
- Reduce steam pressure at point of use .

Important components of steam distribution system include:

1. Pipes
2. Drain points
3. Branch lines
4. Strainers
5. Filters
6. Separators
7. Steam traps
8. Air vents
9. Condensate recovery system
10. Insulation

For efficient steam distribution, an adequate quantity of steam that is dry and free of air must reach the plant at correct pressure; the diameter of steam piping should be optimum, in order to minimize pressure drops and reduce investment & operating costs.

STEAM PIPING: GOOD PRACTICES

- Steam pipes should be laid along the shortest possible distance.
- Provisions need to be made for proper draining of condensate. For example, a 100mm well-lagged pipe 30 meters in length, carrying steam at 7kg/cm^2 pressure, can condense nearly 10kg of water in the pipe in one hour unless it is removed from the pipe through steam traps.
- The pipes should run with a fall (slope) of not less than 12.5mm for every 3 meters distance, in the direction of flow.
- Drain pockets should be provided every 30–50 meters and at any low point in the pipe network.
- Expansion loops are required to take care of the expansion of pipes when they get heated up.
- Automatic air vents should be fixed at the dead ends of steam mains, to allow removal of air, which will tend to accumulate.

STEAM PIPING SIZING & LAYOUT

- Pipe material: carbon steel or copper
- Correct pipeline sizing is important because :

Oversized pipework results in :

- Higher material and installation costs .
- Increased condensate formation .

Undersized pipework results in :

- Lower pressure at point of use .
- Risk of steam starvation .
- Risk of erosion, water hammer and noise .
  - ► Pipeline layout: 1 m fall for every 100 m as shown below in figure.

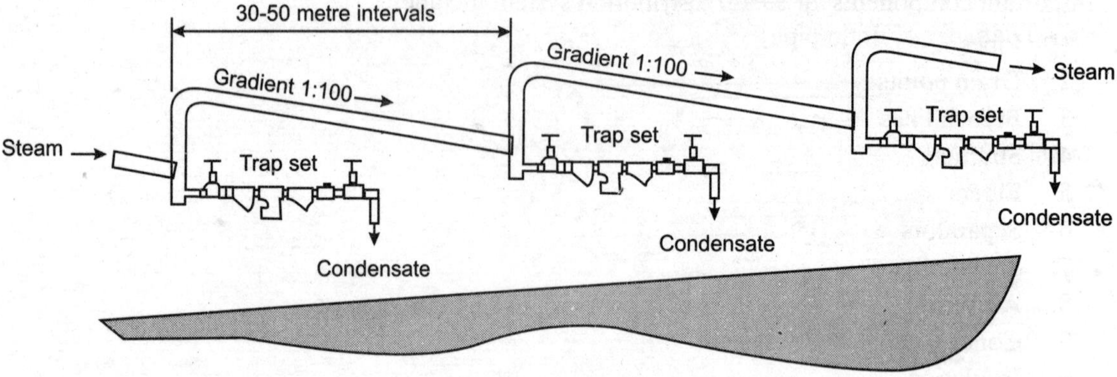

Fig.52.4: Steam pipe layout

Pipe sizing

Proper sizing of steam pipelines helps in minimizing pressure drop. The velocities for various types of steam are:

Superheated: 50–70 m/sec

- Saturated: 30–40 m/sec
- Wet or Exhaust: 20–30 m/sec

The steam piping should be sized based on permissible velocity and the available pressure drop in the line. A higher pipe size will reduce the pressure drop and thus the energy cost. However, higher pipe size will increase the initial installation cost. By use of smaller pipe size, even though the installation cost can be reduced, the energy cost will increase due to higher-pressure drop. Thus, optimum sizing is necessary. (Pressure drop change is inversely proportional to the 5th power of diameter change.)

Pipe Redundancy

Pipe redundancy is to be avoided in steam distribution since it leads to significant losses. All redundant pipelines must be eliminated; this could reduce steam distribution losses significantly.

Drain points

Drain points help in removing water in the pipes due to condensation of steam. The presence of water in steam is undesirable as it causes water hammer, leading to damage to steam pipes and eventually leakages. Steam traps must be provided at the drain points to avoid leakage of steam.

Drain point ensures that condensate can reach steam trap.

Consideration must be given to :

- Design
- Location
- Distance between drain points
- Condensate in steam main at shutdown
- Diameter of drain pipe

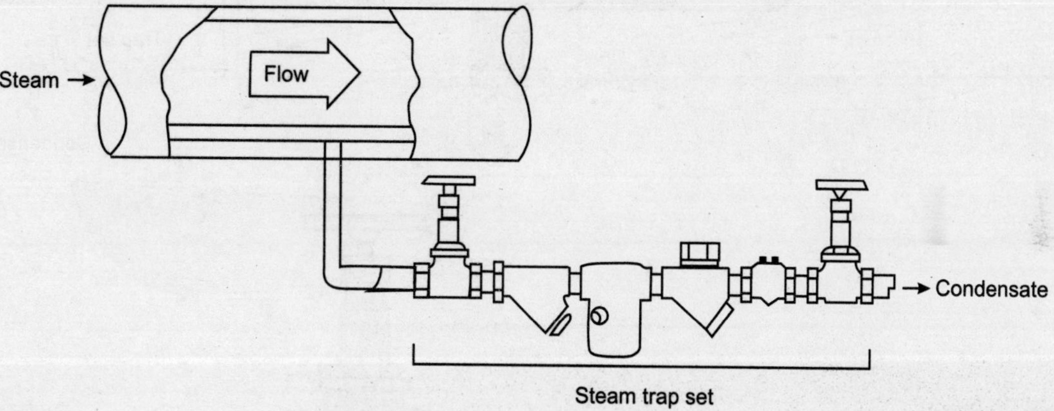

Fig.52.5: Steam condensate drain points

Branch lines :

- Take steam away from steam main .
- Shorter than steam mains .
- Pressure drop no problem if branch line is < 10 m .

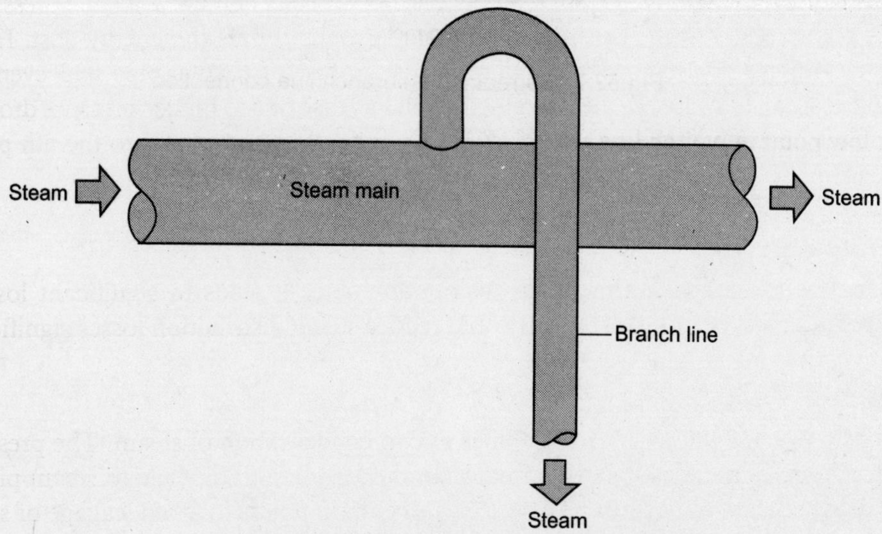

Fig.52.6: Steam branch line connection

Branch line connections :

- Top: driest steam
- Side or bottom: accept condensate and debris

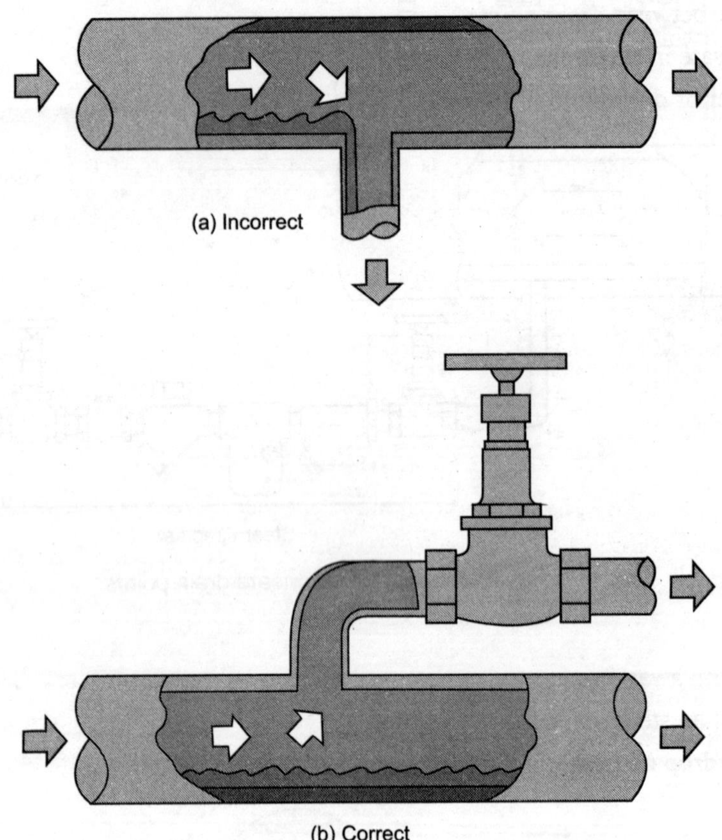

Fig.52.7: Correct steam branch line connection

Drop leg: low point in branch line

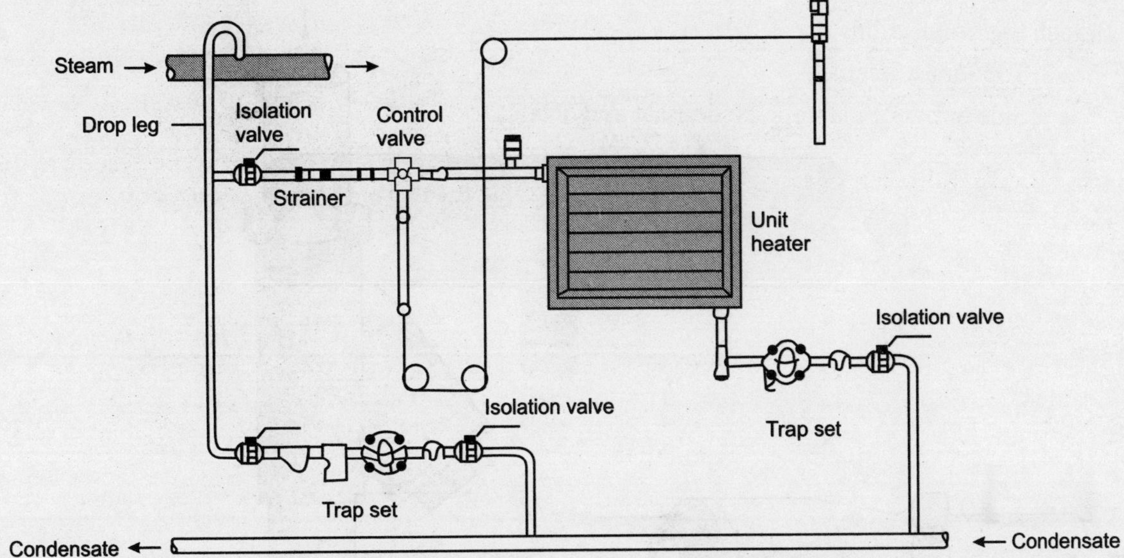

Fig.52.8: Flow direction of steam and condensate

Sometime steam runs across rising ground.Condensate should run against steam flow .

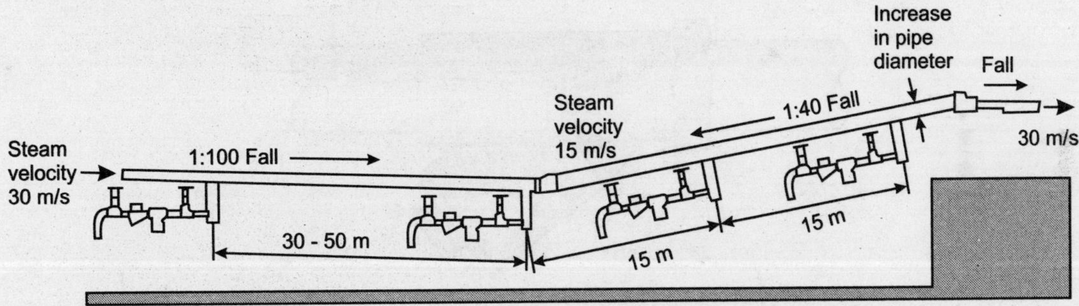

Fig.52.9: Flow of steam across rising ground (reverse gradient)

- Reverse Gradient on Steam Main

Strainers :

- Purpose of strainer is to :
  - Stop scale, dirt and other solids .
  - Protect equipment .
  - Reduce downtime and maintenance .
- Fitted upstream of steam trap, flow meter, control valve .
- Two types of strainer are : Y-type and basket type

Y-Type strainers :

- Handles high pressures .
- Lower dirt holding capacity: more cleaning needed .

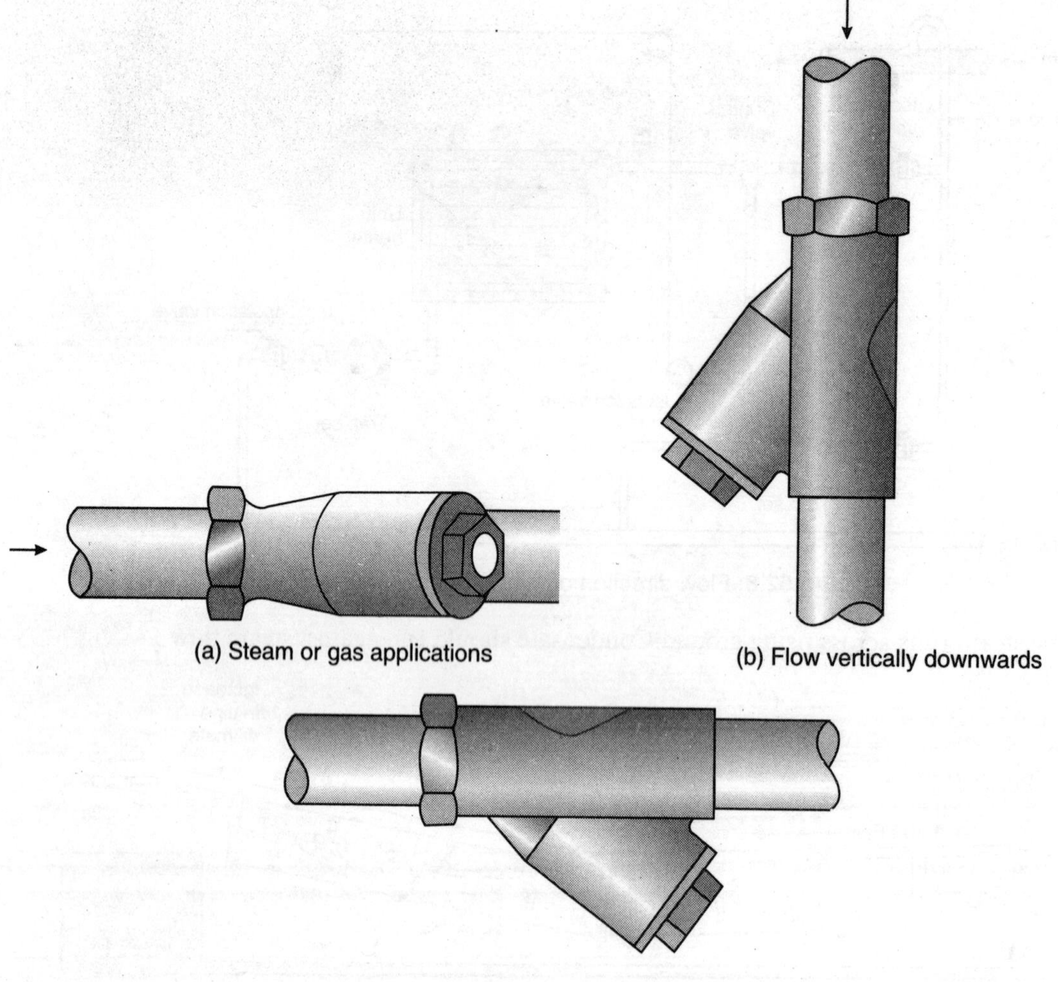

Fig.52.10: Various strainers

Basket type strainers :

- Less pressure drop .
- Larger dirt holding capacity.
- Only for horizontal pipelines.
- Drain plug to remove condensate.

Fig.52.11: Mesh screen strainer

Strainer screens are of two types :

- Perforated screens:
  - Holes punched in flat sheet.
  - Large holes.
  - Removes large debris.
- Mesh screens:
  - Fine wire into mesh arrangement.
  - Small holes.
  - Removes small solids.

Other strainer options include :

- Magnetic inserts: remove iron/steel debris.
- Self cleaning strainers :
  - Mechanical: scraper or brush
  - Backwashing: reverse flow direction
- Temporary strainers: equipment protection during start-ups.

Filters

It Consists of sintered stainless steel filter element.

Removes smallest particles.

- Direct steam injection – e.g. food industry
- Dirty stream may cause product rejection – e.g. paper machines
- Minimal particle emission required from steam humidifiers
- Reduction of steam water content

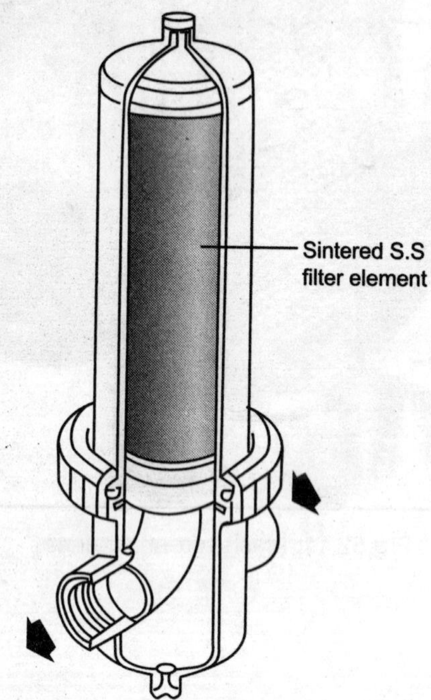

Fig.52.12: Filter

- Choose correct size due to large pressure drop.
- Do not exceed flow rate limits.
- For steam applications.
  - Fit separator upstream to remove condensate.
  - Fit Y-type strainer upstream to remove large particles.
  - Identify when cleaning is needed.

Separators :

- Separators remove suspended water droplets from steam
- Water in steam causes problems
- Water is barrier to heat transfer
- Erosion of valve seals and fittings and corrosion
- Scaling of pipework and heating surfaces from impurities
- Erratic operation and failure of valves and flow meters

Three types of separators are:

– *Baffle type*

- Baffle plates change direction of flow – collect water droplets .
- Cross-sectional area reduces fluid speed – water droplets fall out of suspension.
- Condensate in bottom drained away through steam trap.

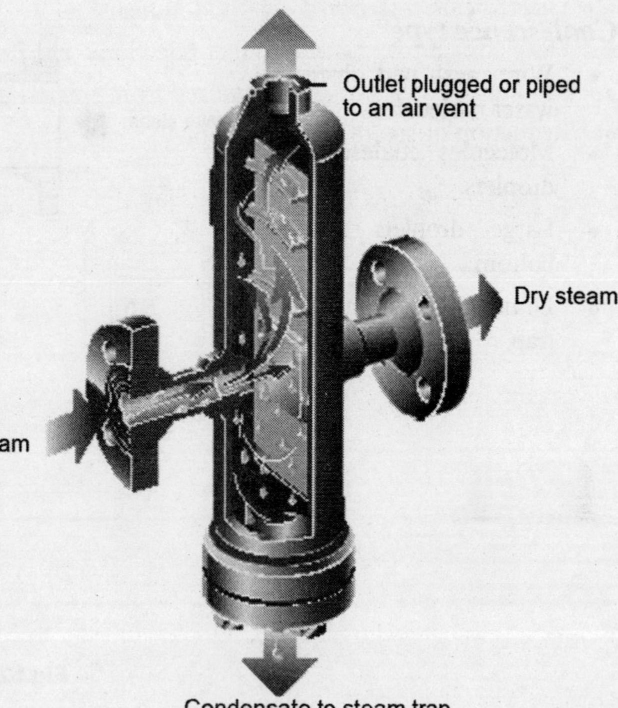

Fig.52.13: Baffle separator

– *Cyclonic type*

- Fins generate cyclonic flow.
- Steam spins around separator body.
- Water thrown to wall.
- Drainage through steam trap.

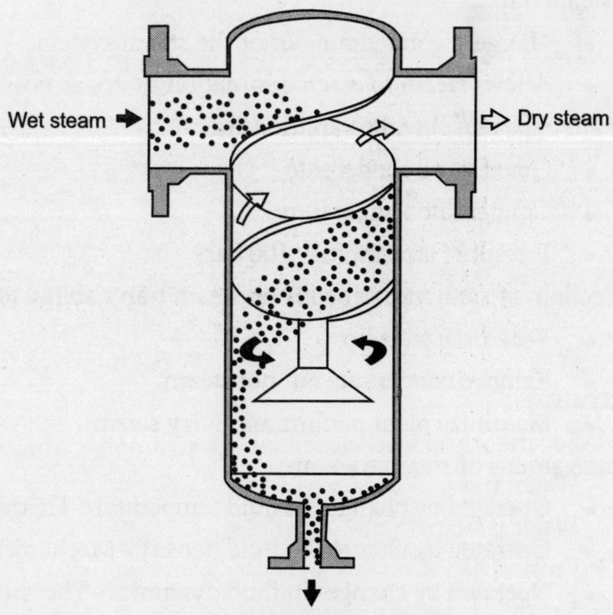

Fig.52.14: Cyclonic type separator

- ***Coalescence type***
  - Wire mesh pad obstructs water molecules.
  - Molecules coalesce into droplets.
  - Large droplets fall to bottom.
  - Drainage through steam trap.

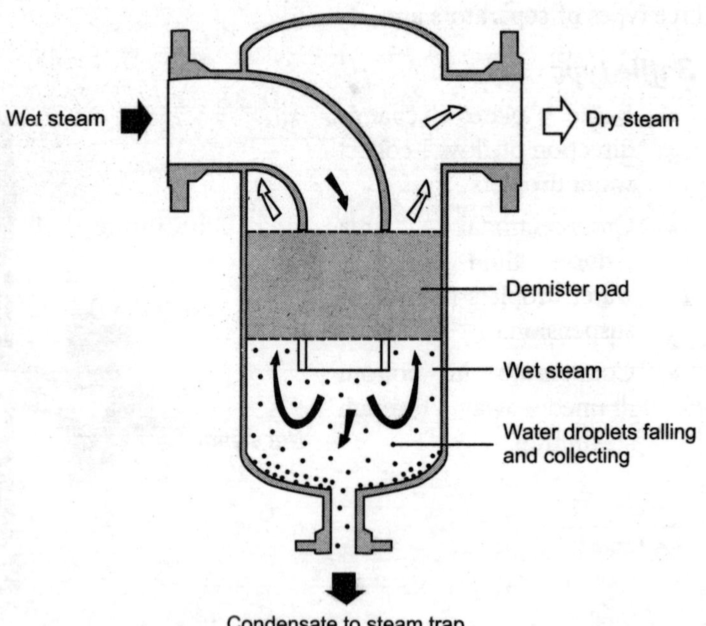

Fig.52.15: Coalescence type separator

Steam traps

A steam trap:

- "Purges" condensate out of the steam system.
- Allows steam to reach destination as dry as possible.

Steam traps must handle variations in :

- Quantity of condensate
- Condensate temperature
- Pressure (vacuum to > 100 bar)

Selection of stem trap depends on steam trap's ability to :

- Vent air at start-up
- Remove condensate but not steam.
- Maximize plant performance: dry steam

Three groups of steam traps are:

- Operated by changes in fluid temperature-Thermostatic type
- Operated by changes in fluid density –Mechanical type
- Operated by changes in fluid dynamics –Thermodynamic type

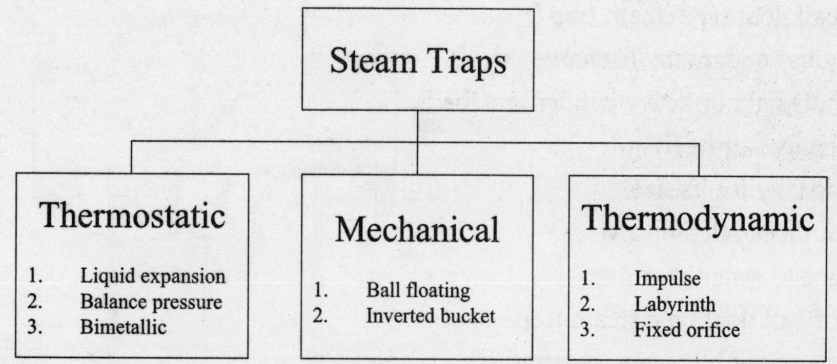

Application	Feature	Suitable trap
Steam mains	▪ Open to atmosphere, small capacity ▪ Frequent change in pressure ▪ Low pressure - high pressure	Thermodynamic, Mechanical: Float
▪ Equipment ▪ Reboiler ▪ Heater ▪ Dryer ▪ Heat exchanger etc.	▪ Large capacity ▪ Variation in pressure and temperature is undesirable ▪ Efficiency of the equipment is a problem	Mechanical: Float Bucket Inverted bucket
▪ Tracer line ▪ Instrumentation	▪ Reliability with no over heating	Thermodynamic, Thermostatic: Bimetallic

– Ball float type steam trap :

- Condensate in trap causes ball float to rise – condensate is released.
- Modern traps use thermostatic air vent to allow initial air to pass.

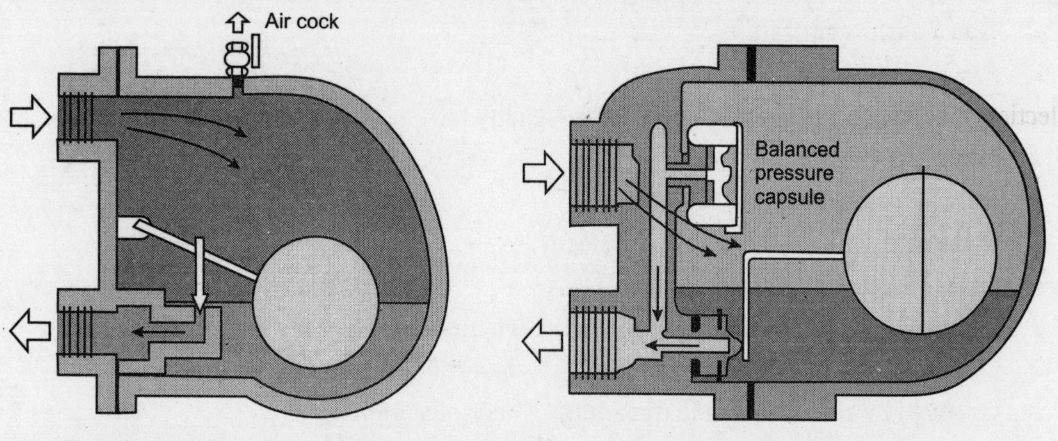

Float trap with air cock vent

Float trap with thermostatic air

Fig.52.16: Ball float type steam trap

Advantages of ball float type steam trap :

- Continuous condensate discharge.
- Can handle light or heavy condensate loads.
- Can discharge air freely.
- Large capacity for its size.
- Has steam lock release valve.
- Resistance to water hammer.

Disadvantages of ball float type steam trap :

- Can be damaged by severe freezing.
- Different internals needed for varying pressures.

– Inverted bucket type steam trap :

- Bucket hangs down.
- Lever pulls off seat.
- Condensate flows under bucket and flows away.
- Steam arrives.
- Bucket rises and shuts outlet.
- Steam in bucket condenses or bubbles through vent hole.
- Main valve opens.
- Condensate is released.

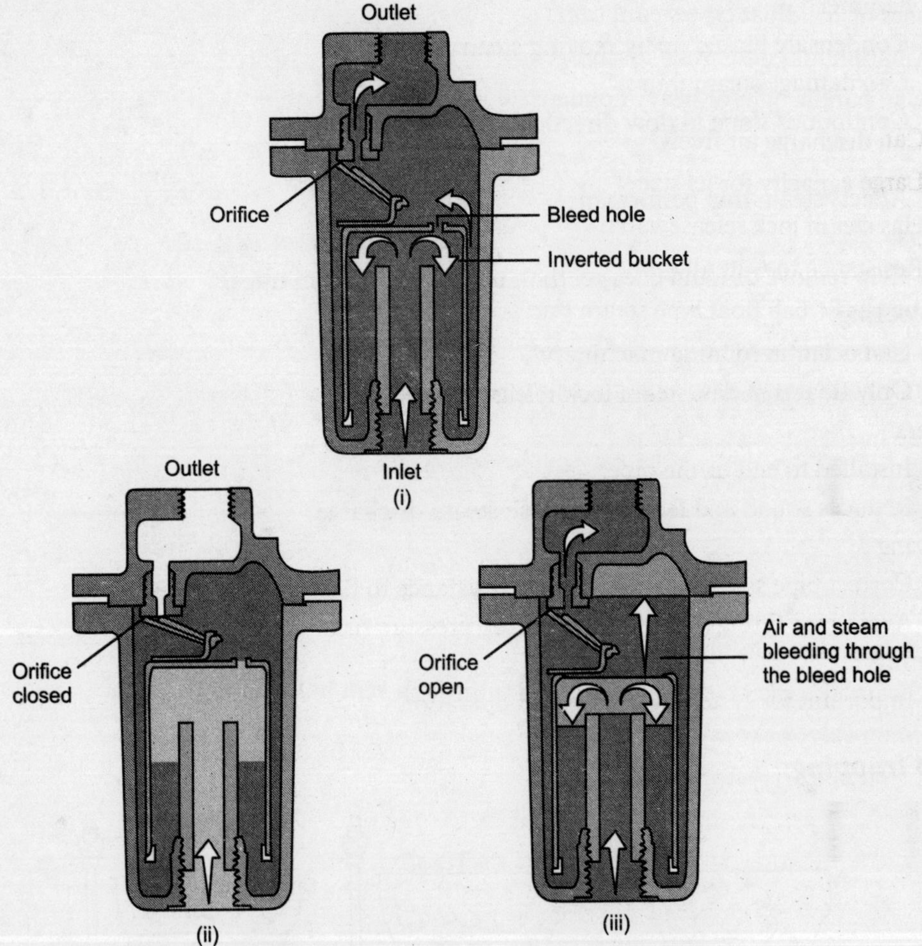

Fig.52.17: Inverted bucket type steam traps

Advantages of inverted bucket type steam trap :

- Can withstand high pressures.
- Tolerates water hammer .
- Suited for superheated steam lines.
- Safer because failure mode is open.

Disadvantages of inverted bucket type steam trap :

- Slow air discharge.
- Trap body must always have enough water.
- Check valve needed if pressure fluctuations.
- Water seal loss by T superheated steam.
- Can be damaged by freezing.

Steam traps considerations include :

Water hammer :

- Condensate picked up by moving steam.
- Can damage steam trap.
- Continuous slope in flow direction reduces this.

Dirt:

- Affects steam trap performance.

Strainers:

- Help remove dirt and cheaper than maintaining steam traps.

Steam locking:

- Can occur in rotating machinery.
- Only float trap has 'steam lock release' valve.

Diffusers:

- Installed to end of the pipe.
- Reduces sound and ferocity of flash steam discharge

Pipe sizing :

- Correct pipe size - traps affected by resistance to flow .
- Avoid pipe fittings close to trap – back pressure risk.

Air venting:

- Important for system warm up and operation

Group trapping:

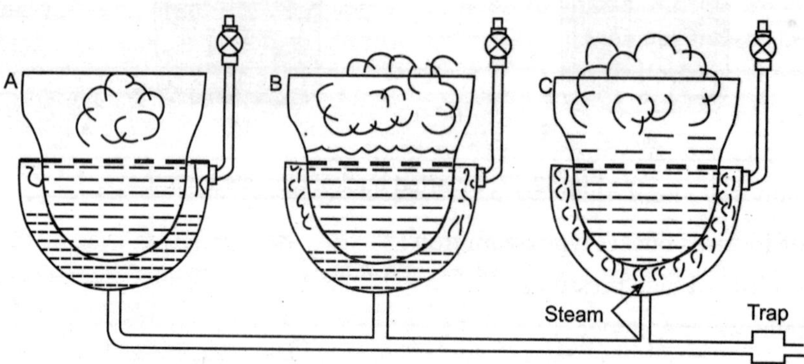

Fig.52.18: Group trapping of steam trap

Drain pocket dimensions

Mains diameter - D	Pocket diameter - d_1	Pocket depth - d_2
Up to 100 mm nb	$d_1 = D$	Minimum $d_2 = 100$ mm
125 - 200 mm nb	$d_1 = 100$ mm	Minimum $d_2 = 150$ mm
200 mm and above	$d_1 \geq D/2$	Minimum $d_2 = D$

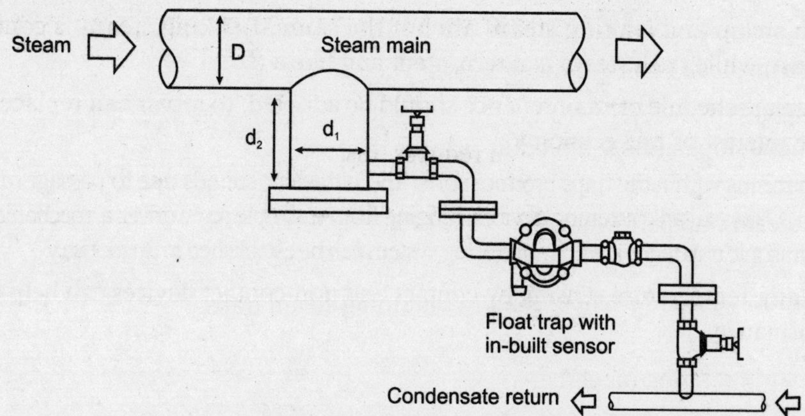

Fig.52.19: Steam drain pocket dimensions

MAINTENANCE OF STEAM TRAPS

The key functions of steam traps are to :

- I discthoarge condensate once it is formed;
- to prevent steam from escaping; and
- to discharge air and non-condensable gases.

Steam trap performance is very important for fuel economy in all steam use systems and is one of the major energy efficiency opportunities .

- Traps that are open, meaning those draining steam & condensate, result in loss of energy and direct loss of condensate (the purest form of water) & significant economic loss as boiler operating costs increase.
- Traps that are closed, or become blocked, result in reduced heating capacity of steam heating equipment, and direct loss of production.

Some tips for steam trap monitoring and maintenance are given below:

- Dirt is one of the most common problem areas. Dirt, scale, or foreign material finds its way to the traps and often damages the valve seat, preventing it from closing properly and leading to leakages.
- Strainers provided for handling dirt and other material before it reaches the traps must be maintained properly.
- Sight glasses, fitted after the traps, permit visual identification of blowing traps and need to be properly taken care of.
- If the trap selected is not the appropriate type and size for the given application, significant losses take place.
- Inverted bucket and thermodynamic traps should have intermittent condensate discharge.
- Float and thermostatic traps should have a continuous condensate discharge.

- Flash steam and leaking steam are not the same. Leaking steam is continuous, like an exhaust, while flash steam is intermittent and like a cloud.
- A regular schedule of maintenance should be adopted, to repair and replace defective traps in the interest of fuel economy.
- Mechanisms within the traps produce sonic and ultrasonic sounds due to passage of condensate and steam. Using suitable listening devices, ranging from a simple screwdriver, a mechanic's stethoscope, or an ultrasonic trap tester, a trap monitoring system can be established in an industry.
- Similarly, temperature sensing by contact and non-contact devices can help to monitor trap performance.

Air vents :

Effect of air on heat transfer

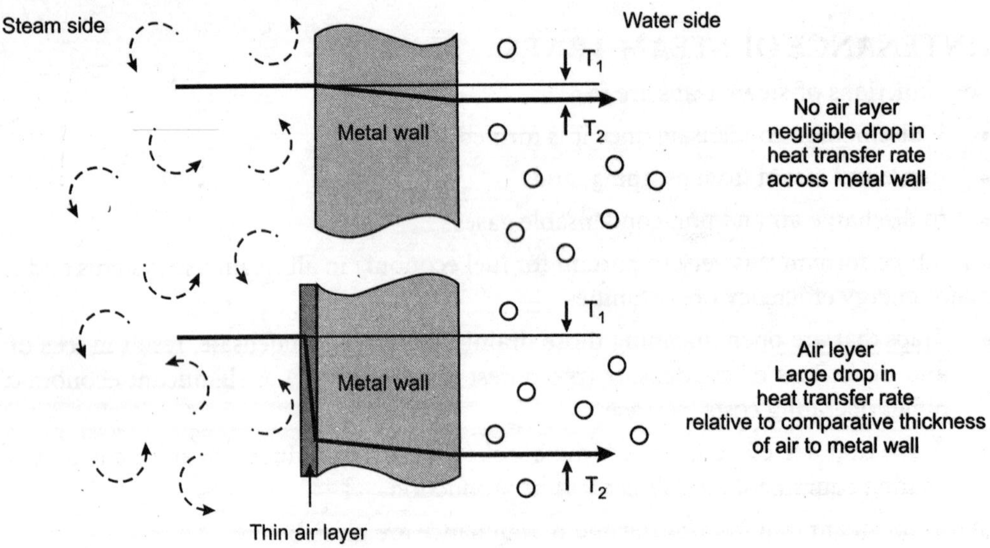

Fig.52.20: Effect of air on heat transfer of steam

Air in the system :

- During start-up
- Condensing steam draws air in pipes
- In solution in the feedwater

Signs of air :

- Gradual fall of output of steam heated equipment
- Air bubbles in the condensate
- Corrosion

Automatic air vent on jacketed pan (vessel)

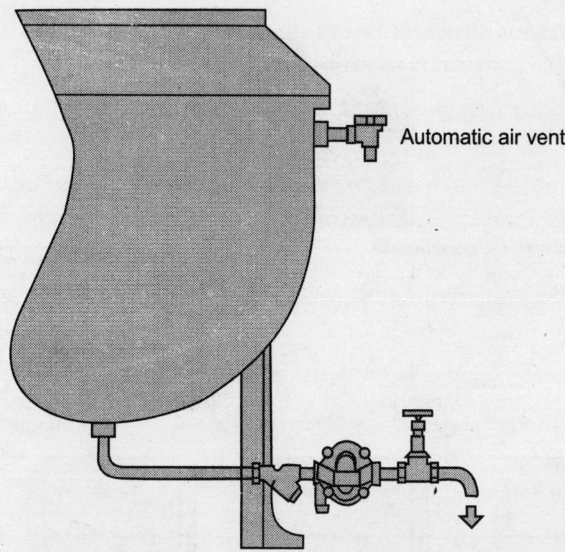

Fig.52.21: Automatic Air Vent on jacketed pan (vessel)

Automatic air vent at the end of steam main

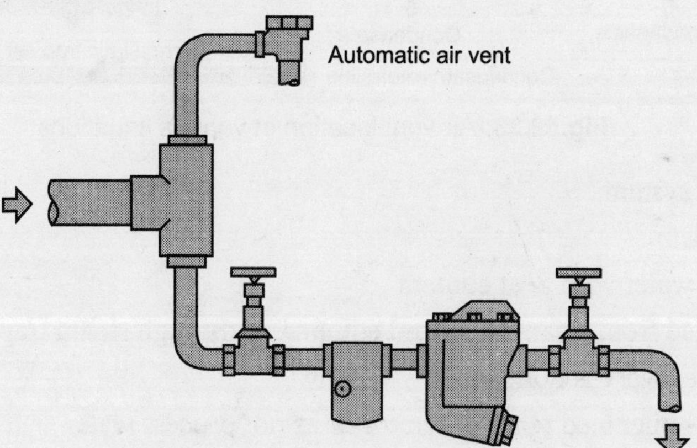

Fig.52.22: Air vent at the end of steam main

Air vent location :

- Within low lying steam trap opposite high level steam inlet .
- Opposite low level steam inlet .
- Opposite end of steam inlet .

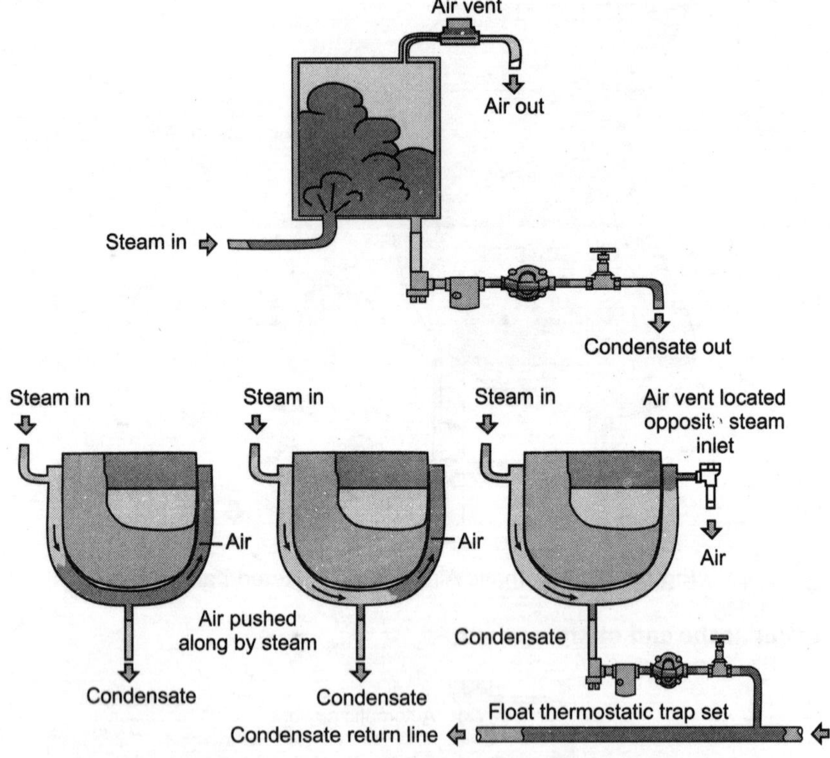

Fig.52.23: Air vent location in various situations

Condensate recovery system :

- Condensate is:
    - Distilled water with heat content .
    - Discharged from steam plant and equipment through steam traps.
- Condensate recovery is done for :
    - Reuse in boiler feed tank, deaerator or as hot process water
    - Heat recovery through heat exchanger

Reasons for condensate recovery include :

- Financial reasons
- Water charges
- Effluent restrictions
- Maximizing boiler output

Typical steam & condensate circuit with condensate recovery is shown below in figure.

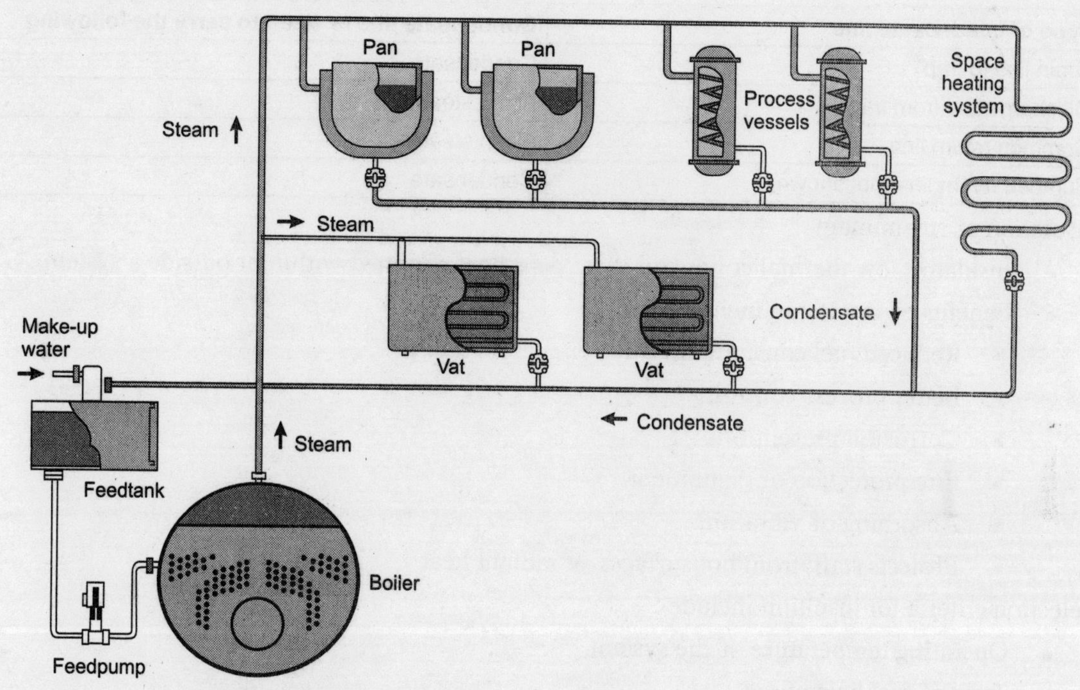

Fig.52.24: Steam & condensate circuit with condensate recovery

Four types of condensate lines are shown below in figure.

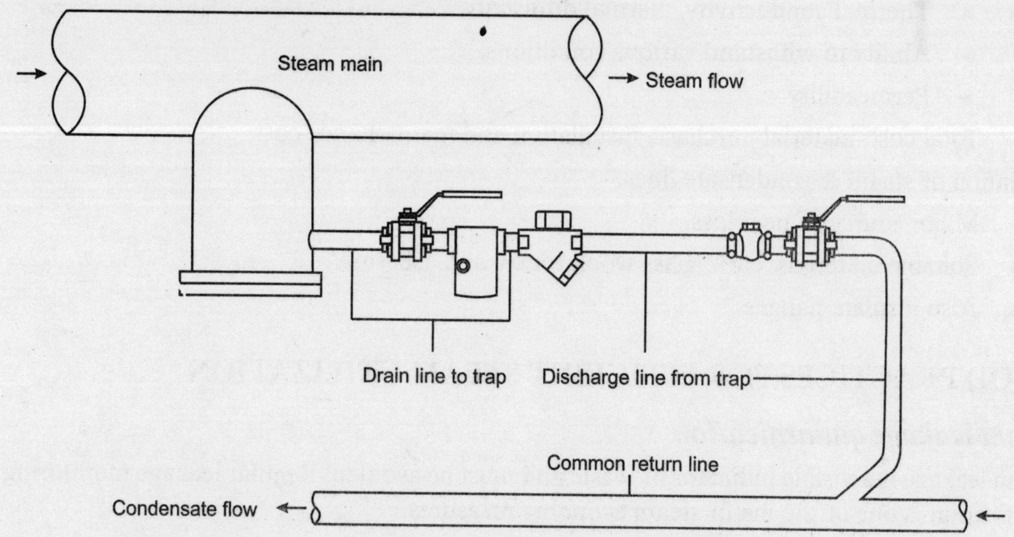

Fig.52.25: Types of condensate lines

Type of condensate line	Condensate line is sized to carry the following
Drain line to trap	Condensate
Discharge line from trap	Flash steam
Common return line	Flash steam
Pumped return line (not shown)	Condensate

Insulation of steam main :

- Insulator: low thermal conductor that keeps heat confined within or outside a system .
- Benefits of insulation include:
  - Reduced fuel consumption.
  - Better process control.
  - Corrosion prevention.
  - Fire protection of equipment.
  - Absorbing of vibration.
  - Protects staff: from hot surfaces & radiant heat

Selection criteria for insultion include :

- Operating temperature of the system.
- Type of fuel being fired.
- Material:
  - Resistance to heat, weather, fire/flames.
  - Thermal conductivity, thermal diffusivity.
  - Ability to withstand various conditions.
  - Permeability
- Total cost: material purchase, installation and maintenance cost.

Insulation of steam & condensate lines:

- Major source of heat loss.
- Suitable materials: cork, glass wool, rock wool, asbestos.
- Also insulate flanges .

GOOD PRACTICES FOR EFFICIENT STEAM UTILIZATION

Steam leakage quantification

Steam leakage is a visible indicator of waste and must be avoided. Regular leakage monitoring and rectification is one of the major steam economy measures.

A graph shown below in figure can be used for estimating steam leakage. The plume is the visible jet of steam issuing from the point of leakage. And, irrespective of pressure, the length of plume helps us to quantify steam loss in kg per hour.

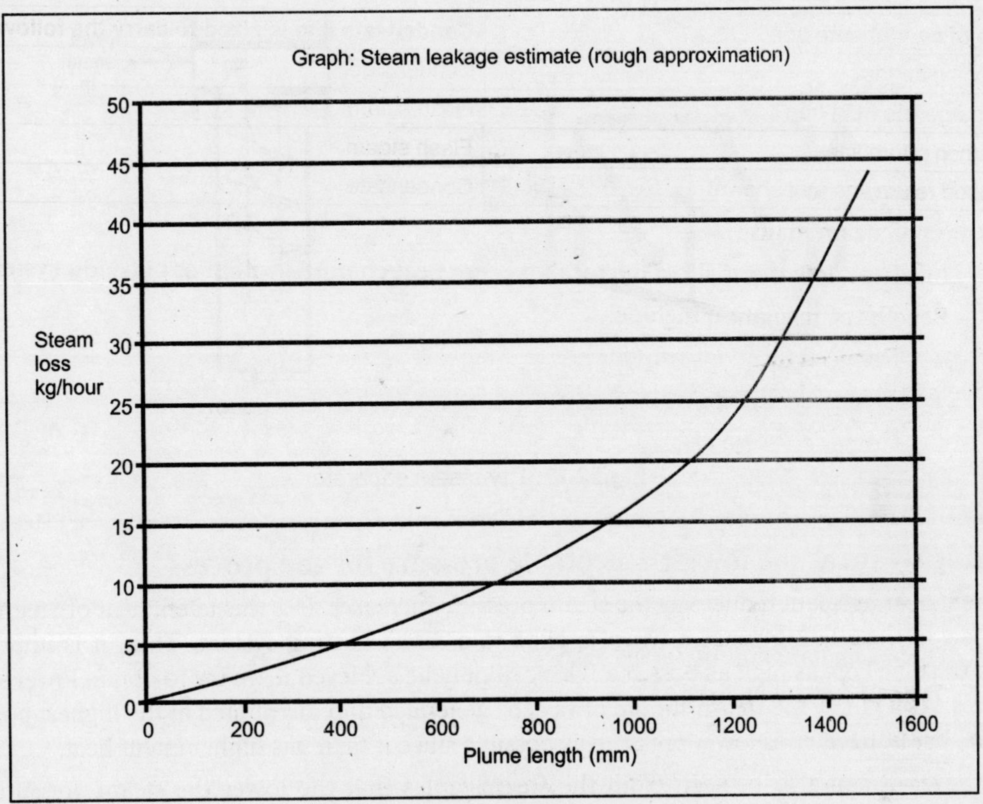

Fig.52.26: Estmating steam leakage

Example: For a plume length of 700 mm, steam loss is 10 kg/hr.

Providing dry steam for the process

Almost all heat transfer processes call for dry steam, and it is necessary to ensure dry steam conditions for efficient use. Wet steam can reduce plant productivity and product quality, and can cause damage to most items of equipment.

Disadvantages of allowing wet steam include lower heat content, extended process time, irregular heating, creation of barriers to heat transfer, and overloading of steam traps. At the same time, superheated steam is not desirable for process heating because of its poor heat transfer coefficient; and it also takes time to give up superheat by conduction. The benefit of using dry steam is that heat transfer is rapid.

To ensure dry steam for processing, steam separators may be fitted appropriately at the steam head and near steam use equipment.

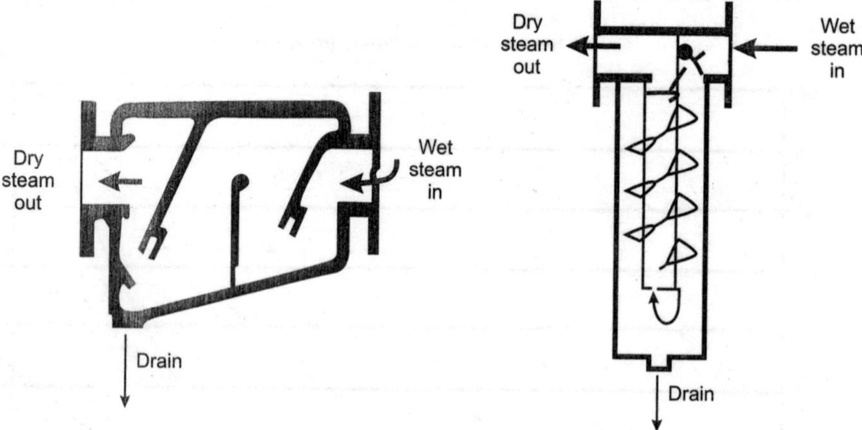

Fig.52.27: Dry steam separator

Utilizing steam at the lowest acceptable pressure for the process

The latent heat in steam reduces as the steam pressure increases. It is this latent heat of steam that takes part in the heating process when applied to a direct heating system. Thus, it is important that its value be kept as high as possible. This can only be achieved if we try to obtain lower steam pressures. As a guide, the steam should always be generated and distributed at the highest possible pressure, but utilized at as low a pressure as possible since it then has higher latent heat.

However, it can also be seen from the steam tables that the lower the steam pressure, the lower will be its temperature. Since temperature is the driving force for the transfer of heat at lower steam pressure, the rate of heat transfer will be slower and the processing time greater. In equipment where fixed losses are high (e.g. big drying cylinders), there may even be an increase in steam consumption at lower pressures due to increased processing time. There are, however, several items of equipment with which one can profitably go in for lower pressures and realize economy in steam consumption without materially affecting production time. Depending on the equipment design, the lowest possible steam pressure with which the equipment can work should be selected without sacrificing either production time or steam consumption.

Insulation of steam pipelines and hot process equipment

Heat can be lost due to radiation from steam pipes. For example, while lagging steam pipes, it is common to see flanges uncovered. An uncovered flange is equivalent to leaving 0.6m of pipeline unlagged. If a steam pipe 0.15m in diameter has 5 uncovered flanges, there would be a loss of heat equivalent to wasting 5 tons of coal or 3000 liters of oil a year. Uncovering is usually done to facilitate checking the condition of the flange, but at the cost of considerable heat loss. The remedy is to provide prefabricated insulation covers, which can be easily removed when necessary. The insulating materials used are cork,glass wool, rock wool, and asbestos.

Table below indicates the heat loss from a hot uninsulated surface to the environment.

Table : Quantity of heat lost at different temperatures

Difference in Temperature between Ambient and Hot Surfaces (°C)	Heat Loss (kCal/m^2 /hr)
50	500
100	1350
200	3790
400	13640

This is based on 35°C ambient temperature, 0.9 emissivity factor, and still wind conditions. The effective insulation of a steam system can bring down the heat losses to less than 75 kCal/m2/hr.

It is desirable to limit hot surface temperature to not more than 20°C above ambient temperature, by insulation, for fuel economy.

Minimizing heat transfer barriers

In indirect heating, a temperature difference is required to transfer heat from steam to the material. The main barrier to heat transfer is the thermal conductivity of the pipe carrying steam. Other impediments include the heat transfer barrier of an air film as well as scaling on the steam side, and scaling as well as a stagnation product film on the product side.

For instance, it has been estimated that air is 1500 times more resistant to heat transfer than steel and 13,000 times more resistant than copper.

Air is the best heat insulator ever known. It is also the most likely material to be trapped in all steam supplies, because when steam condenses, air always tries to take its place, on account of the partial vacuum created. Air is also carried into the steam space by incoming steam during startup. It is, therefore, essential that steam heated equipment should be so designed that all trapped air is pumped out automatically using proper air venting systems. Similarly, scaling of heat transfer areas should be regularly monitored and taken care of by measures like cleaning and descaling.

Condensate recovery

Steam condenses after giving off its latent heat in the heating coil or the jacket of the process equipment. A sizable portion (about 25%) of the total heat contained in the steam leaves the process equipment as condensate (hot water). If this hot condensate is returned to the boiler house, it will reduce the fuel requirements in the boiler.

For every 6°C rise in the feed water temperature, there will be approximately 1% saving of fuel in the boiler.

Further, in most cases, the boiler water has to be chemically treated to prevent or reduce scale formation, whereas condensate is almost entirely pure water, which needs no treatment. With a high share of the condensate returning to the boiler house, the expenses involved for water treatment can therefore be reduced by an appreciable amount.

Flash Steam Recovery

Flash steam is produced when high pressure condensate is flashed in an enclosure to produce low pressure steam, and can be used for heating application. The specially designed enclosure is called a flash vessel, which is available with most steam equipment suppliers.

The flash steam quantity can be calculated by the following formula with the help of steam tables:

Flash steam is calculated in % = $(S_1 - S_2) / L_2$ where,

S_1 is the sensible heat of higher pressure condensate,

S_2 is the sensible heat of steam at lower pressure (at which it has been flashed),

and L_2 is the latent heat of flash steam (at lower pressure)

Thus, higher the steam pressure and lower the flash steam pressure, the greater is the quantity of flash steam that can be generated. In many cases, flash steam from high-pressure equipment is made use of directly by the low- pressure equipment, to reduce the use of fresh steam, through pressure- reducing valves.

Flash steam can also be used on low-pressure applications like direct injection and can replace an equal quantity of fresh steam that would otherwise be required.

Reducing the work to be done by steam

All the equipment should be supplied with dry saturated steam. To reduce the work done by steam, if any product is to be dried, a press could be used to squeeze as much water as possible, before being heated up in a dryer using steam, which is a more efficient method.

Always use the most economical way to remove the bulk of water from the wet material. Steam can then be used to complete the process. For this reason, hydro extractors, spin dryers, squeeze or calendar rolls, presses, and other equipments are initially used in many drying processes to remove the mass of water. The efficiency with which this operation is carried out is most important.

When steam reaches the place where its heating is required, it must be ensured that the steam has no more work to do than is absolutely necessary. Air heater batteries, for example, which provide hot air for drying, will use the same amount of steam whether the plant is fully or partly loaded. So, if the plant is running only at 50% load, it is wasting twice as much steam (or twice as much fuel) as is necessary.

By reducing work done by steam, energy saving can be achieved by the following measures:

- Reduction in operating hours;
- Reduction in steam quantity required per hour;
- Use of more efficient technologies;
- Avoiding part load operations; and
- Minimizing wastage.

Proper utilization of directly injected steam

Where feasible, the heating of a liquid by direct injection of steam is often desirable. The equipment required is relatively simple, cheap, and easy to maintain. No condensate recovery system is necessary. The heating is quick, and the sensible heat of the steam is used up along with the latent heat, making the process thermally efficient. If the dilution of the tank contents and agitation are not acceptable in the process (i.e., if direct steam agitation is not acceptable), indirect steam heating is the only answer.

Ideally, the injected steam should be condensed completely as the bubbles rise through the liquid. This is possible if the inlet steam pressures are kept very low, around 0.5 to 1.0 kg/cm^2. If pressures are high, the velocity of the steam bubbles will also be high, and they will not have sufficient time condense before they reach the surface.

A large number of small-diameter holes (2–5mm), facing downwards, may be drilled on the steam distributor pipe. This will help in dissipating the velocity of bubbles in the liquid. Adopting thermostatic control of steam admitted is highly desirable.

Proper air venting

A 0.25mm thick air film offers the same resistance to heat transfer as a 330mm thick copper wall. The presence of air inside the process equipment will reduce the partial pressure of steam in the steam air mixture, thus dropping the overall temperature of the steam air mixture, which is the heating medium.

However, it is impossible to avoid the entry of air into a steam system that is working intermittently. If the steam condenses during the shutdowns, air enters the process equipment from the steam mains at the time of startup.

Adequate air venting provision should be made at appropriate positions in the pipelines, to purge air as quickly as possible from the equipment, making heat transfer more efficient.

Assessment of Steam Distribution System

Three main areas of assessment include:

- Stream traps
- Heat loss from uninsulated surfaces
- Condensate recovery

Energy Efficiency Opportunities include :

1. Manage steam traps
2. Avoid steam leaks
3. Provide dry steam for process
4. Utilize steam at lowest acceptable pressure
5. Proper utilization of directly injected steam
6. Minimize heat transfer barriers
7. Proper air venting
8. Minimize water hammer
9. Insulate pipelines and equipments
10. Improve condensate recovery

11. Recover flash steam
12. Reuse low pressure steam

Manage steam traps :

- Testing of steam traps:
  - Visual: flow and flow variations
  - Sound: check sound created by flow
  - Temperature: discharge temperature on outlet
  - Integrated: measures conductivity
- Routine maintenance
- Replacement of internal parts
- Replacement of traps

Avoid steam leaks:

- Repair leaks
- Regular leak detection program
- Replace flanged joints by welded joints
- Leakage estimate

Provide dry steam for process:

- Dry saturated steam is best steam because :
  - Wet steam reduces total heat in steam and prevents heat transfer.
  - Superheated steam gives up heat at slower rate.
- Achieve dry steam by:
- Proper Boiler operation
  - Pipeline insulation
  - Separators on steam pipelines

Utilize steam at lowest acceptable pressure :

- Steam should be :
  - Generated & distributed at highest pressure.
  - Utilized at lowest pressure: latent heat highest.
- Select lowest steam pressure without sacrificing:
  - Production time
  - Steam consumption

Proper utilization of directly injected steam :

- Benefits:
  - Equipment simple, cheap and easy to maintain.
  - No condensate recovery system needed.
  - Heating quick and process thermally efficient.
- Only in processes where dilution is not a problem.

Minimize heat transfer barriers :

Temperature gradient across heat transfer barriers is shown below in figure.

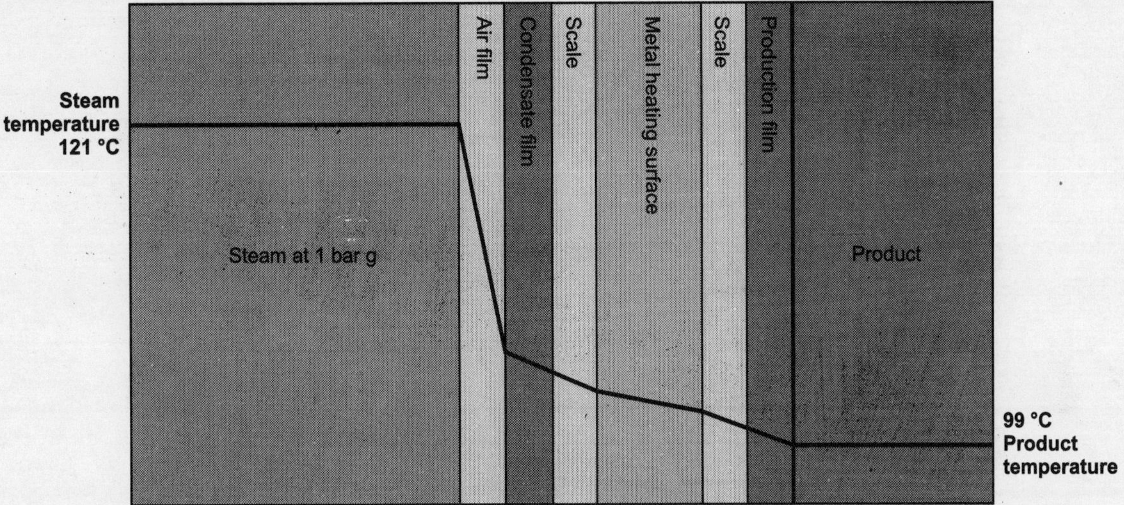

Fig.52.28: Temperature gradient across heat transfer barriers

Possible solutions :

- Stagnant film: product agitation
- Scale :
  - Regular product cleaning
  - Regular surface cleaning on steam side
  - Correct operation of boiler
  - Removal of water droplets with impurities
- Condensation: coat that inhibits wetting
- Air: air venting

Minimize water hammer :

- Banging noise caused by colliding condensate in distribution system
- Sources: low points in the pipework
- Solutions :
  - Steam lines with gradual fall in flow direction
  - Drain points at regular intervals
  - Check valves after all steam traps
  - Opening isolation valves slowly to drain condensate

Economic thickness of insulation:

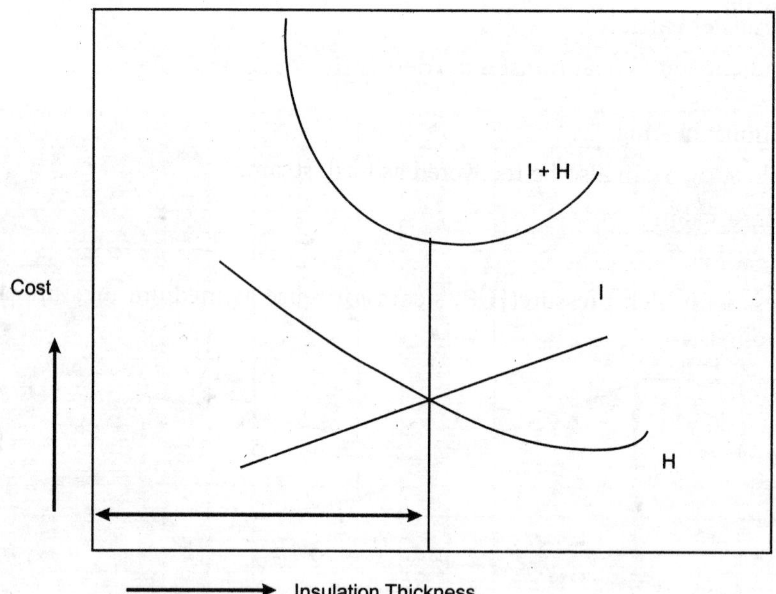

Fig.52.29: Determination of economic thickness of insulation

I+H =Economic thickness of insulation
I = cost of insulation
H = Heat loss savings

Improved condensate recovery :

- Energy in condensate is lower than energy in steam but worth recovering:

Every 6°C rise in the feed water temperature = 1% fuel savings in the boiler

- Energy in condensate lower than energy in steam but worth recovering:

Every 6°C rise in the feed water temperature = 1% fuel savings in the boiler

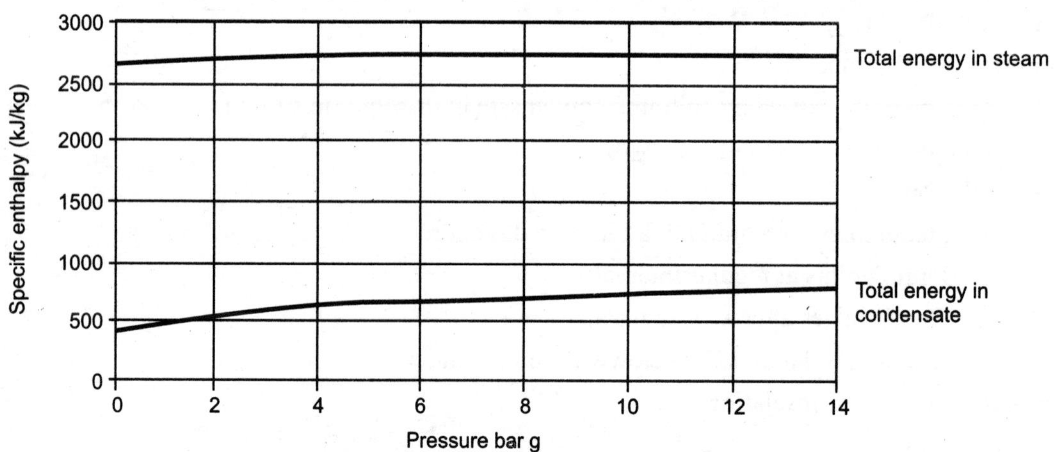

Fig.52.30: Energy in steam vs. that in condensate

Recover flash steam :

- Flash steam released from hot condensate when pressure is reduced.
- Amount available: calculation or tables/charts
- Applications: heating
- Boiler blowdown can also be recovered as flash steam.

Reuse low pressure steam :

- Reuse as water.
- Compress with high pressure(H.P) steam for reuse as medium pressure steam as shown below in figure .

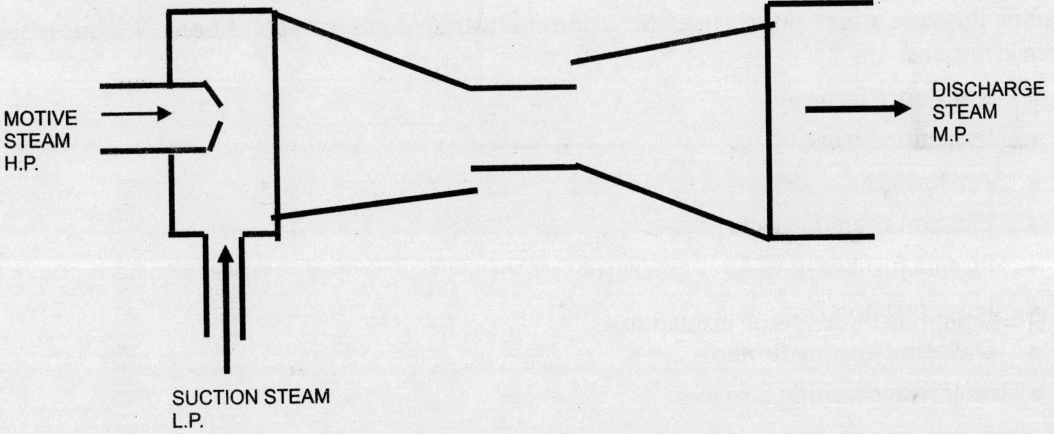

Fig.52.31: Thermo-compressor

53

ELECTRIC FURNACES

Electric furnaces utilize electric heat for various industrial/domestic applications. Various types of electric furnaces are :

- Resistance furnace
- Induction furnace
- Arc furnace
- Electron beam furnaces
- Plasma furnaces
- Laser furnaces
- Dielectric heating furnaces
- Microwave heating furnaces

The characteristics of electric heat are:

1. Precision of the control of the development of heat and of its distribution.
2. The heat development is independent of the nature of the gases surrounding the charge. This atmosphere can be selected at will with reference to the nature of the charge and the chemistry of the heat process. This freedom is a primary reason for the use of electric heat.
3. The maximum temperature is limited only by the nature of the material of the charge. The first two characteristics underlie the design of all electric heating apparatus. The third is utilized in thermal processes for the production of certain materials not obtainable in any other way.

RESISTANCE FURNACE

Basic Principles of Resistance Furnaces

All resistance furnaces operate on the basis of Joule's law, i.e., an electrical conductor emits heat when a current flows through it. The quantity of electric power that is converted into heat may be expressed by:

P =VI _Joule's law_---------------------------------- (1)

where P is the electric power (W), U the voltage (V), and I the current. Ohm's law relates the resistance of the conductor (R in Ω) and the current through it to the applied voltage:

U=RI _Ohm's law_------------------------------------ (2)

Substituting Equation (2) into Equation (1) gives

$$P = I^2 R$$

The amount of energy (heat produced in Joules after t seconds) may be represented by:

$$Q = P\,t = I^2 R\,t$$

The above equations show that the higher the resistance of a conductor or the larger the current, the more electric energy is converted into heat.

The resistance of a conductor depends on the resistivity of the material and on its shape (length and cross-section):

$$R = \sigma l / S$$

where σ is the resistivity ($\Omega\sigma$ cm), l the length, and S the cross-section.

The resistivity is also a function of the temperature:

$$\sigma T = \sigma 273K\ (1+\alpha T)$$

where σT is the resistivity at T Kelvin, $\sigma 273K$ the resistivity at 273 K, T the temperature in K, and α the resistivity temperature coefficient in K-1.

When electricity is passed through the resistance, the resisting element will be heated up and this heat will be utilized to melt the charge.

Advantages of resistance furnace: Accurate temperature control is possible , we can get the accurate temperature and at times it becomes necessary to hold the molten metal at certain temperature for sometime, during which we add some alloys. For that time, for that purpose, we have to hold the molten metal at certain temperature for certain time. That may not be possible in the case of the cupola furnace; that may not be in the case of arc furnace, but that holding the liquid metal is possible in the case of resistance furnace and its applications include melting steel and non-ferrous alloys.

In Resistance Furnaces heat transfer occurs either directly or indirectly according to Ohm's and Joule's laws. They can be operated in an a.c. or d.c. mode.

Applications of resistance furnaces: Furnaces which implement direct resistance include:

- Resistance furnaces for solid-state reactions, e.g., production of graphite and carbides.
- Resistance furnaces for the production of copper, nickel, iron, tin, and zinc or their intermediate products from oxidic and sulfidic materials and slags .
- Electroslag refining (ESR) for the production of clean ferrous or nonferrous metals and alloys such as titanium, steels, and superalloys,
- Furnaces that employ a combination of arc and resistance heating are used for the production of calcium carbide, ferroalloys, pig iron, phosphorus, and silicon compounds by reduction. Electrodes are situated within the charge; reduction and melting occur within the high-temperature zone around the electrodes .
- Fused-salt electrolysis cells for the production of aluminum and alkali metals may also be considered to be electrochemical thermal reactors .
- Salt-bath furnaces for the heat treatment of metals.

Indirect heating is mostly performed in a closed chamber in which heating elements, situated in the walls or suspended in the chamber, transfer their heat via radiation, conduction, and convection to the object to be heated. This type of furnace is applied for heat treatment of metals, ceramics, and for the production of carbides .

In most cases slag acts as the main resistance in furnaces. Direct resistance furnaces exploit the above principles for metallurgical purposes . In direct resistance furnaces the feed or charge is the conductor and its resistance to the current flowing through it creates the heat for melting. The current in these reactors flows between at least two or more electrodes, one of which may be submerged in the material to be heated/melted.

Direct resistance furnaces may be divided into three categories on the basis of the reactions taking place in them:

1. Reduction resistance furnaces
2. Refining resistance furnaces
3. Solid state resistance furnaces

Reduction Resistance Furnaces

Electrothermal reduction furnaces use electric energy to heat the feed to the required operating temperature. In conventional reduction furnaces, heat liberated during the exothermic reaction between, for example, coke and oxygen is utilized to heat the feed. In electrothermal furnaces, coke is only needed for reduction.

Reduction resistance furnaces have a number of advantages over furnaces using fossil fuels:

- Electrothermal processes permit higher productivities due to higher temperatures and energy densities
- Higher grade products can be produced because ash and impurities (e.g., phosphorus and sulfur) derived from the fossil fuel are absent .
- Since the reactors are usually covered, dross formation can be minimized because atmospheric oxygen may be excluded .
- A smaller volume of off-gas volume is produced by resistance reactors with a higher carbon monoxide concentration because only reduction reactions take place.
- Electric furnaces can easily be enclosed, permitting tight control of the off-gas composition
- Metallurgical coke can be replaced by, for example, charcoal which contains much less impurities .
- If the furnace is operated under vacuum conditions, reduction and degassing can take place in a single step .

An obvious disadvantage of electric furnaces are high energy costs. However, where hydropower or inexpensive or excess electric power is available, this is an economically viable method for producing metals.

Refining Resistance Furnaces

The electroslag refining process (ESR) is the classical example for refining operations in a resistance furnace. The high resistance of the slags (CaF_2–CaO–Al_2O_3) is used to produce enough heat to melt and refine high-grade steels .

The main components of the furnace are its consumable electrode (raw material), a water-cooled mold, and a bottom plate, which are connected to an alternating current source . Initially the electrode is submerged in a high-melting liquid slag , the mold and bottom plate contain the slag. At sufficiently high voltages the liquid slag starts to melt the consumable electrode at its surface. The molten metal collects at the tip of the electrode and drops through the slag. During this process it is refined due to its interaction with the slag. The molten metal collects in the mold, and the bottom plate moves downwards at the same rate as that of metal production.

The refined metal is cooled by the water cooled mold and solidifies. The mold and bottom plate are made of copper and water cooled to prevent reactions and to ensure high heat transfer .

Solid-State Resistance Furnaces

The feed material for this type of furnace has a high electric resistance. A current is applied to the feed via two electrodes and the heat generated permits the desired reactions to take place in the solid state.

Two typical applications are the graphite furnace and the silicon carbide furnace .

Graphite Furnace. In this furnace petroleum coke is converted into graphite which is used for electrodes in other electrothermal furnaces. The feed (petroleum coke with tar as a binder) is mixed and usually extruded into the desired shape of the electrodes. The electrodes are heated up to 800 – 1300 °C in fuel heated furnaces to carbonize the binder. Subsequently the electrodes are charged to a resistance furnace ,where they are stacked parallel to the face sides .

The void areas are filled with crushed graphite to ensure that the electrodes are in good electric contact with one another. An electric current flows through the furnace and produces the operating temperatures (2400 – 2800 °C). The amorphous coke is transformed to graphite (hexagonal lattice) within several days .

Silicon Carbide Furnace. In this case two solid materials (sand and coke) are mixed & filled into a furnace which is similar to that used for graphite production. The reaction, $SiO_2+3C = SiC+2CO$ takes place in the solid state. The sand – coke mixture (with additions of sawdust and sodium chloride) is charged into the furnace around a central core of graphitized coke . This central core is connected to electric power and the coke is used as a conductor, which reacts in a similar way as described in the graphitization furnace above.

Silicon carbide formation starts at 1700 °C, at 2400 – 2500 °C a coarse crystalline product is formed. Sodium chloride is added to evaporate undesired elements such as aluminum, iron, and titanium as chlorides. Only 20 –30% of the reaction product is high grade material.

Resistance furnaces may be either the batch or the continuous type. Batch furnaces include box furnaces, elevator furnaces, car-bottom furnaces, and bell furnaces. Continuous furnaces include belt-conveyor furnaces, chain-conveyor furnaces, rotary-hearth furnaces, and roller-hearth furnaces.

Standard resistor furnaces are designed to operate at temperatures within the range 550 to 1,200°C.

The heating chamber of a standard resistance furnace is an enclosure with a refractory lining, a surrounding layer of heat insulation, and an outer casing of steel plate, or for large furnaces an outer layer of brick or tile, as shown below in figure . The hearth of a batch furnace often is constructed of a heat-resisting alloy, made in sections to prevent warping. In some continuous furnaces the conveyor forms the hearth; in others a separate hearth is required.

Insulating firebrics is commonly used for the inner lining of the heating chamber. This material has thermal and physical properties intermediate between those of fire-clay brick and heat-insulating materials. A lining of this kind has less heat-storage capacity than a fire-clay brick lining, and its use accordingly decreases the time periods of heating and cooling the chamber and also decreases the stored-heat loss for a given cycle of operation. Other advantages are its high heat-insulating value and light weight.

The maximum temperature of the inner face of the layer of heat insulation determines the character of material required for the insulation. Practically all resistor furnaces have insulation made of diatomite. Composite wall structures with a 11-cm semi refractory lining and a 23 to 33 cm layer of heat insulation represent general practice for standard furnaces.

Atmospheres. A mixture of air and the gases evolved from the charge constitutes a natural atmosphere in the heating chamber of a resistor furnace. The composition of such an atmosphere in a batch furnace is variable during a heating cycle. A natural atmosphere in the heating chamber of a continuous furnace is mainly air. Natural atmospheres are used where the extent of the action of oxygen on the charge during the heating cycle is not objectionable and for processes where that chemical action is desired.

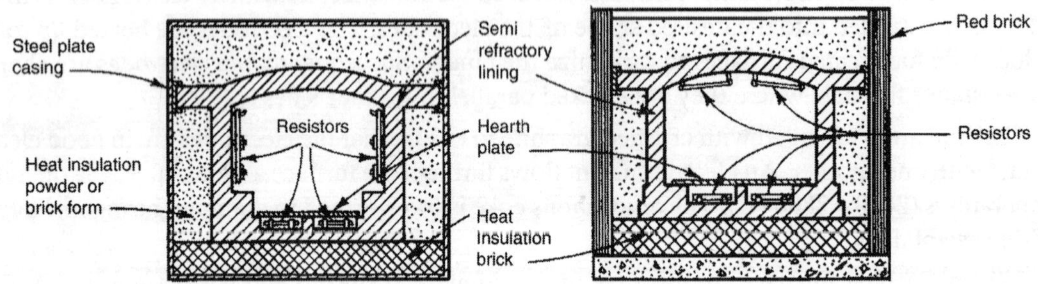

Fig.53.1: Heating chamber with sidewall and hearth resistors.

Fig.53.2: Heating chamber with roof and hearth resistors

High-Temperature Furnaces. Silicon carbide is the basis of a type of non-metallic resistor for heating-chamber temperatures up to about 1,560°C . The material is formed into rods. Resistors of this material do not require protection against oxidation and are operated on constant-voltage circuits.Molybdenum resistors are suitable for temperatures up to 1,670°C . Above that temperature the metal begins to vaporize. A molybdenum resistor cannot be operated in a natural atmosphere, and also it must be protected from reactions with silica and carbon. The metal is immune from reactions with sulfur compounds, nitrogen, and water vapor. Hydrogen is the most common artificial atmosphere used with molybdenum resistors. The difference between the cold and hot resistances of the circuit makes a starting device necessary. Other materials used to some extent for resistors are iron, tungsten, and graphite. These require protection against oxidation.

Temperature Regulation. The temperature of the heating chamber of a resistance furnace is in most cases regulated by a more or less intermittent application of current—the on-and-off method—which is made automatic by instrument control. This method utilizes the heat storage capacity of the inner lining of the heating chamber as a temperature equalizer. The variation from the normal temperature of the chamber can be kept within less than 4°C plus or minus, without undue wear of the temperature-control equipment. Temperature regulation by voltage control is equally applicable to resistor furnaces, and the trend is toward the use of this more accurate method particularly for the more important installations.

Temperature protection for resistor furnaces is obtained by means of a temperature fuse mounted in the heating chamber and connected in the control circuit of the power supply to the furnace.

Multiple-Temperature Control. The resistors of the larger furnaces are divided into two or more circuits. Each circuit can be equipped with individual temperature control. That arrangement provides temperature regulation at more than one location in the heating chamber and is an aid toward maintaining uniform temperature distribution within the chamber. The subdivision of resistor circuits is used also for zone heating— and zone cooling where needed—in continuous furnaces.

Melting Pots. Resistor heating is applied to melting pots for the soft metals and alloys and for lead baths and for salt baths. The immersion heating unit is used for temperatures up to 530 °C . For higher temperatures the metal pot is heated by resistors mounted outside and around the pot. The assembly in each case includes a heat-insulating wall similar to that of a resistor furnace. Another method of heating applicable only to salt baths is the passage of alternating current (of any frequency) between electrodes immersed in the bath.

Tempering Furnaces. The temperature is comparatively low- below 720°C . Electrically heated oil baths and salt baths are used for tempering many kinds of small parts. Another form of tempering furnace is a vertical resistor furnace with the addition of a removable metal cylinder (or basket) to contain the charge and to provide an annular passageway for the circulation of air (by a fan mounted on the furnace) over the resistors and thence through the charge—an application of forced convection heating.

Sizes. The electrical rating is the general method of expressing the size of a resistance furnace. Sizes up to 100 kW predominate, 100 to 500 kW furnaces are common, and others within the range 500 to 1,000 kW are in service.

The losses from a resistance furnace for a given heating chamber temperature are as follows: The open-door loss is a variable depending on the area of the door (or doors) and the percentage of the time that the door is open—from a continuous furnace this loss also varies with the type and speed of the conveyor; with artificial atmospheres, the loss of heat in the gases discharged for atmosphere control; the stored-heat loss, a variable that depends on the extent and frequency of the cooling of the furnace within a given period of operation; the heat dissipated from the outer surfaces of the furnace.

The operating efficiency is expressed as either kg. of material treated or kWh or kWh per ton. Representative values for average service for the heat treatment of steel range from 3-5 kg /kWh. Corresponding values for nonferrous metals and alloys are within the range 5-10 kg /kWh.

Resistance Ovens. The resistor oven is a modification of the resistance furnace to correspond to the low temperatures (<260 °C) of drying and baking processes. The heating chamber is an insulated metal structure with a fresh-air inlet and an exhaust fan for ventilation (for the removal of vapors and gases evolved from the charge). A refractory lining is not required. Ovens may be of the batch type with conventional methods of handling the charge of the continuous type, usually with chain conveyors.

The most common type of electric oven is heated by resistors mounted in a separate compartment of the heating-chamber enclosure. The heat transfer is by forced convection which is accomplished by recirculation of the chamber atmosphere by a motor-driven fan. A resistor oven with filament-type lamps as heating units provides what is generally known as infrared heating. The lamps, usually with self-contained reflectors, surround the charge, and the heat transfer is by radiation, mainly in the infrared portion of the spectrum. This type of oven is best adapted to the continuous heating of charges which present a large surface area in proportion to the mass and which require only surface heating, e.g., baking finishes on sheet products.

Ventilation . The vapors and gases evolved from the charge during baking processes are often flammable, and the continuous discharge of these products from the oven chamber is essential for protection against explosions.

In resistance furnaces of the submerged-arc type, heat is developed by the passage of current from electrode to electrode through the charge. The manufacture of basic products, such as ferroalloys, graphite, calcium carbide, and silicon carbide, is the general service. Alternating current at a standard power frequency is used. An exception is the use of direct current where the product is obtained by electrolytic action in a molten bath, e.g., in the production of aluminum.

SUBMERGED-ARC AND RESISTANCE FURNACES

The resistance furnace is essentially a refractory lined chamber with electrodes—movable or fixed—buried in the charge. Resistance furnaces also are termed submerged-arc furnaces and/or smelting-type furnaces. The only limit on the temperature to which a charge can be heated by this method is the temperature at which the materials of the charge are vaporized. For temperatures beyond the limit of refractory linings, the materials of the charge are used to form a protective layer between the core of the charge (through which the current passes) and the walls of the furnace.

Resistance furnaces with movable electrodes may be either single-phase or polyphase. The materials of the charge are fed more or less continuously, and the product is discharged intermittently or continuously as required. In some cases the product is in the molten state; in others the product is a vapor. The usual method of operation is the use of a single operating voltage and a constant power input. The power is regulated by adjustment of the depths of the electrodes in the charge.The load is fairly uniform and, if polyphase, is kept reasonably well balanced. The resistance furnace with fixed electrodes is designed for heating materials in batches and is usually rectangular in shape with an electrode at each end for single-phase operation. The length and cross-sectional area of the path of the current are proportioned to suit the power characteristics of the charge. Refractory materials have negative temperature-resistance coefficients, and hence to maintain constant power in the furnace circuit the applied voltage must be reduced as the temperature of the charge rises in proportion to the square root of the ratio of the initial resistance of the furnace circuit to the resistance of the furnace circuit at the end of the heat cycle. If the materials of the charge are non conductors at room temperature, a starting circuit is provided by means of a core of carbon—usually coke—placed in the charge. The heat cycles of furnaces of this class generally extend over a period of several days.

Some of the more common uses of the resistance furnace are:

Calcium carbide furnaces are charged continuously with lime and coke. These equipments can be either open or closed top. This type of furnace has been built up to 70,000 kVA in electrical powering— covered and sealed for gas collection.

Ferroalloy furnaces for the production of ferrochrome, ferrosilicon, ferromanganese, etc., are usually three-phase furnaces with movable electrodes and are similar in construction to the three-phase arc furnace. The charge is a mixture of the ore (oxide) of the selected metal, scrap iron, and a reducing agent, generally carbon except for very low carbon content alloys, for which some other reducing agent such as aluminum or silicon is required. Six-electrode furnaces often are used for power inputs of 15,000 kVA and more.

The graphitizing furnace is of the single-phase batch type. Artificial graphite is made by heating amorphous carbon (coal or coke) while shielded from air to a high temperature—around 2,500 °C. The presence of some metallic impurity, such as iron oxide, in thecharge appears to be necessary for the conversion of amorphous carbon to graphitic carbon. The raw material for making bulk graphite constitutes both the charge and the protective layer around the core of the charge. Graphite shapes are made from the corresponding shapes of amorphous carbon which are embedded—between the electrodes—in raw material as noted for the manufacture of bulk graphite.

The silicon carbide furnace is similar to the graphitizing furnace. The charge is a mixture of sand (silica), coke, sawdust, and a small amount of salt. This mixture is packed around a core of granulated coke to form the initial circuit between the electrodes. The sand and coke are the reacting materials. The sawdust serves to make the charge porous so that the gases formed during the heating of the charge can escape freely. The salt vaporizes and removes impurities, such as iron, in the form of chlorides. The temperature of the process is 1,500 to 1,880 °C .

INDUCTION FURNACE

In induction furnaces heat is developed by currents induced in the charge. The service is melting metals and alloys.

Basic Principles of electric induction heating

Electromagnetic induction heating is a direct method for contactless heating with a high power density which implies that the material to be heated (melted) is not contaminated by heating gases or electrode materials. However, contact and reactions with the crucible (graphite or refractories) may take place. Advantages of these types of furnaces are excellent alloying and mixing conditions, good temperature control of the melt, a low dross formation, and low off-gas volume.

The heating principle is as follows:

An a.c. coil induces a potential in an electrical conductor (in induction furnaces solids and/ or melts) situated inside the coil due to the changing magnetic field, which creates eddy currents (Lenz's law). The eddy currents or induced current produce heat according to Joule's law .

The induced current within the conductor is not evenly distributed, but mostly located near the surface (skin effect)

In induction heating, the lateral surface of the charge is exposed to an alternating magnetic flux. The currents thus induced in the charge flow wholly within its mass. The term ''eddy-current heating'' is sometimes applied to the method.

A common assembly, if the charge is to be heated to a temperature below its melting point, is to place the charge within a coil asshown in Figure below . An alternating current in the coil establishes the required alternating magnetic flux around the charge.

A feature of "induction heaters," is the absence of heat insulation; the coil is water-cooled. The charge is heated in open air, or an artificial atmosphere can be used, if the heating assembly is enclosed. This requires rapid heating with heat cycles measured in minutes or seconds.

The frequency required is a function of the electric and magnetic properties of the charge at the temperature specified for the heat application and of the radius, or one-half the thickness, of the charge. This frequency for a given material increases with decrease of the dimension noted. The frequency in any case is not critical. In practice, 480, 960, 3,000, and 9,600 Hz and around 450 kHz suffice for the entire range of induction heating.

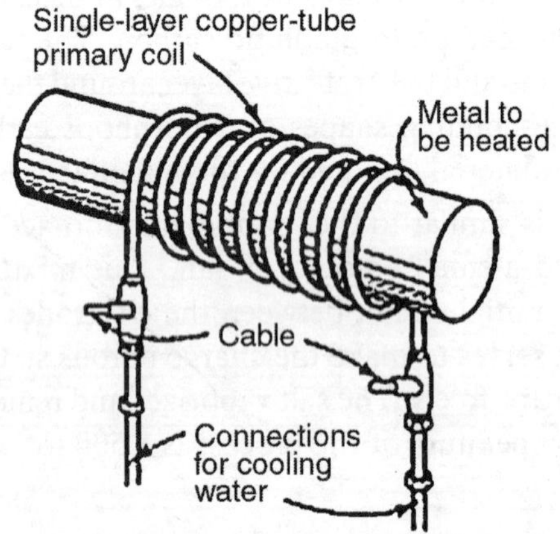

Fig.53.3: Assembly for induction heating

The highest frequencies needed are those for heating steel to temperatures above the Curie point. About 1/2 inch diameter in this case is the lower limit for 9,600 Hz. This limit is decreased for steel heated to temperatures below the Curie point and for all charges of nonferrous materials.

The operation can be either batch heating or continuous heating as required. Applications include heating for forging, for annealing, for hardening steel, for brazing, soldering, and strain relief. As most of the heat is developed within the annular zone of the charge, the method is particularly well adapted to heating steel parts for surface hardening. A recent application of induction heating is the raising in temperature of billet-size ingots for rolling into merchant bars.

Types of induction furnace

There are two basic types of metal-melting induction furnaces:

(1) coreless and (2) core-type. Both types utilize the principle of a transformer. The high-voltage circuit is coupled with that of the low voltage without directly connecting the two circuits. The element responsible for this coupling effect is the magnetic field. Induction heating utilizes the property of the magnetic field, which enables heat to be transferred without direct contact. By correctly disposing the high-voltage winding, which in the case of the induction furnace would be

an induction coil or inductor, the magnetic field is directed so that the metal to be heated or melted is made to absorb energy. The temperature attainable is limited solely by the resistance to heat of the surrounding lining material. Induction heating enables any temperature to be achieved while providing for excellent regulation of temperature and metallurgical properties. Any metal which will conduct electric current can be melted in an induction furnace.

Coreless Induction Furnaces (See Figure below) consists of a crucible, copper coil, and framework on supports arranged for tilting and pouring. The specially designed induction coil acts as the primary of the transformer. The crucible conforms to conventional refractory practice. A rammed crucible is used for furnaces above 50 kW, and preformed crucibles are used on smaller furnaces such as laboratory units.

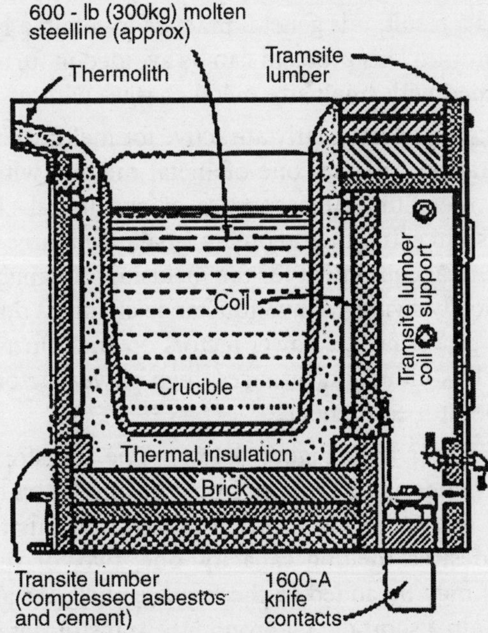

Fig.53.4: Coreless induction furnace.

The principle of operation is essentially the same as that of the induction heater previously described. The initial charge in the furnace is cold scrap metal—pieces of assorted dimensions & shapes and a large percentage of voids. As the power is applied and the heat cycle progresses, the charge changes to a body of molten metal; additional cold metal is added until the molten-metal level is brought to the desired temperature and metallurgical chemistry. The furnace then is tapped. When the metal in the furnace becomes fluid, a certain electromagnetic stirring action will occur. This stirring action is peculiar to the induction furnace and aids in the production of certain types of alloys. The stirring action increases as the frequency is reduced.

There is always an ideal relationship between the size of a coreless furnace and its operating frequency. As a general rule, a small furnace gives best results at high to medium frequencies and large furnaces work best at the lower frequencies. A frequency is suited to a given furnace when it yields good, fast melting with a gentle stirring action. Too high or too low frequencies are accompanied by undesirable side effects. The tabulation below gives the charge weights and frequencies generally to be used in induction furnace :

Charge, weight, lb	Frequency ,Hz
2 – 50	9,600
12 – 500	3,000
200 – 15,000	960
800 – 75,000	60

The coreless induction furnace is usually charged full and tapped empty, although at line frequencies, it may be necessary to retain a certain amount of metal in the furnace to continue the operation, since it is difficult to start the furnace with small metal particles, such as turnings and borings, in a cold crucible. As a result, it is general practice to retain a heel in the furnace of about one-third its molten-metal volume. This problem can be avoided in furnaces of higher frequencies, where start-up can be performed with small-size metal charges without carrying the heel.

Coreless induction furnaces are particularly attractive for melting charges and alloys of known analysis; in essence, the operation becomes one of metal melting with rapidly absorbed electric heat without disturbing the metallurgical properties of the initial charge. These furnaces are supplied from a single-phase source. In order to obtain a balanced three-phase input, it is necessary specifically to design the electrical equipment for the inclusion of capacitors and suitable reactors, which are generally automatically switched (by inductance changes) during the operation in order to provide a reasonably high power factor. Power factors on such furnaces can be kept at or near unity. In high-frequency coreless induction furnaces, high power factors are necessary to prevent overburdening the motor-generator equipment.

Core-Type Induction Furnaces (See Figure below) .The transformer is actually wound to conform to a typical transformer design having an iron core and layers of wire acting as a primary circuit. The melting channel acts as a ring short circuit around this transformer in the melting chamber. According to the desired melting capacity, one, two, or three such transformers (or inductors, as they are called) may be added to the furnace shell. At all times, the channel must hold sufficient metal to maintain a short circuit around the transformer core. Air cooling is used as required to prevent undue heating of the inductor coils and magnetic cores.

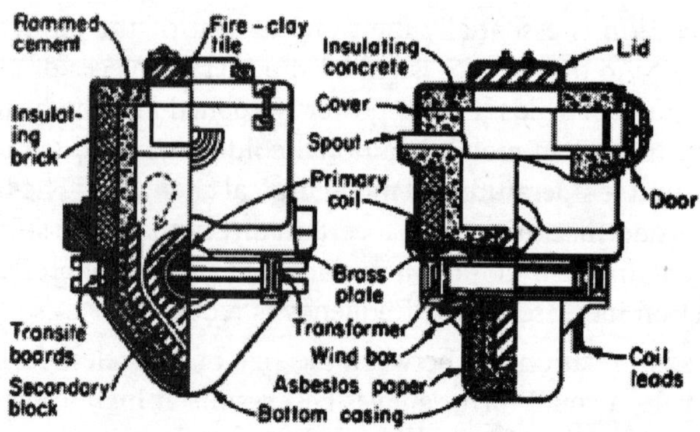

Fig.53.5: Core-type induction furnace.

The melting output is controlled by varying the voltage supplied to the inductors with the aid of a variable-voltage transformer connected to the primary circuit of the supply. Core-type furnaces always use line frequencies. Voltage or power-input regulation, therefore, can be performed by adjusting the tap setting of the transformer feeding the furnace transformer attached to the furnace shell. These transformers are single-phase units, and by using three such units, a balanced three-phase input can be obtained. The current flowing through the primary inductors by transformation causes a much larger current in the metal loop, whose resistance creates heat for melting.

The core-type furnace is the most efficient type of induction furnace because its iron core concentrates magnetic flux in the area of the magnetic loop, ensuring maximum power transfer from primary to secondary. Efficiency in the use of power can be as high as 95 to 98 percent. The essential loop of metal must always be maintained in the core type furnace. If this loop is allowed to freeze by cooling, extreme care is necessary in remelting because the loop may rupture and disrupt the circuit. This could require extensive work in dismantling the coil and restoring the loop. Consequently, core-type furnaces rarely are permitted to cool. This makes alloy changes difficult because a heel of molten metal always is required.

The relatively narrow melting channels must be kept as clean as possible since a high metal temperature exists in this loop. Non metallics or tramps in the charge metal tend to accumulate on the walls in the channel area, restricting the free flow of metal and ultimately closing the passage.

This furnace is particularly useful for melting of nonferrous metals such as aluminum, copper, copper alloys, and zinc.

Average values in kWh/ton of molten metal are as follows: yellow brass, 200 to 350; red brass, 250 to 400; copper, 250 to 400; lead, 30 to 50; steel melting, when making high-quality double-slag basic heats, 650 to 800 .

Induction furnaces are used to melt both ferrous and non-ferrous metals. There are several types of induction furnaces available, but all operate by utilizing a strong magnetic field created by passing an electric current through a coil wrapped around the furnace. The magnetic field in turn creates a voltage across, and subsequently an electric current through, the metal to be melted. The electrical resistance of the metal produces heat, which in turn melts the metal. Induction furnaces are made in a wide range of sizes. Because there is no contact between the charge and the energy-carrier, the induction furnace is suited for the melting of steel, cast iron and nonferrous metals, so long as a suitable lining material can be found.

Preference to Induction Furnaces over Electric Arc Furnaces :

India is using Induction Melting Furnaces for making mild steel. The bulk of structural quality mild steel for long products is manufactured by Induction Melting Furnaces. The EAF (Electric Arc Furnace) units have also installed Induction melting furnaces. There are several reasons for the popularity of Induction Melting Furnaces for making steel. They consume less power compared to EAF . Expenditure on electrode is nil.

They use lesser quantity of refractory. Initial investment is less on plant and equipment. Thus, there are economic advantages in making steel through Induction Furnaces route.

The environmental pollution in case of EAF is much more than Induction furnaces. The only snag of steel production through induction furnace route is of capacity limitation .

Most of the industries are using coreless induction furnaces (IF) as this system is much more efficient as compared to Iron Core furnaces because of efficient melting and long furnace life. The coreless IF is a batch-melting furnace containing a water-cooled copper coil, the inside of which is internally refractory lined. The outside is insulated and enclosed in a steel shell. The furnace body is mounted in a frame equipped with a tilting mechanism. A coreless induction furnace is normally a refractory-lined bucket-shape refractory, the top of which is open for charging and deslagging operations.

The induction furnaces are classified based on their frequency of the current supplied :

High frequency furnaces (200 — 1000 KHz) supplied from valve generators. Medium frequency furnaces (500 — 1000 KHz) supplied from rotary or thyristor convertors.

Low frequency furnaces (50 Hz) which are fed directly from the mains.

Depending on the installed power density and the melting practice the thermal efficiency can exceed 80%, but usually ranges from 60 to 70%.

The coreless induction furnace can be designed to operate at any frequency from 50 Hz upwards. The induction heating of liquid metal causes a stirring effect. The lower the frequency of the primary current, the more intense is the stirring. Therefore, in a mains frequency furnace working at 50 or 60 Hz, the turbulence is greater than in one operating at higher frequency. The frequency of operation affects the current penetration. The higher the frequency, the lesser the penetration depth. This affects the minimal charge piece size and the effective furnace size. 50 Hz furnaces are not practicable at capacities below 750kg. At 10 kHz, charge pieces less than 10 mm in diameter can be heated, so furnaces as small as 5 kg capacity can be used. Frequencies are usually limited to 250 to 350 Hz (in the case of variable frequencies), as at higher frequencies metal homogenisation becomes insufficient. Higher frequencies are used in special cases, such as with very small furnaces.

The increasing demand for high-quality metals and alloys under ecologically clean conditions has increased the application of induction furnaces in foundries for cast iron, steel, and non ferrous metals (mainly Al, Cu, Mg, Ni, precious metals and their alloys). At least 20% of metals produced by electrical furnaces are melted in induction furnaces.

There are two basic types of induction furnace construction: the crucible induction furnace and the channel induction furnace (Figures given below).

Crucible furnaces are used mainly to melt particulate cast iron, stainless steel, base metals, copper, and aluminum. Channel furnaces are applied in the ferrous industry as a holding furnace. In the nonferrous industry they are also used as a melting furnace for aluminum, zinc, and copper.

In vacuum induction furnaces volatile components are removed from less volatile metals and alloys. Examples are the refining of special steels, superalloys & copper. Silver crusts from lead refining are also enriched in silver by volatilization of zinc.

Induction heating is also used during annealing, hardening of steel, brazing, soldering and longitudinal welding .

The application of induction furnaces is mainly limited to melting & refining of metals and alloys. Special applications are zone refining and levitation melting . Inductive heating is also used for brazing and welding, different types of heat treatment (e.g., surface treatment, heating and annealing), metal transport & dosing, and electromagnetic casting (EMC) .

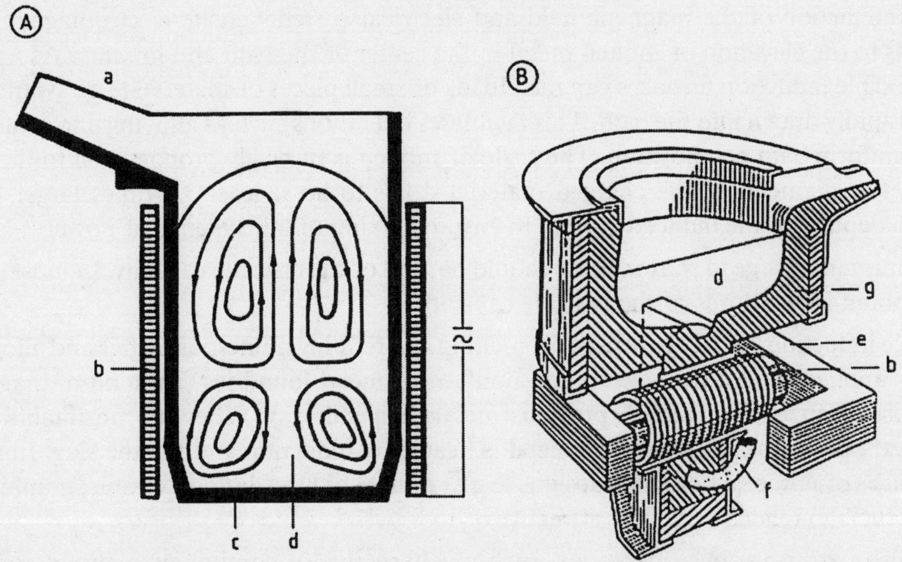

Fig.53.6: Crucible induction (A) and channel induction (B) furnaces

a) Spout; b) Induction coil; c) Crucible; d) Molten metal;
e) Iron core; f) Channel; g) Rammed lining

Crucible Induction Furnace :

Crucible induction furnaces have a cylindrical induction coil consisting of water- cooled copper tubes. The crucible is located inside the coil and usually consists of rammed or brick refractories, graphite, or clay – graphite. In rare cases a more expensive prefabricated ceramic or steel crucible may be used.

The crucible can and should be covered to reduce heat losses. For special applications (metal distillation, refining, melting of reactive or high quality metals and alloys, etc.) the furnace is operated as a vacuum or controlled atmosphere induction furnace .

There are two basic types of vacuum induction furnace, the compact vacuum crucible induction furnace and the vacuum chamber induction furnace . The compact vacuum crucible induction furnace is operated at a low vacuum (10^{-3} bar). It has a simple design and lower investment costs than the vacuum chamber induction furnace, but the load capacities can be higher. The coil, vacuum chamber, and crucible form an integrated unit.

In contrast, the crucible and induction coil of a vacuum chamber induction furnace are located within a sealed,water-cooled chamber. It allows a vacuum of up to 10^{-9} bar and permits the metal to be poured under vacuum within the sealed chamber .

Industrial induction crucible furnaces have different sizes (up to 60 t) and are operated at different frequencies (typically up to 500 Hz), under air or vacuum. Laboratory scale furnaces have capacities down to a few grams and frequencies reaching up to MHz range. Electrical efficiency is >60%.

One of the biggest advantages of a crucible induction furnace is the excellent bath mixing behavior. Interaction of the magnetic field and electrical currents create electromagnetic forces, which leads to the elevation of molten metal in the center of the bath and mixing . As a result of stirring, crucible induction furnaces can melt loads of small pieces of material (e.g., swarfs, chips), which are rapidly drawn into the bath. This facilitates operations such as alloying and refining, and ensures a uniform bath composition. The scale of mixing is inversely proportional to the applied frequency. The diameter of the charged material d has to be at least 3.5 times larger than the penetration depth δ of the induced current to ensure maximum transference of power.

The minimum charge to start melting should be 20% of the crucible capacity. In most cases the starting amount is two-thirds of the crucible capacity.

· Crucible induction furnaces are used for melting and refining numerous metals and alloys. They are usually applied in cast iron, steel, and nonferrous metal foundries (aluminum, magnesium, copper, brass, bronze, zinc, nickel, precious metals, superalloys). They are unsuitable for slag metallurgical operations since only the metal is heated by induction and not the slag. In addition the advantages of numerous crucible designs (e.g., rammed or brick lining; ceramic, graphite, steel, or copper crucible) render them very attractive.

They also permit various operations, such as alternative or continuous melting, continuous melting and pouring, quick alloy changes, and application of vacuum or controlled atmosphere. They are sometimes used as buffer furnaces or grading furnaces.

Line-frequency crucible induction furnaces are mainly employed in cast iron, heavy metal, and aluminum foundries. They are sometimes used in (high-grade) steel foundries. High furnace capacities are possible.

Medium-frequency crucible induction furnaces are mainly used in operations that require frequent alloy changes for melting high-grade steel, alloyed cast iron, copper, aluminum, and precious metals in medium-size furnaces. The low power input required for holding furnaces means that short induction coil furnaces can be used for these applications.

If possible, the crucibles have a rammed lining (acidic monolithic linings) due to their lower costs, which produces significant savings in larger furnaces. They are widely used in cast iron, steel, and nonferrous metal foundries, especially in furnaces with high capacities and little or no metal and alloy changes. The linings are partially infiltrated by the metals/alloys and hence cause contamination during an alloy change. Nonferrous metal foundries, which frequently change alloys often employ more expensive prefabricated crucibles. Precious metal foundries need prefabricated crucibles to minimize metal losses in the linings. Magnesium and its alloys are melted solely in steel crucibles due to their chemical behavior. Alloys that attack acidic linings require basic linings (Al2O3 ,MgO) or graphite crucibles. Non conducting materials (e.g., quartz and glass) are melted in graphite crucibles which heat the charge at low temperature.

Channel Induction Furnace :

In the channel induction furnace the induction coil is located under or beside the crucible and has an iron core. The coil is enclosed in a channel . This construction acts as a transformer with a short-circuit secondary coil producing the best power transfer efficiency (0.95 to 0.98; furnace efficiency amounts to 0.9). The metal in the channel is heated and pumped back to the metal reservoir in the crucible by the repulsive force of the coil and the induced currents.

This means that the furnace must always be charged with a certain amount of molten metal. Special Induction Furnaces Reactions with the crucible or refractory material are undesirable for special high-grade metals and alloys. Use of special induction coil design and suitable frequencies and induction currents, allow metals or alloys to be melted and centered inside the coil by levitation melting. This melting method is only used on a laboratory scale for small amounts of pure metals and alloys.

Induction zone melting is used to refine very pure metals or alloys, especially for the production of semiconductors . Germanium and silicon mono crystals are produced by this technique.

ARC FURNACE

In arc furnaces heat is developed by an arc, or arcs, drawn either to the charge or above the charge. Direct-arc furnaces are those in which the arcs are drawn to the charge itself. In indirect-arc furnaces the arc is drawn between the electrodes and above the charge. A standard power frequency is used in either case. Direct-current (dc) electric power is an alternative source of energy. The general service is melting and refining metals and alloys. The ladle arc furnace is used when a charge of metal is to be processed primarily to refine its chemistry.

In arc furnaces the heating occurs because of the high temperature (>6000 0C) of a gas plasma created by the arc. These furnaces may be operated either in an a.c. or d.c. mode.

Applications of arc furnace :

- Steel scrap and cast iron scrap are melted in arc furnaces in which the arc is situated above the molten slag. The slag acts as a heat distributor and a refining agent for the molten metal
- In vacuum arc remelting (VAR) a consumable electrode, made from the metal to be melted/ refined, is melted in a vacuum to produce high-purity metals. An example is the production of high-purity titanium from titanium sponge .

Two types of arc furnaces are in common use: The three-phase furnace and the single-phase furnace.

The general field of the three phase furnace is the melting and refining of carbon and alloy steels; that of the single-phase furnace is the melting of nonferrous alloys. There is an increasing amount of arc-furnace capacity used for melting and refining various types of iron.

Three-Phase Arc Furnaces :

The general design of this type of furnace is shown in Figure below . In operation, each heat is started by swinging the furnace roof aside and then loading the refractory-lined furnace body with scrap dropped from a crane-handled clamshell charging bucket. Arcs next are drawn between the lower ends of the graphite electrodes and the scrap; melting proceeds under automatic control until the hearth carries the molten metal. This fluidizing stage is effected at about 85 percent thermal efficiency. Several charges usually are needed to build up the bath—particularly in ingot practice.

The furnace tilts forward for pouring; the back tilt serves in the removal of slag and permits the furnace hearth to be kept in proper condition. The slagging door is opposite the pouring spout. Large furnaces frequently also have a side door, known as a working door.

Refractories Furnaces that produce foundry steels operate with acid lining. This means silica brick form the walls; the hearth is of gannister or the equivalent. Silica-brick roofs are the more widely used although, for intermittent operation, clay brick may be preferred. The slags of acid-lining practice remove no phosphorus or sulfur. Essentially all ingot operations are carried on with basic linings. This means magnesite bottom and sidewalls, so that the limey slags employed will not erode them. Entry ports for the electrodes may be of extra-quality refractories to prolong roof life, particularly where the furnace is in continuous operation. In basic practice phosphorus joins the slag readily; sulfur can be removed next by a second slag, when this slag has been made highly reducing. Slag covering the molten bath serves in refining the metal and reduces the heating of wall and roof brick.

In modern arc furnace operations, foamy slag practice is employed, wherein a deep, foamy slag prevents the arcs from damaging the wall and roof linings. Super refractories find application in high-temperature, long-refining operations. Electric irons are made in acid-lined furnaces.

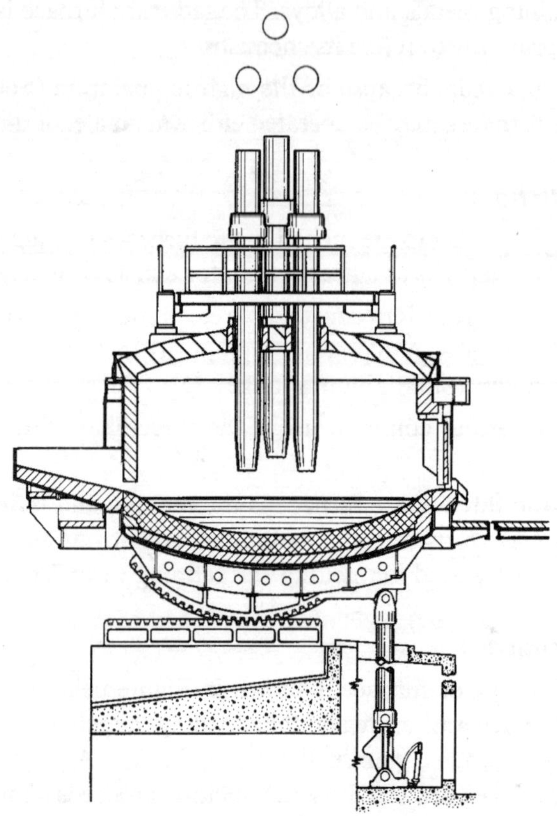

Fig.53.7: Three-phase arc furnace with basic lining.

Temperature. Arcs approximate 3,500°C ; hence operation must be carried out so as to protect the refractories as much as possible. As the top-charge furnace now has supplanted the door-charge furnace in nearly all cold-melt work, the conditions for shielding the refractories during the melt-down stage of each heat are good. With the furnace filled to the top with scrap, electrodes bore down through that scrap, and the heat of the arcs is liberated right in the metallic charge itself. When the charge, and any back charges made, approach the fluid stage it is customary to reduce both the power input and the length of arcs employed. During the finishing stages, roof and sidewalls are protected both by the slag and by the ''umbrella'' effect of the electrodes themselves.

Now a days use of oxygen is being done to gain speed in production, which makes for increasing furnace temperatures. The higher sidewalls of modern furnaces aid in obtaining good roof life. Additionally, water-cooled sidewalls extend refractory life and thus minimize the cost of replacing refractory Charges.

The three-phase arc furnace is primarily a unit for converting scrap charges into steel for pouring into ingots, castings, or a continuous caster. This type of equipment finds increasing use also in the cold melting and duplexing of gray and white irons. One of the main advantages of the top-charge furnace is that the scrap used does not need to be cut to door size, as was the case formerly.

Although first employed only for the more expensive grades of steel, the arc furnace now is used widely in making ingots for rolling into merchant bars and similar grades and supplies liquid metal fed into a continuous caster. The speed of production on this type of working termed single-slag dephosphorizing basic practice can be double that obtained with the same furnace used to make two slag dephosphorized and desulfurized basic alloy steels. Acid working on foundry steels generally approximates the same speed as single-slag dephosphorizing basic practice, and some alloy steels require about half again as much time. While most carbon steel for castings is made on an acid hearth, a basic bottom is regularly used for making manganese steels, for refining nickel & copper, and for the furnacing of many heat-resistant alloys.

In general, approximately 320 kWh at 100 percent thermal efficiency will be needed to melt 1 ton of cold steel scrap. This means about 400 kWh will be needed to fluidize each ton. Additionally, about 100 kWh/ ton will be needed to finish the heat and superheat the bath—this in the case of ordinary plain carbon steels made on single-slag acid or basic practice. Double-slag steel heats will require no more power than others for fluidizing the scrap charge, but the additional power needed for melting new slag, refining, melting added alloys, etc., may require as much as 250 kWh/ton of bath, or even more.

Arcs. The arc in each phase is maintained between the lower end of the electrode and the top of the charge (or bath, after the molten state is reached). Higher voltages can serve for melting as the size of the furnace increases; thus, where a 2.1 m diameter furnace employs 215 V as its highest melting potential, a 4.5 metre diameter furnace would use 290 V or higher as the top tap. For such a furnace constructed with water-cooled sidewall and roof panels, the application of 500 V would be normal.

Small furnaces operate with 600 kVa and even higher powering per ton of charge, whereas in the case of the larger equipments the electrical backing of the furnace normally does not exceed 300 kVA/ton of charge.

Direct Arc Furnaces :

In the Direct Arc Furnaces, electric arcs are formed between the electrodes and the metal being heated, which is thus a component of the electric circuit and is heated by the radiation from the arcs. Direct arc furnaces are used to produce high carbon steels and low alloyed steels.

Submerged Electric Arc Furnaces (SEAF)

In the submerged arc furnaces, the arcs burn under a cover of solid charge which surrounds the electrodes. The charge is heated by the heat generated by the arcs and by joule heat formed on the passage of the current through the charge. SEAF is used for manufacturing ferrosilicon, ferromangenese, ferrochrome and silico-manganese alloys.The SEAFs are open from the top and the raw materials are being fed from the storage bins through chutes.

These storage bins contain different grades of raw materials and are being fed in to the SEAF depending upon the properties of the finished product. In these furnaces, carbon electrodes are used to create the electric arc. These furnaces are continuous in operation but molten metal is being drawn intermittently.

In India, most of the electric arc furnaces are using Direct Arc furnaces.

EAF are principal melting/refining units of mini steel plants producing carbon, alloy, special steel and stainless steel for flat or products having long length applications.

The processes used in these units is divided into following major steps :

- Preparation and charging of raw material
- Melting and Refining of the melt.
- Casting of molten metal into ingots, blooms and billet

The EAF is a batch-melting furnace consisting of a large bowl shaped refractory lined body with a dish shaped hearth. Typically the shell diameter is 2 to 4 m. The shell has a refractory lining inside. The reaction chamber of the furnace is covered from above by a removable roof made of refractory bricks held by a roof ring. The furnace has a main charging door and a tap hole with a tapping spout. It is fed with a three phase alternating current and has three electrodes fastened in electrode clamps which are connected by means of a sleeve led flexible cables and water cooled copper tubes. The furnace rests on two supports sectors which can roll on the furnace stand to tilt ping. The tilting motion being affected by a rack mechanism. The furnace is covered by a refractory roof, which has ports for three graphite electrodes. The electrodes are supported by arms, which allow movement up and down.

D.C. Arc Furnaces

The DC arc furnace is similar to conventional AC arc furnace in metallurgical operation with better stirring of liquid bath. A major problem faced by AC arc furnace operators is that of "flicker", which is dependent on the strength of the local power grid. Consequently in areas with low or insufficient grid strength, it becomes necessary to install costly flicker compensation equipment to mitigate the harmful effects of EAF operation on the utility grid.

The DC arc furnace has single top graphite electrode instead of three graphite electrodes. Roof is similar to conventional water cooled/refractory roof with single hole for electrode entry. In AC arc furnace, there are 8-10 operations of circuit breaker during melting of one heat. The circuit breaker in AC furnace is tripped during short circuiting, charge collapse, tap changing, charging and electrode addition. Whereas in DC arc furnace, circuit breaker is not switched off, instead thyristors are switched off thereby minimizing costly maintenance of high voltage circuit breaker. Bottom electrode is the major component of DC arc furnace through which the liquid bath is continuously stirred. This leads to homogenization of chemistry and temperature of liquid metal bath. In AC arc furnace, temperature distribution on the surface is not uniform due to flaring of arcs and hence cold spots form in the furnace. DC furnace is more effective in terms of bath stirring enabling faster scrap melting with no risk of un-melted scrap and flow of current causes better stirring of liquid metal.

If we compare the AC arc furnaces with DC arc furnaces, following advantages of DC arc furnaces would be observed:

- Decrease in tap to tap time by 5%.
- Decrease in electrical energy consumption by 5-7% .
- Temperature uniformity due to better stirring .
- Low electrode consumption (reduction by 50% e.g. in a 30 ton DC arc furnace, electrode
- consumption is 1.8 kg/t against 3.5-4.5 kg /t in AC arc furance) .
- Considerable reduction in Flicker (by over 50%) .
- One electrode arranged in the centre prevents any local refractory lining over heating .
- Reduced maintenance of various parts like electrode holder, electrode drives, high voltage switchgear etc.
- Less expenditure on pollution control system because of lesser generation of emissions.
- Improved environmental conditions due to less noise during furnace melting and working.

Basic Principles of Arc Furnace

An electric arc can be produced and sustained between two electrodes or an electrode and a liquid melt if the voltage is high enough. The transformation of electric energy to heat takes place through the current in the ionized plasma of the arc, in which the temperatures may reach 6000 0C.

Electric arc furnaces may be classified according to whether transfer of heat from the arc to the furnace feed is indirect or direct. In indirectly heated furnaces , the arc burns between two electrodes without contact with the feed; heat is transferred solely by radiation and convection.

In directly heated furnaces an arc burns between the electrode and the melt . Heat is transferred via the fire point produced by the arc by conduction, radiation, and convection to the melt. Directly heated arc furnaces can be used for melting and refining.

Today one third of the total world steel is produced in electric arc furnaces, this is their principal application .

Arc Furnace Construction :

One-electrode furnaces usually operate in the d.c. mode with the electrode forming the cathode and the bottom of the furnace forming the anode .This kind of furnace may, however, also be operated in an a.c. mode .

Three-electrode furnaces are mostly a.c. furnaces. The electrodes are switched into a three-phase circuit and are usually placed on the vertices of an equilateral triangle in a round furnace. Most furnaces are equipped with removable roofs in order to charge via baskets from the top. The charging of sponge iron, for example, takes place batchwise in baskets or continuously through tube feeding systems. The required atmosphere can be provided inside the furnace which can be tilted during tapping.

Dolomite and magnesite are the refractories mostly used for lining the lower walls and the bottom of the steel shell of the arc furnaces. The upper walls and the roof of the furnaces are water-cooled to provide good resistance against high temperature and temperature changes. The high temperature obtained in arc furnaces allows high production capacities. When scrap is melted down, crater-type holes are burnt into the scrap which protect the furnace roof and walls from overheating.

Modern furnaces may be tilted only 12° during slagging off operations . Steel tapping is done by eccentric bottom tapping. Sometimes the mixing in the furnace is improved by an induction stirring coil underneath the furnace or by bubbling gas through a porous plug.

The advantages of electric arc furnaces (EAF) are listed below:

- High temperatures can be reached within a short time .
- Continuous temperature adjustment is possible via computer control systems which control the level of the arc .
- The atmosphere in the furnace can be tightly controlled .
- In contrast to conventionally heated furnaces, impurities are not brought into the steel by the energy input (e.g., sulfur by coke) .
- The furnace can easily be brought on- or offline with relatively little energy and time, furnace operation may be controlled to meet the production requirements of the cast shops and mills.
- Oxidation, reduction, and alloying can be carried out in one charge with little loss .
- Desulfurization is possible .

Vacuum Arc Refining (VAR) Furnaces

Vacuum arc furnaces are normally used for consolidation and refining reactions. The raw material (sponge or scrap) is pressed into an electrode shape, which is used as a consumable electrode in the arc furnace.

Normally d.c. current is used to provide a stable arc between the consumable electrode and the counter electrode at the bottom of the furnace (Figure given below). After the arc is ignited , the electrode starts to melt, volatile elements & entrapped gases are removed through the high temperature and the vacuum conditions. The refined metal drops to the water or liquid Na/K-cooled copper bottom where it solidifies. Copper is used as the mold material to prevent reactions with refractories.

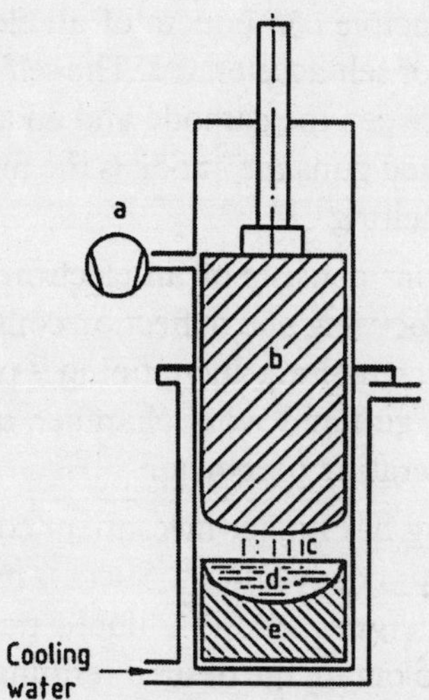

Fig.53.8: Typical vacuum arc refining furnace with consumable electrode

a) Vacuum pump; b) Consumable electrode; c) Arc; d) Molten metal; e) Solid metal

VAR furnaces may also be used with a permanent electrode if the material is fine .This is fed in separately. These furnaces are also used for alloying high-melting metals such as niobium, tantalum, or titanium.

Molten titanium is very reactive, as are molybdenum and zirconium. These metals must therefore be melted in a controlled atmosphere or a vacuum. Vacuum arc refining furnaces are thus suitable for melting and/or recycling these highly reactive metals (also possible in electron-beam furnaces).

As with ESR furnaces, metal sponge and scrap may be mixed and pressed into consumable electrodes and used as such in the VAR furnace. Often refining involves several steps to achieve a pure product, this explains the relatively high energy consumption for titanium (3.8 MW.h per tonne titanium produced).

Vacuum arc furnaces are also used to degas certain steel grades during remelting, energy consumption is 1.0 MW .h per tonne produced metal.

ELECTRON BEAM FURNACE (EBF)

In electron beam furnaces electron guns produce high energy electrons, which impart their energy to the furnace charge to affect its melting.

Applications of electron beam furnace. Electron-beam furnaces are used to melt and/or refine refractory metals such as vanadium, niobium, & tantalum; metals such as molybdenum & tungsten; reactive metals such as zirconium & hafnium; and ceramics such as zirconia, alumina, & uranium carbide.

Basic Principles of EBF .

In electron-beam furnaces the kinetic energy of highly accelerated electrons in a vacuum is transferred to the surface of a material on impact. This leads to local temperature increases and melting of the material.

Furnace Construction. The principle component of an electron-beam furnace is the electron gun which may be work-accelerated or self-accelerated. The self-accelerating gun is most commonly used. Here the voltage is applied between the cathode and an auxiliary hollow anode situated just below the cathode. In work-accelerated guns the anode is the material to be melted, this techniques is used to good advantage in zone melting.

The self-accelerating electron gun consists of an electron emitting cathode , an accelerating hollow anode and electromagnetic focusing and deflection coils . In addition to the gun the furnace comprises of a work chamber in which melting and refining is performed in water-cooled crucibles, vacuum equipment for the electron gun and work chamber, and a computerized control system which controls and monitors the operating parameters.

Although electron-beam melting has higher investment costs per installed kilowatt hour than other melting techniques , it has distinct advantages. Success may be ascribed to its abilities to fuse any low-volatility material that undergoes a solid – liquid transition, and to maintain the fused condition for as long as is required to obtain the desired refining action. Production of high-quality refractory metals is possible, which is not the case with the other methods. Although investment and maintenance costs are high, they are compensated by increased quality.

The refining action in the electron-beam furnace may be classified into:

- Removal of suboxides with high vapor pressures from their respective metals (i.e., NbO can be separated from Nb since their vapor pressure ratio at 2000K is 10 3 , whereas that for TiO/Ti is 1) .
- Removal of oxygen by carbon as carbon monoxide .
- Removal of highly volatile metals from the metal to be melted and refined (e.g., removal of Cr and Fe from Zr)
- Removal of insolubles from the melt material and of inclusions which float to the surface of the melt; since most materials to be processed are usually clean this contribution to refining is small in addition to the usual refractory metals (V, Nb, Ta, Mo, W, Zr, Hf) which are normally melted and/or refined in electron-beam furnaces, beryllium, cobalt, uranium, nickel, and alloys such as special steels have also been melted. For example, the refining action during the melting of hafnium in an electron-beam furnace is far superior to that in vacuum arc melting .

For economic reasons titanium is mostly processed via vacuum arc melting, although electron-beam furnaces are also applied. The impurities oxygen, nitrogen, carbon, and iron remaining after refining lie in similar ranges for both processes, hydrogen values are, however, far lower for electron-beam melting. Ceramics (e.g., ZrO2 and Al2O3), high-quality steels, and carbides of uranium, thorium, & zirconium are also melted in electron-beam furnaces.

Electron beams are also used for metallization, vacuum deposition, centrifugal casting, machining, and welding.

PLASMA FURNACES

In plasma furnaces heating is performed by either transferred or non transferred arc plasma torches. In contrast to arc furnaces, where arcs are produced between a graphite electrode and the charge , plasma furnaces use gas-stabilized plasma arcs produced by water-cooled plasma torches with non consumable electrodes. Consumable hollow carbon electrodes have also been applied.

Applications of plasma furnaces. Plasma furnaces have been employed commercially, e.g., to produce ferro chromium, to melt steel scrap, and to recover valuable metals from steel flue dust .

A plasma is a partially ionized (up to 50 %) gas that contains electrons, ions, energized molecules, dissociated molecules, neutral molecules, and atoms .

The plasma operates at atmospheric pressure and is sufficiently conducting to permit stable transfer of electric power between two or more electrodes. The plasma attains temperatures >2000K (e.g., 10 000 – 20 000K for an argon plasma operated at an arc voltage of 25V and current of 1.5 kA).

Basic Principles and Construction of plasma furnace :

The principal difference between plasma furnaces and arc furnaces is the use of a plasma torch instead of electrodes. In a plasma torch thermionic electrons are emitted from a cathode and accelerated towards the anode.They collide with gas molecules & ionize them.The positively charged gas ions are accelerated in the opposite direction towards the cathode with which they collide, releasing their energy and hence sustaining the thermionic emission. Depending on the type of torch and its construction materials,water cooling may be applied.

Various a.c. or d.c. plasma torches include transferred arc , non transferred arc and superimposed arc.

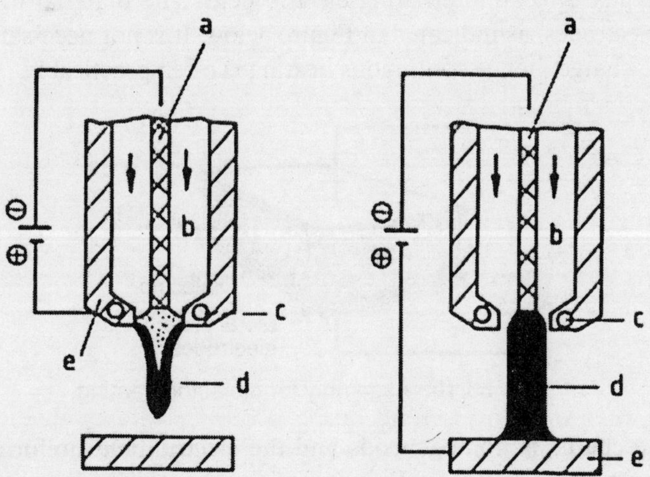

Fig.53.9: Transferred (A) and nontransferred (B) arc torches

a) Cathode; b) Plasma gas; c) Cooling water; d) Plasma jet; e) Anode

Furnace construction is similar to that of arc resistance and arc furnaces. Refractories and other components are selected to suite the specific application.

Applications of plasma furnaces.

Plasma furnaces have only recently been introduced into primary and secondary metallurgy.

Examples include:

- Remelting of steel scrap.
- Production of ferrochromium from chromite fines.
- Recovery of aluminum from aluminum dross.
- Recovery of FeNiCr alloy, slag, and lead and zinc from flue dusts in the stainless steel industry.
- Plasma cold hearth melting is used for melting Ti alloys.
- Several units use plasma torches in research furnaces.

Miscellaneous Types of Electric Heating. In addition to the heating methods discussed above,laser, dielectric, infrared, and microwave heating are also available. But these are not presently used to effect a chemical reaction at high temperature on an industrial scale.

Laser Heating.

Although "laser furnaces" are nonexistent, many applications of laser heating have been reported in , e.g., in cutting, welding, machining, and surface treatment of metals .

DIELECTRIC HEATING

The term relates to the heat developed in dielectric materials, such as rubber, glue, textiles, paper, and plastics, when exposed to an alternating electric field. The material to be heated is placed between plate-form electrodes, as indicated in Figure below. It is not necessary that the electrodes be in contact with the charge; hence continuous heating is often practicable.

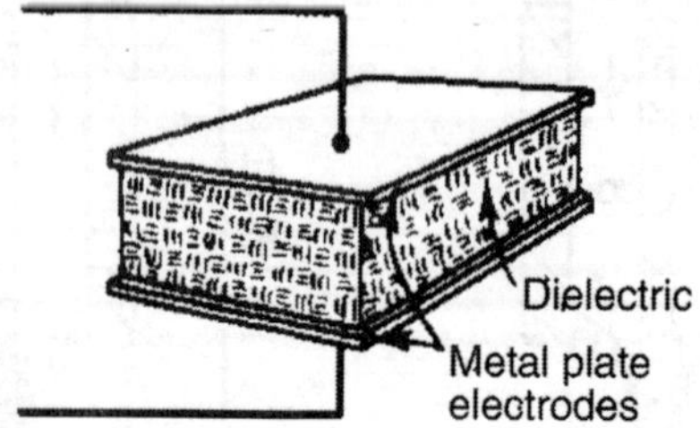

Fig. 53.10: Assembly for dielectric heating.

If the material of the charge is homogeneous and the electric field uniform, heat is developed uniformly and simultaneously throughout the mass of the charge. The thermal conductivity of the material is a negligible factor in the rate of heating. The temperatures and services are within the oven classification (<260 °C).

The frequency and voltage for this class of service depend in each case on the electrical properties of the material of the charge at the temperature specified for the heat application. The frequencies in use range from 2 to 40 MHz; the most common frequencies are from 10 to 30 MHz. It is advisable to select the frequency for heating by trial.

The upper limit of voltage across the electrodes is fixed by the sparkover value and by corona. The permissible voltage gradient depends on the nature of the material of the charge. Values within the range 790 to 2,400 V/cm) are found in practice; the voltage across the electrodes should not exceed 15,000 V.

Applications of dielectric heating include setting glue as in plywood manufacture, curing rubber, drying textiles, and the heat treatment of plastics.

Fused-Salt Electrolysis Cells

Fused electrolysis can be classified both as a resistance electrothermal reactor and as an electrochemical reactor. It is not, however, a purely electrothermal reactor (resistance furnaces).

Due to their low reduction potentials elements such as aluminum, magnesium, and sodium cannot be recovered from aqueous solutions – hydrogen ions would be reduced in preference to these elements to produce hydrogen gas. Recovery of these elements by electrolysis can only be carried out in stable, nonaqueous electrolytes.

In order to illustrate the technology of fused-salt electrolysis,an examples of the recovery of aluminum from alumina is described.

Aluminum Recovery. The feed material for the production of aluminum is alumina (Al_2O_3). Due to alumina's high melting point (2050 °C), fused-salt electrolysis of pure alumina is not economically viable.

The Hall –Heroult process permits electrolysis at 940 – 980 °C, this temperature is produced due to the resistance of the stable salt electrolyte based on the NaF – AlF_3–Al_2O_3 system to the electric current. This electrolyte system has a sufficiently low liquid temperature and a high alumina solubility. Usually about 2–8% Al_2O_3 is dissolved in the cryolite-based (Na_3AlF_6) electrolyte. The density difference between the molten aluminum (ca. 2.3 g/cm^3) and the molten electrolyte (ca. 2.1 g/cm^3) is sufficiently large to ensure good separation of the molten electrolyte and molten aluminum.

Usually the anode gas is a mixture of carbon dioxide and monoxide. Use of carbon anodes lowers the required electrical energy, at the expense of carbon anode consumption (ca. 0.4 kg C/ kg Al).

At the normal current density (0.6– 1.0 A/cm^2) an additional anodic over potential of ca. 0.5V must be added to the theoretical potential of the above reaction (1.06– 1.18 V at 977 °C) as well as a cathodic overpotential of about 0.5 V. This implies that the effective electrolysis potential or polarization potential is ca. 1.7 V. If the voltage drops across the electrodes, then electrolyte, conductors, and anode effects are included, the potential required for aluminum electrolysis lies in the range 3.9 – 4.6 V.

A typical electrochemical cell for the production of aluminum consists of:

- A carbon-lined steel vessel (carbon blocks at the bottom of the furnace 40 cm deep) .
- Two rows containing a total of 16– 24 consumable prebaked carbon anodes .
- Molten salt electrolyte (cryolite-based) .
- A liquid aluminum cathode .
- A layer (20 – 40 cm) of insulation (fire clay refractory) at the bottom of the furnace .
- Anode and cathode bus bars .

In modern cells small amounts of alumina are fed between the two rows of suspended anodes at frequent intervals. Crust breakers between the two rows of anodes break the frozen bath crust, a ventilation shaft removes the 5000 – 8000 m^3/h of anode gas evolved per cell.

Fields of Applications of Electric Furnaces.

Electric furnaces may be used for the production of metals from raw materials by reduction. Electric furnaces may also be applied during metal refining . If cheap electrical energy is available heating in electric furnaces has many advantages over heating with fossil fuels .

Energy Consumption of Various Electric Furnaces are tabulated below.

Process	Type of furnace	lb/kWh
Baking finishes on sheet metal	Batch oven	10 – 18
Baking finishes on sheet metal	Continuous oven	25 – 30
Baking bread	Continuous oven	10 – 12
Annealing brass and copper	Batch furnace	10 – 25
Annealing steel	Batch furnace	5 – 15
Hardening steel	Batch furnace	7 – 11
Tempering steel	Batch furnace	15 – 25
Annealing glass	Continuous furnace	40 – 100
Vitreous enameling, single coat	Batch furnace	5 – 8
Vitreous enameling, single coat	Continuous furnace	10 – 15
Galvanizing	Batch furnace	12 – 20
Melting metals	**Type of furnace**	**kWh/ton**
Lead	Resistor	40 – 50
Solder 50 – 50	Resistor	40 – 50
Tin	Resistor	35 – 50
Zinc	Induction	80 – 100
Brass	Arc and induction	250 – 400
Steel, melting only	Arc and induction	450 – 700
Steel, melting and refining	Arc	600 – 750
Gray iron	Arc and induction	450 – 600
Furnace products	**Type of furnace**	**kWh/ton**
Aluminum	Electrolytic	22,000 – 27,000
Calcium carbide	Resistance	3,000 – 6,000
Ferroalloys	Resistance	4,000 – 8,000
Graphite	Resistance	3,000 – 8,000
Phosphoric acid	Resistance	5,000 – 6,000
Silicon carbide	Resistance	8,500 – 10,000

54

REFRACTORIES

Definition. Refractories are materials which withstand high temperature and do not fuse even at a very high temperature e.g. fire clay, silica, chromite, magnesite etc.

ASTM defines refractories as "non-metallic materials having those chemical & physical properties that make them applicable for structures or as components of systems that are exposed to environments above 538°C.

Refractories are inorganic nonmetallic material which can withstand high temperature without undergoing physical or chemical changes while remaining in contact with molten slag, metal and gases

Refractories perform four basic functions:

1. They act as a thermal barrier between a hot medium (e.g., flue gases, liquid metal, molten slags, and molten salts) and the wall of the containing vessel.
2. They insure a strong physical protection, preventing the erosion of walls by the circulating hot medium.
3. They represent a chemical protective barrier against corrosion.
4. They act as thermal insulation, insuring heat retention.

The principal raw materials used in the production of refractories are: the oxides of silicon, aluminum, magnesium, calcium & zirconium and some non-oxide refractories like carbides, nitrides, borides, silicates & graphite.

What are refractories used for?

- Refractories are used by the metallurgical industry in the internal linings of furnaces, kilns, reactors and other vessels for holding and transporting metal & slag.
- In non-metallurgical industries, the refractories are mostly installed on fired heaters, hydrogen reformers, ammonia primary & secondary reformers, cracking furnaces, utility boilers, catalytic cracking units, coke calciner, sulfur melting furnaces, air heaters, ducting, stacks, etc.
- Majority of these listed equipment operate under high pressure, and operating temperature can vary from very low to very high (approximately 540 to 1600°C).
- The refractory materials are therefore needed to withstand temperatures over and above these temperatures.

- Due to the extremely high melting point of common metals like iron, nickel and copper, metallurgists have to raise furnace temperatures to over 1600°C.

Refractories are used in furnace for the following purposes :

- To protect the steel structure.
- To reduce the heat losses.
- To improve the heating efficiency of the furnace.
- To improve the furnace life.

Classification of refractories

Refractories can be classified on the basis of chemical composition , refractoriness(fusion point) , the methods of manufacture and physical form.

Classification Based on Chemical Composition: Three types of refractory materials namely acidic,basic & neutral refractory exist depending upon their chemical nature as given below in the table.

Table: Classification of Refractory Based on Chemical Composition

Type of refractory material	Examples	Remarks
Acidic refractory (silica is chief constituent) which readily combines with bases.	Fire clay, silica, Quartz, ganister sand, semi-silica, Alumino- silicate etc.	It should not be allowed to come in contact with basic products to avoid the reaction. These are used in areas where slag and atmosphere are acidic.
Basic refractory (consists of basic oxide without free silica)	Bauxite, magnesite dolomite, (lime is not used as refractory due to its dehydrating tendency), Alumina, Zirconia etc.	Should not come in contact with acidic products. They are attacked by acid slags but stable to alkaline slags, dusts & fumes at elevated temperatures.
Neutral refractory which does not combine with acids or bases.	Chromite, graphite carbon refractory, carbide, mullite, Kyanite, pure alumina	Can be used in conjunction with acidic or basic products . They are chemically stable to both acids & bases and are used where slag & atmosphere are either acidic or basic.

Classification based on refractoriness (fusion points).

Refractories are also classified based on their refractoriness as given below :

Table: Classification of Refractories Based on Refractoriness

Type of Refractory	PCE Value (Segar Cones)	Refractoriness/ fusion point, ^{0}C
Low heat duty refractories (LHD)	19—28	1520—1630
Medium/Intermediate heat duty refractories (MHD)	28—30	1630—1670
High heat duty refractories (HHD)	30—33	1670—1730
Super duty refractories (SD)	> 33	> 1730

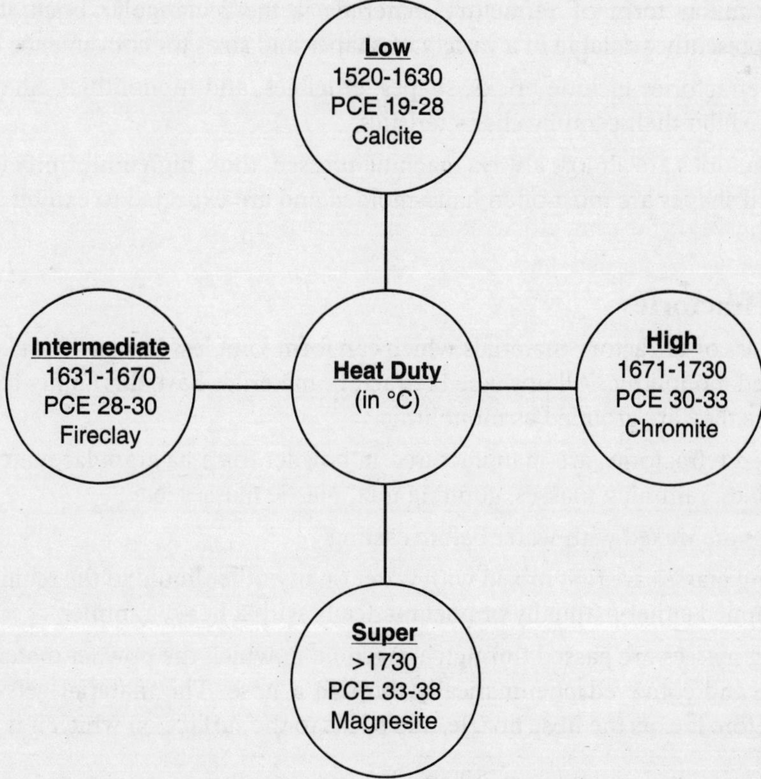

Fig.54.1: Classification of refractories based on refractoriness /fusion points

Classification based on method of refractory manufacture.

Based on the method of manufacture, refractories are classified as :

- Dry pressed
- Fusion cast or electro cast
- Hand moulded
- *Formed* (Normal, fired/burned or chemically bonded)
- *Unformed* (Monolithics-plastics, Ramming mass, Gunning, Castable, Spraying)

Classification Based on Physical Form of refractory

Refractories are classified according to their physical form. These are the shaped and unshaped refractories. The former is commonly known as refractory bricks and the latter as "monolithic" refractories.

Shaped Refractories:

Shaped refractories are those which have fixed shaped when delivered to the user. These are what we call bricks. Brick shapes may be divided into two: standard shapes and special shapes. Standards shapes have dimension that are conformed to by most refractory manufacturers and are generally applicable to kilns and furnaces of the same type. Special shapes are specifically made for particular kilns and furnaces. This may not be applicable to another furnaces or kiln of the same type.

The most familiar form of refractory materials is the rectangular brick shape. However, refractories are presently available in a variety of shapes and sizes for convenience in construction.

Preshaped refractories include bricks, shapes, crucibles, and monolithics. Shaped refractories are pre-fired to exhibit their ceramic characteristics.

Shaped refractories are almost always machine-pressed, thus, high uniformity in properties are expected. Special shapes are most often hand-molded and are expected to exhibit slight variations in properties.

Unshaped Refractories:

There is a class of refractory materials which can form joint less lining. This class of refractory materials is called monolithic. All unshaped refractory materials have this ability to form joint- less lining, and hence they are grouped as monolithic.

- Unshaped refractories are manufactured in powder form as granular material and known as castables, ramming masses, gunning mix, plastic masses, etc.
- Castables are mixed with water before casting.
- Ramming masses are first mixed with water or any other liquid to the required quality, and then rammed either manually or pneumatically with a heavy rammer.
- Gunning masses are passed through a machine in which the powder material is put under pressure and conveyed pneumatically through a hose. The material gets mixed up with water before it exits the hose nozzle, and sticks to the surface on which it is applied to form a lining.
- Plastic masses comprise ready-mix material that is applied manually in the furnace to form a lining.

Unshaped refractories are without definite form and are only given shape upon application. It forms joint less lining and are better known as monolithic refractories. These are categorized as Plastic refractories, ramming mixes, castables, gunning mixes, fettling mixes and mortars.

The manufacture of unshaped refractories differs slightly from shaped refractories. Unshaped refractories typically do not go through a firing process until they reach the final consumer. These unshaped refractories can be installed by spraying, casting, molding, or ramming.

Examples of shaped & unshaped refractories are tabulated below.

Table: Examples of shaped & unshaped refractories

SHAPED REFRACTORIES	UNSHAPED REFRACTORIES
1. Alumino Silicates • High Alumina • Low Alumina **2. Silica** **3. Basic** • Magnesite • Mag-Chrome • Mag-Carbon • Forsterite • Dolomite **4. Carbon** **5. Zirconia** • Zircon (Acidic) • Zirconia (Neutral) **6. Carbides/Nitrides** • Silicon Carbide • Boron Nitride **7. Pure Oxides** • Ferrites • Hofnium Oxides Boron	1. Mortars 2. Castables • LCC • ULCC • NCC • SFC 3. Gunning Mass 4. Ramming Mass 5. Filling Mass 6. Fettling Mass 7. Spraying Compound 8. Patching/Coating Mass

Table below lists each type of shaped & unshaped refractory and a description of its use.

Table: Types and Descriptions of Refractories uses

Types of refractories	Description of its use
Shaped Refractories	
Bricks	Refractories that have shapes and are used to line furnaces, kilns, glass tanks, incinerators, etc.
Insulating firebrick	Low thermal conductivity firebrick.
Unshaped Refractories (Monolithic)	
Mortar	Materials for bonding bricks in a lining. The three types of mortar—heat-setting, air-setting, and hydraulic-setting—have different setting mechanisms.
Castables	Refractories for which raw materials and hydraulic-setting cement are mixed. They are formed by casting and used to line furnaces, kilns, etc.
Plastics	Refractories in which raw materials and plastic materials are mixed with water. Plastic refractories are roughly formed, sometimes with chemical additives.

Gunning mixes	Refractories that are sprayed on the surface by a gun.
Ramming mixes	Granular refractories that are strengthened by gunning formulation of a ceramic bond after heating. Ramming mixes have less plasticity and are installed by an air rammer.
Slinger mixes	Refractories installed by a slinger machine.
Patching materials/ coating materials	Refractories with properties similar to refractory mortar. However, patching materials have controlled grain size for easy patching or coating.
Lightweight castables	Refractories in which porous lightweight materials and hydraulic cement are mixed. They are mixed with water and formed by casting. Lightweight castables are used to line furnaces, kilns, etc.
Fibrous Materials	
Ceramic fiber	Man-made fibrous refractory materials. There are several different types of ceramic fibrous, including blanket, felt, module, vacuum form, rope, loo fiber, etc.

ACID REFRACTORIES

Those refractories which are attacked by basic slags (i.e. bases) are called acidic refractories and include :

(a) **Aluminium-silicate refractories** like fireclay bricks, high alumina bricks, kyanite, silli manite, bauxite, diaspore, mullite, andalusite etc. These refractories vary widely in their physical, chemical and mineralogical characteristics depending on the nature and proportion of silica and alumina present in them. Free silica (SiO_2) should be as little as possible as it lowers the fusion point of these refractories. Addition of grog (ground pre-fired fireclay grains) improves the shrinkage properties of these refractories.

(b) **Silica refractories** consisting mainly of silica (and very low in alkalis and metallic oxide). Raw materials of various grades of silica refractories (containing silica upto 97%, alkalis < 0.3% and alumina < 10%) include natural rock, prepared mass quartz sand (ganister), quartzite, electrically fused quartz etc.

BASIC REFRACTORIES

Those refractories which are attacked by acid slags (acid) are called basic refractories and include :

(a) **Alumina (Al_2O_3)** refractories manufactured from burnt bauxite, electrocast bauxite or electro fused corundum.

(b) **Dolomite refractories** produced by calcining natural dolomite which is mainly a mixture of lime and magnesia.

(c) **Magnesite refractories** produced from dead burnt magnesite or magnesia (MgO).

(d) **Chrome-magnesite and Magnesite-chrome refractories** which are mixture of calcined magnesite and chromium oxide in various proportion necessary to develop required properties in them.

NEUTRAL REFRACTORIES

Those refractories which are not attacked either by basic slag/bases or acid slag/ acids are called neutral refractories and include.

(a) chromite, carbon and graphite refractories

(b) artificially produced refractories like silicon carbide, zirconium carbide, titanium carbide etc.

(c) Metals used as refractory like thorium, tungsten, zirconium, platinum, nickel, iron, molybdenum, copper, vanadium, tantalum etc.

Alumino-silicate refractories are sometimes classified as neutral, refractories but when exposed to basic slag, they exhibit acidic reaction.

MONOLITHIC REFRACTORIES

Monolithic refractory, the name generally given to all unshaped refractory products, are materials installed as some form of suspension that ultimately harden to form a solid mass. Most monolithic formulations consist of large refractory particulates (an aggregate), fine filler materials (which fill the interparticle voids) and a binder phase (that gels the particulates together in the green state).

Monolithic refractories are replacing the conventional type fired refractories at a much faster rate in many applications including those of industrial furnaces.

The main advantages of monolithic refractory are :

- It eliminates joints which is an inherent weakness .
- Method of application is faster and skilled measures in large number are not required.
- Properties can be better than pressed bricks .
- Transportation and handling are simple .
- Offers considerable scope to reduce inventory and eliminate special shapes .
- Offers better scope to reduce downtime for repairs ,
- Has better spalling resistance and volume stability .
- It is a heat saver .
- Ability to install in "Hot Standby" mode .

Monolithic refractories (single piece cast in the shape of equipment such as one for a ladle shown in figure below) are replacing the conventional type fired refractories at a much faster rate in many applications including those of industrial furnaces.

Various means are employed in the placement of monolithics like ramming, casting, gunniting, spraying, sand slinging, etc. Ramming masses are used mostly in cold applications where proper consolidation of the material is important. The same practice can be adopted with both air setting and heat setting materials. Proper ramming tools need to be selected.

Castables by name implies a material of hydraulic setting in nature. Calcium aluminate cement being the binder, it will have to be stored properly to prevent moisture absorption. Further its strength starts deteriorating after a period of 6 to 12 months.

Various means are employed in the placement of monolithic refractories like ramming, casting, gunniting, spraying, sand slinging etc. Ramming masses are used mostly in cold applications where proper consolidation of the material is important. The same practice can be adopted with both air setting and heat setting materials. Proper ramming tools need to be selected.

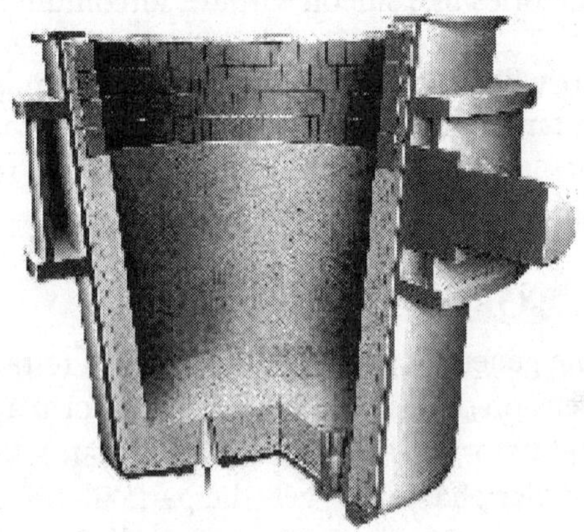

Fig.54.2: Monolithic lining for a ladle

Types of Monolithic Refractories:

Castable Refractories: 'Castable' by name implies a material of hydraulic setting in nature. These are the materials that contain cement binder usually aluminate cement, which imparts hydraulic setting properties when mixed with water. Following the heat-up of the material the binder either transforms or volatilises facilitating the formation of a ceramic bond. The most common binder used in castables is HAC (high alumina cement). Other binders that are often used include hydratable aluminas and colloidal silica. These materials are installed by casting and are also known as refractory concretes.

Insulating Castables: Insulating castables are specialised monolithic refractories that are used on the cold face of applications. There are made from lightweight aggregate materials such as vermiculite, perlite, extend-o-spheres, bubble alumina and expanded clay. Their main function is to provide thermal insulation. They are typically of low density and low thermal conductivity. Insulating refractories have inferior mechanical strength to that of conventional castables.

Plastic Refractories: Plastic refractory is mixtures that is prepared in stiff plastic condition and are delivered in blocks wrapped in polyethylene. During application, the blocks are sliced into pieces and without further preparation, are rammed or poured into place with pneumatic rammer. Plastic are easily rammed to any shape or contour.

Ramming Mixes: Ramming refractory materials are very similar to plastic refractories but are much stiffer. The particle sizes are carefully graded and the final product is usually delivered dry and then mixed with a little amount of water just before application. Other ramming mixes are delivered in wet form and are ready for use immediately upon opening. Application is done with pneumatic rammer.

Patching Refractories: These materials are similar to plastic refractories though have a very soft plasticity allowing them to be pounded into place.

Coating Refractories: This type of product is used to protect refractory linings usually against chemical attack. Coating refractories are normally intended to cover just the working surface of a lining. They tend to be fairly thin layers.

Mortars: Mortars are generally neither classified under refractory brick nor monolithic refractories. These are finely ground refractory materials, which become plastic when mixed with water. These are used to bond the brickwork into solid unit, to provide cushion between the slightly irregular surfaces of the brick, to fill up spaces created by a deformed shell, and to make a wall gas-tight to prevent penetration of slag into the joints.

Gunning Mixes: Gunning mixes are granular refractory materials sprayed on application area using a variety of air placement guns. These are heat setting and are used for patching and maintenance works for kilns and furnaces.

Fettling Mixes: Fettling mixes are also granular refractory materials, similar to gunning mix function, but are applied by shoveling into the furnace needing patching.

INSULATING REFRACTORIES

A refractory suitable for minimising heat losses and thus achieving heat conservation in the furnaces is called insulating refractories. They have high porosity, low thermal conductivity & high thermal insulation properties and are produced from asbestos, fireclay, kieselguhr etc. At low temperature, slagwool, glasswool and vermiculite are also used as insulating materials. For high temperature thermal insulation applications foam ceramics, ceramic fibre and wool are used.

The These have lower densities in range of 65 to 1120 kg/m^3. The three basic types of insulating materials for industrial use are: 1) thin (less than 20 micrometers), low-density (less than 190 kg/m^3) fibers made from organic or inorganic materials; 2) cellular material in closed or open cell form made of organic or inorganic material; and 3) flaked or granular inorganic materials bonded in the desired form. In most cases, glass (silica), mineral wool, high alumina, mulite, or zirconia are the base materials and can be used to temperatures as high as 1600°C. This class of materials has a lower density that varies from 65 to 190 kg/m^3 and offers higher thermal resistance compared to firebricks. In all cases, thermal conductivity of the insulation increases significantly as temperature increases. Insulating materials greatly reduce the heat losses through walls. Insulation is effected by providing a layer of material having a low heat conductivity between the internal hot surface of a furnace and the external surface, thus causing the temperature of the external surface reduced.

The insulating materials may be classified into the following groups :

- Insulating bricks
- Insulating Castables
- Ceramic fibre
- Calcium silicate
- Ceramic coating

Insulating materials owe their low thermal conductivity to their pores while their heat capacity depends on the bulk density and specific heat. Structure of air insulating material consists of minute pores filled with air which have in themselves very low thermal conductivity . Excessive heat affects all insulation material adversely, but the temperatures to which the various materials can be heated before this adverse effect occurs differ widely. The choice of an insulating material depends upon its effectiveness to resist heat conductivity and upon the temperature that it will withstand.

One of the most widely used insulating materials is diatomite, also known as kieselguhr which is made up of a mass of skeletons of minute aquatic plants deposited thousands of years ago on the beds of seas and lakes. Chemically this consists of silica contaminated with clay and organic matter. A wide range of insulating refractories with wide combinations of properties are now available.

The important physical properties of some insulating refractories are shown in the table below:

Table : Physical properties of insulating refractories

Type	Thermal conductivity at 400°C	Maximum safe temperature ,°C	Cold Crushing Strength, Kg/cm²	Porosity, %	Bulk density, Kg/m³
Diatomite Solid Grade	0.025	1000	270	52	1090
Diatomite Porous Grade	0.014	800	110	77	540
Clay	0.030	1500	260	68	560
High Aluminia	0.028	1500-1600	300	66	910
Silica	0.040	1400	400	65	830

Castables and Concretes

Monolithic linings and furnace sections can be built up by casting refractory insulating concretes, and by stamping into place certain light weight aggregates suitably bonded. Other applications include the formation of the bases of tunnel kiln cars used in the ceramic industry. The ingredients are similar to those used for making piece refractories, except that concretes contain some kind of cement, either Portland or a high-alumina cement.

Ceramic fibre

Ceramic fibre is a low thermal mass insulation material, which has revolutionalised the furnace design lining systems.

Ceramic fibre is an alumino silicate material manufactured by blending and melting alumina & silica at temperature of 1800 – 2000°C and breaking the molten stream by blowing compressed air or dropping the melt on spinning disc to form loose or bulk ceramic fibre. The bulk fibre is converted to various products including blanket, strips, veneering and anchored modules, paper, vacuum formed boards and shapes, rope, wet felt, mastic cement etc. for insulation applications.

Fibres are usually produced in two temperature grades based on Al_2O_3 content. A recent addition is $Zr\ O_2$ added alumino silicate fibre, which helps to reduce shrinkage levels thereby rating the fibre for higher temperatures.

These fibres are generally produced in bulk wool form and needled into blanket mass of various densities ranging from 64 to 190 kg/m^3. Converted products and over 40 different forms are made from blankets to suit various requirements.

Properties of Ceramic Fibre

The characteristics of ceramic fibres are a remarkable combination of the properties of refractories and traditional insulation material.

- Lower Thermal Conductivity

The low thermal conductivity – 0.1 kCal/m^3.°C at 600°C for 128 kg/m^3 density ceramic fibre blanket – allows construction of thinner linings with the same thermal efficiency as that of conventional refractories. Hence, for the same outer envelope dimension the furnace volume is much higher. It is 40 % more effective than good quality insulation brick and 2.5 times better than asbestos product. Insulating property of ceramic fibre is better than calcium silicate product.

- Light Weight

Average density of ceramic fibre is 96 kg/m^3. It is one tenth of the weight of insulating brick and one third that of asbestos / calcium silicate boards. For new furnaces structural supports can be reduced by 40%.

- Lower Heat Storage

Ceramic fibre linings absorb less heat because of lower density. Furnace can be heated and cooled at faster rates. Typically the heat stored in a ceramic fibre lining system is in the range of 2700 - 4050 kCal/m^2 as compared to 54200-493900 kCal/m^2 for conventionally lined system.

- Thermal Shock Resistance

Ceramic fibre lining resist thermal shock due to their resilient matrix. Also faster heat up and cool down cycles are possible thereby improving furnace availability & productivity.

- Chemical Resistance

Ceramic fibre resist most of the chemical attack and is unaffected by hydrocarbons, water and steam present in flue gases.

- Mechanical Resilience

This property permits fibre lined furnaces to be shop fabricated and shipped to site in assembled form without fear of damage.

- Low Installation Cost

No special skills are required as application practices are standardised. Fibre linings require no dry out or curing times and can be heated to the capacity of the burners after installation is completed without concern for cracking or spalling.

- Simple Maintenance

In case of physical damage the defective section can be quickly removed and a replacement piece added. Whole panel sections can be pre fabricated for fast installation with minimal down time.

- Ease of Handling

All product forms are easily handled and most can be quickly cut with a knife or scissors. Vacuum formed products may require cutting with a band saw.

- Thermal Efficiency

The low thermal conductivity of ceramic fibre can be advantageously made use of by the lesser lining thickness and reduced furnace volume. The fast response of ceramic fibre lined furnace also allows for more accurate control and uniform temperature distribution within the furnace.

The other advantages offered by ceramic fibre are as follows :

- Light weight furnace
- Simple steel fabrication work
- Low down time
- Increased productivity
- Additional capacity
- Low maintenance cost
- Longer service life
- Higher thermal efficiency
- Faster response

High Emissivity ceramic Coatings

Emissivity, the measure of a material's ability to both absorb and radiate heat, is an inherent physical property which like density, specific heat & thermal conductivity, is not readily amenable to change. However, the development of high emissivity coatings now allows the surface emissivity of materials to be increased, with resultant benefits in heat transfer efficiency and in the service life of heat transfer components. High emissivity coatings are applied in the interior surface of furnaces. The Figure below shows emissivity of various insulating materials including high emissivity coatings. High emissivity coating shows a constant value over varying process temperatures.

The application of high-emissivity coatings in furnace chambers promotes rapid and efficient transfer of heat, uniform heating, and extended life of refractories & metallic components such as radiant tubes & heating elements. For intermittent furnaces or where rapid heating is required, use of such coatings was found to reduce fuel or power to tune of 25-45%. Other benefits are temperature uniformity and increased refractory life.

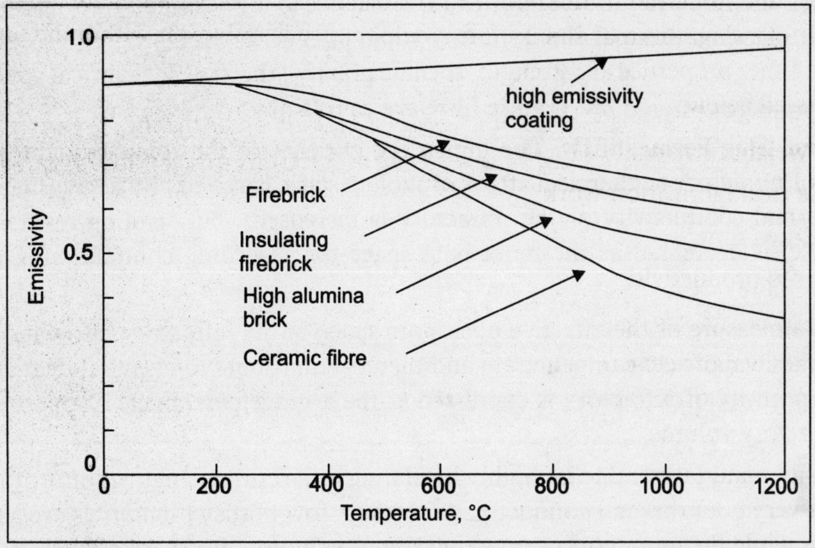

Fig.54.3: Emissivity of Refractory Materials at Different temperatures

Furnaces, which operate at high temperature, have emissivities of 0.3. By using high emissivity coatings this can go upto 0.8 thus effectively increasing the radiative heat transfer.

SPECIAL REFRACTORIES

These are very expensive refractory materials used for making crucibles and furnaces for special/ experimental purposes where cost of refractory is no consideration. They are not very common due to their manufacturing limitation. Special refractories include pure oxides (e.g. magnesia, silica, alumina, beryllia, thoria etc.), borides, nitrides, silicides, carbides etc. Other special refractories are Sialons, zircon, carborundum, alundum (a mixture of fused alumina and clay), sillimanite, electrocast blocks of mullite, magnesia and mixtures of chromite, bauxite and magnesia.

CERMET REFRACTORIES

This refractory is a combination of ceramic materials (e.g. oxide, nitride, carbide, borides etc.) and metallic/metallic alloys materials. It has the combination of good properties of both metallic and ceramic materials e.g. high strength and resistance to high temperature. Cermets are used in nuclear reactors, missiles and space crafts etc.

PROPERTIES OF REFRACTORIES

Important properties of refractories are: chemical composition, bulk density, apparent porosity, apparent specific gravity and strength at atmospheric temperatures. These properties are among those which are used as 'control points' in the manufacturing and quality control process. The chemical composition serves as a basis for classification of refractories and the density, porosity and strength is influenced by many other factors. Among these are type & quality of the raw materials, the size & fit of the particles, moisture content at the time of pressing, pressure at mould, temperature, duration of firing and the rate of cooling.

Other important properties of refractories are resistance to high temperature, heavy load, fluxes, corrosive fluids like slag, thermal shocks due to rapid change in temperature (also called spalling resistance) etc. Other properties are strength, specific gravity , thermal & electrical conductivity etc. They are discussed below.

Porosity and Slag Permeability. The higher the porosity of the refractory, the greater is the susceptibility of refractory to chemical attack of molten slags, gases etc. By decreasing the porosity, strength & thermal conductivity of the refractory is increased ; but spalling resistance (thermal shock resistance) is decreased as the inner pore space for absorbing contraction & expansion is reduced.

Porosity is a measure of the effective open pore space in the refractory into which the molten metal, slag, fluxes, vapors etc can penetrate and thereby contribute to eventual degradation of the structure. The porosity of refractory is expressed as the average percentage of open pore space in the overall refractory volume.

High porosity materials tend to be highly insulating as a result of high volume of air they trap, because air is a very poor thermal conductor. As a result, low porosity materials are generally used in hotter zones, while the more porous materials are usually used for thermal backup.

Such materials, however, do not work with higher temperatures & direct flame impingement, and are likely to shrink when subjected to such conditions. Refractory materials with high porosity are usually NOT chosen when they will be in contact with molten slag because they can be penetrated easily.

Refractoriness or melting/ fusion point. It is the temperature at which the refractory fuses. It should be much higher than the operating temperature of the refractory /furnaces . Fusion point decreases when refractory is under load and this is called refractoriness under load (RUL).

Melting temperatures (melting points) specify the ability of materials to withstand high temperatures without chemical change and physical destruction. The melting point of few elements that constitute refractory composition in the pure state are indicated in the table below:

Table . Melting points of pure refractory compounds

Pure compound	Formula	Melting Temperature , ^0C
Alumina	Al_2O_3	2050
Lime	CaO	2570
Chromite	$FeO.Cr_2O_3$	2180
Chromium Oxide	Cr_2O_3	2275
Magnesia	MgO	2800
Silica	SiO_2	1715
Titania	TiO_2	1850

The melting point serves as a sufficient basis for considering the thermal stability of refractory mixtures and is an important characteristic indicating the maximum temperature of use.

Pyrometric Cone Equivalent (PCE)

Refractories due to their chemical complexity melt progressively over a range of temperature.

Hence refractoriness or fusion point is ideally assessed by the cone fusion method. The equivalent standard cone which melts to the same extent as the test cone is known as the pyrometric cone equivalent (PCE). According to ASTM C24 - 01, PCE is measured by making a cone of the material and firing it until it bends to 3 O'clock.

Representative PCE values for selected refractories include cones 33-34 for super duty fireclay, cones 29-31 for medium duty fire clay and cones 36-37 for a 60% alumina product. The cone values reported for refractories are based on a defined standard time– temperature relationship, so different heating rates will result in different PCE values. PCE can be useful for quality control purposes to detect variations in batch chemistry that result from changes or errors in the raw material formulation.

"Fusibility temperature range" is a more appropriate term for refractory materials as mixture of refractory materials do not have a sharp fusion/melting point. The heat resistance of refractory material is expressed in terms of Pyrometric Cone equivalents (PCE) which is determined by comparing the softening of a test pyramid cone with that of standard pyramid cones of known fusion temperature by heating them under similar conditions.

Temperature at which a refractory will deform under its own weight is known as its softening temperature which is indicated by PCE. Refractories due to their chemical complexity melt progressively over a range of temperature. Hence refractoriness or fusion point is ideally assessed by the cone fusion method. The equivalent standard cone which melts to the same extent as the test cone is known as the pyrometric cone equivalent (PCE).

The pyrometric cone equivalent indicates only the softening temperature. But, in service the refractory is subjected to loads which would deform the refractory at a much lower temperature than that indicated by PCE. With change in the environmental conditions, such as reducing atmosphere, the PCE value changes drastically.

The standard cones are called 'Segar cones' (as Germany standard) and

'Orton cones' (as British standard) and they are numbered according to their refractoriness/ fusion points.

Refractoriness under load (RUL) :

Refractoriness points to the resistance of extreme conditions of heat and corrosion when hot & molten materials are contained while being transported and/or processed.

The ability to withstand exposure to elevated temperatures without undergoing appreciable deformation is measured in terms of refractoriness. The refractoriness under load test (RUL test) gives an indication of the temperature at which the bricks will collapse, in service conditions with similar load. However, under actual service where the bricks are heated only on one face, most of the load is carried by the relatively cooler rigid portion of the bricks. Hence the RUL, test gives only an index of the refractory quality, rather than a figure which can be used in a refractory design. Under service conditions, where the refractory used is heating from all sides such as checkers, partition walls, etc. the RUL test data is quite significant.

The refractoriness under load test (RUL test) gives an indication of the temperature at which the bricks will collapse, in service conditions with similar load.

Refractoriness under load (RUL) of a refractory material is a measure of its failure resistance to the combined action of heat and load. RUL is the softening temperature of a refractory brick (indicated by breaking of the test specimen) when a load of 2 kgf/cm^2 is applied as per the standard test procedure.

Refractory bricks generally having more than two constituent compounds fail under load conditions at much lower temperature (i.e., at RUL) than the fusion point (i.e., refractoriness) of the highest melting and predominant constituent. That signifies that the temperature corresponding to the refractoriness under load of a refractory brick is much lower than its refractoriness (fusion/melting point). A refractory material composed of a single pure compound will have a closer value of its refractoriness and refractoriness under load.

Refractory failure is also influenced by its exposure time to load and heat. Longer exposure time results in failure of refractory under relatively low load & temperature conditions. Similarly refractories will fail at a lower temperature with increased load.

Creep at high temperature:

Creep is a time dependent property which determines the deformation in a given time and at a given temperature by a material under stress. Refractory materials must maintain dimensional stability under extreme temperatures (including repeated thermal cycling) and constant corrosion from very hot liquids and gases. The criterion of acceptance usually adopted is; that compressive creep (deformation at a given time and temperature under stress) for normal working conditions of load and temperature shall not exceed 0.3% in the first 50 hours of the test. This figure has been found to indicate that the creep rate falls by a negligible amount at the end of the stipulated period, and therefore the refractory can be considered safe to use for a much longer time.

Volume stability, expansion and shrinkage at high temperature:

The contraction or expansion of the refractories can take place during service. Such permanent changes in dimensions may be due to:

- The changes in the allotropic forms which cause a change in specific gravity .
- A chemical reaction which produces a new material of altered specific gravity.
- The formation of liquid phase .
- Sintering reactions .
- It may also happen on account of fluxing with dust & slag or by the action of alkalies on fireclay refractories, to form alkali-alumina silicates, causing expansion and disruption. This is an example which is generally observed in blast furnaces.

While it is desirable that all these changes are effected during manufacturing, it is not possible due to economic reasons, as the process is time dependent. Permanent Linear Change (PLC) on reheating and cooling of the bricks give an indication on the volume stability of the product as well as the adequacy of the processing parameters during manufacture. It is particularly significant as a measure of the degree of conversion achieved in the manufacture of silica bricks.

Reversible Thermal Expansion:

Any material when heated expands, and contracts on cooling. The reversible thermal expansion is a reflection on the phase transformations that occur during heating and cooling. The PLC and the reversible thermal expansion are followed in the design of refractory linings for provision of expansion joints. As a general rule, those with a lower thermal expansion co-efficient are less susceptible to thermal spalling.

Strength of the refractory

It is the resistance of the refractory to compressive loads, tension and shear stresses. Refractoriness under load (RUL) is important as some refractories fail at a lower temperature when subjected to load. In taller furnaces, the refractory has to support a heavy load, hence strength under the combined effect of temperature and load is important.

The physical strengths, in both cold and hot conditions, are often characterized as measures of the use of a refractory. Cold strengths indicate the handling and installation of the refractory, whereas hot strengths indicate how the refractory will perform when used at elevated temperatures. Initial strength develops in refractory materials during the forming process.

For shaped refractories, the strengths often develop during the physical processing of the products and sometimes followed by higher temperatures where the refractory goes through a firing process.

For monolithic refractories, the initial strength develops during the installation or forming process (for precast shapes), and the final strength develops while in application. In recent years, more importance has been given to high-temperature strengths of refractories rather than cold strengths since refractories are used at elevated temperatures and not at room temperatures. Strengths of refractories are measured as cold compressive strength (CCS) , cold modulus of rupture, or hot modulus of rupture. Hot modulus of rupture provides the best indication of the performance of a refractory material in use.

Strengths of refractories are usually reported in terms of the three-point bend strength or the flexural strength, frequently called the modulus of rupture (MOR). In case of refractories, there is a standard test for this simple strength measurement.

The flexural strength will depend on sample size, since with increasing size there is an increase in the probability of the existence of a crack-producing flaw, along with a corresponding decrease in the flexural strength.

Characteristic flexural strength values for several ceramic materials are given in table below .

Table : Flexural strength (Modulus of Rupture) of ceramics.

Material	Flexural strength	
	MPa	psi
Silicon nitride (Si_3N_4)	250-1000	35-145
Zirconia[a] (ZrO_2)	800-1500	115-215
Silicon carbide (SiC)	100-820	15-120

Aluminium oxide (Al_2O_3)	275-700	40-100
Glass-ceramic (Pyroceram)	247	36
Mullite ($3Al_2O_3$-$2SiO_2$)	185	27
Spinel ($MgAl_2O_4$)	110-245	16-35.5
Magnesium oxide (MgO)	105[b]	15[b]
Fused silica (SiO_2)	110	16
Soda-lime glass	69	10

[a] Partially stabilized with 3 mol% Y_2O_3.
[b] Sintered and containing approximately 5% porosity.

Cold Crushing strength or cold compressive strength (CCS) :

The cold crushing strength, which is considered to be doubtful relevance as a useful property, other than it reveals little more than the ability to withstand the rigorous of transport. It can be seen as a useful indicator to the adequacy of firing and abrasion resistance in consonance with other properties such as bulk density & porosity.

The cold compressive strength or the cold crushing strength is the capacity of a refractory to provide resistance to a compressive load at room temperature. It is the load, in kg/cm^2, at which the refractory breaks.

The cold compressive strength of a refractory material is an indication of its suitability for use of refractories in construction. It is a combined measure of the refractory for the strength of the grains and also of the bonding system.

Cold Modulus of Rupture (CMOR): (ASTM C-133)

The cold modulus of rupture of a refractory material indicates the flexural strength and its suitability for use in construction. It is indicative of the strength of the bonding system of the refractory product. Since the test is done at room temperature, it can only show its suitability and its use in construction. It provides no indication of how the refractory will behave at elevated temperatures.

Hot Modulus of Rupture (HMOR): (ASTM C-583)

The hot modulus of rupture provides the indication of a refractory material about its flexural strength at elevated temperatures. Since refractories are used at elevated temperatures, the hot modulus of rupture is the true indicator of the suitability and performance of a refractory at high temperatures. Hence, in recent years, the hot modulus of rupture has been prescribed and required by users as the most important test criterion for selection & use of refractories.

Specific gravity. It should be low to produce more number of bricks per unit weight. Also lighter bricks in the lower part of a tall structure will not be subjected to a heavy load. Cost of bricks of higher specific gravity is more than that of lower specific gravity (assuming the price of raw material being same per unit). Increase in bulk density of refractory increases its volume stability, heat capacity and resistance to stag penetration.

Bulk Density:

The bulk density is generally considered in conjunction with apparent porosity. It is a measure of the weight of a given volume of the refractory. For many refractories, the bulk density provides a general indication of the product quality; it is considered that the refractory with higher bulk density (low porosity) will be better in quality.

An increase in bulk density increases the volume stability, the heat capacity, as well as the resistance to abrasion and slag penetration.

Thermal Properties of refractories

(a) Thermal Conductivity:

It is defined as the quantity of heat that will flow through a unit area in direction normal to the surface area in a defined time with a known temperature gradient under steady state conditions.

Thermal conductivity depends upon the chemical & mineralogical compositions as well as the glassy phase contained in the refractory and the application temperature. The conductivity usually changes with rise in temperature. In cases where heat transfer is required though the brick work, for example in regenerators, muffles, etc. the refractory should have high conductivity. Low thermal conductivity is desirable for conservation of heat by providing adequate insulation.

Effect of the structure and the microstructure of refractories on the thermal conductivity:

- The highest conductivity are achieved in the cluttered structure, that in structures consisting of single element, structure made up elements of similar atomic weight, and structure with no extraneous atoms in solid solution. Graphite is good example of single element refractory structure having high thermal conductivity.
- Refractories composed of elements of similar atomic weight and size, have high thermal conductivity. Good examples of this type of refractory are BeO, SiC and B_4C. The refractories having large difference in atomic weight and size have low thermal conductivity as for example UO_2 & ThO_2.
- Impurities, porosity and microcracks exert the main influence on the effective thermal conductivity. Presence of these defects can reduced the thermal conductivity of refractories.

A more porous brick has, in general, lower thermal conductivity. However, the size and number of the pores is also important as heat transfer can take place across the pores by radiation and is prominent at high temperature.

Refractories used in melting furnace should have low thermal conductivity to ensure least heat losses and maximum heat efficiencies whereas recuperates require refractory with high thermal conductivity to ensure maximum heat transfer.

High thermal conductivity refractories are required for some applications where good heat transfer is essential such as coke oven walls, regenerators, muffles and water cooled furnace walls. However, refractories with lower thermal conductivity are preferred in industrial applications, as they help in conserving heat energy.

Porosity is a significant factor in heat flow through refractories. The thermal conductivity of a refractory decreases on increasing its porosity.

Although it is one of the least important properties as far as service performance is concerned, it evidently determines the thickness of brick work.

Ordinary furnaces block should have low thermal conductivity ; but refractory used in coke oven, regenerators, muffle furnace should have high heat conductivity. Thermal conductivity is increased when the porosity of refractory bricks decreases i.e. when the air entrapped in refractory decreases. Insulating bricks (of low thermal conductivity) are maintained at high temperature and hence their life is short.

(b) Thermal Diffusivity

Thermal diffusivity is the thermal conductivity divided by density and specific heat capacity at constant pressure. It measures the ability of a material to conduct thermal energy relative to its ability to store thermal energy. It has the SI unit of m^2/s or m^2/hr.

(c) Thermal Expansion

This is a measure of the refractory about its linear stability when it is exposed to different ranges of high temperatures and then cooled to room temperatures. It is defined as a permanent linear change (PLC) and is measured by the changes in the longest linear dimensions. Most refractory materials expand when heated. Hence, when refractories are installed at room temperatures, the whole structure tightens up when heated.

Thermal expansion of ceramic materials depends on bonding strength between atoms. Ceramic with high degree of ionic bond have high thermal expansion (behavior similar to metals). As the bond strength increases or as the percent of covalent bond increases, the thermal expansion decreases. Both melting temperature & thermal expansion are controlled primarily by bond strength and the magnitude of thermal vibration. As the bond strength increases, the melting temperature increases and thermal expansion coefficient decreases.

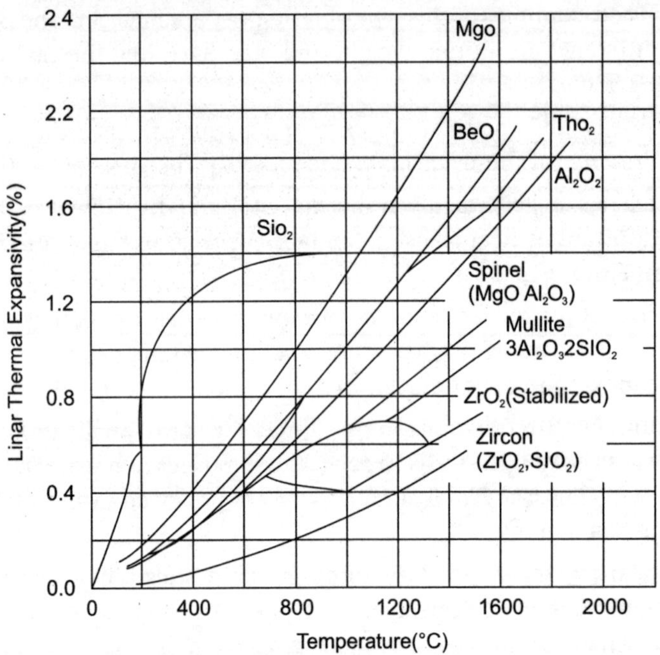

Fig.54.4: Thermal Expansion of Refractory Oxides as Function of Temperature

The expansion characteristic correlates with the nature of the structure, and the atomic bonding. Al_2O_3, TiO_2 and mullite have moderately close packed structures and atomic bond strength. As result, they have moderate thermal expansion. Within the graphite layers, the atomic bonding is very strong (C-C covalent bond) and the coefficient of expansion is low ($1\times10^{-6}/°C$). The atomic bonding between layers is very week (Van der waal's bonds) and the thermal expansion coefficient is high ($27\times10^{-6}/°C$).

Importance of thermal expansion: Many ceramic applications are exposed to a range of varying temperature. Mismatch in thermal expansion behavior between two adjacent materials, can result high enough stresses to fracture. Moderate close thermal expansion match is necessary to minimize thermal stress. Some applications require very low thermal expansion. A well-known domestic application is the use of (LAS) based polycrystalline ceramic ($LiAlSi_2O_6$) for heat-resistant cooking ware.

(d) Thermal Shock

Thermal shock is the fracture of a body resulting from thermal stresses induced by rapid temperature changes. Because ceramic materials are brittle, they are especially susceptible to this type of failure. The thermal shock resistance of many materials is proportional to the fracture strength & thermal conductivity, and inversely proportional to both the modulus of elasticity & the coefficient of thermal expansion.

Thermal shock is an important property for a refractory material. Most high-temperature processes experience heating and cooling. Thermal shock can be indicated by the number of cycles to withstand such temperature fluctuations. Thermal shock resistance can be defined as partial or complete fracture of material due to gradient temperature. The factor that affects the thermal shock resistance includes the shaping methods and the firing temperatures. Thermal shock refers to the thermal stresses that occur in a component as a result of exposure to temperature difference between surface and interior of the component.

As ceramic materials subjected to thermal shock generally fail in tension rather than in shear or compression, the tensile strength is generally used as the criterion for failure rather than the compressive or shear strength. As the thermal shock resistance of a ceramic material is governed mainly by its mechanical & thermal properties, an improvement of the thermal shock resistance of the high Young's modulus of ceramics.

(e) Thermal Spalling

Spalling is defined as the fracture of the refractory brick or block due to any of the following causes:

- A temperature gradient in the brick due to uneven heating or cooling that sets up stresses causing failure (Thermal Spalling). Thermal spalling is the fracture of refractory due to its uneven expansion on account of heat when they are subjected to rapid heating and cooling. For reducing spalling, a refractory should be well fired and its porosity should be more.
- Compression in a structure of refractories due to expansion of the whole from a rise of temperature causing shear failure.
- Variation in coefficient of thermal expansion between the surface layer and the body of the brick, due to the surface slag penetration or due to structural changes in service resulting in shearing off the surface layer.

As a general rule, those with a lower thermal expansion co-efficient are less susceptible to thermal spalling .

Thermal Spalling can be reduced by:

- Using materials with low coefficient of expansion and avoiding sudden temperature fluctuation for example addition of alumina in small quantities decreases expansion to a large extent.
- Using high porosity bricks.
- By designing the furnace such that stress is alienated.

Electrical Conductivity. For electrical furnaces, the electrical conductivity of refractory material should be low. Except, graphite, all other refractories are bad conductors of electricity. The electrical resistivity of the refractory drops rapidly with rise in temperature.

Chemical Composition. The chemical composition of the refractory should be such that the surroundings do not chemically attack the refractory and corrode it. Acidic refractory should not be used in a furnace heating a basic material, otherwise the brick will react with the furnace stock and get eroded and hence its life will be reduced.

Slag Resistance. Slag attacks the refractory particularly at high temperature by forming mixture or new compounds with the refractory substance in contact with. Slag dissolves some of the refractory material and forms a viscous slag film whose melting temperature may be higher or lower. Resistance of refractory to slag action depends on the nature of slag and the refractory. Relative slag resistance of various refractories depend upon temperature, change in slag composition (on combining with refractories) and the nature of contact with the slag i.e. whether the slag is moving, stationary or is carried as fine particles by the gases/flame. A basic slag attacks an acidic refractory while an acid slag attacks a basic refractory very rapidly.

Rate of slag attack is accelerated by the following factors :

- defective joints and cracks in the refractory structures.
- presence of fusible glass of refractories .
- rough and porous surfaces of refractories .
- moving slag as compared to stationary slag.

Rate of slag attack is reduced, if the refractory is dense and well vitrified (i.e. offering a smoother surface) and the slag is stationary.

Abrasion Resistance. A refractory is subjected to wear, when it comes in contact with a moving charge as in case of blast furnace, descending charge materials result in abrasion/ wearing away of the refractory lining material. Abrasion resistance of refractory is increased by using a dense (low porosity), fine grained and wear resistant refractory lining.

Erosion Resistance. Erosion (damage due to mechanical action) occurs when metal or gas carrying dust and slag particles hit against the refractory lining resulting in 'chipping off' of particles from the refractory. Splashing of slag erodes the refractory lining whereas slag particles entrained by flame/gases erode particularly the arches and bends of the structure.

Size , Shape and dimensional stability. It is a part of the design of furnace, since it affects the stability of the furnace structure. Accurate refractory size is important to properly fit the refractory shape inside the furnace and to minimise space between construction joints.

Dimensional accuracy and size is extremely important to enable proper fitting of the refractory shape and to minimize the thickness and joints in construction.

GENERAL METHOD OF MANUFACTURE OF REFRACTORY

Manufacture of refractory consists of the following process :

Grinding. Raw materials are ground so that processes sizes of material are obtained. Ratio of coarse to fine particles should be even. This ratio is normally maintained at 55 : 45. Equipments used for this purpose are various types of crushers, pulverisers, hammer mills, ball mills, shaking tables and screens etc.

Pre-Treatment. The main pre-treatment given to refractory raw material is its firing/ calcination at high temperature for considerable periods of time which bring about the complete mineral conversion and hence stabilises the material. The constituents of the finished refractories produced by reaction between the solid particles of raw materials are rarely completely in equilibrium because it is seldom feasible to actually melt and cast these refractory materials. Natural silica mainly contains quartz which is not stable at high temperatures but undergoes transformation to its allotropic forms tridymite and cristobalite involving high volume changes. Quartz (a stable form of silica up to 870°C is transformed into tridymite on very prolonged heating at this temperature i.e. 870°C. A shorter duration firing between 870 and 1470°C results in transformation of quartz into tridymite. On heating above 1470°C, this is further transformed into cristobalite. Conversion of quartz into tridymite involves 16% increase in volume whereas conversion of tridymite into cristobalite involves a little contraction in volume. Thus raw silica expands considerably when heated to high temperatures and pre stabilization of silica becomes necessary.

Similarly, fireclay which consists of numerous hydrous alumino-silicate minerals, break down to several constituents on heating. And these constituents tend to form stable mullite ($3Al_2O_3$. $2SiO_2$). Mullite formation begins at 1000°C and is accelerated at increasing temperatures. High temperature strength, resistance to slag action and other useful properties of fireclay depend upon the formation of mullite and properties of glassy bonds. Even weathering of mixed fireclay increases its plasticity.

Magnesite is dead burnt to eliminate all the CO_2 from it and to convert it into stable crystalline magnesia (MgO) known as 'Periclase' and involves least shrinkage in refractory shapes on firing.

Dolomite ($CaCO_3$ + $MgCO_3$ mainly) is dead burnt to calcium and periclase. Dolomite as such has the tendency to slaken and is not suitable for refractories and is hence made inactive by converting it into a silicate or some other compound. The slakening tendency of burnt dolomite is eliminated by conversion of calcium to dicalcium or tri-calcium silicate in presence of Fe_2O_3 and surpentine respectively. Presence of Fe_2O_3 and for conversion into dicalcium silicate prevents the formation of y-dicalcium silicate which otherwise results in explosions in service because of about 10% increase in its volume at the working temperature. Calcination of mixture of dolomite and serpentine at 1600°C results in tricalcium silicate and periclase (MgO) which forms a very good refractory material.

Mixing. Ground refractory material is mixed with the binding material in such a way that plastic materials are equally distributed throughout the mass to facilitate easy moulding.

Mixing is usually carried out in pug mills for even distribution of fine and coarse particles in the whole mass. Pre calculated amount of water, additives, binding materials and mineralisers (i.e. grog, stabilisers etc.) are added and the mass is mixed thoroughly to ensure a product of uniform composition and uniform distribution of fine and coarse grains plastic and bond materials etc. for easy moulding and development of useful properties in the fired refractories. To ensure increased plasticity for easy moulding, the mixture is allowed to remain as such for a day or more.

Mixing can be dry, semi-plastic or wet type. 14-20% water is used in wet mixing which are usually used for hand moulding. Semi-plastic mixtures have lower percentages of water to achieve semi-state of plasticity whereas the dry mixture include non-plastic basic mixes and clay mixes containing < 5% water. Since mixing with so little water is difficult, water as a fine spray or mist is used to achieve proper mixing. Due to requirement of high moulding pressures, semi-plastic and dry mixture are machine pressed. Clay mixtures are also obtained in the form of colloidal solutions suitable for moulding by slip casting.

Moulding. It is done mechanically by applying high pressure or by hand to increase density (by deairing) and strength of the refractory. Moulding which follows mixing of refractory materials is done either by hand or machine (by pressing or extrusion). Hand moulding (successful with wet mixtures) is carried out in wooden boxes and is cheaper than machine on a jobbing basis. However, machine moulding has more practical applications and is cheaper for mass production of standard refractory shapes. Machine moulded refractories have higher strength and density than hand moulded refractories. Machine moulding can be used for semi- plastic mixture using moderate moulding pressures. Extrusion is usually used to get the rough shapes of approximate dimensions which are subsequently pressed to exact shapes. Machine moulding of dry mixture requires a pressure of about 1000 kg/cm^2 or more.

At high moulding pressures, air is trapped in the refractories and cause laminations. Therefore de-airing of refractories is done in following ways by :

- pressing the material and then allowed to crack, finally it is repressed to close the voids.
- applying vacuum to the system through the vents in the mould
- passing a gas (such as butane) through the mould to displace air (the gas is absorbed by the clay on the application of pressure)

Density and strength of refractories increase on de-airing due to decrease in laminations and crack. Density increase is also due to elimination of air of a lower density than the refractory material and/or closing of voids.

The method used for moulding refractories of hollow and other special shapes is slip casting in.which a colloidal suspension is poured into a plaster of paris moulds which absorbs the water and causes deposition of uniform layer of clay on the walls of the mould. On deposition of desired thickness of clay (which depends on the time allowed for deposition), the extra mixture in suspension is poured out of the mould resulting in hollow refractories with irregular interval contours.-

Powder pressing is used for moulding refractories involving dry compacting or slightly damp refractory powder mixture in metallic dies using sufficiently high pressure to produce strong and dense refractory shapes. Recent moulding processes used for some typical refractories include reactive hot pressing and isostatic pressing.

Drying. It is done at slow rate to avoid voids and high shrinkage. Drying of moulded refractories increases its green strength by removing moisture and thus making them safe for subsequent handling. Drying is usually carried out under shade (not in direct sunlight) on large drying floors. Floors heated by waste heat from kilns are also used where refractories are laid out in open arrays. Drying is carried out in predetermined set of conditions (of temperature and humidity) and the drying rate is kept at minimum to avoid higher shrinkage and crack formations.

For faster drying of refractory of constant shape and sizes, tunnel kilns are used, where raw refractories are stacked in bogies or placed on belts moving through a tunnel against a stream of hot air. Drying conditions in this case need not be altered (as the shape and sizes of refractories are constant) to ensure drying without damaging the refractories due to high shrinkages and crack formations involved at high rates of drying.

For drying, heavy refractories of intricate shapes, controlled humidity driers are specially used. Drying rate of fireclay bricks should be very low to avoid crack formations as they are liable to maximum shrinkage on drying.

Firing/Burning. Bricks are burnt in Kiln (down draught kiln or tunnel kiln) to remove water of hydration, vitrification and development of stable mineral forms. Shrinkage in volume upto 30% occurs during burning. Burning or firing of refractories which follows their drying facilitates developments of stable mineral forms in them and high crushing strength of finished products.

Degree of firing of silica brick is indicated by its specific gravity. High specific gravity of 2.5 indicates the presence of considerable amount of unconverted quartz in the refractory whereas a specific gravity of 2.32 to 2.37 indicates adequate firing and transformation of quartz to cristobalite and/or tridymite. Rapid heating of silica refractories above transformation temperatures may result in their fracture or spalling due to internal stresses set up owing to large volume changes within the refractories. It can be prevented by following a careful warning of schedule even when fired refractories are finally used in furnaces as some transformation back to quartz takes place on cooling them. Silica bricks are loosely packed in the kiln to meet its expansion requirement during firing whereas fireclay bricks are closely packed (as they shrink on firing) to get maximum production.

Brick firing is done either in down draught, batch type or tunnel kilns. Batch type kiln firing is time consuming due to involvement of three stage operation namely brick charging, firing and cooling. Difficulty in achieving uniform temperature throughout the batch type kiln (as it is heated by coal/coke) results in well or overfired product from the hottest zone and under fired brick from coldest zone. However in a modern kiln (which is divided into different compartments), heating is carried out in one compartment and to a uniform temperature and only after that the fire is advanced to the next compartment. Predetermined quantity of combustion air is preheated through some cooling compartment of a kiln which also accelerates the cooling rate of the fired refractories and thereby reduces the production time. Similarly, hot flue gas from compartments on fire are passed through freshly charged compartments of the kiln to preheat the dried refractories therein which also leads to fuel conservation.

For mass production of refractory bricks however continuous tunnel kilns are more suitable in which dried refractory bricks are charged from one end and fired product at near room temperature is discharged from the other end. Firing is completed in the hottest central zone of the kiln in which the secondary air enters through the discharge end and gets preheated as it cools the fired bricks on their way out of the furnace. Preheated combustion air is used for burning fuel oil (some kilns still use coal/coke as fuel) in the central hottest zone. Air preheating, apart from increasing the temperature in the firing zone also reduces fuel consumption. Hot flue gases while moving towards charging end of the tunnel kiln give away their sensible heat to cold refractories and preheat them as they move towards the firing zone which is the hottest zone.

Firing temperature of some typical common bricks is tabulated in Table 54.1.

Table 54.1: Typical Firing Temperature of Refractories

Refractory Type	Firing temperature, °C
Fireclay bricks	1250 — 1400
High alumina bricks	1450 — 1550
Silica bricks	1450 — 1510
Chrome bricks	1450 — 1650
Magnesite bricks	1450 — 1650
Direct bonded basic bricks	1650 — 1760
Silicon carbide bricks	1370— 1510

Various steps involved in the manufacture of refractories are shown in figure 54.5.

Refractory Manufacturing Process Flow Diagram is shown in Figure 54.6.

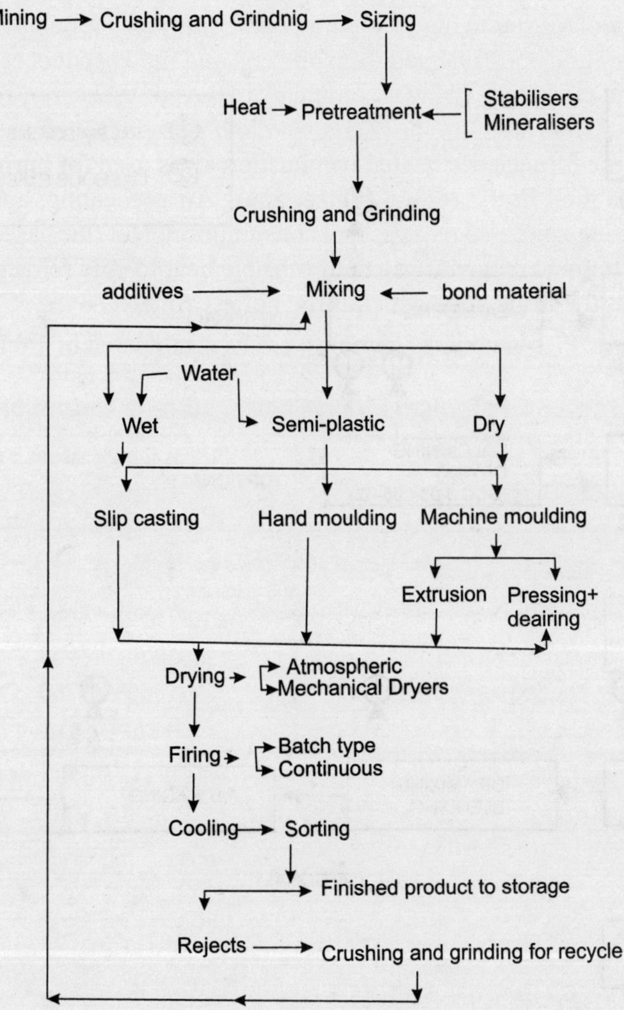

Fig.54.5: Various steps in the manufacture of refractories.

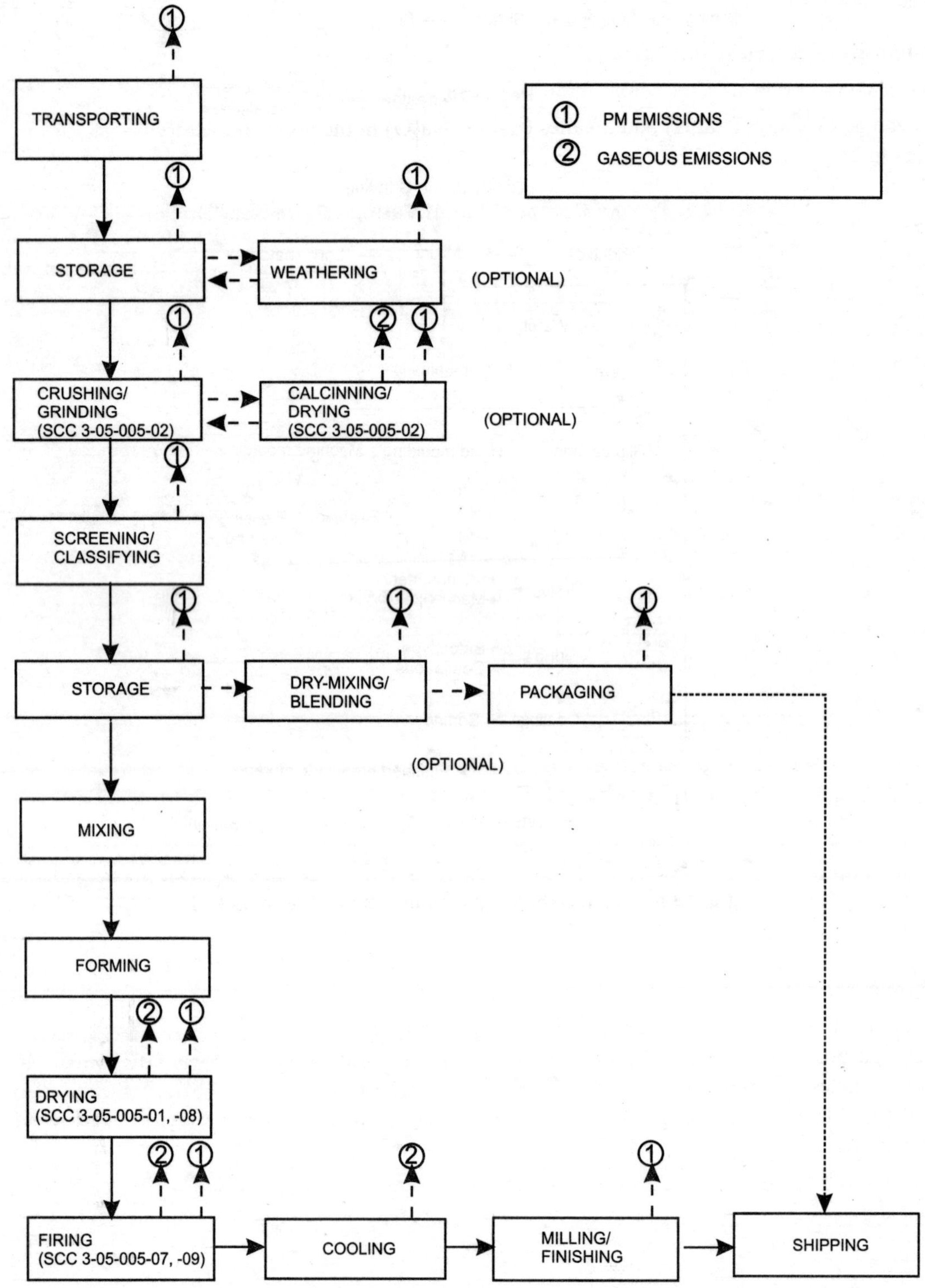

Fig.54.6: Refractory Manufacturing Process Flow Diagram

Composition of Refractory Materials :

A typical composition of refractory material bricks is tabulated in Table 54.2.

Percentage of TiO_2 (Titania) and alkalies ($K_2O + Na_2O$) in the above refractory bricks vary from traces to 2%.

Table 54.2. Typical Composition of Various Refractory Bricks

Principal constituents %	Fire bricks	Aluminious fire bricks	Semi silica bricks	Silica bricks	Sillimanite bricks	*magnesite. Bricks*	Chrome bricks	Chrome magnesite. bricks	Dolomite bricks	Periclase	Foresterite bricks
SiO_2	50—70	50—55	75—90	94—95	25—35	1.5—5	3—10	3—10	12—15	3.0	33.5
Al_2O_3	25—3	38—45	8—15	0.5—2	55—65	1—4	10—25	5—15	2—3	1.0	1.0
Fe_2O_3	2—5	2—4	1—2	0.5—1.5	0.5—1.5	2—8	12—25	8—20	2—4	3.0	0.1
Cr_2O_3	—	—	—	—	—	—	135—45	20—35	—	0.5	0.5
MgO	0.5—1	0.5—1	0.5—1	0.5—1	0.5—1	84—92	15—35	40—60	38—12	98.0	54.5
CaO	0.5—1	0.5—1	0.5—1	2—2.25	0.5—1	2—5	2	1—4	38—42	2.5	1.0

Refractories Product Characteristics

Because the quality & characteristics of refractories vary considerably, consumers employ chemical and physical tests to ensure that the refractories purchased meet their requirements. The American Society for Testing and Materials (ASTM) provides specifications and tests for various kinds and uses of refractory products. Depending on the intended end use, consumers may test refractories for thermal conductivity, resistance to abrasion & corrosion, permeability, oxidation resistance, and other characteristics .

Most refractory products are sold as preformed shapes. However, they are also available in special purpose clays; bonding mortars; and monolithic, plastic refractories; ramming mixes; and gunning mixes. A variety of processed refractory grains and powders are also produced . From the physical form, refractory products can be further classified into oxide bricks, nonoxide bricks, and composites.

Table below lists types of oxide, nonoxide, and composite refractories; their characteristics & their applications.

Table 54.3: General Characteristics & Applications of Various Types of Refractories

Refractory Type	General Characteristics	Application
Oxide Bricks: Silica	High strength at high temperatures, residual expansion, low specific gravity, high expansion coefficient at low temperatures, low expansion coefficient at high temperatures	Glass tank crown, copper refining furnace, electric arc furnace roof
Fused silica	Low thermal expansion coefficient, high thermal shock resistance, low thermal conductivity, low specific gravity, low specific heat	Coke oven, hot stove, soaking pit, glass tank crown
Chamotte (fireclay)	Low thermal expansion coefficient, low thermal conductivity, low specific gravity, low specific heat, low strength at high temperatures, less slag penetration	Ladle, runner, sleeve, coke oven, annealing furnace, blast furnace hot stove, reheating furnace, soaking pit
Alumina	High refractoriness, high mechanical strength, high slag resistance, high specific gravity, relatively high thermal conductivity	Hot stove, stopper head, sleeve, soaking pit cover, reheating furnace, glass tank, high-temperature kiln
High alumina	High refractoriness, high mechanical strength, high slag resistance, high specific gravity, relatively high thermal conductivity	Slide gate, aluminum melting furnace, skid rail, ladle, incinerator, reheating furnace hearth, skid rail, ladle, incinerator
Roseki	Low thermal expansion coefficient, high thermal shock resistance, low thermal conductivity, low specific gravity, low specific heat	Ladle, runner, sleeve, coke oven, annealing furnace, blast furnace hot stove, reheating furnace, soaking pit
Zircon	High thermal shock resistance, high slag resistance, high specific gravity	Ladle, nozzle, stopper head, sleeve
Zirconia	High melting point, low wettability against molten metal, low thermal conductivity, high corrosion resistance, high specific gravity	Nozzle for continuous casting, glass tank, high-temperature furnace, crucible
Alumina zirconia silica	High slag resistance, high corrosion resistance against molten glass	Glass tank, incinerator, ladle, nozzle for continuous casting
Lime	High slag resistance, low hydration resistance	Special refining surface
Magnesia	High refractoriness, relatively low strength at high temperature, high basic slag resistance, low thermal shock resistance, low durability at high humidity	Hot-metal mixer, secondary refining vessel, rotary kiln, checker chamber of glass tank, electric arc furnace
Magnesia- chrome	High refractoriness, high refractoriness under load, high basic slag resistance, relatively good thermal shock resistance (low MgO bricks), high strength at high temperature (direct bonded and fusion cast)	Hot-metal mixer, electric arc furnace, secondary refining vessel, nonferrous refining furnace, rotary cement kiln, lime and dolomite kiln, copper furnace, ladle, checker chamber for glass tank, slag line of electric arc furnace, degasser for copper, nonferrous smelter
Chrome	High refractoriness, low strength at high temperature, low thermal resistance	Buffer brick between acid and basic brick

Dolomite	High refractoriness, high refractoriness under load, high basic slag resistance, low durability in high humidity, high thermal expansion coefficient	Basic oxygen furnace, electric arc furnace, secondary refining vessel, rotary cement kiln
Spinel	High thermal shock resistance, high strength at high temperatures, high slag resistance	Rotary cement kiln, ladle
Non oxide Bricks:		
Carbon	High refractoriness, high slag resistance, low oxidation resistance	Blast furnace hearth, electric arc furnace
Silicon carbide	High refractoriness, high strength at high temperature, high thermal conductivity, high thermal shock resistance, reduced oxidation resistance at high temperature, high slag resistance	Kiln furniture, incinerator, blast furnace
Silicon carbide-graphite	High refractoriness, high strength at high temperature, high thermal conductivity, high thermal shock resistance	Incinerator
Silicon nitride	High strength, high thermal shock resistance, relatively high oxidation resistance	Kiln furniture, blast furnace
Composite bricks:		
Silicon carbide Containing bricks	High corrosion resistance against low iron oxide, high strength at high temperatures, high thermal shock resistance	Ladle, blast furnace, electric arc, torpedo ladle, iron ladle
Magnesia-carbon	High slag resistance, high thermal shock resistance	Basic oxygen furnace, electric arc furnace, ladle
Alumina-carbon	High refractoriness, high thermal shock resistance, high corrosion resistance	Submerged entry nozzle, slide gate

Types & characteristics of raw materials used in various clay & non-clay refractory manufacture is tabulated below.

Table 54.4: Types and Characteristics of Raw Materials used in Refractory Manufacture

Type	Characteristics
Clay Refractories	
Fireclay	Consists of kaolinite ($Al_2O_32SiO_22H_2O$) and minor amount of other clay materials. Fireclay refractories can be low, medium, high, or super-duty based on their resistance to high temperature or refractoriness. Fireclay refractories are used to produce bricks, insulating refractories, and two types of ladle brick.
High-alumina	Composed of bauxite or other raw materials that contain 50 to 87.5 percent alumina. High-alumina refractories are generally multipurpose, offering resistance to chipping and higher volume stability. High-alumina refractories are used to produce brick and insulating refractories.
Nonclay Refractories	
Basic	Produced from a composition of dead-burned magnesite, dolomite, chrome ore, and small amount of other minerals. Basic refractories can be further subdivided into magnesia, dolomite, chrome, and combination bricks. Basic refractories are typically used to make bricks.
Extra-high alumina	Made predominately from bauxite or alumina (Al_2O_3), extra-high alumina refractories contain from 87.5 to 100 percent alumina and offer good volume stability. They are typically poured into special shapes using a fused casting process.
Mullite	Made from kyanite, sillimanite, andalusite, bauxite, or mixtures of alumina silicate materials; mullite refractories are about 70% alumina. They maintain a low level of impurities and high resistance to loading in high temperatures.
Silica	Containing silica, silica refractories are characterized by a high coefficient of thermal expansion between room temperature and 500°C (930°F). Silica brick is available in three grades: super-duty (low alumina and alkali), regular, and coke oven quality. Silica compositions can be used for hot patching, shrouds, and bricks.
Silicon carbide	Produced by the reaction of sand and coke in an electric furnace, silicon carbide refractories are used to make special shapes, such as kiln furniture, to support ceramicware as it is fired in kilns. It has high thermal conductivity, good load bearing characteristics at high temperatures, and good resistance to changes in temperatures.
Zircon	Containing zirconium silicate (ZrO_2SiO_2), zircon refractories maintain good volume stability for extended periods or exposure to high temperatures. Zircon refractories are widely used for glass tank construction.

Clays

Clay is composed mainly of fine particles of hydrous aluminum silicates and other minerals that is plastic when moist but hard when fired . Fireclay is the predominate clay used in firebrick; bentonite, in foundry sand; common clay, in refractory mortar & cement; and kaolin, in calcine, grog, high alumina brick, kiln furniture, and plug, tap, & wad .

Non clays

Nonclay refractories are composed of alumina, mullite, chromite, magnesite, silica, silicon carbide, zircon, and other nonclays.

Typical Refractories used in Industries

The following are the common types of refractories used in various industries :

- **Silica bricks:** Contains at least 92% of silica; bricks of this type are normally bonded with a small amount of lime.
- **Siliceous brick :** Contains from 85-92% of silica, the remainder being essentially alumina.
- Semi-siliceous brick: Contains 77 % of silica, the remainder being essentially alumina.
- **Fire brick:** A brick that in the fired state, consists essentially of alumina-silicates and silica, with <78% of silica and <38% of alumina.
- **Aluminous brick:** Contains from 38-45% of alumina, the remainder being essentially silica.

The names of other types of refractory give a sufficient indication of their general composition. e.g. magnesite brick, chrome brick, chrome magnesite brick, silicon carbide brick, sillimanite brick, etc.

Depending on the area of application such as boilers, furnaces, kilns, ovens etc, temperatures and atmospheres encountered different types of refractories are used.

Fireclay refractories

Fireclay refractories, such as firebricks, siliceous fireclays and aluminous clay refractories consist of aluminium silicates with various amounts of silica ranging from SiO_2 content of less than 78% and containing less than 44% of Al_2O_3.

Table below shows that as the quantity of impurities increases and the amount of Al_2O_3 decreases, the melting point of fireclay brick decreases. Owing to its relative cheapness and widespread location of the raw materials used to manufacture firebricks, this material finds use in most furnaces, kilns, stoves, etc.

Table 54.5: Properties of typical fireclay bricks

Fireclay Brick	Percent SiO_2	Percent Al_2O_3	Other Constituents,%	PCE , °C
Super Duty	49-53	40-44	5-7	1745-1760
High Duty	50-80	35-40	5-9	1690-1745
Intermediate	60-70	26-36	5-9	1640-1680
High Duty (Siliceous)	65-80	18-30	3-8	1620-1680
Low Duty	60-70	23-33	6-10	1520-1595

Firebrick is the most common form of refractory material. It is used extensively in the iron & steel industry, nonferrous metallurgy, glass industry, pottery kilns, cement industry, and by many others.

Fireclays are hydrated aluminum silicates that occur naturally. They are sufficiently pure to serve as raw materials for refractories. The principal mineral in fireclays is kaolinite.

While other clay minerals may be present, the formula $Al_2O_3.2SiO_2.2H_2O$ can usually represent the clay fraction.

Plastic and semi-plastic fireclays, as their names indicate, develop varying degrees of plasticity when they are mixed with water. This is an important factor in the manufacture of fireclay bricks, because the plastic fireclays facilitate the forming process and act as a bonding phase for the raw & calcined flint clays, and they have greater variation in their impurity content.

High alumina refractories

Alumino silicate refractories containing more than 45% alumina are generally termed as high alumina materials. The alumina concentration ranges from 45 to 100%. The refractoriness of high alumina refractories increases with increase in alumina percentage. The applications of high alumina refractories includes the hearth & shaft of blast furnaces, ceramic kilns, cement kilns, glass tanks and crucibles for melting a wide range of metals.

High-alumina bricks serve as a multi-purpose refractory material for severe environments. They are used extensively in the steel industry for such applications as hot metal cars, electric furnace roofs, piers & muffles for a variety of furnaces, and numerous applications where strength at high temperature is an essential requirement. The aluminium and glass industries use high-alumina refractories to keep the melt in the molten state.

Most high-alumina refractories are classified according to their alumina content, which could range from 50 –99%. They are designated as 50%, 60%, 70%, 80%, 85% and 90% alumina.

Two classes of high-alumina refractories are distinguished by a microstructure that is essentially a single, crystalline phase.

These are: (1) mullite refractories and (2) corundum refractories.

Mullite refractories

Mullite is about 72% alumina with 28% silica. The manufacturing procedures are designed to maximize the formation of the compound mullite ($6Al_2O_6.2SiO_2$). A refractory with 71.8% alumina and 28.2% silica will be composed of only mullite ($6Al_2O_6.2SiO_2$) if fired at equilibrium conditions.

However, the extent to which well-developed mullite crystalline form occurs in a refractory depends on the purity of the raw materials used and the manufacturing processes, particularly firing. Therefore, all high-alumina bricks with around 70% alumina may have a well-developed mullite phase.

Mullite refractories have excellent volume stability and strength at high temperatures. They are highly suitable material for electric furnace roofs, blast furnaces and blast furnace stoves, hot metal cars, and the superstructure of glass tank furnaces.

Corundum refractories

The 99% alumina class of refractories is called corundum. These refractories comprise single-phase, polycrystalline, alpha- alumina.

Silica brick

Silica brick is a refractory material containing at least 93% SiO_2. The raw material is quality rocks. Various grades of silica brick have found extensive use in the iron and steel melting furnaces. In addition to high fusion point multi-type refractories, the other important properties are their high resistance to thermal shock (spalling) and their high refractoriness. It finds typical use in glass making and steel industry.

The outstanding property of silica brick is that it does not begin to soften under high loads until its fusion point is approached. This behaviour contrasts with that of many other refractories, for example alumino silicate materials, which begin to fuse and creep at temperatures considerably lower than their fusion points. Other advantages are flux & slag resistance, volume stability and high spalling resistance.

Silica bricks are made from raw materials that are essentially quartz. During the initial firing, alpha-quartz is first converted to beta-quartz, accompanied by an abrupt expansion at 576 °C. Slow rates of heating are required through this temperature range to prevent cracking, as the volume change is about 0.9%. Since the final firing temperature is somewhat over 1026 °C, the brick as it is put into service consists of cristobalite particles (properties of cristobalite are not affected by temperature fluctuation, provided the temperature does not drop below 600°C), with possibly some having residual unconverted quartz cores.

Magnesite Refractories

Magnesite refractories are chemically basic materials, containing at least 85% magnesium oxide. They are made from naturally occurring magnesite ($MgCO_3$). The properties of magnesite refractories depend on the concentration of silicate bond at the operating temperatures. Good quality magnesite usually results from a CaO-SiO_2 ratio of less than 2 with a minimum ferrite concentration, particularly if the furnaces lined with the refractory operate in oxidizing and reducing conditions. The slag resistance is very high particularly to lime and iron rich slags.

Chromite refractories

Here, a distinction must be made between chrome-magnesite refractories and magnesite- chromite-refractories. Chrome magnesite material usually contain 15-35% Cr_2O_3 and 42-50% MgO whereas magnesite-chromite refractories contain at least 60% MgO and 8-18% Cr_2O_3.

Chrome-magnesite refractories are made in a wide range of qualities and are used for building the critical parts of high temperature furnaces. These materials can withstand corrosive slags & gases and have high refractoriness. The magnesite-chromite products are suitable for service at the highest temperatures and in contact with the most basic slags used in steel melting. Magnesite-chromite usually has a better spalling resistance than chrome- magnesite.

Zirconia refractories

Zirconium dioxide (ZrO_2) is a polymorphic, material. There are certain difficulties in its usage and fabrication as a refractory material. It is essential to stabilize it before application as a refractory. This is achieved by incorporating small quantities of calcium, magnesium and cerium oxide, etc. Its properties depend mainly on the degree of stabilization and quantity of stabilizer as well as the quality of the original raw material. Zirconia refractories have a very high strength at room temperature which is maintained upto temperatures as high as 1500°C. They are, therefore, useful as high temperature constructional materials for furnaces and kilns. The thermal conductivity, of zirconium dioxide is found to be much lower than that of most other refractories and the material is therefore used as a high temperature insulating refractory.

Since Zirconia exhibits very low thermal losses and does not react readily with liquid metals, it is particularly useful for making refractory crucibles and other vessels for metallurgical purposes. Zirconia is a useful refractory material for glass furnaces primarily since it is not easily wetted by molten glasses and because of its low reaction with them.

Oxide refractories (Alumina)

Alumina refractory materials which consist of aluminium oxide with little traces of impurities are known as pure alumina. Alumina is one of the most chemically stable oxides known. It is mechanically very strong, insoluble in water & super heated steam, and in most inorganic acids & alkalies. Its properties make it suitable for the shaping of crucibles for fusing sodium carbonate, sodium hydroxide and sodium peroxide. It has a high resistance in oxidizing and reducing atmosphere. Alumina is extensively used in heat processing industries. Highly porous alumina is used for lining furnaces operating up to 1850^0C.

Typical refractory lining pattern in different zones of a steel reheating furnace is given below:

Indicative lining pattern of Furnace Roof:

- **Soaking Zone**: 250 mm thick special shaped bricks made of 60% Alumina Refractories backed by 50 mm thick Insulating castable.
- **Heating Zone**: 250 mm thick special shaped Refractories 60% Alumina quality backed by 50 mm thick Insulating castable.
- **Pre-Heating Zone**: 250 mm thick special shaped Refractories 40% Alumina quality backed by 50 mm thick Insulating castable.

Indicative lining pattern of Furnace side walls and end walls:

- **Discharge end wall**: 230 mm thick 60% Alumina quality refractories backed by 115 mm hot face insulation bricks, 115 mm thick Mica Insulation bricks (lS-2042) and 75 mm thick Calcium silicate blocks insulation and 5 mm thick asbestos sheet.
- **Side walls (Heating & Soaking zones):** 230 mm thick 60% Alumina quality refractory bricks backed by 115 mm thick hot face insulation bricks, 115 mm thick Mica Insulation bricks / Cold face insulation bricks, 75 mm thick Calcium silicate block insulation and 5 mm thick asbestos sheet.
- **Side walls (Preheating Zone):** 230 mm thick 40% Alumina quality (IS-8) refractory bricks backed by 115 mm thick Hot face insulation bricks, 115 mm thick Mica Insulation bricks, 75 mm thick calcium silicate block insulation and 5 mm thick asbestos sheet.
- **End wall (Charging side):** 230 mm thick 40% Alumina quality(IS-8) refractory bricks backed by 115 mm thick Hot face insulation bricks, 115 mm thick Mica Insulation bricks, 75 mm thick calcium silicate block insulation and 5 mm thick asbestos sheet..
- **Flue Off take/down corners**: 150 mm thick ceramic fibre blanket (RT128) to be held in position by heat resisting (Stainless steel) studs and washers in case of flue port at the roof. If the flue line is below the hearth, lining pattern will be same as that of charging side end wall.

- **Soaking Zone of the Hearth**: 150mm thick High Alumina fire bricks (60% Alumina) backed by 75 mm thick 40% Alumina quality firebricks, 230 mm thick cold face insulation bricks and 5 mm thick asbestos sheet.
- **Heating Zone of the Hearth:** 150 mm thick High Alumina quality fire bricks (60% Alumina) backed by 75 mm thick 40% Alumina (IS-8) Firebrick, 115 mm thick Hot face insulation bricks, 115 mm thick cold face insulation bricks and 5 mm thick asbestos sheet.
- **Pre-Heating Zone of the Hearth:** 150 mm thick High Alumina fire bricks (45% Alumina) backed by 75 mm thick 40% Alumina quality firebricks, 115 mm thick Hot face insulation bricks, 115 mm thick cold face insulation bricks and 5 mm thick asbestos sheet.
- **Charging end Doors:** 175 mm thick RT128 grade ceramic fibre module backed by 25 mm thick mineral wool insulation blanket.
- **Discharge end Doors:** 225 mm thick RT128 grade ceramic fibre module backed by 25 mm thick mineral wool insulation blanket.
- **Recuperator area:** 115 mm thick lS-8 quality firebricks backed by 115 mm thick Mica insulating bricks and 50 mm thick calcium silicate block insulation.
- **Flue duct from Recuperator up to chimney:** 115 mm thick IS-6 quality firebricks backed by 115 mm thick Mica insulating bricks.

REQUISITES OF A GOOD REFRACTORY FOR A PARTICULAR JOB

- It should be able to withstand the temperature to which it is liable to be exposed.
- It should not be corroded by the slag produced in the furnace and to which the refractory is to be exposed.
- It should be resistant to the cutting action of flue gases and flames.
- Contraction and expansion of the refractory should not be sudden but should be uniform.
- It should be able to withstand the pressure which is likely to be put on it.
- It should be able to conserve heat and prevent heat loss from the furnace.
- It should be able to withstand sudden changes of temperature.
- It should have the ability to resist the rubbing action of the material that comes in its contact (*i.e.* high abrasion resistant).
- It should be impermeable to liquids and gases {*i.e.* low permeability and porosity)
- It should have ability to promote heat transfer in recuperative type of furnaces (*i.e.* high thermal conductivity)
- It should be able to resist disintegration by undergoing least possible structural changes on cooling and heating (*i.e.* low co-efficient of thermal contraction and expansion)

The general requirements of a refractory material can be summed up as:

- Its ability to withstand high temperatures and trap heat within a limited area like a furnace .
- Its ability to withstand action of molten metal , hot gases and slag erosion etc .
- Its ability to withstand load at service conditions .

- Its ability to resist contamination of the material with which it comes into contact .
- Its ability to maintain sufficient dimensional stability at high temperatures and after/during repeated thermal cycling .
- Its ability to conserve heat.

SELECTION OF REFRACTORIES

The selection of refractories for any particular application is made with a view to achieving the best performance of the equipment, furnace, kiln or boiler and depends on certain properties of the refractories. Further, the choice of a refractory material for a given application will be determined by the type of furnace or heating unit and the prevailing conditions *e.g.* the gaseous atmosphere, the presence of slags, the type of metal charge etc. Therefore, temperature is not the only criterion for selection of refractories.

Any furnace designer or industry should have a clear idea about the service conditions which the refractory is required to face. The furnace manufacturers or users have to consider the following points, before selecting a refractory.

- Area of application.
- Working temperatures.
- Extent of abrasion and impact.
- Structural load of the furnace.
- Stress due to temperature gradient in the structures and temperatures fluctuations.
- Chemical compatibility to the furnace environment.
- Heat transfer and fuel conservation.
- Cost considerations.

It is essential to have an evaluation of the above conditions. The assessment of the desired properties would provide guidelines for selection of the proper refractory materials.

The furnace manufacturer or a user is also concerned with the conservation of energy. Fuel can be saved in two ways, either by insulation or by faster working. Both these methods give low energy cost per ton of product.

There is practically no application where one particular refractory quality can withstand all operating conditions. Usually one would like to have a refractory with low thermal conductivity so that heat may be more effectively contained within a furnace. Sometimes, however refractories and materials having high thermal conductivity are desired; for example, a protective muffle in certain ceramic kilns is designed to prevent combustion gases from reaching the ceramic ware. It must transfer as much heat to the ware as possible, so conductive ceramic materials like *silicon carbide* are often used for muffles.

We must take into account the specific metal we will be melting, the temperatures we will be reaching, the length of melt, how long we will be holding metal in the furnace, how much inductive stirring will take place, what additives or alloying agents we will be using, and furnace relining practices. Therefore refractory selection is based on the most critical factor vis-à-vis the wear out pattern and operating parameters in order to achieve the best possible techno economics.

A right selection will not only gives a longer life but helps in reducing down hours ; thus a better productivity.

The five most important factors governing refractory service are:

- Area of application
- Type of refractory
- Operating conditions
- Quality of refractory
- Workmanship

It is essential to have an objective evaluation of the above considerations. These factors are not necessarily in order of importance. The user needs to adjust them in the order to suit the conditions existing in any particular furnace.

AREA OF APPLICATION

Selecting insulation materials and designing insulation systems requires evaluation of four factors: 1) Thermal performance; 2) Physical properties; 3) Chemical characteristics and 4) Costs

1. Thermal performance- involves temperature limit; melting or fusion temperature; thermal conductivity (or insulation capability); heat capacity or storage; thermal expansion; and thermal shock (spalling) resistance. Operating temperature is an important criterion for selection of refractories. Temperatures exceeding the refractory's rating can soften its surface and cause rapid erosion & failure. If the operating conditions heat or cool the refractory lining beyond its specified range, the resulting thermal shock can damage the integrity of the lining.

- **Design of furnace/vessels** -The performance and stability of the refractory lining depends on the structural design of the furnace and its configuration. Sometimes adequate attention is not given to the refractory lining and its engineering related issues during early plant equipment design. This may lead to several compromises as there is limited scope for change in vessel design, operating conditions and the process to reduce the impact of these factors on lining performance during later stages of detailed engineering.
- The refractory lining failure may occur due to insufficient combustion volume. The heat released within the system is generally more than absorbed by the process and is dissipated through walls or exhausted with flue gases. In such cases, there are possibilities of the lining approaching the flame temperature and causing severe problems. The burner type, its design, location, flame shape, possibilities of flame impingement, flowing pattern of flue gases, etc., may affect the lining. In many cases, limited vessel dimensions, inaccessibility and complicated configuration restricts the best lining practices during initial construction as well as subsequent maintenance & repair.

2. Physical properties- includes density or porosity; abrasion; wear & erosion resistance; electrical resistivity for use in electric heating; mechanical strength; and other structural properties at high temperatures. High mechanical strength is required to resist load, impact, abrasion and erosion in refractory materials.

- **Physical Shock and Mechanical Stress -** Most refractory materials tend to be brittle and weak in tension. Bulky charge material dropped into an empty furnace can easily cause the lining to crack upon impact. If a crack goes unnoticed, molten metal may penetrate, leading to a run-out with the possibility of a water/molten metal explosion.

The sudden or cumulative effects of physical shocks and mechanical stress can lead to failure of refractory lining.

- **Thermal Stress -** The entire refractory lining is subjected to thermal stress which also plays a dominant role in selection. The refractory lining system or design must take care of thermal resistance, expansion relief; prevent cracking and eliminating built-in barriers. Thermo-mechanical FEM analysis is a reliable tool to investigate thermal stresses and develop ways on improving the lining behavior.
- **Refractory Wear -** Refractory linings are subject to regular wear from the scraping of metal on the furnace walls. In theory, refractory wear should be uniform, but in practice this never occurs. The most intense wear occurs at the slag/metal interface; where sidewalls join the floor; and at thin spots caused by poor lining installation.

3. Chemical considerations- It includes uniformity of composition; reactions between base materials & operating environment; and issues such as volatilization of the constituents or binding agents, corrosion, chemical attack, or diffusion and reactions with the product. Chemical attack may occur from gases such as steam, hydrogen, carbon monoxide, acid & alkali slag, sulfurous gases, etc.; these acid gases can initiate various problems in the lining. All of the important operating factors and any other criteria specific to the process under consideration should be verified for their possible effect on the performance of the lining.

- **Chemical v/s Physical Characteristics.** Chemical characteristics of the furnace process usually determine the type of refractory required, although this is not universally true. We have the familiar example of silica brick, the most acidic of refractories, being successfully used in the roofs of basic open hearth steel furnaces. In this case the physical properties of the brick, particularly load bearing strength, are more important than the chemical characteristics. For successful refractory practice the physical & chemical properties of the refractory must be compatible with the furnace operation and nature of the process.

4. Cost / Economics – it means initial installation labor; maintenance, repair; & replacement costs. Selecting materials solely based on price and ease of installation should be avoided. Very often, monolithic lining system is selected in the lining design of critical equipment where brick lining or some other design may be more suitable. The selection may be due to cheap & easy availability of monolithics and easier installation than brick lining and to avoid preparation of too many engineering drawings for the complicated brick shapes & laying details. In many cases, recommendations of refractory manufacturers are biased and based on their available product ranges, which may not be appropriate for the required conditions.

TYPE OF REFRACTORY

Refractory materials are typically quasi brittle at low temperature and have a viscous behavior at high temperature. They are subject to considerable variability in strength, resulting from local variations in their microstructure and their lack of ductility. They are characterized by high-temperature creep or plastic deformation. Their brittleness and their high elastic modulus make them sensible to failure under thermal stresses and shocks.

In order to judge the type of refractory material most likely to give satisfactory service in a certain furnace, one must have knowledge of the limitations as well as the strong points of the principal types of refractories. Very often it is desirable to use different types of refractories in various parts of the same furnace to obtain uniformly good or balanced refractory service.

OPERATING CONDITIONS

Selecting materials should always be based on properties & specifications suitable for the specific application and operating conditions. We frequently hear from users that "the refractories fail prematurely and our operating practice has not been changed in years." This may be true, but very often there can be unintentional changes in furnace conditions long enough in duration to destroy a refractory lining but not long enough to be considered as a change in practice. For example, a burner getting out of adjustment so that the furnace atmosphere becomes highly reducing for a short period of time; possibly the operating temperature of the furnace may become abnormally high and damage the grid-work before it is corrected; or perhaps the chemical composition of raw material charge or the fuel may have changed. Such variations in practice are usually extremely difficult to discover because they leave no record except for their effect on the lining. Another complicating factor is that the failure of the refractory may not become apparent until weeks or even months after the actual damage was done.

Transportation, handling, timely arrival of refractory bricks, mortars, skill of masonry work, proper equipment for application e.g. mixer machine for castable, vibrator for installation, forma etc. are very important.

QUALITY OF REFRACTORY

Quality of refractories can be the controlling factor in the success or failure of thermal equipment as well as the safe and profitable operation of the plant. The performance of any refractory is considered reasonable when similar service lives are achieved on a regular basis. Premature lining failure may be defined as one that does not achieve normal or average performance and service life.

The most common refractory problems are:

- Hot spots (higher casing temperature);
- Excessive cracking;
- Spalling of lining (thermal, mechanical, structural);
- Erosion and thinning of lining;
- Chemical attack/corrosion from process gases (such as hydrogen, carbon monoxide, sulfur dioxide, alkalies), flue gases (sulfur, sodium, vanadium), steam, etc;
- Acid-gas dew-point corrosion of refractory and metallic parts;
- Partial melting and degradation of lining;
- Excessive shrinkage and development of gaps;
- Anchor failure and detachment of lining from wall;
- Failure of metal liner over refractory;

- Explosive spalling during dry out;
- Mechanical damages.

Critical applications, purchasing of materials based on a comparison of product datasheet or catalogue specifications or equivalent principles should be avoided. The actual performance references and records for specific products or brands should be verified against similar applications. Reviews of manufacturing facilities & quality control program, and random inspection & testing of important properties are essential.

WORKMANSHIP

Proper installation of the refractory is as important as selection of the right material. If the material is inadequately compacted during installation, voids or areas of low density may form, creating a weak spot easily attacked by molten metal. Maintaining proper expansion gaps, correct dimension of bricks and monolithics, fixing anchors etc. all are very important to achieve better life of refractories.

Some of the important aspects of refractory construction are:

Bond: Bond in brick work is an arrangement by which the joints between the bricks are staggered in all directions in a pre-determined manner. This ensures stability in construction as well as air tightness. An all header construction is recommended in the hot face of a furnace where heavy refractory wear is anticipated either due to action of molten metal slag or dust laden gas moving at high velocities. Where intermittent repairs to the walls are expected, the working layer is built separately from the main wall.

Wall Thickness: Structural stability should be taken into consideration while deciding the wall thickness. As a good engineering practice, furnace walls lined with 115mm thickness bricks should not be constructed beyond a height of one meter without proper anchorage. From the stability point of view minimum thickness should be around 230mm, which may be suitably increased as the height goes up. Use of metallic sheets with anchoring arrangements, improves the stability of the wall where basic refractories are used. Such anchoring of walls with the furnace structure should have sliding arrangement to take care of the movement of refractories during heating up and cooling down.

Joints: The joints form a small part of the lining of the vessels but they have two important functions.

- To assemble the pieces (bricks or blocks) together .
- To absorb thermal expansion and to limit the stresses generated in the masonry.

The joints may be either mortared joints (i.e. granular material with matrix and voids) or dry (without mortar) joints. In either case, the stress/strain behavior of the joint interface differs considerably. At room temperature, compressibility of the joints is around 20%. At 1200°C, the mortar becomes more and more plastic and its compressibility approaches 50%.

Mortar Joints: Mortar joints improve the stability of construction. It also takes care of slight variation in sizes as well as warpages by offering a uniform bedding surface. Chemical and physical properties of mortar should be identical to that of the bricks which they are joining. Workability of the mortar is important, in the absence of which, good construction is not possible. The excess mortar applied to the brick surface should come out from the edges when one brick is tapped against the other, to ensure that the joints are full and that there is not open gap.

Expansion Joints: Since refractories expand on heating up , suitable provisions should be made at the time of construction for this expansion to take place without creating undue stresses which may result in failure of the refractories. Usually, expansion joints are built in as the construction proceeds. The necessary provision is made depending on the expansion characteristic of the particular refractory at the operating temperature. Care should be taken to check the expansion allowance – in practice just one half of the theoretical expansion is sufficient. The remaining one half will be absorbed by the mortal joints, or even dry joints between the bricks.

Structural Support: Refractories are normally held in position with steel shell or a structural framework by tie rods as failure of structural by other reasons will cause partial failure or limit the total collapse of refractories. Their alignment, shape etc should also be equal concern. Also important is that the structural should not get overheated, and should be adequately insulated wherever required.

Good brickwork is extremely important but there are other factors involved like the proper size & location of expansion joints, the rise of an arch, or the thickness of a wall as properly coming under the head of masonry workmanship. If the refractory lining is improperly cantered or has somehow been distorted during storage or shipment, lining thickness will be uneven and it may fail prematurely before the end of its predicted service life. The refractory manufacturer's procedures for installation, drying and sintering must be followed .

The five most important factors governing refractory service are tabulated below.

Table: Factors governing refractory service

Area Of Application	Type Of Refractory	Operating Conditions	Quality Of Refractory	Workmanship
Selecting insulation materials and designing insulation systems requires evaluation of four factors:	Refractory materials are typically quasi brittle at low temperature and have a viscous behavior at high temperature.	Selecting materials should always be based on properties and specifications suitable for the specific application and operating conditions.	The performance of any refractory is considered reasonable when similar service lives are achieved on a regular basis.	Proper installation of the refractory is as important as selection of the right material.
1) Thermal performance; involves temperature limit; melting or fusion temperature; thermal conductivity , heat capacity or storage; thermal expansion; and spalling resistance	They are subject to considerable variability in strength, resulting from local variations in their microstructure and their lack of ductility.	We frequently hear statements from users that "the refractories fail prematurely and our operating practice has not been changed in years."	Premature lining failure may be defined as one that does not achieve normal or average performance and service life.	If the material is inadequately compacted during installation, voids or areas of low density may form, creating a weak spot easily attacked by molten metal.

2) Physical properties; includes density or porosity; abrasion; wear and erosion resistance; electrical resistivity; mechanical strength; and other structural properties at high temperatures	In order to judge the type of refractory material most likely to give satisfactory service in a certain furnace, one must have knowledge of the limitations as well as the strong points of the principal types of refractories.	This may be true, but very often there can be unintentional changes in furnace conditions long enough in duration to destroy a refractory lining but not long enough to be considered as a change in practice.	The most common refractory problems are: • Hot spots (higher casing temperature); • Excessive cracking; • Spalling of lining (thermal, mechanical, structural); • Erosion and thinning of lining; • Chemical attack/corrosion from process gases (such as hydrogen, carbon monoxide, sulfur dioxide, alkalies), flue gases (sulfur, sodium, vanadium), steam, etc; • Partial melting and degradation of lining; • Excessive shrinkage and development of gaps; • Detachment of lining from wall; • Mechanical damages.	Maintaining proper expansion gaps, correct dimension of bricks and monolithics, fixing anchors etc. all are very important to achieve better life of refractories. Some of the important aspects of refractory construction are:
3) Chemical characteristics; includes uniformity of composition; reactions between base materials and operating environment; and issues such as volatilization of binding agents, corrosion, chemical attack, or diffusion and reactions with the product.	Very often it is desirable to use different types of refractories in various parts of the same furnace to obtain uniformly good or balanced refractory service.	For example, a burner getting out of adjustment so that possibly the operating temperature of the furnace may become abnormally high and damage the grid-work before it is corrected.		Bond: Bond in brick work is an arrangement by which the joints between the bricks are staggered in all directions in a pre- determined manner.
4) Costs; mean initial installation labor; maintenance, repair; and replacement costs.		Such variations in practice are usually extremely difficult to discover because they leave no record except for their effect on the lining.		Wall Thicknes: Joints: Mortar Joints: Mortar joints improve the stability of construction.
		Transportation, handling, timely, skill of masonry work etc is very important.		Expansion Joints: Structural Support: Refractories are normally held in position with steel shell or a structural framework by tie rods

Key refractory Selection Facts:

- Reaction/wear increases exponentially with increasing temperature .
- Temperature cycling will cause cracks, loss of strength,& spalling of material .Stronger is not always better .
- Lining life is controlled by the "weakest link" .

- Shorter service life can be more cost effective .
- Refractory manufacture should use high purity, ultrafine materials .
- Microstructure engineering – control of aggregate shape is important.

Since, no single refractory material meets the above requirements/properties hence refractories with a maximum given set of properties are chosen for a specific purpose to meet the service conditions/requirements. Refractories are expensive and any failure in the refractories results in great loss of production, time, equipment and some time the product itself. The type of refractories also will influence energy consumption and product quality. Therefore, the problem of obtaining refractories best suited to each application is of supreme importance.

FACTORS AFFECTING SELECTION OF REFRACTORIES

Optimum use of refractories is achieved by careful study of furnace design and evaluation of operating conditions prior to selection of refractory products which meet the design and operating requirements.

From the multiple factors listed below, it may appear that the choice of most suitable material would be exceedingly difficult. Sometimes this is true.

Moreover, the best refractory selection often depends on a few requirements so important that other factors play a minor role. In some cases, refractoriness, i.e., maximum service temperature, alone will be the deciding factor, and in others high refractoriness will have to be coupled with resistance to thermal shock. Under other circumstances, resistance to metals, slags, or disintegration by reducing gases may be the governing factors. Sometimes high insulating value is desirable, but in other situations high thermal conductivity may be needed. When selecting refractories, the following major factors must be identified.

OPERATING FACTORS :

- Function of the furnace
- Nature of material being processed
- Rate and continuity of operation
- Range and rapidity of temperature changes
- Chemical attack by metals, slags, ash, etc.
- Fluidity of molten metal or slag
- Velocity of furnace gases
- Abrasion from contained solids or gases
- Impact from charging
- Erosion by molten furnace contents
- Impinging flames or hot spots

FURNACE DESIGN & CONSTRUCTION RELATED FACTORS:

- Type of furnace

- Design and dimensions of walls and arches
- Loads imposed on the lining
- Conditions of heating (one or more sides)
- Amount of insulation
- Air- or water-cooling
- Type of refractory construction (brick or monoliths)
- Methods of bonding or support
- Provision for thermal expansion
- Mechanics of any moving furnace parts

REFRACTORY RELATED FACTORS :

- **Properties at room temperature–**
  - Workmanship and physical strength
  - Density, porosity, permeability
  - Chemical and mineral composition
  - Uniformity
  - Size and design
- **Properties at high temperatures–**
  - Refractoriness or maximum service temperature
  - Reversible thermal expansion
  - Resistance to thermal shock
  - Resistance to chemical attack
  - Resistance to mechanical impact or stress
  - Resistance to abrasion or erosion
  - Permeability to gases or liquids
  - Volume stability (bloating or shrinkage)
  - Resistance to gases and fumes
  - Thermal conductivity
  - Heat capacity
  - Electrical resistance
- **Economic factors** -
  - Delivered cost
  - Cost of installation (brick vs. monoliths)
  - Special shapes or forming required
  - Service life

TESTING OF REFRACTORIES

The durability of refractories is intimately connected with economics and efficiency of the process in they are employed. Refractories are characteristically anisotropic in nature which makes it all the more difficult to judge exactly the durability while it is in use. The furnace designer has to choose the refractories according to the data from different tests available to him. That means in order to fulfill the users or various application requirements and to achieve long service lives, refractory products must be assessed by testing.

The main functions of refractories testing may be classified into three types:

1. Evaluation of new materials before use.
2. Quality control by the manufacturer or user.
3. Post-mortem examination of refractory bricks that behaved differently than the normal.

The testing of refractories can be categorized as:

- Non-destructive type.
- Destructive type.

A list of refractory testing including both these types of testing is given below:

- Bulk density
- Apparent porosity
- Apparent specific gravity
- True specific gravity and true density
- Particle size analysis
- Permeability
- Pore size distribution
- Refractoriness or Pyrometric Cone Equivalent (P.C.E)
- Refractoriness Under Load (R.U.L)
- Cold Crushing Stress (C.C.S)
- Modulus of Rupture (M.O.R)
- Modulus of Elasticity (M.O.E)
- Permanent Linear Change (P.L.C)
- Reversible Thermal Expansion (R.T.E)
- Thermal Conductivity
- Differential Thermal Analysis (DTA)
- Thermo-gravimetric Analysis (TGA)
- Spalling Resistance (water quenching and air spalling)
- Hydration Resistance Test
- Creep in Compression Test
- Abrasion Resistance Test
- Resistance to Carbon Monoxide

- Slag Corrosion Test
- Microstructural analysis under Optical Microscope
- Mineral Phase identification by X-ray Diffractometer (XRD pattern)
- Chemical Analysis

The influence of refractories on fuel economy arises from many causes, among which are the following :

- The stability of the furnace structure is obviously an essential feature in the efficient performance of industrial operation.
- Maintenance of a sound construction free from cracks within the bricks of the joints in essential. Cracks in the brick work may cause short circuiting of flue gases or leakage of live gas into the air or air into the gas, with disastrous effects on temperature uniformity in the furnace, on the amount of fuel used and on the brickwork; this is particularly important in coke ovens and gas retorts.
- The action of slag on furnace brickwork may be such as to destroy it. In extreme circumstances it may give rise to holes through furnace walls, for example. into the heating flues.
- Correct thermal gradient is also important; in many firing operations bricks tend to wear down to a limiting thickness, no matter how thick they may be originally.
- In continuously operated furnaces the longer the periods between cooling down for repairs , the less fuel is needed for heating up the furnace from the cold. On the other hand, planned maintenance may be more economical even though it may mean doing repairs sooner than they are, strictly speaking required.
- In many industrial operational heat has to be transmitted through a wall of refractory material and in all furnaces a great deal of heat is stored within the bricks. Their thermal properties, therefore, affect the quantity of fuel used and are of interest to the furnace operator.

Machineries & equipments used in the refractory production process

Several types of machineries & equipments are used to produce refractories eg : mixing/kneading machines, presses, and kilns.

Mixing/Kneading Machines

Figure below illustrates different machines used to mix or knead refractory products.

There are two types of mixing and kneading machines: fixed vessel and driven vessel. Mixing homogenizes more than two types of bulk materials, and kneading machines make a uniform coating layer. Mixing and kneading machines are equipped with mixing blades or muller wheels. Heating, cooling, or de-airing equipment may also be applied to the vessel. Mixing and kneading machines are used for manufacturing shaped and unshaped refractories. Unshaped refractories, however, are not processed any further .

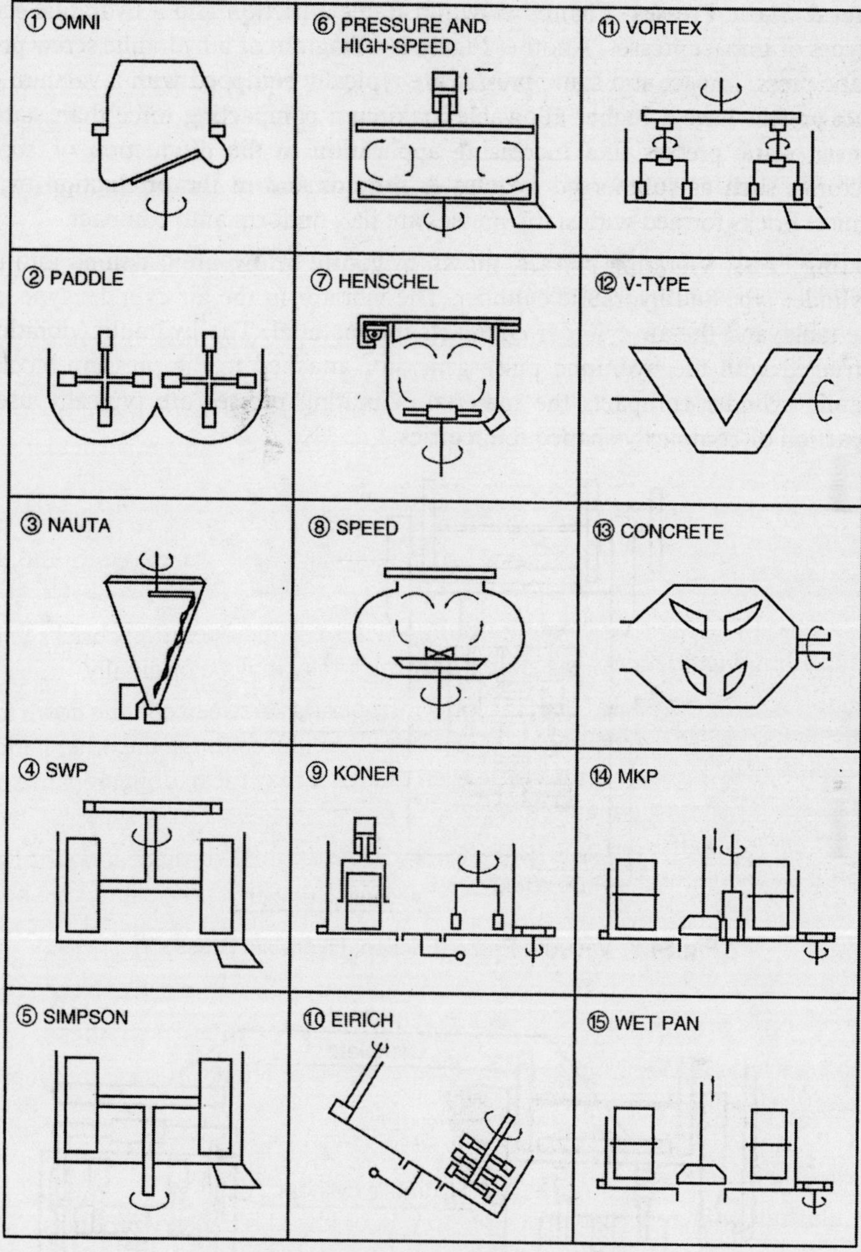

Fig.54.7: Mixing and Kneading Machines

Refractory Presses

Refractory pressing machines are broadly categorized into three groups: impact & static, vibrating, and cold isostatic press. Choosing between the three groups of presses largely depends on the type of raw materials used.

- **Impact & Static Presses**: Figure below illustrates a friction and a hydraulic screw press, two types of impact presses. Another Figure is a diagram of a hydraulic screw press, a type of static press. Impact and static presses are typically equipped with a vacuum deaerator. Impact presses have a higher allowable maximum compacting force than static presses. However, static presses find increasing application in the production of sophisticated refractories such as submerged nozzles & shrouds and in the production of industrial ceramics. Bricks formed with static presses are flat, uniform and, compact.
- **Vibrating Press**: Vibrating presses, shown in Figure below, are classified into two types: air cylinder type and hydraulic cylinder. The vibrator in the air cylinder type is attached to the table, and the air cylinder compacts the material. The hydraulic vibrating press is constructed with the hydraulic pulse generator attached to the pressure block, and the hydraulic cylinder compacts the material. Vibrating presses are typically used for the compaction of complexly shaped refractories .

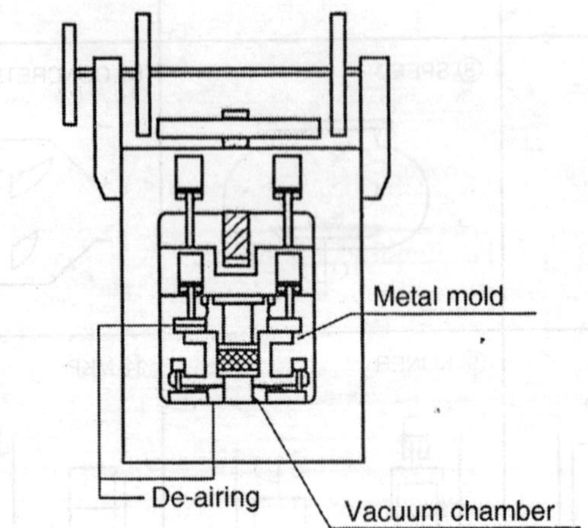

Fig.54.8: Vacuum Press (Friction, Hydraulic Press)

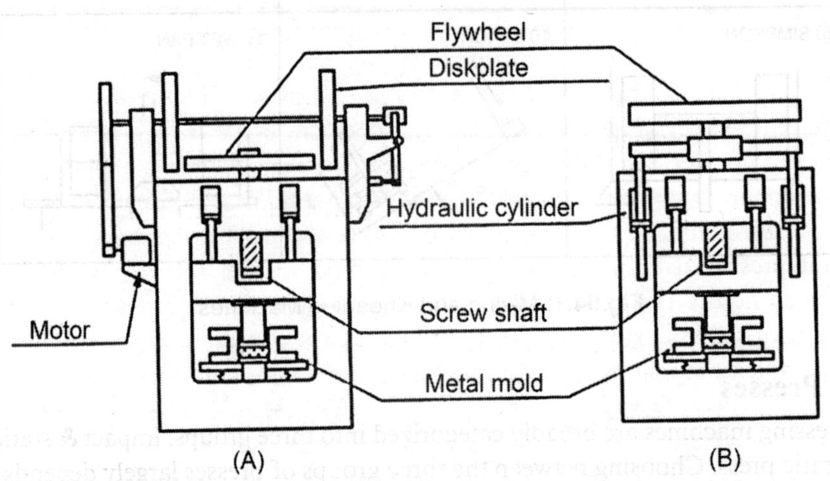

Fig.54.9: Friction Press (A), and Hydraulic Screw Press (B)

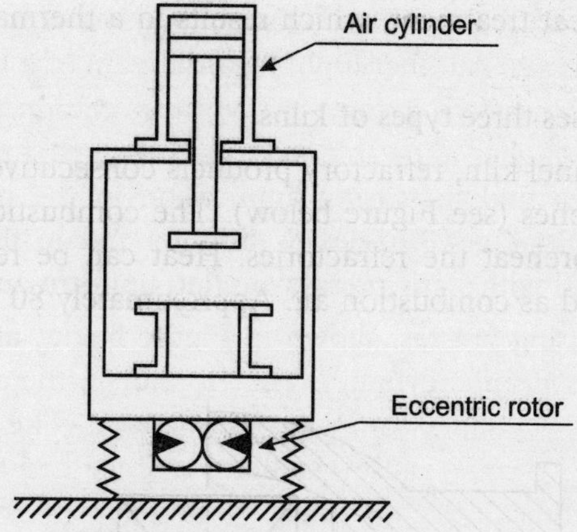

Fig.54.10: Vibrating Press

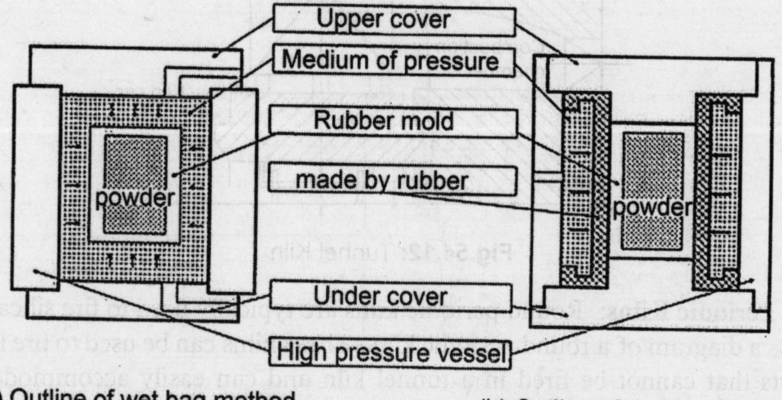

Fig.54.11: Cross Section of CIP (Cold Isostatic Press)

- **Cold Isostatic Press (CIP):** A CIP, illustrated in Figure above, is a molding device that provides homogeneous hydrostatic pressure over the entire surface of a rubber mold filled with powder. This method, also referred to as a hydrostatic press or a rubber press method, is a materials processing technique in which processing in which high fluid pressure is applied to a powder part at ambient temperature to compact it into a predetermined shape. The power part is consolidated into a dense compacted shape. Water or oil is usually used as the presser medium. CIPs are based on either the wet bag method, where the mold is placed in pressurized liquid, or the dry bag method, in which the mold does not touch the pressurized liquid. High pressurized molding provides uniform density, which leads to a reduction of internal stresses, eliminating cracks, strains and laminations, the ability to make complex shapes, and to press more than one shape at the same time .

Refractory firing Kilns

Refractories are fired to develop the materials refractory properties. The unfired ("green") refractories pass through a heat treatment, which results in a thermally stable refractory and/or crystallization.

The refractory industry uses three types of kilns:

- **Tunnel Kiln**: In a tunnel kiln, refractory products consecutively pass through preheating, firing, and cooling zones (see Figure below). The combustion gas from the firing zone is typically used to preheat the refractories. Heat can be recovered from cooling fired refractories and reused as combustion air. Approximately 80 % of shaped refractories are fired in tunnel kilns .

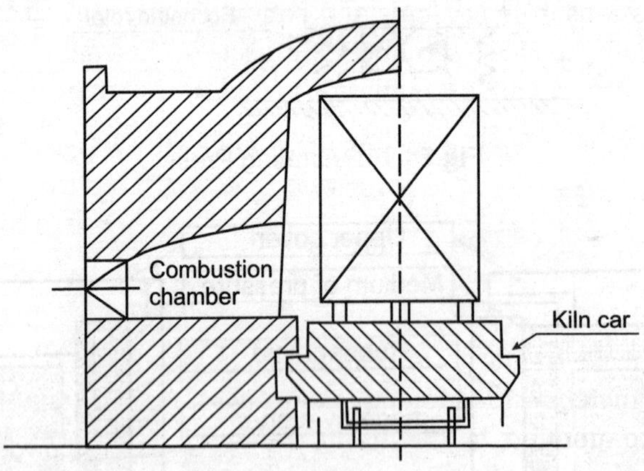

Fig.54.12: Tunnel Kiln

- **Round Periodic Kilns**: Round periodic kilns are typically used to fire silica bricks. Figure below is a diagram of a round periodic kiln. These kilns can be used to fire large refractory products that cannot be fired in a tunnel kiln and can easily accommodate changes in production .
- **Shuttle Kilns**: As illustrated in Figure below, the design of a shuttle kiln resembles the firing zone of a tunnel kiln. Shuttle kilns effectively store heat and are used to fire fireclay and specialty bricks .

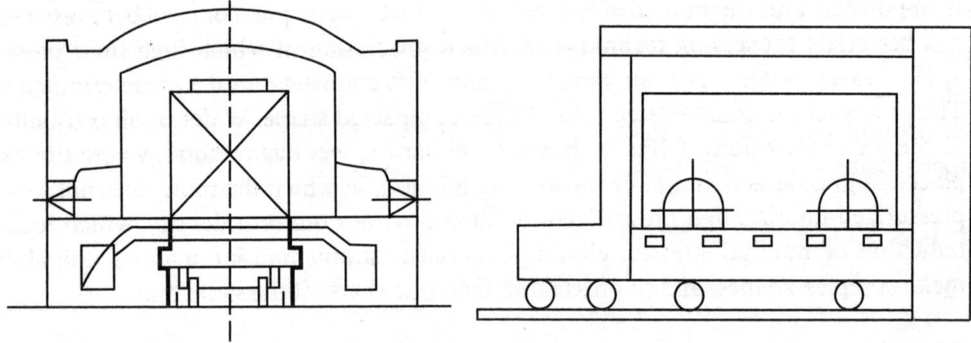

Fig.54.13: Shuttle Kiln

Pollutants Emissions & Controls in Refractory Manufacturing

Refractory production leads to emissions of organic particulate matter (PM); metals; and gaseous pollutants, including sulfur dioxide (SO_2), nitrogen oxides (NO_x), carbon monoxide (CO), carbon dioxide (CO_2), fluorides, & volatile organic compounds (VOCs) .

PM & Metals Emissions

PM and metal emissions occur during the crushing, grinding, calcining, & drying of the raw materials; the drying & firing of unfired refractory bricks and tar & pitch operations; and finishing of the refractories. Emissions from crushing and grinding are readily controlled with fabric filters. Product recovery cyclones and wet scrubbers are used on calciners and dryers to reduce PM emissions. Electric arc furnaces are generally controlled by a baghouse.

The primary pollutant of concern in refractory manufacturing is particulate matter (PM).

Emissions from crushing and grinding operations generally are controlled with fabric filters.

Product recovery cyclones followed by wet scrubbers are used on calciners and dryers to control PM emissions from these sources. The primary sources of PM emissions are the refractory firing kilns and electric arc furnaces. Particulate matter emissions from kilns generally are not controlled.

A number of elements in trace concentrations including aluminum, beryllium, calcium, chromium, iron, lead, mercury, magnesium, manganese, nickel, titanium, vanadium, and zinc also are emitted in trace amounts by the drying, calcining, and firing operations of all types of refractory materials.

Emissions of PM from electric arc furnaces producing fused cast refractory material are controlled with baghouses. The efficiency of the fabric filters often exceeds 99.5 percent. Emissions of PM from the ceramic fiber process also are controlled with fabric filters, at an efficiency similar to that found in the fused cast refractory process. To control blow chamber emissions, a fabric filter is used to remove small pieces of fine threads formed in the fiberization stage. The efficiency of fabric filters in similar control devices exceeds 99 percent. Small particles of ceramic fiber are broken off or separated during the handling and forming of the fiber blankets in the curing oven. An oil is used in this process, and higher molecular weight organics may be emitted. However, these emissions generally are controlled with a fabric filter followed by incineration, at an expected overall efficiency in excess of 95 percent.

Chromium is used in several types of nonclay refractories, including chrome-magnesite, (chromite-magnesite), magnesia-chrome, and chrome-alumina. Chromium compounds are emitted from the ore crushing, grinding, material drying & storage, and brick firing and finishing processes used in producing these types of refractories.

Gaseous & VOCs Emissions

SO_2, NO_x, CO & CO_2 are produced along with refractory products. The source of most of the SO_2 emissions is the fuel used to fire the kiln and dryers. The composition of the clays & plaster added to refractories and the amount of refractories produced affect the amount of SO_2 produced.

Pollutants emitted as a result of combustion in the calcining and kilning processes include sulfur dioxide (SO_2), nitrogen oxides (NOx), carbon monoxide (CO), carbon dioxide (CO_2), and volatile organic compounds (VOCs). The emission of SOx is also a function of the sulfur content of certain clays and the plaster added to refractory materials to induce brick setting. Fluoride emissions occur during the kilning process because of fluorides in the raw materials .

Volatile organic compounds emitted from tar and pitch operations generally are controlled by incineration, when inorganic particulates are not significant. Based on the expected destruction of organic aerosols, a control efficiency in excess of 95 percent can be achieved using incinerators.

APPENDIX-A: TECHNICAL DATA ON FUELS

Table 1. Main conversions used in petroleum industry

S. No.	Items	Conversion Factor
1.	Crude Oil	1 Metric Tonne = 7.33 Barrels =1.165 Cubic Metres (Kilolitres) 1 Barrel = 0.136 Tonnes = 0.159 Cubic Metres (Kilolitres) 1 Cubic Metre = 0.858 Tonnes = 6.289 Barrels 1 Million Tonne = 1.111 Billion Cubic Metres Natural Gas = 39.2 Billion Cubic Feet natural Gas = 0.805 Million tonnes LNG = 40.4 Trillion British Thermal Units

2.	Natural Gas	1 Billion Cubic Metre = 35.3 Billiom Cubic Feet Natural Gas = 0.90 Million tonnes Crude Oil = 0.73 Million Tonnes LNG = 36 Trillion British Thermal Units = 6.29 Million Barrels of Oil Equivalent
3.	LNG (Liquefied Natural Gas)	1 Million Tonne = 1.38 Billion Cubic Metres Natural Gas = 48.7 Billion Cubic Feet Natural Gas = 1.23 Million tonnes Crude Oil = 52 Trillion British thermal Units = 8.68 Million Barrels of Oil Equivalent
4.	CNG (Compressed Natural Gas)	1 Kilogram = 1.244 Standard Cubic Metres Natural Gas = 1.391 Litres of Petrol = 1.399 Litres ofHSDO

Table 2. Conversion factors for petroleum products

S.No.	Products of petroleum refining	TOE /Tonne	Barrel / Tonne
1.	Refinery Gas	1.150	8.00
2.	Ethane	1.130	16.85
3.	LPG	1.130	11.60
4.	Aviation Gasoline	1.070	8.90
5.	Motor Gasoline	1.070	8.53
6.	Jet Gasoline	1.070	7.93
7.	Jet Kerosene	1.065	7.93
8.	Other Kerosene	1.045	7.74
9.	Naphtha	1.075	8.50
10.	Gas/Diesel Oil	1.035	7.46
11.	Heavy Fuel Oil	0.960	6.66
12.	Petroleum Coke	0.740	5.50
13.	White Spirit	0.960	7.00
14.	Lubricants	0.960	7.09
15.	Bitumen	0.960	6.08
16.	Paraffin Waxes	0.960	7.00
17.	Non specified products	0.960	7.00

Note: TOE: Tonnes of Oil Equivalent, 1 barrel = 160 litres

Table 3. Combustion reactions

Combustibles	Combustion reaction	Water in vapour form, kcal	Water in Liquid form,kcal
Carbon	$C+0_2+4N_2=CO_2+4N_2$	+97.6	–
–	$C+ ½O_2+2N_2=CO+2N_2$	+29.4	–
Sulphur	$S_2+2O_2+8N_2=2SO_2+8N_2$	+183.6	–
–	$S_2+3O_2+12N_2=2SO_3+12N_2$	+186.6	–
Hydrogen	$H_2+ ½O_2+2N_2=2N_2+H_2O$	+58.2	(+69.1)
Carbon monoxide	$CO+ ½O_2+2N_2=CO_2+2N_2$	+68.2	–
Methane	$CH_4+2O_2+8N_2=CO_2+8N_2+2H_2O$	+192.5	(+214.3)
Ethylene	$C_2H_4+3O_2+12N_2=2CO_2+12N_2+H_2O$	+319.7	(+341.5)
Acetylene	$C_2H_2+ \frac{5}{2} O_2+10N_2=2CO_2+10N_2+H_2O$	+304.8	(+315.7)
Propane	$C_3H_3+5O_2+20N_2=3CO_2+20N_2+4H_2O$	+485.1	(+528.7)
Butane	$C_4H10 \frac{13}{2} O_2+26N_2=4CO_2+26N_2+5H_2O$	+625.7	(+680.2)
Paraffins	$CnH_{2n}+2+ \frac{3n+1}{2} O_2=nCO_2+(n+1)H_2O$	–	–
Unsaturated hydrocarbons	$C_mH_n+(m+n/4)O_2=mCO_2+nH_2O$	–	–

Table 4. Properties of Gases Concerned in the Combustion of Fuels

Gas	Formulae	Molecular weight	Specific gravity (Air = 1)	Co-efficient of solubility in water at N.T.P.	Density Kcal/ Nm3, (dry)	Calorific value (gross) Kcal/ Nm3 (dry)	Combustion requirements		Products of combustion				
							Unit = Nm^3/Nm^3 = moles/mole						
							O_2	Air	CO_2	h_2o	n_2	Total (dry)	Total (wet)
Air	—	29	1.000	0.029	0.29	—	—	—	—	—	—	—	
Nitrogen	N_2	28	0.967	0.024	1.25	—	—	—	—	—	—	—	
Oxygen	O_2	32	1.104	0.049	1.43	—	—	—	—	—	—	—	
Carbon dioxide	CO_2	44	1.519	1.713	1.96	—	—	—	1.00	—	—	1.00	1.00
Carbon monoxide	CO	28	0.967	0.035	1.25	3036	0.5	2.38	1.00	—	1.88	2.88	2.88
Hydrogen	H_2	2	0.070	0.022	0.90	3054	0.5	2.38	—	1.00	1.88	1.88	2.88
Methane	CH_4	16	0.554	0.056	0.72	9498	2.0	9.52	1.00	2.00	7.52	8.52	10.52
Ethane	C_2H_6	30	1.038	0.099	1.34	16515	3.5	16.66	2.00	3.00	13.16	15.16	18.16

Propane	C_3H_8	44	1.522	—	1.97	23674	5.0	23.80	3.00	4.00	18.86	21.86	25.80
Butane	C_4H_{10}	58	2.006	—	2.59	30690	6.5	30.94	4.00	5.00	24.44	28.44	33.44
Ethylene	C_2H_4	28	0.968	0.226	1.25	14892	3.0	14.28	2.00	2.00	11.28	13.28	15.28
Propylene	C_3H_6	42	1.452	0.446	1.88	21956	4.5	21.42	3.00 ,	3.00	16.92	19.92	22.92
Butylene	C_4H_8	56	1.937	—	2.50	29020	6.0	28.56	4.00]	4.00	22.56	26.56	30.56
Benzene	C_6H_6	78	2.697	—	3.49	35702	7.5	35.70	6.00	3.00	28.20	34.20	37.20
Toluene	C_7H_8	92	3.180	—	4.11	42766	9.0	42.84	7.00	4.00	33.84	40.84	44.84
Water vapor	H_20	18	0.622	—	0.804	—	—	—	—	—	—	—	—
Hydrogen sulphide	H_2S	34	1.177	4.621	1.52	6014	1.5	7.14	1.00 (SO2)	1.00	5.64	6.64	7.64

Table 5. Compostion and Average Properties of Gaseous Fuels

Parameters	B.F. Gas	Coal gas					Producer gas			Water Gas		Oil gas (retort)
		Low temp.	Horizontal retorts	Continuousvertical retorts (No steaming)	Continuous vertical retorts (steaming)	Denbemolised coke oven gas	Hand operated (Coke)	Mechanical (Coal)	Mechanical (Coke)	Blue	Carburetted	
Composition (% by volume)												
O_2	—	0.5	0.4	0.4	0.4	0.4	—	—	—	—	0.4	—
CO_2	11	4	2	3	4	2	7	4	5	4.7	5.6	–
C_mH_m	—	5.5	3.6	3	2	2.6	—	0.4	—	—	7	35
co	27	4	8	9	18	7.4	26	29	29	41	30.5	—
H_2	2	30	F9,	53.6	19.4	54	12	12	11	49	37	25
CH_4	—	52	30	25	20	28	0.5	2.6	0.5	0.8	14	40
N_2	60	4	4	6	6.2	5.6	54.5	52	54.5	4.5	5.5	—
Composition of C_mH_n (assumed)	—	C_4H_8	$C_{2\cdot5}H_5$	C_4H_8	C_4H_8	$C_{2\cdot5}H_5$	—	C_2H_4	—	—	$C_{2\cdot5}H_5$	$C_{2\cdot5}H_5$
Specific gravity (air = 1)	1.01	0.62	0.40	0.43	0.48	0.38	0.90	0.87	0.90	0.55	0.63	0.68
Theoretical air, Nm^3/Nm^3	0.690	8.235	4.908	4.708	4.060	4.572	0.952	1.280	1.00	2.218	4.270	10.936

Theoretical flue gas, Nm^3/Nm^3												
CO_2	0.380	0.950	0.490	0.490	0.500	0.439	0.335	0.364	0.345	0.465	0.690	1.315
H_2O	0.020	1.690	1.210	1.156	0.974	1.165	0.130	0.180	0.120	0.506	0.839	1.965
N_2	1.145	6.545	3.917	3.779	3.269	3.667	1.297	1.531	1.335	1.797	3.428	8.639
Total, wet	1.545	9.185	5.617	5.425	4.743	5.271	1.762	2.075	1.800	2.768	4.957	11.919
Total, dry	1.525	7.495	4.407	4.269	3.769	4.106	1.632	1.895	1.680	2.262	4.118	9.954
Wet flue gas, % by volume:												
CO_2	24.6	10.3	8.7	9.0	10.5	8.3	19.0	17.5	19.2	16.8	13.9	11.0
H_2O	1.3	18.4	21.5	21.3	20.5	22.1	7.4	8.7	6.7	18.3	16.9	16.5
N_2	74.1	71.3	69.8	69.7	69.0	69.6	73.6	73.8	74:1	64.9	69.2	72.5
C.V, (sat.), $Kcal/Nm^3$, Gross	881	8487	5344	5155	4536	5013	1204	1553	1264	2818	4773	11291
Net	872	7680	4767	4604	4071	4457	1142	1467	1207	2577	4373	10354
(CO_2)th in flue gas, %	24.9	12.7	11.1	11.5	13.3	10.7	20.5	19.2	20.5	20.6	16.8	13.2
Dew point of flue gas, °C	11	59	62	62	61	63	40	44	39	59	57	56

Table 6. Mean Specific Heat (Cp) of Gaseous Fuels, in $KCal/Nm^3$ °C

		Coal Gas					Producer Gas			Water Gas		
Temperature, °C	Blast Furnace Gas	Low Temp.	Horizontal Retorts	Continuous vertical retorts (No steaming)	Continuous vertical retorts (steaming)	Debenzodised coke oven gas	Hand operated (Coke)	Mechanical (Coal)	Mechanical (Coke)	Blue	Carburetted	OilGas (retort)
0	0.318	0.385	0.333	0.337	0.331	0.330	0.315	0.315	0.314	0.314	0.337	0.407
100	0.321	0.418	0.346	0.352	0.343	0.342	0.318	0.317	0.316	0.316	0.349	0.449
200	0.324	0.451	0.359	0.367	0.355	0.353	0.320	0.320	0.318	0.318	0.362	0.488
300	0.328	0.483	0.372	0.381	0.366	0.364	0.324	0.323	0.321	0.320	0.373	0.525
400	0.332	0.513	0.384	0.395	0.377	0.375	0.326	0.327	0.323	0.322	0.384	0.560
500	0.336	0.542	0.396	0.409	0.388	0.386	0.330	0.331	0.327	0.325	0.395	0.593

600	0.340	0.570	0.407	0.421	0.398	0.397	0.334	0.335	0.331	0.327	0.405	0.624
700	0.344	0.595	0.417	0.433	0.408	0.406	0.337	0.339	0.334	0.330	0.414	0.653
800	0.348	0.620	0.428	0.446	0.417	0.415	0.340	0.343	0.337	0.332	0.423	0.679
900	0.352	0.644	0.437	0.457	0.426	0.424	0.344	0.346	0.340	0.335	0.431	0.703
1000	0.356	0.665	0.447	0.467	0.434	0.433	0.347	0.350	0.344	0.337	0.439	0.725

*To obtain mean specific heat in kCal/kg.°C or in Btu/lb.°F, divide the tabulated value (given in kCal/Nm3.°C) by the product (1.293 x sp. gravity of gas) [Use sp. gr. w.r.t. Air = 1]

Table 7. Characteristics of Fuel Oils

Properties	Fuel Oils		
	F.O.	L.S.H.S.	L.D.O.
Density (kg/m3 at 15°C)	0.89-0.95	0.88-0.98	0.85-0.87
Flash Point (°C)	66	93	66
Pour Point (°C)	20	72	12 (Winter) 18 (Summer)
G.C.V. (kcal/kg)	10,200	9,500	10,700
Sediment %, Wt. Max.	0.25	0.25	0.1
Sulphur Total %, Wt. Max.	4.0	1.0	1.8
Water Content %, Vol. Max.	1.0	1.0	0.25
Ash % Wt. Max.	0.1	0.1	0.02

Table 8. Typical Composition of Some Fuels

Fuels	Percentage Composition							
	C	H_2	S	N	O_2	H_2O	Ash	GCV
Wood	45.60	3.96	0.07	0.45	37.45	9.33	3.14	4700
Lignite	52.21	3.83	0.88	0.46	17.06	18.39	7.17	5023
Husk	36.14	3.70	0.08	0.46	29.34	8.92	19.40	3364
Bagasse	40.94	4.58	0.04	0.23	36.65	10.53	7.03	3850
F.O.	84.00	11.00	3.50	—	—	1.00	0.50	10200
L.D.O.	85.50	11.50	3.00	—	—	—	—	10300
L.S.H.S.	86.70	11.80	1.00	—	—	0.50	—	10000
L.P.G.	81.50	17.00	—	—	—	1.50	—	11642
Natural Gas	66.00	23.00	—	—	—	1.50	—	10084

Note :

(a) The above compositions can vary widely depending on the mine, location, source etc.

(b) Solid fuels in air dried condition.

† L.P.G. is assumed to be composed of (by volume) 30-40% of Propane and 70-60% of Butane.

Natural Gas is assumed to be composed of (by volume) 76% of Methane, 10% of Ethane, 2% of Propane, 6% of CO_2 and 4% of N_2.

(c) The above figures of G.C.V. are in kcal/kg.

Table 9. Composition and Properties of Some Liquid Fuels

Parameters	Unit	Petrol (motor gaso-line)	Kero-sene	Diesel / gas oil	Mo-tor ben-zol	Light fuel oil	Heavy fuel oil	CTF 200 (PCM)
Composition : (% by volume) :								
— Carbon	%	85.5	86.3	86.3	91.7	86.2	86.2	90.0
— Hydrogen	%	14.4	13.6	12.7	8.0	12.3	11.8	6.0
— Sulphur	%	0.1	0.1	1.0	0.3	1.5	2.0	0.4'
— Nitrogen	%	—	—	—	—	—	—	1.2
— Oxygen	%	—	—	—	—	—	—	2.5
Properties :								
— Specific gravity	at 15°C	0.73	0.79	0.87	0.88	0.89	0.95	1.1
— Kinematic viscosity	Centistokes							
(a) at 20°C		0.75	1.6	5.0	0.72	5	1200	1500
(b) at 100°C		—	0.6	1.2	—	3.5	20	18
— Flash point	°C	-40	39	75	-40	80	110	65
— Net calorific value	Kcal/kg	10460	10390	10290	9600	10100	9890	9000
— Boiling range	°C	40—185	140—280	180	80—170	200	250	200
— Residue at 350°C	%	—	—	15	—	50	60	60

Table 10. Composition (weight %) and hydrogen/carbon ratios for typical fuels and hydrocarbons

S. No.	Fuels	Carbon	Hydrogen	Oxygen	Sulphur	Nitrogen	*H/C ratio
1.	Crude oil	86	11	0.7	1.5	0.5	1.6
2.	Fuel oil	86	13.4	0.2	0.3	0.1	1.8
3.	Anthracite coal	93	3	2.5	0.7	0.8	0.4
4.	Bituminous coal	78	5.7	11.6	3.3	1.0	0.9
5.	Sub-bituminous coal	71.9	6.1	20.2	0.6	1.0	1.0
6.	Lignite	72.3	5.1	19.9	1.6	1.1	0.84

7.	Gasoline	85	15	0.0	0.1	0.0	2.1
8.	Benzene	92.3	7.7	0.0	0.0	0.0	1.0
9.	Naphthalene	93.7	6.3	0.0	0.0	0.0	0.8
10.	Methane	74.9	25.1	0.0	0.0	0.0	4.0

*H/C ratios are atoms of hydrogen/atoms of carbon.

Table 11. Combustion Data for Some Solid and Liquid Fuels

Parameters	Coal	B.F. Grade Coke	Coke Breeze	Heavy Fuel Oil	Tar	Benzol
***(a)* Composition :**						
— % moisture (as received)	9	5	10	—	—	—
— % ash content (dry basis)	9	11	17	—	—	—
— Ultimate analysis (ash free basis)						
% C	90	98.5	98.5	87	94	92
% H	5	0.8	0.8	11	5	7.5
%S	0.5	0.7	0.7	2	1	0.5
% N	1.5	—	—	—	—	—
%O	0.5	—	—	—	—	—
Net Calorific value (dry basis), MJ/kg	31.5	29.5	27.5	40.5	37.0	40.0
(1 cal = 4.18 J), kcal/kg	7535	7057	6579	9689	8851	9570
Theoretical Combustion air required						
— Nm^3/kg (dry basis)	7.9	8.0	7.5	10.6	9.8	10.3
— Nm^3/MJ	0.25	0.27	0.27	0.26	0.26	0.26
Theoretical flue gas volume, Nm^3/kg (dry basis)	8.2	8.0	8.0	11.2	10.1	10.7
Theoretical flue gas analysis from dry fuel :						
$\%CO_2$	18	20	20	14	17	16
$\%H_2O$	6	1	1	11	5	8
$\%N_2$	76	79	79	75	78	76

Table 12. Combustion Data for Some Gaseous Fuels

Parameters	Blast furnace gas	Coke oven gas	Converter gas/ BOF gas	Natural gas
Composition (% by volume) :				
$\%H_2$	3.4	48.4	0.7	—
$\%CH_4$	—	28.0	—	95.3
$\%C_2H_6$	—	2.9	—	3.6
$\%C_4H_{10}$	—	—	—	0.05

%CO	21.7	6.4	67.9	—
$\%CO_2$	20.8	2.2	15.8	0.3
$\%N_2$	54.1	11.3	15.6	0.7
$\%O_2$	—	0.7	—	0.05
Net Calorific value :				
MJ/Nm^3	3.12	17.9	8.7	36.6
(1 Cal = 4.18J) $kcal/Nm^3$	746	4282	2081	8756
Theoretical air requirement:				
Nm^3/Nm^3 gas	0.6	4.3	1.6	9.5
Nm^3/MJ	0.19	0.23	0.19	0.26
Theoretical flue gas volume :				
Nm^3/Nm^3 gas	1.5	5.0	2.3	10.6
Nm^3/MJ	0.47	0.28	0.26	0.29
Theoretical flue gas analysis :				
$\%CO_2$	28.9	8.4	36.6	9.7
$\%H_2O$	2.3	22.5	0.3	19.1
$\%N_2$	68.8	69.1	63.1	71.2

Table 13. Effect of CO_2 % in Flue Gas on Stack Loss for Furnace Oil and Natural Gas

(Flue Gas Temp. = 1000°C)

Fuel	CO_2 % in flue gas	% Excess air	Loss as % of heat in the fuel
Furnace oil	13.0	20	49.0
	11.0	40	56.0
	9.7	60	64.0
	8.5	80	71.5
	7.5	100	79.0
	6.9	120	86.5
Natural gas	10.7	10	41.0
	9.8	20	45.0
	8.3	40	53.0
	7.2	60	60.5
	6.3	80	68.0
	5.7	100	75.5

Table 14. Effect of Excess Air on Flame Temperature of Fuel Oil

% Excess air	Flame temp., °C
0	2080
20	1870
50	1630
80	1450

Table 15. Saving Using Air Preheat for Gaseous Fuels and Furnace Oil

Fuel	Air Preheat Temp., °C	Fuel Saving, %
Gaseous fuels with 10% E.A. (excess air)	100	4.0
	200	8.5
	300	13.5
	400	17.0
	500	20.2
Fuel oil with 20% E.A. (excess air)	100	5.5
	200	11.2
	300	16.2
	400	20.5
	500	24.0

Table 16. Typical % excess air levels used for burning various fuels in practice

Fuels	Combustion appliance used	Typical excess air levels used, %
Coal	Chain grate and travelling grate stokers	15-50
	Underfed stoker	20-50
	Spreader stoker	30-60
	Vibrating-grate stoker (water cooled)	30-60
Palverised coal	Completely water cooled furnace of slag-tap or dry- ash removal type.	15-20
	Partially water cooled furnace of dry-ash removal type.	15-40
Wood /Bagasse	Dutch over (10-23% over grates) and Hofft type	20-25
	All furnaces	25-35
Fuel oil	Oil burners, register type	15-20
	Multifuel burners and flat flame burners	20-30
Natural gas	High pressure burner	5-7
Black liquor of pulp and paper industry	Recovery furnaces for kraft and soda pulping processes.	30-40

Table 17. Typical air requirements and % CO_2 in flue gases for various solid and liquid fuels

Fuels	kg of air Required/ kg of fuel	kg of flue gas / kg of fuel	m^3 of flue gas l/kg of fuel	Theoretical CO_2% in dry flue gas	CO_2% in flue gas achieved in practice
Solid fuels:					
Bituminous coal	10.8	11.7	9.40	18.70	10-13
Lignite	8.4	9.10	6.97	19.40	9-13
Wood	5.8	6.4	4.79	20.3	11-13
Bagasse	3.2	3.43	2.61	20.65	10-12
Paddy husk	4.6	5.63	4.58	19.8	14-15
Liquid fuels:					
Furnace oil	13.90	14.30	11.50	15	9-14
LSHS	14.04	14.63	10.79	15.5	9-14

Table 18. Flame Temperature with Varying Percentage of Excess Air for Fuel Oil

% Excess Air	*% free oxygen in waste gas*	*Theoretical flame temperature, °C*
0	0	2340
10	2.00	2190
20	3.63	2080
30	5.03	1980
40	6.21	1380
50	7.22	1770
60	8.11	1670
70	8.60	1590
80	9.57	1520
90	10.00	1430
100	10.73	1340
110	11.00	1300
120	11.25	1270
130	11.80	1190
140	12.20	1140
150	12.60	1080
160	12.80	1050
170	13.10	1030
180	13.40	980
190	13.70	950
200	14.00	920

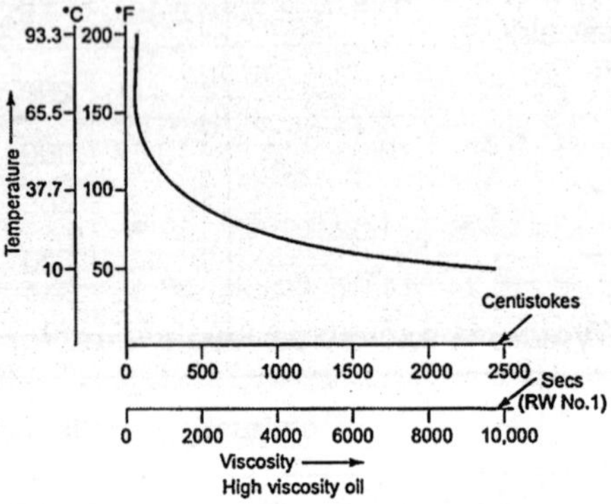

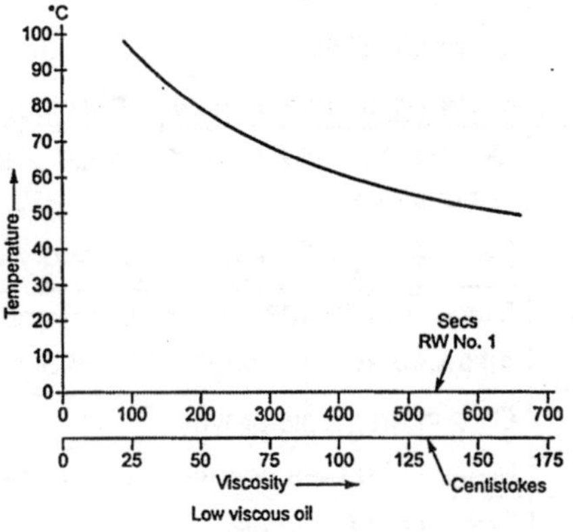

Table 19. Fuel, Air and Flue Gas — Nomogram

Type of Fuel	Air Requirement						Combustion Products					
	Theoretical		Air Volume with different % excess air, Nm^3/ kg				Theoretical		Flue gas volume with- varying % of excess air, Nm^3 / kg			
	kg / kg	Nm^3/ kg	10%	20%	50%	100%	$\frac{kg}{kg}$	$\frac{Nm^3}{kg}$	10%	20%	50%	100%
Solid Fuels												
(a) Coking Coal	9.88	7.35	8.09	9.56	11.03	14.70	10.55	7.62	8.36	9.83	11.30	14.97
(b) Steam Coal	9.31	6.92	7.61	8.99	10.29	13.84	10.00	7.22	7.91	9.29	10.67	14.14
(c) Coal Middlings	7.59	5.64	6.20	7.32	8.44	11.28	8.17	5.88	6.44	7.56	8.68	11.52
(d) Gas Coal	9.91	7.37	8.11	9.59	11.07	14.74	10.64	7.71	8.45	9.93	11.41	15.08
(e) Non-coking Coal	9.77	7.26	7.99	9.45	10.91	14.52	10.48	7.64	8.19	9.65	11.11	14.72
(f) Coke	8.49	6.31	6.94	8.20	9.46	14.62	9.20	6.40	7.03	8.29	9.55	12.72
Liquid Fuels												
(a) Coal Tar	12.28	9.12	10.03	11.86	13.67	18.24	13.22	9.67	10.18	12.00	13.82	18.39
(b) L.S.H.S.	14.04	10.42	11.46	13.54	15.62	20.84	15.01	11.14	12.18	14.26	16.34	21.56
(c) Furnace Oil	13.70	10.24	11.22	13.26	15.30	20.40	14.63	10.79	11.81	13.85	15.89	20.99
Gaseous Fuels	*Volume / Volume*						*Wet /Dry Waste Gas Volume (Wet)*					
(a) C.O. Gas	—	4.20	4.52	5.46	6.30	8.40	4.90	3.84	5.32	6.16	7.00	9.10
(b) B.F. Gas	—	0.70	0.77	0.91	1.05	1.40	1.55	1.51	1.62	1.76	1.90	2.25
(c) Producer Gas	—	1.15	1.27	1.51	1.75	2.30	1.99	1.78	2.10	2.33	2.56	3.14
(d) Mixed Gas: (B.F. + C.O.) Gas)												
(i) With 10% B.F. Gas	—	3.85	4.23	5.00	5.77	7.70	4.56	3.60	4.94	5.71	6.48	8.43
(ii) With 20% B.F gas	—	3.50	3.85	4.55	5.25	7.00	4.23	3.37	4.58	5.28	5.98	7.73
(iii) With 30%B.F gas	—	3.15	3.46	4.09	4.72	6.30	3.89	3.14	4.20	4.83	5.46	7.04
(iv) With 40%B.F gas	—	2.80	3.08	3.64	4.20	5.60	3.57	2.91	3.85	4.41	4.97	6.37
(v) With 50% B.Fgas	—	2.45	2.69	3.18	3.67	4.90	3.22	2.67	3.46	3.95	4.44	5.67
(vi) With 60%B.F gas	—	2.10	2.31	2.73	3.15	4.20	2.91	2.44	3.12	3.54	3.96	5.01

Table 20. Viscometer Conversion Factors

Given Viscosity	Factor for Conversion to Required Viscosity							
	Kinematic viscosity, *centistokes*	**Engler,** *degrees*	**Redwood I,** *seconds*	**Redwood II,** *seconds*	**Saybolt Universal** *seconds*	**Saybolt Furol** *seconds*	**Standard Tar Viscometer** *10 mm, sec.*	**Standard Tar Viscometer** *4 mm, sec.*
Kinematic viscosity, centistokes	1	0.131	4.08	0.408	4.57	0.457	0.0025	0.074
Engler, degree (from 6° upwards)	7.64	1	31.2	3.12	34.9	3.49	0.0191	0.565
Redwood I, sec (from 90 sec (from 90 sec upwards)	0.245	0.0321	1	0.100	1.12	0.112	0.000612	0.0181
Redwood II, sec (from 30 sec upwards)	2.45	0.321	10.0	1	11.2	1.12	0.00612	0.181
Saybolt Universal, sec (from 120 sec upwards)	0.219	0.0287	0.893	0.089	1	0.10	0.000547	0.0162
Saybolt Furol, sec (from 40 sec upwards)	2.19	0.287	8.93	0.893	10.0	1	0.00547	0.162
Standard Tar Viscometer 10 mm cup, sec (from 10 sec upwards)	400	52.4	1630	163	1830	183	1	29.6
Standard Tar Viscometer 4 mm cup, sec (from 100 sec upwards)	13.5	1.77	55.1	5.51	61.6	6.16	0.034	1

Table 21. Table for Viscosity Conversion

Centistokes	Redwood no. 1 seconds at			Saybolt universal seconds at			°Engler
	70°F or 21.1°C	140°F or 60°C	200°F or 93.3°C	100°F or 37.7°C	150°F or 65.5° C	210°F or 98.9°C	
4	35.3	35.9	36.5	39.1	39.2	39.4	1.32
5	37.9	38.5	38.9	42.3	42.4	42.6	1.40
6	40.5	41.1	41.5	45.5	45.6	45.8	1.48
7	43.2	43.7	44.1	48.7	48.8	49.0	1.57

8	46.0	46.3	46.9	52.0	52.1	52.4	1.65
9	48.8	49.1	49.7	55.4	55.5	55.8	1.74
10	51.7	52.0	52.6	58.8	58.9	59.2	1.84
12	57.9	58.1	58.7	65.9	66.0	66.4	2.05
14	64.4	64.6	65.2	73.4	73.5	73.9	2.25
16	71.0	71.4	72.2	81.1	81.5	81.7	2.44
18	77.9	78.5	79.3	89.2	89.4	89.8	2.66
20	85.0	85.8	86.9	97.5	97.7	98.2	2.88
22	92.4	93.3	94.5	106.0	106.2	106.7	3.10
24	99.9	100.2	100.6	114.6	114.8	115.4	3.32
26	107.5	108.6	110.0	123.3	123.5	124.2	3.57
28	115.3	116.5	118.0	132.1	132.4	133.0	3.82
30	123.1	124.4	126.0	140.9	141.2	141.9	4.08
32	131.0	132.3	134.1	149.7	150.0	150.7	4.34
34	138.9	140.2	142.2	158.7	159.0	159.8	4.59
36	146.9	148.2	150.3	167.7	168.0	168.9	4.86
38	155.0	156.2	158.3	176.3	176.7	177.9	5.10
40	163.0	164.3	166.7	185.7	186.1	187.0	5.56
45	183.1	184.5	187.5	208.4	208.8	209.9	6.00
50	203.3	204.7	208.3	231.4	231.8	233.0	6.58
60	243.5	245.3	250.0	277.4	277.9	279.3	7.89
70	283.9	286.0	291.7	323.4	324.0	325.7	9.21
80	323.9	326.6	333.4	369.6	370.3	372.2	10.5
90	364.4	367.4	375.0	415.8	416.8	418.7	11.8
100	404.9	408.2	416.7	462.0	462.9	465.2	13.2
For viscosities above 100 centistokes multiply by							
	4.049	4.082	4.167	4.620	4.629	4.652	0.152

N.B. The redwood no. 2 scale is used to define the viscosity of very thick oil. Redwood no. 1 figure for thick oil is approximately obtained by multiplying the redwood no. 2 figure by a factor of 10. Similarly, there is an alternative saybolt scale known as saybolt furol. Saybolt universal figures is approximately obtained by multiplying the saybolt furol figure by a factor of 10.

Table 22. Gross and Net Calorific values of some common gases

S.N.	Gas	Gross Heating Value		Net Heating Value	
		(Btu/ft^3)	(Btu/lb)	(Btu/ft^3)	(Btu/lb)
1.	Acetylene (ethyne) - C_2H_2	1,498	21,569	1,447	20,837
2.	Benzene	3,741	18,150	3,590	17,418
3.	Blast Furnace gas	92	1,178	92	1,178

4.	Blue water gas	—	6,550	—	—
5.	Butane - C_4H_{10}	3,225	21,640	2,977	19,976
6.	Butylene (Butene)	3,077	20,780	2,876	19,420
7.	Carbon to CO_2	—	14,150	—	14,150
8.	Carbon to CO	—	3,960	—	3,960
9.	Carbon monoxide — CO	323	4,368	323	4,368
10.	Carburetted Water Gas	550	11,440	508	10,566
11.	Coal gas	149	16,500	—	—
12.	Coke Oven Gas	574	17,048	514	15,266
13.	Digester Gas (Sewage or Biogas)	690	11,316	621	10,184
14.	Ethane - C_2H_6	1,783	22,198	1,630	20,295
15.	Ethyl alcohol saturated with water	1,548	12,804	—	—
16.	Ethylene	1,631	21,884	1,530	20,525
17.	Hexane	4,667	20,526	4,315	18,976
18.	Hydrogen (H_2)	325	61,084	275	51,628
19.	Hydrogen Sulphide	672	7,479	—	—
20.	Landfill Gas	476	—	—	—
21.	Methane - CH_4	1,011	23,811	910	21,433
22.	Methyl alcohol saturated with water	818	9,603	—	—
23.	Naphthalene	5,859	17,298	—	—
		950	19,500	850	17,500
24.	Natural Gas (typical)	—	—	—	—
		1,150	22,500	1,050	22,000
25.	Octane saturated with water	6,239	20,542	3,170	10,444
26.	Pentane	3,981	20,908	3,679	19,322
27.	Producer gas	—	2,470	—	—
28.	Propane - C_3H_8	2,572	21,564	2,371	19,834
29.	Propene (Propylene) - C_3H_6	2,332	20,990	2,181	19,630
30.	Propylene	2,336	21,042	2,185	19,683
31.	Sasol	500	14,550	443	13,016
32.	Sulphur		3,940	—	—
33.	Toulene	4.4 us	18,129	4,206	17,301
34.	Water Gas (bituminous)	261	4,881	239	4,469
35.	Xylene	5,155	18,410	—	—

- 1 Btu/ft3 = 8.9 kcal/m3 = 3.73 × 104 J/m3
- 1 Btu/lb = 2,326.1 J/kg = 0.55556 kcal/kg

Table 23. Heat of Combustion and Calorific Values (at 15°C and One Atmosphere Pressure)

Formula	Substance	Heat of Combustion (kcal/kmol)	Gross Calorific Value (G.C.V.), kcal/Nm³ (dry)	Net C.V., kcal/Nm³
C	Carbon	97,000	8,083 (kcal/kg)	—
H_2	Hydrogen	68,317.4	3,050 (33,887.6 kcal/kg)	2580
CO	Carbon monoxide	67,636.1	3,020	3020
CH_4	Methane	2,12,798	9,500	8535
C_2H_6	Ethane	3,72,820	16,644	15140
C_2H_4	Ethylene	3,37,234	15,055	14505
C_2H_2	Acetylene	3,10,615	13,867	13460
C_3H_8	Propane	5,30,605	23,688 (11,947 kcal/kg liquid)	21765
C_4H_{10}	n-Butane	6,87,982	30,714 (11,749 kcal/kg liquid)	28310
C_3H_6	Propene/propylene	4,91,987	21,964 (11,692 kcal/kg liquid)	20885
C_4H_{10}	i-Butane	6,49,757	20,007 (11,581 kcal/kg liquid)	—
C_6H_6	Benzene	7,89,080	35,227 (9,999 kcal/kg liquid)	33900
C_7H_8	Toluene	9,43,580	42,124 (10,143 kcal/kg liquid)	—
H_2S	Hydrogen sulphide	1,34,744	6,015	5465
C_4H_8	B utylene/Butene	—	29030	27145
C_MH_N ($C_{2.5}H_5$)	Unsaturated hydrocarbons	—	18150	17000

Table 24. Lower heating value for some organic compounds (at 25°C)

Organic Fuel Compounds	Net C.V., MJ/kg
Paraffins	
Methane	50.009
Ethane	47.794
Propane	46.357
Butane	45.752
Pentane	45.357
Hexane	44.752
Heptane	44.566
Octane	44.427
Nonane	44.311
Decane	44.240
Undecane	44.194

Dodecane	44.147
Isoparaffins	
Isobutane	45.613
Isopentane	45.241
2-Methylpentane	44.682
2,3-Dimethylbutane	44.659
2,3-Dimethylpentane -	44.496
2,2,4-Timethylpentane	44.310
Naphthenes	
Cyclopentane	44.636
Methylcyclopentane	44.636
Cyclohexane	43.450
Methylcyclohexane	43.380
Monoolefins	
Ethylene	47.195
Propylene	45.799
1-Butene	45.334
cis-2-Butene	45.194
trans-2-Butene	45.124
Isobutene	45.055
1-Pentene	45.031
2-Methyl- 1-pentene	44.799
1-Hexene	44.426
Diolefins	
1,3-Butadiene	44.613
Isoprene	44.078
Nitrous derivated	
Nitromethane	10.513
Nitropropane	20.693

Organic Fuel Compounds	Net C.V., MJ/kg
Acetylenes	
Acetylene	48.241
Methylacetylene	46.194
1-Butyne	45.590
1-Pentyne	45.217
Aromatics	
Benzene	40.170

Toluene	40.589
o-Xylene	40.961
m-Xylene	40.961
p-Xylene	40.798
Ethylbenzene	40.938
1,2,4-Trimethylbenzene	40.984
Propylbenzene	41.193
Cumene	41.217
Alcohols	
Methnol	—
Ethanol	28.865
n-propanol	30.680
Isopropanol	30.447
n-Butanol	33.075
Isobutanol	32.959
Tertiobutanol	32.587
n-Pentanol	34.727
Ethers	
Methoxymethane	28.703
Ethoxyethane	33.867
Propoxypropane	36.355
Butoxybutane	37.798
Aldehydes and ketones	
Methanal	17.259
Ethanal	24.156
Propionaldehyde	28.889
Butyr aldehyde	31.610
Acetone	28.548
Other species	
Carbon (graphite)	32.808
Hydrogen	12.0971
Carbon monoxide	10.112
Ammonia	18.646
Sulfur (solid)	9.163

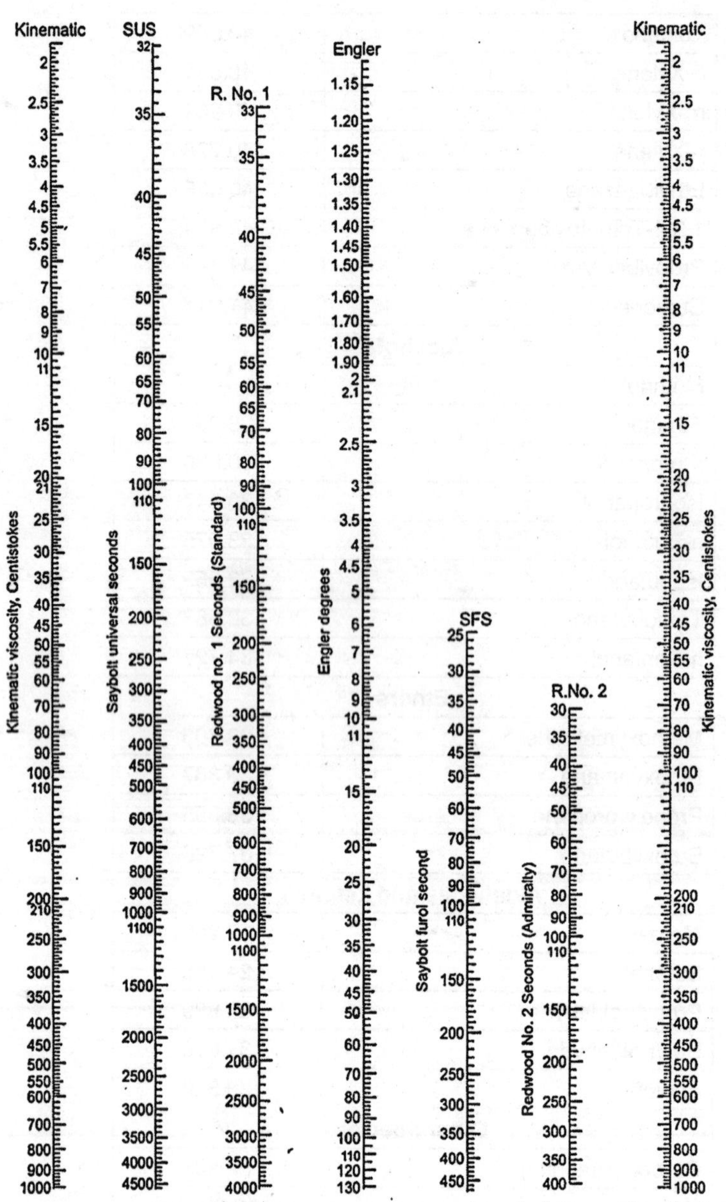

Fig.A .I : Viscosity conversion nomogram

To make use of it, line up the straight edge so that Centistokes value on both Kinematic scales is the same. Viscosities at the same temperature on all scales are then equivalent. To extend range of only the Kinematic, Saybolt Universal, Redwood No. 1, and Engler scales : Multiply by 10, the viscosities on these scales between 100 and 1000 centistokes on the Kinematic scale and the corresponding viscosities on the other three scales. For further extension, multiply these scales as above by 100 or a higher power of 10. (Example : 1500 centistokes = 150 x 10 CS = 695 x 10 SUS = 6950 SUS).

Table 25. Spontaneous Ignition Temperature in Air and Oxygen

Substance	In Air, °C	In Oxygen, °C	Substance	In Air, °C	In Oxygen, °C
Acetone	561	485	Hexane	248	—
Acetylene	305	—	Hexane, iso	—	284
Ammonia	651	—	Hydrocyanic acid	538	—
Amyl benzene	—	255	Hydrogen	572	560
Amylene	273	—	Hydrogen sulphide	292	220
Anthracene	472	—	Isododecane	500	—
Benzene	580	566	Isoprene	440	—
1, 3-Butadiene	418	335	Kerosene	254	—
Butane	408	283	Methane	632	556
Butylene	443	—	Methyl alcohol	470	461
Carbon disulphide	120	107	2-Methyl butane		
Carbon monoxide	609	588	(isopentane)	420	294
Cetane	235	—	Methyl chloride	632	—
Creosote	356	—	a-Methyl naphthalene	566	—
Cresol, 0	599	—	2-Methyl propane	462	319
Cresol, m	626	—	Naphtha	277	—
Cyanogen	850	—	Naphthalene	587	560
Cyclohexane	270	296	Nonane	285	—
Cyclopropane	498	454	Octane	218	208
Cymene	466	—	Pentane	290	258
Decalin	262	—	Petroleum ether	329	—
Decane, n	250	—	Phenol	715	500
Dimethyl ether	350	252	Propane	493	468
Dodecane	534	—	Propylene	458	—
Ethane	472	—	Styrene	490	450
Ether, diethyl	193	182	Toluene	552	516
Ethyl alcohol	- 392	—	Turpentine	252	—
Ethyl benzene	477	468	Xylene, O	496	—
2-Ethyl butane	—	273	Wood	300	—
Ethylene	490	485	Wood charcoal	200—350	—
Gasoline (regular)	280	—	Peat	220—250	—
Gasoline (73 octane)	299	—	Lignite	250—350	—
Gasoline (92 octane)	390	—	Butuminous coal	380—420	270—290
Gasoline (100 Octane)	429	—	Anthracite	500	340
Heptane	230	214	Soft coke	425-500	—
Gobar gas	650	—	Hard coke	500—650	—

Table 26. Limits of Inflammability of Gases/Vapours in Air (at Ordinary Temperature and Pressures)

(The figures are for upward propagation of flame in large vessels, open at their lower ends.)

Gas/Vapour	% by volume		Stoichiometric mixture, volume%	Gas/Vapour	% by Volume		Stoichiometric mixture, volume%
	Lower Limit L_1	Upper Limit L_2			Lower Limit L_1	Upper Limit L_2	
Inorganic Substances				**Esters**			
Amonia, NH_3	15.5	26.6	—				
Hydrogen, H_2	4.0	75	29.5	Methyl formate	5.9	20	—
Hydrogen sulphide, H_2S	4.3	45.5	12.24				
Carbon disulphide, CS_2	1.25	44	6.52	Ethyl formate	2.7	13.5	—
Hydrogen cyanide, HCN	5.6	40	—				
Carbon monoxide, CO	12.5	74	29.5	Ethyl acetate	2.5	9.0	4.02
Hydrocarbons				Vinyl acetate	2.6	13.4	—
Methane, CH_4	5.3	15	9.47	Propyl acetate	2.0	—	—
Ethane, C_2H_6	3.0	12.5	5.64	Butyl acetate	1.7	—	—
Ethylene, C_2H_4	3,1	32	6.52	**Bases**			—
Acetylene, C_2H_2	2.5	80.0	7.72	Methyl amine	4.9	20.7	—
Propane, C_3H_8	2.2	9.5	4.02	Dimethyl amine	2.8	14.4	—
Propylene, C_3H_6	2.4	10.3	4.44	Trimethyl amine	2.0	11.6	—
Cyclopropane, C_3H_8	2.4	10.4	4.44	Ethyl amine	3.5	14.0	—
Butane, C_4H_{10}	1.9	8.5	3.12	Diethyl amine	1.8	10.1	—
Butene-1, C_4H_8	1.6	9.3	3.37	Propyl amine	2.0	10.4	—
Butene-2, C_4H_8	1.8	9.7	3.37	**Halogen Compounds**			
Butadiene —	2.0	11.5	3.66	Methylchloride, $CH_3.Cl$	10.7	17.4	—
Pentane, C_5H_{12}	1.5	7.8	2.55	Ethyl chloride, C_2H_5 .Cl	3.8	15.4	—
Hexane, C_6H_{14}	1.2	7.5	2.16	**Hydrocarbon Fuels** Naphtha	1.1	6.0	
Cyclohexane, C_6H_{14}	1.3	8	2.27	Petrol (gasoline)	1.4	7.6	—
Benzene, C_6H_6	1.4	7.1	2.71	Natural gas	4.8	13.5	—
Toluene, $C_6H_5CH_3$	1.4	6.8	2.27	**Fuel Gases**			
Ethyl benzene, $C_6H_5C_2H_5$	1.0	—	1.95	Coke-oven gas	5	32	—
Alcohols				Oil gas	4.7	33	—

Methyl, CH_3OH	7.3	36.5	12.24	Coal gas	5.3	32	—
Ethyl, C_2H_5OH	4.3	19.0	6.52	Carburetted water gas	5.5	36	—
Ethers				Water gas	7.0	72	—
Methyl	3.4	14	—	Producer gas	17	70	—
Ethyl,	1.9	48	—	Blast-furnace gas	37	71	—
Ethyl n-propyl	1.9	24	—	LPG	1.8	8.5	—
Ethylene oxide	3.0	80	7.72	L.D. converter gas	20	70	—
Aldehydes & Ketones				Gobar gas	5	15	—
Acetaldehyde	4.1	55	—				
Butaldehyde	2.5	—	—				
Acetone	3.0	11	4.97				

Table 27. Inflammability limits of gaseous organic vapors in air (% gas in gas-air mixture)

Gaseous organic vapors *(1)*	Lower inflammability limit (% by volume) *(2)*	Upper inflammability limit (% by volume) *(3)*
Acetaldehyde	4.0	57
Acetone	2.5	12.8
Acetylene	2.5	80
Allyl alcohol	2.5	—
Ammonia	15.5	26.6
Amyl acetate	1.0	7.5
Amylene	1.6	7.7
Benzene (benzol)	1.3	6.8
Benzlyl chloride	1.1	—
Butene	1.8	8.4
Butyl acetate	1.4	15.0
Butyl alcohol	1.7	—
Butyl cellosolve	—	—
Carbon disulfide	1.2	50
Carbon monoxide	12.5	74.2
Chlorobenzene	1.3	7.1
Cottonseed oil	—	—
Cresol meta or para	1.1	—
Crotonaldehyde	2.1	15.5
Cyclohexane	1.3	8.4
Cyclohexanone	1.1	—
Cyclopropane	2.4	10.5

Cymene	0.7	—
Dichlorobenzene	2.2	9.2
Dichloroethylene (1,2)	9.7	12.8
Diethyl selenide	2.5	—
Dimenthyl formamide	2.2	—
Dioxane	2.0	22.2
Ethane	3.1	15.5
Ether (diethyl)	1.8	36.5
Ethyl acetate	2.2	11.5
Ethyl alcohol	3.3	19.0
Ethyl bromide	6.7	11.3
Ethyl collosolve	2.6	15.7
Ethyl chloride	4.0	14.8
Ethyl ether	1.9	48
Ethyl lactate	1.5	—
Etylene	2.7	28.6
Ethylene dichloride	6.2	15.9
Ethyl formate	2.7	16.5
Ethyl nitrite	3.0	50
Ethylene oxide	3.0	80
Furfural	2.1	—
(1)	*(2)*	*(3)*
Gasoline (variable)	1.4-1.5	7.4-7.6
Heptane	1.0	6.0
Hexane	1.2	6.9
Hydrogen cyanide	5.6	40.0
Hydrogen	4.0	74.2
Hydrogen sulfide	4.3	45.5
Illuminating gas (coal gas)	5.3	33.0
Isobutyl alcohol	1.7	—
Isopentane	1.3	—
Isopropyl acetate	1.8	7.8
Isopropyl alcohol	2.0	—
Kerosene	0.7	5
Linseed oil	—	
Methane	5.0	15.0
Methyl acetate	3.1	15.5
Methyl alcohol	6.7	36.5

Methyl bromide	13.5	14.5
Methyl butyl ketone	1.2	8.0
Methyl chloride	8.2	18.7
Methyl cyclohexane	1.1	—
Methyl ether	3.4	18
Methyl ethyl ether	2.0	10.1
Methyl ether ketone	1.8	9.5
Methyl formate	5.0	22.7
Methyl propyl ketone	1.5	8.2
Mineral spirits No. 10	0.8	—
Napthalene	0.9	—
Nitrobenzene	1.8	—
Nitroethane	4.0	—
Nitromethane	7.3	—
Nonane	0.83	2.9
Octane	0.95	3.2
Paraldehyde	1.3	—
Paraffin oil	—	—
Pentane	1.4	7.8
Propane	2.1	10.1
Propyl acetate	1.8	8.0
Propyl alcohol	2.1	13.5
Propylene	2.0	11.1
Propylene dichloride	3.4	14.5
Propylene oxide	2.0	22.0
Pyridine	1.8	12.4
Rosin Oil	—	
Toluene (toluol)	1.3	7.0
Turpentine	0.8	—
Vinyl ether	1.7	27.0
Vinyl chloride	4.0	21.7
Water gas (variable)	6.0	70
Xylene (xylol)	1.0	6.0

Table 28. Mean Specific Heat of Gases at Constant Pressure (Cp) between 0°C and T°C (Mass Basis, kcal/kg.°C)

Values for the Undissociated Gases

T°C	O_2	n_2	Air	CO	H_2	CO_2	H_2O Va-pour	Meth-ane	Ethylene	Ethane	Ben-zene Vapour
0	0.218	0.248	0.240	0.249	3.40	0.195	0.443	0.496	0.342	0.397	0.224
100	0.220	0.248	0.240	0.249	3.44	0.208	0.445	0.545	0.387	0.450	0.273
200	0.223	0.249	0.242	0.250	3.45	0.219	0.450	0.590	0.431	0.501	0.316
300	0.227	0.250	0.243	0.252	3.46	0.228	0.456	0.634	0.470	0.548	0.356
400	0.230	0.252	0.246	0.254		0.236	0.462	0.677	0.508	0.594	0.390
500	0.234	0.254	0.248	0.257		0.244	0.470	0.719	0.542	0.636	0.422
600	0.237	0.256	0.250	0.259		0.250	0.477	0.758	0.574	0.676	0.448
700	0.240	0.258	0.253	0.262		0.256	0.485	0.794	0.602	0.713	0.474
800	0.243	0.261	0.256	0.265		0.261	0.494	0.830	0.628	0.748	0.494
900	0.245	0.264	0.258	0.268		0.266	0.501	0.864	0.651	0.780	0.514
1000	0.247	0.266	0.260	0.270	3.53	0.270	0.510	0.896	0.671	0.810	0.534
1100	0.249	0.268	0.263	0.273	3.55	0.274	0.517				
1200	0.251	0.271	0.265	0.275	3.57	0.277	0.525				
1300	0.253	0.272	0.267	0.277	3.59	0.280	0.531				
1400	0.255	0.275	0.269	0.278	3.61	0.283	0.540				
1500	0.256	0.276	0.271	0.280	3.63	0.285	0.546				
1600	0.258	0.278	0.272	0.282	3.65	0.288	0.552				
1700	0.258	0.280	0.273	0.283	3.67	0.290	0.560				
1800	0.260	0.281	0.274	0.285	3.69	0.292	0.566				
1900	0.262	0.282	0.276	0.286	3.71	0.293	0.572				
2000	0.263	0.284	0.277	0.287	3.74	0.295	0.577				
2100	0.264	0.285	0.278	0.289	3.76	0.297	0.583				
2200	0.265	0.286	0.279	0.290	3.78	0.298	0.588				
2300	0.266	0.288	0.280	0.291	3.80	0.300	0.595				
2400	0.267	0.288	0.281	0.292	3.82	0.301	0.598				
2500	0.268	0.289	0.282	0.293	3.84	0.302	0.603				
2600	0.269	0.291	0.283	0.294	3.86	0.303	0.608				
2700	0.270	0.291	0.284	0.295	3.87	0.304	0.612				
2800	0.271	0.292	0.284	0.295	3.89	0.305	0.617				
2900	0.272	0.293	0.285	0.296	3.91	0.306	0.621				
3000	0.274	0.294	0.286	0.297	3.93	0.307	0.623				

T°C	Pro-pane	Butane
0	0.359	0.342
100	0.407	0.390
200	0.456	0.438
300	0.504	0.486
400	0.553	0.534
500	0.601	0.582
600	0.650	0.630
700	0.698	0.678
800	0.747	0.726
900	0.795	0.774
1000	0.844	0.822

N.B. At temperatures above 1500°C dissociation must be taken into account.

Table 29. Mean Specific Heat of Gases at Constant Pressure (Cp) Between 0°C and T°C (Volume Basis kcal/Nm^3.°C)

Values for the Undissociated Gases

T°C	*O_2*	*N_2*	*Air*	*CO*	*H_2*	*CO_2*	*H_2O Vapour*	*Methane*	*Ethylene*	*Ethane*	*Benzene Vapour*
0	0.312	0.311	0.310	0.311	0.306	0.384	0.356	0.355	0.428	0.532	0.78
100	0.315	0.311	0.311	0.311	0.309	0.409	0.358	0.390	0.485	0.603	0.95
200	0.319	0.312	0.313	0.312	0.310	0.431	0.362	0.422	0.539	0.671	1.10
300	0.324	0.314	0.315	0.315	0.311	0.449	0.367	0.454	0.589	0.735	1.24
400	0.329	0.316	0.318	0.318	0.311	0.464	0.372	0.484	0.635	0.796	1.36
500	0.334	0.319	0.321	0.321	0.312	0.480	0.378	0.514	0.678	0.853	1.47
600	0.339	0.322	0.324	0.324	0.313	0.492	0.384	0.542	0.718	0.906	1.56
700	0.343	0.324	0.328	0.328	0.314	0.503	0.390	0.568	0.754	0.956	1.65
800	0.347	0.328	0.331	0.332	0.315	0.513	0.397	0.594	0.786	1.003	1.72
900	0.350	0.331	0.334	0.335	0.316	0.522	0.403	0.618	0.815	1.046	1.79
1000	0.353	0.334	0.337	0.338	0.318	0.530	0.410	0.641	0.840	1.085	1.86
1100	0.356	0.337	0.340	0.341	0.319	0.538	0.416				
1200	0.359	0.340	0.343	0.344	0.321	0.544	0.422				
1300	0.362	0.342	0.345	0.346	0.323	0.550	0.427				
1400	0.364	0.345	0.347	0.348	0.325	0.556	0.434				
1500	0.366	0.347	0.350	0.350	0.327	0.560	0.439				
1600	0.368	0.349	0.352	0.352	0.328	0.565	0.444				
1700	0.369	0.351	0.353	0.354	0.330	0.569	0.450				
1800	0.372	0.353	0.355	0.356	0.332	0.573	0.455				
1900	0.374	0.354	0.357	0.358	0.334	0.576	0.460				
2000	0.376	0.356	0.358	0.359	0.336	0.580	0.464				
2100	0.377	0.357	0.360	0.361	0.338	0.583	0.469				
2200	0.378	0.359	0.361	0.362	0.340	0.586	0.473				
2300	0.380	0.361	0.362	0.364	0.342	0.589	0.478				
2400	0.382	0.361	0.363	0.365	0.344	0.591	0.481				
2500	0.383	0.363	0.365	0.366	0.345	0.593	0.485				
2600	0.385	0.365	0.366	0.368	0.347	0.595	0.489				
2700	0.386	0.365	0.367	0.369	0.348	0.597	0.492				
2800	0.387	0.367	0.368	0.369	0.350	0.599	0.496				
2900	0.389	0.368	0.369	0.370	0.352	0.601	0.499				
3000	0.391	0.369	0.370	0.371	0.353	0.603	0.501				

T°C	Propane	Butane
0	0.707	0.886
100	0.802	1.011
200	0.898	1.135
300	0.993	1.260
400	1.089	1.384
500	1.184	1.569
600	1.280	1.633
700	1.375	1.758
800	1.471	1.882
900	1.566	2.007
1000	1.662	2.131

N.B. At temperatures above 1500°C, dissociation must be taken into account.

Table 30. Viscosity Data A. Viscosity of Water at Various Temperatures

Temperature, °C	Kinematic Viscosity, centistoke	Absolute Viscosity, centipoise
0	1.792	1.792
10	1.308	1.308
15	1.141	1.140
20	1.007	1.005
30	0.804	0.801
50	0.556	0.549
70	0.415	0.406
100	0.296	0.284

B. Viscosity (Centistokes) of Some Hydrocarbon Fuels at Various Temperatures

Tempera-ture, °C	Gasoline	Diesel	Motor Benzol	Heavy fuel oil	Power alcohol
0	0.80	10.8	—	20400	—
10	0.74	7.00	0.83	5000	1.5
20	0.68	4.90	0.72	1200	1.2
30	0.62	4.00	0.66	500	1.0

C. Viscosity of Gases (at Atmospheric Pressure) at Various Temperatures

Tem-pera-ture, °C	Kinematic Viscosity, Centistokes (Cst) and Absolute Viscosity, Centipoise (Cp)													
	Air		Steam		Coke Oven gas		Blast Furnace gas		Water gas		Producer gas with fuel as			
											Coal		Coke	
	Cst	*Cp*	*Cst*	*Cp*	*Cst*	*Cp*	*Cst*	*Cp*	*Cst*	*Cp*	*Cst*	*Cp*	*Cst*	*Cp*
	13.1	0.0170	10.9	0.0088	23.3	0.0140	12.3	0.0160	21.7	0.0153	14.4	0.0163	14.3	0.0166
20	15.2	0.0180	12.8	0.0095	26.4	0.0121	14.1	0.0170	24.6	0.0162	16.4	0.0172	16.2	0.0176
40	17.2	0.0190	14.7	0.0103	29.8	0.0128	15.9	0.0179	27.6	0.0171	18.4	0.0181	18.2	0.0185
80	21.2	0.0209	19.1	0.0118	37.1	0.0141	19.7	D.0196	34.2	0.0187	22.8	0.0198	22.6	0.0203
120	25.4	0.0225	24.0	0.0135	44.9	0.0153	23.8	0.0214	41.2	0.0203	27.5	0.0215	27.2	0.0220
160	30.0	0.0242	29.5	0.0152	53.5	0.0166	28.2	0.0230	48.8	0.0217	32.5	0.0231	32.2	0.0236
200	34.6	0.0258	35.6	0.0167	62.2	0.0177	32.8	0.0245	56.7	0.0231	37.8	0.0248	37.4	0.0252
250	41.3	0.0276	43.6	0.0186	74.0	0.0190	39.0	0.0263	67.2	0.0248	44.8	0.0263	44.4	0.0270
300	48.0	0.0295	52.8	0.0203	86.6	0.0203	45.5	0.0280	78.3	0.0264	52.3	0.0280	51.7	0.0287
400	62.8	0.0329	73.0	0.0238	113.6	0.0227	59.5	0.0312	102.3	0.0293	68.1	0.0312	67.6	0.0319
500	78.7	0.0359	96.3	0.0272	143.7	0.0249	74.9	0.0342	128.0	0.0321	85.7	0.0341	85.0	0.0349
Density at 0°C, 760 mm Hg in Kg/Nm^3	1.293		0.804		0.49		1.3		0.712		1.13		1.16	

Table 31. Properties of Liquid Fuels

Properties	Density, gm/c.c	Chemical composition (Weight %)							Heating value (kcal/kg)		The-retical amount of air nm_3/kg	Theoretical amount of waste gas Nm^3/kg					Maxi-mum theo-retical dry CO2 in flue gas %
Type of fuel		C	h	O	n	S	h20	Ash con-tent	Gross	Net ..		CO_2	H_2O	SO_2	n_2	T	
Kerosene	0.78-0.83	85.7	14.0	—	—	0.5	0.0	0.0	11,100	10,400	11.4	1.59	1.56	0.00	9.02	12.17	15.0
Light gas oil	0.81-0.84	85.9	13.6	—	—	0.5	0.0	0.0	,10,£60	10,280	11.2	1.60	1.57	0.00	8.87	11.98	15.3
Heavy fuel oil (No. 1)	0.83-0.89	84.6	11.8	0.7	0.5	2.0	0.3	0.05	10,780	10,110	10.8	1.58	1.32	0.01	8.53	11.44	15.5
Heavy fuel oil (No. 2)	0.90-0.93	84.5	11.3	0.4	0.4	3.0	0.4	0.05	10,470	9,860	10.7	1.58	1.27	0.02	8.44	11.31	15.7
Heavy fuel oil (No. 3)	0.92-1.00	82.7	10.9	0.5	0.4	3.5	2.0	0.1	10,4210	9,840	10.3	1.54	1.21	0.02	8.18	10.96	15.8
Gasoline	0.70-0.74	85.0	15.0	—	—		0.0	0.0	11,310	10,540	11.6	1.59	1.68	0.00	9.18	12.45	14.7

Table 32. Theoretical Air Requirement & % CO_2 Evolved in Flue Gas for Different Liquid Fuels

Fuels	*Typical density at 15°c*	*Theoretical amount of air required for complete combustion*		*Theoretical CO_2 % in exhaust Gas*
		in Nm³/kg fuel	*in kg air/kg fuel*	
Naphtha	0.72	11.6	15.0	12.77
Gasoline	0.72	11.6.	15.0	12.77
Kerosene	0.805	11.4	14.74	13.07
High speed diesel (HSD)	0.825	11.2	14.48	13.36
Light diesel oil (LDO)	0.86	10.9	14.09	13.85
Furnace oil, heavy	0.95	10.4	13.45	14.25
LPG (0°C, 1 atm), air = 1	17.65	27.4 (Nm³/Nm³)	—	12.32

Table 33. Equations for Calculation of Theoretical Air Requirement and Theoretical Flue Gas Volumes for Various Types of Fuels Based on their Net C.V.

(A) Rosin's Equation :	**(H_1 = Net heating value, of the fuel)**	**Units for C.V : kcal/kg — for solid & liquid fuels ,kcal/Nm³— for gaseous fuels**
Fuels :	***Theoretical flue gas volume***	***Theoretical air requirement***
(a) Solid fuels	$\frac{0.89\,H_1}{1,000}+165$	$\frac{1.01\,H_1}{1,000}+0.5$
(b) Liquid fuels	$\frac{1.11\,H_1}{1,000}$	$\frac{0.85\,H_1}{1,000}+2.0$
(c) Low c.v. fuel gas (500 - 3000 kcal/Nm³)	$\frac{0.725\,H_1}{1,000}+1.0$	$\frac{0.75\,H_1}{1,000}$
(d) High c.v. fuel gas (4000 - 7000 kcal/Nm³)	$\frac{1.14H_1}{1,000}+0.25$	$\frac{1.09H_1}{1,000}+0.25$
(B) Boie's Equation :		
(a) Solid fuels	$\frac{0.905(H_1+550)}{1,000}+1.17$	$\frac{1.01(H_1+550)}{1,000}$
(b) Liquid fuels	$\frac{15.75(H_1-1,100)}{10,000}+0.$	$\frac{12.38(H_1-1,100)}{10,000}$
(c) Gaseous fuels (Net C.V. > 3500 kcal/Nm³)	$\frac{11.9H_1}{10,000}+0.5$	$\frac{11.05H_1}{10,000}+0.2$
Units for theoretical flue gas volume & theoretical air requirement : (a) Nm³/kg for solid and liquid fuels, (b) Nm³/Nm³ for gaseous fuels.		

Table 34. Composition and Properties of Some Typical Commonly Used Refractory Materials

Parameters	Units	Ordinary Firebricks	Aluminous Firebricks	Semi-silica	Silica	Stabilised Dolomite	Magnesite	Chrome-Magnesite	Silicon Carbide	Sillimanite	Ceramic Fibre
		(1)	(2)	(3)	(4)	(5)	(6)	(7)	(8)	(9)	(10)
Composition (%)											
Al_2O_3	%	36.5	43.0	12.0	1.0	2.0	2.0	10.5	1.5	61.5	43-47
SiO_2	%	57.5	51.0	85.0	96.0	14.5	3.0	4.5	4.5	34.0	50-54
Cr_2O_3	%	—	—	—	—	—	—	30.5	—	—	—
Fe_2O_3	%	2.5	2.0	1.0	0.5	4.0	4.5	12.5	1.0	0.5	1-1.5
CaO	%	1.0	0.5	0.5	2.0	39.0	2.5	1.0	—	1.0	0.1-1.0
MgO	%	0.5	0.5	0.5	0.2	40.5	88.0	41.0	—	0.5	Traces
TiO_2	%	1.0	2.0	0.5	0.1	—	—	—	—	2.0	1.5-3.5
Alkalies% 1.0			1.0	0.5	0.2	—	—	—	—	0.5	0.5-2.0
B_2O_3	%	—	—	—	—	—	—		—	—	0.5-1.0
SiC	%	—	—	—	—	—	—	—	93.0	—	—
Properties :											
Bulk density	kg/m^3	1920-2080	1920-2080	1840-2000	1680-1840	2480-2640	2640-2880	2800-3360	2400-2640	2000-2160	60-130
Cold crushing strength (CCS)	kgf/cm^2	170-380	170-380	170-380	140-380	410-550	340-550	210-340	> 550	340-480	—
Refractoriness	Segar cone no.	28-31	32-34	29-31	31-33	38	38	38	38	35-36	Max. temp., 1260°C
Refractoriness	°C	1630-1700	1710-1760	1650-1690	1690-1730	1850	1850	1850	1850	1770-1790	
Refractoriness under load (RUL), at 2 kg/cm^2	°C	1440-1520	1550-1620	1550-1600	1660-1700	1550-1650	1500-1630	1500-1700	1700	1700	—
Thermal conductivity at 1000°C	$W.m^{-1}.°C^{-1}$	1.1	1.2	1.4	1.7	2.2	3.6	1.6	11.7	1.6	0.07-0.22

		1	2	3	4	5	6	7	8	9	10
Maximum temperature of use	°C	1300-1450	—	1300-1400	1550-1700	—	1700-1800	1700-1750	1500-1600	1600-1800	1260
Resistance to basic slag		low		low	alkaline vapors low		very high	very high	very low	high especially to alkaline vapors	
Resistance to acid slag	—	moderate	—	moderate	high	—	low	moderate	low	high	—
Resistance to molten metal		high upto 1300°C		high upto 1300°C	high	—	high in absence of carbides and oxides	high to Fe, Cu, Sn, Pb, Bi, Sb	low to vapors, high to Zn, Cu	high	
Liability to spall	—	small or nil	—	small	high<600°C low >600°C	—	high	small	very low	very small	—
Resistance to oxidising atmosphere		very high		very high	very high		very high	very high	decomposes at 1100°Cbut protective film forms, can be used upto 1500°C	very high	
Resistance to reducing atmosphere	—	high upto 1350°C	—	high upto 1400°C	high upto 1700°C	—	high	moderate	high	high	—

Table 35. Thermal Conductivities of Some Important Refractories at Various Temperatures (in CGS Units)

Refractories	Thermal Conductivity (cal/cm. sec. °C) at			
	30°C	300°C	700°C	1100°C
Asbestos	0.00008	0.0001	0.0001	—
Slagwool	—	0.0001	0.0001	—
Kieselguhr	0.0001	0.0003	0.0004	—
Insulating firebrick	0.0007	0.0006	0.0008	0.001
Firebrick	0.002	0.002	0.002	0.003
Sillimanite	—	0.003	0.004	0.004
Silica	—	0.003	0.004	0.005
Chrome	—	0.004	0.004	0.0G5
Chrome-magnesite	—	0.005	0.005	0.004
Magnesite	—	0.011	0.008	0.007
Carborundum	—	—	0.03	0.02
Carbon	0.1	—	0.005	0.008
Graphite	—	0.21	0.12	0.08

Table 36. Volumetric Composition of Dry Air

Constituents of dry air	% (by Volume)
Nitrogen	78.03
Oxygen	20.99
Argon	0.94
Carbon dioxide	0.03
Hydrogen	0.01
Neon	0.00123
Helium	0.0004
Krypton	0.00005
Xenon	0.000006
Total	100.00

Table 37. Emissivity (e) of Various Materials/Surfaces (within 0-1000°C range)

Material / Surface	Emissivity	Material / Surface	Emissivity
Aluminium		Asbestos	0.95
Polished aluminium	0.1	Concrete	0.7
Heavily acidised aluminium	0.25	Marble	0.9
Aluminium oxide at 260°C	0.6	Carborundum	0.85
Aluminium oxide at 800°C	0.3	Plaster	0.9
Aluminium oxide, various	0.1-0.25	Alumina, fine grain	0.25
Copper		Alumina, coarse grain	0.45
Polished	0.05	Silica, coarse grain	0.55
Plate oxidised	0.8	Zirconium silicate upto 500°C	0.85
Molten	0.15	Zirconium silicate at 800°C	0.6
Lead		Quartz rough	0.9
Pure	0.1	Carbon graphite	0.75
Oxidised at 25°C	0.3	Carbon lamp black soot	0.95
Oxidised, heated at 200°C	0.6	Glass	0.95
Iron & Steel		Timber	0.8-0.9
Cast iron, polished	0.2	**Miscellaneous Materials :**	
Cast iron, turned at 100°C	0.45	Enamels white fused on iron	0.90
Cast iron, turned at 1000°C	0.6-0.7	Paint, aluminium	0.27-0.67
Steel, ground sheet	0.6	Paint flat black lacquar	0.97
Mild Steel	0.3-0.5	Polished sheet	0.07
Steel plate, oxidised	0.9	Sheet plate, rough	0.95
Iron plate, rusted	0.7-0.85	Enamel (any colour)	0.9
Cast iron (rough) rusted	0.95	Oil paint (any colour)	0.95
Rough ingot iron	0.9	Lacquer	0.9
Molten cast iron	0.3	Matt Black paint	0.95-0.98
Molten mild steel	0.3-0.4	Aluminium laquer	0.5
Stainless steel, polished	0.1	Water	0.98
Stainless steel, various	0.2-0.6	Rubber smooth	0.9
Brass		Rubber rough	0.98
Polished	0.1	Plastic various solid	0.8-95
Roughened surface	0.2	Plastic film (0.5 mm thick)	0.5-0.95
Oxidised	0.6	Paper + Card board	0.9
Nickel & Its Alloys		Silicon polished (0.3 mm thick)	0.7
Pure nickel	0.1	Zinc oxidised	0.1
Ni plate oxidised	0.4-0.5	Galvanised Iron	0.3

Nichrome	0.7	Tin plate steel	0.1
Nichorome, oxidised	0.95	Gold, polished	0.1
Refractory & Building Materials :		Silver, polished	0.1
Red brick, rough	0.75-0.9	Chromium, polished	0.1
Fireclay	0.75		

Table 38. Absorptivity of Various Surfaces for Sun-rays

Materials	Absorptivity
Iron polished	0.45
Iron oxidised rusted	0.74
Galvanised iron dirty	0.89
Galvanised iron (new)	0.66
Aluminium polished	0.26
Copper polished	0.26
Black paint	0.97-0.99
White paint	0.12-0.26
Brick red	0.70-0.77
Asphalt	0.89
Paper, white	0.27
Tile, red and brown	0.65-0.74

APPENDIX-B: SHORT QUESTIONS & ANSWERS ON FUEL TECHNOLOGY

Q. How are coal generally classified ? Describe the grading of Indian coal based on its calorific value(C.V) range .

Ans. Coal is classified into three major types namely anthracite, bituminous, and lignite. However there is no clear demarcation between them and coal is also further classified as semi- anthracite, semi-bituminous, and sub-bituminous. Anthracite is the oldest (most matured) coal from geological perspective. It is a hard coal composed mainly of carbon with little volatile content and practically no moisture. Lignite is the youngest (most immature) coal from geological perspective. It is a soft coal composed mainly of volatile matter and moisture content with low fixed carbon. Fixed carbon refers to carbon in its free state, not combined with other elements. Volatile matter refers to those combustible constituents of coal that vaporize when coal is heated.

The common coals used in Indian industry are bituminous and sub-bituminous coal. The gradation of Indian coal based on its calorific value is as follows:

Grade of coal	Calorific Value Range (in kCal/kg)
A	> 6200
B	5600 – 6200
C	4940 – 5600
D	4200 – 4940
E	3360 – 4200
F	2400 – 3360
G	1300 – 2400

Q. Describe briefly various Natural solid fuels.

Ans. Various natural solid fuels are briefly described below.

Wood: It consists of mainly carbon and hydrogen. The wood is converted into coal when burnt in the absence of air. The average calorific value of the wood is 19,700 kJ/kg.

Peat: It is a spongy humid substance found in boggy land. It may be regarded as the first stage in the formation of coal. It has a large amount of water contents (up to 30%) and therefore has to be dried before use. It has a characteristic odour at the time of burning, and has a smoky flame. Its average calorific value is 23,000 kJ/kg.

Lignite or brown coal: It represents the next stage of peat in the coal formation, and is an intermediate variety between bituminous coal and peat. It contains nearly 40% moisture and 60% of carbon. When dried, it crumbles and hence does not store well. Due to its brittleness, it is converted into briquettes, which can be handled easily. Its average calorific value is 25,000 kJ/kg.

Bituminous coal: It represents the next stage of lignite in the coal formation and contains very little moisture (4 to 6%) and 75 to 90% of carbon. It is weather resistant and burns with a yellow flame. The average calorific value of bituminous coal is 30,000 kJ/kg.

Anthracite coal: It represents the final stage in the coal formation, and contains 90% or more carbon with a very little volatile matter. It is thus obvious, that the anthracite coal is comparatively smokeless, and has very little flame. It possesses a high Calorific value of about 36,000 kJ/kg .

Wood charcoal: It is made by heating wood with a limited supply of air to a temperature not less than 280°C. It is a good prepared solid fuel, and is used for various metallurgical processes.

Coke: It is produced when coal is strongly heated continuously for 42 to 48 hours in the absence of air in a closed vessel. This process is known as carbonization of coal. Coke is dull black in colour, porous and smokeless. It has high carbon content (80 %) and has a higher calorific value than coal.

If the carbonisation of coal is carried out at 700°C, the resulting coke is called low temperature coke or soft coke. It is used as a domestic fuel. The coke produced by carbonisation of coal at 1,100 °C, is known as hard coke. The hard coke is mostly used as a blast furnace fuel for extracting pig iron from iron ores, and to some extent as a fuel in cupola furnace for producing cast iron.

Briquetted coal: It is produced from the finely ground coal by moulding under pressure with or without a binding material. The binding materials usually used are pitch, coal tar, crude oil , clay etc. The briquetted coal has the advantage of having, practically, no loss of fuel through grate openings and thus it increases the heating value of the fuel.

Pulverised coal: The low grade coal with high ash content, is powdered to produce pulverised coal. The coal is first dried and then crushed into a fine powder by pulverising machines. The pulverised coal is widely used in the cement industry , boilers and also in metallurgical processes.

Q. Discuss the merits and demerits of liquid fuels over solid fuels .

Ans.

Merits:

- Higher calorific value.
- Lower storage capacity required.
- Better economy in handling.
- Better control of consumption by using valves.
- Better cleanliness and freedom from dust.
- Practically no ashes.
- Non-deterioration in storage.
- Non-corrosion of boiler plates.
- Higher efficiency.

Demerits:

- Higher cost.
- Greater risk of fire.
- Costly containers are required for storage and transport.

Q. Discuss the Merits and demerits of gaseous fuels .

Ans.

Merits:

- The supply of fuel gas, and hence the temperature of furnace is easily and
- accurately controlled.
- The high temperature is obtained at a moderate cost by pre-heating gas and
- air with heat of waste gases of combustion.
- They are directly used in internal combustion engines.
- They are free from solid and liquid impurities.
- They do not produce ash or smoke.
- They undergo complete combustion with minimum excess air supply.

Demerits:

- They are readily inflammable.
- They require large storage capacity.

Q. What is calorific value of fuels ? What is the difference between Gross & Net calorific values ?

Ans. The calorific value (C.V.) or heat value of a fuel is defined as the amount of heat given out by the complete combustion of 1 kg of fuel. It is expressed in terms of kJ/kg or kcal/kg of solid & liquid fuels .The calorific value of gaseous fuels is, however, expressed in terms of kcal/m3 or kJ/m3 at a specified temperature and pressure.

Following are the two types of the calorific value of fuels:

- Gross or higher calorific value, and
- Net or lower calorific value.

Gross or Higher Calorific Value (H.C.V) :

All fuels, usually, contain some percentage of hydrogen. When a given quantity of a fuel is burnt, some heat is produced. Moreover, some hot flue gases are also produced. The water, which takes up some of the heat evolved, is converted into steam. If the heat, taken away by the hot flue gases and the steam is taken into consideration, i.e. if the heat is recovered from flue gases and steam is condensed back to water at room temperature (15°C), then the amount of total heat produced per kg is known as gross or higher calorific value of fuel. In other words, the amount of heat obtained by the complete combustion of 1kg of a fuel, when the products of its combustion are cooled down to the temperature of supplied air (usually taken as 15°C), is called the gross or higher calorific value of fuel. It is briefly written as H.C.V.

If the chemical analysis of a fuel is available, then the higher calorific value of the fuel is determined by the following formula, known as Dulong's formula:

H.C.V. = 33,800 C + 1,44,000 H2 + 9,270 S kJ/kg .... (i)

where C, H2 and S represent the mass of carbon, hydrogen and sulphur in 1 kg of fuel, and the numerical values indicate their respective calorific values. If the fuel contains oxygen (O2), then it is assumed that the whole amount is combined with hydrogen having mass equal to 1/8th of that of oxygen. Therefore, while finding the calorific value of fuel, this amount of hydrogen should be subtracted.

H.C.V. = 33,800 C + 1,44,000 [H2-O2/8]+ 9,270 S kJ/kg K ....(ii)

Net or Lower Calorific Value (LCV) :

When the heat absorbed or carried away by the products of combustion is not recovered (which is the case in actual practice), and the steam formed during combustion is not condensed, then the amount of heat obtained per kg of the fuel is known as net or lower calorific value. It is briefly written as L.C.V. If the higher calorific value is known, then the lower calorific value may be obtained by subtracting the amount of heat carried away by products of combustion (especially steam) from H.C.V.

L.C.V = H.C.V. - Heat of steam formed during combustion

Let ms = Mass of steam formed in kg per kg of fuel = 9 H_2 Since the amount of heat per kg of steam is the latent heat of vaporisation of water corresponding to a standard temperature of 15°C, is 2,466 kJ/kg, therefore

L.CV. = H.C.V. – ms X 2466

= H.C.V. – 9 H_2 × 2,466 kJ/kg (since, ms = 9 H_2)

Q. Describe the procedure of experimental determination of higher calorific value of solid & liquid fuels by a Bomb Calorimeter .

Ans. The most satisfactory method of obtaining the calorific value of a fuel is by actual experiment. In all these experimental methods, a known mass of fuel is burnt in a suitable calorimeter, and the heat so evolved is found by measuring the rise in temperature of the surrounding water. The calorimeters used for finding the calorific value of fuels are known as fuel calorimeters. Bomb calorimeter and Boy's gas calorimeter are used to determine the calorific values of the solid & liquid and gaseous fuels respectively .

Bomb Calorimeter:

It is used for finding the higher calorific value of solid and liquid fuels. In this calorimeter , the fuel is burnt at a constant volume and under a high pressure in a closed vessel called bomb.

The bomb is made of acid-resisting stainless steel, machined from the solid metal, which is capable of withstanding high pressure (upto 100 bar), heat and corrosion. The cover or head of the bomb carries the oxygen valve for admitting oxygen and a release valve for exhaust gases. A cradle or carrier ring, carried by the ignition rods, supports the silica crucible, which in turn holds the sample of fuel under test. There is an ignition wire of platinum or nichrome which dips into the crucible. It is connected to a battery, kept outside, and can be sufficiently heated by passing current through it so as to ignite the fuel.

The bomb is completely immersed in a measured quantity of water. The heat, liberated by the combustion of fuel, is absorbed by this water, the bomb and copper vessel. The rise in the temperature of water is measured by a precise thermometer, known as Beckmann thermometer which reads upto 0.01 °C.

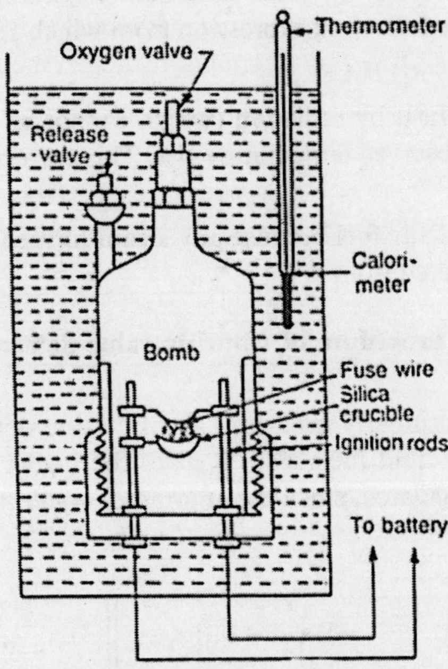

Fig.56.1: Bomb calorimeter.

Experimental Procedure:

A weighed sample of the fuel (usually one gram or so) is placed in the crucible. Pure oxygen is then admitted through the oxygen valve, till pressure inside the bomb rises to 30 atmosphere. The bomb is then completely submerged in a known quantity of water contained in a large copper vessel. This vessel is placed within a large insulated copper vessel to reduce loss of heat by radiation.

When the bomb and its contents have reached steady temperature (this temperature is noted), fuse wire is heated up electrically. The fuel ignites, and continues to burn till whole of it is burnt. The heat released during combustion is absorbed by the surrounding water and the apparatus itself. The rise in temperature of water is noted.

Let ,

m_f = Mass of fuel sample burnt in the bomb in kg

H.C.V. = Higher calorific value of the fuel sample in kJ/kg

m_w = Mass of water filled in the calorimeter in kg

m_e = Water equivalent of apparatus in kg

C_w = Specific heat of water, 4.2 KJ/kg. K

t_1 = Initial temperature of water and apparatus in °C,

t_2 = Final temperature of water and apparatus in °C

We know that heat liberated by fuel = $m_f \cdot$ H.C.V ....(1)

and heat absorbed by water and apparatus = $(m_w + m_e)\, C_w\, (t2 - t1)$ ....(2)

Since the heat liberated is equal to the heat absorbed (neglecting losses), therefore equating equations (1) and (2), we get the following expression from which H.C.V can be calculated.

$m_f \cdot$ H.C.V. = $(m_w + m_e)\, C_w\, (t_2 - t_1)$ H.C.V.

To compensate for the loss of heat by radiation (which cannot be totally eliminated), a cooling correction (tc) is added to the observed temperature rise. This corrected temperature rise is used in the above expression.

This calorimeter gives H.C.V. of the fuel because any steam formed is condensed (since it cannot escape) and hence heat is recovered from it.

Q. Describe the experimental procedure of calorific value determination of gaseous fuels by Boy's Gas Calorimeter .

Ans. Boy's gas calorimeter is primarily used for calorific value determination of gaseous fuels, though it can be modified for liquid fuels also. It gives the higher calorific value only. But the lower calorific value may be calculated, since the amount of water produced can be collected and measured.

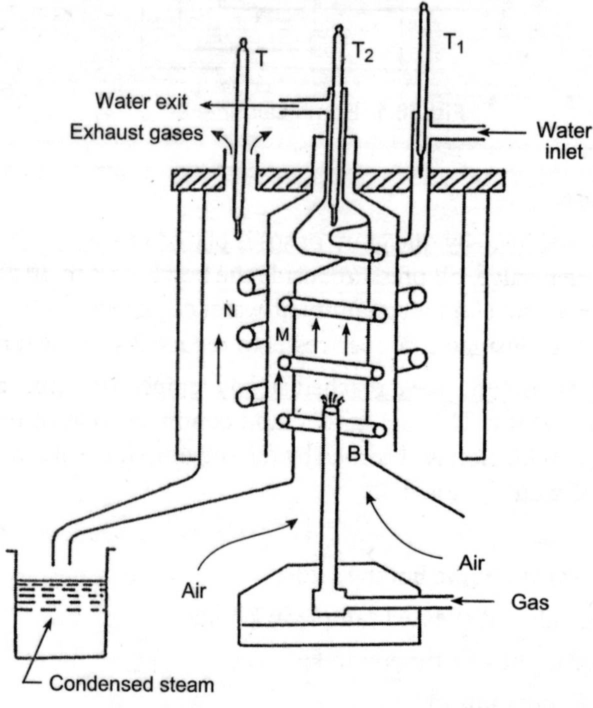

Fig.56.2: Boy's Gas Calorimeter.

It consists of a suitable gas burner at B, in which a known volume of gas at a known pressure is burnt. The hot gases, produced by combustion, rise up in the copper chimney or combustion chamber, which is surrounded by a double metal tubing through which a continuous flow of water under a constant head is maintained. From the top of inner chamber, hot gases are deflected downwards through the space containing the inner water tubes M. From here the gases are deflected upwards through the space containing the outer water tubes N. Then the gases escape into the atmosphere from the top and their temperature is recorded just before their exit. During this process of playing up and down the water tubes, the gases give out, practically, whole of their heat so that any steam formed during combustion is condensed back into water.

The temperature of the circulating water is measured at inlet and outlet by thermometers T1 and T2 respectively. After an initial warming up period during which conditions are established, simultaneous readings are taken of:

- The volume of gas burnt in a certain time.
- The quantity of water passing through the tube during the same time, and
- The rise in temperature of water.

The volume of the gas consumed is reduced to some standard conditions of pressure and temperature (15°C and 760 mm of Hg), the normal temperature and pressure (N.T.P.) being 0°C and 760 mm of Hg.

Let , v = Volume of gas burnt at standard temperature and pressure (S.T.P.) in m^3

m_w = Mass of cooling water used in kg,

H.C.V. = Higher calorific value of the fuel in kJ/m^3,

t_1 = Temperature of water at inlet, in °C, and

t_2 = Temperature of water at outlet in °C.

We know that heat produced by the combustion of fuel = v · H.C.V. .... (1)

and heat absorbed by circulating water = m_w cw $(t_2 - t_1)$ .... (2)

Since the heat produced by the combustion of fuel is equal to the heat absorbed by the circulating water (neglecting losses), therefore equating equations (1) and (2), we get

$v \times H.C.V. = m_w c_w (t_2 - t_1)$

H.C.V. can be calculated from the above expression.

The lower calorific value may be calculated if the water produced during the combustion is drained off from the bottom of the calorimeter, collected and weighed. If this mass is m, then

L.C.V. = H.C.V. – (m × 2,466) kJ/kg

The modification of this gas calorimeter is known as Junker's calorimeter.

Q. Describe briefly the procedure of Flue Gas Analysis by Orsat Apparatus .

Ans. To check the combustion efficiency of furnaces/boilers, it is essential to determine the constituents of the flue gases. Such an analysis is carried out with the help of Orsat apparatus as shown in Figure below . It consists of a graduated measuring glass tube (known as eudiometer tube) and three flasks A, B and C, each containing different chemicals for absorbing carbon dioxide, carbon monoxide and oxygen. An aspirator bottle containing water is connected to the bottom of the eudiometer tube by means of a rubber tube. It can be moved up and down, at will, for producing a suction or pressure effect on the sample of the flue gas.

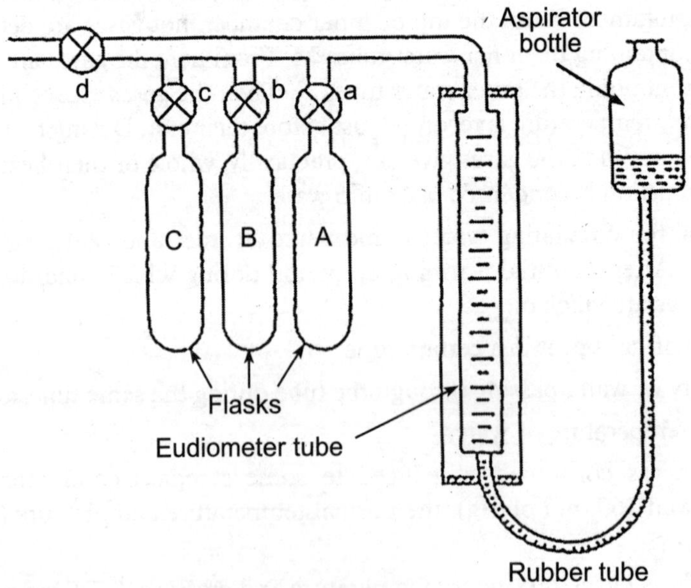

Fig.56.3: Orsat apparatus

The flask-A contains caustic potash (KOH) and is used for absorbing carbon dioxide in the sample of the flue gas. The flask-B contains caustic potash (KOH) and pyrogallic acid, which absorbs oxygen from the sample of the flue gas. The flak-C contains a solution of cuprous chloride (Cu2 Cl2) in hydrochloric acid (HCl). It absorbs carbon monoxide (CO) from the sample of the flue gas. Each of the three flasks has stop cock 'a', 'b' and 'c' respectively and a three-way cock 'd'

The sample of the flue gas to be analysed is first sucked in the eudiometer tube, and its volume is noted. It can take usually 100 cm3 of gas. By manipulating the level of aspirator bottle, the flue gas can, in turn, be forced into either of the flasks A, B and C by opening the respective cocks a, b and c. The flue gas is left there for sometime and then sucked back into eudiometer tube. The chemicals, in the three flasks absorb carbon dioxide, oxygen and carbon monoxide and the resulting contraction in volume enables the percentage of each gas present in the sample to be read on the eudiometer tube.

Since the flue is collected over water in the tube, any steam present in it will be condensed. Similarly, the sulphur dioxide present in it will also be absorbed. Hence the percentages of dry flue gases are only obtained by Orsat apparatus.

Q. Discuss the portable and online continuous Flue Gas Analysis instruments used for monitoring the optimum use of excess air in the furnace .

Ans. Microprocessor based Flue gas analyser used for the purpose are portable and widely used today. Depending on the type of sensor fitted in the instrument, it displays Oxygen & C02 levels in percentage and CO in ppm. The sampling pump fitted in the instrument draws the sample of flue gas from stack. The flue gas after passing through the sensors is finally ejected from instrument to atmosphere. The sensor needs a voltage to the microprocessor, based on the concentration of Oxygen, CO_2 & CO in the flue gas. The microprocessor calculates & displays the Oxygen %, CO_2 % and CO in ppm, based on the input voltage from these sensors.

Some instruments have additional features for measuring temperature and excess air also and even display the combustion efficiency based on standard formulae & calorific value of the particular fuel, already programmed into the microprocessor.

On line Oxygen, CO_2, CO measurement facilities are also available which work on same principle. Many systems are now available which tune the burner automatically after sensing Oxygen, CO_2 & CO levels in the stack gas .

Q. What are the safe storage and handling procedure for coal ?

Ans. Uncertainty in the availability and transportation of fuel necessitates storage and subsequent handling. Stocking of coal has its own disadvantages like build-up of inventory, space constraints, deterioration in quality and potential fire hazards. Other minor losses associated with the storage of coal include oxidation, wind and carpet loss. A 1% oxidation of coal has the same effect as 1% ash in coal, wind losses may account for nearly 0.5 – 1.0% of the total loss.

The main goal of good coal storage is to minimise carpet loss and the loss due to spontaneous combustion. Formation of a soft carpet, comprising of coal dust and soil causes carpet loss. On the other hand, gradual temperature builds up in a coal heap, on account of oxidation may lead to spontaneous combustion of coal in storage.

In industry, modes of coal handling range from manual to conveyor systems. It is advisable to minimize the handling of coal so that further generation of fine coal dust is avoided.

Measures to minimize spontaneous oxidation & combustion include:

- Releasing stagnant trapped air in stacked coal by inserting perforated pipes in coal heaps. Sprinkling water occasionally as needed, especially in the hot summer months .
- Adopting "first in, first out" principles for coal use and avoiding long durations of coal stacking/storage.

Measures to reduce carpet losses of coal include:

- Preparing a hard surface for coal to be stacked upon .
- Preparing standard storage bays made of concrete and brick .
- Controlling the height of coal heaps in storage.

Q. Describe the various stages in Preparation of coal for its industrial uses .

Ans. Preparation of coal prior to feeding into the boiler/furnace is an important step for achieving good combustion. Burning of large and irregular lumps of coal may cause the following problems:

- Poor combustion conditions and inadequate furnace temperature.
- Higher excess air resulting in higher stack loss.
- Increase of unburnts in the ash.

- Low thermal efficiency.

Various stages in coal preparation are as follows ;

Sizing of coal

Sizing, with specific relevance to the type of coal firing system, helps towards even burning, reduced ash losses and better fuel combustion efficiency.

Coal is reduced in size by crushing and pulverizing. Pre-crushed coal can be economical for smaller units, especially those which are stoker fired. In a coal handling system, crushing is limited to a top size of 6 or 4 mm. The devices most commonly used for crushing are the rotary breaker, the roll crusher and the hammer mill.

It is necessary to screen the coal before crushing, so that only oversized coal is fed to the crusher. This helps to reduce power consumption in the crusher .

The Table below gives the proper size of coal for various types of firing systems .

Table : Proper Size of Coal for Various Types of coal Firing System

S. No.	Types of Firing System	Size (in mm)
1.	**Hand Firing**	
	(a) Natural draft	25-75
	(b) Forced draft	25-40
2.	**Stoker Firing**	
	(a) Chain grate	25-40
	i) Natural draft	15-25
	ii) Forced draft	15-25
	(b) Spreader Stoker	
3.	Pulverized Fuel Fired	75% below 75 micron*
4	Fluidized bed boiler	< 10 mm

Proper coal sizing, to match the specifications of the type of combustion system, helps to maintain even burning, reduce ash losses, and achieve better combustion efficiency. Proper maintenance and upkeep of equipment like crushers, screens & mills is very important to ensure proper coal sizing in accordance with manufacturer specifications.

Using undersized coal will lead to increased loss through unburnt coal, and using oversized coal can lead to clinker formation losses apart from increases in unburnt losses.

Conditioning of coal

Segregation of fine particles from larger coal pieces can be reduced to a great extent by conditioning coal with water. Water helps fine particles to stick to the bigger lumps due to surface tension of the moisture, thus stopping fines from falling through grate bars or being carried away by the furnace draft.

While tempering the coal, care should be taken to ensure that moisture addition is uniform and preferably done in a moving or falling stream of coal. \

If the percentage of fines in the coal is very high, wetting of coal can decrease the percentage of unburnt carbon and the excess air level required to be supplied for combustion. Table below shows the extent of coal wetting, depending on the percentage of fines in coal.

Table : Extent of Wetting: Fines Vs Surface Moisture in Coal	
Fines (%)	Surface Moisture (%)
10 - 15	4 - 5
15 - 20	5 - 6
20 - 25	6 - 7
25 - 30	7 - 8

Blending of coal

When coal lots have excessive fines, it is advisable to blend the predominantly lumped coal with lots containing excessive fines. Coal blending may thus help to limit the proportion of fines in coal being fired to 25% or less. Blending of different qualities of coal may also help to supply a uniform coal feed to the boiler.

In case of carbonization of coal in coke ovens, prime coking coal,medium coking coal & low coking coal are blended in various proportion to produce hard coke of desired quality specifications for use in blast furnace.

Q. Describe the significance of important constituents of coal .

Ans. Significance of various coal constituents are described below :

Fixed carbon

Fixed carbon is the solid fuel left in the furnace after volatile matter is distilled off. It consists mostly of carbon but also contains some hydrogen, oxygen, sulphur and nitrogen not driven off with the gases. Fixed carbon gives a rough estimate of heating value of coal . Higher carbon means higher calorific value.

Volatile matter

Volatile matters are the methane, hydrocarbons, hydrogen & carbon monoxide, and incombustible gases like carbon dioxide & nitrogen found in coal. Thus volatile matter is an index of the gaseous fuels present in coal . The typical proportion of volatile matter in coal is 20 to 35%. The presence of volatile matter in coal :

- Proportionately increases flame length and helps in easier ignition of coal;
- Sets minimum limits on the furnace height and volume;
- Influences secondary air requirement and distribution aspects.
- Influences secondary oil support

Ash content

Ash is inorganic material that will not burn. The typical proportion of ash found in coal is 5 to 40% . Ash presence in coal has several detrimental effects such as :

- It reduces handling and burning capacity.
- It Increases handling costs.
- It affects flame temperature, heat transfer, combustion efficiency, and boiler efficiency.
- It causes clinkering and slagging, affecting availability.

It increases auxiliary power consumption.

Moisture content

The presence of moisture decreases the C.V of coal, typically by amounts ranging from 0.5 to 10%. It also increases heat loss through stack gases, due to evaporation and superheating. A small quantity, however, at times helps in binding fines together.

Moisture in coal must be transported, handled and stored. Since it replaces combustible matter, it decreases the heat content per kg of coal.

Moisture in coal :

- Increases heat loss, due to evaporation and superheating of vapour .
- Helps, to a limit, in binding fines.
- Aids radiation heat transfer.

Sulphur content

The typical amount of sulphur found in coal is 0.5 to 5%. An excess of it affects clinkering & slagging tendencies, corrodes mild steel chimneys & other equipment such as air heaters & economizers, limits the exit flue gas temperature due to acid dew point, and affects efficiency.

Q. What is meant by proximate & ultimate analysis of coal . What is their significance & how are they related ?

Ans.

Proximate Analysis :

Proximate analysis indicates the percentage by weight of the Fixed Carbon, Volatile matter, Ash, and Moisture Content in coal. The amounts of fixed carbon and volatile combustible matter directly contribute to the heating value of coal. Fixed carbon acts as a main heat generator during burning. High volatile matter content indicates easy ignition of fuel. The ash content is important in the design of the furnace grate, combustion volume, pollution control equipment and ash handling systems of a furnace. A typical proximate analysis of various coal is given below in the Table .

Table : Typical Proximate Analysis Of Various Coals (In Percentage By Weight)

Parameter	Indian Coal	Imported Coal	
Moisture	5.98	9.43	
Ash	38.63	13.99	
Volatile matter	20.70	29.79	
Fixed Carbon	34.69	46.79	

Ultimate Analysis:

The ultimate analysis indicates the various elemental chemical constituents of coal such as Carbon, Hydrogen, Oxygen, Sulphur, etc. It is useful in determining the quantity of air required for combustion and the volume & composition of the combustion gases. This information is required for the calculation of flame temperature and the flue duct design etc. Typical ultimate analyses of various coals are given below in the Table.

Table : Typical Ultimate Analyses of Coals

Parameter	Indian Coal, %	Imported Coal, %
Moisture	5.98	9.43
Mineral Matter (1.1 x Ash)	38.63	13.99
Carbon	41.11	58.96
Hydrogen	2.76	4.16
Nitrogen	1.22	1.02
Sulphur	0.41	0.56
Oxygen	9.89	11.88

Table : Relationship Between Ultimate Analysis and Proximate Analysis

%C	=	0.97C+ 0.7(VM - 0.1A) - M(0.6-0.01M)
%H	=	0.036C + 0.086 (VM -0.1xA) - 0.0035M2 (1-0.02M)
$\%N_2$	=	2.10 -0.020 VM
		where ,
C	=	% of fixed carbon
A	=	% of ash
VM	=	% of volatile matter
M	=	% of moisture

Q. How are various coal constituents of its proximate analysis determined in the laboratory ?

Ans. There are two methods of coal analysis : ultimate analysis and proximate analysis. The ultimate analysis determines all coal component elements, solid or gaseous and the proximate analysis determines only the fixed carbon, volatile matter, moisture and ash percentages. The ultimate analysis is determined in a properly equipped laboratory by a skilled chemist, while proximate analysis can be determined with a simple apparatus.

Determination of various constituents in its proximate analysis is described below.

Determination of Moisture content in coal

Determination of moisture is carried out by placing a sample of powdered raw coal of size 200-micron size in an uncovered crucible and it is placed in the oven kept at 108+2 oC along with the lid. Then the sample is cooled to room temperature and weighed again. The loss in weight represents moisture content in coal .

Determination of Volatile Matter content in coal

Fresh sample of crushed coal is weighed, placed in a covered crucible,and heated in a furnace at 900 + 15 oC . The sample is cooled and weighed. Loss of weight represents moisture and volatile matter. The remainder is coke (fixed carbon and ash).

Determination of Fixed Carbon and Ash content in coal

The cover from the crucible used in the last test is removed and the crucible is heated over the Bunsen burner until all the carbon is burned. The residue is weighed, which is the incombustible ash. The difference in weight from the previous weighing is the fixed carbon. In actual practice Fixed Carbon derived by subtracting from 100 the value of moisture, volatile matter and ash.

Q.Describe the salient Features of coal combustion .

Ans. 1 kg of coal will require around 8 – 10 kg of air depending upon the carbon, hydrogen, nitrogen, oxygen and sulphur content for complete combustion. This air is also known as theoretical or stoichiometric air.

If for any reason the air supplied is inadequate, the combustion will be incomplete. The result is poor generation of heat with some portions of carbon remaining unburnt (black smoke) and forming carbon monoxide instead of carbon dioxides.

As in the case of oil, coal cannot be burnt with stoichiometric quantity of air. Complete combustion is not achieved unless an excess of air is supplied.

The excess air required for coal combustion depends on the type of coal firing equipment. Hand fired boilers use large lumps of coal and hence need very high excess air. Stoker fired boilers as shown in the Figure below use sized coal and hence require less excess air.

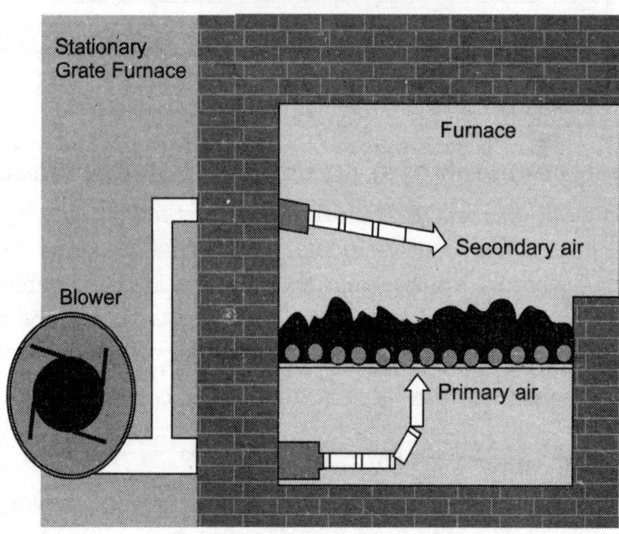

Fig.56.4: Coal Combustion in stoker fired boiler

In these systems , primary air is supplied below the grate and secondary air is supplied over the grate to ensure complete combustion.

Fluidised bed combustion in which turbulence is created leads to intimate mixing of air and fuel resulting in further reduction of excess air. The pulverized fuel firing in which powdered coal is fired has the minimum excess air due to high surface area of coal ensuring complete combustion.

Q. What is the detrimental effect of clinker formation in combustion of coal ?

Ans. Clinker is a mass of rough, hard, slag-like material formed during combustion of coal due to low fusion temperature of ash present in coal. Presence of silica, calcium oxide, magnesium oxides etc. in ash lead to a low fusion temperature. Typically Indian coals contain ash fusion temperature as low as 1100oC. Once clinker is formed, it has a tendency to grow. Clinker will stick to a hot surface rather than a cold one and to a rough surface rather than a smooth one thereby reducing heat transfer in the furnace .

Q. Describe The Basic Oil Refinery Processes.

Ans. Present day refineries produce a variety of products including many required as feedstock for the petrochemical industry.The basic refinery processes are as follows :

Distillation Processes. The first refinery, in 1861, produced kerosene by simple atmospheric distillation. Its by-products included tar and naphtha. It was soon discovered that high-quality lubricating oils could be produced by distilling petroleum under vacuum. However, for the next 30 years kerosene was the product consumers wanted. Two significant events changed this situation: (1) invention of the electric light decreased the demand for kerosene, and (2)invention of the internal combustion engine created a demand for diesel fuel and gasoline (naphtha).

Thermal Cracking Processes. With the advent of mass production and World War I, the number of gasoline-powered vehicles increased dramatically and the demand for gasoline grew accordingly. However, distillation processes produced only a certain amount of gasoline from crude oil. In 1913, the thermal cracking process was developed, which subjected heavy fuels to both pressure and intense heat, physically breaking the large molecules into smaller ones to produce additional gasoline and distillate fuels. Visbreaking, another form of thermal cracking, was developed in the late 1930's to produce more desirable and valuable products.

Catalytic Processes. Higher compression ratio gasoline engines required higher octane gasoline with better antiknock characteristics. The introduction of catalytic cracking and polymerization processes in the mid 1930's met the demand by providing improved gasoline yields and higher octane numbers.

Alkylation, another catalytic process developed in the early 1940's, produced more high-octane aviation gasoline and petrochemical feedstock for explosives and synthetic rubber. Subsequently, catalytic isomerization was developed to convert hydrocarbons to produce increased quantities of alkylation feedstock.

Improved catalysts and process methods such as hydrocracking and reforming were developed throughout the 1960's to increase gasoline yields and improve antiknock characteristics. These catalytic processes also produced hydrocarbon molecules with a double bond (alkenes) and formed the basis of the modern petrochemical industry.

Treatment Processes. Throughout the history of refining, various treatment methods have been used to remove nonhydrocarbons, impurities, and other constituents that adversely affect the properties of finished products or reduce the efficiency of the conversion processes. Treating can involve chemical reaction and/or physical separation. Typical examples of treating are chemical sweetening, acid treating, clay contacting, caustic washing, hydrotreating, drying, solvent extraction, and solvent dewaxing. Sweetening compounds and acids desulfurize crude oil before processing and treat products during and after processing.

Following the Second World War, various reforming processes improved gasoline quality & yield and produced higher quality products. Some of these involved the use of catalysts and/or hydrogen to change molecules and remove sulfur.

Q. Explain how crude quality affects the topology of refinery configuration?

Ans: Usually, refinery crudes are characterized as light, moderate and heavy crudes. Light and moderate crudes are typically targeted for gas, naphtha, diesel, light and heavy gas oil fractions. Heavy crudes are targeted for coke and residue product streams. Therefore, the crude quality does affect the topology of the refinery configuration. According to the choice of the crude available, refineries are classified into four types namely :

(a) Those that target fuels. This is very prominent in a country like India .
(b) Those that target coke. This is very much targeted for refineries that supply coke as an important raw materials to other industries such as steel, catalysts etc.
(c) Lubricants.
(d) Petrochemicals.

According to the desired product pallete, the refinery configuration and hence topology is affected with the crude quality.

Q. What is the basic difference between a chemical and a refinery process?

Ans: A chemical process essentially involves streams whose composition is fairly known. As far as refinery processes are concerned, their chemical constituents are not exactly known but are estimated as functions of various measurable properties such as viscosity, cetane number, octane number, flash point, TBP/ASTM distillation etc. Therefore, refinery process technology should accommodate the details pertaining to these issues in addition to the technology issues.

Q. Why are refinery process flow sheets very complex?

Ans.: Large refineries involve the production of about 30 to 40 refinery products with diverse specifications and needs of the consumers. And necessarily all these products are derived from the crude oil in a complex way. Due to complicated physical & chemical processes that are sequentially applied for various refinery process streams, refinery process flow sheets are very complex.

Flowsheet of a modern refinery is shown below.

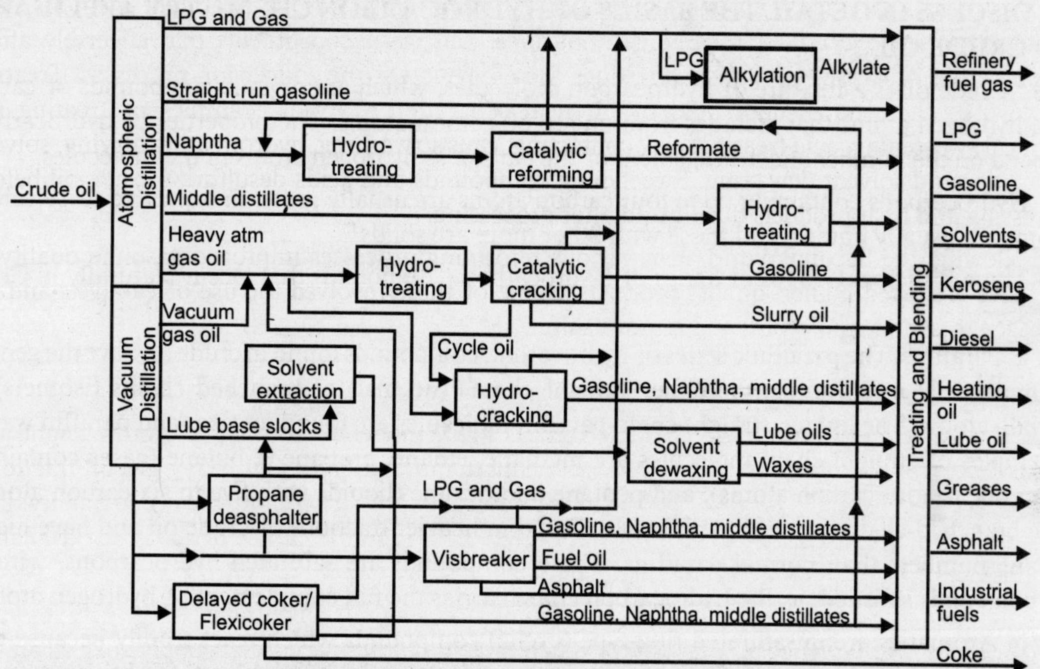

Fig.56.5: Flowsheet of a modern oil refinery

Q. DESCRIBE THE NATURE & CHARACTERISTICS OF CRUDE OIL.

Ans. Crude oils are complex mixtures containing many different hydrocarbon compounds that vary in appearance and composition from one oil field to another. Crude oils range in consistency from water to tar-like solids, and in color from clear to black. An "average" crude oil contains about 84% carbon, 14% hydrogen, 1%-3% sulfur, and less than 1% each of nitrogen, oxygen, metals & salts. Crude oils are generally classified as paraffinic,naphthenic,or aromatic,based on the predominant proportion of similar hydrocarbon molecules. Mixed-base crudes have varying amounts of each type of hydrocarbon. Refinery crude base stocks usually consist of mixtures of two or more different crude oils.

Relatively simple crude oil assays are used to classify crude oils as paraffinic, naphthenic, aromatic, or mixed. One assay method is based on distillation, and another method is based on gravity & boiling points. More comprehensive crude assays determine the value of the crude (i.e., its yield and quality of useful products) and processing parameters. Crude oils are usually grouped according to yield structure.

Crude oils are also defined in terms of API (American Petroleum Institute) gravity. The higher the API gravity, the lighter the crude. For example, light crude oils have high API gravities and low specific gravities. Crude oils with low carbon, high hydrogen, and high API gravity are usually rich in paraffins and tend to yield greater proportions of gasoline and light petroleum products; those with high carbon, low hydrogen & low API gravities are usually rich in aromatics. Crude oils that contain appreciable quantities of hydrogen sulfide or other reactive sulfur compounds are called "sour." Those with less sulfur are called "sweet."

Q. DISCUSS IN DETAIL THE BASICS OF HYDROCARBON CHEMISTRY APPLICABLE TO CRUDE OIL.

Ans. Crude oil is a mixture of hydrocarbon molecules, which are organic compounds of carbon and hydrogen atoms that includes from one to 60 carbon atoms. The properties of hydrocarbons depend on the number and arrangement of the carbon & hydrogen atoms in the molecules.

Hydrocarbons containing up to four carbon atoms are usually gases, those with 5 to 19 carbon atoms are usually liquids, and those with 20 or more are solids.

Three Principal Groups or Series of Hydrocarbon Compounds that Occur Naturally in Crude Oil are :

a. Paraffins. The paraffinic series of hydrocarbon compounds found in crude oil have the general formula CnH_2n+2 and can be either straight chains (normal) or branched chains (isomers) of carbon atoms. The lighter, straight-chain paraffin molecules are found in gases and paraffin waxes. Examples of straight-chain molecules are methane, ethane, propane & butane (gases containing from one to four carbon atoms), and pentane and hexane (liquids with five to six carbon atoms). The branched-chain (isomer) paraffins are found in heavier fractions of crude oil and have higher octane numbers than normal paraffins. These compounds are saturated hydrocarbons, with all carbon bonds satisfied, ie, the hydrocarbon chain carries the full complement of hydrogen atoms.

b. Aromatics are unsaturated ring-type (cyclic) compounds which react readily because they have carbon atoms that are deficient in hydrogen. All aromatics have at least one benzene ring (a single-ring compound characterized by three double bonds alternating with three single bonds between six carbon atoms) as part of their molecular structure. Naphthalenes are fused double-ring aromatic compounds. The most complex aromatics, polynuclears (three or more fused aromatic rings), are found in heavier fractions of crude oil.

c. Naphthenes are saturated hydrocarbon groupings with the general formula CnH_2n, arranged in the form of closed rings (cyclic) and found in all fractions of crude oil except the very lightest. Single-ring naphthenes (monocycloparaffins) with five and six carbon atoms predominate, with two-ring naphthenes (dicycloparaffins) found in the heavier ends of naphtha.

Other Hydrocarbons.

a. Alkenes are mono-olefins with the general formula CnH2n and contain only one carbon-carbon double bond in the chain. The simplest alkene is ethylene, with two carbon atoms joined by a double bond and four hydrogen atoms. Olefins are usually formed by thermal & catalytic cracking and rarely occur naturally in unprocessed crude oil.

b. Dienes and Alkynes. Dienes, also known as diolefins, have two carboncarbon double bonds. The alkynes, another class of unsaturated hydrocarbons, have a carbon-carbon triple bond within the molecule. Both these series of hydrocarbons have the general formula CnH2n-2. Diolefins such as 1,2-butadiene & 1,3-butadiene, and alkynes such as acetylene, occur in C5 & lighter fractions from cracking. The olefins, diolefins, and alkynes are unsaturated because they contain less than the amount of hydrogen necessary to saturate all the valences of the carbon atoms. These compounds are more reactive than paraffins or naphthenes and readily combine with other elements such as hydrogen, chlorine, and bromine.

Q. Describe the Non hydrocarbons constituents of crude oil .

Ans. Non hydrocarbon components of crude oil are as follows :

a. Sulfur Compounds. Sulfur may be present in crude oil as hydrogen sulfide(H2S), as compounds (e.g. mercaptans, sulfides, disulfides, thiophenes, etc.) or as elemental sulfur. Each crude oil has different amounts and types of sulfur compounds, but the proportion, stability, and complexity of the compounds are greater in heavier crude-oil fractions. Hydrogen sulfide , elemental sulfur and mercaptans are primary contributor to corrosion in refinery processing units. The corrosive sulfur compounds have an obnoxious odor.

Pyrophoric iron sulfide results from the corrosive action of sulfur compounds on the iron and steel used in refinery process equipment, piping & tanks. The combustion of petroleum products containing sulfur compounds produces undesirables such as sulfuric acid and sulfur dioxide. Catalytic hydrotreating processes such as hydrodesulfurization remove sulfur compounds from refinery product streams. Sweetening processes either remove the obnoxious sulfur compounds or convert them to odorless disulfides, as in the case of mercaptans.

b. Oxygen Compounds. Oxygen compounds such as phenols, ketones, and carboxylic acids occur in crude oils in varying amounts.

c. Nitrogen Compounds. Nitrogen is found in lighter fractions of crude oil as basic compounds, and in heavier fractions of crude oil as nonbasic compounds that may also include trace metals such as copper, vanadium, and/or nickel. Nitrogen oxides can form in process furnaces. The decomposition of nitrogen compounds in catalytic cracking and hydrocracking processes forms ammonia and cyanides that can cause corrosion.

d. Trace Metals. Metals, including nickel, iron & vanadium are often found in crude oils in small quantities and are removed during the refining process. Burning heavy fuel oils in refinery furnaces & boilers can leave deposits of vanadium oxide & nickel oxide in furnace boxes, ducts, & tubes. It is also desirable to remove trace amounts of arsenic, vanadium, and nickel prior to processing as they can poison certain catalysts.

e. Salts. Crude oils often contain inorganic salts such as sodium chloride, magnesium chloride, and calcium chloride in suspension or dissolved in entrained water (brine). These salts must be removed or neutralized before processing to prevent catalyst poisoning, equipment corrosion, and fouling. Salt corrosion is caused by the hydrolysis of some metal chlorides to hydrogen chloride (HCl) and the subsequent formation of hydrochloric acid when crude is heated.Hydrogen chloride may also combine with ammonia to form ammonium chloride (NH4Cl), which causes fouling and corrosion.

f. Carbon Dioxide. Carbon dioxide may result from the decomposition of bicarbonates present in or added to crude, or from steam used in the distillation process.

g. Naphthenic Acids. Some crude oils contain naphthenic (organic) acids, which may become corrosive at elevated temperature .

Q. What are the major major oil refinery products ?

Ans. Major refinery products are described below :

Gasoline(Petrol) . The most important refinery product is motor gasoline, a blend of hydrocarbons with boiling ranges from ambient temperatures to about 2050C. The important qualities for gasoline are octane number (antiknock), volatility(starting & vapor lock), and vapor pressure (environmental pollution control). Additives are used to enhance performance & provide protection against oxidation and rust formation.

Kerosene. Kerosene is a refined middle-distillate petroleum product used as a jet fuel and in cooking and space heating. When used as a jet fuel, some of the critical qualities are freeze point, flash point, and smoke point. Commercial jet fuel has a boiling range of about 190-275° C, and military jet fuel 55-290° C. Kerosene, with less-critical specifications, is used for lighting, heating, solvents, and blending into diesel fuel.

Liquified Petroleum Gas (LPG). LPG, which consists principally of propane & butane,is produced for use as fuel and is an intermediate material in the manufacture of petrochemicals. specifications for proper performance include vapor pressure and control of contaminants.

Distillate Fuels. Diesel fuels and domestic heating oils have boiling ranges of about 205-370° C. The desirable qualities required for distillate fuels include controlled flash & pour points, clean burning, no deposit formation in storage tanks, and a proper diesel fuel cetane rating for good starting & combustion.

Residual Fuels. Many marine vessels, power plants, commercial buildings and industrial facilities use residual fuels or combinations of residual & distillate fuels for heating and processing. Two most critical specifications of residual fuels are viscosity &low sulfur content for environmental control.

Coke and Asphalt. Coke is almost pure carbon with a variety of uses from electrodes to charcoal briquets. Asphalt, used for roads & roofing materials, must be inert to most chemicals and weather conditions.

Solvents. A variety of products, whose boiling points and hydrocarbon composition are closely controlled, are produced for use as solvents. These include benzene, toluene, and xylene.

Petrochemicals. Many products derived from crude oil refining, such as ethylene, propylene, butylene, and isobutylene, are primarily intended for use as petrochemical feedstock in the production of plastics, synthetic fibers, synthetic rubbers, and other products.

Lubricants. Special refining processes produce lubricating oil base stocks. Additives such as demulsifiers, antioxidants, and viscosity improvers are blended into the base stocks to provide the characteristics required for motor oils, industrial greases, lubricants, and cutting oils. The most critical quality for lubricating oil base stock is a high viscosity index, which provides for greater consistency under varying temperature conditions.

Q. Briefly Describe Various Petroleum Refining Operations.

Ans. Petroleum refining begins with the distillation, or fractionation, of crude oils into separate hydrocarbon groups. The resultant products are directly related to the characteristics of the crude processed. Most distillation products are further converted into more usable products by changing the size & structure of the hydrocarbon molecules through cracking, reforming, and other conversion processes . These converted products are then subjected to various treatment & separation processes such as extraction, hydrotreating, and sweetening to remove undesirable constituents & improve product quality.Integrated refineries incorporate fractionation,conversion, treatment & blending operations and may also include petrochemical processing plants .

Q. What are the basic areas of oil refining processes and operations ?

Ans. Petroleum refining processes & operations can be separated into following five basic areas:

1. Fractionation (distillation) is the separation of crude oil in atmospheric and vacuum distillation towers into groups of hydrocarbon compounds of differing boiling-point ranges called "fractions" or "cuts."

2. Conversion processes change the size and/or structure of hydrocarbon molecules. These processes include:

- Decomposition (dividing) by thermal & catalytic cracking;
- Unification (combining) through alkylation & polymerization; and
- Alteration (rearranging) with isomerization & catalytic reforming.

3. Treatment processes are intended to prepare hydrocarbon streams for additional processing and to prepare finished products. Treatment may include the removal or separation of aromatics and naphthenes as well as impurities & undesirable contaminants. Treatment may involve chemical or physical separation such as dissolving, absorption, or precipitation using a variety and combination of processes including desalting, drying, hydrodesulfurizing, solvent refining, sweetening, solvent extraction, and solvent dewaxing.

4. Formulating and Blending is the process of mixing and combining hydrocarbon fractions, additives, and other components to produce finished products with specific performance properties.

5. Other Refining Operations include: light-ends recovery; sour-water stripping; solid waste & wastewater treatment; process-water treatment & cooling; storage & handling; product movement; hydrogen production; acid & tail-gas treatment; and sulfur recovery. Auxiliary operations and facilities include: steam and power generation; process & fire water systems; flares & relief systems; furnaces & heaters; pumps & valves; supply of steam, air, nitrogen & other plant gases; alarms & sensors; noise & pollution controls; sampling, testing,& inspecting; and laboratory, control room, maintenance & administrative facilities.

Q. Describe the crude oil pretreatment (desalting) process .

Ans. Crude oil contains water, inorganic salts, suspended solids, and water soluble trace metals. As a first step in the refining process, to reduce corrosion, plugging & fouling of equipments and to prevent poisoning of the catalysts in processing units, these contaminants must be removed by desalting (dehydration).

The two typical methods of crude-oil desalting, chemical and electrostatic separation, use hot water as the extraction agent.In chemical desalting,water & chemical surfactant (demulsifiers) are added to the crude, heated so that salts & other impurities dissolve into the water or attach to the water, and then held in a tank where they settle out. Electrical desalting involves the application of high voltage electrostatic charges to concentrate suspended water globules in

the bottom of the settling tank. Surfactants are added only when the crude has a large amount of suspended solids. Both methods of desalting are continuous. A third and less common process involves filtering heated crude using diatomaceous earth.

The feedstock crude oil is heated to between 65 and 175°C to reduce viscosity & surface tension for easier mixing and separation of the water. The temperature is limited by the vapor pressure of the crude oil feedstock. In both methods other chemicals may be added. Ammonia is used to reduce corrosion. Caustic or acid may be added to adjust the pH of the water wash. Wastewater and contaminants are discharged from the bottom of the settling tank to the wastewater treatment facility. The desalted crude is continuously drawn from the top of the settling tanks and sent to the crude distillation (fractionating) tower.

Q. Describe the various crude oil distillation (fractionation) practices followed in petroleum refineries.

Ans. The first step in the refining process is the separation of crude oil into various fractions or straight run cuts by distillation in atmospheric and vacuum distillation towers.The main fractions or "cuts" obtained have specific boiling point ranges and can be classified in order of decreasing volatility into gases, light distillates, middle distillates, gas oils, and residuum.

Atmospheric Distillation Tower. At the refinery, the desalted crude feedstock is preheated using recovered process heat. The feedstock then flows to a direct fired crude charge heater where it is fed into the vertical distillation column just above the bottom, at pressures slightly above atmospheric and at temperatures ranging from 340 to 370° C (heating crude oil above these temperatures may cause undesirable thermal cracking). All but the heaviest fractions flash into vapor. As the hot vapor rises in the tower, its temperature is reduced. Heavy fuel oil or asphalt residue is taken from the bottom. At successively higher points on the tower, the various major products including lubricating oil, heating oil, kerosene, gasoline, and uncondensed gases (which condense at lower temperatures) are drawn off.

The fractionating tower, a steel cylinder about 35 metres high, contains horizontal steel trays for separating and collecting the liquids. At each tray, vapors from below enter perforations and bubble caps. They permit the vapors to bubble through the liquid on the tray, causing some condensation at the temperature of that tray. An overflow pipe drains the condensed liquids from each tray back to the tray below, where the higher temperature causes re evaporation. The evaporation, condensing, and scrubbing operation is repeated many times until the desired degree of product purity is reached. Then side streams from certain trays are taken off to obtain the desired fractions. Products ranging from uncondensed fixed gases at the top to heavy fuel oils at the bottom can be taken continuously from a fractionating tower. Steam is often used in towers to lower the vapor pressure and create a partial vacuum. The distillation process separates the major constituents of crude oil into so called straight run products. Sometimes crude oil is "topped" by distilling off only the lighter fractions, leaving a heavy residue.

Vacuum Distillation Tower. In order to further distill the residuum or topped crude from the atmospheric tower at higher temperatures, reduced pressure is required to prevent thermal cracking. The process takes place in one or more vacuum distillation towers. The principles of vacuum distillation resemble those of fractional distillation and, except that larger diameter columns are used to maintain comparable vapor velocities at the reduced pressures, the equipment is also similar. The internal designs of some vacuum towers are different from atmospheric towers in that random packing and demister pads are used instead of trays. A typical first phase vacuum tower may produce gas oils, lubricating oil base stocks, and heavy residual for propane deasphalting. A second-phase tower operating at lower vacuum may distill surplus residuum from the atmospheric tower, which is not used for lube stock processing, and surplus residuum from the first vacuum tower not used for deasphalting. Vacuum towers are typically used to separate catalytic cracking feedstock from surplus residuum , often distilled further under high vacuum.

Other Distillation Towers. Within refineries there are numerous other, smaller distillation towers called columns, designed to separate specific and unique products which all work on the same principles as the towers described above. For example, a depropanizer is a small column designed to separate propane and lighter gases from butane and heavier components. Another larger column is used to separate ethyl benzene and xylene. Small "bubble" towers called strippers use steam to remove trace amounts of light products from heavier product streams.

Q. What similarities are there for the atmospheric crude distillation unit(CDU)with vacuum distillation unit(VDU)?

Ans: Both crude and vacuum distillation units have similar architecture of the main and secondary columns i.e., both have complex stream circuitries with pump arounds, heat exchanger networks and utilization of steam. Only basic difference is that while we operate the VDU at lower pressure (30 – 40 mm Hg), the operating temperatures will be lower than those in the CDU. Otherwise, the basic principles remain the same.

Q. What primary disadvantage exists by using live steam in the CDU columns?

Ans.

- Live steam once it enters the column does not condense anywhere, as we don't want any condensation to happen.
- When live steam is used, vapor load increases significantly in the column.
- This increases the diameter of the column at various sections. There will be of course a section that has maximum vapor load and this section will have the maximum diameter.

Q. Can we use steam instead of reboiler in the fractionator?

Ans: Steam utilization instead of reboilers is usually carried out for heavier hydrocarbons but not for the light hydrocarbons such as naphtha and gas oil. This is to ensure that lighter hydrocarbons are easily stripped from the heavier desulphurized product and good product quality is obtained in the top and bottom products of the fractionators.

Q. Do we produce only gas when steam stripping facilitiated in the fractionators?

Ans: No, we produce three products from the partial condenser of the fractionator. These are (a) sour water (steam condenses to form water), unstabilized naphtha (this will again enter a stabilization unit) and a gas stream.

Q. What is the purpose of thermal cracking ?

Ans. Because the simple distillation of crude oil produces amounts and types of products that are not consistent with those required by the marketplace,subsequent refinery processes change the product mix by altering the molecular structure of the hydrocarbons. One of the ways of accomplishing this change is through "cracking," a process that breaks or cracks the heavier, higher boiling point petroleum fractions into more valuable products such as gasoline, fuel oil, and gas oils. The two basic types of cracking are thermal cracking, using heat & pressure, and catalytic cracking.

The first thermal cracking process was developed around 1913. Distillate fuels and heavy oils were heated under pressure in large drums until they cracked into smaller molecules with better antiknock characteristics. However, this method produced large amounts of solid,unwanted coke.

Q. Describe the various thermal cracking processes used in petroleum refineries .

Ans. The processes that has evolved into the applications of thermal cracking include: visbreaking, steam cracking, and coking.

- Visbreaking Process. Visbreaking, a mild form of thermal cracking, significantly lowers the viscosity of heavy crude oil residue without affecting the boiling point range. Residual from the atmospheric distillation tower is heated (425-510° C) at atmospheric pressure & mildly cracked in a heater. It is then quenched with cool gas oil to control overcracking, and flashed in a

distillation tower. Visbreaking is used to reduce the pour point of waxy residues and reduce the viscosity of residues used for blending with lighter fuel oils. Middle distillates may also be produced, depending on product demand. The thermally cracked residue tar, which accumulates in the bottom of the fractionation tower, is vacuum flashed in a stripper and the distillate is recycled.

- Steam Cracking Process. It is a petrochemical process sometimes used in refineries to produce olefinic raw materials (e.g., ethylene) from various feedstock for petrochemicals manufacture. The feedstock range from ethane to vacuum gas oil, with heavier feeds giving higher yields of byproducts such as naphtha. The most common feeds are ethane, butane, and naphtha. Steam cracking is carried out at 815-870°C, and at pressures slightly above atmospheric. Naphtha produced from steam cracking contains benzene, which is extracted prior to hydrotreating. Residual from steam cracking is sometimes blended into heavy fuels.
- Coking Processes. Coking is a severe method of thermal cracking used to upgrade heavy residuals into lighter products or distillates. Coking produces straight run gasoline (coker naphtha) and various middle-distillate fractions used as catalytic cracking feedstock. The process so completely reduces hydrogen that the residue is a form of carbon called "coke." The two most common processes are delayed coking and continuous (contact or fluid) coking. Three typical types of coke are obtained (sponge coke, honeycomb coke, and needle coke) depending upon the reaction mechanism, time, temperature, and the crude feedstock.

a. Delayed Coking. In delayed coking the heated charge (typically residuum from atmospheric distillation towers) is transferred to large coke drums which provide the long residence time needed to allow the cracking reactions to proceed to completion. Initially the heavy feedstock is fed to a furnace which heats the residuum to high temperatures (480-510° C) at low pressures (1.7- 2 kg/ cm^2) and is designed & controlled to prevent premature coking in the heater tubes. The mixture is passed from the heater to one or more coker drums where the hot material is held approximately for 24 hours (delayed) at pressures of 1.7-5 kg/cm^2, until it cracks into lighter products. Vapors from the drums are returned to a fractionator where gas, naphtha, and gas oils are separated out. The heavier hydrocarbons produced in the fractionator are recycled through the furnace.

After the coke reaches a predetermined level in one drum, the flow is diverted to another drum to maintain continuous operation. The full drum is steamed to strip out uncracked hydrocarbons, cooled by water injection, and decoked by mechanical or hydraulic methods. The coke is mechanically removed by an auger rising from the bottom of the drum. Hydraulic decoking consists of fracturing the coke bed with high-pressure water ejected from a rotating cutter.

b. Continuous Coking. Continuous (contact or fluid) coking is a moving-bed process that operates at temperatures higher than delayed coking. In continuous coking, thermal cracking occurs by using heat transferred from hot, recycled coke particles to feedstock in a radial mixer, called a reactor, at a pressure of 3.5 kg/cm^2. Gases and vapors are taken from the reactor, quenched to stop any further reaction, and fractionated. The reacted coke enters a surge drum and is lifted to a feeder and classifier where the larger coke particles are removed as product. The remaining coke is dropped into the preheater for recycling with feedstock. Coking occurs both in the reactor and in the surge drum. The process is automatic in that there is a continuous flow of coke and feedstock.

Q. Are there any heat integration opportunities existent for the visbreaking operation?

Ans: No, the reason is that the vapour from the fractionators is a hot stream and the quenching stream is also a hot stream. We need atleast one cold stream and hot stream to enable heat integration.

Q. Do opportunities exist in coking operation for heat integration?

Ans: Yes, they exist very much. The hot products from the complex distillation unit as well as the vapors generated from the coke drums are the hot streams. The cold stream is the feed which is subjected to pre-heated. Since we have a cold stream and good number of hot streams, we can have a good heat integration opportunity in the coking operation.

Q. Why thermal cracking is most preferable for vacuum residue than catalytic cracking?

Ans: Since heavier compounds are present in the vacuum residue, they can totally poison the catalyst with coke deposition. Therefore, thermal cracking but not catalytic cracking is preferred.

Q. What is flexicoking?

Ans: In flexi coking, coke is fed to a gasifier to react with air and steam. Eventually, the coke produces hydrogen, CO, CO_2, N_2, H_2O and H_2S. The gas thus obtained is used as a fuel gas in the refinery itself for various furnace operations. Therefore, flexicoking refers to flexibly alter the coking process to produce additionally gas using an extra gasification unit.

Q. What is catalytic cracking ? In what ways it is different from thermal cracking ?

Ans. Catalytic cracking breaks complex hydrocarbons into simpler molecules in order to increase the quality & quantity of lighter, more desirable products and decrease the amount of residuals. This process rearranges the molecular structure of hydrocarbon compounds to convert heavy hydrocarbon feedstock into lighter fractions such as kerosene, gasoline, LPG, heating oil, and petrochemical feedstock.

Catalytic cracking is similar to thermal cracking except that catalysts facilitate the conversion of the heavier molecules into lighter products. Use of a catalyst in the cracking reaction increases the yield of improved quality products under much less severe operating conditions than in thermal cracking. Typical temperatures are from 470-510° C at much lower pressures of 0.7-1.5 kg/cm^2. The catalysts used in refinery cracking units are typically solid materials (zeolite, aluminum hydrosilicate, treated bentonite clay, fuller's earth, bauxite, and silica-alumina)that come in the form of powders, beads, pellets or shaped materials called extrudites.

Q. What are the basic functions & types of catalytic cracking processes ?

Ans. There are three basic functions in the catalytic cracking process:

1. Reaction: Feedstock reacts with catalyst and cracks into different hydrocarbons;
2. Regeneration: Catalyst is reactivated by burning off coke; and
3. Fractionation: Cracked hydrocarbon stream is separated into various products.

The three types of catalytic cracking processes are :

Fluid catalytic cracking (FCC), moving-bed catalytic cracking, and Thermofor catalytic cracking (TCC). The catalytic cracking process is very flexible, and operating parameters can be adjusted to meet changing product demand.

Q. What are the main advantages of catalytic cracking over thermal cracking?

Ans.

- Good product quality .
- Less coke formation .
- Lower operating Temperature & pressure and uniformity in operation .

Q. Describe the features of various catalytic cracking processes .

Ans. Essential features of various catalytic cracking processes have been described below :

Fluid Catalytic Cracking (FCC) Process

FCC is the most common process , in which the oil is cracked in the presence of a finely divided catalyst which is maintained in an aerated or fluidized state by the oil vapors. The fluid cracker consists of a catalyst section and a fractionating section that operate together as an integrated processing unit. The catalyst section contains the reactor and regenerator, which, with the standpipe & riser, forms the catalyst circulation unit. The fluid catalyst is continuously circulated between the reactor and the regenerator using air, oil vapors & steam as the conveying media.

A typical FCC process involves mixing a preheated hydrocarbon charge with hot, regenerated catalyst as it enters the riser leading to the reactor. The charge is combined with a recycle stream within the riser, vaporized, and raised to reactor temperature (480-535° C) by the hot catalyst. As the mixture travels up the riser, the charge is cracked at 0.7-2 kg/cm2. In the more modern FCC units, all cracking takes place in the riser. The "reactor" no longer functions as a reactor; it merely serves as a holding vessel for the cyclones. This cracking continues until the oil vapors are separated from the catalyst in the reactor cyclones. The resultant product stream (cracked product) is then charged to a fractionating column where it is separated into fractions, and some of the heavy oil is recycled to the riser.

Spent catalyst is regenerated to get rid of coke that collects on the catalyst during the process. Spent catalyst flows through the catalyst stripper to the regenerator, where most of the coke deposits burn off at the bottom where preheated air and spent catalyst are mixed. Fresh catalyst is added and worn-out catalyst removed to optimize the cracking process.

Moving Bed Catalytic Cracking Process.

The moving-bed catalytic cracking process is similar to the FCC process. The catalyst is in the form of pellets that are moved continuously to the top of the unit by conveyor or pneumatic lift tubes to a storage hopper, then flow downward by gravity through the reactor, and finally to a regenerator. The regenerator and hopper are isolated from the reactor by steam seals. The cracked product is separated into recycle gas, oil, clarified oil, distillate, naphtha, and wet gas.

Thermofer Catalytic Cracking Process

In a typical thermofor catalytic cracking unit, the preheated feedstock flows by gravity through the catalytic reactor bed. The vapors are separated from the catalyst and sent to a fractionating tower. The spent catalyst is regenerated, cooled, and recycled. The flue gas from regeneration is sent to a carbonmonoxide boiler for heat recovery.

Q. Explain the reason behind recycling of intermediate process streams in fluid coker?

Ans: An essential issue in fluidization is to maintain a uniform temperature throughout the fluidized bed. To do so, partial recycle of intermediate vapour streams is carried out so that uniform temperatures are maintained.

Q. What is hydrocracking ? What is the role of hydrogen in hydrocracking ?

Ans. Hydro cracking is a two stage process combining catalytic cracking and hydrogenation, wherein heavier feed stocks are cracked in the presence of hydrogen to produce more desirable products. The process employs high pressure, high temperature, a catalyst, and hydrogen. Hydro cracking is used for feedstock that are difficult to process by either catalytic cracking or reforming, since these feed stocks are characterized usually by a high polycyclic aromatic content and/or high concentrations of the two principal catalyst poisons, sulfur & nitrogen compounds.

The hydro cracking process largely depends on the nature of the feedstock and the relative rates of the two competing reactions, hydrogenation and cracking. Heavy aromatic feedstock is converted into lighter products under a wide range of very high pressures (70-140 kg/cm2) and fairly high temperatures (400-815° C), in the.presence of hydrogen and special catalysts. When the feedstock has a high paraffinic content, the primary function of hydrogen is to prevent the formation of polycyclic aromatic compounds. Another important role of hydrogen in the hydrocracking process is to reduce tar formation and prevent buildup of coke on the catalyst. Hydrogenation also serves to convert sulfur and nitrogen compounds present in the feedstock to hydrogen sulfide and ammonia.

Q. Describe the Hydrocracking Process.

Ans. In the first stage, preheated feedstock is mixed with recycled hydrogen and sent to the first-stage reactor, where catalysts convert sulfur and nitrogen compounds to hydrogen sulfide and ammonia. Limited hydrocracking also occurs. After the hydrocarbon leaves the first stage, it is cooled and liquefied and run through a hydrocarbon separator. The hydrogen is recycled to the feedstock. The liquid is charged to a fractionator. Depending on the products desired (gasoline components, jet fuel, and gas oil), the fractionator is run to cut out some portion of the first stage reactor out turn. Kerosene range material can be taken as a separate side draw product or included in the fractionator bottoms with the gas oil.

The fractionator bottoms are again mixed with a hydrogen stream and charged to the second stage. Since this material has already been subjected to some hydrogenation, cracking, and reforming in the first stage, the operations of the second stage are more severe (higher temperatures and pressures). Like the outturn of the first stage, the second stage product is separated from the hydrogen and charged to the fractionator. Hydrocracking produces relatively large amounts of isobutane for alkylation feedstock.Hydrocracking also performs isomerization for pour-point control and smoke point control, both of which are important in high quality jet fuel.

Q. Comment upon hydrogen requirements for various hydrotreating feedstocks?

Ans: Depending upon the feedstocks, hydrotreating feedstocks would be fed with variant hydrogen requirements. For straight run naphtha, the H2/Naphtha feed ratio is 1.4 – 5.7 Nm^3/bbl of feed. For Gas oil, it is 5.5 – 16 Nm3/bbl feed. For kerosene, it is 2.7 – 11.2 Nm3/bbl feed.

Q. Why two phase separators are used for the reactor product in hydrocracking ?

Ans: The first phase separator provides three streams based on its temperature and pressure. However, when the temperature of the gas stream is again reduced, again two phase streams would emanate. This is due to the heat removal and due to this reason, adiabatic flash conditions would exist and two streams would be generated. The second phase separator could be optional as well, but it will affect the catalyst performance. Long term performance studies along with process simulation studies provide such minor changes in the process flow sheet which are important to eventually phase out long term profitability of the hydrocracking flow sheet.

Q. What is catalytic reforming & its purpose ?

Ans. Catalytic reforming is a process used to convert low octane naphthas into High octane gasoline blending components called reformates. Reforming represents the total effect of numerous reactions such as cracking, polymerization, dehydrogenation, and isomerization taking place simultaneously. Depending on the properties of the naphtha feedstock (as measured by the paraffin, olefin, naphthene,& aromatic content) and catalysts used, reformates can be produced with very high concentrations of

toluene, benzene, xylene, and other aromatics useful in gasoline blending & petrochemical processing. Hydrogen, a significant by product, is separated from the reformate for recycling and use in other processes.

A catalytic reformer comprises a reactor section and a product recovery section. More or less standard is a feed preparation section in which, by combination of hydrotreatment and distillation, the feedstock is prepared to specification. Most processes use platinum as the active catalyst. Sometimes platinum is combined with a second catalyst (bimetallic catalyst) such as rhenium or another noble metal.

Q. What are various commercial catalytic reforming processes .

Ans. There are many different commercial catalytic reforming processes including platforming, powerforming, ultraforming, and Thermofor catalytic reforming(TCR). In the platforming process, the first step is preparation of the naphtha feed to remove impurities from the naphtha and reduce catalyst degradation. The naphtha feedstock is then mixed with hydrogen, vaporized, and passed through a series of alternating furnace & fixed-bed reactors containing a platinum catalyst. The effluent from the last reactor is cooled and sent to a separator to permit removal of the hydrogen rich gas stream from the top of the separator for recycling. The liquid product from the bottom of the separator is sent to a fractionator called a stabilizer (butanizer). It makes a bottom product calledreformate; butanes and lighter go overhead and are sent to the saturated gas plant.

Some catalytic reformers operate at low pressure (3.5-14 kg/cm^2), and others operate at high pressures (up to 70 kg/cm^2). Some catalytic reforming systems continuously regenerate the catalyst while in other systems , one reactor at a time is taken off stream for catalyst regeneration, and some facilities regenerate all of the reactors during turnarounds.

Q. Why is hydrogen used in the reforming reaction?

Ans: Hydrogen reduces the coke formation on the catalyst and therefore increases the shell life of the highly expensive platinum catalyst.

Q. Why is reforming a very important process in the refinery?

Ans: The reforming is one of the most important operation in the refinery to enhance product quality. Further, the total hydrogen requirement in the refinery for various hydrotreaters is produced in the reforming process. Therefore, the reformer contributes to both most important requirements of the refinery.

Q. What is catalytic hydrotreating process and its purpose ?

Ans. Catalytic hydrotreating is a hydrogenation process used to remove about 90% of contaminants such as nitrogen, sulfur, oxygen, and metals from liquid petroleum fractions. These contaminants, if not removed from the petroleum fractions as they travelthrough the refinery processing units, can have detrimental effects on the equipment, the catalysts, and the quality of the finished product.

Typically, hydrotreating is done prior to processes such as catalytic reforming so that the catalyst is not contaminated by untreated feedstock. Hydrotreating is also used prior to catalytic cracking to reduce sulfur & improve product yields, and to upgrade middle distillate petroleum fractions into finished kerosene, diesel fuel, and heating fuel oils. In addition, hydrotreating converts olefins and aromatics to saturated compounds.

Q. Describe the Catalytic Hydrodesulfurization Process.

Ans. Hydrotreating for sulfur removal is called hydrodesulfurization. In a typical catalytic hydrodesulfurization unit, the feedstock is deaerated and mixed with hydrogen, preheated in a fired heater (315-425°C) and then charged under pressure (up to 70 kg/cm^2) through a fixed bed catalytic reactor.

In the reactor, the sulfur & nitrogen compounds in the feedstock are converted into H_2S and NH3.

The reaction products leave the reactor and after cooling to a low temperature enter a liquid/gas separator. The hydrogen rich gas from the high pressure separation is recycled to combine with the feedstock, and the low-pressure gas stream rich in H_2S is sent to a gas treating unit where H_2S is removed. The clean gas is then suitable as fuel for the refinery furnaces. The liquid stream is the product from hydrotreating and is normally sent to a stripping column for removal of H_2S and other undesirable components. In cases where steam is used for stripping, the product is sent to a vacuum drier for removal of water. Hydrodesulfurized products are blended or used as catalytic reforming feedstock.

Q. Describe the features of some Other Hydrotreating Processes.

Ans. Hydrotreating processes differ depending upon the feedstock available and catalysts used. Hydrotreating can be used to improve the burning characteristics of distillates such as kerosene. Hydrotreatment of a kerosene fraction can convert aromatics into naphthenes, which are cleaner-burning compounds.

Lube-oil hydrotreating uses catalytic treatment of the oil with hydrogen to improve product quality. The objectives in mild lube hydrotreating include saturation of olefins and improvements in color, odor & acidic nature of the oil. Mild lube hydrotreating also may be used following solvent processing. Operating temperatures are usually below315° C and operating pressures below 55 kg/cm2. Severe lube hydrotreating, at temperatures in the 315-400° C range and hydrogen pressures up to 210 kg/cm2, is capable of saturating aromatic rings, along with sulfur & nitrogen removal, to impart specific properties not achieved at mild conditions.

Hydrotreating also can be employed to improve the quality of pyrolysis gasoline (pygas), a by-product from the manufacture of ethylene. Traditionally, the outlet for pygas has been for motor gasoline blending, a suitable route in view of its high octane number. However, only small portions can be blended untreated owing to the unacceptable odor, color,& gum forming tendencies of this material. The quality of pygas, which is high in diolefin content, can be satisfactorily improved by hydrotreating, whereby conversion of diolefins into mono olefins provides an acceptable product for motor gas blending

Q. What is the purpose of isomerisation in petroleum refining operations ?

Ans. Isomerization converts n-butane, n-pentane and n-hexane into their respective isoparaffins of substantially higher octane number. The straight chain paraffins are converted to their branched chain counterparts whose component atoms are the same but are arranged in a different geometric structure. Isomerization is important for the conversion of n-butane into isobutane, to provide additional feedstock for alkylation units, and the conversion of normal pentanes & hexanes into higher branched isomers for gasoline blending. Isomerization is similar to catalytic reforming in that the hydrocarbon molecules are rearranged, but unlike catalytic reforming, isomerization just converts normal paraffins to isoparaffins.

Q. Describe the features of various isomerisation processes.

Ans.There are two distinct isomerization processes, butane (C4) and pentane/hexane (C5/C6) isomerisation . Butane isomerization produces feedstock for alkylation. Aluminum chloride catalyst plus hydrogen chloride are universally used for the low-temperature processes.

Platinum or another metal catalyst is used for the higher-temperature processes. In a typical low temperature process, the feed to the isomerization plant is n-butane or mixed butanes mixed with hydrogen (to inhibit olefin formation) and passed to the reactor at 110-170° C and 14-20 kg/cm2. Hydrogen is flashed off in a high pressure separator and the hydrogen chloride removed in a stripper column. The resultant butane mixture is sent to a fractionator (deisobutanizer) to separate n-butane from the isobutane product.

Pentane/hexane isomerization increases the octane number of the light gasoline Components , n-pentane & n-hexane, which are found in abundance in straight run gasoline. In a typical C_5/C_6 isomerization process, dried & desulfurized feedstock is mixed with a small amount of organic chloride & recycled hydrogen, and then heated to reactor temperature. It is then passed over supported metal catalyst in the first reactor where benzene and olefins are hydrogenated. The feed next goes to the isomerization reactor where the paraffins are catalytically isomerized to isoparaffins. The reactor effluent is then cooled and subsequently separated in the product separator into two streams: a liquid product (isomerate) and a recycle hydrogen-gas stream. The isomerate is washed (caustic and water), acid stripped, and stabilized before going to storage.

Q. Looking at the process of the reforming and isomerisation process, what insights you gain towards the evolution?

Ans: Both reforming and isomerisation flow sheets are typical examples of reactor-separator-recycle systems that are extensively studied in process synthesis and evaluation studies. These flow sheets are candidate examples to illustrate the process development required for other non-conventional and novel technologies such as bio-fuel processing etc.

Q. Why is make up $AlCl_3$ required in the isomerisation process?

Ans: A part of the $AlCl_3$ gets converted to the HCl and therefore gets lost. The make up $AlCl_3$ therefore replaces the lost HCl in the process.

Q. What is alkylation in petroleum refinery & its purpose ?

Ans. Alkylation process combines low molecular weight olefins (primarily a mixture of propylene and butylene) with isobutene in the presence of a catalyst, either sulfuric acid or hydrofluoric acid. The product is called alkylate and is composed of a mixture of high octane, branched chain paraffinic hydrocarbons. Alkylate is a premium blending stock for gasoline because it has exceptional antiknock properties and is clean burning. The octane number of the alkylate depends mainly upon the kind of olefins used and upon operating conditions.

Q. Describe the features of Sulfuric Acid Alkylation Process.

Ans. In cascade type sulfuric acid (H_2SO_4) alkylation units, the feedstock (propylene, butylene, amylene, and fresh isobutane) enters the reactor and contacts the oncentrated sulfuric acid catalyst (in concentrations of 85% to 95% for good operation and to minimize corrosion). The reactor is divided into zones, with olefins fed through distributors to each zone, and the sulfuric acid and isobutanes flowing over baffles from zone to zone. The reactor effluent is separated into hydrocarbon and acid phases in a settler, and the acid is returned to the reactor. The hydrocarbon phase is hot-water washed with caustic for pH control before being successively depropanized, deisobutanized, and debutanized. The alkylate obtained from the deisobutanizer can then go directly to motor fuel blending or be rerun to produce aviation grade blending stock. The isobutane is recycled to the feed.

Q. Describe the features of Hydrofluoric Acid Alkylation Process.

Ans. Phillips and UOP are the two common types of hydrofluoric acid alkylation processes in use. In the Phillips process, olefin and isobutane feedstock are dried and fed to a combination reactor/ settler system. Upon leaving the reaction zone, the reactor effluent flows to a settler (separating vessel) where the acid separates from the hydrocarbons. The acid layer at the bottom of the separating vessel is recycled. The top layer of hydrocarbons (hydrocarbon phase), consisting of propane, normal butane, alkylate, and excess (recycle) isobutane, is charged to the main fractionator, the bottom product of which is motor alkylate. The main fractionator overhead, consisting mainly of propane, isobutane, and HF, goes to a depropanizer. Propane with trace amount of HF goes to an HF stripper for HF removal and is then catalytically defluorinated, treated, and sent to storage. Isobutane is withdrawn from the main fractionator and recycled to the reactor/settler, and alkylate from the bottom of the main fractionator is sent to product blending.

The UOP process uses two reactors with separate settlers. Half of the dried feedstock is charged to the first reactor, along with recycle and makeup isobutane. The reactor effluent then goes to its settler, where the acid is recycled and the hydrocarbon charged to the second reactor. The other half of the feedstock also goes to the second reactor, with the settler acid being recycled and the hydrocarbons charged to the main fractionator. Subsequent processing is similar to the Phillips process. Overhead from the main fractionator goes to a depropanizer. Isobutane is recycled to the reaction zone and alkylate is sent to product blending.

Q. Why isobutane is fed to the deisobutanizer unit in the sulphuric acid flow sheet?

Ans: The isobutane feed consists of n-butane and it should be separated so that the alkylation reactor is fed with only isobutane. Now, the issue is to use a separate distillation column for separating isobutane feed into isobutane and butane. This will incur additional costs. Therefore, instead of two separate columns, both separations are carried out in the same distillation column. Of course, costs will be saved by this concept in processing.

Q. What are the advantages of H_2SO_4 process when compared to the HF process?

Ans.

- Additonal equipment is not need in this process. This is not the case in HF process where HF recovery or neutralization requires additional processes.
- HF can form a toxic vapour cloud therefore, safety parameters need to be more effectively considered.
- Higher capital costs in HF process Good quality product is produced in sulphuric acid process but not the HF process.

Q. Why is deisobutanizer placed before the depropanizer?

Ans: Isobutane is used in excess. If propane is removed first then both the columns have high flow rates. Since column costs are proportional to the flow rates, deisobutanizer is placed before depropoanizer so that the depropanizer can process less feed and hence have lower diameters and cost.

Q. What additional units are there in the HF process flow sheet when compared to the sulphuric acid flow sheet?

Ans.

a. Drier

b. Deflourinator

c. Safety instrumentation and ace

Q. What is the purpose of polymerization process in petroleum refinery ?

Ans. Polymerization in the petroleum industry is the process of converting light olefin gases including ethylene, propylene & butylene into hydrocarbons of higher molecular weight and higher octane number that can be used as gasoline blending stocks. Polymerization combines two or more identical olefin molecules to form a single molecule with the same elements in the same proportions as the original molecules. Polymerization may be accomplished thermally or in the presence of a catalyst at lower temperatures.

In the petroleum industry, polymerization is used to indicate the production of gasoline components, hence the term "polymer" gasoline. Furthermore, it is not essential that only one type of monomer be involved. If unlike olefin molecules are combined, the process is referred to as "copolymerization." Polymerization in the true sense of the word is normally prevented, and all attempts are made to terminate the reaction at the dimer or trimer (three monomers joined together) stage. However, in the petrochemical section of a refinery, polymerization, which results in the production of, for instance, polyethylene, is allowed to proceed until materials of the required high molecular weight have been produced.

Q. Describe the features of polymerization process.

Ans. The olefin feedstock is pretreated to remove sulfur and other undesirable compounds.In the catalytic process , the feedstock is either passed over a solid phosphoric acid catalyst or comes in contact with liquid phosphoric acid, where an exothermic polymeric reaction occurs. This reaction requires cooling water and the injection of cold feedstock into the reactor to control temperatures between 150 and 230° C at pressures from 15 to 85 kg/cm^2.The reaction products leaving the reactor are sent to stabilization and/or fractionator systems to separate saturated and unreacted gases from the polymer gasoline product.

Q. What modification you can think in the polymerization reactor process instead of using a shell and tube design?

Ans: If we analyze the shell and tube design, we want to achieve simultaneous cooling of the olefin stream in the tube. Instead of cooling, we can also try cold shot technique or diluent in the stream. When diluents is used, then the diluents or inert carrier needs to be separated and its an additional task in the process. Therefore, the best possibility is to target either a cold shot feed or a cold shot product with recycle. Since conversions are low, doing cold shot with feed is not beneficial. Therefore, cold shot product will be the optimal arrangement to replace the shell and tube design. In that case, the product should be cooled and send back to the reactor.

Q. Can optionally propane be fed to the polymerization reactor?

Ans: Few process flowsheets involve propane as well fed to the polymerization reactor, but conversions are expected to be significantly low for propane. Therefore, as such much gain will not be there by feeding propane to the polymerization reactor.

Q. What is the purpose of sweetening and treating processes in a petroleum refinery ?.

Ans. Treating is a means by which contaminants such as organic compounds containing sulfur, nitrogen & oxygen; dissolved metals & inorganic salts; and soluble salts dissolved in emulsified water are removed from petroleum fractions or streams. Petroleum refiners have a choice of several different treating processes, but the primary purpose of the majority of them is the elimination of unwanted sulfur compounds. A variety of intermediate and finished products, including middle distillates, gasoline, kerosene, jet fuel & sour gases are dried and sweetened. Sweetening, a major refinery treatment of gasoline, treats sulfur compounds (hydrogen sulfide, thiophene & mercaptan) to improve color, odour, and oxidation stability. Sweetening also reduces concentrations of carbon dioxide.

Treating can be accomplished at an intermediate stage in the refining process, or just before sending the finished product to storage. Choices of a treating method depend on the nature of the petroleum fractions, amount & type of impurities in the fractions to be treated, the extent to which the process removes the impurities, and end product specifications. Treating materials include acids, solvents, alkalis, oxidizing, and adsorption agents.

Q. Describe the process of Acid, Caustic, or Clay Treating.

Ans. Sulfuric acid is the most commonly used acid treating process.Sulfuric acid treating results in partial or complete removal of unsaturated hydrocarbons, sulfur, nitrogen,& oxygen compounds, and resinous & asphaltic compounds. It is used to improve the odor, color, stability, carbon residue, and other properties of the oil. Clay/lime treatment of acid-refined oil removes traces of asphaltic materials and other compounds improving product color, odor, and stability. Caustic treating with sodium (or potassium) hydroxide is used to improve odor and color by removing organic acids (naphthenic acids, phenols) and sulfur compounds (mercaptans, H2S) by a caustic wash. By combining caustic soda solution with various solubility promoters (e.g., methyl alcohol and cresols), up to 99% of all mercaptans as well as oxygen and nitrogen compounds can be dissolved from petroleum fractions.

Q. Describe the process of Drying and Sweetening.

Ans. Feedstocks from various refinery units are sent to gas treating plants where butanes and butenes are removed for use as alkylation feedstock, heavier components are sent to gasoline blending, propane is recovered for LPG, and propylene is removed for use in petrochemicals. Some mercaptans are removed by water-soluble chemicals that react with the mercaptans. Caustic liquid (sodium hydroxide), amine compounds (diethanolamine) or fixed-bed catalyst sweetening also may be used. Drying is accomplished by the use of water absorption or adsorption agents to remove water from the products. Some processes simultaneously dry and sweeten by adsorption on molecular sieves.

Q. How Sulfur Recovery is done in refineries ?.

Ans. Sulfur recovery unit converts hydrogen sulfide in sour gases and hydrocarbon streams to elemental sulfur. The most widely used recovery system is the Claus process, which uses both thermal and catalytic conversion reactions. A typical process produces elemental sulfur by burning hydrogen sulfide under controlled conditions. Knockout pots are used to remove water and hydrocarbons from feed gas streams. The gases are then exposed to a catalyst to recover additional sulfur. Sulfur vapor from burning & conversion is condensed and recovered.

Q. Explain how using three reactors in series will enhance equilibrium conversion of the second conversion reaction i.e, H_2S to S in Claus process?

Ans: Equilibrium conversion is a function of temperature and pressure. There are two types of equilibrium reactions namely exothermic and endothermic reactions. For exothermic reactions, the conversion increases with decreasing temperature. Therefore, after first converter, the product is cooled and sulphur is also removed. The reduction in temperature and elimination of sulphur moves the process conditions to a shift equilibrium to the right hand side. Therefore, series of chemical reactors followed with product cooling will enhance equilibrium conversion. By the by, the same concept is also used in hot and cold shot reactors. The only difference between this arrangement and the shot reactor arrangement is that there direct heat transfer by quenching or heating is allowed where as in the shown flow sheet, indirect heat transfer is facilitated only.

Q. How Hydrogen Sulfide Scrubbing is done ?

Ans. Hydrogen sulfide scrubbing is a common treating process in which the hydrocarbon feedstock is first scrubbed to prevent catalyst poisoning. Depending on the feedstock and the nature of contaminants, desulfurization methods vary from ambient temperature activated charcoal adsorption to high temperature catalytic hydrogenation followed by zinc oxide treating.

Q. Describe the purpose of solvent extraction and dewaxing process used in petroleum refineries .

Ans. Solvent treating is a widely used method of refining lubricating oils as well as a host of other refinery stocks. Since distillation (fractionation) separates petroleum products into groups only by their boiling-point ranges,impurities may remain.These include organic compounds containing sulfur,nitrogen & oxygen; inorganic salts & dissolved metals; and soluble salts that were present in the crude feedstock. In addition, kerosene & distillates may have trace amounts of aromatics & naphthenes, and lubricating oil base-stocks may contain wax. Solvent refining processes like solvent extraction and solvent dewaxing usually remove these undesirables at intermediate refining stages or just before sending the product to storage.

Solvent Extraction

The purpose of solvent extraction is to prevent corrosion, protect catalyst in subsequent processes, and improve finished products by removing unsaturated, aromatic hydrocarbons from lubricant and grease stocks. The solvent extraction process separates aromatics, naphthenes, and impurities from the product stream by dissolving or precipitation. The feedstock is first dried and then treated using a continuous countercurrent solvent treatment operation. In one type of process, the feedstock is washed with a liquid in which the substances to be removed are more soluble than in the desired resultant product. In another process, selected solvents are added to cause impurities to precipitate out of the product. In the adsorption process, highly porous solid materials collect liquid molecules on their surfaces.

The solvent is separated from the product stream by heating, evaporation, or fractionation, and residual trace amounts are subsequently removed from the raffinate by steam stripping or vacuum flashing. Electric precipitation may be used for separation of inorganic compounds. The solvent is then regenerated to be used again in the process.

The most widely used extraction solvents are phenol, furfural, and cresylic acid. Other solvents less frequently used are liquid sulfur dioxide, nitrobenzene, and 2,2'-dichloroethyl ether. The selection of specific processes and chemical agents depends on the nature of the feedstock being treated, the contaminants present, and the finished product requirements.

Solvent dewaxing

It is used to remove wax from either distillate or residual basestocks at any stage in the refining process. There are several processes in use for solvent dewaxing, but all have the same general steps, which are: mixing the feedstock with a solvent, precipitating the wax from the mixture by chilling, and recovering the solvent from the wax and dewaxed oil for recycling by distillation and steam stripping. Usually two solvents are used: toluene, which dissolves the oil & maintains fluidity at low temperatures, and methyl ethyl ketone (MEK), which dissolves little wax at low temperatures and acts as a wax precipitating agent. Other solvents that are sometimes used include benzene, methyl isobutyl ketone, propane, petroleum naphtha, ethylene dichloride, methylene chloride, and sulfur dioxide. In addition, there is a catalytic process used as an alternate to solvent dewaxing.

Q. Describe the process of asphalt production in a refinery ?

Ans. Asphalt is a portion of the residual fraction that remains after primary distillation operations. It is further processed to impart characteristics required by its final use. In vacuum distillation, generally used to produce road tar asphalt, the residual is heated to about 400 °C and charged to a column where vacuum is applied to prevent cracking.

Asphalt for roofing materials is produced by air blowing. Residual is heated in a pipe still almost to its flash point and charged to a blowing tower where hot air is injected for a predetermined time. The dehydrogenization of the asphalt forms hydrogen sulfide, and the oxidation creates sulfur dioxide. Steam, used to blanket the top of the tower to entrain the various contaminants, is then passed through a scrubber to condense the hydrocarbons.

Another process used to produce asphalt is solvent deasphalting. In this extraction process, which uses propane (or hexane) as a solvent, heavy oil fractions are separated to produce heavy lubricating oil, catalytic cracking feedstock, and asphalt. Feedstock and liquid propane are pumped to an extraction tower at precisely controlled mixtures, temperatures (65-120 °C), and pressures of 25-40 kg/cm^2. Separation occurs in a rotating disc contactor, based on differences in solubility.

The products are then evaporated and steam stripped to recover the propane, which is recycled. Deasphalting also removes some sulfur & nitrogen compounds, metals, carbon residues, and paraffins from the feedstock.

Q. Describe the lubricant, wax, & grease manufacturing processes in a refinery .

Ans. Lubricating oils and waxes are refined from the residual fractions of atmospheric and vacuum distillation. The primary objective of the various lubricating oil refinery processes is to remove asphalts, sulfonated aromatics, and paraffinic & isoparaffinic waxes from residual fractions. Reduced crude from the vacuum unit is deasphalted and combined with straight run lubricating oil feedstock, preheated, and solvent extracted (usually with phenol or furfural) to produce raffinate.

Wax Manufacturing Process. Raffinate from the extraction unit contains a considerable amount of wax that must be removed by solvent extraction and crystallization. The raffinate is mixed with a solvent (propane) and precooled in heat exchangers. The crystallization temperature is attained by the evaporation of propane in the chiller & filter feed tanks. The wax is continuously removed by filters and cold solvent washed to recover retained oil. The solvent is recovered from the oil by flashing and steam stripping. The wax is then heated with hot solvent, chilled, filtered, and given a final wash to remove all oil.

Lubricating Oil Manufacturing Process. The dewaxed raffinate is blended with other distillate fractions and further treated for viscosity index, color, stability, carbon residue, sulfur, additive response, and oxidation stability in extremely selective extraction processes using solvents (furfural, phenol, etc.). In a typical phenol unit, the raffinate is mixed with phenol in the treating section at temperatures below 205°C. Phenol is then separated from the treated oil and recycled. The treated lube oil base stocks are then mixed and/or compounded with additives to meet the required physical & chemical characteristics of motor oils, industrial lubricants, and metal working oils.

Grease Compounding. Grease is made by blending metallic soaps (salts of long chained fatty acids) and additives into a lubricating oil medium at temperatures of 205-315° C. Grease may be either batch produced or continuously compounded. The characteristics of the grease depend to a great extent on the metallic element (calcium, sodium, aluminum, lithium, etc.) in the soap and the additives used.

Q. What is the purpose of blending ?

Ans. Blending is the physical mixture of a number of different liquid hydrocarbons to produce a finished product with certain desired characteristics. Products can be blended in-line through a manifold system, or batch blended in tanks and vessels. In-line blending of gasoline, distillates, jet fuel, and kerosene is accomplished by injecting proportionate amounts of each component into the main stream where turbulence promotes thorough mixing. Additives including octane enhancers, metal deactivators, anti-oxidants, anti-knock agents, gum & rust inhibitors, detergents, etc. are added during and/or after blending to provide specific properties not inherent in hydrocarbons.

Q. What are the common refinery chemicals ?

Ans. Most common refinery chemicals are as follows :

Leaded Gasoline Additives. Tetraethyl lead (TEL) and tetramethyl lead (TML)are additives formerly used to improve gasoline octane ratings but are no longer in common use except in aviation gasoline.

Oxygenates. Ethyl tertiary butyl ether (ETBE), methyl tertiary butyl ether (MTBE), tertiary amyl methyl ether (TAME), and other oxygenates improve gasoline octane ratings and reduce carbon monoxide emissions.

Caustics. Caustics are added to desalting water to neutralize acids and reduce corrosion. They are also added to desalted crude in order to reduce the amount of corrosive chlorides in the tower overheads. They are used in some refinery treating processes to remove contaminants from hydrocarbon streams.

Sulfuric Acid and Hydrofluoric Acid. Sulfuric acid and hydrofluoric acid are used primarily as catalysts in alkylation processes. Sulfuric acid is also used in some treatment processes.

Q. What is the function of unsaturated gas plants in petroleum refineries ?

Ans. Unsaturated gas plants recover light hydrocarbons (C3 and C4 olefins) from wet gas streams from the FCC, TCC, and delayed coker overhead accumulators or fractionation receivers. In a typical unsaturated gas plant, the gases are compressed and treated with amine to remove hydrogen sulfide either before or after they are sent to a fractionating absorber where they are mixed into a concurrent flow of debutanized gasoline. The light fractions are separated by heat in a reboiler, the offgas is sent to a sponge absorber, and the bottoms are sent to a debutanizer. A portion of the debutanized hydrocarbon is recycled, with the balance sent to the splitter for separation. The overhead gases go to a depropanizer for use as alkylation unit feedstock.

Q. What is the function of amine plants in refineries ?

Ans. Amine plants remove acid contaminants from sour gas and hydrocarbon streams. In amine plants, gas & liquid hydrocarbon streams containing carbon dioxide and/or hydrogen sulfide are charged to a gas absorption tower or liquid contactor where the acid contaminants are absorbed by counterflowing amine solutions (i.e., MEA, DEA, MDEA). The stripped gas or liquid is removed overhead, and the amine is sent to a regenerator. In the regenerator, the acidic components are stripped by heat & reboiling action and disposed of, and the amine is recycled.

Q. What is the purpose of saturate gas plants in a petroleum refinery ?

Ans. Saturate (sat) gas plants separate refinery gas components including butanes for alkylation, pentanes for gasoline blending, LPG's for fuel, and ethane for petrochemicals. Because sat gas processes depend on the feedstock and product demand, each refinery uses different systems, usually absorption-fractionation or straight fractionation. In absorption-fractionation, gases and liquids from various refinery units are fed to an absorber-deethanizer

where C2 and lighter fractions are separated from heavier fractions by lean oil absorption and removed for use as fuel gas or petrochemical feed. The heavier fractions are stripped and sent to a debutanizer, and the lean oil is recycled back to the absorber-deethanizer. C3/C4 is separated from pentanes in the debutanizer, scrubbed to remove hydrogen sulfide, and fed to a splitter where propane and butane are separated. In fractionation sat gas plants, the absorption stage is eliminated.

Q. Describe the process of hydrogen production in a refinery to meet its requirements in various operations .

Ans. High-purity hydrogen (95%-99%) is required in a refinery for hydrodesulfurization, hydrogenation, hydrocracking, and petrochemical processes. Hydrogen, produced as a by-product of refinery processes (principally hydrogen recovery from catalytic reformer product gases), often is not enough to meet the total refinery requirements, necessitating the manufacturing of additional hydrogen or obtaining supply from external sources.

In steam methane reforming,desulfurized gases are mixed with superheated steam(590-870° C) and reformed in tubes containing a nickel base catalyst. The reformed gas, which consists of steam, hydrogen, carbon monoxide, and carbon dioxide,is cooled and passed through converters containing an iron catalyst where the carbon monoxide reacts with steam to form carbon dioxide and more hydrogen. The carbon dioxide is removed by amine washing. Any remaining carbon monoxide in the product stream is converted to methane.

Steam-naphtha reforming is a continuous process for the production of hydrogen from liquid hydrocarbons and is, similar to steam methane reforming. A variety of naphthas in the gasoline boiling range may be employed, including fuel containing up to 35% aromatics. Following pretreatment to remove sulfur compounds, the feedstock is mixed with steam and taken to the reforming furnace (675-815° C) where hydrogen is produced.

Q. How good is conversion in the steam methane reforming reaction?

Ans: By maintaining good process conditions, the conversion of the methane steam reforming reaction can be achieved to about 60 – 80 % in a single furnace reactor. Therefore, a single furnace reactor is sufficient to achieve the desired conversion.

Q. Why methanation cannot be avoided in the hydrogen production process?

Ans: The trace gases CO and CO_2 are reactive towards the catalysts due to the presence of oxygen molecule. Therefore, H2 sent to various other units should be bereft of the CO and CO_2 which can poison the catalyst due to the reactive oxygen molecule.

Q. Why CO_2 bulk removal is targeted in the hydrogen production process?

Ans: CO_2 bulk production if avoided will affect the product quality. The methanation catalyst is not competent to fully convert CO_2 back to methane due to the reversibility of the reaction. Therefore, CO_2 removal is required.

Q. What oppurtunites exist for heat integration in the hydrogen production process?

Ans: The feed to the process needs pre-heating therefore, is a cold steam. Now this stream can be integrated with the furnace itself where the combustion gases can be used to pre-heat the stream. Other than this the furnace product, high and low temperature shift reactor products are also available. The choice of heat integration and steam generation at various sites in the process is a complex optimization problem and has to be addressed by solving separate problems to finally arrive at the one optimal solution. Mostly, the furnace combustion gases will be integrated for energy recovery with the feed stream and steam generation is effected from the high and low temperature shift reactors.

Q. How various heating & cooling operations are carried out in refineries ?

Ans. Heating Operations. Process heaters and heat exchangers preheat feedstock in distillation towers and in refinery processes to reaction temperatures. Heat exchangers use either steam or hot hydrocarbon transferred from some other section of the process for heat input. The heaters are usually designed for specific process operations, and most are of cylindrical vertical or box-type designs. The major portion of heat provided to process units comes from fired heaters fueled by refinery or natural gas, distillate, and residual oils.Fired heaters are found on crude and reformer preheaters, coker heaters, and large column reboilers.

Cooling Operations. Heat also may be removed from some processes by air & water exchangers, fin fans, gas & liquid coolers, and overhead condensers, or by transferring heat to other systems. The basic mechanical vapor-compression refrigeration system, which may serve one or more process units, includes an evaporator, compressor, condenser, controls, and piping. Common coolants are water, alcohol/water mixtures, or various glycol solutions.

Q. How steam is generated in petroleum refineries to meet its process needs ?

Ans. Steam is generated in main generation plants, and/or at various process units using heat from flue gas or other sources. Heaters (furnaces) include burners and a combustion air system, the boiler enclosure in which heat transfer takes place, a draft or pressure system to remove flue gas from the furnace, soot blowers, and compressed air systems that seal openings to prevent the escape of flue gas. Boilers consist of a number of tubes that carry the water-steam mixture through the furnace for maximum heat transfer. These tubes run between steam distribution drums at the top of the boiler and water collecting drums at the bottom of the boiler. Steam flows from the steam drum to the superheater before entering the steam distribution system.

Heaters may use any one or combination of fuels including refinery gas, natural gas, fuel oil, and powdered coal. Refinery off-gas is collected from process units and combined with LPG in a fuel gas balance drum which provides constant system pressure, fairly stable calorific value fuel, and automatic separation of suspended liquids in gas vapors, and it prevents carryover of large slugs of condensate into the distribution system.

Fuel oil is typically a mix of refinery crude oil with straight run and cracked residues and other products. The fuel oil system delivers fuel to process unit heaters and steam generators at required temperatures and pressures. The fuel oil is heated to pumping temperature, sucked through a coarse suction strainer, pumped to a temperature control heater, and then pumped through a fine mesh strainer before being burned.

In one example of process unit heat generation, carbon monoxide boilers recover heat in catalytic cracking units as carbon monoxide in flue gas is burned to complete combustion. In other processes, waste heat recovery units use heat from the flue gas to make steam.

Q. How steam is distributed in refineries to meet its process needs ?

Ans. The steam distribution system consists of valves, fittings, piping, and connections suitable for the pressure of the steam transported. Steam leaves the boilers at the highest pressure required by the process units or electrical power generation. The steam pressure is then reduced in turbines that drive process pumps and compressors. Most steam used in the refinery is condensed to water in various types of heat exchangers. The condensate is reused as boiler feedwater or discharged to wastewater treatment. When refinery steam is also used to drive steam turbine generators to produce electricity, the steam must be produced at much higher pressure than required for process steam. Steam typically is generated by heaters (furnaces) and boilers combined in one unit.

Q. Describe the process of production of boiler feedwater.

Ans. Water used in steam generation must be free of contaminants including minerals and dissolved impurities that can damage the system or affect its operation. Suspended materials such as silt, sewage, & oil, which form scale and sludge, must be coagulated or filtered out of the water. Dissolved gases, particularly carbon dioxide & oxygen, cause boiler corrosion and are removed by deaeration and treatment. Dissolved minerals including metallic salts, calcium, carbonates, etc., that cause scale, corrosion, and turbine blade deposits are treated with lime or soda ash to precipitate them from the water. Recirculated cooling water must also be treated for hydrocarbons and other contaminants.

Depending on the characteristics of raw boiler feedwater, some or all of the following six stages of treatment are applicable:

- Clarification;
- Sedimentation;
- Filtration;
- Ion exchange;
- Deaeration; and
- Internal treatment.

Q. Describe the process of wastewater treatment in petroleum refinery .

Ans. Wastewater treatment is used for process, runoff, and sewerage water prior to discharge or recycling. Wastewater typically contains hydrocarbons, dissolved materials, suspended solids, phenols, ammonia, sulfides, and other compounds. Wastewater includes condensed steam, stripping water, spent caustic solutions, cooling tower and boiler blowdown, wash water,alkaline & acid waste neutralization water, and other process associated water.

Pretreatment Operations. Pretreatment is the separation of hydrocarbons and solids from wastewater. API separators, interceptor plates, and settling ponds remove suspended hydrocarbons, oily sludge,& solids by gravity separation, skimming, and filtration. Some oil-in-water emulsions must be heated to assist in separating the oil and water. Gravity separation depends on the specific gravity differences between water and immiscible oil globules and allows free oil to be skimmed off the surface of the wastewater. Acidic wastewater is neutralized using ammonia, lime, or soda ash. Alkaline wastewater is treated with sulfuric acid, hydrochloric acid, carbon dioxide rich flue gas, or sulfur.

Secondary Treatment Operations. After pretreatment, suspended solids are removed by sedimentation or air flotation. Wastewater with low levels of solids may be screened or filtered. Flocculation agents are sometimes added to help separation. Secondary treatment processes biologically degrade and oxidize soluble organic matter by the use of activated sludge, unaerated or aerated lagoons, trickling filter methods, or anaerobic treatments. Materials with high adsorption characteristics are used in fixed bed filters or added to the wastewater to form a slurry which is removed by sedimentation or filtration. Additional treatment methods are used to remove oils and chemicals from wastewater. Stripping is used on wastewater containing sulfides and/or ammonia, and solvent extraction is used to remove phenols.

Tertiary Treatment Operations. Tertiary treatments remove specific pollutants to meet regulatory discharge requirements. These treatments include chlorination, ozonation, ion exchange, reverse osmosis, activated carbon adsorption, etc. Compressed oxygen is diffused into wastewater streams to oxidize certain chemicals or to satisfy regulatory oxygen content requirements. Wastewater that is to be recycled may require cooling to remove heat and/or oxidation by spraying or air stripping to remove any remaining phenols, nitrates, and ammonia.

Q. Discuss the role of COOLING TOWERS in a refinery operation.

Ans. Cooling towers remove heat from process water by evaporation and latent heat transfer between hot water & air. The two types of cooling towers are crossflow and counterflow. Crossflow towers introduce the airflow at right angles to the water flow throughout the structure. In counterflow cooling towers, hot process water is pumped to the uppermost plenum and allowed to fall through the tower. Numerous slats or spray nozzles located throughout the length of the tower disperse the water and help in cooling. Air enters at the tower bottom and flows upward against the water. When the fans or blowers are at the air inlet, the air is considered to be forced draft. Induced draft is when the fans are at the air outlet.

Recirculated cooling water must be treated to remove impurities and dissolved hydrocarbons. Because the water is saturated with oxygen from being cooled with air, the chances of corrosion are increased. One means of corrosion prevention is the addition of a material to the cooling water that forms a protective film on pipes and other metal surfaces.

Q. How the electrical power requirements of a petroleum refinery is met ?

Ans. Petroleum Refineries may receive electricity from outside sources or produce their own power with generators driven by steam turbines or gas engines. Electrical substations receive power from the utility or power plant for distribution throughout the facility. They are usually located in nonclassified areas, away from sources of vapor or cooling tower water spray. Transformers, circuit breakers, and feed circuit switches are usually located in substations. Substations feed power to distribution stations within the process unit areas. Distribution stations can be located in classified areas, provided that classification requirements are met. Distribution stations usually have a liquid filled transformer and an oil filled or air break disconnect device.

Q. What types of gas & air compressors are used in petroleum refineries ?

Ans. Both reciprocating and centrifugal compressors are used throughout the refinery for gas & compressed air. Air compressor systems include compressors, coolers, air receivers, air dryers, controls, and distribution piping. Blowers are used to provide air to certain processes. Plant air is provided for the operation of air powered tools, catalyst regeneration, process heaters, steam-air decoking, sour water oxidation, gasoline sweetening, asphalt blowing, and other uses. Instrument air is provided for use in pneumatic instruments & controls, air motors and purge connections.

Q. What equipments are used in refineries for loading , unloading & storage of petroleum products ?

Ans. Facilities for loading liquid hydrocarbons into tank cars, tank trucks, and marine vessels & barges are usually part of the refinery operations. Product characteristics, distribution needs, shipping requirements, and operating criteria are important when designing loading & unloading facilities. Tank trucks and rail tank cars are either top or bottom loaded, and vapor recovery systems may be provided where required. Loading and unloading liquefied petroleum gas (LPG) require special considerations in addition to those for liquid hydrocarbons.

Atmospheric storage tanks and pressure storage tanks are used throughout the refinery for storage of crudes, intermediate hydrocarbons (during the process), and finished products. Tanks are also provided for fire water, process and treatment water, acids, additives, and other chemicals. The type, construction, capacity and location of tanks depends on their use and materials stored.

Q. What types of pumps, piping and valves are used in petroleum refineries ?

Ans. Centrifugal and reciprocating pumps are used to move hydrocarbons, process water, fire water, and wastewater through piping within the refinery. Pumps are driven by electric motors, steam turbines, or internal combustion engines. The pump type, capacity, and construction materials depend on the service for which it is used.

Process and utility piping distribute hydrocarbons, steam, water, and other products throughout the facility. Their size and construction depend on the type of service, pressure, temperature, and nature of the products. Vent, drain, and sample connections are provided on piping, as well as provisions for blanking.

Different types of valves are used depending on their operating purpose. These include gate valves, bypass valves, globe & ball valves, plug valves, block & bleed valves, and check valves. Valves can be manually or automatically operated.

Q. Discuss the general criteria for selecting a fuel .

Ans. Different types of fuels, (solid, liquid & gaseous, are available for firing in domestic ovens,automobiles and industrial boilers & furnaces. General criteria for selecting a fuel include the following:

Availability: Availability is one of the important criteria in the selection of a fuel. It is desirable to select a fuel that is indigenous and is locally available.

Ease of Storage and handling: It is desirable to select a fuel that is easy to handle; and, for storage, adequate space is required so that the fuel can be stored safely.

Cost of fuel: Cost is a key deciding factor for any fuel.

Fuel properties: The constituent properties of fuels are important from the point of view of storage, handling, preparation, and combustion equipment performance.

Calorific value: The calorific value of a fuel is the measure of energy content; reported as either gross calorific value (GCV) or net calorific value (NCV), units being kCal/kg . The difference between the two calorific values is the latent heat of condensation of the water vapor produced during the combustion process. Efficiency is reported both ways (GCV as well as NCV basis). It is important to note that the efficiency value reported with NCV basis would be higher than the value reported with GCV basis.

Sulphur content in fuel: The presence of sulphur in fuel is related to cold end corrosion in cool parts of the chimney/stack, air pre-heating equipments , and economic factors as the sulphur content determines the minimum stack temperature for avoiding acid dewpoint corrosion in a stack. This means higher losses from the stack when sulphur content in the fuel is higher. Besides, it causes atmospheric air pollution after combustion.

Analysis (composition) of fuel: Two types of analysis, proximate and ultimate analysis, are used. Ultimate analysis is at the elemental level, and includes carbon, hydrogen, nitrogen, sulphur, etc, and is often useful in design, while proximate analysis gives the broader composition of carbon, volatile matter, moisture, and ash.

Q. Describe various factors for safe storage of fuel oil .

Ans. Furnace oil, that is delivered, may contain dust, water and other contaminants.

The sizing of storage tank facility is very important. A recommended storage estimate is to provide for at least 10 days of normal consumption. Industrial heating fuel storage tanks are generally vertical mild steel tanks mounted above ground. It is prudent for safety and environmental reasons to build bund walls around tanks to contain accidental spillages.

As a certain amount of settlement of solids & sludge will occur in tanks over time, cleaning should be carried out at regular intervals-annually for heavy fuels and every two years for light fuels. A little care should be taken when oil is decanted from the tanker to storage tank. All leaks from joints, flanges and pipelines must be liquidated . Fuel oil should be free from possible contaminants such as dirt, sludge and water before it is fed to the combustion system.

Furnace oil arrives at the factory site either in tank lorries by road or by rail. Oil is then decanted into the main storage tank. To prevent contaminants such as rags, cotton waste, loose nuts or bolts or screws entering the system and damaging the pump, coarse strainer of 10 mesh size (not more than 3 holes per linear inch) is positioned on the entry pipe to the storage tanks.

Progressively finer strainers should be provided at various points in the oil supply system to filter away finer contaminants such as external dust and dirt, sludge or free carbon. It is advisable to provide these filters in duplicate to enable one filter to be cleaned while oil supply is maintained through the other.

Various factors for safe storage of fuel oil are listed below :

- It is hazardous to store furnace oil in barrels.
- It is preferable to store oil in cylindrical tanks either above or below the ground.
- Storage capacity should be at least 10 days of normal consumption.
- Bund walls should be built around tanks.
- Periodical cleaning of tanks is necessary: annually for heavy fuels and every two years for light fuels.
- Leaks from joints, flanges, and pipelines must be attended to immediately, and should be avoided at all costs.

— Loss of even one drop of oil every second can cost over 4000 liters a year.

— Fuel oil should be free from contaminants such as dirt, sludge, and water before it is fed to a combustion system.

Q. What are various pumping choices for fuel oil ?

Ans. Heavy fuel oils are best pumped using positive displacement pumps, as they are able to get fuel moving when it is cold. A circulation gear pump running on LDO should give between 7000-10000 hours of service. Diaphragm pumps have a shorter service life, but are easier and less expensive to repair. A centrifugal pump is not recommended, because as the oil viscosity increases, the efficiency of pump drops sharply and the horsepower required increases. Light fuels are best pumped with centrifugal or turbine pumps. When higher pressures are required, piston or diaphragm pumps should be used.

The viscosity of furnace oil and LSHS increases with decrease in temperature, which makes it difficult to pump the oil. At low ambient temperatures (below 25o C), furnace oil is not easily pumpable. To circumvent this, preheating of oil is accomplished in two ways:

a) the entire tank may be preheated. In this form of bulk heating, steam coils are placed at the bottom of the tank, which is fully insulated;

b) the oil can be heated as it flows out with an outflow heater. To reduce steam requirements, it is advisable to insulate tanks where bulk heating is used.

Bulk heating may be necessary if flow rates are high enough to make outflow heaters of adequate capacity impractical, or when a fuel such as Low Sulphur Heavy Stock (LSHS) is used. In the case of outflow heating, only the oil, which leaves the tank, is heated to the pumping temperature. The outflow heater is essentially a heat exchanger with steam or electricity as the heating medium.

Temperature control of the oil is necessary to prevent overheating, especially when oil flow is reduced or stopped. This is particularly important for electric heaters, since oil may get carbonized when there is no flow and the heater is on. Thermostats should be provided at a region where the oil flows freely into the suction pipe. The temperature at which oil can readily be pumped depends on the grade of oil being handled. Oil should never be stored at a temperature above that necessary for pumping as this leads to higher energy consumption.

It is advisable to use a positive displacement pump such as a gear pump, for pumping furnace oil. Sometimes no oil gets transferred through the pipes because of excessive pressure drop and cavitation at the pump. A centrifugal pump is therefore, not recommended, as the oil viscosity increases, the efficiency of the pump drops sharply and the horse power required increases.

It is also necessary to size the pipelines properly so that there is no excessive pressure drop and the required oil flow rates are obtained. On expansion of facilities, the adequacy of pipe sizes for transporting larger quantities of oil should be reviewed. It is preferable to lay the oil pipelines over ground as any oil leaks would then be visible and could be quickly attended to.

Fuel oil has to be pumped from the main storage tank to the day service tank. Depending upon the desired quantity of oil to be pumped, and the distance between the storage tank and the service tanks, adequate pumping capacity have to be provided.

The following are good pumping choices for fuel oils:

- Positive displacement pumps are best for heavy fuel oils.
- Gear pumps are best for light diesel oils.
- Diaphragm pumps have a shorter service life, but are easier and less expensive to repair.

- Light fuel oils are best pumped with centrifugal or turbine pumps. When higher pressures are required, piston or diaphragm pumps should be used.

Q. What are different important properties/Parameters in liquid fuel combustion .

***Ans.* Various important properties are discussed below.**

Density

Density is the ratio of mass of fuel to the volume of fuel. Density is measured by an instrument called hydrometer. The knowledge of density is useful for quantity calculations and assessing ignition quality.Reference temperature is typically 15°C, and units are (kg/m^3.)

Specific gravity

The density of liquid fuel, relative to water, is called its specific gravity. The specific gravity of water is defined as 1. Since specific gravity is a ratio, it has no unitsSpecific gravity is the ratio of weight of a given volume of oil to the weight of the same volume of water at a given temperature. The measurement of specific gravity is generally made by a hydrometer. Higher specific gravity normally means higher heating value . Specific gravity is used in calculations involving weights and volumes.

Sp.gr. of some liquid fuels are as follows ;

- Light oil = 0.85–0.87
- Furnace oil = 0.89–0.95
- L.S.H.S. = 0.88–0.98

Viscosity

Viscosity is internal resistance to flow of fluid. Viscosity depends on temperature and decreases as the temperature increases. Any numerical value for viscosity has no meaning unless the temperature is also specified. Viscosity is measured in Stokes / Centistokes. Sometimes viscosity is also quoted in Engler, Saybolt or Redwood. Each type of oil has its own temperature - viscosity relationship. The measurement of viscosity is made with an instrument called Viscometer.

Viscosity is the most important characteristic in the storage and use of fuel oil. It influences the degree of pre-heat required for handling, storage and satisfactory atomization. If the oil is too viscous, it may become difficult to pump, hard to light the burner, and tough to operate. Poor atomization may result in the formation of carbon deposits on the burner tips or on the furnace walls. Therefore pre-heating is necessary for proper atomization.

Specific heat

The specific heat of a fuel oil is the amount of energy (kcals) needed to raise the temperature of 1 kg of oil by 1°C. The unit of specific heat is kcal/kg .°C and denotes heat capacity, and is useful in designing of heating systems. It varies from 0.22 to 0.28 depending on the oil-specific gravity. It also helps to quantify how much steam or electrical energy is required for preheating.

Flash point

Flash point is the lowest temperature at which the fuel vapors released flash momentarily when an open flame is passed over them. The flash point for furnace oil is 66°C. It affects safe storage conditions.

Pour point

Pour point is the lowest temperature at which fuel oil can be poured/can flow under prescribed conditions. It is a very rough indication of the lowest temperature at which fuel oil can be readily pumped.

Water content

Water content of fuel oil is normally very low as the product is handled hot at the refinery site and a maximum limit of 1% is specified in the standard, but water may be present in free or emulsified form. Water can cause damage to the inside furnace surfaces during combustion, especially if it contains dissolved salts. It can also cause spluttering of the flame at the burner tip, possibly extinguishing the flame and reducing the flame temperature or lengthening the flame.

Calorific Value

The calorific value of fuel is the measurement of heat produced, and is measured either as gross calorific value or net calorific value. The difference being the latent heat of condensation of the water vapour produced during the combustion process. Gross calorific value (GCV) assumes all vapour produced during the combustion process is fully condensed. Net calorific value (NCV) assumes the water leaves with the combustion products without fully being condensed. Fuels should be compared based on the net calorific value.

The calorific value of coal varies considerably depending on the ash, moisture content and the type of coal while calorific value of fuel oils are much more consistent. The typical Gross Calorific Values of some of the commonly used liquid fuels are given below:

Liquid Fuel		GCV (kCal/kg)
Kerosene	-	11,100
Diesel Oil	-	10,800
L.D.O	-	10,700
Furnace Oil	-	10,500
LSHS	-	10,600

Sulphur

The amount of sulphur in the fuel oil depends mainly on the source of the crude oil and to a lesser extent on the refining process. The normal sulfur content for the residual furnace oil is of the order of 2-4 %.

Typical figures of Sulphur % in some liquid fuels are:

Liquid Fuel		**% of Sulphur**
Kerosene	-	0.05—0.2
Diesel Oil	-	0.05 – 0.25

L.D.O(Light Diesel Oil) - 0.5 – 1.8

Furnace Oil - 2.0 – 4.0

LSHS(Low Sulphur Heavy Stock) < 0.5

The main disadvantage of sulphur is the risk of corrosion by sulphuric acid formed during & after combustion, and condensing in cool parts of the chimney or stack, air pre heater & economiser.

Ash Content

The ash content of liquid fuel is related to the inorganic material in the fuel oil. The ash levels of distillate fuels are negligible. Residual fuels have more of the ash-forming constituents. These salts may be compounds of sodium, vanadium, calcium, magnesium, silicon, iron, aluminum, nickel, etc.

Typically, the ash value is in the range 0.03-0.07 %. Excessive ash in liquid fuels can cause fouling deposits in the combustion equipment. Ash has erosive effect on the burner tips, causes damage to the refractories at high temperatures and gives rise to high temperature corrosion & fouling of equipments.

Carbon Residue

Carbon residue indicates the tendency of oil to deposit a carbonaceous solid residue on a hot surface, such as a burner or injection nozzle, when its vaporisable constituents evaporate. Residual oil contains carbon residue ranging from 1 percent or more.

Q Why preheating of fuel oil is necessary ? What is the pre-heating requirement of fuel oil for its combustion ?

Ans. Preheating of oil in storage is needed to make it less viscous and easily pumpable. Heating an entire oil tank or outflow heater may be needed to heat oil which is to be pumped away. Outflow heaters with steam or electricity are used to ensure smooth pumping.

Furnace oil increases in viscosity as temperature drops, which makes handling a little difficult, particularly in winter. At low ambient temperatures (below 25°C), furnace oil is not easily pumpable. The oil supplying companies should be approached to ascertain the correct- pre-heating temperature for easy pumping of furnace oil, as its viscosities are subject to seasonal changes.

The preheating of oil in the storage tank to render it pumpable can be accomplished in two ways. Either the entire tank can be maintained at the pumping temperature (bulk heating) or an outflow heater may be fitted to heat the oil as it flows out. Bulk heating, which uses steam coils positioned at the floor of the tank, requires additional heat to compensate for the heat losses from the tank surface. To reduce steam requirements, it is advisable to insulate tanks where bulk heating is used.

In the case of outflow heating, only the oil, which leaves the tank, is heated to the pumping temperature. It is preferred choice for storage tank with capacity less than 70,000 litters.

While the outflow heater is the preferred choice, bulk heating may be necessary where flow rates are so high that outflow heaters of adequate capacity can not be easily installed. or when using special fuels such as Low Sulphur Heavy Stock (LSHS) / RFO (Residual Fuel Oil)/ HPS (Heavy Petroleum Stock).

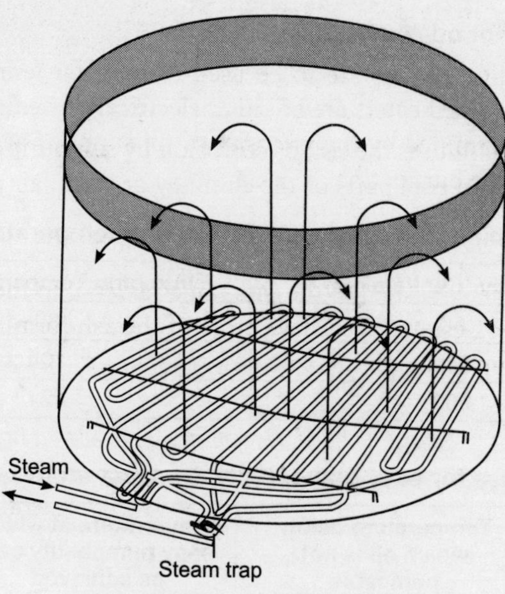

Fig.56.6: Oil storage tank

The out flow heater is essentially a heat exchanger, with steam or electricity as the heating medium. Thermostatic temperature control of the oil is necessary to prevent over heating especially when oil flow is reduced or stopped. This is particularly important for electric heaters, since oil may get carbonised when there is no flow and the heater is on.

Thermostats should be provided at a region where the oil flows freely into the suction pipe.

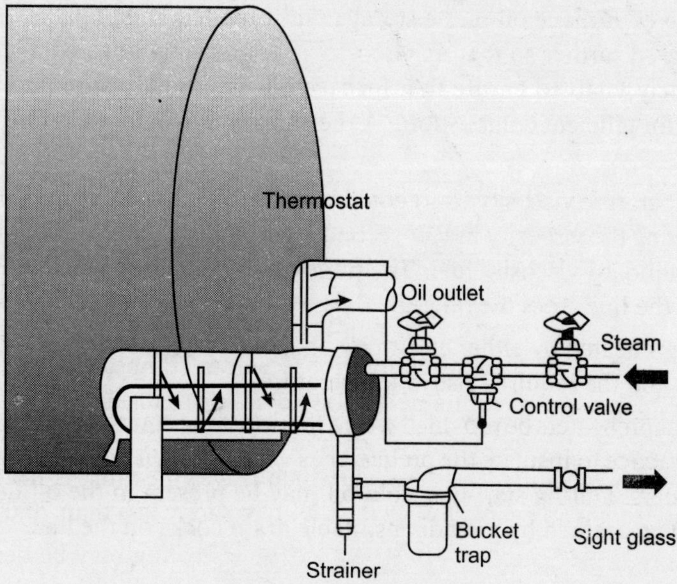

Fig.56.7: Oil heater

Pre-heating requirements for oil combustion

For pre-heating of oil, in-line heaters are to be used to raise oil temperature from pumping to combustion temperature. In-line heaters are of either electrical or steam tracing type. The pre-heat temperatures of oil are determined by the viscosity of oil supplied at room temperature and the recommended viscosity at the burner tip.

Table : Viscosity and oil pumping temperature

Viscosity (Centistokes)	Pumping Temperature, °C
50	60
230	104
900	121

Table : Temperature for easy pumpability & proper atomization of furnace oil

Viscosity of furnace oil in Red Wood No.1 @ 37.8°C(100°F) (Seconds)	Temperature below which oil is not pumpable (°C)	Temperature at which easy pumpbality can be achieved (°C)	Temperature for proper atomization (°C)
600	4.5	11 -22	71°C-87°C
1500	15.5	25 -37	88 -87
3000	26	34 -46	99 -115
LSHS	60	70 -80	120 -130

Q. What are the various steps involved in Preparation of Furnace Oil for Combustion .

Ans. The preheating of furnace oil in the storage tanks renders it pumpable, however, furnace oil generally has to heated further so that its viscosity is appreciably lowered to suit the burner. This operation is extremely vital to ensure that furnace oil is properly atomised by the burner - an essential condition for efficient combustion. A thermometer should, therefore, be provided on the oil line.

A standard temperature viscosity chart could be used to obtain the correct preheat temperature. Any fuel after knowing the viscosity in RW second No.1 is plotted on viscosity - Temperature chart parallel to line of standard viscosity fuel. The pumpability and atomising temperature will be the temperature where the line cross the range.

The oil can be preheated by either electricity or steam. The capacity of the preheater depends on the oil flow rate and the required rise in temperature.

It is essential that preheaters be provided with a thermostatic control to prevent oil carbonisation. It would be good practice to insulate the preheater as well as the oil delivery line from the preheater right up to the burners. During start-ups, cold oil may be present in the oil lines upstream of the preheater. It should be drained by providing suitable drain cocks on the line.

Important Points to note :

- Spilt oil is irretrievable. Plug all leakages.

- Impurities in furnace oil affect combustion. Filter oil in stages.
- Oil has to be preheated to obtain the right viscosity to the burner. Provide adequate preheater capacity.

The reasons for various problems encountered in oil pumping and combustion along with their reasons are tabulated below.

Oil not pumpable	• Viscosity too high
	• Blocked lines and filters
	• Sludge in oil
	• Leak in oil suction
	• Vent pipe choked
Blocking of strainers	• Sludge or wax in oil
	• Heavy precipitated compounds in oil
	• Rust or scale in tank
	• Carbonisation of oil due to excessive heating
Excessive water in oil	• Water delivered along with oil
	• Leaking manhole
	• Seepage in case of under tank
	• Ingress of moisture form vent pipe
	• Leaking heater steam coils
Pipe line plugged	• Sludge in oil
	• High viscosity oil
	• Foreign materials such as rags, scale and wood splinters in line
	• Carbonisation of oil

Q What are the requisite parameters for achieving good efficiency in fuel oil combustion ?

Ans. Following points should be noted for achieving efficient fuel oil combustion .

- Aaintain viscosity of oil to 100 Redwood sec I at burner tip by proper setting of pre-heat temperature.
- Provide atomizing air at 1–3 kg/cm2 (about 2% of total air requirement)
- About 14 kg of air/kg of fuel oil is required for complete combustion. Optimum efficiency occurs with around 10% excess air.
- To control for excess air, flue gases should be continuously analyzed for CO_2 or O_2.
- Acid formation occurs at around 160°C, and cold end corrosion should be avoided by maintaining stack temperature at above 160°C.
- Slightest damage to the burner tip may increase fuel consumption by 10– 15%, and hence any wornout burner tips should be replaced immediately.
- Oil pressure at the burner tip for pressure jet burners should be 17–20 kg/cm2.

- Correct flame is normally short. Flame impingement on walls and tubes is to be avoided since it causes carbon/soot deposits.
- Too sThort a flame indicates high excess air, and air supply to burners should be adjusted for a light hazy brown color smoke at the chimney.

Q. List out the statutory and safety requirements for storage of petroleum/ petroleum products.

Ans. Various statutory and safety requirements for storage of petroleum / petroleum products are listed below:

- Every tank or receptacle for storage of petroleum/petroleum products in bulk (other than a well-head tank) shall be constructed of iron or steel in accordance with the relevant Indian Standards or specifications approved by the chief controller of explosives .

However, API-R21615-1987 recommends that underground petroleum storage tanks may also be manufactured from Fibreglass Reinforced Plastic (FRP).

- When 2 or more storage tanks are installed in one enclosure, the total capacity of the tanks in the enclosure shall not exceed 60,000 Kl in case of conventional fixed roof tanks and 1,20,000 Kl in the case of floating roof type tanks.
- A combination of fixed and floating roof tanks can be also located in the same enclosure/ premises but the total capacity in this case shall not exceed 60,000 Kl.
- CI (Cast Iron) valves are not permitted in oil installations.
- Height of storage tanks shall not exceed one and a half times their diameter or 20 in whichever is less.
- An air space of not less than 5% of the total storage capacity shall be considered while designing oil storage tank capacity.
- All tanks for the storage of petroleum in bulk installed above or below ground shall be protected against corrosion by use of protective coatings and cathodic protection.
- Tanks after erection or repair shall be tested with water pressure. The water used for testing shall be free from petroleum and shall not be passed through any pipe or pump used for the conveyance of petroleum.
- Ministry of Petroleum prescribes that for large industries either a full rake of 76 tank wagons or half rake of 38 tank wagons shall be procured and unloading points shall be accordingly provided.
- Every storage and handling facility shall, at all times, maintain the prescribed distances from any other facility .
- Grouping of petroleum products for storage purposes shall be based on the product classification. Class- A and /or Class-B petroleum can be stored in the same dyked enclosure. Class-C should preferably be stored in separate enclosure. However, where class-C is stored in a common dyke along with Class- A and/or Class—B, all safety stipulations applicable for class A or B should apply.
- No fire water/foam ring main shall pass through dyked enclosure.
- Piping from/to any tank located in a single dyked enclosure should not pass through any other dyked enclosure. Piping within the dyked enclosure should be minimised.

- Tanks shall be arranged in maximum two rows so that each tank is approachable from the road/area surrounding the enclosure. However, tank having capacity of 50, 000 m^3 and above shall be laid in a single row.
- Any tank having a diameter more than 30 m should be separated with fire walls from other tanks.
- Firewalls should be constructed by limiting the aggregate capacity of each group of tanks within 20,000 m^3.
- The minimum distance between a tank shell and the inside of the dyke wall shall not be less than half of the tank diameter.

Q. Describe the Explosives Regulations On The Storage of Petroleum Products.

Ans. Various provisions under the the Explosives Regulations On The Storage of Petroleum Products are discussed below .

a) Petroleum products are classified as under as per the Explosive Regulations.

Flash Point	Classification	Products
Below 23°C	A	Petrol, Naphtha
23°C — less than 65°C	B	Kerosene Mineral Turpentine High Speed Diesel Oil
65°C — less than 93°C	C	Light Diesel Oil, Furnace Oil
93°C and above	Excluded	LSHS, Lubricating Oils & Greases

b) If any two classes of petroleum products are stored together, the storage rules applicable for more dangerous product between the two shall be applicable

e.g. if furnace oil and LSHS are stored adjacent to each other, even the storage of LSHS which is normally excluded from the purview of the Explosive Regulations should conform to those applicable to furnace oil. The whole installation in such case will be deemed to be that of furnace oil alone.

c) If LSHS alone is being stored in the premises, any amount of it could be stored without either the approval or a license from the Dept. of Explosives since it falls under exempted category.

d) Minimum safety distance to be observed in the storage of petroleum products.

i) When the total quantity of Class 'C' products stored does not exceed 45 KL:	
• Between retaining wall and the face	= 3m
• Between to center of lorry platform and the fence	= 3m
ii) When the total quantity of Class 'C' product stored exceeds 45 KL:	
• Between Class C fuel tank and storage or filling shed of Class A Fuel	= 9m
• Between Class C fuel tank and storage or filling shed of Class B Fuel	= 0.5 Diameter of the Class C fuel tank
• Between Class C fuel tank and Class A fuel tank lorry/wagon loading or unloading point	= 9m

• Between Class C fuel tank and Class B fuel tank lorry/wagon loading or unloading point	=	4.5m
• Between Class C fuel tank and office building/stores	=	0.5 D or 3m whichever is more
• Between Class C fuel tank and boundary fencing around the installations	=	0.5 D or 3m whichever is more

Storage of a quantity of more than 45 KL of Class C fuel become a licensable installation for which application should be made in Form VIII, as per Petroleum Rules, 1976.

Q. Describe various Fittings requirements of Storage Tanks & Service Tanks for various petroleum products.

Ans. Each of the following fittings provided on storage tanks has a definite role to play. All tanks do not have some of them.

The outlet or draw-off pipe should be located at the higher point of the tank bottom slope to facilitate filling by venting air and oil vapour during the filling operation. It should be at least 5 cms in diameter and provided with a strainer at the outlet.

The fill-pipe could be located either at the top or at the bottom of the tank. If located at the top, it must be ensured that the pipe descends almost to the bottom of the tank. When located near the bottom, it is important to position a non-return valve on the line to avoid furnace oil re-entry into the tank-truck on account of back pressure. To avoid leakage of oil during decanting, it is advisable to have a camlock coupling at the filling end so that it fits snugly on the delivery hose.

The drain pipe, which is fitted at the lower tank end bottom, is used to periodically remove the accumulated water. It is a difficult task to drain the water and sludge from underground storage tanks. The water and sludge which settle down at the bottom should, however, be removed by a hand pump. It would be advantageous to have two suction pipes — one at a lower level and the other at a higher level — so that in case the sludge accumulation is up to the lower point and the suction line chokes up, the alternative line could be put into operation and adequate time would be available to decide upon a time schedule for cleaning the tank.

Furnace oil as delivered may contain dust, dirt, gummy matter and water. During storage for prolonged periods, these tend to form sludge, which accumulates in the various 'dead spaces' of the tank, where the oil is not agitated during normal operations of the tank. This results in unsteady flames and clogging of filters. To avoid this, some means of agitation of oil in the tank may be necessary.

Cleaning of storage tanks about once a year, will help in removing any accumulated sludge. Besides cleaning, periodic inspection and maintenance of the storage tank & auxiliaries is important.

Oil Level Indicator . A calibrated dip stick measuring tape should be used for determining the quantity of oil in the tank. A dipping hole with non ferrous cap should be located on a dip pipe in the oil tank.

For bigger size storage tanks a float (metallic) and level indicator can be installed to know the level of oil.

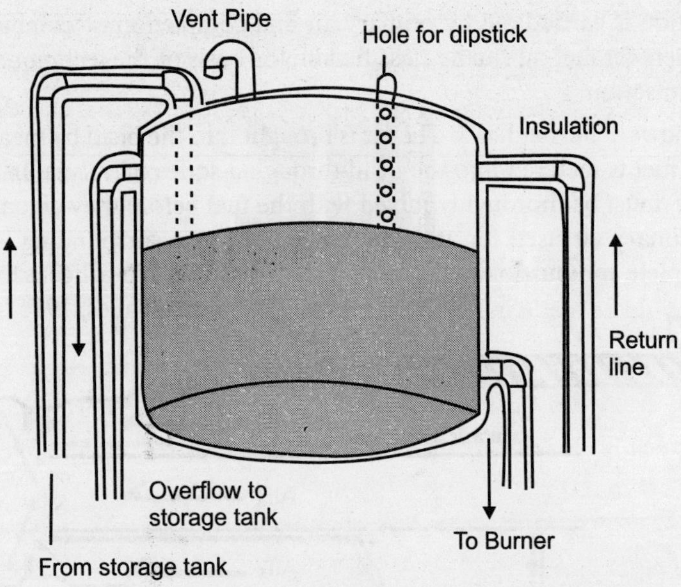

Fig.56.8: Various fittings of oil storáge tank & service tank

The Service Tank

In addition to the main storage tank(s), industries may have one or a number of day service tanks located in proximity to the units consuming oil.

The principles for the maintenance of main storage tank apply here too. The service tank may be sized for a shift's consumption or daily consumption — the choice depends on capacity required and convenience of location of tank near the shop.

Pumping capacity, be taken with respect to service tanks:

(a) Service tanks should not be located very near or above furnace, hot air or process ducts, etc.

(b) (i) They may be provided-with overflow pipes which lead below to a sump. As a further refinement, limit switches could be provided to avoid the possibility of overflow.

(ii) Where service tank is located at a higher elevation than the top of storage tank the overflow pipe may be directly connected to storage tank.'

(c) In gravity feed systems, the difference in height between the service tank and the burners should be at least 10 metres. In case pump is installed after service tank to facilitate pumping fuel oil to burner, the service tank should be kept at 3 to 5ft above ground level for ensuring, positive suction to the pump.

(d) A float and level indicator is to be installed to keep track of the level of oil in the tank.

Q. Describe the salient features of liquid fuel/ fuel Oil Firing Burners .

Ans. The burner is the principal device for the firing of fuel. The primary function of burner is to atomise fuel to millions of small droplets so that the surface area of the fuel is increased enabling intimate contact with oxygen in air. The finer the fuel dróplets are atomised, more readily will the particles come in contact with the oxygen in the air and burn.

Normally, atomisation is carried out by primary air and completion of combustion is ensured by secondary air. Burners for fuel oil can be classified on the basis of the technique to prepare the fuel for burning i.e. atomisation.

Figure below shows a burner head. The air is brought into the head by means of a forced draft blower or fan. The fuel is metered into the head through a series of valves. In order to get proper combustion, the air must be thoroughly mixed with the fuel before they actually burn. The air in the center is the primary air used for atomization and the one surrounding is the secondary air which ensures complete combustion.

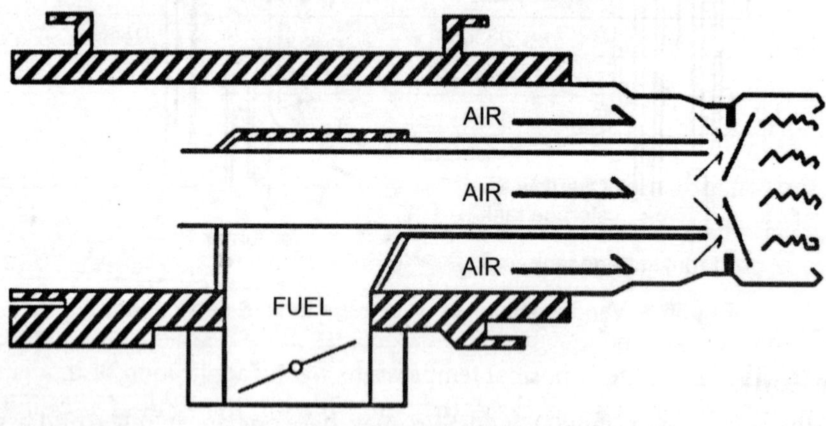

Fig.56.9: Burner Head

The mixing of liquid fuel with air is achieved by burner parts designed to create high turbulence. If sufficient turbulence is not produced by the burner, the combustion will be incomplete and samples taken at the stack will reveal presence of carbon monoxide .

An important factor in selection of burner is turndown ratio. Turndown ratio is the relationship between the maximum and minimum fuel input without affecting the excess air level. For example, a burner whose maximum input is 250,000 Kcals and minimum rate is 50,000 Kcals, has a 'Turn-Down Ratio' of 5 to 1.

Since the velocity of air affects the turbulence, it becomes difficult to get good fuel & air mixing at higher turndown ratios since the air amount is reduced. Towards the highest turndown ratios of any burner, it becomes necessary to increase the excess air amounts to obtain enough turbulence to get proper mixing. The better burner design will be one that is able to properly mix the air and fuel at the lowest possible air flow or excess air.

Q. Describe important properties of gaseous fuels including calorific value

Ans. Commonly used gaseous fuels include liquefied petroleum gases (LPG), Natural gas, producer gas, blast furnace gas, coke oven gas etc. The calorific value of gaseous fuel is expressed in kilocalories per normal cubic meter (Kcal/Nm3) i.e. at normal temperature (15oC) and pressure (760 mm Hg) .

Calorific Value

Since most gas combustion appliances cannot utlilize the heat content of the water vapour, gross calorific value of fuel is of little interest. Fuel should be compared based on the net calorific value. This is especially true for natural gas, since increased hydrogen content results in high water formation during combustion.

Typical physical and chemical properties of various gaseous fuels are given in Table below .

Table : Typical Physical and Chemical Properties of Various Gaseous Fuels.

Fuel Gas	Relative Density	Gross calorific value, kcal/ Nm^3	Air/Fuel ratio- m^3 of air to m^3 of Fuel	Flame Temp. °C	Flame Speed, m/s
Natural Gas	0.6	9350	10	1954	0.290
Propane	1.52	22200	25	1967	0.460
Butane	1.96	28500	32	1973	0.870

Q. What are the constituents & salient features of Liquefied Petroleum Gas(LPG) as a fuel ?

Ans. LPG is a predominant mixture of propane(65%) and butane(30%) with a small percentage of unsaturated hydrocarbons (Propylene & Butylene) and some lighter C2 as well as heavier C5 fractions.

LPG are those hydrocarbons, which are gaseous at normal atmospheric pressure, but may be condensed to the liquid state at normal temperature, by the application of moderate pressures. Although they are normally used as gases, they are stored & transported in cylinders/bullets / pipelines as liquids under pressure for convenience and ease of handling. Liquid LPG evaporates to produce about 250 times volume of gas.

LPG vapour is denser than air: butane is about twice as heavy as air and propane about one & a half times as heavy as air. Consequently, the vapour may flow along the ground and into drains sinking to the lowest level of the surroundings and be ignited at a considerable distance from the source of leakage. In still air vapour will disperse slowly. Escape of even small quantities of the liquefied gas can give rise to large volumes of vapour / air mixture and thus cause considerable hazard. To aid in the detection of atmospheric leaks, all LPG's are odorized by addition of mercaptans (RSH) in small quantity . There should be adequate ground level ventilation where LPG is stored. For this very reason LPG cylinders should not be stored in cellars or basements, which have no ventilation at ground level.

Q. Discuss the constituents & salient features of natural gas as a fuel.

Ans. Methane is the main constituent of Natural gas accounting for about 90 % of the total volume. Other components are: Ethane, Propane, Butane, Pentane, Nitrogen, Carbon Dioxide, and traces of other gases. Very small amounts of sulphur compounds are also present. Since methane is the largest component of natural gas, generally properties of methane are used when comparing the properties of natural gas to other fuels.

Natural gas is a high calorific value fuel requiring no storage facilities. It mixes with air readily and does not produce smoke or soot. It has practically no sulphur content. It is lighter than air and disperses into air easily in case of leak.

A typical composition of elemental constituents in fuel oil, coal and natural gas is given below in the table .

Table : Comparison of Chemical Composition of Various Fuels

Constituents	Fuel Oil Coal		Natural Gas
Carbon	84	41.11	74
Hydrogen	12	2.76	25
Sulphur	3	0.41	-
Oxygen	1	9.89	Trace
Nitrogen	Trace	1.22	0.75
Ash	Trace	38.63	-
Water	Trace	5.98	-

Q. What are the general principles & basic chemical reactions involved in combustion of fuels?

Ans. Combustion refers to the rapid oxidation of fuel accompanied by the production of heat, or heat & light. Complete combustion of a fuel is possible only in the presence of an adequate supply of oxygen which makes up 21 % of air. Solid or liquid fuels must be changed to a gas before they will burn. Usually heat is required to change liquids or solids into gases. Fuel gases will burn in their normal state if enough air is present.

Air contains 79 % of air, with traces of other elements. Nitrogen is considered to be a temperature reducing dilutant . Nitrogen reduces combustion efficiency by absorbing heat from the combustion of fuels and diluting the flue gases. This reduces the heat available for transfer through the heat exchange surfaces. It also increases the volume of combustion by-products, which then have to travel through the heat exchanger and up the stack faster to allow the introduction of additional fuel air mixture. This nitrogen also can combine with oxygen (particularly at high flame temperatures) to produce oxides of nitrogen (NOx),which are toxic air pollutants.

Carbon, hydrogen & sulphur in the fuel combine with oxygen in the air to form carbon dioxide, water vapour & sulphur dioxide, releasing 8084 kcals, 28922 kcals & 2224 kcals of heat respectively. Under certain conditions, Carbon may also combine with Oxygen to form Carbon Monoxide, which results in the release of a smaller quantity of heat (2430 kcals/kg of carbon) than carbon burned to CO2 which produces more heat per kg. of fuel than when CO or smoke are produced.

Combustion is a high-speed, high-temperature chemical reaction in which the rapid union of a combustible element / compound with oxygen liberates heat .

Basic chemical reactions involved in combustion of fuels are as follows :

$C + O_2 \rightarrow CO_2$ + 8084 kCals/kg of Carbon

$2C + O_2 \rightarrow 2\,CO$ + 2430 kCals/kg of Carbon

$2H_2 + O_2 \rightarrow 2H_2O$ + 28,922 kCals/kg of Hydrogen

$S + O_2 \rightarrow SO_2$ + 2,224 kCals/kg of Sulphur

Each kilogram of CO formed means a loss of 5654 kCal of heat (8084-2430=5654).

Q.What is the significance of the three T's (Time, Temperature, Turbulence) involved in combustion ?

Ans. The objective of good combustion is to release all of the heat in the fuel. This is accomplished by controlling the "three T's" of combustion which are :

- Temperature high enough to ignite and maintain ignition of the fuel . The temperature around the fuel release area at the burner tip must be must be greater than ignition temperature, or else incomplete combustion results.
- Turbulence or intimate mixing of the fuel and oxygen . Proper turbulence close to the flame area helps to bring fuel & air into intimate contact and provides ideal conditions for complete combustion.
- Time sufficient for complete combustion of fuel . All combustion reactions require sufficient time (residence time) for completion, or else incomplete and inefficient combustion results.

Q..Describe three types of fuel combustion .

Ans. Three types of fuel combustion are perfect combustion, complete combustion and incomplete combustion of fuel .

- **Perfect combustion** is achieved when all the fuel is burnt, using only the theoretically ideal amount of air, but perfect combustion can rarely be achieved in practice.
- **Complete combustion** is achieved when all the fuel is burnt, using the minimal amount of excess air (over and above the theoretically ideal amount of air needed to burn the fuel). Complete combustion with minimum excess air is always our goal since heat losses due to high excess air in flue gases are unaffordable and unacceptable from the fuel efficiency point of view.
- **Incomplete combustion** occurs when the fuel is not completely burnt and some escapes as CO in flue gases or as unburnts in refuse, both of which result in high losses and low efficiency.

Flue gas analysis of combustion is important, as it helps us to achieve efficient combustion conditions by controlling excess air and reducing CO in flue gases.

Q. What are different types of draft systems used in combustion of fuels in furnaces ?

Ans. The function of a draft in a combustion systems is to exhaust the products of combustion (flue gas) into the atmosphere through the chimney . The draft can be classified into two types namely Natural and Mechanical Draft.

Natural draft:

A natural draft is the draft produced by a chimney alone. It is caused by the difference in weight between the column of hot gas inside the chimney and the column of outside air at the same height & cross-section. Being much lighter than outside air, chimney flue gas tends to rise, and the heavier outside air flows in through the ash pit to take its place. It is usually controlled by hand-operated dampers in the chimney . Here no fans or blowers are used. The products of combustion are discharged through chimney at such a height that it will not be a nuisance to the surrounding community.

Mechanical draft:

A mechanical draft is a draft artificially produced by fans. (There are three basic types.)

(a) Forced draft

A forced draft system uses a fan to deliver the combustion air into the furnace, & to force combustion products to pass through the boiler/furnace and up the stack.

(b) Induced draft

An induced draft system provides suction at the chimney side, to evacuate products of combustion all through the furnace. The furnace is kept at a slight negative pressure below the atmospheric pressure due to suction ..

(c) Balanced draft

A balanced draft is a combination of a forced & an induced draft system where a forced-draft fan pushes combustion air into the furnace and an induced-draft fan draws gases into the chimney, thereby providing a draft to remove the gases from the furnace . Here the pressure is maintained between 1.25 to 2.5 mm of water gauge below atmospheric pressure in the case of boilers and slightly positive pressure for reheating & heat treatment furnaces.

In all the above mechanical draft systems, electric power is used for fan requirements.

Q. What are the Advantages and Disadvantages of Mechanical Draught in a boiler/ furnace .

Ans. These days the mechanical draught is widely used in large boiler plants.

Advantages of mechanical draught :

- It is more economical.
- It is better in control.
- The flow of air through the grate and furnace is uniform.
- It produces more draught.
- Its rate of combustion is very high.
- Low grade fuel can be used.
- The air flow can be regulated according to the changing requirements.
- It is not affected by the atmospheric temperature.
- It reduces the amount of smoke.
- It reduces height of chimney.
- It increases efficiency of the plant.
- It reduces the fuel consumption. About 15% of fuel is saved for the same amount of work.

Disadvantages of mechanical draught :

- Its initial cost is high.
- Its running cost is also high.
- It has increased maintenance cost.

Q. Discuss the salient comparative features of Forced Draught and induced draught .

The following gives the comparison between forced draught and induced draught.

Forced draught :

- The fan in placed before the fire grate.
- The pressure inside the furnace is above the atmospheric pressure.
- It forces fresh air into the combustion chamber.
- It requires less power as the fan has to handle cold air only. Moreover, volume of air handled is less because of low temperature of the cold air.
- The flow of air through grate and furnace is more uniform.
- As the leakages are outward, therefore there is a serious danger of blow out when the fire doors are opened and the fan is working.

Induced draught:

The fan is placed after the fire grate.

— The pressure inside the furnace is below the atmospheric pressure. It sucks hot gases from the combustion chamber, and forced them into the chimney.

— It requires more power as the fan has to handle hot air and flue gases. Moreover, volume of air gases is more because of high temperature of the air and gases.

— The flow of air through grate and furnace is less uniform.

— As the leakages are inward, therefore there is no danger of blow out. But if the fire doors are opened and the fan is working there will be a heavy air infiltration.

Q. Describe the salient features of Combustion of Liquid Fuel / Oil .

***Ans.* Salient features of liquid fuel/oil combustion are as follows :**

Heating Oil to Correct Viscosity

When atomizing oil, it is necessary to heat it enough to get the desired viscosity. This temperature varies slightly for each grade of oil. The lighter oils do not usually require pre-heating. Typical viscosity at the burner tip for furnace oil should be 100 Redwood seconds-1,which would require heating the oil to about 105°C.

Ensuring Stoichiometric Combustion

The efficiency of a boiler or furnace depends on efficiency of the combustion system. The amount of air required for complete combustion of the fuel depends on the elemental constituents of the fuel that is Carbon, Hydrogen, Sulphur etc. This amount of air is called stoichiometric air. For ideal combustion process for burning one kg of a typical fuel oil containing 86% Carbon, 12% Hydrogen, 2% Sulphur, theoretically required quantity of air is around 14 kg. This is the minimum air that would be required if mixing of fuel & air by the burner and combustion is perfect. The combustion products are primarily Carbon Dioxide (CO_2), water vapor (H_2O) and Sulphur Dioxide (SO_2), which pass through the chimney along with the Nitrogen (N_2) in the air,.

After surrendering useful heat in the heat absorption area of a furnace or boiler, the combustion products or flue gases leave the system through the chimney, carrying away a significant quantity of heat with them.

Golden rules for combustion of oil are :

- Atomize the oil completely to produce a fine uniform spray .
- Mix the air and fuel thoroughly .
- Introduce enough air for combustion, but limit the excess air to a maximum of 15% .
- Keep the burners in good operating condition .

Q. Describe the key result areas for improved combustion efficiency of fuel.

Ans. For energy efficiency in combustion systems, irrespective of the fuel used, the following are the key result areas to be pursued: nsuEng proper fuel storage, handling, and preparation, for achieving good combustion conditions.

- Avoiding partial load operations of combustion equipment.
- Operating with minimum excess air for fuel economy.
- Operating with lowest the possible stack temperature for fuel economy.
- Operating with minimum amounts of combustibles in refuse (ash) and CO in flue gases,
- Operating with variable speed options for fan motors, if capacity control isneeded (rather than inefficient damper control operations) in order to achieve power savings.

Q. Describe the basic components of a furnace with a neat diagram.

Ans. A furnace is a piece of equipment used to melt metal for casting or to heat materials in order to change their shape (rolling, forging, etc.) or their properties (heat treatment).

A furnace should be designed so that in a given time, as much of the material as possible can be heated to as uniform a temperature as possible with the least possible fuel and labor.

The flue gases directly contact the stock, hence the type of fuel chosen is of importance. For example, some materials will not tolerate sulphur in the fuel. In addition, use of solid fuels will generate particulate matter, which may affect the stock placed inside the furnace. Hence, many furnaces use liquid fuel, gaseous fuel, or electricity as energy input. Melting furnaces for steel and cast iron use electricity in induction & arc furnaces. Non-ferrous melting furnaces utilize oil as fuel.

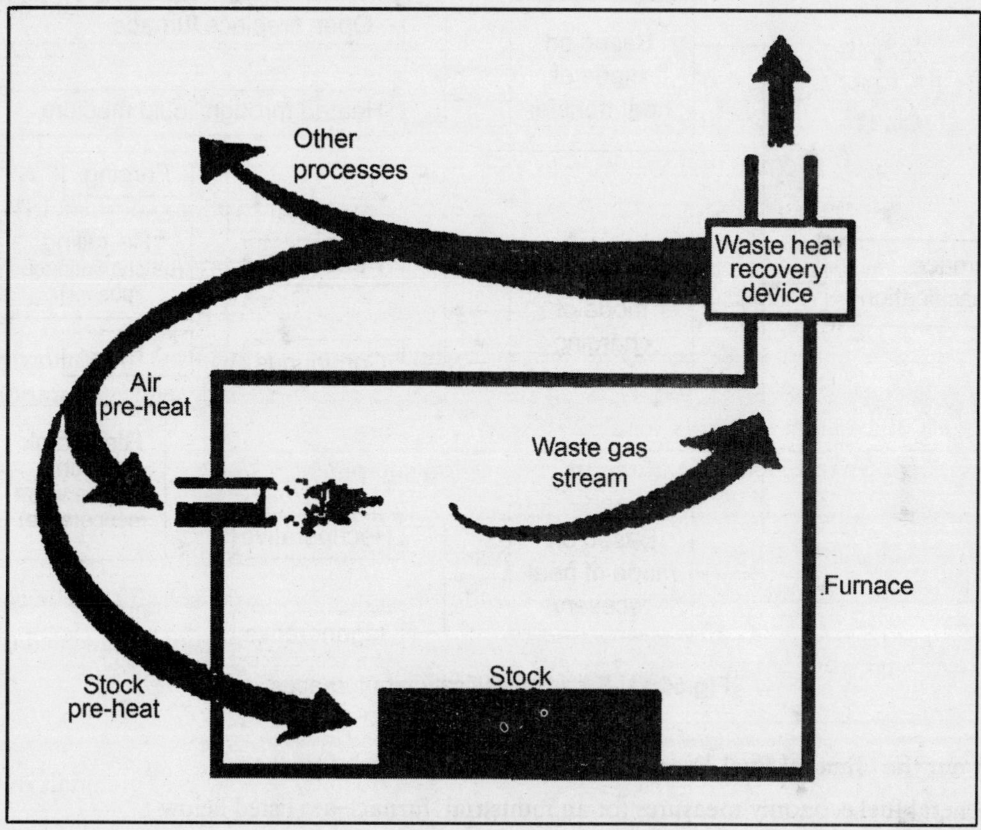

Fig.56.10: Basic components of a furnace

Q. Describe various types and classification of furnaces.

Ans. Based on the method of generating heat, furnaces are broadly classified into two types namely combustion type (using fuels) and electric type. In case of combustion type furnace, depending upon the kind of combustion, it can be broadly classified as oil fired, coal fired or gas fired.

Based on the mode of charging of material furnaces can be classified as (i) Intermittent or Batch type furnace or Periodical furnace and (ii) Continuous furnace.

Based on mode of waste heat recovery furnaces are classified as recuperative and regenerative furnaces.

Another type of furnace classification is made based on mode of heat transfer, mode of charging and mode of heat recovery as shown in the Figure below.

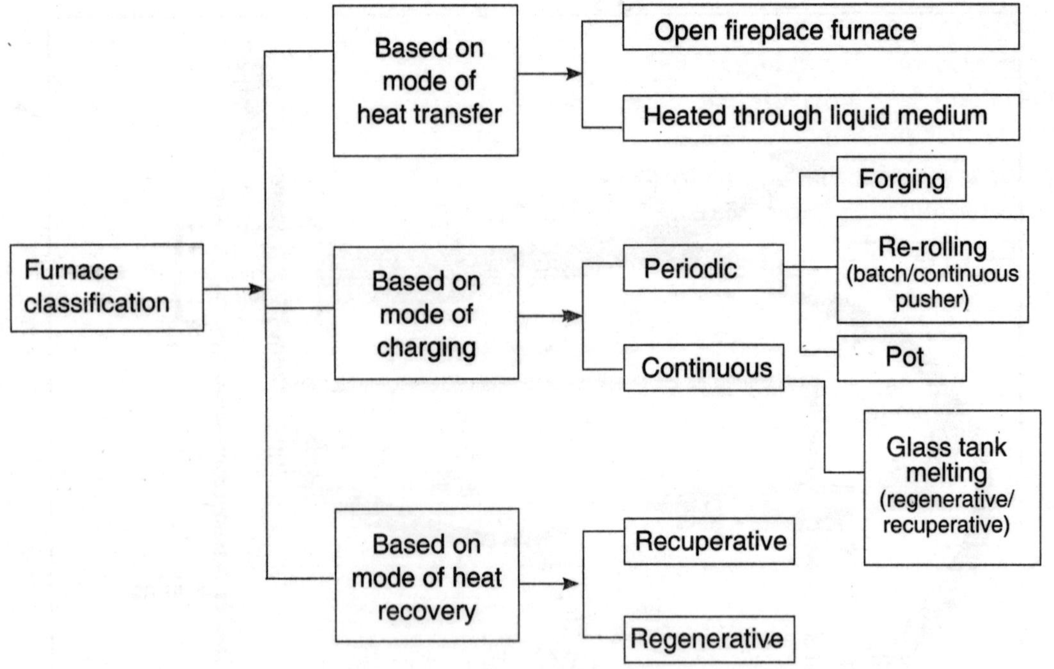

Fig.56.11: Broad classification of furnaces

Q. List out the General Fuel Economy Measures in industrial Furnaces .

Ans. General fuel economy measures for an industrial furnace are listed below :

- Complete combustion with minimum excess air .
- Correct heat distribution in the furnace.
- Operating the furnace at the desired temperature .
- Reducing heat losses from furnace openings .
- Maintaining correct amount of furnace draught .
- Ensuring optimum capacity utilization of the furnace.
- Ensuring waste heat recovery from the flue gases of furnace .
- Ensuring minimum refractory losses in the furnace.
- Use of Ceramic Coatings in the furnace .

Q. What are the benefits of ensuring Complete Combustion of fuels with Minimum Excess Air?

Ans. The amount of heat lost in the flue gases (stack losses) depends upon amount of excess air.

To obtain complete combustion of fuel with the minimum amount of air, it is necessary to control air infiltration, maintain pressure of combustion air, fuel quality and excess air monitoring.

Higher excess air will reduce flame temperature, furnace temperature and heating rate. On the other hand, if the excess air is less, then unburnt components in flue gases will increase and would be carried away in the flue gases through stack .Therefore the optimization of combustion air is the most economical measure for fuel economy . The impact of this measure is higher when the temperature of the furnace is high. More excess air also means more scale losses, which is equally a big loss in terms of money.

Air ratio is the value that is given by dividing the actual air amount by the theoretical combustion air amount, and it represents the extent of excess of air.

The figure below indicates relation between air ratio and exhaust gas loss.

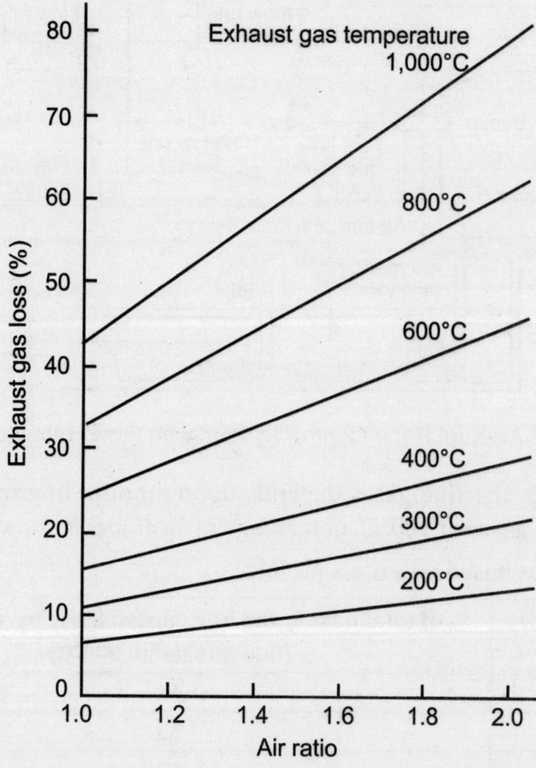

Fig.56.12: Relation Between Air Ratio and Exhaust Gas Loss

A simple equation to calculate excess air from the percentage of oxygen in flue gases for any fuel in a combustion system is:

$$\% \text{ excess air} = (\% \text{ Oxygen})*100/ (21 - \% \text{ Oxygen}).$$

If a furnace is not equipped with an automatic air/fuel ratio controller, it is necessary to periodically sample gas in the furnace and measure its oxygen contents by a gas analyzer. The Figure below shows a typical example of a reheating furnace equipped with an automatic air/fuel ratio controller.

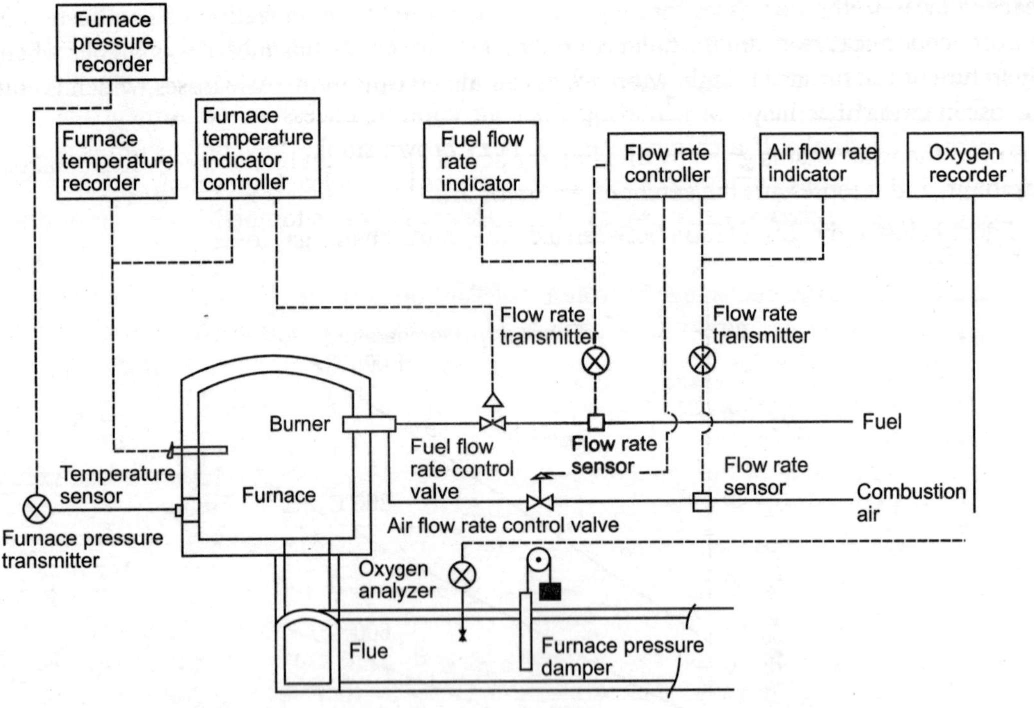

Fig.56.13: Air/Fuel Ratio Control System with Flow Rate Controller

The amount of heat lost in the flue gases depends upon amount of excess air. In the case of a furnace carrying away flue gases at 900°C, percentage of heat lost is shown below in Table .

Table : Heat loss in flue gas based on excess air level

% Excess Air	% of total heat in the fuel carried away by waste gases (flue gas temp. 900°C)
25	48
50	55
75	63
100	71

Q. Why it is important to optimize the Excess Air used for Combustion .How can it be done by excess air analysis ?

Ans. We know that we need 14 kg of air for every kg of fuel oil for complete combustion. This is the minimum air that would be theoretically needed if mixing of fuel and air by the burner is perfect. In actual practice, since mixing is never perfect, a certain amount of excess air is needed to complete the combustion and ensure release of the entire heat contained in fuel oil. If too much air than what is required for completing combustion is allowed to enter, additional heat would be lost in heating the surplus air to the chimney temperature. Less air would lead to the incomplete combustion and smoke. More air would lead to increased stack losses. Hence, there is an optimum excess air level for each type of fuel. Fuel savings are the most at the optimum excess air level.

Excess Air Analysis

An experienced operator relies on the smoke emitted by the chimney or the colour of the flame to help him adjust air at the optimum level. However, a chemical analysis of flue gases is a more objective method that helps in achieving finer air control. Excess air adjustment has a direct relationship with smoke. If a chimney emits a hazy brown smoke, the combustion is considered proper, but if the chimney emits black smoke, it indicates incomplete combustion and available fuel wastage. Similarly, a chimney that is clear & smokeless indicates too high excess air and wasted fuel.

By measuring either the percentage by volume of Carbon Dioxide (CO_2) or Oxygen (O_2) in flue gases, it is possible to estimate the percentage of excess air in flue gases. Continuous on line recording instruments, as well as cheaper portable instruments are available for this purpose.

The excess air ratio (EAR) can be calculated by empirical equation, EAR = $21/21\text{-}O_2$. As excess air increases, the concentration of these gases (CO_2 & SO_2) decreases and the concentration of Oxygen steadily increase in the flue gases. This forms the basis for determination of excess air though chemical analysis. The CO_2% varies with type of fuel oil at same Oxygen % and excess air. So, measurement of Oxygen percentage in flue gas for evaluation of excess air is a more accurate method and is presently followed everywhere.

Presence of carbon monoxide (CO) in the flue gas is the indication of improper /incomplete combustion . It is possible to operate installations with Zero Stack Carbon Monoxide, but in general, to do so would result in the requirement of higher than necessary excess air operation resulting in additional heat loss.

In most commercially available boilers and furnaces, combustion at the theoretical stoichiometric point cannot be achieved. The result is the presence of Carbon Monoxide in stack gases, coincident with the presence of Oxygen. The appliance must therefore, be set with a minimum Carbon Monoxide level coincident with .a minimum Oxygen level.

In boilers it is considered a good practice to maintain CO at 300 ppm and in furnances at 500 ppm. Paradoxically Carbon Monoxide may also be present along side a high degree of excess air. This may be due to a faulty nozzle, diffuser plate or badly maintained burners.

Q. What measures should be taken for proper heat distribution in furnaces to achieve fuel economy?

Ans. Furnace design should be such that in a given time, as much of the stock could be heated uniformly to a desired temperature with minimum fuel firing rate.

For ensuring proper heat distribution in furnaces , the following measures should be taken:

(a) Prevent flame impingement by aligning the burner properly to avoid touching the material. The flame should not touch any solid object and should propagate clear of any solid object. Any obstruction will disturb the atomized fuel particles, thus affecting combustion and creating black smoke. If flame impinges on the stock, there will be an increase in scale losses, which are hidden in nature but precious in value.

Correct heat distribution & burner alignment are shown below in figure.

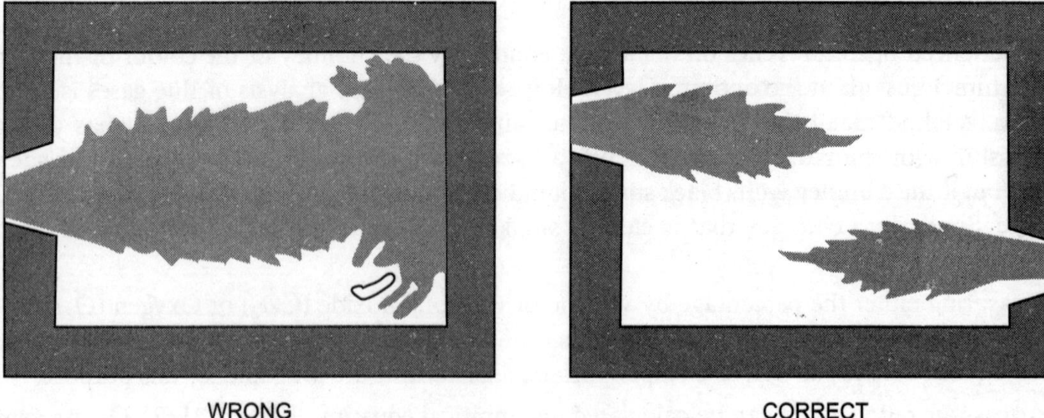

Fig.56.14: Heat distribution in a furnace

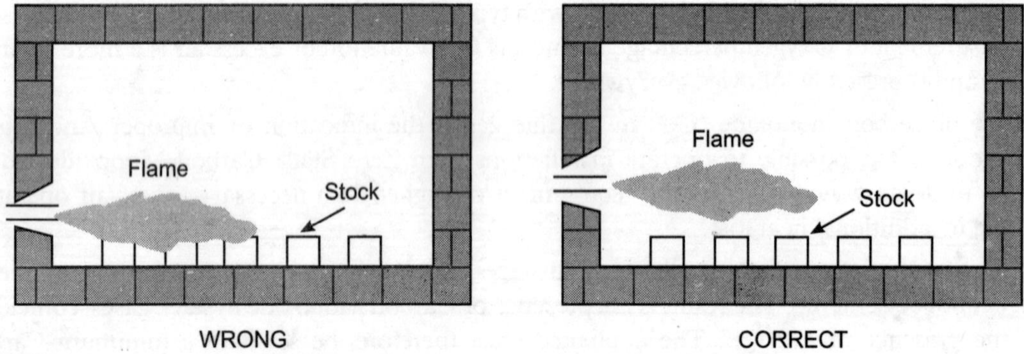

Fig.56.15: Burner alignment in a furnace

(b) If the flames impinge on refractories, the incomplete combustion products can settle and react with the refractory constituents at high flame temperatures.

(c)The flames of different burners in the furnace should stay clear of each other. If they intersect, inefficient combustion will occur. It is desirable to stagger the burners on the opposite sides.

(d)The burner flame has a tendency to travel freely in the combustion space just above the material. In small reheating furnaces, the axis of the burner is never placed parallel to the hearth but always at an upward angle. Flames should not hit the roof.

(e)Designs with few large-capacity burners produce a long flame, which may be difficult to contain within the furnace walls. More burners of less capacity give a better distribution of heat in the furnace and also increase furnace life.

(f)For small reheating furnaces, it is desirable to have a long flame with golden yellow color while firing furnace oil for uniform heating. The flame should not be so long that it enters the chimney and comes out through the top or through the doors. This will happen if excess oil is fired. In such cases, a major portion of additional fuel is carried away from the furnace.

Q. Why it is important to maintain Optimum Operating Temperature of Furnaces ?

Ans. It is important to operate the furnace at optimum temperature. Operating at too high temperatures than optimum causes heat loss, excessive oxidation, de-carbonization as well as over-stressing of the refractories. These controls are normally left to operator judgment, which is not desirable. To avoid human error, on/off controls should be provided.

Q. Why Prevention of Heat Loss through Furnace Openings is important ?

Ans. Heat loss through openings consists of the heat loss by direct radiation through openings and the heat loss caused by combustion gas that leaks through openings.

If the furnace pressure is slightly higher than outside air pressure (as in case of reheating furnace) during its operation, the combustion gas inside may blow off through openings and heat is lost with that. But damage is more, if outside air intrudes into the furnace, making temperature distribution uneven and oxidizing steel slabs/billets. This heat loss is about 1% of the total quantity of heat generated in the furnace, if furnace pressure is controlled properly.

Q. Why proper control of furnace draft is necessary during its operation ?

Ans. If negative pressures exist in the furnace, air infiltration will occur through the cracks & openings thereby affecting air-fuel ratio control. Not maintaining proper furnace pressure could result in cold metal & non-uniform metal temperatures, which could affect subsequent metal operations like forging & rolling and result in increased fuel consumption. For optimum fuel consumption, slight positive pressure should be maintained in the furnace as shown below in Figure .

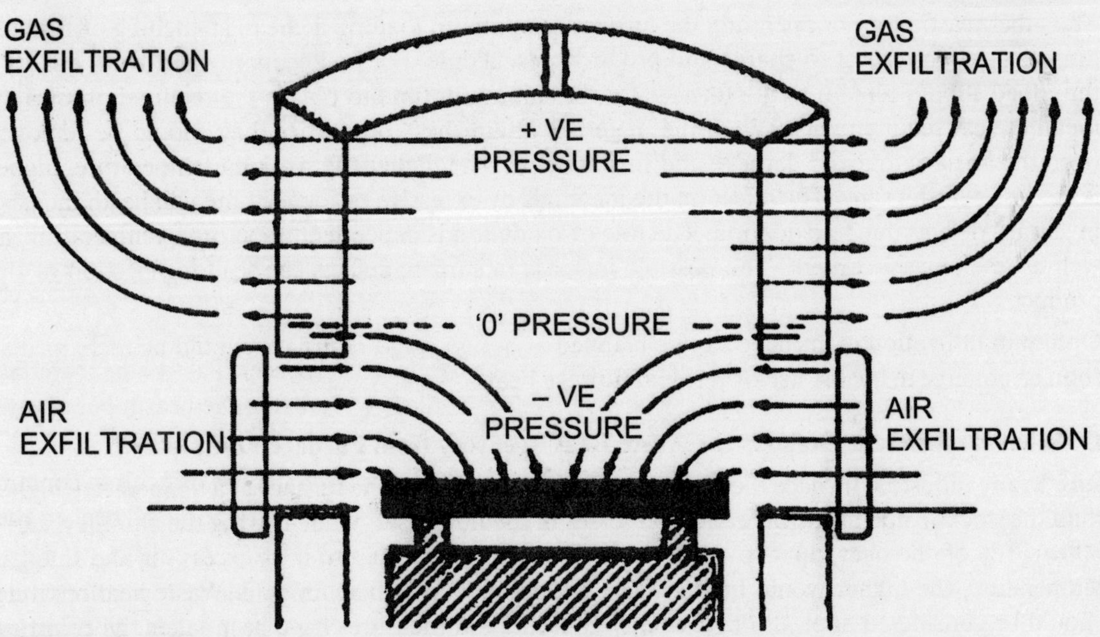

Fig.56.16: Effect of Pressure on the Location of Zero Level and Infiltration of Air

Ex-filtration is less serious than infiltration. Some of the associated problems with ex filtration are leaping out of flames, overheating of the furnace refractories leading to reduced brick life, increased furnace maintenance, burning out of ducts & equipments attached to the furnace, etc. In addition to the proper control on furnace pressure, it is important to keep the openings as small as possible and to seal them in order to prevent the release of high temperature gas & intrusion of outside air through openings such as the charging inlet, extracting outlet and peephole on furnace walls or the ceiling.

Q. How Optimum Capacity Utilization of furnaces is important for its fuel economy?

Ans. One of the most important factors affecting thermal efficiency of a furnace is its loading. There is a particular loading at which the furnace will operate at maximum thermal efficiency. If the furnace is under loaded a smaller fraction of the available heat in the working chamber will be taken up by the load and therefore efficiency will be low.

Furnaces should be loaded at the rate associated with optimum .The loading of the charge on the furnace hearth should be arranged so that :

- It receives the maximum amount of radiation from the hot surfaces of the heating chambers and the flames produced.
- The hot gases are efficiently circulated around the heat receiving surfaces .
- Stock should not be placed in the following position :
- In the direct path of the burners or where flame impingement is likely to occur.
- In an area which is likely to cause a blockage or restriction of the flue system of the furnace.
- Close to any door openings where cold spots are likely to develop.

The other reason for not operating the furnace at optimum loading is the mismatching of furnace dimension with respect to charge and production schedule. For fuel economy and work quality , the load should remain in the furnace for the minimum time to obtain the required physical & metallurgical requirements. When the materials attain these properties they should be removed from the furnace to avoid damage and fuel wastage. The higher the working temperature, higher is the loss per unit time. The effect on the materials by excessive residence time will be an increase in surface defects due to oxidation. The rate of oxidation is dependent upon time, temperature, as well as free oxygen content. The possible increase in surface defects can lead to rejection of the product.

Optimum utilization of furnace can be planned at design stage. Furnace should be recharged as soon as possible to enable use of residual furnace heat.

Q. What are different methods for Waste Heat Recovery from Furnace Flue Gases ?

Ans. In any industrial furnace the products of combustion leave the furnace at a temperature higher than the stock temperature. Sensible heat losses in the flue gases, while leaving the chimney, carry 30 to 50 % of the heat input to the furnace. The higher the quantum of excess air and flue gas temperature, the higher would be the waste heat availability for recovery .. Waste heat recovery should be considered after all other energy conservation measures have been taken. Minimizing the generation of waste heat should be the primary objective. The sensible heat in flue gases can be generally recovered by the following methods :

- Charge (stock) preheating,

- Preheating of combustion air,
- Utilizing waste heat for other process (to generate steam by a waste heat boiler)

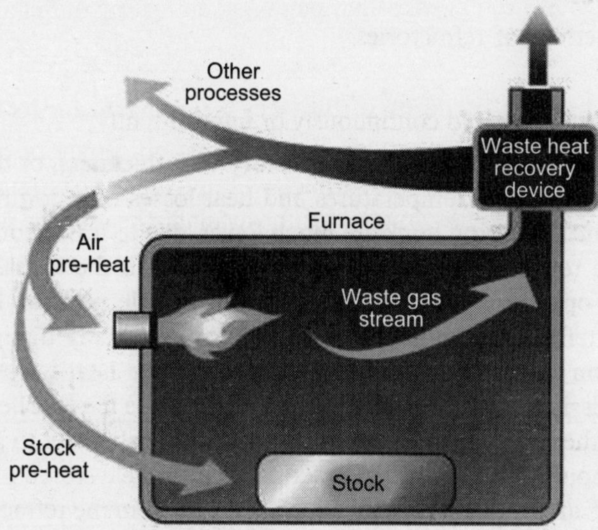

Fig.56.17: Methods of Waste heat recovery from a furnace flue gas

Charge Pre-heating

When raw materials are preheated by exhaust gases before being placed in a heating furnace, the amount of fuel necessary to heat them in the furnace is reduced. Since raw materials are usually at room temperature, they can be heated sufficiently using high-temperature gas to reduce fuel consumption rate.

Preheating of Combustion Air

For a long time, the preheating of combustion air using heat from exhaust gas was not used except for large boilers, metal-heating furnaces and high-temperature kilns. This method is now being employed in compact boilers & industrial furnaces as well.

The energy contained in the exhaust gases can be recycled by using it to pre-heat the combustion air. A variety of equipment is available; external recuperators are common, but other techniques are now available such as self-recuperative burners.

For example, with a furnace exhaust gas temperature of 1,000°C, a modern recuperator can pre-heat the combustion air to over 500 0C, giving energy savings compared with cold air of up to 30%

Q. Describe the factors affecting the heat losses from furnace walls. How can these losses be minimized ?

Ans. About 30-40% of the fuel input to the furnace generally goes to make up for heat losses in intermittent or continuous furnaces. The appropriate choice of refractory and insulation materials goes a long way in achieving high fuel savings in industrial furnaces.

The heat losses from furnace walls affect the fuel economy considerably. The extent of furnace wall heat losses depend on:

- Emissivity of wall
- Thermal conductivity of refractories
- Wall thickness
- Whether furnace is operated continuously or intermittently

Heat losses can be reduced by increasing the refractory wall thickness, or through the application of insulating bricks. Outside wall temperatures and heat losses of a composite wall of a certain thickness of firebrick and insulation brick are much lower, due to lesser conductivity of insulating brick as compared to a refractory brick of similar thickness. In the actual operation in most of the small furnaces the operating periods alternate with the idle periods. During the off period, the heat stored in the refractories during the on period is gradually dissipated, mainly through radiation and convection from the cold face. In addition, some heat is abstracted by air flowing through the furnace. Dissipation of stored heat is a loss, because the lost heat is again imparted to the refractories during the heat "on" period, thus consuming extra fuel to generate that heat. If a furnace is operated 24 hours, every third day, practically all the heat stored in the refractories is lost. But if the furnace is operated 8 hours per day all the heat stored in the refractories is not dissipated. For a furnace with a firebrick wall of 350 mm thickness, it is estimated that 55 % of the heat stored in the refractories is dissipated from the cold surface during the 16 hours idle period. Furnace walls built of insulating refractories and cased in a shell reduce the flow of heat to the surroundings.

Q. What is furnace efficiency ? How it is determined ?

Ans. The efficiency of a furnace is the ratio of useful heat output to heat input. The furnace efficiency can be determined by both a direct and an indirect method.

Direct method of furnace efficiency determination

The efficiency of the furnace can be directly computed by measuring the amount of fuel consumed per unit weight of material produced from the furnace.

$$\text{Thermal efficiency of the furnace} = \frac{\text{Heat in the stock}}{\text{Heat in the fuel consumed}}$$

The quantity of heat to be imparted (Q) to the stock can be found from the formula :

$$Q = m x C_p x (t_2 - t_1)$$

Where ,

Q = Quantity of heat in kCal

m = Weight of the material in kg

C_p = Mean specific heat, in kCal/kg°C

t_2 = Final temperature desired, in °C

t_1 = Initial temperature of the charge before it enters the furnace, in °C

Indirect method of furnace efficiency determination

Furnace efficiency can also be calculated by an indirect method. Furnace efficiency is calculated after subtracting sensible heat loss in flue gas, loss due to moisture in flue gas, heat loss due to openings of the furnace, heat loss through the furnace skin, and other unaccounted losses from the input to the furnace.

The parameters that must be taken into account in order to calculate furnace efficiency using the indirect method include hourly fuel consumption, material output, excess air quantity, temperature of flue gas, temperature of the furnace at various zones, skin temperature and hot combustion air temperature. Efficiency is determined by subtracting all the heat losses from 100.

The following measurements should be made to calculate the energy balance in heating furnaces :

i. Weight of stock.

ii. Temperature of furnace walls, roof, etc.

iii. Flue gas temperature.

iv. Flue gas analysis.

v. Fuel oil consumption.

Instruments like infrared thermometer, fuel consumption monitor, surface thermocouple and other measuring devices are required to measure the above parameters.

Q. Show with a neat sketch the various components of heat balance in furnace . List out and describe the normal furnace losses.

Ans. A furnace heat balance has the components as shown in Figure below.

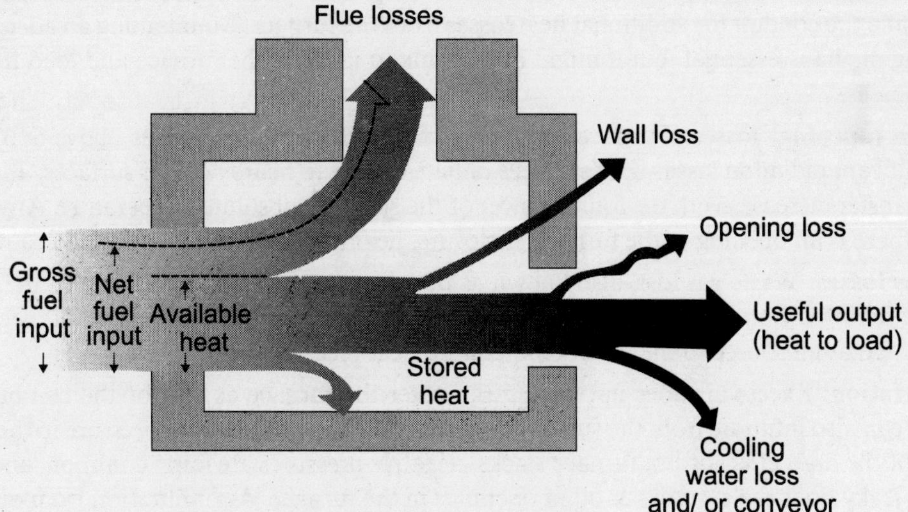

Fig.56.18: Furnace heat balance

Normal furnace losses include the following:

- Heat storage in the furnace structure.
- Losses from the furnace outside walls or structure.
- Heat transported out of the furnace by the load conveyors, fixtures, trays.
- Radiation losses from openings, hot exposed parts, etc.

- Heat carried by cold air infiltration into the furnace.
- Heat carried by the excess air used in the burners.

Stored Heat Loss: First, the metal structure and insulation of the furnace must be heated so their interior surfaces are about the same temperature as the product they contain. This stored heat is held in the structure until the furnace shuts down, then it leaks out into the surrounding area. The more frequently the furnace is cycled from cold to hot and back to cold again, the more frequently this stored heat must be replaced. Fuel is consumed with no useful output.

Wall losses: Additional heat losses take place while the furnace is in operation. Wall or transmission losses are caused by the conduction of heat through the walls, roof, and floor of the heating device . Once that heat reaches the outer skin of the furnace and radiates to the surrounding area or is carried away by air currents, it must be replaced by an equal amount taken from the combustion gases. This process continues as long as the furnace is at an elevated temperature.

Material handling losses: Many furnaces use equipment to convey the work into and out of the heating chamber, and this can also lead to heat losses. Conveyor belts or product hangers that enter the heating chamber cold and leave it at higher temperatures drain energy from the combustion gases. In car bottom furnaces, the hot car structure gives off heat to the room each time it rolls out of the furnace to load or remove work. This lost energy must be replaced when the car is returned to the furnace.

Cooling media losses: Water or air cooling protects rolls, bearings, and doors in hot furnace environments, but at the cost of lost energy. These components and their cooling media (water, air etc.) become the conduit for additional heat losses from the furnace. Maintaining an adequate flow of cooling media is essential, but it might be possible to insulate the furnace and load from some of these losses.

Radiation (opening) losses: Furnaces and ovens operating at temperatures above 540°C might have significant radiation losses. Hot surfaces radiate energy to nearby colder surfaces, and the rate of heat transfer increases with the fourth power of the surface's absolute temperature. Anywhere or anytime there is an opening in the furnace enclosure, heat is lost by radiation, often at a rapid rate.

Waste gas losses: Waste gas loss, also known as flue gas or stack loss, is made up of the heat that cannot be removed from the combustion gases inside the furnace. The reason is heat flows from the higher temperature source to the lower temperature heat receiver.

Air infiltration: Excess air does not necessarily enter the furnace as part of the combustion air supply. It can also infiltrate from the surrounding room if there is a negative pressure in the furnace. Because of the draft effect of hot furnace stacks, negative pressures are fairly common, and cold air slips past leaky door seals, cracks & other openings in the furnace. Air infiltration from outside the furnace occure every time the door is opened,and a considerable amount of heat is lost.

Economy in fuel consumption can be achieved if the total heat that can be passed on to the stock is as large as possible.

It can be seen that some of these losses do not.exist in equipment like boilers, which is the reason why boilers are often much more efficient than industrial furnaces.

Q. Why waste heat recovery from furnaces is important ? Show with a neat sketch a simple heat exchanger type of waste heat recovery equipment .

Ans. Since furnaces operate in high temperature domains; the stack temperatures are very high, and waste heat recovery is a major energy efficiency opportunity in furnaces. The recovered waste heat is conventionally used for either preheating of combustion air or to heat the stock material itself.

Waste heat diagram of a furnace is shown below .

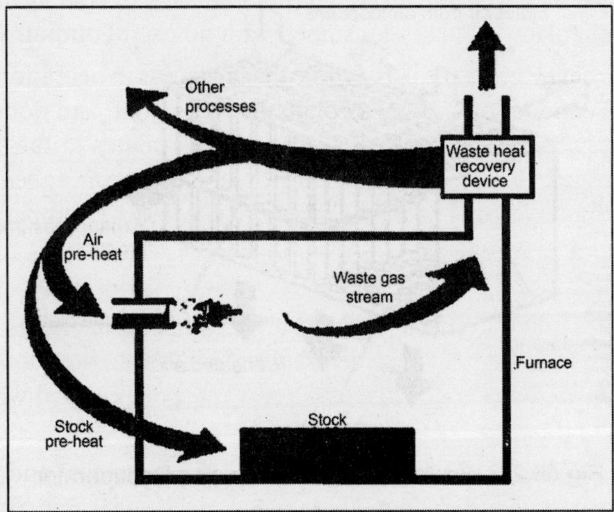

Fig.56.19: Waste heat diagram of a furnace

A common waste heat recovery system is a simple double pipe type heat exchanger (also called a recuperator) shown below in figure.

These generally take the form of concentric cylinders, in which the combustion air passes through the annulus and the exhaust gases from the furnace pass through the center. Such recuperators are very cheap to make, are suitable for use with dirty gases, have a negligible resistance to flow, and can replace the flue or chimney if space is limited.

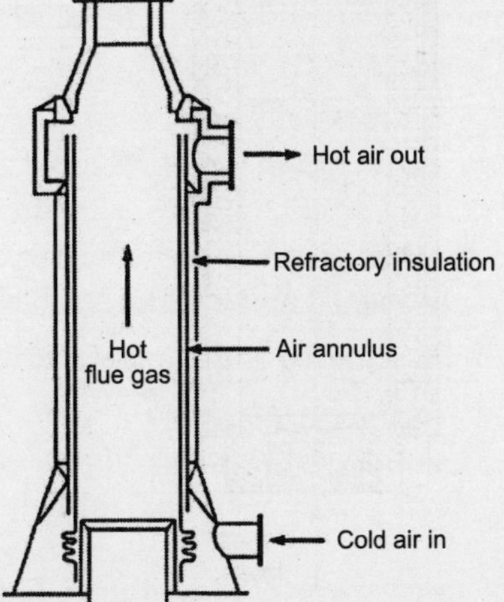

Fig.56.20: A simple double pipe heat exchanger

Q. Describe the operating principle of a recuperator with a neat sketch.

Ans. In a recuperator, heat exchange takes place between the flue gases and the air through metallic or ceramic walls. Ducts or tubes carry the air for combustion to be preheated ; the other side contains the waste heat stream. A recuperator for recovering waste heat from flue gases is shown below in Figure.

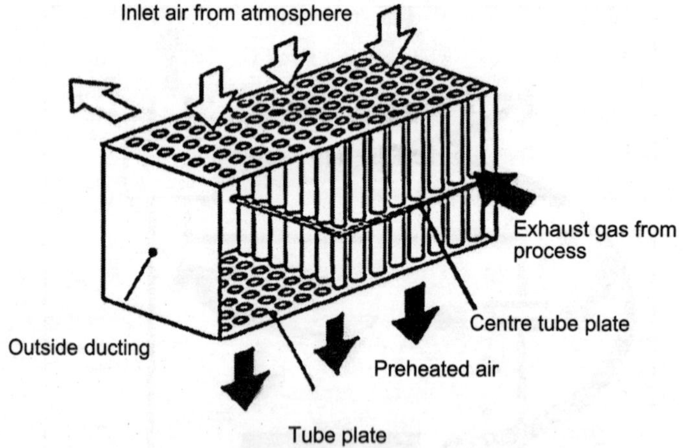

Fig.56.21: Waste Heat Recovery using Recuperator

Q. Describe the functioning of a metallic radiation recuperator .

Ans. The simplest configuration for a recuperator is the metallic radiation recuperator, which consists of two concentric lengths of metal tubing as shown below in Figure .

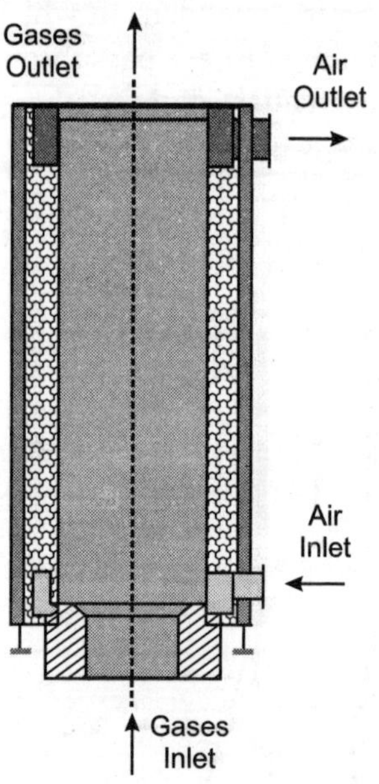

Fig.56.22: Metallic Radiation Recuperator

The inner tube carries the hot exhaust gases while the external annulus carries the combustion air from the atmosphere to the air inlets of the furnace burners. The hot gases are cooled by the incoming combustion air, which now carries additional energy into the combustion chamber. This is the energy, which does not have to be supplied by the fuel; consequently, less fuel is burned for a given furnace loading. The saving in fuel also means a decrease in combustion air and therefore, stack losses are decreased not only by lowering the stack gas temperatures but also by discharging smaller quantities of exhaust gas. The radiation recuperator gets its name from the fact that a substantial portion of the heat transfer from the hot gases to the surface of the inner tube takes place by radiative heat transfer. The cold air in the annuals, however, is almost transparent to infrared radiation so that only convection heat transfer takes place to the incoming air. As shown in the diagram, the two gas flows are usually parallel, although the configuration would be simpler and the heat transfer would be more efficient if the flows were opposed in direction (or counterflow).

The reason for the use of parallel flow is that recuperators frequently serve the additional function of cooling the duct carrying away the exhaust gases and consequently extending its service life.

Q. Describe the operating principle and usefulness of a Convection recuperator as a waste heat recovery equipment .

Ans. A second common configuration for recuperators is called the tube type or convective recuperator. As seen in the figure below, the hot gases are carried through a number of parallel small diameter tubes, while the incoming air to be heated enters a shell surrounding the tubes and passes over the hot tubes one or more times in the direction normal to their axes.

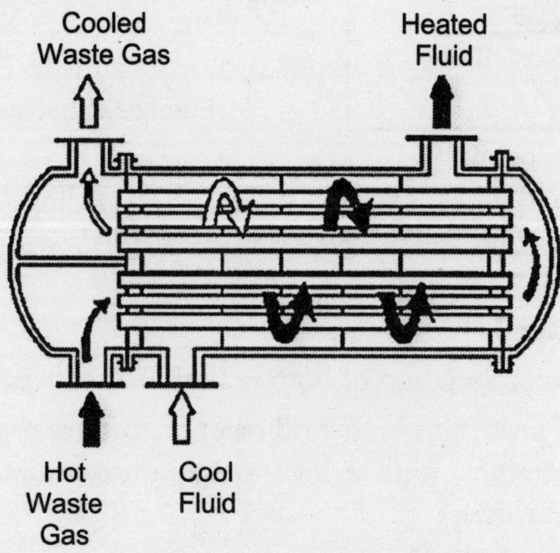

Fig.56.23: Operating principle of Convective Recuperator

If the tubes are baffled to allow the gas to pass over them twice, the heat exchanger is termed a two-pass recuperator; if two baffles are used, a three-pass recuperator, etc. Although baffling increases both the cost of the exchanger and the pressure drop in the combustion air path, it increases the effectiveness of heat exchange. Shell and tube type recuperators are generally more compact and have a higher effectiveness than radiation recuperators, because of the larger heat transfer area made possible through the use of multiple tubes and multiple passes of the gases.

Convection recuperators consist essentially of bundles of drawn or cast tubes. Internal and/or external fins can be added to assist with heat transfer. The combustion air normally passes through the tubes and the exhaust gases outside the tubes, but there are some applications where this is reversed. For example, with dirty gases, it is easier to keep the tubes clean if the air flows on the outside. Design variations include 'U' tube and double pass systems. Convection recuperators are more suitable for exhaust gas temperatures of less than about 900°C. Beyond 900°C, ceramic recuperators made of silicon carbide can be used which can withstand higher temperatures.

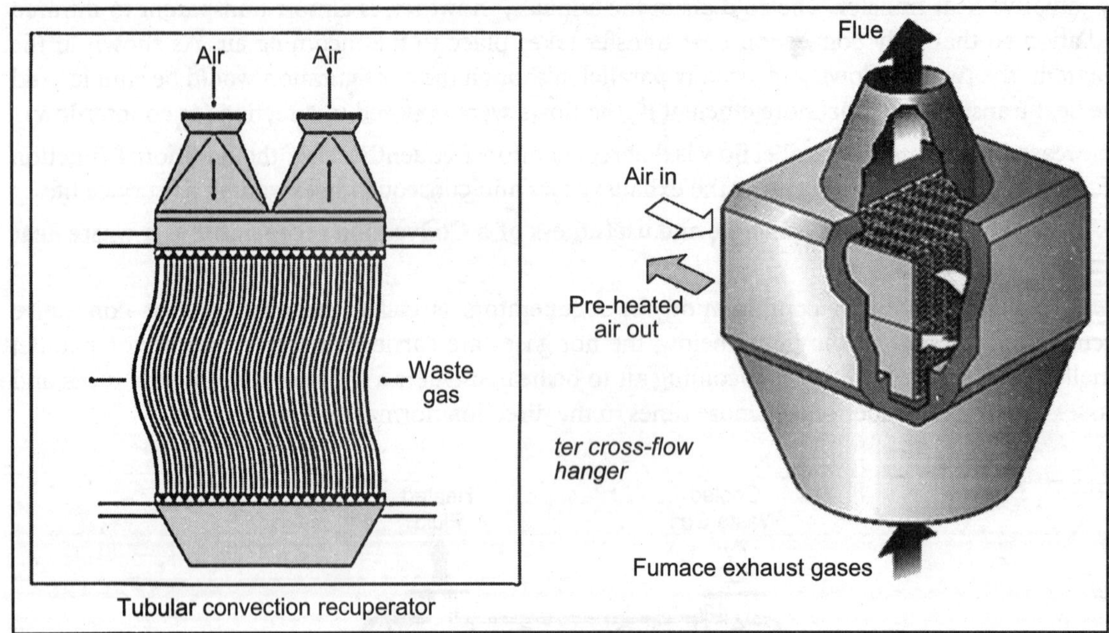

Fig.56.24: Convection recuperator

Q. What is a hybrid recuperator ?

Ans. A hybrid recuperator is a combination of both radiation & convection recuperator .

For maximum effectiveness of heat transfer, hybrid recuperators are used. These are combinations of radiation and convective designs, with a high-temperature radiation section followed a by convective section (see Figure below).

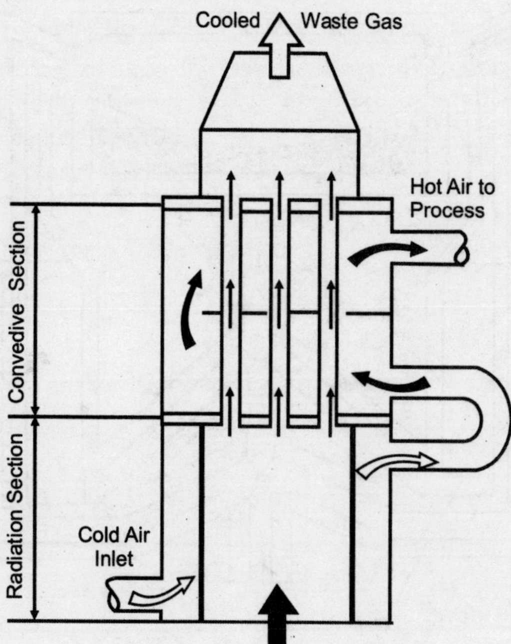

Fig.56.25: Hybrid Recuperator

Hybrid recuperator are more expensive than simple metallic radiation recuperators, but are less bulky.

Q. What is a ceramic recuperator ?

Ans. The principal limitation on the heat recovery of metal recuperators is the reduced life of the liner at inlet flue gastemperatures exceeding11000C. To overcome the temperature limitations of metallic recuperators, ceramic tube recuperators have been developed whose material of construction allow operation on the flue gas side to be at 1550 0C and on the preheated air side to be 815 0C .

Early ceramic recuperators were built of tile and joined with furnace cement, and thermal cycling caused cracking of joints & rapid deterioration of the tubes. Later developments introduced various kinds of short silicon carbide tubes, which can be joined by flexible seals located in the air headers. Earlier ceramic recuperator designs had experienced leakage rates from 10 to 60 % . The new designs are with air preheat temperatures as high as 700 0C, with much lower leakage rates.

Q. Discuss the factors which affect the effectiveness of a regenerator as a waste heat recovery equipment.

Ans. Regenerators are suitable for large capacities and have been widely used in glass & steel melting furnaces. Important relations exist between the sizes of the regenerator, time between reversals, thickness of brick, conductivity of brick and heat storage ratio of the brick.

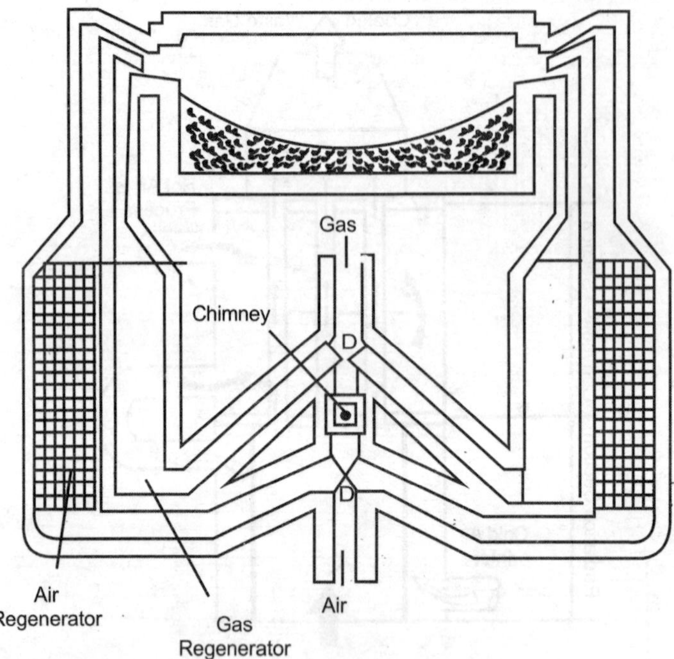

Fig.56.26: Regenerator

In a regenerator, the time between the reversals is an important aspect. Long periods would mean higher thermal storage and hence higher cost. Also long periods of reversal result in lower average temperature of preheat and consequently reduction in the fuel economy. Accumulation of dust and slagging on the surfaces reduce efficiency of the heat transfer as the furnace becomes old. Heat losses from the walls of the regenerator and air in- leaks during the gas period and out- leaks during the air period also reduces the heat transfer.

Q. What is a Heat Wheel ? Describe its areas of applications.

Ans. A heat wheel is finding increasing applications in low to medium temperature waste heat recovery systems.

It is a sizable porous disk, fabricated with material having a fairly high heat capacity, which rotates between two side-by-side ducts: one is a cold gas duct, the other a hot gas duct. The axis of the disk is located parallel and on the partition between the two ducts. As the disk slowly rotates, sensible heat (moisture that contains latent heat) is transferred to the disk by the hot air and, as the disk rotates, from the disk to the cold air. The overall efficiency of sensible heat transfer for this kind of regenerator can be as high as 85 % . Heat wheels have been built as large as 21 meters in diameter with air capacities up to 1130 m^3/min.

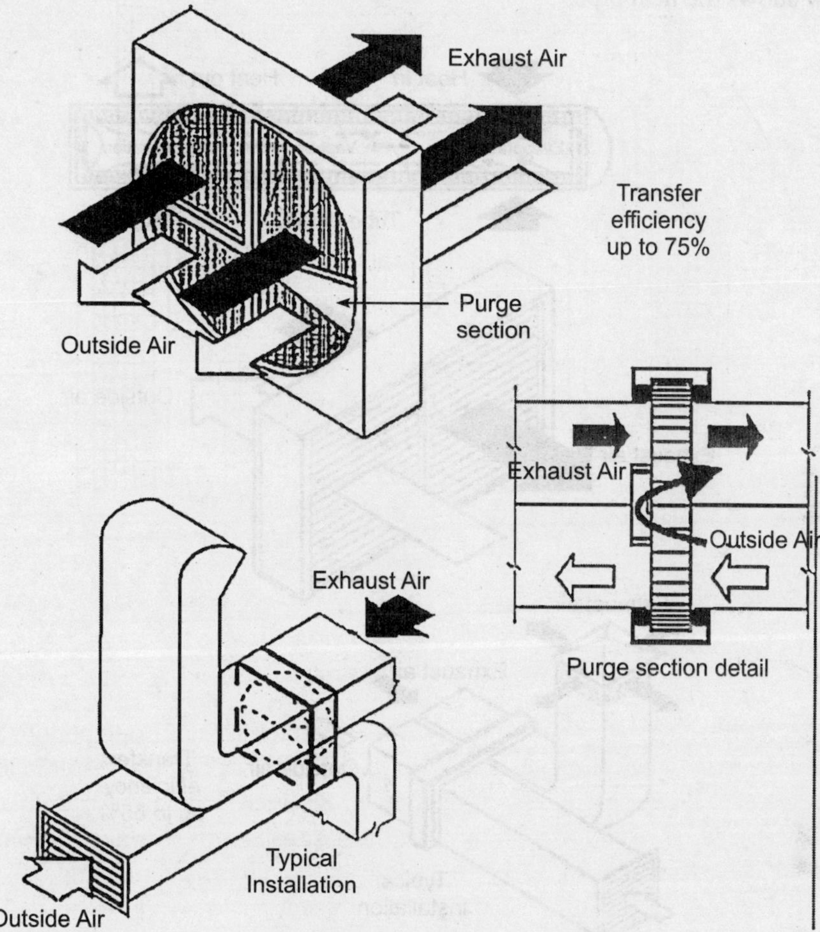

Fig.56.27: Heat Wheel

A variation of the heat wheel is the rotary regenerator where the matrix is in a cylinder rotating across the waste flue gas and air streams. The heat or energy recovery wheel is a rotary gas heat regenerator, which can transfer heat from exhaust to incoming gases.

Its main area of application is where heat is exchanged between large masses of air having small temperature differences. Heating & ventilation systems and recovery of heat from dryer exhaust air are typical applications.

Q. What is a heat pipe? Explain its functioning and applications with a neat sketch .

Ans. The heat pipe exchanger (HPHE) is a lightweight compact heat recovery system. It virtually does not need any mechanical maintenance, as there are no moving parts that wear out. It does not need input power for its operation and is free from cooling water & lubrication systems. It also lowers the fan horsepower requirement and increases the overall thermal efficiency of the system. The heat pipe heat recovery systems are capable of operating at 315°C with 60% to 80% heat recovery capability.

Figure below shows the heat pipe.

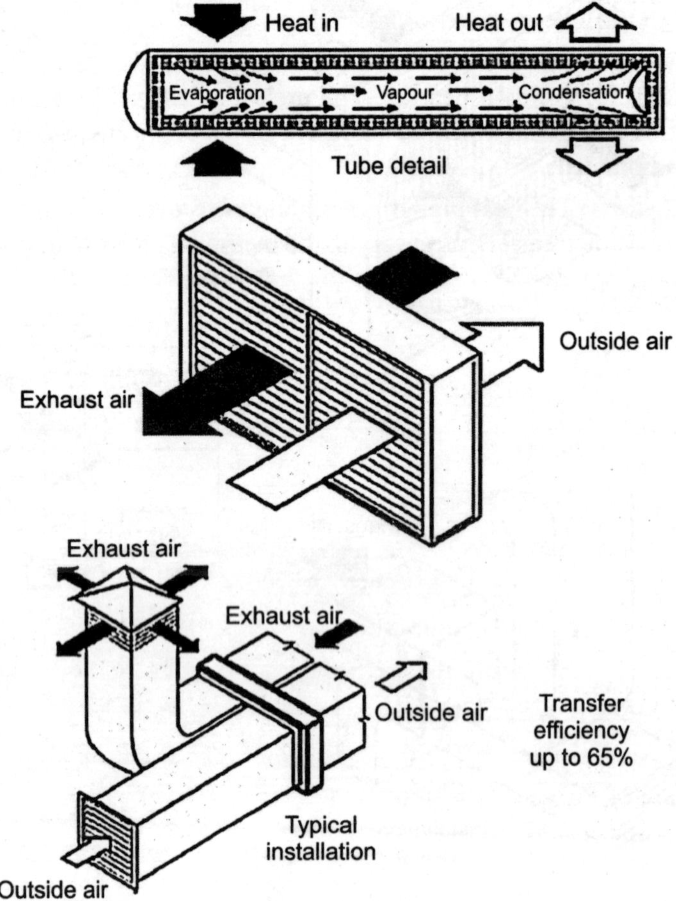

Fig.56.28: Heat Pipe

The heat pipe comprises of three elements – a sealed container, a capillary wick structure and a working fluid. The capillary wick structure is integrally fabricated into the interior surface of the container tube and sealed under vacuum. Thermal energy applied to the external surface of the heat pipe is in equilibrium with its own vapour as the container tube is sealed under vacuum. Thermal energy applied to the external surface of the heat pipe causes the working fluid near the surface to evaporate instantaneously. Vapour thus formed absorbs the latent heat of vaporization and this part of the heat pipe becomes an evaporator region. The vapour then travels to the other end the pipe where the thermal energy is removed causing the vapour to condense into liquid again, thereby giving up the latent heat of the condensation. This part of the heat pipe works as the condenser region. The condensed liquid then flows back to the evaporated region.

A heat pipe can transfer up to 100 times more thermal energy than copper, the best-known conductor. Heat pipe is a thermal energy absorbing and transferring system having no moving parts and hence requires minimal maintenance.

Applications of heat pipe

Heat pipes are used in the following industrial applications:

- **Process of Space Heating:** The heat pipe heat exchanger transfers the thermal energy from process exhaust for building heating. The preheated air can be blended if required. The requirement of additional heating equipment to deliver heated make up air is drastically reduced or eliminated.
- **Process to Process:** The heat pipe heat exchangers recover thermal energy waste from the exhaust process and transfer this energy to the incoming process air. The incoming air thus becomes warm and can be used either for the same process/other processes and hence, reduce process energy consumption.
- **HVAC Applications:**
  - Cooling: Heat pipe heat exchangers pre-cools the building make up air in summer and thus reduces the total tonnes of refrigeration, apart from the operational saving of the cooling system. Thermal energy supply is recovered from the cool exhaust and transferred to the hot supply make up air.
  - Heating: The above process is reversed during winter to preheat the make up air.
- Other applications of heat pipe in industries are:
  - Preheating of boiler combustion air
  - Recovery of Waste heat from furnaces
  - Reheating of fresh air for hot air driers
  - Recovery of waste heat from catalytic deodorizing equipment
  - Reuse of Furnace waste heat as heat source for other oven
  - Cooling of closed rooms with outside air
  - Preheating of boiler feed water with waste heat recovery from flue gases in the heat pipe economizers.
  - Drying, curing and baking ovens
  - Waste steam reclamation
  - Brick kilns (secondary recovery)
  - Reverberatory furnaces (secondary recovery)
  - Heating, ventilating and air-conditioning systems

Q. Describe the functioning of an economizer used in boiler system for feed water preheating.

Ans. In the case of boiler systems, an economizer is provided to utilize the flue gas heat for preheating the boiler feed water. On the other hand, in an air pre-heater, the waste heat is used to heat combustion air. In both the cases, there is a corresponding reduction in the fuel requirements of the boiler.

Functioning of an economizer is shown below with the help of figure.

For every 220°C reduction in flue gas temperature by passing through an economizer or air preheater, there is 1% saving of fuel in the boiler. In other words, for every 60°C rise in feed water temperature through an economizer, or 20°C rise in combustion air temperature through an air preheater, there is 1% saving of fuel in the boiler.

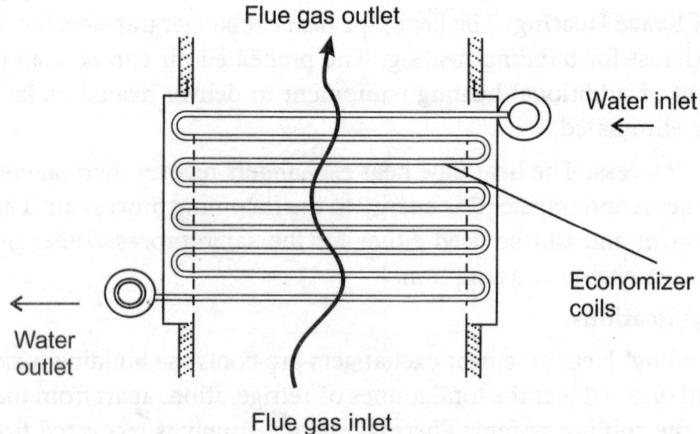

Fig.56.29: Economizer

Q. Describe the functioning of a Waste Heat Recovery Boiler with a neat sketch.

Ans. Waste heat boilers are ordinarily water tube boilers in which the hot exhaust gases from gas turbines, incinerators, industrial furnaces etc., pass over a number of parallel tubes containing water. The water is vaporized in the tubes and collected in a steam drum from which it is drawn out for use as heating or processing steam.

Because the exhaust gases are usually in the medium temperature range and in order to conserve space, a more compact boiler can be produced if the water tubes are finned in order to increase the effective heat transfer area on the gas side.

Figure below shows a mud drum, a set of tubes over which the hot gases make a double pass, and a steam drum which collects the steam generated above the water surface. The pressure at which the steam is generated and the rate at which steam is produced depend on the temperature of waste heat. The pressure of a pure vapor in the presence of its liquid is a function of the temperature of the liquid from which it is evaporated. If the waste heat in the exhaust gases is insufficient for generating the required amount of process steam, auxiliary burners, which burn fuel in the waste heat boiler or an after-burner in which the exhaust gases flue are added. Waste heat boilers are built in capacities from 25 almost 30,000 m^3 /minute of exhaust gas.

Heat Pumps

In the various commercial options previously disc

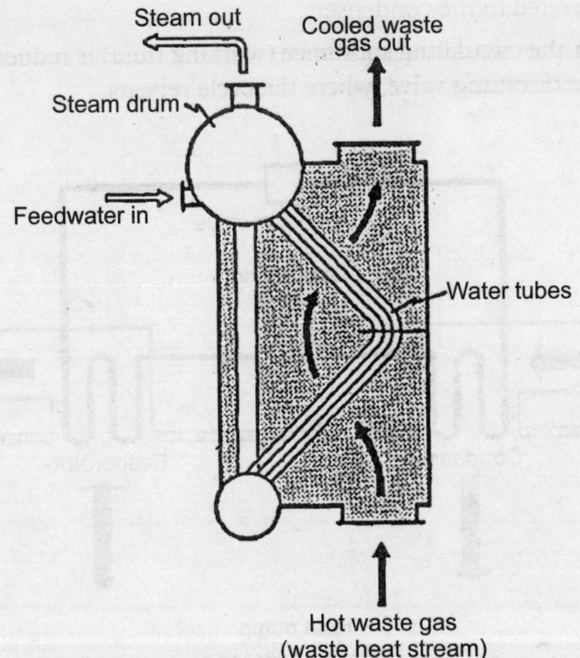

Fig.56.30: Two-Pass Water Tube Waste Heat Recovery Boiler

Q. What is a heat pump ? Discuss its function and application with a neat sketch.

Ans. In the various commercial waste heat recovery equipments , waste heat is transferred from a hot fluid to a fluid at a lower temperature. Heat must flow spontaneously "downhill", that is from a system at high temperature to one at a lower temperature. When energy is repeatedly transferred or transformed, it becomes less and less available for use. Eventually, energy has such low intensity (resides in a medium at such low temperature) that it is no longer available to function.

In industrial operations fluids with temperatures less than 120°C (or,150°C to provide a safe margin), are set as the limit for waste heat recovery because of the risk of condensation of corrosive liquids. However, as fuel costs continue to rise, even such waste heat can be used economically for space heating and other low temperature applications. It is possible to reverse the direction of spontaneous energy flow by the use of a thermodynamic system known as heat pump.

The majority of heat pumps work on the principle of the vapour compression cycle. In this cycle, the circulating substance is physically separated from the source (waste heat, with a temperature of Tin) and user (heat to be used in the process, Tout) streams, and is re-used in a cyclical fashion, therefore being called 'closed cycle'.

In the heat pump, the following processes take place:

- In the evaporator, the heat is extracted from the heat source to boil the circulating substance;
- The compressor compresses the circulating substance, thereby raising its pressure and temperature. The low temperature vapor is compressed by a compressor, which requires external work. The work done on the vapor raises its pressure and temperature to a level where its energy becomes available for use.

- The heat is delivered to the condenser;
- The pressure of the circulating substance (working fluid) is reduced back to the evaporator condition in the throttling valve, where the cycle repeats.

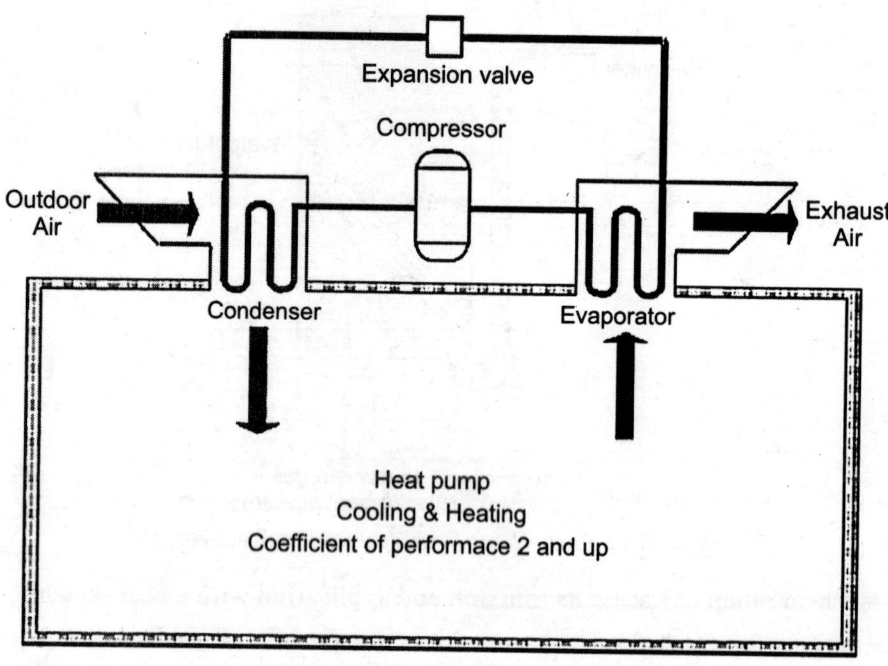

Fig.56.31: Heat Pump Arrangement

The heat pump was developed as a space heating system where low temperature energy from the ambient air, water, or earth is raised to heating system temperatures by doing compression work with an electric motor-driven compressor. The arrangement of a heat pump is shown above in figure .

The heat pumps have the ability to upgrade heat to a value more than twice the energy consumed by the device. The potential for application of heat pumps is growing and a growing number of industries have been benefited by recovering low grade waste heat by upgrading it and using it in the main process stream.

Heat pump applications are most promising when both the heating and cooling capabilities can be used in combination. One such example of this is a plastics factory where chilled water from a heat is used to cool injection-moulding machines, whilst the heat output from the heat pump is used to provide factory or office heating. Other examples of heat pump installation include product drying, maintaining dry atmosphere for storage and drying compressed air.

Q. What is a thermocompressor ? What is its use ?

Ans. In many cases, very low-pressure steam is reused as water after condensation for lack of any better option of reuse. In many cases , it becomes feasible to compress this low pressure steam by very high pressure steam and reuse it as a medium pressure steam. The major energy in steam is in its latent heat value, and thus thermo compressing would give a big improvement in waste heat recovery.

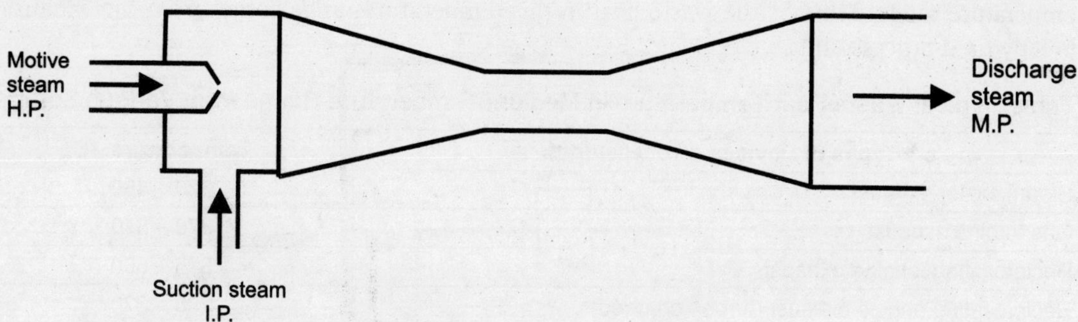

Fig.56.32: Thermo compressor

The thermo-compressor is a simple equipment with a nozzle where HP steam is accelerated into a high velocity fluid. This entrains the LP steam by momentum transfer and then recompresses in a divergent venturi. A figure of thermo compressor is shown in figure above .

It is typically used in evaporators where the boiling steam is recompressed and used as heating steam.

Q. List out the temperatures of waste gases from industrial process equipments/furnaces in the high temperature range.

Ans. Table below gives the temperatures of waste gases from industrial process equipment in the high temperature range. All of these results are from direct fuel fired processes.

Table: Typical Waste Heat Temperature in High Temperature Range from Various Sources

Types of Devices/furnaces	Temperature , °C
Nickel refining furnace	1370 – 1650
Aluminium refining furnace	650 –760
Zinc refining furnace	760 – 1100
Copper refining furnace	760 – 815
Steel heating furnace	925 – 1050
Copper reverberatory furnace	900 – 1100
Open hearth furnace	650 – 700
Cement kiln (Dry process)	620 – 730
Glass melting furnace	1000 – 1550
Hydrogen plants	650 – 1000
Solid waste incinerators	650 – 1000
Fume incinerators	650 – 1450

Q. List out the temperatures of waste gases from process equipment in the medium temperature range.

Ans. Table below gives the temperatures of waste gases from process equipment in the medium temperature range. Most of the waste heat in this temperature range comes from the exhaust of directly fired process units.

Table :Typical Waste Heat Temperature in Medium Temperature Range from Various Sources

Types of Devices /Process units	Temperature ,°C
Steam boiler exhaust	230 – 480
Gas turbine exhaust	370 – 540
Reciprocating engine exhaust	315 – 600
Reciprocating engine exhaust (turbo charged)	230 – 370
Heat treatment furnace	425 – 650
Drying & baking ovens	230 – 600
Catalytic crackers	425 – 650
Annealing furnace cooling systems	425 – 650

Q. List out Waste Heat Temperature in Low Temperature Range from Various Sources .

Ans. Table below lists some heat sources in the low temperature range. In this range, it is usually not practical to extract work from the source, though steam production may not be completely excluded if there is a need for low-pressure steam. Low temperature waste heat may be useful in a supplementary way for preheating purposes.

Table : Typical Waste Heat Temperature in Low Temperature Range from Various Sources

Source	Temperature , °C
Process steam condensate	55-88
Cooling water from: Furnace doors	32-55
Bearings	32-88
Welding machines	32-88
Injection molding machines	32-88
Annealing furnaces	66-230
Forming dies	27-88
Air compressors	27-50
Pumps	27-88
Internal combustion engines	66-120
Air conditioning and refrigeration condensers	32–43
Liquid still condensers	32-88
Drying, baking and curing ovens	93-230
Hot processed liquids	32-232
Hot processed solids	93-232

Q. List out the options to recover waste heat to maximize energy efficiency.

Ans. The most important options to maximize energy efficiency when applying waste heat recovery are :

- Recover heat from flue gas, engine cooling water, engine exhaust, low pressure
- waste steam, drying oven exhaust, boiler blowdown, etc.
- Recover heat from incinerator off gas.
- Use waste heat for fuel oil heating, boiler feedwater heating, outside air heating, etc.
- Use chiller waste heat to preheat hot water.
- Use heat pumps.
- Use absorption refrigeration.
- Use thermal wheels, run-around systems, heat pipe systems, and air-to-air exchangers.

Q. Explain the functioning of a regenerative burner for waste heat recovery with a neat diagram.

Ans. In regenerative burners, the heat of the gases is stored before they exit and incoming combustion air is preheated from this stored heat, switching between two chambers alternately. The cycle time is accurately controlled for switchover. Options include regenerative burners and regenerative air pre- heaters for heat recovery. Figure below shows a regenerative burner.

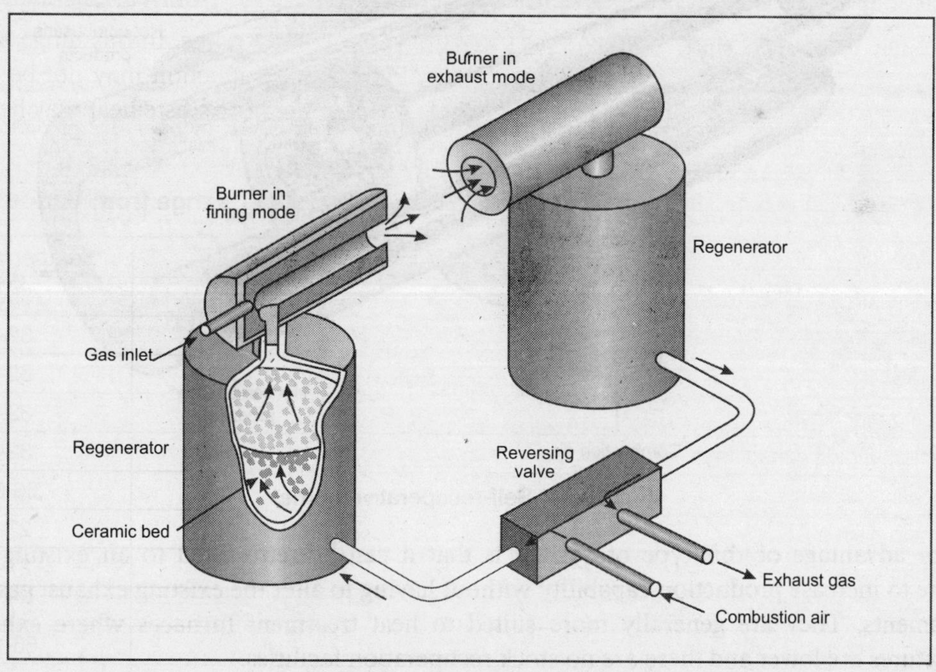

Fig.56.33: Regenerative burner

Q. What is a radiation recuperator ? Describe a self recuperative burner used for preheating the combustion air .

Ans. Radiation recuperator is in the form of concentric cylinders, in which the combustion air passes through the annulus and the hot exhaust gases from the furnace pass through the centre. The simple construction means that such recuperators are suitable for use with dirty gases, have a negligible resistance to flow, and can replace the flue or chimney if space is limited. The annulus can be replaced by a ring of vertical tubes, but this design is more difficult to install and maintain. Radiation recuperators rely on radiation from high temperature exhaust gases and should not he employed with exhaust gases at less than about 800°C.

Self-recuperative burners (SRBs) are based on traditional heat recovery techniques in that the products of combustion are drawn through a concentric tube recuperator around the burner body and used to pre-heat the combustion air as shown below in figure.

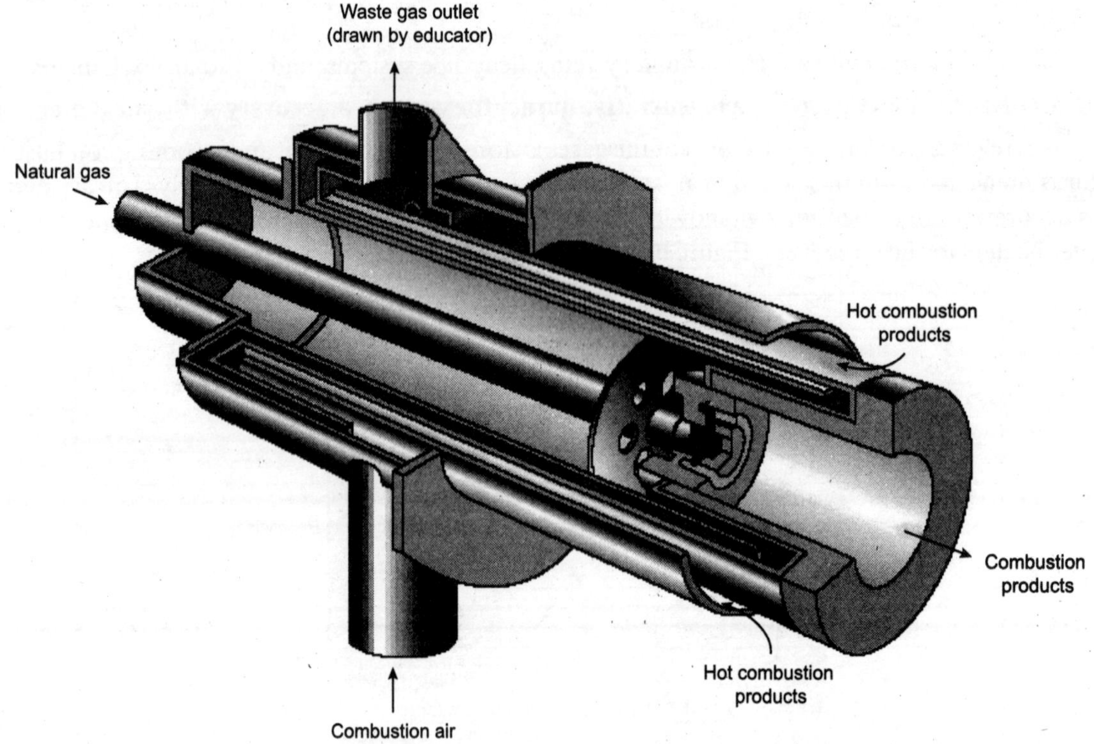

Fig.56.34: Self-recuperative burner

A major advantage of this type of system is that it can be retro-fitted to an existing furnace structure to increase production capability without having to alter the existing exhaust gas ducting arrangements. They are generally more suited to heat treatment furnaces where exhaust gas temperatures are lower and there are no stock recuperation facilities.

Q. List out the working temperature range, sources and applications of various waste heat recovery devices .

Ans. Matrix of Waste Heat Recovery Devices And Their Applications are tabulated below.(H-High temperature, M- Medium temperature, L- Low temperature)

Waste Heat Recovery Devices	Temp. Range	Typical Sources	Typical Uses
Radiation Recuperator	H	Incinerator or boiler exhaust	Combustion air preheat
Convective Recuperator	M-H	Soaking or annealing ovens, melting furnaces, afterburners, gas incinerators, radiant tube burners, reheat furnaces	Combustion air preheat
Furnace Regenerator	H	Glass and Steel melting furnaces	Combustion air preheat
Metallic Heat Wheel	L-M	Curing and drying ovens, boiler exhaust	Combustion air preheat, Space preheat
Ceramic Heat Wheel	M-H	Large exhaust boiler or Incinerator	Combustion air preheat
Finned tube Regenerator	L-M	Boiler Exhaust	Boiler makeup water preheat
Shell & tube Regenerator	L	Refrigeration condensates, waste steam, distillation condensates, coolants from engines, air compressors, bearings and lubricants	Liquid flows requiring heating
Heat Pipes	L-M	Drying, curing and baking ovens, Waste steam, air dryers, Kilns and Reverberatory furnaces	Combustion air preheat, boiler makeup water preheat,Steam generation, domestic hot water, space heating
Waste heat boiler	M-H	Exhaust from gas turbines, Reciprocating engines, incinerators and furnaces	Hot water or steam generation

Q. Describe various parameters/actions necessary for improving the fuel economy/efficiency in furnaces .

Ans. Various actions for improving fuel economy of the furnaces are listed below.

- Prevent infiltration of air, using doors or air curtains.
- Monitor O_2/CO_2/CO in flue gas and control excess air level.
- Ensure that the furnace combustion chamber is under slight positive pressure.
- Use ceramic fiber linings in the case of batch operations.
- Match the load to the furnace capacity.
- Investigate cycle times and avoid extended hours of run time and excess heating.
- Provide automatic temperature controllers in the furnace .
- Ensure that the flame does not touch the stock.
- Recover maximum heat from flue gases by waste heat recovery equipment .

- Ensure upkeep of heating surfaces by regular cleaning.
- Repair damaged insulation.
- Use an infrared gun to check for hot wall areas during hot weather.
- Ensure that all insulated surfaces are clad with aluminum lining.
- Insulate all flanges, valves, and couplings.
- Establish a management information system on loading, efficiency, and specific fuel consumption of the furnace .

Q. Describe briefly the various types of Combustion Control Systems .

Ans. Combustion controls assist the burner in regulation of fuel supply, air supply, (ie, fuel to air ratio), and removal of product gases of combustion to achieve optimum boiler / furnace efficiency. The amount of fuel supplied to the burner must be in proportion to the steam pressure and the quantity of steam required. The combustion controls are also necessary as safety device to ensure that the boiler operates safely.

Various types of combustion controls systems in use are:

On/Off Control

The simplest control, ON/OFF control means that either the burner is firing at full rate or it is OFF. This type of control is limited to small furnaces / boilers.

High/Low/Off Control

Slightly more complex is HIGH/LOW/OFF combustion control system where the burner has two firing rates. The burner operates at slower firing rate and then switches to full firing as needed. Burner can also revert to low firing position at reduced load. This control is fitted to medium sized boilers.

Modulating Control

The modulating control operates on the principle of matching the steam pressure demand by altering the firing rate over the entire operating range of the boiler. Modulating motors use conventional mechanical linkage or electric valves to regulate the primary air, secondary air, and fuel supplied to the burner. Full modulation means that boiler keeps firing, and fuel & air are carefully matched over the whole firing range to maximize thermal efficiency.

Q. What are the benefits of ceramic coatings of inside of a furnace ?

Ans. Ceramic coatings in furnace chamber promote rapid & efficient transfer of heat, uniform heating and extended life of refractories. The emissivity of conventional refractories decreases with increase in temperature whereas for ceramic coatings it increases. This outstanding property has been exploited for use in hot face insulation of furnaces .

Ceramic coatings are high emissivity coatings which when applied has a long life at temperatures up to 1350°C. The coatings fall into two general categories-those used for coating metal substrates, and those used for coating refractory substrates. The coatings are non-toxic, non-flammable and water based. Applied at room temperatures, they are sprayed and air dried in less than five minutes.

The coatings allow the substrate to maintain its designed metallurgical properties and mechanical strength. Installation is quick and can be completed during furnace shut down. Energy savings of the order of 10-20% have been reported depending on the type of furnace and operating conditions.

Heat transfer in a furnace depends mainly on radiation, which in turn depends on emissivity. At higher temperatures, the emissivity decreases and is of the order of 0.3. High emissivity refractory coating, if applied on the internal surface of the furnace, increases the emissivity to 0.8, thus contributing to increased heat transfer. The application of high emissivity coatings in furnace chambers promotes rapid & efficient transfer of heat, uniform heating, and extended life of refractories & metallic components such as radiant tubes & heating elements. For intermittent furnaces or where rapid heating is required, use of such coatings is found to reduce fuel consumption up to 10%. Other benefits are temperature uniformity and increased refractory life.

Q. How is ceramic fiber lining of furnace inner is beneficial ?

Ans . Ceramic fibre is a low thermal mass refractory used in the hot face of the furnace and fastened to the refractory walls. Due to its low thermal mass the storage losses are minimized. This results in faster heating up of furnace and also faster cooling. Energy savings by this application is possible only in intermittent furnaces.

Thermal mass in a furnace contributes significantly to storage heat loss, especially in batch furnaces with huge refractory brickwork. Every time the furnace goes through a heating cycle, the whole refractory mass needs to be heated all over. Introduction of ceramic fiber to replace conventional fire brick lining helps to reduce thermal mass to almost 15% of the original, helping to lower fuel consumption and batch time reduction as well.

A furnace lined with ceramic fibre is shown below.

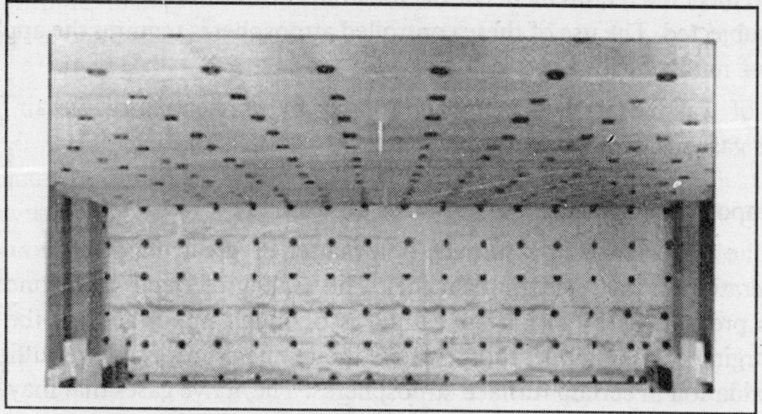

Fig.56.35: Use of ceramic fiber in a furnace

Q. Describe the various factors to be considered in the selection of furnaces .

Ans. In selecting a furnace, the purpose for which it is required and the fuel consumption are of primary importance. Its capacity should be related closely to the probable output since one of the main causes of heat loss lies in incompletely filled furnaces.

The cost of the fuel is a continuous charge on the heating operation, usually for outweighing the capital charges. Consequently the use of mechanical aids to operation, control instruments, and means of heat recovery should not be excluded on grounds of capital costs. An apparently high capital cost spread over the life of the furnace may in the long run produce a considerable economy.

Other factors for special attention in selection of furnaces are as follows :

- Robustness of the furnace structure: Besides a mechanically stable framework, good quality refractories and attention to efficient laying of the bricks are essential.
- Ease of control: All valves, damper controls and instruments should be situated in a convenient, and as far as possible in a central position to give ready accessibility for control purposes. Provided that over-elaboration is avoided, automatic control is an instrument of efficiency.
- Ease of maintenance: This implies accessibility, the use of readily replaceable parts, and means for inspection & cleaning.
- Insulation: This must be regarded as an essential feature of almost any type of furnace.
- Provision for adequate draught (preferably balanced) : A chimney is not necessarily the most efficient means of providing draught. Fans should be carefully rated for the duty intended. Flues, dampers & sight holes should be so arranged and constructed as to reduce the possibility of air in leakage to a minimum.
- Furnace atmosphere: It may be necessary to heat the goods in the furnace in an atmosphere having a composition that will not injure them. The controlled atmospheres are provided for the heating of materials in which a specific quality of surface must be preserved. In such cases the controlled atmosphere may consist of cracked ammonia, or other gaseous mixtures prepared from town gas, or from hydrocarbon gases, the composition of which is adjusted to meet the requirements of the materials heated and of the process to which they are to be subjected. The use of these controlled atmospheres requires the application of the principle of muffle heating.
- Recovery of waste heat: This is applied mainly by pre-heating of the air, or of the fuel where it is gaseous, or of the furnace charge.

Q. What is the importance of control of furnaces atmosphere ?

Ans. Control of the atmosphere in a furnace is a matter of great importance and may be the dominant consideration in the selection of a fuel. The scaling of steels, as distinct from surface tarnishing that is preventable by bright annealing, is of much importance in the re-heating of material before forging or rolling operations, where the temperatures used are sufficiently high to cause excessive oxidation in certain furnace atmospheres. The active gases that may be present are oxygen, carbon dioxide and steam, usually diluted with nitrogen and possibly containing reducing & sulphurous gases. The degree of scaling is determined by the temperature of treatment (which is the most important factor), the time at that temperature, and the composition of the furnace atmosphere.

Scaling is diffusion phenomenon and is governed by diffusion laws. With specific alloying constituents in a metal, protective layers of scale may be formed; these reduces the rate of scaling. The scaling losses in oxidizing gases are substantially decreased by adding small percentages of a reducing gas, but with rising temperature higher concentrations of the reducing gas are required to effect a corresponding reduction in scaling. Thus, in a carbon dioxide atmosphere the addition of 2 % of carbon monoxide at 930°C will reduce the scaling losses to the same extent as the addition of 10 % at 1,040°C. The addition of hydrogen to carbon dioxide will, at low concentrations, give an increase in the rate of scaling due to the formation of water vapour, which is more strongly oxidizing than is carbon dioxide.

As a practical consideration in the heating of ferrous metals the condition of the metal surface beneath the scale and the character of the scale may be more important than merely the amount of scale formed. This applies particularly to decarburization, which may follow the oxidation process and is a cause of ultimate failure of the metal in use. In producing decarburization in a combustion atmosphere moist hydrogen has by far the greatest effect, followed by carbon dioxide; air and steam are less effective.

Sulphur dioxide has a much greater effect on scaling than other sulphurous gases . In addition to increasing the scaling losses it will also promote the absorption of sulphur by the steel, the attack being localized and proceeding by way of the grain boundaries. Both sulpur dioxide and hydrogen sulphide react with the oxides of iron at high temperatures to form an oxysulphide eutectic, which has a comparatively low melting point-about 950°C - and is, therefore, molten at the heating temperatures usual for hot working. It is found below the layer of scale, and in its presence the extent of oxidation is two to there times as great as that obtained in a similar desulpherized atmosphere.

To counteract the effect of the active gases on the oxidation of mild steel, for example, the atmosphere must contain carbon monoxide and hydrogen, firstly in sufficient quantity to ensure that the stable sulphurous gases are converted to hydrogen sulphide, and secondly in concentrations, relative to carbon dioxide and water vapour, such that oxidation is reduced to a minimum. At high temperatures such necessary concentrations may be so high as to make the process entirely uneconomical in an internally heated furnace chamber.

Q. Discuss the role of Heat Transmission and Gas Flow in relation to Furnace Design .

Ans. The successful performance of any furnace depends upon factors like : behaviour of the fuel during combustion (particularly in respect of mode of heat release), the character of the flow of the hot gases through the furnace system and the mode of heat transmission in the furnace structure & the charge.

Heat Input

All the factors relating to the determination of the heat input, the properties & heating values of the fuels, sensible heat of air & fuel pre-heat, the heat content of the gases at any point in the furnace system, provided the temperature is known, are ascertainable with accuracy.

Flow of Gases

Within a heated furnace chamber the influence of turbulence-an important operative factor-is still incompletely understood with respect to heat transmission, though useful qualitative guidance as to the change of the gaseous flow may be derived from the use of models on the basis of dimensional analysis.

Heat Release from the Flame

The heat release from the flame involves the phenomena of both luminous and non-luminous radiation. The effective emissivity in coal flames, including both luminous & non-luminous radiation, is related to the excess air used in combustion by a logarithmic relationship, the radiation being greater in the presence of less excess air. The fundamental factors involved for a particular fuel are the air/fuel ratio, and the aerodynamical conditions produced by the combustion appliance.

Q. Explain with neat sketch underfeed and overfeed stoker principle of firing coal.

Ans.

Underfeed stoker firing:

Under stoker firing is the process of combustion in which the new coal is heated by radiation in the presence of air and located under ignited fuel bed. The heating of coal is running less rapidly and release volatile matter combine with air, so generate low smoke.

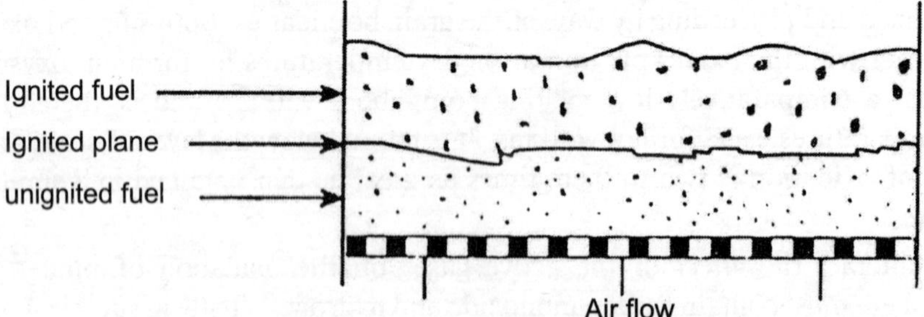

Fig.56.36: Underfeed stoker firing in steam boiler

Overfeed stoker firing:

Overfeed stoker firing is the process of combustion in which the unignited fuel or incoming coal is located above ignited fuel bed. The ignited fuel transfer heat to the incoming coal by radiation. Moreover coal is heated convection from hot gases that has been through the combustion. Secondary air is added to perform complete combustion unless steam boiler will produce more smoke because the hot gases contain little oxygen.

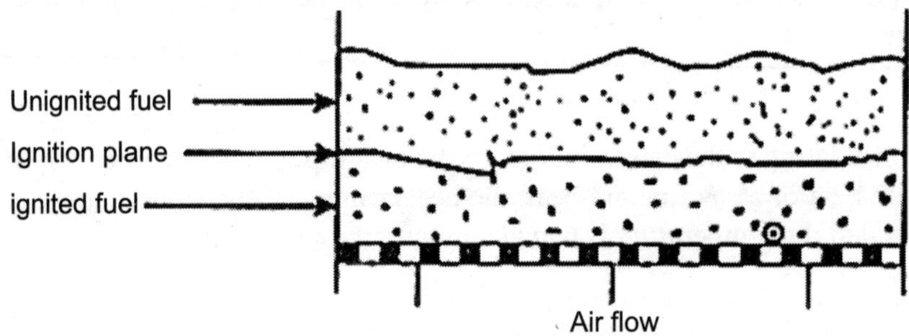

Fig.56.37: Overfeed stoker firing in steam boiler

Q. List the requirements of pulverized coal burners. Explain a cyclone furnace with its advantages and disadvantages.

Ans. The main requirements of pulverized fuel burners are:

- It should mix thoroughly primary air with coal particles and secondary air.
- It should create turbulence and maintain stable combustion.
- It should control the flame shape and it travel in the furnace.
- The velocity of primary air and coal particles should be same as that of flame velocity to avoid flash.
- The burner should have ability to withstand overheated due internal fires and excessive abrasive wear.

Cyclone Burner

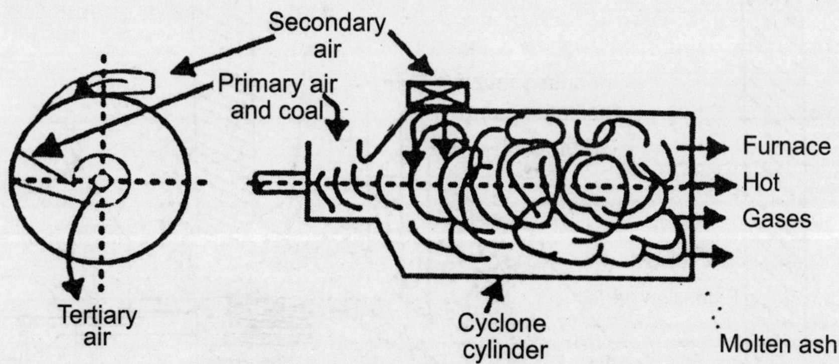

Fig.56.38: Cyclone furnace

It consists of horizontal cylinder of water cooled construction 2 to 3 meters in diameter and 2.5 m in length. The horizontal axis of the burner is slightly deflected downward towards the boiler. These burners are externally attached to the furnace. The cyclone burner receives pulverized coal carried by the primary air tangentially to the cylinder at outer and creates stroke and highly turbulent Vortex. Secondary air enters in to the cylinder tangentially to complete the combustion. These burners can be + or – 30 degree up and down ; it helps in controlling the super heater temperature. The fuel supplied burns quickly with high heat liberation rates . The ash forms the molten film over the inner wall surface and molten ash flows to an ash disposal system. The cyclone burners gives best results with the graded fuel.

Advantages:

- Crushed coal can be used instead of costly pulverized coal.
- It can burn low grades of coal.
- Percentage of excess air required is less.
- Combustion rates can be easily controlled by varying fuel and air supply.
- Combustion efficiency is high.
- High furnace temperature can be obtained .
- Disadvantages:

- Consumes high power.
- Produces more oxides of nitrogen.

Q. Describe with a neat sketch the working of multi retort stoker .

Ans. Multi retort stoker is shown below in figure. it consists of series of alternate retorts and tuyers boxes for supply of air. Each retorts is fitted with reciprocating ram for feeding pusher plates for uniform distribution of coal. Coal falling from hopper is pushed forward during inward stoke of the stoker arm. Then distributing ram pushes the entire coal down to length of the stoker. The ash formed is collected at the end as shown below in the figure. The number of retorts may be vary from 2 to 20 with burning capacity varying from 200 kg to 2000 kg/hr/retort.

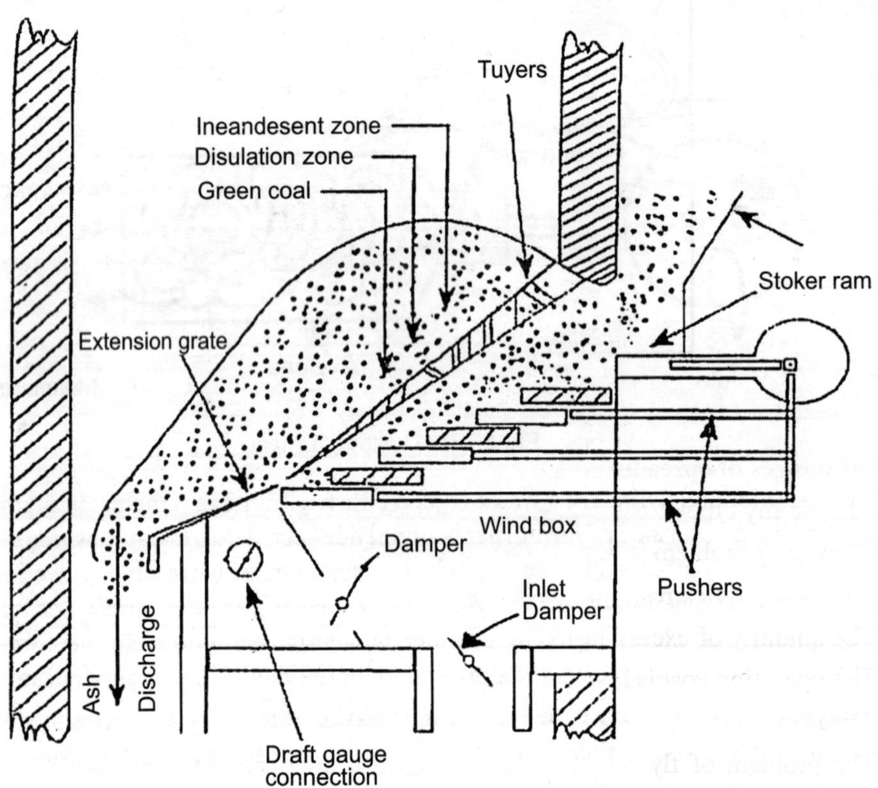

Fig.56.39: Multi Retort Stoker

Q. With the help of a neat diagram explain the working of spreader stoker. State its limitations.

Ans. The coal from hopper is fed by a rotating feeder is fed by a rotating feeder, drum fitted with short blades on its surfaces, to the spreader or distributor below which projects the coal particles on to grate holding an ignited fuel bed. The finer particles burns in suspension and the coarse particles are consumed on the grate.

The speed of the feeder is directly proportional to the steam output of the boiler. The secondary air helps in creating turbulence and completing combustion. High capacity boilers may have travelling grate stoker in addition to spreader stoker. The grate consists of cast iron links underneath . The grate connect all the bars to a lever that makes the ash fall through to the ash pit below. Spreader stokers are capable of burning any type of coal.

Spreader stoker is shown below in figure.

Ans.

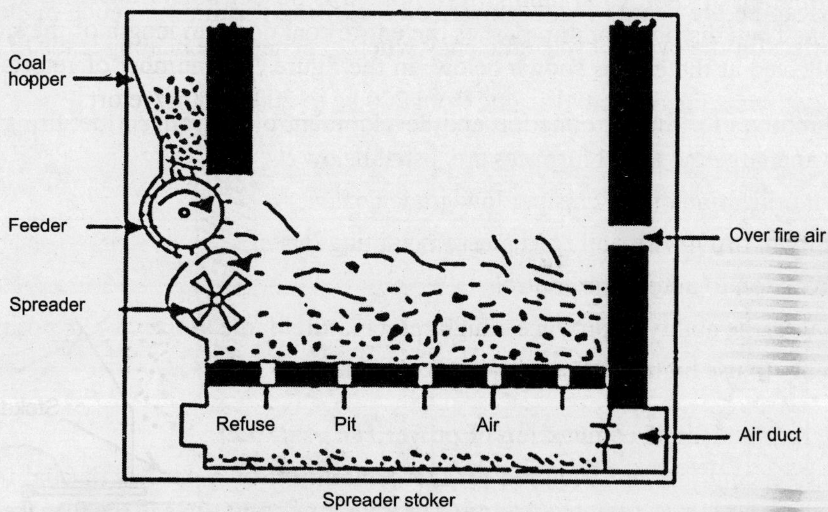

Fig.56.40: Spreader stoker

Advantages of spreader stoker :

- Almost any type of the coal can be burnt.
- Clinkering problem is less.
- It is having quick response to varying load.
- The quantity of excess air required is less.
- The operation cost is low .

Disadvantages :

- The problem of fly ash is high. It requires a dust collector to prevent the environment pollution.
- Coal particles trapping mechanism is necessary to prevent their escape with excess air.
- Its operating efficiency decreases with varying sizes of coal.

Q. What are the advantages of pulverized fuel (P.F) firing in the furnace/boiler ?

Ans. Pulverised fuel firing is a method whereby the finely crushed fuel, generally reduced to a fineness such that 75-85 % , passes a 200 B.S. sieve, is carried forward by air through pipes directly to burners or to storage bins from where it is passed to burners. When discharged into the combustion chamber the mixture of air and fuel burns in a manner very similar to that of the combustion of a gas.

The use of pulverized fuel has increased rapidly and constitutes the principal form in which coal is used.

The increasing proportion of fines in the coal raised since machine-mining became general has rendered pulverized fuel the most satisfactory way of using fine, low-grade coal.

There is less likelihood of bonded deposits accumulating on the heating surfaces than with stoker firing. In addition, while with stoker firing ,there is an upper limit to the temperature to which the combustion air can be pre-heated ; with pulverized fuel firing this limit is much higher. This has an important effect on the overall cycle efficiency where regenerative heating leads to a high feed temperature .

The economic motives for the introduction and development of pulverized fuel firing in thermal power stations and large industrial furnaces are listed below :

- Efficient utilization of the cheaper low-grade coal.
- Flexibility in firing with ability to meet fluctuating loads.
- Better reaction to automatic control.
- In furnaces, the ability to produce a high-temperature flame at the correct position.
- The ability to use high combustion air temperatures.

Q. Discuss the mechanism of combustion of pulverized coal/fuel.

Ans. Factors influencing the time required for the combustion of a particle of coal include coal properties such as rank & size, temperature, and aerodynamic conditions in the furnace into which the particle is injected.

Studies of the ignition and combustion of particles of coal injected into a hot furnace in a stream of air indicate that combustion takes place in three stages:

(a) pre-ignition during which changes in shape & size (depending on the rank of the coal) and some evolution of volatile matter occurs,

(b) ignition and combustion of the volatile matter and

(c) combustion of the residual carbon.

During combustion a particle is surrounded by an "atmosphere" of combustion products through which the oxygen has to diffuse to react with the carbon. Experiments have shown that the time for the combustion of a 200-mesh (74μ)

particle of coal containing 30-40 % volatile matter is approximately 0.3 second, and that the time required for combustion varies approximately as d1.5, 'd' being the diameter of the particle. Higher burning rates can be attained with relative motion between the air and the burning particle to increase the rate of diffusion to the carbon surface. With small particles however, this is difficult to achieve as the particle rapidly achieves the velocity of the air stream into which it is introduced.

Since the solid residue from low volatile coals and anthracites is relatively unreative, these fuels are more difficult to burn than coals of lower rank. Consequently they should be more finely ground and used with less excess air, whilst heat losses should be avoided by providing refractory walls backing the ignition zone. These precautions are essential to secure satisfactory operation.

Q. What are the functions and requirement of a burner ?

Ans. The main requirements of a burner are that it shall ensure rapid & intimate mixing of fuel & air, and maintain an adequate supply of oxygen surrounding each fuel particle until its combustion is completed. The need for intimate and rapid mixing of fuel & air within the furnace dominates the design of the burner. According to whether a short or a long flame is required the burner is designed for turbulent or parallel flow.

Primary air that carries the powdered coal from the ball mill to the furnace is only about 20% of the total air needed for combustion. Before the coal enters the furnace, it must be mixed with additional air, known as secondary air, in burners mounted in the furnace wall.

In addition to the prime function of mixing, burners must also maintain stable ignition of fuel-air mix and control the flame shape & travel in the furnace. Ignition depends on the rate of flame propagation. To prevent flash back into the burner, the coal-air mixture must move away from the burner at a rate equal to flame front travel. Too much secondary air can cool the mixture and prevent its heating to ignition temperature.

The requirements of a burner can be summarised as follows:

- The coal and air should be so handled that there is stability of ignition.
- The combustion is complete.
- In the flame , the heat is uniformly developed avoiding any superheat spots.
- Adequate protection against overheating, internal fires and excessive abrasive wear.

Q. How are the pulverised fuel burners classified :

Ans. Pulverised fuel burners may be classified as follows:

- Long flame burners
- Turbulent burners or short flame burner
- Tangential burners
- Cyclone burners.

Q. Describe the salient features of long flame burners used for pulverized fuel/coal firing .

Ans. Long flame burners are also called U-flame or steamlined burners. In this type of burner , coal is floated on a portion of air supply (primary air) and supplied to the burner in one stream. Secondary and tertiary air supplies are maintained as shown in Figure below . The length of flame is increased in the combustion chamber by downward initial flow of the flame. The flame produced is stable, long and intense but it can be made short & intense by adding much secondary air. Tertiary air enters through the burner and forms an envelope around the primary air & fuel and provides better mixing. Furnaces for low volatile coal are equipped with such burners to give a long flame path.

Long flame burner is shown below.

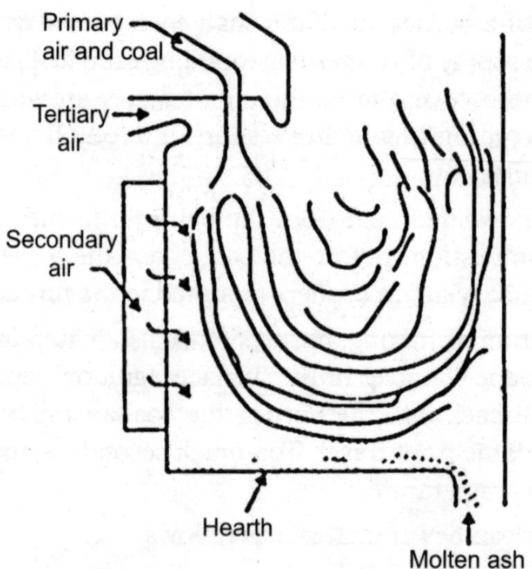

Fig.56.41: Long flame burner

Q. Describe the salient features of short flame / turbulent burners used for pulverized fuel/ coal firing .

Ans. A turbulent burner is also called a short flame burner. As shown in Figure below , these burners can fire horizontally or at some inclinations by adjustment. The fuel-air mixture and secondary hot air are arranged to pass through the burner in such a way that there is good mixing and the mixture is projected in highly turbulent form in the furnace. Due to high turbulence created before entering the furnace, the mixture burns intensely and combustion is completed in short distance.

This burner gives high rate of combustion compared with other types. This is generally preferred for high volatile coals. All modern plants use this type of burner.

A short flame turbulent burner is shown below.

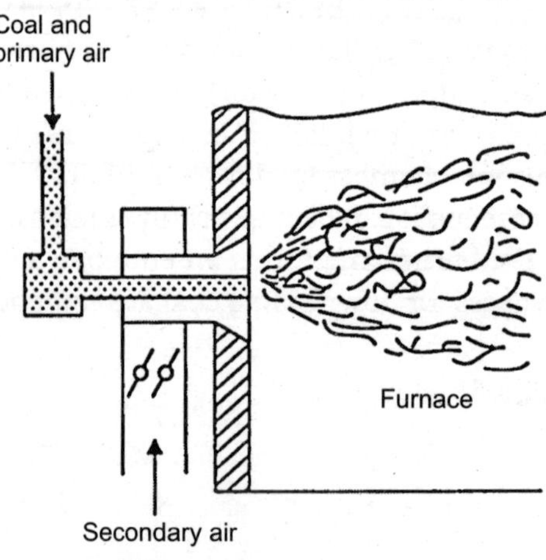

Fig.56.42: Turbulent burner

Q. Describe the salient features of tangential burners used for pulverized fuel/coal firing .

Ans. These burners are shown below in Figure . In this case four burner are located in the four corners of the furnace and are fired in such a way that the four flames are tangential to an imaginary circle formed at the centre. The swirling action produces adequate turbulence in the furnace to complete the combustion in a short period and avoids the necessity of producing high turbulence at the burner itself.

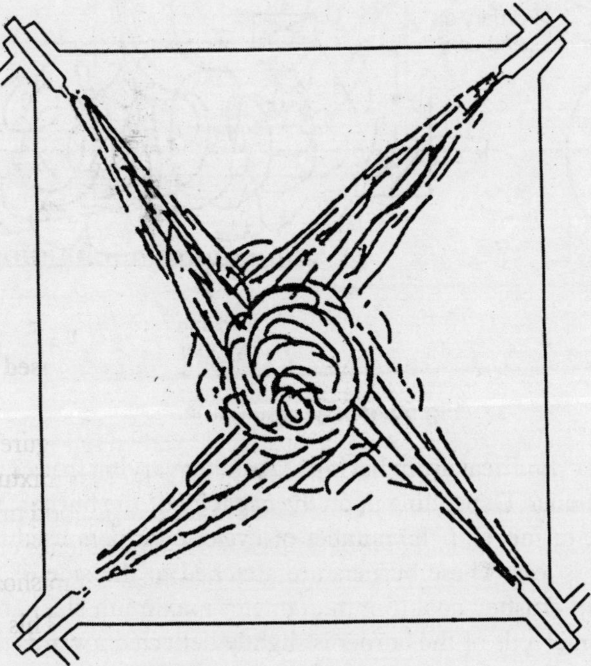

Fig.56.43: Tangential burners.

The tangential burners have the following advantages:

- The parts of the burner are well-protected by furnace wall tubes.
- The operation of these burners is very simple.
- High heat release with complete and effective utilisation of furnace volume is possible.
- The completeness of combustion is exceptionally good and a maximum degree of turbulence exists throughout the furnace.
- Liquid, gaseous and pulverised fuels can be readily burned either separately or in combination.

Q. Describe the salient features of cyclone burners used for pulverized fuel/coal firing .

Ans. Two main disadvantages associated with pulverised coal firing are:

- High cost.
- 70% of the ash escapes as fly-ash , which requires expensive dust collectors in the flue gas path.

These disadvantages are offset by a cyclone burner. In a cyclone burner coal is crushed to a Maximum size of 6 mm and blows into a cylindrical cyclone furnace. The fuel is quickly consumed and liberated ash forms a molten film flowing over the inner wall of the cylinder. Owing to inclination of the furnace, the molten ash flows to an appropriate ash disposal system.

The description of a cyclone burner shown below in figure is given below:

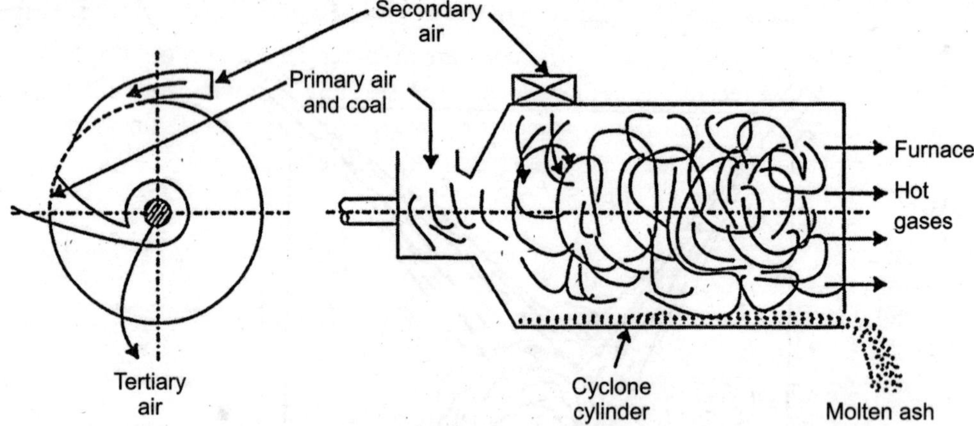

Fig.56.44: Cyclone burner

It consists of a horizontal cylindrical drum having a diameter varying from 2 to 4 meter, depending upon the capacity of the boiler. Depending upon the capacity of the burner the number of cyclone burners used may be one or more. If the number of cyclone burners used is more than one, the diameter of each burner is less. These burners are attached to the side of the furnace wall and have vents for primary air, crushed coal (6 mm diameter maximum size) and secondary air. It is water-cooled. The horizontal axis of the burner is slightly deflected towards the boiler.The cyclone burner receives crushed coal carried in primary air (at 800 mm water gage pressure) at the left end. Tangential entry of the coal throws it to the surface of the cylindrical furnace. Secondary air enters the furnace through tangential ports at the upper edge at high speed and creates a strong & highly turbulent vortex. Extremely high heat liberation rate and use of preheated air cause high temperatures to the tune of 2000°C in the cyclone. The fuel supplied is quickly consumed and liberated ash forms a molten film flowing over the inner wall of the cylinder.

Due to horizontal axis of the burner being tilted , the molten ash flows to an appropriate ash disposal system. The cyclone furnace gives best results with low grade fuels.

Advantages of cyclone burners :

- High furnace temperatures are obtained.
- Simplified coal existing equipments can be used instead of costly pulverised mills.
- The cyclone burners reduce the percentage of excess air used.
- It can burn poorer and cheaper grades of coal.
- As the swirling effect and consequently the mixing of air & crushed coal is better, it provides for a higher furnace capacity & efficiency.

- Boiler efficiency is increased.
- The cost of milling in cyclone fire is less as the finer particles are not required in this case.
- Combustion rates can be controlled by simultaneous manual adjustment of fuel feed & air flow and response in firing rate changes is comparable to that of pulverised coal firing.

Q. Describe the various factors to be considered in the design of a pulverized fuel/coal burner .

Ans. In designing burners it is assumed that the coal/air ratio can be correlated with the rate of flame propagation. Any appreciable lack of balance between these two will cause either loss of ignition or back-firing. The speed of flame propagation varies with various fuel/air mixtures and with different types of coal. The maximum rate of flame propagation, for low volatile coals is at a point where the air/coal ratio is low, and this ratio must be still further reduced for anthracite.

To a major degree the volatile content of the fuel controls the stage of ignition, which-is the point where combustion becomes self-supporting. The temperature of ignition of bituminous coal is significantly lower than that of anthracite and this is another factor of which account must be taken in burner design. To overcome this difficulty, arrangements have been made in certain installations to extract part of the air from the fuel/air mixture at the burner by the use of an "air separation" burner, and to control the ignition by regulating the quantity of air extracted, the vented air being subsequently introduced into the combustion chamber at a suitable point as secondary air.

In direct-fired systems , the air/coal ratio is important since the amount of air required for sweeping the mill and conveying the pulverized fuel may affect the composition of the air/fuel mixture at the burner. Some systems incorporate an additional air control which enables the composition of the fuel/air mixture to be adjusted and reduces the need to relate the burner design closely to the air- flow characteristics of the pulverizer.

FUEL DISTRIBUTION

It is important to secure uniform distribution of fuel to the several burners feeding a furnace and to avoid local differences in concentration ;attempts have been made to achieve this by situating the exhaust fans and fuel piping symmetrically in relation to the burners. These attempts have not been completely successful and mechanical distributors have been found to be necessary.

AIR PRE-HEATING

The importance of high furnace temperature to secure a high rate of radiant heat transmission from burning fuel to the receiving surfaces is clear. Air preheat will help to heat the fuel more quickly and will thus enable the processes of distillation and combustion to be completed in a shorter time. The main practical advantage of preheating air, however, is to reduce the flue gas loss.

It is impossible to prescribe exact temperatures for either the primary air or the secondary air. If conveying air is used for drying coal and also, as secondary air the ultimate temperature will depend on the extent of drying & upper limit of hot drying air. Limit of the temperature of that air when used as primary air being about 65 0C . The use of a much higher temperature for secondary air is, of course, desirable, but here also there are limits according to the nature of the combustion cycle. An unduly high temperature cannot be used in the case of high volatile bituminous coals owing to the risk of fire ; with lower volatile coals & anthracites, of course, much higher temperatures are possible, i.e. up to 315 to 380 0C.

One of the main problems in pulverized-fuel firing is to shorten the time required for combustion. The development of combustion chamber design has been to burn the maximum amount of coal dust in the minimum space and in minimum time. To this end, means were sought for grinding the coal to a very fine sub division at an economic cost. But for a given fineness, shortening of the combustion time is mainly the aerodynamic problem of bringing air and fuel together within the furnace as quickly as possible. This means that the air, waste gas and fuel particles must be in a state of violent motion, or turbulence, relative to one another. Early designers sought to achieve this through the use of burners causing a turbulent or whirling motion, skillfully introducing primary, secondary and sometimes tertiary air. Such burners promote quick ignition but it is a characteristic of pulverized-fuel firing, in distinction to stoker-firing, that it is extremely difficult to influence the tail of combustion. This cannot be done by burner agitation or by introducing secondary air at points farther along in the path of the burning dust.

Q. Describe the principle of fuel oil firing . How are oil burners classified ?

Ans.

Principle of oil firing. The functions of an oil burner are to mix the fuel & air in proper proportion and to prepare the fuel for combustion. Figure below shows the principle of oil firing.

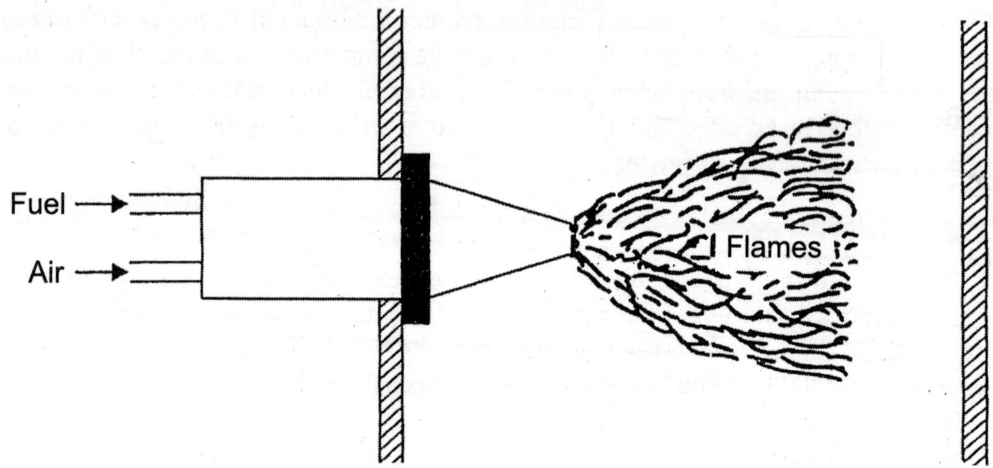

Fig.56.45: Principle of oil firing

Classification of oil burners:

The oil burners may be classified as:

(a) Vapourising oil burners:

- Atmospheric pressure atomising burner
- Rotating cup burner
- Recirculation burner
- Wick type burner

(b) Atomising oil burners:

- Mechanical or oil pressure atomising burner
- Steam or high pressure air atomising burner
- Low pressure air atomising burner.

The broad classification of oil burners based on the principle used for atomisation is given below.

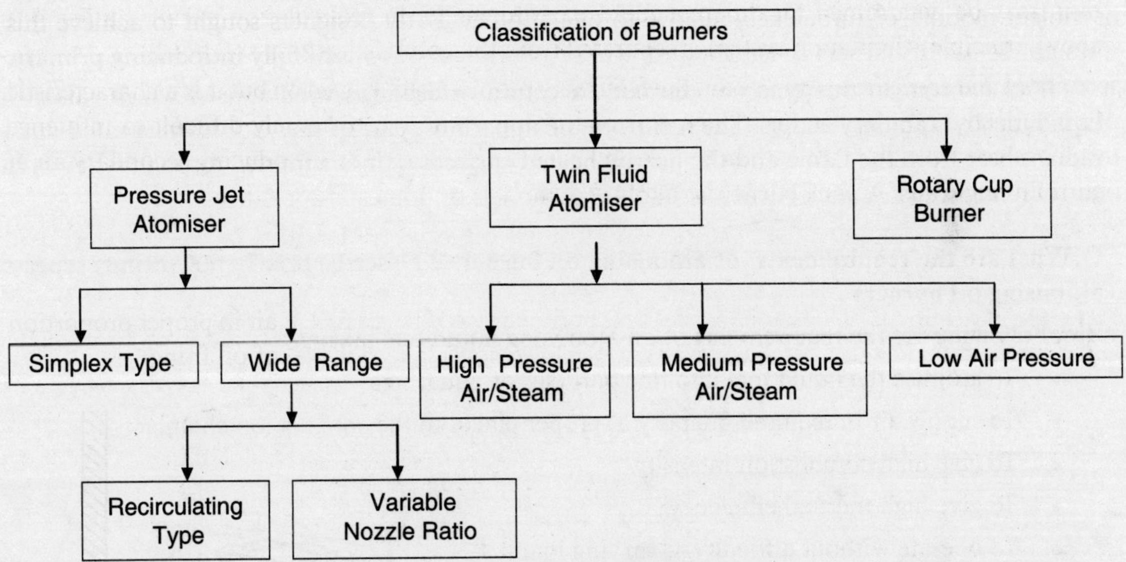

Q. What are the requirements of vapourising oil burners ? Describe briefly the various types of vapourising oil burners .

Ans. Following are the requirements of a vapourising oil burners :

- To vapourise the liquid fuel before ignition.
- To mix the vapourised fuel thoroughly with the air.
- To minimise the soot formation.
- To give high heat release by burning large quantity of oil per hour.
- To allow for efficient combustion of fuel at part load operation.

Various vapourising oil burners are described below :

Atmospheric pressure oil atomising burner . This burner makes use of highly volatile liquid fuels such as naphtha, volatile gasoline etc. Here the fuel at low pressure is passed through a tube adjacent to the flame before being released through an orifice. While passing through the hot tube, most of the fuel is vapourised so that the fluid ejected from the orifice is more or less a vapour. The required quantity of primary air is supplied to burn the vapour stream in a cylindrical tube.

Rotating cup burner. In this type of burner, the fuel oil flows through a tube in the hollow shaft of the burner and into the cup at the furnace end. An electric motor or an air turbine runs the shaft and the cup at high, speeds (3000 to 10000 r.p.m.). As a result of centrifugal force fuel is split into small droplets. About 10 to 15 percent of air is supplied, as primary air. This air is supplied from a blower surrounding the cup. The shape of the flame is governed by the sharp edge of the cup and the position of air nozzle. These burners are used on low as well as medium capacity boilers.

Recirculating burner. A part of the combustion products may be recirculated in order to heat up the incoming stream of fuel and air. Low ratio of the mass of recirculated combustion products to the mass of un burnt fuel-air mixture results in less temperature rise of the mixture, whereas, high ratio may extinguish the flame due to increased proportions of circulated products. An optimum ratio may be determined for different fuels experimentally. In recirculation burner (utilising the above principle) circulation system is separated from the combustion by a solid wall.

Wick burners. In this type of burner , a cotton or asbestos wick is used which raises the liquid fuel by capillary action. The fuel from the uppermost part of the wick is evaporated due to radiant heat from the flame and the nearby heated surfaces. Air is admitted through holes in the surrounding walls. A wick burner is suitable for models or domestic appliances.

Q. What are the requirements of atomising oil burners ? Describe briefly the various types of atomising oil burners .

Ans. Following are the requirements of an atomising liquid fuel burner:

- To atomise the liquid fuel into fine particles of equal size.
- To supply air in required quantity at proper places in the combustion chamber.
- To give high combustion intensity.
- To give high thermal efficiency.
- To operate without difficulty at varying loads.
- To create necessary turbulence inside the combustion chamber for proper combustion of fuel.
- To minimise soot formation and carbon deposit, particularly on the burner nozzle.

Various atomising liquid fuel burners are described below :

(i) Mechanical atomising burners. A mechanical atomising oil burner consists of the following four principal parts:

(1) Atomiser (2) Air register (3) Diffuser (4) Burner throat opening.

Atomiser breaks up the oil mechanically into a fine uniform spray that will burn with minimum of excess air when projected into the furnace. The spray is produced by using relatively high pressure to force oil at high velocity through small tangential passages of sprayer plate.

(ii) Steam atomising burners. This method may, however, absorb some 4 to 5% of the total amount of steam generated. These burners may be divided into two categories:

a. The outside mixing type

b. The inside mixing type

In case of outside mixing [Figure (b)] type burners, oil is rejected through one side of the holes and is blasted by a high velocity jet of steam issuing from other holes. Mixing, however, occurs outside the burner.

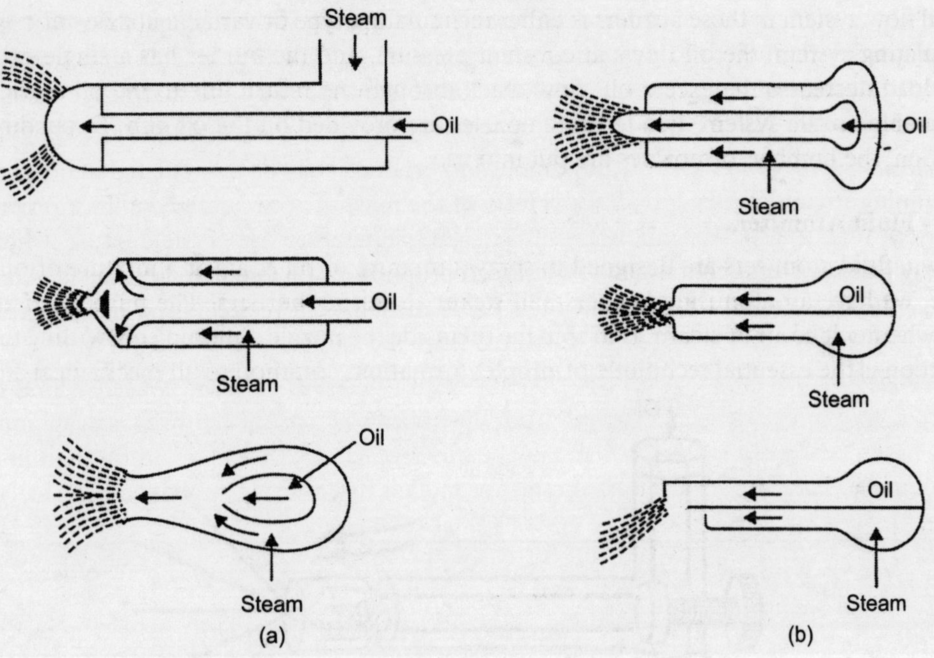

Fig.56.46: (a) Inside mixing burner (b) Outside mixing burner

In case of inside mixing type burners steam and oil are mixed inside the burner before the mixture is projected in the furnace. These burners provide high efficiency at the high firing rates and flexible flame shape.

(iii) Low pressure air atomising burners. In this case air pressure required ranges from 0.015 bar to 0.15 bar.

Q. Describe the salient features of a pressure jet atomizing oil burner .

Ans. The key component of pressure Jet Burners is the high pressure nozzle that consists primarily of two components. The outer body or tip of the nozzle, functions like an orifice for the discharge of the oil and the inner section termed as the swirler, snugly screws fits into the tip, leaving a space in between called the swirl chamber upstream of the orifice. The swirler has several tangential slots that divide the oil stream when it flows through the swirler under pressure. This makes the oil spin in the swirl chamber with high velocity before being spouted through the orifice into the combustion space, as an exploding mist in the shape of a cone, either solid spray or with a hollow cone.pattern, depending upon the nozzle. Nozzles are available with a range of spray angles and patterns.

The oil flow rate can be reduced in the simple design merely by reducing the oil pressure. Efficient operation at varying loads is usually carried out by using nozzles of various jet sizes. The oil pressure in this burner varies from 7 to 20 kg/cm^2.

Pressure jet burners are relatively simpler and inexpensive. This burner is generally used in Boilers, Thermospacs, Aluminum melting furnace etc. where short and wide angled flame is desired. The disadvantage of this type of burner is its poor turn down ratio and high maintenance cost.

The oil flow system in these burners is either recirculating type or variable nozzle ratio type. In the recirculating system, the oil flows at constant pressure, and the burner has a single nozzle. Thus when load decreases the excess oil flow back through the return line to the oil service tank. In the variable nozzle system, two to three nozzles are provided on the oil gun. Depending on load condition, the number of nozzles are put into use.

Twin - Fluid Atomisers

The twin fluid atomisers are designed to spray a mixture of oil & air or a mixture of oil & steam. Hence, we have air-atomising burners and steam atomising burners. The purpose of the second fluid, whether it be air or steam, is to spin the oil inside the nozzle. Spinning or swirling the oil prior to ejection is the essential technique of droplet formation, common to all mechanical atomizers.

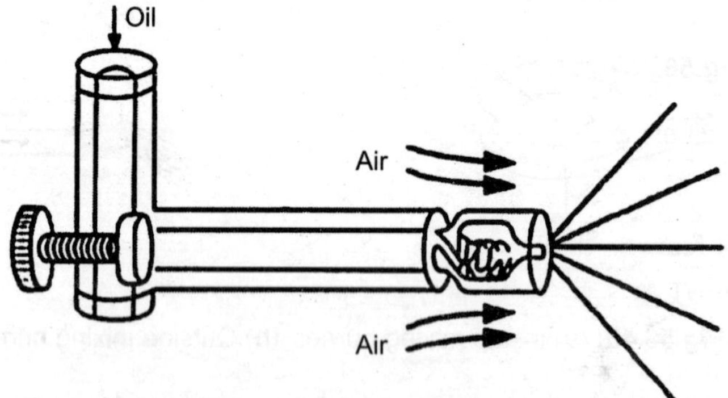

Fig.56.47: Pressure jet atomizing oil burner

Q. Describe the saient features of Medium & High Air Pressure Oil Atomiser Burners .

Ans. Air for atomising is provided by an air compressor at high pressures. When thermal loads change, the quantity of atomising air is not changed but only the secondary air entering the system is regulated. Therefore, these burners have better efficiencies at even low loads. The compressed air may be replaced by steam as the atomising medium in these burners. Steam assists in the breaking up of oil droplets in the combustion zone and hence steam assisted atomisation is superior, especially in the case of burning of heavy fuel oils and LSHS. Steam pressure burners should be maintained carefully, since a slight increase in the nozzle size can lead to considerable wastage of steam or electric power besides creating a distortion in the spray pattern. Atomising air forms a lower proportion of total combustion air with higher atomising air velocities the combustion intensity increases proportionately. For these reasons, these burners are more popular in high temperature furnace applications, where combustion air can be preheated considerably.

As air at 300°C can easily crack the oil in the burner,hence only colder air (a small percentage of total air) is fed to the burner for atomisation. Hot air is separately fed in to the combustion chambers through secondary air-ports as secondary air. Steam or compressed air, ranging from 0.5 to 10 kg/cm^2 is used as the atomising medium. Steam consumption may vary from 1 to 15 pound per gallon of oil and air consumption from 22 to 100 cft of STP air per gallon of oil, depending on the design & size of the atomiser and oil viscosity. For external oil mixing burner types , oil pressure at 0.5 to 1 kg/cm^2 is recommended at the burner but for emulsion type (oil and steam or air mix inside the unit) the oil pressure must be about the same as the steam or air pressure.

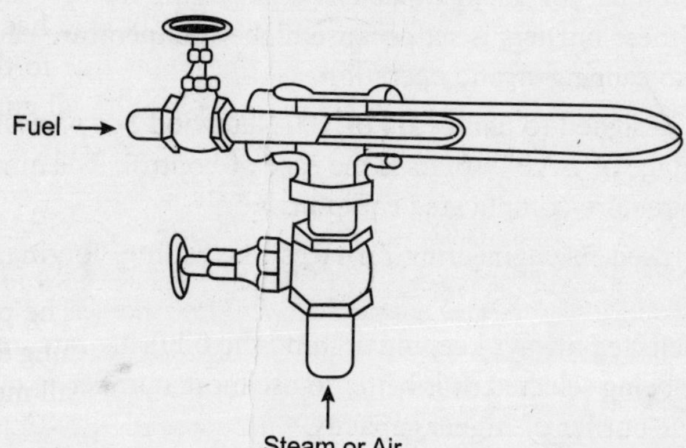

Fig.56.48: Medium & high air pressure oil atomiser burner

Medium air pressure burners are used in higher capacity Boilers/Thermopacs etc. High pressure burners are widely used in Kilns.

Q. Describe the salient features of Low Air Pressure (LAP) Oil Atomisation Burner.

Ans. In the low air pressure(LAP) oil atomisation burner, air at low pressure is provided by a blower and acts as the atomising medium. Air pressure recommended at the burner tip for light distillate oil is 600-700 mm WG and for residual oil is 850-1000 mm WG. The fuel oil pressure at burner tip is maintained at 0.5-0.8 kg/cm^2 just enough for positive delivery and flow control. The oil and air streams are brought in to atomiser through different routes, made to interact with each other and leave through a common orifice. The amount of aeration in the mixing process is not much although fine oil droplet formation is promoted through colloidal dispension. Most of the designs are based on pipe within a pipe nozzle line. These burners work efficiently at the designated oil flow rate but the efficiency drops sharply as the oil flow rate is reduced at low loads. It is hence necessary to avoid selection of over sized burner of this type. Air and oil flow rates are regulated by means of valves on the respective supply lines. In the self-proportioning type, it is possible to regulate air and oil flow simultaneously by operation of a single lever. Few manufacturers use primary air for atomisation and secondary air for combustion purpose. The turn down for burner increases in case secondary air is used but it is effective only for burner working at high firing rate.

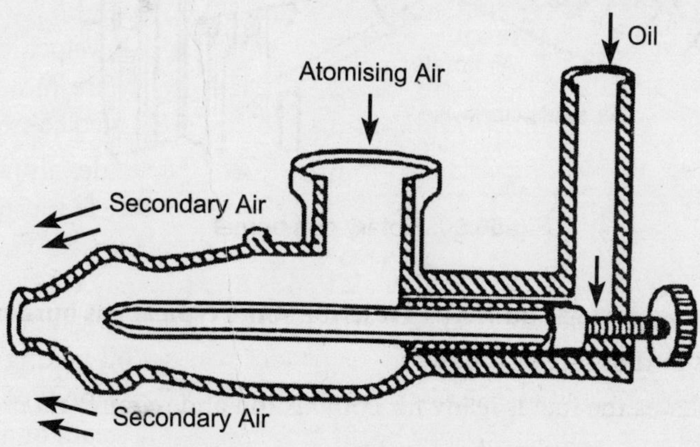

Fig.56.49: Low pressure air atomiser

LAP burners are more suitable for small capacity ranging from 10 to 100 lt/hr. The limit upto 250°C on air preheat on these burners is set because higher temperature cause the oil to vaporise and crack in the burner, so causing erratic operation.

LAP burners are usually designed to handle oil of 100 Redwood Sec viscosity at the burner inlet. One of the biggest advantage of LAP burners is the ease of control. The air/oil ratio can be easily maintained and does not require complicated equipment.

LAP burners are widely used in engineering furnaces like rolling, forging, melting galvanizing, annealing furnaces.

LAP burners should be selected always keeping in mind the oil firing rate, required in the furnace, for which the burners are being selected. It is better to use more number of burners of smaller sizes (if space permits) than one burner of higher capacity.

Q. Describe the salient features of Rotary Cup Oil Atomising Burners.

Ans. In this burner, a thin film of oil is formed by injecting oil into a rapidly spinning cup: This film is converted into minute droplets as it leaves the cup, by the counter action of the atomising air, which is fed by the burner.

The rotary-cup burner functions better with heavy oil than with light or distillate type oils. High maintenance cost and lubrication problems of revolving parts are the main problems which discourage the use of this type of burner.

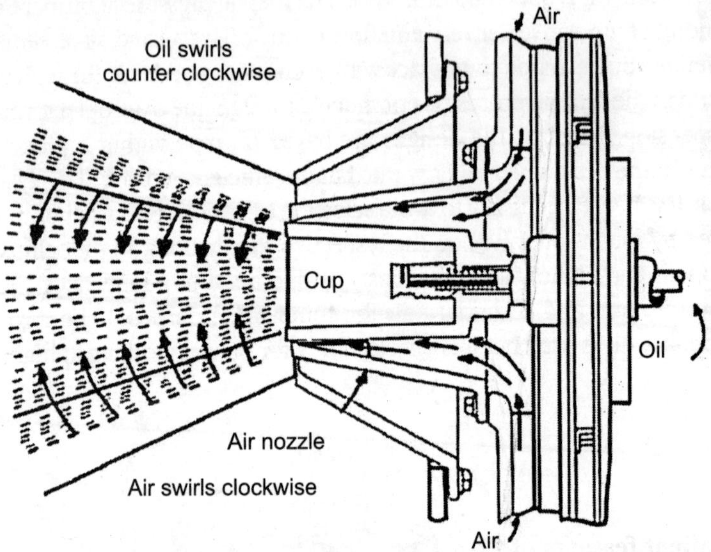

Fig.56.50: Rotary cup burner

Q. What are the advantages of gas burners ? Describe some typical gas burners.

Ans. Advantages of gas burners are as follows :

- It is much simpler as the fuel is ready for combustion and requires no preparation.
- Furnace temperature can be easily controlled.

- A long slow burning flame with uniform and gradual heat liberation can be produced.
- Clean operation is ensured.
- High chimney is not required.
- No ash removal is required.

For generation of steam, natural gas is invariably used in the following cases:

- Gas producing areas.
- Areas served by gas transmission lines.
- Where coal is costlier.

Typical gas burners used are shown below in Figure .

In this burner [Figure (a)], the mixing is poor and a fairly long flame results.

This is a ring type burner [Figure (b)] in which a short flame is obtained.

This arrangement [Figure (C)] is used when both gas and air are under pressure.

In order to prevent the flame from turning back, the velocity of the gas should be more than the "rate of flame propagation".

Typical gas burners are shown below.

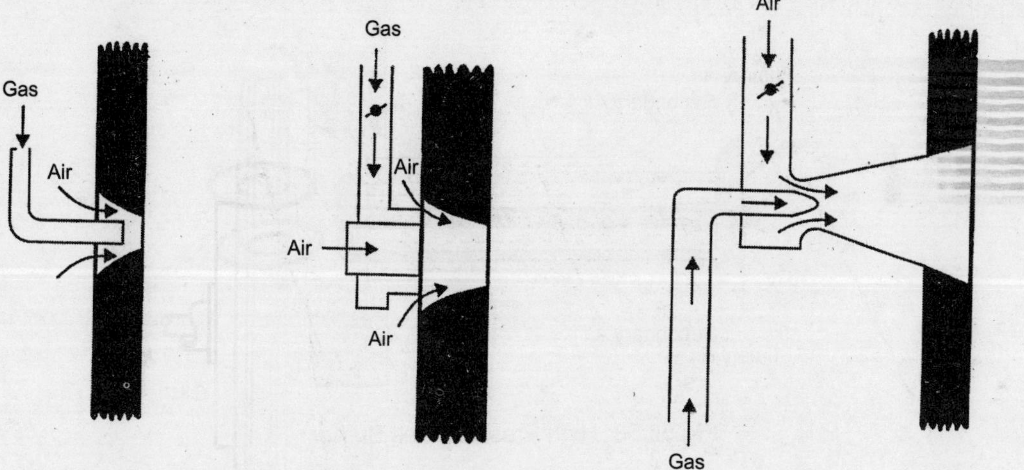

Fig.56.51: Typical gas burners.

Q. Describe the salient features of Low And High Pressure Gas Burners.

Ans. The important thing in all gaseous fuel burning devices is a correct air-and-gas mixture at the burner tip. Low-pressure burners (shown below in figure), using gas at a pressure less than 0.15 kg/cm2 , are usually of the multi-jet type, in which gas from a manifold is supplied to a number of small single jets, or circular rows of small jets, centered in or discharging around the inner circumference of circular air openings in a block of some heat-resisting material. The whole is encased in a rectangular cast-iron box, built into the boiler setting and having louver doors front to regulate the air supply. Draft may be natural, induced, or forced.

Fig.56.52: Low Pressure Gas Burner

In a high-pressure gas mixer/burner (shown below in figure), the energy of the gas jet draws air into the mixing chamber and delivers a correctly proportioned gas-air mixture to the burner. When the regulating valve is opened, gas flows through a small nozzle into a venturi tube (a tube with a contracted section). Entrainment of air with high-velocity gas in the narrow venturi section draws air in through large openings in the end. The gas-air mixture is piped to a burner.

The gas-burner tip may be in a variety of forms. In a sealed-in tip type, the proper gas-air mixture is piped to the burner, and no additional air is drawn in around the burner tip. Size of the air openings in the venturi tube end is increased or decreased by turning a revolving shutter, which can be locked in any desired position. Excess air levels in natural gas burner is in the order of 5%.

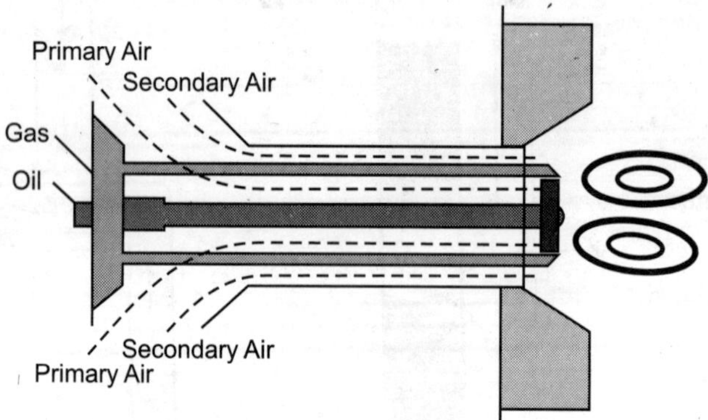

Fig.56.53: High Pressure Gas Burner

Q. What are the direct and indirect benefits of waste heat recovery?

Ans. Direct Benefits:

Recovery of waste heat has a direct effect on the efficiency of the process. This is reflected by reduction in the utility consumption & costs, and process cost.

Indirect Benefits:

Reduction in pollution: A number of toxic combustible wastes such as carbon monoxide gas, sour gas, carbon black off gases, oil sludge, Acrylonitrile and other plastic chemicals etc, releasing to atmosphere if/when burnt in the incinerators serves dual purpose i.e. recovers heat and reduces the environmental pollution levels.

Reduction in equipment sizes: Waste heat recovery reduces the fuel consumption, which leads to reduction in the flue gas produced. This results in reduction in equipment sizes of all flue gas handling equipments such as fans, stacks, ducts, burners, etc.

Reduction in auxiliary energy consumption: Reduction in equipment sizes gives additional benefits in the form of reduction in auxiliary energy consumption like electricity for fans, pumps etc.

Q. How the coal is pulverized, and what are the advantages of pulverized coal firing systems ?

Ans. It is a two stage process :

Stage I.

Raw and lump coal is crushed to a particle size not more than 15-25 mm in the crusher.

Stage II.

Crushed coal is delivered into raw coal bunkers and from here it is transferred to grinding mills that grind the feed into the final particles of 200 mesh size. During grinding hot air is blown through the fuel to dry it to impart good fluidity of the coal dust.

Advantages of pulverized fuels are as follows:

- Pulverization brings about a large increase in surface area per unit mass of solid fuel. Since combustion is a surface reaction, greater the extent of coal surface available, higher will be the rate of combustion. Herein lies the success of pulverized coal fired systems.
- Less excess air is required for complete combustion because of greater surface area of fuel exposed.
- Higher combustion air temperatures ensure higher cycle efficiency.
- A good range of coal right from anthracite to peat can be successfully burnt.
- Better combustion control enables the system to respond quickly to extensive load variation.
- Better response to instrument control on auto.
- Large amount of heat release makes it very suitable for super thermal power stations where the rate of steam generation is as high as 2000 t/h.
- Slagging and clinkering problems are low.
- Carry over of unburnt fuel to ash is practically nil.
- Ash handling problem low.
- Can operate successfully in combination with gas and oil fired systems.
- Cold start-up of boilers is very rapid and efficient.
- Less furnace volume is required.
- Low banking loss.

Q. What is draft in combustion system? Explain briefly the different types of draft.

Ans. The function of draft in a combustion system is to exhaust the products of combustion into the atmosphere. The draft can be classified into two types namely Natural and Mechanical Draft.

Natural Draft : It is the draft produced by a chimney alone. It is caused by the difference in weight between the column of hot gas inside the chimney and column of outside air of the same height & cross section.

Mechanical Draft : It is draft artificially produced by fans. Three basic types of mechanical drafts that are applied are :

Induced Draft: An induced-draft fan draws enough draft for flow into the furnace, causing the products of combustion to discharge to atmosphere

Forced Draft: The Forced draft system uses a fan to deliver the air to the furnace, forcing combustion products to flow through the unit and up the stack.

Balanced Draft: Forced-draft (F-D) fan (blower) pushes air into the furnace and an induced-draft (I-D) fan draws gases into the chimney thereby providing draft to remove the gases from the boiler.

Q. Explain why furnaces should operate at slightly positive pressure?

Ans. It is important to operate furnace at a slightly positive pressure. Negative pressure lead to air ex-filtration affecting air fuel ratio and furnace temperature thus increasing fuel consumption. Excessive positive pressure leads to infiltration resulting in leaking out of flames, overheating of furnace refractories, reduced brick life and other associated problems.

Q. Discuss need of providing excess air for combustion. What is the disadvantage of providing too much of excess air ?

Ans. In practice, fuel-air mixing is never perfect, hence a certain amount of excess air is needed to complete combustion which ensures the release of the entire heat contained in fuel . If too much air than what is required for completing combustion are allowed to enter, additional heat would be lost in heating the surplus air to the chimney temperature. This would result in increased stack losses. Less air would lead to the incomplete combustion and smoke. Hence, there is an optimum excess air level for each type of fuel. For example ,complete combustion of one kg of fuel oil requires 14 kg of air .

Q. Enumerate reasons for incomplete combustions of solid, liquid and gaseous fuels.

Ans. Incomplete combustion can arise from a shortage of air or surplus of fuel or poor distribution of fuel. It is usually obvious from the colour of smoke, and must be corrected immediately. In the case of oil and gas fired systems, CO or smoke (for oil fired systems only) with normal or high excess air indicates burner system problems. A more frequent cause of incomplete combustion is the poor mixing of fuel and air at the burner. Poor oil fires can result from improper viscosity, worn burner tips, carbonization on tips and deterioration of diffusers or spinner plates.

With coal firing, unburned carbon can comprise a big loss. It occurs as grit carry-over or carbon-in-ash and may amount to more than 2% of the heat supplied to the boiler.

Non-uniform fuel size could be one of the reasons for incomplete combustion. In chain grate stokers, large lumps will not burn out completely, while small pieces and fines may block the air passage, thus causing poor air distribution. In sprinkler stokers, stoker grate condition, fuel distributors, wind box air regulation and over-fire systems can affect carbon loss. Increase in the fines in pulverized coal also increases carbon loss.

Q. Explain the significance of the ultimate analysis of the fuels from the combustion point of view.

Ans. Ultimate analysis of the fuel besides its moisture and ash content gives percentage of various elemental chemical constituents of the fuel such as carbon, hydrogen, oxygen, sulpher etc. It helps in estimating the quantity of stoichiometric air required for combustion and volume & composition of combustion gases. This also helps in estimating the flame temperature, flue gas duct design and calorific value of the fuel.

Q. Why should LPG cylinders not be stored in basements or cellars ?

Ans: LPG is a predominant mixture of propane and butane. Both propane and butane are denser than air. Consequently, the vapour flows along the ground into drains and sinks to the lowest level of the surroundings and gets ignited at a considerable distance from the source of leakage. Escape of even small quantities of LPG can give rise to large volume of vapour mixture and can cause considerable hazard. Hence there should be adequate ground level ventilation where LPG cylinders are stored. For this reason LPG cylinders should not be stored in cellars or basements, which have no ventilation at ground levels.

Q. Explain the working principles and applications of 'Heat pump'.

Ans. Heat must flow spontaneously "downhill", that is from a system at high temperature to one at a lower temperature. It is possible to reverse the direction of spontaneous energy flow by the use of a thermodynamic system known as a heat pump. This device consists of two heat exchangers, a compressor and an expansion device. A liquid or a mixture of liquid and vapor of a pure chemical species flows through an evaporator, where it absorbs heat at low temperature and, in doing so, is completely vaporized. The low temperature vapor is compressed by a compressor, which requires external work. The work done on the vapor raises its pressure and temperature to a level where its energy becomes available for use. The vapor flows through a condenser where it gives up its energy as it condenses to a liquid. The liquid is then expanded through an expansion valve back to the evaporator where the cycle repeats. The heat pump was developed as a space heating system where low temperature energy from the ambient air, water, or earth is raised to heating system temperatures by doing compression work with an electric motor-driven compressor.

The heat pumps have the ability to upgrade heat to a value more than twice that of the energy consumed by the device. The potential for application of heat pump is growing and number of industries have been benefited by recovering low grade waste heat by upgrading it and using it in the main process stream.

Heat pumps applications are most promising for following situations :

(i) (a) space heating system, (b) use in plastic factory where chilled water is used to cool injection moulding machines, (c) drying applications such as maintaining dry atmosphere in storage and drying compressed air .

(ii) In a situation when both the cooling and heating capabilities of the cycle can be used in combination .

A heat pump arrangement is shown below in figure .

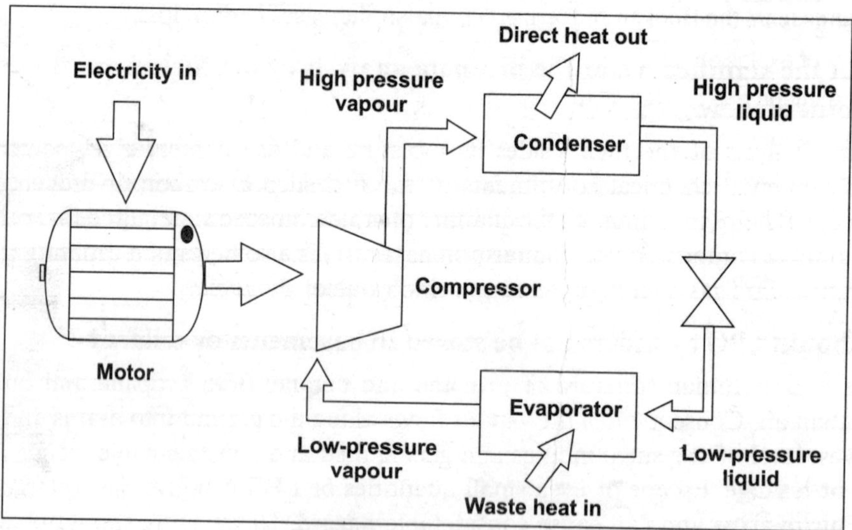

Fig.56.54: Heat pump arrangement

Working principle of heat pump arrangement is as follows;

Step 1: in the evaporator the heat is extracted to boil the circulating working fluid .

Step 2: the evaporated working fluid is compressed in a compressor rising working fluid temperature and pressure..

Step 3: the heat is delivered to the condenser .

Step 4: the pressure of the working fluid is reduced in a throttling valveand condensate returned to the compressor .

Q. How water hammer is produced in a steam piping system and how it can be eliminated?

Ans. A water hammer in a steam system is caused by condensate collection in the plant or pipe work picked up by the fast moving steam and carried along with it. When this collection hits obstructions such as bends, valves, steam traps or some other pipe fittings, it is likely to cause severe damage to fittings & equipment and result in leaking pipe joints. The problem of water hammer can be eliminated by positioning the pipes so that there is a continuous slope in the direction of flow. A slope of at least 12mm in every 3 metres is necessary, as also an adequate number of drain points every 30 to 50 metres.

Q. What are the comparative advantages and disadvantages of fire tube and water tube boilers?

Ans. Comparative features are tabulated below.

S.No.	Parameters	Fire tube boilers	Water tube boilers
a.	Rate of steam generation	Less rapid	More rapid
b.	Suitability for power plants	Unsuitable	Suitable. All major power plants are based on these
c.	Operating steam pressure	Limited to 25kgf/cm^2	Can well exceed 125 kgf/cm^2
d.	Chances of explosion	Less	More
e.	Risk of damage due to explosion	Much more	Much less
f.	Water treatment	Not very necessary as minor scaling would not go far enough to cause overheating and tube bursting	Required as scaling will lead to tube bursting
g..	Floor space requirement	Much	Less
h.	Cost & construction problem	Higher	Much less
i.	Transportation	Inconvenient due to large size of the shell	Comparatively easier
j.	Skill required for efficient operation	Less	More

Q. Describe the mechanism of Fluidised Bed Combustion (FBC). What are different types of FBC boilers ?

Ans. When an evenly distributed air or gas is passed upward through a finely divided bed of solid particles such as sand supported on a fine mesh, the particles are undisturbed at low velocity. As air velocity is gradually increased, a stage is reached when the individual particles are suspended in the air stream – the bed is called "fluidized bed ".

With further increase in air velocity, there is bubble formation, vigorous turbulence, rapid mixing and formation of dense defined bed surface. The bed of solid particles exhibits the properties of a boiling liquid and assumes the appearance of a fluid – the bed is then called "bubbling fluidized bed".

At still higher velocities, bubbles disappear, and particles are blown out of the bed. Therefore, some amounts of particles have to be recirculated to maintain a stable system – this is then called "circulating fluidised bed".

Fluidization depends largely on the particle size and the air velocity. The mean solids velocity increases at a slower rate than does the gas velocity . The difference between the mean solid velocity and mean gas velocity is called as slip velocity. Maximum slip velocity between the solids & the gas is desirable for good heat transfer and intimate contact.

If sand particles in a fluidized state is heated to the ignition temperatures of coal, and coal is injected continuously into the bed, the coal will burn rapidly and bed attains a uniform temperature. The fluidized bed combustion (FBC) takes place at about 800 to 900 °C. Since this temperature is much below the ash fusion temperature, melting of ash and associated problems are avoided.

The lower combustion temperature is achieved because of high coefficient of heat transfer due to rapid mixing in the fluidized bed and effective extraction of heat from the bed through in-bed heat transfer tubes & walls of the bed. The gas velocity is maintained between minimum fluidisation velocity and particle entrainment velocity. This ensures stable operation of the bed and avoids particle entrainment in the gas stream.

This principle of fluidisation is illustrated in Figure below.

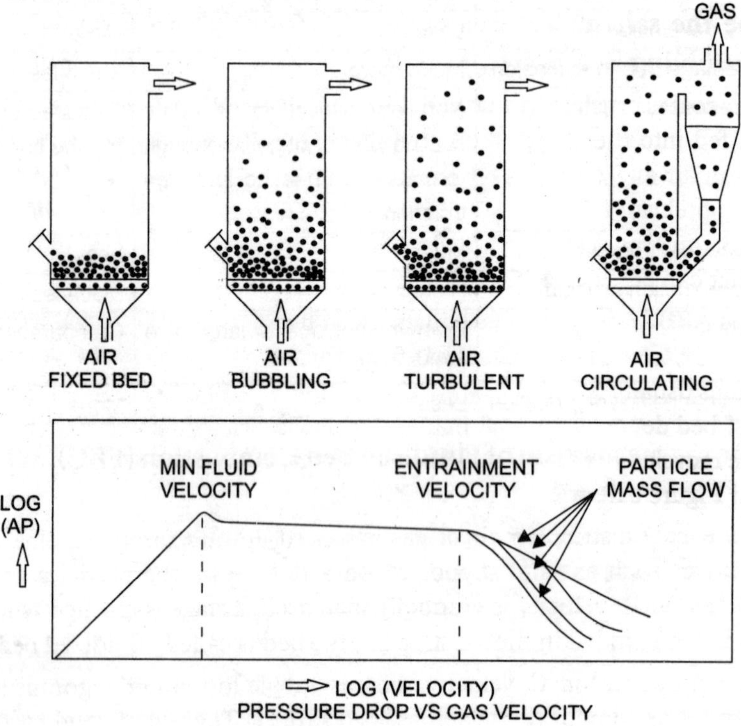

Fig.56.55: Principle of fluidisation

Combustion process requires the three "T"s that is Time, Temperature and Turbulence. In FBC, turbulence is promoted by fluidisation. Improved mixing generates evenly distributed heat at lower temperature. Residence time is many times greater than conventional grate firing. Thus an FBC system releases heat more efficiently at lower temperatures.

Since limestone is used as particle bed, control of sulfur dioxide and nitrogen oxide emissions in the combustion chamber is achieved without any additional control equipment. This is one of the major advantages over conventional fired boilers.

Since the major portion of the coal available in India is of low quality, high ash content and low calorific value the traditional grate fuel firing systems have got limitations and are techno-economically unviable to meet the challenges of future.

Fluidized bed combustion has emerged as a viable alternative and has significant advantages over conventional firing system and offers multiple benefits eg; compact boiler design, fuel flexibility, higher combustion efficiency and reduced emission of noxious pollutants such as SOx & NOx. The fuels burnt in these boilers include coal, washery rejects, rice husk, bagasse & other agricultural wastes. The fluidized bed boilers have a wide capacity range- 0.5 T/hr to over 100 T/hr.

There are three basic types of fluidised bed combustion boilers:

- Atmospheric Fluidised Bed Combustion System (AFBC)
- Atmospheric circulating (fast) Fluidised Bed Combustion system(CFBC)
- Pressurised Fluidised Bed Combustion System (PFBC).

Q. Describe the salient features of AFBC (Atmospheric Fluidised Bed Combustion) / Bubbling Bed Boiler.

Ans. In AFBC, coal is crushed to a size of 1-10 mm depending on the rank of coal, type of fuel feed and fed into the combustion chamber. The atmospheric air, which acts as both the fluidization air and combustion air, is delivered at a pressure and flows through the bed after being preheated by the exhaust flue gases. The velocity of fluidising air is in the range of 1.2 to 3.7 m /sec. The rate at which air is blown through the bed determines the amount of fuel that can be reacted.

Almost all AFBC/ bubbling bed boilers use in-bed evaporator tubes in the bed of limestone, sand and fuel for extracting the heat from the bed to maintain the bed temperature. The bed depth is usually 0.9 m to 1.5 m and the pressure drop averages about 2.5 cm of water per cm of bed depth. Very little material leaves the bubbling bed – only about 2 to 4 kg of solids are recycled per ton of fuel burned. Typical fluidized bed combustors of this type is shown in Figure below

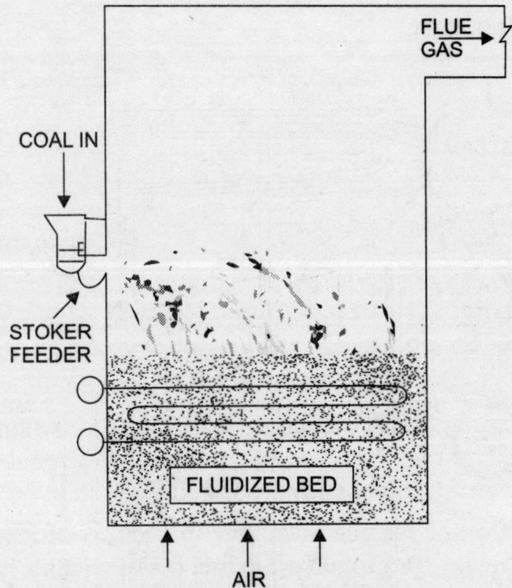

Fig.56.56: AFBC/ Bubbling Bed Boiler

Atmospheric fluidised bed boiler can operate at near atmospheric or elevated pressure and have these essential features:

- Distribution plate through which air is blown for fluidizing.
- Immersed steam-raising or water heating tubes which extract heat directly from the bed.

- Tubes above the bed which extract heat from hot combustion gas before it enters the flue duct.

The combustion gases pass over the super heater sections of the boiler, flow past the economizer, the dust collectors and the air preheaters before being exhausted to atmosphere.

The special feature of atmospheric fluidised bed combustion is the constraint imposed by the relatively narrow temperature range within which the bed must be operated. With coal, there is risk of clinker formation in the bed if the temperature exceeds 950°C and loss of combustion efficiency , if the temperature falls below 800°C. For efficient sulphur retention, the temperature should be in the range of 800 to 850 °C.

Q. Discuss the general arrangements of systems of AFBC / Bubbling Bed Boiler.

Ans. AFBC boilers comprise of following systems:

- Fuel feeding system
- Air Distributor
- Bed & In-bed heat transfer surface
- Ash handling system

Many of these are common to all types of FBC boilers .

The general arrangements of systems of AFBC / Bubbling Bed Boiler is shown below in figure .

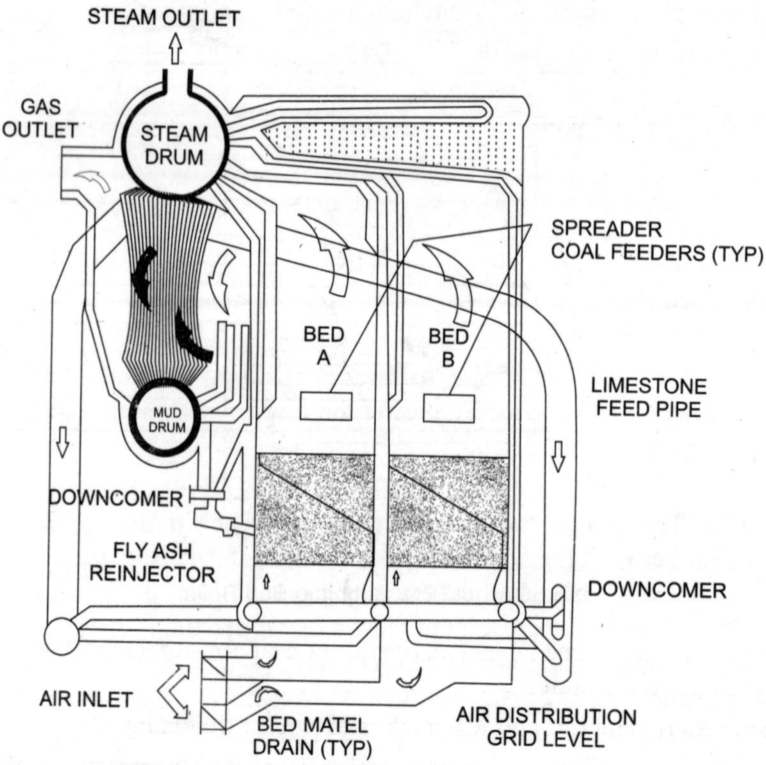

Fig.56.57: Systems of AFBC/Bubbling Bed Boiler

- **Fuel Feeding system**

For feeding fuel and sorbents like limestone or dolomite, usually two methods are followed: under bed pneumatic feeding and over bed feeding.

Under Bed Pneumatic Feeding : If the fuel is coal, it is crushed to 1-6 mm size and pneumatically transported from feed hopper to the combustor through a feed pipe piercing the distributor. Based on the capacity of the boiler, the number of feed points is increased, as it is necessary to distribute the fuel into the bed uniformly.

Over-Bed Feeding : The crushed coal, 6-10 mm size is conveyed from coal bunker to a spreader by a screw conveyor. The spreader distributes the coal over the surface of the bed uniformly. This type of fuel feeding system accepts over size fuel also and eliminates transport lines, when compared to under bed feeding system.

- **Air Distributor**

The purpose of the distributor is to introduce the fluidizing air evenly through the bed cross section thereby keeping the solid particles in constant motion, and preventing the formation of defluidization zones within the bed. The distributor, which forms the furnace floor, is normally constructed from metal plate with a number of perforations in a definite geometric pattern. The perforations may be located in simple nozzles or nozzles with bubble caps, which serve to prevent solid particles from flowing back into the space below the distributor.

The distributor plate is protected from high temperature of the furnace by:

i) Refractory Lining

ii) A Static Layer of the Bed Material or

iii) Water Cooled Tubes.

- **Bed & In-Bed Heat Transfer Surface**

a) **Bed :** The bed material can be sand, ash, crushed refractory or limestone, with an average size of about 1 mm. Depending on the bed height these are of two types: shallow bed and deep bed.

At the same fluidizing velocity, the two ends fluidise differently, thus affecting the heat transfer to an immersed heat transfer surfaces. A shallow bed offers a lower bed resistance and hence a lower pressure drop and lower fan power consumption. In the case of deep bed, the pressure drop is more and this increases the effective gas velocity and also the fan power.

b) In-Bed Heat Transfer Surface : In a fluidized in-bed heat transfer process, it is necessary to transfer heat between the bed material and an immersed surface, which could be that of a tube bundle, or a coil. The heat exchanger orientation can be horizontal, vertical or inclined. From a pressure drop point of view, a horizontal bundle in a shallow bed is more attractive than a vertical bundle in a deep bed. Also, the heat transfer in the bed depends on number of parameters like (i) bed pressure (ii) bed temperature (iii) superficial gas velocity (iv) particle size (v) Heat exchanger design and (vi) gas distributor plate design.

- **Ash Handling System**

a) **Bottom ash removal :** In the AFBC boilers, the bottom ash constitutes about 30 - 40 % of the total ash, the rest being the fly ash. The bed ash is removed by continuous over flow to maintain bed height and also by intermittent flow from the bottom to remove over size particles, avoid accumulation & consequent defluidization. While firing high ash coal such as washery rejects, the bed ash overflow drain quantity is considerable so special care has to be taken.

b) ***Fly ash removal :*** The amount of fly ash to be handled in AFBC boiler is relatively very high, when compared to conventional boilers. This is due to elutriation of particles at high velocities. Fly ash carried away by the flue gas is removed in number of stages; firstly in convection section, then from the bottom of air preheater/economizer and finally a major portion is removed in dust collectors.

The types of dust collectors used are cyclone, bag filters, electrostatic precipitators (ESP's) or some combination of all of these. To increase the combustion efficiency, recycling of fly ash is practiced in some of the units.

Q. Describe the salient features of Circulating Fluidised Bed Combustion (CFBC).

Ans. Circulating Fluidised Bed Combustion (CFBC) technology has evolved from conventional bubbling bed combustion as a means to overcome some of the drawbacks associated with conventional bubbling bed combustion.

CFBC technology utilizes the fluidized bed principle in which crushed (6-12 mm size) fuel and limestone are injected into the furnace. The particles are suspended in a stream of upwardly flowing air (60-70% of the total air), which enters the bottom of the furnace through air distribution nozzles. The fluidising velocity in circulating beds ranges from 3.7 to 9 m/sec. The balance of combustion air is admitted above the bottom of the furnace as secondary air. The combustion takes place at 850-900°C, and the fine particles (<450 microns) are elutriated out of the furnace with flue gas velocity of 4-6 m/s. The particles are then collected by the solids separators and circulated back into the furnace. Solid recycle is about 50-100 kg per kg of fuel burnt.

There are no steam generation tubes immersed in the bed. The circulating bed is designed to move a lot more solids out of the furnace area and to achieve most of the heat transfer outside the combustion zone - convection section, water walls, and at the exit of the riser. Some circulating bed units even have external heat exchanges.

The particles circulation provides efficient heat transfer to the furnace walls and longer residence time for carbon and limestone utilization. Similar to Pulverized Coal (PC) firing, the controlling parameters in the CFB combustion process are temperature, residence time and turbulence.

For large units, the taller furnace characteristics of CFBC boiler offers better space utilization, greater fuel particle & sorbent residence time for efficient combustion and SO_2 capture, and easier application of staged combustion techniques for NOx control than AFBC generators. CFBC boilers achieve better calcium to sulphur utilization – 1.5 to 1 vs. 3.2 to 1 for the AFBC boilers, although the furnace temperatures are almost the same.

CFBC boilers are generally more economical than AFBC boilers for industrial application requiring more than 75 – 100 T/hr of steam.

CFBC requires huge mechanical cyclones to capture and recycle the large amount of bed material, which requires a tall boiler.

A Circulating Fluidised Bed Boiler(CFBC) is shown below in figure.

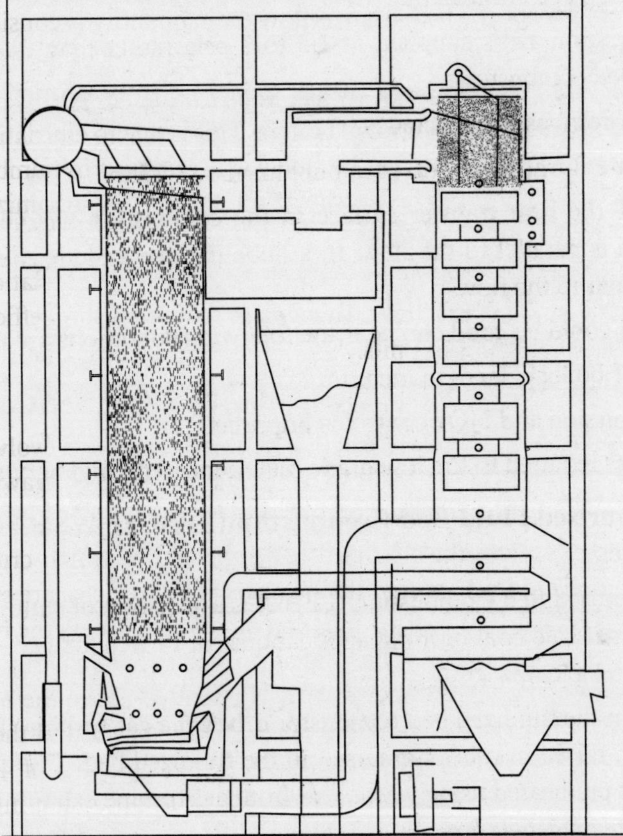

Fig.56.58: Circulating Fluidised Bed Boiler(CFBC)

At high fluidizing gas velocities in which a fast recycling bed of fine material is superimposed on a bubbling bed of larger particles , the combustion temperature is controlled by rate of recycling of fine material. Hot fine material is separated from the flue gas by a cyclone and is partially cooled in a separate low velocity fluidized bed heat exchanger, where the heat is given up to the steam. The cooler fine material is then recycled to the dense bed.

Q. Describe the performance features of the circulating fluidized bed system . List out the conditions under which CFBC could be a good choice.

Ans. Main performance features of the circulating fluidized bed system are as follows:

- It has a high processing capacity because of the high gas velocity through the system.
- The temperature of about 870 °C is reasonably constant throughout the process because of the high turbulence and circulation of solids. The low combustion temperature also results in minimal NOx formation.

- Sulfur present in the fuel is retained in the circulating solids in the form of calcium sulphate and removed in solid form. The use of limestone or dolomite sorbents allows a higher sulfur retention rate, and limestone requirements are substantially less than with bubbling bed combustor.
- The combustion air is supplied at 1.5 to 2 psig rather than 3-5 psig as required by bubbling bed combustors.
- It has high combustion efficiency.
- It has a better turndown ratio than bubbling bed systems.
- Erosion of the heat transfer surface in the combustion chamber is reduced, since the surface is parallel to the flow. In a bubbling bed system, the surface generally is perpendicular to the flow.

A CFBC boiler could be good choice if the following conditions are met.

- Capacity of boiler is large to medium .
- Sulphur emission and NOx control is important .
- The boiler is required to fire low-grade fuel or fuel with highly fluctuating fuel quality.

Q. What is Pressurised Fluid Bed Combustion(PFBC) ? What are its applications and advantages .

Ans. Pressurised Fluidised Bed Combustion (PFBC) is a variation of fluid bed technology that is meant for large-scale coal burning applications. In PFBC, the bed vessel is operated at pressure upto 16 kg/cm2 .

The off-gas from the fluidized bed combustor drives the gas turbine. The steam turbine is driven by steam raised in tubes immersed in the fluidized bed. The condensate from the steam turbine is pre-heated using waste heat from gas turbine exhaust and is then taken as feed water for steam generation.

The PFBC system can be used for cogeneration or combined cycle power generation. By combining the gas and steam turbines in this way, electricity is generated more efficiently than in conventional system. The overall conversion efficiency is higher by 5 to 8%.

A PFBC Boiler for co-generation is shown below in figure.

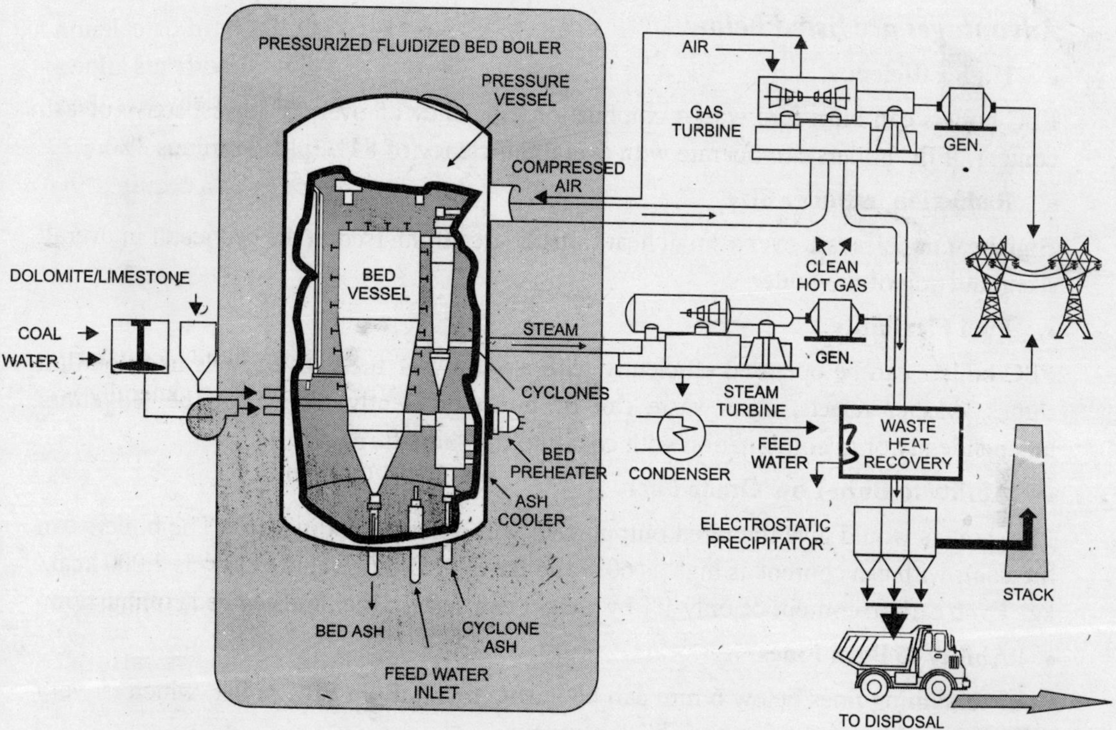

Fig.56.59: PFBC Boiler for Cogeneration

At elevated pressure, the potential reduction in boiler size is considerable due to increased amount of combustion in pressurized mode and high heat flux through in-bed tubes.

Q. What is the principle of CFBC (circulating fluidized bed combustion) boiler?

Ans. CFBC technology utilizes the fluidized bed principle in which crushed (6 –12 mm size) fuel and limestone are injected into the furnace or combustor. The particles are suspended in a stream of upwardly flowing air (60-70% of the total air), which enters the bottom of the furnace through air distribution nozzles. The balance of combustion air is admitted above the bottom of the furnace as secondary air. While combustion takes place at 840-900 0C, the fine particles (<450 microns) are elutriated out of the furnace with flue gas velocity of 4-6 m/s. The particles are then collected by the solids separators and circulated back into the furnace. This combustion process is called circulating fluidized bed (CFB).

Q. Write down in detail about major advantages of FBC(Fluidised Bed Combustion) boiler system.

Ans. Fluidized bed combustion has emerged as a viable alternative and has significant advantages over conventional fuel firing system and offers multiple benefits – compact boiler design, fuel flexibility, higher combustion efficiency and reduced emission of noxious pollutants such as SOx & NOx. The fuels burnt in these boilers include coal, washery rejects, rice husk, bagass and other agricultural wastes.

Advantages are listed below.

- **High Efficiency**

FBC boilers can burn fuel with a combustion efficiency of over 95% irrespective of ash content. FBC boilers can operate with overall efficiency of 84% (plus or minus 2%).

- **Reduction in Boiler Size**

High heat transfer rate over a small heat transfer area immersed in the bed result in overall size reduction of the boiler.

- **Fuel Flexibility**

FBC boilers can be operated efficiently with a variety of fuels. Even fuels like flotation slimes, washer rejects, agro waste can be burnt efficiently. These can be fed either independently or in combination with coal into the same furnace.

- **Ability to Burn Low Grade Fuel**

FBC boilers would give the rated output even with inferior quality fuel. The boilers can fire coals with ash content as high as 60% and having calorific value as low as 2,000 kcal/kg. Even carbon content of only 1% by weight can sustain the fluidised bed combustion.

- **Ability to Burn Fines**

Coal containing fines below 6 mm can be burnt efficiently in FBC boiler, which is very difficult to achieve in conventional firing system.

- **Pollution Control**

SO_2 formation can be greatly minimised by addition of limestone or dolomite for high sulphur coals. 3% limestone is required for every 1% sulphur in the coal feed. Low combustion temperature eliminates NO_x formation.

- Low Corrosion and Erosion The corrosion and erosion effects are less due to lower combustion temperature, softness of ash and low particle velocity (of the order of 1 m/sec).

- **Easier Ash Removal – No Clinker Formation**

Since the temperature of the furnace is in the range of 800 – 900° C in FBC boilers, even coal of low ash fusion temperature can be burnt without clinker formation. Ash removal is easier as the ash flows like liquid from the combustion chamber. Hence less manpower is required for ash handling.

- **Less Excess Air – Higher CO_2 in Flue Gas**

The CO_2 in the flue gases will be of the order of 14 – 15% at full load. Hence, the FBC boiler can operate at low excess air - only 10 – 15%.

- **Simple Operation, Quick Start-Up**

High turbulence of the bed facilitates quick start up and shut down. Full automation of start up and operation using reliable equipment is possible.

Q. Briefly list the advantages of fluidized bed combustion boiler over fixed grate boiler.

Ans. The advantages of fluidized bed combustion boiler over fixed grate boiler are listed below;

- efficiency is not so much affected when firing higher ash fuels
- better fuel flexibility
- burns low grade fuels
- burns fines
- reduces NO_x formation
- may reduce SO_x formation through additives
- no clinker formation
- no soot blowing
- quicker response to changing demand

Q. List the advantages of CFBC boilers over AFBC boilers.

Ans. The advantages of CFBC boilers over AFBC boilers are listed below :

- Higher processing temperature because of high gas velocity through the system.
- Lower combustion temperature of about 870 ^{0}C can be achieved constantly, which results in minimal NOx formation.
- The combustion air is supplied at 1.5 to 2 psig rather than 3 to 5 psig as required by bubbling bed combustors.
- Higher combustion efficiency.
- Better turndown ratio.
- Erosion of heat transfer surface in the combustion chamber is reduced, since the surface is parallel to the flow. In AFBC system, the surface is generally perpendicular to the flow .

Q. List down few points that need attention while selecting biomass fed fluidised bed combustion (FBC) boiler .

Ans. The following areas need attention with biomass fed FBC boilers:

- Uneven spreading of biomass fuel on boiler grate can lead to secondary combustion in the super-heater zone, resulting in overheating of super heater tubes and fluctuations in steam pressure.
- Frequent erosion of super-heater and economizer coils can occur, due to high silica content in the biomass, especially in rice husk.
- High extraneous matter in biomass (sand & mud) causes tube fouling and fluidized bed to be drained more frequently, with resultant heat loss.
- Carbon and dust coating of boiler tubes results in lowering of steam temperatures, especially during soot blowing.
- Presence of Pesticides (used during farming) adds to tube failure frequencies; mainly due to potassium constituents.

- Corrosive constituents in biomass adversely affect boiler internals, especially the superheater tubes. Chloride content in certain types of biomass (like cotton stalk, 8–9%) can combine with sodium and potassium in high temperature regime to aggravate the corrosion process.
- Some boilers which use red Gram husk/twigs as fuel pose corrosion problems at the cold end (i.e., secondary super-heater and economizer tubes), due to the sulfur content.
- The biomass fuel mix fed to the boiler, in quite a few cases, contains a combination of 6 to 7 biomass types. Each biomass has a separate air to fuel ratio, and it is difficult to set a workable air to fuel ratio.
- High moisture content in the biomass causes frequent jamming of the fuel in feeders, leading to fluctuations in steam pressure and temperature.
- High moisture content in the biomass also leads to plugging and choking of closely spaced heating surfaces. This situation is further aggravated by the super-heater tube coil with very close spacing, often the result of a desire to achieve a compact design.
- Due to biomass fuel size variation, occurence of unburnts in flue gases & bottom ash is high, resulting in lower efficiency and also variation in steam pressure and temperature.
- Absence of biomass feed rate measurement mechanism leaves little room for accurate assessment of heat rate/efficiency. Providing a weighing mechanism is difficult on account of different biomass fuel combinations being used, with different (and low) bulk densities.
- Degradation of biomass during storage in exposed ambient wet atmosphere leads to loss of heat value. Loss of material due to windage & carpet loss, coupled with loss of heat value on account of decay (inherent biomass characteristics), can cause an error in assessment of input fuel energy (as the input heat is customarily evaluated based on received biomass quantities & GCV).

Q. Describe various consideration involved in retrofitting of FBC systems to conventional boiler system.

Ans. Retrofitting fluidised bed coal fired combustion systems to conventional boilers has been carried out successfully both in India and abroad.

The important aspects to be considered in retrofit projects are:

- Water/steam circulation design
- Furnace bottom-grate clearance
- Type of particulate control device
- Fan capacity
- Availability of space.

Retrofitting of a fluidised bed combustor to a conventional stoker fired water tube boiler may involve:

- The replacement of grate by a distributor plate with short stand pipes for admitting air from the wind box located underneath.

- Installations of stand pipes to remove ash from the bed.
- Provision of horizontal hair pin tubes in the bed with a pump for forced circulation from the boiler drum.
- Modification of crusher to size the coal/limestone mixture for pneumatic under bed injection of the mixture.

The conversion of a conventional coal fired system to a fluidised bed combustion system can be accomplished without effecting major changes, after making a cost-benefit analysis. Oil fired boilers can also be converted to coal fired fluidised bed combustion systems. However it has to be examined on a case-to-case basis.

Q. Why should the stack temperature of furnace oil fired boilers not be maintained below 160 °C ?

Ans. For fuels containing sulphur, low temperatures (below 160 °C) of stack can lead to sulphur dew point corrosion. The main disadvantage of sulphur is the risk of corrosion by sulphuric acid formed during & after combustion, and condensing in cool parts of the chimney or stack, air preheater and economiser.

Q. Explain why natural gas requires less amount of excess air compared to solid/liquid fuels ?

Ans. Natural gas requires less amount of excess air compared to solid/liquid fuels .Natural gas is in the gaseous form and lighter than air, it mixes with air readily [intimate mixing of air (oxygen) and fuel takes place] and aids to complete combustion with less amount of excess air. It does not produce smoke or soot. It has no sulphur content. It is lighter than air and disperses into air easily in case of leak. Natural gas is also free of ash. Solid or liquid fuels must be changed to a gaseous form before they will burn. Hence it requires more excess air compared to natural gas for complete combustion. Solid fuels need to be pulverized properly to get uniform sizes and liquid fuels need to be preheated & atomized properly for intimate mixing with air to ensure complete combustion. Hence more excess air is provided for solid and liquid fuels.

Q. Why steam is to be used at the lowest practicable pressure for indirect process heating ?

Ans. Steam is to be used at the lowest practicable pressure for indirect process heating .A study of the steam tables indicates that the latent heat in steam reduces as the steam pressure increases. It is only the latent heat of steam, which takes part in the heating process when applied to an indirect heating system. Thus, it is important that its value be kept as high as possible. This can only be achieved ; if we go in for lower steam pressures. As a guide, the steam should always be generated and distributed at the highest possible pressure, but utilized at as low a pressure as possible', since it then has higher latent heat. However, it may also be seen from the steam tables that the lower the steam pressure, the lower will be its temperature. Since temperature is the driving force for the transfer of heat at lower steam pressures, the rate of heat transfer will be slower and the processing time greater. In equipment where fixed losses are high (e.g. big drying cylinders), there may even be an increase in steam consumption at lower pressures due to increased processing time. There are, however, several equipment in certain industries where one can profitably go in for lower pressures and realize economy in steam consumption without materially affecting production time. Therefore, there is a limit to the reduction of steam pressure. Depending on the equipment design, the lowest possible steam pressure with which the equipment can work should be selected without sacrificing either on production time or on steam consumption.

Q. What are the Benefits of Waste Heat Recovery (WHR)devices implementation in Industry? List down few WHR sources in a typical industry.

Ans. Benefits of waste heat recovery include :

- Reduction in the utility consumption & costs, and process cost.
- **Reduction in pollution:** A number of toxic combustible wastes such as carbon monoxide gas, sour gas, carbon black off gases, oil sludge,
- **Reduction in equipment sizes:** Waste heat recovery reduces the fuel consumption, which leads to reduction in the flue gas produced. This results in reduction in equipment sizes of all flue gas handling equipments such as fans, stacks, ducts, burners, etc.
- **Reduction inauxiliary energy consumption:** Reduction in equipment sizes gives additional benefits in the form of reduction in auxiliary energy consumption like electricity for fans, pumps etc..

WHR sources in a typical industry are :

- Heat in flue gases.
- Heat in vapour streams..
- Convective and radiant heat Low grade
- Heat losses in cooling water.

The criteria for development of WHR are :

Identify the following before selection of WHR devices :

- Sources and use of waste heat
- Upset conditions if any (back pressure)
- Availability of space
- Acid due point problem , if any

Review the process flow, Layout Piping, Instrumentation & Control.

Economic analyzer of WHR Such as Investment, depreciation, pay back period, Commercial availability, etc

Q. What are waste heat recovery boilers? Explain their need and benefits?

Ans. Waste heat boilers are ordinarily water tube boilers in which the hot exhaust gases from gas turbines, incinerators, etc., pass over a number of parallel tubes containing water. The water is vaporized in the tubes and collected in a steam drum from which it is drawn off for use as heating or processing steam. Because the exhaust gases are usually in the medium temperature range and in order to conserve space, a more compact boiler can be produced if the water tubes are finned in order to increase the effective heat transfer area on the gas side. The pressure at which the steam is generated and the rate of steam production depends on the temperature of waste heat. The pressure of a pure vapor in the presence of its liquid is a function of the temperature of the liquid from which it is evaporated. If the waste heat in the exhaust gases is insufficient for generating the required amount of process steam, auxiliary burners which burn fuel in the waste heat boiler or an after-burner in the exhaust gases flue are added. Waste heat boilers are built in capacities from 25 to 30,000 m3 / min. of exhaust gas.

Typical applications of waste heat boilers are to recover energy from the exhausts of gas turbines, reciprocating engines, incinerators, and furnaces.

Q. List down major advantages of waste heat recovery in rolling mill furnace .

Ans. Advantages are listed below:

- Fuel economy
- Improved combustion/less excess air/reduction in stack losses
- Increased output
- Reduction in scale losses
- Uniform temperature across the material
- . Reduced pollution
- Less auxiliary energy consumption
- Reduction in equipment sizes

Q. Write short notes on factors affecting wall heat losses in batch type reheating furnace .

Ans. Factors affecting wall heat losses in batch type reheating furnaces are as follows :

(a) **Emissivity of walls :** Emissivity of fire brick refractory should be high . Emissivity of most of the refractory bricks decreases with increase in temperature. High emissivity coatings whose emissivity increases with temperature can be used to increase emissivity and decrease wall losses.

(b) **Conductivity of refractories :** The refractory and insulating bricks should have low thermal conductivity. Chosing low thermal conductivity bricks will reduce wall heat losses. Conductivity rises with rise in temperature. Batch type furnaces can use ceramic fibre to reduce storage losses.

(c) Wall thickness of batch furnaces : Heat losses can be reduced by increasing the wall thickness, or through the application of insulating bricks. Outside wall temperature and heat losses for a composite wall of a certain thickness of firebrick and insulation brick are much lower due to lesser conductivity of may be worked out to reduce the heat storage.

Q. Describe with a sketch about continuous steel reheating furnace.

Ans. The main function of a reheating furnace is to raise the temperature of a piece of steel, typically to between 900 and 1250°C, until it is plastic enough to be rolled to the desired section, size or shape. The furnace must also meet specific requirements & and objectives in terms of stock heating rates for metallurgical and productivity reasons. In continuous reheating, the steel stock forms a continuous flow of material and is heated to the desired temperature as it travels through the furnace.

All furnaces possess the features as shown below in Figure

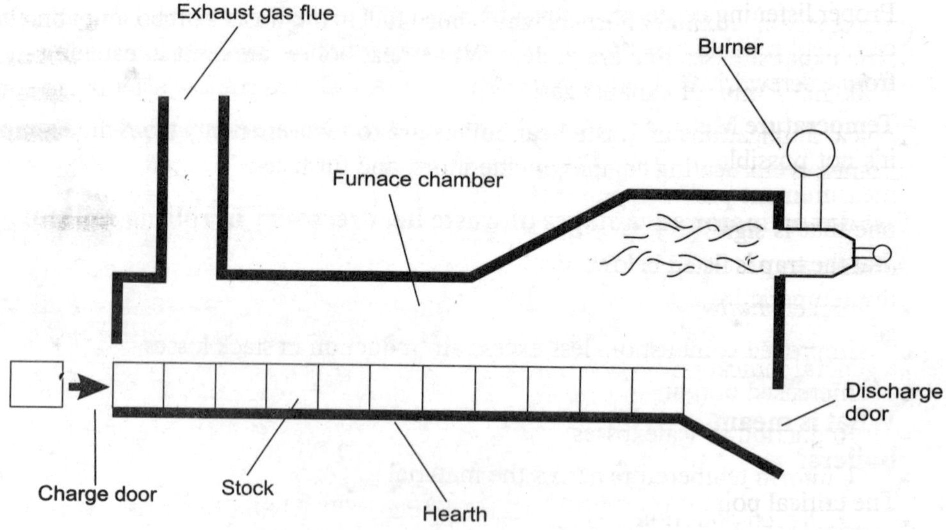

Fig.56.60: Furance features

- A refractory chamber constructed of insulating materials for retaining heat at the high operating temperatures.
- A hearth to support or carry the steel. This can consist of refractory materials or an arrangement of metallic supports that may be water cooled.
- Burners that use liquid or gaseous fuels to raise and maintain the temperature in the chamber. Coal or electricity can be used for reheating. A method of removing the combustion exhaust gases from the chamber.
- A method of introducing and removing the steel from the chamber.
- These facilities depend on the size and type of furnace, the shape and size of the steel being processed, and the general layout of the rolling mill.
- Common systems include roller tables, conveyors, charging machines and furnace pushers.

Q. What are the functions of a steam trap ?

Ans. The three important functions of steam traps are:

- • To discharge condensate as soon as it is formed in the steam line / pipes.
- • Not to allow steam to escape.
- • To be capable of discharging air and other incondensable gases from the steam pipe.

Q. Explain any two proven methods of testing steam traps?

Ans. There are two proven methods of testing of steam traps: - Sound method and Temperature method.

Sound Method : Mechanisms within steam traps and the flow of steam and condensate through steam traps generate sonic (audible to the human ear) and supersonic sounds. Proper listening equipment, coupled with the knowledge of normal and abnormal sounds, can yield reliable assessments of steam trap working condition. Listening devices range from a screwdriver or simple mechanic's stethoscope that allow listening to sonic sounds.

Temperature Method : Saturated steam and condensate exist at the same temperature. So it's not possible to distinguish between the two based on temperature. Still, temperature measurement provides important information for evaluation purposes. A cold trap (i.e., one that is significantly cooler than the expected saturated steam temperature) indicates that the trap is flooded with condensate, assuming the trap is in service. On the other hand, the temperature downstream of the trap will be nearly constant if significant steam is getting past the trap. At the low-end, spitting on the trap and watching the sizzle provides a general indication of temperature.

Q. What is meant by critical point of steam ? Give two advantages of Super Critical boilers?

Ans: The critical point is the highest temperature at which liquid can exist. Any compression at constant temperature above the critical point will not produce a phase change. Compression at constant temperature below the critical point however, will result in liquefaction of the vapour as it passes from the superheated region into the wet steam region. The critical point occurs at 374.15°C and 221.2 bar (a) for steam and at this point the latent heat of steam is zero.

Advantages of super critical boiler are as follows :

- Higher Heat Transfer Rate
- More Flexible in accepting load variations
- Greater ease of operation
- High thermal efficiency of 40 %
- The absence of two phase mixture minimize the problem of erosion and corrosion
- Steadier Pressure level

Q. List down energy conservation measures in steam system .

Ans: Energy conservation measures in steam system are listed below.

- Fix steam leaks and condensate leaks.
- Use back pressure steam turbines to produce lower steam pressures.
- Use dry steam for indirect process heating.
- Ensure process temperatures are correctly controlled.
- Maintain lowest acceptable process steam pressures.

- Remove or blank off all redundant steam piping.
- Ensure condensate is returned or re-used in the process.
- Preheat boiler feed-water.
- Recover boiler blowdown heat .
- Check operation of steam traps.
- Remove air from indirect steam using equipment .
- Inspect steam traps regularly and repair malfunctioning traps promptly.
- Consider recovery of flash steam .
- Use waste steam for water heating.
- Reduce work done by steam

Q. How are Stoker classified ? Describe chain grate stoker and spreader stoker fired boiler.

Ans. Stokers are classified according to the method of feeding fuel to the furnace and by the type of grate. The main classifications are:

- Chain-grate or traveling-grate stoker
- Spreader stoker

Chain-Grate or Traveling-Grate Stoker Boiler

Coal is fed onto one end of a moving steel chain grate. As grate moves along the length of the furnace, the coal burns before dropping off at the end as ash. Some degree of skill is required, particularly when setting up the grate, air dampers and baffles, to ensure clean combustion leaving minimum of unburnt carbon in the ash.

The coal-feed hopper runs along the entire coal-feed end of the furnace. A coal grate is used to control the rate at which coal is fed into the furnace, and to control the thickness of the coal bed and speed of the grate. Coal must be uniform in size, as large lumps will not burn out completely by the time they reach the end of the grate. As the bed thickness decreases from coal-feed end to rear end, different amounts of air are required- more quantity at coal-feed end and less at rear end .

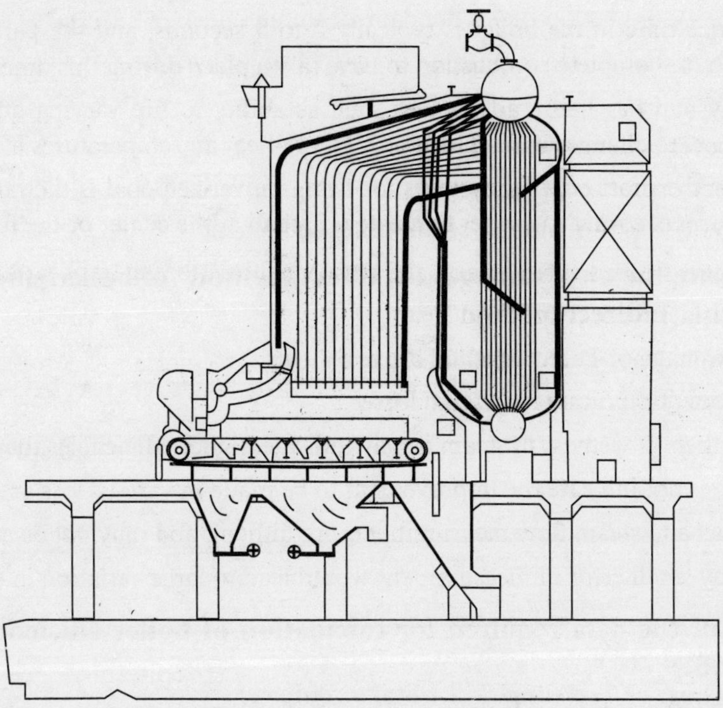

Fig.56.61: Chain Grate Stoker

Spreader Stoker Boiler

Spreader stokers utilize a combination of suspension burning and grate burning. The coal is continually fed into the furnace above a burning bed of coal. The coal fines are burned in suspension; the larger particles fall to the grate, where they are burned in a thin, fast-burning coal bed. This method of firing provides good flexibility to meet load fluctuations, since ignition is almost instantaneous when firing rate is increased. Hence, the spreader stoker is favored over other types of stokers in many industrial applications.

Q. Describe the salient features of a Pulverized Fuel Fired Boiler .

Ans. Most coal-fired power station boilers use pulverized coal, and many of the larger industrial water-tube boilers also use this pulverized fuel. This technology is well developed, and there are thousands of units around the world, accounting for well over 90% of coal-fired capacity.

The coal is ground (pulverised) to a fine powder, so that less than 2% is +300 micro metre (μm) and 80 % is below 74 microns (200 mesh) , for a bituminous coal. Too fine a powder is wasteful of grinding mill power. On the other hand, too coarse a powder does not burn completely in the combustion chamber and results in higher unburnt losses.

The pulverised coal is blown with part of the combustion air into the boiler plant through a series of burner nozzles. Secondary and tertiary air may also be added. Combustion takes place at temperatures from 1300-1700°C, depending largely on coal grade. Particle residence time in the boiler is typically 2 to 5 seconds, and the particles must be small enough for complete combustion to have taken place during this time.

This system has many advantages such as ability to fire varying quality of coal, quick responses to changes in load, use of high pre-heat air temperatures etc.

One of the most popular systems for firing pulverized coal is the tangential firing using four burners corner to corner to create a fireball at the center of the furnace.

Q. What are the disadvantages of "direct method" of boiler efficiency calculation over the "indirect method"?

Ans: Disadvantage of Direct Method are as follows ;

- Does not indicate individual losses .
- If there is wetness in steam it may indicate higher efficiencies than actual .
- Does not indicate the improvement to be made in various loss areas .
- Fuel and steam flow measurements are difficult and may not be accurate .
- Any small error in measurement would lead to large variation in efficiency levels .

Q. List out the data required for calculation of boiler efficiency using 'indirect method'.

Ans. The data required for calculation of boiler efficiency using indirect method are:

- Ultimate analysis of fuel (H2, O2, S, C, moisture content, ash content)
- Percentage of Oxygen or CO2 in the flue gas
- Flue gas temperature in °C (Tf)
- Ambient temperature in °C (Ta) & humidity of air in kg/kg of dry air.
- GCV of fuel in kcal/kg
- Percentage combustible in ash (in case of solid fuels)
- GCV of ash in kcal/kg (in case of solid fuels)

Q. List out various energy conservation opportunities in a boiler.

Ans. Various energy conservation opportunities in a boiler are listed below.

- Heat recovery from flue gas through air preheating.
- Heat recovery from flue gas through an economizer for feed water heating.
- Excess Air Control.
- Ensurance of complete combustion
- Reduction in Radiation and Convection losses from boiler shell surface.
- Reduction in scale and soot losses.
- Reduction in boiler steam pressure.
- Variable speed control of fans, blowers and pumps
- Automatic Blow down control.

- Proper Boiler scheduling.
- Replacement of inefficient boiler.

Q. List few energy saving measures in a steam distribution & utilization system.

Ans. Energy saving measures in a steam distribution & utilization system are listed below :

- Monitoring Steam Traps
- Avoiding Steam Leakage
- Providing Dry Steam for Process
- Proper Utilisation of Directly Injected Steam
- Miminising Heat Transfer Barrier
- Proper Air Venting
- Condensate Recovery
- Insulation of steam pipe lines and hot process equipments
- Flash Steam Recovery
- Reducing the work to be done by steam

Q. Explain briefly with a neat sketch the following heat recovery devices.

a) Heat Pipe

b) Metallic radiation recuperator

c) Plate heat exchanger

Ans. **(a) Heat pipe**

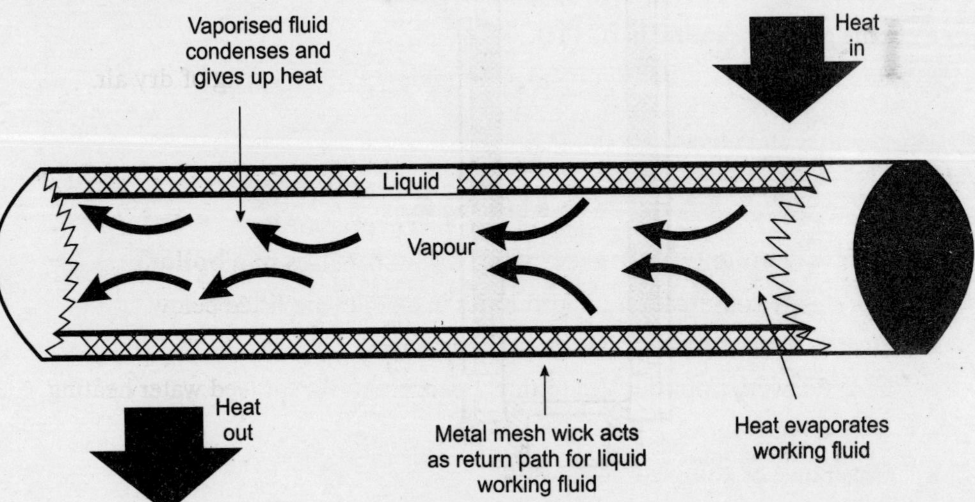

Fig.56.62: Heat pipe

The Heat Pipe comprises of three elements – a sealed container, a capillary wick structure and a working fluid. The capillary wick structure is integrally fabricated into the interior surface of the container tube and sealed under vacuum. Thermal energy applied to the external surface of the heat pipe is in equilibrium with its own vapour as the container tube is sealed under vacuum. Thermal energy applied to the external surface of the heat pipe causes the working fluid near the surface to evaporate instantaneously. Vapour thus formed absorbs the latent heat of vapourisation and this part of the heat pipe becomes an evaporator region. The vapour then travels to the other end the pipe where the thermal energy is removed causing the vapour to condense into liquid again, thereby giving up the latent heat of the condensation. This part of the heat pipe works as the condenser region. The condensed liquid then flows back to the evaporated region.

Applications of heat pipe in industries are:

- Preheating of boiler combustion air
- Recovery of Waste heat from furnaces
- Reheating of fresh air for hot air driers
- Recovery of waste heat from catalytic deodorizing equipment
- Reuse of Furnace waste heat as heat source for other oven
- Cooling of closed rooms with outside air
- Preheating of boiler feed water with waste heat recovery from flue gases
- Drying, curing and baking oven

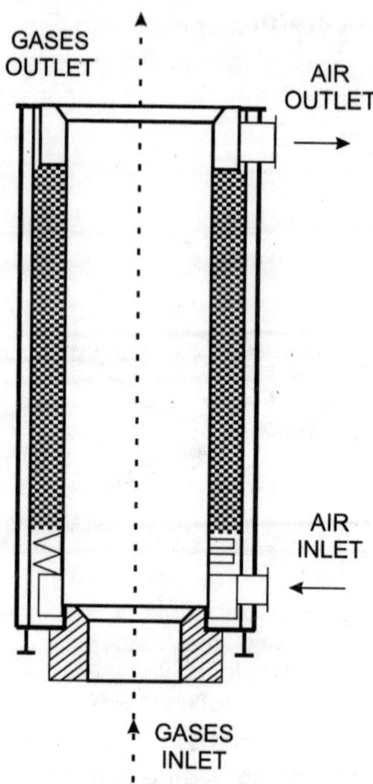

Fig.56.63: Metallic radiation repuperator

(b) Metallic Radiation Repuperator

A metallic radiation recuperator consists of two concentric lengths of metal tubing. The inner tube carries the hot exhaust gases while the external annuius carries the combustion air from the atmosphere to the air inlets of the furnace burners. The hot gases are cooled by the incoming combustion air which now carries additional energy into the combustion chamber. Radiation rècuperator gets its name from the fact that a substantial portion of the heat transfer from the hot gases to the surface of the inner tube takes place by radiative heat transfer.

(c) Plate heat exchanger

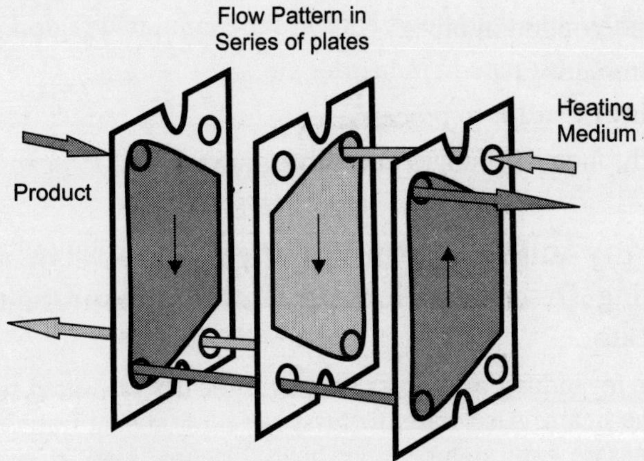

Fig.56.64: Plate heat exchanger

A plate type heat exchanger consists of a series of separate parallel plates forming thin flow pass. Each plate is separated from the next by gaskets and the hot stream passes in parallel through alternative plates whilst the liquid to be heated passes in parallel between the hot plates. To improve heat transfer the plates are corrugated.

Hot liquid passing through a bottom port in the head is permitted to pass upwards between every second plate while cold liquid at the top of the head is permitted to pass downwards between the odd plates. When the directions of hot & cold fluids are opposite, the arrangement is described as counter current. The plate heat exchanger is a potential heat recovery device to recover heat from hot liquids and hot effluents.

Q. Briefly explain the principle of 'thermo compression'.

Ans. In many cases, very low pressure steam is reused as water after condensation for lack of any better option of reuse. In many cases it becomes feasible to compress this low pressure steam by very high pressure steam and reuse it as a medium pressure steam. The major energy in steam is in its latent heat value and thus thermo compressing would give a large improvement in waste heat recovery.

The thermo compressor is simple equipment with a nozzle where HP steam is accelerated into a high velocity fluid. This entrains the LP steam by momentum transfer and then recompresses in a divergent venturi. It is typically used in evaporators where the boiling steam is recompressed and used as heating steam.

Q. Explain why dry saturated steam is preferred over wet or superheated steam for industrial process heating ?

Ans Dry saturated steam is the preferred choice because:

- Superheated steam gives up heat at a slower rate than saturated steam.
- Dry steam alone condenses quickly, thereby providing a higher heat transfer rate.

Q. List five disadvantages if the steam pipe line is not insulated .

Ans : Disadvantages are listed below.

- Heat Loss from bare surfaces .
- Steam condensation in pipes.
- Water Hammering.
- Poor quality of steam to process.
- Leakage through the flanges and valves due to vibration caused by water hammering effect.

Q. Explain why dry saturated steam is preferred over wet or super heated steam for process heating .Describe various precaution/measures one must take to supply saturated steam.

Ans. The best steam for industrial process heating is the dry saturated steam. Wet steam reduces total heat in the steam. Also water forms a wet film on heat transfer surface and overloads traps & condensate equipment. Super heated steam is not desirable for process heating because it gives up heat at a rate slower than the condensation heat transfer of saturated steam.

It must be remembered that a boiler without a superheater cannot deliver perfectly dry saturated steam. At best, it can deliver only 95% dry steam. The dryness fraction of steam depends on various factors, such as the level of water to be a part of the steam. Indeed, even as simple a thing as improper boiler water treatment can become a cause for wet steam.

As steam flows through the pipelines, it undergoes progressive condensation due to the loss of heat to the colder surroundings, The extent of the condensation depends on the effectiveness of the lagging. For example, with poor lagging, the steam can become excessively wet.

Since dry saturated steam is required for process equipment, due attention must be paid to the boiler operation and lagging of the pipelines.

Wet steam can reduce plant productivity and product quality, and can cause damage to most items of plant & equipment. Whilst careful drainage and trapping can remove most of the water, it will not deal with the water droplets suspended in the steam. To remove these suspended water droplets, separators are installed in steam pipelines.

The steam produced in a boiler designed to generate saturated steam is inherently wet. Although the dryness fraction will vary according to the type of boiler, most shell type steam boilers will produce steam with a dryness fraction of between 95 and 98%. The water content of the steam produced by the boiler is further increased if priming and carryover occur.

A steam separator may be installed on the steam main as well as on the branch lines to reduce wetness in steam and improve the quality of the steam going to the units. By change of direction of steam, steam seperators causes the entrained water particles to be separated out and delivered to a point where they can be drained away as condensate through a conventional steam trap.

Q. List the characteristic of steam which make it most popular for carrying energy.

Ans. The characteristics are :

bm. Highest specific heat and latent heat.

bn. Highest heat transfer coefficient both at generation and use point.

bo. Easy to control and distribute.

bp. Cheap and Inert.

bq. Source (water) is easily available in the quantities required anywhere any time.

Q. List out the important properties of ceramic fibre.

Ans. Important properties of ceramic fibre are listed below .

- **Lower Thermal Conductivity:** This allows construction of thinner linings.
- **Lower Heat Storage :** Less heat absorption due to lower density.
- **Good thermal shock resistance:** Allows faster heating and cooling.
- **Mechanical Resilience:** This allows pre fabrication.
- Low installation Cost.
- Simple Maintenance.
- **Higher Thermal Efficiency:** Allows more accurate and uniform temperature.

Q. How does ceramic coatings help in reducing energy consumption in a furnace?

Ans. Ceramic coatings in furnace chamber promote rapid and efficient transfer of heat, uniform heating and extended life of refractories. The emissivity of ceramic-coated refractories increases with increase in temperature. Energy savings to an order of 8 to 10% can be achieved for high temperature applications.

Q. List advantages of ceramic fibre insulation?

Ans. The advantages are :

- Low Down time
- Increased productivity
- Low maintenance cost
- Longer service life
- Higher thermal efficiency
- Faster response

Q. Classify the various types of thermal insulation as per temperature and give an example for each type.

Ans. The Insulation can be classified into three groups according to the temperature ranges for which they are used.

- **Low Temperature Insulations (up to 90 °C):** This range covers insulating materials for refrigerators, cold and hot water systems, storage tanks, etc. The commonly used materials are Cork, Wood, 85% magnesia, Mineral Fibers, Polyurethane and expanded Polystyrene, etc.

- **Medium Temperature Insulations (90 – 325 °C):**Insulators in this range are used in low temperature, heating and steam raising equipment, steam lines, flue ducts etc.The types of materials used in this temperatures range include 85% Magnesia, Asbestos, Calcium Silicate and Mineral Fibers etc.
- **High Temperature Insulations (> 325 °C):**Typical uses of such materials are super heated steam system, oven dryer and furnaces etc. The most extensively used materials in this range are Asbestos, Calcium Silicate, Mineral Fibre, Mica and Vermiculite based insulation, Fireclay or Silica based insulation and Ceramic Fibre.

Q. Describe the properties & applications of commonly used insulation materials .

Ans. Insulation materials can be organic and inorganic types. Organic insulations are based on hydrocarbon polymers, which can be expanded to obtain high void structures eg; Thermocol (Expanded Polystyrene) and Poly Urethane Form(PUF). Inorganic insulation is based on Siliceous/Aluminous/Calcium materials in fibrous, granular or powder forms eg; Mineral wool, Calcium silicate etc.

Properties & applications of common insulating materials are as under:

Calcium Silicate: Used in industrial process plant piping where high service temperature and compressive strength are needed. Temperature ranges of application varies from 40 to 950 0C.

Glass mineral wool: These are available in flexible forms, rigid slabs and preformed pipe work sections. Good for thermal and acoustic insulation for heating & chilling system pipelines. Temperature range of application is –10 to 500 °C .

Thermocol: These are mainly used as cold insulation for piping and cold storage construction.

Expanded nitrile rubber: This is a flexible material that forms a closed cell integral vapour barrier. Originally developed for condensation control in refrigeration pipe work and chilled water lines; now-a-days also used for ducting insulation for air conditioning.

Rock mineral wool: This is available in a range of forms from light weight rolled products to heavy rigid slabs including preformed pipe sections. In addition to good thermal insulation properties, it can also provide acoustic insulation and is fire retardant.

Lagging materials can be obtained in bulk, in the form of moulded sections; semi cylindrical for pipes, slabs for vessels, flanges, valves etc. The main advantage of the moulded sections is the ease of application and replacement when undertaking repairs for damaged lagging.

APPENDIX-C: MULTICHOICE QUESTIONS & ANSWERS ON FUEL & COMBUSTION TECHNOLOGY

1. In flue gas analysis by Orsat's apparatus, carbon monoxide is absorbed by
   (a) cuprous chloride
   (b) potassium hydroxide
   (c) alkaline pyrogallol solution
   (d) none of these.
2. Calorific value as determined by bomb calorimeter is the
   (a) higher calorific value at constant volume.
   (b) gross calorific value at constant pressure.
   (c) lower calorific value at constant pressure.
   (d) net calorific value at constant volume.
3. Incomplete combustion of a fuel is characterised by the high..................in the flue gas.
   (a) smoke *(b)* temperature
   (c) oxygen *(d)* carbon monoxide
4. For maximum discharge through a chimney, its height should be
   (a) 200 meters.
   (b) infinitely long.
   (c) more than 105.7 metres.
   (d) equal to the height of the hot gas column producing draught.
5. Oxygen percentage (by weight) in atmospheric air is
   (a) 19 (b) 21
   (c) 23 (d) 29
6. Stack (chimney) height in a big thermal power plant is dictated by the
   (a) pollution control aspect.
   (b) draught to be created.
   (c) limitation of constructional facilities.
   (d) none of these.
7. As time passes, the calorific value of stored coal
   (a) decreases
   (b) increases
   (c) remains unaltered
   (d) may increase or decrease (depends on the method of storage).
8. Emission of dense white smoke out of the chimney of a thermal power plant is an indication of the use of
   (a) less air for combustion.
   (b) correct amount of air for combustion.
   (c) too much air for combustion.
   (d) pulverised coal in boilers.
9. Use of pulverised coal in boiler furnace provides
   (a) high calorific value.
   (b) better combustion.
   (c) smokeless burning.
   (d) less erosion on furnace walls
10. Which of the following fuels is the best for burning on chain grate stoker ?

ANSWERS

1.*(a)* 2.*(a)* 3.*(d)* 4.*(d)* 5.*(c)* 6.*(b)* 7.*(a)* 8.*(c)* 9.*(b)* 10.*(a)*

(a) Non-caking coal
(b) Caking coal
(c) Coking coal
(d) Pulverised coal.

11. Which of the following accounts for maximum energy loss in a boiler ?
(a) Flue gases.
(b) Ash content in the fuel.
(c) Incomplete combustion.
(d) Unburnt carbon in flue gases.

12. "Overfire burning" in a furnace is a phenomenon characterised by the
(a) supply of excess fuel.
(b) supply of excess air.
(c) burning of carbon monoxide and other incombustibles in upper zone of furnace by supplying more air.
(d) none of these.

13. When steam is passed over coal resulting in the endothermic reaction, it is represented by, $C + H_2O = CO + H_2$, it is called the..............of coal.
(a) carbonisation (b) oxidation
(c) coalification (d) gasification

14. Fischer-Tropsch method aims at the
(a) gasification of coal.
(b) synthesis of gasoline (from water gas).
(c) hydrogenation of coal to produce gasoline.
(d) none of these.

15. Commercial production of petrol from coal (as practised in a factory at Sasol in South Africa) is done by the...............of coal.
(a) hydrogenation (b) gasification
(c) carbonisation (d) none of these

16. In case of pulverised coal fired steam boiler, the secondary air serves the main purpose of
(a) transportation of coal.
(b) drying of coal.
(c) combustion of coal by supplying it around the burner.
(d) preheating the primary air.

17. The main function of primary air in pulverised coal fired burner is to
(a) burn CO to CO_2.
(b) dry and transport the coal.
(c) have proper combustion by supplying it around the burner.
(d) preheat the tertiary air used for complete combustion of CO to CO_2.

18. The advantage of firing pulverised coal in the furnace lies in the fact, that it
(a) permits the use of high ash content coal,
(b) permits the use of low fusion point ash coal.
(c) accelerates the burning rate and economises on fuel combustion.
(d) all (a), (b) and (c).

19. Orsat apparatus is meant for
(a) gravimetric analysis of flue gas.
(b) finding out combustion efficiency.
(c) direct determination of nitrogen in flue gas by absorbing it in ammoniacal cuprous chloride.
(d) none of these.

20. Pick out the wrong statement.
(a) Theoretical flame temperature is the temperature attained by the products of combustion, when the fuel is burned without loss or gain of heat.
(b) Burning the fuel with theoretically required amount of pure oxygen results in attainment of maximum adiabatic flame temperature.
(c) Burning the fuel with excess pure oxygen results in maximum theoretical flame temperature.
(d) Adiabatic flame temperatures of actual combustions are always less than the maximum values.

21. The maximum adiabatic flame temperature in air as compared to that in pure oxygen is
(a) much lower.
(b) much higher.
(c) same.
(d) either lower or higher, depends on the type of fuel.

22. Adiabatic flame temperature of a fuel is dependent on the initial temperature of
(a) fuel
(b) air
(c) both (a) & (b)
(d) neither (a) nor (b)

23. Preheating of..................the flame temperature.
(a) the gaseous fuel before combustion decreases
(b) combustion air decreases
(c) either the fuel or the air or both, increases
(d) either the fuel or the air does not affect

24. A fuel containing carbon and carbon monoxide (but containing no hydrogen or its compounds) is burnt in pure oxygen at constant pressure. Its

ANSWERS

11.(a) 12.(c) 13.(d) 14.(b) 15.(a) 16.(c) 17.(b) 18.(c) 19.(d) 20.(c) 21.(a) 22.(c)
23.(c) 24.(c)

gross calorific value as compared to net calorific value will be
(a) more
(b) less
(c) same
(d) data insufficient; can't be predicted.

25. Removal of hydrogen from coke oven gas
(a) increases its calorific value.
(b) decreases its calorific value.
(c) does not alter its calorific value.
(d) is not possible on commerical scale.

26. With increase in calorific value of fuels, their adiabatic flame temperatures
(a) increase
(b) decrease
(c) remain unchanged
(d) may increase or decrease ; depends on the quantity of products of combustion

27. Grindability index of a coal is 100. It implies that the
(a) coal can be pulverised with great difficulty.
(b) coal can't be pulverised.
(c) coal can be easily pulverised.
(d) power consumption in grinding the coal will be very high.

28. If the specific heat of gaseous products of combustion of a fuel is high, the abiabatic flame temperature will be
(a) low
(b) high
(c) very high, if the fuel is of low calorific value
(d) none of these

29. 'Calorific value of dry wood may be around.......................Kcal/kg.
(a) 5 (b) 50
(c) 500 *(d)* 5000

30. A coal containing high amount of volatile matter will have
(a) low ignition temperature
(b) very little ash content
(c) high fusion point of its ash
(d) low adiabatic flame temperature

31. Steam is intermittently admitted into the fuel bed during the production of producer gas to
(a) convert CO to CO_2.
(b) increase the combustion rate.
(c) increase the gas production rate.
(d) minimise the chances of clinker formation.

32. Which of the following is called "blue gas" ?
(a) coke oven gas *(b)* water gas
(c) natural gas *(d)* producer gas

33. During combustion of gaseous fuels, deficiency of air
(a) lengthens the flame.
(b) tends to shorten the flame.
(c) does not affect the flame length.
(d) increases the flame temperature.

34. Use of excess of combustion air in the combustion of fuels results in
(a) heat losses.
(b) long flame.
(c) condensation of water vapour from the fuel gas.
(d) none of these.

35. A gaseous fuel in order to develop luminousity on burning, must contain
(a) carbon monoxide *(b)* hydrocarbons
(c) hydrogen *(d)* oxygen

36. Preheating of a gaseous fuel results in increased
(a) flame length
(b) flame temperature
(c) quantity of flue gas
(d) ignition temperature

37. Quantity of coke oven gas produced by high temperature carbonisation of one ton of dry coal may be around.................Nm^3 .
(a) 30 (b) 300
(c) 3000 (d) 30,000

38. Coke oven gas is a better fuel than blast furnace gas, because of its higher
(a) calorific value, cleanliness and relatively low distribution cost (due to its low specific gravity).
(b) adiabatic flame temperature.
(c) heat release rate (thus requiring smaller combustion chamber).
(d) all *(a)*, *(b)* and *(c)*.

39. Which of the following constituents of a fuel does not contribute to its calorific value on combustion ?
(a) Hydrogen *(b)* Sulphur
(c) Carbon *(d)* None of these

40. Combustion of pulverised coal as compared to that of lump coal
(a) develops a non-luminous flame.

ANSWERS

25.*(a)* 26.*(d)* 27.*(c)* 28.*(a)* 29.*(a)* 30.*(a)* 31.*(d)* 32.*(b)* 33.*(a)* 34.*(a)* 35.*(b)* 36.*(b)*
37.*(b)* 38.*(d)* 39.*(d)* 40.*(c)*

(*b*) develops a low temperature flame.
(*c*) can be done with less excess air.
(*d*) provides a lower rate of heat release.

41. Which of the following is the most important deterrents to an extended use of pulverised coal in boiler firing ?
(*a*) Ash disposal problem.
(*b*) Excessive fly-ash discharge from the stack.
(*c*) High power consumption in its transportation.
(*d*) Erosion of induced draft fan blades.

42. Pulverised coal used in boiler firing need not have
(a) less moisture content
(b) high fusion point of its ash
(c) high bulk density
(d) lower ash content

43. Dry air required to burn 1 kg of carbon completely may be around...............kg.
(a) 11 (b) 2
(c) 20 (d) 38

44. Dry air requirement for burning 1 Nm^3 of CO to CO_2 may be around..............Nm^3.
(*a*) 2.4 (*b*) 1.75
(c) 0.87 (*d*) 11.4

45. Washing of coal
(*a*) reduces its sulphur and ash content.
(*b*) controls its ash fusibility and increases its calorific value.
(*c*) improves its coking properties.
(*d*) all (*a*), (*b*) and (*c*).

46. Which of the following has the highest gross calorific value ?
(*a*) Blast furnace gas
(*b*) Coke oven gas
(*c*) Carburetted water gas
(*d*) Oil refinery gas

47. Most of the coking coals are
(a) anthracite (*b*) bituminous
(*c*) lignite (*d*) none of these

48. High excess air in combustion of fuels results in
(a) increased fuel consumption
(*b*) incomplete combustion
(*a*) smoky flame
(*d*) none of these

49. Atomising steam to fuel oil ratio in a burner should be around
(*a*) 0.5 (*b*) 1.5
(c) 2.5 (*d*) 3.5

50. For every 10% increase in the excess air ; the fuel consumption increases by.............percent.
(*a*) 0.1 (*b*) 2
(c) 5 (*d*) 10

51. Ash content in the coke produced from a coking coal having 20% ash may be around ....percent.
(a) 6 (b) 12
(c) 18 (d) *24*

52. Out of the following fuels, the difference between the net and gross calorific value is maximum in case of
(a) pitch (b) fuel oil
(c) blast furnace gas (*d*) bituminous coal

53. Spontaneous combustion of coal on storage results due to
(a) inadequate ventilation.
(b) low temperature oxidation.
(c) storage in large heaps with small surface to volume ratio.
(*d*) all (*a*), (*b*) and (*c*).

54. Low temperature oxidation of coal resulting from bad storage conditions does not decrease its
(a) caking power (b) calorific value
(c) hydrogen content (*d*) oxygen content

55. Efficient burning of anthracite coal requires
(*a*) low preheat of air.
(*b*) fine grinding.
(*c*) high excess air.
(*d*) all (*a*), (*b*) and (*c*).

56. Rate of low temperature oxidation of coal due to bad storage conditions
(*a*) decreases with increase in surface area.
(*b*) does not vary with increase in surface area.
(*c*) is more for low volatile Coal compared to high volatile coal.
(*d*) is accelerated by storage in large heaps with small surface to volume ratio.

57. Fusion point of coal ash increases with increase in its.............content.
(a) iron sulphate
(b) iron silicate
(c) lime and magnesia
(d) all (*a*), (*b*) and (*c*)

58. Which of the following accentuates clinkering trouble on furnace grate burning coal ?

ANSWERS

41.(*a*) 42.(*b*) 43.(*a*) 44.(*a*) 45.(*d*) 46.(*d*) 47.(*b*) 48.(*a*) 49.(*a*) 50.(*a*) 51.(*d*) 52.(*b*)
53.(*d*) 54.(*d*) 55.(*b*) 56.(*d*) 57.(*d*) 58.(*d*)

(a) Low reactivity of carbonised residue containing high proportions of iron & sulphur.
(b) Low forced draught and fuel bed temperature.
(c) Thick fire bed and preheated primary air.
(d) All (a), (b) and (c).

59. A coal having high ratio of volatile matter to fixed carbon as compared to a coal having low ratio of volatile matter to fixed carbon
(a) is less liable to spontaneous combustion on storage.
(b) is more difficult to ignite and produces a shorter flame.
(c) requires smaller combustion space and less secondary air.
(d) none of these.

60. Which of the following can be made into briquettes without the use of a binder ?
(a) Lignite
(b) Bituminous coal
(c) Anthracite coal
(d) None of these

61. Presence of free moisture in coal is most disadvantageous during
(a) its pulverisation (as it requires more power).
(b) combustion of fire slacks on the grates.
(c) handling (*e.g.* when emptying wagons).
(d) none of these.

62. Which of the following is not a binder for coal briquetting ?
(a) Coal tar (b) Bitumen
(c) Molasses (d) None of these

63. Highly caking coals
(a) produce weak coke.
(b) produce strong coke.
(c) may damage the coke oven walls during carbonisation.
(d) both (b) and (c).

64. Presence of free moisture in coal during its high temperature carbonisation
(a) reduces the coking time.
(b) protects the volatile products from pyrolysis (cracking) in the presence of hot coke and hot oven walls.
(c) increases the loss of fine coal dust from the ovens when charging.
(d) none of these.

65. Higher percentage of ash in coal meant for the production of metallurgical grade coke
(a) decreases the hardness of coke.
(b) decreases the abrasion resistance of coke.
(c) causes brittleness in steel.
(d) none of these.

66. Increase in ash content of blast furnace coke
(a) reduces its consumption in the furnace.
(b) increases its consumption in the furnace.
(c) does not affect its consumption in the furnace.
(d) decreases its hardness and abrasion resistance.

67. High sulphur content in a fuel................of the flue gases.
(a) decreases the dew point
(b) increases the dew point
(c) reduces the combustion efficiency by limiting the permissible temperature reduction
(d) both (b) and (c)

68. With increase in carbonisation temperature
(a) coke even gas yield increases.
(b) tar yield increases.
(c) hydrogen percentage in the coke oven gas decreases.
(d) methane percentage in the coke oven gas increases.

69. With increase in the time of carbonisation at a particular temperature (say 1000°C), the... percentage in coke oven gas increases.
(a) hydrogen
(b) methane
(c) unsaturated hydrocarbons
(d) all (a), (b) and (c)

70. Producer gas containing least amount of tar is produced by the
(a) partial combustion of coal.
(b) partial combustion of large size (< 50 mm) coke.
(c) high pressure gasification of coal (e.g. in Lurgi gasifier).
(d) atmospheric pressure gasification of coal (*e.g.*, in Kopper-Totzek gasifier).

71. High temperature in gasification of coal favours
(a) high production of CO_2.
(b) low production of CO_2.
(c) high production of CO.
(d) both (b) and (c).

ANSWERS

59.(d) 60.(a) 61.(a) 62.(d) 63.(d) 64.(b) 65.(d) 66.(b) 67.(d) 68.(a) 69.(a) 70.(b) 71.(d)

72. Lurgi gasifier (high pressure gasifier) as compared to Kopper Totzek gasifier (atmospheric pressure gasifier) produces
(a) higher amount of methane.
(b) lower amount of hydrogen.
(c) both *(a)* and *(b)*.
(d) higher amount of both methane and hydrogen.

73. Which of the following constituents of coal is the most important in the production of coke?
(a) Moisture (b) Ash
(c) Volatiles *(d)* Carbon

74. Which of the following is the most reactive (as regards the formation of $CO + H_2$ from $C+H_2O$) ?
(a) Blast furnace coke
(b) Low temperature coke
(c) Anthracite coal
(d) Sub-bituminous coal

75. A coal that softens and fuses on heating is
(a) classified *(b)* carbonised
(c) caking *(d)* non-caking

76. Purity of coke means that, it is high in carbon and low in
(a) volatiles (b) ash
(c) iron *(d)* moisture

77. Calorific value of coke oven gas is around........... Kcal/Nm^3.
(a) 900 (b) 4200
(c) 7500 *(d)* 2000

78. Calorific value of blast furnace gas is around................KCal/Nm^3.
(a) 850 *(b)* 1800
(c) 4200 *(d)* 6500

79. Blast furnace gas is a very poisonous gas because of its predominantly high...........content.
(a) H_2O (b) CO_2
(c) CO *(d)* CH_4

80. Percentage of carbon monoxide in blast furnace gas may be around
(a) 8 *(b)* 14
(c) 22 *(d) 52*

81. Percentage of hydrogen in coke oven gas may be around
(a) 10 *(b)* 25
(c) 45 *(d)* 60

82. Percentage of methane in coke oven gas may be around
(a) 5 (b) 15
(c) 25 *(d)* 50

83. Coke oven gas compared to blast furnace gas is
(a) more explosive and inflammable.
(b) less poisonous.
(c) lighter.
(d) all *(a)*, *(b)* and *(c)*.

84. Which of the following fuel gases will require maximum amount of air for combustion of 1 Nm^3 gas ?
(a) Blast furnace gas *(b)* Natural gas
(c) Producer gas *(d)* Water gas

85. Which of the following will generate maximum volume of product of complete combustion (Nm^3/Nm^3 of fuel) ?
(a) Carburetted water gas
(b) Blast furnace gas
(c) Natural gas
(d) Producer gas

86. Which is the heaviest fuel gas out of the following ?
(a) Blast furnace gas
(b) Coke oven gas
(b) Water gas
(c) Carburetted water gas

87. 'Fuel' can be defined as a substance which produces heat by
(a) combustion.
(b) nuclear fission.
(c) nuclear fusion.
(d) all *(a)*, *(b)* & *(c)*.

88. Catalyst used in Fischer-Tropsch process is
(a) nickel *(b)* zinc oxide
(c) alumina *(d)* thorium oxide

89. In flue gas analysis by Orsat's apparatus, oxygen is absorbed by
(a) potassium hydroxide.
(b) cuprous chloride.
(c) alkaline pyragllol solution.
(d) none of these.

90. In flue gas analysis by Orsat's apparatus, carbon dioxide is absorbed by
(a) potassium hydroxide.
(b) dilute potassium carbonate.
(c) cuporus chloride.
(d) alkaline pyragllol solution.

ANSWERS

72.*(c)* 73.*(d)* 74.*(b)* 75.*(c)* 76.*(a)* 77.*(b)* 78.*(a)* 79.*(c)* 80.*(c)* 81.*(d)* 82.*(c)* 83.*(d)*
84.*(b)* 85.*(c)* 86.*(a)* 87.*(d)* 88.*(a)* 89.*(c)* 90.*(a)*

91. The sequence of absorption in flue gas analysis by Orsat's apparatus is respectively
(a) CO_2, O_2, CO (*b*) CO, O_2, CO_2
(c) CO_2, CO, O_2 (*d*) O_2, CO_2, CO

92. Which of the following has the highest calorific value ?
(a) Lignite
(b) Sub-bituminous coal
(c) Anthracite
(*d*) Peat

93. The main product of high temperature carbonisation of coal is
(a) coke (b) ammonia
(c) tar (*d*) coke oven gas

94. High temperature carbonisation of coal produces
(a) inferior coke compared to low temperature carbonisation.
(b) less of gases compared to liquid products.
(c) large quantity of tar compared to low temperature carbonisation.
(*d*) none of these.

95. High temperature carbonisation of coal takes place at..............°C.
(*a*) 2000 (*b*) 600
(c) 1100 (*d*) 1600

96. Low temperature carbonisation of coal takes place at..............°C.
(a) 300 (b) 1100
(c) 700 (d) 150

97. Proximate analysis of coal determines its......... content.
(*a*) moisture, ash, sulphur & volatile matter
(*b*) moisture, volatile matter, ash & fixed carbon
(*c*) moisture, sulphur, nitrogen & fixed carbon
(*d*) none of these

98. Ultimate analysis of coal determines its........content.
(*a*) carbon, hydrogen, nitrogen & sulphur
(*b*) carbon, ash, sulphur & nitrogen
(*c*) carbon, sulphur, volatile matter & ash
(*d*) carbon, volatile matter, ash & moisture

99. Washing of coal is done to reduce the
(*a*) inherent impurities
(b) adhering impurities
(c) mineral matter
(d) both (*b*) and (*c*)

100. A good quality coal should have
(a) low fusion point of ash
(*b*) high ash content
(c) high sulphur
(*d*) none of these

101. Coke oven gas consists mainly of
(a) H_2, & CH_4 (b) CO, & CO_2
(c) H_2, &CO (*d*) CH_4, & CO

102. Lurgi coal gasifier is a pressurised...........bed reactor.
(a) moving (*b*) fixed
(c) fluidised (*d*) entrained

103. In Lurgi coal gasifier,
(a) coking coals cannot be used.
(b) low carbon conversion efficiency is achieved.
(c) entrainment of solids is higher.
(d) larger quantity of coal can be processed.

104. The catalyst used in shift converter is
(a) nickel (*b*) vanadium
(c) silica gel (*d*) alumina

105. The gasification reaction represented by, $C+H_2O=CO+H_2$, is
(a) exothermic (b) endothermic
(c) catalytic (d) autocatalytic

106. The combustion reaction, $C+O_2=CO_2$, is
(a) exothermic (b) endothermic
(c) autocatalytic (d) none of these

107. Kopper-Totzek coal gasifier
(a) can give ammonia synthesis gas (H_2+N_2).
(b) is a moving bed reactor.
(c) cannot use coking coal.
(d) operate at very high pressure.

108. A good metallurgical coke
(a) should have high porosity.
(b) should be brittle.
(c) must contain moderate quantities of ash, moisture, sulphur and volatile matters.
(*d*) should have low fusion point of its ash.

109. Gross heating value of coal is..........the net heating value.
(*a*) higher than (*b*) lower than
(c) same as (*d*) none of these

110. The maximum adiabatic flame temperature in air is...........the maximum flame temperature in pure oxygen.

ANSWERS

91.(*a*) 92.(*c*) 93.(*a*) 94.(*d*) 95.(*c*) 96.(*c*) 97.(*b*) 98.(*a*) 99.(*d*) 100.(*d*) 101.(*a*) 102.(*a*)
103.(*d*) 104.(*a*) 105.(*b*) 106.(*a*) 107.(*a*) 108.(*a*) 109.(*a*) 110.(*a*)

(a) lower than (b) higher than
(c) same as (d) not related to

111. Junkar's calorimeter is used to determine the calorific value of
(a) pulverised coal (b) gaseous fuels
(c) fuel oil (d) none of these

112. Preheating of combustion air is done to
(a) increase the adiabatic flame temperature.
(b) increase the calorific value of the fuel.
(c) complete the combustion of fuel.
(d) reduce its requirement for effecting the complete combustion.

113. The average molecular weight of a flue gas having the composition by volume as CO_2 = 25%, O_2 = 25%, N_2 = 50% will be
(a) 27.6 (b) 23
(c) 47.3 (d) 42.9

114. Which of the following is a poisonous fuel gas ?
(a) Coke oven gas
(b) Blast furnace gas
(c) Natural gas
(d) None of these

115. Which of the following fuel gases is heavier than air ?
(a) Blast furnace gas (b) Coke oven gas
(c) Natural gas (d) Water gas

116. A fuel with high heat release rate will
(a) require smaller combustion chamber.
(b) have high calorific value.
(c) have high adiabatic flame temperature.
(d) none of these.

117. Initial pressure of oxygen introduced into the 'bomb' of the bomb calorimeter for determination of calorific value of coal/fuel oil may be around.................atm.
(a) 3–5 (b) 25–30
(c) 60–65 (d) 95–100

118. Net calorific value is the gross calorific value less the................heat of water in the product of combustion when cooled to 15°C.
(a) sensible
(b) latent
(c) sensible and latent
(d) none of these

119. Presence of nitrogen in combustion air does not reduce the
(a) amount of heat liberated.
(b) flame temperature.
(c) flue gas temperature.
(d) any of the above.

120. A liquid/gaseous fuel containing hydrocarbons and high amount of sulphur is burnt with 40% excess air. The flue gas still contains large amount of carbon monoxide. This may be due to the
(a) presence of large quantity of hydrogen in the fuel.
(b) low calorific value of the fuel.
(c) high sulphur content in the fuel.
(d) lack of thorough mixing of fuel with air.

121. Which of the following would require least amount of secondary air for combustion ?
(a) Coke breeze containing 25% ash and 2% volatile matter.
(b) Anthracite containing 10% volatile matter and 8% ash.
(c) Bituminous coal containing 20% ash and 25% volatile matter.
(d) Semi-bituminous coal containing 25% ash and 20% volatile matter.

122. A coal gasifier opearating at 20 atm. (e.g. Lugri gasifier) as compared to one operating at atmospheric pressure (*e.g.* Kopper- Totzek or Winkler gasifier) will produce a gas having
(a) higher methane content and thus higher calorific value.
(b) higher carbon monoxide content.
(c) lower carbon dioxide content.
(d) none of these

123. Which of the following coke has the least percentage of ash ?
(a) Petroleum coke
(b) Beehive coke
(c) Foundary coke
(d) Metallurgical coke

124. Blast furnace gas burns with a bluish flame, because of the presence of
(a) CO (b) CH_4
(c) CO_2 (d) S

125. Coke oven gas burns with an yellowish flame, because of the presence of
(a) CO_2 (b) CH_4
(c) H_2 (d) NH_3

126. Fixed carbon in coal is defined as
(a) that present in volatile matters.

ANSWERS

111.(b) 112.(a) 113.(b) 114.(b) 115.(a) 116.(a) 117.(b) 118.(c) 119.(a) 120.(d) 121.(a) 122.(a) 123.(a) 124.(a) 125.(b) 126.(c)

(*b*) the total quantity of carbon present in the coal.
(*c*) hundred minus the percentage of volatile matter, ash and moisture.
(*d*) the one which is present in the residue after combustion.

127. Natural draught produced by a chimney depends upon the
(*a*) density of the chimney gases.
(*b*) height of the chimney.
(*c*) both (*a*) and (*b*).
(*d*) neither (*a*) nor (*b*).

128. Artificial draught produced by a fan can be controlled by the
(a) speed of the fan.
(b) damper.
(c) variation in the pitch of the fan blades.
(d) all (*a*), (*b*) and (*c*).

129. The liquid used for the washing of coal in an industrial coal washery is a mixture of water and
(*a*) carbon tetrachloride.
(*b*) sand (40%).
(*c*) mineral oil of high viscosity & specific gravity (1.6)
(*d*) none of these.

130. White flue gas (resembling steam) coming out of the chimney of a thermal power plant indicates that the fuel used in the boiler furnace is
(*a*) tar (*b*) coke oven gas
(c) pitch (*d*) pulverised coke

131. Flue gas discharge velocity through chimney of a big thermal power plant may be around................m/sec.
(a) 0.5 (*b*) 10
(c) 50 (*d*) 500

132. The optimum percentage of excess air for combustion depends upon the................of the fuel.
(a) type (solid, liquid or gaseous)
(b) calorific value
(*b*) sulphur content
(*d*) ignition temperature

133. Percentage of nitrogen in blast furnace gas may be around
(a) 5 (b) 25
(c) 55 (*d*) 80

134. Out of the following fuels, the difference between gross and net calorific value will be minimum in case of
(a) coke oven gas
(*b*) water gas
(c) natural gas
(*d*) blast furnace gas

135. Coalification means the
(*a*) process of conversion of lignite into anthracite.
(*b*) underground gasification of coal.
(*c*) complete combustion of coal.
(*d*) direct hydrogenation of coal.

136. Calorific value of bituminous coal may be around................Kcal/kg.
(a) 500 (*b*) 1500
(c) 6500 (*d*) 20000

137. Blast furnace gas constitutes mainly of
(a) N_2 & CH_4 (*b*) N_2 & H_2
(c) N_2 & CO (*d*) CH_4 & CO_2

138. The maximum percentage of CO_2 in a flue gas (from a carbonaceous fuel) can be
(a) 21 (b) 77
(c) 79 (*d*) 29

139. Water gas constitutes mainly of
(a) CO & H_2 (b) CO & N_2
(c) CO_2 & H_2 (*d*) CH_4 & H_2

140. Gobar gas constitutes mainly of
(a) CH_4 & CO_2 (*b*) CO & CO_2
(c) CH_4 & N_2 (d) CO & N_2

141. Gray-King Assay and Roga Index of coal is a measure of its
(*a*) abradability.
(*b*) grindability.
(*c*) weathering properties.
(*d*) caking and swelling properties.

142. Which of the following would require maximum amount of % excess air for complete combustion?
(*a*) Coke oven gas
(b) Furnace oil
(c) Pulverised coal
(*d*) Lump coal (fired on chain grate).

143. Efficiency of the combustion of a fuel is judged by the..................the flue gas.
(a) % of CO_2 in (*b*) % of O_2 in
(c) temperature of (*d*) colour of

144. A carbonaceous fuel (containing no H_2 or hydrocarbons) is burnt and the resulting flue gas contains 21% CO_2. It means that

ANSWERS

127.(*c*) 128.(*d*) 129.(*b*) 130.(*b*) 131.(*b*) 132.(*a*) 133.(*c*) 134.(*d*) 135.(*a*) 136.(*c*) 137.(*c*) 138.(*a*)
139.(*a*) 140.(*a*) 141.(*d*) 142.(*d*) 143.(*b*) 144.(*c*)

(a) 21% excess air has been used for combustion.
(b) 21% excess oxygen has been used for combustion.
(c) complete combustion of fuel has taken place.
(d) no excess air has been used for combustion.

145. Oxygen content in a flue gas was found to be 4%. It implies that excess air used for combustion was around................percent.
(a) 4 (b) 96
(c) 20 (d) 40

146. Quantity of coke produced from metallurgical coal may be around.................percent.
(a) 30 (b) 50
(c) 75 (d) 95

147. Volatile matter content of metallurgical coke may be around................percent.
(a) 1–2 (b) 10–15
(c) 22–26 (d) 30–33

148. Lignite is
(a) a high rank coal.
(b) a coking coal.
(c) can be used for gasification and steam generation.
(d) a black banded coal which burns with a non-smoky yellowish flame.

149. Anthracite coal
(a) contains more volatile matter than bituminous coal.
(b) ignites more easily than bituminous coal.
(c) is essentially a coking coal.
(d) burns with short, bluish, yellow-tipped flame producing very little smoke.

150. Bituminous coal
(a) ignites less easily than anthracite.
(b) is generally coking.
(c) burns with smoky yellow flame.
(d) both (b) and (c).

151. Calorific value of a typical dry anthracite coal may be around...........Kcal/kg.
(a) 1000 (b) 4000
(c) 8000 (d) 15000

152. Which of the following petrological constitutes is responsible for bright and lustrous black band of bituminous coal ?
(a) Clarain (b) Vitrain
(c) Durain (d) Fussain

153. Fussain
(a) is friable, charcoal like substance.
(b) has highest fixed carbon and lowest volatile matter content of all the four banded components of coal.
(c) is non-coking, but when blended with highly coking coal, controls its swelling and produces high strength coke on carbonisation.
(d) all (a), (b) and (c).

154. Which of the following is coking ?
(a) Vitrain
(b) Fussain
(c) both (a) & (b)
(d) Neither (a) nor (b)

155. Bright coal
(a) contains more than 90% durain.
(b) contains more than 90% fussain.
(c) contains mainly vitrain & clarain and is generally coking.
(d) is non-coking.

156. A coal containing very high percentage of durain is called...............coal.
(a) bright (b) splint
(c) non-banded (d) boghead

157. Cannel coal
(a) is non-coking.
(b) has a high volatile matter content and burns with a luminous smoky flame.
(c) is a non-banded coal which can be ignited easily with a match stick or candle flame, hence is so named.
(d) all (a), (b) and (c).

158. To avoid fire by spontaneous combustion of coal due to its low temperature oxidation, it should be stored in
(a) shallow and small piles.
(b) fine sizes without the presence of any lump.
(c) closed space without any ventilation facility.
(d) large heaps with small surface to volume ratio.

159. Low temperature oxidation of stored coal results in the
(a) decrease in its caking power & calorific value.
(b) decrease in its carbon & hydrogen content.
(c) increase in its oxygen content.
(d) all (a), (b) and (c).

ANSWERS

145.(c) 146.(c) 147.(a) 148.(c) 149.(d) 150.(d) 151.(c) 152.(a) 153.(d) 154.(a) 155.(c) 156.(b) 157.(d) 158.(a) 159.(d)

160. A coal having high amount of volatile matter
(*a*) would require smaller combustion chamber.
(*b*) produces very little of tar and gas on carbonisation.
(*c*) ignites easily and burns with long smoky flame.
(*d*) all (*a*), (*b*) and (*c*).

161. A coal with high.............content, would ignite most easily.
(*a*) fixed carbon (b) volatile matter
(c) ash (*d*) oxygen

162. A coal with high ash content is undesirable, as
(a) it is abrasive to the coal pulveriser (*e.g.* ball mill) and the combustion chamber.
(b) the ash in molten condition gets absorbed in the pores of the refractory lining of the furnace and causes its spalling due to different co-efficient of expansion/contraction of the refractory and the ash.
(c) the ash retains the sulphur & phosphorus and thus affects the quality of products in metallurgical furnace apart from increasing the slag volume. Besides, it may fuse and stick to the boiler tubes thereby reducing the heat transfer.
(d) all (*a*), (*b*) and (*c*).

163. Which of the following has the highest flame speed ?
(*a*) CO (b) H_2
(c) CH_4 (*d*) C_2H_6

164. Gross calorific value will be equal to the net calorific value for
(a) H_2 (b) C_2H_2
(c) CO (*d*) C_2H_6

165. Fuel gases containing hydrocarbons (*e.g.* coke oven gas) are not preheated before burning, mainly because
(*a*) the hydrocarbons crack thereby choking and fouling the heat transfer surface by carbon soot.
(*b*) it reduces its calorific value tremendously.
(*c*) it reduces its flame temperature tremendously.
(*d*) there are chances of explosion during preheating.

166. Sometimes water is sprayed during coal charging in the coke oven, which helps in
(*a*) prevention of clinker formation in the oven.
(*b*) controlling the dust nuisance while charging the coal.
(*c*) reducing the cracking of hydrocarbons in the gas and increasing the percentage of CO & H2 in the gas (due to the endothermic reaction represented by
$C + H_2O = CO + H_2$)
(*d*) all (*a*), (*b*) and (*c*).

167. Caking coal is desirable for
(a) burning on travelling grate.
(b) coke making.
(c) burning on firebars.
(*d*) all (*a*), (*b*) and (*c*).

168. As the equilibrium moisture content of the coal increases, its
(a) calorific value increases.
(b) swelling number decreases.
(c) swelling number increases.
(*d*) bulk density decreases.

169. Froth floatation is used for
(a) washing fine coal dust (< 0.5 mm size).
(b) washing lump coal (> 80 mm size).
(c) removing ash from the coal based on difference in specific gravity of coal and ash.
(*d*) all (*a*), (*b*) and (*c*).

170. Sulphur in metallurgical coal
(*a*) contributes to its heating value.
(*b*) affects the quality of steel produced as cracks develop on the surface while rolling the steel.
(*c*) both (*a*) and (*b*).
(*d*) neither (*a*) nor (*b*).

171. Presence of phosphorous in metallurgical coal
(*a*) is not undersirable.
(*b*) reduces its calorific value.
(*c*) badly affects the quality of steel.
(*d*) increases its caking power.

172. Nitrogen in coal
(a) is present upto 1–2%.
(b) comes from protein in parent vegetable matter.
(c) is recovered as ammonia during its carbonisation.
(d) all (*a*), (*b*) and (*c*).

173. Pick out the correct statement.
(*a*) Oxygen content decreases from lignite to bituminous coal as the coalification increases.

ANSWERS

160.(*d*) 161.(*b*) 162.(*d*) 163.(*b*) 164.(*c*) 165.(*a*) 166.(*d*) 167.(*b*) 168.(*b*) 169.(*a*) 170.(*c*) 171.(*c*) 172.(*d*) 173.(*d*)

(b) The less the oxygen content, better is the coa l, as it reduces the calorific value.
(c) With increase in oxygen content, moisture holding capacity of coal increases and the caking power decreases.
(d) all (a), (b) and (c)

174. Pick out the wrong statement.
(a) Pulverised fuel can be completely burnt with less percentage of excess air compared to lump coal.
(b) Low grade coal can be used, but generally high volatile matter coals are more suitable for making pulverised fuel.
(c) Regulation of furnace temperature and atmosphere (oxidising or reducing) is easily possible with pulverised fuel firing.
(d) none of these.

175. Combustion of pulverised fuel
(a) requires larger combustion chamber than rich gaseous fuel.
(b) in a, metallurgical furnace does not contaminate the product in the furnace by ash from fuel.
(c) suppresses the discharge of fine dust and grit into atmosphere
(d) all (a), (b) and (c)

176. Anthracite can be used for
(a) recarbonising steel.
(b) making carbon electrodes.
(c) blending with highly coking coal to check its swelling which helps in saving coke even walls from damage and to produce high strength coke.
(d) all (a), (b) and (c).

177. In high temperature carbonisation of coal compared to low temperature carbonisation
(a) yield of ammonia is less.
(b) aromatic content of tar is low.
(c) H_2 content in the coke oven gas is more.
(d) calorific value of the coke oven gas is lower.

178. Low temperature carbonisation
(a) is mainly for producing the smokeless domestic coke.
(b) is meant for the production of 'metallurgical coke'.
(c) produces higher quantity of gas than high temperature carbonisation.
(d) produces less quantity of tar than high temperature carbonisation.

179. During the carbonisation of coal
(a) all tar is evolved at < 600°C.
(b) evolution of H_2 and formation of methane and aromatics occur at > 700°C.
(c) hard semi-coke starts shrinking at 600°C.
(d) all (a), (b) and (c).

180. Width of the coke oven towards coke side is slightly more than that on pusher side to
(a) facilitate easy discharging of coke as it swells during carbonisation.
(b) facilitate uniform heating of the oven.
(c) increase the output of the coke.
(d) none of these.

181. Traces of tar fog present in the coke oven gas is removed by
(a) cyclone separator.
(b) wet packed scrubber.
(c) electrostatic precipitator.
(d) washing with monoethanolamine.

182. Which is the most matured coal ?
(a) Lignite (b) Bituminous
(c) Semi-anthracite (d) Anthracite

183. With increase in moisture content of coal, its
(a) calorific value increases.
(b) caking properties diminish.
(c) swelling during carbonisation becomes excessive.
(d) none of these.

184. Which will have the least volatile matter and hence will be the most difficult to ignite?
(a) Bituminous coal
(b) Coke
(c) Semi-bituminous coal
(d) Anthracite

185. High ash containing coke
(a) produces more slag when used in the blast furnace.
(b) has poor strength and abrasion resistance.
(c) is desirable in producer gas manufacture.
(d) none of these.

186. High amount of sulphur and phosphorous in coke causes
(a) decrease in its calorific value.
(b) increase in its strength.
(c) brittleness of steel made by using it.
(d) none of these.

ANSWERS

174.(d) 175.(a) 176.(d) 177.(c) 178.(a) 179.(d) 180.(a) 181.(c) 182.(d) 183.(b) 184.(b) 185.(a) 186.(c)

187. High rate of heating of coke ovens
(a) may damage its walls due to abrupt excessive swelling of coal.
(b) produces larger size coke.
(c) increases the time of carbonisation.
(d) none of these.

188. The cooling medium used in dry quenching of coke is
(a) nitrogen *(b)* air
(c) phenolic water *(d)* carbon dioxide

189. Coke having higher porosity has
(a) lower bulk density.
(b) lower strength.
(c) higher reactivity.
(d) all *(a)*, *(b)* and *(c)*.

190. Preheating of coal charge for the coke ovens reduces the
(a) time of carbonisation.
(b) yield of gas and tar.
(c) fuel consumption in coking.
(d) all *(a)*, *(b)* and *(c)*.

191. Narrow coke ovens as compared to wider coke ovens
(a) produce smaller coke.
(b) produce stronger coke.
(c) require less time of carbonisation.
(d) all *(a)*, *(b)* and *(c)*.

192. Fuel consumption in coke making can be reduced by
(a) preheating, blending and de-ashing the coal.
(b) stamped charging of coal.
(c) dry quenching of coke.
(d) all *(a)*, *(b)* and *(c)*.

193. A good coking coal should have high.........content.
(a) ash
(b) sulphur & phosphorus
(c) moisture
(d) none of these

194. A coal having high volatile matter content will
(a) give less yield of tar and gas on carbonisation.
(b) burn with a small non-smoky flame.
(c) have a very high calorific value.
(d) none of these.

195. Tar yield in the low temperature and high temperature carbonisation of dry coal may be respectively...............percent.
(a) 3 & 10 (b) 10 & 3
(c) 10 & 20 *(d)* 15 & 8

196. Coking time in narrow by-product coke ovens is around..........hours.
(a) 18 *(b)* 48
(c) 8 *(d)* 80

197. Beehive coke oven
(a) facilitates by-products recovery.
(b) takes 2-3 days for coking of coal but requires no external fuel for heating.
(c) gives larger yield of coke (around 90 %) as compared to by-product ovens.
(d) produces coke with very poor strength.

198. Size of blast furnace grade coke is............mm.
(a) 25–80 *(b)* 0–15
(c) 15–25 *(d)* > 100

199. Caking index of the coal blend used for the manufacture of metallurgical coke should be around
(a) 5 (b) 21
(c) 40 *(d)* 48

200. Critical Air Blast (CAB) value of coke is a direct measure of its
(a) reactivity (b) hardness
(c) Strength *(d)* none of these

201. "Micum Index" of a coke is a measure of its
(a) reactivity.
(b) porosity.
(c) bulk density.
(d) hardness & strength.

202. Desirable "Micum Index" values of metallurical coke are
(a) M_{40} > 78% and M_{10} < 10%.
(b) M_{40} > 4% and M_{10} < 80%.
(c) M_{40} > 10% and M_{10} < 78%.
(d) M_{40} > 98% and M_{10} < 2%.

203. Prime coking coal is always blended with medium or non- coking coal before earbonisation
(a) to check against its excessive swelling during heating, which may exert high pressure and damage coke oven walls.
(b) because, it alone produces unreactive coke.
(c) both *(a)* and *(b)*.
(d) neither *(a)* nor *(b)*.

204. High ash coals
(a) are soft & friable (poor strength and size stability).
(b) require longer time of carbonisation as ash offers resistance to heat transfer.

ANSWERS

187.*(a)* 188.*(a)* 189.*(d)* 190.*(d)* 191.*(d)* 192.*(d)* 193.*(d)* 194.*(d)* 195.*(b)* 196.*(a)* 197.*(b)* 198.*(a)*
199.*(b)* 200.*(a)* 201.*(d)* 202.*(a)* 203.*(c)* 204.*(b)*

(c) produce larger quantity of coke oven gas.
(d) none of these.

205. In high temperature carbonisation (as compared to low temperature carbonisation) of coal
(a) coke oven gas yield is more.
(b) tar yield is less but free carbon in tar is more.
(c) calorific value of coke oven gas is less.
(d) all *(a)*, *(b)*, and *(c)*.

206. Shatter index of the coke is a measure of its
(a) strength *(b)* bulk density
(c) reactivity *(d)* none of these

207. Heat penetration rate in narrow coke ovens in high temperature carbonisation of coal is around..................cm/hr.
(a) 2.5 (b) 0.5
(c) 10 (d) 20

208. Shatter index of metallurgical coke on 2 inches and 0.5 inch screen should be respectively around................percent.
(a) 80 and 97 (b) 95 and 100
(c) 40 and 70 (d) 25 and 97

209. Abrasion index of blast furnace coke should be around.................percent.
(a) 20 (b) 35
(c) 55 (d) 80

210. Gray King Assay and Swelling Index of a coal is a measure of its
(a) swelling characteristics.
(b) abradability.
(c) agglutinating properties.
(d) resistance to impact breakage.

211. Height of coke oven is limited (say maximum upto 7 metres) mainly by the
(a) problem of uniform heating along its height.
(b) structural strength of silica bricks.
(c) problem in door cleaning.
(d) buckling of ram of pusher car at the time of coke pushing.

212. Caking index of a coal is a measure of its
(a) abradability
(b) reactivity
(c) agglutinating (binding) properties
(d) porosity

213. With increase in the temperature of carbonisation of coal
(a) hydrogen content of coke oven gas increases due to cracking of hydrocarbons.
(b) methane content in the coke oven gas decreases and carbon monoxide content increases.
(c) calorific value of the coke oven gas decreases due to cracking of hydrocarbons which is not compensated by increase in CO & H_2 content.
(d) all *(a)*, *(b)* and *(c)*

214. Ionisation potential applied across the electrodes of electrostatic tar precipitator is around
(a) 230 V AC (b) 60 KV AC
(c) 230 V DC *(d)* 60 KV DC

215. Main constituents of benzol are
(a) benzene, toluene & xylene.
(b) tar creosote.
(c) ammonia & phenol.
(d) anthracene & phenol.

216. Which of the following fertilisers is produced in the by-products recovery (from coke oven gas) plant attached to an integrated steel plant ?
(a) Ammonium sulphate
(b) Ammonium nitrate
(c) Ammonium phosphate
(d) None of these

217. Naphthalene is removed from coke oven gas by
(a) scrubbing with wash oil (a petroleum product).
(b) adsorbing on bog iron bed.
(c) absorbing in vetro-coke solution.
(d) none of these.

218. Coke oven gas is stripped of its H_2S and mercaptans content by
(a) absorption in wash oil.
(b) adsorption on bog iron bed.
(c) bubbling it through dilute sulphuric acid.
(d) adsorption on silica gel.

219. Which of the following is not a product of tar distillation ?
(a) Phenol & naphthalene
(b) Benzol & pitch
(c) Anthracene & creosote
(d) None of these

220. Benzene is used
(a) as a motor fuel.
(b) as an explosive.
(c) for making insecticides (*e.g.*, DDT, BHC etc.), detergent & rubber (SBR).
(d) as a perfume.

ANSWERS

205.*(d)* 206.*(a)* 207.*(a)* 208.*(a)* 209.*(d)* 210.*(a)* 211.*(a)* 212.*(c)* 213.*(d)* 214.*(d)* 215.*(a)* 216.*(a)* 217.*(a)* 218.*(b)* 219.*(d)* 220.*(c)*

221. Which of the following is used for making the explosive TNT' ?
(a) Benzol (b) Toluene
(c) Pyridine *(d)* Cerosote

222. Hard pitch is used for making
(a) pulverised fuel
(b) perfumes
(c) insecticides
(d) plastics

223. Benzol is used
(a) as a motor fuel blend.
(b) for producing benzene, tolune & xylene by its distillation.
(c) both *(a)* & *(b)*.
(d) neither *(a)* nor *(b)*.

224. Yield of pitch from distillation of high temperature tar is around..................percent.
(a) 10 (b) 25
(c) 65 *(d)* 90

225. Naphthalene is used for making
(a) insecticides (*e.g.* moth balls).
(b) unsaturated polyesters.
(c) drug intermediates *e.g.* β–naphthol.
(d) all *(a)*, *(b)* and *(c)*.

226. Calorific value of coal middling generated in coal washeries during washing of coal may be around.................Kcal//kg.
(a) 1000 (b) 4000
(c) 6000 *(d)* 8000

227. Volatile matter content in coking coal may be about.................percent.
(a) 1 *(b)* 7
(c) 22 *(d)* 46

228. Calorific value of tar is about................Kcal/kg.
(a) 8800 *(b)* 12000
(c) 5000 *(d)* 16000

229. Calorific value of furnace oil is about............ Kcal/kg.
(a) 10000 *(b)* 15000
(c) 5000 *(d)* 20000

230. Tar is a better fuel than furnace oil, because of its
(a) higher calorific value.
(b) lower sulphur content.
(c) higher emissivity (0.8 - 0.9) resulting in higher radiation heat transfer rate.
(d) both (a) and (b).

231. Washing of coal
(a) reduces its ash & sulphur content.
(b) improves its coking properties.
(c) increase the fusion point of its ash by removing chlorine compounds.
(d) all *(a)*, *(b)* and *(c)*.

232. Gross and net calorific value of a fuel will be the same
(a) if its ash content is zero.
(b) if its carbon content is very low.
(c) if its hydrogen/hydrogen compound content is zero.
(d) under no circumstances.

233. With increase in moisture content of coal, its
(a) calorific value increases sometimes.
(b) bulk density always decreases.
(c) clinkering tendency during combustion increases.
(d) none of these.

234. Low temperature oxidation and spontaneous combustion of freshly mined coal is accentuated, if
(a) it contains large amount of volatile matter.
(b) it is stored in tall heaps.
(c) smaller fines are stored in large quantity.
(d) all *(a)*, *(b)* and *(c)*.

235. With increase in the oxygen content of the coal, its................decreases.
(a) calorific value
(b) caking power
(c) both *(a)* & *(b)*
(d) neither *(a)* nor *(b)*

236. Yield of blast furnace gas is about............Nm^3/ton of pig iron.
(a) 300 *(b)* 2000
(c) 5000 *(d)* 10000

237. Largest constituent of blast furnace gas is
(a) N_2 *(b)* CO
(c) CO_2 (d) H_2

238. Largest constituent of coke oven gas is
(a) N_2 *(b)* H_2
(c) CH_4 *(d)* CO_2

239. Blast furnace gas compared to coke oven gas has
(a) lower ignition temperature.
(b) narrower limit of inflammability.
(c) higher calorific value.
(d) lower theoretical flame temperature.

ANSWERS

221.*(b)* 222.*(a)* 223.*(c)* 224.*(c)* 225.*(d)* 226.*(b)* 227.*(c)* 228.*(a)* 229.*(a)* 230.*(d)* 231.*(d)* 232.*(c)*
233.*(d)* 234.*(d)* 235.*(c)* 236.*(b)* 237.*(a)* 238.*(b)* 239.*(d)*

240. Tolerable concentration of toxic carbon monoxide in atmospheric air is about................PPm.
(a) 50 (b) 1000
(c) 5000 (d) 10000

241. L.D. converter gas (produced in steel plant) comprises mainly of
(a) CO (upto 65%) & CO_2
(b) CO_2 & H_2
(c) CO & O_2
(d) CO_2 & O_2

242. The calorific value of L.D. converter gas is about..............Kcal/Nm^3.
(a) 1800 (b) 800
(c) 4500 (d) 10000

243. Which of the following is the most poisonous gas ?
(a) Coke oven gas
(b) Producer gas
(c) Blast furnace gas
(d) L.D. converter gas

244. The calorific value of producer gas is around...............kcal/Nm^3
(a) 1300 (b) 500
(c) 4500 (d) 9000

245. Main constituent of natural gas is
(a) CH_4 (upto 90%) (b) C_2H_6
(c) C_3H_8 (d) H_2

246. The calorific value of natural gas is about................kcal/Nm^3.
(a) 10,000 (b) 2500
(c) 25,000 (d) 35,000

247. The calorific value of'LPG' (50% propane + 50% butane) is about................kcal/Nm^3.
(a) 5000 (b) 25,000
(c) 25,000 (d) 15,000

248. Ignition temperature decreases progressively from anthracite to lignite, because
(a) volatile matter content increases.
(b) carbon content decreases.
(c) moisture content increases.
(d) ash content increases.

249. Coke compared to the coal from which it has been made, contains
(a) less volatile matter.
(b) more carbon.
(c) greater percentage of ash.
(d) all (a), (b) and (c).

250. Which of the following is not a by-product fuel ?
(a) Producer gas
(b) Blast furnace gas
(c) Coke oven gas
(d) Refinery gas

251. Which of the following is not a manufactured fuel ?
(a) Furnace oil (b) Bagasse
(c) Semi-coke (d) Kerosene

252. The calorific value of a gas of composition CO_2 = 20%, H_2 = 50%, CH_4 = 30%, is y. If the composition of gas is changed to H_2 = 50%, CH_4 = 30%, CO_2 = 10%, N_2 = 10%, then the calorific value will be
(a) y (b) 0.95 y
(c) 1.05 y (d) 1.8 y

253. The calorific value is the highest out of the following for
(a) producer gas
(b) water gas
(c) coke oven gas
(d) blast furnace gas

254. In Orsat's apparatus, ammoniacal cuprous chloride is used for selectivily absorbing
(a) CO (b) CO_2
(c) O_2 (d) H_2O

255. Eschka mixture, which is used for the determination of sulphur in coal, is a mixture of
(a) MgO & Na_2CO_3 (b) $MgSO_4$ & $BaCl_2$
(c) $BaSO_4$ & NaCl (d) $MgCO_3$ & NaCl

256. Abel apparatus is used for those oils, whose flash points are..................°F.
(a) <120 (b) >120
(c) <280 (d) 300–600

257. Pensky-Marten apparatus is used for those oils, whose flash points are...................°F.
(a) < 120 (b) > 120
(c) < 90 (d) 90–110

258. Mott and Wheeler test is conducted on coke to find its
(a) reactivity with O_2.
(b) abradability.
(c) phosphorus content.
(d) volatile matter content.

ANSWERS

240.(a) 241.(a) 242.(a) 243.(d) 244.(a) 245.(a) 246.(a) 247.(b) 248.(a) 249.(d) 250.(a) 251.(b) 252.(a) 253.(c) 254.(a) 255.(a) 256.(a) 257.(b) 258.(a)

259. Pick out the wrong statement.
(a) Cokes of high reactivity are obtained from weakly coking coals.
(b) Cokes of high reactivity are obtained from strongly coking coals.
(c) Reactivity of coke is inversely proportional to its absolute density.
(d) Abrasion index of the coke is a measure of its hardness.

260. Which of the following is not endothermic?
(a) Cracking (b) Reforming
(c) Gasification (d) Partial oxidation

261. Ratio of primary air to secondary air increases with increase in the rank of coal, because the
(a) high rank coals have higher amount of volatile matter.
(b) ratio of fixed carbon to volatile matter increases.
(c) oxygen content progressively decreases.
(d) calorific value of the coal increases.

262. 'Fat' coal means a coal having
(a) low calorific value.
(b) high volatile matter.
(c) low ash content.
(d) non smoking tendency.

263. Method of maintaining fires in furnace during standby periods without undue consumption of fuel is called
(a) back draughting.
(b) banking.
(c) under pressurising.
(d) none of these.

264. Which of the following will be unsuitable for dust cleaning from flue gas at 400°C from a pulverised coal fired boiler ?
(a) Multicyclones (b) Bag filter
(c) Wet scrubber (d) Hydrocyclones.

265. For long flame and easy ignition, the coal used should have
(a) high volatile matter
(b) low ash
(c) high calorific value
(d) high ash

266. Mineral matter, 'M' and ash percentage 'A' in coal are roughly related as
(a) $M = 1.8A$ (b) $M = 0.5A$
(c) $M = 1.1A$ (d) $M = A$

267. Coal tar (produced by high temperature carbonisation) is the main source of
(a) aromatic compounds
(b) alphatic compounds
(c) paraffins
(d) olefins

268. The pyragallol solution used in orsat apparatus can absorb
(a) only O_2
(b) both O_2 and CO_2
(c) both O_2 and CO
(d) all CO, CO_2 and O_2

269. The cuprous chloride used in orsat apparatus can absorb
(a) only CO
(b) both CO and CO_2
(c) both CO and O_2
(d) all CO, CO_2, and O_2.

270. The internal energy of the combustion products is..................compared to that of reactants.
(a) more
(b) less
(c) equal
(d) more or less ; (depends on the state of fuel.

271. A sooty flame can be made non-luminous by
(a) admitting more gas.
(b) shutting gas supply.
(c) admitting more air.
(d) shutting air supply.

272. The heat of combustion of a fuel
(a) is equal to the heat of formation.
(b) is always negative.
(c) can't be known without calculating it.
(d) is always positive.

273. The hottest part of the flame lies in its......... zone.
(a) non-luminous (b) luminous
(c) yellow (d) unburnt gases

274. As the C/H ratio of the fuel increases, the amount of CO_2 formed on combustion..................for the same percentage of excess air.
(a) decreases
(b) increases
(c) remains same
(d) may increase or decrease depending on the type of fuel

275. Calorific value of
(a) light paraffinic fuel oils is equal to that of equivalent olefins.

ANSWERS

259.(b) 260.(d) 261.(b) 262.(b) 263.(b) 264.(b) 265.(a) 266.(c) 267.(a) 268.(b) 269.(d) 270.(b)
271.(c) 272.(b) 273.(a) 274.(b) 275.(c)

(b) n-paraffin is lower than that of iso-paraffins of the same compounds.
(c) light paraffmic fuel oils is higher than that of equivalent olefins.
(d) light paraffinic fuel oils is lower than that of equivalent olefins.

276. KOH solution used in orsat apparatus absorbs
(a) CO_2
(b) SO_2
(c) both *(a)* & *(b)*
(d) neither *(a)* nor *(b)*

277. If CO_2 is not fully absorbed by the KOH solution meant for its absorption in the orsat apparatus, it will appear as
(a) O_2 (b) CO
(c) N_2 *(d)* SO_2

278. With increase in C/H ratio of a fuel, the dew point of flue gases
(a) increases
(b) decreases
(c) remains same
(d) may increase or decrease ; depends on the type of fuel

279. The maximum adiabatic flame temperature is attained, when the fuel is burnt with
(a) theoretically required amount of air.
(b) more than theoretically required amount of air.
(c) less than theoretically required amount of air.
(d) theoretically required amount of oxygen.

280. Too much of excess air in combustion results in high
(a) fuel consumption for the same heat load.
(b) stack gas temperature.
(c) percentage of oxygen in flue gases.
(d) all *(a)*, *(b)* and *(c)*.

281. Oxygen required for theoretically complete combustion of 1 Nm^3 methane is............Nm^3.
(a) 2 (b) 4
(c) 6 (d) 1

282. One kg of carbon for theoretically complete combustion requires................kg of air.
(a) 11.5 (b) 0.5
(c) 23 (d) 12

283. When the coal is heated in absence of air, it is called its
(a) deoxidation (b) gasification
(c) coalification *(d)* carbonisation

284. Ammonia content in raw coke oven gas is about...................gm/Nm^3.
(a) 1 (b) 15
(c) 50 *(d)* 100

285. Calorific value of coke even gas produced by low temperature carbonisation of coal is about..................Kcal/Nm^3.
(a) 4000 (b) 2500
(c) 6500 *(d)* 10000

286. Yield of coke oven gas in low temperature carbonisation of coal is about.................Nm^3/ton of dry coal.
(a) 60 (b) 160
(c) 500 (d) 750

287. Coke ovens in steel plant are heated by
(a) electricity (b) coke oven gas
(c) mixed gas (d) both *(b)* & *(c)*

288. Coking time in a commercial high temperature coal carbonisation plant is about...........hours.
(a) 6 (b) 18
(c) 28 (d) 36

289. Coking time in beehive coke oven is about
(a) 12 hours (b) 2–3 days
(c) one week *(d)* two weeks

290. Fuel combustion is never cent per cent efficient due to
(a) incomplete combustion
(b) dry gas/stack gas loss
(c) moisture loss
(d) all *(a)*, *(b)* and *(c)*

291. When incomplete combustion loss is high, the flue gas analysis shows large amount of
(a) CO_2 (b) CO
(c) O_2 (d) C

292. If oxygen content in the flue gas rises too high, fuel is being wasted by
(a) incomplete combustion.
(b) dry gas/stack gas loss.
(c) moisture loss.
(d) none of these.

293. The ratio of % total carbon obtained in the ultimate analysis of coke and % fixed carbon obtained in the proximate analysis is always
(a) 1 *(b)* < 1
(c) > 1 *(d)* unpredictable

ANSWERS

276.*(c)* 277.*(d)* 278.*(b)* 279.*(d)* 280.*(d)* 281.*(a)* 282.*(a)* 283.*(d)* 284.*(b)* 285.*(c)* 286.*(b)* 287.*(d)* 288.*(b)* 289.*(b)* 290.*(d)* 291.*(b)* 292.*(b)* 293.*(c)*

294. C.V. (kcal/Nm^3) of gaseous fuels..............with increase in molecular weight.
(a) decreases
(b) increases
(c) remains constant
(d) may increase or decrease, depends on combustibles.

295. Actual flame temperature is always lower than the adiabatic flame temperature, because there is
(a) no possibility of obtaining complete combustion at high temperature.
(b) always loss of heat from the flame.
(c) both *(a)* and *(b)*.
(d) neither *(a)* nor *(b)*.

296. The ratio of maximum adiabatic flame temperature in air to that in pure oxygen is always
(a) 1 *(b)* < 1
(c) > 1 *(d)* unpredicatable

297. The bright glow of a combustion process is the characteristic of..................temperature oxidation of coal.
(a) fast & high (b) slow & low
(c) slow & high *(d)* fast & slow

298. C/H ratio is the maximum in case of
(a) coal *(b)* furnace oil
(c) natural gas *(d)* naphtha

299. C/H ratio is the minimum in case of
(a) furnace oil (b) natural gas
(c) coal *(d)* naphtha

300. Which of the following is not a 'manufactured' fuel?
(a) LPG *(b)* Coal briquettes
(c) Tar (d) Colloidal fuels

301. Which of the following is not a by-product fuel?
(a) Pitch *(b)* Blast furnace gas
(c) Petrol *(d)* Refinery gas

302. Carbon Content by weight in air dried wood may be about..................percent.
(a) 10 (b) 25
(c) 50 *(d)* 80

303. Yield of charcoal in high temperature carbonisation of wood is about.....................percent.
(a) 10 *(b)* 25
(c) 50 *(d)* 75

304. The most matured coal out of the following is
(a) lignite *(b)* semi-anthracite
(c) sub-bituminous (d) bituminous

305. Coking coals are invariably
(a) lignites
(b) bituminous coals
(c) semi-anthracites
(d) anthracites

306. Cannel coal and boghead coal are the examples of
(a) humic coals (b) liptobiolites
(c) sapropelic coals *(d)* none of these

307. A coal having higher volatile matter content, has lower
(a) smoking tendency on burning.
(b) coke oven gas yield on carbonisation.
(c) chance of catching fire during storage in open space.
(d) ignition temperature.

308. A coal with high ash content
(a) has higher calorific value.
(b) is harder and stronger.
(c) is not subjected to washing.
(d) has low quantity of mineral matter.

309. The difference between total carbon and fixed carbon of coal will be minimum in case of
(a) lignite
(b) bituminous coal
(c) anthracite
(d) high temperature coke (V.M < 0.5%)

310. Assam coals suffers mainly from the disadvantage of high...........content.
(a) ash *(b)* volatile matter
(c) sulphur (4-6%) *(d)* moisture

311. Which of the following petrographic constituents of coal is non-coking ?
(a) Vitrain *(b)* Clarain
(c) Durain *(d)* Fussain

312. Washing of coal does not reduce its
(a) S & P content.
(b) heating value.
(c) fusion point of ash.
(d) none of these.

313. Low temperature oxidation of coal during storage does not decrease its
(a) oxygen content *(b)* caking power
(c) calorific value *(d)* none of these

314. Low temperature oxidation of coal is accelerated by the
(a) storage in large heaps.

ANSWERS

294.*(d)* 295.*(c)* 296.*(b)* 297.*(a)* 298.*(a)* 299.*(b)* 300.*(c)* 301.*(c)* 302.*(c)* 303.*(b)* 304.*(b)* 305.*(b)*
306.*(c)* 307.*(d)* 308.*(b)* 309.*(d)* 310.*(c)* 311.*(d)* 312.*(d)* 313.*(d)* 314.*(a)*

(*b*) storage in compressed piles.
(*c*) absence of porous or friable particles.
(*d*) all (*a*), (*b*) and (*c*).

315. Mineral matter content (M) and ash content (A) in coal are approximately related as
(a) $M = 1.1A$ (*b*) $M = A$
(c) M = 1.5 A (*d*) M = 2 A

316. During combustion of coal on grate, clinker formation is increased by the
(a) use of thick fire bed.
(b) low fusion point of ash (< 1100°C).
(c) use of preheated primary air.
(d) all *(a), (b) & (c).*

317. Weathering of coal during storage causes
(a) reduction in coal size.
(b) increase in its friability.
(c) decrease in its caking capacity.
(d) all *(a), (b)* and *(c).*

318. Hardgrove grindability index of four coal samples are given below. Which of them is the easiest to grind ?
(a) 50 (*b*) 70
(c) 85 (d) 100

319. Caking index of coal blend used for blast furnace coke making is about
(a) 8 (b) 12
(c) 22 (*d*) 36

320. 'Fat coals' are those coals which have very high
(*a*) caking capacity.
(*b*) volatile matter content.
(*c*) fusion point of its ash.
(*d*) inherent moisture content.

321. Bulk density of pulverised coal may be about..................kg/m^3.
(a) 100 (b) 500
(c) 1000 (*d*) 1500

322. In low temperature carbonisation of coal, the
(*a*) yield of coke oven gas is 290 Nm^3/ton dry coal.
(*b*) volatile matter in coke is zero,
(c) temperature maintained is 700°C.
(*d*) yield of tar is about 3% of dry coal.

323. Combustion reaction of fuels is a/an.............reaction.
(a) auto catalytic (*b*) exothermic
(*c*) endothermic (*d*) none of these

324. Which of the following gaseous fuels has the highest net calorific value (kcal/Nm^3)?
(a) Producer gas
(b) Carburetted water gas
(c) Natural gas
(*d*) Liquefied petroleum gas

325. ....has the widest inflammability limit (explosion limit) of all the gases.
(a) Hydrogen (*b*) Carbon monoxide
(c) Acetylene (*d*) Methane

326. Gobar gas is produced by the...................of 'gobar' (cow dung).
(a) hydrolysis (b) fermentation
(c) oxidation (*d*) dehydration

327. Theoretical flame temperature of a fuel is that temperature which is attained, when the fuel is completely burnt 'using theoretical amount of
(a) air
(b) oxygen
(c) either *(a)* or *(b)*
(*d*) either *(a)* or (*b*) without gain or loss of heat

328. 'Wobbe index' is a characteristic of
(*a*) solid fuels (*b*) gaseous fuels
(c) liquid fuels (*d*) fat coals

329. Calorific value of gobar gas (containing CH_4 = 60% , H_2 = 10% ; CO_2 = 30%) may be about Kcal/Nm^3.
(a) 1800 (b) 3200
(c) 5400 (d) 10200

330. Calorific value of wood gas is about............ kcal/Nm^3.
(a) 1500 (b) 3300
(c) 5400 (d) 8500

331. Main component of sewage gas produced during aneorobic decomposition of organic waste (by suitable bacteria) during sewage disposal is
(a) H_2 (*b*) CH_4
(c) CO_2 (d) N_2

332. Which of the following has the highest heat of combustion ?
(a) H_2 (b) CO
(c) CH_4 (d) C_2H_6

333. Flue gas obtained on complete combustion of pure acetylene (a hydrocarbon fuel) will contain
(a) CO (b) CO_2
(c) H_2O (d) both *(b) & (c)*

334. About..................Nm^3 of air will be required for the complete combustion of 2 Nm^3 of CO.
(a) 2 (b) 3
(c) 5 (d) 1

ANSWERS

315.*(a)* 316.*(d)* 317.*(d)* 318.*(d)* 319.*(c)* 320.*(a)* 321.*(b)* 322.*(c)* 323.*(b)* 324.*(d)* 325.*(c)* 326.*(b)*
327.*(d)* 328.*(b)* 329.*(c)* 330.*(b)* 331.*(b)* 332.*(d)* 333.*(d)* 334.*(c)*

335. Pick out the wrong statement.
(a) High concentration of oxygen in flue gas means high stack loss.
(b) Gaseous fuels require the least % excess air for complete combustion.
(c) The ratio of fixed carbon to volatile matter percentage in coal is called its 'fuel ratio'.
(d) Calorific value of natural gas is more than that of LPG.

336. Pick out the wrong statement.
(a) Main constituents of LPG are propane and butane.
(b) C.V. of natural gas is about 10000 KCal/Nm^3.
(c) C.V. of LPG is about 26000 kcal/Nm^3 (11500 kcal/kg).
(d) L.P.G. is lighter than air.

337. Which of the following is unsuitable fuel for producer gas manufacture ?
(a) Coke.
(b) Anthracite.
(c) Coal having low fusion point of its ash.
(d) Coal having high fusion point of its ash.

338. Which of the following has the least calorific value (kcal/Nm^3) ?
(a) Blast furnace gas (b) Coke oven gas
(c) Sewage gas (d) Natural gas

339. Which of the following has the highest calorific value (kcal/Nm^3) ?
(a) Carburetted water gas
(b) Gobar gas
(c) Natural gas
(d) LPG

340. Which of the following has the highest theoretical flame temperature ?
(a) Blast furnace gas (b) Hydrogen
(c) Acetylene (d) Coke oven gas

341. Supply of excess air for complete combustion of fuel is necessitated to facilitate
(a) its thorough mixing with air.
(b) attainment of chemical equilibrium.
(c) attainment of high temperature.
(d) none of these.

342. Main constituents of purified Lurgi gas are
(a) H_2, C_mH_n & CO_2 (b) CO, H_2 & CH_4
(c) CO_2, O_2 & CO (d) N_2, H_2 & CO_2

343. The function of secondary combustion air is to
(a) pneumatically convey the pulverised coal.
(b) completely burn the volatile matter.
(c) burn the lumpy coal.
(d) none of these.

344. The main advantage of forced draft over natural draft is that
(a) combustion of fuel is complete.
(b) smaller height chimney can be used.
(c) furnace control is easier.
(d) all *(a)*, *(b)* & *(c)*.

345. Short/intense flame is produced during combustion of gaseous fuel by using
(a) high amount/current of combustion air.
(b) low amount/current of combustion air.
(c) preheated secondary air.
(d) very little excess air.

346. Ignition temperature of a fuel in air is............... that in pure oxygen.
(a) more than
(b) less than
(c) equal to
(d) either more or less; depends on the type of fuel

347. Actual flame temperature of fuels depends upon the
(a) C.V. of the fuel.
(b) volume & specific heats of flue gas.
(c) endothermic dissociation of the CO_2 & H_2O present in flue gas.
(d) all *(a)*, *(b)* & *(c)*.

348. Explosion limit of blast furnace gas is 37 to 71% gas in gas-air mixture. It means that the blast furnace gas will explode when burnt in a confined space, if its concentration in the gas-air mixture is...............percent.
(a) < 37
(b) > 71
(c) in between 37 & 71
(d) none of these

349. In general, the limit of inflammability/explosion limit of fuel gases is widened by the
(a) increase in gas pressure.
(b) increase in temperature (*i.e.* preheating).
(c) use of pure oxygen for combustion instead of air.
(d) all *(a)*, *(b)* & *(c)*.

350. Deficiency of combustion air during combustion of a gaseous fuel
(a) lengthens the flame.

ANSWERS

335.*(d)* 336.*(d)* 337.*(c)* 338.*(a)* 339.*(d)* 340.*(c)* 341.*(a)* 342.*(b)* 343.*(b)* 344.*(d)* 345.*(a)* 346.*(a)* 347.*(d)* 348.*(c)* 349.*(d)* 350.*(c)*

(b) causes heat loss of fuel by its partial combustion.
(c) both (a) & (b).
(d) shortens the flame.

351. Combustion of pulverised coal compared to the lumpy coal
(a) provides better control of furnace temperature.
(b) facilitates combustion with lower excess air.
(c) provides higher thermal efficiency & flame temperature.
(d) all (a), (b) & (c).

352. Turndown ratio of a gas burner is equal to
(a) maximum to minimum heat input ratio.
(b) maximum to minimum permissible gas flow rate.
(c) both (a) & (b).
(d) minimum to maximum heat input ratio.

353. Bunsen burner is an example of a/an.......... burner.
(a) inside mixing/premix type
(b) outside mixing/diffusion flame/nozzle mix type
(c) rotary cup
(d) submerged combustion

354. Orsat apparatus
(a) gives flue gas analysis on dry basis.
(b) determines N_2 in flue gas indirectly.
(c) can't determine CO in flue gas, if it is <0.2%.
(d) all (a), (b) & (c).

355. Advantages of fluidised bed combustion are
(a) reduced NO_x formation in flue gas.
(b) lower furnace operating temperature.
(c) high heat transfer rate.
(d) all (a), (b) & (c).

356. Main use of soft coke is as................fuel.
(a) domestic (b) blast furnace
(c) foundary (d) none of these

357. ....process is meant for direct hydrogenation of coal to produce liquid fuel.
(a) Fischer-Tropsch (b) Bergius
(c) Lurgi (d) None of these

358. Calorific value of both the solid & liquid fuels can be determined by using ....calorimeter.
(a) Junker's (b) Bomb
(c) Boy's (d) none of these

359. Air/gas ratio for complete combustion will be the highest for
(a) LPG
(b) gobar gas
(c) coke oven gas
(d) carburetted water gas

360. Nitrogen present in the flue gas is determined in the orsat apparatus by absorbing it in
(a) KOH
(b) ammoniacal cuprous chloride
(c) pyrogallol solution
(d) none of these

361. A travelling grate stoker is meant for the efficient burning of..................coal.
(a) caking (b) pulverised
(c) non-caking (d) high ash

362. Volumetric ratio of N_2 to O_2 in dry atmospheric air is
(a) 0.21 (b) 3.76
(c) 4.76 (d) 0.79

363. Which of the following is the maximum coal producing state in India?
(a) Orissa (b) West Bengal
(c) Jharkhand (d) Assam

364. During coal carbonisation process, the conversion of semi-coke to coke is accompanied by an increase in the.................of the mass.
(a) density
(b) porosity
(c) electrical resistivity
(d) none of these

365. During combustion of a gaseous fuel, the presence of a non-luminous flame is an indication of the
(a) incomplete combustion.
(b) complete combustion.
(c) high oxygen in flue gas.
(d) none of these.

366. Bomb calorimeter can be used to determine the...................of the coal.
(a) sulphur content
(b) calorific value
(c) both (a) & (b)
(d) neither (a) nor (b)

367. Pick out the wrong statement. :
(a) Carburetted water gas is also called blue gas.

ANSWERS

351.(d) 352.(c) 353.(a) 354.(d) 355.(d) 356.(a) 357.(b) 358.(b) 359.(a) 360.(d) 361.(c) 362.(b) 363.(c) 364.(a) 365.(b) 366.(c) 367.(a)

(*b*) Coals are divided in four species according to their carbon content in Seylor's classification.
(*c*) Carbonisation time in a by-product coke oven is about 16 hours.
(*d*) Gross and net calorific value of a fuel is the same, if it does not contain hydrogen or hydrocarbons.

368. Coal is pulverised before burning in large capacity boiler furnaces mainly to
(*a*) ensure its complete combustion.
(*b*) facilitate easy ash removal.
(*c*) enhance its calorific value.
(*d*) provide trouble free operation.

369. Function of secondary air in pulverised coal firing is to
(a) transport the coal to the burner.
(b) dry the coal.
(c) ensure efficient burning of coal around the burner.
(*d*) reduce primary air requirement.

370. Which of the following coal gasification processes will produce gas having maximum methane content ?
(*a*) Winkler process.
(*b*) Lurgi process.
(*c*) Kopper-Totzek process.
(*d*) All can produce same methane content.

371. The main reason for making the copper calorimeter (used in bomb calorimeter) silvery white and shining/polished is to
(*a*) minimise its corrosion.
(*b*) avoid radiation heat loss.
(*c*) make it look attractive.
(*d*) none of these.

372. High pressure coal gasification is employed in....................process.
(*a*) Lurgi (*b*) Kopper-Totzek
(c) Winkler (*d*) none of these

373. By-products are..................recovered in 'Beehive ovens'.
(a) fully (b) partially
(c) not at all (*d*) negligibly

374. ....of the coal is the basis for Seylor's coal classification.
(*a*) Proximate analysis
(*b*) Ultimate analysis
(*c*) Caking index
(*d*) Calorific value

375. By-products recovery process from coal carbonisation is termed as direct, indirect and semi direct process depending upon the method of recovery of
(a) tar (*b*) benzol
(c) ammonia (*d*) none of these

376. Out of the following, which is the most important parameter for the blast furnace grade coke?
(a) CSR & CRI
(b) Ash content
(c) Moisture content
(*d*) Volatile matter content

377. In high temperature carbonisation (as compared to low temperature carbonisation) of coal, the
(a) gas yield is less.
(b) tar yield is more.
(c) ignition temperature of coke produced is less.
(*d*) aromatic content of tar produced is more.

378. Coke oven gas produced by high temperature carbonisation of coal (as compared to that produced by low temperature carbonisation), has
(a) higher calorific value.
(b) lower hydrogen content.
(c) both *(a)* and *(b)*.
(d) neither *(a)* nor *(b)*.

379. In low temperature carbonisation (as compared to high temperature carbonisation) of coal
(a) ammonia yield is more.
(b) aliphatic tar is produced.
(c) free carbon in tar is more.
(*d*) all *(a)*, *(b)* and *(c)*.

380. Flue gas from the heating chamber of by-product coke ovens is removed by
(*a*) forced draft fan
(*b*) induced draft fan
(*c*) steam ejector
(*d*) natural draft

381. Undercharging of coal in the by-product coke ovens results in
(a) decrease in the c.v. of coke oven gas.
(b) increase in its throughput.
(c) increase in the c.v. of coke oven gas.
(*d*) no change in the c.v. of coke oven gas.

382. Coke made in narrower by-product coke ovens (as compared to wider ovens) is
(a) less reactive (b) stronger
(c) smaller in size (*d*) all *(a)*, *(b)* & *(c)*

ANSWERS

368.*(a)* 369.*(c)* 370.*(b)* 371.*(b)* 372.*(a)* 373.*(c)* 374.*(b)* 375.*(c)* 376.*(a)* 377.*(d)* 378.*(d)* 379.*(b)* 380.*(d)* 381.*(a)* 382.*(d)*

383. Which of the following fuels is generally not used in thermal power plant boiler firing?
(a) Furnace oil, light diesel oil and tar/PCM.
(b) Pulverised coking coal.
(c) Coal middling from washeries.
(d) B.F. gas, coke oven gas and L.D. converter gas.

384. Optimum preheating temperature for pitch creosote mixture (PCM) which is also termed as C.T.F-200, before atomisation through burners is
(a) 200 °F (b) 200 °C
(c) 200 °K (d) 200 °R

385. Pitch creosote mixture (PCM) as compared to furnace oil is a better fuel, because its
(a) emissivity factor is higher.
(b) sulphur content is lower.
(c) flue gas has lower dew point thereby facilitating more waste heat recovery.
(d) all *(a)*, *(b)* and *(c)*.

386. Ash content in the middling coal (a by-product of coal washeries) in India ranges between................percent.
(a) 15 to 25 *(b)* 35 to 45
(c) 55 to 75 *(d)* 75

387. Calorific value of pitch creosote mixture (PCM) *i.e.*, C.T.F.–200 is about
(a) 8800 kcal/m^3
(b) 8800 kcal/kg
(c) same as that of coal middling
(d) 25000 kcal/kg

388. Which adsorbant is used for removing sulphur compounds, (SO_2, H_2S, RSH etc.) removal from coke oven gas in by-products recovery plant ?
(a) Silica gel
(b) Diatomaceous earth
(c) Basalt
(d) Bog iron (*i.e.*, moist ferric hydroxide)

389. Of the total tar present in raw coke oven gas, the tar recovered in primary cooler is about...............percent.
(a) 5 *(b)* 25
(c) 55 *(d)* 75

390. Temperature of coke oven gas just before entering into saturator (for recovery of NH_3) is about................°C.
(a) *30* *(b)* 65
(c) 120 *(d)* 180

391. 'Mixed gas' used in steel plants is a mixture of
(a) B.F. gas and coke oven gas.
(b) coke oven gas and converter gas.
(c) coke oven gas and L.P.G.
(d) blast furnace gas and naphtha vapor.

392. With increase in C/H ratio of a fuel, the amount of CO_2 formed on its complete combustion
(a) increases
(b) decreases
(c) remains same
(d) either *(a)* or *(b)*, depends on other factors

393. Stack heat losses can be minimised by
(a) controlling the excess air.
(b) oxygen enrichment of combustion air.
(c) using low c.v. fuels.
(d) maintaining proper draft in the furnace.

394. Gas yield in the Kopper-Totzek coal gasifier is about..................Nm^3/ton coal (ash = 35%).
(a) 150 *(b)* 1500
(c) 3500 *(d)* 5000

395. Higher fuel combustion efficiency can not be achieved by
(a) preheating of fuel gases & combustion air.
(b) reducing sulphur content in the fuel.
(c) adopting proper fuel firing technique & fuel preparation.
(d) supplying correct amount of combustion air.

396. Safe condition for storage of high V.M. bituminous coal is that
(a) height of the coal heap should be < 3 metres.
(b) maximum 200 tons should be stored in a heap.
(c) both *(a)* and *(b)*.
(d) neither *(a)* nor *(b)*.

397. Which of the following is not applicable to fluidised bed combustion of coal?
(a) It can not burn low grade coal.
(b) It achieves higher fuel combustion efficiency.
(c) Less heat transfer surface area is required in boilers.
(d) None of these.

398. Higher efficiency in the combustion of solid fuel can not be achieved by
(a) proper fuel preparation.
(b) keeping the flue gas exhaust temperature very high.

ANSWERS

383.*(b)* 384.*(a)* 385.*(d)* 386.*(b)* 387.*(b)* 388.*(d)* 389.*(b)* 390.*(b)* 391.*(a)* 392.*(a)* 393.*(a)* 394.*(b)* 395.*(b)* 396.*(c)* 397.*(a)* 398.*(b)*

(c) adopting efficient fuel firing technique & equipment.
(d) supplying correct quantity of combustion air.

399. 'Wind loss' resulting from unscientific storage of coal may be the order of about.............percent.
(a) 2.5 (b) 7.5
(c) 10.5 (d) 14.5

400. The main drawback of supplying more excess air in the combustion of fuel is the
(a) excessive power requirement of air blower.
(b) enhanced sensible heat loss in the flue gas.
(c) intermittant and uncontrolled combustion of the fuel.
(d) high exit flue gas temperature from the furnace.

401. A certain thickness of the coal, if stored unscientifically on soft (katcha) ground having no metallic/concrete flooring gets sunked into the ground, which is termed as the 'carpet loss'. The carpet loss may be of the order of.............cms.
(a) 1 to 2 (b) 3 to 4
(c) 5 to 15 (d) 20 to 40

402. Furnace oil consumption in a furnace for a given duty employing preheated combination air (at 300°) may be reduced by about.................percent as compared to the use of atmospheric combustion air.
(a) 5 (b) 10
(c) 20 (d) 35

403. Correct viscosity of furnace oil at the burner tip for proper atomisation is about 25 centistokes. To reduce the viscosity of high viscosity furnace oil (250 centistokes) to the correct atomisation viscosity (*i.e.* 25cst), it should be preheated to about............°C.
(a) 70 (b) 85
(c) 105 (d) 145

404. Presence of.................in a dry gaseous fuel does not contribute to its calorific value.
(a) sulphur (b) oxygen
(c) hydrogen (d) carbon

405. Which of the following oil gasification processes is non-catalytic?
(a) Semet-Solvay process
(b) Segas process
(c) Onia-Gegi process
(d) Steam reforming of naphtha

406. Number of macrocomponents present in coal according to Stopes are
(a) four (b) five
(c) six (d) two

407. Which of the following is not a micro component present in coal?
(a) Micrinite (b) Clarain
(c) Fusinite (d) Liptinite

408. ....present in coal is not determined in its ultimate analysis.
(a) Fixed carbon (b) Total carbon
(c) Hydrogen (d) Nitrogen

409. Use of preheated air for combustion of fuel in the furnace, increases the
(a) scale losses of the furnace stock.
(b) calorific value of the fuel.
(c) flame temperature.
(d) none of these.

410. A coal having higher volatile matter content will necessarily have lower
(a) ash fusion temperature.
(b) calorific value.
(c) ignition temperature
(d) caking index.

411. Improper storage condition results in the weathering of coal and spontaneous combustion, which increases its
(a) caking index.
(b) yield of carbonised products.
(c) friability & oxygen content.
(d) calorific value.

412. Washing of coal decreases its
(a) caking index.
(b) mineral matter content.
(c) ash content.
(d) both (b) and (c).

413. High sulphur (4-6%) coal in India are found in
(a) Assam (b) Andhra Pradesh
(c) Bengal (d) Madhya Pradesh

414. Coal is heated in presence of air to a temperature of about.............. °C, while determining its ash content for proximate analysis.
(a) 500 (b) 750
(c) 950 (d) 1100

415. During its calorific value determination by bomb calorimeter, coal is combusted by
(a) air (b) oxygen
(c) oxygen enriched air (d) none of these

ANSWERS

399.(a) 400.(b) 401.(c) 402.(c) 403.(c) 404.(b) 405.(a) 406.(a) 407.(b) 408.(a) 409.(c) 410.(c) 411.(c) 412.(b) 413.(a) 414.(b) 415.(b)

416. Which of the following is a measure of the agglutinating *(i.e.,* binding) property of coal?
(a) Thickness of plastic layer
(b) Caking index
(c) Swelling index
(d) Gray-king index

417. While the first commercial low temperature coal carbonisation plant is located at Neyveli (in Tamilnadu), the second such plant has been built at
(a) Dankuni (West Bengal)
(b) Dhanbad (Jharkhand)
(c) Singreni (A.P.)
(d) Raniganj (W.B.)

418. Which of the following is a single stage, fixed bed high pressure coal gasification process ?
(a) Winkler process
(b) Kopper-Totzek process
(c) Lurgi Process
(d) None of these

419. Soft coke is not
(a) produced by low temperature carbonisation of coal.
(b) a domestic fuel.
(c) used in blast furnaces.
(d) none of these.

420. Which of the following is not a by-product recovered in a high temperature coal carbonisation plant?
(a) Benzol
(b) Pitch-cresosote mixture (PCM)
(c) Napthalene
(d) Ethylene

421. Coal is heated in absence of air for 7 minutes at a temperature of about................ °C, during its volatile matter determination for proximate analysis.
(a) 550 (b) 750
(c) 950 *(d)* 1150

422. Pick out the wrong statement.
(a) Percentage of ash in coke produced from medium coking coal is more than that in coal.
(b) The calorific value (kcal/Nm^3) of coke oven gas reduces on removal of hydrogen from it by cryogenic method.
(c) Ash is normally removed as 'flyash' in kopper-Totzek process of coal gasification.
(d) Coal based fertiliser plants in India at Talcher (Orissa) and Ramagundam (A.P) employ Kopper-Totzek process of coal gasification.

423. Insitu theory and drift theory are related to the
(a) origin of petroleum oil.
(b) origin of coal.
(c) coalification.
(d) variation of coal quality with depth.

424. Coal containing large quantity of................are difficult to wash.
(a) ash
(b) inherent mineral matter
(c) free impurities
(d) volatile matter

425. Main use of hard coke produced by high temperature carbonisation is in the
(a) iron blast furnace
(b) cupola in foundaries
(c) sinter making
(d) domestic ovens

426. Compound coke ovens are those which can be heated by
(a) both steam and electrical power.
(b) lean gas *(e.g.,* B.F. gas).
(c) rich gas *(e.g.,* coke oven gas).
(d) both *(b)* and *(c)*.

427. Degree of carbonisation of coal during coke making can be roughly judged by the...........of the coke produced.
(a) colour *(b)* moisture content
(c) ash content *(d)* volatile matter

428. Highly reactive coke have lower
(a) porosity.
(b) coke reactivity index (CRI).
(c) Critical air blast (CAB) value.
(d) both *(b)* and *(c)*.

429. The temperature at which plastic layer ' formation takes place during carbonisation of coal varies from...............°C.
(a) 100 to 150 *(b)* 350 to 450
(c) 550 to 650 (d) 700 to 850

430. Pick out the wrong statement.
(a) Indian coals on an average contain 25- 30% ash as against 10-12% ash in imported coking coal.

ANSWERS

416.*(b)* 417.*(a)* 418.*(c)* 419.*(c)* 420.*(d)* 421.*(c)* 422.*(c)* 423.*(b)* 424.*(b)* 425.*(a)* 426.*(d)* 427.*(d)* 428.*(c)* 429.*(b)* 430.*(c)*

(b) Ammonia is recovered in the form of ammonium sulphate in direct process of by-product recovery.
(c) A high swelling index number of coking coal indicates poor caking properties of coal.
(d) Wash oil is used for scrubbing benzol from coke oven gas.

431. Which of the following combustibles is absent in blast furnace gas?
(a) H_2 *(b)* CH_4
(c) CO *(d)* none of these

432. Which of the following is a lean fuel gas?
(a) Coke oven gas *(b)* LPG
(c) Blast furnace gas *(d)* Natural gas

433. Which of the following is a rich fuel gas?
(a) Producer gas.
(b) Coal gas from underground gasification of coal.
(c) Refinery gases.
(d) None of these.

434. The reaction, $C + CO_2 \leftrightarrow 2CO$, taking place during coal gasification is called the............reaction.
(a) Neumann reversal *(b)* shift conversion
(c) Boudouard *(d)* reduction

435. Which of the following reactions occuring during coal gasification is called the Neumann reversal reaction ?
(a) $2CO \leftrightarrow C + CO_2$
(b) $CO + H_2O \leftrightarrow CO_2 + H_2$
(c) $C + H_2O \leftrightarrow CO + H_2$
(d) $C + 2H_2O \leftrightarrow CO_2 + 2H_2$

436. The shift conversion reaction taking place during water gas manufacture is given by
(a) $C + H_2O \leftrightarrow CO + H_2$
(b) $C + 2H_2O \leftrightarrow CO_2 + 2H_2$
(c) $CO + H_2O \leftrightarrow CO_2 + H_2$
(d) none of these

437. Bomb calorimeter is used for the determination of calorific value of the............fuels.
(a) gaseous (b) solid
(c) liquid *(d)* both *(b)* and *(c)*

438. Bomb calorimeter is used to determine the
(a) GCV at constant pressure.
(b) GCV at constant volume.
(c) NCV at constant pressure.
(d) NCV at constant volume,

where, GCV – Gross Calorific Value.
NCV – Net Calorific Value.

439. Octane number of 2, 2, 4 – trimethyl pentane is
(a) 0
(b) 100
(c) in between 0 and 100
(d) more than 100

440. Which of the following is a primary fuel?
(a) Blast furnace coke *(b)* Gasoline
(c) Natural gas *(d)* Wood charcoal

441. ....is used as jet engine fuel.
(a) Petrol (b) Diesel
(c) Kerosene (d) LPG

442. Laboratory gas is obtained by the cracking of
(a) gasoline (b) diesel
(c) fuel oil *(d)* kerosene

443. Stoichiometric combustion of 12kg of carbon requires................of oxygen.
(a) 1 kg mole *(b)* 22.4 Nm^3
(c) 32 kg *(d)* all *(a)*, *(b)* and *(c)*

444. Presence of free moisture in coal is most disadvantageous during its
(a) transportation and handling
(b) washing
(c) pulverisation
(d) storage

445. Wood charcoal is obtained by the destructive distillation of wood. It is used in the production of activated carbon, which is not used for the
(a) decolorisation of sugar.
(b) solvent recovery from air and gases
(c) absorption of gases and vapor.
(d) electrode manufacture.

446. Which of the following is not used as a binder in coal briquetting?
(a) *Tar* *(b)* Molasses
(c) Pitch *(d)* Lime

447. Gasification of the solid fuel converts its organic part into combustible by interaction with air/ oxygen and steam so as to obtain a secondary gaseous fuel of high calorific value having no ash. Gasification reactions are normally carried out at about................°C.
(a) 400-500 *(b)* 900-1000
(c) 1400-1500 *(d)* 1700-1800

ANSWERS

431.*(b)* 432.*(c)* 433.*(c)* 434.*(c)* 435.*(a)* 436.*(c)* 437.*(d)* 438.*(b)* 439.*(b)* 440.*(c)* 441.*(c)* 442.*(d)*
443.*(d)* 444.*(c)* 445.*(d)* 446.*(d)* 447.*(b)*

448. Pick out the wrong statement.
(a) LPG is also used as fuel for automobiles & small furnaces and for cutting & welding of metals.
(b) The minimum temperature, at which a petroleum oil vapor catches fire and continues to burn, is called its flash point.
(c) Each ton of petroleum oil on distillation produces about 30-50 Nm³ of gas.
(d) Maximum yield of naphthalene is obtained on distillation of crude oil.

449. Light oil whose major component is benzol, is obtained by the distillation of crude tar in the temperature range of..................°C.
(a) 80-170 *(b)* 200-300
(c) 250-270 (d) 280-300

450. The weathering index of a coal
(a) gives an idea of the fusion temperature of ash.
(b) is related to its calorific value.
(c) is a measure of its size stability, when stored & exposed to weather.
(d) is a measure of its caking tendency.

451. Softening temperature of coal ash is a measure of the.................of coal.
(a) caking tendency
(b) coking tendency
(c) clinkering tendency
(d) size stability

452. Fossil fuels mean
(a) solid fuels
(b) liquid fuels
(c) those fuels which are found in the crust of earth.
(d) premature fuels with low calorific value.

453. A good metallurigical coke should have very low
(a) sulphur & phosphorous content (b) porosity
(c) fusion point of its ash
(d) hardness & strength

454. Kopper-Totzek coal gasifier (installed in a coal based nitrogeneous fertiliser plant) employs a/an.................bed gasifier.
(a) entrained *(b)* moving
(c) fixed *(d)* fluidised

455. Calorific values of both the solid as well as the liquid fuels can be determined by.................calorimeter.
(a) bomb (b) Boy's
(c) Junkar's (d) none of these

456. Gobar gas is produced by the........of cow dung.
(a) fermentation (b) oxidation
(c) hydrogenation *(d)* none of these

457. Main constituent of the gas produced from a gobar gas plant is
(a) CO_2 (b) CH_4
(c) H_2 *(d)* CO

458. The lowest temperature, at which a solid fuel produces enough vapors to support continuous combustion, is called
(a) fire point
(b) smoke point
(c) burning temperature
(d) kindling temperature

459. Which of the following is not a by-product fuel?
(a) Sewage gas *(b)* Refinery gas
(c) Producer gas *(d)* Bagasse

460. Which of the following is not a manufactured fuel?
(a) Coke breeze *(b)* Soft coke
(c) Colloidal fuels *(d)* Charcoal

461. Pick the odd man out.
(a) Briquettes
(b) Wood
(c) Oil gas
(d) Pitch creosote mixture

462. Hard coke is manufactured from
(a) lignite *(b)* bituminous coal
(c) semi-anthracite (d) anthracite

463. Temperature of preheated air used for the transportation of pulverised coal through pipes to the burner of a boiler furnace is restricted to a maximum limit of about 300° C to avoid the
(a) requirement of large volume combustion chamber.
(b) risk of explosion.
(c) chances of clinker formation.
(d) incomplete combustion of coal.

464. Coals used for the generation of producer gas should have
(a) high caking index.
(b) low fusion point of ash.
(c) high volatile matter content.
(d) very low ash content (8-10%).

ANSWERS

448.*(b)* 449.*(a)* 450.*(c)* 451.*(c)* 452.*(c)* 453.*(a)* 454.*(a)* 455.*(a)* 456.*(a)* 457.*(b)* 458.*(c)* 459.*(c)*
460.*(a)* 461.*(b)* 462.*(b)* 463.*(b)* 464.*(c)*

465. Fusion point of coal ash generally varies from 1000 to 1700° C. Ash having fusion point less than.................°C is liable to form clinker.

(a) 1100 (b) 1250
(c) 1350 (d) 1400

466. Very.................. coals are completely devoid of cokability *i.e,* it is non coking.

(a) young (b) mature
(c) both 'a' & 'b' (d) neither 'a' nor 'b'

467. In low temperature carbonisation of coal as compared to high temperature carbonisation ....produced is less.

(a) difference in gross & net calorific value of the coke oven gas
(b) free carbon content in tar
(c) yield percentage of coke
(d) yield of ammonia present in coke oven gas

468. Pick out the wrong statement.

(a) Angle of repose of coal increases with its increasing size.
(b) Specific gravity of coal increases with its increasing maturity.
(c) Refractive index and reflectance of coal increases with the increasing rank of coal.
(d) Specific heat of coal decreases with increases in the volatile matter and decrease in the C/H ratio of coal.

469. Naphthalene recovered from coke oven gas is used

(a) as moisture proof coating on fibres.
(b) as moth balls (insecticides).
(c) as a fuel in furnaces.
(d) for making electrodes.

470. Coal tar fuel - 200 (CTF 200) is a mixture of

(a) pitch and creosote oil.
(b) light oil and pitch.
(c) anthracene and crecosote oil.
(d) solar oil and pitch.

471. Coal tar fuels (CTF) as compared to petroleum based fuel oils have higher

(a) calorific value
(b) higher C/H ratio
(c) sulphur content
(d) difference in gross & net calorific value

472. ....prohibits the use of alcohols directly in petrol engines.

(a) Low octane number
(b) High cost & availability
(c) Low flash point
(d) Low calorific value

473. Power alcohol as compared to straight run gasoline has lower

(a) calorific value (b) octane number
(c) specific gravity (d) viscosity

474. Blast furnace coke is made from coal by

(a) low temperature carbonisation.
(b) high temperature carbonisation.
(c) medium temperature carbonisation.
(d) heating the coal in an oven in presence of air.

475. Dust content in raw blast furnace gas is about...................gm/Nm^3, hence it is cleaned to the dust level of about < 10mg/Nm^3 before use.

(a) 15 - 30 (b) 1 - 2
(c) 70 - 100 (d) 150 - 200

476. Which of the following is not increased by using preheated combustion air in place of ordinary air at room temperature ?

(a) Calorific value of the fuel.
(b) Flame temperature achieved.
(c) Speed of combustion of fuel.
(d) Heat transfer rate to the stock in the furnace.

477. Which of the following gaseous fuels has the lowest calorific value?

(a) Gobar gas
(b) Refinery gas
(c) Converter gas
(d) Blast furnace gas

478. Which of the following fuel gases has the highest calorific value?

(a) Natural gas.
(b) Coal bed methane (CBM).
(c) Liquefied petroleum gas (LPG).
(d) Sewage gas.

479. Gross & net calorific value is the same for

(a) blast furnace gas
(b) coke oven gas
(c) L.D. converter gas
(d) none of these

480. For which pair of the fuel gases, calorific value (C.V.) of one fuel is almost double that of the other on volume basis (*i.e.,* kcal /Nm^3), while the C.V. is same on weight basis (*i.e.,* kcal/kg) ?

ANSWERS

465.(a) 466.(c) 467.(a) 468.(d) 469.(b) 470.(a) 471.(b) 472.(b) 473.(a) 474.(b) 475.(a) 476.(a)
477.(d) 478.(c) 479.(c) 480.(a)

(a) Propane and acetylene
(b) Propane and LPG
(c) Sewage gas and gobar gas
(d) B.F. gas and coke oven gas

481. Combustion process occuring at a subsonic velocity with slow pressure build up is termed as
(a) explosion (b) detonation
(c) deflagration (d) none of these

482. Which of the following gaseous fuels requires the highest amount of theoretical combustion air?
(a) LPG (b) Coke oven gas
(c) Natural gas (d) Lurgi gas

483. Differences of gross calorific value and net calorific value of a fuel indicates the............content in the fuel.
(a) carbon (b) hydrogen
(c) moisture (d) volatile matter

484. Suitable atomising viscosity of fuel oil for use in burner is..................Redwood-I seconds.
(a) 100 (b) 400
(c) 800 (d) 1200

485. The limiting temperature to which the flue gases can be cooled is influenced by the percentage of.................in the fuel.
(a) hydrogen (b) carbon
(c) sulphur (d) ash

486. Which parameter of coal gives a rough estimate of its calorific value?
(a) Ash (b) Hydrogen
(c) Fixed carbon (d) Total carbon

487. Gross calorific value of fuel oil, diesel & petrol is of the order of..................MJ/kg.
(a) 20 (b) 30
(c) 40 (d) 50

488. Gross calorific value of a good coal is 26000 MJ/ton, which is equivalent to.................kcal/kg.
(a) 621 (b) 6210
(c) 62100 (d) 621000

489. Specific heat of fuel oil is in the range of................kcal/kg.°C.
(a) 0.1-0.15 (b) 0.22-0.28
(c) 0.34-0.42 (d) 0.45-0.50

490. In a flue gas, the theoretical CO_2 is 15.5% and measured CO_2 is 11% by the volume. The percentage of excess air will be
(a) 40.9 (b) 20.5
(c) 80.8 (d) 145.2

491. In a oil fired furnace, preheating of combustion air by 20°C will save about...............percent of fuel.
(a) 0.2 (b) 1
(c) 5 (d) 10

492. In an oil fired boiler, increase in stack gas temperature of 22°C due to tube fouling will increase oil consumption by about..........per cent.
(a) 1 (b) 3.5
(c) 7.5 (d) 10

493. Which of the following is not measured in ultimate analysis of coal?
(a) Sulphur (b) Nitrogen
(c) Ash (d) Total carbon

494. With increase in the percentage of excess air for combustion of fuel oil, the percentage of CO_2 in flue gas.
(a) increases (b) decreases
(c) remain same (d) unpredictable

495. Which of the following requires the least amount of oxygen/Kg for complete combustion?
(a) Hydrogen (b) Methane
(c) Sulphur (d) Carbon

496. Which of the following fuels has the highest calorific value (in kcal/kg) ?
(a) Petrol (b) Hydrogen
(c) LPG (d) Natural gas

497. Out of the following, the lowest excess air will be required by the...............burner.
(a) coal
(b) high pressure oil
(c) high pressure gas
(d) low pressure oil

498. The purpose of oil atomisation in oil fired burner is to
(a) increase the surface area of oil.
(b) increase the excess air.
(c) reduce the power consumption.
(d) reduce the flue gas temperature.

499. Match the percentage of sulphur (List II) present in various petrofuels (List I).

List I	ListII
A. Kerosene	I. < 0.5
B. Light diesel oil (LDO)	II. 2.0 - 4.0

ANSWERS

481.(c) 482.(a) 483.(c) 484.(a) 485.(c) 486.(d) 487.(c) 488.(b) 489.(b) 490.(a) 491.(b) 492.(a) 493.(c) 494.(b) 495.(c) 496.(c) 497.(c) 498.(a) 499.(a)

C. Furnace oil III.0.5 - 1.8
D. LSHS IV.0.05-0.20
(a) A-IV, B-III, C-II, D-I
(b) A-III, B-II, C-I, D-IV
(c) A-II, B-III, C-I, D-II
(d) A-I, B-II, C-III, D-IV

500. Pick out the wrong statement.
(a) Proximate analysis of coal has no connection with the word "approximate".
(b) Volatile matter content of coal influences the secondary air requirement for its combustion.
(c) Moisture present in coal aids in radiation heat transfer.
(d) High sulphur Assam coals contain a maximum of 2% sulphur.

501. Losses associated with the storage of coal include oxidation loss, wind loss and carpet loss. 1% oxidation of coal has the same effect as.................% ash in coal.
(a) 0.1 *(b)* 1
(c) 3 *(d)* 5

502. Combustion of large size & irregular lump coal does not result in higher
(a) excess air requirement.
(b) stack loss.
(c) unburnts in the ash.
(d) thermal efficiency.

503. Match the proper size of coal to be used (List II) for various coal firing systems (List I).

List I
A Fluidiscd bed boiler
B Pulverised fuel fired boiler
C Natural draft hand firing
D Spreader stoker firing

List II
I. 80% — 200 mesh (74 micron)
II. >10 mm
III. 15 – 25 mm
IV. 25 – 75 mm
(a) A-II, B-I, C-IV, D-III
(b) A-I, B-II, C-IV, D-III
(c) A-II, B-I, C-III, D-IV
(d) A-I, B-II, C-III, D-IV

504. Natural gas
(a) is heavier than air.
(b) produces heavy smoke & soot on burning.
(c) practically contains no sulphur.
(d) contains about 10% hydrogen.

505. When 1 kg of carbon burns completely to CO_2 ($C + O_2 = CO_2$), it releases 8084 kcal of heat; while 1 kg of carbon when burns incompletely to produce CO ($C + 1/2\, O_2 = CO$), it releases much less quantity (kcal) of heat. Each kg of CO thus formed means a loss of...............kcal of heat.
(a) 2430 (b) 5654
(c) 4042 (d) 7044

506. Turn down ratio of a burner is the ratio of the maximum fuel input to minimum fuel input of the burner without affecting the excess air level. The optimum turn down ratio of an oil burner is
(a) 1:2 *(b)* 1:5
(c) 1:10 *(d)* 1:15

507. Typically, Indian coals have ash fusion temperature as low as 1100°C, which results in clinker formation. Once clinker is formed, it has a tendency to
(a) stick to a hot surface rather than a cold one.
(b) grow.
(c) stick to a rough surface rather than a smooth one.
(d) all 'a, 'b' & 'c'.

508. Pick out the wrong statement about combustion of fuels.
(a) Gaseous fuels require the least % excess air for combustion out of solid, liquid & gaseous fuels.
(b) Spreader stoker grate firing of coal is thermally more efficient than pulverised coal firing in boilers.
(c) Combustion of 1 kg of bituminous coal requires about 11 kg of theoretical air producing about 11 kg of flue gas.
(d) Combustion of 1 kg of fuel oil requires about 14 kg of theoretical air producing about 14 kg of flue gas.

509. Out of the following fuels, differences between gross calorific value (GCV) and net calorific value (NCV) will be maximum in case of
(a) natural gas (b) coal
(c) furnace oil *(d)* rice husk

510. Preheating of combustion air by 100°C will save about..................per cent of fuel.
(a) 1 *(b)* 5
(c) 15 *(d)* 25

511. The heat loss dependent on the chemical characteristics of liquid fuel is

ANSWERS

500.*(d)* 501.*(b)* 502.*(d)* 503.*(a)* 504.*(c)* 505.*(b)* 506.*(b)* 507.*(d)* 508.*(b)* 509.*(a)* 510.*(b)* 511.*(c)*

(a) radiation loss (b) blowdown loss
(c) hydrogen loss (d) unburnt loss

512. Excess air can be derived by measuring the % of O_2 in flue gases. In a Flue gas, the theoretical CO_2 is 15% and the measured CO_2 is 12% by volume. The % excess air is
(a) 25 (b) 50
(c) 100 (d) 150

513. Fluidised bed combustion takes place in the temperature range of................°C.
(a) 500-600 (b) 800-900
(c) > 1000 (d) < 5000

514. Complete combustion of 8 kg of methane produces...............Kg of CO_2.
(a) 16 (b) 22
(c) 44 (d) 56

515. When 100 kg of fuel containing 60% carbon in burnt with theoretical air, the weight of CO_2 produced will be................kg.
(a) 220 (b) 320
(c) 440 (d) 4400

516. Which of the following fuels will have maximum viscosity at the same temperature?
(a) Light diesel oil (LDO).
(b) High speed diesel (HSD).
(c) Furnace oil.
(d) LSHS (low sulphur heavy stock).

517. The lowest limit of exit flue gas temperature of furnace oil fired system is dependent on.................content in fuel.
(a) carbon (b) sulphur
(c) ash (d) moisture

518. Which of the following releases the maximum energy per kg on complete combustion?
(a) Carbon (b) Hydrogen
(c) Sulphur (d) Magnesium

519. Which of the following fuels requires maximum theoretical/stoichiometric combustion air?
(a) Coal (b) Furnace oil
(c) Bagasse (d) Hydrogen

520. The viscosity of LSHS (low sulphur heavy stock) will be maximum at which of the following temperature?
(a) 35 (b) 65
(c) 95 (d) 115

521. LSHS contains about..................per cent sulphur.
(a) 0.5 (b) 2
(c) 4 (d) 5.5

522. When pure hydrogen is burnt with stoichiometric air, percentage nitrogen on volume basis in flue gas on dry basis will be
(a) 21 (b) 79
(c) 100 (d) < 21

523. The country which has the largest share of coal reserve is
(a) India (b) China
(c) U.S.A (d) Russia

524. A factory substitutes expensive fuel oil and starts firing rice husk in a boiler. The following scenario is most likely to happen.
(a) Higher boiler efficiency.
(b) Higher energy consumption.
(c) Higher fuel cost.
(d) Less particulate emission in the chimney.

525. Changing from furnace oil firing to natural gas firing in a boiler furnace will result in
(a) decreased % of wet flue gas loss.
(b) decreased SO_2 emission.
(c) increased CO_2 emission.
(d) none of these.

526. With increase in excess air for combustion, which of the following will result in flue gas?
(a) % CO_2 decreases.
(b) % O_2 decreases.
(c) Both *'a'* & *'b'*.
(d) Neither *'a'* nor *'b'*.

527. Sulphur percentage is furnace oil
(a) improves viscosity.
(b) does not add to heat value.
(c) forms soot.
(d) sets lower flue gas exit temperature limit.

528. Theoretical combustion air ratio for burning 1 kg of carbon is
(a) 3.76 (b) 4.76
(c) 11.6 (d) 44

529. Excess air can be derived by measuring percentage of...................in flue gases.
(a) CO (b) CO_2
(c) NO_2 (d) N_2

530. Carpet loss occurs in
(a) coal combustion.
(b) coal storage.
(c) furnace operation.
(d) oil atomisation.

ANSWERS

512.(a) 513.(b) 514.(b) 515.(a) 516.(d) 517.(d) 518.(b) 519.(d) 520.(a) 521.(a) 522.(c) 523.(c) 524.(b) 525.(b) 526.(a) 527.(d) 528.(c) 529.(b) 530.(b)

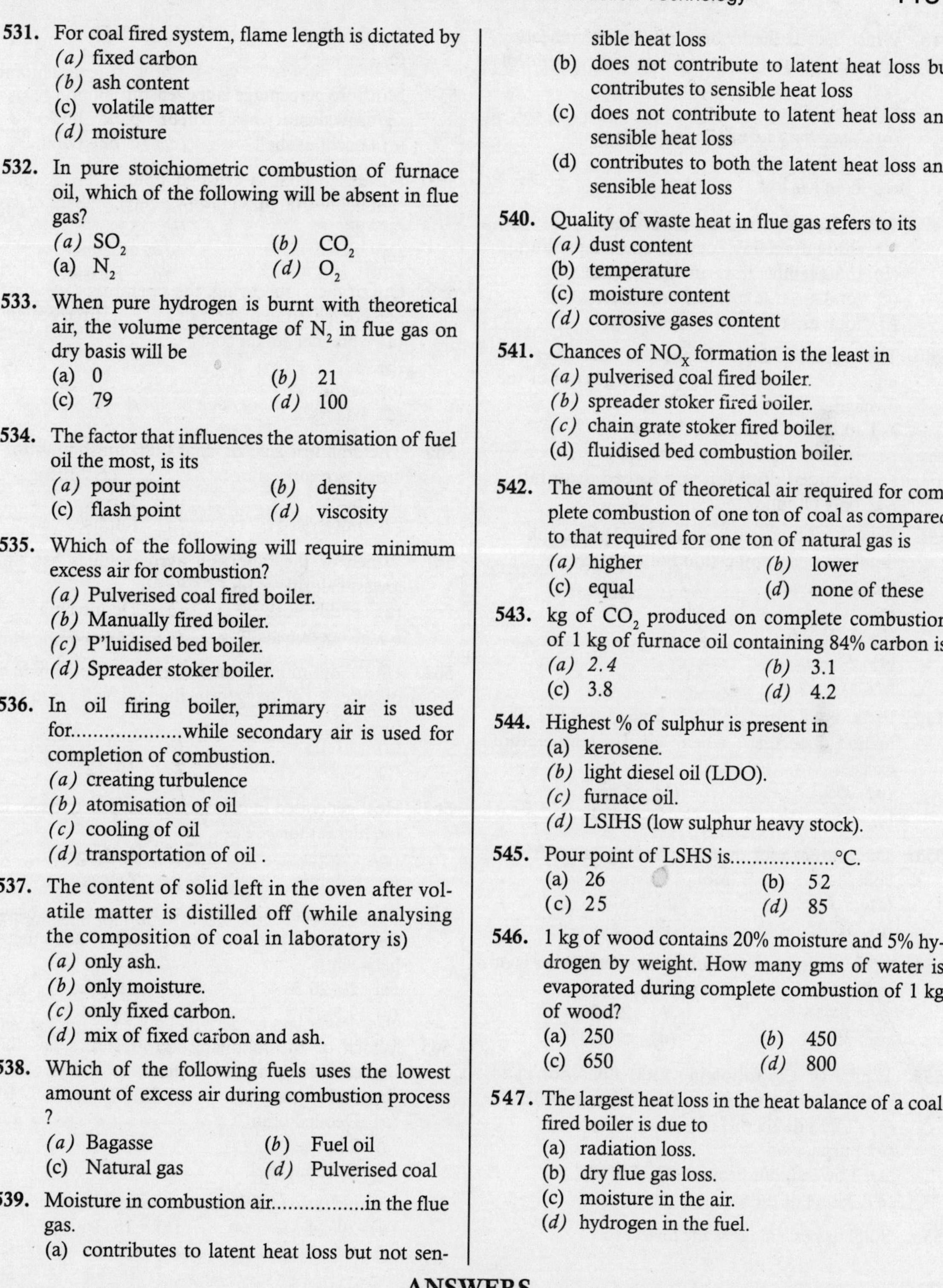

531. For coal fired system, flame length is dictated by
(a) fixed carbon
(b) ash content
(c) volatile matter
(d) moisture

532. In pure stoichiometric combustion of furnace oil, which of the following will be absent in flue gas?
(a) SO_2 (b) CO_2
(a) N_2 (d) O_2

533. When pure hydrogen is burnt with theoretical air, the volume percentage of N_2 in flue gas on dry basis will be
(a) 0 (b) 21
(c) 79 (d) 100

534. The factor that influences the atomisation of fuel oil the most, is its
(a) pour point (b) density
(c) flash point (d) viscosity

535. Which of the following will require minimum excess air for combustion?
(a) Pulverised coal fired boiler.
(b) Manually fired boiler.
(c) P'luidised bed boiler.
(d) Spreader stoker boiler.

536. In oil firing boiler, primary air is used for..................while secondary air is used for completion of combustion.
(a) creating turbulence
(b) atomisation of oil
(c) cooling of oil
(d) transportation of oil .

537. The content of solid left in the oven after volatile matter is distilled off (while analysing the composition of coal in laboratory is)
(a) only ash.
(b) only moisture.
(c) only fixed carbon.
(d) mix of fixed carbon and ash.

538. Which of the following fuels uses the lowest amount of excess air during combustion process ?
(a) Bagasse (b) Fuel oil
(c) Natural gas (d) Pulverised coal

539. Moisture in combustion air................in the flue gas.
(a) contributes to latent heat loss but not sensible heat loss
(b) does not contribute to latent heat loss but contributes to sensible heat loss
(c) does not contribute to latent heat loss and sensible heat loss
(d) contributes to both the latent heat loss and sensible heat loss

540. Quality of waste heat in flue gas refers to its
(a) dust content
(b) temperature
(c) moisture content
(d) corrosive gases content

541. Chances of NO_X formation is the least in
(a) pulverised coal fired boiler.
(b) spreader stoker fired boiler.
(c) chain grate stoker fired boiler.
(d) fluidised bed combustion boiler.

542. The amount of theoretical air required for complete combustion of one ton of coal as compared to that required for one ton of natural gas is
(a) higher (b) lower
(c) equal (d) none of these

543. kg of CO_2 produced on complete combustion of 1 kg of furnace oil containing 84% carbon is
(a) 2.4 (b) 3.1
(c) 3.8 (d) 4.2

544. Highest % of sulphur is present in
(a) kerosene.
(b) light diesel oil (LDO).
(c) furnace oil.
(d) LSIHS (low sulphur heavy stock).

545. Pour point of LSHS is..................°C.
(a) 26 (b) 52
(c) 25 (d) 85

546. 1 kg of wood contains 20% moisture and 5% hydrogen by weight. How many gms of water is evaporated during complete combustion of 1 kg of wood?
(a) 250 (b) 450
(c) 650 (d) 800

547. The largest heat loss in the heat balance of a coal fired boiler is due to
(a) radiation loss.
(b) dry flue gas loss.
(c) moisture in the air.
(d) hydrogen in the fuel.

ANSWERS

531.(c) 532.(d) 533.(d) 534.(d) 535.(a) 536.(b) 537.(d) 538.(c) 539.(b) 540.(b) 541.(d) 542.(b)
543.(b) 544.(c) 545.(c) 546.(c) 547.(b)

548. Which loss is the highest in a typical reheating furnace operating at a furnace temperature of 1250°C?
(a) Wall losses.
(b) Necessary furnace opening losses.
(c) Cooling water loss.
(d) Flue gas loss.

549. Three 'T's for good combustion are
(a) through mixing, total air and temperature.
(b) time, temperature and turbulence.
(c) total air, true fuel and turbulence.
(d) total air, time and temperature.

550. The balance draft furnace is the one using..................for exiting the flue gas out of the furnace.
(a) forced draft fan and chimney
(b) induced draft fan and chimney
(c) induced draft fan and forced draft fan
(d) only chimney

551. Velocity of fluidised air in a AFBC (Atmospheric fluidised bed combustion boiler) is.................. metre/second,
(a) 1.2 – 3.7
(b) 3.7 – 9.5
(c) 9.5 – 12
(d) 12 – 15.5

552. In a coal fired AFBC boiler, clinker formation occurs when bed temperature exceeds.....................°C.
(a) 750 (b) 850
(c) 950 (d) 1050

553. The coal size for over bed feeding in a AFBC . boiler is.................mm.
(a) 3 (b) 6–10
(c) 10–15 (d) 15

554. Excess air quantity required in a FBC boiler is of the order of.................percent.
(a) 10–15 (b) 20–25
(c) 35–40 (d) 5–10

555. Which of the following fuels does not need preheating for pumping from storage tank?
(a) Light diesel oil (LDO).
(b) Furnace oil.
(c) Low sulphur heavy stock (LSHS).
(d) None of these.

556. Sulphur percentage is the lowest in
(a) LSHS (b) LDO
(c) furnace oil (d) kerosene

557. Moisture percentage is the least in
(a) saw dust (b) paddy husk
(c) coconut shell (d) deoiled bran

558. Pulverised coal is crushed upto................percent below 200 mesh (74 microns).
(a) 45 (b) 55
(c) 75 (d) 95

559. Out of the following, the operating excess air percentage is the highest in.............fired boiler.
(a) spreader stoker coal
(b) oil
(c) wood
(d) bagasses

560. The smallest size of coal (200 mesh= 74 micron) is required in...................coal firing.
(a) hand (b) stoker
(c) pulverised (d) fluidised bed

561. Which of the following agro residues has the lowest calorific value?
(a) Paddy husk (b) Deoiled bran
(c) Coconut shell (d) Saw dust

562. Under optimum combustion condition for fuel oil, the % of oxygen in flue gas will be in the range of
(a) 0.1–05 (b) 2–3
(c) 5–8 (d) 10–12

563. Heat generated in combustion on per kg basis, is the highest for
(a) carbon (b) hydrogen
(c) sulphur (d) oxygen

564. Theoretical air fuel ratio (AFR) on volume basis for complete combustion of natural gas ranges between
(a) 20–20.5 (b) 15–15.5
(c) 9.5–10 (d) 5.5–6

565. Which of the following agro-residues has the highest volatile matter content?
(a) Paddy husk
(b) Deoiled bran
(c) Saw dust
(d) Coconut shell

566. Air fuel ratio (AFR) for coal ranges between
(a) 20–24 (b) 15–16
(c) 8 – 10 (d) 2 – 4

ANSWERS

548.(d) 549.(b) 550.(c) 551.(a) 552.(c) 553.(b) 554.(b) 555.(a) 556.(d) 557.(d) 558.(c) 559.(a) 560.(c) 561.(b) 562.(b) 563.(b) 564.(b) 565.(c) 566.(b)

567. Which of the following releases the maximum energy per kg on complete combustion?
(a) Carbon *(b)* Magnesium
(c) Hydrogen *(d)* Sulphur

568. An oil fired boiler is operating at 5% oxygen in the flue gas. The percentage excess air supplied to the boiler is approximately....................per cent.
(a) 11 *(b)* 20
(c) 31 *(d)* 44

569. Controlled wetting of coal (during coal preparation) prior to its combustion would result in the
(a) decrease in the percentage of unburnt carbon.
(b) reduction in the flue gas exit temperature.
(c) increase in the fines of coal.
(d) improper combustion.

570. By higher octane number of SI fuel, it is meant that the fuel has
(a) higher heating value
(b) higher flash point
(c) lower volatility
(d) longer ignition delay

571. The presence of nitrogen in the products of combustion ensures that:
(a) complete combustion of fuel takes place.
(b) incomplete combustion of fuel occurs.
(c) dry products of combustion are analysed.
(d) air is used for the combustion.

572. Consider the following statements:
The maximum temperature produced by the combination of a unit mass of fuel depends upon
1. LCV
2. Ash content
3. Mass of air supplied
4. Pressure in the furnace

of these statements:
(a) 1 alone is correct
(b) 1 and 3 are correct
(c) 2 and 4 are correct
(d) 3 and 4 are correct

573. The amount of CO_2 produced by 1 kg of carbon on complete combustion is................kg.
(a) 3/11 *(b)* 3/8
(c) 8/3 *(d)* 11/3

574. Consider the following statements:
1. Pulverised fuel gives up high and controlled burning rate.
2. Insufficient air causes excessive smoking of exhaust.
3. Excess air is provided to control the fuel gas temperature.
4. Effect of sulphur in fuel is to give high heat transfer rate.

Which of the following statements are correct?
(a) 3 and 4 (b) 2 and 3
(c) 1 and 2 (d) 1 and 4

575. Consider the following statements:
1. Octane rating of gasoline is based on iso-octane and iso- haptane fuels, which are paraffins.
2. Tetraethyl lead is added to gasoline to increase octane number.
3. Ethylene dibromide is added as scavenging agent to remove lead deposits on spark plugs.
4. Surface ignition need not necessarily cause knocking.

Which of the following statements are correct?
(a) 1, 2, 3 and 4 *(b)* 2, 3 and 4
(c) 1 and 4 *(d)* 1, 2 and 3

576. Consider the following statements:
1. For the combustion of pulverised coal, 5 to 10% excess air is required.
2. Air contains 21% oxygen by weight.
3. The flue gases from a coal fired furnace contain around 70% nitrogen by volume.
4. In the combustion of liquid fuels, the number of moles of the products are invariably greater than the number of moles of reactants.

of these statements:
(a) 1, 2 and 4 are correct
(b) 1, 3 and 4 are correct
(c) 2, 3 and 4 are correct
(d) 1 and 3 are correct

ANSWERS

567.*(c)* 568.*(c)* 569.*(a)* 570.*(d)* 571.*(d)* 572.*(b)* 573.*(d)* 574.*(b)* 575.*(c)* 576.*(a)*

APPENDIX-D: GLOSSARY OF TERMS RELATED TO FUELS & COMBUSTION TECHNOLOGY

A

Abel tester: A closed-cup flash point tester for kerosene and other oils

Absolute Viscosity: A property of a fluid that determines the internal resistance to flow without regard to its density. It is also called dynamic viscosity which is expressed as a product of kinematic viscosity and density of the fluid.

Acid Treatment: A process in which unfinished petroleum products such as gasoline, kerosene, and lubricating oil stocks are treated with sulfuric acid to improve color, odor, and other properties.

Aerodynamics of the Furnace: A study of fluid flow and its pressure as well as velocity distribution in the furnace, flue duct, etc. Fluid pressure impaired by a fan/injector, buoyancy, and different flow resistances, such as friction loss in the flow path, pressure drop at bends and changes of section, resistance of beds of solid particles, and resistance due to regenerators, recuperators, etc, are considered for the aerodynamics of a furnace.

Aeration: Addition and mixing of air.

Agglomerating Characteristics: The tendency of fine particles to bind together into a larger mass.

Air/Fuel Ratio: The ratio of the air flow rate to the fuel flow rate, measured in the same units under the same conditions.

Air Heater or Air Preheater: Heat-transfer equipment through which air is passed and heated by a medium of higher temperature, such as the products of combustion or steam.

- **Regenerative air preheater:** An air heater in which heat is first stored up in the structure itself by the passage of the products of combustion, and which then gives up the heat so stored to the subsequent passage of air.
- **Recuperative air heater:** An air heater in which the heat from products of combustion passes through a partition, which separates the products from the air.

(a) Tubular air heater. An air heater containing a group of tubular elements through the walls of which heat is transferred from a flowing heating medium to an air stream.

(b) Plate air heater. An air heater containing passages formed by spaced plates through which heat is transferred from a flowing heating medium to an air stream.

Air Purge: The removal of undesired matter by replacement with air.

Air Register: A type of burner mounting that can admit secondary air to the combustion space through the openings around the burner.

Air Shutter: An adjustable shutter on a burner air register by means of which the amount of air induced into the furnace through the register can be controlled.

Air-Swept Pulverizers: A pulverizer through which air flows and from which pulverized fuel is removed by the stream of air.

Alkylation: A Process using sulfuric or hydrofluoric acid as a catalyst to combine olefins (usually butylene) and isobutane to produce a high octane product known as alkylate.

Aniline Point: The aniline point of a petroleum product is the minimum equilibrium solution temperature with an equal volume of freshly distilled aniline.

Anthracite: A hard, compact coal characterized by high luster and low sulphur content. Anthracite is comprised of less volatile matter than bituminous coal, which provides for its nearly non-luminous flame. ANTHRACITE : ASTM coal classification by rank: Dry fixed carbon 92 % or more and less than 98 %; and dry volatile matter 8 % or less and more than 2 % on a mineral-matter-free basis.

API Gravity: The specific gravity of a liquid expressed as API, a scale adopted by the American Petroleum Institute. As the gravity readings according to this scale are higher for less dense liquids, the API gravity for a liquid increases as its temperature rises.

Astm: American Society for Testing and Materials. The ASTM provides a system for classifying coal as well as other testing parameters.

Atomising Air: The air supplied through a burner which is used to break the oil stream into a spray of tiny droplets and subsequently takes part in the combustion process.

Atomizing Media: A supplementary medium, such as steam or air, which assists in breaking the fuel oil into a fine spray.

Ash: Non-combustible mineral matter left out after a fuel is combusted. In residual fuel oils, ash consists mainly of inorganic oxides and chlorides.

Ash Fusion Temperature: The lowest temperature at which the ash can melt and blend into a vitrified substance or clinker.

Atomization: The process of breaking a liquid into a spray of tiny droplets.

Auger Mining: A method of mining following a surface mining operation where the overburden becomes too costly to remove. The auger usually operates on a horizontal plane, comes in sections, and the diameter depends on the height of the coal.

Automatic Fuel Shut-off Valve: A valve for stopping the flow of fuel automatically when an abnormal or a dangerous situation develops.

Available Heat: The quantity of heat available for useful purposes from the gross quantity of heat released within a combustion space taking into account the dry flue gas loss and the moisture loss.

B

Barrel: A unit of volume measurement used for petroleum and its products. 1 barrel = 42 U.S. gallons or 35 British gallons or 160 litres.

Batch Type Furnace: A furnace in which charging and discharging operations are carried out after each heating/heat treatment cycle.

Bbl: Abbreviation for barrel.

Beehive oven: A periodic oven with a circular hearth and dome like roof (crown).

Bituminous Coal: A hard, black, combustible substance formed from decomposed vegetable matter subjected to pressure, temperature, and moisture for millions of years. Varying conditions of formation result in diverse coal chemistries and heat contents. Bituminous coal is one of several phases in an evolutionary development process which includes (from least developed to most developed): peat, lignite (sometimes called brown coal), sub-bituminous, bituminous, anthracite (sometimes called hard coal) and graphite. The latter phases have generally experienced greater extremes of pressure, temperature and time.

Blast (Pressure) Burner: A burner delivering a combustible mixture under pressure, normally above 10 mm WC to the combustion zone.

Blast Furnace Gas: A low calorific value gas recovered from blast furnace as a by-product and used as a fuel.

Blower: Equipment for moving large volumes of gas against low pressure heads.

Boiler: A closed vessel in which water is heated, steam is generated, steam is superheated, or any combination thereof, under pressure or vacuum by the application of heat from combustible fuels, electricity, or nuclear energy. The term does not include such facilities of an integral part of a continuous processing unit but does include fired units of heating or vaporizing liquids other than water where these units are separate from processing systems and are complete within themselves.

Breeching: A duct for the transport of the products of combustion between parts of a steam generating unit or the stack.

Bridgewall: A wall in a furnace over which the products of combustion pass.

Blue Water Gas: A fuel gas manufactured by passing steam over a bed of incandescent coke. It mainly contains hydrogen & carbon monoxide.

Burner: A device through which predetermined quantities of fuel and air are brought in contact with each other for continuous combustion. Flame propagation as well as heat release are controlled to produce the desired effect at a given location .

Burner Assembly A burner that is factory built as a single assembly or as two or more sub assemblies which include all parts necessary for its normal function when installed as intended.

Burner, Atmosheric: A gas burner in which all air for combustion is supplied by natural draft, the inspiriting force being created by gas velocity.

Burner, Automatically Lighted: A burner in which fuel to the main burner is normally turned on and ignited automatically.

Burner Block: A special refractory block with a suitable hole at its centre for mounting the burner.

Burner, Natural Draft Type: A burner which depends primarily on the natural draft created in the flue to induce the air required for combustion into the burner.

Burner, Power: A burner in which all air for combustion is supplied by a power driver fan that overcomes the resistance through the burner to deliver the quantity of air required for combustion.

Burner Wind Box: A plenum chamber around a burner in which an air pressure is maintained to ensure proper distribution and discharge of secondary air.

C

Caking: Property of certain coals to become plastic when heated and form large masses of coke.

Calorie: The amount of heat required to raise the temperature of 1 gm of water through 1°C. As calorie is very small, kilocalorie (1 kcal = 1000 cal) is more commonly employed in engineering calculations.

Calorific Value: The amount of heat generated by the complete combustion of a unit quantity of fuel measured in calories. It is also called heating value or heat of combustion of a fuel.

Calorific Value, Gross: The quantity of heat which is evolved when the products of combustion of a unit quantity of fuel are cooled to 15.5°C and the water vapour produced from the hydrogen is completely condensed to liquid water at the temperature. It is also called as higher heating value of a fuel and expressed as kcal/kg or kcal/m3 (J/kg , or J/m^3) as applicable.

Calorific Value, Net: A gross calorific value without the latent heat of vaporisation of the water vapour formed by the combustion of hydrogen content of the fuel. It is also known as lower heating value.

Carbon/Hydrogen Ratio: The weight of carbon in a hydrocarbon fuel divided by the weight of hydrogen.

Carbon Residue: The amount of carbonaceous material left from a sample of oil heated or distilled in the absence of air. An indication of carbon/hydrogen ratio.

Catalytic Cracking: The process of breaking up heavier hydrocarbon molecules into lighter hydrocarbon fractions by use of heat and catalysts.

Centipoise: A unit for measurement of absolute viscosity. Centipoise is equal to one hundredth of a poise which is the metric unit of absolute viscosity. One centipoise is equal to 10-2 Ns/m which is SI unit of viscosity.

Centistoke: A unit for kinematic viscosity in metric system.Centistoke is equal to one hundredth of a stoke. One centistoke is equal to 10^{-2} m^2/s in SI units.

Centrifugal Atomising Oil Burner: A burner in which oil is thrown into air stream by centrifugal force from a rotating cup causing the oil to break into a spray of tiny droplets.

Cetane number: An empirical measure of ignition quality. Defined as the percentage by volume of cetane in a mixture of cetane and alpha methyl naphthalene which has the same ignition quality when used in an engine as a fuel under test.

Checker Work: An arrangement of alternately spaced brick in a furnace with openings through which air or gas flows.

Check Valve: A valve to prevent reverse flow,usually closed by reverse movement of the fluid.

Cinder: Particles of partially burned fuel from which volatile gases have been driven off, which are carried from the furnace by the products of combustion.

Circulating Loop: The main loop in which the oil is circulated from the oil storage tanks to the branch circuits and then back to the storage tank.

Closed Circuit Oil System: A fuel oil supply system in which oil is pumped from the storage tank through a circulating loop and back into the storage tanks.

Cloud Point: Temperature at which wax begins to crystallize from a distillate fuel.

Coal: Includes all ranks of coal – anthracite, bituminous coal, sub-bituminous coal and lignite.

Coal Seam: A "bed" or stratum of coal. The term "seam" usually is applied to a large deposit of coal.

Coal Washing: The process of separating coal of various sizes, densities and shapes by allowing them to settle in a fluid.

Coal Tar: A viscous mixture of organic compounds. A byproduct of coke production. Limited fuel use.

Coke: The solid residue, mainly carbon, resulting from destructive distillation of coal or other carbonaceous materials in an oven or close chamber in the absence of air.

Coke Breeze: Following a screening process, the smallest coke particles remaining in the form of a residue.

Coke Button: A button-sized piece of coke produced during laboratory testing of the coking or free swelling characteristics of coal. Numerical references of one through nine are assigned to coke buttons based on the degree of swelling during the coking process.

Coke Oven Gas: The volatile matter from coal which comes out of coke ovens in the form of a gas having a high calorific value during the process of coking. This may be used as a fuel.

Combination Burner: A dual purpose burner capable of firing either gas or oil.

Combustibles: Substances which may be combusted.

Combustion: A rapid combination of oxygen with a fuel resulting in release of heat.

Combustion Air: Air supplied through a burner for combustion of a fuel. It excludes atomising air.

Combustion Intensity: Volumetric rate of combustion. The ratio of fuel energy input to the flame volume.

Combustion Products: Flue gases, water vapour and ash which are formed as a result of combustion of fuel substance.

Compressed Air Atomising Oil Burner: A burner in which compressed air is used to atomize the oil stream into a spray of droplets.

Condensable Hydrocarbon Content: The fraction of a gas that may condense as the gas is cooled. Natural Gasoline is a condensable hydrocarbon present in raw natural gas.

Continuous Furnace: Furnaces operated on an uninterrupted cycle in which the charge material is added constantly, moved through and removed.

Continuous Miner: A mechanical mining machine consisting of a cutting head, a coal- gathering device, a chain conveyor with flexible loading boom, and a crawler equipped chassis. Its function is to excavate the mineral and to load it onto shuttle cars or continuous haulage systems. It is electrically powered, with a hydraulic subsystem for auxiliary functions. Power is supplied through a trailing cable.

Continuous Regenerator: A waste heat recovery device in which continuous transfer of heat through a rotating heat absorption media take place.

Contour Stripping or Surface Mining: The removal of overburden and mining from a coal seam that outcrops or approaches the surface at approximately the same elevation, in steep or mountainous areas.

Controller: A device that detects a change in a process variable and then automatically uses an external source of power to amplify the detected signal and to energise a mechanism that' will correct the deviation in the process variable until it returns to a pre-set value.

Control Zone: That section of a furnace within which the temperature is controlled by one temperature sensing device and one set of valves or regulators (one each for air and fuel).

Copper Strip Corrosion: An oil test standardized by ASTM as an indication of sulphur in the oil. Generally applied only to light distillates and gasoline.

Corner Firing: A method of firing liquid, gaseous, or pulverized fuel in which the burners are located in the corners of the furnace. Also called tangential firing.

Counter Current Recuperator: A recuperator in which the hot gases and air/gas to be preheated flow in opposite directions, therefore, these conditions favour the maximum preheat, the greatest possible cooling of the waste gases and the highest efficiency.

Cracking: The process of breaking hydrocarbon molecules so that they recombine into both lighter and heavier molecules.

Creosote: A light fuel produced by the distillation of coal tar.

Cross Fired Furnace: A furnace with fuel supplied from the side.

Cross Current Recuperator: A recuperator in which the air and the gases flow through passages at right angles to one another.

Crown: A furnace roof, especially one which is dome shaped. The highest point of an arch.

Crude Oil: A naturally occurring mixture of hydrocarbons that usually includes small quantities of sulfur, nitrogen, and oxygen derivatives of hydrocarbons as well as trace metals.

Cutter stock: A petroleum stock which is used to reduce the viscosity of a heavier residual stock by dilution.

D

Damper: A mechanically controlled construction in flue duct to pressurize the furnace. A device for introducing a variable resistance of regulating the volumetric flow of gas or air.

- ***Butterfly type***. A single-blade damper pivoted about its center.
- ***Curtain type***. A damper consisting of one or more blades, each pivoted about one edge.
- ***Flap type***. A damper consisting of one or more blades, each pivoted about one edge.
- ***Louvre type***. A damper consisting of several blades, each pivoted about its center and linked together for simultaneous operation.
- ***Slide type***. A damper consisting of a single blade which moves substantially normal to the flow.

Dead End Oil System: An oil supply in which there is no return line to the storage tank.

Demulsibility: The resistance of an oil to emulsification, or the ability of an oil to separate from any water with which it is mixed. The better the demulsibility rating, the more quickly the oil separates from water

Deficiency of Air: A supply of air which is inadequate for complete combustion of fuel. This is same as an excess of fuel.

Delayed Mixing: A process in which the fuel and air leave the burner nozzle unmixed and thereafter mix relatively slow primarily through diffusion, this results in long luminous flame called a diffusion flame, luminous flame or long flame.

Desalting: Removal of mineral salts (most chlorides, e.g., magnesium chloride and sodium chloride) from crude oil.

Desulfurization: A chemical treatment to remove sulfur or sulfur compounds from hydrocarbons.

Detonation: A violent explosion involving high-velocity pressure waves; in a gasoline engine, the spontaneous combustion of part of the fuel compresses charge after spark occurs. Detonation usually produces a characteristic metallic sound, or knock.

Dewaxing: The removal of wax from petroleum products (usually lubricating oils and distillate fuels) by solvent absorption, chilling, and filtering.

Dew Point: The combination of temperature and pressure at which a condensable vapour produces the first small amount of liquid in equilibrium with the vapour .

Diffusion Flame: A long luminous flame that results when the fuel and air leaving the burner nozzle unmixed and mix relatively slowly through diffusion.

Dissociation: This usually occurs at high temperatures when the products of combustion break into combustibles and oxygen, with absorption of heat, and act as one of the limiting factors to the maximum temperature of a flame.

Distillate Oil: Oil separated from the crude by fractional distillation.

Distillation: The process of heating a liquid to its boiling point and condensing and collecting the vapors .

Dowtherm: An organic chemical with an exceedingly high boiling point, sometimes used in special types of boilers for high-temperature service. It is composed of diphenyl and diphenyloxide.

Dragline: An overburden removal machine used in stripping coal seams because of its great reach and its ability to cast spoil farther from the pit – used for the recovery of moderate depth coal seams.

Dried Coal: Moisture,an inherent component of coal,requires heat for its evaporation & release in the products of combustion. Surface moisture in cold climates increases handling problems. Normal atmospheric drying of coal is accelerated by passing hot combustion gases over or through beds of coal to be dried (usually metallurgical coal).

Dry Bottom Furnace: A pulverized-fuel furnace in which the ash particles are deposited on the furnace bottom in a dry, non adherent condition.

Dry Steam: Steam containing no moisture. Commercially dry steam containing not more than 0.5 percent moisture.

Duplex Pump: A pump having two cylinders operated in parallel to provide a more even flow of liquid. Also refers to composite fuel oil pumping units having two sets of oil pumps, one to operate and the other to remain as standby.

E

Economizer: A series of tubes located in the path of flue gases. Feed water is pumped through these tubes on its way to the boiler in order to absorb waste heat from the flue gas.

Effective Chimney Height: The height that is necessary to overcome the pressure drop caused by the gas flow.

Efficiency: The ratio of useful heat output to the gross heat input of a furnace expressed as percentage.

Electric Furnace: A furnace used for the refinement of high-grade steel.

Electrostatic Precipitator: A device for collecting dust, mist, or fume from a gas stream, by placing an electric charge on the particle and removing that particle onto a colleting electrode.

Emissivity: The ability of a surface to radiate energy as compared with that of a black body which emits radiation at the maximum rate possible at any given temperature and which has an emissivity of unity.

Engler viscosity: A viscosity obtained by dividing the out flow time in seconds for 200 ml. of the material being tested, by the time in seconds for 200 ml. of water at 68°F (20°C) to flow out of an Engler viscometer.

Excess Air: The amount of air supplied over and above the theoretically required quantity of air for complete combustion.

Exit Temperature: The temperature of combustion gases as they leave a furnace.

Explosion Door: A door in a furnace or boiler setting designed to be opened by a pre determined gas pressure.

Explosion Fireside : Combustion which proceeds so rapidly that a high pressure is generated suddenly.

Explosion Flap/Rupture Disc: A mettallic sheet at the zone header end of a gas air line which gets ruptured during certain increase of pressure in pipe line. It is used as safety measure.

F

Fan Mixer: An air blower in which the fuel gas is admitted to the inlet to be mixed with air.

Fire Box: A combustion chamber. If external to the furnace or kiln, also called a Dutch Oven or Dog House .

Fire Point: The lowest temperature at which an oil vaporizes rapidly enough to burn for at least 5 seconds after ignition, under standard conditions.

Fire-Tube Boiler: A boiler that has water on the outside of the tubes with the heat and product of combustion flowing inside the tube. Heat through convection is transmitted through the inside of the tube to the water that is on the outside of the tube. It is the reverse of a water tube boiler.

Firing Rate: The rate at which air, fuel or a fuel air mixture is supplied to furnace for combustion. It may be expressed in volume, weight or heat units supplied per unit time.

Firing Zone: That portion of a furnace or kiln, usually a continuous
furnace or kiln, through which the load passes and which is at or near the process temperature.

Fixed Carbon: The combustible residue remaining after volatile matter has been driven off by heating of coal.

Flame: A hot, luminous reaction front (or wave) in a gaseous medium into which the reactants flow and out of which the product flow.

Flame Blow/Lift Off: Extinguishing of flame when velocity of fuel mixture leaving burner is greater than velocity of flame propagation.

Flame Detector: A device which indicates if fuel, such as liquid, gaseous, or pulverized, is burning or if ignition has been lost. The indication may be transmitted to a signal or to a control system.

Flame Front: An area of very rapid chemical reaction. It is a boundary between burnt and unburnt gas that continually moves towards the unburnt gas.

Flame Propagation/Profile: The spread of a flame in a combustible environment outward from the point at which the combustion started.

Flame Monitoring Device -Flame 'Sensor', Flame Scanner: A device for flame supervision eg,ultra violet detector,flame rod, flicker detector, infra-red detector, photocell, thermopile, bimetal warp switch, etc.

Flame Retaining Nozzle: Any burner nozzle with built in features to hold the flame at high fuel-air mixture pressures.

Flame Retention Burner: A burner whose nozzle is surrounded with small ports that act as pilots to relight the main flame if it blows off. The velocity through the small ports is less, so the flame almost never blows off. Also called a "Stick-tight nozzle".

Flame Speed: The rate at which the flame propagates through a combustible mixture.

Flame Temperature: Measure of heat release by the flame due to net heating value of the fuel.

Flame Temperature(Theoretical or Calculated): Equilibrium temperature at which the rising flame temperature comes to a halt due to the effect of dissociation which is the maximum flame temperature.

Flame Temperature (Actual): Theoretical flame temperature minus the temperature drop caused by heat transfer to surroundings before the heat release process is completed.

Flame Velocity: The speed at which the flame progresses into a mixture relative to the speed of the mixture. Also called flame speed , ignition velocity, rate of flame propagation.

Flammability Limits: The maximum & minimum percentage of a fuel in an air-fuel mixture for combustion. This is also known as inflammability limits or explosion limits .

Flare Back: A burst of flame from a furnace in a direction opposed to the normal flow, usually caused by the ignition of an accumulation of combustible gases.

Flash Back: When the flame velocity exceeds the velocity with which the air-fuel mixture leaves the burner, then the flame travels back through the burner nozzle. This tendency of a flame is termed flash back or back-fire.

Flash Point: The temperature at which an oil gives off just sufficient vapour to create, in the air space above it, an explosive mixture that may 'flash' if brought into contact with a flame. It is a measure of the fire risk of an oil in bulk.

Flue Gas: All gases including gaseous products of combustion, water vapour, excess air and nitrogen, which are formed after a fuel is burnt and leave the furnace to the chimney through flue passage.

Flue Gas Analysis: A statement of the quantities of the various components of a sample of flue gas, usually expressed in percentages by volume .

Flue Gas Loss (or Stack Loss): The sensible and latent heat carried away by the flue gases to the chimney.

Fluidized Bed Combustion: A coal burning system which received attention because of its ability to produce low sulfur dioxide and nitrogen oxide emissions along with a high overall efficiency. Coal is burned as it is mixed with other materials and suspended in space by an upward moving stream of air and combustion gases. This bed of suspended particles moves much as a boiling liquid.

Fly Ash: The finely divided particles of ash entrained in gases resulting from the combustion of fuel.

Foot Valve: A check valve at the bottom end of a suction line in a tank to prevent the oil system from draining back into the tank when the pump is not operating.

Forced Draft: Air under positive pressure produced by fans at the point where air or gases enter a unit, such as a combustion unit.

Forced-Draft Fan: A fan supplying air under pressure to the fuel burning equipment.

Fossil Fuel: Any naturally occurring fuel of an organic nature, such as coal, crude oil or natural gas.

Friability: The tendency of a coal to crumble or break into small pieces.

Fuel: Any substance used for combustion as a heat source.

Fuel Oil: A liquid petroleum product used as a fuel . The following terms are used to describe its properties:

- ***Flash point***. The flash point of a fuel oil is an indication of the maximum temperature at which it can be stored and handled without serious fire hazard.
- ***Pour point***. The pour point is an indication of the lowest temperature at which a fuel oil can be stored and still be capable of flowing under very low forces.
- ***Viscosity***. The viscosity of an oil is a measure of its resistance to flow. In fuel oil it is highly significant since it indicates both the relative ease with which the oil will flow or may be pumped and the ease of atomization.

Fuel Shut Off Valve: Any valve for stopping flow of fuel. The term is frequently used to mean an automatic fuel shut off valve which is spring loaded and tripped by de-energising an electric or pneumatic hold open mechanism when any connected interlock that senses a dangerous condition.

Fuel Train: The fuel handling system between the source of fuel and the burner. It may Include regulators, shut off valves, pressure switches, flow meters, control valves.

Furnace: An enclosed place in which heat is produced by the combustion of a fuel.

Furnace Pressure: The pressure exerted by flue gases within a furnace combustion space is called furnace pressure. It is said to be positive , if greater than atmospheric pressure, negative if less than atmospheric pressure and neutral if equal to atmospheric pressure.

G

Gas Booster: Gas tight blower used to increase gas pressure.

Gasification: The production of gaseous fuels from coal for residential, commercial and industrial users. The type of conversion process determines the heating values of the produced gas.

Gas Mixer: Any device for mixing gas and air such as a fan aspirator or inspirator.

Gas Oil: Middle distillate petroleum fraction with a boiling range of about 175-400°C ; usually includes diesel fuel, kerosene, heating oil, and light fuel oil.

Gas Oil Burner: A burner designed to burn gas and oil simultaneously.

Gasoline (Petrol): A blend of napthas and other refinery products with sufficiently high octane and other desirable characteristics to be suitable for use as fuel in internal combustion engines.

Gauge Pressure: The difference between atmospheric pressure and the absolute pressure.

Grades of Fuel Oil: Voluntary commercial standards recommended by the ASTM for different classification of fuel oils based on various characteristics among which the most important are specific gravity and viscosity.

Grindability: It is the characteristic of coal representing its ease of pulverizing and is one of the factors used in determining the capacity of a pulverizer. The index is relative, with the large values, such as 100, representing coals easy to pulverize and smaller values such as 40 representing coals difficult to pulverize.

Grindability Index: Indicates the ease of pulverizing a coal in comparison to a reference coal. This index is helpful in estimating mill capacity. The two most common methods for determining this index are the Hardgrove Grindability Method and Ball Mill Grindability Method. Coals with a low index are more difficult to pulverize.

H

Hardgrove Grindability Index: A measure of the relative ease with which coal may be pulverized. On a one to 100 scale, the lower the value the harder the coal.

Heat Balance: Its primary object is to determine the overall thermal efficiency of a furnace. Heat balance is an attempt to balance the total energy entering the system against that leaving which is indicative of the directions in which the heat supplied is utilized or wasted and if so to what extent.

Heat Content: The sum total of latent & sensible heat stored in a substance minus that contained at an arbitrary set of conditions chosen as the base; usually expressed as J/kg or J/m^3.

Heat Exchanger: Any device for transferring heat from one fluid to another without allowing the fluids to mix.

Heat Flux: The rate of heat flow through a unit area.

Heat of Combustion: The heat released by combustion of a unit quantity of a fuel measured in kcal.

Heat Release: The total quantity of thermal energy above a fixed datum introduced into a furnace by the fuel, the product of the hourly fuel rate and its higher heating value, expressed in Kcal / hour /m3 of furnace volume or m2 of heating surface.

Heat Transfer: The flow of heat by conduction, convection and radiation which depend upon a difference of temperature between the bodies emitting the heat and those receiving it.

Heavy Oil: A term denoting residual oil, as opposed to light or distillate oil.

Higher Heating Value(HHV)/Gross Heating Value: The total heat obtained from combustion of specified amount of fuel and its stoichiometrically correct amount of air, both being at 15.5 °C when combustion starts and the combustion products being cooled to 15.5 °C before the heat released is measured.

High Pressure Switch: A device to monitor liquid, steam or gas pressure and arranged to shut-down the burner at a preset high pressure.

Hot-mix Temperature: A theoretical average temperature of all products of combustion of a fuel with air, including excess air assuming a plug flow reaction with no heat loss.

Hydrocracking: A process used to convert heavier feedstock into lower boiling, higher value products. The process employs high pressure, high temperature, a catalyst, and hydrogen.

Hydrodesulfurization: A catalytic process in which the principal purpose in to remove sulfur from petroleum fractions in the presence of hydrogen.

Hydrofinishing: A catalytic treating process carried out in the presence of hydrogen to improve the properties of low viscosity index naphthenic and medium viscosity index napthenic oils. It is also applied to paraffin waxes and microcrystalline waxes for the removal of undesirable components. This process consumes hydrogen and is used in lieu of acid treating.

Hydroforming: Catalytic reforming of naphtha at elevated temperatures and moderate pressures in the presence of hydrogen to form high octane BTX aromatics for motor fuel or chemical manufacture. This process results in a net production of hydrogen and has rendered thermal reforming some what obsolete. It represents the total effect of numerous simultaneous reactions such as cracking, polymerization, dehydrogenation, and isomerization.

I

Igniter: A burner smaller than the main burner, which is ignited by a spark or other independent & stable ignition source and which provides proven ignition energy required to immediately light off the main burner.

Ignition: A system in which the fuel to a main burner or gas or oil pilot is ignited directly either by an automatically energized spark or glow coil or by a gas or oil pilot.

Ignition Temperature: The lowest temperature in a combustion process at which chemical reaction proceeds without heat being supplied from an external source. No substance will burn until it has been raised to a temperature at which its reaction with oxygen issufficiently rapid. Visible ignition is preceded by a temperature range in which slow combustion occurs. The ignition temperature does not necessarily correspond with the fuel starting to flow or with the first appearance of flame.

Ignition Velocity: Same as Flame Velocity .

Incomplete Combustion: Partial combustion of fuel in a combustion process. Formation of carbon monoxide in a combustion process is a good example of incomplete combustion.

Induced Air: Air that flows into a furnace through openings because the furnace pressure is less than the atmospheric pressure. Also refers to air brought into a furnace by entrainment in a high velocity stream.

Induced Draft: A mechanical draft produced by suction stream jets or fans at the point where air or gases leave a unit.

Induced Draft Fan: A fan or blower that produces a negative pressure in the combustion chamber or in a chimney(stack) either by taking its suction from the combustion chamber or by induction means (ejector principle).

Industrial Furnaces: These include the furnaces in which heat is imparted for the purpose of heating, heat treatment, etc, without effecting any substantial chemical change.

Inerts: Non-combustible constituents of a fuel.

Infiltrating Air: The air which forces itself into a furnace through various openings when the furnace pressure is negative.

Inherent Moisture: That moisture remaining in the coal after natural drying in the air. In general, the moisture that is present in the coal in the bed.

Insulation: A material which is relatively bad conductor of heat used to reduce heat loss from a given space.

Interlock: An electrical, pneumatic or mechanical connection between elements of a control system that verifies conditions satisfactory to a proper operating sequence and which commands a shutdown of the system when a dangerous or unwanted condition develops. Example: Excess furnace temperature limits, low fuel or air pressure limits, etc.

Intermittent Furnace: Furnaces which require to be cooled to a greater or lesser degree between charges,and again heated to full temperature.

Isomerization: A reaction that catalytically converts straight-chain hydrocarbon molecules into branched-chain molecules of substantially higher octane number. The reaction rearranges the carbon skeleton of a molecule without adding or removing anything from the original material.

K

Kerosene: A light liquid petroleum fuel, mostly Nonane . A constituent in some of the light fuel oils and jet propulsion fuels.

K-Factor: Thermal conductivity of a material:

Kiln: A heated chamber used for burning, hardening and/or drying stage of an industrial (usually ceramic) process.

Kiio-calorie (kcal): The heat energy required to raise the temperature of one kilogram of water by 1°C

Kilowatt-hour (kWh): A unit of work or heat, equivalent to that resulting from the use of electricity at the rate of 1 kW for one hour .

Kinematic Viscosity: It is a measure of the tendency of a fluid to resist flow. The kinematic viscosity is expressed as the absolute viscosity divided by the density.

L

Laminar Flow, Streamline Flow, Viscous Flow: Flow of viscous fluids in which the particles of a fluid move in straight lines parallel to the direction of flow.

Large Port Burner: A burner characterised by a single large discharge opening or nozzle.

Latent Heat: Heat absorbed or given off by a substance without changing its temperature, as when melting, solidifying, evaporating, condensing, changing crystalline structure.

Latent Heat of Fusion: Heat given off by a unit weight of a liquid freezing to a solid or gained by a solid melting to a liquid, without a change in temperature .

Latent Heat of Evaporation: Heat given off by a unit weight of a vapour condensing to a liquid or gained by a liquid evaporating to a vapour without a change in temperature.

Lean Mixture: A mixture of fuel and air in a premix burner system in which an excess of air is supplied in relation to the amount needed for complete combustion.

Light Crude: Crude oil with a low specific gravity and high API gravity due to the presence of a high proportion of light hydrocarbon fractions and low metallic compound.

Light Ends: The more volatile products of petroleum refining; eg. butane, propane, gasoline.

Light Fuel Oil: A term designating a distillate fuel oil.

Light Up: The process of igniting a burner or system of burners.

Lignite: A brownish-black woody-structured coal, lower in fixed carbon and higher in volatile matter and oxygen than either anthracite or bituminous coal.

Liquefied Petroleum Gas (LPG): It is a mixture of butane and propane obtained during fractional distillation of petroleum or other compounds with which they occur. They are stored under pressure in liquid form. LPG is used as a fuel or as a feed stock for generation of protective gas for atmospheric control as in bright annealing.

Long Flame Burner: A burner used for producing a long luminous flame by uneven mixing of fuel and air.

Longwall Mining: A mining method in which a large rectangular section of coal is removed in one continuous operation. Equipment is installed along one side of the section (longwall face) and the coal is removed in slices 2 to 4 feet thick & lengths of up to 200 meters.The roof is supported by hydraulic props; a revolving cutting wheel or plow cuts & breaks the coal, and the coal is moved from the mining face with a rugged chain conveyor.

Low Pressure Air Atomising Oil Burner: A burner that purposely has poor mixing of fuel and air so that the fuel takes more time to burn thus producing a long, luminous flame.

Low Pressure Switch: A device to monitor liquid, steam or gas pressure and arranged to shut down the burner if the pressure falls below a pre-set low pressure limit.

Luminous Flame Burner, Long Flame Burner: A burner that purposely has poor mixing of fuel and air so that the fuel takes more time to burn, thus producing a long, luminous flame.

M

Main Circulating Loop: The section of an oil handling system that delivers the oil from storage to the branch circuits and returns the unused oil to the storage tank.

Manifold: A header to which a number of incoming/outgoing branch pipes are connected.

Manual Ignition: Lighting up of a burner by use of a manually applied portable gas torch or gas pilot. The term does not necessarily imply that the pilot is hand held, however defines that the burner is not automatically ignited.

Manual Reset Automatic Fuel Shut Off Valve: A shut-off valve that automatically closes by spring action when its hold open mechanism is electrically or pneumatically tripped by any connected interlock sensing a dangerous or unwanted condition. It must, however, be reopened by hand after the dangerous or unwanted condition is rectified and the hold open mechanism is re energised.

Manufactured Gas: Any gas made artificially on large scale for use as fuel.

Mean Stack Temperature: The average temperature of flue gases in a chiminey. Sometimes approximated by measuring the temperature of the flue gas at a point mid way between the flue gas inlet of chimney base and the top of the chimney.

Mechanical Draft: A pressure differential produced by mechanical means, such as fans, blowers, etc.

Mechanical Pressure Atomising Burner: It is a burner in which liquid fuel is atomized into a spray of fine droplets by passing it under pressure through a small orifice and allowing it to expand.

Mechanical Stoker: A device consisting of a mechanically operated fuel feeding mechanism and a grate; is used for the purpose of feeding solid fuel into a furnace, distributing it over the grate, admitting air to the fuel for the purpose of combustion, and providing a means for removal or discharge of refuse.

Metallurgical Coal: The various grades of coal suitable for carbonization to make coke for steel manufacture. Also known as "met" coal, it contains four important qualities: volatility, which affects coke yield; the level of impurities, which affects coke quality; composition, which affects coke strength; and basic characteristics, which affect coke oven safety.

Micron: A unit of length. One millionth of a meter or one thousandth of a millimeter. One micron equals 0.00004 of an inch.

Middle Distillate: Term applied to hydrocarbons in the so-called "middle range" of refinery distillation. Examples: heating oil, diesel fuels, and kerosene.

Minimum Firing Rate: The lowest input rate for a burner or a process.

Mining, Strip: Strip mining involves the use of large equipment which removes overburden (rock, dirt, etc.), exposing the seam of coal. This method is usually used when the coal seam lies close to the ground surface and the overburden can be removed economically.

Mining, Underground:

- **Drift Mine:** Driven into the side of the hill where the coal seam outcrops. The seam is above the valley floor or water level.

- **Slope Mine:** Driven into the ground above the coal seam level. The seam usually has no outcrop and is situated below the valley floor level.
- **Shaft Mine:** Driven into the ground vertically to the seam of coal. Most often used for deep seams.

Modulating Control: A system of automatic control which provides as infinite number of control positions, as opposed to system with a finite number of positions.Modulating control is generally used to adjust the fuel and air rates to correct for the difference between the actual temperature in a furnace and the temperature desired.

Moisture:

- **Surface Moisture:** wetness of coal from external sources such as weather or coal washing processes.
- **Inherent Moisture:** water which comprises an integral portion of the composition of coal. Moisture percentage measured in proximate analysis of coal reflects inherent moisture only.

Monolithic Lining: A furnace lining without joints formed of material which is rammed, cast, gunned or sintered into place.

Motor Gasoline: A complex mixture of relatively volatile hydrocarbons with or without small quantities of additives, that have been blended to form a fuel suitable for use in spark-ignition engines.

Muffle: It is a chamber or enclosure within a furnace where the charge being heated is kept out of direct contact with the combustion products.

Multifuel Burner: A burner by means of which more than on fuel can be burned either separately or simultaneously, such as pulverized fuel, oil or gas.

Multiple Retort Stoker: An underfeed stoker consisting of two or more retorts, parallel & adjacent to one another, but separated by a line of tuyeres and arranged so that the refuse is discharged at the ends of the retorts.

N

Naphtha: A general term used for low boiling hydrocarbon fractions that are a major component of gasoline. Aliphatic naphtha refers to those naphthas containing less than 0.1% benzene and with carbon numbers from C_3 through C_{16}. Aromatic naphthas have carbon numbers from C_6 through C_{16} and contain significant quantities of aromatic hydrocarbons such as benzene (> 0.1%), toluene, and xylene.

Natural Convection, Free Convection: Transfer of heat by currents set up in fluids by differences of density resulting from differences in temperature.

Natural Draft: It is a measure of the difference in pressure heads of hot flue gases passing through the chimney and an equivalent column of cold atmospheric air around it.

Natural Gas: Any gas found in the crust of the earth as opposed to manufactured gas.

Net Heating Value, Lower Heating value: Gross heating value minus the latent heat of vaporisation. of water vapaur formed by the combustion of hydrogen in the fuel. For a fuel with no hydrogen in it, the net and gross heating values are the same .

Neutral Atmosphere: An atmosphere condition in firing a furnace or kiln that is neither oxidising nor reducing.

Nozzle Mixing Burner: A burner in which gas and air are kept separate until discharged from the burner into the combustion chamber or tunnel. Generally used with low pressure gas (up to 3 kPa) and low pressure air (up to 35 kPa).

O

Octane Number: A number indicating the relative antiknock chacacteristics of gasoline.

Oil Burner: A burner that atomizes fuel oil and blows it into the combustion chamber in the form of a fine mist or vapor. Steam or mechanical motion plus air may be used as the operating medium.

Oil Filters: A device to separate suspended impurities from fuel oil to avoid or minimise choking and corrosion of burner, pipelines and other combustion equipment.

Oil Preheating Temperature: A temperature to which medium and heavy grades of fuel **oil are heated to attain:** (a) better pumpability, and (b) to lower viscosity if required down to the limit as stipulated by burner design,

Oil to Gas Converter: A device used to vaporise distillate oils. The vaporised oil and air mixture can then be burned through pre-mix gas burners. The system consists of an oil fired air heater, atomisers and ratio controller.

Once Through Boiler: A steam-generating unit usually operated above the critical pressure in which there is no recirculation of the working fluid in any part of the unit. In the case of a supercritical steam generator, there is a constant increase in temperature and enthalpy from inlet to outlet.

On-Off Control: A control which turns the input on or off but does no proportioning or throttling as is the case with modulating control.

Open-Pit Mining: Surface mining, a type of mining in which the overburden is removed from the product being mined and is dumped back after mining; or may refer specifically to an area from which overburden has been removed, which has not been filled.

Open Port Burner/Open Type Burner/Open Burner/Open Register Burner: A burner that fires across a gap into an opening in the furnace or combustion chamber wall and is not sealed into the wall. Sometimes the burner is surrounded by an air register through which controlled secondary air or tertiary air can enter the combustion chamber.

Optical Pyrometer: An instrument used to measure the temperature of a hot body (such as furnace) by matching a standard source of light (from the filament of an electric lamp) with the light emitted from the hot body. The current flow through the lamp for this condition is measured in terms of temperature by an electrical instrument.

Optimum Combustion Air: The quantity of air which shall give greatest thermal efficiency under actual conditions.

Orsat Analyzer: An apparatus used to determine the percentage (by volume) of carbon dioxide, oxygen, hydrogen, methane and carbon monoxide in gases on dry basis.

Overburden: The earth, rock and other materials that lie above the coal seam.

Over Firing: A heat treatment that causes deformation or bloating of ceramic products or damages any product.

Oxidising Atmosphere: The presence of flue gases inside a furnace chamber with excess oxygen content in them, thereby rendering the furnace atmosphere oxidising.

Oxidising Flame: Flame resulting from combustion of a mixture containing too much air and too little fuel.

Oxygen Analyser: A device for measuring oxygen content of a gas.

P

Percent Excess Air: The percentage of air supplied in excess of that required for complete combustion.

Perfect Combustion (Theoretical Combustion): The combustion obtained by mixing and burning chemically correct proportions of fuel and oxygen so that no fuel or oxygen is left over.

Petrochemical: An intermediate chemical derived from petroleum, hydrocarbon liquids or natural gas, such as: ethylene, propylene, benzene, toluene and xylene.

Petroleum: A generic name for hydrocarbons, including crude oil, natural gas liquids, natural gas and their products.

Petroleum Coke: A residual by-product from the oil refining industry. Following heat processing or thermal cracking that extracts all the more valuable gases & oil liquids from the crude input, a basic hydrocarbon is left which does have a number of important properties & applications to industry . There are essentially three types of petroleum coke: specialty or needle coke; calcined or anode coke; and green, raw or fuel coke.

Photocell Flame Detector: A device that generates or rectifies an electric current while exposed to the light from a flame.Failure of the current or lack of rectification may be used to close an automatic fuel shutoff valve.

PI Transducer: A device to convert a pressure signal to an electrical current.

Pilot, Interrupted Spark-Ignited: A small flame that is electrically ignited automatically each time there is a call for heat and is cut-off automatically by the programming flame failure controls at the end of the trial for ignition period of the main burner.

Pipe Burner: Any burner made in the form of the tube or pipe with ports or trips spaced over its length.

Post Purge: An acceptable method for scavenging the combustion chamber & connected pipelines, boiler passes and breeching to remove all combustible gases after flame failure controls have sensed pilot and main burner shutdown and fuel shut-off valves are closed.

Pour Point: The lowest temperature of an oil at which it flows under test conditions.

Ppm (Parts per million): The unit commonly used to designate the concentration of a substance in a wastewater in terms of weight ie. one kg per million kg , etc. PPM is synonymous with the more commonly used term mg/l (milligrams per liter).

Preheated Air/Gas: Air/Gas heated prior to combustion by utilizing the sensible heat of outgoing flue gases in order to achieve higher thermal efficiency.

Pre ignition Purge/Pre-Purge: An acceptable method for scavenging the combustion chamber, connected pipelines, boilers passes, and breeching to remove all combustible gases before the ignition system can be energised.

Premix Burner: A burner within which the gas and air are mixed before they reach its nozzle.

Pressure Burner/Blast Burner: A premix burner delivering a combustible mixture under pressure normally above 10 mm water column to the combustion zone.

Pressure Control: A system for control of air/fuel ratio by proportioning the pressures in the air and fuel supply lines.

Pressure Regulator: A device used to maintain a constant pressure in a fuel or air supply line regardless of the flow.

Primary Air: The air which is mixed with the fuel at (or in) the burner.

Producer Gas: An artificial fuel gas of low calorific value made by passing a mixture of air and steam through a hot bed of coke or coal.

Proximate Analysis: A percentage measurement of the physical properties of coal including moisture, volatile matter, fixed carbon and ash. Proximate analysis is usually accompanied by a statement of sulphur content, Kcal/kg , ash fusion temperature and grindability.

Pulverized Fuel: Solid fuel reduced to a fine size, such that 80 % passes through a 200-mesh screen.

Pulverizer: A machine which reduces a solid fuel to a fineness suitable for burning in suspension. Pulveriser types used are:

- **High speed (over 800 rpm)**
  - (a) **Impact pulverizer.** A machine in which the major portion of the reduction in particle size of the fuel to be pulverized is affected by fracture of larger sizes by sudden shock, impingement, or collision of the fuel with rotating members and casing.
  - (b) **Attribution pulverizer.** A machine in which the major portion of the reduction of particle size is by abrasion, either by pulverizer parts on coal or by coal on coal.
- **Medium speed (between 70 and 300 rpm)**
  - (a) **Roller pulverizer**. A machine having grinding elements consisting of conical or cylindrical rolls and a bowl, bull-ring mating rings or table, any of which may be the rotating member, the fuel to be pulverized being reduced in size by crushing and attrition between the rolls & the rings.
  - (b) **Ball pulverizer.** A machine in which the grinding elements consist of one or more circular rows of metal balls arranged in suitable raceways, wherein the fuel to be pulverized is reduced in size by crushing and attrition between the balls & raceways.
- **Low speed (under 70 rpm)**
  - (a) **Ball or tube pulverizer.** A machine having a rotating cylindrical or conical casing charged with metal ball or slugs and the fuel to be pulverized, with reduction in particle size being affected by crushing & attrition resulting from continuous relative movement of the charge on rotation of the casing.

Purging: The process of removal of undesirable gases from a system by means of suitable media like steam, compressed air, inert/protective gas, etc.

Pyrolysis Gasoline: A by-product from the manufacture of ethylene by steam cracking of hydrocarbon fractions such as naphtha or gas oil.

Q

Quarl: A refractory throat around a burner port. It helps direct air into the flame, and radiation from the hot refractory. Also assists efficient combustion.

Quenching of Flame: A process in which the reactants in a flame are rapidly cooled. This usually prevents the reaction from reaching completion in the localised area where quenching occurs. Large scale quenching may result into incomplete combustion.

R

Radiation: A process of heat transfer in which energy is transmitted from a hot body to another body separated from it by a medium (such as air) transparent to the radiation.The emission of heat from a furnace to their surroundings is a good example of radiation. All matter-solids, liquids and gases may radiate.

Radiation/Radiant Burner: A burner designed to transfer a significant part of the combustion heat in the form of radiation from surfaces of various refractory shapes of alloy screens.

Radiant Tube Burner: A burner designed to operate inside a tube which heats the load indirectly by radiation or by air or gas convection.

Ratio Regulator: An automatic device used to maintain proportional flow of fuel and air for designed combustion conditions.

Raw Coal (Run-of-Mine Coal): Coal extracted from the seam and not processed, washed, crushed or sized.

Recirculation Type Oil System: A closed circuit oil system in which the oil not used at the burner is returned to the storage tank.

Recuperator: A continuous heat exchanger in which heat from combustion products is transferred to incoming air or gas through metal or refractory wall.

Reduced Crude Oil: Crude oil that has undergone at least one distillation process to separate some of the lighter hydrocarbons. Reducing crude lowers its API gravity, but increases the handling safety by raising the flash point.

Reducing Atmosphere: A furnace atmosphere which tends to remove oxygen from substances placed in the furnace. It may be produced by supplying inadequate air to the burners, thus intentionally making the combustion incomplete.

Reducing Flame: A flame resulting from combustion of a mixture containing too much of fuel and too little air.

Redwood viscosity: The number of seconds required for 50 ml. of an oil to flow out of a standard Redwood viscosimeter at a definite temperature; British viscosity standard.

Reforming: The thermal or catalytic conversion of petroleum naphtha into more volatile products of higher octane number. It represents the total effect of numerous simultaneous reactions such as cracking, polymerization, dehydrogenation, and isomerization.

Refractory Block/Refractory Tile/Burner Tile/Burner Block/Burner Refractory: A piece of refractory material moulded with a conical or cylindrical hole through its center through which the flame propagates from a burner when mounted at one end concentric to the block. Helps maintain continuous combustion, reduces the possibility of flash back & blow-off from pre mix burners.

Refuse: The ash and unburnt fuel remaining after combustion of a solid fuel.

Regeneration: In a catalytic process the reactivation of the catalyst, sometimes done by burning off the coke deposits under carefully controlled conditions of temperature and oxygen content of the regeneration gas stream.

Regenerator: Arrangement (chambers filled with a checker work of refractory bricks) for receiving and then utilizing heat from hot spent gas from a furnace.

Residual Oil: The heavy portion of the crude oil remaining after distillation and cracking.

Return Line: The low pressure side of main circulating oil loop or branch circuit through which the unused oil flows back to the storage tank after having passed the burner.

Rich Mixture: A mixture of fuel and air in system in which an excess of fuel is present relation to the amount of air needed for combustion.

Rotary Cup Oil Burner: Cetrifugal Atomising Burner which throws oil from a rotating cup into an air stream, breaking the oil into a fine conical spray.

Rotary Dryer/Kiln: An inclined rotating drum, usually refractory lined and fired with a burner at the lower end for drying or calcining loose material as they roll through by gravity sometimes lifted, mixed and exposed by flights or shelves attached to the inner wall.

Rotary Hearth furnace: A continuous type round shaped furnace for heating/reheating/heat treatment in which the hearth, made in the form of a circular disc/plate/annular ring, rotates either continuously or stepwise within the stationary inner chamber of the circular furnace.

S

Safety Shut Off Valve: A valve which automatically and completely shuts off the supply of fuel to a burner or set of burners in case of any irregularity which may lead to an unsafe situation.

Saturated Air: Air containing all the maximum water vapour it can normally hold under specified conditions.

Saturated Steam: Steam in contact with water at a given temperature is said to be saturated when it is of maximum density for that temperature. Steam in contact with water is saturated steam.

Saybolt Universal (SSU): A scale used for measuring the viscosity of oil, expressed in seconds required for a specified amount of oil to flow through an orifice.

Sealed-in Burner: A burner mounted on a furnace in an air tight fashion so that there can be no inflow of secondary air around the burner.

Seam: A deposit or bed of coal which can vary in thickness from less than one inch to over fifty feet; from several acres to thousands of square kms. in size; and from several metre in depth to thousands of metre below ground level.

Secondary Air: Air brought in around the burner or through other openings in the furnace wall.

Secondary Combustion: Combustion which occurs as a result of ignition at a point beyond the furnace. Also called delayed combustion.

Semi-Muffle Furnace: A furnace with a partial muffle in which the pro-ducts of combustion can eventually come in contact with the load, but the temperature uniformity is improved by protecting the load from intense direct heat transfer from the flame.

Sensible Heat: Heat, the addition or removal of heat which results in a change in temperature of a given system.

Shaft Kiln: A furnace for heating lump material, consists of a vertical refractory lined shaft.

Shortwall: A mining method with a panel layout similar to longwall but at a panel width reduced by approximately 50%. It uses continuous-mining equipment for coal cutting and haulage and a specially designed hydraulic roof support.

Shutdown: The total process of terminating operation of a combustion system.

Shuttle Kiln: A type of periodic kiln in which the ware is loaded on the cars rather than directly on the hearth.

Single Port Burner: A burner having only one discharge opening or port.

SIT(Spontaneous Ignition Temperature): The temperature at which an oil ignites of its own in the presence of air or oxygen under standard conditions.

Slag: Coal ash that is or has been in a molten state.

Slagging of Refractories: Destructive chemical action between refractories and external agencies at high temperatures resulting in the formation of a viscous liquid.

Slag-Tap Furnace: A pulverized-fuel-fired furnace in which the ash particles are deposited & retained on the floor in molten condition, and from which molten ash is removed by tapping either continuously or intermittently.

Sludge: Heavy materials found at the bottom of fuel oil storage tanks including oil-water emulsion, heavy chemicals, oxidation products, dirt, etc.

Slurry: A suspension of finely pulverised solid material inwater.

Smoke: Small gas-borne particles of carbon or soot, less than 1 μm in size, resulting form incomplete combustion of carbonaceous materials and of sufficient number to be observable.

Sour Gas: Natural gas that contains corrosive, sulfur bearing compounds such as hydrogen sulfide and mercaptans.

Soak (Soaking): To hold the load in a furnace or kiln at a particular temperature for a time to allow equalization of temperature throughout the load.

Soot: A black substance, consisting mostly of very small particles of carbon, which appears in smoke resulting from incomplete combustion.

Sour Gas: Gaseous fuel which contains a relatively large proportion of sulphur or sulphur compounds.

Space Heating: Heating of a large volume of air to desired temperature for human occupancy or for storage.

Spalling of Refractories: The cracking or fracture of a refractory product caused by differential expansion due to thermal shock, a steep temperature gradient,a crystalline inversion or a change in composition near the hot face.

Specific Gravity of Gases: The ratio of the density of a gas to the density of dry air at standard conditions of temperature and pressure.

Specific Gravity of Liquids: A ratio of the density of a liquid to the density of water when both the liquid and the water are at 27°C.

Specific Heat: The amount of heat required to raise the temperature of unit mass/volume of a substance through one degree Celsius. It is measured in kcal/ kg °C or kcal/m^3 °C as applicable.

Specific Volume: The volume occupied by 1 kg of a substance under any specified conditions of temperature and pressure.

Spontaneous Combustion: Ignition of combustible material following slow oxidation without the application of high temperature from an external source.

Spray Nozzle: A nozzle form which a liquid fuel is discharged in the form of a spray.

Stabilization: A process for separating the gaseous and more volatile liquid hydrocarbons from crude petroleum or gasoline and leaving a stable (less volatile) liquid so that it can be handled or stored with less change in composition.

Stack Loss/Flue Gas Loss: The sensible heat carried away by the dry flue gas plus the sensible and latent heat carried away by the water vapour in the flue gas.

Standard Air/Standard Atmosphere/Standard Barometer/Standard Pressure: Air at standard temperature and pressure Viz. 25 °C and 760 mm Hg of mercury.

Static Pressure: The force per unit area exerted by a fluid upon a surface at rest relative to the fluid. For fluid flowing in a pipe static pressure is measured by a gauge connected to the side of the pipe, perpendicular to the direction of flow.

Stability (Combustion): The quality of a burner enabling it to remain. lighted over a wide range of fuel/air ratio and input rates without the help of a pilot or spark .

Steam Atomising Burner: A burner in which high pressure steam is used to break the oil stream into a spray of tiny droplets or vapour.

Steam Generators – Furnace Design: The furnace design of steam generators may be classified into two categories by type: dry bottom and wet bottom.

The dry-bottom units are fired with pulverized fuel. About 10 to 20% of the fired ash goes to the bottom of the furnace in a dry form and is removed continually, while the remainder of the ash (80 to 90%) passes up through the furnace, through the boiler convection banks, and air heater into the dust collector.

Wet-bottom units may be fired by crushed coal, such as is fed to cyclone furnaces, or by pulverized coal. The pulverized coal-fired wet-bottom unit traps 50% or less of the ash as molten slag on the furnace walls; this slag runs down to the furnace floor and taps down into the slag tanks for removal. Cyclone furnace boilers are the wet-bottom type because they make slag that flows out on the furnace walls, then to the floor, and into slag tanks. Wet-bottom type boilers firing is no longer being used because of NOx emission considerations.

Stoichiometric Ratio: A chemically correct mixture of fuel and air capable of perfect combustion theoretically.

STP: Standard Temperature (25 °C) and Pressure (760 mm Mercury).

Straight Run Gasoline: Gasoline produced by the primary distillation of crude oil. It contains no cracked, polymerized, alkylated, reformed, or visbroken stock.

Strainer: A fine mesh screen used to separate foreign particles from an oil stream.

Strip Mine: Refers to a procedure of mining that entails the complete removal of all material from over the product to be mined in a series of rows or strips; also referred to as "open cut," "open pit," or "surface mine."

Sub-bituminous Coal: A glossy black weathering and non agglomerating coal which is lower in fixed carbon than bituminous coal, with more volatile matter & oxygen.

Sulphur: A non-metallic chemical element comprising varying degrees of coal's composition. Sulphur burns off when coal is heated but promotes clinkering, slagging and corrosion. Coals with high sulphur contents are susceptible to spontaneous combustion in storage piles. Sulphur present in stack emissions is regulated .

Sulphur Classifications:

- **Pyritic:** The only inherent form of sulphur in coal which can be removed by washing.
- **Organic:** Inherent and generally cannot be removed from coal.
- **Sulphates:** Oxidation products, present on the surface of fresh coal in small amounts. Sulphates can be removed conveniently through standard cleaning processes.

Superheated Steam: If dry saturated steam is heated when not in contact with water, its temperature is raised and its density is diminished or its pressure is raised.The steam is then said to be superheated . Steam heated to a temperature higher than that corresponding to the temperature equivalent to the pressure.

Superheater: A series of tubes exposed to high-temperature gases or to radiant heat. Steam from the boiler passes through these tubes to attain a higher temperature than would be possible otherwise. This superheated steam ensures dryness.

Surface Coefficient/Radiation Heat Transfer Coefficient: The rate of heat flow per unit area (heat flux) for each degree of temperature difference between the emitting surface and the absorbing surface.

Surface Mine: A coal-producing mine that is usually within a few hundred metres of the surface. Earth above or around the coal (overburden) is removed to expose the coalbed, which is then mined with surface excavation equipment such as draglines, power shovels, bulldozers, loaders, and augers. It is also known as an area, contour, open-pit, strip or auger mine.

Sweetening: Processes that either remove obnoxious sulfur compounds (primarily hydrogen sulfide, mercaptans & thiophens) from petroleum fractions or streams, or convert them, as in the case of mercaptans, to odorless disulfides to improve odor, color,& oxidation stability.

T

Tangential Firing: A method of firing by which a number of fuel nozzles are located in the furnace walls so that the centerlines of the nozzles are tangential to a horizontal circle. Corner firing is usually included in this type.

Tap Hole: An opening for the removal of slag from a slag tap furnace.

Tertiary Air: A third supply of combustion air introduced downstream from the secondary air.

Theoretical Air: The amount of air required for making chemically correct mixture with a given amount of a particular fuel for complete combustion.

Theoretical Draft: The calculated difference between the pressure in a furnace and the atmospheric pressure, which is caused by the buoyancy of the hot gases in the furnace and its chimney.

Thermal Conductivity: The amount of heat conducted through a body per unit time per unit area per unit temperature gradient.

Thermal Cracking: The breaking up of heavy oil molecules into lighter fractions by the use of high temperature without the aid of catalysts.

Thermal Diffusivity: Thermal conductivity divided by the product of specific heat and density of a material is thermal diffusivity. It is an index which indicates the ease with which heat penetrates within the body.

Thermal Shock: A sudden temperature change.

Thermopile: Many thermocouples assembled in series to provide a single strong enough to actuate a valve without amplification by an external power source .

Thermocouple: Two dissimilar metal wires welded together at one end, used for measuring temperature. The welded end, when exposed to elevated temperature while the other ends are maintained at a steady low temperature, generates an emf forming the basis of temperature measurement. Potential difference between the unwelded ends gives a measure of temperatue at the welded end.

Throttling Valve: A valve used to control flow rate of a fluid. A throttling valve does not necessarily provide tight shut-off.

Transducer: An element or device which receives information in the form of one physical quantity and converts the information to the same or another physical quantity. For example, T/I Transducer is a device to convert a temperature signal to electric current.

Travelling-Grate Stoker: A stoker similar to a chain-grate stoker with the exception that the grate is separate from but is supported on and driven by chains. Only enough chain strands are used as may be required to support and drive the grate.

Tunnel Burner: A burner sealed into a furnace wall in which combustion takes place mostly in a refractory tunnel or tuyer which is a part of the burner .

Tunnel Kiln: A tunnel-shaped furnace through which the ware is pushed on cars passing through preheating, firing and cooling zones consecutively, enabling continuous operation.

Turndown Ratio: The ratio of maximum to minimum input rates, within the range of controllability measurement.

Tuyers: A refractory shape containing one or more holes through which air and other gases are introduced into a furnace.

U

Ultimate Analysis: A statement of the quantities of various elements of which a substance is composed, usually expressed in percentages by weight.

Ultimate Percent, CO_2: The percentage by volume of carbon dioxide that appears in the dry flue gases when a fuel is completely burnt with chemically correct air/fuel ratio.

Underground Mine: A mine where coal is produced by tunneling into the earth to the coalbed, which is then mined with underground mining equipment such as cutting machines and continuous, longwall, and shortwall mining machines. Also called deep mines, underground mines are classified according to the type of opening used to reach the coal; i.e., drift (level tunnel), slope (inclined tunnel), or shaft (vertical tunnel).

Upper Limit of Flammability (Formally Inflammability): The maximum percentage of gaseous fuel in an air-fuel mixture which can be ignited. Above this percentage, the mixture will be too rich to burn.

V

Vacuum Distillation: The distillation of petroleum under vacuum which reduces the boiling temperature sufficiently to prevent cracking or decomposition of the feedstock.

Valve Control: A system for control of air-fuel ratio by mechanical linkage of valves having the same flow characteristics.

Valve, Safety Shutoff: A valve automatically closed by the safety control system or by an emergency device to completely shut off fuel supply to the burner.

Vane: A fixed or adjustable plate inserted in a gas or air stream used to change the direction of flow.

Vane Control: A set of movable vanes in the inlet of a fan to provide regulation of air flow.

Vane Guide: A set of stationary vanes to govern direction, velocity, and distribution of air or gas flow.

Vapour: A gas that is near the condensing state, where molecules are so close together that the forces between them significantly affect their behavior so that P_1V_1/T_1 , is not equal to P_2V_2/T_2.

Vapourising Oil Burner: A burner in which the oil is vapourised in a single step by direct heating of the liquid.

Vapour Lock: An obstruction to the flow of a liquid in a pipe caused by vapour from the liquid or by air.

Vapour Pressure: The pressure of the vapour of a liquid or solid in equilibrium with the liquid or solid.

Vent: A small hole or opening in a system for the escape of a fluid.

Venturi: A section in a pipe or passageway that converges to a narrow constriction and then smoothly flares out (Named after the Italian Physicist G. B. Venturi who first noted the effect of a constriction).

Visbreaking: Viscosity breaking is a low-temperature cracking process used to reduce the viscosity or pour point of straight run residuum.

Viscometer: A device for determining the viscosity of oil. There are several methods or devices in general use. Basically, a fixed quantity of oil is allowed to pass through a fixed orifice at a specified temperature over a measured time span and then compared to a standard liquid such as a calibration oil or water.

Viscosity: A measure of fluids tendency to resist flow. Very significant in the design and selection of oil burners and oil handling systems.

Viscosity (Absolute), Dynamic Viscosity: A measure of fluid's tendency to resist flow without regard of its density. Absolute viscosity is equal to kinematic viscosity multiplied by density or specific gravity.

Viscosity, Kinematic: The relative tendency of a fluid to resist flow including the effect of fluid's density. Kinematic viscosity is equal to absoulte viscosity divided by the density or specific gravity.

Volatile: Normally refers to the more easily vapourized component of a liquid.

Volatile Matter: Portion of coal which is driven off in a gaseous form when the coal is heated. Volatility is classified as follows:

	Composition by Weight
high volatile	over 31%
medium volatile	22% to 31%
low volatile	under 22%

Volumetric Analysis: A statement of the various components of a substance (usually applied to gases only), expressed in percentage by volume.

W

Wall Loss: The heat lost from a furnace through its walls to the surroundings by radiation or convection,

Warm Up (Bring Up) Time: The time required to bring a furnace and its charge, if any, up to a prefixed operating temperature.

Water Tube Boiler: A boiler in which the water or other fluid flows through the tubes and the products of combustion surround the tubes. This kind of boiler is mainly used for high pressure steam but also can be used to produce low pressure steam.

Water-Wall: A row of water tubes lining a furnace or combustion chamber, exposed to the radiant heat of the fire; used to protect refractory and to increase capacity of the boiler.

Wobbe Index/Wobbe Number: An index used to show fuel interchangeability . Wobbe Index is equal to gross heating value divided by the square root of its specific gravity.

Z

Zone Control: The section of a furnace within which the temperature is controlled by one temperature measurement (and usually with one control valve).